MODERN
POWER STATION PRACTICE

Third Edition

(in 12 volumes)

Incorporating Modern Power System Practice

Main Editorial Panel

D. J. Littler, BSc, PhD, ARCS, CPhys, FInstP, CEng, FIEE (Chairman)

Professor E. J. Davies, DSc, PhD, CEng, FIEE

H. E. Johnson

F. Kirkby, BSc, CEng, MIMechE, AMIEE

P. B. Myerscough, CEng, FIMechE, FINucE

W. Wright, MSc, ARCST, CEng, FIEE, FIMechE, FInstE, FBIM

Volume Consulting Editor

Professor E. J. Davies, DSc, PhD, CEng, FIEE

Volume Advisory Editor

F. Kirkby, BSc, CEng, MIMechE, AMIEE

Authors

Chapter 1 **F. Kirkby**, BSc, CEng, MIMechE, AMIEE

Chapter 2 **F. Kirkby**, BSc, CEng, MIMechE, AMIEE
F. J. Trotter, CEng, MIMechE, MIEE

Chapter 3 **I. Fisher**, BSc, CEng, MIEE

Chapter 4 **P. M. Randle**, CEng, MIMechE, MBIM, MINucE

Chapter 5 **P. Bellamy**, BSc, MIMechE, MBIM

Chapter 6 **R. T. Stevens**, CEng, MIEE

Series Production

Managing Editor P. M. Reynolds

Production Editor H. E. Johnson

Resources and Co-ordination T. A. Dolling
J. R. Jackson

MODERN POWER STATION PRACTICE

Third Edition

Incorporating Modern Power System Practice

British Electricity International, London

Volume H

Station Commissioning

PERGAMON PRESS

Member of Maxwell Macmillan Pergamon Publishing Corporation

OXFORD · NEW YORK · BEIJING · FRANKFURT
SÃO PAULO · SYDNEY · TOKYO · TORONTO

U.K.	Pergamon Press plc., Headington Hill Hall, Oxford OX3 0BW, England
U.S.A.	Pergamon Press, Inc., Maxwell House, Fairview Park, Elmsford, New York 10523, U.S.A.
PEOPLE'S REPUBLIC OF CHINA	Pergamon Press, Room 4037, Qianmen Hotel, Beijing, People's Republic of China
FEDERAL REPUBLIC OF GERMANY	Pergamon Press GmbH, Hammerweg 6, D-6242 Kronberg, Federal Republic of Germany
BRAZIL	Pergamon Editora Ltda, Rua Eça de Queiros, 346, CEP 04011, Paraiso, São Paulo, Brazil
AUSTRALIA	Pergamon Press Australia Pty Ltd., P.O. Box 544, Potts Point, N.S.W. 2011, Australia
JAPAN	Pergamon Press, 5th Floor, Matsuoka Central Building, 1-7-1 Nishishinjuku, Shinjuku-ku, Tokyo 160, Japan
CANADA	Pergamon Press Canada Ltd., Suite No. 271, 253 College Street, Toronto, Ontario, Canada M5T 1R5

First edition 1963

Second edition 1971

Third edition 1990

Library of Congress Cataloging in Publication Data

Modern power station practice: incorporating modern power system practice/British Electricity International.—
3rd ed. p. cm.
Includes index.
1. Electric power-plants. I. British Electricity International.
TK1191.M49 1990
62.31'21 — dc20 90-43748

British Library Cataloguing in Publication Data

British Electricity International
Modern power station practice.—3rd. ed.
1. Electric power-plants. Design and construction
I. Title II. Central Electricity Generating Board
621.3121.

ISBN 0-08-040510-X (12 Volume Set)
ISBN 0-08-040518-5 (Volume H)

ISBN 0-08-042248-9 (Flexicover)

Printed in the Republic of Singapore by Singapore National Printers Ltd

Contents

Foreword

G. A. W. Blackman, CBE, FEng

Chairman, Central Electricity Generating Board
and Chairman, British Electricity International Ltd

For over thirty years, since its formation in 1958, the Central Electricity Generating Board (CEGB) has been at the forefront of technological advances in the design, construction, operation, and maintenance of power plant and transmission systems. During this time capacity increased almost fivefold, involving the introduction of thermal and nuclear generating units of 500 MW and 660 MW, to supply one of the largest integrated power systems in the world. In fulfilling its statutory responsibility to ensure continuity of a safe and economic supply of electricity, the CEGB built up a powerful engineering and scientific capability, and accumulated a wealth of experience in the operation and maintenance of power plant and systems. With the privatisation of the CEGB this experience and capability is being carried forward by its four successor companies — National Power, PowerGen, Nuclear Electric and National Grid.

At the heart of the CEGB's success has been an awareness of the need to sustain and improve the skills and knowledge of its engineering and technical staff. This was achieved through formal and on-job training, aided by a series of textbooks covering the theory and practice for the whole range of technology to be found on a modern power station. A second edition of the series, known as Modern Power Station Practice, was produced in the early 1970s, and it was sold throughout the world to provide electricity undertakings, engineers and students with an account of the CEGB's practices and hard-won experience. The edition had substantial worldwide sales and achieved recognition as the authoritative reference work on power generation.

A completely revised and enlarged (third) edition has now been produced which updates the relevant information in the earlier edition together with a comprehensive account of the solutions to the many engineering and environmental challenges encountered, and which puts on record the achievements of the CEGB during its lifetime as one of the world's leading public electricity utilities.

In producing this third edition, the opportunity has been taken to restructure the information in the original eight volumes to provide a more logical and detailed exposition of the technical content. The series has also been extended to include three new volumes on 'Station Commissioning', 'EHV Transmission' and 'System Operation'. Each of the eleven subject volumes had an Advisory Editor for the technical validation of the many contributions by individual authors, all of whom are recognised as authorities in their particular field of technology.

All subject volumes carry their own index and a twelfth volume provides a consolidated index for the series overall. Particular attention has been paid to the production of draft material, with text refined through a number of technical and language editorial stages and complemented by a large number of high quality illustrations. The result is a high standard of presentation designed to appeal to a wide international readership.

It is with much pleasure therefore that I introduce this new series, which has been attributed to British Electricity International on behalf of the CEGB and its successor companies. I have been closely associated with its production and have no doubt that it will be invaluable to engineers worldwide who are engaged in the design, construction, commissioning, operation and maintenance of modern power stations and systems.

March 1990

Preface

THIS VOLUME reflects the changes in the approach to commissioning which have taken place during my working life. I readily accepted the offer to become the Advisory Editor (and an author) for this volume because of my recent involvement with commissioning power stations.

My first experience of commissioning was in the era of the early 500 MW units in 1964-8 and was followed by commissioning 500 and 660 MW oil-fired units between 1975 and 1984. Returning to the commissioning scene in 1975, I was struck by the different, more professional approach then adopted by Regionally-based, full-time commissioning teams compared with that which was seen previously as a 'fill in' job for operations engineers. The whole scene had changed to a more structured approach, with better documentation and an insistence on demonstrating that quality standards had been achieved. I became an enthusiast for the new method of working and, in consequence, being asked to describe it was a pleasure.

The authors are colleagues of many years, standing who have experienced commissioning activities over the same timescale as myself. They entered into the task with great enthusiasm and I am indebted to them for their contributions and the manner in which they have tackled a daunting task.

When discussing this volume in the formative stages, the authors agreed that they would attempt to produce a practical book. It was not intended to describe exactly how to carry out every stage of commissioning, but to give sufficient detail to enable the readers to adapt our methods to their plant. It was considered important to indicate why certain activities were performed in order that the implications of ignoring, or including them, could be evaluated. Where possible, quantities of fuel oil, chemicals and other commodities are stated for guidance. Similarly, the times taken for many activities and the manpower involved are also stated. Again, these are related to particular types of plant to assist the readers in translating these items to their particular circumstances.

The volume contains six chapters, the first two dealing with the organisation and planning of commissioning and the next three with the pre-steam to turbine commissioning of the major plant areas. The final chapter deals with commissioning from steam to turbine through to post-commissioning testing.

The three plant chapters deal with common plant, boiler and turbine-generator respectively. The common plant chapter is extensive, ranging from large motors and high voltage switchgear through to small pumps and valves. The boiler and turbine chapters cover all commissioning aspects of these areas from activities such as oil and water systems flushing, through to electronic governor, automatic voltage regulator and condensate system commissioning. The last chapter covers such items as running the turbine for the first time, open-circuit and short-circuit testing, load rejections, performance testing and flexibility tests.

In each, the objectives and background to the test are discussed and a procedure with typical timescales and results is given. Although the review of the first two chapters is left to the end, this does not imply their lack of importance. In many respects they are the foundation on which the approach to the other chapters is based. It is extremely important early in the commissioning phase to be clear as to how this phase is to be approached, recognising any limitations which are imposed and their implications. The methods adopted over a number of years in the CEGB are outlined along with the reasons for change. The structures which have been used to commission power stations successfully are described along with reasons for particular deployments.

The final point is that commissioning does not finish when the plant is fully in service with all tests completed. Plant is taken out of service from time to time for minor and major works and much of the commissioning documentation can be used for the return to service from these activities. It imposes a discipline on the staff and gives a logical, progressive method of returning the plant to service which will save time and money in the longer term. It also produces further benefits from the heavy investment in the original commissioning process.

F. KIRKBY
Advisory Editor — Volume H

Contents of All Volumes

Volume L — System Operation

Volume M — Index

CHAPTER 1

Introduction

1 Readership

This volume is primarily aimed at people who are involved in the commissioning of plant. It has a further role in presenting information to others who may be interested in this aspect of CEGB activities. Throughout, it assumes a basic technical knowledge and will rarely cover technical points in detail unless there is a specific need. It assumes either that the reader is of sufficient technical awareness for explanation to be superfluous, or that the technical explanation is covered in one of the companion volumes or in the standard works of reference. For example, the theory of thermocouples for measuring temperature will not be covered in this volume, but frequent reference will be made to this method of measurement.

2 Basis of experience

All the authors responsible for this volume are employed by the CEGB and have been for many years. So the experience reflected must of necessity be the CEGB side of the commissioning activity. Also, the period dealt with by the authors is approximately the last 20–25 years, covering the change to 500 MW turbine-generators and ultimately to 660 MW turbine-generators. During this interval (1960–1984), the CEGB approach to commissioning has changed significantly. These changes will be covered later, as will the need for such changes.

It is interesting to reflect that the experience of commissioning described in this volume has taken place in circumstances constantly changing plant. Very little of the plant has been an exact replica of what went before, which is not surprising when one considers the technical developments that took place on equipment during the period.

3 Commissioning practice

The practices reflected in this volume are those adopted by the CEGB in conjunction with its consulting engineers and contractors during the last 20–25 years, but with obvious emphasis on the recent past. They have been carried out against quite different contractual backgrounds. Many of the early years were dominated by fixed price contracts and the later years by reimbursable contracts. It is not proposed to deal with the differing merits of these two contractual approaches here — that would fill a separate volume! From a commissioning viewpoint, other changes have been more significant:

- Acceptance of the need for a separate Commissioning Organisation
- Better quality control prior to, and during, commissioning
- Preparation of detailed commissioning procedures prior to commissioning
- Provision of detailed operating and maintenance instructions prior to operation

As a result of the above changes, and others, it was recognised that a significant increase in manpower was needed to commission successfully. This also meant that commissioning was recognised as the means of obtaining sound longer-term commercial operation.

Within this volume, practices in the following areas will be covered:

- Site organisation relationships (Chapter 2, Section 4.1)
- Site commissioning organisation (Chapter 2, Section 4)
- Commissioning documentation (Chapter 2, Section 5)

- Commissioning procedures for all areas of plant (Chapters 3, 4, 5, 6)
- Commissioning safety (Chapter 2, Section 6)

This volume does not explain in detail how every specific piece of plant is commissioned. It does, however, deal with many generic items of plant and systems and indicates clearly the type of approach needed to commission them successfully. The commissioning programme cannot cater for errors in design but, if the approach outlined is adhered to, it is possible to detect some of them before catastrophic failure occurs. The use of these procedures as a discipline has often led to aspects of the design being queried and, in certain cases, to major modifications being carried out. Modifications during construction are undesirable as they often cause delays. It is always a difficult decision to introduce modifications unless the predicted problems are so severe and obvious that there is no alternative. Once the potential problems are recognised, consideration has to be given to proceeding with the existing erection programme and completing the modifications later. This may also involve all parties in recognising either that the plant will not achieve full output, or that it is achieved but with some longer-term loss of life. A typical example of this is the recognition that certain parts of the boiler pressure circuit will run at a higher temperature than they were designed for, thus the achievement of full-load will result in a shorter life for these components.

4 Practical needs

As far as possible, the contents of the volume are arranged so that the reader with a specific interest can find his topic in one place, For example, the person who has the responsibility for organising a site Commissioning Organisation goes to Chapter 2, or the person who is responsible for carrying out a boiler chemical clean to Chapter 4. The needs of each are different, but certain common factors emerge, such as:

- How many people will be needed?
- What background and experience will they have?
- How will they be deployed?
- What other resources are required?

Where appropriate, facts such as numbers of personnel and quantities of documentation produced, fuel oil burned and treated water consumed have been itemised. These have usually been quoted against a plant size and for particular circumstances. Although, from experience it is realised that other people's conditions never exactly fit your own, it is often useful to have some guidelines with which to work. The risk of quoting figures is that it enables other groups to feel superior, e.g., they use fewer people, consume less oil, etc. If that is true, then it is obvious that something should be learned from that group, and the aims of commissioning realigned.

5 Commissioning team development

To appreciate where we are at present, and why, it is necessary to look back over the last 20 years and consider the stages passed through.

At one time commissioning on site was the responsibility of the construction organisation, assisted by the main contractors and the Station Manager's staff. This meant that the main contractors were to put the plant in service at the request of the construction organisation and at some later date the station staff would take over the plant in a running state. In practice, once plant had been initially operated, the station staff became involved in its operation on a regular basis, allowing the contractors' staff to progress with further plant commissioning. This resulted in station staff running all plant outside the main plant area before main plant commissioning.

This approach gradually resulted in an informal takeover of plant commissioning by the Station Manager's staff, with the main plant contractors still accepting the obligation to direct initial operation. The role of the Station Manager's staff in commissioning was recognised by the Board and this was reflected in the Plant Commissioning Procedure (*Green Book*, 1964). This involvement became stronger with the insistence of the application of the Board's Safety Rules to completed and, in some cases, partially completed plant. Normally the only people authorised to carry out the full duties under the Safety Rules were the Station Manager's staff. Consequently they became intimately involved at the stage of completion of erection and virtually controlled operation from then onwards.

This state of affairs persisted into the mid 1960s with station operation staff carrying out most of the plant commissioning duties, assisted by Construction Division colleagues and contractors. It usually took the form of the operational engineer agreeing to apply the working fluid if he was satisfied that certain conditions had been met. These conditions were usually provided verbally by the construction organisation and/or the contractor.

As plant became more complicated, and because a step change in plant size had occurred, the system began to show its weakness. One which had been there all the time started to come to the forefront. This was the conflict which occurred when the first unit was commissioned and other commissioning activities were required to be carried out. It often happened that the running unit took priority over the other commissioning work or that this work was done to a standard lower than required. A further complication was that the people who carried out the commissioning were shift workers and in consequence liaison was extremely difficult. The result of this was the setting up of small commissioning teams, usually from seconded operation staff but not always so.

5.1 The first commissioning teams

The benefits of these small teams quickly became apparent. These were that:

- A dedicated resource for commissioning was accessible on a regular basis
- The number in each team was small enough for the right working relationships to be developed with Construction Division colleagues and contractors' staff
- Consistent commissioning policies could be developed and implemented
- Operations staff were left free to concentrate on optimising plant operation

By using these teams, the early 500 MW unit stations completed their erection and commissioning programmes. During this period many design problems were identified over a wide range of plant and the commissioning teams in one form or another assisted in pursuing these problems to successful conclusions.

5.2 The regional commissioning teams

At this time the organisation of the CEGB was changing and the concept of individual accountability and the need to have a strong engineering function was recognised. This resulted in the centralisation of all power station construction and the setting up of engineering departments within each Region, among a number of organisational changes.

While the role of the Regional engineering function was being considered, it was recognised that a Commissioning Organisation based on the Region had certain advantages. These were that:

- Personnel could be recruited with career prospects in commissioning
- Experience was retained and was transferable
- It was possible to develop common commissioning policies across the Region
- The engineering well being of the Region could be improved if efforts were made to commission plant to predetermined standards

At the same time, the Construction Division was attempting to develop a Commissioning Organisation. This was mainly aimed at standardising documentation and procedures on a national basis, although there were some initial attempts to lead commissioning at site level. As a result of these moves those Regions which had conventional plant either under construction or planned set up Region-based Commissioning Organisations. As the Directors-General are autonomous within their Regions, the organisations set up were not identical but the differences were sufficiently small to be ignored. Figure 1.1 shows a typical Regional organisation. At the same time as these commissioning organisations were set up, it was realised that their

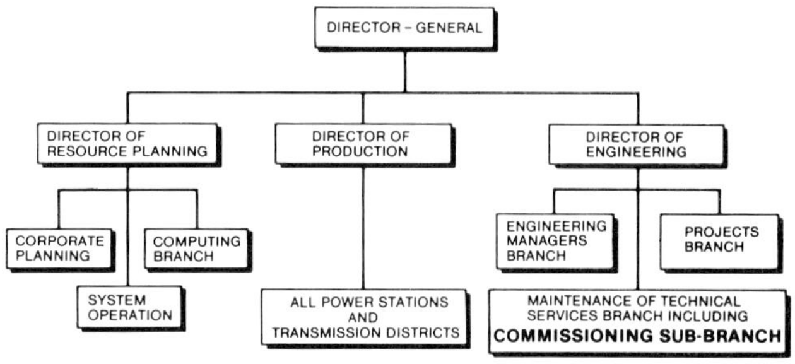

Fig. 1.1 Regional organisation featuring commissioning

involvement would be greater than the small teams, originally station based, for the following reasons:

- Plant had become more complicated
- The teams were to be largely self-contained
- They would undertake all commissioning duties and take plant up to final commissioning stage
- They would undertake all liaison duties with the Construction Division and contractors
- They would prepare all detailed commissioning documentation

In consequence the simple teams of about five people, who were mainly of an operational background, became teams of between 15 and 40 people, depending upon the exact role of the team. Within these larger teams were specialists in mechanical and electrical plant, computers, control and instrumentation and chemistry. With a Regional organisation there was also the opportunity to vary the mix of people, depending upon the workload. The team of 40 people mentioned was associated with the commissioning of the second half of a large generating station and, in addition to the duties listed above, was required to:

- Write all plant operating instructions
- Write all plant maintenance instructions
- Order all plant spares
- Ensure that all the above was consistent with existing station policy
- Minimise any loss of output on the existing station due to commissioning activities on the new station

An example of a station commissioning organisation, with numbers of staff, is given in Fig 1.2.

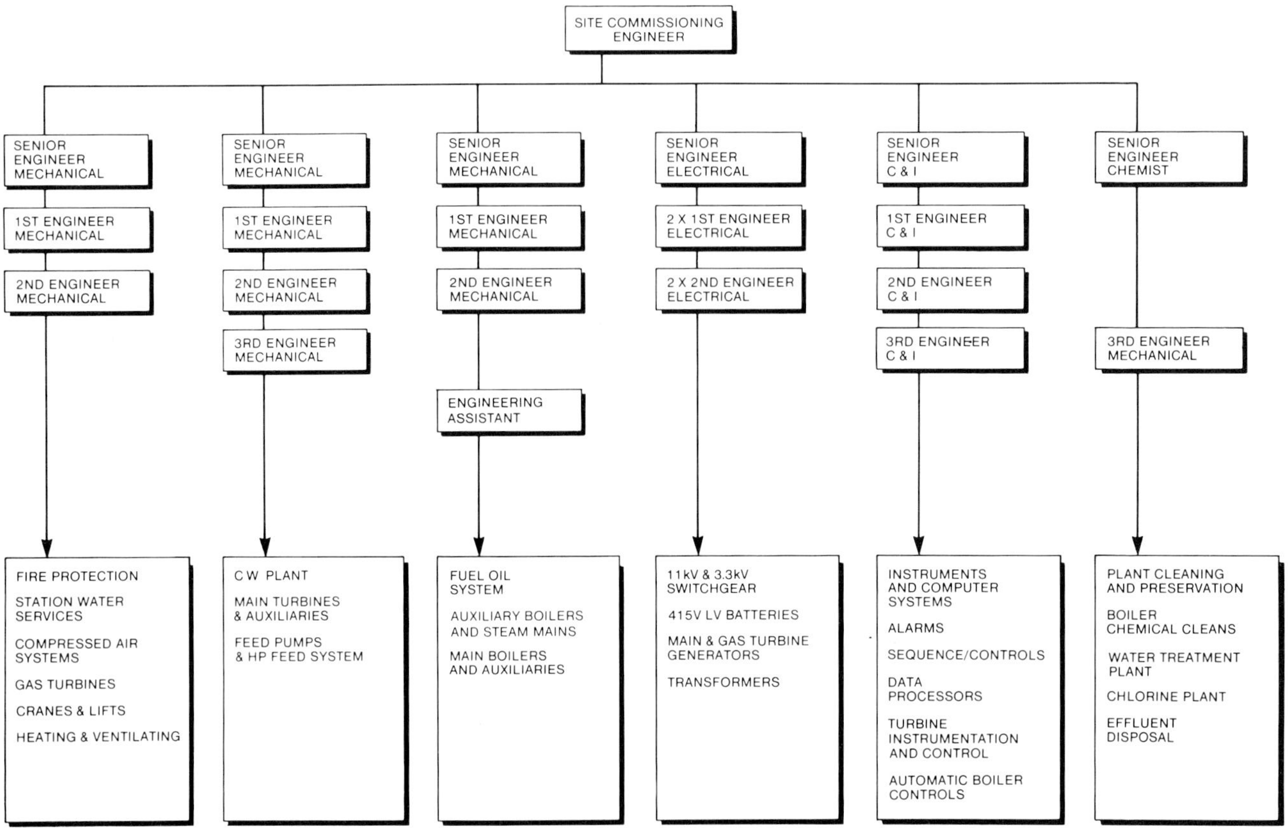

FIG. 1.2 Regional commissioning team organisation at a power station

Within the CEGB's latest commissioning procedure *Plant Completion and Station Commissioning — Procedure for New Power Stations* (1979) the Station Manager is still responsible for ensuring that the station is commissioned properly. In consequence, even though the team is Regionally employed it is usually responsible to the Station Manager.

CHAPTER 2

Principles of commissioning

This chapter will cover the objectives of commissioning, an outline of the organisation for commissioning and the basic building blocks of commissioning.

1 Objectives

The term 'Commissioning' is used in this volume to mean 'the process of taking plant from an agreed state of physical completion to the point where it has demonstrated its ability to meet its design and commercial requirements'.

It is important to avoid the impression that the aim of commissioning is simply 'putting plant to work'. Whereas that might be the initial intention, it is only part of the story. Virtually all installed plant is purchased to a specification, so it should be tested over the full range of that specification. It is essential to do this as early as possible, so that a chain reaction does not set in. For example, if auxiliary cooling systems underperform and this goes unchecked, it could mean that the main plant cannot be fully utilised or tested, with the consequent risk of hidden problems that are not found until much later in the plant life.

1.1 The agreed programme

'Of course one must have an agreed programme', states the constructor, 'which is totally linked with my construction programme'. In the initial stages of the project, time is allowed within the programme for commissioning activities. These times are often unrealistic, being the very best that were achieved on the fourth unit of a previous four-unit station. At an early stage, it is imperative that the Commissioning Organisation has sufficient influence on project management to be able to insist on realistic times from recent experience and, even more important, to be able to insist that these times will still be needed, even if the construction programme slips. Because commissioning activities are at the end of a construction programme, they come under enormous pressure to be shortened if construction has been delayed. The attempt to shorten the commissioning activity must be considered realistically. It is sometimes possible to do this with combinations of increased resources and working long hours. The latter aspect has to be reviewed carefully as it is at this stage that mistakes can be most costly. Resources at the later stage of commissioning involve many specialists and these are not readily duplicated or amenable, for example, to shift working. If the commissioning programme has been properly constructed, it is unlikely that significant reductions can be achieved.

The other factors to take into account when agreeing a programme are the number and build-up of staff, the safety policy prevailing and the standard of commissioning required.

1.2 Commercial operation

The plant must be brought into commercial operation as soon as technically possible, the CEGB definition of this being 'the completion of a 72-hour, full-load run, with all required tests fully completed'. There must also be joint agreement on site that this state has been reached.

Difficulties arise from this objective if the need to have the plant on the system overrides other factors. A typical example of this is the unit which can only reach full load by exceeding metal temperature limits on the boiler. In the past, the extensive definition of boiler metal temperatures, available today, did not exist. In consequence, plant was often operated in a manner which caused premature failure of boiler tubes, so commercial operation was achieved in the short term, at the expense of long-term life. It is therefore important that it is absolutely clearly agreed under what terms commercial operation is to be achieved and that, if limitations exist, they should be written down.

1.3 Plant checks

It is important for the CEGB to know that the plant has been manufactured and erected correctly. The initial reaction to this objective by the constructor or supplier is one of 'it's no concern of the Commissioning Organisation' and, in essence, they are correct. The manufacturer is paid to make plant to given standards and is subjected to quality assurance checks. Similarly the erector's work is supervised for quality. All the Commissioning Organisation is concerned about is that these quality standards have been adhered to. This will take the form of seeing evidence that vital checks, such as alignment figures, clearances, and settings on pipe hangers' have been properly carried out.

These checks are an attempt to ensure that there is a reasonable chance that plant will function without catastrophic failure as working fluids are applied to equipment. It also acts as a back-up should any of these items have been forgotten.

Certain areas where commissioning-oriented checks have been found useful are:

- Flow checking boiler tubes during flushing to detect debris
- Percentage sampling on boiler tubing and headers for use of correct material
- Checking items in storage and awaiting erection to ensure correct preservation

1.4 Special proving tests

Commissioning should establish the flexibility of operation and assess the inherent reliability of the plant. The operational flexibility of plant is usually defined in the purchasing specification, and virtually all fossil-fuel-fired plant over the last 20 years has had the defined requirement to operate in a two-shift mode. Nevertheless, it is important that the commissioning programme covers all the known operating conditions, including those which are not defined within the contract but may be required for internal use, either within the location or by the CEGB. Examples are:

- Start-up capability from hot, warm or cold starts.
- Ability to remain at synchronous speed after being run up from cold or tripped from a fully heat-soaked condition.
- Ability to be started from any temperature condition, using station emergency supplies — 'dead-station start'.
- Ability to follow system frequency changes.
- Use of all modes of governor operation.
- Use of all modes of computer control.

The assessment of the reliability of the plant is more difficult, but clearly if the plant does not get used it is impossible to assess its reliability. This means that when systems are commissioned it is important to use them fully over as large a range as possible to detect shortcomings as early as possible. This, on occasion, will demand a degree of flexible thinking as to how an auxiliary system can be tested without its main item of plant being available, without affecting erection work and without costing too much. Examples of this are:

- Providing an auxiliary steam dump to test an auxiliary boiler prior to main plant completion
- Insisting on having early completion of a single condenser pass to ensure that cooling water pump systems can be tested
- Manufacturing a 'black box' to simulate plant inputs for checking sequence cubicles and computer functions prior to plant completion
- Simulating plant items on an auxiliary cooling system by means of orifice plates to ensure that the pumps and cooling tower systems will work

This approach allows the detection of auxiliary system faults and their correction prior to main plant operation. In that way, one only has to overcome the faults in the main plant! Assessment of the potential reliability of plant during the commissioning phase can rarely go much further than the first part of the classic 'bathtub' curve for reliability. However, if modifications can be identified soon enough, then it is possible to incorporate them in succeeding units before completion of construction.

1.5 Recording and feedback

It is important that any differences between the plant as conceived and the plant as designed, manufactured, constructed and commissioned are identified and fed back to the organisations who need to use such information for future designs. The role of the Commissioning Organisation will generally be restricted to certain areas of this work. For instance, differences in manufacture will usually be detected, agreed and recorded by the Construction Division, but if these differences have an influence on plant operation then the Commissioning Organisation must be made aware of them and a system must be established for this to take place effectively both on site and at the design centre. It will also involve staff employing self-discipline to adhere to such a system and not to initiate or agree to the modifications without following the correct procedures.

1.6 Specification and guarantees

To establish that individual suppliers have met the specifications, it is necessary to be clear which items of plant have contractual guarantees and what are the means for assessing plant against its guarantees. For major items of plant, there is a test code or national standard and it is usually sufficient to ensure that there is time in the commissioning programme for the contractual tests to be done. Nearer to the time when the tests are to occur, a Working Group is set up to ensure they are correctly carried out (see Section 2 of this chapter).

Where either an individual item of plant or a system has to be checked against specification, it is necessary to specify a test to ensure that the item meets its requirements. This can be complicated when items or systems on which the item itself is dependent are not functioning correctly. Under these circumstances, either agreed correction factors have to be applied and the performance assessed, or the performance 'as operated' accepted. This can lead to the acceptance of an unsatisfactory system because it cannot be operated under design conditions. This re-emphasises the need to try to operate plant as early as possible, and so eliminate faults.

1.7 Training

The role of the commissioning staff is one of putting the plant to work and not of operating it on a long term commercial basis. With that in mind the Commissioning Organisation must make provision both in time and manpower to assist in the training of a nucleus of station staff to carry out the routine operation during the commissioning phase and the longer term commercial operation.

This can take the form of lectures, of practical demonstrations on the plant, and of ensuring that relevant operating instructions are available. It is good policy to introduce the operators to the plant as early as possible while providing them with sufficient work of the right type to keep them occupied and interested.

1.8 Engineering data

Engineering data should be produced during commissioning to enable the design organisation to judge which plant improvements are necessary to increase the profitability of the investment and assess likely future operating costs. This has some similarity to recording and feedback, but in fact goes further in certain areas. It is an attempt to highlight the costs involved in problems or in areas where improvements are possible. The nature of the costs incurred may vary from those associated with shortfall in plant output to those due to a change in legislation. It is also an attempt, at an early stage, to indicate to the power station management the likely costs associated with reduced life.

1.9 Plant parameters

There are often occasions when other responsible departments within the CEGB may wish to know about specific parameters of plant. The questions during commissioning are whether the request is outside the contractual requirements, and what the implications are of carrying out special tests. Examples of this are:

- Forms of stored energy testing
- Simulated electrical faults of the main transmission system which subject the generator to stress
- Running a turbine-generator as a synchronous compensator

If the CEGB has sound and substantial reasons for wanting information, then it is usually possible to get agreement to do such tests, even during the early stages of commissioning. This will sometimes involve the Board granting waivers in the case of damage being sustained to certain parts of the plant. The main function of the Commissioning Organisation is to ensure that these requests are rigorously examined and that the appropriate agreements are obtained. They are then responsible for ensuring that a properly conducted and controlled test is carried out.

2 Commissioning organisation

In this section the following areas are considered:

- Responsibilities of Generation Development and Construction Division (GDCD), Transmission and Technical Services Division (TTSD), the Region and the Contractor.
- Logistics of the commissioning organisation.
- Site organisation.

2.1 Responsibilities of GDCD, TTSD, the region and the contractor

In Section 1 of this chapter, the objectives of commissioning were defined and, whereas they are not exhaustive, they cover the majority of commissioning circumstances met within the CEGB. It is therefore important to describe an organisation which is constructed to meet these objectives, which will have to:

- Provide sufficient staff with the right skills, including clerical back-up.
- Develop documentation schemes which monitor progress of all activities relative to commissioning.
- Be involved in, and where necessary draw up, a detailed commissioning programme.
- Ensure clear definition of commissioning areas and responsibilities.
- Define responsibilities between all concerned parties.
- Ensure formal and informal liaison between all concerned parties.

2.1.1 GDCD responsibilities

(a) To ensure that the design and operation of the plant installed meets the requirements of the CEGB Station Development Particulars.

(b) To manage contracts, and to oversee all activities necessary to ensure that contractual obligations and commitments are met in respect of setting the plant to work, testing, commissioning and the meeting of performance guarantees.

(c) To establish the Plant Completion and Station Commissioning Organisation with the Region and TTSD.

(d) To ensure that commissioning Papers of Principle for the main plant and station systems are prepared in conjunction with the Region.

(e) To ensure that contractors' documentation and programmes are available to meet the station programme.

(f) To provide suitable, experienced, technical staff to co-ordinate and plan the work, and to participate in plant completion and station commissioning.

(g) To take over plant which has fulfilled the contractual obligations and to hand it over to the Region.

(h) To ensure that a plant history record system is set up in agreement with the Region, and is maintained during manufacture, construction and commissioning.

(j) To ensure that the contractor is instructed in the use and requirements of the CEGB Safety Rules and the need to comply with them at all times.

(k) To ensure that the contractor complies with all statutory requirements in respect of plant and personnel safety and maintains the necessary records.

(l) To establish the certification procedures for plant energising, operation and taking over by the CEGB.

2.1.2 TTSD responsibilities

(a) To ensure that the design and operation of the transmission plant meets those requirements of the CEGB Transmission Reinforcement Instruction (TRI) which are applicable to the station.

(b) To manage contracts and to ensure that contractual obligations and commitments are met in respect of setting the plant to work, testing, commissioning and meeting performance guarantees for the transmission substation plant.

(c) To establish the Plant Completion and Station Commissioning Organisation for the transmission substation with GDCD and the Region.

(d) To prepare a commissioning Paper of Principle for transmission substation plant in conjunction with GDCD and the Region.

(e) To ensure that the contractors' documentation and programmes are available to meet the substation programme.

(f) To ensure that technical staff are available to co-ordinate and plan the setting-to-work, testing and commissioning of transmission substation plant, and to participate in plant completion and station commissioning.

(g) To take over plant which has fulfilled the contractual obligations and to hand it over to the Region.

(h) To ensure that a plant history records system is set up in agreement with the Region and maintained during manufacture, construction and commissioning.

(j) To ensure that the contractor is instructed in the use and requirements in respect of plant and personnel safety and maintains the necessary records.

2.1.3 Responsibilities of the region

(a) To establish the Plant Completion and Station Commissioning Organisation for station plant and transmission substation plant with GDCD and TTSD.

(b) To establish the Regional requirements for inclusion in commissioning Papers of Principle and in Plant Completion and Station Commissioning Procedures.

(c) To participate in the preparation and approval of Papers of Principle and of Plant Completion and Station Commissioning Procedures with GDCD and TTSD.

(d) To provide operational and technical staff to operate and service the plant during commissioning to meet the requirements of the commissioning programme. Also, to participate in plant completion and station commissioning activities.

(e) To accept, from the appropriate Construction Division, plant which has been satisfactorily tested and commissioned.

(f) To maintain plant history records during the contractual maintenance period and to report to the appropriate Construction Division the behaviour of plant which has been cleared for station operation.

(g) To ensure that the requirements of the CEGB Safety Rules are complied with and to establish and operate the necessary permit-for-work arrangements.

2.1.4 Responsibilities of the contractor

(a) To take all measures necessary for setting plant to work, testing and commissioning and for undertaking maintenance obligations under the contract.

(b) To provide suitably qualified and experienced technical staff to carry out the work and to supervise station operational staff as required.

(c) To provide any other services, equipment or facilities required under the contract for this stage of the work.

(d) To provide procedures, test schedules and programmes for the work to the approval of the Board.

(e) To provide plant description, operation and maintenance instructions and other necessary information for Station Manuals to agreed time scales.

(f) To comply with the requirements of the CEGB Safety Rules.

(g) To comply with all statutory requirements in respect of plant and personnel safety and to maintain the necessary records.

(h) To ensure and maintain the well-being of plant until taken over by the CEGB.

2.2 Logistics of the commissioning organisation

2.2.1 Commissioning staff

Commissioning staff have special problems:

- They have probably moved their family to the area, with all the resulting problems.
- They have to develop new relationships with new colleagues.
- They have to develop new relationships with various manufacturers' representatives.
- They may be under pressure to produce documentation, programmes, etc.
- They may be working in poor surroundings.
- They may be involved in 'power struggles' as certain of the organisations on site endeavour to impose their will.
- They may have difficulty in establishing their identity and imposing control on site.

Later, as the project develops, new circumstances arise:

- The commissioning staff can be seen as 'controlling everything'
- Relationships with colleagues on the station staff can deteriorate as they are seen as doing 'the job' of the station staff

- Plant handed over by them develops problems which the station staff feel that they should have foreseen
- The project is starting to wind down and they become anxious about their next job/position

All these pressures are what individual members of the Commissioning Organisation will feel to a greater or lesser degree during their stay on site. So how does one choose members of a Commissioning Organisation to ensure that, even if these problems are met, commissioning is not affected? Certain points must be considered when selecting individuals, bearing in mind the special needs of a commissioning team:

- People who lead these teams must be able to establish working relationships with their equals to minimise friction at lower levels.
- Members of the teams should be selected for ability to work within a team. This would usually be indicated by their previous record of working in groups. There may be exceptions to this if certain types of specialists are required but, even so, the best specialist is not much use if he cannot work with his colleagues.
- Members of the teams should establish credibility in their own sphere of operations. This may be due to previous direct experience or extensive training in the area of work.

It is therefore useful to consider the type of background from which commissioning staff are selected. Within the CEGB most of the staff have had experience of plant operation, but this varies from large modern plant to considerably older and smaller stations. It is not really the size and date of the plant which is important, but more the staff's attitude to, and understanding of, plant operation. If, however, the commissioning team is to carry plant operational commissioning through to an advanced stage, it is imperative that it has a certain number of people with the background and confidence that is only achieved through direct experience at a senior level.

A commissioning team does use staff with backgrounds other than plant operation, e.g., much of the initial work is directly related to electrical, instrument and mechanical maintenance. It is, however, when the stage is reached of putting systems to work that an operational background is most useful as it incorporates the ability to consider a picture that encompasses all disciplines. As with most staff selection, there is no magic formula; it is important to try to recognise some of the problems and special difficulties that commissioning staff may encounter.

2.2.2 Facilities

If the commissioning organisation is to function efficiently, certain facilities are required. They are:

- Office accommodation — this includes area(s) for meetings
- Administrative back-up
- Canteen facilities — with provision of food for unusual working hours
- Good communications with all groups

At first sight, this list sounds formidable for what is, after all, a transitory group of people. However, in recent years the standard of hired portable accommodation has become very high indeed, and comfortable surroundings can be obtained in a very short time and removed equally quickly.

(a) *Office accommodation* The office accommodation will obviously depend upon the size of the team involved but should comprise a mixture of individual and group offices and meeting rooms. This is important as the group offices tend to foster the team spirit, if grouping by plant area or discipline can take place. An example of an office layout is shown in Fig 2.1. The individual offices

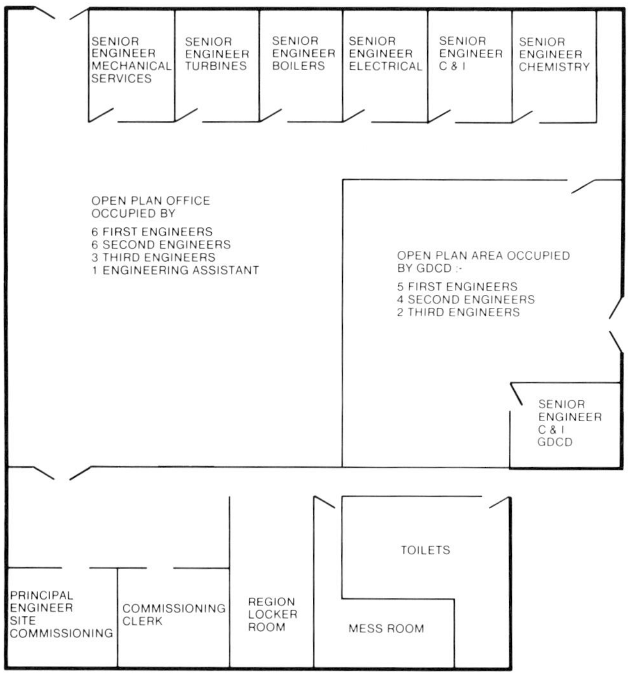

Fig. 2.1 Typical layout of commissioning team offices

are for section leaders who may need a quieter environment to prepare documentation, carry out across-boundary liaison and generally discuss matters with their staff. The provision of meeting room(s) is necessary, as much of the commissioning work is organised on a team basis and regular team meetings are necessary. There is also a need for a meeting room when the wider aspects of commissioning work are to be discussed with groups outside the Commissioning Organisation.

(b) *Administrative back-up* Some people are needed who are not directly involved in the commissioning work. Some simple examples are as follows:

- The commissioning system is largely based on documentation which has to be produced, needing typists and reprographic equipment.
- The complete documentation needs to be kept in an orderly manner, which implies clerical staff.
- A system needs to exist to record the arrival, and dispatch, and possibly the circulation of both on-site and off-site documentation. Again, this is a function of clerical staff.
- Where a group of staff is located, there are certain returns to be made on a monthly, quarterly and yearly basis.

It has been argued that all these tasks are well within the competence of most engineering staff. It is common experience that most engineers are bad typists and, whereas they can keep reasonable records of their own work, across-boundary work is usually poorly treated. It is also false economy to use their skills for this purpose: the provision of reasonable clerical back-up, sometimes on a joint basis, eases that part of the workload where engineering enthusiasm is low and allows greater concentration on the more relevant areas. As guidance, with a commissioning team in the range 15–20, who are to prepare all the commissioning documentation and to provide adequate storage and retrieval facilities, three or four clerical staff will be required. At least two of these must be competent typists, and of the others one should be able to type to overcome peak workloads and absences. With a small clerical team, flexibility is essential.

With this administrative back-up must go the provision of equipment. This will normally include typewriters (usually electric or electronic), reprographic equipment (including the ability to collate and bind documents) and filing cabinets. The last often occupy a lot of space and serious consideration should be given to microfilming, with appropriate retrieval systems. The space occupied by completed commissioning documentation at a 5 × 660 MW oil-fired power station is typically 40 metres of shelving and involves 600 volumes.

(c) *Canteen facilities* Canteen facilities do not present problems on most sites as they are normally provided for the construction labour force and are more than adequate for commissioning staff. If a facility to obtain food for out-of-hours working can also be set up, then this can improve the conditions under which the commissioning staff work. This will usually be in the form of sandwiches, where work has to be undertaken over normal meal times and other provisions during weekend and/or night working.

(d) *Communication system* Good communications is something that is widely quoted today as the cause or the cure of all our ills in society. Here, the term 'communications' is used in its most direct sense, i.e., means of communication must exist between all the parties concerned in the commissioning activity. This will normally take the form of a site-based private automatic branch exchange (PABX) system. The usual development of this system is:

- Initial site PABX installed by the construction organisation covering all on-site offices, with the facility to have direct contact with construction design offices.
- Installation of station PABX systems which then have to be linked to the construction PABX, so that, as construction is completed, plant areas lose their site PABX and gain the station PABX.
- If construction work is protracted, or if certain contractors remain on site, they can be included in the station PABX system, thus minimising the tie-line traffic and consequent risk of bottlenecks.

In addition to the PABX system, most organisations on site will have external means of communication via the national telecommunications network. With the increasing use of telex and facsimile transmissions it is essential that provision is made for this type of installation. The use of computers or computer terminals on site which are directly involved in construction and commissioning activities is widespread and their special requirements with regard to accommodation and personnel have to be considered.

For many commissioning activities, direct point-to-point communication is important, and for this UHF radio telephones are required. This service may utilise the permanent installation for the station, the emergency system installed by the construction organisation or a separate system purchased by the commissioning organisation. If the best use of personnel is to be obtained it may also be necessary to have a staff alerter or 'bleeper' system, operated via the PABX system.

3 Basic steps in commissioning

In this section will be covered some of the basic points of commissioning and how they are integrated into the overall programme. They are discussed in the sense of why they are chosen as either discrete identifiable activities or points from which other activities start. They are given in the order which they normally occur.

3.1 Completion of erection

This is normally considered to be an easily identified point and, when reached, implies that further activities may take place. However, is it so easily identified? Interpretation of 'completion of erection' of even a simple item like a pump/motor unit can vary considerably. Consider these two situations:

(a) Pump and motor mechanically complete. Electrical supplies available from temporary supplies. Working fluid available through previously commissioned pipework. Access poor.

(b) Pump and motor mechanically complete. All labels fitted. Connected to permanent supplies with all protection tested. Working fluid as in (a) and permanent access fitted.

It is probably fair to say that the pump/motor unit would operate equally well in both circumstances, although with some increased risk of trip and personal injury in case (a). The important point illustrated is that it is no use just stating 'completion of erection'; it must be clearly and unambiguously defined.

3.2 Establishing cleanliness

Establishing cleanliness can be described as an erection activity and in part that is true. It is current practice to indicate in what state pipework, motors, pressure vessels, etc., will arrive on site. Depending upon proposed and achieved erection dates, the items may require degreasing, or to have protective coatings stripped off, and in some cases re-applied. These activities should be viewed in the overall context of achieving rapid commissioning; then if the plant is cleaned before erection and is left clean during erection, the amount of post-erection cleaning can be kept to a minimum. For this reason, the Commissioning Organisation must be involved in monitoring prior to erection. A typical example was lubricating-oil pipework which was cleaned prior to delivery to site but was left exposed during construction delays. The programme to clean it up *in situ* using flushing techniques was long and, in certain cases, the standard required was impossible to achieve. This led to the pipework being stripped out, sent away for re-cleaning, and re-installed. All this occurred at a time when 'erection' was complete and the delay was attributed to commissioning!

The Commissioning Organisation must have an involvement with the site quality assurance section to know which systems will require extensive flushing and which should be inherently clean in erection. It is also incumbent upon the Commissioning Organisation to have decided in advance how systems are to be flushed and, if special equipment is required, to ensure that it is obtained in good time. On the major items of plant, the manufacturers are usually quite specific on the flushing and cleaning required, but in the case of auxiliary pipework, compressed air systems, town main services, etc., the Commissioning Organisation will have to decide the methods and acceptance criteria.

3.3 Static checks

These checks are carried out before systems and items of plant are put into service. Examples of these are:

- Physical examination of the plant — to see if it is all there!
- Stimulated tests on sequence, computer and alarm systems.
- Manual operation of plant, such as valves, dampers, etc., to check physical free movement and whether other equipment, like limit switches, operates.
- Examination or checks on clearances and alignment.

It is normal practice for these checks to be specified in the associated commissioning documentation. The checks, plus others, will be stated and will normally require confirmation of exact state before a decision is taken to proceed.

3.4 Admission of the working fluid

The term 'fluid' is used here in a wide context. It can mean gas, steam, water, air or electricity. This stage implies that the equipment has had all agreed checking carried out to a state where it is considered safe to admit the working fluid. The documentation should be specific about what should have been carried out prior to this stage being reached.

Sometimes the admission of the working fluid may do nothing more that raise the pressure or temperature, or both. Sometimes, the equipment may start to move and, in other instances, the risk of flammable or toxic gases escaping may increase. It is therefore important to have considered the following points:

- What is the source of the working fluid and how is it controlled?
- What are the implications of admitting the working fluid (a) if everything is normal and (b) if circumstances are abnormal?
- How do you monitor the admission of the working fluid to detect abnormalities?

It is obviously important to try to admit the fluid for the first time through a positive control device, e.g., valve or switch. If problems occur, quick removal of the fluid can be important. Time spent considering the effects of admitting the working fluid under both normal and abnormal circumstances is well spent because it can prevent major incidents occurring from minor lapses. This means that the provision of fire fighting equipment, anti-flooding equipment and communications has to be considered. The monitoring of the admission of the working fluid can easily be covered by visual observation for leaks in simple cases, progressing up to 24-hour (and longer) pressure-decay tests in more demanding environments or the provision of protective devices operated by flow, current or pressure.

3.5 Application of the CEGB Safety Rules

There is a time when equipment is ready for use and it is decided that it must come under the CEGB Safety Rules. The great debate is about when this point has been reached. The constructors usually see the imposition of the Safety Rules as a hindrance to their progress and, in certain areas, they are very difficult to implement. The implementers of the Safety Rules are involved in a great deal of effort if the plant comes under the Safety Rules too early.

A general simple rule is that the equipment must be covered by the Safety Rules before it is energised or the working fluid is admitted for the first time, but certain problems arise from this. Take, for example, the need to flush a lubricating oil system on an item of plant which is mechanically complete but has not yet been cabled. To progress work, it is not unusual for temporary supplies to be applied to the motors for the flushing. In these circumstances, one either insists that the constructors take the basic precautions about developing 'safe systems of work' and allow them to progress the job, or the plant is brought under the CEGB Safety Rules. The disadvantage of the latter approach is that quite often temporary switchgear may not have the type of facilities demanded by the CEGB Safety Rules, so it is not always possible to meet the CEGB requirements. A further simple rule is that if the motor is fed from its normal supply, then it should come under the Safety Rules. The pipework/tank, etc., may then be treated separately. It would be left for the constructor to use his own 'safe system of work', which would include direct liaison with implementor(s) of the CEGB Safety Rules.

To put systems or part-systems under the CEGB Safety Rules does imply that the necessary isolating points and equipment exist. This must be checked during erection and remedied, if necessary.

3.6 Initial operation

This action normally takes place after the equipment has come under CEGB Safety Rules and it is witnessed by representatives of all interested parties. It is usually the result of accumulated efforts and is the opportunity of assessing the system or item for compliance with the contract.

It is important *before this stage is reached* to have agreed under what circumstances initial operation will take place. For example, it may be acceptable for a small motor-driven auxiliary to be operated for the first time with no remote alarms, but this could be totally unacceptable for a large system like the initial run on an electrically-driven boiler feed pump.

As usual, compromises have to be reached on what is essential and what is desirable. It is usually sensible, when drawing up a 'state of plant requirement' before a particular piece of plant is run, to decide which items are essential and that the plant will not be run, under any circumstances, unless the requirements for these items are met. It is also essential to aim higher than that as much of the other equipment, such as alarms and indicators, has not been provided just for amusement. Once again, this tends to be a pressure area when the constructors can see that the plant is almost ready to run and is only being held up for such alarms or indicators. If the essential requirements have been decided properly, then there can be room to compromise on some of the other less essential areas. For example, it may be acceptable to use a local observer at a remote point to signal back should that be necessary and no remote indicator available. At the same time it should be pointed out that the plant will not be operated unless that observer is present until the remote facility is provided.

3.7 System checks

When a system or sub-system has completed its initial operation, it has then to be checked to see if it has the correct response to operational circumstances. Sometimes, e.g., where main plant operation is required, it may be necessary to simulate inputs to cause an initial response of the system. A typical example is a control valve when it is first operated in its working fluid. Prior to that time all the static checks will have been carried out to check its travel, freedom of movement, initial settings on controllers, etc. If its working fluid is admitted under normal working conditions, then adjustments can be carried out until the correct operation has been achieved. If the fluid is not at normal working conditions, then certain simulations may have to be carried out to enable appropriate settings to be obtained. This will normally shorten the final commissioning time on this item.

Frequently a system is part of a much larger system, e.g., items of automatic boiler control; then, as the other parts of the overall system become available, simulated checks are replaced by 'live' checks. Much useful information on control systems has been obtained by computer simulations of plant and the checking of control systems against known plant responses. This has enabled settings to be obtained for plant in advance of testing.

3.8 Performance testing

The decision as to whether the plant meets its major contractual and design criteria is taken on the results of performance tests. On the major items of plant, these are well defined in standard test codes and it is only necessary to agree the logistics of carrying out such tests. In most cases the CEGB duplicates the manufacturer's instrumentation and virtually carries out an independent test while witnessing the manufacturer's test. On the major acceptance tests, it is essential to have personnel involved who are test 'professionals'. This usually implies people who regularly conduct these large-scale tests and are thoroughly familiar with test code procedures. The debate after such tests can be long and it is not unknown for repeat tests to be called for.

It is not a good policy to restrict testing only to that required for contractual matters. It is also important that overall plant operation and performance is established in as many modes of operation as possible, typical examples being:

- Operation with HP feed heaters in and out of service.
- Operation with electric feed pumps only in service.
- Operation, where possible, with only one condenser in service.
- Operation with different combinations of boiler fans in service.

The reason for some of these modes may not be obvious but it is a fact that, throughout the life of a station, many of these unusual combinations will occur. If these tests have taken place, limits will have already been established, as will the cost penalties. This information can be useful in establishing alternative courses of action.

Performance testing can also be useful in establishing such matters as transient performance of cooling water systems employing multiple pumps, vacuum raising equipment, water recovery systems and auxiliary cooling systems. As an example, there can often be some debate about poor vacuum on a turbine, but if figures for cold vacuum pull on completion of erection have been established and the capability of the vacuum raising plant is known, it is then fairly easy to decide whether the poor vacuum is due to low vacuum raising equipment performance and/or air leaks on the system.

4 Site organisation for commissioning

Having defined the objectives of commissioning, the steps that a Commissioning Organisation will need to take to meet these objectives, and the detailed responsibilities of the four main parties involved in the commissioning activity, it is now necessary to describe the organisation on site to put all these into effect.

The basic staff tree used for a station-based Regional commissioning team was shown in Fig 1.2. Similar staff trees for station and GDCD site organisations are shown in Figs 2.2 and 2.3. The latter organisation, however, is not just for commissioning but will include numbers of staff for construction purposes, although some of these staff will have commissioning responsibilities. In the case of the Regional staff, they will have responsibilities for taking over plant, maintaining it, operating it and generally ensuring its commercial use. With these four organisations on site it is clear that regulation of activities is required and this is carried out by three 'commissioning committees' and a 'construction meeting'. The relationships, obviously simplified, are shown in Figs 2.4 to 2.7; they will be examined first, followed later by discussion of the defined roles of the committees and the need for such committees.

4.1 Relationships via committees

It may be asked 'why do we need a committee structure to commission power plant?' If it was only a limited operation on relatively uncomplicated plant, it would probably not be needed and an informal gathering of the interested parties would probably suffice. However, the commissioning of power stations involves many actions over a wide range of plant and equipment with a very real need to co-ordinate these activities. This co-ordination alone implies that the people involved, or their representatives, should meet together at regular intervals. Having decided that there is a need for people to get together to attempt to co-ordinate their activities, why should three committees or panels be required? (Figs 2.4 to 2.7). First one has to consider the responsibilities outlined in Section 2.1 of this chapter. These imply in the broadest terms that it is GDCD responsibility, along with the contractor, to build the plant to a time and a price. The Region ensures that it is commissioned correctly and it is the responsibility of all three parties to ensure that the plant performs correctly. With this broad separation of responsibilities, the provision of separate committees is the most effective way of fulfilling them. When the detailed terms of reference are considered later, it will be realised that the Testing and Commissioning Panel is really the implementer of the instructions of both the Plant Completion Committee and the Station Commissioning Committee.

From the relationships shown in Figs 2.4 to 2.7, the Site Progress Meeting (Fig 2.4) is the main forum for the contractor. At this meeting the details of construction progress are considered. Here, the role of the Region and TTSD is generally one of being updated on construction progress in order that they may be able to assess forthcoming commissioning requirements. In practice, this meeting can be used as a clearing house for difficulties foreseen by the commissioning staff in relation to the construction work or work force.

At the other meetings, the contractor is not excluded and can be involved at the discretion of the committee members. This approach is adopted because the contract to carry out the work is between the contractor

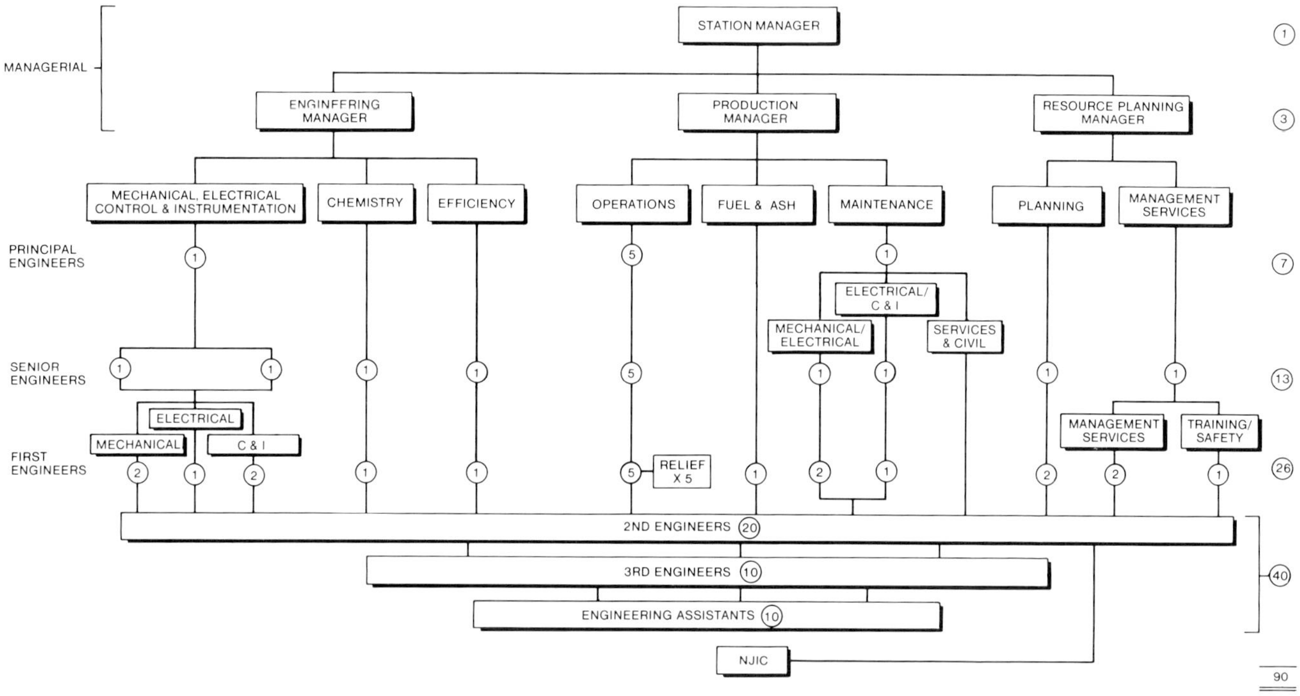

Fig. 2.2 Power station staff tree

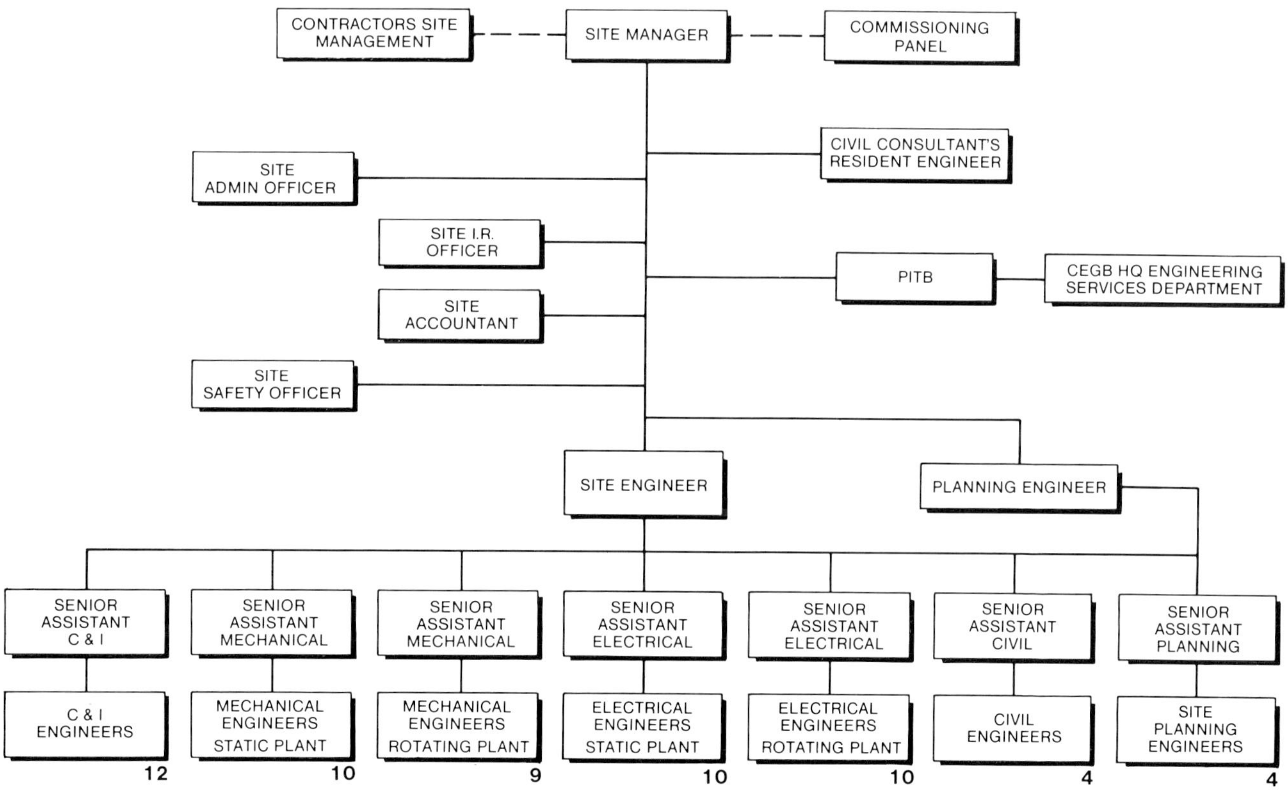

Fig. 2.3 GDCD site staff tree

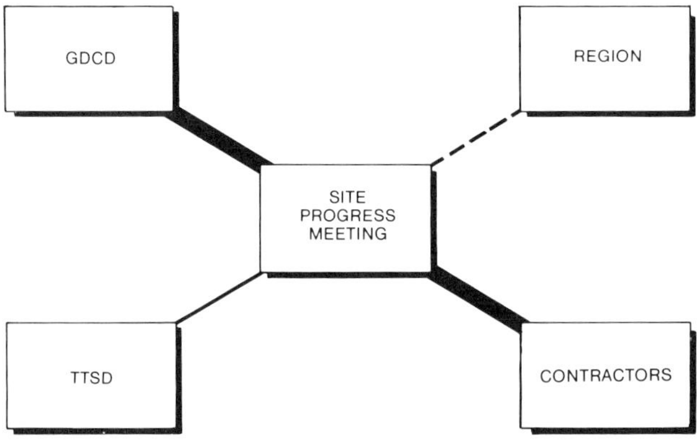

FIG. 2.4 Interaction for site progress meeting

and GDCD, or TTSD, and it is desirable to avoid circumstances where another group or committee may become involved in discussions which could have contractual implications for which they have no authority. There is, however, a great deal of informal discussion with the contractor and his staff on a wide range of topics concerning the commissioning of the plant. In their areas they have considerable experience and they make a valuable contribution to any Commissioning Organisation.

This volume is primarily concerned with commissioning, so it is not intended to discuss further the site progress meeting as this is essentially a construction activity. The two committees and the panel will now be discussed in more detail.

4.2 Plant completion committee

This committee (Fig 2.5) meets under the chairmanship of the GDCD Site Manager and will have the following membership:

- Site Engineer (GDCD)
- Project Manager (GDCD)
- Station Manager (Region)

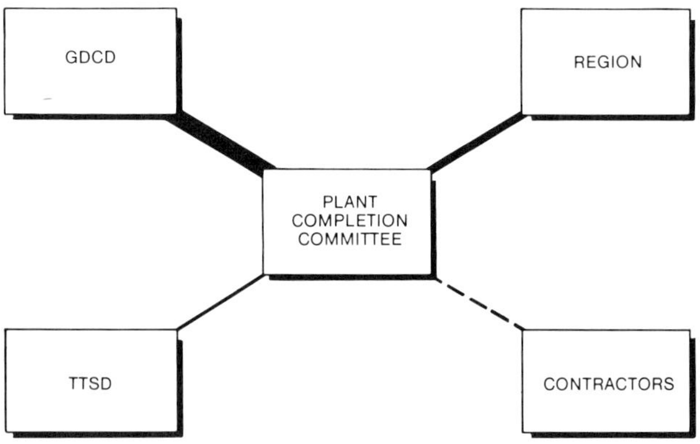

FIG. 2.5 Interaction for plant completion committee

- Station Commissioning Engineer (Region)
- Project Manager (TTSD)

In addition it has the power to co-opt other CEGB staff or contractors as appropriate. Its terms of reference are:

(a) To call for and accept programmes, schedules and instructions for the carrying out, completion and recording of the necessary mechanical, electrical and functional tests on each plant item and to co-operate in these activities with the Station Commissioning Committee and the Site Progress Meeting.

(For this purpose, mechanical, electrical and functional tests are defined as 'those tests, both static and running, on a plant item to prove satisfactory manufacture, installation or construction, so that the plant may be considered available for commissioning tests in association with other plant, or for operation').

(b) To ensure that information, materials, services and personnel are available as required, so that the work is carried out stage by stage under complete control and without hindrance.

(c) To ensure that at all times during the work the procedures being carried out will secure the safety of personnel and plant in compliance with the CEGB Safety Rules and all relevant statutory requirements.

(d) To ensure that information arising from plant completion activities is recorded, correlated and issued to the CEGB departments concerned and appropriate contractors.

In order to fulfil these terms of reference, it is necessary to divide the plant into packages. These packages initially will be large, e.g., the boiler. But this subsequently has to be divided into systems and ultimately into sub-systems in order to define individual test requirements. A list of main plant areas is:

- Unit
- Boiler and auxiliary systems
- Turbine condensate and feed heating plant
- Generator, unit and generator transformers
- Fuel handling and storage
- Water treatment plant
- Main cooling water and auxiliary cooling water systems
- Station auxiliary systems
- Gas-turbines
- Electrical power and services
- Control and instrumentation
- Civil, site and building services

A list of sub-sections of boiler and auxiliary systems is as follows:

- Induced draught and forced draught fans
- Air heaters
- Burner and combustion equipment
- Coal milling plant
- Sootblower system
- Chemical dosing and sampling
- High pressure pipework (steam and feed)
- Automatic boiler controls
- Gas treatment plant
- Boiler hydraulic tests
- Boiler air flow and leakage test
- Initial fill, fire and chemical clean
- Steam purge and float safety valves
- Protection and inter-trip equipment

To enable the Plant Completion Committee to function, Working Parties are set up to cover the main plant areas. The mix of people in these Working Parties and their terms of reference is discussed in Section 4.5.1 of this chapter. The various Working Parties draw up a schedule of tests to be completed either prior to first running the plant or before the system can be handed over for inclusion in a further test. An example of the latter is the testing of alarm and indication systems prior to the running of a main auxiliary.

It is important that the Plant Completion Committee is carefully integrated with the construction programme and that the sequence of testing fits in with these activities. This committee therefore puts on a formal basis the responsibility of the Site Manager to take notice of and agree to the work that needs to be done before plant can be operated. This may sound obvious, and many site managers would claim that this is what they always did, but it was not unknown for completion of plant to take precedence over any prior testing which was required. What is also important is that it recognises that resources have to be made available to carry out this work. It is no use getting the Site Manager to adopt this approach, only to see his construction programme jeopardised because there are not enough people to carry out the work.

The presence of Regional members on this committee is important because ultimately the Station Manager is responsible for commissioning. This committee functioning correctly should ensure that plant is handed over to the 'Commissioners' in a tested or known state. The Regional members have a strong interest in reaching agreement with the Site Manager on which activities will be included in 'Plant Completion' and which are 'Commissioning'. The category itself is arbitrary, but what is important is that the state of the plant is formally known and recorded.

Further important requirements of this committee are to ensure that the testing work is carried out safely and where appropriate under the CEGB Safety Rules. In addition, the information gathered from testing has to be made available to a wide range of people to check designs, to remedy mistakes, and to avoid problems on one system being passed on to another.

4.3 Station commissioning committee

This committee (Fig 2.6) meets under the chairmanship of the Regional Station Manager and has the following membership:

- Power Station Departmental Head } Region
- Station Commissioning Engineer } Region
- Transmission District Manager } Region
- Site Manager } GDCD
- Site Engineer } GDCD
- Project Manager } GDCD
- Project Engineer } GDCD

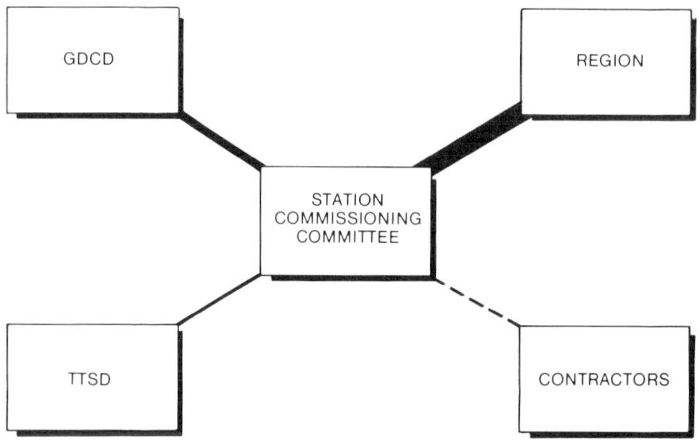

Fig. 2.6 Interaction for station commissioning committee

In addition, it has the power to co-opt other CEGB staff or contractors, as appropriate. Its terms of reference are:

(a) To call for and accept programmes, schedules and instructions for the commissioning of the complete station in co-operation with the Plant Completion Committee. For each plant item, the completion of the mechanical, electrical and functional tests is taken as the starting point.

(b) To ensure that information, materials, services and personnel are available, so that the work is carried out stage by stage under complete control.

(c) To ensure that at all times the procedure being carried out will secure the safety of plant and personnel in compliance with the CEGB Safety Rules and all relevant statutory requirements.

(d) To ensure that information, arising from station commissioning activities, is recorded, correlated and issued to CEGB departments concerned and appropriate contractors.

(e) To approve organisational staff trees, working party representation, and details of day staff and staff engaged on permanent or temporary shift patterns of working. Also, to ensure that the qualifications, experience and numbers of contracts staff are adequate for commissioning main items of plant, as decided by the committee.

Terms of reference (a) to (d) are similar to those for the Plant Completion Committee. This does not mean that one committee should do both jobs, but rather that it recognises the real differences between the prime responsibilities of the GDCD Site Manager and the Regional Station Manager. This similarity in terms of reference also indicates how interdependent are the two activities of plant completion and commissioning.

It would be possible to combine the functions of both committees if one of the organisations was totally responsible for plant completion and commissioning, but this would then require total rethinking on how we arrange to build and operate power stations. This approach has not been adopted within the CEGB, but it does not mean that such an arrangement would not work under a different organisation.

The last term of reference (e) is important because it clearly requires this committee not only to decide how commissioning is to be organised overall but also to indicate the need for shift work and day work. It also requires them to ensure that contractors' staff are of the calibre necessary to perform the tasks allocated to them, which can sometimes cause difficulties. Questions of qualifications and suitability are best approached through the GDCD organisation, as the contract is between GDCD and the contractor. There is normally a clause in the contract obliging the contractor to supply staff to meet this. It is then the duty of the contractor to demonstrate that the people he will provide:

- Are sufficient in number
- Have the right technical background
- Have the correct experience

It is usually in his interest to do all these things in order to ensure that his plant is commissioned safely. However, experience suggests that he will normally only be able to supply large numbers of staff for short periods owing to the associated costs; thus a predictable commissioning timetable is very much to his liking, as it enables him to control costs in this area. The other problem that the contractor often faces is the demand for his more specialised resources to be in several places at the same time, especially with electronic governor, automatic voltage regulator and turbovisory experts. If movement of staff occurs frequently to and from the site, then working relationships and continuity can be damaged. (This changing of staff on site applies equally to all no matter who is employing them.) It is, however, a fact that CEGB staff are there for a longer period owing to their continuing commitment, whereas often the contractor's staff may only be on site for certain specific tasks, with gaps between. It is not unreasonable for these people to be away from the site during the gaps, as long as they can be recalled to fulfil their obligations. So the planning of commissioning means scheduling not only plant, but people as well. It is no use expecting a man to appear at short notice, as it has to be assumed that in the gaps his employer wishes to gain the benefit of his services in a profitable way. So be warned: if you leave the contact too late, do not be surprised if the man you want is in Hong Kong or South Africa, and unavailable.

This raises the important general issue of understudies: staff involved in a particular area should be encouraged to be aware of related work undertaken by their colleagues. In this manner, the loss of one person has only a temporary effect on the output and effectiveness of a team.

The need for Working Parties as outlined in the discussion on the Plant Completion Committee is equally valid for this committee. Figure 2.8 shows the relationship between both committees and the Working Parties. It is seen that the Working Parties are the same for both and in practice they have responsibilities for producing both test and commissioning procedures. This is reasonable as much of this work has to be fitted together to meet the complete commissioning objective. The details of Working Party representation are dealt with in Section 4.5.1 of this chapter.

4.4 Test and commissioning panel

The reason for designating this as a 'Panel' (Fig 2.7) is that it has no executive authority but exists to implement the day-to-day decisions of both the Plant

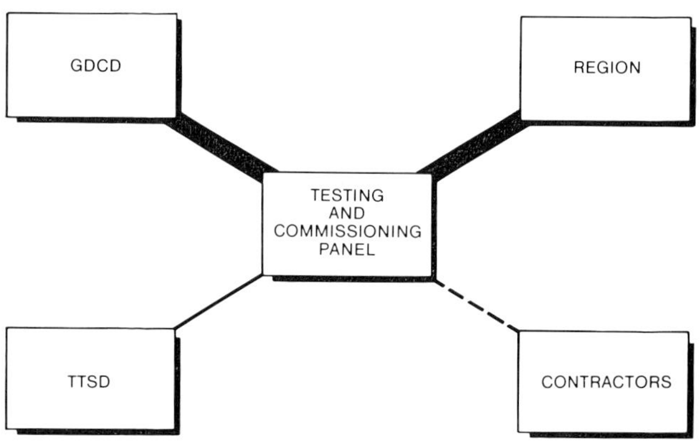

FIG. 2.7 Interaction for testing and commissioning panel

Completion and the Station Commissioning Committees. As the two committees are not in regular session, probably meeting no more frequently than monthly, it is necessary to have a locally-based meeting of interested parties to ensure that the actions placed upon them by the committees are discharged on a day-to-day basis. Figure 2.8 illustrates the relationships. The panel is not empowered to change, other than in minor detail, anything which has been agreed by the two committees. It normally consists of:

- Station Manager
- Site Manager
- Station Commissioning Engineer
- Site Engineer

In practice most of the business is transacted between the Station Commissioning Engineer and the Site Engineer, with the other two members of the Panel being involved only on important issues. The terms of reference are:

(a) To ensure that the setting to work, testing and commissioning is carried out in accordance with the programme, methods and approved procedures of the Plant Completion Committee and the Station Commissioning Committee.

(b) To report progress, results and problems to the appropriate committee.

(c) To authorise any detailed changes necessary to the approved procedures to enable the work to continue. Where changes to principles are involved, these shall be referred for approval to the Plant Completion Committee or to the Station Commissioning Committee before implementation.

(d) To nominate Working Party leaders and membership of Working Parties for each plant area to the Station Commissioning Committee for approval.

(e) To approve nominations from the Working Parties of test team leaders and membership of test teams for each test or series of tests.

From the terms of reference, the reporting of progress, results and problems to the appropriate committee can be a time consuming and sometimes frustrating task. Two main approaches have been tried:

(a) The Site Commissioning Engineer summarises progress, results and problems over all the Working Party areas. This is done in the form of either a written or a verbal report to the Committee. *Or*

(b) The Working Party leaders are invited to the committees to report in their area.

The latter has the advantage that first hand experience is injected into the deliberations of the committee. However, it suffers from the severe disadvantage of it not being possible to predict exactly when people are required to attend the meeting, which can sometimes result in a queue of people waiting their turn to brief the committee. It is possible to control this by agreeing time bands.

Using a spokesman to summarise all Working Party activities can mean that certain of the detail is lost, but the Working Party leaders are allowed to continue working. On balance, it is preferable to use the spokesman approach, but to supplement it if there are particular problems in certain areas.

The Panel also has the role of recommending Working Party leaders and Working Party membership to the Station Commissioning Committee. As the Working

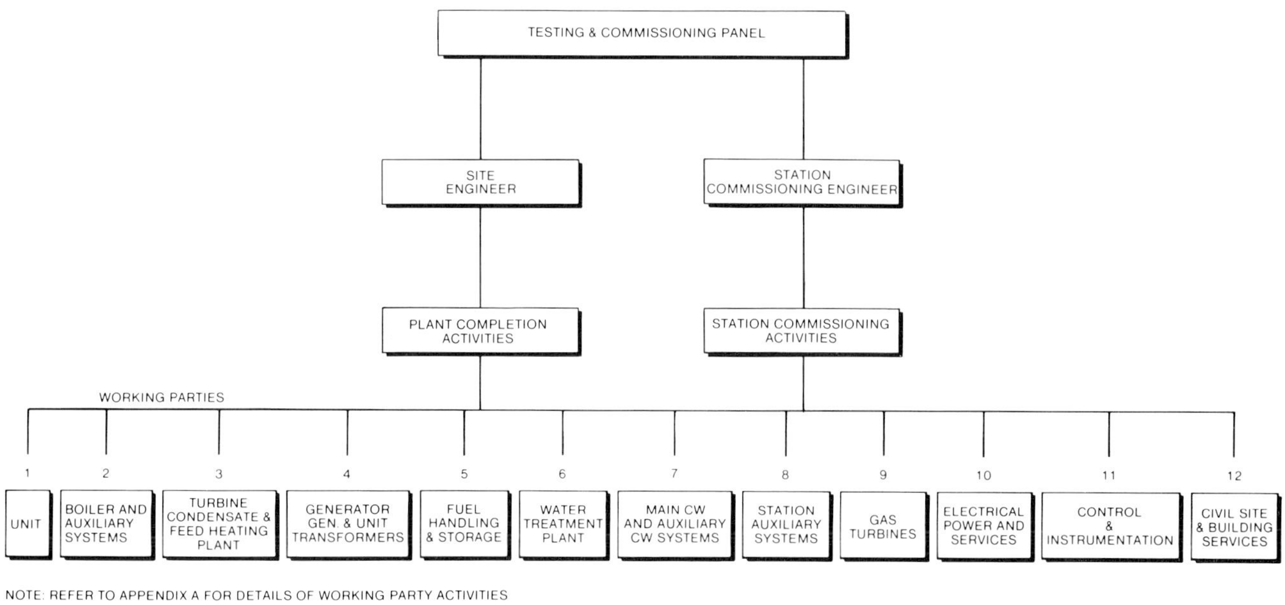

FIG. 2.8 Commissioning management and co-ordination

Parties are the same for both Plant Completion and Station Commissioning Committees, the needs of both groups have to be carefully assessed. There is usually some bargaining to ensure a reasonably even spread of duties between both GDCD and Regional staff.

4.5 Working parties and test teams

4.5.1 Working Parties

Working Parties are established by the Station Commissioning Committee based on recommendations made by the Testing and Commissioning Panel (Fig 2.9). The terms of reference and representation of Working Parties are as follows:

(a) To develop the necessary Plant Completion and Station Commissioning Procedures (see Section 5.4 of this chapter) to meet the requirements of the Papers of Principle. These procedures are approved by the Plant Completion Committee or the Station Commissioning Committee, as appropriate.

(b) To plan and progress the work and ensure that setting-to-work, testing and commissioning is carried out in accordance with these procedures.

(c) To nominate test team leaders and membership of test teams for each section of the work to the Testing and Commissioning Panel for approval.

(d) To report progress, results and problems to the Testing and Commissioning Panel.

(e) To prepare the necessary reports.

The representation on Working Parties is as follows, dependent upon local circumstances:

- *GDCD* Usually a site-based engineer, but on occasions he can be supplemented by a GDCD engineer.
- *Region* Usually an engineer from the Regional Commissioning Team, but he can be an engineer from the station production department.
- *Contractor* Usually a site commissioning engineer or site agent.
- *TTSD* A Regional transmission district engineer.

Working parties with this representation can be seen to be very useful working groups, as they bring together all four main participants on site. They have to function correctly if progress is to be made and the selection of people for them is an extremely important and sometimes delicate task. If there is one area where team work counts, this is it. From time to time, mistakes are made in selection, but any such mistakes *must* be remedied quickly if damage to relationships between the main parties is to be minimised. As stated earlier, in personnel matters there are no magic formulae — only judgement and experience.

The representative from GDCD may be the most senior man for a major plant area, or somebody more junior for a lesser plant area. While it can be argued that all plant areas are important, the main role of GDCD is completing erection and they have to balance staff numbers on Working Parties against their other needs. A senior member of GDCD staff allocated to a working party will usually be nominated as chairman.

Representation from the Region may be a compromise but, if there is a strong Regional Commissioning

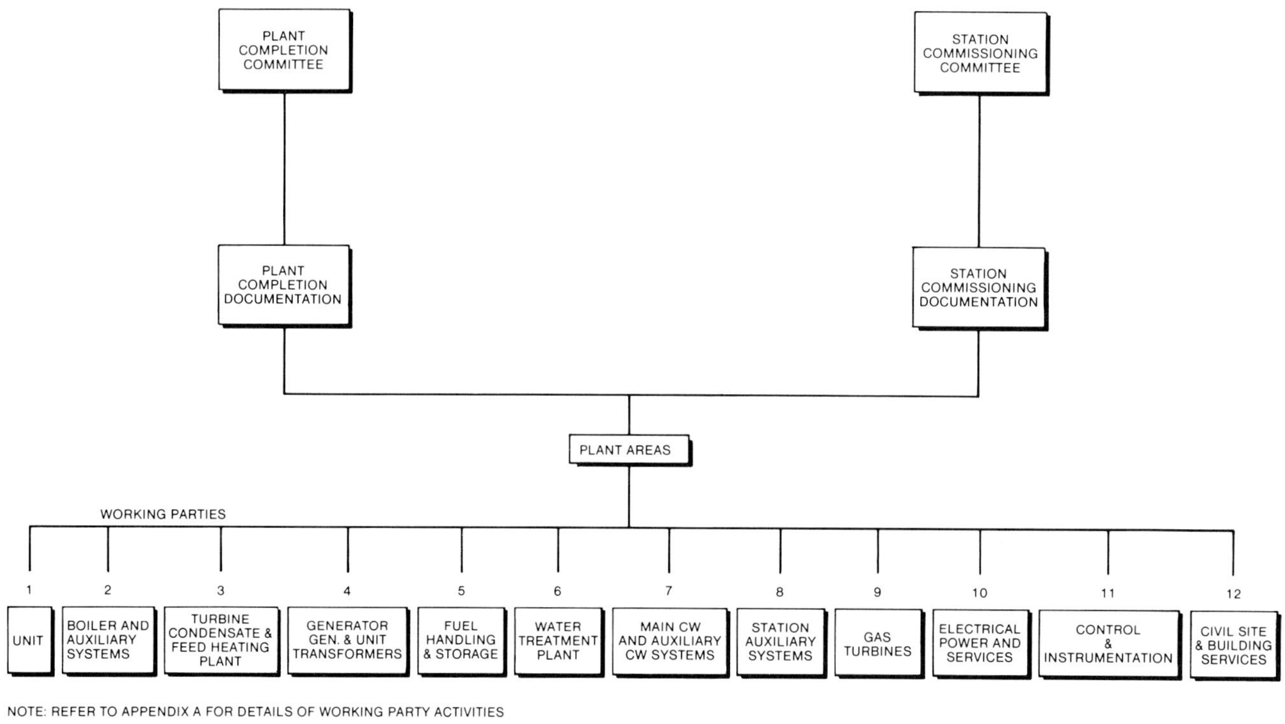

FIG. 2.9 Plant area and committee responsibilities

Team, then they usually tend to represent the Region on Working Parties. Depending upon previous experience, they will often lead Working Parties, particularly where operational knowledge or experience is required. It is normal for most of the large plant area contractors to employ their own specialist commissioning staff. In certain of the smaller areas this can be left to erection staff or delegated to one of the main contractors if the plant is a sub-contract.

TTSD usually ask the Transmission District staff to represent them on working parties, as the Transmission District staff normally commission the equipment and will ultimately be responsible for its maintenance.

The schedules in Appendix A give an indication of how the activities of the Working Parties are divided for any plant area. These schedules are drawn up in the CEGB *Plant Completion and Station Commissioning Procedure for New Power Stations.*

The division between Plant Completion Committee and Station Commissioning Committee activities is illustrated in an attempt to guide those committees when they are set up at new sites. The schedules are intended to guide but not rigidly define, as it is important that the right relationships are established as soon as possible on site. This will require flexibility on the part of all parties involved in the commissioning process, and the sooner this is demonstrated the easier will be the path to agreement on the initial important issues. Although the schedules are intended as a guide, it is unlikely that there will be major variations from them. If a major variation is being considered the reasons for such action must be examined by the main parties concerned. An example of this was if the floating of safety valves was being considered to be a Plant Completion Committee matter instead of an activity covered by the Station Commissioning Committee. This may be acceptable if, for instance, the boiler was going to be completed considerably in advance of the turbine but could not be used in commercial operation. The agreement may then be that the construction organisation assumed responsibility for its preservation until such time as it was required. There would be an agreed procedure to bring it from this state to that required for further commissioning activity, which may include refloating the safety valves.

The range of documentation which has to be produced directly by Working Parties, or with their assistance, includes:

- Papers of Principle
- Plant completion and station commissioning procedures
- Check lists and test procedures
- Design change notifications
- Programmes
- Plant completion and station commissioning reports
- Plant assessment reports

In addition, they will be involved in certain aspects of plant certification, including:

- Safety rules clearance certificates
- Record of initial operation certificate
- Clearance for station operation certificate
- Internal taking-over certificate
- Final internal control clearance certificate

All this documentation is covered in detail in Section 5 of this chapter. In addition, there is other documentation produced for use by the station staff. This includes detailed maintenance and descriptive manuals, along with operating instructions. Depending upon the division of responsibilities betwen station staff and Regional commissioning staff, they may be required to check the accuracy of submissions or, in certain cases, to write sections. The role that this type of documentation plays in commissioning is discussed in Section 5 of this chapter.

The Working Parties have the responsibility for drawing up programmes for their individual areas and for ensuring that they are compatible with the main construction and overall commissioning programmes. As GDCD normally have a programming section on site they are usually charged with producing the overall commissioning programme, but the Regional Commissioning Organisation will normally have the capability of producing the smaller system or sub-system programmes. Examples of a system and overall unit programmes are shown in Figs 2.10 and 2.11. They have been simplified to assist clarity.

4.5.2 Test teams and test team leaders

These are established by the Testing and Commissioning Panel based on nominations from the Working Parties. The representation on test teams is as follows:

- *GDCD* Usually a site-based engineer
- *Region* A member of the Regional Commissioning Team
- *Contractor* An engineer from the site team who in certain instances will be from their specialist commissioning organisation.

The terms of reference of the test teams are:

(a) To carry out their activities in accordance with the previously agreed procedure.

(b) To ensure that the activities carried out comply with safety standards either defined in the procedures or considered to be more generally applicable.

(c) To raise and complete all the documents defined in their procedures.

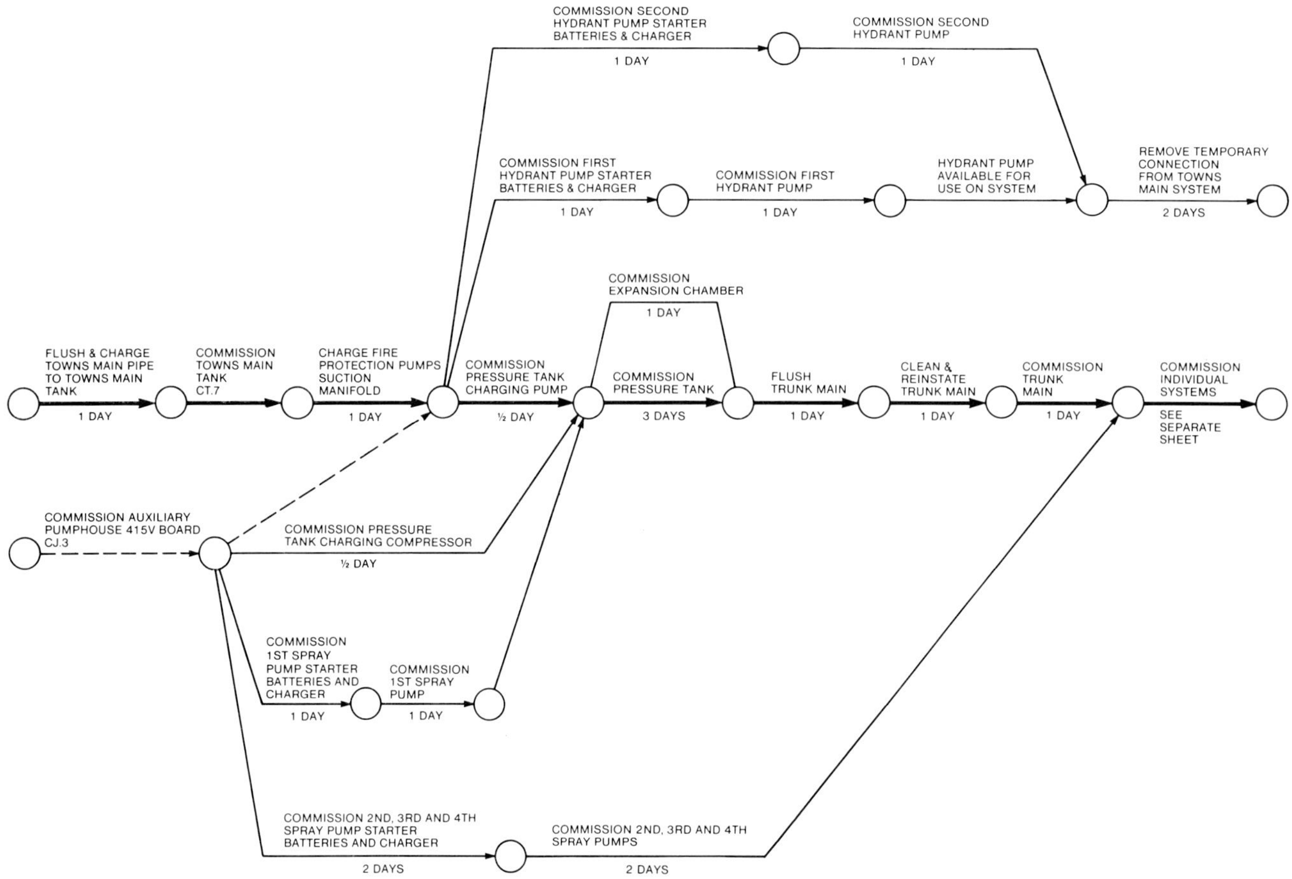

FIG. 2.10 Example of a system programme

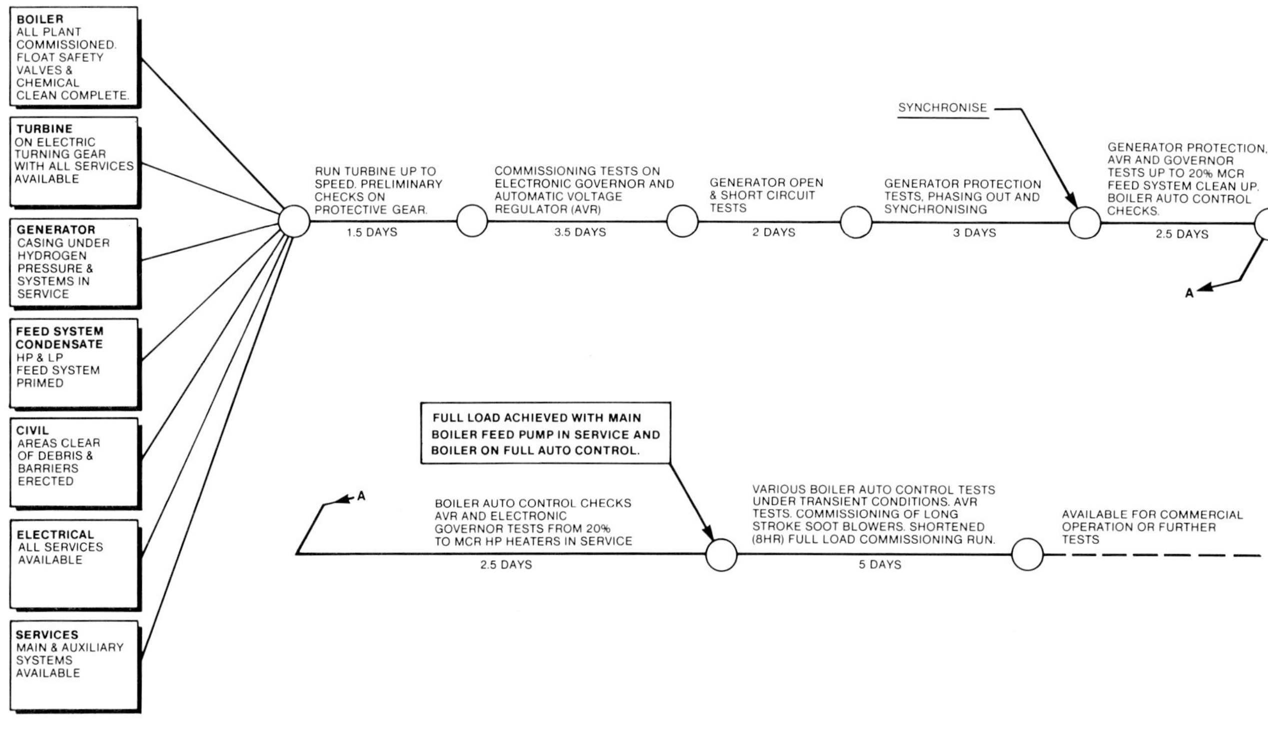

FIG. 2.11 Example of an overall unit programme

(d) To report any incident, accident or near accident which occurs as a result of their activities.

(e) To prepare Commissioning Progress Reports for the Working Party and to indicate any difficulties in implementing the previously agreed procedure.

The test team's function is one of taking a particular section of plant, within a working party area and ultimately bringing it up to an agreed state of operation. As indicated in Section 4.2 of this chapter, the working party area might be the boiler, but within that there are many sub-sections some of which will require test teams to bring the plant to a certain operational state. An important role within the test team is that of the test team leader. This person is responsible for co-ordinating the activities which take place, but does not undertake the contractor's responsibility for commissioning the plant. This relationship requires careful handling, if contractual responsibilities are not to be set aside.

It is crucial when plant is being commissioned that the responsibilities of the individuals concerned are clear but also that there is a clear chain of command. For example, when steam is being applied to the turbines for the first time there will be a leader of the test team responsible for this activity, but he will be co-ordinating the actions of other parties to implement the agreed schedule, plus any detailed requirements of the turbine contractor's representative. If, during this activity, problems develop in other plant areas which impinge on it, he has no authority to change their conditions but can only consider the best course of action in the circumstances. It is normally left to the individuals concerned, who each have their separate responsibilities and authorities, to decide the best course of action, bearing in mind the need to:

(a) Protect personnel from danger

(b) Protect plant from danger

(c) Progress commissioning

But, to remove any uncertainty, it is normal to nominate one person who has overriding control to protect all plant and personnel if, in his opinion, circumstances are developing which are unexpected or cannot be explained. This, in most circumstances where major activities are being undertaken, will be the shift charge engineer. It must be emphasised, however, that within the specific activity being undertaken the test team leader has the authority to co-ordinate actions to achieve the objectives.

5 Documentation systems

5.1 Introduction

To ensure satisfactory execution of the commissioning process as described in this volume, it must be defined, regulated, monitored and recorded. The system of documentation acts as the framework for the process of commissioning as a whole and also provides details of individual actions within that framework.

The documentation allows communication between the various organisations involved. It also encourages a maximum amount of deliberation and planning to be applied to each stage and reduces the reliance on individuals deciding what to do as things progress.

On completion of commissioning, the detail of the procedures and check lists forms a base of information for the station history system and for comparison of plant performance between the 'present' and the 'as new' state.

Any commissioning activity, great or small, uses some documentation and certification, depending on the need.

Within the CEGB, the biggest commissioning job is the complete commissioning of a new power station. The large capital involvement in a new station and the need to have it operating at maximum availability and efficiency as soon as possible may appear to conflict with the desire to commission thoroughly and effectively. However, it has been demonstrated that the gains of rapid setting to work of plant, caused by load demands on the system, are only short term. The required amount of discipline must be applied to ensure that the laid-down procedures, tests and checks are carried out and that the resulting *defects* and *omissions* work is completed, which will result in higher levels of availability, flexibility and efficiency. It is usually flexibility which is the loser in the event of over-rapid commissioning of plant, as the automatic and remote controls do not receive sufficient attention. For example, where a sequence fault is easily bypassed by having someone perform the job manually, this tends to become accepted practice and the fault is never attended to.

In large commissioning projects, the amount of administrative effort required to manage the documentation system and certification procedures should not be underestimated, and responsibilities for this task should be clearly laid down. Also, the natural aversion of most engineers to preparing and completing documents must be discouraged. The correct approach to this aspect of commissioning is at least as important as the physical actions of setting plant to work.

In any contractual agreement, a commitment to the commissioning activity must be obtained, including the preparation of commissioning documents. Agreement must be reached between the plant supplier and the customer or his agents on the content of commissioning documentation.

This section will concentrate on the documentation systems for commissioning new power station plant, using current CEGB practice. The principles embodied in the systems can be applied to any commissioning task, whether for large or small jobs, new plant or the return to service of older plant.

Documentation systems will be described in the following sections.

5.2 Location commissioning procedure document

At each location where the commissioning activity is going to take place, a document should be produced to describe the overall approach to the activity, the scope of the commissioning, the detail of the documentation and the certification procedures to be used. For a new CEGB power station, the location document will be based on the *Plant Completion and Station Commissioning Procedure for New Power Stations* (22 March 1979). The location document will contain all the requirements of the CEGB procedures, as well as the detailed local application.

A typical summary of contents would be as follows:

- *Statement of objectives* This would contain, for example, references to contractual obligations, safety and engineering standards.
- *Organisation* A description of the commissioning organisation, clearly defining the interaction of the various parts of the organisation and the roles of each group within it.
- *Responsibilities* Responsibilities of all interested parties to be set out clearly.
- *Documentation and certification* A detailed description of the documentation system which will be applied, describing the function of each document and the responsibility for its production and completion. Documents will generally detail the procedure to be followed and the checks to be made, each stage being authorised by the responsible person, when completed. Certificates, on the other hand, are formally completed at certain stages in the commissioning process, indicating when significant points have been reached.
- *Safety* The responsibilities for safety are described, both before and after the application of the CEGB Safety Rules. The process of applying the rules to a new piece of plant is described, including the issue of the Safety Rules Clearance Certificate.
- *Membership of committees and plant area responsibilities* The document details the membership of the Plant Completion Committee, the Station Commissioning Committee, and the Testing and Commissioning Panel. When the decisions have been made by the Station Commissioning Committee, the actual composition of working parties will be detailed, along with the plant areas for which they are responsible.
- *Location details* The document will include any other important points relevant to the commissioning process, such as the physical location of commissioning staff offices, the storage of documents and certificates, and the hours of the day during which commissioning may take place, this being particularly relevant for boiler safety valve floating and steam purging, where noise could be a problem.

5.3 Papers of Principle

Papers of Principle are the first documents to be produced for the commissioning of individual plant systems and will be written by GDCD or TTSD, as appropriate. They are prepared in conjunction with the Region and main plant contractors, for consideration and approval by the Station Commissioning Committee. Papers of Principle set out the testing and commissioning principles and objectives for both plant completion and station commissioning work, and schedule the necessary tests and commissioning procedures.

From the outset, it is important to ensure that the coverage of the various Papers of Principle lines up with the Working Party plant areas. The subsequent Plant Completion and Station Commissioning Procedures relate to the coverage of the Papers of Principle. A typical breakdown of documentation for a Working Party is shown in Fig 2.12. Depending on the plant system, it is possible that the Paper of Principle and the Location Commissioning Procedure Document will have the same coverage.

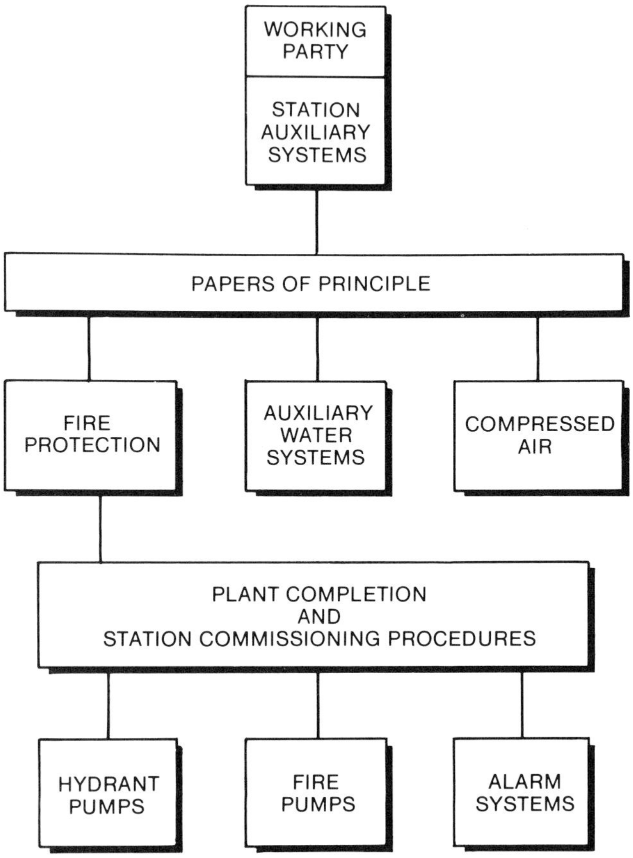

FIG. 2.12 Breakdown of commissioning documentation for a working party

The Papers of Principle need to give sufficient information about the design, function and performance of the plant or system, for the Station Commissioning Committee to judge easily the suitability of the testing and commissioning proposals. They should also contain adequate information for the committee to be satisfied about personnel and plant safety. Each paper should contain sections covering the following headings, and should also reflect the general requirement for commissioning, as dealt with in the Location Commissioning Procedure Document:

- *Scope of the paper* The plant area covered by the paper is defined and the state of the plant at the start of the commissioning activity described, including relevant tests or checks which have already been carried out in the erection process. This section also summarises the contents of the Paper of Principle.
- *Description of plant and/or system* A description should be included of sufficient detail to enable the commissioning principles to be understood. Information concerning the items in the plant system, relevant design features, and the function of the system should all be included. Also, details of the final expected performance will be covered.
- *Objectives of commissioning activities* The overall objective is to establish that the plant is capable of safe and reliable operation and control and that it meets its design specification and performance guarantees. Each Paper of Principle will contain statements of objectives relevant to the particular plant or system being covered.
- *Commissioning and test proposals* This section summarises the various commissioning activities and tests needed to satisfy the general objectives, when applied to each individual piece of major plant or a plant system. For each of the activities or tests, the objectives are defined. The more important stages will be mentioned, particularly those which interact with tests in other plant areas.
- *Safety* For the purposes of the Paper of Principle, this section is general rather than specific and refers to the CEGB *Procedure for Commissioning New Power Stations*. The major safety concerns are centred around the initial energising or pressurising and the issue of the Safety Rules Clearance Certificate (SRCC).

  Any particular major hazards which affect the manner or order in which commissioning tests are carried out should also be mentioned. For example, work during commissioning involving access to a hydrogen-cooled generator frame requires that the CO_2 system and air purge system are commissioned along with the hydrogen system.

  The responsibilities for safety prior to the SRCC issue and during commissioning will be covered in principle only.
- *Programme* This section covers the programming of the commissioning tests and activities, putting them in the correct order and sharing possible concurrent activities. An approximate duration of the activities and the timing of any interaction with the commissioning of other systems will be included. This information allows the programming of the overall station commissioning activities at the earliest possible time and can be used to decide priorities for the completion of erection of plant.
- *Commissioning and test documentation* This section deals with the requirements for Plant Completion and Station Commissioning Procedures with respect to the particular Paper of Principle. There may be only one procedure for each paper, or a number, depending on the size and complexity of the plant area.
- *Reports* The number and types of reports to be written are specified.
- *Staffing and resource requirements* The requirements for staff numbers and types are stated and, if possible, individuals named. Normally, the minimum complement will be one engineer from Region, one from GDCD and a contractor's representative.

  The type of engineer to act as test team leader, i.e., Region or GDCD, will be specified and, if possible, the individual named. Since individuals are nominated by the Working Party, it will not be possible to fit names to tasks in those Papers of Principle which are written early in the commissioning phase.

  Any particular equipment, test gear, materials, fluids, temporary supplies or any other resources, the need for which is known at the time of writing the Papers of Principle, should be included.

  If it can be foreseen that other test teams will need to be present to operate their plant to allow commissioning to proceed, then this should be mentioned, along with the period of their involvement.

5.4 Plant Completion and Station Commissioning Procedures

5.4.1 Definition of plant completion and station commissioning activities

Each Paper of Principle for a commissioning plant area or system normally gives rise to a number of Plant Completion Procedures (PCP) and Station Commissioning Procedures (SCP). Both types of document are produced by the Working Parties, for approval by the Plant Completion Committee or Station Commissioning Committee, as appropriate.

The basis of the decision whether a PCP or a SCP is applicable to a given commissioning activity is as follows:

- Plant Completion Procedures cover the carrying out, completion and recording of the necessary mechanical and electrical functional tests on each plant item. These tests are those, both static and running, that prove satisfactory manufacture, installation or construction, so that the plant may be considered available for commissioning tests in association with other plant, or for operation. So, the emphasis in a PCP is on the completion of erection checks and individual tests on items of plant. If the item of plant does not constitute part of a system, then it may be suitable for immediate operation by the station staff, following the plant completion activity.

- Station Commissioning Procedures are intended to continue the activities of plant completion into commissioning of systems, and eventually of the complete power station. They normally cover commissioning tests on systems which are made up of plant items, all previously checked under individual PCPs.

  In practice, the Plant Completion Committee and the Station Commissioning Committee initially have the same membership, so it is likely that initial agreement of the breakdown of documentation into PCPs and SCPs will be decided by some joint body or representatives of the interested parties.

  A typical breakdown of documentation for water treatment plant is shown in Fig 2.13.

PAPER OF PRINCIPLE	PLANT COMPLETION PROCEDURE	STATION COMMISSIONING PROCEDURE
WATER TREATMENT PLANT	RAW WATER DELIVERY, STORAGE AND PUMPING PLANT	PERFORMANCE AND ACCEPTANCE TESTS
	MIXED BED TREATMENT PLANT	
	REGENERATION PLANT	
	TREATED WATER STORAGE AND PUMPING PLANT	
	CHEMICAL CONTROL EQUIPMENT	

FIG. 2.13 Breakdown of documentation for water treatment plant commissioning

5.4.2 Procedure contents

Both PCPs and SCPs have similar subject coverage, but with different levels of detail and with emphasis placed in different areas. The procedures should contain information and instructions along the following lines:

- A description is included of the plant item or system being commissioned, giving information as to its purpose, design features, performance and specification. The description will include the smaller items of equipment, such as instruments or switchgear, which may be commissioned at the same time. The extent of the commissioning or plant completion activity is also defined.
- A statement of the objectives of the procedure is given, describing the expected state of the plant at the start of the activity and what is to be achieved. In a PCP, this might be the completion of erection checks on a pump, including its ancillary equipment, operation of associated valves, switchgear, rotation check on the motor and a coupled run. In an SCP, this means an outline of the items of plant included in the system being commissioned, the tests to be carried out, and the main performance parameters to be monitored.
- The manpower and equipment/services resources needed to carry out the commissioning are detailed. The test team members are named, including the test team leader; also the responsibility for supplying labour for particular jobs during the activity will be clarified. For example, operators from the Regional staff may be required to operate and monitor plant that is being commissioned or equipment which must be running to service that plant item. Craftsmen may be needed during the activity at unspecified times to repair leaks, to carry out fault investigation, to change over connections on motor supplies, or for any other activity which is an integral part of the commissioning process.
- The equipment and services required prior to and during the commissioning activity should be included in the procedure, allowing preparations to be made prior to commissioning, so minimising the actual commissioning period. Information on equipment and services should cover such items as:

Cooling water supplies	Test equipment
Electrical supplies and switchgear	Communication systems
Plant drainage arrangements	Tools
Control air supplies	Lighting and heating
Remote controls and instruments	

- Any arrangements for routine commissioning which do not require the complete team to be present are agreed and included in this section.
- The test team comprises individuals from different organisations, with different allegiances, so their individual responsibilities within the test team should be described. This description covers the roles of the test team leader, the GDCD engineer, the CEGB Region engineer and contractors' representatives.
- It is important to state clearly any particular precautions or instructions relevant to the safety standards applicable to the commissioning activity. For any PCP or SCP, there will be a number of common safety aspects, such as application of the Health and Safety at Work Act, the Factories Act,

the CEGB Safety Rules, plant identification, access, fire precautions, lifting equipment and personal safety equipment. Also, for every commissioning activity there will be safety considerations which are of particular importance. These might be fire risks in a system containing oil or hydrogen, risks from injury due to handling corrosive chemicals, possible damage to hearing during purging exercises, and so on.

- The procedures must also contain instructions on what to do in emergencies. Some instructions apply to all commissioning activities, based on the power station site procedures for such items as first aid, fire, oil spillages, and flooding, and there will also be instructions relating to the particular commissioning activity. Both should be stated simply and clearly, and be practicable.

The major part of a PCP or SCP will be the detailed procedure for the commissioning activity. This should cover all the requirements of the following headings:

(a) Narrative procedure

(b) Programme

(c) Initial state of plant

(d) Procedure detail

(e) Final state of plant

The *Narrative procedure* (a) summarises the activities in chronological order, suitable for reading by people both directly and indirectly involved. For those with direct involvement, it introduces the procedure and describes the basic steps; for people not directly involved, the narrative helps to co-ordinate different commissioning activities.

The *Programme* (b) is a clear statement of the sequence of events, showing any interactions or critical paths and the expected times for the major events. Normally, a network will assist in this area; it could be backed up with a bar chart, if necessary, to show the activities on a time base. An example of a commissioning network for a unit condensate system is shown in Fig 2.14.

The *Initial state of the plant* (c) again assists those either directly or indirectly involved. It informs those responsible for co-ordinating erection activities with plant completion activities of the expected state of the plant at the interface and is of particular importance in co-ordinating plant completion and station commissioning activities.

This section describes the access requirements, the position of valves, the availability of electrical supplies, the state of checks on instruments, actuators and motors, and details the running state of other systems which have to be in service to allow commissioning to proceed.

A *detailed step-by-step procedure* (d) of the commissioning work must be drawn up, using the narrative commissioning procedure and any relevant network as a skeleton. This document will be followed by the test team and retained on completion of commissioning as a record of what was done. The procedure sheets leave room to record actual measurements taken and show any change in the procedure which arose during the commissioning: they allow for signatures of the test team who carried out or witnessed the activities to be recorded.

The detailed procedure may refer to other commissioning procedures or to operating instructions, but otherwise it contains all the necessary detail for actual commissioning. The use of correct nomenclature in all plant references is essential to avoid confusion and, in particular, the plant identification must be to a high standard. Valve identification is a problem where fitting of labels is the last step in the erection process,

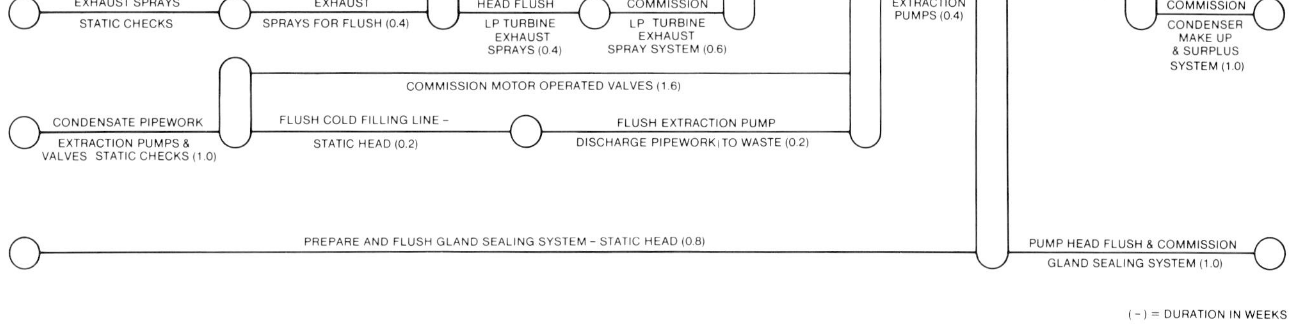

FIG. 2.14 Unit condensate system — commissioning network

but is essential to ensure safe and effective commissioning.

The interim use of contractors' valve numbers should be ruled out and the CEGB nomenclature used from the start, both on the valves themselves and in the procedure. An extract from a typical procedure is shown in Fig 2.15.

CENTRAL ELECTRICITY GENERATING BOARD
POWER STATION

COMMISSIONING SCHEDULE No. ....

PLANT .... CONDENSATE SYSTEM

SECTION .... DETAILED COMMISSIONING PROCEDURE

ITEM	DESCRIPTION	RESPONSIBILITY	WITNESSED BY: SUPPLY CONTR.	WITNESSED BY: G.D. C.D.	WITNESSED BY: CEGB REGION	DATE
15.	Run Extraction Pumps – Flush Extraction Pump Discharge Pipework and L.P. Heater Feed Pipework (CAP A/2282/3)					
15.1	Check that the following system valves are closed:					
	WC-73 (V616) Condr./RFW Tanks Condst. Outsurge Isol.					
	WC-75 (V614) Condr./RFW Tanks Condst. Outsurge Isol.					
	WC-74Y (V617) Condr./RFW Tanks Condst. Outsurge Bypass					
	WC-V93 (V529) Condr./LP Heater 5-1 Air Release Isol.					
	WC-96 (V546) Condenser Hotwell 5A Condst. Recirc. Isol.					
	WC-99 (V550) Condr./Anti-void Tank 5 Make-up Isol.					
	WC-107 (V551) LP Htr. Lift Pumps 5A and 5B Gland Sealing Isol.					
	WC-110 (V547) Condr. to WTP Condst. Dumping Isol.					
15.2	Run pump 'A'					
15.2.1	Put extraction pump 'A' MSD into the service position and on local.					
15.2.2	Carry out a pre-start check on Extraction Pump 'A'. (Refer to plant operating instructions Job No. 3.6.1.4/5 Extraction Pump Pre-start checks). Monitor the remote, or temporary local ammeter. Post personnel around the pump and discharge point and establish radio contact with the switchroom.					
15.2.3	Open:					
	WC-56 (V510) Condr./LP Heater 1 Condst. Discharge Isol.					
15.2.4	Making up to condenser as necessary, start extraction pump 'A' against closed discharge valves, monitor the starting current, running current and the time that max current is sustained:					
	Starting current					
	Running current					
	Time of Max current					

FIG. 2.15 Extract from typical commissioning procedure

Just as important as the initial state of the plant is the *final state of the plant* (e) on completion of the commissioning activity. This information will be needed for co-ordination purposes, either between plant items within a working party area or for pulling together systems, for say, the final commissioning of the complete generating unit. The state of plant after the activity will also define any preservation requirements, particularly where the system or item will not be used for a period while awaiting the completion of other systems. This might include filling with dosed water, draining and drying out, nitrogen sealing, lubrication and the use of heaters for electric motors.

5.4.3 Standard Check Lists and Test Procedures

A large number of plant items in a modern power station are common to many plant systems. For example, the type of electric actuator used on a project will be applied to boiler dampers and turbine valves alike, and used in large numbers. It, therefore, makes sense to prepare commissioning procedures for these common items, which can then be used by various working parties and test teams, as appropriate. Also, the Plant Completion Procedure can refer to these standard procedures and reduce the volume of paperwork for each procedure document. It also ensures that a common standard of commissioning is applied to the same items across the whole power station.

Examples of the use of Standard Check Lists and Test Procedures are:

Manually-operated valves	415 V switchgear
Coupling alignment	Temperature control loops
Small AC motors	Pressure switches
Low pressure pipework	

Standard Check Lists are quoted by number only and referred to in schedules within a commissioning procedure document. Examples of a schedule and a Standard Check List are shown in Figs 2.16 and 2.17.

5.5 Design Change Notification

The design process for a conventional power station tends to be prolonged well into the erection and commissioning phase, as a result of experience at existing power stations or of reappraisal of original designs. There may also be a small amount of redesign work arising from examination of the plant systems by the site staff engaged in the construction work. The number of changes to the designs should be kept to a minimum, for many reasons, but inevitably there will be changes, and these must be covered within the Commissioning Organisation.

To ensure that any changes to the test or commissioning activities necessitated by changes in design are catered for, a procedure for the particular location should be established. The procedure would probably require GDCD to notify the Plant Completion or Station Commissioning Committee of design changes made after the approval of the Paper of Principle.

On occasions, the commissioning activity itself will bring about changes of a minor nature, but still important as far as the operation of the plant is concerned. A typical example would be the alteration of pressure switch settings from the original design levels in a turbine oil system when the plant characteristics do not exactly match design.

It is very important to feed back any changes of this nature to the design staff at GDCD, to avoid the station staff putting incorrect settings on instruments in future recalibrations by following schedules and design settings which have not been updated.

COMMISSIONING SCHEDULE CONDENSATE SYSTEM

SECTION ________________ **VALVE SCHEDULE No.** ________________ **CONDENSATE PIPEWORK SYSTEM**

Valve Nameplate Inscription				Contractors Valve Number	Valve Type	Standard Check List	Sign when SCL Completed
CEGB Valve No.	Valve Title						
TG5-WC-51A	Condr. Ext. Pump 5A	Condst. Suction	Isol	V500			
TG5-WC-51B	Condr. Ext. Pump 5B	Condst. Suction	Isol	V501			
TG5-WC-52A	Condr. Ext. Pump 5A	Condst. Discharge	Non-return	V502			
TG5-WC-52B	Condr. Ext. Pump 5B	Condst. Discharge	Non-return	V505			
TG5-WC-53A	Condr. Ext. Pump 5A	Condst. Discharge	Isol	V503			
TG5-WC-53B	Condr. Ext. Pump 5B	Condst. Discharge	Isol	V506			
TG5-WC-53AY	Condr. Ext. Pump 5A	Condst. Discharge	Bypass	V504			
TG5-WC-53BY	Condr. Ext. Pump 5B	Condst. Discharge	Reg.	V507			
TG5-WC-54	Condr./LP Heater 5-1	Condst. Control	Reg.	V509			
TG5-WC-55	Condr./LP Heater 5-1	Condst. Discharge	Non-return	V548			
TG5-WC-56	Condr./LP Heater 5-1	Condst. Discharge	Isol	V510			
TG5-WC-68A	Condr. Ext. Pump 5A	Gland Sealing	Isol	V884			
TG5-WC-68B	Condr. Ext. Pump 5B	Gland Sealing	Isol	V885			
TG5-WC-D69A	Condr. Ext. Pump 5A	Gland Sealing Drain	Isol	V888			
TG5-WC-D69B	Condr. Ext. Pump 5B	Gland Sealing Drain	Isol	V889			
TG5-WC-70A	Condr. Ext. Pump 5A	Gland Sealing	Isol	V896			
TG5-WC-70B	Condr. Ext. Pump 5B	Gland Sealing	Isol	V897			

FIG. 2.16 Typical valve schedule

SCL No. M101
Sheet 1 of 1
Ref. No. ....

.... Power Station

Standard Check List – Valves

Valve Identification Number ....
Valve Identification Title ....
Valve Schedule Reference No. ....
System Drawing Number ....
Manufacturer's Name ....
Figure No. .... Serial No. .... Type ....
Contract Reference No. ....
Actuator Type & Make ....

ITEM	DESCRIPTION	RESP	WITNESSED			DATE
			CONT	UTILITY		
1.	Check that valve identification agrees with details on schedule a) Number b) Title					
2.	Check that access is safe for operational purposes.					
3.	Check receipts of manufacturer's test certificate and record ref. no.					
4.	Check that valve is correctly fitted into pipeline for a) direction of flow b) tightness of bolts c) adequacy of support d) correct joint and gland packings.					
5.	Check that valve is correctly lubricated.					
6.	Check that locking facility is fitted.					
7.	Check that local position indication is correct.					
8.	Check and set limit switches (if fitted).					
9.	Check free operation with actuator fitted. Note actuator SC ref. number					

FIG. 2.17 Typical standard check list

5.6 Plant completion and test reports

During the commissioning period of a power station, various reports and test results are produced. These arise from Working Party activities and are prepared by working parties in line with Plant Completion and Station Commissioning Committee requirements.

These reports are approved by the relevant committee with recommendations for any necessary action and for their distribution.

Each committee determines the method and form of recording test results. In some cases, the test log sheets will be a sufficient record, whilst in others test results may require careful assessment against design data, corrections may be applied, and a formal final test report will be issued at some later time.

5.6.1 Test Certificates

The testing and insurance inspection of all lifting equipment will need to be phased in during the commissioning process. This may be to allow further erection work to continue or for early maintenance work to commence. Also, the manufacturer's test certificates for pressure vessels and relief valves will be issued. These will be followed by the insurance inspection and issue of the appropriate certificates. Plant

Completion and Station Commissioning Procedures detail the stages at which test and insurance certificates should be available. The appropriate storage system and a procedure for subsequent handover to the permanent station staff should be decided as early as possible.

As part of the plant completion process, calibration checks will be carried out on various instruments, such as pressure gauges, pressure switches and transducers, temperature indicators and thermocouples. A record of the results will initially be retained with the plant completion documents, but will be transferred into the recording system of the permanent station organisation at the appropriate time.

5.6.2 Works tests

Where a contractor has to demonstrate the performance and capability of individual items of plant, such as pumps, fans or motors, this is often done at the contractor's works. These tests are carried out with Production Inspection and Test Field Services (PITFS) present to witness the tests on behalf of the CEGB. Test reports are produced by PITFS and are passed via the Commissioning Organisation to the station staff for inclusion in the station technical files. The Commissioning Organisation should make provision for the handling of such reports and their transfer to the station organisation.

5.6.3 Performance tests and acceptance tests

These tests are normally included in the Station Commissioning Procedures. They will cover areas such as boiler/turbine acceptance tests, cooling towers, gas turbines, auxiliary boilers and electrostatic precipitators.

Completed systems are tested to demonstrate that they are satisfactory in terms of output, efficiency and reliability. Any other parameters which were part of the plant specification are checked.

5.6.4 Supplementary tests

In order to confirm certain of the CEGB Station Development Particulars, tests will be carried out which are not the responsibility of one contractor, or even of one particular working party. They may cover the operation of a number of plant systems together to demonstrate a certain capability. Examples of such tests are:

- Station start-up with no auxiliary supplies available
- CW system surge tests
- Unit load rejection tests
- Station capacity tests

These tests are specified by the Station Commissioning Committee and separate groups may need to be set up from interested Working Parties to perform particular tests. Reports of the tests will be prepared by these groups and submitted to the Station Commissioning Committee.

6 Safety, operation and certification in commissioning

The application of the CEGB Safety Rules is an integral part of the commissioning process, as specified in the procedure documents. This section describes how the Safety Rules are applied and the manner in which the plant is operated to cater for the resulting change in status. It also deals with the formal certification process from the issue of the Safety Rules Clearance Certificate to the Final Internal Contract Clearance Certificate.

6.1 Application of the CEGB Safety Rules

The application of the Safety Rules has been mentioned in the section dealing with basic steps in commissioning, as it is a major step in the erection/commissioning process. Following the application of the Safety Rules, the behaviour of the erection contractor has to be modified to cater for the need for documents before commencing work. This sudden change in safety requirements is difficult for the contractor to understand, and consequently an appreciation by all concerned is needed for effective application of the rules.

The stage at which the Safety Rules are applied is stated in general terms in the *Commissioning Procedures for New Power Stations*. Before any item of electrical plant is energised or put into service for the first time, or any item of mechanical plant is pressurised with air, gas, water, steam, oil or other working fluid, a Safety Rules Clearance Certificate (SRCC) must be issued. This certificate states that the plant is to be regarded as alive, or in service, from a specified date and time, and that all future work on the plant will be subjected to the Safety Rules.

This definition, although comprehensive, still requires interpretation at each location. Difficulties arise with plant being subjected to the Safety Rules too early in the erection/commissioning process. The Plant Completion Committee and Station Commissioning Committee have to produce guidelines on the application of the Safety Rules within the Station Commissioning Procedure document described in Section 5 of this chapter. The following points should be considered and suitable guidance given:

- Premature issue of an SRCC could result in erection progress being hindered by the need for documentation, when there is no danger to the person in charge of the work from the system.

- Pressure testing of plant is necessarily carried out some time before erection is complete and before the start of the commissioning activity. A typical example is a hydraulic test on a boiler, which is carried out with temporary water supplies and pumps. This activity, on the face of it, comes under the definition quoted above, but to issue an SRCC at this stage in the boiler erection process would be wrong.
- Flushing activities are often part of the erection/commissioning procedure for pipework systems. These activities vary in the level of danger which might arise, and their position in the overall erection process. The decision to issue an SRCC will be made after considering these two factors.
- There is no doubt that an SRCC must be issued when plant is energised electrically from its normal permanent point of supply. If, however, temporary supplies are used in low voltage systems early in the erection process, it could be premature to apply the Safety Rules.
- When the Safety Rules are applied, there must be effective means of isolation, by lockable valves and/or by lockable electric isolation.

It is important that the decision to issue an SRCC is taken after consideration of these points and is not confused with a division between plant completion and station commissioning activities. The issue of an SRCC could occur at any time, but most probably at the start of, or during, the plant completion process.

6.2 SRCC and issue of safety documentation

6.2.1 The SRCC

The energising and operation of plant for the first time is carried out only with the agreement of the Site Manager and Station Manager, and in the presence of the contractor, GDCD and the Region. As already stated, a Safety Rules Clearance Certificate (SRCC) must be issued prior to this event. A specimen copy of an SRCC is shown in Fig 2.18.

- The certificate is initiated by the Site Manager in agreement with the Station Manager. All the details of the plant, the time and date for it to come into force, and the contractors to which it applies, must be completed.
- Each contractor acknowledges notification by signing the certificate and retaining a copy.
- The completed master certificate is then issued by the Site Manager to the Station Manager.
- Just before the SRCC comes into force, the Region staff, who are responsible for administering the Safety Rules, apply the necessary isolations. Safety documents are then prepared and issued on the time and date specified on the SRCC.

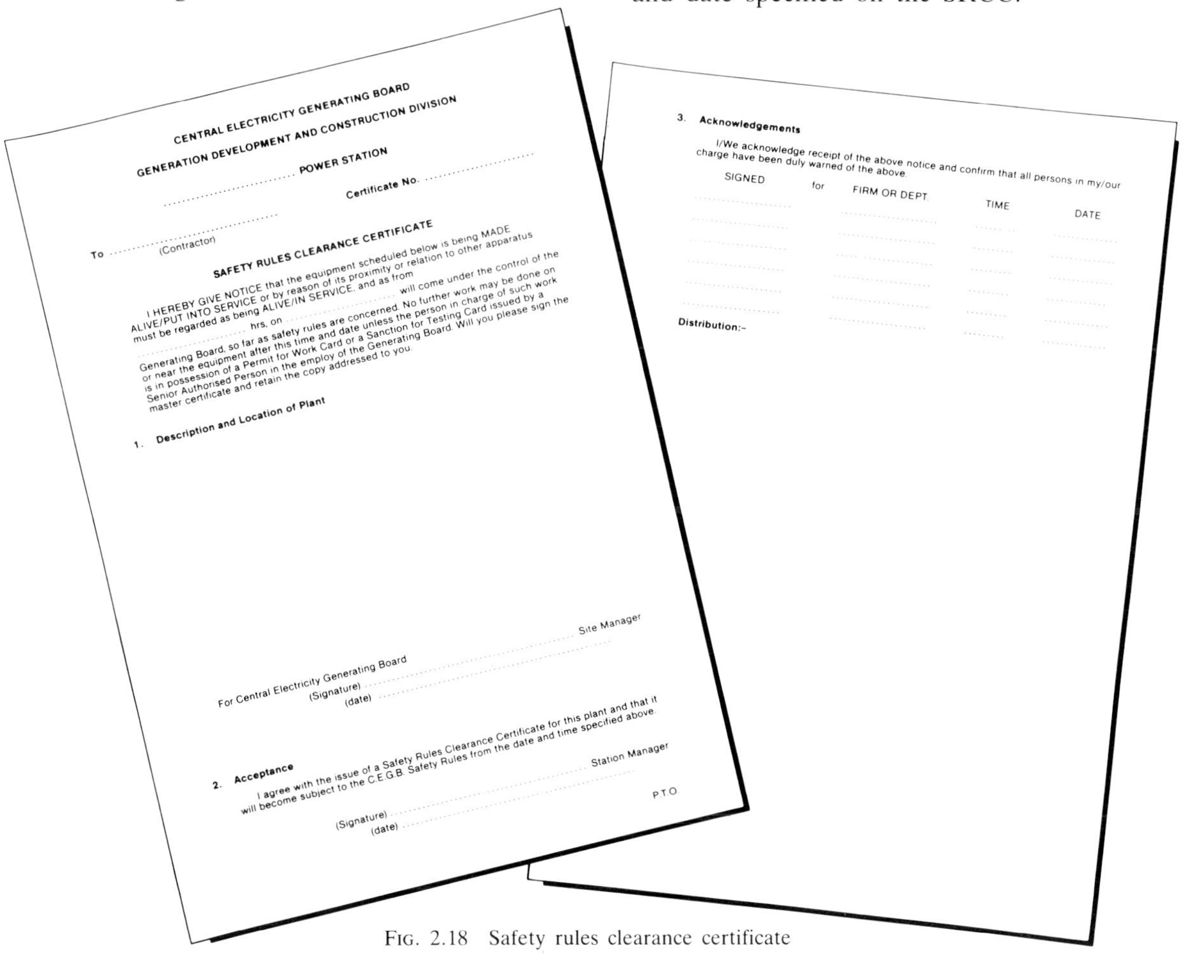

CENTRAL ELECTRICITY GENERATING BOARD
GENERATION DEVELOPMENT AND CONSTRUCTION DIVISION

.... POWER STATION

Certificate No. ....

To .... (Contractor)

SAFETY RULES CLEARANCE CERTIFICATE

I HEREBY GIVE NOTICE that the equipment scheduled below is being MADE ALIVE/PUT INTO SERVICE or by reason of its proximity or relation to other apparatus must be regarded as being ALIVE/IN SERVICE, and as from .... hrs, on .... will come under the control of the Generating Board, so far as safety rules are concerned. No further work may be done on or near the equipment after this time and date unless the person in charge of such work is in possession of a Permit for Work Card or a Sanction for Testing Card issued by a Senior Authorised Person in the employ of the Generating Board. Will you please sign the master certificate and retain the copy addressed to you.

1. Description and Location of Plant

For Central Electricity Generating Board
(Signature) .... Site Manager
(date) ....

2. Acceptance

I agree with the issue of a Safety Rules Clearance Certificate for this plant and that it will become subject to the C.E.G.B. Safety Rules from the date and time specified above.

(Signature) .... Station Manager
(date) ....

P.T.O.

3. Acknowledgements

I/We acknowledge receipt of the above notice and confirm that all persons in my/our charge have been duly warned of the above.

SIGNED	for	FIRM OR DEPT	TIME	DATE
....		....	....	....

Distribution:–

FIG. 2.18 Safety rules clearance certificate

6.2.2 Plant isolation

Following the issue of an SRCC for a plant item and during the plant completion and commissioning activities, there will be a need for contractors to work on that plant. A system for co-ordinating the need to work and the preparation of safety documents has to be agreed between the Site Manager and the Station Manager. It has to be borne in mind that this system has to work within the station's own organisation, as more and more plant is maintained and operated by station staff.

The agreed system should contain the following elements:

- Work on plant under an SRCC is to be considered by a Senior Authorised Person and any safety document requirements identified.
- The effect on the commissioning programme of releasing the plant item for isolation is to be taken into account.
- A standard form is necessary to request the issue of safety documentation. This form is completed by a representative of the contractor, and countersigned by the Site Manager's representative.
- The standard form is to be issued on a date satisfying an agreed period of notice, giving an estimation of the duration of the work.
- In requesting the issue of safety documentation, the plant should be correctly identified by name and location. The nature of the work to be carried out and the time and date for commencement should also be specified.

6.3 Plant operation

The operation of plant during the commissioning of a power station involves a process of transfer of responsibility from the contractors (in the early stages) to the station staff (on completion of commissioning). In practice, station staff are used to operate plant under the direction of the contractor from the earliest practicable time. This gives the station staff as much time as possible to gain experience in operating the plant and reduces the need for contractor labour.

On completed plant, which is being operated by station staff, any isolation of plant for maintenance work is carried out by locking-off at sources of supply, whether electrical or mechanical.

6.3.1 Mechanical plant

During the process of bringing mechanical plant into the CEGB Safety Rules, boundary terminal points are established. These separate those parts of a system which are covered by SRCC, and therefore liable to filling or pressurising, from the remainder of the system. Any work being carried out on the remainder of the system is controlled, from a safety point of view, by GDCD and commissioning staff.

A typical method of controlling the keys to boundary terminal points is to safeguard them in an 'uncommissioned plant' key safe; access to this safe will only be possible using two keys, one each in the control of a GDCD engineer and a Regional commissioning engineer.

When additional plant is added as the system is extended, then an additional SRCC will state this. The boundary terminal points will be extended to include the new plant, which is now covered by the additional SRCC, and placed under CEGB Safety Rules.

6.3.2 Electrical plant

The electrical system of a power station is progressively constructed and commissioned ahead of the need for electrical supplies for plant commissioning purposes.

- When electrical switchboards or distribution boards are to be energised for the first time, an SRCC is issued to cover the incoming feeders, busbars and switchgear. Suitable isolation and earthing arrangements must be available as part of the equipment.
- All outgoing circuits should be isolated, circuit shutters locked off, fuses and links removed and keys placed in an 'uncommissioned plant' key safe, as with mechanical plant. These outgoing circuits become subject to the CEGB Safety Rules as each is individually covered by the issue of an SRCC, and subsequently energised.
- The operation of switchgear and energising of boards is only carried out by authorised persons under the CEGB Safety Rules. These will normally be the Region's commissioning staff or the station staff. Other switching, defined as 'routine switching' within the Safety Rules, such as individual motor circuits off switchboards, may be carried out by the test teams for testing purposes to an agreed programme.

6.3.3 Initial operation

The initial operation of an item of plant or a complete plant system is normally the first time that equipment is energised and operated in the completed state. For example, an initial operation of a boiler feed pump would be the first time that the motor was started when driving the pump, which itself was primed with water. Prior to the initial operation, all sub-systems — such as the lubricating oil systems, the leak-off control system and the water cooling system — would have been checked out as far as possible without the pump itself being run.

The initial operation would be carried out by the commissioning test team with as much involvement as possible from the station staff. All checks and operations needed for the commissioning procedure will be carried out, and the procedure document signed, by the appropriate representatives on the test team.

Immediately following the initial operation, a *Record of Initial Operation Certificate* is prepared, signed by representative witnesses. Also, a list of 'defects' and 'omissions' is prepared by the test team, which contains all the faults and instances of incomplete erection that became obvious during the initial operation. This list could range from incomplete plant labelling, poor access to a valve, or oil leaks on the lubricating oil system, to major faults such as the feed pump vibrating heavily or not achieving its full load.

The Record of Initial Operation Certificate does not signify any change in the responsibilities for the operation and maintenance of the plant, which still rests with the contractor.

At the time of issue of the SRCC, or between such time and the initial operation, it should be made clear, by using suitable notices, that the plant item is liable to be energised or rotated at any time. This is of particular importance on switchboards, when the busbars will be energised first, followed by individual circuits as commissioning proceeds.

6.4 Plant certification and status

To ensure that satisfactory procedures are established and that adequate records of the whole commissioning process are maintained, a system of plant certification is used. This provides records for the initial energising and operation of plant, the subsequent take-over from the contractor and the reporting of defects. A diagrammatic summary of the certification process is shown in Fig 2.19.

ISSUE / RECEIPT	CERTIFICATE	COMMENTS
BY – SITE MANAGER TO – CONTRACTORS	SAFETY RULES CLEARANCE CERTIFICATE	ISSUED WHEN PLANT IS TO BE MADE ALIVE OR PUT IN SERVICE FOR THE FIRST TIME. RECEIPT IS ACKNOWLEDGED BY EACH CONTRACTOR LISTED.
	↓	FROM NOW ONWARDS THE CEGB SAFETY RULES APPLY. COMMISSIONING PROGRESSES TOWARDS THE INITIAL OPERATION.
BY – TEST TEAM TO – SITE MANAGER	RECORD OF INITIAL OPERATION CERTIFICATE	ISSUED BY THE TEST TEAM FOLLOWING A SATISFACTORY INITIAL OPERATION AND INCLUDES THE DEFECTS AND OMISSIONS LIST.
	↓	COMMISSIONING CONTINUES ALSO RECTIFICATION OF DEFECTS AND OMISSIONS TO A LEVEL WHERE THE NEXT CERTIFICATE CAN BE ISSUED.
BY – SITE MANAGER TO – STATION MANAGER	CLEARANCE FOR STATION OPERATION CERTIFICATE	THIS AUTHORISES STATION TO OPERATE THE PLANT WITHIN SPECIFIED LIMITS AND TO CARRY OUT ROUTINE MAINTENANCE. THE STATION MANAGER SIGNS HIS ACCEPTANCE OF THE AUTHORITY.
	↓	THE RECTIFICATION OF DEFECTS AND OMISSIONS PROGRESSES AND COMMISSIONING WORK IS COMPLETED. LIMITED OPERATION AND ROUTINE MAINTENANCE IS CONTINUED BY THE STATION STAFF.
BY – SITE MANAGER TO – STATION MANAGER	INTERNAL TAKING-OVER CERTIFICATE	THIS NOTIFIES THE STATION MANAGER OF THE INTENTION TO ISSUE THE CONTRACTUAL TAKING OVER CERTIFICATE AND THE EFFECTIVE DATE. THE STATION MANAGER MAY AGREE OR COMMENT ON THE INTENTION.
	↓	LIMITED OPERATION AND ROUTINE MAINTENANCE IS CONTINUED BY THE STATION STAFF. GD AND CD PROGRESSES THE RECTIFICATION OF THE CONTRACTUAL DEFECTS AND OMISSIONS WITH THE CONTRACTOR.
BY – PROJECT MANAGER TO – STATION MANAGER	FINAL INTERNAL CONTRACT CLEARANCE CERTIFICATE	ISSUED ONE MONTH BEFORE EXPIRY OF MAINTENANCE PERIOD AND LISTS ANY OUTSTANDING DEFECTS AND OMISSIONS. RETURNED TO SITE MANAGER AFTER COMPLETION BY STATION MANAGER. CONTRACTUAL FINAL CERTIFICATE THEN ISSUED.

FIG. 2.19 Summary of commissioning certification process

During the final commissioning of the plant, various dates will be agreed for defined stages in the process, up to the full commissioning date.

6.4.1 Plant safety, operational and internal taking-over certificates

(a) *Safety Rules Clearance Certificate* The application of the CEGB Safety Rules has been discussed in Section 6.1 of this chapter. The Safety Rules Clearance Certificate is the first stage in the certification process and is used to indicate to all relevant persons the date and time from which the Safety Rules will apply. Each relevant contractor acknowledges notification by signing a master certificate and retaining a copy. When all relevant contractors have signed, the completed master certificate is issued by the Site Manager to the Station Manager for the necessary actions to be taken.

SRCCs should be issued in their completed state at least two days before the effective date, so that copies can be circulated to the Station Staff responsible for administering the Safety Rules. Also, the Site Manager must supply a list of 'Competent Persons' from his own and the contractors' staff for the receipt and clearance of safety documentation.

(b) *Record of Initial Operation Certificate* This certificate (Fig 2.20) records the date and time of the initial operation of the plant, and any defects and omissions identified at the time.

(c) *Clearance for Station Operation Certificate* Following the initial operation of the plant, defects and omissions are rectified to the point when the plant can be operated by the station staff. A 'Clearance for Station Operation Certificate' (Fig 2.21) is then issued, recording any limitation to the operation of plant without the contractor being present. In the case of a generating unit, there may be a limitation requiring the contractor to be present for unit start-up operation or turbine overspeed testing. This certificate has to contain the signed agreement of the Station Manager for its issue to be valid. As

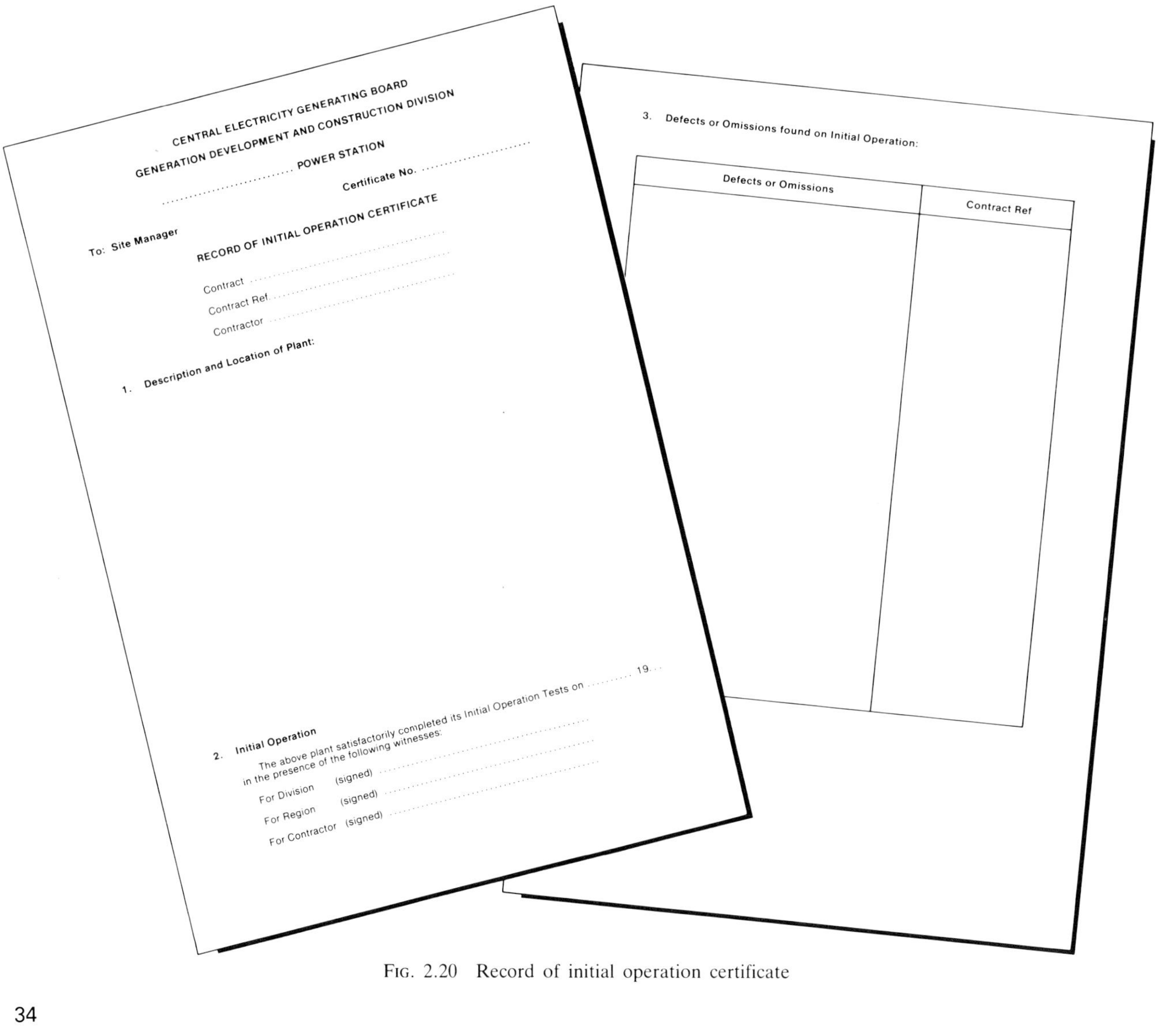

CENTRAL ELECTRICITY GENERATING BOARD
GENERATION DEVELOPMENT AND CONSTRUCTION DIVISION

.... POWER STATION

Certificate No. ....

To: Site Manager

RECORD OF INITIAL OPERATION CERTIFICATE

Contract ....
Contract Ref ....
Contractor ....

1. Description and Location of Plant:

2. Initial Operation
The above plant satisfactorily completed its Initial Operation Tests on .... 19...
in the presence of the following witnesses:
For Division (signed) ....
For Region (signed) ....
For Contractor (signed) ....

3. Defects or Omissions found on Initial Operation:

Defects or Omissions	Contract Ref

FIG. 2.20 Record of initial operation certificate

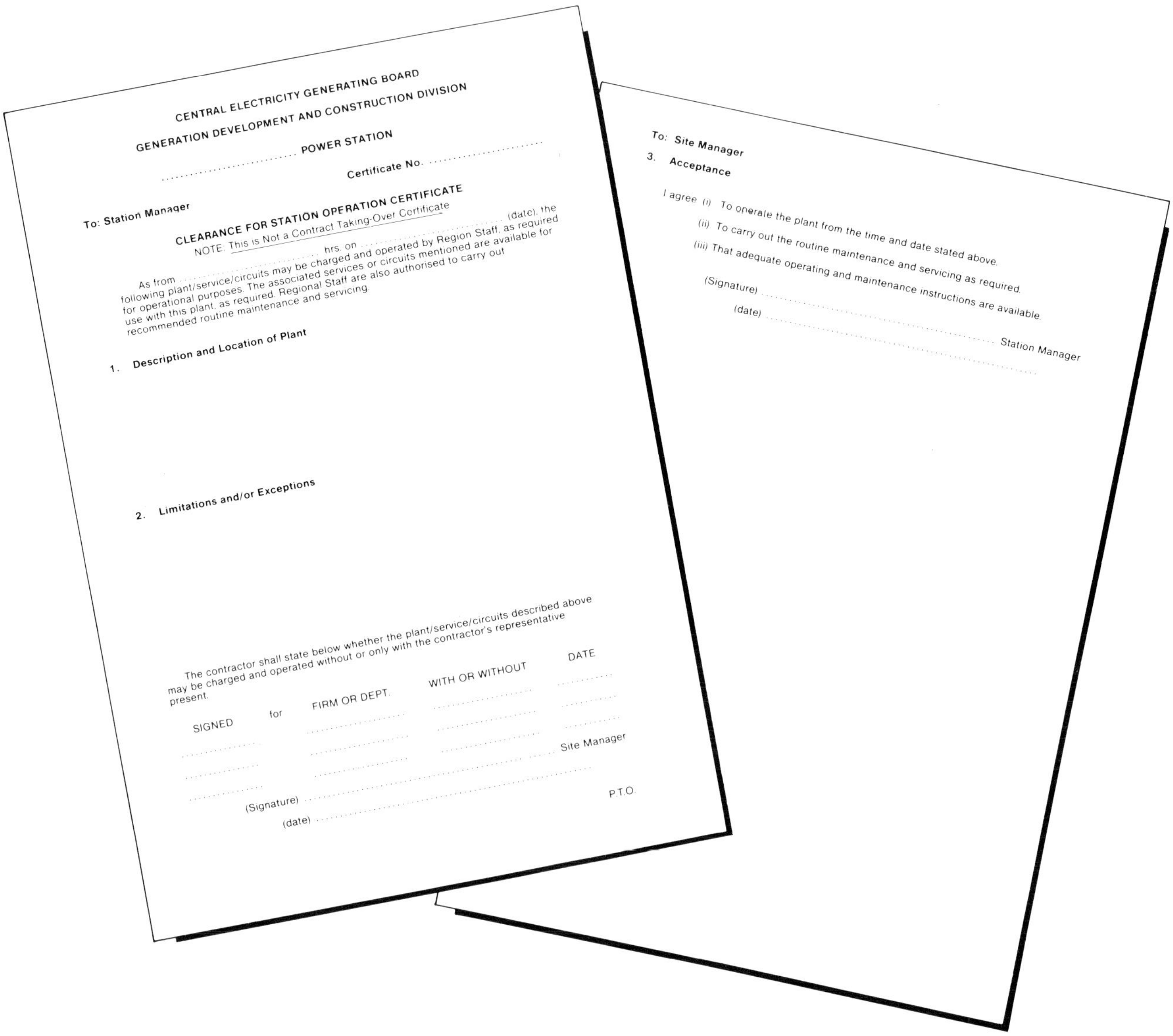

CENTRAL ELECTRICITY GENERATING BOARD

GENERATION DEVELOPMENT AND CONSTRUCTION DIVISION

.... POWER STATION

Certificate No. ....

To: Station Manager

CLEARANCE FOR STATION OPERATION CERTIFICATE

NOTE: This is Not a Contract Taking-Over Certificate

As from .... hrs. on .... (date), the following plant/service/circuits may be charged and operated by Region Staff, as required for operational purposes. The associated services or circuits mentioned are available for use with this plant, as required. Regional Staff are also authorised to carry out recommended routine maintenance and servicing.

1. Description and Location of Plant

2. Limitations and/or Exceptions

The contractor shall state below whether the plant/service/circuits described above may be charged and operated without or only with the contractor's representative present.

SIGNED	for	FIRM OR DEPT.	WITH OR WITHOUT	DATE
....		....	....	....
....		....	....	....
....		....	....	....

.... Site Manager

(Signature) ....

(date) ....

P.T.O.

To: Site Manager

3. Acceptance

I agree (i) To operate the plant from the time and date stated above.

(ii) To carry out the routine maintenance and servicing as required.

(iii) That adequate operating and maintenance instructions are available.

(Signature) ....

(date) .... Station Manager

FIG. 2.21 Clearance for station operation certificate

well as accepting the plant for operation, the Station Manager agrees to carry out routine maintenance and servicing, and confirms that adequate operating and maintenance instructions are available.

The issue of this certificate does not signify a contractual taking-over, the responsibility for the plant still being with the contractor at this stage.

(d) *Internal Taking-over Certificate* This certificate (Fig 2.22) is issued by the Site Manager to inform the Station Manager of the intention to take over the plant from the contractor, in compliance with the appropriate clauses of the conditions of contract and specification requirements. It carries the proposed date of issue of the Taking-over Certificate and indicates that this date will be the start of the maintenance period.

On the Internal Taking-over Certificate, the Station Manager states any observations he may have on the proposal to issue the contractual Taking-over Certificate.

The Project Manager issues the Taking-over Certificate when he is satisfied that the conditions of contract have been met. Any exceptions to the acceptance of the plant are recorded with the certificate. A standard exception is the current contractual defects and omissions list.

During the maintenance period, Regional staff record defects and omissions which come to their notice. Those which are considered to be contractual are indicated to the Site Manager on the Internal Defect Report (see Fig 2.23). The Site Manager, upon receipt of this report, notifies the contractor concerned by issuing him with a Defect Report (see Fig 2.24). When the defect has been rectified, the Site Manager will be notified to this effect by the contractor, and will then issue the Clearance of Internal Defect Report (see Fig 2.25) to the Station Manager for acceptance and record purposes.

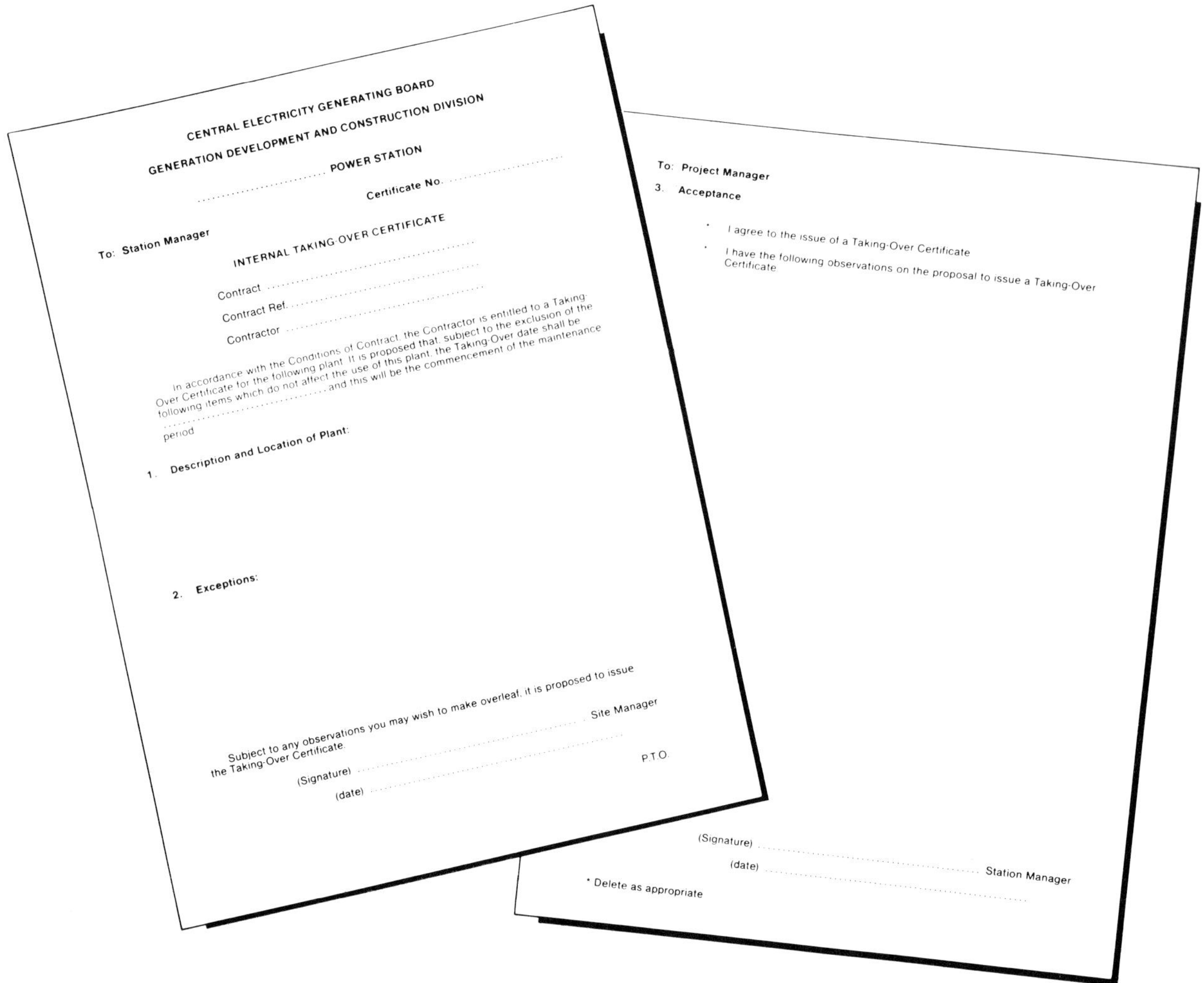
CENTRAL ELECTRICITY GENERATING BOARD
GENERATION DEVELOPMENT AND CONSTRUCTION DIVISION
.... POWER STATION
Certificate No. ....
To: Station Manager
INTERNAL TAKING-OVER CERTIFICATE
Contract ....
Contract Ref. ....
Contractor ....
In accordance with the Conditions of Contract, the Contractor is entitled to a Taking-Over Certificate for the following plant. It is proposed that, subject to the exclusion of the following items which do not affect the use of this plant, the Taking-Over date shall be .... and this will be the commencement of the maintenance period.
1. Description and Location of Plant:
2. Exceptions:
Subject to any observations you may wish to make overleaf, it is proposed to issue the Taking-Over Certificate.
(Signature) .... Site Manager
(date) ....
P.T.O.
To: Project Manager
3. Acceptance
* I agree to the issue of a Taking-Over Certificate
* I have the following observations on the proposal to issue a Taking-Over Certificate
(Signature) ....
(date) .... Station Manager
* Delete as appropriate

Fig. 2.22 Internal taking-over certificate

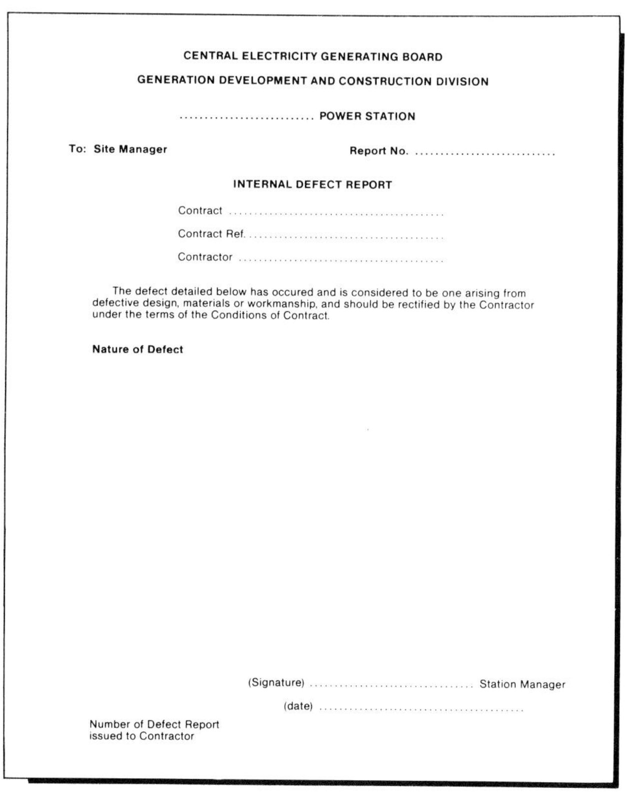
CENTRAL ELECTRICITY GENERATING BOARD
GENERATION DEVELOPMENT AND CONSTRUCTION DIVISION
.... POWER STATION
To: Site Manager Report No. ....
INTERNAL DEFECT REPORT
Contract ....
Contract Ref. ....
Contractor ....
The defect detailed below has occured and is considered to be one arising from defective design, materials or workmanship, and should be rectified by the Contractor under the terms of the Conditions of Contract.
Nature of Defect
(Signature) .... Station Manager
(date) ....
Number of Defect Report issued to Contractor

Fig. 2.23 Internal defect report

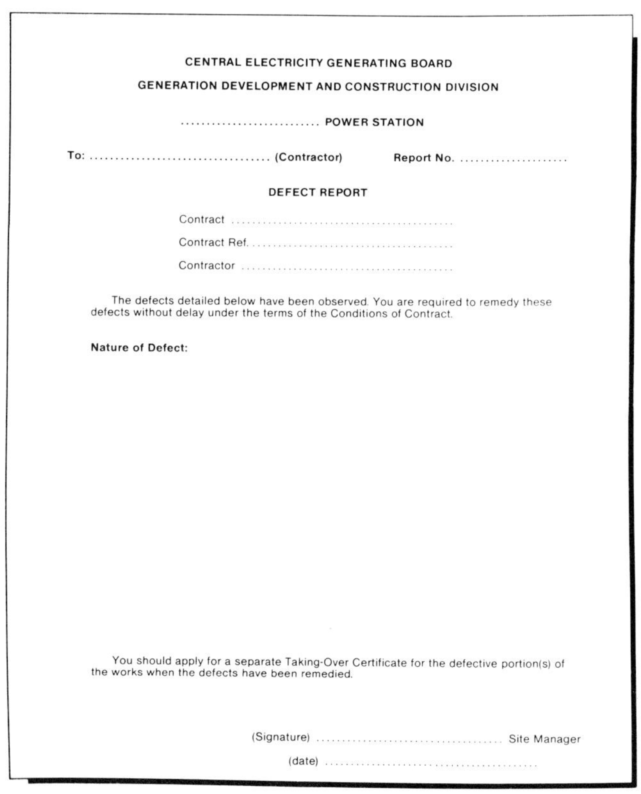
CENTRAL ELECTRICITY GENERATING BOARD
GENERATION DEVELOPMENT AND CONSTRUCTION DIVISION
.... POWER STATION
To: .... (Contractor) Report No. ....
DEFECT REPORT
Contract ....
Contract Ref. ....
Contractor ....
The defects detailed below have been observed. You are required to remedy these defects without delay under the terms of the Conditions of Contract.
Nature of Defect:
You should apply for a separate Taking-Over Certificate for the defective portion(s) of the works when the defects have been remedied.
(Signature) .... Site Manager
(date) ....

Fig. 2.24 Defect report

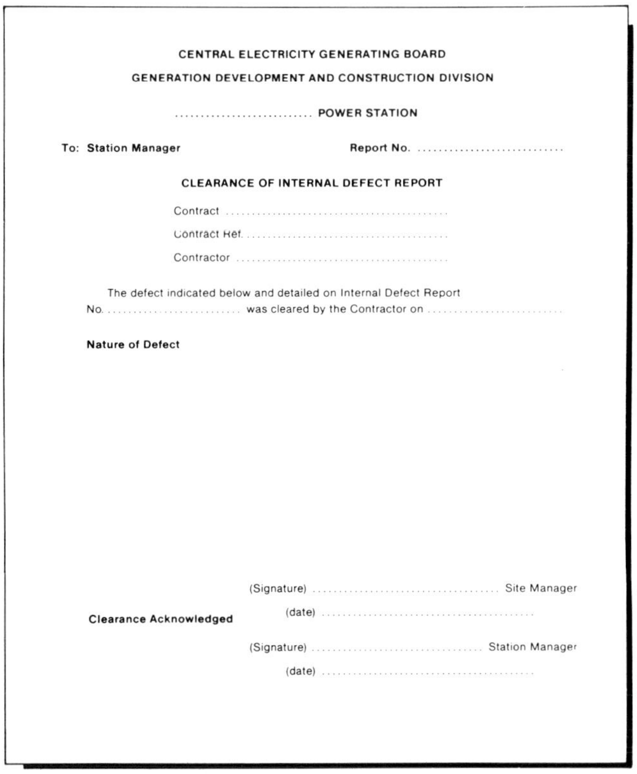
CENTRAL ELECTRICITY GENERATING BOARD
GENERATION DEVELOPMENT AND CONSTRUCTION DIVISION

.... POWER STATION

To: Station Manager Report No. ....

CLEARANCE OF INTERNAL DEFECT REPORT

Contract ....
Contract Ref. ....
Contractor ....

The defect indicated below and detailed on Internal Defect Report No. .... was cleared by the Contractor on ....

Nature of Defect

(Signature) .... Site Manager
(date) ....

Clearance Acknowledged

(Signature) .... Station Manager
(date) ....

FIG. 2.25 Clearance of internal defect report

(e) *Final Internal Contract Clearance Certificate* This certificate (Fig 2.26) is initiated by the Project Manager approximately one month before the expiry of the maintenance period. It lists outstanding defects and omissions which still require attention by the contractor. At this stage, defects and omissions should be concerned with minor points only, and these should not affect the use of the plant. The use of the plant could be said to be affected if:

- Full control to start, stop or regulate the plant is not available.
- Sufficient instruments and protective devices to avoid plant abuse or danger are not available.
- Safe access is not available or there has been a failure to observe all statutory requirements.
- Satisfactory operating instructions are not available.
- The plant is not clearly identified and labelled.
- Its condition involves regular over-manning.

The completed certificate is forwarded to the Project Manager before issue of the contractual Final Certificate.

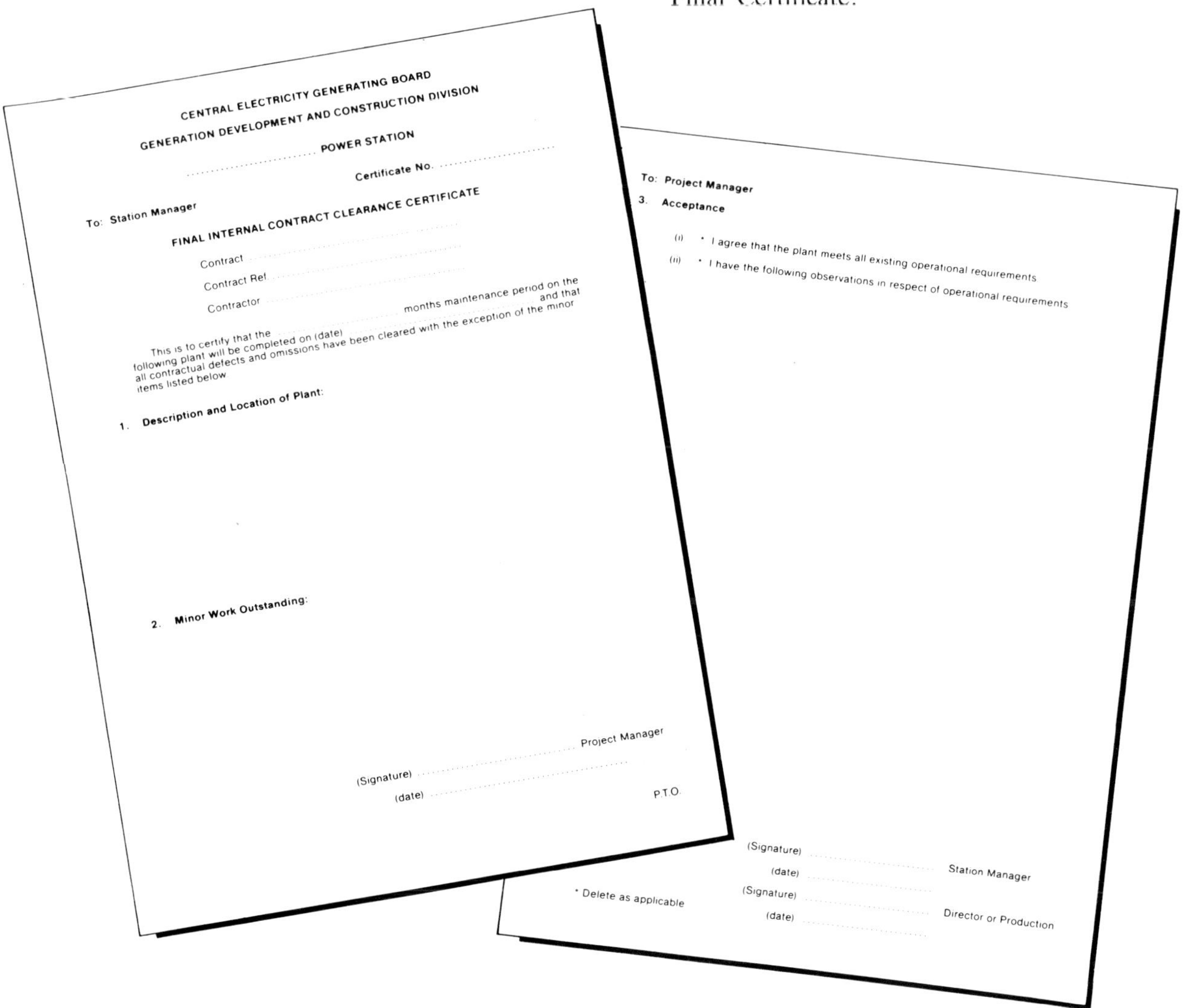
CENTRAL ELECTRICITY GENERATING BOARD
GENERATION DEVELOPMENT AND CONSTRUCTION DIVISION

.... POWER STATION

Certificate No. ....

To: Station Manager

FINAL INTERNAL CONTRACT CLEARANCE CERTIFICATE

Contract ....
Contract Ref ....
Contractor ....

This is to certify that the .... months maintenance period on the following plant will be completed on (date) .... and that all contractual defects and omissions have been cleared with the exception of the minor items listed below

1. Description and Location of Plant:

2. Minor Work Outstanding:

(Signature) .... Project Manager
(date) ....

P.T.O.

To: Project Manager

3. Acceptance

(i) * I agree that the plant meets all existing operational requirements
(ii) * I have the following observations in respect of operational requirements

(Signature) .... Station Manager
(date) ....
(Signature) .... Director of Production
(date) ....

* Delete as applicable

FIG. 2.26 Final internal contract clearance certificate

6.4.2 Plant status

As the commissioning of power station plant proceeds, certain key points are reached, as follows:

(a) *Synchronising date of new generating plant* The synchronising date of a new generating set is the date when it is connected to the system and is deemed by the Station Commissioning Committee to have been synchronised. For steam plant, this will normally be when the generator is first synchronised, using steam raised by its associated main boiler. Early synchronising of generators on steam from auxiliary or temporary boilers, before the main boilers are available, would not qualify.

(b) *Full commissioning date of a boiler* This is the date, agreed between the Director-General of GDCD and the Regional Director-General, on which a boiler has satisfactorily completed all necessary tests and has achieved a continuous 72-hour run. During this run, it must have satisfactorily supplied steam to operate its associated turbine generator at full load and is deemed by the Station Commissioning Committee to have been commissioned.

(c) *Full commissioning date of a turbine-generator* This is the date agreed between the Director-General of GDCD and the Regional Director-General, on which a turbine-generator has satisfactorily completed all necessary tests, including a continuous run of 72 hours at the specified rating, and is deemed by the Station Commissioning Committee to have been commissioned.

(d) *Date of commissioning* The date of commissioning of a generating set or unit is either the date of full commissioning or, if it has not achieved its specified rating, the date when it is first taken into the CEGB *Record of Total Plant Capacity in Service* at an interim rating.

(e) *Interim rating of new plant* Interim ratings will only be declared for newly-commissioned plant where this has not achieved its specified rating, but has successfully completed a continuous run of 72 hours at an average load normally not less than 60% of that rating. The first interim rating will normally be agreed between the Director-General of GDCD and the Regional Director-General on the recommendation of the Station Commissioning Committee.

Appendix A

Schedule of working party activities for new power station plant

General engineering requirements *All plant areas*

Component/sub-system	PCC activity	SCC activity
Mechanical General mechanical plant items including pumps, fans, hydraulic drives, valves, dampers, actuators, pipework, etc.	Cleaning, pressure and leak testing of plant and pipework. Initial proving runs for correct rotation, clearances, lubrication, cooling, heating, balancing, proving of controls and protection.	System operational testing of the plant as part of a sub or main system.
Electrical General electrical plant associated with mechanical plant items including motors, switchgear and cables.	Relevant PCC tests as indicated in 'Electrical power and services' plant area 10.	Relevant SCC testing as in 'Electrical power and services' plant area 10.
C and I General C and I equipment associated with mechanical plant items or systems including indicating and controls, instrumentation, alarms sequence and interlocks instrumentation pipework and wiring, transmitters etc.	Relevant PCC tests as indicated in 'control and instrumentation' plant area 11.	Relevant SCC testing as in 'Control and Instrumentation' plant area 11.

Unit *Plant area 1*

Component/sub-system	PCC activity	SCC activity
Unit Boilers/turbine-generator condensate and feed heating plant	All necessary PCC activity tests and checks completed satisfactorily to their relevant plant areas.	Initial steam to set, synchronising and loading tests. Unit plant loading, operation and control tests. Boiler, reactor and turbine acceptance tests.

Boilers and auxiliary systems *Plant area 2*

Component/sub-system	PCC activity	SCC activity
Pressure parts	Hydraulic test Chemical clean Post chemical clean storage	Float safety valves. Operation of boiler and auxiliaries for STS. Synchronising and unit testing.
Draught plant	Initial runs on motors and fans including fan balancing. Initial operation of dampers including controls, alarms and limits. Initial runs of air heaters to check cold clearances and to set seals.	Initial operational runs of whole system to carry out furnace and ductwork leak tests, venturi calibrations and to establish damper settings. Initial light-up trials on oil to check burner operation and carry out flame and ignition tests. Burner trials on coal including furnace purge tests.
Combustion plant	Simulated burner insertion and withdrawal tests and ABC control tests without fuel. Setting of air registers and primary air dampers and air calibration tests.	
PF milling plant	Initial and independent runs on feeders, mills, classifiers Exhauster balancing runs. Speed control tests of regulators on fluid drives.	Initial runs of all plant with coal. Feeder calibrations, tests for mill throughput PF transport velocities and distribution.
Sootblowing plant	Insertion and retraction tests and full functional tests without air or steam. Testing of automatic controls, sequence and interlocks.	Tests with air or steam to prove system operation, adequacy of coverage, air or steam usage and sootblowing routine and frequencies.
Precipitators	Initial mechanical runs on rotating rectifier equipment, rapper gear and vibration equipment. Initial HT energising and testing.	Flow distribution tests. Functional tests on-load. Precipitator efficiency test.
Dust handling plant	Initial runs on pumps, vacuum plant and compressors, filters and vibrators including tests on control equipment, dampers and valves. Bunker and air slide dry-out using dry air.	Testing of plant using imported ash or dust. Operational tests with boiler on load. Performance tests to establish capabilities and capacities both for handling and disposal including main bunkering equipment.
Ash handling plant	Initial runs on pumps, crushers and mechanical handling plant such as cranes, grabbing equipment and transporters.	

Plant area	PCC activity	SCC activity
Boilers and auxiliary systems	No PCC activity requirements	72-hour reliability run. (Unit test). Boiler efficiency test. (Unit test).

Turbine and condensate system *Plant area 3.1*

Component/sub-system	PCC activity	SCC activity
HP/IP and LP turbine governor and valve gear	Steam purge of HP pipework. Governor and turbine valve tests (cold). Initial turning on barring gear with generator. Initial raise vacuum tests in cold state (air leakage).	Unit SCC testing for load control.
Bearing, jacking and governor oil systems	Cleaning and flushing of each system and initial oil fill for commissioning. Initial circulation of lubricating oil through bearing and jacking oil systems and governor tests without steam.	Operation and control tests on start-up and on load.
Barring gear	Motor tests and interlock checks. Engagement, disengagement and operational trials without steam.	Tests with machine taking steam.
Gland steam system	System cleaning and steam purging of pipework. Tests to set up desuperheater controls using auxiliary steam.	Operational and control checks on-load. Live steam and spindle leak-off tests.
Condensate and condensate extraction plant	Condenser and condensate cleaning and leak testing. Initial condenser filling and make up capacity tests. Initial runs of extraction pumps and pumping tests through LP plant to deaerator.	Unit SCC testing.
Turbine and condensate system	No PCC requirement.	Vacuum raising with glands sealed. Initial plant proving and pre-synchronising tests to verify plant operation and control. Turbine overspeed tests, synchronising and on load operational tests (Unit test). Load rejection tests (Unit test). 72-hour reliability run (Unit test). Turbine heat rate test (Unit test).

Feed heating system *Plant area 3.2*

Component/sub-system	PCC activity	SCC activity
LP and HP feed heating plant including deaerator	System cleaning and leak tests including heater drains system. Simulated testing of heater steam and water side protection including flooding and HP heater bypass operation. Initial filling of deaerator and steam coil and electrical heating tests. Tests of feed regulating valves to prove control and operation.	Pumping trials and operation tests of HP and LP systems. On-load testing of feed regulating valves. Heater drains control tests.
Start and standby BFPs (motor driven)	Pump pressure test. Speed regulator plant proving tests. Initial runs of pump motor uncoupled for rotation and speed control tests.	Initial operation on leak-off. On-load speed control tests. Flow control tests.
Boiler feed pump turbine	Oil systems flushing and cleaning and initial lubricating oil pumping and circulation tests. Governor and turbine valve tests (cold) and initial turning on barring gear. Steam purge of HP pipework.	Initial steaming runs and overspeed tests uncoupled. Initial coupled pumping runs. Speed/load control tests.

Plant area	PCC activity	SCC activity
Feed heating	No PCC activity requirements.	Operational control tests of plant without steam. LP and HP heater tests on steam. Feed heater efficiency test (Unit test).

Generator/generator and unit transformers *Plant area 4*

Component/sub-system	PCC activity	SCC activity
Main generator and exciter	Electrical current injection or voltage pressure tests for establishing and/or checking the insulation resistance, polarisation index, CT and VT ratios, magnetisation curves, circuit and winding polarities. Hot water or warm air dry-out. Stator HV pressure test. Electrical and mechanical testing of rectifiers, main and pilot exciters, AVR and associated control and switching equipment.	Excitation tests and setting up of AVR with machine at speed/load. Windage dry out with machine at speed. Open circuit and short circuit tests.
Seal oil system	System oil flush and leak tests. Initial runs on pumps and tests to set up seal flows and controls with machine at rest and on barring.	Check with turbine at speed/load, to check temperature setting, pressure differentials and plant operation.
Stator coolant system (Hydrogen and stator water)	Leak and pressure testing of gas and water side systems. Initial runs of pumps and circulation checks. Initial purging of hydrogen system with N_2 or CO_2.	Stator temperature control tests with machine on-load. Initial fill of H_2 and consumption tests on load.
Main connections	Electrical injection or pressure tests for establishing CT and VT ratios, mag curve and polarity, conductor IR, HV and joint resistances.	Heat run.
Generator transformer	Electrical injection and pressure tests for establishing and/or checking core and winding ratios, IRs, mag currents and vector groupings. Functional tests on tap changing equipment, oil cooler and pumps.	Cooling system tests on load.
Generator, generator and unit transformer	Protection tests with machine at rest.	Protection tests with machine at speed and on-load (Unit test).

Fuel handling and storage *Plant area 5*

Component/sub-system	PCC activity	SCC activity
	Coal handling and storage	
Merry-go-round railway sidings and signalling	Testing of merry-go-round sensing devices, train control and marshalling equipment, wagon hopper door-operating devices. 'Dry runs' on paddle feeders.	Operational testing of railway marshalling, signalling and controls. Loading tests. Flow control tests. Sequence and interlock control tests. Emergency and routine safety control system tests. On-load environmental testing of ventilation systems.
Stocking and reclaiming plant	Conveyor belt tensioning and alignment. Testing of feeders, chute flap valves, magnetic separators, fuel sampling equipment and safety devices.	
Bunker equipment	Testing of shuttle conveyors, coal distributors, ventilation equipment.	
	Oil handling and storage	
Jetties and unloading equipment	Testing of unloading arms. Testing of valves. Pressure testing and cleaning of pipework. Testing of trace heating and cathodic protection.	Operation and equipment control tests for oil unloading, transfer and circulation. On-load pressure and control tests.
Tanks and tank heating	Hydrostatic testing of tanks. Cleaning and leak test of steam heating system. Civil foundation settlement tests during filling.	
Pumping and heating plant	System cleaning and leak testing of plant and pipework. Initial runs of pumps. Pipework trace heating tests.	

Water treatment plant *Plant area 6*

Component/sub-system	PCC activity	SCC activity
Raw water delivery and storage plant	Plant and pipework cleaning, leak and pressure testing.	Initial delivery and storage of regeneration and neutralising chemicals.
Pre-treatment plant	Anion, cation and mixed bed initial flow and control tests.	Plant reliability runs.
Main plant	Pre-coat plant tests.	Regeneration tests.
Regeneration plant	Effluent system plant initial operation runs.	Main plant throughput tests.
Polishing plant	Overall plant control equipment tests.	Polishing plant tests at temperature and load.
Treated water storage plant		Acceptance tests for water quality, throughput, resin life, efficiency.
Acid/caustic handling and storage plant		

NB: All tests using acids or alkalies, including initial delivery and storage, to be classed as an SCC activity.

Circulating water and auxiliary circulating water system *Plant area 7*

Component/sub-system	PCC activity	SCC activity
Civil works, culverts and main CW valves	Leak and pressure testing of culverts, valves, forebay and tunnels, etc. Pump discharge valve tests to optimise settings.	Initial flooding of system including condensers. Initial runs on main inlet screens. Initial CW pump runs. Surge tests. Loss of single CW pump. Stability and load sharing tests.
CW pumps, motors and auxiliaries	Motor rotation checks, gland sealing system and bearing lubrication system cleaning and testing.	
Screens	Initial dry runs on screens. Testing of flood protection equipment.	
Dewatering plant	PCC activity tests as for general engineering requirements.	Dewatering tests of selected sections of system.
ACW plant	PCC activity tests as for general engineering requirements.	Flooding and proving of ACW system and cooler tests.
Chlorination plant	PCC activity tests as for general engineering requirements.	System injection tests.

Plant area	PCC activity	SCC activity
CW system	No PCC activity requirements.	System surge tests. Recirculation tests. Acceptance tests.

NB: All tanks using chlorine to be classed as an SCC activity.

Station auxiliary services *Plant area 8*

Component/sub-system	PCC activity	SCC activity
Air systems plant	Tests as for general, mechanical, electrical and C and I plant and equipment.	Individual system testing to prove functional capability of each system, overall operational control, output, plant changeover and loss of main auxiliaries.
Gas systems plant		
Water systems plant		
Fire detection and protection systems		Deluge tests of selected plant items or areas. Simulated tests to prove fire detection equipment, operation and controls.
Auxiliary boilers and protection systems	Plant cleaning and pipework steam purge. Plant testing of individual items. Manual and auto control equipment tests. Float safety valves.	Feed system and chemical control tests. Boiler manual and auto control tests. Auxiliary steam system loading and performance.
Cranes and lifts	Plant operational control tests. Testing of emergency braking systems and/or arrester gear. Proof load testing.	No SCC activity testing requirements.

Gas-turbines — not applicable to stand alone GT stations *Plant area 9*

Component/sub-system	PCC activity	SCC activity
Fuel handling and storage	Storage tank hydrostatic tests. Pipework oil flush, leak and pressure tests. Tests on pumps, motors, filters and heaters without fuel oil.	Operational tests of plant with fuel oil.
Gas generator and power turbine	Lubricating oil system flush, leak and pressure test. Tests on pumps, motors, compressors, oil heaters.	Initial start-up tests and engine balancing. Turbine auxiliaries functional tests.
Main generator, exciter and AVR	Electrical injection and pressure tests for IR, polarisation index and rotor resistance measurements. AVR, intertripping and protection tests with machine at rest.	Open circuit and short circuit tests. Synchronising tests. AVR, tests with machine at speed/load.

Plant area	PCC activity	SCC activity
Gas turbine	Overall protection and tripping tests.	Initial synchronising and loading tests. Operational tests of control equipment. Auto synchronising tests and load pick-up. Dead station start tests. Performance and reliability trials. Dead station start load pick-up and load transfer on loss of grid. Parallel operation tests.

Electrical power and services *Plant area 10*

Component/sub-system	PCC activity	SCC activity
Transformers	Core and winding IR tests. Ratio, magnetising current and vector group tests. Cooling system functional tests. Winding temperature and buchholz tests. Tap changer operation tests.	On-load temperature rise tests. On-load tap changer operation. Noise and vibration checks. System loading and power flow checks.
Switchgear	HV and ductor tests. CT, wiring, protection and operation tests.	No SCC tests other than in combination with switchboard or plant tests.
Cables	IR and HV tests. Partial discharge tests.	System loading temperature tests. Ambient conditions checks.
Batteries and chargers	IR and charger operation and battery load tests. Charter output and voltage variation tests.	Operational loading and battery room atmospheric conditions checks.
Motors and control equipment	IR and continuity tests. Rotation checks. Initial runs to prove plant items.	Operational loading checks. Low voltage starting tests.
Protection	Primary and secondary injection tests. Relay and intertripping tests.	Discrimination proving tests. Stability tests.

Plant area	PCC activity	SCC activity
Station electrical system	No PCC activity requirement.	Prove automatic changeover, emergency essential supplies. Overall system loading and regulation tests.

Control and instrumentation *Plant area 11*

Component/sub-system	PCC activity	SCC activity
Display, recording and control instrumentation	Calibration, range and functional tests.	Tests to be included in plant test programme.
Computers and data processors	Computer and peripheral equipment tests, including reliability tests, using manufacturer's software. Testing and control of necessary functions using operational software and simulated plant inputs.	Functional tests using operational software with actual plant inputs.
Control system	Control circuit logic and loop tests including switching and relay equipment (dry runs). Loss of 'supply' tests (electrical and mechanical) with plant inputs simulated.	Tests with plant operational to be included in plant test programme.
Sequence, supervisory and interlock equipment	Logic and functional tests of equipment using simulated inputs. Reliability testing.	Functional tests with plant operational.
Alarm equipment	'End to end' tests using simulated inputs. Reliability testing.	Functional tests to be included in plant test programme.
Communications systems	Transmission and reception tests for coverage and interference.	No SCC testing.
Station metering equipment	No PCC testing.	Metering accuracy tests.
Instrument supplies	Tests for voltage and frequency variations. Loss and change over of supplies tests, with plant inputs simulated where necessary.	No SCC testing.

Site Civil and building services *Plant area 12*

Component/sub-system	PCC activity	SCC activity
Site drainage	Leak tests and inspections of open and closed drainage systems. Continuity and segregation tests.	No SCC activity testing requirement.
Site fire mains	Pressure and leak testing of pipework, hydrants and valves. Hydrant discharge tests.	No SCC activity testing requirement.
Domestic services	Pressure and leak testing of cold and hot water service pipework and valves. Drinking water fluorescence tests for segregation and identification.	No SCC activity testing requirement.
Street lighting	Electrical supply tests for phase loading and control. Illumination level checks.	No SCC activity testing requirement.
Heating and ventilation plant	Tests as for general, mechanical, electrical and C and I.	Overall system testing to optimise individual room or building controls. Performance testing.
Lagoons and reservoirs	Inspections and clearance for initial and stage filling.	Initial filling and monitoring of levels. Tests on clean effluent discharge.

CHAPTER 3

Common equipment and station plant commissioning

Introduction

Principles for the organisation and documentation of commissioning are outlined in Chapter 2. This chapter deals with the practical approach to commissioning of common equipment and station plant in modern fossil-fuelled power stations in the UK. Commissioning of main boiler/turbine-generator units is covered in Chapters 4, 5 and 6.

Scope

The first four sections of this chapter outline commissioning checks for some of the more common items of equipment installed throughout a power station site. Valves, dampers, actuators, instruments, electric motors, small pumps and fans come into this category. Section 5 deals with different types of station lifting equipment, much of which is usually commissioned relatively early in the construction programme to provide additional facilities for plant erection. Though commissioning of data processing, alarm and common control systems is discussed in Sections 6, 7 and 8, on-line proving of each will be found in the appropriate sections of subsequent chapters. The remainder of this chapter is devoted to the commissioning of various individual systems, classed as station plant, necessary to support commissioning and operation of the main generating units.

Content

As a general rule, each section has been subdivided into the following headings:

System description, to give the reader sufficient detail to be able to understand the commissioning activities.

More detailed information will be found in companion volumes in this series, other published works, or manufacturers' literature.

Commissioning objectives, which outline the objectives specifically applicable to the system involved, though in each case the following general objectives also apply:

- Make plant available for commercial operation as soon as possible, consistent with contractual requirements and statutory obligations, particularly safety.
- Establish that the plant meets the requirements expressed in the Station Development Particulars.
- Demonstrate that the plant is capable of safe and reliable operation and control and that it meets its design specification and performance guarantees.
- Identify and establish the reasons for any limitations of either performance or flexibility of operation.
- Prove the validity of the plant operating instructions.

Safety considerations, where specific hazards associated with the particular system are highlighted. It is important that the following general aspects are also taken into account:

- A Safety Rules Clearance Certificate (SRCC) must be in force and all terminal point isolations secured before charging or energising any plant.
- All plant and equipment must be clearly identified with its agreed title. The contents must also be stated where chemicals are involved.
- Lighting must be adequate to promote a safe working environment. Where practicable, emergency lighting also should be available.
- Safe access to and around all plant being commissioned must be maintained at all times. Clearly identified escape routes will be necessary in some instances.
- Access to be restricted to personnel necessarily involved with testing. Temporary barriers may be required to inhibit access of unauthorised personnel.
- Permanent fire protection equipment has to be in service, augmented by additional staff and/or temporary extinguishers where necessary.
- Plant being commissioned should be clearly marked with durable, easily-observed, self-adhesive labels to warn personnel which sections of the plant are subject to the Safety Rules, e.g. COMMISSIONED, LIVE, PFW REQUIRED, UNCOMMISSIONED CIRCUIT, etc.

Resources, to identify any additional services, including staffing, needed to progress commissioning of the particular system. Anticipated programme times are given where appropriate.

Commissioning, in which the basic principles of the activities necessary to set the plant to work are outlined. Simplified commissioning networks are also given where appropriate. Figure 3.1 shows a typical overall programme for commissioning of the various systems encompassed within this chapter in order to meet key commissioning events for the first main unit.

It is self-evident that standards of commissioning for station plant have to be similar to those for the main units, if continued availability of the latter is to be assured.

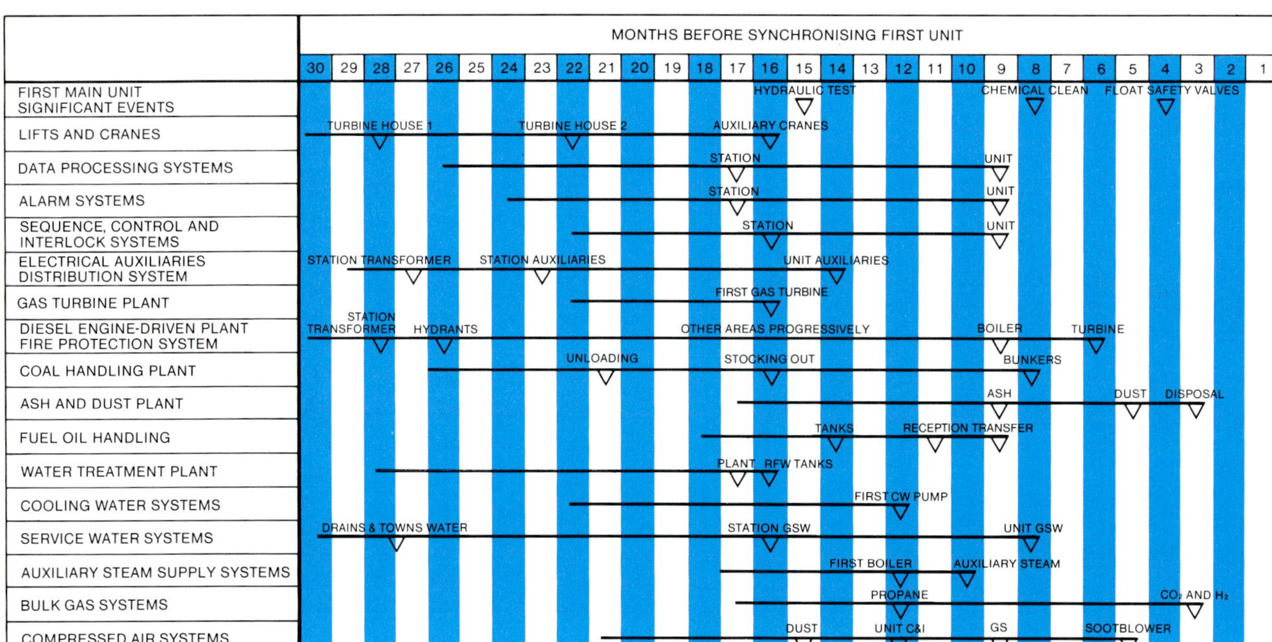

FIG. 3.1 Overall commissioning programme

1 Valves and dampers

This section outlines basic commissioning checks for various types of valves and dampers fitted to modern plant. All valves and dampers employed for process fluid regulation, as well as many used for isolating purposes, are fitted with an electric, pneumatic or hydraulic actuator. Most of these actuators are arranged for operation from a remote control position. Commissioning of these associated electric, pneumatic and hydraulic actuators is covered in Section 2 of this chapter.

1.1 Valves

One Standard Check List (SCL) should be adequate for the common types of valve. Others may have to be produced for some of the more specialised designs of valve, e.g., modulating control valves, boiler safety valves, etc.

Figure 3.2 shows a typical SCL for simple valves of the following types:

- Parallel-slide
- Wedge-gate
- Globe
- Non-return, (check or reflux)
- Butterfly
- Ball
- Diaphragm
- Relief.

For a detailed description of the construction and operation of each type, reference should be made to the appropriate volume in this series.

It is necessary to have comprehensive Valve Schedules, which contain detailed manufacturing information about each valve, to cross-reference with completed check lists.

The following points in connection with Fig 3.2 are worthy of further explanation.

- *Responsibility* This column is used to denote the contractor or organisation responsible for conducting the particular check.
- *Witnessed* Provision is normally made for representatives of the contractor, the power station manager and the construction division (GDCD) to record their acceptance that the check has been carried out satisfactorily.
- *Safe access* (item 2) This check ensures not only that personnel have safe access to operate the valve, but also that their hands are not liable to injury whilst doing so.
- *Direction of flow* (item 4) This is particularly important in the case of globe and non-return valves. It is customary to recheck this aspect during system commissioning.
- *Locking facilities* (item 6) Manual valves are often used for isolation purposes, so it must be possible to apply padlocks to secure the points of isolation before the issue of safety documentation.

.... POWER STATION

SCL No. M101
Sheet 1 of 1
Ref. No. ....

STANDARD CHECK LIST – VALVES

Valve Identification Number .... Valve Schedule Reference No. ....
Valve Agreed Plant Title ....
System Drawing Number .... Contract Reference No. ....
Manufacturer's Name ....
Figure No. .... Type .... Serial No. ....
Actuator Make .... Type .... Serial No. ....
Output Speed ....

ITEM	DESCRIPTION	RESP.	WITNESSED CONT.	STATION	GDCD	DATE
1.	Check that valve identification label agrees with details on schedule					
2.	Check that access is safe for operational purposes					
3.	Check receipt of manufacturer's test certificate. Record ref. no.					
4.	Check that valve is correctly fitted into pipeline for:- a) direction of flow b) adequacy of supports c) tightness of bolts d) correct joint and gland packings					
5.	Check that valve is correctly lubricated and is free to operate over full travel.					
6.	Check that locking facility is fitted					
7.	Check that local position indication is correct					
8.	Check and set limit switches (if fitted)					
9.	Check free operation with actuator fitted. Note actuator SCL ref. number					
10.	Check that set operating pressure of relief valve agrees with test certificate					

FIG. 3.2 Typical standard check list (SCL)

- *Limit switches* (item 8) Normally these are integral with the actuator. Some manual valves have limit switches for remote position indication, or for plant interlocking. This check ensures that each switch is set to operate within about a quarter of a turn of the valve handwheel reaching its extreme position.
- *Relief valves* (item 10) Generally this covers the small safety valves fitted to compressed air, oil fuel, low pressure steam and other systems. Each valve should be bench-tested to check that the actual setting agrees with that stamped on the valve body. It is sometimes practicable to subsequently carry out an *in-situ* proof test using the system process fluid.

Specialised designs of valve

As mentioned earlier, additional checks may be required for some specialised types of valve. These will depend upon the function and design of the valve. Examples of the approach are given below.

Modulating control valves Figure 3.3 shows one design of a typical balanced double-seat modulating control valve. The figure also gives typical characteristic curves for different types of control valve. The characteristic of a control valve is the relationship

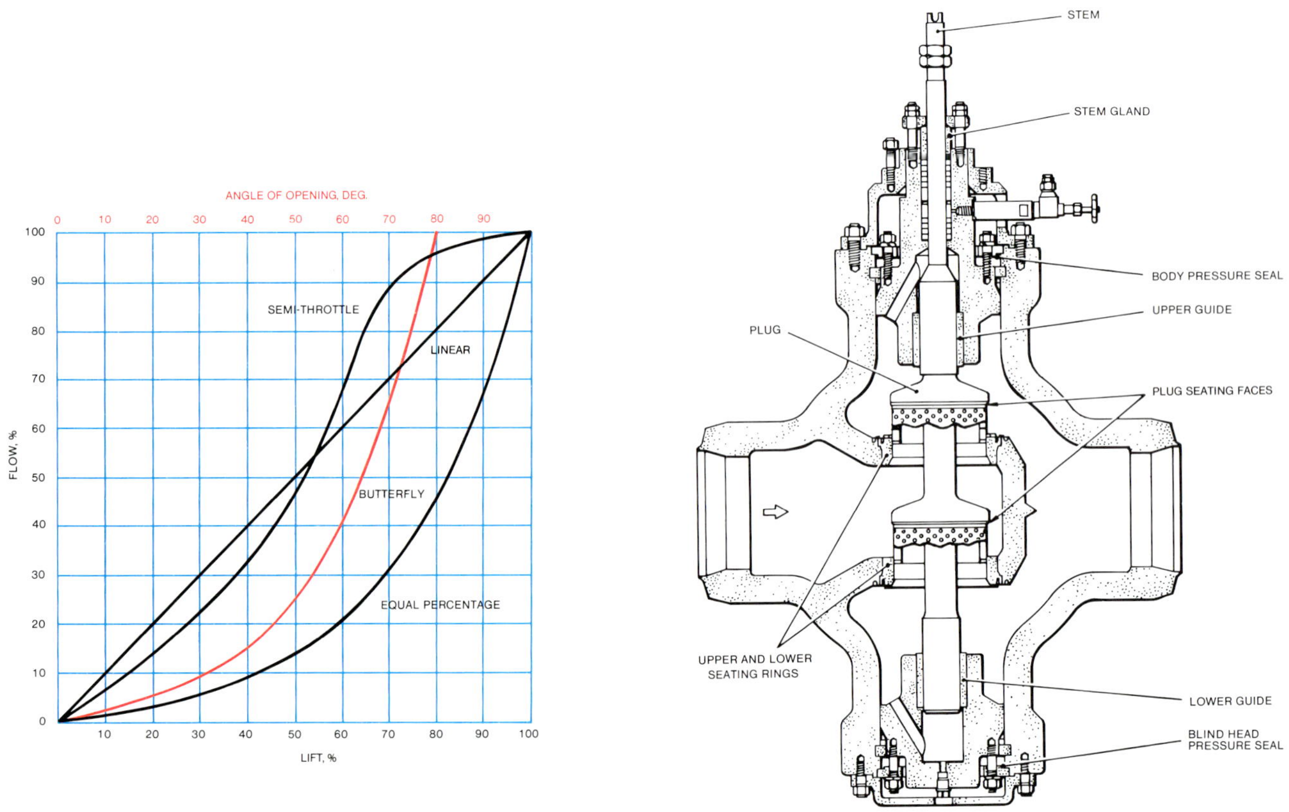

FIG. 3.3 Typical control valve and flow characteristics

between the rate of fluid flow and the lift of the plug at a constant pressure drop. Three basic forms of valve plug, sometimes known as valve trims, are called:

- Quick-opening or semi-throttle, where the flow increases rapidly with plug lift and approaches its maximum value after only 75% of valve movement.
- Linear, where the flow is directly proportional to valve lift.
- Equal percentage, where the flow changes by a constant percentage with unit change in plug lift.

One application of a valve which has an onerous duty to perform is that for the regulation of feedwater flow to a boiler. Because such valves are usually machined to relatively fine clearances, the valve plugs are not fitted until after all debris has been flushed out of the HP feed system. Typical checks are:

- Valve plug (or trim) has the specified lift/flow characteristic.
- Seats and plug are of the correct material and are free to move without binding after installation.
- Valve stem is located to prevent rotation about its axis, since even balanced valves can be subject to rotational forces.
- Plug shuts firmly without shock, and is pressure-tight when closed.
- Valve stem gland and body pressure seals are correctly set before the valve is subjected to system pressure, and then readjusted afterwards where appropriate.

CW pump discharge valves A large diameter butterfly valve, often fitted with an inflatable seal, is employed for this duty. The construction is shown in Fig 3.4, which illustrates the arrangement for inflating the circumferential sealing hose. The hose is normally inflated by means of the system internal water pressure, but it can be inflated by air or water pressure applied via shaft drillings, shuttle valves and interconnecting tubing. The shuttle valves are connected so that sealing water pressure can be obtained from either the curved or the flat face of the valve dependent upon which is at the higher pressure. For maintenance isolation purposes, the pressure is usually applied through the shaft, the shuttle valves automatically sealing against any water ingress from the CW system.

The characteristic for a butterfly valve of this type is also shown in Fig 3.3.

Typical additional checks are:

- Confirmation that the circumferential sealing hose is undamaged and that it has has been correctly fitted and secured to prevent inadvertent displacement or damage as the valve operates.
- Pressurising tubing and shuttle valves to be correctly mounted and proved free from leaks. Where practicable, the tubing should be blown clear to ensure that no debris, likely to inhibit shuttle valve operation, remains.

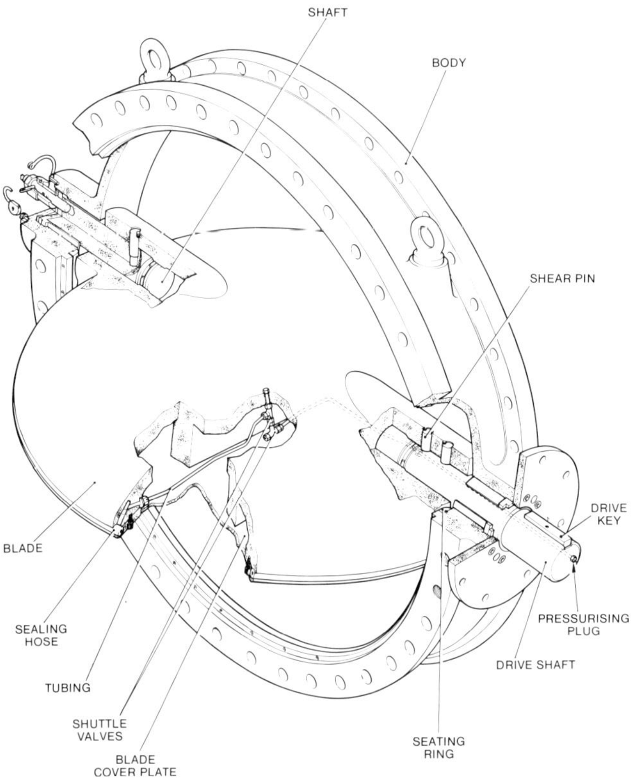

FIG. 3.4 Large butterfly valve (with inflatable seal)

- Arrangements made, involving temporary connections where appropriate, to demonstrate correct inflation of seal from each source of pressure. It is important to prove that the sealing hose deflates as the pressure is released and before the valve starts to open.

Safety valves By custom and practice this term tends to be applied to the more specialised designs of relief valve used mainly on steam boilers. In common with many relief valves, the proof setting has to be witnessed by an authorised representative of an approved insurance company to comply with statutory regulations. Detailed test procedures have to be developed, since safety valves and their escape pipes are always proof tested by overpressurising the boiler. This aspect is more fully covered in Chapter 4.

1.2 Dampers

Though dampers are normally identified with boiler plant, many are to be found in heating and ventilating (H and V) systems. Because of the large cross-sectional area of gas and air ducts in modern boilers, most dampers have multiple blades, or leaves, interconnected by external linkages.

Most of the items on the valve check list (Fig 3.2) are equally applicable to dampers. The following additional features are especially relevant:

- Check correct assembly of damper blades and linkage.
- Check free movement over full travel.
- Confirm that indication marked on damper shaft truly agrees with the actual blade alignment.
- Check that the blades are parallel with gas flow in the fully open position.
- Check that damper blades seal effectively in the closed position.
- Confirm that the environmental location is unlikely to cause operational problems, e.g., bearing seizure, blades fouling each other or the ducting, dust build-up, etc.
- Ensure that each damper can be secured (and locked) at the fully-open and fully-closed positions. This is important where access to certain sections of ducting may have to be achieved by climbing through the open damper. Usually, however, dampers are used for safety isolation when in the shut position.
- Check that limit switches operate some 3 to 5 degrees before the damper reaches its extreme limits of travel. If an actuator is fitted, then the actuator limits must operate just before the damper reaches the mechanical travel stops, with due allowance for any overtravel of the linkage.

In the case of H and V ducting, fire dampers or automatic shutters are fitted at strategic points and particularly where ducts pass through fire-break walls. These dampers are designed to automatically close when fire occurs in order to prevent the passage of smoke, heat or flame. The damper may be held open either by a heat sensitive device in the duct, e.g., a fusible link, or by an electromagnetic latch mechanism released by a heat or smoke detector in an adjacent room. Closure is then effected by gravity or by a spring. Each section of ductwork incorporating a fire damper must be provided with a readily accessible removable panel to facilitate maintenance and testing. The mounting has to be checked to confirm that the normal direction of airflow acts to assist closure of the damper. Fire dampers must be proved to close smartly and seal effectively when the restraining link is released. Random testing of selected fusible links by the application of heat from a portable electric hot-air blower is advisable. Typically fusible links release at a temperature of about 74°C.

Another type of fire damper is in the form of a honeycomb matrix of paper or aluminium coated with an intumescent paint. When subjected to temperatures above about 200°C the paint expands and carbonises to form a complete seal across the duct. Other than ensuring that the matrix is undamaged and is properly sealed to the duct walls no specific checks are possible.

2 Actuators

Where continuously variable (modulating) position control is required, a pneumatic or a hydraulic actuator is usually fitted. As a general rule, electric actuators are fitted where the driven device has an isolating or non-modulating role. However, the application of thyristor controls for modulation of conventional electric actuators is meeting with considerable success.

All types of actuator can be arranged for remote manual control and/or automatic operation in conjunction with centralised control systems, as outlined in Section 8 of this chapter.

This section deals with commissioning of the three main classes of actuator usually fitted to valves, dampers, etc.:

- Electric
- Pneumatic
- Hydraulic

Because of the diversity of actuator types in use, only the basic philosophy, citing some typical examples, can be outlined here.

2.1 Electric actuators

2.1.1 Description

An electric actuator is essentially a motor-driven, speed-reducing gearbox, the output shaft of which actuates the valve, damper or other device. An auxiliary shaft operates limit and torque switches for both OPEN and CLOSE movements. Usually this shaft also drives local and remote position indicators, plus banks of additional limit switches for other interlocking and control functions. Figure 3.5 shows the general arrangement of a typical electric actuator.

Each actuator is provided with OPEN-STOP-CLOSE pushbuttons and is capable of manual operation by means of an integral handwheel. When the electric motor is energised, the mechanism first disconnects the manual handwheel drive automatically. All of the components are housed in a weatherproof enclosure, capable of operation under adverse environmental conditions of steam, dust and temperature.

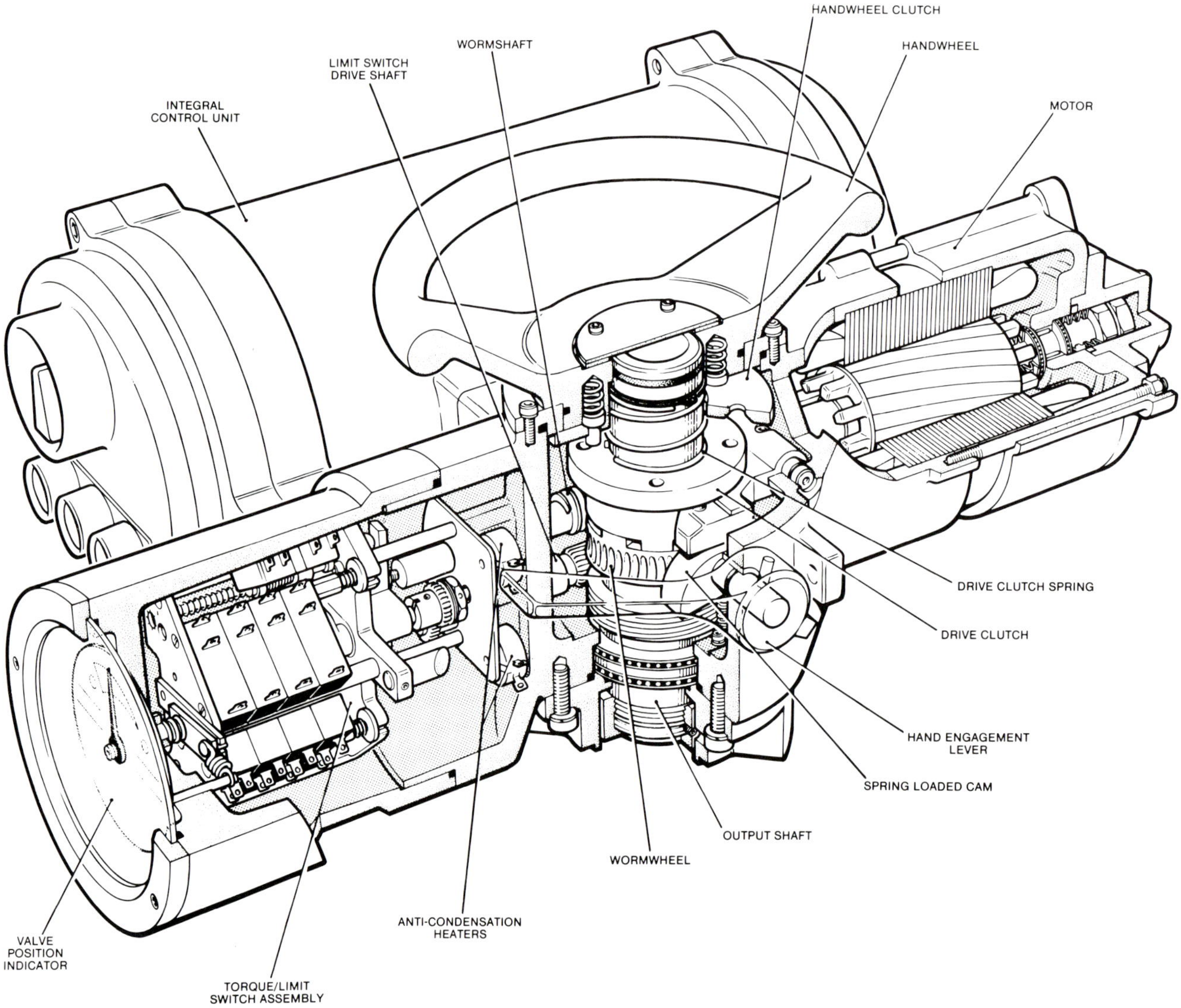

Fig. 3.5 General arrangement of electric actuator

Power supply is generally at 415 V AC three-phase through a locally-mounted lockable isolating switch which may incorporate the OPEN-STOP-CLOSE pushbutton controls for LOCAL TO PLANT operation.

Dependent upon the station design philosophy, the control equipment may be located in one of three places:

- In strategically located valve group-starter boards.
- Integral with the actuator enclosure.
- In a locally-mounted cabinet.

Control equipment usually comprises:

- 110 V AC electrically-held reversing contactors.
- Overload, single-phasing and undervoltage protection.
- 415/110 V control transformer.
- Control selector switch for REMOTE, LOCAL TO PLANT, or SWITCHGEAR operations.
- Pushbuttons for SWITCHGEAR operation.
- Interposing relays for control from the REMOTE (e.g., central control room) location.

For effective positioning of electric actuators when being used for a modulating duty, the reversing contactors have to be replaced by a thyristor controlled reversing switch module. Such modules pulse the actuator to its desired position then bring the motor to a standstill by DC braking. Precise position control is achieved by using relatively short pulses, typically of 50 ms duration, giving much shorter positioning times (e.g., 10 to 15 s) than reversing contactors can. Many actuator motors now have thermistors embedded in the stator windings. These are arranged to trip the power supply contactor if the sensing elements detect a temperature in excess of 120°C to 140°C. For further details, refer to Section 9 of this chapter.

Actuators should be capable of the following operational sequences, without overheating:

- Six consecutive starts against full-load torque.
- Five consecutive 'open-close-open' operations when coupled to the driven device.

Contactors should be able to close smartly at 80% nominal control supply voltage and remain closed down to 50% nominal volts.

2.1.2 Commissioning

Before commissioning any actuator, it is important to be fully conversant with the manufacturer's setting-up instructions. Commissioning engineers should be in possession of a comprehensive Actuator Schedule, to cross-reference with completed standard check lists (SCLs).

A check list, with a format similar to Fig 3.2, must be prepared to cover all of the following static and operational checks.

Static checks

- Correct identification in accordance with actuator schedule.
- Correct motor rating and actuator output speed for the driven device.
- Safe access for operation and maintenance.
- Lubricated as detailed on lubrication schedule.
- Correct cable termination and earth bonding.
- Compatibility of overload protection with motor rating (refer to Section 4 of this chapter).
- Insulation resistance of motor, cables and control equipment all tested at 500 V DC.

The characteristics of the driven device determine whether the actuator travel is halted by limit switches, or by adjustable torque switches. With parallel slide and ball valves, the torque switches are usually set to their maximum, so that they will only operate if the limits are overrun. Wedge-gate, globe, butterfly and diaphragm valves (and also dampers) generally stop on minimum torque setting in the close direction but on limit switch in the open direction. The CLOSED limit switch then acts as a back-stop, and the OPEN torque switch safeguards against the OPEN limit being overrun. Figure 3.6 shows two common arrangements for adjusting the setting of torque switches.

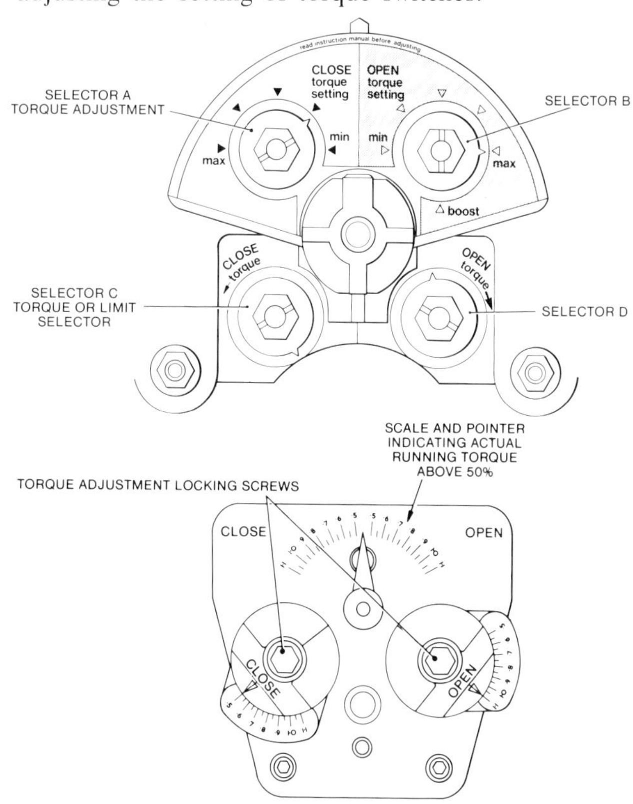

FIG. 3.6 Torque switch setting selector arrangement

Where control equipment is mounted in a valve group-starter board, correct operation of the switchgear is usually proved as part of switchboard commissioning. However, if the equipment is located at the actuator, then control function checks are better completed before the actuator drive is energised.

Operational checks

- Manually operate actuator to the fully-open and fully-closed positions to adjust and set the limit and torque switches, as required.
- With the actuator at an intermediate position, energise the motor and check for correct direction of motion, using the local control pushbuttons.
- Confirm that manual handwheel disengages automatically when the motor is first energised.
- Run actuator to each extreme position to check local indication and correct operation of limit and torque switches, adjusting the settings if necessary.
- Recheck operation from, and indication at, each control location.
- Check correct operation of any interlocks with associated valves, e.g., main and bypass.
- Run actuator for five consecutive operations over full travel to confirm repeatability of switch settings; check for motor overheating.

2.2 Pneumatic actuators

2.2.1 Description

Each pneumatic actuator has some combination of air receiver, filters, pressure and flow regulators, auto/manual changeover valves, solenoid valves and/or signal converter associated with it. The equipment is fed with dry air from the control and instrumentation compressed air system (see Section 21 of this chapter). Unless pneumatic actuators are equipped with a handwheel for manual operation, compressed air supplies must be available before the equipment can be operated manually. An actuator may open, close or remain in its last position when the operating compressed air supply fails. Before installing an actuator, it is necessary to confirm whether the driven device should open or close in the event of air failure. Pneumatic valves which open on air failure are usually called *direct acting*; those which close on air failure are *reverse acting*. It is very important that the correct type of action is used, otherwise the process system fail-safe philosophy may be invalidated in the event of air failure.

Non-modulating actuators

Figure 3.7 shows a typical arrangement of the pneumatic system for a piston-type non-modulating actuator. This type is often fitted to the larger bled-steam isolating valves in feed heating systems. Solenoid valves control the position of the cylinder control valve, and hence the direction of travel of the piston in the cylinder. The speed of movement is determined by adjustment of the regulators, which restrict the rate of air flow to the actuator. Air exhausted from the other side of the piston has free passage to the exhaust filters, via non-return valves which bypass the flow regulators.

Single-acting actuators operate by compressed air acting on one side of a piston or diaphragm. Movement of the actuator spindle by application of pressure is usually resisted by adjustable loading springs. Expensive cumbersome locking mechanisms are necessary if

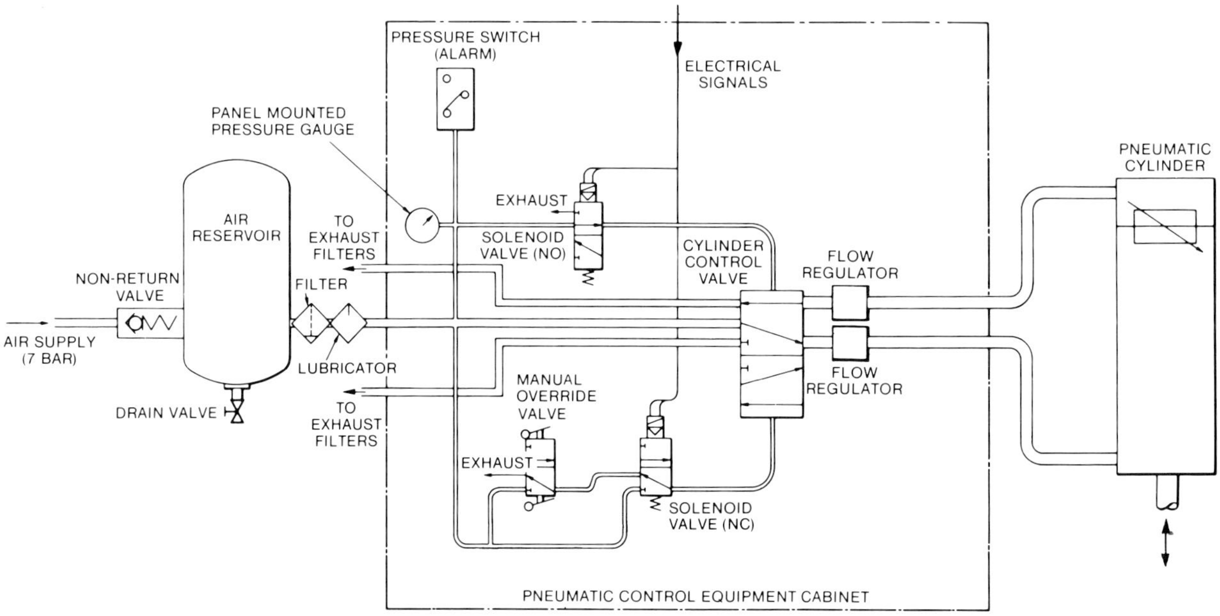

FIG. 3.7 Pneumatic system for non-modulating actuator

single-acting actuators have to lock in position in the event of air failure. In general, simple diaphragm-actuated valves can be converted from direct-to-reverse action by inverting the valve plug and the seating rings.

Double-acting actuators require compressed air applied to the appropriate side of the piston or diaphragm to create the desired direction of movement. In the event of air failure, the pneumatic system may allow the actuator to drift slowly to its limit of travel, though in some cases the actuator locks in position until it is manually released.

Another form of actuator, used where only 90° of stem movement are required, is shown in Fig 3.8. The one shown is a single-acting device which can be converted to double-acting by removal of the springs followed by valved reversal of the air supply and vent connections to achieve the desired direction of motion, as shown by the inset detail.

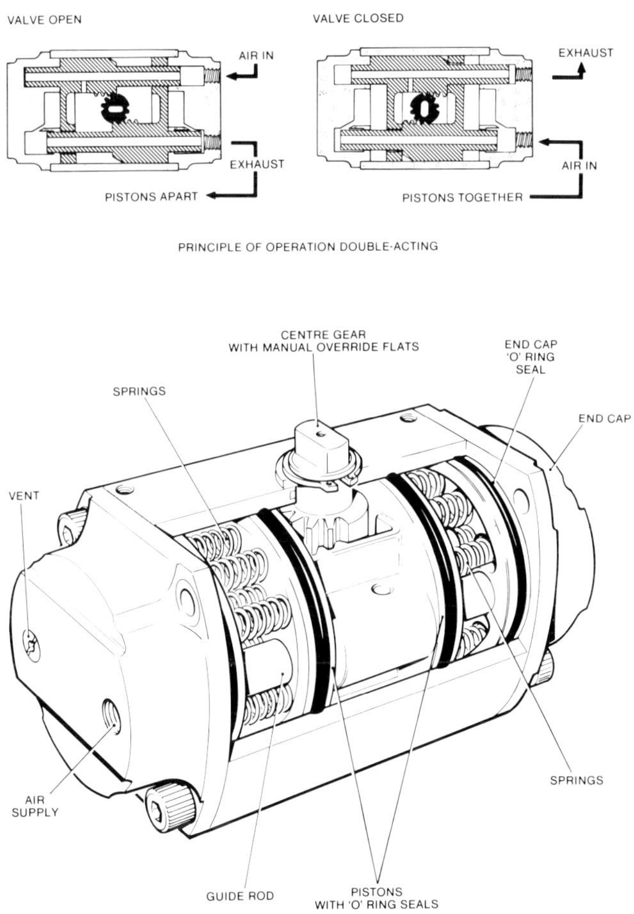

FIG. 3.8 Rotary non-modulating pneumatic actuator

Modulating actuators

The most common type of modulating actuator is the single-acting diaphragm-type mentioned above, where some form of pneumatic servo-positioner is fitted to regulate the air pressure applied to the diaphragm. Operation of the actuator and its driven device in the event of air failure has to be carefully assessed, since the action assigned to a valve, for example, may conflict with that of its modulating actuator. Typically actuator action depends upon whether an increasing control input either extends the actuator operating rod (direct-acting) or retracts it (reverse-acting). Figure 3.9 shows one form of modulating actuator, where a

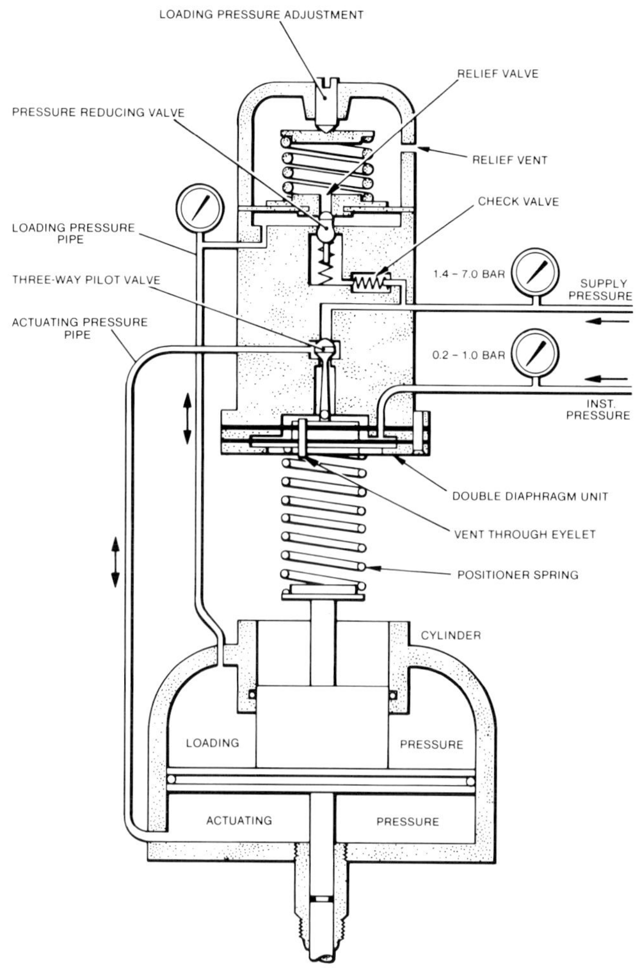

FIG. 3.9 Modulating pneumatic actuator

constant preset air-loading pressure is maintained on one side of the piston. This is, in effect, the counter-balancing spring-loading of a single-acting diaphragm actuator. A signal converter turns a variable electrical input signal, e.g., 4–20 mA DC, into a proportional 0.2–1.0 bar pneumatic output which flexes the double diaphragm unit. The diaphragm unit then positions a three-way pilot valve which modifies the actuating pressure on the other side of the piston. Movement of the piston then resets the diaphragm unit, via the positioner spring, to maintain the actuator at its desired position. Remote indication of valve position is derived from a linear variable displacement transducer (LVDT) connected to, and driven by, the actuator output shaft.

2.2.2 Commissioning

The commissioning engineer must be familiar with the manufacturer's setting-up instructions for each piece of equipment. It is likely that a separate SCL will be needed for each item of pneumatic equipment. These are then integrated by a simple test procedure for the pneumatic loop. All test results and setting values should be clearly recorded for future reference on a test results record, similar to that shown by Fig 3.10.

SCL No. A48
Sheet 4 of 4
Ref No. ....

.... POWER STATION

ELECTRO-PNEUMATIC ACTUATOR TEST RESULTS

Function .... Stroke ....
Location .... Traverse Time..................................
Schedule No. .... Serial No. ....
Manufacturer .... Type ....

Manual filter/regulator pressure .... bar
Auto filter regulator pressure .... bar
S-P Converter output pressure .... bar

Interlock	Operates at (bar)	Resets at (bar)

Number of turns of servo motor shaft for full travel of actuator ....

Turns of Servo Motor Shaft	Ram Position mm		LVDT Output		Pot. Output		S-P Converter Output	
	Extend	Retract	Extend	Retract	Extend	Retract	Extend	Retract

Max. Deviation from Linear: (a) Ram .... (b) LVDT .... (c) Pot ....
(d) S-P Converter ....

Max. Hysteresis (a) Ram .... (b) LVDT .... (c) Pot ....
(d) S-P Converter ....

Stroke Length .... Deadband (Open) ....
(Close) ....

FIG. 3.10 Electro-pneumatic actuator test results record

General principles for commissioning pneumatic actuators are given below. These are based on a type which relies on an electric servo-motor to vary the output from a servo-pneumatic (S-P) converter to regulate actuator position. The servo-motor usually is operated from a control station or module in the central control room which has the facility for AUTO or MANUAL control selection. In the auto position the process variable automatically positions the actuator to achieve a preset desired value. When selected to manual, the operator positions the actuator to achieve a desired value applicable to the operational requirements at that point in time. The control module also has facilities for bumpless transfer from manual to auto and vice versa.

Initially, the actuator and control equipment are checked to ensure that:

- Identification and all equipment details are correct and in accordance with the actuator schedule.
- There is safe access to all equipment.
- Erection is complete and all pipework correctly connected.
- Valve (or damper, etc.) is available for operation.

The compressed air supply line is disconnected and purged to atmosphere to displace any construction debris, water, etc., before being recoupled to the control equipment (refer also to Section 21 of this chapter). With the air supply valve open, the filter regulators are set to the specified air pressure values; then any air leaks are identified and rectified. The following checks are then carried out:

- The auto/manual control selector is set to manual, and then the actuator is traversed, or stroked, over full travel in each direction to prove that the movement is smooth and free from stiction or obstruction. Local manual control can be conveniently achieved by manual operation of the servo-pneumatic converter.
- The position indication loop is set up, followed by recording of the calibration figures.
- Servo-motor drive from the remote control station is checked to confirm operation of the valve or damper over full travel. The operating times are recorded at this stage.
- Control is selected to auto to check that operation of the manual control has no effect.
- Finally, the action in the event of air supply failure is demonstrated to be correct. The minimum air pressure to operate the air-fail interlocks is adjusted, if necessary, then recorded for future maintenance history.

2.3 Hydraulic actuators

2.3.1 Description

Actuators are powered by electro-hydraulic pumping units, either integral with the actuator or mounted in an adjacent power frame. One power frame may supply high pressure oil for up to four separate actuators. Each main boiler/turbine unit may be equipped with over 50 hydraulic actuators associated with modulating valves, dampers and motor speed regulators.

Modern steam turbine and main boiler servo-hydraulic systems often use a special synthetic phosphate-ester fire-resistant fluid (FRF) instead of a mineral oil. Filtration of hydraulic oil to an extremely fine particle size is necessary to ensure trouble-free operation of the equipment. Hydraulic pumping units now have suction and discharge filters fitted for control of particulate matter. It is well known that a pump supplied with fluid carrying particulate contamination will, with this contaminant, rapidly produce more contamination. Beyond a particular level and size of particle contamination, the rate of production becomes exponential. It is also very important that hydraulic systems, particularly those employing phosphate-esters, be kept dry. Phosphate-esters break down in the presence of water to form an acid which rapidly attacks vulnerable components. A typical limit of total acidity would be not more than 0.3 mg potassium hydroxide per gram of fluid.

Self-contained hydraulic actuators

A typical self-contained hydraulic power pack servo-actuator is shown in Fig 3.11, with a diagrammatic representation of the hydraulic system in Fig 3.12. This unit is designed to operate reciprocating-stem and rotary-shaft control valves etc., using standard electronic control signals, e.g., 4–20 mA DC. The components comprise a motor-driven gear type pump, a transducer and a positioner, all completely self-contained in one enclosure. Typically, the motor would be rated at 400 W and be connected to a standard 415 V three-phase supply. The electronic control unit processes analogue input and feedback signals to operate an hydraulic directional control valve. Hydraulic pressure, generated at 70 bar by the actuator pump unit, is applied to the cylinder ram via the directional control valve. Input signal failure is detected by the control unit which automatically causes the actuator either to lock in position, or to fail-safe by moving to one or other of the extreme stroke positions. Electrical supply failure or hydraulic malfunction usually causes the actuator to lock in position. Remote position indication, as with the pneumatic actuator, is derived from a linear variable displacement transducer (LVDT) mounted on the actuator drive shaft.

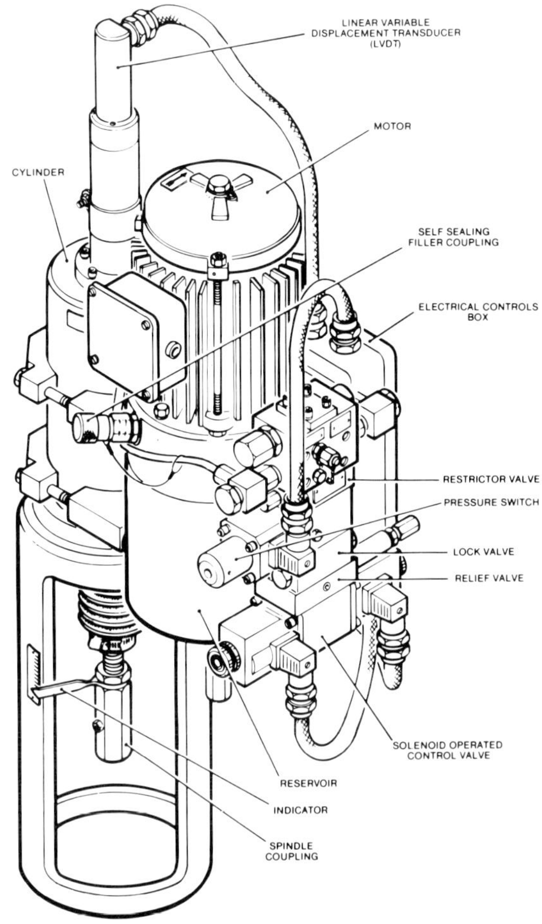

FIG. 3.11 Self-contained hydraulic actuator arrangement

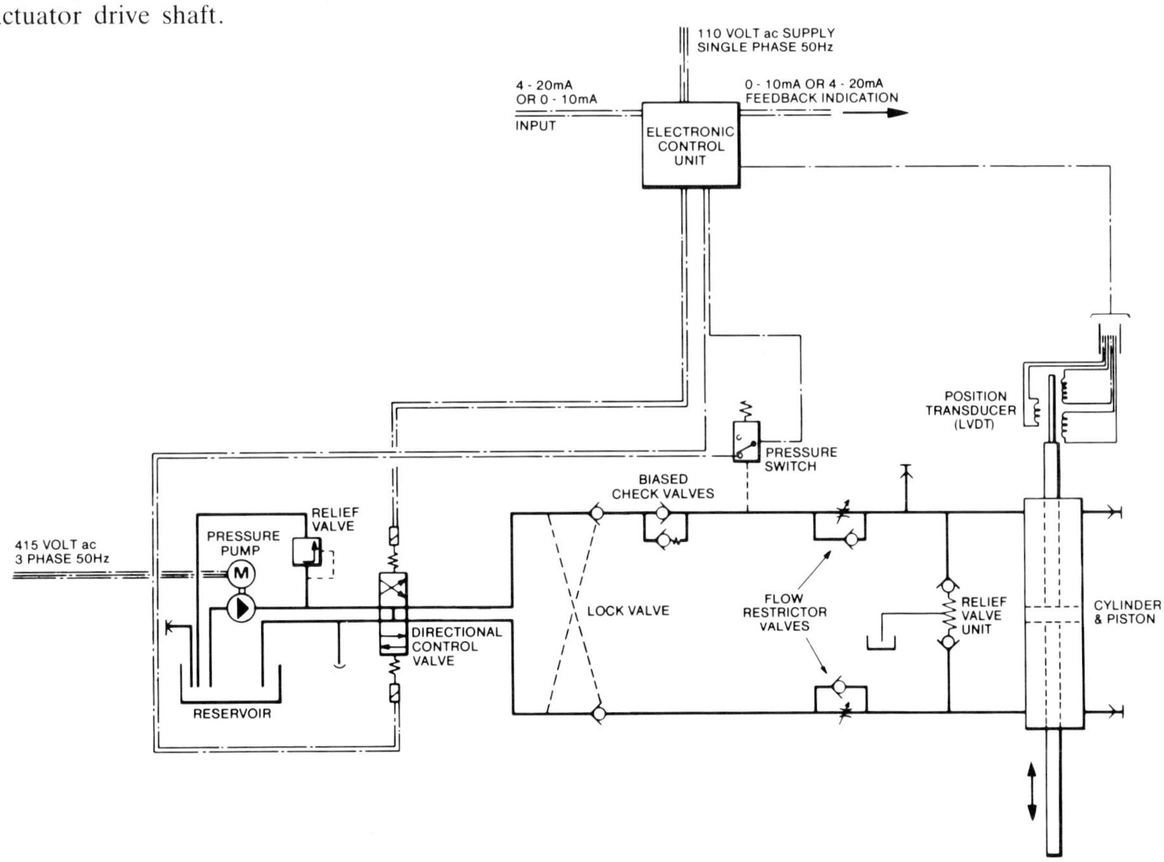

FIG. 3.12 Hydraulic schematic diagram for self-contained actuator

Hydraulic power frames

During the normal operation of power frame hydraulic systems, there can be large transfers of fluid into and out of the storage tank. To cater for this, the vapour space is allowed to breathe via water-absorbent filters, e.g., silica gel. Some modern systems are equipped with flexible, rather than rigid, tanks which almost eliminate excessive breathing of atmospheric air. Hydraulic oils have to be kept warm, i.e., within a band of 30°C to 50°C, and it is customary to fit tank heaters for use during shutdown periods. Air-blast coolers are installed for on-load temperature control when the heat input from the pressure pumps justifies it. Hydraulic pipelines are often electrically trace-heated and thermally insulated to prevent the formation of condensation if their length exceeds a few metres. The need is determined from consideration of anticipated lowest ambient temperatures, plant load factor, pipeline location, etc.

The hydraulic system for a typical power frame is shown in Fig 3.13. Each frame comprises an electrically-heated reservoir supplying oil to a hand pump and a pair of motor-driven gear or axial-piston-type pressure pumps. One pressure pump runs continuously whilst the other is on standby. The duty pump is automatically loaded and unloaded by changes in system pressure, via pilot-operated relief valves on the pump discharge main. A nitrogen filled accumulator is fitted to smooth pressure pulsations in the fluid. The accumulator is precharged to 180 bar; the system then operates in a pressure band of 190 to 200 bar.

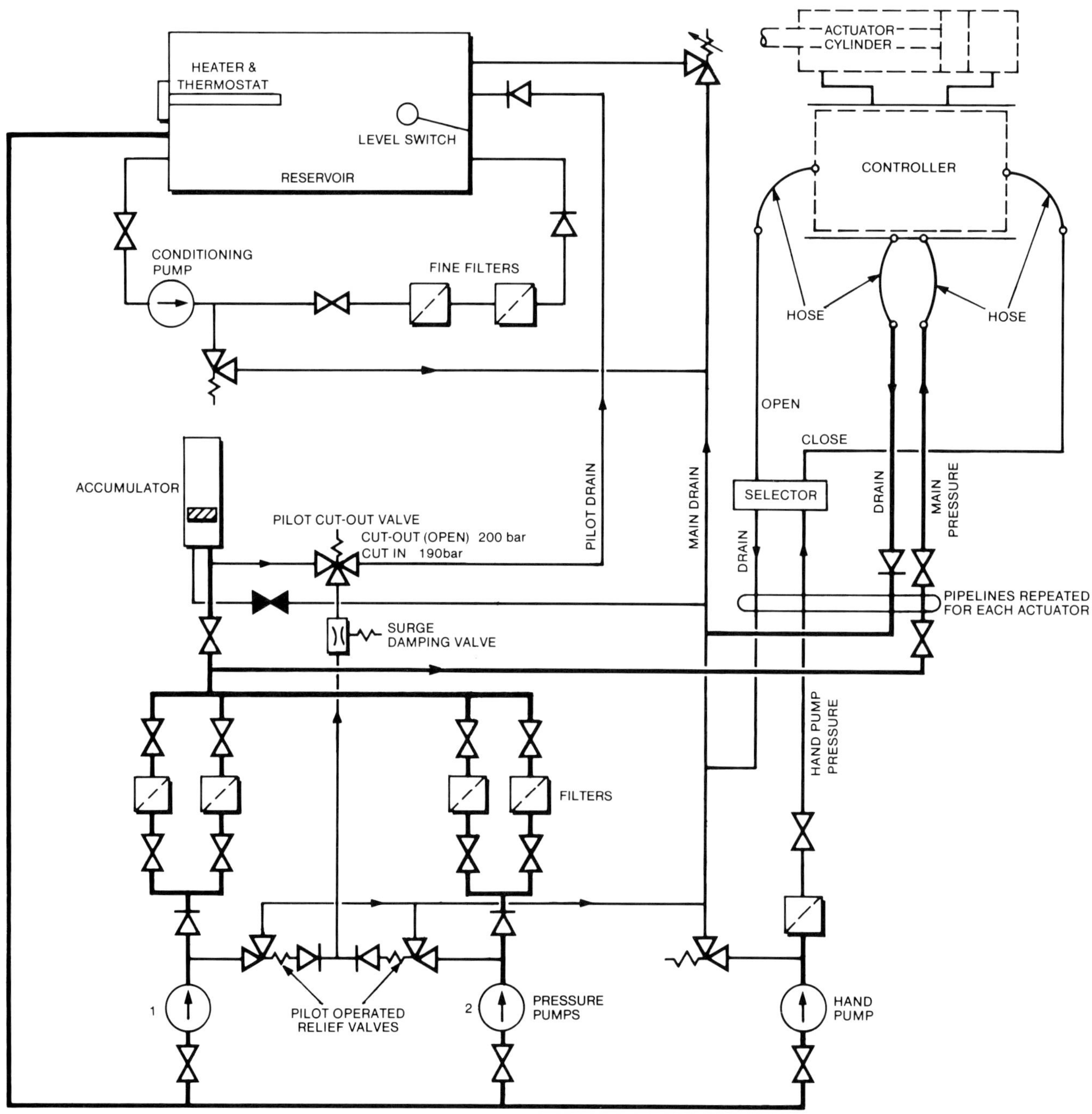

FIG. 3.13 Hydraulic power pack system

A small conditioning pump continuously recirculates a small proportion (10% of the contents per hour) of the oil in the reservoir through a pair of fine filters.

Steel piping, terminating with flexible hoses, conveys hydraulic fluid to and from each actuator. On this particular design the hand pump also has separate flow and return lines. All four lines are led through a valve selector block at the power frame. At the actuator, an electrically-driven servo-positioner spool valve directs hydraulic fluid to one or other side of the piston in the hydraulic cylinder. Depending upon the side to which the fluid is directed, the actuator will extend or retract. Should there be a loss of hydraulic pressure, the actuator locks in position and cannot be freed until the hydraulic lines have been isolated and the pressure re-established to reset the hydraulic locks.

Another form of power frame, often fitted to coal-handling plant, contains an axial-piston pump with servo-control of its swash plate. The swash plate angle determines the magnitude of the oil flow and it can be angled to either side of a neutral position to provide flow in either direction to an actuator. For this type of application, e.g., for crane slewing duties, a rotary actuator, comprising a radial-piston hydraulic motor, is employed. The actuator output shaft is fitted with a spur gear which engages with the main drive-gear ring of the slewing mechanism. In most other respects, this system is similar to that of the power pack described earlier.

Similar, less sophisticated, individual power frames (refer to Fig 3.14) are employed on station services plant, such as CW pump discharge valves of the type shown earlier by Fig 3.4. These generally use a mineral oil as a hydraulic fluid, because there is not the same risk of fire. Hydraulic actuators for such functions usually do not modulate. Instead, oil from the pressure pump drives the actuator to the open position, and then the pump shuts down. Non-return valves in the hydraulic circuit maintain the actuating ram under pressure to hold the valve open. Should the hydraulic pressure leak, the valve would drift closed under the action of its counterbalance weight. After a few degrees of movement, the 'open' limit switch resets and restarts the hydraulic pump to 'kick back' the valve to the fully-open position. When required, closure of the valve is effected by energising a solenoid valve which releases the trapped hydraulic pressure. The valve then closes under the influence of the counterbalance weight. Automatic starting of the hydraulic pump is inhibited whenever the normal or the emergency closing solenoid valves are energised.

2.3.2 Commissioning

A single SCL may be adequate for commissioning each type of hydraulic actuator. The test procedure for a power frame will cross refer to several SCLs for individual items of equipment associated with the hydraulic loop. The more important checks are as detailed below.

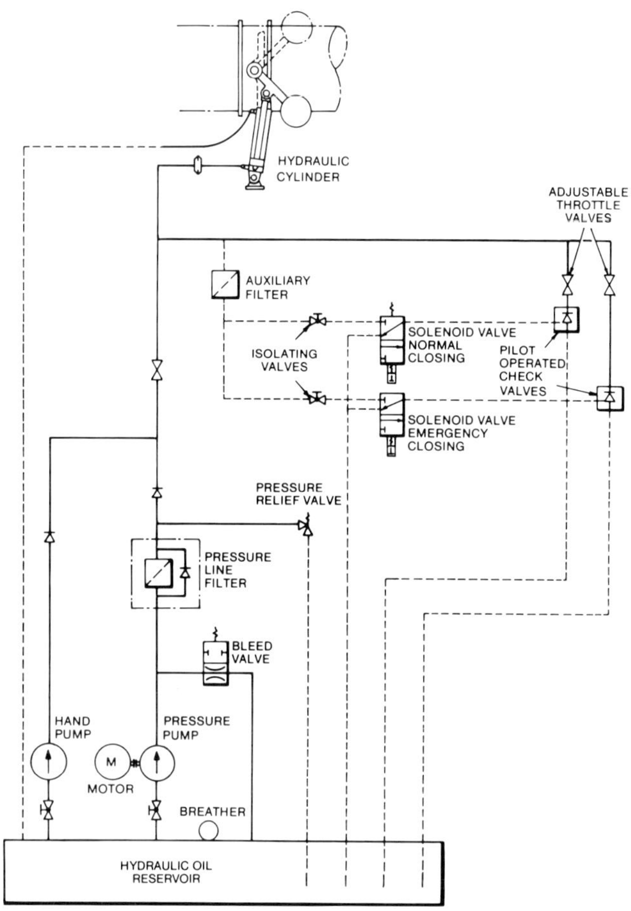

FIG. 3.14 Self-closing valve hydraulic system

It must be emphasised that all personnel have to be thoroughly familiar with the safety procedures for handling of phosphate-ester hydraulic fluids. Appropriate protective clothing must be worn, and any spillages must be controlled and then recovered for disposal in strict accordance with current legislation. If fluids come into contact with very hot surfaces, toxic fumes may be released.

Self-contained hydraulic actuator

After carrying out the customary checks for safe access, correct identification, etc., the following tests are carried out:

- Check that the actuator unit is cleaned down and that the internal surface of the oil reservoir is also clean.
- Fill reservoir with correct quantity of the appropriate hydraulic oil. Typically about 12 litres will be necessary.
- Energise the oil pump motor, ensuring that its direction of rotation is correct, and allow all of the oil pressure lines to fill.
- Ensure that the system is fully primed by venting off any accumulated air. If necessary, add oil to the reservoir to restore the correct running level.

- Confirm that the actuator and driven shaft are correctly connected together, then stroke the actuator over full travel, using the manual handwheel (*NB* The oil pump bypass valve may have to be opened to balance the hydraulic pressures before the handwheel can be rotated.)
- Disengage manual drive and, with the pump running, set the zero position at one end of the stroke to correspond with minimum input signal of, say, 4 mA.
- Increase the input signal to maximum (say, 20 mA) and adjust the actuator span control to give the desired length of travel.
- When both zero and span positions have been correctly set, the speed of travel may have to be adjusted. Normally, it is only possible to adjust speed for one direction of travel.
- Recheck traversing of actuator a further five or six times to confirm repeatability of settings, and also to circulate some oil through the pipelines.
- Check, clean or replace the oil filter as appropriate and leave the actuator available for further plant commissioning.

Hydraulic power frames

The principles outlined below are based on the type of power frame shown in Fig 3.12. They are readily adapted to suit different hydraulic systems. For practical purposes, commissioning splits into three main activities:

- Commission power-frame-mounted auxiliaries.
- Flush all hydraulic pipework between power frame and the associated actuators.
- Commission each hydraulic actuator.

Initially all equipment, including pipework up to the actuators, is checked to the appropriate SCL. After checking that the correct filters are in position, the reservoir is filled with the specified hydraulic fluid. A sample of the fluid is analysed for:

Conductivity resistivity greater than 50 MΩ m at 20°C.

Moisture less than 1000 PPM.

Particulate contamination not more than 10 000 particles of a size greater than 5 μm per 100 ml of fluid.

The following activities are then carried out:

- Reservoir heater commissioned and thermostat proved to control oil temperature at 50°C.
- All three pumps primed and rotation checked, leaving the conditioning system pump in service. The direction of rotation must be correct, otherwise the pump may be damaged.
- Charge the nitrogen accumulator, check for leaks and put it on line.
- Check the lifting pressure of all relief valves by operating one of the pumps and carefully throttle the drain to the tank. Rectify any oil leaks found during this exercise.
- Check the integrity of all electrical trace-heating tapes and energise all circuits prior to flushing of the pipelines.

At one actuator, disconnect a pair (flow and return) of hoses. Couple them together, remembering to blank the open orifices on the actuator, and flush the pipework, as follows:

- Run each pressure pump in turn for half an hour, flushing through the hydraulic loop back to the reservoir, keeping the filter differential pressures under close observation in case the filters have to be replaced.
- After a further hour, the return oil should be sampled and analysed as before.
- Continue the flushing exercise until the chemical standard of that loop is considered acceptable.
- Hoses are then rearranged at both ends of the line, so that the pressure pumps can flush the hand pump pipelines.
- The whole procedure is repeated until all hydraulic lines associated with the power frame have attained an acceptable standard of cleanliness.
- Assuming that all hydraulic oil leaks have been cured, thermal insulation has to be applied to the trace heated pipelines.

It may be prudent to impart mechanical shocks to the pipework during the flushing exercise. The need for this depends upon the standard of cleanliness achieved during erection. One method is to include a solenoid-operated shut-off valve in the loop to impart hydraulic shocks each time the valve is actuated by some form of automatic timing device.

Actuator commissioning should be a relatively simple task after all the pipelines have been flushed to an acceptable standard. Preliminary adjustments of stroke length and limit-switch settings are usually carried out on a workshop test rig, before the actuator is erected in position. Once erected, the hand pump is used to traverse the actuator to the fully-open and fully-closed positions. This is followed by rotating the servo-motor drive shaft and checking actuator movements, in a similar manner to the pneumatic actuator discussed earlier.

Operation of the actuator with the servo-motor energised is checked. If the servo-drive unit incorporates a stepper motor, rather than a highly-geared conventional DC motor, it is important to count the pulses sent out from the controller and then check that the actuator has traversed the appropriate length and in the correct direction of travel. Finally, application and release of the hydraulic lock is proved.

3 Instrumentation

There is such a wide range of commercially available instrumentation that it would be impossible to give detailed descriptions and commissioning procedures in these pages. Instead, this section outlines only the broad principles involved in testing the main types of measuring and indication equipment.

3.1 Description

Instrumentation systems include sensors or transducers fitted to plant to provide signals that are processed and displayed, or provide measured variables for interlocks, protection and closed loop modulating controls.

The main types of measurement are:

- Temperature and pressure
- Flow and level
- Analytical (gas, oil, water) and moisture
- Electrical and supervisory

Figure 3.15 shows some typical measurement loops applied to a process fluid pipeline as an example. Each item in a loop has its own individual specified accuracy which, when combined, results in a typical overall loop accuracy of ±2%, referred to the full-scale indication value.

Many indicators are mounted at the point of measurement, i.e., local to plant. To facilitate remote operation of systems, most measurements are transmitted for display on centralised control panels or data processing VDUs (see Section 6 of this chapter). Instrumentation transmitters are usually grouped together in local equipment housings (LEHs) strategically placed about the plant. Usually these comprise angle-iron-framed open, but roofed, structures provided with panel lighting and having 110 V AC instrumentation supplies to power the transmitter units. Each transmitter, which incorporates its own stabilised power supply unit, converts the measured value into either 0–10 mA or 4–20 mA DC current output for transmission by a two-wire system to the central control room (CCR).

Modern pressure transmitters are usually in the form of a four-arm, active, unbonded, strain gauge bridge configuration. These give an extremely accurate indication with little hysteresis, when compared with bellows or Bourdon tube mechanisms.

The largest single type of instrumentation is usually that employed for temperature measurement. Platinum resistance thermometer detectors (RTDs) are used for all measurements needing a high level of accuracy, e.g., for plant performance monitoring. However, they have

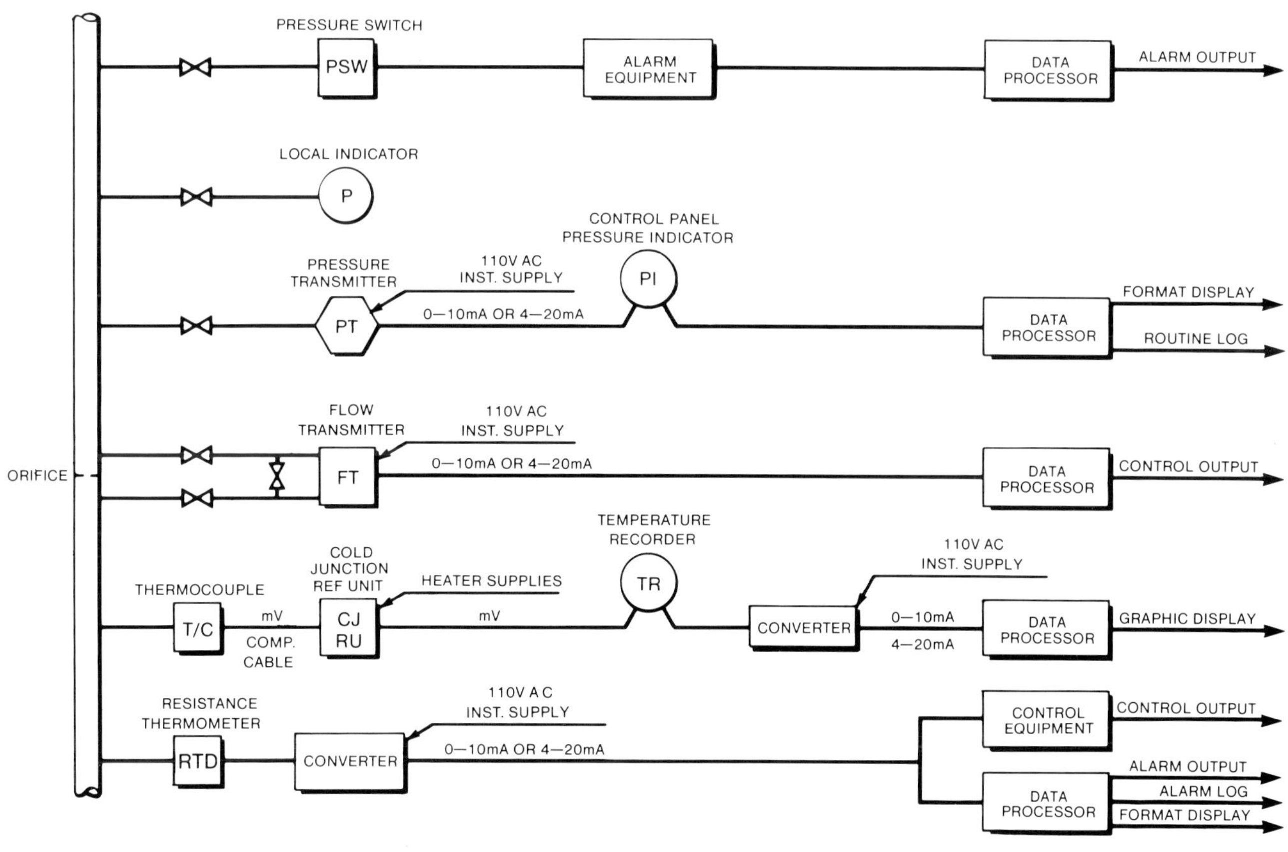

FIG. 3.15 Typical measurement loop arrangements

a relatively narrower range of operating temperature than thermocouples (T/C). Typically, this may be up to 650°C, as against 1750°C for a thermocouple. Resistance thermometers usually have a nominal resistance of 100 Ω at 0°C. For higher temperature applications (i.e., over 600°C) 10 Ω detectors are often preferred for reliable service.

Depending upon the required accuracy, RTDs can be connected as two-, three- or four-wire elements into a bridge network in a transmitter which needs a 110 V AC instrumentation supply source to power it. Thermocouples used in power stations are usually of type K, i.e., nickel-chromium/nickel-aluminium (Chromel/Alumel), fitted into a stainless steel sheath, with mineral electrical insulation. The output terminals are usually connected to some form of temperature reference unit by means of special compensating cable. Compensating cable is of different materials to the thermocouple, but has similar thermal/EMF characteristics up to ambient temperatures of approximately 80°C. If incorrect readings are to be avoided it is very important that the correct connection polarity is observed.

Typical thermal characteristic of each device is as follows:

Resistance thermometers 0.38 Ω increase per °C rise in temperature.

Thermocouples (type K) 40.9 mV increase per °C rise in temperature.

For precise determination of temperature for a given output, reference must be made to appropriate data tables for each type.

Temperature reference units form the 'cold' junction and the temperature measuring junction is the 'hot' junction. Thus cold junction reference units (CJRU) form a marshalling point for large numbers of thermocouples, which share the same reference point. Some designs of reference unit maintain the cold junction at 0°C by means of special cooling equipment. Other forms are based on a fixed reference temperature (typically 60°C), closely maintained by means of an electric heating element; they are used extensively in power stations. Lately, with the advent of digital computing techniques, there has been a marked shift toward isothermal reference units. The isothermal 'cold' junction temperature is permitted to vary in response to ambient temperature changes. Individual thermocouple outputs are transmitted to the data processing computer, along with a measurement of the 'cold' junction temperature. The computer then compensates each measured value for the actual reference temperature, and relates them to 0°C for display.

3.2 Commissioning

Before an instrumentation commissioning exercise is started, the extent and importance of the task should be fully appreciated. Sufficient resources must be available to ensure that plant completion is not delayed due to lack of calibrated instruments.

Ideally all instrumentation equipment should be transferred from the plant to a centralised facility for witnessed calibration tests. Staff in this facility need to be specially trained and allocated solely to this work. Typically, some six people are necessary to meet the workload at a 2000 MW power station. Accurate standard test equipment should be kept within this facility. All test equipment must be recalibrated at frequent intervals against standards which are traceable to national and international standards of accuracy.

3.2.1 Test equipment

Typically, a deadweight tester would be the 'site standard' for pressure measurement. Standard test gauges, in conjunction with a comparison test pump, would be used for calibration of industrial grade instruments and transmitters. For pressures less than 2 bar, the comparison test pump is usually replaced by manometric test equipment, or by a specially calibrated transmitter with a digital readout. When using the comparator, it is important to ensure that the test fluid is compatible with the system into which the measuring device being calibrated is to be connected.

Standard test equipment for each of the more common measurements may comprise:

Temperature Decade resistance boxes; resistance thermometer simulators; thermocouple calibrators, capable of measurement and injection suitable for the types of thermocouples in use; millivolt source; oil, water or sand temperature reference baths and possibly a thermocouple furnace.

Pressure Deadweight tester; comparison test pumps; full range of standard test gauges to an accuracy of better than ±0.25%; liquid and solid state digital manometers; low range pressure source.

Note Current developments with digital pressure gauges will soon enable wholly-digital calibration systems to be adopted.

Flow Differential pressure types need low range pressure gauges and sources. Inferential types require the use of signal generators to simulate the pulse outputs. A pulse counter is also needed to check the actual output from the flowmeter.

Level In general, differential pressure test equipment capable of producing low range pressures is needed.

Analytical Appropriate certified test gases and test solutions are necessary for each particular application.

Moisture Laboratory calibration facilities are required.

Electrical Test quality voltmeters, ammeters, wattmeters, etc., have to be used.

Supervisory Specialist electronic test equipment is usually necessary to simulate each particular quantity.

3.2.2 Testing procedures

Testing must be implemented to an agreed priority order which is compatible with the overall plant commissioning programme. Items of instrumentation equipment must be tested in accordance with the individual manufacturer's operating and maintenance instructions.

It is good practice for each item in an instrumentation loop to be calibrated separately. Items are then connected together in their correct locations and a loop test carried out, by either simulating or creating plant conditions, to compare indicated readings with the input value. If necessary, adjustments are made to any items which give rise to unacceptable loop-accuracy errors. All test results and adjustments, including details of the test equipment used, must be fully recorded for future maintenance purposes.

A separate SCL may need to be prepared for each individual item forming the instrumentation loop. Check lists and test records could follow the format of those depicted earlier by Figs 3.2 and 3.9 respectively. Instrumentation check lists need to be cross-referenced to a comprehensive Instrumentation Schedule.

Another item of documentation, which is of inestimable value to station engineering staff, is a comprehensive Settings Schedule which lists the approved switching set-points for all pressure, vacuum, temperature, level and other switches. One nominated location should be responsible for maintaining the master copy up-to-date. Naturally the value of this depends upon a disciplined approach from commissioning engineers to ensure that the master copy is updated with any alterations to the previously approved settings.

Detailed test procedures for putting instrumentation systems on-line when plant goes into service will also be needed. A typical example of an on-line commissioning procedure is given later. Furthermore, an instrument-loop routing-sheet should be prepared for each measurement loop. This could take the form of that shown by Fig 3.16; this is for a four-wire platinum resistance thermometer, measuring a fan bearing temperature for display on a switched digital panel meter (DPM) in the CCR.

Standard checks (likely to appear on each SCL):

- Safe access.
- Clear identification.
- Correct mounting.
- Electrical connections correct to drawing.
- Equipment bonded to earth, where applicable.
- Signal (impulse) pipework complete (nominally 'horizontal' runs installed with the correct 1 in 20 fall toward the measuring device).
- Impulse pipework blown through with dry compressed air, positively identified to follow the correct route, and then pressure tested.
- Head corrections applied, where appropriate, to cater for difference in height between point of measurement and the measuring device. Sometimes it is necessary to relate the measured value to some particular datum, e.g., basement-mounted oil pres-

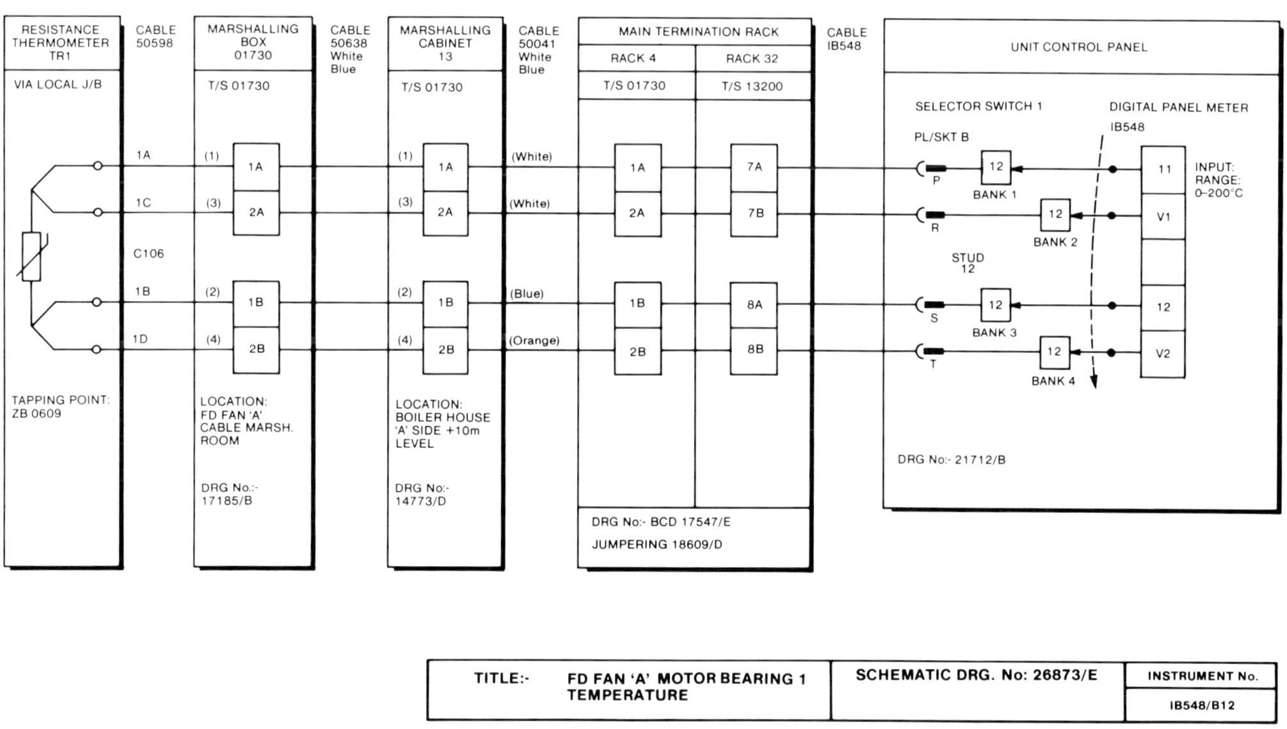

FIG. 3.16 Instrumentation loop routing sheet

sure transmitters, whose indication is often referred to turbine shaft centre line.

- Calibration of analogue indicators carried out at 20% scale increments for increasing and decreasing measured values. For switches, the set point is adjusted, then it and the reset value are recorded for increasing and decreasing measured values as appropriate.
- Manufacturer's test certificate available.

Specific measurement checks

Temperature:

- Detector pocket to be cleaned out; and, if appropriate, filled with a heat transfer medium.
- Thermocouple and compensating cable polarity checked. Insulation resistance tested at 100 V DC giving better than 50 MΩ.
- Cold junction reference unit checked to control at specified temperature (typically 60°C) within one hour of switch-on. High and low temperature alarms checked. The stability of reference temperature control over a 24-hour period is recorded.
- Resistance thermometers are very delicate and should be handled with care. Correct sequence of terminations checked to be suitable for the type of circuit, i.e., two-, three- or four-wire connections. Element continuity is checked with a low voltage source and the resistance value noted. Insulation resistance is checked at 100 V DC between each terminal and sheath to be at least 100 MΩ.

Pressure:

- Differential pressure switches, transmitters and gauges should all be capable of calibration whilst subjected to an ambient static pressure, as well as retaining their calibration at the normal operating static pressure.

Flow:

- Pipelines to be flushed before fitting of orifice plates or inferential flowmeter measuring elements.
- Flow switches to be checked with the system fluid flowing. It is difficult to create accurately calibrated flows under site testing conditions, particularly for the wide range of switches fitted.
- Integrating flowmeters to be injected with pulses, at 20% increments of range, over specific periods of time.
- Orifice plate sizes and orientation (in respect of direction of pipeline flow) to be checked. Tappings correctly piped to the transmitter.

Analytical:

- Equipment for measuring relative acidity/alkalinity is normally tested with three different standard buffer solutions whose pH lies within the range of the measurement loop. The measuring electrode must not be allowed to dry out.
- Silica measuring loops are usually tested at 80% of range with a specially prepared solution.
- Other measurement loops are checked, at several points over the range, with known solutions or certified test gases.

Electrical:

- Transducers to be tested with appropriate 110 V AC, 5 A AC or 1 A AC inputs from special injection equipment normally used as part of switchgear commissioning (see Section 9 of this chapter). Indicators are calibrated at 20% increments of full scale, including the normal running value.

Supervisory:

- Transducers should be tested at three or five points within their normal range. Service indicators are compared with indications from specialist portable equipment.
- Tachometers are checked at three to five points over the full range by means of a stroboscope directed at a convenient part of the rotating shaft.

Recorders:

- Check that chart drive operates with acceptable speed and accuracy over at least a 24-hour period.
- Inking and printing mechanisms proved for each channel at 20% increments of span.
- Induce a 50% step change of input from zero and record the time for indication to stabilise.
- Check setting and operation of any alarm/control switches.

Radio frequency interference (RFI) checks

Modern instrumentation equipment is often susceptible to interference from mobile radio transceivers used for commissioning and operational messages. It is prudent to test for RFI whilst the loops are undergoing functional or operational tests. Checks similar to the following should be conducted adjacent to the cubicles housing the instrumentation electronic equipment:

- Operate radio transceivers within 1.5 m of cubicles with all doors shut.
- Repeat the test with transducers 300 mm away from cubicles with doors shut, and again with the doors open.

Any positive response, particularly where auto controls or protection circuits are involved, should be carefully noted. The appropriate action to be taken will depend upon local conditions and the degree of risk to plant safety.

In-service checks

As each section of plant is put into service, instrumentation loops should be put on line in accordance with approved test procedures. A typical on-line commissioning procedure for a transmitter measuring steam or water flow would be (referring to Fig 3.17 for valve identification) as follows:

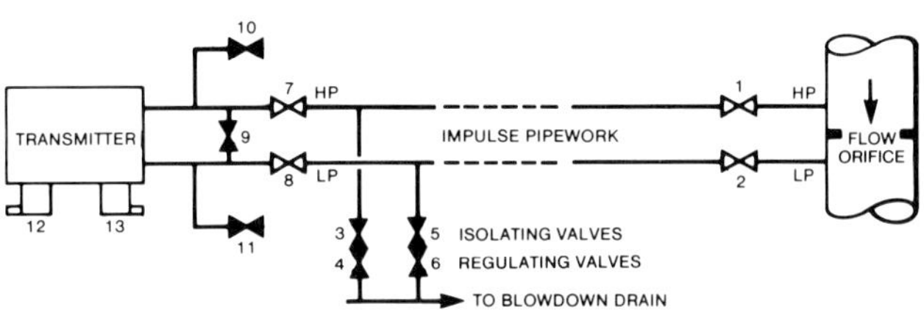

FIG. 3.17 Flow transmitter impulse pipework system

- Check that all valves are in closed position.
- Slowly open both tapping point isolating valves (1 and 2).
- Blow down the HP line by opening isolating valve (3) then open the regulating valve (4). Close the regulating valve before the isolating valve, when the line has been blown clear.
- Blow down the LP line (using valves 5 and 6) in a similar manner.
- Allow sufficient time for the lines to fill with condensate (in the case of steam measurement) and to cool to a safe temperature for the transmitter before proceeding any further.
- Slowly open equalising valve (9) and HP isolating valve (7). Check that transmitter output is zero.
- Slowly bleed the HP line by removing the blanking plug then cracking open the test valve (10) until clean water flows. Close test valve, replace blanking plug. Repeat for LP line (11).
- Slowly open HP bleed screw (12) on transmitter pressure chamber to remove any air or gas which has gathered.
- Close the HP bleed screw and repeat for the LP side (13). Close equalising valve (9) and slowly open the LP isolating valve (8).
- Check the system for leaks.
- Check that central control room (CCR) and any other indicators are giving expected readings.

Indications from all associated displays should be noted, then cross-checked against other measured parameters, to assess if the indicated value is what could be reasonably expected at that particular state of plant. This correlation exercise is repeated for all main unit measurements at discrete intervals of load, usually at the time when the response of various control loops is being tuned. Where extensive data logging facilities are available, it is advantageous to print out a hard copy of all analogue input values for subsequent analysis.

4 Motors and auxiliary plant

The first part of this section is devoted to commissioning of various types of electric motor commonly used in modern power plants. For ease of reference, they have been grouped into three classes:

- High voltage (HV) AC motors, i.e., supplied at 11 kV and 3.3 kV.
- Low voltage (LV) AC motors, i.e., connected at 415 V three-phase and also at 240 V and 110 V single-phase.
- Direct current (DC) motors, usually supplied at 240 V and 110 V DC.

Commissioning of small pumps and fans, generally driven by LV AC and DC motors up to 150 kW rating, is covered in the second part. Larger pumps and fans are covered in the appropriate plant section of this chapter, and also in Chapters 4, 5 and 6.

4.1 Motors

4.1.1 Description

Present designs for the electrical auxiliaries distribution system (see Section 9 of this chapter) of modern power plants are based on the three standard voltage levels of 11 kV, 3.3 kV and 415 V. Systems using 6.6 kV as an intermediate voltage are no longer considered. Motors rated at over 1500 kW are connected at 11 kV and, with a few exceptions, motors between 150 and 1500 kW are supplied at 3.3 kV. All motors of less than 150 kW rating are connected to LV AC and DC supplies.

The construction and testing of motors at works is governed by BS4999 and BS5000. As a general rule, most motors used in modern power stations are single-speed and arranged for direct-on-line (DOL) starting. Where there is a requirement for variable-speed drives, slipring induction motors may be used. Squirrel-cage induction motors with scoop-controlled fluid couplings are also specified in some cases. Commissioning of fluid couplings is covered in Chapter 4, since variable-speed drives for boiler auxiliary plant is their main application. The use of thyristor speed-controlled motors has not been widely adopted to date, other than for actuators and some specialised low power rated drives. High voltage, high power solid-state converter/inverter variable frequency controllers, suitable for application to standard HV induction motors, are becoming available. Because of insufficient operating experience, commissioning of such units will not be covered in this chapter.

High voltage (HV) motors fall into the following categories:

- Squirrel-cage induction motors, which may be of the double-wound stator (i.e., two-speed) type or may be of twin-cage rotor construction. Most, however, are of conventional single-stator winding, single-cage rotor construction.
- Slipring induction motors, with liquid regulator control of speed.
- Commutator motors with induction regulator control of speed.
- Specialised wet stator cage-rotor motors.

Two-speed motors require a separate circuit-breaker for each stator winding. Safety interlocks are fitted to the circuit-breakers to prevent both being closed simultaneously, and also to prevent the high speed winding being energised with the motor at rest.

Most of the HV motors are air cooled by means of shaft-mounted double-impeller fans. One impeller circulates air in a closed circuit through the motor windings and an air/air heat exchanger. The other blasts atmospheric air through the heat exchanger tubes. This cooling arrangement is identified as CACA. Boiler feed pumps, however, generally use water from a service water system (Section 18 of this chapter) as a secondary cooling medium: CACW is the classification for this cooling arrangement. CW pump motors are usually open-ventilated machines, where the warmed air is discharged into the pumphouse. Often special arrangements are necessary to ventilate the pumphouse, particularly during summer months.

Figure 3.18 shows simplified arrangements of speed regulators for slipring and commutator motors. The liquid speed regulator requires pumps to circulate electrolyte through heat exchangers which are then cooled by water from a service water system. Induction regulator cooling is usually of the CACA arrangement with integral cooling fans.

Larger motors usually have both bearings insulated from the general earthed mass of the motor. This prevents induced shaft voltages from causing circulating currents likely to damage the bearing surfaces. One of the bearings is usually fitted with a removable earth-bonding strap to facilitate testing of the bearing insulation resistance. Some designs of motors are provided with one or two shaft earthing brushes to discharge induced voltages in a controlled manner. In this case both bearings would be fully insulated from earth at all times.

Oil-fed journal/thrust bearings for 11 kV motors are often supplied from a common pumped-lubrication system for the motor and its fan, pump, etc. Bearings for 3.3 kV motors are usually grease-filled ball and roller, though in some cases oil-lubricated journal bearings are fitted. Lubricating oil may be fan cooled, or water cooled from a service water system.

Some modern boiler plants are provided with wet stator pumps to assist the natural thermo-syphonic circulation through the furnace water wall tubes. Figure 3.19 shows a typical boiler circulating pump which will be discussed in greater detail in Chapter 4. These motors are subjected to boiler pressure and have a

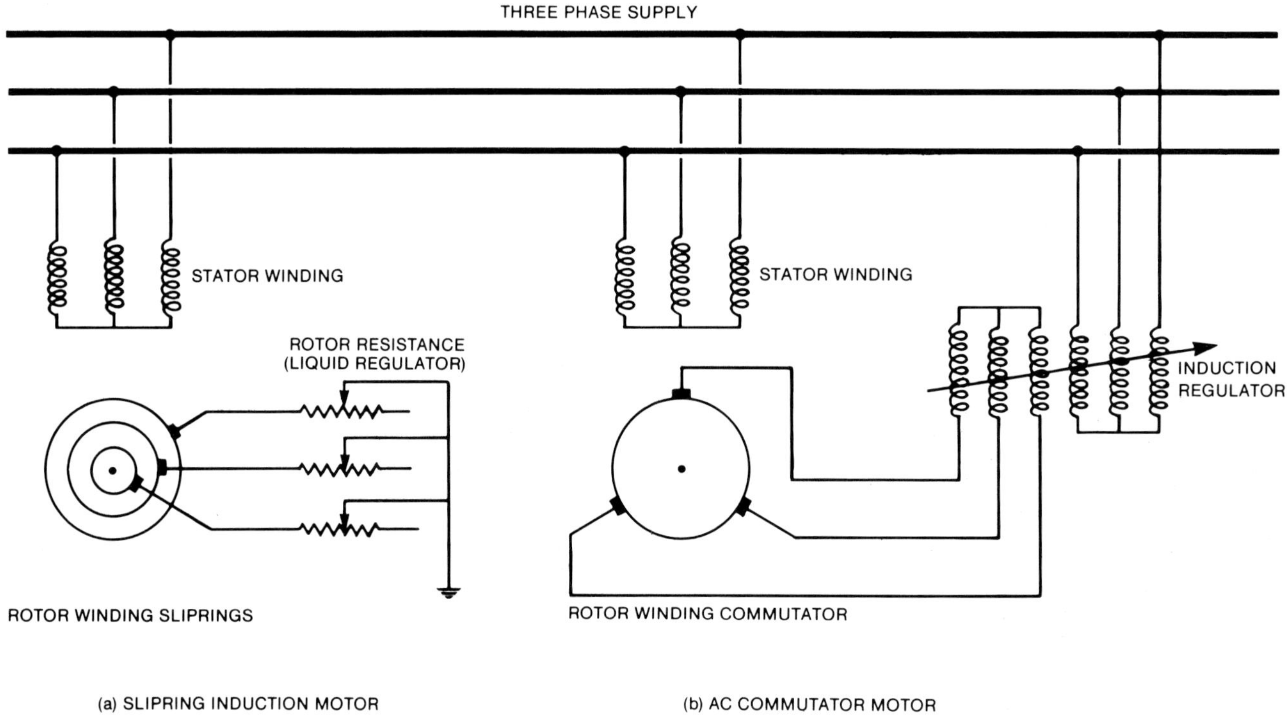

FIG. 3.18 Diagrammatic arrangement of speed regulators

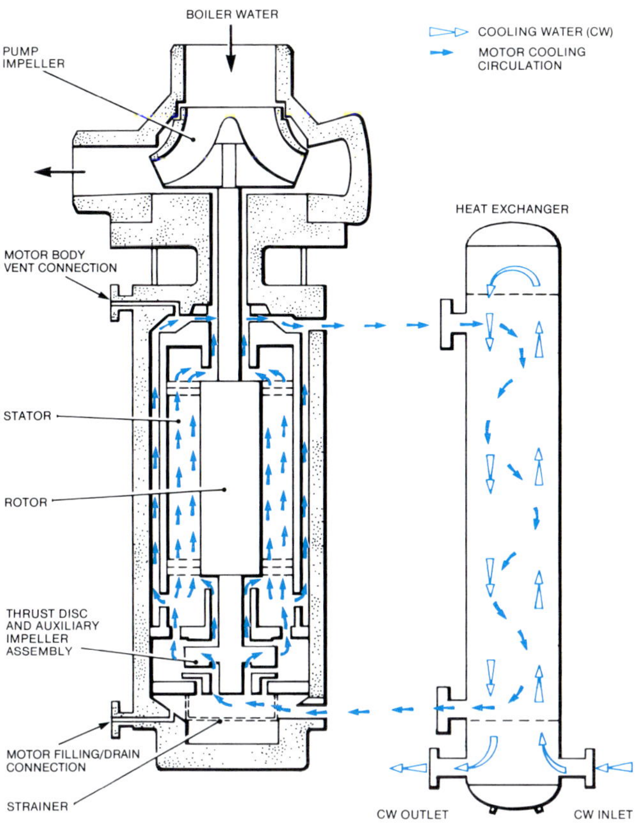

FIG. 3.19 General arrangement of wet stator motor

cooling circuit which uses a motor-shaft-driven impeller to circulate the coolant through a heat exchanger cooled by service water. Some boiler plants are fitted with wet stator motors connected at 415 V; but, apart from that difference, the units are almost identical to the HV ones.

The group classed as LV AC motors includes most of the motors installed to drive pumps, fans, electric actuators, etc., with power ratings of less than 150 kW. Almost all are squirrel-cage induction motors connected for a DOL start. For some specific duties, commutator motors with induction regulator speed control are still used. Generally the motor cooling arrangement is totally-enclosed fan-cooled (TEFC). Bearings are usually of grease-packed ball and roller combinations to accommodate end thrust.

DC motors are usually fitted to emergency pumps for circulating lubricating-oil and main generator stator cooling water. Typically the motors would be rated at up to 25 kW at 240 V DC. Starters usually incorporate two or three stages of armature resistance, automatically shorted-out on a fixed time basis. Motors at 110 V DC are often used for control equipment drives. Other specialised lower voltage designs are used for modulating actuator positioning duties.

4.1.2 Commissioning

One of the more outstanding developments in power stations has been the rapid increase in size and numbers of motors employed. Because of this, commissioning staff are involved in setting to work a variety of motors, which requires a methodical technique equally applicable for diverse items ranging from large CW pump motors to small drainage sump pumps. It may be convenient to produce a separate standard check list (SCL) or a test procedure for each type of motor, but many of the static checks are common to all. The following checks are based on commissioning HV squirrel-cage induction motors. Differences of approach for other types are highlighted later.

Squirrel-cage induction motor

It is convenient to think of the commissioning checks as being subdivided into two sections. The first, often called *static checks*, are as follows:

- Foundation bolts grouted in and tightened down.
- Alignment complete and bearing pedestals dowelled.
- Motor air gaps checked to be correct within 10% of nominal value.
- Bearing and cooling fan clearances recorded.
- Bearings cleaned, flushed if appropriate, and charged with correct grade of lubricant.
- Ball and roller bearings greased, if necessary.
- Shaft proved to be free to rotate, uncoupled from its driven member.
- Plant clearly identified with safe access for commissioning.
- Pedestal insulation resistance checked with 100 V or 500 V DC tester. Readings of up to 50 MΩ should be possible with a new, clean system. It is important to test the insulation resistance of all insulated flange joints in bearing oil pipelines.
- Stator winding phase resistance balance measured.
- Stator winding insulation resistance and polarisation index measured, using 5 kV DC tester for 11 kV motors and 2.5 kV DC tester for 3.3 kV motors. Values after one minute should be better than 1000 and 500 MΩ, respectively. With modern epoxy-resin insulation systems, insulation resistances of 20 000 MΩ are likely. The polarisation index (PI) is a ratio of the ten-minute reading to the one-minute reading, and a PI of better than 1.5 would be acceptable, providing that the insulation resistance results were also acceptable.

  Note All stator winding thermocouples must be short-circuited to earth during these tests.
- 'Tan delta' test carried out as a further check of the integrity of the winding insulation.
- Motor steelwork bonded to earth and the bearing pedestal earth continuity link fitted, where appropriate.

- Motor supply switchgear all checked and available for use.
- All cabling completed and HV test carried out on main supply cable.
- Motor heater insulation resistance checked at 500 V DC, then energised and proved to be working. It is strongly recommended that the motor heaters should be energised from temporary site electrical supplies during storage and the early erection phase.
- Motor cable terminations completed, then taped to the specified standard. Terminal box explosion diaphragm and dessicator (if fitted) checked to be satisfactory, then terminal box cover secured.
- All control, alarm, indication and protective devices checked to the appropriate SCL.
- Motor nameplate details recorded and electrical protection confirmed to be correct for motor rating, as outlined later.

When all of the preceding static checks have been satisfactorily completed, the following *running checks* should be considered:

- Temporary start-stop button provided adjacent to HV switchgear, so that personnel can operate the circuit-breaker from a safe distance, yet have it in line of sight.
- Motor half-coupling guard supported clear of shaft; temporary safety barrier put around motor half-coupling.
- With the HV circuit-breaker isolated, but simulated to be in the service position, all protective devices are proved to be operational. Generally this condition is referred to as having the circuit-breaker 'on jumpers'.
- Set up temporary instrumentation for recording of starting current, run-up time and no-load running current.
- The circuit-breaker should then be restored to the normal service position. Start motor and confirm correct direction of rotation. Take due regard of any limitations on frequency of motor starting. Typically, this may be three evenly-spaced starts per hour, where a start is assumed to be a run-up to normal speed.
- Check that motor runs with its magnetic centre as marked on the shaft. If not, then the true magnetic centre should be inscribed.
- Allow motor to run for up to four hours until bearing temperatures stabilise. Note indication of all instrumentation for future use.
- Carry out bearing vibration datum, or fingerprint, record.
- Compare all local and remote indications of current, temperature, oil pressure, etc.

At the end of the run, shut down and record the run-down time, then release the motor for coupling to the driven unit. Commissioning then becomes part of the associated plant system.

Slipring induction motors

For slipring induction motors with liquid regulator speed control, the following checks have to be added at the appropriate stage:

- Rotor winding resistance balance measured.
- Rotor winding insulation rèsistance checked at 1000 V DC.
- Rotor slipring surface cleaned to an acceptable finish for good current collection. Brushboxes correctly aligned to sliprings, with brushes of correct grade, all properly bedded to the curvature of the sliprings. Brushes also checked to be free to slide in the boxes and with the correct spring tension applied.
- Electrolyte system filled with clean demineralised water and checked to be free from leaks. System circulated to flush any debris into temporary fine mesh filters placed in the path of water returning into the tank. Flushing water drained to waste. Tank and resistance chambers inspected, cleaned out if necessary, then refilled with clean demineralised water.
- Electrolyte strength adjusted, by mixing sodium bicarbonate with water whilst it is circulating, to obtain the initial value of resistance required.
- Check that the resistance of each of the three phase electrodes is balanced at 20% increments of electrode travel. The test is carried out using an AC current injection test set. Results for each phase must be similar and be within the manufacturer's specified tolerance.
- Instrumentation calibrated and interlocks checked. Electrolyte pump auto/standby control checked.
- Rotor-to-regulator cabling all positively identified at both sets of terminations, and electrical tests completed.
- Regulator set to the minimum speed position, after proving that the speed control actuator drive gives the correct stroke. Control from remote position checked to be operable. It is important to check that the fixed and moving electrodes cannot come into contact with each other. Generally 5 mm insulated buffers prevent this, so that there is always a very small resistance in the rotor circuit when the regulator is at the maximum speed position.

Careful note of the starting current, run-up time and electrolyte temperature must be taken. If the starting current is more than 70% full-load current with an electrolyte temperature of 20°C, then the electrolyte resistance should be increased. When the electrolyte is hot, i.e., at 60°C, the starting current must not exceed 1.5 times full-load current. It should be appreciated that a 40°C increase in electrolyte temperature virtually doubles the rotor starting current. Hot starts are an infrequent occurrence, which is why a limit of 1.5 times is acceptable. High starting currents are likely to result in damage to the stator winding.

If practicable, the regulator should be operated over its full travel to record the range of speed in relation to regulator position. An acceptable combination of electrolyte conductivity, electrode position, starting current and motor speed takes patience and a lot of adjustment to achieve. This must be carried out when the pump or fan is coupled and able to be put on load. Rotor circuit resistance is only a means to an end, as it is the top speed that matters, subject to the starting current being acceptable.

HV commutator motors

Induction regulator speed-controlled commutator motor checks should also include:

- Brushgear checks, similar to those for the slipring motor.
- Induction regulator stator and rotor windings have to be electrically tested, as for a slipring induction motor.
- The interconnecting cables and all motor/regulator terminations must be thoroughly checked from end to end to be absolutely certain that every connection is correct.
- Cooling fans, anti-condensation heaters and instruments have to be checked and commissioned.
- Regulator position control proved and left at minimum speed setting.

When the drive motor is started, the regulator must be kept under close observation for any undue increase in temperature. If practicable, the regulator should be operated over its full travel to record the range of speeds attained in relation to regulator position.

Wet stator HV motors

It is customary to store these units with the shaft vertical. The motor and pump impeller are kept filled with a corrosion-inhibited water solution during storage. After mounting and connection of the motor cooler pipework, the motor water-circuit should be filled with demineralised water through the motor drain connection. The circuit then should be flushed out through the pump shaft neck-bush and also out of the motor body vent connection to remove all traces of the storage fluid. A maximum filling rate of 2 l/min must be observed to ensure that all air is driven out of the stator windings. Some 50 minutes should be allowed for the motor and heat exchanger to fill. Flushing, interspersed with several hours of soaking, should continue until the outlet conductivity is less than 200 μS/m.

Stator winding insulation resistance and PI tests for 3.3 kV motors are carried out using either a 2.5 kV or a 1000 V DC tester. Motors connected at 415 V would be tested with a 500 V DC tester. A target insulation resistance of better than 2000 MΩ should be achievable. The motor must be full of clean water for this test.

The only positive rotation check is to ensure that each phase of the motor supply cables is correctly terminated in accordance with the manufacturer's instructions. It may be practicable to give each motor a very short (e.g., 5 s) rotational spin with suction and discharge valves shut, providing the pump bowl is fully primed. Running tests have to be deferred until the boiler is filled. These generally comprise a check on the motor load current with cold boiler water, plus noting the differential pressure generated by the pump. These checks are repeated when the boiler is at working temperature. Incorrect rotation will give rise to approximately 40% lower differential pressure, accompanied by a 10% increase in power input.

Other vital checks in the case of boiler circulating pumps are:

- To ensure that the motor supply cables are adequately protected and have sufficient flexibility to accommodate a downward movement of some 250 mm as the boiler is filled then expands with temperature.
- To ensure that warm boiler water (i.e., above 70°C) is not permitted to flow into the motor cooling circuit at any time. All motor drain valves should be locked shut, or blanked off, to prevent this happening otherwise the stator winding insulation will be destroyed.

Low voltage AC motors

Commissioning checks are likely to be set out on a single SCL to cater for all sizes of motor. They will include the following:

- Plant clearly identified with safe access for commissioning.
- Motor anti-condensation heater insulation resistance checked with 500 V DC tester and heater energised as soon as possible.
- Stator winding resistance phase-balance checked, then insulation resistance checked with 500 V DC tester. A reading in excess of 50 MΩ should be achievable.
- Motor frame securely bonded to station earth bar.
- Motor supply cable correctly and securely terminated, then taped up, where appropriate.
- Motor protection confirmed to be correct for motor rating as described later.

- Switchgear all checked, ready for energising, with all control functions proved to operate correctly.
- Motor uncoupled from driven unit (if practicable) and checked to be free to rotate.
- Rotation check carried out and larger motors (say above 50 kW) allowed to run for 30 minutes or so. Starting current, run-up time and no-load current to be recorded.
- Motor coupled to its pump or fan, after which commissioning becomes part of the system into which it is fitted.

DC motors

Commissioning checks generally are of a similar pattern to the LV AC motor, with the inclusion of the following:

- Insulation resistance check of armature and field windings at 500 V DC.
- Field and armature circuit resistances measured and recorded.
- Armature starting resistance short-circuiting contactor timers given preliminary settings, until the starting current can be measured and final settings deduced.
- Brushgear alignment checked. All brushes checked to be of the correct grade, properly bedded to the commutator and free to slide in the brush box with the correct spring tension.
- Brushgear neutral position rechecked by DC injection of shunt field winding to confirm marked position.

Motor electrical protection settings

Electrical protection is covered in greater detail in Section 9 of this chapter. Protection settings applied during switchboard commissioning tests are usually based on scheduled information rather than motor nameplate details. However, the responsibility for ensuring that motor protection settings are correct often rests with the engineer involved with commissioning of the motor. Though it is a relatively simple matter to check the applied settings, no adjustments should be made without reference to the switchgear test team.

As one example, consider a typical 11 kV CW pump motor rated at 2600 kW for which the following data can be obtained by observation:

Motor full-load rated current (I_F): 172 A

Protection current transformer (CT) ratio: 200/1

Protection relay rated current (I_R): 1 A

Relay tapping set to: 90%

Thermal overload set at: 100% rated current

Instantaneous overcurrent setting: 8 times full-load

Note The maximum continuous-running current that most modern designs of motor can safely withstand is only 5% above the rated full-load current. Hence:

Motor maximum continuous current (I_H): 1.05 I_F = 180 A for the above motor

In order to circulate rated current through the relay when the motor is at rated load the relay tappings have to be adjusted to a ratio of I_F and the CT rated primary current, i.e., 172/200 = 0.86, or 86%. Since the closest relay tapping is 90%, this is the one selected. In effect, this means that a primary current of 180 A will correspond to secondary current of 1 A, i.e., 100% relay rated current. Because the calculated maximum thermal rating of this motor (I_H) just happens to be 180 A, then the thermal protection has to be set to 100%. When the motor is carrying rated current the '% load current' indicator will then read just over 95%. Large motors, started direct on line, normally have a starting current of up to 6 times full-load current. To avoid spurious tripping during starting, the instantaneous overcurrent protection elements are usually set at 8 times full-load current (i.e., 8 × 180 or 1440 A in this case) for 11 kV motors. It is not uncommon for 3.3 kV motors to be set to 10, and sometimes 12, times full-load current.

At the other end of the scale, a 415 V drainage sump pump motor, with a full-load current of perhaps 5.2 A, will have the contactor unit isolator switch fused at perhaps 20 A, the nearest standard fuse rating to 3.5 times full-load current. If the motor is protected by a simple thermal overload unit, then the heater code would be selected such that it would have a range of 4.8 A to 5.8 A. The tripping point of the relay would be set at about mid-point to correspond to a motor full-load current of 5.2 A, with an inherent allowance for 5% overcurrent before the relay operates.

Drying out

With modern insulation systems, there should be little need to dry out motor windings to achieve acceptable values of insulation resistance. It would be unwise to state here the absolute minimum values of insulation resistance that are acceptable. This is a matter for careful judgement based on a knowledge of many factors concerning each particular motor and its application. The method adopted for drying out will depend very much on the size, type and voltage rating of the machine and the ease or difficulty of conducting the operation anyway. In all cases, circulation of heated air, rather than the application of direct heat, should be the rule. An air temperature limit of no more than 80°C to 90°C at the inlet to the machine should be observed. It should be noted that insulation resistance values are very sensitive to winding temperature; low readings may cause unnecessary alarm unless this is appreciated. Typically, the insulation resistance with the winding at a temperature of 90°C may be only a tenth of that at normal ambient temperature levels of 20°C.

4.2 Pumps and fans

This group includes pumps and fans associated with smaller auxiliaries, heat exchanger circuits, heating and ventilation plants, etc. A separate SCL is likely to be required for fans and another for pumps.

4.2.1 Pumps

Centrifugal pumps are the most widely used type. Positive displacement pumps are used for low flow, high pressure duties, e.g., chemical injection.

All pumps must be fully primed before they are started. Centrifugal pumps must be started with the discharge valve closed. They should not be run for longer than a few minutes without establishing a flow of water through the pump. Opening of the discharge valve must be carried out carefully, particularly when pumping into an empty system, so that the pump unit is loaded gradually, and does not become overloaded. Positive displacement pumps must always be started and run with the discharge valve at least partially, but preferably fully, open, otherwise over-pressurisation may occur.

Most small pump shafts are fitted with conventional packing glands. Mechanical seals are being increasingly used, especially where hot or corrosive liquids are being handled. Frequently, larger pumps have a water supply to the mechanical seals and also for bearing cooling purposes. The water supply is usually from either the non-potable main or a service water system (see Section 18 of this chapter).

Commissioning checks would include:

- Plant clearly identified with safe access for commissioning.
- Bearings correctly lubricated.
- Rotating element free in its bearings.
- Motor checked, and rotation proved to appropriate SCL.
- Instrumentation and protection devices checked to appropriate SCL.
- Coupling alignment accepted.
- Holding-down bolts and dowels fitted securely.
- Drain and vent valves accessible, free to operate, and led to convenient drainage point; also drains from pump-shaft gland-leakage collection bowl piped away.
- Glands or mechanical seals correctly fitted.
- Gland sealing and bearing cooling pipework flushed clear of debris and proved to be leak free.
- Motor and pump coupled and, if appropriate, the coupling greased.
- Prime pump, confirm no leaks and adjust packing glands, where appropriate.

Subsequent operation of the plant is carried out in accordance with the requirements for the particular system.

4.2.2 Fans

Axial flow fans are generally used for oil or water/air heat-exchanger duty, but centrifugal fans are almost universally employed for heating and ventilating plant air circulation.

There are seldom any starting problems with axial fans, but centrifugal fans must have the outlet damper closed before being started. In many cases, small fans can be operated for a considerable period of time without any air flow through the ductwork.

Commissioning checks include those appropriate from the list for pumps: the first six items are particularly relevant.

Axial fans are usually mounted directly on the motor shaft and an uncoupled rotation check is impracticable. For centrifugal fans, the main check comprises making up the coupling or, in most cases, fitting, lining up and adjusting the drive belts to the correct tension. Belt-driven fans must have the belt tension checked and readjusted, if necessary, after the first 50 hours operation, again after 100, and then after 300 hours of continuous running, to prevent premature belt failure.

5 Lifts and cranes

The first part of this section deals with commissioning of power station goods and passenger lifts. Escalators are seldom fitted in power stations; however, the principles of commissioning are broadly similar to those for lifts. The second part considers the principles involved in commissioning various cranes, hoists and runway beams installed to facilitate plant installation and maintenance.

It is common practice to commission much of the equipment early in the plant construction programme as an aid to plant erection. Because of this, electrical connections usually have to be made to the construction site supplies system (see Section 9 of this chapter). At some convenient stage in the programme, the permanent supply cabling has to be connected, ensuring the correct phase rotation is maintained.

Usually an arrangement is reached whereby contractors carry out routine maintenance and testing and then refurbish the equipment to an 'as new' condition prior to handing it over for station use. Lifts, for example, are subject to misuse, so a temporary protective lining is fitted in the car. A competent person is employed as a lift car attendant, with responsibility for completing the daily routine maintenance checks.

The standards for construction, testing and use of lifts and cranes must be in accordance with all current statutory requirements.

5.1 Lifts

5.1.1 System description

Several goods lifts, with a load capacity of some 3000 kg, are installed for transporting personnel and materials to all major levels of the boiler house. Passenger lifts, generally of up to 600 kg capacity, are installed in administration and control buildings, though some installations may also be provided in the boiler house. At some locations one lift, providing it complies with the necessary requirements as regards size, capacity and speed, will be fitted with a 'fire control' switch, usually at ground floor level. Operation of the switch overrides all other control stations and calls the car immediately to the ground floor where it stops with the doors parked open. All safety devices, including maintenance switches, remain operative but sole control is then exercised from the car control station.

Each lift is also provided with a control station on top of the car to facilitate examination and maintenance of the lifting rope, runners, sheaves, guides, counterbalance weight and safety gear. This overrides all other control stations and also incorporates a 'dead man's handle' feature. An emergency stop button is fitted to allow the car to be stopped at any position. It is good practice to provide telephone communication between car top and the lift motor room, in addition to the mandatory telephone fitted inside the car.

A special key is required to release the landing door safety-lock at any one floor to give access on to the car top or into the base of the lift shaft. When the car is at its upper/lower limit of travel there must be a prescribed safety clearance above/below the car for safety of maintenance personnel. Shock-absorbing buffers are fitted at the base of the shaft to cushion overtravel of the car (lift car down) or counterbalance weight (lift car up). The mass of the counterbalance weight overcompensates for the weight of the empty car, but is less than that of the fully-laden car, so four loading conditions are chosen for testing lifts, i.e., empty, balanced and full, as well as overload.

Usually all of the control equipment is mounted in a sheet steel cubicle, located in the lift motor room at the top of the lift shaft. The motor room and the control gear must be kept warm and dry by adequate space heating and ventilation. The design of the ventilation system should ensure that the motor room temperature is kept below an absolute maximum of 40°C.

5.1.2 Commissioning objectives

The following specific objectives are additional to the general objectives highlighted at the beginning of this chapter:

- To prove that the mechanical and electrical installation is in accordance with the specification.
- To check that all indicators, controls and protective devices are functioning correctly, including 'fire control', if fitted.
- To prove that the lift will raise and lower the contract load and that the contract speed and acceleration can be achieved.
- To prove that the brakes will support a load equivalent to 125% contract load.
- To prove that the safety gear will arrest the contract load within the specified slide tolerance.
- To prove that the lift motor thermal performance is within specification.
- To obtain statutory certification for each lift.

5.1.3 Safety considerations

The primary danger is the possibility of personnel inadvertently gaining access into and/or falling down a lift shaft. To maintain a safe system of work, the following should be considered in addition to the points outlined at the beginning of this chapter:

- Approved procedures must be available for gaining access to the lift car top, and into the base of the shaft.
- Special procedures must be prepared for maintenance control of the lift when personnel are on top of the lift car. These will include issue of safety documents, with restoration of motive power (ROMP) for work which involves personnel riding on top of the car whilst carrying out testing or specific maintenance tasks. Particular attention must be paid to training of personnel to act as authorised 'competent persons' for this work.
- Whilst maintenance work is in progress, warning notices must be affixed at each landing, preferably mounted on secured temporary barriers placed across the landing doorway.
- Access to motor room must be restricted to authorised personnel only. The door must be kept locked at all other times.

5.1.4 Resources

Services

A special key for opening landing doors, and tools for hand winding the car in an emergency, must be available. Portable barriers with warning notices for placing across landing access doors are to be ready for use. Telephone communication between lift motor room, car top and inside the car must be in service. The motor room lifting beam and tackle must be certified for use and be properly identified. Portable BCF fire extinguishers will be needed in the motor room and temporarily in the car.

Staffing

A team of three, including the authorised test engineer from the lift contractor, will require three to five days to commission and test each lift.

5.1.5 Commissioning

Commissioning of lifts is a specialised function carried out by the supplier, in accordance with BS2655 and BS5655. He also has the responsibility for issuing the statutory certificate.

The commissioning engineer has to be satisfied, mainly by documentary evidence, that erection has been correctly completed. The lifting rope test certificate must be available, all specified lubricants applied and all equipment adequately bonded to earth. The insulation resistance of the electrical wiring should be tested at 500 V DC to earth, the minimum acceptable result being 0.5 MΩ, though, on a new installation, it should be possible to achieve almost 'infinity'. Using the hand winding gear, all control equipment, protective devices and indicators should be checked for correct operation, then the following running tests carried out:

- Prove that operation of the 'fire control' switch routes the lift immediately to the ground floor and stops it with the doors open, irrespective of its direction of travel or location.
- Confirm that sole control of the lift car is exercised from the car control buttons when under 'fire control'. Check that the doors remain open whilst the car is at any selected floor.
- Functionally test the motor, brake, control equipment and door locking devices over full travel with the lift unloaded.
- Repeat the test, but with the lift under balanced loading.
- Repeat the test, with the lift carrying the contract load, to determine speed, acceleration, motor current, levelling accuracy and repeatability and also that the specified starting frequency can be achieved without overheating motor or brake.
- Demonstrate the operation of the safety gear while the car, containing its contract load, is descending at the contract speed.
- Static overload test at 125% contract load, followed by one full traverse, usually with halts at the upper and lower levels *only*, to prove that the brake will arrest and hold the overload.

Upon satisfactory completion of these tests, a statutory certificate will be issued. During subsequent service, the motor room temperature should be monitored to ensure that it does not become excessive as a result of high ambient temperatures created by operation of adjacent plant, e.g., the main boiler.

5.2 Cranes and hoists

5.2.1 System description

A typical modern power station contains two main overhead cranes and at least ten auxiliary overhead cranes, plus a number of power-operated hoists and a variety of runway beams with portable lifting tackle. In addition, some specially designed lifting aids are supplied by plant manufacturers for future station maintenance work.

The two main cranes are located in the turbine house. Each crane typically has a span of 50 m and is provided with main and auxiliary hoists having a safe working load (SWL) of 120 t and 30 t, respectively. The two cranes can be mechanically and electrically coupled as 'master' and 'slave', then operated in tandem from one of the control cabins for very heavy lifts using a special lifting beam slung on both main hoists.

Turbine house crane main hoists are equipped with special creep-speed devices, e.g., Microsen, which usually operate when the controller is selected to the first two stages of speed control. Generally, both brakes on the main hoist drive are of the hydraulic thruster type. These comprise a self-contained hydraulic pump unit, which holds the brake off as long as the motor is energised. De-energising the pump motor collapses the hydraulic pressure, allowing strong springs to apply the brake.

Infra-red or radio remote control features are not as widely adopted in power stations as they are in other industries. If remote controls of this type are fitted, then their application is normally to turbine house cranes only. BS5744 'Code of Practice for Safe Use of Cranes' specifies that remote controls must have a very limited operating range. No crane can be moved, nor any of the other drives energised, unless the operator is within a preset range for safe operation of that crane. This feature is relatively easy to achieve with low power infra-red controls which depend upon a line of sight between transmitter and receiver. Radio controls utilise the electromagnetic component of radio waves, the frequency of which typically lies between 200 and 400 kHz. The generally accepted range for safe operation of cranes is usually between 15 and 35 m, a distance which is less that 10% of the wavelength corresponding to that range of frequencies. The receiver is then tuned to accept signals from the transmitter only while it is within the designated 'controlled range' for safe operation.

The Station Manager has to be in possession of the Induction Communications Licence (rather than a Radio Communications Licence) before any radio control equipment is used. Like all forms of crane control, remote facilities are designed on the 'dead-man's handle' principle and incorporate a special permissive key-operated control isolating switch.

Some cranes are now being fitted with automatic load-limiting control and indication systems. These usually depend upon signals from a strain gauge transducer bolted to the main beam of the crane to warn the operator of an impending overload. If corrective action is not taken, the equipment usually trips the main hoist when a predetermined strain is reached.

Auxiliary cranes, with SWL between 2 t and 30 t, are fitted in other plant areas, including workshops and heavy stores. Larger capacity cranes will be equipped with power operation for all directions of motion. Cranes with a power-operated traverse drive are equipped with an audible warning device, which may be either automatically or manually sounded as the crane is moved. Smaller capacity cranes may be fitted with manually-operated long and/or cross travel motions. Usually all are provided with a pendant-operated electrical control station for personnel use at ground level. Control for all power-operated motions is on the 'dead-man's handle' principle, in that each pushbutton has to be manually depressed to energise the appropriate motors, which stop and brake as soon as the button is released.

Coal-fired power plants are provided with special purpose cranes for handling coal deliveries and for ash disposal. These may be either traversing gantry cranes, or jib cranes arranged for luffing and slewing as well as traversing. Wire-rope-operated grabs are usually fitted in place of a lifting hook for these applications. At some sites, lifting and lowering of the grab is by wire ropes, but opening and closing of the grab is carried out by hydraulic actuator.

Hoists usually are installed at various places in the boiler house and control building. Lifting capacities generally range up to 5 t. At least one hoist is designed to lift equipment on to the boiler house roof level. They have a very restricted field of movement, but are power operated for all motions. Control is from interlocked pendant stations located at each working floor level. Hundreds of runway beams, with a wide range of safe working loads, are fitted over plant, where fixed cranes or hoists do not have clear access. Electrical supplies are all from a 415 V, three-phase and earthed-neutral system, via a local isolating switch. Where the crane has a long traverse, i.e., more than 10 m or so, sheathed power conductors, or downshop leads, are fastened along the side of the crane rail. Shielded pick-up shoes on the crane engage with the downshop leads. The 415 V isolator used to supply downshop leads is arranged so that it can be selected into a position which first isolates and then connects all three phase conductors to earth. For shorter traverses, or in locations where the environment is harsh, special flexible cable systems are used for power supplies and control purposes. One form comprises loops of cable suspended from a series of wheeled runners which slide along either a plastic-coated steel-stretched wire or a rail as the cable is extended or retracted. On turbine house and ash plant cranes, a form of cable reel is used to transfer power and control signals to the travelling crab containing the hoist motors, etc. Coal and ash plant cranes are normally fitted with power-operated cable reeling drums to hold their power supply cable (see Section 13 of this chapter) while traversing along the rails.

5.2.2 Commissioning objectives

The following specific objectives supplement the general objectives outlined at the beginning of this chapter:

- To confirm that all electrical equipment is correctly installed and bonded to earth. Confirm that crane steelwork and crane rails are effectively bonded to earth.
- To prove that the downshop lead isolator switch will connect all three phase conductors to earth when selected to the 'earthed' position.
- To energise the downshop leads, prove that all warning lamps are operational, then ensure that all controls function correctly.
- To complete motor rotation checks, couple the shafts and then wind (or reeve) hoist ropes on to the drum.
- To demonstrate the performance of each crane or hoist in respect of control, speeds and loads up to SWL, including emergency lowering of main hoist.
- To carry out an overload test at 125% SWL.
- To prove that access to downshop lead current collection equipment is prevented unless the conductors are connected to earth. Check that safe maintenance access is then possible.
- To prove that alignment of the running rails is within the specified tolerance.
- To obtain statutory certification for each appliance.
- To prove correct and safe operation of any infra-red or radio remote control facilities, including consistent repeatability of the 'controlled range' feature.

5.2.3 Safety considerations

The requirement for the thorough examination of all lifting equipment, cranes, ropes, chains, etc., is governed by formidable legislation under the Health and Safety at Work etc. Act 1974. Figure 3.20 is an example of how specific legislation affects the examination and testing frequency of power station lifting equipment. During the construction phase, the Construction Regulations 1961 apply to all work being carried out, and there are specific Construction (Lifting Operations) Regulations 1961 which apply to lifting tackle and lifting operations. Once the first prime mover generates electricity, then most of the activities within the physical area of the plant operation are classed as being in a 'factory'. When the 'client' assumes responsibility for work in the area then the requirements of the Factories Act 1961 apply, particularly Sections 22 to 27. All lifting equipment at jetties is governed by the requirements of The Docks Regulations 1934.

	TYPE OF EQUIPMENT		
	Cranes, Crabs, Winches, Pulley Blocks, Transporters, Runways, etc.	Chains, Ropes, Rings, Hooks, Shackles, Swivels, etc.	Mechanically clamped wire ropes, Chains, Rings, Hooks, Shackles, Swivels, Cranes, Crabs, Winches, Pulley Blocks, Transporters, Runways, etc.
REGULATIONS	**Examinations**		**Testing**
Factories Act, 1961	After testing and every 14 months	After testing and every 6 months	Before being taken into use and after any repair
The Construction (Lifting Operations) Regulations, 1961	After testing and every 14 months Inspect in position every week	After testing and every 6 months. Items not in regular use need only be examined when necessary.	Before being taken into use and after any repair. In the case of cranes, crabs and winches after every 4 years, also in the case of cranes after adjustment or re-erection involving anchorages or ballasting. (In the case of blocks, only applicable when raising or lowering a load of one ton or more.)
Dock Regulations, 1934	After testing and every 12 months. Also must be inspected before use unless inspection has taken place in preceding 3 months	After testing and every 12 months. Also must be inspected before use unless inspection has taken place in preceding 3 months.	Before being taken into use and after any repair

FIG. 3.20 Lifting equipment regulations

Valid test certificates must be available for all lifting tackle before it is accepted into use. It is most important that a register of all lifting equipment be maintained, along with all current test certificates. Each piece of tackle should be identified with a unique number and be clearly marked with its safe working load (SWL).

It is good practice to mark the direction of movement for long and cross travel of all power-operated cranes with four cardinal points of the compass on the crane beam and at all control positions. The SWL and compass point indication should be clearly visible both from the control positions and from ground level.

The main hazard to personnel is contact from inadvertent movement of the crane or hoist without their knowledge. Other than the driver, no personnel should be permitted on to a live crane, unless special safety precautions have been taken. Access to any section of the crane rail should be prevented until full safety precautions have been taken and a safety document issued. Whilst load tests are being carried out, barriers should be erected and all personnel not involved should be cleared out of the delineated area. Should any equipment show signs of distress whilst the test weights are suspended, the weight must be slowly lowered to the ground by manual operation of the brake where practicable. Special care is required before energising extensions to the downshop leads or permitting movement of any crane into other areas as construction work progresses. Temporary rail stops, warning signals and safety barrier isolations may be necessary to define the agreed working area.

A crane driver emergency escape device must be available in the cab and checked to ensure that it will work correctly when required. BS5744 gives guidance on the current intervals between testing of the 'controlled range' feature of infra-red or radio controls for cranes or hoists.

Each of the general safety considerations highlighted at the beginning of this chapter should be applied in addition to those above.

5.2.4 Resources

Services

Specialist equipment will be needed for checking alignment of the crane rails and for measuring deflection of the crane beam when it is loaded. Test weights must be available to cater for 25% above the SWL of each lifting device.

Staffing

Commissioning will be carried out by a team of three or four people, augmented by an authorised test engineer during load testing. Approximately three days will be required for each crane or hoist, and up to one day for each runway beam.

5.2.5 Commissioning

In the following pages commissioning of a turbine house crane has been chosen to outline the principles for commissioning of various types of lifting equipment. Individual components should be tested in accordance with standard check lists (SCL). Where appropriate, test procedures have to be prepared to integrate various groups of components. A summary of the main activities is as follows:

- Using optical or laser equipment, the gauge and level of the running rails in relation to fixed datums is checked to be within the designed acceptable tolerance.

- After proving that access for operation and maintenance is safe and that all equipment has been correctly lubricated, cables terminated and steelwork bonded to earth, the 415 V supply to the crane rail conductors, i.e., downshop leads, is energised.
- Providing that the warning lamps installed along the track are functioning correctly, the crane control circuits can be energised.
- All control functions and protective devices are then proved.
- The drive motors are checked for rotation, and the couplings made up.

The next operation is to wind (or reeve) the hoist ropes on to the drums and the hook, after ensuring that the rope and hook test certificates are valid:

- The raise and lower overtravel limits should be set.
- Checks are carried out to confirm that at least two turns of wire remain on the drum at the lower limit and that there are still at least two empty grooves on the drum at the raise limit.
- Operation of the main hoist emergency-lowering technique must be demonstrated before any test loads are suspended from the hook.
- The crab is operated to prove the settings of the cross travel 'in' and 'out' limits, confirming that the hook does not foul on any supporting steelwork.
- Similarly, the long travel limits and anti-collision device are proved by traversing the crane along the available length of running rail. This will prove that joints in the rail do not cause the crane to jerk, that the wheel/rail alignment is within acceptable limits, and that the crane does not foul structural steelwork.

With the crab in mid-span and the test safe working load hung on the hook, the following tests are conducted:

- The raise and lower speeds and motor current are noted for every position of the controller.
- The deflection of the main beam is recorded, first with the load on and then with it off, to confirm that there is no permanent set.
- These tests are repeated with 125% SWL overload on the hook.
- Whilst the overload is still suspended, the hook should be lowered at top speed and the effectiveness of the brake proved. Both normal and standby brakes are tested independently whilst lowering the SWL at the fastest speed.
- The SWL is hoisted up clear of all obstructions, then the crane traversed along the track with the load suspended at mid-span.
- The load is repositioned at each extreme position of the cross travel in turn and the crane traversed along the entire length of available rail. Rail/wheel alignment is rechecked during each traverse.

Providing all of the tests have yielded results within acceptable tolerances, the test engineer will issue a certificate of compliance. During subsequent service, the running rails will need to be periodically re-examined for tightness of fastenings and wheel/rail alignment.

Commissioning of either infra-red or radio control facilities is usually carried out by the equipment supplier's specialist staff. Normally this is undertaken after completion of all statutory tests, when it is a relatively simple matter to prove the remote control equipment interface with already proven control systems. Radio controls, in particular, need special additional attention to ensure that the equipment is immune to interference from any other on-site equipment employing similar radio frequencies. Equally, the radio control should not cause interference with the other radio-controlled equipment within its range of transmission.

Before any commissioning of radio control equipment takes place, the Station Manager has to be in possession of a valid Induction Communications Licence. The maximum 'controlled range' for safe operation of the particular crane has to be agreed, then the receiver adjusted to this limit. Once safe correct operation of the remote control features has been proved then consistent repeatability of the 'controlled range' limit has to be confirmed. Providing that the results lie in a band of ±5% of the preset limit they are usually acceptable. Whenever the limit is exceeded, confirm that all drives stop safely and that they cannot be restarted until the operator with the remote control station has been repositioned within the safe range.

6 Data processing systems

This section outlines the principles involved in commissioning a typical computer system on a modern power station. Software development is usually completed off site, then the programs are loaded and tested on site. Proving of the system under operational conditions will be found in the appropriate sections of Chapters 4, 5 and particularly 6.

6.1 System description

6.1.1 General

Computer systems provide data processing and plant alarm annunciating facilities, plus control of certain plant operations. The advantages of direct digital control (DDC) over conventional analogue control are now being exploited in closed loop control systems (see Section 8 of this chapter). Modern computer-based alarm displays present the operator with an overview of alarms in various areas, laid out in a specific pattern on the visual display unit (VDU). Codes displayed with this overview enable the operator to call up other more detailed displays. Computers also permit sophisticated alarm analysis which can present the operator with the prime cause of a group of related alarms. Alarm analysis is a feature seldom provided at conventional power plants; however, the additional cost and complexity of the software can be justified for nuclear stations. Comprehensive computer systems, particularly in nuclear power stations, utilise on-line digital computers to replace many of the conventional indicators, recorders and analogue controllers.

Figure 3.21 shows a simplified block diagram of a typical modern installation comprising several independent data processing (DP) systems coupled by data links to form a distributed network. DP systems are provided for the following functions:

- Each main unit
- Station services
- Maintenance and standby
- Data amendment
- Local HV transmission substations

If a separate DP system could not be justified for the HV transmission substation, then the information would be processed by the station services computer.

Each system usually consists of a main central computer controlling several data centres, VDUs and line printers. The main computers are usually located adjacent to the central control room (CCR) in separate air-conditioned rooms, often equipped with Halon 1301 fire protection schemes (see Section 12 of this chapter).

Data centres are strategically located in plant switchrooms or close to the CCR. Their purpose is to collect and process data from the plant and to transmit control commands to plant. Specific data centres also interface with the microcomputer systems for automatic boiler control and plant sequence control equipments (see Section 8 of this chapter).

Analogue inputs from plant-mounted devices are converted into binary numbers representing engineering units, before being multiplexed for transmission by a single data link from each data centre to the main computer. This results in economies of cabling, and the plant signals are less susceptible to electromagnetic interference.

Data centres consist of a scanner, controlled by a microcomputer which linearises and scales the data before passing it to the main computer. Digitised analogue values are compared with preset high or low

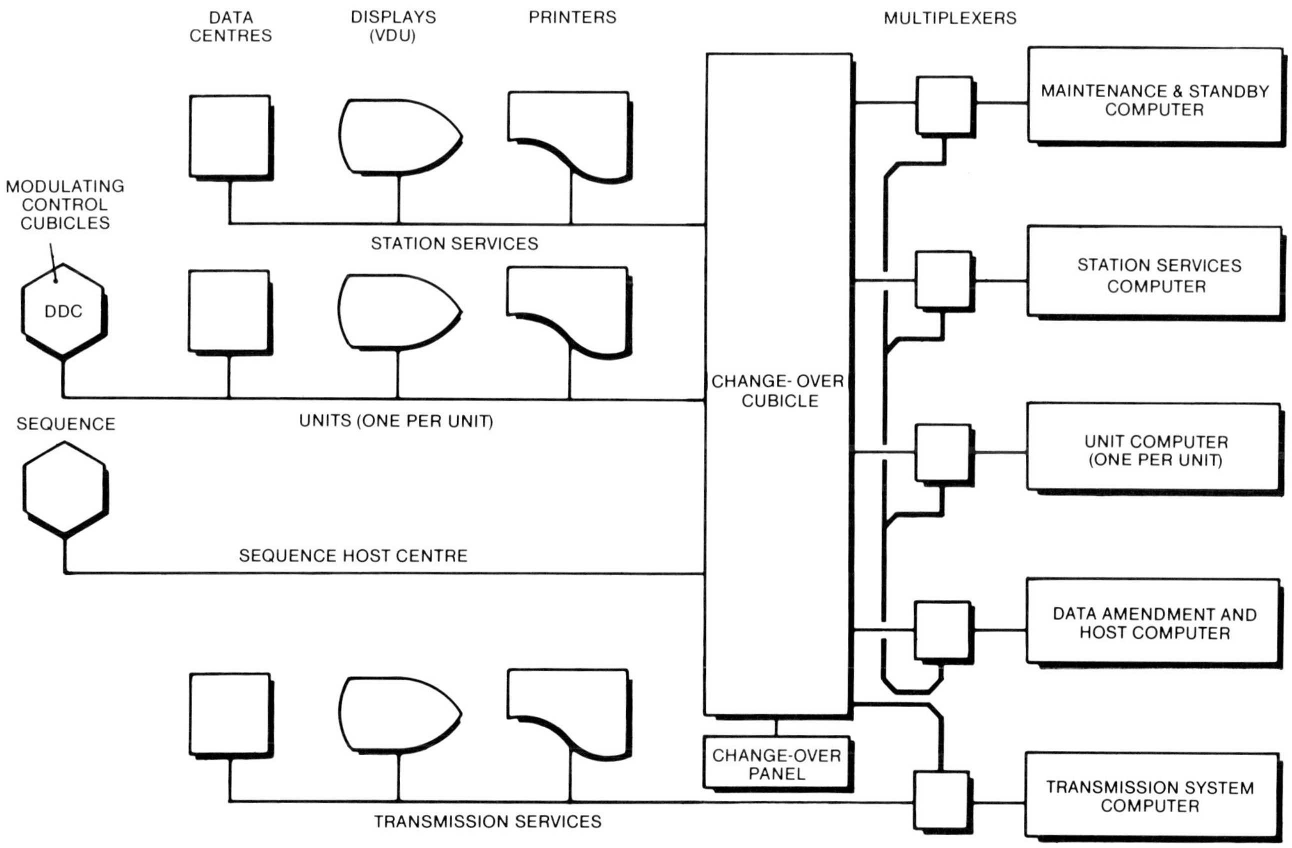

FIG. 3.21 Computer system single-line diagram

alarm limits: if either of the alarm levels has been exceeded for more than one scan, the alarm condition is passed to the main computer.

Power station computer systems are usually connected to duplicated 110 V AC single-phase guaranteed power supplies from the instrumentation supplies system (see Section 9 of this chapter).

6.1.2 Unit DP system

The unit computer system will have two central processors and perhaps six data centres. One processor, known as the 'A' processor, carries out all DP tasks, while the 'B' processor is concerned with supervising the control programs for specific plant microcomputers. A VDU and a special function keyboard are provided at the unit control desk. Separate conventional alarm facias are fitted to annunciate computer system fault alarms in the CCR.

Each computer system usually has the capability of dealing with over 1000 analogue and 1600 digital inputs. Functions available from the system are:

- Data display and logging
- Alarm display and logging
- Post-incident and routine logs
- Performance monitoring and efficiency calculations
- Boiler header life factor determination
- Turbine run-up and loading
- Unit load control

6.1.3 Station services DP system

Generally, only one processor and perhaps two data centres are fitted. A conventional alarm facia and a VDU are located in the station services control panels. This system may have a capability of only 40 analogue and 800 digital inputs to perform the following functions:

- Data display and logging.
- Alarm display and logging.
- Display and logging of certain conditions common to the main units.
- Display and logging of selected HV substation information.

6.1.4 Maintenance and standby computer

This has two central processors only. It functions similarly to the main unit computers, and performs the following functions:

- Acts as a main computer replacement for any of the above systems, when manually selected to do so.
- Provides test centre facilities.
- Maintains and tests hardware.

6.1.5 Host and data amendment computer

One main computer performs these functions:

- Down-line loads each of the other computer systems.
- Down-line loads, via the appropriate main computers, the data centres and specific plant control microcomputers.
- Creates, alters and integrates new programs and routines into the existing software systems.
- Inspects and modifies plant parameters.

6.1.6 HV transmission DP system

This computer system as, for example, at Drax Power Station, comprises one central processor and seven data centres. It is complete with its own special function keyboard, VDU and conventional alarm facia. This system deals with 170 analogue and over 2000 digital inputs. The following functions are available:

- Data display and logging
- Alarm display and logging
- Operation of the local transmission substation from the CCR

6.1.7 Software

Computer software generally falls into one of two categories, i.e., either operating system software or application software, known collectively as the 'system software'. The former is necessary to enable the computer system hardware to run: it includes executive device drivers for the input/output facilities, peripherals and also communication data links. Application software defines the use to which the system is put, e.g., details of formats, etc. The modern trend is to adopt a high level language, which is independent of the computer type, then to use a compiler to enable it to work on the particular computer system. It is generally accepted that software for DDC systems is more complex than that for a centralised computer system, but there are other compensations for the expansion of DDC in modern process plant control.

Within the UK electricity supply industry most software is now written in the CEGB high level engineer-orientated language CUTLASS. This has been specifically developed for power station use and has been applied in several modern plants. Another language in use is CORAL.

Special procedures have to be implemented to ensure security of system software throughout commissioning and subsequent operation (see Section 6.5.7 of this chapter).

6.1.8 Facilities available to the operator

Data Analogue and digital information, calculated values, etc., are available upon demand by selecting any one of a number of specific plant-group data formats. Modern systems make full use of colour displays in mimic, tabular, graphical or histogram form on the control desk VDU. For effective operator comprehension, the number of lines of alphanumeric detail is usually limited to 20 per format. Typically, up to 200 formats, including mimics, etc., may be prepared to suit local conditions. The operator can usually request a printed copy of most formats.

Alarms These are displayed on a VDU in order of receipt from either of two sources. One source is digital inputs, either from the conventional alarm system or direct from plant-mounted devices. The other is from digitised analogue measurements, which are then compared with alarm limits preset by computer program. Generally, all alarm messages are displayed and logged, but the program provides for the allocation of selected alarms to be logged only. The operator usually also has the facility to suppress (or shelve) a relatively small number of either 'standing' or defective alarms.

Post-incident and routine logs The computer continuously monitors and stores typically 100 preselected analogue input measurements. Normally these, along with all other analogues, are scanned at 3-, 5- or 30-second intervals, dependent upon the relative importance of each. However, for the purposes of the post-incident log (PIL), the selected inputs are also checked every minute against the value last recorded. In event of a specific plant incident, e.g., unit trip, all of the values that have shown a change during the preceding 25 minutes and subsequent 5 minutes are printed out automatically in a timed sequence. This is known as a 30-minute PIL. Provisions usually exist for a 3-minute PIL, i.e., 2 minutes before and 1 minute after the event.

For a routine plant log, up to 60 inputs can be selected for printing automatically at, say, hourly intervals.

Turbine run-up and load control Facilities are provided to enable the unit operator to preset the rate and then initiate an automatic run-up of the main turbine or steam-turbine-driven feed pump. The computer can also be selected to load the unit at predetermined rates, using ramp or boiler pressure functions to control turbine governor set point. Furthermore, the unit load control program will maintain the load at any selected value, in accordance with either a preselected governor droop characteristic by adjustment of the governor or the total fuel controller. The governor characteristic (droop) is capable of manual selection over a very narrow load/frequency ratio.

6.2 Commissioning objectives

Apart from the general objectives outlined at the beginning of this chapter, the following are specific to this system:

- Make each DP system and associated data centres available for plant monitoring and control, consistent with plant commissioning programmes.
- Establish that failures or program corruptions do not occur as a result of disturbances on the instrumentation supplies system which feeds the computers.
- Demonstrate that the standby computer can be selected to, and will drive, each of the unit or station services DP systems.
- Prove that programs, once loaded on site, function together without corruption. Also, prove that modifications or additions to the software are compatible.
- Test the display, logging, alarm and data amendment facilities of each system.
- Confirm that alarm message phraseology is correct and that it is compatible with that of any associated conventional alarm display.
- Check that the alarm state is not displayed when the plant is in a healthy state.
- Demonstrate that the alarm analysis programs function correctly, by using different combinations of inputs.
- Check that all analogue inputs display correct values, when the plant is on load.
- Confirm that correct inputs have been used and allocated to their designated positions on data formats, etc.
- Confirm that the turbine run-up, boiler feed pump run-up, plant loading control and unit load control programs operate the plant effectively within the specified parameters.
- Demonstrate that the immunity to RF interference, proved at works, has not been degraded by connection of permanent input cabling.
- Check that the automatic power fail/restart procedure operates correctly.

6.3 Safety considerations

The following must be considered in conjunction with those points outlined at the beginning of this chapter:

- Access to the computer equipment must be restricted to authorised personnel only at all times.

- All personnel must be familiar with the action to be taken in event of operation of the fire protection equipment.
- There is a danger to personnel terminating other cables in marshalling boxes through which computer input or output cables may be energised at 50 V DC or lower voltages. These voltages are normally considered safe; nevertheless, harmful electric shocks may be experienced dependent upon local conditions.
- It may be necessary to work inside equipment cubicles while the plant is in service. Extreme care must be taken to ensure that all work, including any temporary connections necessary in furtherance of testing, is to a safe standard.
- Another problem, from the plant operating staff point of view, is to maintain the integrity of the displayed information, as different inputs are commissioned.

6.4 Resources

6.4.1 Services

- Air conditioning plant available and running.
- Fire protection system in service.
- Instrumentation supplies energised.
- Timing pulses available from the station master clock.
- Conventional alarms system ready to accept inputs.
- Control desks and panels completed, or temporary mountings installed for VDUs and keyboards, etc., using permanent cable connections as far as practicable.
- Software security system set up and all necessary equipment in position.
- Computer Input/Output Schedule and analogue Alarm Setting Schedules prepared for use.
- Diagnostic test programs for fault finding on system hardware.

6.4.2 Staffing

A team of four engineers is necessary for checking and commissioning the hardware. A number of teams, each of two or three people, are needed for checking, testing and proving of all analogue and digital input/output channels between computer system and plant. Several software specialists will be needed throughout the system and plant commissioning phases.

The commissioning programme is likely to occupy more than fifteen months, dependent upon the extent of the system. Input channel commissioning should be progressed in parallel with the hardware tests. The bar chart in Fig 3.22 indicates an approximate timescale for commissioning to meet main plant requirements.

6.5 Commissioning

6.5.1 General

Relevant hardware and software of the computer systems must be proved and available for the associated plant to be commissioned. Subsequent hardware/software commissioning must not interfere with plant testing.

Computer systems are subjected to tests to prove their reliability, accuracy, speed of operation and

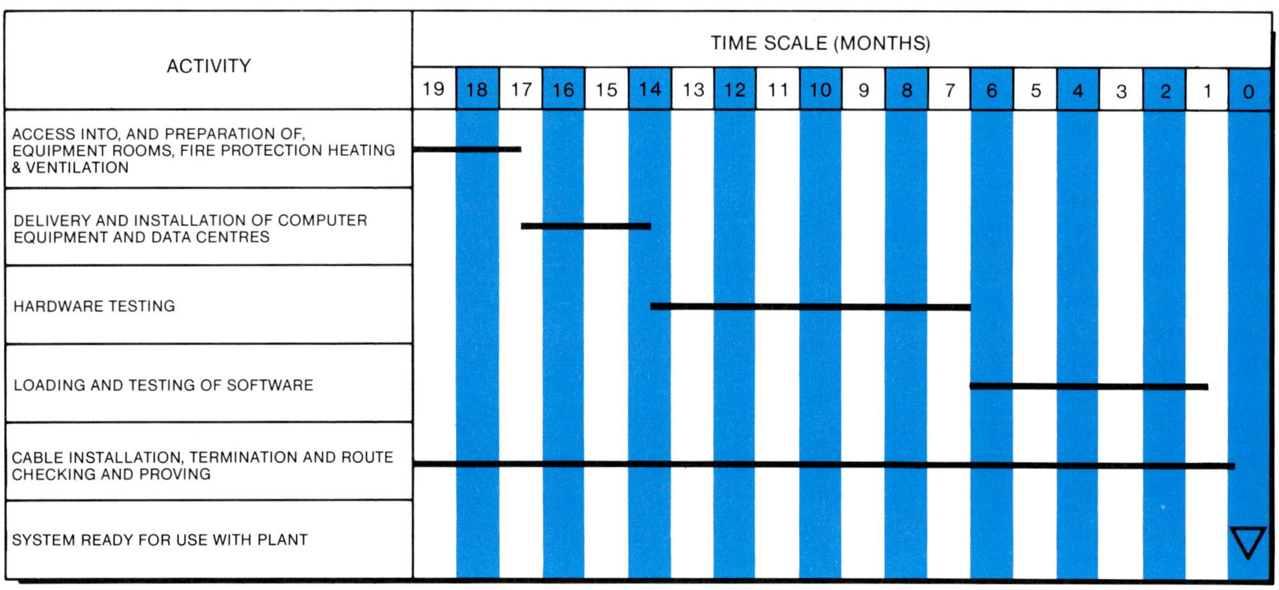

FIG. 3.22 Computer system commissioning bar chart

correct activation. BS5887 is a relevant reference document for testing of computer-based systems. Basically, the tests are in two parts:

(a) Works tests to prove hardware reliability and accuracy.

(b) Site tests, which demonstrate both hardware and software reliability and accuracy. These also establish the integrity and speed of operation of the assembled system.

The commissioning activities can be grouped into four distinct sections:

- Hardware testing, using test programs.
- Software loading and testing of hardware.
- Input/output testing by simulation of plant parameters.
- Computer system proving, with the main plant operational.

The final section will not be covered in any detail here, since it is largely a function of putting the main plant, rather than station services, to work. Such testing will be dealt with in later chapters, particularly in Chapter 6.

6.5.2 Hardware testing

Generally, all equipment modules are tested individually at the manufacturer's works, using stand-alone test programs designed to test all the facilities of particular items. After installation on site, the equipment is again subjected to a test programme to ensure that it is in the same satisfactory state in which it left the works. An off-line suite of diagnostic programs is used on site to check for interaction and lack of interference between sections of the system. These diagnostic programs are capable of being run individually or grouped into a repetitive sequence for a system test to fault-find on the hardware. Following satisfactory completion of a 1000-hour test programme, the system software is loaded.

The erection checks comprise the following main items:

- Each equipment room is complete, cleaned out and painted, with all access doors fitted.
- Air conditioning and fire protection equipment is commissioned where appropriate.
- All equipment within computer area is installed, and correctly identified.
- All peripheral equipment is installed in desks, panels, switchrooms, etc., with cabling completed.
- Earth bonding is completed to the station instrumentation-system earth bar.

Next, the power supplies are energised and the system operation proved thus:

- Energise each module, checking direction of rotation of fans, drums, discs, etc., and prove that each operates satisfactorily.
- Fully test all modules with other modules.
- Check the computer supplies for voltage and current, and confirm that the voltage drop to remote equipment is within acceptable limits.
- Run an interactive hardware system acceptance test program. This simultaneously exercises all hardware blocks with the distributed data processing configuration for the 1000-hour test.
- Test the links to other plant microcomputers, if the relevant equipment is available.
- Finally, demonstrate that the standby computer system can be switched to replace any of the other available systems, and that it functions correctly.

Throughout all of these tests, it is customary to record the following environmental parameters:

- System supply voltage and frequency
- Room temperature and humidity

6.5.3 Software testing

Software production is very vulnerable to human errors, which may not be revealed until certain functions are exercised. Such errors may have serious consequences. By using special techniques, it is practicable to achieve a satisfactory performance as, for example, by:

- Rigorous testing, using plant simulation, as comprehensively as practicable, e.g., model testing or computer simulation of a turbine.
- Exemplary software management documentation, with close administrative control of modifications.
- Comprehensive testing of the system after any modifications.

The main test activities are as follows:

- Demonstrate that all system programs can be loaded into all the computer systems from the data amendment machine, and also from local loading sources where appropriate. Confirm that the computer systems operate satisfactorily.
- With the system running, prove that repeated operation, plus disconnection and reconnection of peripheral equipment, has no detrimental effect on system operation.
- Show that the response of the system to display and logging demands is satisfactory, when the load on the system is onerous.

- Test all non-control programs by checking the correct operation and speed of response of each function in turn.
- Initiate a large number of inputs in a short time and check that corruption of software or overloading of the processors does not occur.
- Check that the automatic power fail and restart sequence operates correctly.
- Test each of the listed control programs by checking the logic, action levels of each input, control loop response and correct digital outputs for turbine and boiler feed pump run-up, turbine loading and unit load control. It is also necessary to prove that control parameters within the program can be modified, using the appropriate level of access for data amendment (see Section 6.5.7 of this chapter).
- Analogue and digital input testing, by simulation at the plant end of cables, to prove correctness of indication on VDU and/or printer. See Section 6.5.4 of this chapter for further details of this aspect.
- Test application-addressing features by confirming that the control software outputs to, and receives inputs from, the correct devices.
- Carry out RF interference test to confirm that the system still has the same immunity that it had at the works.
- Prove that failure of a data centre does not affect the operation of any of the specific plant microcomputer systems.

As mentioned earlier, but it justifies repetition, the current state of the DP system has to be carefully recorded at all times. In particular, the software currently loaded, and its state of modification, should be clearly identified (see Section 6.5.7 of this chapter).

6.5.4 Input/output testing

Every input and output is allocated a unique data location number (DLN) which is listed on a comprehensive Input/Output Schedule. Personnel involved in commissioning must be in possession of up-to-date schedules which also detail initiating device information.

The cable core route from the computer marshalling cubicles to every remote initiating device (or nearest junction box) is checked to be complete. It is customary to prepare a loop schematic, similar to those shown by Figs 3.16, 3.24 and 3.28, in conjunction with this check, as a future maintenance aid. The insulation resistance of the loop is checked with a 100 V DC tester to be at least 50 MΩ. These checks are carried out in parallel with, but independently of, most of the computer system testing. However, the system software has to be loaded and proved before most of the following testing can be started:

- Each digital input has to be initiated in turn at the computer cable marshalling cubicles. The message received on the VDU and/or the printers is checked for correctness.
- Selected inputs should have the highest specified ON resistance and lowest specified OFF resistance imposed on them and a check made that the appropriate plant state is displayed.
- Each analogue input has to be injected in turn at the computer cable marshalling cubicles. The message and calibration received at the VDU and/or printer is checked for correctness. Testing of the alarm settings and system response to out-of-limit signals will be carried out at the same time. The analogue input schedule should be marked up at this stage to show the current alarm levels of all inputs.
- The systems are checked to ensure that they can scan all inputs at the correct speed.
- Each analogue and digital input is injected or initiated in turn from the terminals of the plant device and the computer messages rechecked for correctness. At this stage, it is advisable to re-open the plant cable isolation links until each input is ready to be proved under actual service conditions.
- Characterisation tests are carried out to determine stroking non-linearities, hysteresis, backlash, etc., for all actuators driven from the software.

6.5.5 System proving

In conjunction with the plant test teams, the plant cable-isolation links are closed and each previously calibrated initiating device operated under normal service conditions. Correct receipt of alarms, indications and logged messages, etc., should be verified, then the computer input channel is validated and the status record brought up to date.

6.5.6 Control program testing

Either an analogue computer model or digital computer simulation of the particular plant to be controlled is used to test the control programs. This entails checking that each of the facilities available to the operator functions correctly in accordance with the design criteria. It may be necessary to make adjustments to the software control parameters when the actual plant becomes available. The extent of these adjustments will be a function of the accuracy of the modelling/simulation exercise. Subsequent on-line proving of the control programs is carried out in conjunction with the main unit test teams as outlined later, particularly in Chapter 6.

6.5.7 Security of software

Security procedures are needed to control access to computer systems and for administration of system software, tapes, listings, etc. During commissioning, security is usually vested in the computer test team, using a procedure compatible with that to be adopted by station staff. Responsibility transfers to the Station Manager when the Board officially takes over the computer system.

Control of access into the computer system is usually on the basis of three levels of security. Direct access via a teletype is customary for Level 1; but for Levels 2 and 3, individual keys have to be used in conjunction with the teletype before data amendment can be carried out. Nominated members of staff have to be authorised for each level of access. Typical functions that can be performed at each level may be as follows:

Level 1

- Set and adjust the system time.
- Commence, or cease, printing of specific logs.
- Change the printing frequency of certain logs.
- Display current status (valid or invalid) and the value of any selected input.

Level 2

- Make any input either 'valid' or 'invalid'.
- Modify or set alarm levels of analogue measurements.
- Change alarm message phraseology and select each for 'log and display' or 'log' only.

Level 3

- Create and alter various logging and calculation routines.
- Modify data format content and layout.
- Make changes to constants, etc., in plant control programs.

Security of system software is usually based on there being three copies of each item, with no two copies kept in the same area. Generally, a master copy is kept in a fire-resistant safe, the keys to which may be kept in the station safe. A second copy is often held in a secure locked cabinet in, for example, the station library. The third (working) copy is retained in locked cabinets located in, or adjacent to, the computer rooms. It is customary to maintain an emergency (or back-up) copy of application software, which lags the master by one update. In the event of a catastrophe, the back-up can be used to restart the computer system. Access to the various cabinets is restricted to selected personnel only.

Meticulous operating and software maintenance logs must be kept of all actions carried out by the computer user, including details of software amendment, patches, tests, etc. A master log, kept in the fire-resistant safe, should be updated periodically from information contained in the operating and maintenance logs.

7 Alarm systems

Alarm systems are installed for annunciating abnormal plant conditions to operating staff. The two main types of alarm system are:

Conventional Where the alarm messages appear on facias containing engraved legends each illuminated by a lamp unit.

Computer Where alarm messages are displayed on VDU driven by centralised DP systems.

This section deals with commissioning of alarm systems from the initiating devices to the associated facia and computer VDU displays. Alarm system commissioning interacts with commissioning of the DP system which was dealt with in Section 6 of this chapter.

7.1 System description

A centralised control policy has been adopted for most modern power stations. This means that alarms, along with other controls and indicators, tend to be concentrated on desks and panels located in a central control room (CCR). In modern designs, the alarm system is integrated with the DP system, which permits mounting of a smaller number of alarm facias in the control desks. Facia lamp disposition patterns are more easily recognised by the operator as being due to certain plant conditions, whereas the DP system allows ease of alarm handling, logging, etc. The general philosophy is for conventional alarms to provide sufficient coverage, either to maintain steady load or to enable the plant to be shut down safely, should the computer fail whilst the main unit is on load.

Computer-displayed alarms are initiated from three sources:

- Digital alarms repeated from interposing relays within the conventional alarm system cubicles.
- Digital alarms direct from plant-mounted devices.
- Digitised analogue values, which are compared with preset high or low alarm limits within the computer system. If either of the alarm levels is exceeded for more than one scan, an alarm is displayed.

Computers also permit sophisticated alarm analysis, utilising selected groups of direct inputs, which are normally unconnected with the conventional alarm system. In general, however, this form of alarm presentation is limited to nuclear power stations.

Conventional alarm facias and the controls for groups of plant are usually located close to each other on the same section of the control desk or panel. Computer alarms may warn of an abnormal plant condition or give supporting details in event of a critical alarm. A typical example may be turbine bearing temperature alarms, where the facia will warn of a 'high bearing temperature' but the VDU will identify which bearing is at fault and may also display the actual temperature.

The VDU may also display plant states which might not represent abnormal plant conditions, e.g., 'turbine shaft on barring'. The computer system also logs each alarm message and plant-state change, along with its time of initiation, on a line printer.

Conventional alarm facias are also located on local control panels (LCPs) associated with station services plant groups. Alarms from these facias are repeated into the CCR conventional and computer alarm systems. Typical locations for out-stationed local alarm systems would be:

- Coal plant
- Ash and dust plants
- Centralised compressor systems
- Water treatment plant
- Auxiliary boiler system
- Gas turbines
- Hydrogen production plant
- Main unit generator coolant control panels

Historically, alarm equipment has generally operated on the basis of a plant contact closing to initiate an alarm condition. Modern designs now specify that plant contacts be normally closed for a healthy condition. Centralised alarm systems are usually based on reed relays mounted on printed circuit boards (PCBs) contained within floor-mounted cabinets or cubicles. Local alarm systems now employ microprocessor-based logic to drive the conventional facias on each LCP. They are also arranged for self-monitoring of internal faults, e.g., 'system unhealthy'.

Intelligent microprocessor-based local alarm systems are now provided with a serial data link to the centralised DP system which allows the CCR operator to interrogate the status of local alarms. A typical 128-way facia can then be displayed on a VDU in the CCR to show the state of all associated alarm inputs, i.e., alarmed, reset or accepted. Such local alarm systems are also provided with remote accept/reset facilities to permit alarms to be accepted and reset, via the main DP computer system.

Conventional and computer alarm systems are usually powered from separate guaranteed supply sources. It is impossible to keep the two systems completely electrically separate, since in general they use the same initiating devices. Most alarm schemes operate at either 50 V or 24 V DC derived from the 50 V DC distribution or 110 V AC instrumentation supply systems.

Inputs from initiating devices are connected to the alarm equipment cubicles through combined terminal blocks and isolating links (see Fig 3.23). Each alarm circuit can be individually isolated for testing, independently of the alarm equipment. Interposing relays are provided, so that conventional alarm inputs can be grouped for display and also individually repeated to the digital marshalling cubicles of the DP computer. Incoming alarm cables to the computer system are also fitted with isolation links. Figure 3.23 is a simplified single-line diagram of a typical alarm scheme.

Power station alarm systems are usually divided into three separate identifiable plant area alarm schemes:

- Unit alarms, with a separate system for each main unit.
- Station services alarms, which may be grouped to cover two or three different plant areas, including out-stationed local alarm schemes.
- Fire alarms.

The unit and station services conventional alarms schemes may display up to 500 alarm messages each. Most of these are of the general/abnormal (or grouped) category, with detailed individual alarms displayed on the VDU via interposing and repeat relays (see Fig 3.23). Each plant area alarm scheme is provided with a unique audible warning and an amber flashing beacon in the CCR. Most facias are engraved with white letters on a black background. In some cases, coloured engravings are utilised to group the alarms into relative order of priority. Fire alarms are usually treated separately with a wholly conventional scheme; the facias usually have black lettering on a red background and are surmounted by a red flashing beacon supported by a distinctive fire alarm.

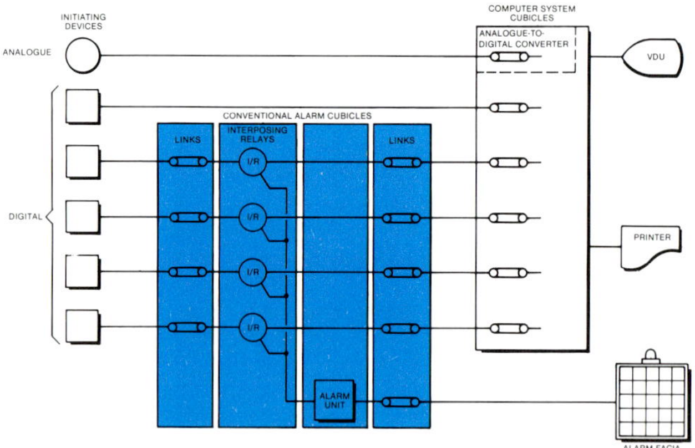

Fig. 3.23 Alarm system single-line diagram

7.2 Commissioning objectives

Apart from the general objectives listed earlier in this chapter, the following are relevant to this section:

- To check that each facia and its associated alarm-cubicle-mounted equipment operates correctly by simulation at the cubicle terminal racks.
- To prove each alarm channel by simulation from as close as possible to the initiating device.
- In conjunction with the computer test team, confirm that each VDU alarm message is correctly worded and that it is not displayed when the plant is in a healthy state.
- Confirm that each local alarm system connected to the station DP system has the correct address, and that, when the appropriate local alarm format is selected, the displayed information is correct. Check also that the remote control dialogue, i.e., accept/reset command, is with the appropriate local alarm system.
- In conjunction with plant area test teams, ensure that each inititating device has been calibrated and that it operates correctly into the alarm scheme.
- Carry out a six-month reliability run.

7.3 Safety considerations

In addition to the following, the general safety considerations highlighted at the beginning of the chapter should be taken into account:

- The main danger is that the plant alarm input cables may be energised at 50 V DC or lower voltage, whilst personnel are terminating other cables in the same cable-marshalling boxes. These voltages normally are considered safe to personnel; nevertheless, harmful electric shocks may be experienced dependent upon local conditions.
- Another problem for plant operating staff is to maintain the integrity of the alarm scheme in a continuously changing commissioning environment.
- Where appropriate, personnel must be aware of the action to be taken in event of release of Halon fire protection gas.

7.4 Resources

7.4.1 Services

Other than appropriate electrical supplies and calibrated initiating devices, no special services are required. Two-way radio communication will be necessary during later stages of testing.

7.4.2 Staffing

Teams of two, occasionally three, people are needed to check all equipment and cabling and to prove the correct operation of the scheme. Typically, each channel may take from four to eight hours just to check the cable route and prepare a cable core routing sheet as shown later in Fig 3.24.

7.5 Commissioning

Availability of comprehensive conventional and computer Alarm Schedules and a procedure for status marking of commissioned alarms are prerequisites for this activity. Each schedule should identify the location and engraving details for every facia window and also list relevant information about the initiating devices and their operating settings, plus identification of alarm repeats to the DP computer.

One status marker system that has been used successfully in several power stations makes use of small coloured self-adhesive spots that are affixed to each facia window. The presence of a green spot indicates that the alarm channel has been checked. This is superseded by a red spot when the channel has been proved under service conditions: the operator then knows that all 'red spotted' alarms should be considered as being operational.

Where a single window is initiated from several inputs, either a differently-coloured or a shaped marker has to be used until all inputs are in service. The status marker system must be backed up by a comprehensive up-to-date recording system which has to be available for reference by shift operations staff at all times.

In the case of computer alarms, it is customary to make all inputs invalid after completion of the data processing equipment tests. Each input is then made valid at the time the initiating device is proved in service. Again, a comprehensive recording system must be adopted in the interests of maintaining the integrity of the indications.

System commissioning tests, be they for a small locally-mounted self-contained unit or a full-scale main unit alarm suite, follow a similar pattern. For simplicity, they can be divided into the following stages:

- Alarm cubicle checks and facia commissioning.
- Channel proving by simulation at initiating device.
- Live testing of each channel with plant in service.

Since most alarm schemes are designed to continue operating satisfactorily with a specified minimum supply voltage, it is prudent to carry out the first stage checks at this minimum. Typically, 50 V DC equipment would be tested at a supply voltage of 46 V. Generally, this is achieved by variable resistance banks temporarily connected between the supply and each suite of alarm cubicles.

Cubicles and facia commissioning

A single standard check list (SCL) will probably be adequate to record all of the following static and powered checks:

- Cubicles identified, bonded to earth and safe access available.
- Relays, PCB and power supply units correctly installed, and identified.
- All plant cabling correctly terminated.
- Facia engravings correct and fitted to the appropriate window location.
- All plant cable input links, and also output links to the computer, open.
- Indication lamps of the appropriate voltage rating for their duty. (Generally 60 V lamps are used in a 50 V system.)

The power supply is then energised and equipment proved to function by:

- Checking power supply units to confirm that the output voltage has the correct value and polarity.
- Checking cubicle power supply failure warning alarms.
- Operation of 'lamp test' facility for each facia.
- Simulating plant contact changes of sense at the cubicle terminals, then confirming correct operation of the audible alarm, flashing beacon and flashing facia. After proving that the audible alarm circuit is functioning correctly it is prudent to replace the sounder with a lamp display to reduce the nuisance caused by frequent initiation of the audible signal.
- Proving correct operation of the facia 'accept' and 'reset' buttons for both fleeting and non-fleeting inputs.
- Testing each input to prove that correct window illuminates and also that the correct repeat relay to the computer operates, by checking at the appropriate outgoing terminals.
- Confirming that the window engraving and computer alarm message phraseology is consistent.

To complete this stage of commissioning, the temporary voltage dropping resistor is removed and the permanent supply reinstated. The facia status marker is applied and the records updated.

Channel proving The cable core route from the alarm cubicle termination racks to the remote initiating device (or nearest junction box) is checked to be complete. It is customary to prepare an alarm routing sheet, similar to Fig 3.24, which can be used as an SCL for this activity and also serves as a future maintenance aid. The insulation resistance of the loop is checked with a 500 V DC tester to be at least 50 MΩ. If this is satisfactory, the alarm equipment-cubicle incoming cable isolation links are closed.

Each channel is then proved, if necessary by simulation of the correct sense of the plant contact, from the terminals of the initiating device. Correct receipt of the appropriate alarm, including computer display and logging, is confirmed. It should be noted that some alarm sources have numerous contacts in parallel and that documenting of the proving checks for each contact requires close attention.

Live testing In conjunction with the relevant plant test team, each initiating device, having been calibrated

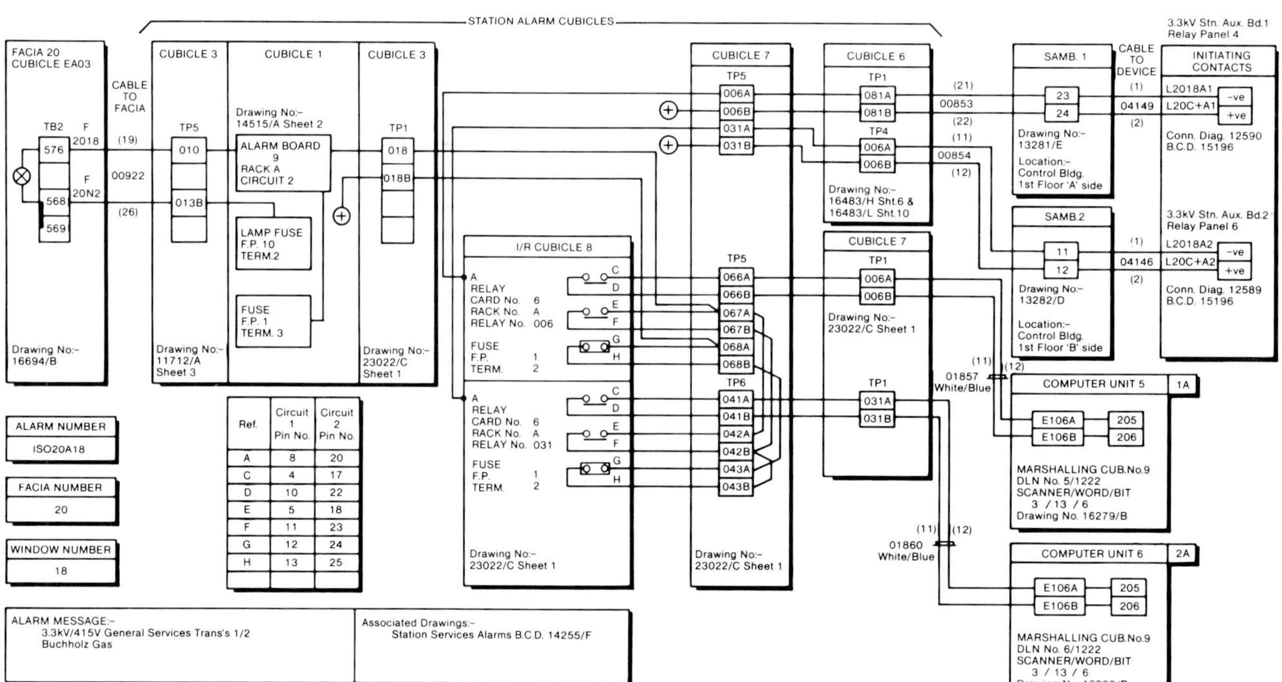

FIG. 3.24 Typical alarm routing sheet

as outlined in Section 3 of this chapter, should be caused to operate under normal service fault conditions. Correct receipt of alarms, both conventional and computer, are verified, then the status marker and recording systems are brought up-to-date. Where appropriate, the computer alarm channel is made 'valid'.

Reliability monitoring After all windows in each facia have been put into service, the alarm system reliability should be monitored over a six-month period. Clear records of all failures, plus details of any rectification work, must be maintained.

8 Sequence, control and interlock systems

The majority of controls and indicators for most modern power plants are concentrated in a central control room (CCR) as a result of the adoption of a centralised control policy. Most of the systems covered by this section are associated with control of main boiler/turbine units. Generally there is only a limited application on station services plant. It has been estimated that over 700 individual operations are necessary during a hot start of one unit. Application of sequence, control and interlock equipment can reduce that by more than half.

This section discusses commissioning of equipment enclosed within suites of various control cubicles. System testing, necessarily carried out in conjunction with plant commissioning, is outlined for completeness. Details of such plant tests are to be found in subsequent chapters of this volume.

For ease of reference, this section is subdivided into the three main subject headings to deal with control equipment defined as follows:

Sequence Equipment installed to start and stop individual or selected groups of plant items from a single control station. It is customary to arrange for automatic starting of standby plant should the duty plant fail.

Remote/manual control System of control applied to plant items which are not included within sequence or modulating control loops, yet need to be operated from a central location. A typical example would be a single control station on the unit desk from which the operator would initiate opening of by-pass valves for four boiler stop valves. When satisfied with plant conditions, he would again operate the control to open the four main stop valves.

Modulating control Equipment which continuously adjusts the plant to maintain operating set points or desired values. In its simplest form, this could be a closed loop that controls lubricating oil temperature by regulation of the flow of cooling water through a heat exchanger.

Interlocks Automatic schemes employed to secure basic plant safety, since it is usually impracticable to provide sufficient hardware to guarantee automatic foolproof operation of the plant. Considering the boiler stop valves mentioned earlier, interlocks would ensure that the main valves could not be opened unless the associated by-pass valves were fully open. General logic control would automatically reclose the by-pass valves after the main valves reached the fully open position.

8.1 Sequence systems

8.1.1 System description

The function of sequence systems is to ensure correct and safe start-up and shutdown of selected groups of associated plant items. A sequence includes operation of standby plant when necessary, e.g., automatically starting a condenser air extraction pump, if one of the running pumps trip, as distinct from being purposely shut down. Sequence equipment is entirely separate from plant control and interlocking equipment, though all three interact with each other.

A sequence consists of a number of steps comprising commands and checks, carried out by the sequence control equipment. Figure 3.25 shows a typical sequence flow chart for starting a gas air heater drive and setting up a standby lubricating oil pump. Each sequence is initiated by operation of a single control switch. The sequence continues to completion, without further operator intervention, provided all commands to, and checks from, the plant are completed in a preset time period.

In general, a step initiates a command signal to an item of plant within the plant group under control. The command signal is usually pulsed for one or two seconds to avoid damage to control equipment. This might arise if a continuous start signal were conflicting with a protection circuit trip signal to the same plant item.

Each check completes when the plant item has obeyed the command, in order for the sequence to continue. If any one check is unsatisfactory, progression of the sequence is halted. When the sequence overall time delay operates, a 'timed-out' alarm is annunciated in the CCR. The basic cause of the fault is determined by inspection of a coded display, plus cross-reference to the sequence flow charts. The coded display is sometimes located on the control equipment modules but, on modern plants, is often placed beside the operator's control switch. Some systems allow the operator to override certain plant checks after being satisfied that the appropriate input is faulty.

Typical coverage of individual sequences for a boiler/turbine unit may be as follows:

- Induced and forced draught fans
- Gas air heaters
- Individual PF milling groups

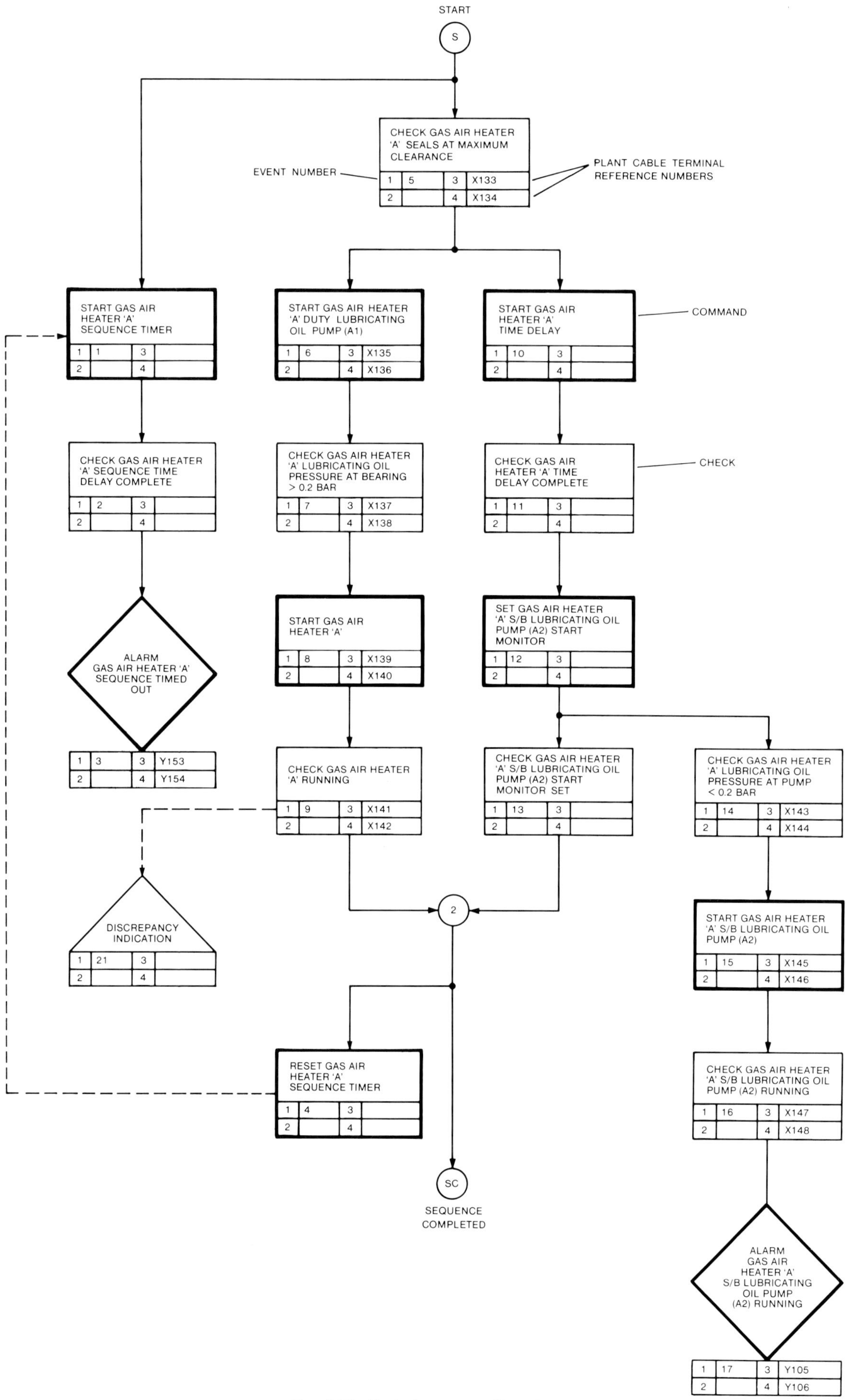

FIG. 3.25 Typical sequence flow chart

- Electrically-driven boiler feed pumps
- Turbine gland steam and vacuum raising
- Fuel oil pumping plant
- Automatic chemical control and monitoring
- Boiler furnace purge
- Steam air-heaters (auxiliary and bled steam controls)

Station services sequences would, for example, cover:

- Individual cooling-water system pumps
- Ash and dust plant blowers
- Compressor systems

Most modern sequence control systems comprise either reed relay or microprocessor based equipments, dependent upon the function and location of the controlled plant. The equipment is either grouped together in suites of cubicles located close to the CCR, or in sequence centres strategically out-stationed in plant electrical switchrooms. Sequences for various plant groups are selectively allocated amongst the cubicles or sequence centres, so as to maintain system integrity in the event of an equipment fault.

Each suite incorporates separate termination and equipment cubicles, the latter being provided with self-testing and monitoring facilities. The termination cubicle contains plant cabling isolating links, to facilitate checking and testing. Power supplies are usually at 50 V DC and/or 110 V AC from guaranteed supplies systems. Indication of fuse failure, supply failure, and equipment fault are provided, with the alarms being grouped and repeated to the conventional alarm system.

Reed-relay-based systems are designed on a modular basis, using a limited number of different printed circuit boards (PCBs) to achieve the desired functions. Each type of PCB has a specific function, which is performed by means of various relays, logic circuits, etc. Typical PCB functions would be step, control, timer, check, monitor and clock.

Microprocessor-based systems typically comprise a microcomputer with memory, watch dog timer, analogue and digital inputs, and digital outputs. Program loading of sequence centres is carried out individually by means of a tape cassette unit. Frequently, the data amendment computer of the centralised data processing (DP) system is used for down-line loading of the sequence centres, via a sequence minimal host centre. See Section 6 of this chapter for further details of this facility.

8.1.2 Commissioning objectives

In addition to the general objectives outlined at the beginning of this chapter, the following are specifically for sequence systems:

- Make each system available to meet the programmes of the plant commissioning teams.
- Check correct operation of the equipment, using simulated plant inputs.
- Prove overall operation of the sequence to run plant group auxiliaries, with the main plant item disconnected from its power source.
- Confirm that computer programs, once loaded on site, function together without corruption and that any modifications to the software do not dilute their compatibility.
- In conjunction with other plant commissioning teams, operate all of the various sequences under live operational conditions to confirm safe control of plant.

8.1.3 Safety considerations

Refer to the general safety considerations listed at the start of this chapter. In addition, the following are more specific to sequence control equipment:

- Most of the equipment is self-contained and the risks to personnel arise from low voltage signals being routed through cable marshalling boxes while other termination work is still in progress.
- Computer systems must not be left unattended, unless the alarm systems are fully operational.
- Only authorised personnel should have access to the control equipment, particularly as commissioning progresses.
- Operations staff should have access to a concise record of the status of the sequences which are available for their use: this record has to be supported by some form of status marking system (see Section 7.5 of this chapter). It may be acceptable to mark up a copy of appropriate sequence flow charts to highlight any checks, etc., that have not been proven, even though the sequence is operational.
- Where appropriate, personnel must be aware of the action to be taken in event of release of Halon fire protection gas (see Section 12 of this chapter).

8.1.4 Resources

Services

Apart from electrical supplies at 50 V DC or 110 V AC for control equipment, 240 V AC will be required for cubicle illumination and anti-condensation heating.

Cabling to plant, control stations, interlock equipment, alarm systems and the centralised DP system will need to be substantially complete. A portable VDU/tape cassette unit will be needed for program loading of microprocessor systems.

Some manufacturers provide portable test units for checking the sequence logic of their own equipment. However, it may be advantageous to manufacture suitable test units on site, since each is, in effect, a portable desk or panel structure on which are mounted switches, relays and indication lamps to represent plant items which will eventually be operated by the sequence control equipment. For this reason it is good practice to use relays similar to those which will be operated by the sequence control equipment, so that the correct resistive/inductive load is applied to the contacts of the sequence equipment under test. The surface of the board should be lined out with some form of mimic diagram showing which switch, relay or lamp is wired to the appropriate terminals. Flying leads bearing these terminal numbers are used to couple the termination blocks of the test unit to those of the cubicle under test. Switches then represent relay contacts, switchgear auxiliary contacts or limit switches, whereas the lamps indicate relay operation or outputs. Temporary labels are used to identify each component with the correct title or mnemonic applicable for the circuit being tested. Where appropriate, power supplies for the test unit are derived from the cubicle under test.

Staffing

Teams of three or four people will be needed for testing of each suite of sequence centres and also for checking associated plant cabling. Each suite may require a week for proving tests, but the total programme will spread over many months, dependent upon availability of plant for overall testing. Microprocessor-based systems may need a proportionately longer time span for control program proving/modifications. They also require the attendance of software specialist staff right through the plant commissioning period.

8.1.5 Commissioning

Irrespective of the type of sequence control system, it is convenient to subdivide the activities into four phases or stages, as shown in Fig 3.26. The basic approach is similar to that often adopted for commissioning of other control and alarm systems. The general principles discussed are based on a relay logic system, but reference to microprocessor-based systems is made, where appropriate.

Reed-relay-based systems

Phase 1 These are classed as cubicle checks. Here completion of installation is checked, plus termination of the power supply and alarm cabling. All equipment is checked to be correctly fitted and bonded to a power system earth bar. The cubicles are energised with plant cable-isolation links open to prove all the common features and alarm initiations. Non-interchangeability of different types of printed circuit boards has to be proved as well. The portable test unit is connected to command output terminals. Operation of the cubicle-mounted equipment is proved by means of the self-testing facilities. There should be no command outputs during this exercise. The self-test facility allows all timers to be by-passed for sequence checking; then individual timer settings can be checked and set to their design values.

Phase 2 Each individual cable loop between cubicles and plant devices is checked for completeness. It is prudent to prepare a simple loop schematic diagram in conjunction with this test, as a future maintenance fault

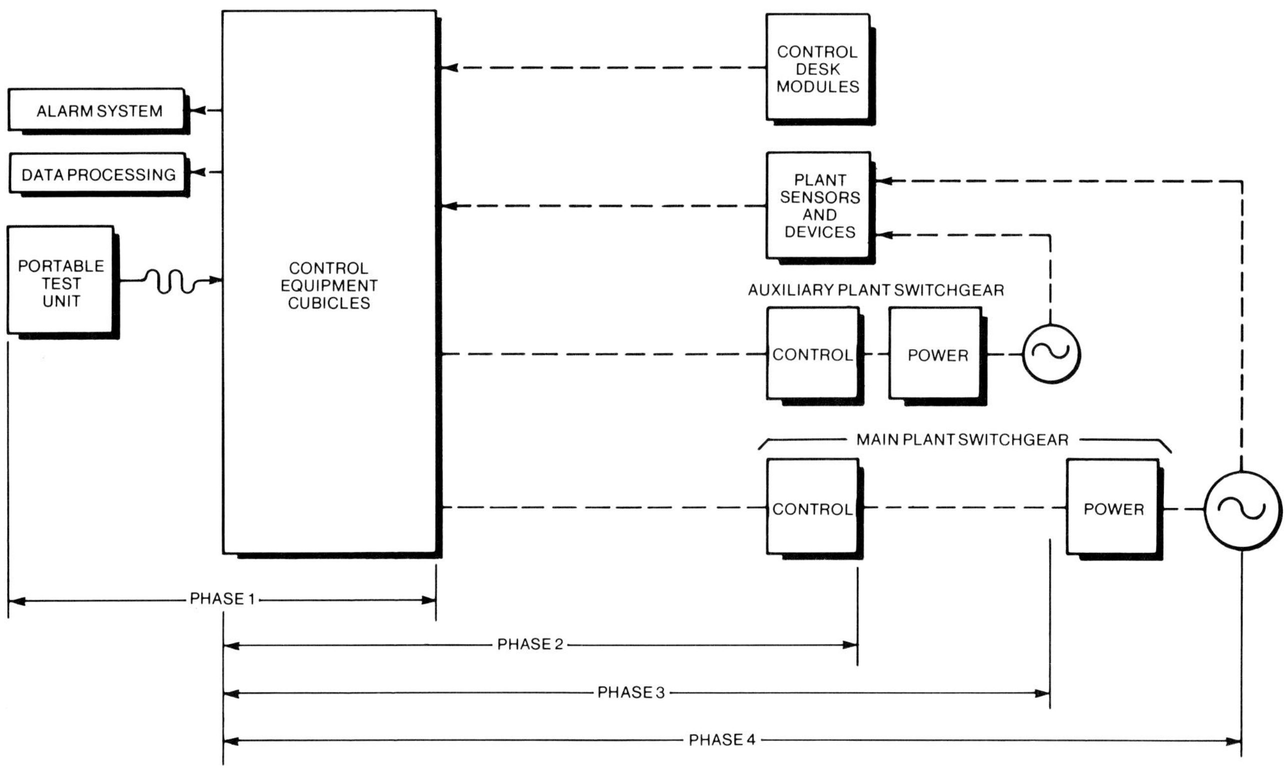

FIG. 3.26 The four phases of control systems commissioning

diagnosis aid. Refer to Figs 3.16, 3.24 and 3.28 for typical examples of loop routing sheets.

Using a 50 V DC source, command outputs are sent to interposing relays in switchgear, valves or other interlock and control equipment. Correct operation of each relay in turn is confirmed by observation at the switchgear. Digital inputs are proved by applying a short-circuit to the plant device terminals, or at the nearest available junction box, if plant cabling is incomplete. Confirmation is checked at the sequence equipment cable-isolation links.

Phase 3 For this stage, the overall sequence operation is proved by running all available group auxiliaries under sequence control. For plant groups which interact with interlock systems, it is often convenient to conduct Phase 3 of both systems as a parallel exercise, in conjunction with other plant test teams. However, the switchgear for the main plant item is kept isolated from its primary connections but with all control circuits energised, just as if it were in the 'service' position. This is often referred to as having the switchgear 'on jumpers'. Before commencing this phase of the testing programme, it is necessary to inhibit operation of the overall time delay units until all of the checks have proved satisfactory. It is also necessary to ensure that the plant cabling isolation links for incoming check signals are all in the 'open' position.

Providing that any check signals necessary to begin the sequence are correct, a command signal is initiated from the sequence equipment. Correct operation of the plant item is confirmed by receipt of a check signal, then the appropriate check input links are closed to permit initiation of the next command. The process is repeated until all steps and parallel paths have been proved. As part of this test, it may be necessary to revise the settings of individual timers, if valves or dampers take longer to travel than at first thought. Even at this stage, some check inputs may have to be simulated, for example, the establishment of oil pressure from the shaft-driven lubricating oil pump of a main plant item. Eventually, the entire sequence is run through without any manual intervention to check and, if necessary, adjust the design setting of the overall timer. Correct automatic starting of standby plant, by simulation of a duty plant failure, is also proved at this stage.

Phase 4 This is a full live sequence run of the entire plant group controlled by each separate sequence. It is carried out in conjunction with the main plant test team, as part of their commissioning activities. Details of such tests will be found in the appropriate sections elsewhere in this volume.

Microprocessor-based systems

Microprocessor-based sequence centres are tested with the appropriate diagnostic program, which is loaded via a portable tape cassette unit. Generally this is run for up to 1000 hours to exercise all the hardware in the system. It is then necessary to demonstrate that the system programs can be loaded into their respective computers from the portable terminal and, where appropriate, from the host computer associated with the unit or station data processing systems.

Sequence control schemes need to be exhaustively tested by simulation of different failure conditions. This ensures that the program does not enter a loop from which it cannot exit, as a result of some peculiar combination of plant inputs.

Other general tests applicable to computer-based systems are outlined in Section 6 of this chapter.

Commissioning of the system in conjunction with inputs and outputs between the equipment, CCR control modules and the plant is carried out generally as described for reed-relay-based sequence systems.

8.2 Control systems

8.2.1 System description

Equipment considered under this heading covers the facilities provided on centralised control panels for operation of specific sections of plant. Most of the controls are located on desks and panels for operation of the main boiler/turbine units, plus a selection of station plant functions from the CCR. The controls operate relay-based equipment, housed in cubicles adjacent to the control location.

Other centralised control panels may be outstationed in the following plant areas:

- Coal plant
- Ash plant
- Auxiliary boiler house
- Gas turbine house
- Water treatment plant
- Compressor house

Two basic types of control, viz. remote/manual and modulating, are described below:

Remote/manual control This provides the operator with facilities for direct control of open/close or start/stop operation of specific switchgear, valves, dampers, etc., or for selection of duty/standby modes for pumps, fans, compressors, etc. In many cases, these controls interface with sequence control equipments for plant start-up/shutdown, and also with modulating controls for automatic/manual operation and changeover from one mode to the other. A typical arrangement is shown later on Fig 3.28.

The most common form of operator control is the discrepancy control switch (DCS). There may be over 200 switches for station services and more than 100 for each main unit. A DCS is a miniature control switch containing several microswitch contacts and an integral lamp unit that illuminates the centre of the control knob. The switch is selected or 'dressed' to the desired position by rotating it through 90°. When the knob is depressed and then overthrown a further 30° against a spring, microswitches activate the control equipment. When it is released the switch returns to the selected position.

The integral lamp flashes if the switch selected position is in disagreement with the state of the plant. In event of the plant state changing without operation of the DCS, warning is also given by a flashing beacon on top of the appropriate unit control panel, as well as by a distinctively-toned audible warning. The warnings are cancelled when the DCS is redressed to be in agreement with the plant state.

Discrepancy control equipment is often based on reed relays mounted on PCBs located within sheet steel cubicles. The cubicles are usually located relatively close to the associated control panel. Electrical supplies at 50 V DC are required for control and alarm functions. Each suite of cubicles is fitted with indication of fuse failure, supply failure, and equipment fault. These alarms are grouped before being repeated to the station alarms system as a general alarm for the suite.

A slightly different form of control is provided by valve selection equipment. This is usually applied to groups of turbine steam pipework drain valves. Each group has its own control plaque, comprising a valve selector switch, a temperature indicator, one set of open/close pushbuttons and a common indicator for valve state (i.e., open or shut). The operator has the facility to select any valve in the group and check its drain temperature, then either to open or to close the valve from the pushbuttons. He can also select all valves in the group to open or close from a single initiation of the pushbutton. When selected to 'auto', each valve closes when its condensate drainage temperature reaches a predetermined value. This equipment is similar in many respects to that for the discrepancy controls.

Modulating control Each main unit, and selected sections of station plant, is provided with a number of individual control loops, each designed to achieve continuous control of a small section of the plant. Many of the loops form an integrated system, operable in either automatic or manual control modes, with indication for safe changeover between each state. Figure 3.27 describes a simplified block diagram of a typical loop for boiler feedwater control.

Typically there can be more than twenty closed loops, classed as automatic boiler controls (ABC), for each main boiler unit. These cover the following functions, which are described in detail in Chapters 4 and 6:

- Boiler steam pressure control
- Combustion controls
- Superheat and reheat steam temperature controls

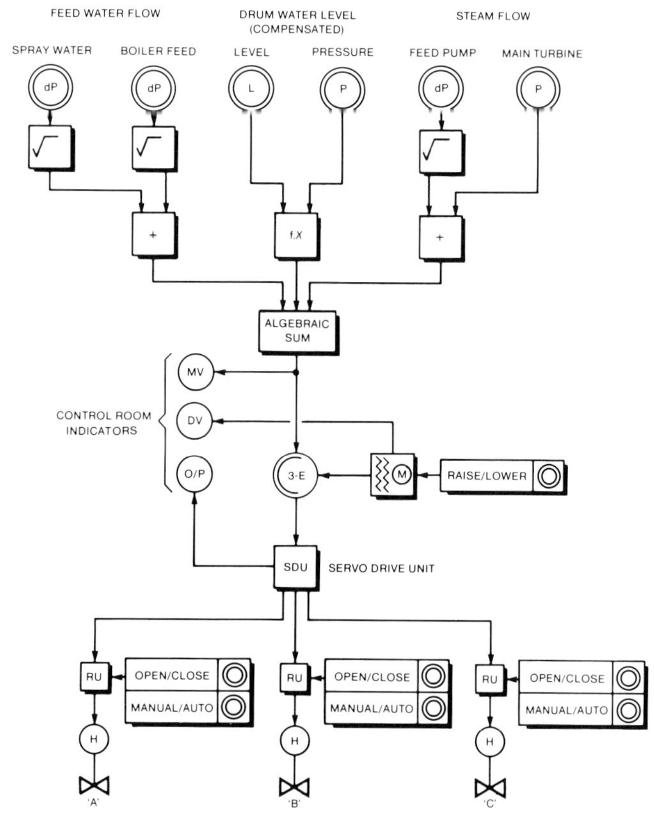

FIG. 3.27 Modulating control simplified block diagram

This is usually referred to as the *three-element feedwater control* loop. Feedwater flow to the boiler is balanced against the steam flow from the boiler to maintain a desired water level in the boiler drum.

Signals for feedwater and desuperheating spraywater flows are derived from differential pressure measurements by square root extractors. Main and feed pump turbine steam flow signals are derived from turbine nozzle box pressure and a square rooted differential pressure measurement respectively. The drum water level signal is compensated for density changes by a drum pressure measurement.

An algebraic summation of the drum level, feed flow and steam flow signals is applied to the three-element (3E) controller as the *measured value* (MV). The *desired value* (DV), manually adjusted by the operator via a motorised potentiometer, is also applied to the 3E controller. The controller output (O/P) positions the servo drive unit (SDU) which drives, via amplifiers and relay units, servo-converters to position the main feedwater regulating valves. Operator controls for auto/manual selection and for manual operation of each valve are also shown.

- Feedwater controls
- PF milling group controls

Similar, though much less extensive, controls are fitted to auxiliary boilers. There are also many other independent control loops around the plant for localised control of pressure, temperature, flow or level. The main turbine condensate and feedheating control systems, for example, are usually of the pneumatic 'Leveltrol' type which require pauses in the plant loading programme to optimise response tuning.

Main boiler modulating control equipment is designed on a modular basis, housed in sheet steel cubicles, and connected to the 110 V AC instrumentation supplies system. Each suite of cubicles is fitted with supply failure, fuse failure, and equipment fault indicators, which are grouped and repeated to the unit or station alarm system. Usually comprehensive self-testing and monitoring facilities are provided on each module.

Modern modulating control systems are becoming increasingly microprocessor based. Such systems operate on a distributed direct digital control (DDC) basis, with a high level data link to the centralised data processing (DP) installation for processing of control information and down-line loading of control programs. Each of the functional control groups outlined earlier has its own dedicated microcomputer outstationed at some convenient position in the plant. Integration and overall control of these distributed control systems is exercised either manually or automatically by means of the central DP system computers.

8.2.2 Commissioning objectives

The following are specific to control systems, but they should be read in conjunction with the general objectives outlined at the beginning of this chapter:

- Make the control systems available in time to meet the programmes of plant commissioning teams.
- Ensure completion of erection and check that all components are installed correctly and free from damage.
- Prove that the operational and control features of each system are satisfactory and to design, first by simulation and then by running auxiliary plant.
- Demonstrate all safety interlocks, protection and alarm facilities.
- Identify and record any cause for a shortfall in performance of the equipment.
- Produce sufficient engineering data to allow a full assessment of any equipment improvements needed.

8.2.3 Safety considerations

The following should be read in conjunction with those described at the beginning of this chapter:

- The main danger to personnel is the risk of low voltage signals being routed through cable marshalling boxes while other termination work is still in progress.
- There is also a risk of inadvertent operation of plant, unless adequate safeguards are taken to prevent it happening.
- Access to control equipment must be restricted to authorised personnel only at all times.
- A concise record of commissioning progress must be maintained and kept up-to-date for reference by operating staff at all times.
- A clearly understood commissioning-status marker system must be adopted for prevention of plant maloperation during the commissioning period. The use of small self-adhesive coloured spots as status markers (see Section 7.5 of this chapter) has been effective at several locations.

8.2.4 Resources

Services

Apart from the appropriate control power supplies at 50 V DC or 110 V AC, cubicle heating and lighting supplies at 240 V AC will be needed.

For discrepancy equipment, a special discrepancy test set similar to the portable test unit outlined in Section 8.1.4 of this chapter is required. Modulating controls require the availability of a range of instrumentation test equipment to create or simulate the plant conditions necessary to set up the control loops. Refer to Section 3 of this chapter for examples of test equipment.

Cabling must be substantially completed, initially at the various control panels and later to plant-mounted equipment. All instrumentation should be calibrated and the cabling terminated. If the modulating controls are based on direct digital control (DDC), then the centralised DP system, complete with interconnecting cabling, has to be available for software loading and commissioning.

Staffing

Several teams of two or three people are deployed across the whole field of control equipment commissioning. Programme times are fairly lengthy, probably extending from three months for remote/manual controls to ten or twelve months for modulating controls, dependent upon plant availability. In each case, cubicle checks (Phase 1) may occupy up to a month per suite.

Software specialists have to be involved from early commissioning right through to final completion of plant proving when DDC schemes are involved.

8.2.5 Commissioning

The activities are usually conveniently subdivided into four stages or phases, as outlined below (refer also to Fig 3.26). The final two phases are conducted in conjunction with, and usually under the direct control of, the relevant plant test teams. These tests will be found in the appropriate sections of Chapters 4, 5 and 6 of this volume. The following notes give a generalised approach to commissioning of centralised discrepancy control and modulating control systems. Differences in approach for DDC and self-contained modulating controls are highlighted later, where applicable.

Phase 1 This deals with checking all of the cubicle-mounted equipment as follows:

- Completion of cubicle erection, plus bonding to appropriate earth bar.
- Correct termination of power supply and alarm cables, with other plant cable-isolation links left open.
- Cubicle equipment all installed and correctly identified.
- Interchangeability of similar modules and PCB confirmed. This includes checking that dissimilar modules cannot be interchanged.
- Power supplies energised and all common features and alarm initiations proved.
- Appropriate test equipment to simulate plant items is connected to the cubicle terminals, then all internal equipment is proved to function correctly (see Section 8.1.4 of this chapter). Equipment self-testing features are also proved to operate correctly, with confirmation that signals to the plant are inhibited during the self-test.

Phase 2 For this phase, the plant cabling is checked and the test unit used to simulate plant conditions, using as much of the cabling as possible. Again it is convenient to prepare loop schematic diagrams for future fault diagnosis work at this stage. Figure 3.28 is an example for a discrepancy control. The reader should refer to Figs 3.16 and 3.24 for other examples of the approach. All functions from the desk or panel controls into the cubicle-mounted equipment are then proved to the satisfaction of the test team.

Phase 3 The control equipment is now used to operate auxiliary plant, with the main plant switchgear 'on jumpers' in a similar manner to that adopted for sequence control commissioning. Satisfactory completion of this stage is identified by application of the green status marker adjacent to the appropriate control station.

Phase 4 These are the integrated system tests, where the equipment is proved with the main plant in service; then the correct status marker is applied. Modulating control loops are optimised loop-by-loop, with the appropriate plant conditions, at designated stages in the plant commissioning programme.

Other modulating controls

Microprocessor-based modulating control system commissioning follows the basic pattern outlined above.

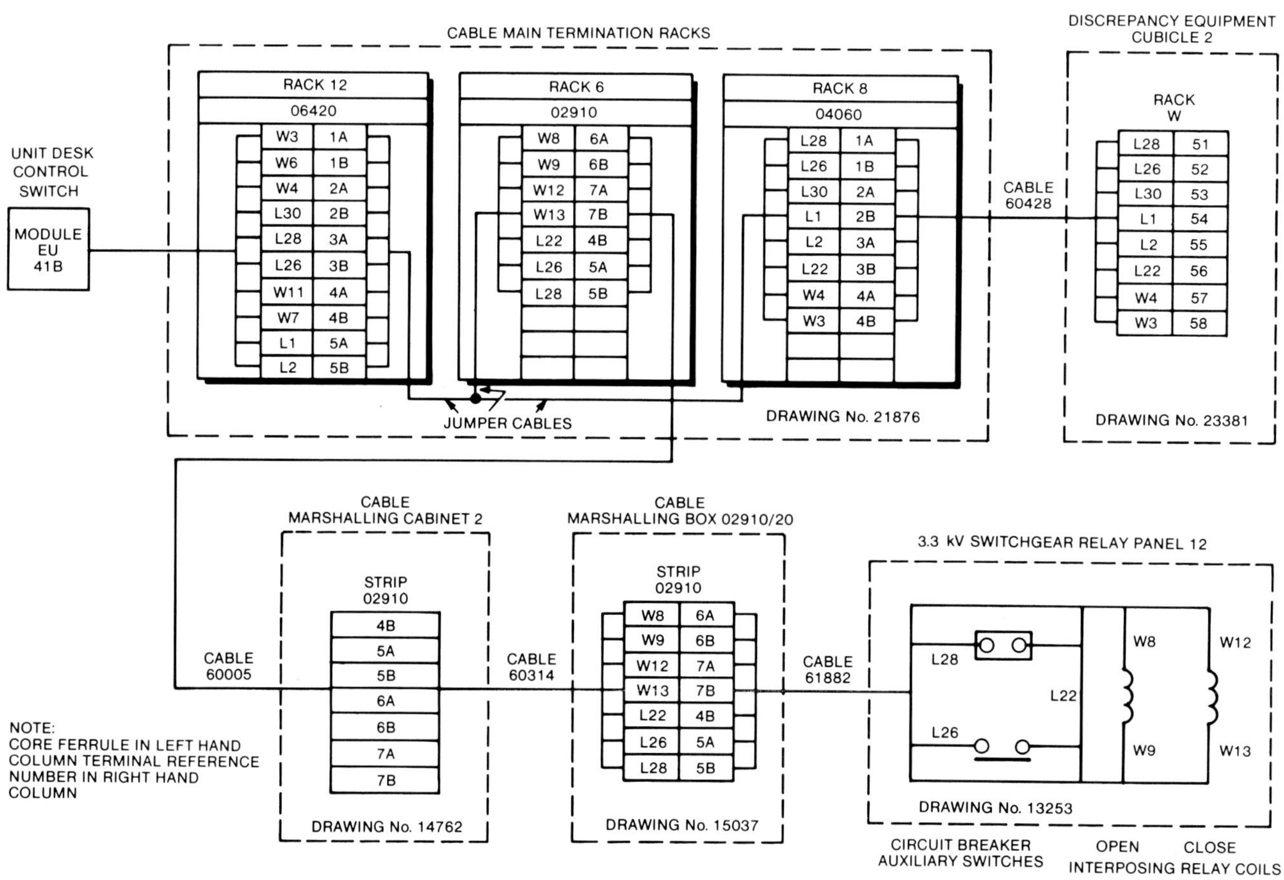

FIG. 3.28 Discrepancy control routing sheet

However, Phase 2 incorporates a repeat of the works test for 200 hours, followed by a further 800-hour test, but using the application programs in their respective DDC centres. Testing of software has to be proved completely satisfactorily before the DDC equipment is integrated with the plant systems in Phase 3.

Local control loops are commissioned, as outlined earlier in Sections 2 and 3 of this chapter, so that they are available, when required, for commissioning of the system of which they form part. Fine tuning of the control response is carried out at designated stages in the plant commissioning programme.

8.3 Interlock systems

8.3.1 System description

Interlock schemes are installed to ensure basic plant safety, independently of sequence or any other control systems. The equipment performs the functions of:

- Essential interlocking and intertripping of certain plant groups.
- General logic control of plant not forming part of a sequence control system.
- Repeating of plant states to other control or protection schemes.

Sequence control and interlock systems are interrelated, in that signals are exchanged between each to enable plant to be operated safely. Because of this, commissioning of both systems has to be interleaved.

An interlock is a basic safety control function which is generally applied directly into switchgear control circuits. Such interlocks may be of a 'permit-start' or 'permit-run' type. 'Starting' interlocks have to be satisfied in order to start the plant safely but have no subsequent tripping action. 'Running' interlocks have to be satisfied before the plant is started and must remain satisfied while it is in operation, otherwise a plant trip will result.

A typical 'permit-start' interlock may ensure that the speed regulator for a feed pump is in the minimum speed position, prior to energising the drive motor. Once the motor starts, the regulator has to be repositioned to obtain the desired speed. Establishing and maintaining lubricating oil to the bearings would be classed as a running interlock.

Figure 3.29 shows a simplified logic diagram for an interlock chain in the control circuit of a main boiler induced draught (ID) fan. Each relay is usually identified by a unique mnemonic related to its function in the chain.

Intertripping signals are those which are transmitted between plant groups to ensure plant protection. A typical condition would arise if a boiler ID fan were tripped by operation of its electrical protection. The intertripping system would ensure that the associated forced draught (FD) fan, plus selected PF mill groups, would also trip. This action would assist the boiler

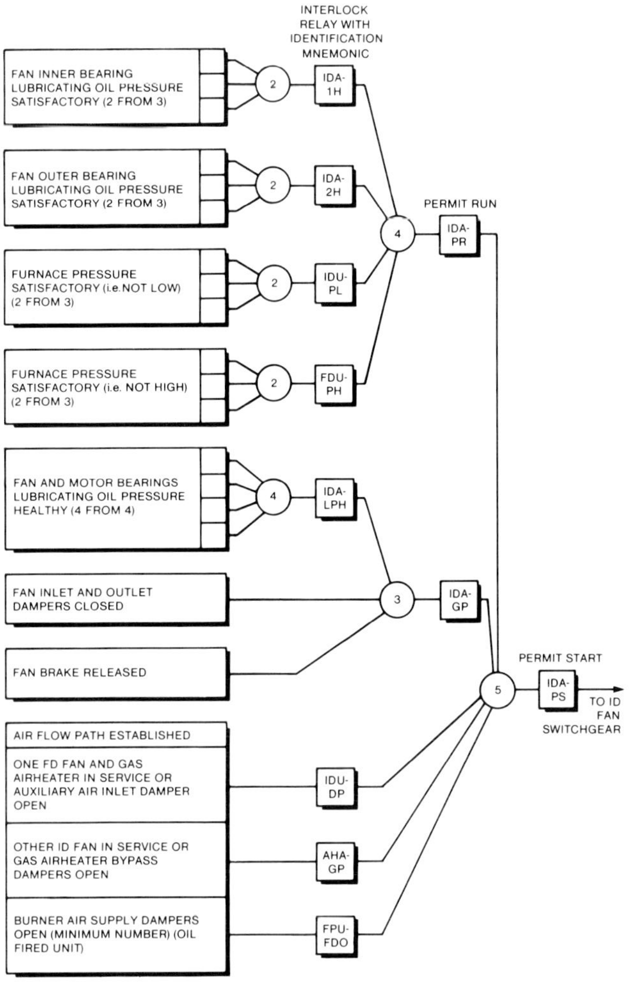

FIG. 3.29 Interlocking system single-line diagram

modulating controls to maintain safe combustion conditions, despite the reduction in gas flow arising from the loss of a fan group.

Starting interlocks are often arranged on a '1 out of 1' logic basis, whereas running interlocks are of either a '1 out of 2' or a '2 out of 3' and occasionally a '2 out of 4' coincidence.

Interlocking equipments usually employ a combination of either mechanically-latched or electrically-held relays, with contacts arranged to close to create a plant trip. All relays are fitted with light-emitting diodes (LED), arranged to be lit when the associated relay is in the healthy, energised, or SET condition.

Dependent upon the design of the scheme, relay coils may be operated from either 50 V or 110 V DC supplies. The supply for the LED displays is usually from a common internal 110 V AC to 5 V DC power supply unit connected to the instrumentation supplies system.

Microprocessor-based interlocking schemes are gradually being adopted, which means that there are no hard-wired relays because all of the interlocks are written into the logic of the programmable logic controllers (PLCs). They have the advantages that fault diagnosis is simpler and changing the logic is quicker and cheaper than before. For convenience this section is based on hard-wired relay logic which continues to be installed in most of the power stations in the UK.

The equipment is mounted in sheet steel cubicles strategically located in switchrooms around the plant. Each suite of cubicles is complete with its own supply-fail, fuse-fail and equipment-fault alarm indications, which are also grouped and repeated to the unit or station alarms system. To enhance the integrity of the protective schemes, termination racks generally do not have an isolation type of cable link fitted. It is standard practice to use Category 1 connections for this application, i.e., either stud connectors, with nuts and special washers, or spring-loaded clamps on to hooked terminations.

Typical application of interlocking equipments may be as follows:

- Induced and forced draught fans
- Gas air heaters
- PF mills, plus associated coal feeders and PA fans
- Coal conveyor systems
- Auxiliary boiler auxiliaries
- General service water pumps
- Ash and dust pumps
- Ash plant conveyor systems

8.3.2 Commissioning objectives

The general objectives outlined at the beginning of this chapter should be read in conjunction with the following, which are more specific to interlocking systems:

- Ensure that each part of the system is available to meet the programmed requirements of main plant and sequence control test teams.
- Prove the correct operation of the cubicle-mounted equipment, using simulated inputs, and then carry out live functional tests in conjunction with plant test teams.
- Confirm satisfactory performance of the system under abnormal operating conditions, e.g., loss of cubicle control supplies.
- Determine any operating philosophy errors at an early stage, and prove the correctness of any modifications arising therefrom.
- Establish that the systems give reliability and performance compatible with the plant protection requirements.

8.3.3 Safety considerations

The following are specific to interlocking equipments and should be read in conjunction with those outlined at the start of this chapter:

- Most of the equipment is self-contained and the risks to personnel arise from low voltage signals being routed through cable marshalling boxes in which termination work is still in progress.
- Access into the cubicles must be kept under secure control at all times.
- Concise records of the current state of commissioning should be readily available for reference by operational personnel.
- Inadvertent energising of plant is unlikely to be a risk, since interlocking equipment generally provides a service to other control schemes, which in turn operate the plant.

8.3.4 Resources

Services

Apart from control supplies at 110 V AC, 110 V DC and 50 V DC, cubicle illumination and anti-condensation heating supplies at 240 V AC will be necessary. Cabling to the sequence control equipment, to plant and to the station alarm systems will need to be substantially complete.

A portable test unit, to prove the correctness of logic by simulation of plant contacts as well as relay inputs and outputs, will be needed. This would be similar to the unit used for sequence system commissioning outlined in Section 8.1.4 of this chapter.

Staffing

Teams of two to four people will be needed for testing each suite of cubicles and also for checking cabling to plant devices. One or two weeks will be necessary for proving each suite of cubicles. The total programme will spread over many months, dependent upon the availability of plant and sequence control systems for overall system testing.

8.3.5 Commissioning

As with commissioning of sequence and control equipment, it is convenient to subdivide the activities for interlock systems into four phases (refer also to Fig 3.26). Generally, for plant groups where sequence and interlock systems interact, the programme is arranged so that Phase 3 of both systems can be progressed in parallel.

Phase 1 Usually these are classed as cubicle checks, where the following are observed:

- Completion of cubicle erection and bonding to station earth bar.
- Correct termination of power supply and alarm cabling. Plant cabling may be either left unconnected, or subsequently disconnected then reconnected as part of testing.
- Equipment all installed and correctly identified.

transformer to supply the unit electrical auxiliary load, via the unit transformer, when its associated generator is shut down. The generator switch is then used for synchronising the main generator. In some cases, the unit transformers have an increased rating and are connected as station transformers to supply a proportion of the station load. When the generator switch is open, the generator/unit transformer combination usually has no neutral point earth. This is overcome by connecting an earthing transformer to the generator transformer LV connections. The neutral point of the earthing transformer is then earthed through either a liquid resistor or a high impedance distribution-type transformer.

The auxiliary system main voltage level is determined by the required rating of the station transformers. Three distribution voltage levels have been adopted at modern 500 MW and 660 MW unit power stations. These are 11 kV, 3.3 kV and 415 V three-phase AC; with sub-distribution at 415 V, 240 V and 110 V single-phase AC. The selection of 3.3 kV (rather than 6.6 kV) as the intermediate level is based on technical and economic grounds. As a general rule, all motor drives rated between 1500 kW and 150 kW are connected at 3.3 kV.

Control and supervision of the electrical auxiliaries system is exercised from a control panel in the central control room (CCR). The control panel displays the system in mimic form and permits the control room supervisor to carry out selective switching operations, including synchronising.

Comprehensive interlocking is provided at all voltage levels to prevent operation of the system with prospective fault levels in excess of switchgear rating. Unless operational flexibility demands electrical interlocking, then locally-operated mechanical key interlocks are fitted. Microprocessor-based interlocking schemes are sometimes applied to the more complex 11 kV busbar systems. These also display the prospective fault levels with different combinations of selected switching operation, so that the supervisor can assess the situation before carrying out any particular operation.

Battery systems are installed for essential power, control, alarm and emergency lighting functions. The standard voltage levels adopted are 240 V, 110 V and 50 V DC. Instrumentation supplies at 110 V AC are derived from the battery systems.

9.2 Commissioning objectives

The following are relevant to the plant covered by this section, in addition to the general objectives outlined earlier in this chapter:

- Check, test and energise individual items of equipment to a priority order which ensures that plant commissioning programmes are not delayed.
- Prove satisfactory and safe operation of all interlocking schemes.
- Confirm correct operation and setting of all protective devices.
- Ensure that electrical protection schemes give correct discrimination and stability.
- Demonstrate in-built protection relay test facilities for HV switchboards.
- Test switchgear operation with reduced control voltages.
- Ensure that no transformer is energised until its fire protection is in service.
- Conduct noise and vibration checks on transformers and motors.
- Check loading and temperature rise of various major items of plant.
- Carry out system loading, power flow and voltage level checks.
- Confirm that the station earthing system complies with BS CP1003 and 1013.
- Test and record satisfactory installation of the lightning protection schemes in accordance with BS CP326.
- Prove the integrity of each battery by a 10-hour discharge test in accordance with BS440.
- Confirm that each battery charger is capable of delivering its rated output without any distress.
- Demonstrate that instrumentation supplies are within voltage, frequency, power factor and harmonic content tolerances under all modes of operation.

9.3 Safety considerations

Hazards raised by commissioning of the station electrical system are usually fully appreciated by those involved. However, the general mass of employees on a construction site have to be adequately supervised to ensure that they are not inadvertently exposed to danger when working in close proximity to live plant and equipment.

The following items should be given appropriate consideration in addition to the general aspects of safety outlined at the beginning of this chapter:

Environmental safety

- TREATMENT FOR ELECTRIC SHOCK notices must be erected in advance of energising any equipment.
- Combustible materials must be kept clear of all switchgear rooms and cableways. Ducts and floor openings should be sealed and fire barriers established as soon as practicable.

- Permanent fire detection and fire protection schemes must be in service before cables and transformers are energised. Portable temporary fire appliances should be to hand to augment the permanent equipment for initial energising.
- There must be strict control over work in cable tunnels and also around switchboards and transformers.
- Battery room ventilation must be in service at all times.
- NO SMOKING/NO NAKED LIGHTS notices, water supplies and eyewash bottles must be maintained in all battery rooms.
- Absorbent material must be available to mop up dilute acid spillages.

Operational safety

- Switchrooms should be clear of personnel whenever switchboards are being first energised. Preferably, each HV circuit should be energised using a circuit breaker that has previously been energised. Provision for remote operation of the circuit breaker should be made.
- HV switching must be carried out to an agreed written procedure and be conducted by word-of-mouth commands, rather than by signs or time-lapse intervals.
- Electrical systems should be energised in small manageable sections rather than as a large system in cascade.
- Lightning system testing must be suspended if there is any possibility of a lightning discharge in the surrounding area.

9.4 Resources

9.4.1 Services

It is self-evident that the appropriate electrical supplies must be available before any commissioning work can be undertaken. This includes all lighting, heating, test, control and alarm supplies. Reliable telephone and two-way radio communications will also be necessary.

Availability of standard items of electrical test equipment is assumed. Typically these may include the following:

- Accurate multimeters
- Insulation testers for 500 V and 1000 V DC
- Mains-powered insulation tester for 2.5 kV and 5 kV DC
- 30 kV high voltage test set
- HV potential testing and phasing-out equipment
- 0–500 V high impedance voltmeter
- Phase rotation meter
- Current transformer-fed ammeters (1 A and 5 A input)
- Portable current transformers with 1 A and 5 A secondary tappings
- Multirange clip-on ammeters (AC/DC)
- Portable digital $\mu\Omega$ meter
- Primary and secondary injection test sets
- Single- and three-phase variable autotransformer units (Variac)
- Portable oscilloscope
- Stop watches
- 0–100°C thermometers
- Portable hydrometer
- Schering bridge and variable-voltage discharge-free test transformer

The following services will be necessary for specific functions:

- Heated oil bath for winding temperature indicator calibration.
- Inert gas cylinder or 'test bomb' for testing the Buchholz relay.
- Inclinometer for checking alignment of Buchholz relay.
- Torque spanners for checking bolted terminations.
- Approved portable leads for earthing of HV switchgear.
- Supply of demineralised water and sodium carbonate for liquid neutral earthing resistor (LNER) calibration.
- Special earth-loop-impedance testing equipment.
- Portable 415 V three-phase diesel generator (say 200 kVA rating), plus up to 2000 m of 2.5 mm flexible cable for earthing system checks.
- Availability of a remote zero-potential earth point and conducting path to it over telephone lines for earthing system checks.
- Supply of electrolyte and demineralised water for battery topping-up, including hydrometer and protective clothing.
- Water supply and absorbent material to dilute and mop up battery spillages.
- Battery room ventilation in service
- Permanent fire protection systems in service and supply of temporary portable fire extinguishers, where necessary.

- Resistive variable loading bank for testing of batteries and instrumentation supplies conversion equipment.

9.4.2 Staffing

The scale of the exercise is such that several teams, comprising two or three people, would be concurrently engaged in each area. The numbers would need to be doubled for some of the more extensive activities, particularly when energising transformers or HV switchboards.

Overall programme times and sequencing of the work in each area is difficult to assess, as so much depends upon the particular site programme for the release of plant for commissioning. Testing of the cabling and the earthing system, in particular, is an exercise spread over many years as construction proceeds. Typical duration of testing and commissioning activities for specific sections of plant may be:

Transformers Perhaps three weeks for large transformers, but five to seven days should be sufficient for each auxiliary transformer.

HV switchboards Three to six weeks per switchboard, dependent upon the number of circuits and the complexity of interlocking and controls.

LV switchboards Due to the larger number of circuits per switchboard, these may take from four to twelve weeks each.

Protection Secondary injection of protection relays could take up to one day per relay but, on average, perhaps two relays per day is likely.

Battery systems Up to five days per battery/charger combination, inclusive of discharge and load testing. Outgoing circuits to main distribution boards may take half a day each.

Instrument supplies Each inverter/by-pass combination may take up to three days for load testing, exclusive of battery commissioning, plus a further day for every 110 V AC load distribution board.

9.5 Commissioning

9.5.1 General

Commissioning the electrical auxiliaries distribution system within a power station is one of the first activities to start and is almost the last to finish. Though construction site supplies are not normally considered part of the station electrical system, commissioning checks have been included for completeness.

Figure 3.31 depicts a typical related programme for commissioning of the main distribution network. For convenience, the description and commissioning of the following plant areas is dealt with in turn:

- Transformers

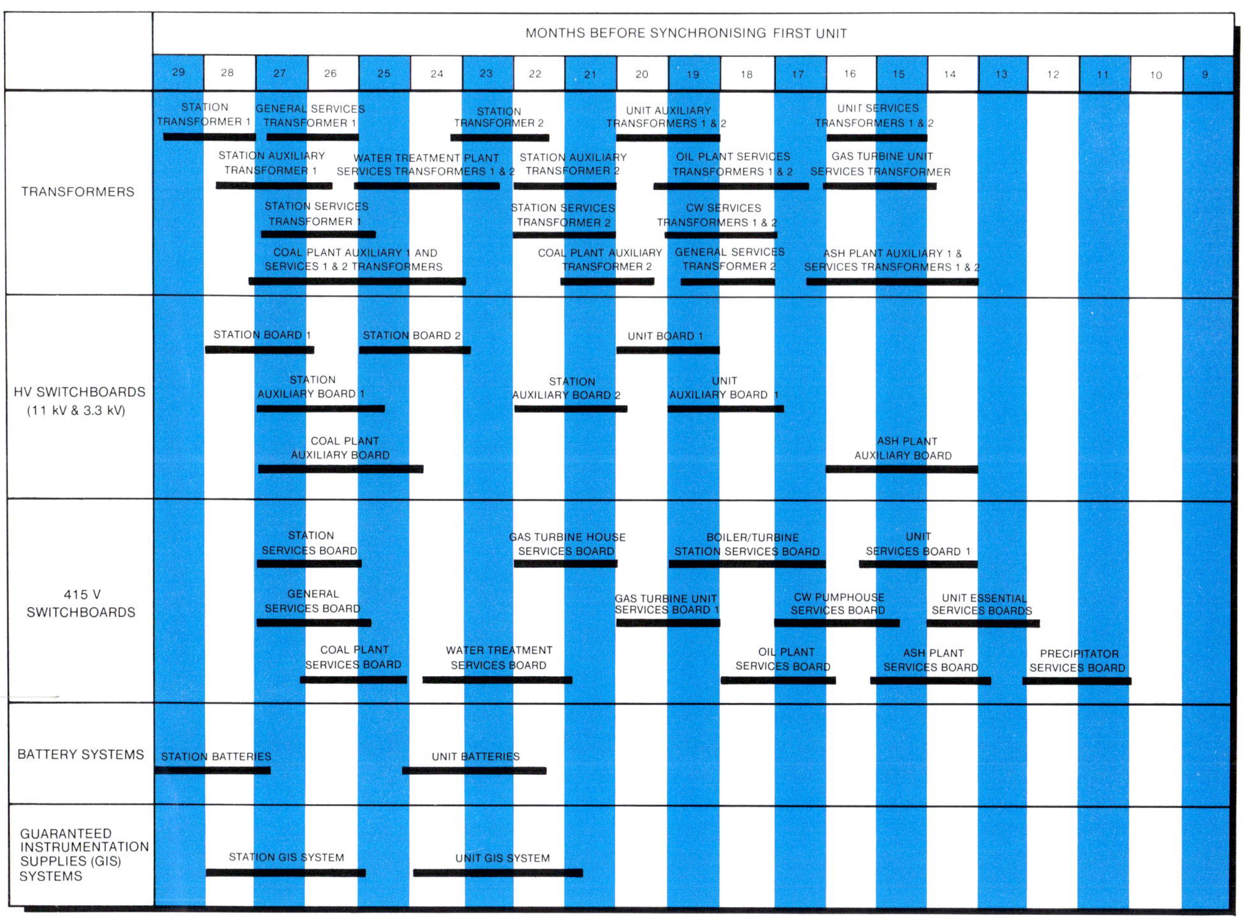

FIG. 3.31 Commissioning programme for electrical auxiliaries distribution system

- High voltage (HV) switchboards
- Low voltage (LV) switchboards
- Electrical protection
- Cabling, earthing, lightning protection
- Battery systems
- Guaranteed instrumentation supplies (GIS)
- Construction site supplies

Testing of electrical equipment is widely covered by appropriate British Standards, which are quoted where applicable. Extensive use should be made of standard check lists (SCL) and test procedures based on nationally accepted documents for commissioning electrical equipment. Because of the close interrelationship between transformers, switchgear and protection, it may be advantageous to combine test procedures into composite plant completion documents.

To avoid unnecessary repetition, these general statements apply to the commissioning of each plant group:

- All relevant safety considerations applied
- Services necessary for commissioning available when required
- Terminal point isolations secured concurrent with the issue of Safety Rules Clearance Certificates (SRCCs)

9.6 Transformers

9.6.1 System description

Power station transformers are manufactured in accordance with BS171. Generally they are oil-immersed and of double-wound construction, contained within a single steel tank arranged for outdoor mounting. To compensate for thermal cycling of the oil volume, a conservator tank is mounted above, and connected to, the highest point of the transformer tank. A gas/oil (Buchholz) protection relay is located in the inclined section of the interconnecting pipe. Moisture ingress during thermal cycling is prevented by allowing the conservator air space to breathe, via a filter charged with silica gel crystals.

Larger transformers are now fitted with a refrigerant-type breather, which utilises the Peltier effect to keep the conservator air space extremely dry. Such breathers, e.g., Drycol, need a separate electrical supply to operate. Figure 3.32 shows the general

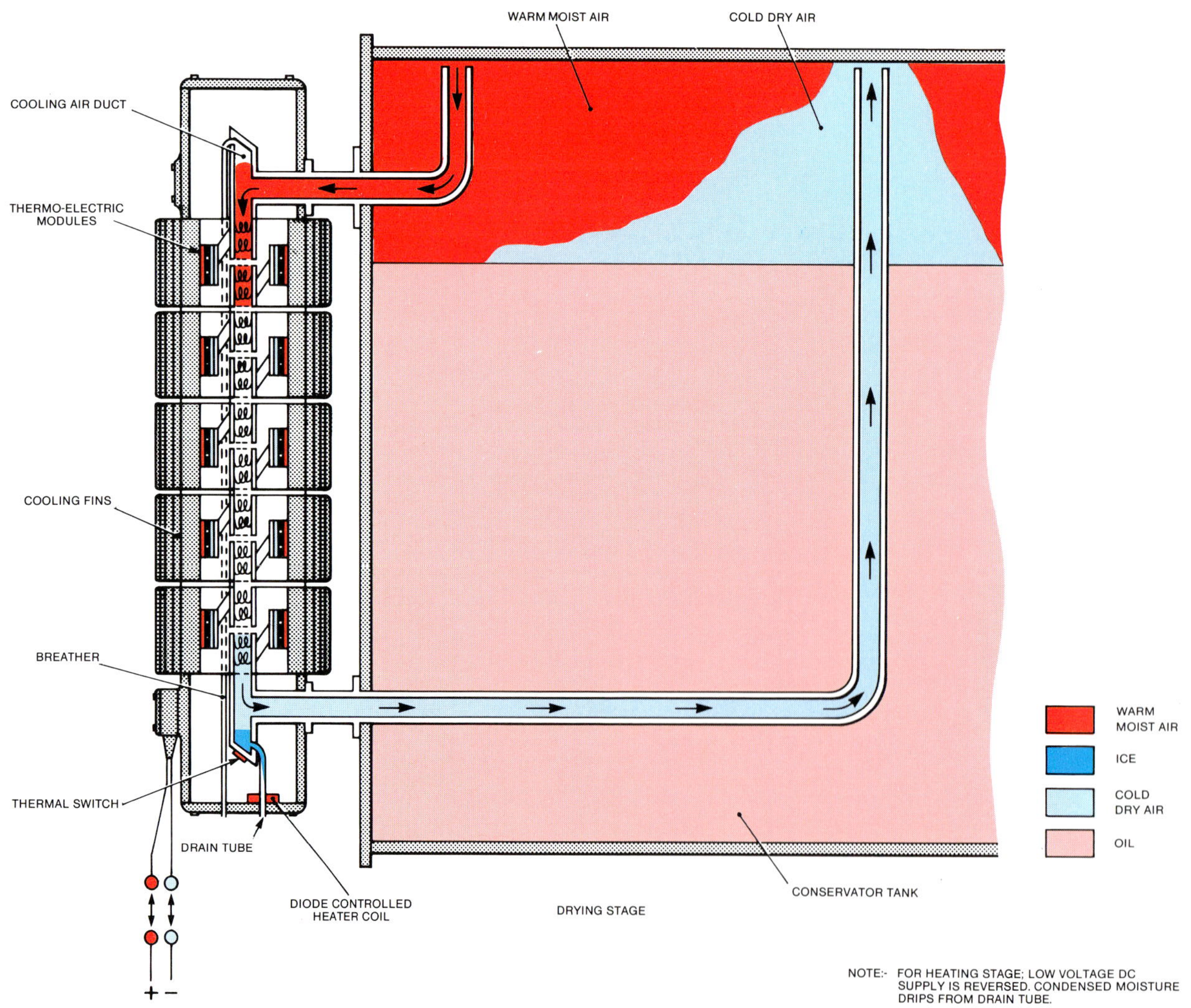

FIG. 3.32 Drycol breather functional diagram

arrangement of a Drycol breather, the principle of operation of which is as follows.

Air in the conservator circulates through the breather by thermo-syphonic action. As air in the breather is refrigerated, it passes downwards, condensed moisture is deposited in the breather, and relatively dry air is returned to the conservator. After a preset freezing period, the thermal cycle is reversed, the frozen condensed moisture melts and drains through the water outlet. When the water has drained off, the breather reverts to air refrigeration and the cycle repeats automatically. Typically, a freezing period of between 5 and 8 hours is followed by a defrost period of some 15 minutes. It is important that the breather and radiating fins are kept painted white to minimise the heating effect of the sun. The polarity of the low voltage DC supply from the control cabinet to the breather must be checked and rechecked to be absolutely correct, otherwise the unit will not operate.

Transformers rated above 2 MVA are fitted with an automatic self-sealing spring-operated pressure relief device (e.g., Qualitrol), typically set to operate when the pressure differential across the device reaches 0.69 bar. Most of the smaller auxiliary transformers have an oil-resistant replaceable diaphragm, mounted on a vent pipe above the normal oil level, to relieve any pressure surges. Figure 3.33 shows the construction of a Qualitrol device.

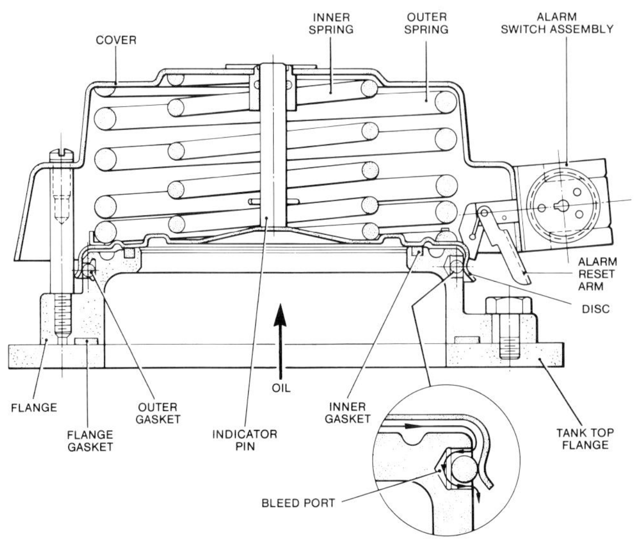

FIG. 3.33 Qualitrol pressure surge relief device

Fixed fire protection is fitted only where a transformer fire could pose a fire risk to adjacent plant. When outdoor location is not practicable, either air cooled or fire-resistant synthetic liquid-filled transformers are mounted in the switchhouses or are integral with the 415 V switchboards.

Transformer winding configuration is usually delta/star, but other phasor groupings are used to achieve the desired phase relationship between the distribution voltage levels. As a general rule, the phasors for the 415 V, 11 kV and generator transformer HV busbars voltages are all in phase.

The neutral point of star-connected windings is usually earthed through liquid resistors (LNER) which limit the fault current to approximately the full-load rated current of the transformer. The LNER is usually rated to carry this current for up to 30 seconds. Thermostatically-controlled anti-frost heaters maintain the electrolyte minimum temperature between 7°C and 4°C. A blanket of plastic balls is often floated on the electrolyte surface to reduce heat loss by convection.

The tapping range, open-circuit voltage and impedance of auxiliary transformers are such that rated voltage appears at the busbars under full-load conditions without causing unacceptably high voltages at no-load. In order to minimise the effects of transient voltage depressions on other equipment, e.g., discharge lighting, the design aim is to keep busbar voltage depressions within 15% during motor starts.

Generator transformers

Modern 660 MW units are connected to generator transformers rated at 800 MVA 23.5/400 kV. These comprise three single-phase units connected in star on the HV side and delta on the LV side, i.e., a YNd phasor grouping. Each phase tank of the transformer has two 100% duty oil pumps, connected to separate sources of supply, which automatically change over upon failure of the duty pump. As a general rule, the cooling type is OFWF supplied with water from the station CW system. These transformers have virtually no natural cooling capacity, so it is practice to initiate a unit trip upon failure of both oil pumps for any one phase tank.

Generator transformers for 500 MW units are usually rated at 600 MVA 22/400 kV or 22/275 kV. Usually they are of single-tank design, OFWF cooling, with three 50% duty oil pumps, each connected to a different supply source. At power stations where the cooling water quality is aggressive, the oil may be cooled by air-blast fans, i.e., an OFAF type of cooling. Typically five or six fan coolers would be fitted, though only four would be needed for full rating: the others automatically start at a selected oil temperature. Total loss of oil pumps or fans usually initiates an alarm, rather than causing a unit trip.

The neutral point of the HV winding is brought out and connected solidly to the station and grid substation earth bonding system. High speed on-load tapchangers are usually fitted to the neutral end of the HV windings. Tapchangers are capable of manual operation in an emergency, but are usually power-operated from the CCR.

Unit transformers

Ratings of unit transformers vary from 28 MVA for an oil-fired 500 MW unit to 48 MVA for a coal-fired 660 MW unit. They are single-tank units arranged for OFWF cooling and have off-load manual tapchangers, giving a plus and minus 7.5% range. Voltage ratios are either 22/11 kV or 23.5/11 kV and the winding arrangement is delta/star, i.e., a Dyn vector group. Three 50% duty oil pumps are fitted and loss of all three usually initiates an alarm only.

Station transformers

As a general rule, four station transformers are necessary to give security of supply for a 4–6 unit station, but two are adequate for two or three unit power stations. The rating of station transformers is determined by the auxiliary load, which in turn determines the auxiliary system main voltage. At 11 kV, the current is limited by the maximum rating of the switchgear available, i.e., 3000 A. This equates to a rating of 57 MVA, so station transformers for 500 MW and 660 MW unit plants are rated at 60 MVA. The primary voltage is dependent upon that of the local grid network and 400/11 kV, 275/11 kV, and 132/11 kV ratios are all fairly common. High speed on-load tapchangers, operated from the CCR, are fitted at the neutral end of the HV windings.

Winding arrangements are usually star/star with a delta tertiary winding (i.e., YNynd). The HV neutral connection is bonded directly to earth, whilst that of the LV winding is earthed through a liquid resistor.

Cooling is usually OFAF, but the station transformer usually has a natural cooling (ONAN) rating equivalent to 50% of its full-load OFAF rating. Two 100% or three 50% duty oil pumps and four fans are fitted. Loss of all pumps initiates an alarm only. Pairs of fans automatically start at selected oil temperatures.

Auxiliary transformers

Ratings are selected from manufacturers' standard sizes, and range from 12.5 MVA down to 150 kVA to suit the application. No-load voltage ratios are typically 11/3.45 kV, 11 kV/433 V or 3.3 kV/433 V. Off-load tap selector switches are fitted to give a range of either ±7.5% or ±5% in 2.5% increments, dependent upon the rating of the transformer. All are mounted in single oil-filled tanks arranged for outdoor location with ONAN cooling, unless the situation dictates that air cooled or synthetic liquid-filled transformers be used instead.

Winding arrangements are usually delta/star (Dyn) for 11/3.45 kV and 3.3 kV/433 V transformers. Delta/zigzag (Dzn) or star/star (YNyn) connections are necessary for the 11 kV/433 V transformers to maintain the correct busbar voltage phase relationship. The neutrals of 433 V windings are solidly earthed to pass sufficient current to operate the earth fault protection.

Low flammability non-toxic biodegradable synthetic insulating liquids (e.g., MIDEL) are used in transformers located indoors. This type of transformer is contained within a hermetically-sealed tank, which has neither conservator nor breather, but does have an automatic self-sealing pressure relief device. An inert-gas-filled expansion space ensures that the internal pressure does not exceed 0.35 bar above or below atmospheric pressure over a liquid temperature range of −10°C to +90°C. Contemporary designs of sealed distribution transformers are using a non-flammable liquid (FORMEL NF) which has a slightly toxic characteristic. In most respects, the design characteristics are similar to those of the MIDEL-filled transformer. However, the pressure relief device is non-resettable and is designed to operate at a pressure greater than 1 bar gauge. The expansion space in this type of transformer is evacuated at works to reduce dissolved air, and is delivered with an internal pressure equivalent to 0.4 bar absolute at 20°C. The increase in fluid temperature with load current may result in an internal pressure of 0.15 bar gauge at rated output, rising to 0.6 bar gauge at 50% overload.

Smaller auxiliary transformers may be provided with either an oil temperature alarm or an indicator only, rather than a load-current-sensitive winding temperature alarm and/or trip device.

It is customary to include the word 'auxiliary' in the title of 3.3 kV switchboards and their incoming transformers. Transformers supplying 415 V switchboards, as well as the associated LV boards, generally have the word 'services' in the title.

9.6.2 Commissioning

Figure 3.34 shows a simplified commissioning network for a typical transformer circuit.

Load testing of generator and unit transformers will be found in Chapter 6. The general principles for commissioning them, as well as for station and auxiliary transformers, are as follows:

- Check that transformer bunded area is complete, with all cable ducts sealed. Drainage system should be tested and available for use, including proof that the bund wall will retain water.
- All cabling to be terminated but links to transformer HV and LV connections left disconnected until the transformer tests have been completed.
- Electrical tests on transformer to check insulation resistance of each winding to earth and to each other. Core-to-tank insulation resistance checked and then earth link fitted.
- Using a 415 V three-phase supply, check the transformer voltage ratio and magnetising current at each tap position, then complete vector group checks.
- Sample transformer oil from conservator, Buchholz relay and tank base; test each for breakdown, moisture and acidity in accordance with BS148. No attempt must be made to obtain a fluid sample from sealed synthetic-liquid-filled transformers.
- Check that inclination of Buchholz relay is within acceptable limits, then prove operation of the gas alarm and gas surge trip functions.
- Calibrate the winding/oil temperature indicators at 20% increments of scale, then set and prove correct

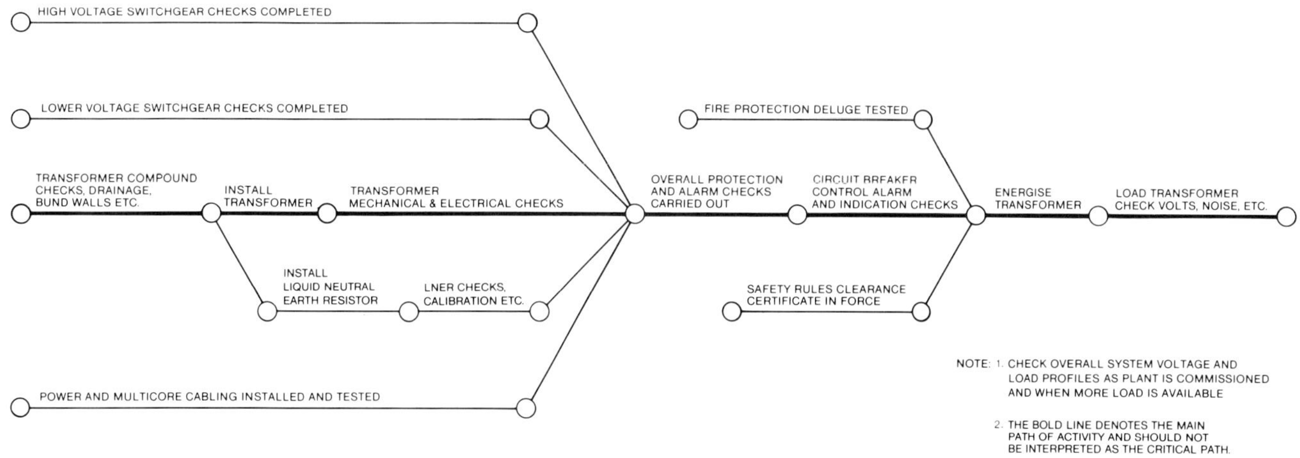

FIG. 3.34 Simplified transformer commissioning network

operation of the switches for cooler control, alarm and trip, where appropriate.

- Confirm that pressure-relief diaphragm is undamaged.
- Check correct installation of the self-sealing (Qualitrol) relief device where appropriate.
- Energise and check correct operation of the refrigerant drier unit on the station transformer. Ensure that there is a fresh charge of silica gel in the breathers of lower rated transformers.
- Charge LNER with electrolyte and adjust the conductivity to achieve the desired resistance when related to 15°C. Ensure that anti-frost protection heater and thermostat are operating correctly. Check that an adequate number of plastic balls have been loaded on to the surface of the electrolyte to form a heat loss reduction blanket.
- Commission oil pumps and cooling fans, checking correct operation of oil-flow fail alarms and auto-start flow switches on station transformers.
- Prove correct local and remote operation and indication of on-load tapchanger. Check manually-operated off-load tap position selector switch.

After checking that the following items have been completed satisfactorily, the fire protection system should be commissioned in preparation for initial energising of the plant:

- Transformer and LNER terminal chambers, cubicles, etc., all secured and water tight.
- Switchgear checks, including protection trips and intertrips, all complete.
- Electrical protection equipment all tested and correctly set (see Section 9.9.2 of this chapter).
- All inputs to station alarm and data processing (DP) systems checked satisfactorily.

Immediately prior to energising a transformer, the cable and transformer winding insulation resistance is rechecked and compared with previous results. The lower voltage circuit-breaker and its voltage transformer or voltage indicator fuses are isolated. The HV circuit-breaker is operated in the 'isolated' position to prove that controls and protective gear trips are working. It is a matter of personal preference whether or not to set down electrical protection settings. If one has sufficient confidence in the integrity of the checks carried out earlier, then, within power stations, there is little merit in interfering with the prescribed settings.

All personnel are cleared out of the commissioning/energising plant areas, then the transformer is energised from the HV side using a circuit-breaker that has already been proved in service. Providing there are no signs of distress, then the HV circuit-breaker controlling the transformer can be used to de-energise, then re-energise, the circuit. Voltage levels and phase rotation are then checked at the lower voltage circuit-breaker. The plant is then made available for energising the lower voltage switchboard in conjunction with the test team concerned.

It is difficult to create full-load conditions on auxiliary transformers until the main plant is available to run. So, apart from an initial assessment of the no-load noise levels, tests for power flow, voltage drop, temperature rise and full-load noise levels have to be conducted when conditions are suitable at a much later date in the overall station programme. As more plant is commissioned, it will be necessary to re-examine the voltages experienced at different parts of the system. It is highly likely that many of the auxiliary transformer tap selector switches will have to be adjusted. A balance may have to be struck between no-load and full-load voltage regulation.

9.7 High voltage switchboards

9.7.1 System description

Switchboards are located in heated switchrooms positioned in close proximity to the load centres to minimise the length of cable runs. Switchgear currently employed in modern switchboards is of the metal-clad indoor type, air insulated, single busbar arrangement. The air-break circuit-breakers (ACB) are of the truck type, arranged for horizontal isolation. Each circuit-breaker is equipped for integral earthing of its associated cable circuit. Selected outgoing circuit breakers, normally two per busbar section, are also capable of being used for busbar earthing.

The switchgear fault ratings adopted are:

11 kV 900 MVA make, 750 MVA break.

3.3 kV 150 MVA make and break.

Electrical protection relays, control switches and indicating instruments are mounted on separate relay panels which are located opposite the associated switchboard where they may act as a screen for personnel carrying out local operation.

Where the incoming and busbar sectioning circuit-breakers are operated from a remote position, e.g., CCR, electrical control circuit interlocks are fitted. Otherwise mechanical key interlocks prevent operation with prospective fault levels in excess of the switchgear rating. Mechanical key interlocks, unique to each circuit, are fitted to ensure that selection of busbar or cable earthing is carried out in a safe manner. In the case of transformer circuits, the interlocks usually require the HV circuit-breaker to be isolated before the lower voltage circuit-breaker can be selected and closed on cable earth. An interlock key is then released to permit earthing of the HV cable end. These features do not apply to every transformer circuit, since earthing facilities are not normally fitted to 415 V circuit-breakers. Key exchange interlocking prevents busbars from being earthed until all of the circuit-breakers controlling possible power infeeds have been opened and isolated.

Voltage transformers (VT) and straight-bar primary current transformers (CT) are fitted to the switchgear for protection, metering, instrumentation and (at 11 kV) for synchronising purposes. Each VT is mounted on a withdrawable carriage to give both isolation of, and access to, the VT high voltage fuses. These fuses are certified to match the switchgear fault rating. The majority of VT are single-phase, but at least one three-phase VT per busbar is connected to the switchboard bus-wiring for synchronising and efficiency metering purposes.

All 11 kV, and the larger 3.3 kV (i.e., over 1000 kW), motor drives are switched by circuit-breakers. Lower rated 3.3 kV motors are usually switched by motor switching devices (MSD) which are of the air break, horizontal isolation, truck-mounted design, arranged in two tiers. The MSD is purpose built for motor control and comprises a circuit-breaker of limited fault capacity (typically 50 MVA) backed up by a three-phase set of 350 A high blowing capacity (HBC) striker pin fuses. The fuses give protection against short-circuits up to the 3.3 kV system fault level of 150 MVA. When the fuse blows, operation of the striker pin automatically trips the MSD.

The majority of modern HV switchgear is solenoid-closed with a shunt trip coil, though some motor-wound spring-closing equipment is used. Control supplies are all at 110 V DC for closing and tripping functions, with the proviso that the 11 kV switchgear closing solenoid is often 240 V DC, powered via a 110 V DC contactor unit. A supply of 50 V DC is used for the interposing control relays operated from the CCR but mounted in the switchgear relay panels.

Switchgear and motor switching devices incorporating vacuum interrupters have been available for some time but have not seen much service in modern high merit power plant.

9.7.2 Commissioning

It is customary to commission each switchboard as a separate exercise, generally as indicated by the simple network shown in Fig 3.35. These activities are summarised in the following paragraph with the more important aspects detailed later.

On completion of switchgear erection, a full inspection is carried out, then integral equipment static checks are completed. All auxiliary electrical supplies for controls, heating and panel illumination etc. are energised. Functional checks of each ACB and MSD

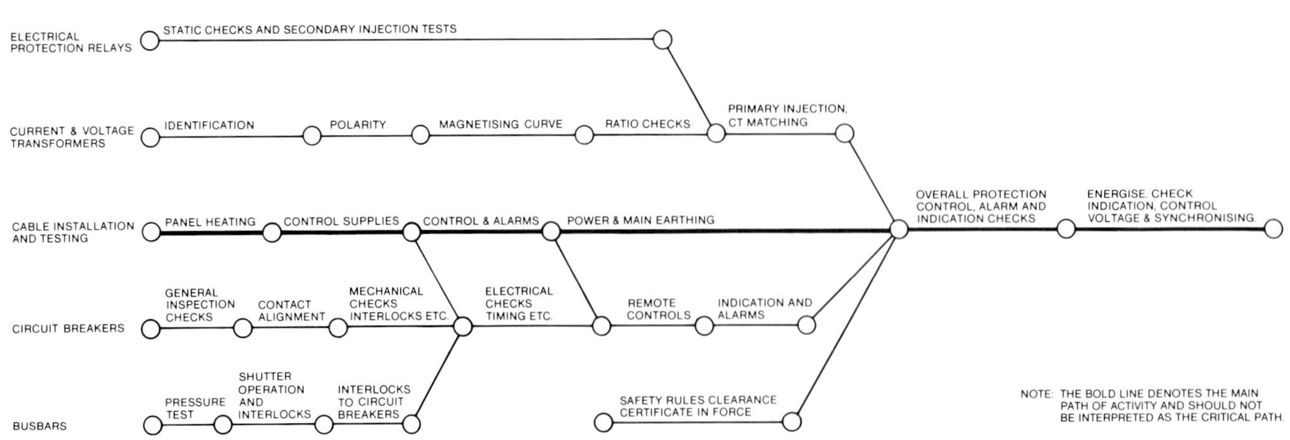

FIG. 3.35 Switchboard commissioning network

are conducted, proving correct operation of all local and remote controls and interlocks with the busbars dead. Electrical protection equipment is tested, settings are applied, then switchgear tripping operations are proved. Finally, the busbar and incoming ACB insulation resistance is rechecked. The switchboard is then energised, voltage and phase rotation are noted and, where appropriate, the undervoltage protection scheme is commissioned.

More important points to be considered are:

- Switchgear and switchroom to be kept dry, heated, dust-free and secure from unauthorised entry.
- Anti-condensation heaters energised at all times.
- Circuit-breaker truck alignment during withdrawal and re-insertion to be smooth and straight with floor absolutely level.
- Busbars cleaned, resistance of all site-made joints checked to be less than 5 $\mu\Omega$. High voltage test and insulation resistance checks completed.
- Alignment of switchgear contacts, contact pressures, conductivity of line and earth connections all recorded.
- Mechanical operation of ACB and MSD mechanisms proved, followed by functional check of all mechanical key interlocking schemes. This will include selection of circuit and busbar earthing and will involve having all switchgear racked on to DEAD busbars. If it is impracticable to isolate the busbars in order to carry out operational checking of MSDs, the line fuses can be replaced with substantial strips of electrical insulating board. This will maintain mechanical stability of the contacts and operating mechanism during testing.
- Isolation keys and padlocks for each circuit and function proved to be unique and correctly labelled. Human error symbols on switchgear, i.e., on cable box, front door, truck, in cubicle and on front and rear of relay panel, to be checked for consistency.
- Insulation resistance checks of all secondary wiring for controls, interlocks, alarms and protection circuits. Current and voltage transformer checks, including indicating instruments and transducers for remote instruments, plus electrical protection relays.
- Switchgear control function tests at 80% voltage for closing and 50% voltage for tripping. Switchgear timing checks carried out and results recorded.
- ACB and MSD arc chute splitter-plate insulation resistance (IR) checked. Arc chutes dried out, if necessary, to achieve an IR of at least 3000 MΩ at 10 kV DC.
- Insulation resistance and HV tests of each ACB and MSD satisfactorily completed.
- Final tripping and intertripping tests, plus proving of electrical interlocking functions, carried out.

Prior to energising the switchboard for the first time, all circuits except the incomer should be isolated along with any associated VT. If the incoming transformer has already been commissioned, then the HV ACB of the circuit should be used to energise the lower voltage busbars via an already closed incomer ACB. As a final check before energising, the insulation resistance of the busbars is retested and compared with previous results. Personnel are cleared out of the switchroom and the busbars energised. The incoming circuit-breaker operation is then proved followed by voltage and phase rotation checks. Where appropriate, operation of the undervoltage protection scheme can be proved.

9.8 Low voltage switchboards

9.8.1 System description

This part covers the whole range of LV AC switchgear, particularly air-break 415 V circuit-breakers, fuse switches and contactors mounted in sheet steel cubicles. The circuit-breakers are usually of the horizontal isolation design, fitted with 110 V DC solenoid closing and a shunt trip coil.

Modern 415 V switchboards have a multi-tier contactor unit arrangement. Dependent upon the individual ratings, there may be between four and eight units per tier. Contactor gear usually has 110 V AC control circuits supplied from individual 415/110 V control transformers for each circuit. Certain applications require the use of a mechanically-latched contactor; these generally have closing and tripping coils supplied at 110 V DC. Contactors rated over 40 kW often have 415 V single-phase AC closing coils, though the control circuit will be at 110 V.

Mechanical key interlocks are fitted to prevent operation of 415 V incomer and bus section circuit-breakers with prospective fault levels in excess of the switchgear rating of 31 MVA. If the circuit breakers are arranged for remote operation, then electrical control circuit interlocking is applied instead.

Auxiliaries needed to shut down a boiler/turbine unit safely are usually connected to an Essential Services Board, which has two different sources of supply. The normal source is from the unit system; the alternative source is from a station services board. In the event of loss of unit supplies, the incoming supply contactors change over automatically to the station source, but have to be manually reset to normal when unit supplies are restored. The supply contactors are mechanically, as well as electrically, interlocked to ensure that the two incoming supplies cannot be connected in parallel.

Electrical protection gear, along with remote control interposing relays, is mounted in the cubicle for each individual circuit. Distribution of 415 V supplies, not required for motor drives, is from multi-way fuse distribution boards strategically located around the plant. Such distribution is at 415 V three-phase for power, etc., or 240 V single-phase AC for heating and lighting. Lower voltage power supplies are distributed from 110 V AC distribution boards served by 415/110 V AC single-phase transformers, with the centre point of the LV winding earthed to give 55–0–55 V between poles.

9.8.2 Commissioning

Every switchboard is commissioned as a separate exercise, with similar pre-energising checks carried out on each ACB, fuse switch or contactor circuit. The initial erection check is conducted using the manufacturer's equipment schedule, which lists the type, duty, rating, etc., of every device fitted to each cubicle.

Insulation resistance checks of 415 V busbars and line equipment are conducted at 1000 V DC. All other circuits are checked using a 500 V DC insulation tester. The completed busbar system is tested at 2 kV AC for 1 minute to prove its integrity. Insulation resistance testing must be carefully considered in the light of the increased usage of semiconductor devices, which must not be subjected to test voltages in excess of 100 V DC.

When appropriate, the auxiliary electrical supplies for closing, tripping, heating, lighting and other control purposes are energised. At a later stage, it will be necessary to prove that each item of equipment will operate satisfactorily with reduced control voltages. Typically these are:

80% volts for DC closing

50% volts for DC tripping

85% volts for AC closing

Checks conducted on each circuit comprise those appropriate from this list:

- Protection equipment tested and settings applied in accordance with available schedules.
- Fuse ratings confirmed to be correct for the particular duty.
- Local and remote indication of voltage and current proved.
- Mechanical and electrical switching interlocks proved.
- All live terminals that are accessible when cubicles are open for maintenance are to be fully shrouded for personnel safety.
- Integrity of isolation locking devices proved.
- Interlocks between isolating switch and cubicle access door checked for integrity.
- Correct functioning of all control and alarm features proved.
- This may involve extensive simulation of plant contacts, etc., since plant and cabling completion may not be particularly well advanced before the switchboard is energised.

Immediately prior to energising a switchboard, the insulation resistance of the busbars and incoming circuit-breaker is rechecked. Operation of all local and remote controls and electrical protection for the incoming ACB is also rechecked. The supply cable is energised, followed by energising of the busbars. Indicators are checked and the phase rotation recorded. The protection against loss of supply must be proved to perform its specified functions.

When or where appropriate, each source of supply to a section of busbars is energised in turn and the busbar voltage phase rotation checked to be consistent. On switchboards where it is permissible to parallel supplies, both supplies have to be phased out relative to each other.

Though the above is based on commissioning a 415 V switchboard, similar principles can be adopted for all lower voltage AC or DC switchboards or distribution fuseboards.

9.9 Electrical protection

9.9.1 System description

Theory and practice of electrical protection schemes and the associated relays are fully covered in other volumes, publications or manufacturers' literature. This summary is only intended to highlight the more important aspects relevant to modern power station auxiliary plant.

It is important that protection schemes should clear an electrical fault as quickly as possible. At the same time they should discriminate, such that only the plant affected by the fault is disconnected. Maximum fault clearance times depend upon whether the fault current is to be interrupted by a circuit-breaker or a fuse. Discrimination is achieved by using relays with variable operating current and time settings. The system protection is time graded, so that the protection nearest the fault operates first. Where discrimination by time is adopted, relays having an inverse definite minimum time (IDMT) characteristic are used.

As a general guide, the following protection schemes may be installed on the various types of circuit.

Transformers

Overcurrent and earth fault protection is usually applied to both windings, using IDMT relays. Station transformers are supplemented by a biased differential protection scheme. Buchholz gas alarm and trip functions are provided, but winding temperature is usually arranged to alarm only. Figure 3.36 depicts the arrangement of current and voltage transformers for a typical auxiliary transformer circuit.

Because transformers commonly have delta-connected HV windings, earth faults on the lower voltage windings are not detected by the protection on the HV side, so a low voltage winding restricted or

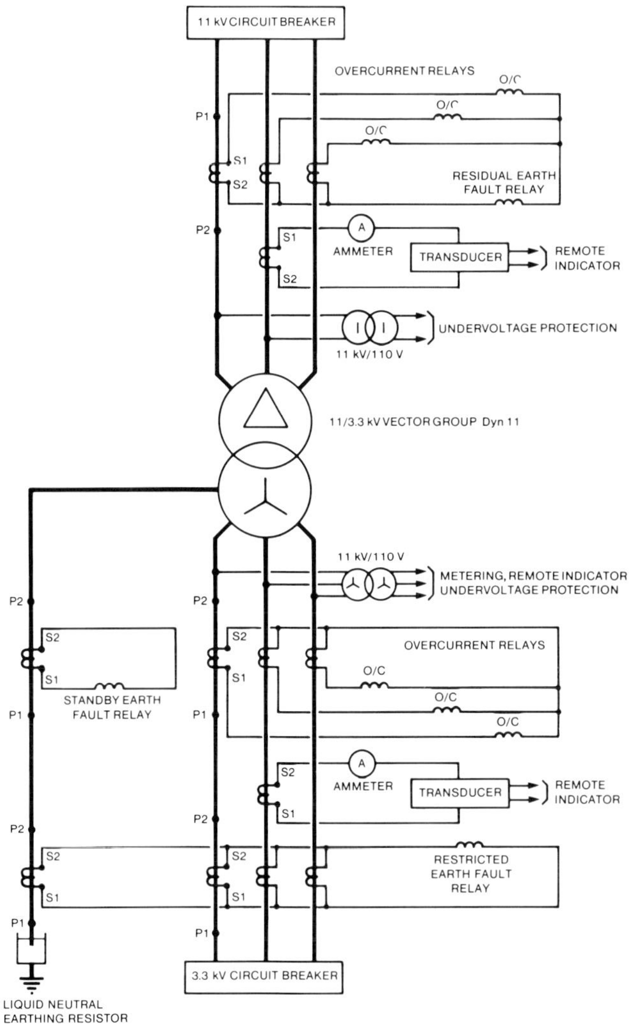

FIG. 3.36 Auxiliary transformer CT and VT arrangement

balanced earth fault scheme, utilising instantaneous relays, is fitted. Transformers with a 415 V secondary winding are often fitted with a HV winding high-set instantaneous overcurrent protection scheme as well.

Interconnectors and busbars

Earlier designs of electrical auxiliary systems relied on overcurrent and earth fault protection applied to interconnectors to give busbar protection. The interconnector cables were protected by a circulating current scheme. Modern HV busbar systems are now protected by a complete busbar protection scheme.

Gas turbine generator

Stator faults are covered by an overall biased differential scheme, which is backed up by IDMT overcurrent and earth fault protection. The gas turbine unit services transformer is fitted with Buchholz gas alarm and trip, plus local indication of oil temperature only. Figure 3.53 illustrates a typical arrangement of a gas turbine-generator electrical protection scheme.

Motors

Thermal overload, high-set instantaneous overcurrent, instantaneous earth fault and single-phasing protection is fitted to all drives controlled by circuit-breakers or MSDs. All of these features are now incorporated into a single motor protection relay. In order to achieve close thermal protection of the windings of large motors during starting conditions, it is necessary to fit a relay, the heating curve of which closely matches the motor heating curve. Thermal protection is usually set to operate at 5% over motor full-load current (FLC). Instantaneous overcurrent elements for 11 kV motors tend to be set at 8 × FLC, whereas those for 3.3 kV motors are set at 10 or 12 × FLC dependent upon the mechanical inertia of the driven member.

Many modern HV motors are also fitted with stalling protection relays. Should a motor fail to run up, it will continue to draw locked rotor current from the supply. It is difficult to differentiate between a stalled condition and a normal start-up purely on a current basis, since even a healthy motor will maintain normal starting current for 80 to 90% of the total run-up time. The starting period for most induction motors is less than 10 s, whereas the permissible stall time to avoid deterioration of the motor insulation is usually of the order of 20 s. It is generally accepted that if the running load current of the motor should rise to approximately 3 × FLC then the motor will stall, and the load current will quickly rise to the locked rotor value. A stalling relay therefore employs a current-sensitive contactor, which operates at a value corresponding to 3 × FLC. When the contactor operates, it puts a thermal relay into service. The heating curve of the thermal relay is carefully matched to the motor heating characteristic under stalled conditions. This characteristic, when compared with that of a normal thermal relay designed for, say, 105% overload operation, gives a much faster clearance time for currents in excess of 3 × FLC.

Motors at 415 V, rated above 50 kW, are protected by fuses supplemented with high grade relays, which give thermal overload, single-phase and instantaneous earth fault protection. Motors at 415 V, rated below 50 kW, are protected by fuses in conjunction with simple thermal overload and single-phase protection relays. Simple thermal overload units now have scale settings which incorporate a 5% overcurrent allowance, so that they can be set to the motor FLC marking. The combination of fuse/contactor/protection relay is designed to ensure that the motor and its supply cable are effectively provided with earth fault protection. Fuse ratings are generally about 2.5 to 3 times the motor FLC, up to a maximum of 4 × FLC. In some cases, where the motor starting current is up to 6 × FLC, special motor protection fuses may have to be fitted.

Positive temperature coefficient (PTC) thermistors are embedded into each phase of the stator windings of many electric motors to give full protection against excessive temperature rise. Generally the three thermistors are connected in series to a transistorised detector circuit fed from a low voltage AC source. The circuit detects a large increase in thermistor resistance (see Fig 3.41) to de-energise an auxiliary trip relay at the design temperature. The circuit resistance at this temperature, typically 120°C, is of the order of 3000 Ω. Instead of thermistors, actuator motors often incorporate a bi-metallic strip temperature switch to give over-temperature protection. This switch is self-resetting and is connected directly into the actuator control circuit.

Undervoltage protection

Undervoltage protection schemes are applied to selected switchboards. To ensure integrity of the undervoltage protection schemes applied to the HV switchboards, switchgear auxiliary contacts and VT-operated relays are integrated into a relatively complex logic, all of which must be satisfied before any plant is caused to trip. Typically the voltage sensing relays operate at between 50 and 70% of nominal voltage and, if conditions are correct, a common timing relay (set between 5 and 10 seconds) has to operate before the trip relays are initiated. In the case of LV switchboards, if at least two of the three voltage elements sense a voltage of less than 70% of the nominal value which persists for longer than 5 seconds, selected plant motors will be tripped.

At some power stations, the protection is arranged to trip the ACB of all HV motors connected to the affected switchboard. This protection is also applied to all plant associated with firing equipment of either main or auxiliary boilers and may include fuel-oil transfer pump drives.

Earth leakage protection

Socket outlets for portable equipment supplied at 415 V three-phase or 240 V single-phase are fitted with earth leakage protection based on residual current devices (RCD). Typical applications are welding and bolt-heating transformers, portable pumps, steam cleaners, etc. Even for steam cleaners operating from a 110 V AC centre-tapped earth source, it is prudent to fit RCD protection.

9.9.2 Commissioning

This can be a very involved subject and reference to manufacturers' literature relating to the particular equipment is necessary. It is important that the commissioning staff be in possession of an accurate schedule, listing the various components and detailing the recommended settings for all protection relays.

The following notes refer primarily to electrical protection equipment. Testing of gas/oil (Buchholz) relays and electric motor thermistor protection is mentioned later. After checking that all connections have been terminated correctly to the latest wiring diagrams, the following tests are conducted:

- Insulation resistance check of all wiring
- Current transformer (CT) and voltage transformer (VT) tests
- Check and set protection relays by secondary injection
- Prove operation of the schemes by primary injection

Insulation resistance

Generally this is carried out by means of a 500 V DC tester after ensuring that all transistorised equipment has been disconnected. The insulation resistance must be measured between electrically separate circuits, in addition to the customary circuit-to-earth measurement. Acceptable values will depend upon the extent of the circuit involved and on environmental conditions, but the absolute minimum should be 2 MΩ. If the battery system for the protection supplies does not have any of the poles connected to earth, it will be necessary to apply a high resistance earth-return for this test. As far as practicable, all auxiliary contacts in the circuit under test should be closed to ensure that all wiring is being tested.

CT tests

Each current transformer must be subjected to a polarity test to prove that the primary (P1/P2) and secondary (S1/S2) terminals are correctly identified. For this test, a centre zero moving-coil ammeter is connected across the secondary winding, with the meter positive terminal coupled to S1. A low voltage battery, with series-connected switch, is then connected to the primary terminals, such that the battery positive terminal is coupled to P1. When the battery is switched on, the ammeter should flick in the positive direction. The ammeter will flick negatively when the switch is re-opened. Figure 3.37 shows the method of connection.

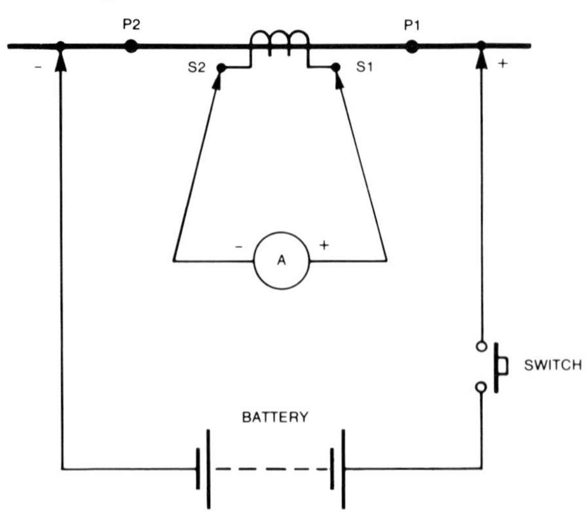

FIG. 3.37 Current transformer polarity check

The magnetising curve of each CT is checked by energising the secondary winding from a variable-voltage AC mains source, whilst the primary remains open-circuited. It is customary to measure the applied voltage and the magnetising current at five points up to 25% above the CT rated secondary current. The test results, when plotted, resemble a *B–H* curve, as shown by Fig 3.38, but represent the properties of the CT as a whole rather than just the properties of the core material. The CT is said to be approaching the 'knee' or saturation point when a 10% increase of applied voltage results in a 50% increase in magnetising current. For a protection class CT, the knee point may not occur until approximately 10-times rated secondary current is attained. In order to obtain the voltage required to carry out a magnetising test for a 1 A secondary CT, a step-up transformer may have to be used.

Finally, a ratio check has to be carried out, as shown by Fig 3.39. It is customary to conduct this in conjunction with the primary injection tests mentioned later. Current is passed through the primary circuit with the secondary winding connected to its normal circuit load. The ratio of the primary to the secondary currents should approximate closely to that stamped on the CT identification plate.

VT tests

Voltage transformer polarity is checked in the same manner as for a CT, taking care to ensure that the battery is connected to the primary winding.

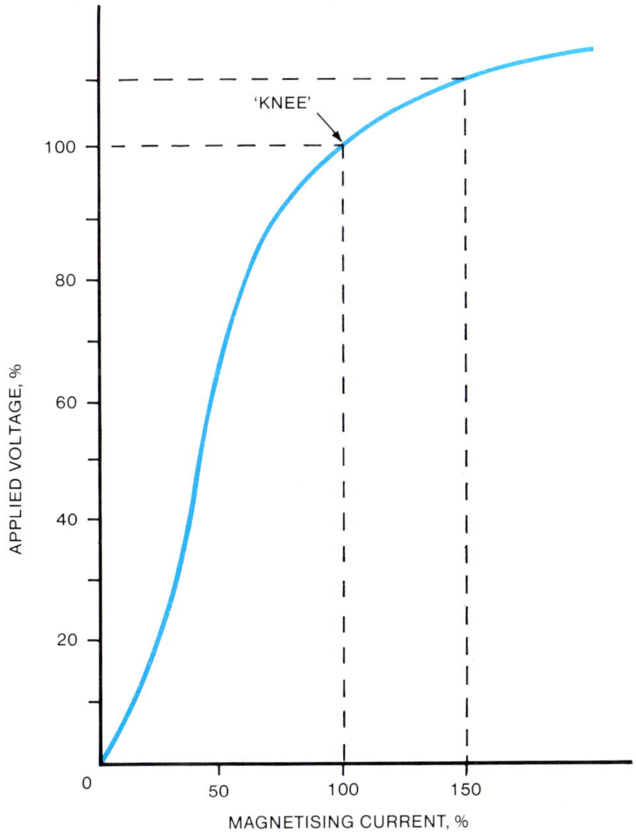

FIG. 3.38 Current transformer magnetising curve

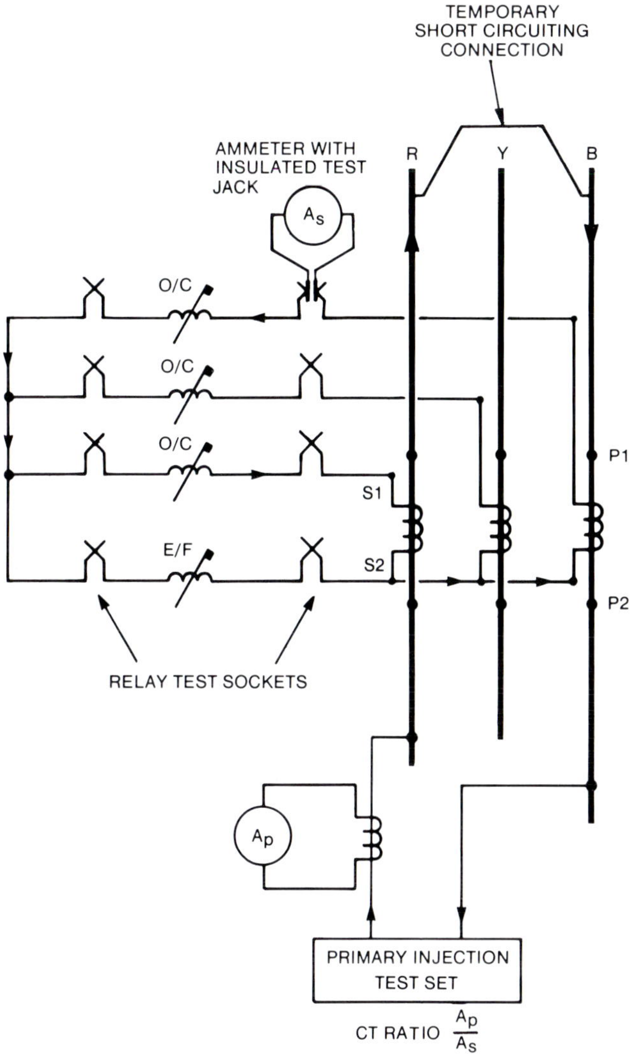

FIG. 3.39 Current transformer ratio check

Ratio and phasor group tests are conducted using a low voltage, i.e., mains, source. A three-phase 415 V supply is applied to the primary winding with the secondary winding open-circuited, then a high impedance voltmeter is used to measure the secondary voltage to determine the voltage ratio. For the phasor group check, one terminal of the primary and secondary windings (typically Al and al) are strapped together. The three-phase 415 V supply is connected to the primary winding, then voltages are measured between each terminal and to the neutral, from which the phasor diagram can be constructed. If it is a star/star winding configuration, care has to be taken to ensure that the primary and secondary neutral points are not also connected together, e.g., by a common earth bond.

Alternatively, the ratio check can be carried out when the VT is in service, by direct comparison of the secondary voltage with that of a VT already connected to the same set of busbars. It is necessary to verify that the phase relationship of a three-phase VT is correct. This is achieved with the VT in service. The secondary voltage between phases and neutral are measured and then a phase rotation meter is connected across the three-phase terminals.

If a three-phase VT has a broken delta tertiary winding, then the output voltage from the tertiary winding should be measured. With the VT primary coupled to a balanced supply, and the tertiary rated burden connected, this voltage should be less than 5 V.

Secondary injection

This involves passing current through protection relays, i.e., simulating CT output currents, and confirming correct operation at two points, in accordance with the published characteristic for each particular relay. The relay then has its designated settings applied and the operation is rechecked to confirm the set values of current and, where appropriate, time. Modern relays, particularly those mounted in modular cases, can be tested *in situ* by insertion of a special plug connected to the secondary injection test set. However, there are many instances where it is advantageous to withdraw relays from their cases and secondary inject them at a central test facility. Testing of small thermal overload units fitted to contactor switchgear is usually carried out by connecting all three thermal elements in series to prove the overcurrent features. Proving of the single-phasing characteristic is achieved by circulating current through any two of the three thermal elements.

Primary injection

This type of test involves each protection circuit, including all interconnecting wiring. It is usually carried out as a final proving test to ensure that all terminations are correct and that all CT are mounted correctly. Typically, a primary injection test set of up to 10 kVA output should be adequate for the majority of circuits encountered in power stations. For primary currents of over 3000 A, a 40 kVA set may be necessary.

There is insufficient space to detail testing of all types of relays here. This information is readily available from manufacturers' literature and other published works. Figure 3.39 showed the connection arrangement for testing current transformers. By rearranging the connections, the sensitivity of different protective schemes can be tested.

Gas/oil (Buchholz) relay

Three standard sizes of relay are generally fitted to transformers provided with conservator tanks. The size selected is related to the diameter of the pipeline into which they are inserted, which in turn is determined by the rating of the transformer. Figure 3.40 shows a typical Buchholz relay and includes some typical test results. Commissioning checks are carried out to confirm that the relays have been installed with the correct upward slope of 5 degrees to allow gas to rise from the transformer into the conservator. The relay must also

TRANSFORMER RATING MVA	PIPE SIZE MILLIMETRE	ALARM VOLUME MILLILITRE	MINIMUM TRIP VELOCITY RANGE METRE/SECOND
< 1	25	110	0.70/1.30
1 TO 10	50	200	0.75/1.40
10 TO 100	75	250	0.90/1.60
> 100	75	250	1.50/2.50

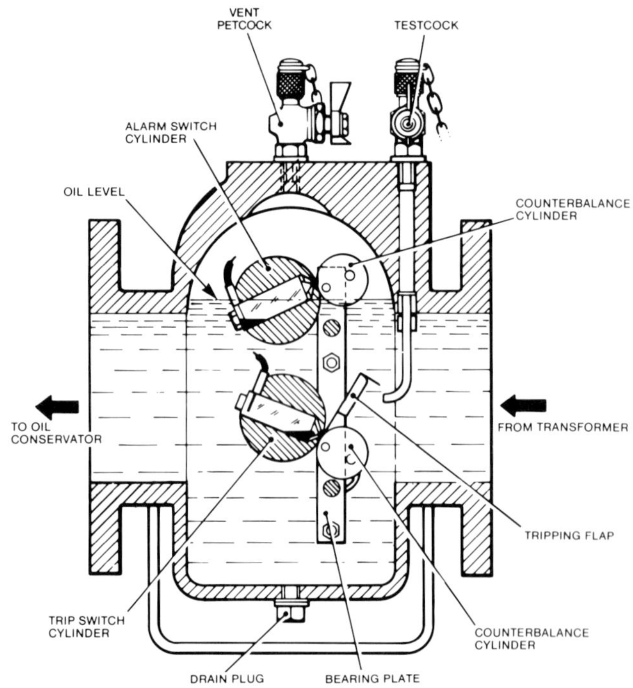

FIG. 3.40 Gas/oil (Buchholz) relay and test results

be located within 5 degrees of the vertical in the transverse direction. Correct operation is proved using compressed air from a portable test cylinder, charged by means of a small footpump. The gas alarm is tested by charging the cylinder to a pressure of 0.7 bar, then allowing the air to trickle into the relay, via the test cock, until the alarm contacts close. The indicated contents of air enclosed in the relay are recorded from the graduated scale on the viewing glass. Air is then vented until the relay is again full of oil.

The test cylinder is charged to approximately 2.8 bar, then the test cock is opened and shut rapidly to simulate gas surges and operate the trip contacts. After venting the relay, the test is repeated at successively reducing pressures in 400 mbar increments until the trip contact fails to operate. This pressure is also recorded for future reference.

It is not unknown for there to be a surge of oil in the conservator pipeline when oil circulation pumps start, so it is necessary to confirm stability of the relay during a pump start.

Motor thermistor protection

Operation of the protection has to be tested with the transistorised detector circuit energised, but with a calibrated (decade) resistor connected to simulate the thermistor resistance. Reference to the thermistor manufacturer's characteristic curves in relation to the particular class of stator insulation will give the desired tripping temperature. Figure 3.41 shows a typical curve.

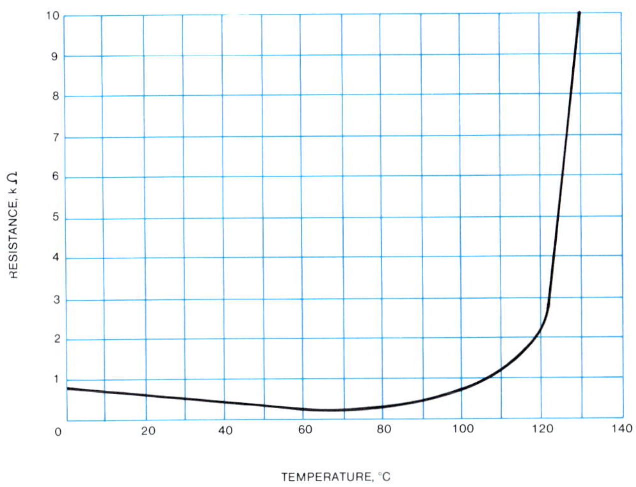

FIG. 3.41 Typical characteristic for PTC thermistor

9.10 Cabling, earthing, lightning protection

9.10.1 System description

Cabling

The co-ordinated design of cable systems is now carried out by computer programs. Main cable routes are subdivided into segments, each of which lie between pairs of uniquely numbered points or nodes. Each cable is allocated a unique five-digit number, where the first digit identifies the main function of the cable, i.e., 0 for station, 1 for unit 1, etc. The remaining four digits identify each cable, the numbers for similar cables on different units being identical.

Cable segregation is carried out, as far as practicable, to ensure that:

- Power cables to main and standby plant follow different routes.
- Power and control cables are not placed together on the same cable tray.
- Power cables for station and unit services are placed on different cable trays, even if they follow the same route.

Fire-resistant barriers are fitted at strategic points along cable runs and cable risers to prevent the spread of fire. Main cable routes close to fire risk areas, e.g., around the main boilers, are often enclosed by fire-resistant sheeting. In areas susceptible to accumulations of pulverised coal, flat horizontal cable trays are avoided wherever possible. Section 12 of this chapter deals with methods of fire detection and the sprinkler fire protection systems which are applied to many of the main cable routes.

Cable sheaths are now generally manufactured from reduced-propagation or flame-retardant black PVC compounds. Single core high-current-rated cables normally have XLP or EPR conductor insulation, whereas cables with two or more cores have PVC insulation. Conductor sizes of 6 mm^2 and below are manufactured from copper. All other conductors are manufactured from either stranded or solid aluminium. Cables with aluminium conductors may be either steel or aluminium wire armoured, but those with copper conductors are steel-wire armoured. Certain types of control cable have either a steel or an aluminium tape armouring instead of wire. Modern developments with short-time rated plastic-insulated cables has resulted in a marked reduction in the use of mineral- insulated copper-clad (MICC) cables other than for a few specialised applications.

Control cables have sequentially numbered cores of 2.5 mm^2 copper conductors, whereas multi-pair signal cables have colour coded, single-strand 0.6 mm^2 conductors laid in twisted pair configurations. Twisted pair cables are usually sheathed with grey-coloured PVC to facilitate function identification.

Multi-core composite cables are often used to link low power drives, such as electric actuators, to their motor control centres or valve group starter boards. One design of cable comprises four power cores and ten pairs of screened control cores within the one PVC sheath which has PVC insulation and a steel tape armour.

All control cables between the plant and the CCR are grouped together in plant-mounted marshalling cabinets from which trunk multi-cored cables are routed to main termination racks (MTR) close to the control room area. The incoming and outgoing cable terminals are then interconnected by jumper leads. The computer programs which co-ordinate the cabling generally specify the jumpering details as well.

Cables used for earth bonding have conductors of solid aluminium, with the exception of 500 mm^2 and above which are stranded aluminium, and sizes below 6 mm^2 which are stranded copper. All cables used for earth bonding functions are sheathed with a green-coloured PVC.

It is now common practice for 11 kV cables to be run as three single cores, rather than a three-core cable, and also laid in a flat formation to reduce the likelihood and effects of phase-to-phase faults.

Earthing

Power station earthing (or grounding) system design is based on BS Code of Practice CP1013. Bonding of equipment in hazardous areas, e.g., hydrogen and propane plants, is also governed by the requirements of BS Code of Practice CP1003.

The basic requirement of an earthing system is to bond the following to earth:

- The neutral points of each separate supply voltage system.
- Non-current-carrying metalwork of apparatus framework, e.g., transformer tanks, equipment cubicles, cable armouring, etc.
- Extraneous metalwork not directly associated with the power system, e.g., boundary and transformer compound fencing.

In the event of earth fault currents flowing, the potential difference between any two normally earthed systems that may be touched simultaneously by one person should not exceed 55 V.

The design of the earthing system depends upon the resistivity of the soil surrounding the plant superstructure to achieve a resistance to earth at any earth electrode of no greater than 0.5 Ω. The earth bonding system must be capable of safely withstanding fault currents appropriate to the rating of the switchgear bonded to it. Rectangular aluminium bars and earth cables within all of the station buildings are interconnected to form a closely-knit equipotential earthed network. These are all connected to one or more earth electrodes to achieve the desired minimum resistance to earth. Cross-sectional areas of bar and cable are chosen to suit the anticipated thermal rating. Though the earth connection voltage drop is not readily calculable, the size of conductor necessary for thermal considerations during fault current flow is likely to keep the voltage drops within acceptable limits.

Earth electrodes at some sites are formed into mats by interconnection of groups of copper rods or sheet steel piles driven into the ground. At other sites the piled foundations for structures, e.g., cooling towers, double as earth electrodes.

The HV (grid) substation area is normally given separate treatment in that the earth mat generally covers all of the ground within the substation boundary fence. Copper earth rods, interconnected by copper tapes, are driven in at predetermined spacings. This earth system is usually bonded to the station main earthing system, via removable links to facilitate periodic testing of earth loop resistance.

Lightning protection

All buildings and structures at power stations have to be given individual consideration for protection against lightning strike. Construction and testing of the installation is carried out in accordance with the requirements of BS Code of Practice CP326 (soon to be superseded by BS6651).

Discharge from a thundercloud towards a structure is diverted to, then safely conducted to earth by, an air termination and downconductor respectively. At ground level, the downconductor has an earth termination which is then connected to earth electrodes. Re-entrant loops in downconductors, particularly over personnel access doors, must be avoided since they can allow a lightning discharge to jump across the open side of the loop. The distance between an earth termination and earth electrode must be as short as possible to reduce the voltage gradient along the ground surface should there be a lightning strike. Figure 3.42 illustrates the salient points mentioned in this section.

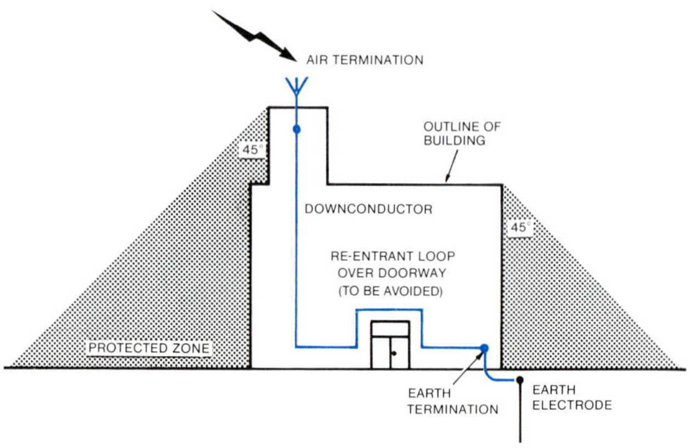

FIG. 3.42 Diagrammatic arrangement of lightning protection system

Metal, steel-reinforced-concrete and metalclad structures are, to a degree, self-protecting against lightning strikes, providing they are adequately bonded to the station main earth system. Building support columns are provided with an earth termination near their base for this purpose. The degree of self-protection is improved, where necessary, by bonding all metal parts together, then adding air terminations and downconductors.

Any building within a 45° shadow of a protected building is normally considered to be protected against lightning strikes. Very tall structures, like chimneys and cooling towers, are not so effective over large horizontal distances, so buildings within only a 30° shadow or cone can be considered as being protected by them. Other buildings, less than 15 m in height, normally do not need separate protection.

The main chimney reinforcing steelwork is specially bonded together to provide a low resistance downconductor to which the aircraft warning lamp units are bonded as well. At ground level an extensive earth electrode system is usually fitted to supplement that provided by reinforcing steel in the base supporting structure. This earth electrode is normally also bonded to the station main earth system.

Oil storage tanks and flammable gas stores have to be given special consideration. Generally oil tanks are provided with several earth terminations round the periphery of the base. These are connected to earth rods driven into the ground close to the base. The rods are interconnected and also bonded to the station main earth. In this case, the shell of the oil tank acts as its own air termination and downconductor. All access ladders, etc., attached to each tank must be securely bonded to it.

Flammable gas stores, if not within the zone of protection afforded by adjacent structures, may require the erection of one or more earthed masts with an earthed catenary suspended over the plant to give the required protection. This is in addition to ensuring that all metalwork, vessels, fencing, etc., are securely bonded together and to the station main earth.

9.10.2 Commissioning

Cables

Testing of cables is a very extensive exercise and the resources required should never be underestimated. Typically, there may be up to 40 000 cables, perhaps of 150 different types, with a total length in excess of 2000 km on a 2000 MW power station. Fortunately, the bulk of the effort is usually provided by the cable installation contractor, who must follow a meticulous system for recording the scheduling, installation and testing of every cable.

The essential reference document for commissioning is a computerised cable schedule, which details a wide range of information of value to commissioning personnel. Installation and commissioning checks blend together with an indistinct demarcation line. Normally, the cable contractor has a separate team to carry out most of the undernoted tests and also rectify minor errors discovered in the process. They furnish commissioning staff with partially completed standard check lists (SCL) for each cable. It is then up to the commissioning engineer to complete each SCL for the cables in his plant area and forward the details to the centralised records office.

These checks are typical of what is needed for each cable:

- Confirmation that the cable is securely fastened along its specified route.
- Both ends glanded and earth bonded to the correct specification.
- Cable positively identified to run between the specified 'A' and 'Z' end locations, through the designated node points, and also to have the correct cable number attached to each end.
- Cable cores at both ends fitted with the specified type of termination and connected in accordance with current drawings.
- Cable record card available and up-to-date.
- Insulation resistance, and HV tests where necessary, carried out with the appropriate testing equipment.

HV cables have their insulation resistance checked, then an HV pressure test carried out, generally in accordance with BS6346 or BS6480 dependent upon the type of cable. The insulation resistance is rechecked after the pressure test. The pressure test is normally a DC test, during which the voltage is gradually increased to the specified test value, held there for 15 minutes, then gradually decreased to zero. The leakage current is recorded for future reference. LV cables are normally subjected to insulation resistance checks only. Mineral-insulated copper-clad cables have their insulation resistance checked at 1000 V DC, 24 hours after the cable end seals have been completed. Figure 3.43 lists test voltages for new PVC cables as an example.

SYSTEM VOLTAGE	HV PRESSURE TEST VOLTAGE		INSULATION TEST VOLTAGE
	CORE-TO-EARTH	CORE-TO-CORE	
11 kV	18 kV DC	30 kV DC	5 kV DC
3.3 kV	5 kV DC	9 kV DC	1 kV DC

FIG. 3.43 PVC cables — typical test voltages

Other important aspects of cable system commissioning that are difficult to record but which must be kept under close observation are as follows:

- Fire barriers and fire breaks should be fitted as soon as practicable after installation of cables. Special safeguards may be necessary in order to permit subsequent installation of other cables in the same location.
- Wherever possible, fire protection and detection systems should be operational before cable systems are energised.

Earthing

Proof that the earthing system is complete is achieved by testing every path to each earth bar and thence to the earth electrodes, in accordance with standard check lists. Testing of the earth electrodes, and of the overall earthing system, is normally carried out by specialists in accordance with BS CP1013. Typically, the earth resistance at each earth electrode should be less than 1 Ω and the resistance to earth of the total interconnected system should be less than 0.5 Ω from any point.

Before commencing tests, it is vitally important that there be a detailed drawing of the main earth matrix, with every joint allocated a unique reference number. Each joint must be made as specified to achieve the minimum electrical resistance with the highest resistance to in-service corrosion. There should also be a schedule listing torque loading of fastenings and resistivity values for earthing bars and cables.

To establish that each section of earth bar is satisfactorily connected, testing is commenced at the earth electrode and then radiates outward, checking every joint in turn. It may be advantageous to use a status marker to identify the integrity of each visible section. The standard check list covers the following typical tests:

- Confirm that dimensions and cross-sectional area of conductor are as specified.

- Check that joints have been correctly prepared and that appropriate high tensile steel fastenings have been used.
- Torque test all bolted joints and record results.
- Record electrical resistance across every joint to ensure that the resistivity is no greater than 110% that of the same length of parent bar.
- Apply a blue-coloured marker to denote that the joint has been satisfactorily checked.

It is also necessary to identify that successive sections of earth bar are continuously connected to an earth electrode. A green-coloured marker can be applied at intervals, working outwards from the earth electrode. Thus, personnel commissioning plant and equipment need only check their earth bonding to the nearest section of green-marked earth bar.

Overall earthing system test procedures depend upon the nature of the site and its location. A typical test carried out at one power station site utilised a 415 V AC portable generator located several kilometres distant. Full-load current from one phase of the generator was conducted to the site over a de-energised HV overhead line which had been connected to the site earth electrode. A remote earth was made available over the public telephone network: then the voltage between this earth and a portable ground electrode was measured. The portable electrode was repositioned at 100 m intervals in a straight line radiating from the main earth electrode over a total distance of 1000 m. A series of tests were conducted with different power system frequencies and levels of current.

Lightning protection

Testing is carried out in accordance with BS Code of Practice CP326 (refer also to BS6651). The basic requirement is that the overall resistance from air termination to earth should be less than 10 Ω. Generally this is measured in two parts, i.e., resistance of downconductor and resistance of earth electrode, which are then summed. Care must be taken to ensure that all interconnections to the station earthing system are disconnected before the lightning protection earth resistance is measured. After testing, the interconnecting link must be reinstated.

It is important that there should be scale drawings of the installation and that all major components are uniquely identified. Records of all tests must be maintained, including cross-references to the system diagram.

9.11 Battery systems

9.11.1 System description

A number of battery systems are installed to furnish 240 V, 110 V and 50 V DC supplies throughout the power station. These are based on lead-acid cells of the high performance Plante type, manufactured in accordance with BS440. Each battery is provided with a full-duty charger, which supplies the load and keeps the battery fully charged, by float or trickle charging. On mains failure, the battery immediately takes over to supply the whole of the connected load, without any break in continuity. The charger and its associated battery are connected to a DC switchboard from which supplies are taken to load distribution fuseboards for various services. Facilities for interconnecting adjacent battery switchboards are usually provided. Figure 3.44 shows a single-line diagram for a typical battery system.

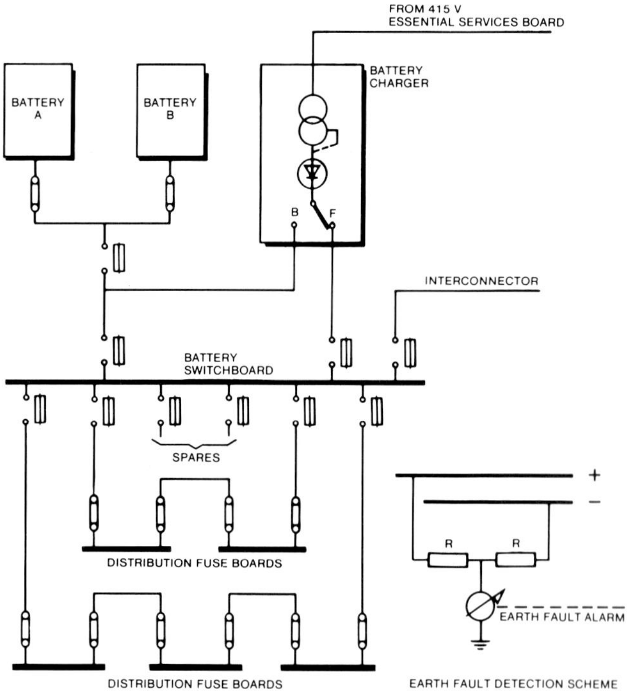

FIG. 3.44 Typical battery system

Contemporary designs provide for separate three-phase 415 V or single-phase 240 V AC -fed chargers for float and boost charging of 240 V and 110 V batteries, but with combined float and boost chargers for 50 V batteries only. Mechanical key interlocks are fitted to ensure that each battery is off-line before it can be boost charged. Normally, the 50 V battery charger circuit facilitates boost charging with the battery still in service, though some do have off-load mechanical key interlocking. Usually both poles of the 240 V and 110 V batteries are insulated from earth. An earth fault detection system comprising a centre-tapped resistor connected across the poles, with a detection relay wired to the tapping, warns of earth faults. The positive pole of 50 V batteries is solidly connected to earth, at both the battery and its distribution board.

Batteries are located in fresh air fan-ventilated rooms provided with an acid-resistant floor covering. Illumination is by sealed fittings because of the possible release of hydrogen from the electrolyte, particularly during boost charging operations. Chargers and distribution switchboards are located in adjacent rooms.

The capacity and number of batteries installed varies widely from plant to plant. Typically, there could be up to 30 different batteries ranging in capacity from over 2000 Ah to as low as 15 Ah. The principal duty of each battery system is summarised in the following paragraphs.

240 V system

Batteries are usually arranged on a station basis with there being one battery for every main generating unit. The batteries are used for supplies to:

- Emergency auxiliary pumps etc.
- Emergency lighting
- HV circuit-breaker closing
- AVR and excitation rectifier controls
- Guaranteed instrumentation supplies (GIS) conversion equipment
- Emergency valve operation

Some designs of the GIS conversion equipment are provided with separate battery systems on a unit basis.

110 V system

Individual batteries for each unit, plus others at the coal, ash, dust and CW plants, to provide supplies for:

- HV and LV switchgear operation
- Control equipment supplies
- Interlocks and protection
- Emergency lighting in areas not covered by 240 V DC lighting

Additional battery systems are installed for:

- Gas turbine controls
- Gas turbine emergency auxiliaries
- Gas turbine starter motors

50 V system

Separate batteries are often provided for station and unit functions within the main plant buildings. Supplies for out-stationed plant such as coal, ash, CW, etc., may be derived from the GIS system by transformer/rectifier units. They are used for:

- Alarms and indicators
- Sequence and control systems
- Telecommunications
- Grid substation control, indication and communication

Some recent power station designs have provision for only minimal 50 V DC battery capacity, preferring to derive the unit 50 V supplies at various load centres around the main plant from the GIS system, via transformer/rectifier/distribution units.

9.11.2 Commissioning

As soon as practicable after installation, batteries should be fully charged and then left on float charge until required. Frequently the construction site electricity supplies are used for this purpose. When the permanent cables are connected, outstanding commissioning tests have to be conducted to prove the integrity of the battery/charger system.

The following series of checks have to be completed irrespective of the source of the electrical supply to the charger. Personnel handling lead-acid cells should be familiar with the requirements of BS6133:

- Battery and charger rooms to be clean, well lit and ventilated.
- Equipment correctly installed. Battery stands level and rigid. All cells carefully lined up, with the interconnecting links fitted so that the pillars are not subjected to any strain. All fastenings tight and coated with a protective jelly.
- Battery capacity checked to be correct. Details engraved on each cell identifies the number of plates and the battery rating, typically 7.5, 25 or 100 Ah.
- Cells all topped up to correct level with distilled water or sulphuric acid in compliance with BS3031.
- Emergency eyewash bottles; NO SMOKING/NO NAKED LIGHTS, and TREATMENT FOR ELECTRIC SHOCK notices all erected. Supply of fresh water available at battery room sink.
- Switchgear operational checks carried out, including functioning of mechanical interlocks for boost charging.
- Insulation resistance of all cables, panel wiring, switchgear, busbars, etc., tested at 500 V DC (after isolation of semiconductor devices).
- Energise auxiliary control, indication, panel heating and lighting supplies, then prove functioning of alarms. At this stage it is unlikely that inputs to the station alarm scheme will be cabled.
- Energise charger on open circuit, then check that polarity of charger output and battery input are identical.
- Connect battery to charger on float control until required for further proving tests. Record the cell voltage and specific gravity of every cell for future reference.

The next series of tests is aimed at proving the ability of the battery and then the charger to meet the specified load. For these tests, an adjustable load resistor bank, located in a safe area outside the battery room, is connected to the DC switchboard busbars. When testing large capacity batteries, it may be necessary to provide an air-blast cooling fan to assist with heat dissipation from the loading resistor.

Prior to conducting the battery discharge test, the battery must be charged fully, then left on open-circuit for at least 12 hours but not more than 18 hours. The load resistor is switched in and the load current adjusted to 10% of the standard rating. Throughout the test, the resistor is trimmed to maintain the load current reasonably constant at its specified value.

When conducting a discharge test on a battery containing a large number of cells, the following may be a more practical set of details to record instead of those listed in BS440:

- Select a pilot cell at random, perhaps in the centre of the battery, and record cell volts, load current, specific gravity and cell temperature every half hour.
- Specific gravity of every cell at the start and finish of the test. Typically, this will drop from 1210 to 1155 at 15°C as the cell discharges.
- Cell voltage of every cell at the start and at every half hour for the final hour, i.e., three final readings.
- Cell voltage of every fifth cell at half hour intervals throughout the test.

The prevailing battery voltage, load current and ambient temperature are also recorded. If practical, it is advantageous to try to maintain the ambient temperature close to 15°C to minimise the correction factors to be applied.

The test is terminated either when the voltage of any one cell has reduced from 2.2 V to 1.85 V, or at the end of the tenth hour, whichever is sooner. The calculated capacity is then compared with the specified rating.

The charger is reinstated to recharge the battery fully on boost charge over some 12 to 16 hours, after which the float charging characteristic is checked.

When convenient, the battery is taken off-line and the load resistor switched onto the charger output. The load current is adjusted to that specified for the charger and a test conducted over at least four hours. During this period, the temperature rise of the charger transformer and rectifier is recorded along with measurement of ripple in the DC output wave form.

Outgoing feeders and their distribution fuseboards fed from the battery switchboard are commissioned in a similar manner to other LV supplies. Most DC distribution fuseboards have duplicated supplies and obviously the polarity of each supply must be identical. All cabling etc. is subjected to an insulation resistance check at 500 V DC before energising.

Outgoing circuits from DC distribution boards will be checked and energised by members of the various plant commissioning test teams. As more plant is energised, the load distribution and voltage regulation at different locations have to be checked and kept under review to ensure security of supplies.

9.12 Guaranteed instrumentation supplies (GIS)

9.12.1 System description

The GIS system provides secure power supplies at 110 V AC single-phase 50 Hz, with a close tolerance on frequency and voltage, to essential instrumentation and control equipment. Voltage variation between ±3% and a frequency range of 49.5 to 50.5 Hz are normally acceptable. Instrumentation supplied from this system must be capable of tolerating an interruption of supply of up to 100 ms when alternative sources are being automatically switched.

GIS systems are normally supplied from a 240 V battery via conversion equipment producing a regulated single-phase 50 Hz output. During normal operation, input power is derived from the supply mains, via the battery charger. In the event of mains failure, the conversion equipment is supplied directly from the battery, which should be capable of maintaining the load for at least 30 minutes before the battery terminal volts drop to approximately 90% of their nominal value. Should the converter fail, an unregulated single-phase standby supply direct from a 415 V Essential Services Board is automatically switched in.

Two types of conversion systems are in use at present. One comprises a motor-generator set with an interlocked 415/110 V standby transformer, as shown by Fig 3.45, connected to a 110 V AC main switchboard. This is duplicated for each main unit, with a further duplicated installation for station services instrumentation. Typical ratings for the motor-generators and standby transformers would be 30 kVA. Facilities for selective synchronising of motor-generators to each other and to the standby transformers are provided.

The other, more recent, design utilises 30 kVA static inverters to provide a regulated 415 V single-phase 50 Hz output to a main switchboard. In this case, the standby supply is at 415 V single-phase, connected in to the switchboard as shown by Fig 3.46. Controlled

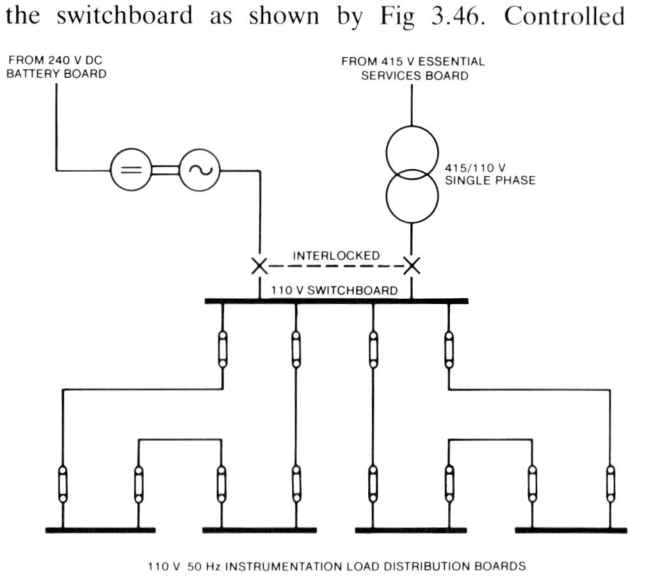

FIG. 3.45 Instrumentation supplies — motor generator system

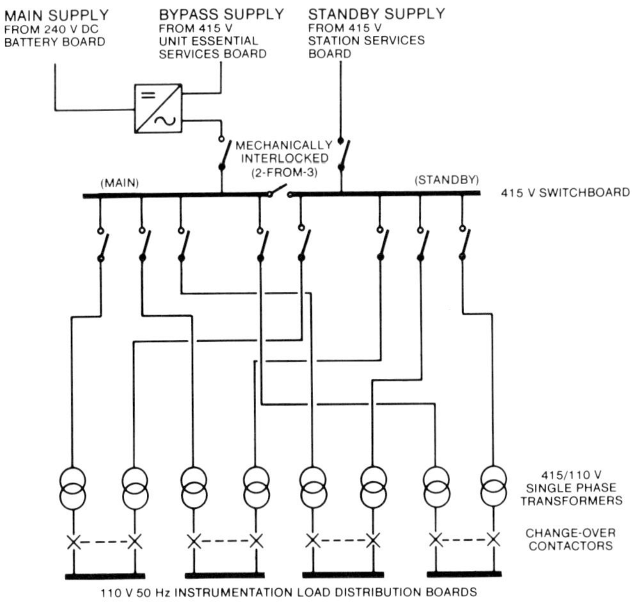

FIG. 3.46 Instrumentation supplies — inverter system

synchronising of each inverter with its by-pass supply is provided for maintenance purposes. Instrumentation supplies are distributed to load centres at 415 V single-phase, where it is transformed to 110 V single-phase for local distribution. When the inverter is on-load, it is kept in synchronism with the by-pass supply by means of an internal control loop. In the event of an inverter fault, the by-pass supply automatically takes over the load without interruption, providing that the inverter and by-pass voltages are still in synchronism. Load transfer from the by-pass supply to the inverter is completed automatically without interruption after the inverter has maintained a stable synchronised output for at least 10 seconds. Should the inverter output voltage fall below 80% for longer than 10 ms, then automatic changeover from the 'main' to the 'standby' supply occurs at each distribution board. This results in an interruption of supply for between 60 and 100 ms. Transfer back from 'standby' to 'main' is carried out manually, following restoration of the main supply. This has to be carried out in stages to avoid excessive inverter overload, arising from magnetising inrush current on the 415/110 V step-down transformers.

9.12.2 Commissioning

Commissioning of batteries and chargers has been covered in preceding paragraphs. The following notes refer to the conversion equipment necessary to produce the 110 V 50 Hz single-phase secure supply for instrumentation.

Details of the test procedures will depend upon the plant arrangement and the specified performance for the installation. Typical tests are outlined to cover each of the following:

- Motor-generator set equipment
- Static inverter equipment
- Instrumentation supplies load distribution centres

Motor generator (MG)

The initial checks follow a similar pattern to that for other types of small rotating machines and electrical control equipment. Typically these may be:

- Mechanical checks of coupling alignment and security of foundations.
- Electrical checks of DC drive motor and AC generator, including all cabling.
- Testing of motor and generator protection, instrumentation, alarms and automatic control loops.

Energising and no-load checks follow, and include:

- Direction of rotation, run-up time, no-load current, commutation, plus bearing temperature after a few hours of running.
- Prove manual, then automatic, control of generator output voltage and frequency at no-load.
- Demonstrate manual and automatic synchronising features.
- Carry out automatic starting tests and prove that, if a pair of machines is started simultaneously, one synchronises whilst the other remains locked out.

Finally, the following loading tests are conducted with a temporary load bank connected to the busbars:

- Operation of plant under full-load conditions for an extended period, then under selected overload conditions and finally with changing (transient) loads.
- Prove that the control equipment has the ability to transfer the load on to the standby supply automatically in the event of a failure of the MG set. Manual transfer of load from one MG set to the other is also demonstrated.
- The ability of the control equipment to maintain the output frequency within acceptable time/error tolerance, when related to the station standard clock pulses, is checked.

Static inverter

The control equipment is all checked for completeness in accordance with the latest connection diagrams. The insulation resistance of the wiring and cables is tested at 500 V DC, noting that the semiconductors must first be isolated. Auxiliary supplies are energised to prove control and alarm functions.

Each inverter is energised on open-circuit to prove the controls and to provide the auxiliary supplies for the integral cooling fans. Various part-load and overload cycle tests are carried out with the temporary load bank connected to the output busbars.

Typical tests are as follows:

- Checks for harmonic content of inverter output voltage.
- Assessment of battery current ripple with inverter in service.
- Demonstrate manual and automatic load transfer to and from the by-pass supply by simulating various modes of inverter fault.
- Check system response to charger failures.
- Prove inverter thermal performance during full-load heat run for 6 hours.

Distribution system

Each load distribution centre is energised in turn from the main supply switchboard. First of all the transformers, changeover contactors, undervoltage protection, indicators, alarms and interlocks have all to be checked. Correct functioning of the control features may have to be proved by simulation at this stage.

Following cable identification and insulation resistance tests of the items comprising the distribution centre at 500 V DC, each incoming supply is energised in turn. The polarity (line and neutral) is confirmed to be identical and the automatic changeover from 'main' to 'standby' supply is proved. This includes timing checks for operation of the changeover contactors. Typical limit is less than 10 ms for all control voltages between 75% and 110% of nominal 110 V. Manual reinstatement of the 'main' supply has to be demonstrated too.

Outgoing circuits from the 110 V AC distribution boards are checked and energised, as required by various plant commissioning test teams. At a later stage, once sufficient instrumentation load is available, load distribution and voltage regulation has to be assessed.

9.13 Construction site supplies

9.13.1 System description

Construction site supplies are made available almost right at the start of the site works. A main substation, housing an 11 kV switchboard provided with duplicate supplies from the local electricity undertaking, is established at a central point. Self-contained skid-mounted primary substations, each comprising an 11 kV ring main isolation switch unit, an 11 kV/415 V 1000 kVA transformer and a 415 V distribution board, are strategically located around the periphery of the site works. The primary substations are usually interconnected into two or three 11 kV rings with the opposite ends of each ring connected to different busbar sections of the main switchboard. Sub-distribution at 415 V, 240 V and 110 V AC single-phase is achieved by 415 V portable distribution units (PDU) located within the plant areas but connected to the primary substations. The selection and use of low voltage equipment on construction sites should be as outlined in BS4363 and BS CP1017. Figure 3.47 illustrates a single-line diagram of a typical network.

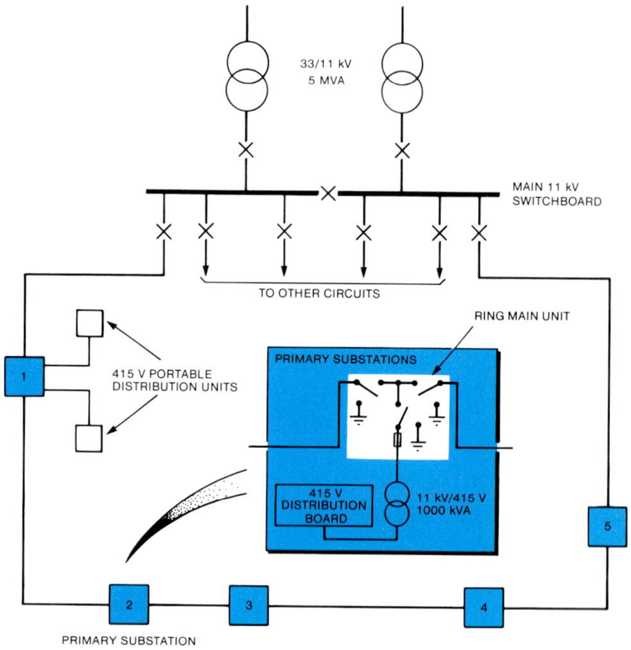

FIG. 3.47 Construction site supplies system

During the early stages of plant erection, several items of station equipment, including some switchboards, may have to be connected to the site supplies system to maintain programme dates. At a later stage, they have to be re-energised from their permanent electrical supply source. It is important to ensure that identical phase sequence is maintained after the changeover.

9.13.2 Commissioning

The main supply switchboard and the primary substations are commissioned generally as outlined earlier in the appropriate parts of this section. At present oil-break switchgear is used for the ring main isolating switches and the oil must be subjected to the same stringent tests as transformer oil. These switches are designed for load-breaking, but fault-making when related to a 250 MVA prospective fault level system. The primary substations are not normally fitted with interlocks to prevent paralleling of supplies from both sections of the main switchboard.

Again, because the substations are normally interconnected in a ring (or mesh) formation, particular attention must be given to phasing out of both 11 kV supplies at every substation. Continuity of the system earth conductor has to be assured at all times and the earth loop impedance checked frequently.

Security of all portable substations and temporary cables must be strictly observed to ensure that construction personnel are not inadvertently put at risk by contact with live cables or equipment. This is particularly relevant when 415 V distribution units are located in the midst of plant being erected, where there is an ever-present danger of physical damage.

10 Gas turbine plant

Gas turbine plant is installed at all major fossil-fuelled power stations. The plant is housed in a purpose built building and is, in effect, a self-contained power plant within a power station.

The grid (HV transmission) system operational requirements set the following objectives for gas turbine plant:

- To provide system generation for normal merit-order running.
- To supply reserve capacity for system frequency regulation.
- To improve security of supply to the station auxiliaries under system emergency conditions.
- To permit station start-up without external supplies.

Gas turbine-generators comprise one or more gas generators exhausting to a power turbine. Each gas generator is basically an aircraft jet engine derated for operation nominally at sea level. The power turbine is directly coupled to a two-pole electrical generator. Output from this generator is connected to an appropriate 11 kV Unit Board.

The CEGB has a 70 MW industrial gas turbine power plant in service, comprising a compressor, a power turbine and an electrical generator, all coupled into a single-shaft unit. A diesel engine is used to accelerate the shaft from rest up to 1600 r/min, at which point the gas turbine is self-sufficient and the diesel automatically shuts down. Propane gas is used for ignition of the fuel during the starting sequence. A clutch allows the AC generator to be disconnected from the power turbine and run as a synchronous motor for grid system voltage control. Most of the auxiliary plant, sequence and controls are similar to those for the aero-engine powered units.

This section outlines the activities concerned with commissioning two 25 MW twin-Avon gas turbine-generators, as a typical example. Industrial gas turbine commissioning is a similar, but perhaps slightly simpler, exercise. Figure 3.48 shows a general arrangement of the plant chosen for this example.

10.1 System description

10.1.1 General

It is customary to provide one gas turbine-generator for each main boiler/turbine unit. The general philosophy is that total gas turbine output should match the maximum station auxiliary load when all main units are generating.

Early power station gas-turbine plants were of 17.5 MW capacity, powered by a single Olympus jet engine. Later installations used twin Avon jet engines to drive a common power turbine/25 MW generator unit. Current designs of 35 MW generator are based on twin Olympus jet engines as a power source.

Individual gas turbine power stations, connected directly into the local electricity distribution network, utilise four gas generators, with various designs of power turbines, to drive 50 MW and 70 MW generators. The most common layout is a double-ended twin gas generator arrangement, where the AC generator can be declutched from both power turbines and used for synchronous compensation.

The Olympus gas generator is a twin-spool design, in which separate turbines drive the HP and LP compressors, via concentric shafts. The Avon engine is a single-spool machine, where a three-stage axial turbine drives a 17-stage axial compressor. In each case, the exhaust gases drive a two-stage power turbine overhung from its own combined journal and thrust bearing pedestal, which also houses the coupling to the electrical rotor shaft. Figures 3.49 and 3.50 show the general arrangements of Olympus and Avon gas generators, respectively.

10.1.2 Fuel oil system

Figure 3.51 is a schematic diagram of a typical fuel-oil supply system. Diesel oil to BS2869 Class D is used as gas turbine fuel. Other designs of fuel-oil system accept deliveries into a 'dirty' storage tank, where it is allowed to settle for a few days. The oil is then passed through a bank of electrically-driven centrifuges before being discharged into a 'clean' storage tank, whence it passes to the transfer pumps. The quality of delivered oil has improved to the point where the centrifuges can be bypassed and the settled fuel pumped direct into the 'clean' tank ready for use. Stations where coal and boiler fuel-oil is delivered by rail generally also have rail reception facilities for gas-oil. Provision for delivery by 23 000-litre road tankers is often made as well.

Storage tanks are mounted in individual reinforced-concrete bunds, each of sufficient capacity to hold the entire contents of a ruptured tank. Accumulations of storm water are usually pumped out of the bunded areas by manually-controlled pumps. This avoids oil contamination of site drains which could occur if automatic pumps were fitted and there were an oil spillage.

Storage tanks are usually fitted with an articulated floating suction line, the mouth of which is supported by a group of floats, such that the inlet is held about one metre below the oil surface. When an empty tank is being filled for the first time, it is not unknown for the

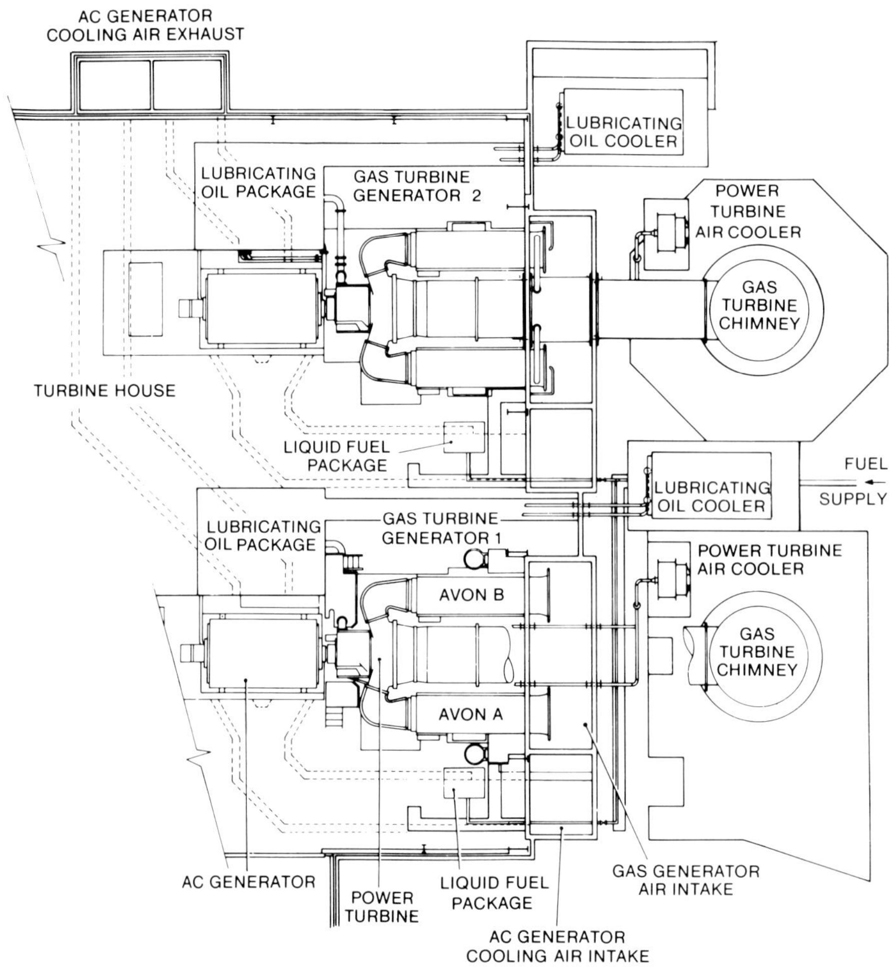

FIG. 3.48 Layout of gas turbine plant

articulated joints to become buoyant before the supporting floats do. The suction line then remains air-locked and is unable to rise in the correct attitude to release the trapped air. Before commissioning the transfer pumps, it is necessary to induce a reverse flow of oil through the transfer pump suction line into the storage tank. This should displace the trapped air in the articulated joint and allow the floating suction to resume its normal attitude.

Fuel for the gas generators comes from the oil tanks via a common fire-trip valve into the transfer pump house. Each transfer pump is capable of supplying the total fuel requirements of the plant. Two of the pumps have 415 V AC motors and the third a 110 V DC motor.

The transfer pumps discharge fuel oil to the liquid fuel packages associated with each gas turbine set. Pipework from the fuel tanks to the liquid fuel packages is covered with thermal insulation. Exposed sections of piping are also electrically trace heated to facilitate gas generator light-up under adverse weather conditions. A small-bore recirculation line is led from a

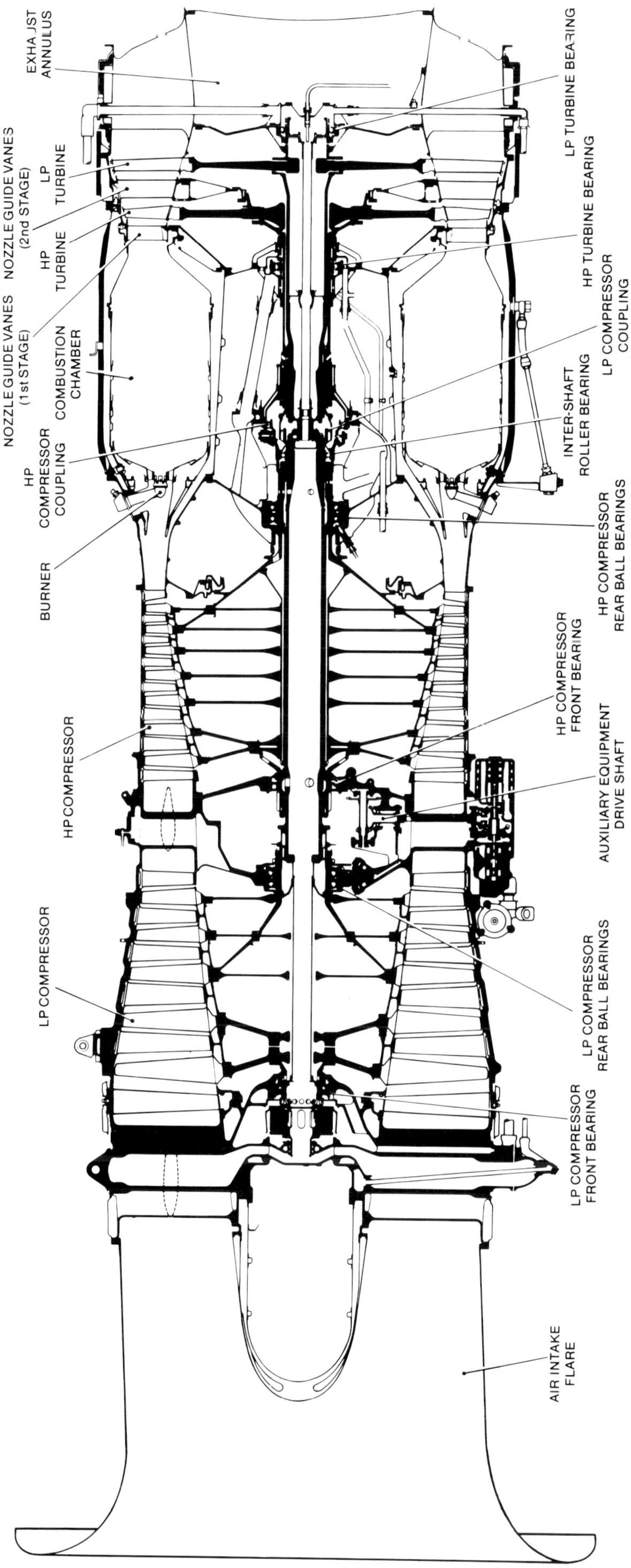

FIG. 3.49 General arrangement of Olympus gas generator

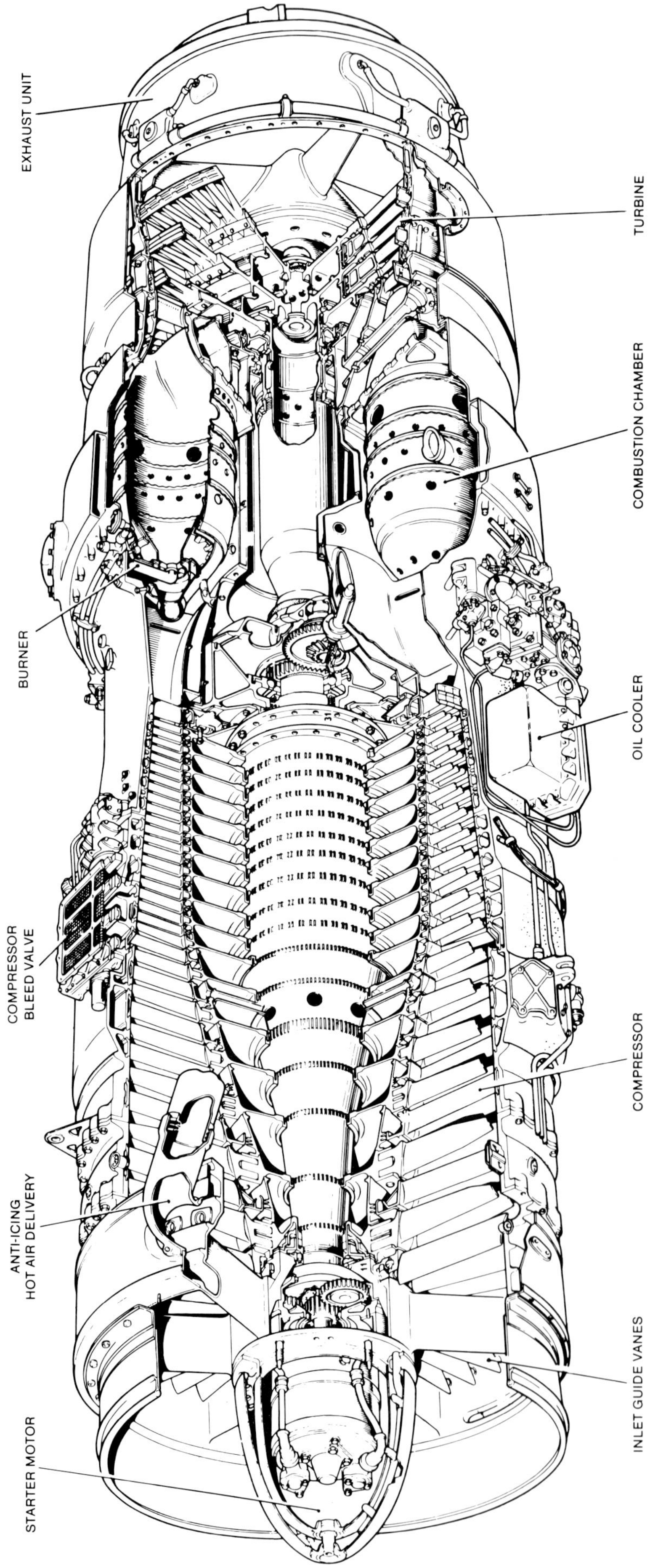

FIG. 3.50 General arrangement of Avon gas generator

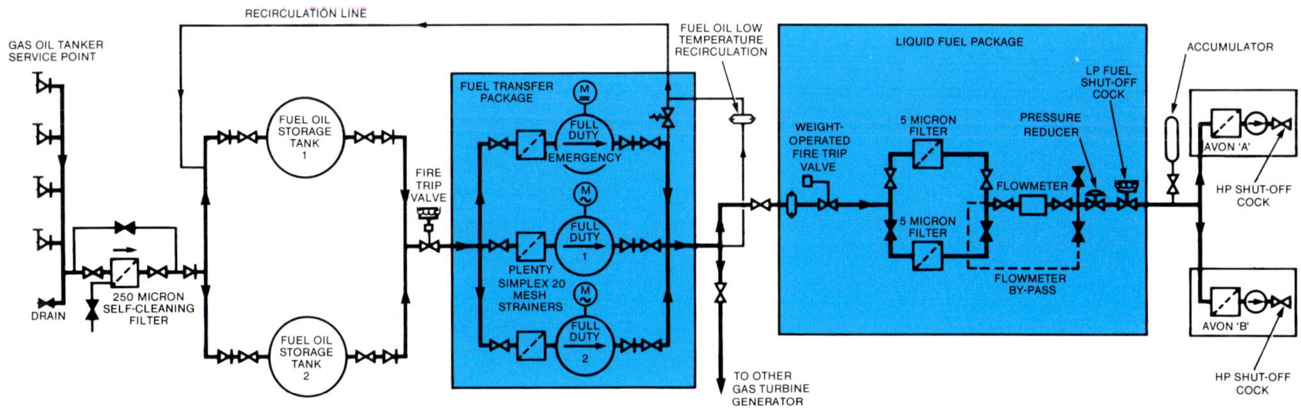

FIG. 3.51 Gas turbine fuel oil supply system

point close to the liquid fuel packages back to the storage tanks. When the fuel-oil temperature drops to about 5°C, a thermostat automatically starts the duty fuel oil transfer pump to establish a recirculation flow through the external pipework until the oil temperature rises to about 7°C.

Initial filtration of the fuel oil is effected by a motor-operated self-cleaning 250 μm filter in the storage tank supply line. Each transfer pump is equipped with a 20 μm suction filter. Further filtration is carried out by duplicate 5 μm filters at each liquid fuel package.

Fuel entering the liquid fuel package passes through a fire-trip valve which is activated by the Avon fire detection equipment. After being metered, the fuel is piped to the fuel control unit located on each gas generator.

10.1.3 Air system

Combustion air for the gas generators is taken from the atmosphere, via filters and silencers mounted at the top of the building. Filter elements may be of the fixed 'Vee' bag form, or of the automatic roll-on design. The roll-type of filter comprises a long length of filtering medium wound on a wide reel, which is then stretched across the air intake and attached to a similar (empty) reel. As the differential pressure across the filtering medium increases due to contamination, a control sequence rotates the reels so that a fresh section of material is moved into the air stream. Protective devices prevent all of the material being wound off one reel on to the other. The filters are fitted with automatic by-pass doors which open if the pressure drop exceeds approximately 6 mbar. Cooling air for the electrical generator is drawn through another section of the same filter pack. Typically an air flow rate of 66 m^3/s is needed for the gas generators and 21 m^3/s for the electrical generator.

Exhaust gas from the power turbine is led to atmosphere, via a silencer and a steel chimney. Sometimes the chimneys for all gas-turbine units are enclosed in the one reinforced-concrete windshield.

Heated cooling air from the electrical generator is discharged to atmosphere through louvred vents in the wall of the plant house.

10.1.4 Gas generators

Each gas generator is enclosed in a fire-resistant, sound-absorbent enclosure equipped with an automatic fire protection scheme. The fire detection system is based on heat-sensitive cables led all over the Avon. Duplicate systems are fitted and both have to be activated before automatic discharge of the extinguishing gas takes place. The gas turbine plant upon which this section is based uses CO_2 gas, but Halon 1301 is the preferred extinguishant on more recent installations (see also Section 12 of this chapter).

The gas generator is a single-shaft open-cycle unit, as shown earlier by Fig 3.50. It has a three-stage axial-flow turbine driving a 17-stage axial-flow compressor. Atomising fuel burners are mounted in eight interconnected radially-disposed combustion chambers. The power output of each gas generator is controlled by regulating the quantity of fuel supplied to the burners. An electronic governor unit controls a servo valve which regulates the fuel input by adjustment of the swash plate angle of the HP fuel oil pump.

A fuel oil control unit mounted on each gas generator incorporates a filter, an engine-driven HP fuel pump and an HP shut-off cock. This cock is held open by an electromagnetic clutch, but will close in about 200 ms if the clutch is de-energised. The common fuel oil supply line from the liquid fuel package incorporates a low pressure shut-off cock and a nitrogen-filled accumulator. Pressure surges created by rapid closure of the HP shut-off cock are suppressed by the accumulator.

Gas generator lubrication systems are mounted on, and driven from, the engine. A special synthetic lubricant is used. It is cooled by incoming fuel in oil/oil heat exchangers, which preheat the fuel oil to the burners. A small tank collects fuel drains from several points on both gas generators, transition duct, power turbine and the liquid fuel package. Drains from the main lubricating oil system filter are also collected in this tank. Periodically, the oil has to be manually pumped out and disposed of by burning.

Air is bled off from the intake for the engine and led into the gas generator enclosure for ventilation and cooling, then exhausted into the power turbine chimney via a venturi ejector and an automatic flap-valve. Power air for the ejector is derived from the compressor discharge stage. The flap-valve is tripped closed when the gas generator fire protection operates, to avoid dilution of the fire extinguishing gas.

A proportion of the compressor airflow is bled off for control purposes and also for cooling the power turbine stator and rotor blading discs. The power turbine casing shaft-seal is pressurised with air tapped off the disc cooling air system. The cooling air from the compressor tapping is first passed through air-blast heat exchangers before entering the power turbine cooling ducts.

Each gas generator is fitted with a 33 kW 120 V DC series-wound starter motor and duplicate electric spark igniters. The igniters are supplied at 2 kV from a 110 V DC power pack capable of generating up to 60 sparks per minute.

Gas-generator performance is adversely affected by deposits of dirt, salt particles, etc., on the compressor blades. Two separate cleaning systems are often provided, namely carbo-blast and crank-soak-wash.

The carbo-blast system is used with the gas generator in service. It entails feeding a soft blast material into the compressor intake. The soft blast material removes atmospheric deposits from the blades, then passes harmlessly through the combustion chambers to the exhaust.

The crank-soak-wash is an off-load process comprising two washing cycles, each followed by a 15-minute soaking period. A special fluid is sprayed as a fine mist into the compressor intake while the shaft is being rotated by the starter motor. During the soaking period, with the shaft at rest, excess fluid drains out of drillings in the compressor casing into prepositioned drip trays.

10.1.5 Power turbine and electrical generator

Hot gases leaving the gas generators are turned through 180° and led through a common transition duct into the two-stage power turbine. The power turbine is overhung from a pair of journal bearings, separated by a thrust bearing, enclosed within a common pedestal. A flexible coupling, e.g., Turboflex, is used to connect the power turbine and AC generator rotor shafts. The main lubricating oil pump is gear-driven from the power turbine shaft.

Electronic and mechanical overspeed protection devices are also driven from the power turbine shaft. Operation of either device, or the hand trip, shuts the high pressure shut-off cocks for both gas generators.

The 31.8 MVA air-cooled, two-pole, brushless generator is supported in two force-lubricated journal bearings. A revolving armature main exciter and a permanent magnet pilot exciter are mounted on the outboard extension of the main generator rotor shaft. Output from the generator is cabled to the appropriate 11 kV Unit Board. The neutral of the generator winding is connected to earth, via a harmonic rejector and a liquid resistor. The harmonic rejector suppresses the circulation of third-harmonic currents, when the generators operate in parallel.

Connected on to the generator terminals is a 11 kV/415 V Gas Turbine Unit Services Transformer to supply gas turbine auxiliary plant.

10.1.6 Main lubricating oil system

The main lubrication system, shown in Fig 3.52, supplies a standard grade turbine oil to the power turbine and electrical generator bearings only. Two lubricating oil pumps are mounted on an oil storage tank. One pump is a 415 V AC full-duty standby unit and the other is a 110 V DC 50% duty emergency unit. The main shaft-driven oil pump on the power turbine bearing pedestal supplies lubricating oil requirements when the gas turbine is running. Duplicated air-blast oil coolers are mounted outside the plant building, along with those for the power turbine cooling air. A thermostatic valve, connected in parallel with the oil coolers, proportions cooled and uncooled oil to maintain a steady oil temperature to the bearings. Drains from the lubricating oil filter unit are led to the gas generator fuel-oil drains tank.

10.1.7 Control system

A fully automatic starting sequence is installed, so that each gas turbine unit can be run up, automatically synchronised and loaded from a single control initiation. Control is normally from the central control room (CCR). The same facilities are also provided on the local control panel (LCP). All the control and alarm indication equipment is fitted in the LCP. Manual control for testing purposes is only provided at the LCP. Four modes of control are as follows:

No-load The unit starts and automatically runs up to the minimum governor speed — generally about 2800 r/min. The operator then has to synchronise and load the generator manually.

Peak The unit starts, runs up to speed and synchronises, then loads to full-load (25 MW) over a four-minute period by automatic sequence. At some power stations, the 'peak' selection causes a 25 MW machine to be loaded to 29 MW. If this facility is fitted, it is augmented by a 'base' selection which limits the load to only 23 MW. The lower base-load output results in an extension of the gas generator service life between overhauls.

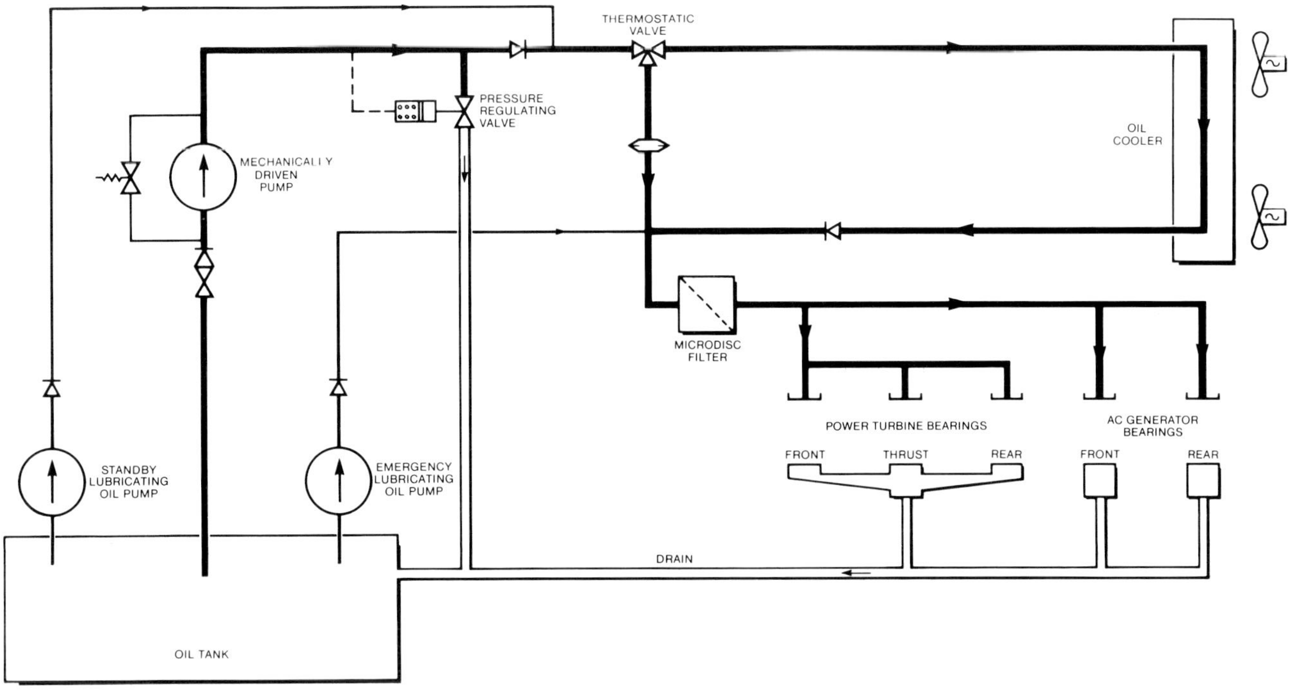

FIG. 3.52 Gas turbine lubricating oil system

Standby If the 11 kV Unit Board system frequency falls below 49.7 Hz, the gas turbine starts, synchronises and loads automatically. The loading cycle is completed within 22 seconds of synchronisation. Should the 11 kV Unit Board lose its supply, the undervoltage protection automatically starts the gas turbine and runs it up to synchronous speed. The operator has to override the synchronising check relay to close the circuit breaker, and then load manually to the desired level.

Shutdown The unit will automatically de-load, disconnect from the 11 kV Unit Board and shut down in a sequential manner.

As mentioned earlier, the fuel supply to the burners is regulated by a servo-valve. The servo-valve control system is based on a 'least gate' logic circuit which allows only the signal that commands the minimum fuel input for control of the power turbine speed to regulate the servo-valve. All inputs are inhibited until the power turbine speed is greater than 450 r/min. Thereafter, any regulation of fuel flow is related to the compressor discharge pressure (or P2) which thus forms a basis for gas generator control. One of the governing parameters for operation of twin gas generator units is the relative balance between the P2 pressures of the two engines.

10.1.8 Fire protection

Separate fire protection systems serve the following risk areas:

(a) CO_2 (or Halon) for protection of both gas generators.

(b) Foam for direct injection into the fuel oil storage tanks.

(c) Waterspray for:

- Fuel oil storage tanks
- Fuel oil unloading area
- Fuel oil transfer pump package
- Liquid fuel package
- Power turbine/generator lubricating oil tank and pipework within plant building
- Lubricating oil coolers
- Gas turbine unit services transformer

All waterspray systems are of the automatic dry-deluge type as described in Section 12 of this chapter.

Usually, operation of any of the systems results in an automatic trip of the gas-turbine unit. Fires within the plant building also trip the roof ventilating fans. A fireman's switch is fitted to override the fan controls and establish ventilation for release of smoke and heat, when required.

10.1.9 Electrical services

Figure 3.53 shows a simplified line diagram of the electrical system. All of the auxiliary plant directly associated with each gas turbine is connected to a 415 V Gas Turbine Unit Services Board. This board has two sources of supply, which automatically change over when the gas turbine 11 kV circuit-breaker closes. One supply is derived from a 415 V Station Services Board, via a 415 V Gas Turbine House Services Board, which also provides supplies to associated plant not directly concerned with gas turbine operation. The second

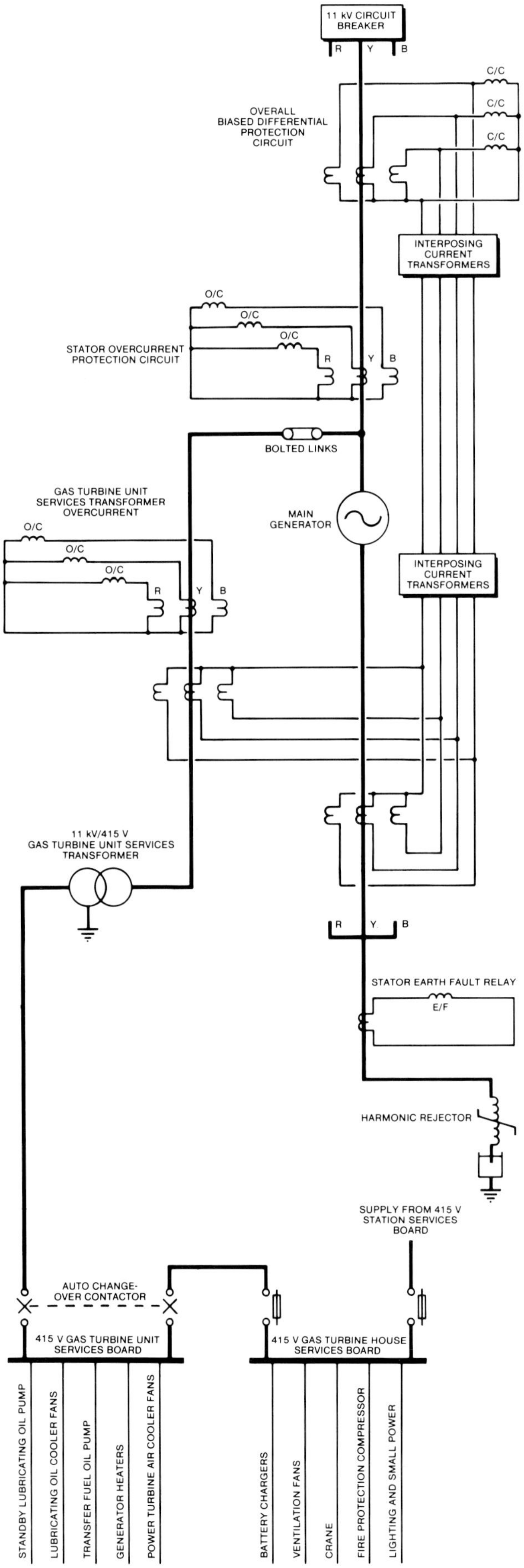

FIG. 3.53 Arrangement of gas turbine-generator electrical services and protection

supply is obtained from the 11 kV/415 V Gas Turbine Unit Services Transformer.

Since the plant must be able to operate in the event of a total loss of electrical supplies to the power station, an extensive battery installation is needed. A 120 V DC starter battery provides power for the gas generator starter motors. The standby fuel and lubricating oil pumps are connected to a 110 V DC auxiliaries battery. All other functions needed to run up and synchronise the gas turbine are derived from a 110 V DC controls battery. When 415 V AC supplies are restored upon closure of the gas turbine 11 kV circuit-breaker, the standby pumps shut down.

10.2 Commissioning objectives

In addition to the general objectives outlined at the beginning of the chapter, the following are specific to the gas turbine plant.

- Check that the plant has been erected correctly.
- Ensure that the fuel oil storage tanks are clean and oil-tight.
- Prove that the fuel oil transfer system will meet its design specification.
- Demonstrate that all protective devices function safely and correctly.
- Confirm that the automatic control features operate correctly.
- Conduct a heat run on the generator main connections busbars.
- Demonstrate that the automatic voltage regulator and auto-synchroniser operate correctly over the specified range of voltage and frequency.
- Show that the plant is capable of completing the agreed reliability test programme.
- Establish all necessary operational and efficiency parameters at 25% increments of load.
- Prove that the plant will provide security of supply to main unit auxiliaries, particularly during *system islanding*. (This is where one or a group of power stations becomes disconnected from the main HV transmission system. There will be a sudden change in both frequency and load, particularly if the group was exporting to the system.)
- Demonstrate that the plant is capable of permitting a station start-up without external supplies.
- Confirm, by regular monitoring, that noise levels emanating from the plant are within specification.
- Identify any differences in intent between the original concept and the completed plant. Produce sufficient engineering data to progress necessary plant improvements.

10.3 Safety considerations

Commissioning of gas turbine plant is a relatively major activity. It can generate a number of safety hazards, some of which may be new to personnel on a construction site. The following points are worthy of consideration in addition to those highlighted at the beginning of this chapter:

- Access to the commissioning area to be restricted to authorised personnel only. The area must be defined by barriers. If necessary, NO SMOKING/NO NAKED LIGHTS rules to be enforced.
- Ear defenders to be worn in areas of expected high noise levels until measurements define the type of hearing protection needed.
- Access doors to gas generator air intake, filter chamber, electrical generator cooling and main connection chambers kept locked shut.
- No fuel oil should be accepted until the permanent fire protection systems are in operation.
- Each oil delivery should be tested for flashpoint before off-loading begins. (It is not unknown for methanol or petrol to be delivered in error.)
- Fuel oil tank filling to be conducted slowly to observe tank settlement.
- The synthetic oil used for gas generator lubrication presents particular health hazards. Gloves and goggles must be worn and personnel must be aware of the hazards.
- Temporary piping used for oil flushing must be fabricated from rigid, rather than flexible, pipework. This is particularly important for heated lubricating oil, since pipe failure may cause fire.
- Frequency of operation of the starter motors must be kept under strict control. Typically, three starts in 15 minutes, followed by a 20-minute cooling period, would be acceptable.
- Special precautions are necessary during extended operation of the gas generators with the sound absorbent enclosure open. The fire protection must be available on manual control, so that it can be released in the event of fire. The enclosure ventilation inlet ducts should be temporarily blanked off to prevent the reverse flow of air conducting debris into the engine intake. Clear notices warning that 'blanks are fitted' must be posted.
- The current state of plant controls and limitations must be logged and kept available for reference by the operations and commissioning staff.
- In the event of a discharge of CO_2 or Halon gas, care must be exercised in case oxygen levels have been depleted in low-lying areas of the plant.

10.4 Resources

10.4.1 Services

- Gas turbine 415 V AC switchboards available for use.
- Gas turbine batteries charged and ready for use.
- Waterspray, foam and gas fire protection systems commissioned and in service.
- Temporary pipework, filters and oil heaters for system flushing.
- Nitrogen cylinder for accumulator charging.
- Portable fire fighting equipment, including mobile foam tender.
- Plentiful supply of oil-absorbent materials for mopping up spillages.
- Supply of flushing oils, including empty road tanker for reception of flushing oil at various stages (say 35 000 litres).
- Supply of service fuel and lubricating oils when required.
- Cabling to control and alarm systems in CCR complete and available.
- Range of instrumentation and electrical test equipment, including a low power variable-frequency testing source.
- Fuel oil flashpoint testing equipment (e.g., SETA-FLASH)
- Temporary supply for electrical generator anti-condensation heaters, if permanent supply is not available.
- Gas turbine house ventilation system ready for use.

10.4.2 Staffing

A team of four to six engineers will be required, supplemented by station operational staff as required. The part-time services of specialist staff from the following organisations will be needed:

- Gas generator manufacturer to test and set up the gas generator control systems.
- Transmission district staff for generator protection and auto-synchroniser commissioning.
- Performance section staff to conduct the agreed test programme.
- Scientific services personnel to monitor noise levels around the site.

Commissioning of the fuel oil system may take up to three weeks. To complete commissioning of each gas turbine, including the reliability tests, some 20 to 24 weeks need to be allocated.

10.5 Commissioning

Figure 3.54 depicts a simplified commissioning network for the common services and one 25 MW unit of a typical gas turbine plant. This particular installation accepts gas-oil by road tanker and filters it only before storage and use. Furthermore, the air intake filter is a static 'Vee' bag design, rather than the automatic roll-on type.

For simplicity, commissioning of the plant is best carried out in distinct phases, allowing proving of individual systems prior to co-ordinated testing of the complete installation.

Typical sub-division of the testing activities is as follows:

- Fuel oil reception, storage and transfer system
- Air filters and ducting

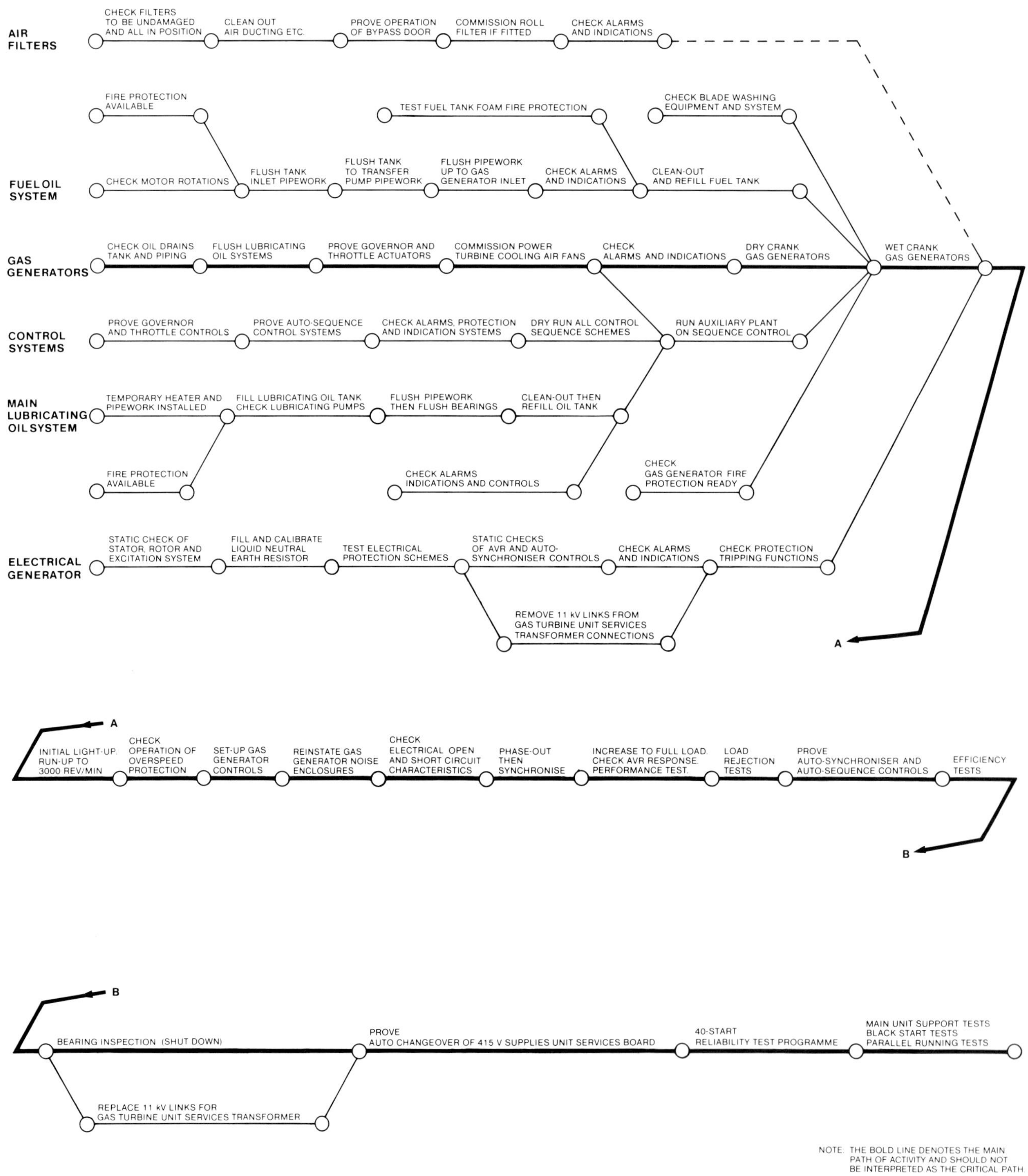

FIG. 3.54 Gas turbine-generator commissioning network

- Gas generators
- Main lubrication system
- Generator, exciter, automatic voltage regulator (AVR) and auto-synchronising unit (ASU)
- Sequence and controls
- Initial running tests, including proving of overspeed protection
- Loading and efficiency tests
- Bearing inspection
- Reliability proving tests
- Type tests, e.g., black start and main unit support

The principles involved with each are considered in turn on the following pages. In each case, it is assumed that all relevant safety considerations have been fully implemented.

10.5.1 Fuel oil system

This summary covers the entire fuel oil system from the tanker off-loading points, right through the liquid fuel packages associated with each gas turbine, to temporary blanks fitted at the take-off points for each gas generator. The pipework system is flushed in defined sections until temporary filters are clean. A tanker of gas-oil is needed to supply flushing oil and to receive the initial flushings. Thereafter, one fuel oil tank is partially filled for flushing the remainder of the systems. Upon completion of the exercise, the stored oil will be checked to see if it is fit for use. If not, it is drained and despatched off-site for cleaning. Before this is done, it will be advantageous to prove operation of the gas-oil tank foam fire protection system (see Section 12 of this chapter). Both fuel oil tanks are then filled with clean oil ready for gas turbine commissioning activities.

Preliminary checks are carried out to ensure that the plant is in a fit state to be charged with fuel oil. The more important checks are:

- Confirm that oil tank retaining bunds are water tight.
- Check that all tanks, steelwork, electrical equipment, etc., are securely bonded to earth.
- Confirm that fire protection equipment is functional.
- Check that all construction blanks have been removed and that all pipework and plant erection is complete.
- Confirm that all instruments, protective and alarm devices have been calibrated and all cabling has been completed.
- Check that all motor drive cables are complete and that electrical protection is set correctly for the motor rating.
- Check that pipework, thermal insulation and trace heating is complete and operational.
- Prove correct operation of all fire trip-valves.
- Temporary elements fitted to all normal service filters for the flushing exercise. Also additional temporary filter element at the tanker filling connection.
- Check that the fuel oil tank hydrostatic test has been completed and the tanks cleaned out afterwards.
- Tank contents calibration survey carried out.
- Calibration certificates for relief valves and flowmeters. Works test certificates for accumulators and pressure regulating valves.
- Flowmeters temporarily replaced with distance pieces fabricated from sections of pipework.

The fuel system is flushed in three stages, as follows:

- Tank filling pipework.
- Fuel oil pipework from tanks to transfer pumps.
- From transfer pumps to gas generator fuel inlets.

During flushing, the integrity of piping, pumps, filters, tanks, instrumentation and alarms is proved. Operation of the floating suction in each fuel oil tank is checked at the appropriate time.

Before any oil is accepted into the system, it must be checked to confirm that the tanker contains only gas-oil. A sample is removed from each separate compartment and visually checked for colour (pink), clarity and lack of suspended matter. The sample is also tested to see whether its smell is similar to gas-oil. Next, the flash point of the sample is tested, e.g., by using SETA-FLASH equipment. This test involves heating approximately 2 ml of the gas-oil to 60°C in a closed vessel, then applying a test flame to the vapour released from the oil. If the vapour ignites, then the oil must be rejected as unsuitable, pending an independent check. Though BS2869 specifies a minimum flash point of 56°C, in practice most gas turbine fuels have a flash point well above the 60°C selected as the go/no-go datum.

Initially, fuel oil is circulated from a road tanker through the filling manifold up to the blanked inlet branches of the fuel oil tanks. Temporary pipework conveys the oil back to the tanker from the inlet valve bodies. This process continues, with the 250 μm filter by-passed, until the temporary filter at the filling manifold is clean. Each fill point is treated similarly, then the 250 μm filter put into service. Flushing continues until the temporary inlet filter stays clean, after which the temporary pipework is disconnected.

One fuel oil tank is then filled with approximately 35 000 litres of clean oil. Temporary pipework is led from the fuel oil transfer pump recirculation branch to the tank-filling manifold via a temporary filter. Temporary arrangements have to be made, so that the floating suction unit can be back-filled from the transfer pump suction main. This is to prime the articulated section of the pipeline, so that it sinks below the surface of the oil to release any pockets of air trapped in the pipe. The system is primed up to the inlet valves of each liquid fuel package. All transfer pumps are run in turn to flush oil around the system, using the fuel oil tank as a reservoir.

Whilst each transfer pump is running, the non-return valves of the other pumps are checked to be tight.

Temporary pipework is connected to the blanks on the branches to the gas generators of each gas turbine in turn and led back to the tank filling manifold via a temporary filter.

Each of the normal-duty transfer pumps is run to circulate oil around each gas turbine liquid fuel package in turn until the return line filter is clear. Automatic operation of the transfer pumps along with alarms and indicators are all proved during this stage of the flushing progress. Operation of the standby fuel-oil pump with reduced supply voltage is also checked.

Upon completion of flushing, all temporary pipework is removed and temporary filters are replaced with the normal service elements. The fuel oil system is reinstated to its normal status by fitting the flowmeters and by charging the accumulators and pressure regulating valves with nitrogen.

It may be prudent at this stage to commission the gas-oil tank foam fire protection. This is carried out generally as given in Section 12 of this chapter, whereby foam is discharged into one tank only, as a type-test. Before doing so, it is wise to consider what effect, if any, the application of foaming compounds will have on the characteristics of the oil remaining in the tank.

Providing the oil in the fuel tank is satisfactory for further use, supplies of gas-oil are ordered. Whilst each tank is being filled, the following items are carried out:

- Monitor tanks and joints for leaks.
- Confirm that level switches operate correctly.
- Compare accuracy of contents gauges by comparison with tanker invoices and by ullaging with a dip tape.
- Check that the floating suction unit is functioning correctly by visual observation through an access hatch. Note that it may be necessary to back-fill the floating suction unit of the second tank when it is being filled.

Finally, the fuel oil system is reprimed up to the LP shut-off cock at each liquid fuel package, to await gas turbine commissioning.

10.5.2 Air filters and ducting

It is vitally important to check that all air intake shafts on the 'clean' side of the filters are absolutely free of any debris, no matter how small. The concrete surfaces should be smoothed and sealed, rather than painted, to prevent dust nuisance. This applies equally to the gas generator and AC-generator cooling air. All instrumentation has to be calibrated and alarm channels proved. Securing nuts and screws for filters and silencer assemblies have to be tight and locked. Access doors must be lockable and kept locked after the final internal inspection.

The load calculated to operate the air filter by-pass doors should be applied by means of a spring balance to confirm that the doors will operate and will initiate the appropriate control and alarm actions, when required.

Cooling air outlet ducting from the generator and the exhaust from the power turbine should also be clear of all debris. The appropriate filters and silencers must be fitted and secured.

Finally, temporary blanks have to be fitted to the gas generator enclosure cooling air inlet ports to prevent a reverse flow of air to the gas generator intake, whilst the enclosure is opened up for commissioning tests. Warning notices to this effect must be posted.

10.5.3 Gas generators

Most of the activities associated with the gas generators are usually carried out by the manufacturer's specialist engineer. The main points for consideration are as follows:

- Prove all drains tank pipework, correct functioning of alarms, and operation of the drains flame trap.
- Fill and flush the gas generator lubrication oil system with the specified synthetic oil. Flushing requires a temporary pump and pipework recirculating to the lubricating oil tank via the system suction strainer.
- Turn the compressor shaft by hand from the intake flare to check freedom of movement.
- Calibrate all alarm, indication and protective devices.
- Check operation of the electrical controls to LP and HP shut-off cocks, throttle actuators, etc.
- Check operation of fire protection equipment manually. Repeat the test by simulation of a signal from the sensors on the gas generators.
- Commission air-blast coolers for power turbine cooling air system.
- Using the starter motor, dry-crank each gas generator with the fuel supply and the igniters isolated. Check engine speed, starter current, lubricating oil pressure and run-down time.
- Charge the fuel oil system and carry out a wet-crank, using the starter motor with the ignition circuit isolated. Check that there are no leaks and confirm recovery of fuel drains into drains tank.

  Note Each wet-crank sequence must be followed by a dry-crank to purge all fuel from the engine.

- Flush compressor wash system and set it up for normal use.

10.5.4 Main lubrication system

The system should be flushed, using a flushing charge of the specified lubricating oil. A temporary heater will be needed to raise the oil temperature to about 75°C to:

- Reduce the oil viscosity.
- Increase the flow rate.
- Provide the differential expansion effects necessary to promote release of debris within the system.

As far as practicable, all restrictions should be removed from the pipework. A temporary gauze filter is placed in the bearing oil return line strainer-basket to monitor the efficacy of the exercise.

When the oil tank has been filled, the standby and emergency lubricating oil pumps are commissioned as are the lubricating oil air-blast coolers. For the first stage of flushing, the power turbine and generator bearings are by-passed, the lubricating oil strainer internals are removed and the main oil pump inlet is blanked off. The next stage entails replacing the strainer internals and bringing the lower half-bearings into the circuit, then flushing until the system is clean. The flushing charge is then removed, the oil tank cleaned out, if necessary, and the normal running charge added. The lubricating oil system is reinstated to normal. Sequence control of the lubricating oil pumps and air-blast coolers is proved, along with checking of any outstanding instrumentation and alarms. Adequacy of oil return flows from each bearing is checked, particularly when the emergency lubricating oil pump is running on a reduced voltage test.

10.5.5 Generator, exciter, AVR and ASU

The generator anti-condensation heaters should be in service as soon as the machine arrives on site. The heaters may have to be connected to the construction site supplies system until normal station electrical supplies are commissioned.

During the erection programme, regular checks should be carried out to monitor the integrity of the stator winding insulation resistance.

Note All stator winding thermocouples must be short-circuited to earth during these tests.

Pre-running commissioning checks will include:

- Rotor winding impedance and resistance check for comparison with works test results.
- Rotor insulation resistance check with 500 V DC motorised tester. Acceptable value greater than 100 MΩ.
- Stator insulation resistance and polarisation index taken with 5 kV DC tester. Acceptable values would be greater than 1000 MΩ and 2, respectively.
- Recurrent surge oscillogram (RSO) of rotor winding and 'tan delta' test of stator winding insulation carried out.
- With rotating rectifier diodes removed, the pilot exciter and main exciter insulation resistances are checked at 500 V DC to be better than 200 MΩ.
- Bearing pedestal insulation checked with 500 V DC tester. A minimum of 1 MΩ would be acceptable. Shaft earthing brush checked to be effectively bonding the shaft to earth.
- Check that all earth bonding is completed and that the generator stator casing earth bond joints have been torque and continuity tested back to the station main earth bar.
- Exciter field diodes checked for polarity and resistance, and compared with works test results.
- High voltage test on 11 kV cables and on the stator winding carried out.
- Neutral earth resistor calibrated.
- All electrical protection tests completed, including all inter-tripping signals into the gas generator control system. (Refer to Section 9 of this chapter for details of electrical protection commissioning.)
- Torque and conductivity test all generator main connection busbar joints, noting that the connecting links to the 11 kV/415 V gas turbine unit services transformer should be left disconnected until later in the commissioning programme.
- Attach temperature sensitive indicator strips to selected busbar joints to measure temperature rise during subsequent heat run.
- Complete static checks for AVR and ASU equipment to the current issue of the circuit diagrams. Refer to the appropriate sections of Chapters 5 and 6 for details of main unit AVR commissioning checks.

10.5.6 Sequence and controls

The preliminary checks comprise confirming that erection is complete, that all earth bonding is satisfactory, and that all cabling is correctly glanded and terminated. Following a check that all equipment is installed to drawing, that fuses are correctly rated and that every item is clearly identified, the various control supplies are energised.

Alarm systems will be tested by simulation or functional testing, where practicable, and repeats to the CCR and to the centralised data processing (DP) system proved. The approved status marker system should be adopted, as outlined in Section 7 of this chapter. Governor controls will be set up and statically checked, using simulated inputs where necessary.

Sequence control checks will be carried out, using the current issue of the schematic diagram. As each path is completed it will be marked with a distinctive colour, e.g., red. Any constraints or disconnections necessary to facilitate testing are noted on a separate log sheet.

Finally, overall sequence control 'dry runs' will be completed, followed by live operation of the auxiliary plant until confidence is established that all checks are satisfactory.

10.5.7 Initial running tests

Prior to conducting the first light-up of the gas generators, it is prudent to confirm the following items:

- Air-intake shafts clear and access doors locked. Chimney-access doors secured.
- Generator main-connections chamber clear of debris and all loose equipment removed or secured in place. Gas turbine unit services transformer 11 kV links disconnected. Access door secured in shut position.
- Gas turbine 11 kV circuit-breaker and VT, and the incomer from the gas turbine unit services transformer, isolated and locked off. Automatic changeover feature for 415 V supplies to be inhibited until much later in the programme.
- Excitation controls at minimum, with the AVR on hand control and the field control switch open.
- Lubrication systems all at working levels.
- All auxiliary plant checked to operate and left with controls correctly selected.
- Governor control at minimum speed position.
- All unauthorised personnel cleared out of the commissioning area.
- Fire protection equipment checked to be in operational state. Additional temporary fire fighting equipment in position, where necessary.

Both gas generators are started on manual from the LCP after completing a dry-crank sequence. The gas generators should run up to idling speed of approximately 3400 r/min. The speed at which the starter motors shut down should be noted. Checks for speed, lubricating oil pressures, exhaust gas temperatures, vibration levels, expansion and freedom from leaks will be carried out.

The gas generator speeds are gradually increased until the power turbine starts to rotate, i.e., breaks away, when oil flow rates and absence of rubbing noises will be checked. Gas generator speeds will then be steadily increased until the power turbine is at normal running speed (3000 r/min). The hand trip at the power turbine pedestal bearing is then operated to prove correct operation of the shutdown features, after which the plant is run back up to normal speed.

Providing that all parameters are satisfactory, all personnel will be cleared out of the plant room so that the power turbine mechanical and electronic overspeed protection can be proved to operate at 3300 $\pm$100 r/min.

Setting-up of the gas generator controls, if necessary, should be completed, including auto/manual drive of the governor. When these tests have all been satisfactorily concluded, the gas generator enclosures are refitted, the enclosure ventilation inlet blanks removed and the fire protection restored to full automatic control.

10.5.8 Loading and efficiency tests

Automatic changeover of 415 V auxiliary supplies when the 11 kV current breaker closes has to be inhibited and remain so until the links to the 11 kV/415 V gas turbine unit services transformer are refitted.

The following programme of tests should be carried out, in accordance with the terms of the contract document. Typically these would be:

Open-circuit test Determine the open-circuit characteristic, first with excitation on manual control and then using the AVR controls.

Short-circuit test Measure the generator short-circuit characteristic with a temporary short-circuiting link fitted at the 11 kV circuit-breaker (ACB).

Phasing-out and synchronising Check the generator for correct phase sequence and manually 'synchronise' on to a dead busbar to energise the busbar (running volts) VT. After a recheck of the voltage transformer phase relationship, the generator will be synchronised to a live busbar.

Loading tests The load will be increased gradually to full-load over approximately one hour, pausing at 25% increments to carry out checks on the AVR response characteristics.

At full-load, the gas generator control settings will be tuned with due correction for ambient pressure and temperature. The full-load rating is usually based on atmospheric conditions of 1 bar absolute at 15°C.

Next, a full-load run at rated output and power factor will be conducted over a six-hour period. This also tests the generator main connections for temperature rise, deduced from subsequent examination of the temperature sensitive indicator strips attached earlier.

Load rejection tests Load will be rejected by manually opening the 11 kV ACB with the unit output at 25%, 50%, 75% and 100% MCR to prove that the transient speed rise is controlled by the governing system, without operating the overspeed protective devices.

AVR tests The AVR will be adjusted and then tested to prove that it is capable of control over the specified ranges of terminal voltage and frequency in both manual and automatic modes. The response to overfluxing conditions (for the unit services transformer), leading power factor operation and overvoltage in event of load rejection will be proved.

ASU tests Five consecutive satisfactory dead busbar synchronising operations will be made using the auto-synchroniser: the auto-synchroniser has to be supplied with a simulated busbar voltage by temporary cabling from a live voltage transformer for these tests. This is followed by a further five satisfactory closes on to live busbars, all at system nominal frequency.

By means of a variable-frequency testing source, the behaviour of the ASU is then checked, using dead busbars, for the various specified frequency conditions.

Control system tests This series of tests is designed to prove satisfactory operation of the automatic sequences. Start-ups are initiated from either LCP or CCR for different system voltage and frequency conditions to test each of the selectable operational modes. Perhaps a total of twelve separate tests is involved in this group alone. Automatic shutdown sequence is also proved.

Efficiency test This performance test is carried out in accordance with BS3135 when the contractor agrees that the plant is acceptable (from his point of view). Additional personnel and specialist instrumentation are required for this. Fitting of the test equipment may require a short outage, which is usually combined with the contractual bearing inspection (see next sub-section).

10.5.9 Bearing inspection

A short outage is usually arranged for completion of a range of remedial work, adjustment, etc., deemed necessary before the reliability proving programme commences. During this outage, some of the more important tasks are:

- Remake the 11 kV connections to the 11 kV/415 V gas turbine unit services transformer.
- Inspect and record results from the temperature sensitive indication strips on the generator main connections.
- Open up and inspect the power turbine and electrical generator bearings.

When the unit is next run up to speed, the relative phasing between the two supplies to the 415 V Gas Turbine Unit Services Board has to be checked. The automatic changeover then has to be proved to operate satisfactorily on start-up and shutdown. In each case, the changeover operation should be so rapid that there is no disturbance, either to running plant or to the control sequence equipment.

10.5.10 Reliability proving tests

The basic statement governing this aspect of the test programme is usually:

The reliability of the plant will be demonstrated by achieving 40 uninterrupted automatic starts, run-ups and shutdowns, with not less than 40 hours and not more than 60 hours of generation, at loads up to the rated output.

It is considered acceptable for some of the 40 starts to have taken place during the earlier testing phases — but this has to be carefully considered. Two plant failures may be acceptable within the 40 starts, dependent upon the action necessary to correct the cause of the failures.

10.5.11 Type tests

These tend to fall into three categories; they are not all conducted on every gas turbine unit.

Parallel operation The reactive load-sharing capabilities of each pair of machines is checked. At least two units need to be synchronised to the same station auxiliary load, but preferably isolated from the main transmission system.

Main unit support The plant is expected to function as a power supply to the main unit and a series of tests needs to be formulated to demonstrate the capability under different emergency conditions.

Black start Sometimes referred to as *dead station start*. The gas turbine plant is required to start-up, be 'synchronised' and must then restart the necessary main unit and station services auxiliary plant, following a total loss of electrical supplies. In order to prove this capability, a series of tests is necessary. The initial test should demonstrate that the gas turbine is capable of being started, run-up to speed and excited to 11 kV using DC auxiliary supplies only. Following synchronising onto a live 11 kV Unit Board, all of the gas turbine 415 V auxiliary plant should start automatically and ensure continued operation of the gas turbine unit. At a later stage the exercise has to be repeated, but this time the gas turbine-generator is used to energise the 11 kV Unit Board. The station and unit supplies are restored in a controlled manner (under test conditions) up to the full rating of the generator. It is of particular importance to monitor the governor and AVR response as large motors are started, e.g., boiler fans, feed pumps and CW pumps.

Details of these three type-tests will be found in the appropriate section of Chapter 6.

11 Diesel engine driven plant

Diesel generators have been supplanted by gas turbine-generators for emergency electrical supplies at modern fossil-fuelled power stations. As a consequence, fixed diesel engine installations are used solely for fire-protection purposes. Nuclear and pumped-storage power plants, however, usually have emergency diesel generators in addition to fire protection diesels.

This section concentrates on activities associated with commissioning fire pump diesel engines. Important points concerned with diesel generator commissioning are highlighted, where appropriate.

11.1 System description

11.1.1 Fire protection diesels

A typical fire protection installation may comprise four 280 kW waterspray 5, with one and sometimes two 120 kW hydrant, diesel engine driven pumps. These are of V6 and V8 cylinder arrangement, run at approximately 1800 r/min with 24 V DC starter motors. Figure 3.55 shows the general arrangement of a typical fire-protection pump unit.

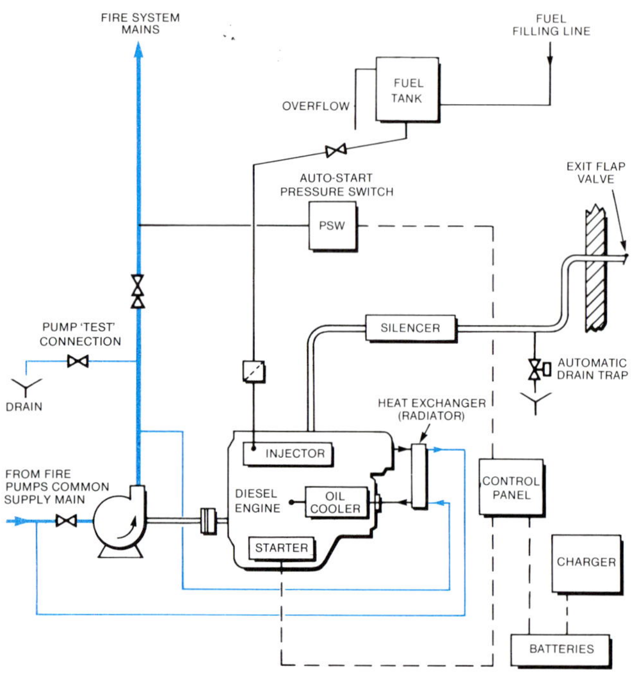

FIG. 3.55 General arrangement of diesel fire pump

Each engine is equipped with a fuel storage tank, usually capable of holding sufficient diesel oil to BS2869 Class A, for 24 hours operation. Fuel flows by gravity to an engine-mounted injector pump. The engine and fuel tank are covered by a waterspray fire protection scheme (see Section 12 of this chapter).

Engine-driven auxiliary pumps circulate lubricating oil and engine cooling water. The cooling water, dosed with antifreeze, flows in a closed circuit, comprising engine jacket, oil cooler and a water/water heat exchanger. Secondary cooling water is derived from the fire pump discharge branch and returned to the suction of an adjacent pump. In some cases, the heat exchanger is air cooled by an engine-driven fan. A 50 mm test connection on the pump discharge pipeline is used to maintain flow of water through the pump whilst it is being run manually for testing purposes.

Control equipment is located in a weatherproof cabinet adjacent to the engine. Duplicate 24 V starter batteries are connected in parallel for control circuit supplies. One battery is dedicated to automatic starting duty, whilst the other is used for local/manual starting of the starter motor. A dual-output battery charger meets the trickle or boost charging requirement of both batteries.

Starting of the engine is generally initiated automatically by pressure switches detecting a fall in the fire protection system water pressure. Diesel fire pumps are designed to continue running until they are stopped by manual isolation of the fuel supply at the injector.

Silenced exhaust gases are led through the outside wall of the building where they can readily disperse into the atmosphere. The exhaust line is often fitted with an exit flap to prevent ingress of rain. An automatic drain is sometimes fitted at a low point between the engine and it exhausts to atmosphere to release accumulated condensation. Thermal insulation is applied for personnel protection, where necessary.

11.1.2 Diesel generators

Emergency diesel generators may be rated at 5000 kW or more and generate at 3.3 kV. Typically, such engines have V12 or V16 cylinder arrangements and run at approximately 600 r/min. A dedicated common 27 bar compressed-air installation would be provided for diesel generator starting purposes.

Emergency generators usually have separate electrically-driven pumps for fuel, lubricating oil and cooling water. The cooling water system is much more extensive and several air cooled heat exchangers may be employed. An electric-motor-driven slow-speed shaft turning gear is often fitted too. Diesel generator operation must be completely independent of the normal power supplies in the station, and the vital auxiliaries are powered by emergency DC motors. Since generators are required to operate at full-load shortly after starting, electric immersion heaters are usually installed to maintain the oil and cooling water temperatures near to their normal operating values while the plant is in a shutdown state.

11.2 Commissioning objectives

In addition to the main objectives outlined at the beginning of this chapter, the following are specific to this system:

- To demonstrate that filling of fuel oil storage tanks can be accomplished easily and safely.
- To ensure that all pipework systems are leak and pressure tested, and then flushed.
- To set, calibrate and functionally test all instrumentation devices, as appropriate.
- To check that all major plant items are effectively bonded to the station main earthing system.
- To test all alarm channels by simulation and then by overall loop tests.

- To check all engine controls, interlocks and monitoring systems, and functionally test them by simulation.
- To carry out the diesel engine initial runs and prove all associated systems and protective devices.
- To check vibration levels at various loads.

11.3 Safety considerations

The main dangers associated with diesel engine driven plant are:

- Fuel oil spillage
- Risk of fire
- Excessive noise
- Crankcase explosions
- Asphyxiation by leaking exhaust fumes

These must be considered in conjunction with the general safety aspects outlined at the beginning of this chapter.

When it is suspected that any internal damage or overheating has occurred, the engine must be allowed to cool before any crankcase doors are removed. This is to avoid admitting air, which could mix with hot oil vapour, risking fire or an explosion in the presence of an overheated engine component.

11.4 Resources

11.4.1 Services

- Supplies of diesel fuel to BS2869 Class 'A' (typically 5000 litres)
- Lubricating oil
- Fresh water and antifreeze for cooling systems
- Waterspray fire protection
- Portable fire extinguishers
- Oil-absorbent material

11.4.2 Staffing

Depending upon the type of installation, a team of three to six people will need from two days to two weeks to commission each diesel engine and its associated equipment.

11.5 Commissioning

The steps listed below are based on commissioning a diesel-engine-driven waterspray fire protection pump. These interface with the appropriate paragraphs in Section 12 of this chapter. The simplified commissioning network shown in Fig 3.56 links the main activities.

Before running tests commence, the following basic requirements must be met:

- All relevant safety considerations implemented.
- All plant items checked or calibrated to the appropriate Standard Check List (SCL).
- Pipework systems proved leak-free and flushed clean.
- Services available whenever they are needed.

The engine must be correctly lined-out in relation to its driven member and all holding-down bolts grouted in. The fuel supply line to the tank has to be flushed into the tank, then the fuel tank cleaned out. If practicable, it may be advantageous to fit a temporary gauze filter at the inlet to the tank for the flushing exercise. Providing

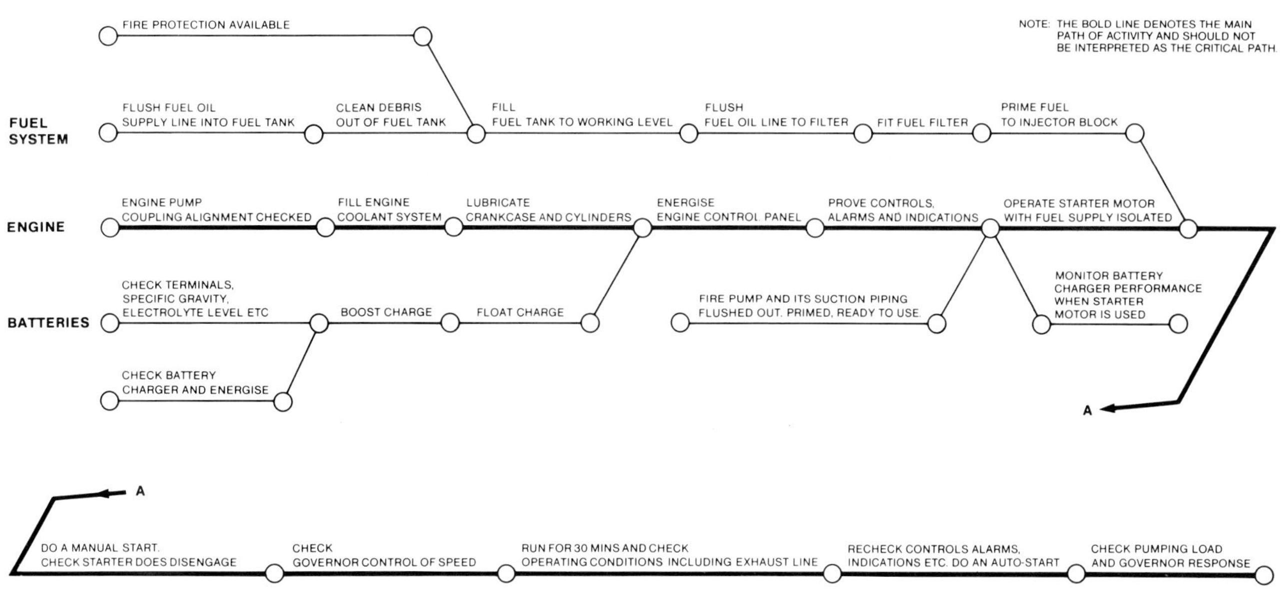

Fig. 3.56 Diesel-driven fire pump commissioning network

the fire protection is in service, fuel is transferred into the tank slowly, so that the contents gauge reading can be verified. Once the tank is full, the overflow line is checked to be clear by slowly raising the oil level, taking care to collect the spillage in a suitable receptacle.

The electrolyte level of each starter battery is checked and topped up with distilled water. The specific gravity of every cell is noted to commence regular plant history records. The battery charger is energised and then used to boost charge both batteries to the fully charged condition, before being left on trickle charge duty. Typically, the specific gravity at 15°C would range from 1270 for a new fully-charged cell to 1190, if the cell was in a discharged state.

Charge the engine cooling system with clean fresh water, adding antifreeze as appropriate, then check for leaks. Ensure that the crankcase is filled to the indicated level with the correct lubricant. Add a small quantity of oil into each cylinder to assist with lubrication during initial start-up. Prime from fuel tank to fuel filter, draining any debris into a suitable receptacle, then fit the filter element. Providing that the fire pumps common water supply main has been flushed and is ready for use, the fire pump and the diesel engine secondary cooling circuit are primed and vented. The 50 mm test connection from the pump discharge branch to the station drains is also primed.

Energise the engine control panel and prove all controls and alarms. After rotating the engine by hand to ensure that it is free to turn, the starter motor is operated with the fuel system isolated.

The fuel system is primed through to the injectors and all final prestart checks carried out, including a check of the exhaust line drain and flap valves, where fitted. Initiate a manual start, confirm that the starter motor disengages as the engine runs up to speed, then check that the governor holds the speed constant at its nominal value. Ensure that the TEST connection is open to maintain a minimum flow of water through the pump.

The engine is allowed to run for at least 30 minutes while its general condition is monitored. Engine cooling water outlet temperature should be of the order of 75°C and the lubricating oil pressure about 3.2 bar. Check round for hot spots and evidence of excessive vibration. During this run, it is advisable to check the exhaust line for evidence of gas leakage or inadequacy of the supports to cater for thermal expansion. Check also that exhausted gases do lift and disperse into the surrounding atmosphere quickly. Confirm that the silencer is effective, particularly if the plant is relatively close to domestic premises.

Finally, reprove all alarms and remote controls (if fitted) and simulate automatic operation of each pump comprising the installation. For example, progressive decay of waterspray system pressure would initiate successive automatic starting of each of four pumps at approximately 9.5, 9.1, 8.7 and 8.3 bar, respectively, for a nominal system pressure of 10 bar. Confirm that each battery charger is operating correctly to replace the energy used to start pumps during the testing programme.

On-load tests will have to be carried out in conjunction with the fire protection commissioning test team when they are in a position to create a large enough water demand. This will allow assessment of governor response to changes in load.

Commissioning of a diesel generator follows a similar pattern, but the following items need additional consideration:

- Record crankshaft deflections during lining out.
- Check the rotation of all the auxiliary motors and couple them to their fans, pumps, etc. (see also Section 4 of this chapter).
- Each water and oil system will need to be flushed, using temporary strainers to retain construction debris.
- Commission compressed-air plant generally, as shown in Section 21 of this chapter.
- Operate engine at slow speed and gradually work up to normal service speed over a period of several hours. Crankcase inspections will need to be carried out at prescribed intervals during the programme.
- Demonstrate governor response, then prove that the overspeed trip operates at 15% over nominal speed.

Once the diesel engine is operating satisfactorily with all alarms, controls, etc., functional, the generator has to be commissioned. This is tackled generally in accordance with the procedure for gas-turbine-driven generators given in Section 10 of this chapter. Finally, prove that the plant can be successfully run up to speed and put on load, without normal station supplies to auxiliaries.

12 Fire protection systems

Modern power station design philosophy is based on prevention of fire by segregation of fire risk areas, plus installation of comprehensive fire detection and protection systems.

This section considers the following:

- High pressure waterspray and sprinkler installations
- Hydrant systems
- Foam generation equipment
- Carbon dioxide and Halon gas discharge schemes
- Detection and alarm systems

Generally, the risk of fire is far greater during plant commissioning than in normal station operation, because of the almost inevitable accumulations of combustible construction debris at various locations.

Fire detection and protection schemes for main boiler gas airheaters and pulverised coal milling plant are covered in the appropriate sections of Chapter 4.

12.1 System description

12.1.1 High pressure waterspray and sprinkler installations

A typical waterspray system may comprise four diesel-engine-driven pumps discharging into a galvanised steel trunk main from which branch mains distribute high pressure water to the various risk areas, as shown by Fig 3.57. Each pump is provided with a discharge test connection so that a minimum flow of water can be maintained when the pump is being test run, without having to deluge a risk area. The pumps are designed for automatic start-up in succession at 9.5, 9.1, 8.7 and 8.3 bar, as the trunk main pressure falls from its nominal value of 10 bar.

A 45 m^3 pressure tank, connected to the trunk main, contains approximately 15 000 litres of water pressurised with air at 10 bar. This water reserve supplies the initial demand until the fire pumps run up to speed. The pressure tank also serves to provide a supply of water for diesel engine fire protection in the extremely unlikely event of a diesel engine fire disabling the fire pumps. A small charging pump is fitted to fill the pressure tank to the prescribed level and then to maintain that level should there be any leakage from the waterspray system. A compressor maintains the air pressure at about 10 bar. The compressor and the charging pump are interlocked to ensure that the compressor will only run if the water level is correct. Losses of pressure due to system leakage have, in the first instance, to be restored by returning the water level to normal. A 'low air pressure' alarm is raised if the tank pressure falls to 9.6 bar.

A variety of deluge and section control valves, spray projectors, sprinklers and detectors are installed to suit each particular type of risk. The various types of waterspray systems are either 'automatic or manual'. *Automatic* systems are of either wet or dry section types to suit the location of the risk. Generally speaking, wet systems are employed within buildings, whilst the dry systems are used outdoors to obviate the risk of frost damage to the protection equipment. *Manual* systems are employed where the associated risks require visual assessment before the application of a fire extinguishant. The most common application of a manual system is the main turbine-generator fire protection system, where personnel operating the valves controlling water supplies are shielded by fire-resistant barriers.

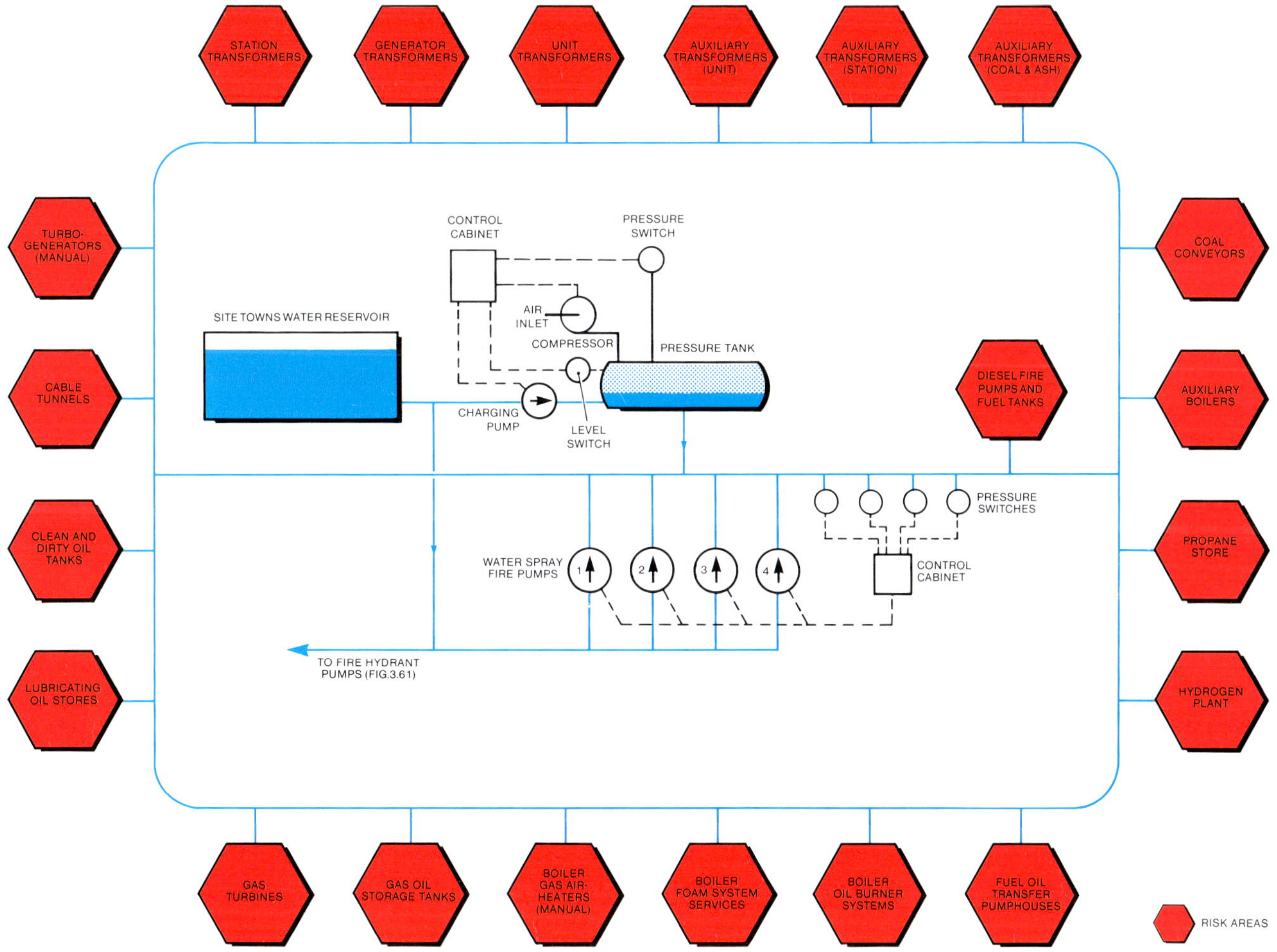

FIG. 3.57 Diagrammatic arrangement of waterspray system

Wet section In this system, the distribution pipework is permanently charged with water at normal system pressure. Water is contained within the piping by quartzoid bulbs fitted either to the projector nozzles (sprinklers) or to detector control valves, as shown later by Fig 3.63. The flow of water which results from failure of the bulb causes the section deluge valve to lift off its seat and supply a continuous flow of water to the affected projectors. The section of pipework downstream of the detector control valve is normally dry and is fitted with open projectors covering the risk area. Figure 3.58 depicts two common forms of waterspray projector. When the deluge valve unseats, it allows water pressure into a recess to operate a local audible alarm and a remote 'fire' or 'water applied' alarm.

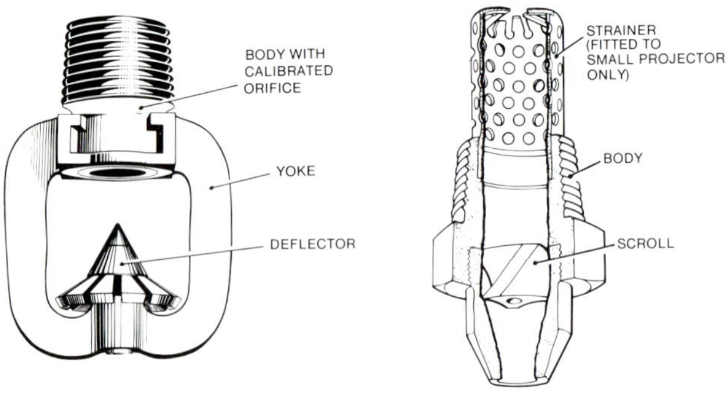

FIG. 3.58 Typical waterspray protectors

Dry section The distribution pipework, which has open projectors spaced around the risk area, is normally empty and at atmospheric pressure. A separate detection system, using compressed air trapped in a system of pipework by quartzoid bulbs, acts on the diaphragm of the associated automatic deluge valve, as shown by Fig 3.59. Loss of air pressure to the diaphragm releases a mechanical latch to allow the deluge valve to unseat and pass high pressure water to all of the open projectors. Operation of the deluge valve initiates alarms in a similar manner to the wet section.

For convenience, some systems are arranged into subsidiary sections, each being controlled by its own air detection system, which operates a simple pneumatically-actuated butterfly valve. Only one automatic deluge valve is fitted to feed all of the associated sections, as shown by Fig 3.60. Loss of air pressure within a section opens that subsidiary deluge valve (and initiates a zonal alarm) and also opens the automatic deluge valve to apply water to the affected section only.

Fire hydrant outlets are provided from the waterspray system for water supplies to:

- Oil-fired boiler foam-generation systems
- Pulverised fuel mill fire fighting (Chapter 4)

Fire protection systems almost always use fresh water, which is drawn from either the CW forebay or the site towns water reservoir (see Sections 17 and 18 of this chapter).

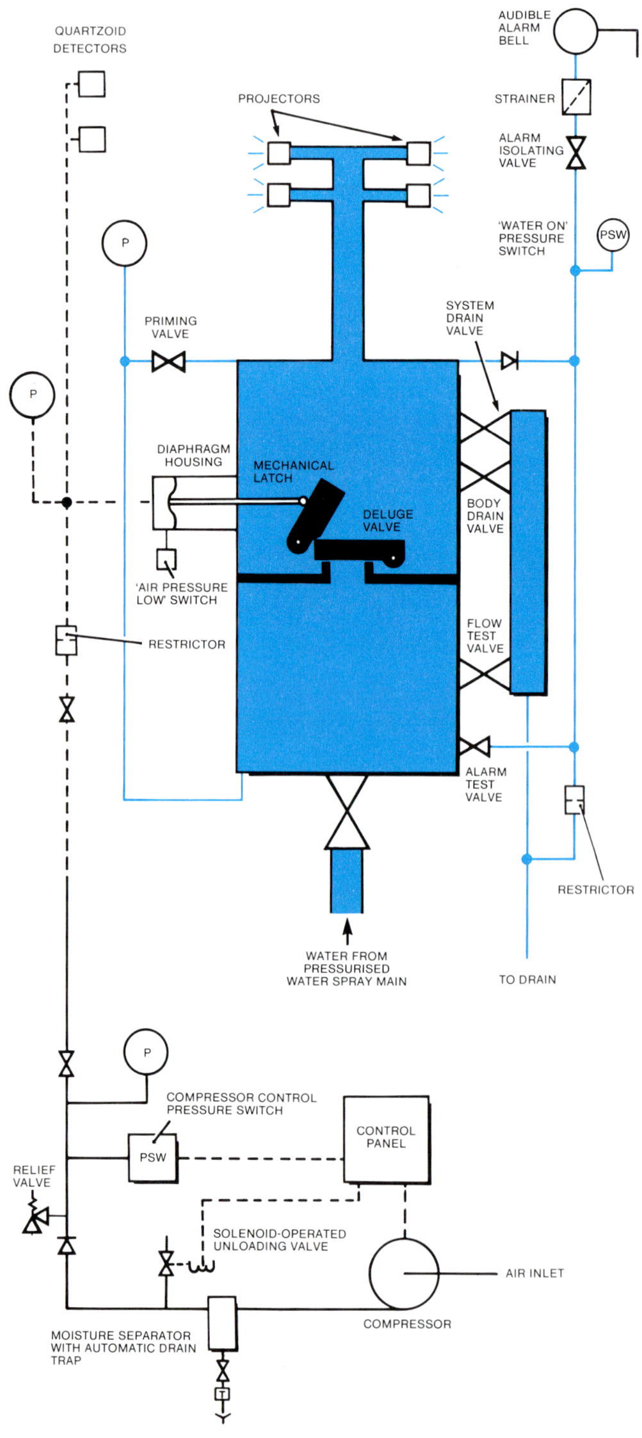

FIG. 3.59 Typical automatic deluge valve scheme

Thermal insulation, supplemented where necessary by electrical trace heating, is applied to valves and sections of pipework that may be liable to damage from freezing.

12.1.2 Hydrant systems

For areas not protected by the waterspray system, and also as a second line of protection in the turbine and boiler house, the site is provided with a system of internal and external hydrant ring mains.

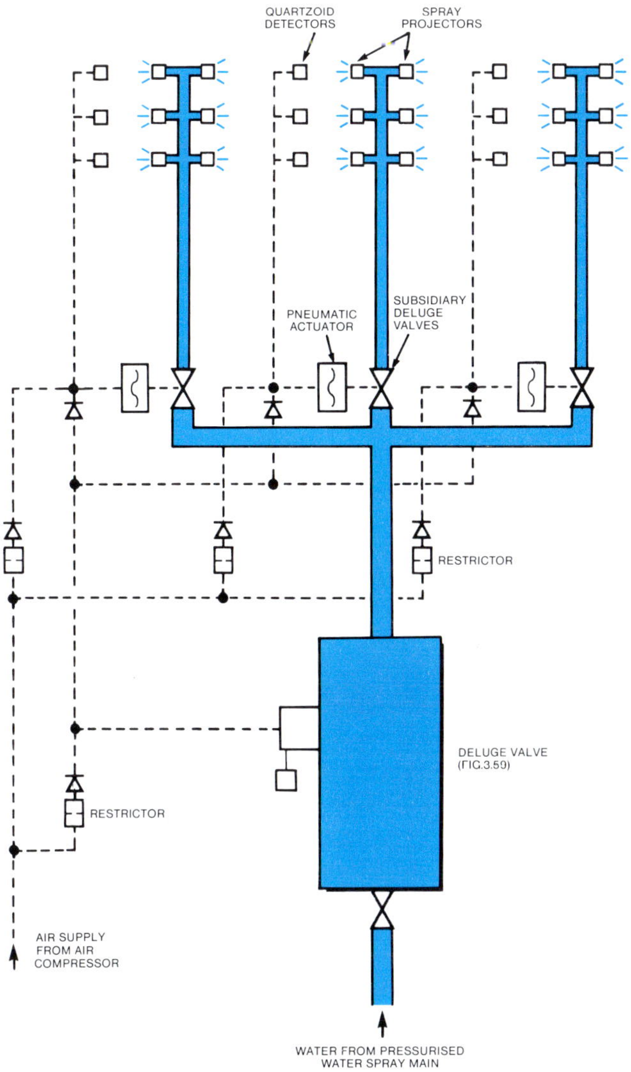

FIG. 3.60 Typical automatic subsidiary deluge valve scheme

Figure 3.61 shows a typical arrangement of fire hydrant pumping facilities. The external hydrant system is usually supplied with water from a separate electrically-driven fire pump, which can be started from the central control room (CCR) when required. Two fire pumps are provided for the internal system; one is usually electrically-driven and the other diesel-engine-driven. The internal hydrant system at some sites is kept pressurised to 8 bar by a 20% capacity jockey pump to compensate for system leakage. The jockey pump is automatically started and stopped by pressure switches, which respond to the system pressure. A progressive fall in internal hydrant main water pressure causes the associated fire hydrant pumps to start automatically in sequence. The diesel pump has to be shut down manually by cutting off the fuel supply at the injectors.

Dry risers (Fig 3.62), fitted with pumping-in connections at ground level, serve the turbine house roof and higher levels of the boiler house. The external and internal hydrant systems are also fitted with pumping-in facilities which, like those on the dry risers, are compatible with the local fire brigade equipment.

Supplies of fresh water are derived from the same source as for the waterspray pumps. Electric trace heating and thermal insulation is applied where there is a risk of frost damage.

12.1.3 Foam generation equipment

Foam is usually applied to oil fires by means of wheeled portable units. Each comprises a storage tank for low expansion foam-making concentrate, plus hoses, branches and an inductor for mixing the concentrate with fire hydrant water.

Gas-oil storage tanks are usually fitted with external fixed foam-transfer branch pipes connected to pouring nozzles around the inner surface of the tank top. One end of the fixed pipework terminates with instantaneous hose couplings near ground level for attachment of a mobile foam mixing and pumping trailer unit. At the inlet to the tank foam-pouring nozzles, the branches are fitted with externally mounted vapour seal boxes, containing rupturing diaphragms to act as a gas-oil vapour seal. This prevents volatiles escaping from the tank through the fire protection pipeline. When foam is applied, the diaphragms fracture and have to be replaced later.

Oil-fired boilers are provided with several fixed tanks of foam concentrate, each complete with an inductor, located close to the oil burner assemblies. The end connections of the inductor accept standard fire hoses used in conjunction with either the waterspray system hydrant outlets or the fire hydrant system to apply either low-expansion foam or water to the fire. The foam system augments the fixed waterspray installations provided at each burner.

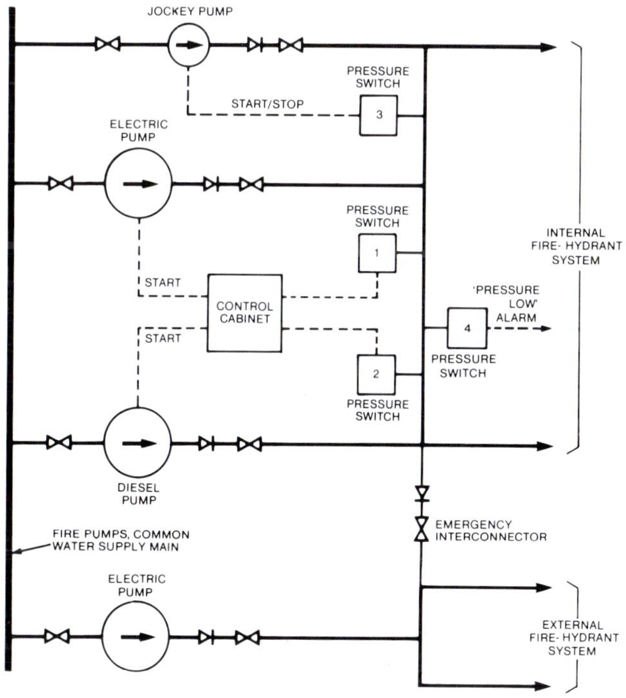

FIG. 3.61 Arrangement of fire hydrant pumps

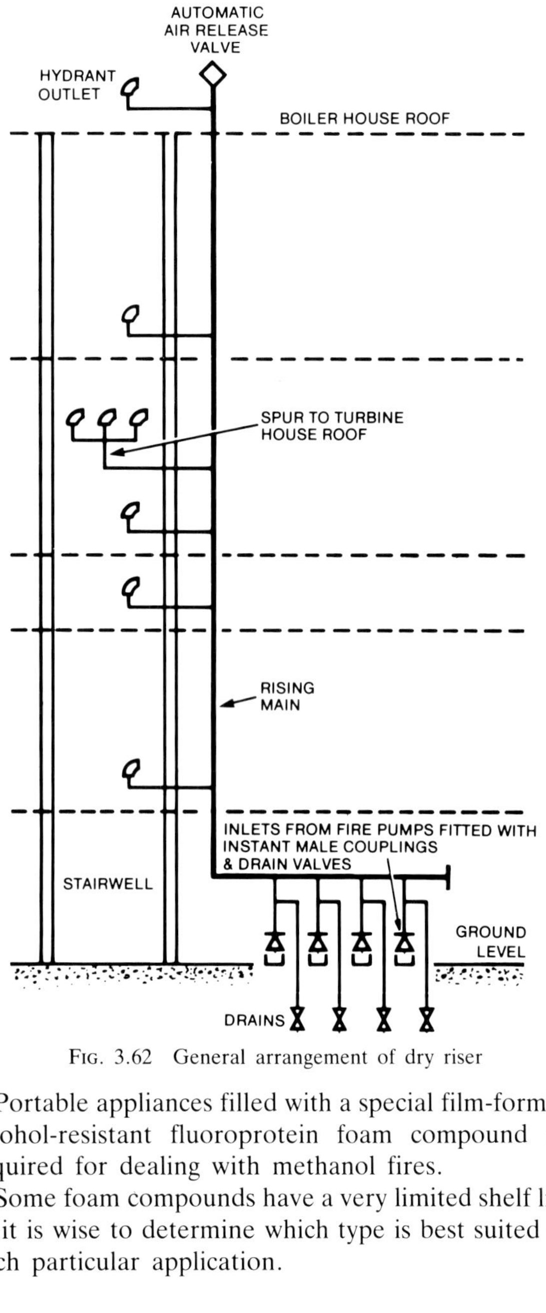

FIG. 3.62 General arrangement of dry riser

Portable appliances filled with a special film-forming alcohol-resistant fluoroprotein foam compound are required for dealing with methanol fires.

Some foam compounds have a very limited shelf life, so it is wise to determine which type is best suited for each particular application.

12.1.4 Carbon dioxide (CO_2) and Halon gas schemes

Halon 1211, bromochlorodifluoromethane (BCF), and Halon 1301, bromotrifluoromethane (BTM), are almost universally employed for total flood systems. They have numerous advantages over CO_2 as a fire extinguishing gas in power stations. Particular applications are where personnel are normally present or where there is delicate electrical or computing equipment. Another application is for extinguishing fires in pulverised-fuel mills; the affected mill has to be taken out of service and the dampers closed to achieve satisfactory concentrations of extinguishing gas. Halon installations are usually designed to discharge within four seconds and maintain a 5% concentration for up to ten minutes.

Smoke and/or heat detection systems are used to raise a fire alarm and/or automatically release the gas. With Halon 1301 (BTM), there is a 10-second delay after sounding of the alarm, to allow personnel to vacate the area before automatic discharge. Before CO_2 or Halon 1211 (BCF) are released, all personnel must be out of the area. Ventilating equipment has to be isolated, either by tripping the fans or by closing dampers, before the release of fire extinguishing gas into an area.

One design of Halon 1301 system uses small modular units mounted inside small enclosed areas, e.g., equipment cubicles. Larger Halon and CO_2 installations incorporate banks of cylinders connected to distribution pipework which leads to nozzles located in the risk area. Sometimes a second bank of manually-released cylinders is provided as a standby.

12.1.5 Detection systems

Detection systems are employed primarily to warn personnel where there is a fire, or conditions conducive to one, on the plant. In most cases, the detection system also initiates operation of an automatic fire protection system. On modern power stations, the most common system is the compressed-air detector system, used in conjunction with high pressure waterspray fire protection schemes (see Fig 3.59). This nominally 2.75 bar detection system is often arranged in up to 20 separate groups, each with a 1.1 kW motor-driven air compressor covering several risk areas. Reduction in air pressure normally initiates a 'low air' alarm at 2.0 bar and warns of 'fire' at about 1.4 bar; this is concurrent with operation of automatic deluge valves. Almost all compressed-air detector systems rely on a quartzoid bulb to release air pressure in the event of fire. Some use a rate-of-rise detector which operates a solenoid valve to release air from the deluge valve control system. Both forms of detector are detailed below, together with a range of other detectors that normally initiate alarms only.

The most common type of detector is the *quartzoid bulb* (QB). Five different temperature ratings covering the 60°C to 140°C range are used and it is not unusual to find up to 2500 bulbs at a site. When subjected to a specific temperature, the bulb shatters to release either compressed air (detector) or high pressure water (sprinkler). Figure 3.63 show two common locations for QB units.

Another type is the *rate of temperature rise detector* (ROR) (Fig 3.64) of which there could be over 300 on a site. This is a mercury switch which operates an alarm and/or initiates release of compressed air, Halon gas or high pressure water dependent upon the system. Two interconnected inert-gas-filled bulbs, one shielded from direct heat, contain a flow restricting orifice and a column of mercury. In the event of a fire, the heat creates a differential pressure between the two bulbs, which transfers the mercury column to change the state of the mercury switch.

Heat-sensitive cables are suspended along cable trays, as shown by Fig 3.65, to detect an increase in ambient

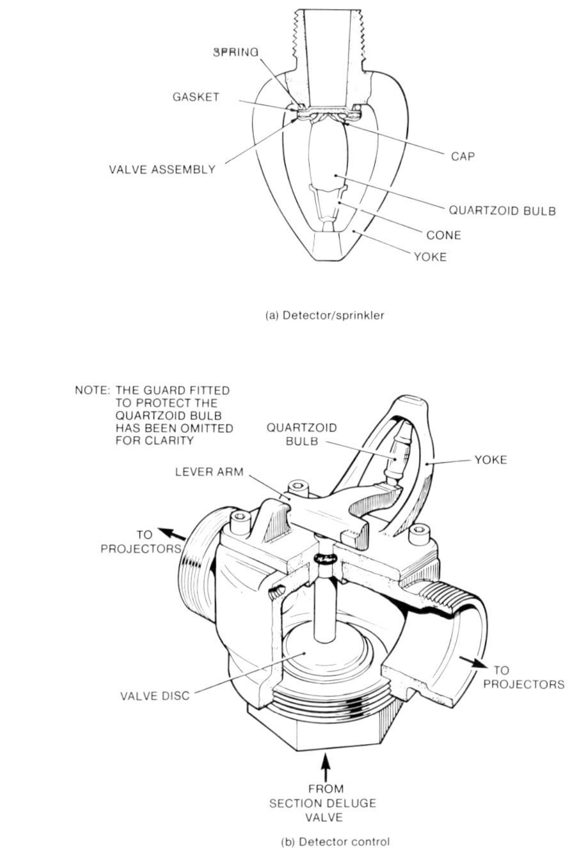

FIG. 3.63 Quartzoid bulb detectors

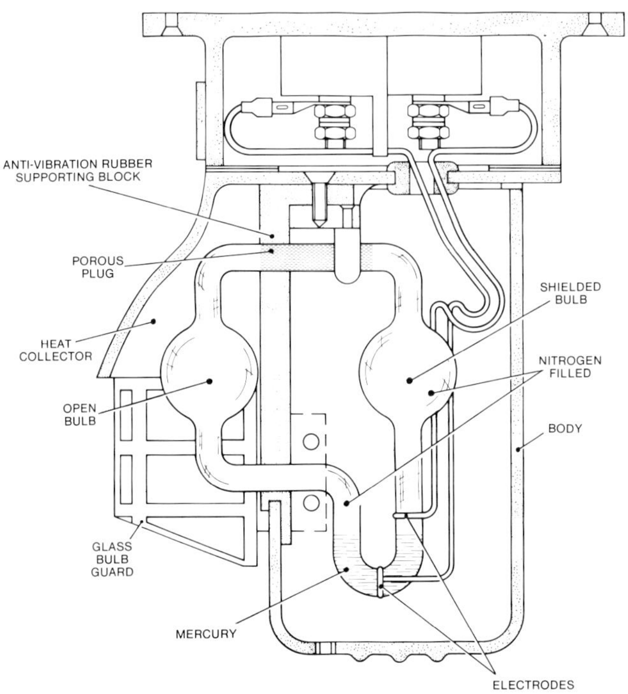

FIG. 3.64 Rate-of-rise detector

temperature. The detector cable consists of conducting cores insulated from each other by a heat-sensitive negative temperature coefficient dielectric whose insulation resistance reduces rapidly with increasing temperature. The outer sheath is usually made from a distinctively-coloured high temperature PVC compound to safely withstand localised fire. One form of

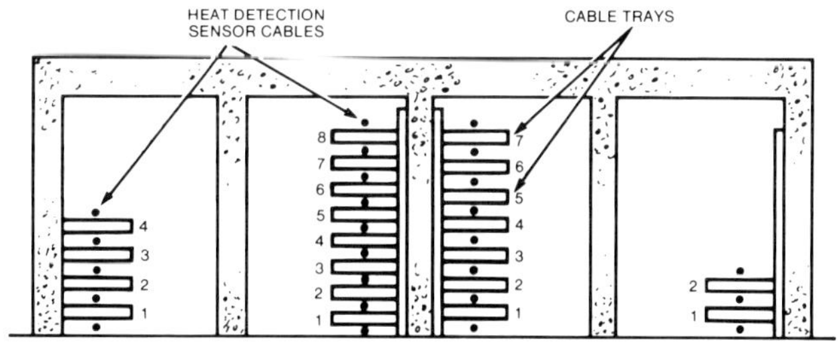

FIG. 3.65 Location of cable tray heat detectors

cable exhibits a response to temperature according to the percentage of its total length exposed to heat. Typically the temperature rise, above a specified minimum ambient temperature of 38°C, to initiate an alarm condition ranges from 21°C over the entire length of the cable to 76°C over only a one metre length. Thus, it is not normally practicable to extend or shorten standard lengths of cable, since each is matched to its monitoring unit. As a general rule, cable heat detection schemes are used to initiate alarms only. These are displayed at the monitoring unit, with a general alarm being repeated to the CCR.

Smoke and heat detectors, located on ceilings of risk areas, often have neon indicators to identify which device has operated. Smoke detectors may be of either the ionisation type (Fig 3.66) or the photo-electric cell type. Occasionally laser or light beam designs are adopted where there is a relatively long unobstructed clear space. Heat detectors are usually bimetallic strips,

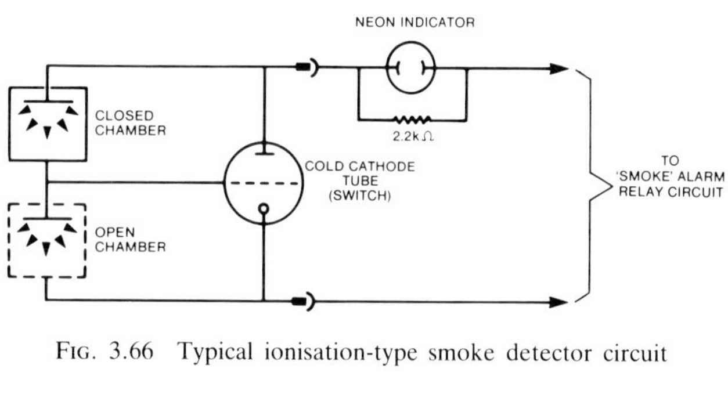

FIG. 3.66 Typical ionisation-type smoke detector circuit

though thermistor and electrical fusible-link types are in general use. Generally, such detectors initiate alarms only, though they may initiate a discharge of Halon gas in computer suites. Mechanical fusible-links are often employed to operate fire dampers in ventilation ducting (see Section 1 of this chapter).

Heat-sensitive *fire detection canopies*, mounted over selected coal conveyors, detect a rate-of-rise in temperature; they then stop the conveyor, so that any fire will activate the sprinkler fire protection system. Optical methods of detecting conveyor fires are prone to spurious operation caused by clouds of coal dust created when dry coal is being handled. In one design of heat canopy, a group of compensated electrical detectors, incorporating exposed and shielded elements, is used. The shielded element is covered by a silicone compound, which has a larger thermal mass than the exposed element. In the event of fire, the hot gases trapped under the canopy, together with the radiated heat, warm each exposed element to create a small voltage difference between the shielded and exposed element. Since the group of elements is connected in series, the resultant voltage is used to actuate the detection system.

12.2 Commissioning objectives

The following are specific for the fire protection system, but should be read in conjunction with the general objectives outlined at the beginning of this chapter:

- Ensure that each section of fire protection equipment is available to protect the main plant, before plant commissioning is due to start.
- Demonstrate that each system is capable of initiating alarms and, where appropriate, discharging fire extinguishant into the correct risk area.
- Check that the design criteria for water flow, pressure and coverage are met.
- Establish a system to ensure that the correct temperature rating of QB detector is fitted to the appropriate risk area.
- Confirm that the fire hydrant pumping-in facilities are compatible with local fire brigade equipment.
- Prove that the CO_2 and Halon systems achieve and maintain the specified concentrations necessary to extinguish fire.

12.3 Safety considerations

The main aspects of safety to be considered are as follows. They have to be considered in conjunction with the general safety aspects outlined at the beginning of this chapter:

- Possibility of flooding during flushing or testing of fire hydrants or waterspray systems, if drainage channels are not clear or are inadequate for the flow of water (i.e., up to 0.6 m^3/s from the largest risk area).
- Risk of asphyxiation arising from CO_2 or Halon 1211 discharge into confined spaces. Checks of oxygen levels to be carried out in adjacent lower-lying areas, in case there has been a seepage of the gas.
- Following discharge of Halon 1301, the atmosphere must be checked to ensure that a 5% concentration has not depleted the atmospheric oxygen to less than 20%.
- Vulnerable sections of plant must be adequately protected by temporary sheeting before deluge testing is carried out.
- Ionisation-type smoke detectors contain a low-energy radioactive source and large numbers should not be stored together before installation.

12.4 Resources

12.4.1 Services

Apart from electrical supplies and copious quantities of water, other services needed are:

- Supply of diesel fuel to BS2869 Class A for the diesel pumps.
- Mobile foam trailer for commissioning of gas-oil tank foam system.
- Fixed fire fighting waterspray to be available before diesel fire pump commissioning.
- Portable fire pump, hoses and branches for proving dry risers and fire hydrants.
- Replacement supplies of Halon and CO_2 gases.
- Approximately 2000 m^2 of temporary sheeting, plus hoses, for deluge testing of boiler and turbine risk areas.
- Temporary supply of compressed air for purging air detection pipework.
- Temporary flowmeter for insertion into fire pump discharge main to measure the rate of flow to each risk area.

12.4.2 Staffing

The commissioning period is likely to extend over a significant part of the main plant erection and commissioning programme. It is likely that several teams, each of three or four people, will be involved concurrently in different areas.

Application of temporary sheeting for either a turbine or a boiler oil-burner front deluge test may take over a week for a half-hour test.

12.5 Commissioning

Commissioning of the system should be carried out in a logical sequence, to ensure that each section is available before main plant commissioning commences. For commissioning purposes, the installation can be subdivided into some 100 distinctive stages. Due to the repetitive nature of many of the stages, relatively few test procedures need to be prepared to cover the different types of stage. These procedures should make full use of Standard Check Lists (SCLs) to detail all functional testing of controls, indicators, alarms, etc., including a live operational test of the system.

Typical coverage of standard test procedures may be:

- Trunk main sections

- Pressure tank, plus its auxiliaries
- Fire pumps (see Section 11 of this chapter)
- Automatic deluge valve installation
- Subsidiary deluge valve installation
- Wet section detection control valve installation
- Manual spray system
- Sprinkler system
- Air compressor and associated compressed-air detector system
- Heat or smoke detection scheme
- ROR detector scheme
- Foam generation system
- Hydrant/dry riser system

12.5.1 High pressure waterspray and sprinkler installations

As a general rule, all waterspray pipework is subjected to a hydraulic test at 15 bar, with no detectable leaks over a minimum period of 2 hours. All water pipework is flushed to waste through system drains and then through the projector nozzles. Generally, the smaller nozzles with integral strainers are removed first to prevent impact damage from any construction debris which may have been left in the pipework.

Compressed-air detector pipework is pneumatically tested at 3 bar also for 2 hours with no detectable leakage. Using a temporary compressor to create a high velocity air flow, detector pipework is purged to atmosphere to clear construction debris before fitting of the quartzoid bulbs.

Deluge testing philosophy should be based on conducting a full functional test of every risk area. Special arrangements may have to be made for sensitive areas. For example, nozzles for oil burner equipments and steam turbine fire protection should first be proved clear with temporary hoses covering each projector. The hoses check that the projectors are clear by allowing water to be safely conducted to waste without causing any plant damage. A full deluge test without the hoses is carried out once the plant has been protected against water damage by temporary sheeting.

In the case of main turbine fire protection, with perhaps eight separate zones, each is deluged independently to facilitate easier observation of the coverage. Later, it is advantageous to initiate all of the zones simultaneously, to confirm that the total flow is within the rated capacity of the fire pumps. The pressure available at the most remote projector of each zone should be recorded for future reference.

The commissioning procedure should follow these general guidelines as outlined in Fig 3.67:

- Flush suction pipework to fire pumps, then charge fire pumps and first section of discharge trunk main.
- Commission pressure tank and its auxiliaries. This involves commissioning of the charging pump and the air compressor, and then proving the interlocks between the two auxiliaries. The pressure tank is then filled to the correct level and correct automatic operation of the charging pump proved. Next, the compressor is used to charge the pressure tank to 10 bar, in increments of 3 bar, to enable quick leak checks to be conducted. Once the pressure has been established, correct operation of the compressor is checked by relieving the air pressure. All interlocks, controls and alarms are thus proved; then the system is left on test for 24 hours, whilst automatic operation of the auxiliaries is monitored.
- Prove operation of the fire protection for the diesel engines and their fuel tanks.
- Commission at least two of the waterspray fire pumps. Refer to Section 11 of this chapter for commissioning of the diesel engine and its auxiliaries. Automatic operation of the available pumps in response to falling waterspray system pressures has to be proved. This will also cover proving of all alarms including the 'fire pump failed to start' alarm sequence.
- Charge and flush available sections of the trunk main, followed by fitting of the temporary flowmeter.
- Charge, flush and deluge each risk area. This includes setting up each type of deluge valve system and subsystem and proving of all alarms. Compressed-air detection compressors and pipework systems have to be set up and auto-controls proved. A second deluge may be necessary, if some of the projectors are found to be misaligned or become blocked during the first deluge.

When a sufficient number of risk areas have been put into service, the capability of the fire pumps should be proved. This is achieved by selecting a combination of risk areas to induce the greatest anticipated flow rate through the most onerous lengths of pipeline. As with deluge testing, nozzle output and coverage with different combinations of pumps should be closely monitored. It is important that this test be repeated with the pressure tank isolated from the system to assess the response of the fire pumps and their ability to deliver the required flow quickly. As commissioning of risk areas progresses, operation of the pressure tank charging pump and compressor should be monitored. Normally the charging pump should only run at infrequent intervals to make up for system water losses. The compressor is required to make up air losses from the pressure tank, and thus should need to be run only occasionally.

12.5.2 Hydrant systems

The fire pump suction pipework is flushed to waste, then the fire pumps charged. Fire protection for the

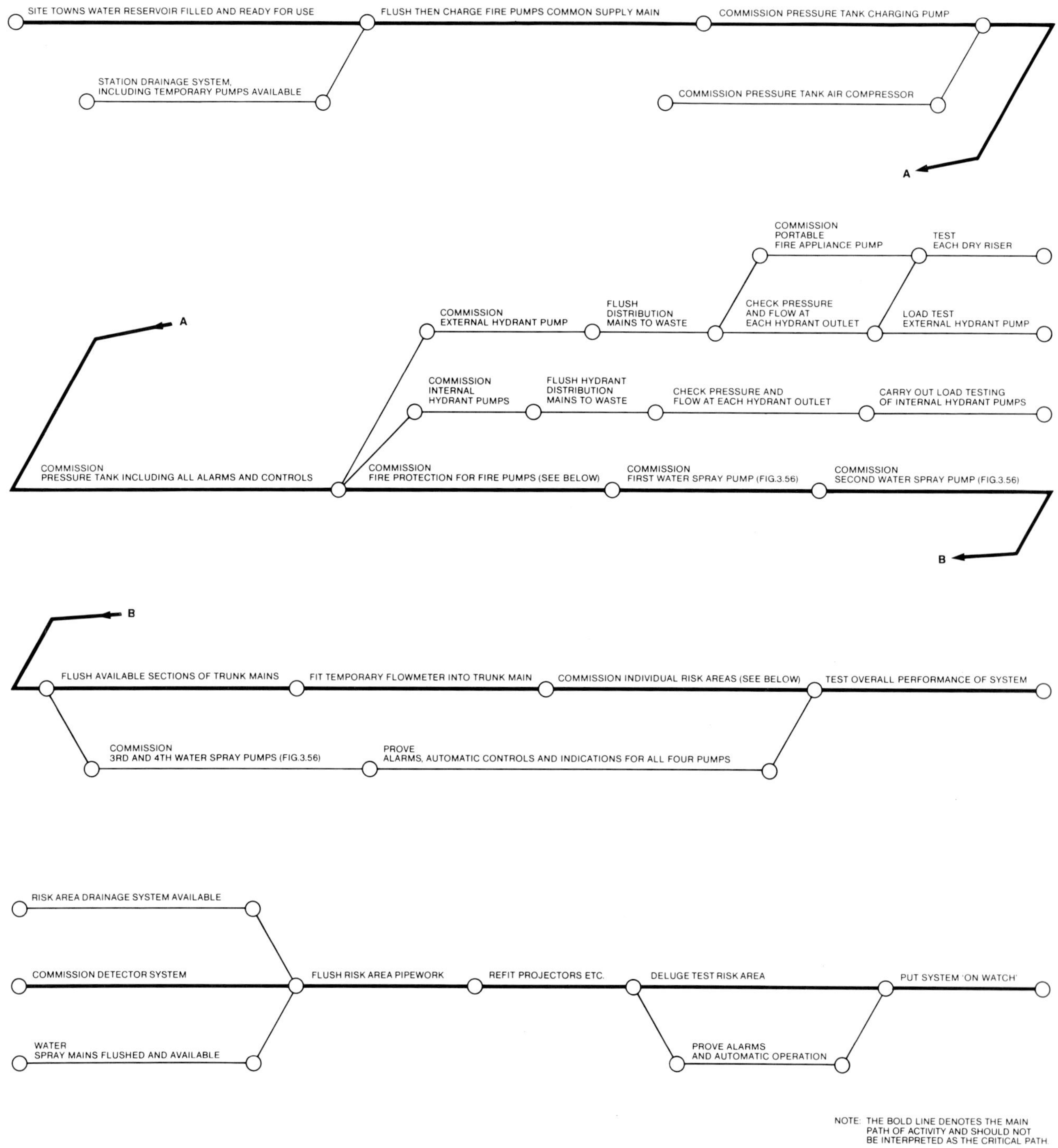

FIG. 3.67 Waterspray fire protection system commissioning network

diesel-driven fire pump has to be put into service before that pump is commissioned. However, usually an electrically-driven hydrant pump is used for system proving. This pump may well be commissioned relatively early in the construction programme to provide plant erection fire protection coverage and the electrical supplies would be derived temporarily from the construction site supplies system (see Section 9 of this chapter).

Each hydrant system is charged, using an electric pump, ensuring that automatic air release valves operate correctly. Sections of pipework are subjected to a hydraulic test to prove that there are no leaks. Water is then flushed through an open-ended standpipe at each hydrant outlet in turn, before coupling up hoses and a branch to measure the pressure and coverage, thence deducing the line pressure drop.

Finally, a flow test is carried out from the most remote part of the system. Hoses are led from sufficient outlets to create a demand equivalent to the design conditions of one pump, which should be able to run for half an hour without any overheating.

Dry risers are flushed, automatic air release valves proved and branch pressures recorded in a similar manner. The portable fire pump should be run for half an hour, discharging water from the highest branch, to prove that a consistently adequate flow of water can be maintained.

12.5.3 Foam generation equipment

The foam generating appliance is filled with foam concentrate, ensuring that the pressurising filler cap is sealed. Water supply hoses are connected and foam discharge hoses run out to a nearby clear, safe, barriered area. The equipment is then put into service and the regulator adjusted to produce foam of acceptable consistency. After the test, the hoses must be thoroughly flushed with clean water before being hung up and allowed to dry.

For the gas oil tanks, this procedure continues with the connection of the foam hoses on to the tank injection pipework. After confirming that the level of gas-oil is acceptable for this test, foam is injected into the tank. The consistency and coverage of foam is closely monitored from the roof hatch. When the test has been satisfactorily concluded, the foam injector vapour-seal diaphragms are replaced; then the injection pipeline (and the oil tank, if necessary) is cleaned out to remove all traces of foam. It is customary to carry out injection to only one gas-oil tank to prove the adequacy of the design. Subsequent tanks are tested by diverting the foam nozzles, so that they discharge down the outer wall of the tank. The foam is relatively easily cleaned out of the bunded area.

12.5.4 CO_2 and Halon gas schemes

For such systems, most of the testing is carried out with the gas discharge valve either disconnected or securely inhibited to prevent inadvertent release. Full functional testing of all detection, control, indication, alarm and test facilities, including inter-tripping signals to other plant, is carried out.

Finally, each area should be cleared of personnel and the system operated to confirm that the specified gas concentrations can be achieved. Oxygen concentration tests must be carried out in rooms adjacent to a CO_2 or Halon discharge, in case there has been any seepage or accumulation into low-lying areas.

12.5.5 Detection systems

Compressed-air detector systems are usually commissioned in conjunction with the high pressure water system, since their main application is to deluge valve installations. As mentioned earlier, sections of the detector pipework are pneumatically tested at 3 bar. There should be no detectable loss in pressure over a 2-hour period. The next phase of activity is the blowing-out of the detector lines to remove any construction debris. It is advantageous to use a temporary compressor for this purpose, since the system compressors are unlikely to be able to promote sufficient velocity for transporting debris. The compressor for each section, probably covering several risk areas, has to be commissioned. This involves pressurising the discharge pipework to a value which proves the correct setting of the compressor relief valve, followed by checking the operation of the compressor control pressure switch. The next part of the exercise is carried out in conjunction with deluge valve testing, where operation of the detector system and its attendant alarms is checked, whilst deluge valves are proved to trip at the appropriate pressure. Air detector system commissioning must ensure consistency of pressure settings throughout the entire installation.

Typical values may be:

- Compressor relief valve proved to lift at: 3.5 bar
- Compressor auto start/stop at: 2.3/3.0 bar
- System 'air pressure low' alarm: 2.0 bar
- 'Fire' alarm (typically at): 1.4 bar

Quartzoid-bulb testing methods depend upon the type of protection system. In some cases, it is practicable to use either a blow torch or a portable electric hot-air blower to shatter specific bulbs. Detector control valves can be replaced temporarily with a special test unit, though this does involve having to disturb the pipework installation twice. Live operation of the QB also requires disturbing the pipework to replace the whole detector control valve unit, since refitting the QB is usually carried out by the manufacturer of the valve.

ROR detectors are best tested with a portable electric hot-air blower played on to the exposed bulb. This unit needs to detect a rate-of-change of temperature of approximately 8°C/min to operate.

Smoke and heat detector systems need to be thoroughly checked by simulation of fault conditions, to prove that all the correct alarms, indications, etc., are produced. Often, self-testing facilities permit simulation of fault conditions at the detector system monitoring units. As a final test, gentle heat or a special test gas should be applied to each appropriate detector to prove correct operation. Electrically-powered hot-air blowers are best suited for application of heat in a controlled manner without damaging the detection sensor.

13 Coal handling plant

This section outlines the principal activities involved in commissioning the plant needed to receive, unload and then transfer coal, either to stock or into the bunkers for the main boilers.

The information within this section is based on the coal handling plant installed at Drax, a 4000 MW power station, where delivery of coal is by rail. However, reference to seaborne coal handling is made, where relevant.

13.1 General description

Coal handling plant layout design depends to a large extent upon whether coal is delivered to the station by either rail, sea or canal. Delivery of the entire load by road is not normally a practicable long-term proposition even for 2000 MW power stations. A simplified block diagram (Fig 3.68) illustrates the essential components of coal handling plant. Coal unloaded from ships, barges or railwagons is conveyed to a junction house, where it is selected for routing to the coal stock or screened before being transferred to the boiler bunkers. Reclaimed coal is conveyed to the same junction house and thence to the boiler bunkers. In order to make full use of the coal stocking area, the stocking-out/reclaiming equipment may have to be supplemented by large capacity mobile plant.

The coal consumption of a 4000 MW power station is of the order of 39 000 t per day. Coal conveyor systems for stations of 2000 MW and over have been standardised at a rating of approximately 3000 t per hour. Reclaiming conveyors are usually rated at 1500 t/h. Modern coal handling plant is usually designed to accept a wide range of washed and unwashed coals, having a size of up to 75 mm cube, with a maximum total moisture content of 25%. Average coal density is of the order of 820 kg/m^3.

Operation of the plant is usually exercised from a panel in a strategically located control room. The coal plant control room is often shared with the one for the ash and dust plant. The panel normally has a mimic diagram from which the operator can select a route for the coal, as well as giving limited manual control of some of the plant. Another panel, complete with mimic diagram, is provided for control of all rail traffic movements within the site. The control facilities are supplemented by comprehensive sequence, interlocking and alarm schemes.

Electrical supplies are, at minimum, duplicated to ensure that one stream of conveyors can be kept in service in event of a failure of any one supply source. The majority of the plant is driven by 415 V three-phase AC motors, but the larger drives use 3.3 kV motors.

Since many of the coal plant civil engineering structures have rooms below ground level, extensive pumping arrangements are needed to control accumulations of storm water. Submersible pumps, fitted into sumps at the base of each main junction house and in the railway unloading complex, discharge water into the station drainage system. Special arrangements have to be made to collect drains from the coal plant transformer bays: refer to Section 18 of this chapter for details.

Selected coal conveyors are fitted with an overhead heat-detector canopy, the heat sensors of which are arranged to trip the conveyor. Enclosed conveyors are protected from fire damage by a dry section sprinkler system over the length of the belt. Cable routes and transformers are protected in a similar manner to those for the main power station plant. Refer to Section 12 of this chapter for further details of the fire detection equipment.

13.2 Commissioning objectives

In addition to the general objectives listed at the beginning of this chapter, the following are specific to the coal handling plant:

- Ensure that the plant is commissioned in sufficient time to provide coal supplies for main unit commissioning.

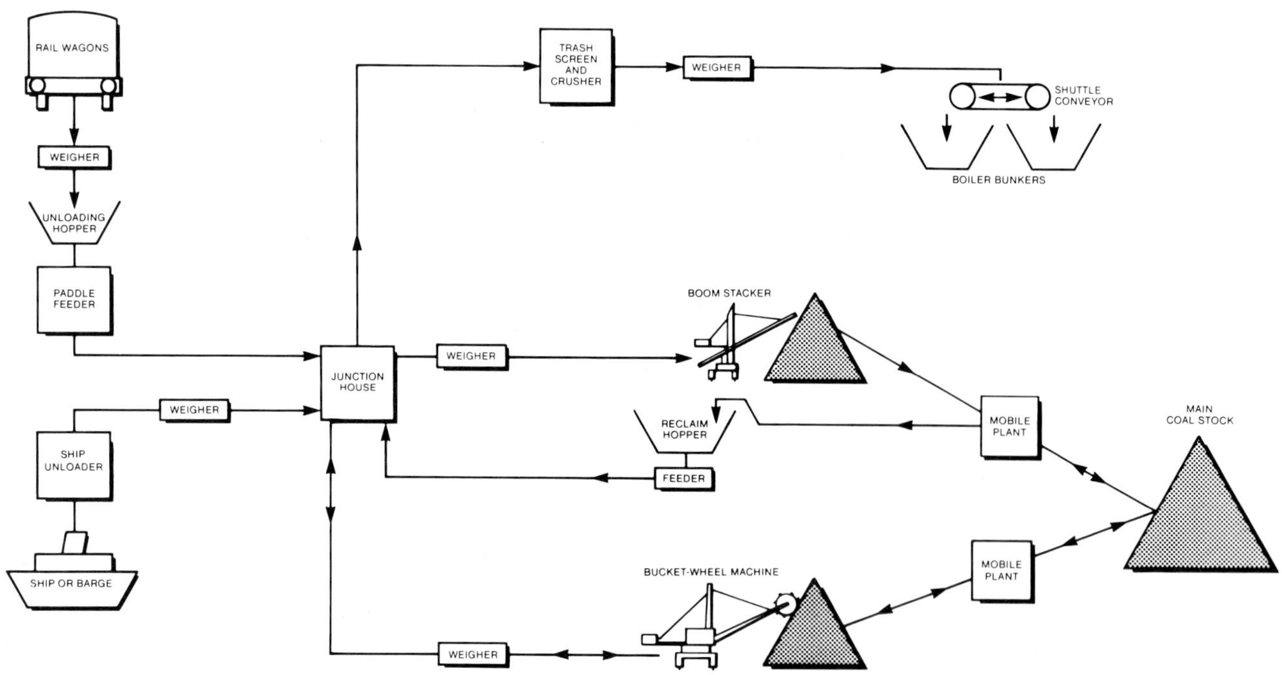

Fig. 3.68 Coal handling plant — simplified block diagram

- Demonstrate that all plant is checked and set to work in a safe manner, with all interlocks and tripping devices functioning correctly.
- Test lineside equipment for effective unloading of wagons, including hopper level-sensing devices and emergency trip wires.
- Prove that the lineside equipment air-compressor plant is capable of meeting demands imposed upon it.
- Demonstrate that the continuous ship unloaders are stable under all conditions of operation and especially when subjected to adverse weather conditions.
- Prove that grabbing cranes and ship unloaders are capable of maintaining the specified discharging rate.
- Check that all coal weighing devices give the specified accuracy to the satisfaction of the Department of Metrology inspector.
- Conduct performance tests to prove the capacities of conveyors, boom-stackers or bucket-wheel machines while stocking-out and reclaiming.
- Confirm that all conveyors can be restarted when fully laden with wet coal, without belt or coal slippage.

13.3 Safety considerations

The main risk to plant is from conveyor belt fires, usually as a result of careless usage of flame cutting or welding equipment. Even if fire does not occur, there may be sufficient physical damage to the belt to cause failure in service. Conveyor belting must be adequately protected by temporary metal sheeting in areas where burning or welding is being carried out. Propane or oxyacetylene cylinders must be kept well away from conveyors. Additional temporary fire fighting equipment should be readily available.

If proper care is not exercised, personnel can be exposed to the following risks during early operation of the coal handling plant:

- Accidental contact with railway traffic or marine unloading plant.
- Being trapped by moving machinery.
- Getting caught up by conveyor belting.
- Falling into unguarded hoppers, bunkers and through access hatches for underground chambers.

The following items should be given proper consideration, in addition to the general safety aspects detailed at the beginning of this chapter:

- Access restricted to authorised personnel only, using temporary barriers if necessary.
- Safety trips, including conveyor pull-wires, to be operational before any equipment is run.
- Coal handling plant safety-guarding all complete.
- Conveyor belts all inspected for damage and freedom from obstruction before initial running, then kept under direct supervision during first few hours of operation.
- Adequate telephone and two-way radio communications to be available before commissioning streams of conveyors.
- Work in hoppers bunkers, etc., to be in accordance with the appropriate safety rules.
- Handling of all marine and rail-borne traffic to be given extra special attention to comply with appropriate statutory regulations.
- Statutory test certificates to be available for all lifting equipment, including boom and bucket conveyors. Personnel escape safety gear to be available in elevated control cabins.

Since early operation of the plant is labour intensive, station operational staff will need to be involved at an early stage to augment contractors' resources. Extensive training will be necessary, particularly for the ship unloaders and stocking-out/reclaiming machines. This aspect needs appropriate attention, if safety of plant and personnel is to be assured.

13.4 Resources

13.4.1 Services

In addition to having the electrical power and control supplies systems commissioned and ready for use, the following will be needed:

- Two-way radio communications.
- PAX telephone system for various positions along the conveyor routes.
- Fixed fire fighting system, including fire hydrants.
- Plentiful supply of portable fire extinguishers.
- Station drains system for storm and foul water.
- Domestic water supply for ablutions and fire hose reels.
- Locomotive and empty wagons (mixture of long and short wheelbase types) to test signal interlocking, weighbridges and for initial adjustment of lineside equipment.
- Locomotive with load of full wagons to prove lineside equipment and check weighbridge accuracy.
- A barge or ship load of coal for initial testing of ship unloaders.

- Continuous supply of coal to test conveyors and belt-weighers; initially to stock, then to bunkers.
- Supply of appropriate hydraulic oil for ship unloaders, bucket-wheel machines and paddle feeders.
- General service compressed air system for bunker coal flow stimulators.

13.4.2 Staffing

Several teams of three will be needed to work concurrently on ship unloaders, railway sidings and line-side equipment, conveyors and stocking-out/reclaiming plant, plus alarm, sequence and control systems. Specialist staff from the railway company and the Department of Metrology will be needed for certain activities.

Personnel drawn from the station permanent staff will be needed to operate plant under the contractors' direction from very early stages of the commissioning activity. Progressive involvement of a supervisor and up to eight men several months before reception of the first coal delivery would be advantageous.

The overall commissioning programme may occupy a six to twelve months period, dependent upon the plant erection plan for the particular station. Typical times for testing some of the individual plant may be:

- Railway signalling — 3 weeks
- Lineside equipment — 4 weeks
- Paddle feeder — 2 weeks
- Conveyor — 5 days
- Dust extraction equipment — 5 days
- Coal sampling system — 6 days
- Weighbridge — 3 days
- Belt weigher — 3 days
- Control equipment — 10 weeks
- Bucket-wheel machine — 8 weeks
- Ship unloader — 10 weeks

Plant proving, with coal available, could well occupy a further three or four months of effort, inclusive of personnel training.

13.5 Commissioning

13.5.1 General

Figure 3.69 shows a typical diagram of the commissioning activities necessary to achieve minimum plant availability for transferring coal to boiler bunkers. Commissioning of plant downstream of the bunker outlets, i.e., coalfeeders, mills, etc., will be dealt with in the appropriate sections of Chapters 4 and 6. In practice, the programme is usually arranged so that coal-to-stock precedes coal-to-bunkers by a few weeks. This has the advantage of proving much of the plant well in advance of requiring coal for main boiler firing.

Irrespective of how the coal is delivered to site, commissioning of the plant from the first set of conveyors is similar for most power stations. Though the following notes are based on the rail-borne coal handling plant for Drax Power Station, the reader should be able to identify the basic principles for other arrangements of coal handling systems.

The plant has been subdivided into these areas, for both description and commissioning, on the following pages:

- Seaborne coal handling
- Rail-borne coal handling
- Paddle feeders
- Conveyors
- Ancillary plant
- Stocking-out/reclaiming plant
- Control systems

13.6 Seaborne coal handling

13.6.1 System description

Delivery is either by specially designed bulk-carrier ships or by barges equipped with large hatchways to maximise mechanised discharging of the load. Generally, vessels have to carry more than 7500 t to be practicable for large power stations. Modern bulk carriers may have a capacity of up to 120 000 t.

Early power plants used various designs of grabbing crane for coal unloading, usually from relatively small colliers of up to 3000 t capacity. This method is not really suitable for handling the volume of coal involved with multiple 500 MW unit plants. A typical grabbing crane is capable of handling up to 600 t/h. Higher-capacity continuous ship unloaders (CSU) have been developed to overcome the limitations of grabbing cranes. Typically these have the capability of discharging at a rate of over 1000 t/h.

Grabbing cranes

Some power stations are provided with gantry-type grabbing cranes similar to those employed for ash handling (see Section 14 of this chapter). Part of the gantry which overhangs the ship's hold has to be hinged upwards to permit free passage of deck fittings, masts, etc., while the ship is being manoeuvred into position.

The most common type is a jib crane mounted on a portal frame which traverses along rail tracks on the jetty as shown by Fig 3.70. The portal frame also carries a hopper into which the grab empties its load. Vibrating feeders at the hopper outlet direct coal via short transfer conveyors and then on to one of two conveyor belts, usually located alongside the crane rail. Each crane has traversing, luffing, slewing and lifting

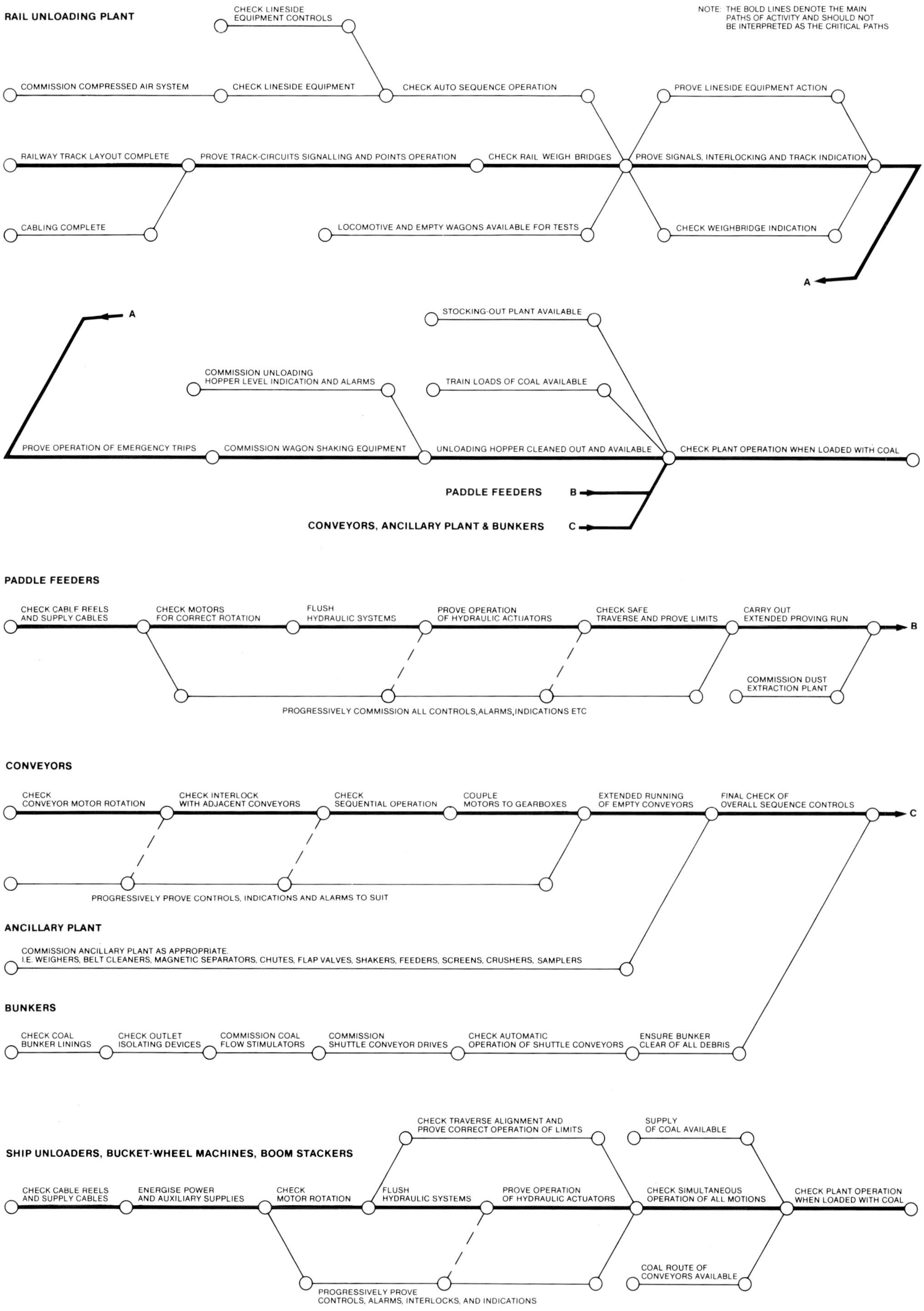

FIG. 3.69 Coal handling plant commissioning network

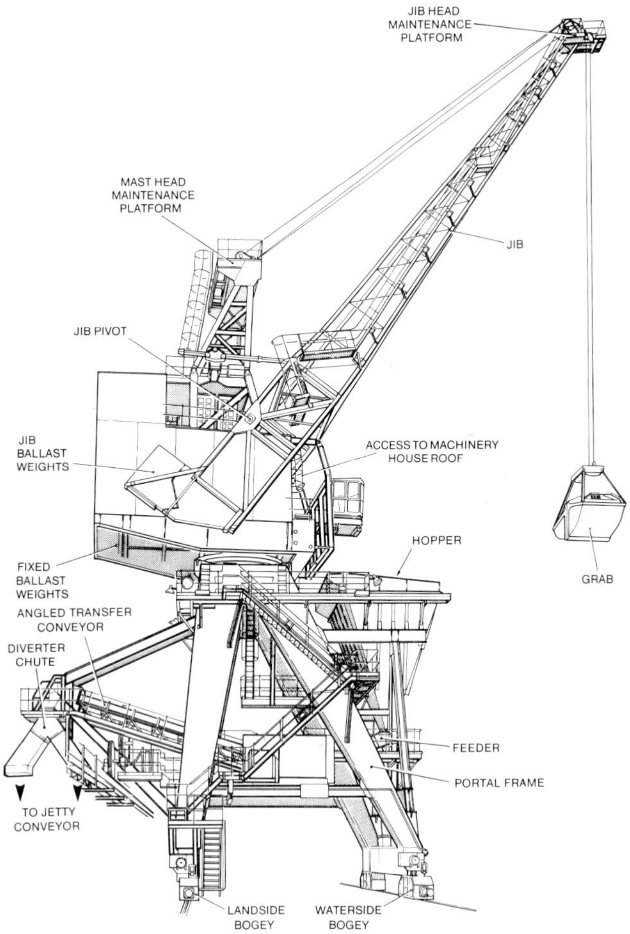

FIG. 3.70 Typical grabbing crane

motions, all driven by individual reversible variable speed induction motors. Power supply at 415 V is usually via a four-core flexible cable wound on a cable-reeling drum. The grab winch on some modern cranes is of the four-rope type wound onto two drums, each of which is provided with its own drive motor and electromagnetic brake to provide the lifting/lowering and open/close functions. Other designs have a small electro-hydraulic actuator mounted on the grab for the open/close function, with a conventional wire rope drive for lifting and lowering.

Screw-type continuous ship unloader (CSU)

One example of a modern CSU is the unit installed at Kingsnorth Power Station, manufactured by the Swedish company Siwertell. The unloader has a capacity of some 1100 t/h, traverses along rail tracks in the jetty and is operated in the slewing, luffing and pendulum modes during the unloading process. Figure 3.71 shows a general arrangement of this unloader.

Unloading of coal is carried out by a vertical Archimedean-screw conveyor which, in conjunction with a counter-rotating inlet device, lifts coal from the ship's hold. The vertical conveyor discharges from its upper end into a horizontal screw-conveyor secured to the luffing beam. Coal discharged from the landward end of the horizontal conveyor drops down a vertical chute in the centre of the slewing tower to be picked up by the portal screw conveyor. This conveyor, positioned on top of the main portal frame, is cantilevered over the duplicated belt conveyors running along the

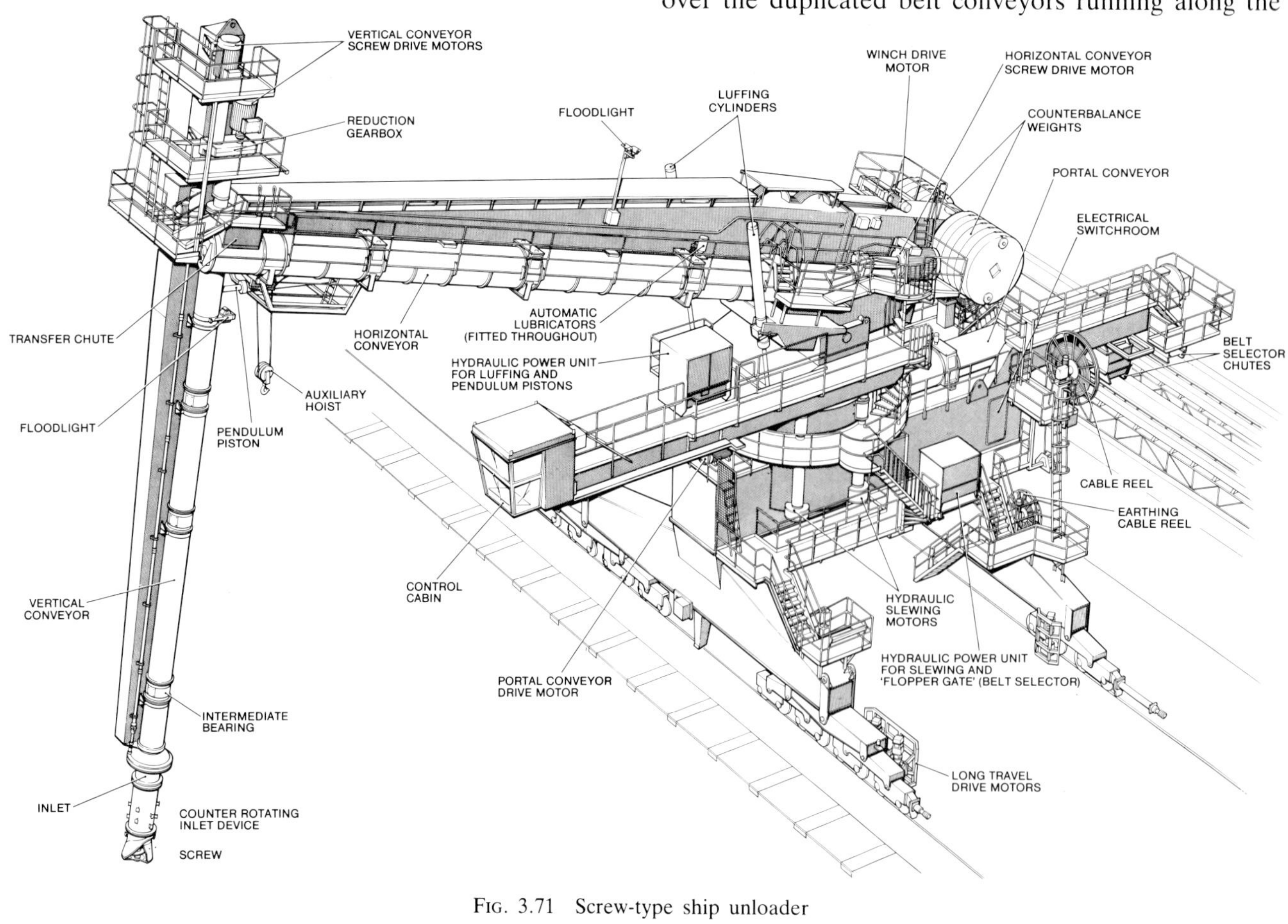

FIG. 3.71 Screw-type ship unloader

length of the jetty. The unloader weighs some 224 t and the luffing beam has a 100 t counterbalance weight.

Two control positions are provided. One is from a cabin mounted on the outboard end of an extension of the slewing turret, such that the operator has a clear view down into the ship's hold. The other is a portable control station which can be operated on board ship should better visibility be required. The unloader is equipped with a 12.5 t auxiliary rope hoist, so that a front-loading wheeled tractor shovel or calf-dozer can be lifted into the hold to assist with final clearing out.

The vertical screw conveyor is driven by two 315 kW thyristor-controlled induction motors and the other conveyors by one each. The 110 kW counter-rotating inlet device motor and the eight 3 kW traverse motors are all thyristor speed-controlled DC motors. Two duplicated 18.5 kW hydraulic pump sets discharging at 210 bar are fitted. One set is for operating the rams for the pendulum and luffing motions, the other is for driving the slewing-motion hydraulic motors.

The speed of the counter-rotating inlet device regulates filling, and thus loading, of the vertical conveyor. The device is fitted with a series of cutting knives to assist with feeding coal into the vertical conveyor. Auxiliary fans, driven by 415 V motors, are fitted to force-cool the main conveyor drive gearboxes.

Three power-operated cable-reeling drums are fitted. One is for the 3.3 kV main supply and another for the 415 V auxiliary control supplies. The third, smaller, cable-reeling drum carries an earthing cable which is only used to maintain earth bonding of the unloader during maintenance work on the 3.3 kV main supply cable. The power and auxiliary control cable reels are arranged so that the power-operated drive rotates the drum only when the cable is being wound in. When the cable is reeling off the drum, the natural tension in the cable causes the clutch in the gear housing to slip. Should the cable tension detector sense a loss of tension, the motor operates to re-impose a slight tension in the cable. The earthing cable drum is motor-operated, in both directions of rotation, from a local control station. All traversing bogies and each moving joint are strapped over by substantial flexible earth bonding cables. The entire unloader is bonded to the station main earth system via the fourth core in the 3.3 kV main supply cable.

The 3.3 kV main power supply is transformed by a 1600 kVA 3.3 kV/415 V transformer, mounted on the portal frame, connected to a 415 V motor control switchboard. The incoming 415 V auxiliary control supply is transformed to various voltages, some of which are also rectified, for controls, lighting, heating, safety equipment, interlocks, etc.

Comprehensive interlocks are fitted to safeguard plant and personnel. The main aspects of these are:

- Inhibit rotation of all three Archimedean screws and the inlet device whilst the vertical arm is over the jetty area.
- Limit pendulum motion to ±30° to the vertical.
- Restrict the luffing motion to +20° to the horizontal when working (this can be by-passed to permit unloaded luffing up to +45° to clear the ship hatchways).
- Limit slewing to an arc of 200°.
- Prevent operation of the screw conveyors, unless at least one of the jetty belt conveyors is running; and also to ensure correct sequential operation of the three screws and the inlet device.
- Prevent traversing of the unloader, if the traverse bogie carriage storm lock is applied.
- Prevent operation of the unloader, if wind speed is in excess of 20 m/s.
- Prevent the slewing motors creating excessive sideways forces on the pendulum beam from the digging action.

Interlocks are supported by the more familiar hydraulic oil tank level, oil filter differential pressure, hydraulic oil pressure and temperature, motor temperature and overload protection and alarm systems. The 350 litre oil tank associated with each hydraulic pump set is fitted with a heater to inhibit condensation.

13.6.2 Commissioning

Grabbing cranes

In most respects, the commissioning checks follow the principles laid out in Section 5 of this chapter for overhead cranes and hoists. Rather than repeat the details here, the reader is asked to refer to that section.

Unlike continuous ship unloaders, operation of grabbing cranes is not interlocked with the jetty coal conveyors, but it is still necessary to check that the coal-feeder chute attached to the crane portal frame maintains its alignment with the jetty conveyor.

Application of the traverse bogie storm locks, including the interlocking with the traverse motor controls, needs to be proved. The coal plough fitted to the leading rail wheel bogie has to be adjusted and checked for correct alignment throughout the entire length of traverse. It also has to be demonstrated that the plough will clear accumulations of spilled coal out of the rail track. Any interlocks which prevent the jibs of adjacent cranes fouling each other have to be carefully proved. The operating temperature of the grab opening electro-hydraulic actuator system has to be monitored over an extensive period of operation to ensure that it does not become excessive, even with high ambient temperatures.

Continuous ship unloaders

In many respects, commissioning tests follow a similar pattern to those for coal reclaiming bucket-wheel machines. The extent of the tests will be a function of the amount of site erection involved, as distinct from factory tested modules. A ship unloader is usually classed as a lifting device, which means that much of the testing has to be to the same standard as for cranes.

Initial checks are often carried out with the machine mechanically secured to the traversing rails. These generally comprise:

- Statutory certification in order.
- Electrical installation checked to latest wiring diagrams, special attention being given to earth bonding checks and lightning protection, where appropriate.
- Hydraulic pipework systems completed, apart from having temporary short-circuiting connections fitted at hydraulic actuators.
- Temporary strainers fitted for flushing the system, which is then filled with the correct fluid.
- Cable reels are checked, then power and control supplies are energised to facilitate checking of control sequences. Motor rotation checks and hydraulic oil tank heater tests conducted.
- Checks that safe access from the operator's control cabin is available, even under adverse weather conditions.
- Confirm adequate visibility from the operating position.
- Prove correct operation of the emergency escape equipment.
- Ensure that the communications channel to the nearest manned control room is operational.
- Demonstrate that the klaxon for warning shore-based personnel is functioning before any machine movements are made.
- Commission the hydraulic pumping units and flush the pipework for each actuator separately until all debris has been removed. Refer to Section 2 of this chapter for more detailed information.
- Remove temporary strainers, include hydraulic actuators into system and individually prove correct operation of each function, i.e., luffing, slewing, pendulum motion, elevator drive, bucket-wheel rotation, etc., as appropriate.
- Operate all drives sequentially and together to confirm that unloader remains physically stable and is controllable. Prove that the hydraulic pump sets adequately cope with the simultaneous demand for oil from all actuators.
- Confirm correct operation of storm locks, then energise the traversing drives and operate the machine over the entire length of the rails. Check that the rail alignment is maintained and that there are no bumps or other discontinuities.
- Observe operation of cable reels to ensure that cables pay out and take up smoothly without any snatching or kinking of the cable.
- Confirm correct operation of the end-of-travel limit switches.
- Ensure that unloader coal chute remains aligned over the jetty conveyor belts at all points.
- Check that the bogie-mounted coal ploughs are correctly aligned and prove that they do clear coal spillage from the rail tracks.
- Operate all control motions while the machine is being traversed to check that stability is maintained. Simulate plant failures to confirm that response of controls is always safe.
- Check sequential interlocks for the group of conveyors and also the interlocks with the jetty conveyors.
- When coal is available, carry out loading checks, rates of flow, etc. Prove also that conveyors can be restarted, following a trip when fully loaded with coal. Confirm stability of machine in winds of up to 20 m/s when handling coal at various rates of flow.

13.7 Rail-borne coal handling

13.7.1 System description

Specially designed hopper-type rail wagons, fitted with three pairs of bottom-opening doors, are used for coal deliveries. These wagons are of 32 t capacity, usually made up into 1100 t train loads. Some 50 trains per day are needed, five days a week, to deliver the coal supplies for a 4000 MW power station.

The rail track layout is arranged on a merry-go-round principle in one of the variations shown in Figs 3.72 and 3.73, and also by Fig 3.75.

- Loop system — this may be of the form of an open-ended loop (Fig 3.72), where locomotive and wagons enter then leave from the same end of the site, which is the most common layout. Alternatively, it may be a full loop with crossover, where the reception and departure junctions are at different locations.
- Run-round system — this is used where space precludes a loop. Figure 3.73 shows the arrangement, where the locomotive transfers or 'runs round' to the other end of the train before approaching the unloading house. At some sites it is more convenient to design for this 'run-round' to take place after the train has discharged its load.

Locomotive and permanently-coupled wagons approach the unloading house over a weighbridge which automatically weighs each loaded wagon. The train then passes over underground unloading hoppers where lineside equipment automatically unloads the wagons. Empty wagons are automatically reweighed,

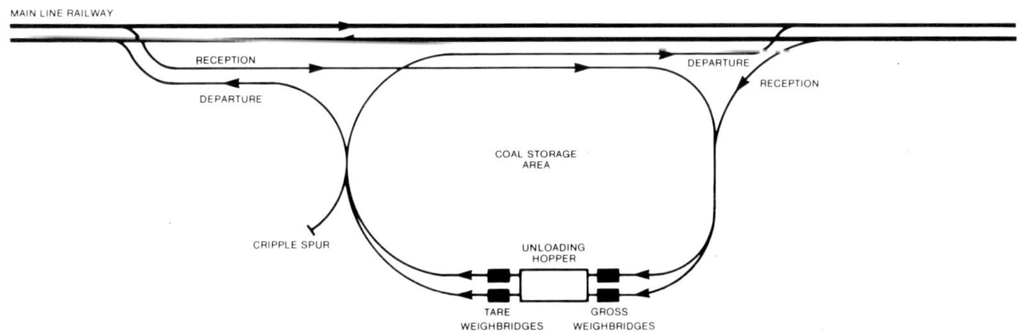

FIG. 3.72 Open-ended loop system

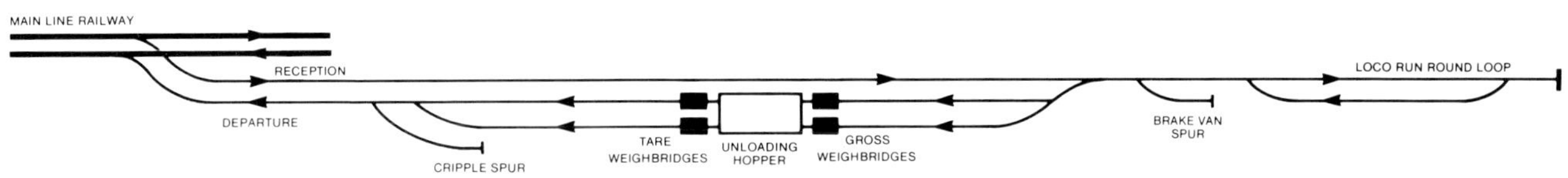

FIG. 3.73 Run-round system

so that the weight of coal discharged can be deduced. Throughout the actual weighing and unloading operations, the train moves at a steady speed of 0.8 km/h. As wagons exit from the unloading house, they are checked by inspectors employed by the railway company. Any wagons considered unfit to re-enter the railway system are designated as 'cripples'; they are separated from the train and shunted into a special siding to await repair.

Modern computerised weighbridges weigh each axle independently, then compute the total weight. Axle detectors are employed to differentiate between wagon wheelbase centres and those of adjacent wagons or locomotives. The detectors also activate the 'weigh' action when the axle is in the centre of the weighbridge. Display, plus printing, of gross and tare weights, including the total tonnage of coal discharged per train, is available in the coal plant control room. Usually only a display is provided at the weighbridge.

The rail sidings are usually capable of holding six trains waiting to unload and a further four waiting clearance to re-enter the public railway system. Another rail loop is provided for handling trainloads of boiler and/or gas turbine fuel oils (see Fig 3.75).

Figure 3.74 details the lineside (or trackside) equipment within the unloading house. Each rail track is fitted with pneumatically-operated sequence controlled units, which carry out the following functions as the rake of wagons moves across the unloading hopper:

- Unlatch the wagon door mechanism safety catches (two sets).
- Open the wagon doors at preselected positions over the unloading hopper (four to eight sets).
- Reclose the wagon doors after emptying (two sets).
- Relatch the door mechanism safety catch (two sets).

Lineside equipment is controlled from an elevated cabin at the exit end of the unloading house. The operator usually has the facility either for remote manual control of individual units or for automatic sequence of all units for any one track. Each unit is also provided with local manual control, as well as an escape mechanism to cope with jammed hopper doors, etc., and wagon run-back. Wagon wheel proximity detectors are used for control of the automatic sequencing equipment. These are designed to differentiate between the wheelbase axle centres of wagons, typically 5.56 m, and the axle centres of a pair of coupled wagons (3.49 m) or locomotive wheels. Duplicate 7 bar pressure compressor/receiver/drier sets are fitted to provide the motive air for operating the lineside equipment. Typically, these may have 1.5 kW motors and be rated at 6.5 litres/s. Compressed air pipework is frequently thermally insulated and is also provided with a filter/lubricator at each item of lineside equipment.

If any wagon fails to discharge fully, the operator can set all rail signals to 'stop' so that a wagon shaker (see Fig 3.76), suspended from a power-operated hoist running over the track centre line, can be lowered on to the wagon. The wagon shaker comprises an eccentric-weight, belt-driven from an electric motor, mounted in a frame which straddles the wagon sides at specially strengthened positions.

A system of emergency trip wires, comprising red PVC-coated pull wires operating manually-reset electrical switches, is fitted for personnel and plant protection. These are located along the handrails of both sides

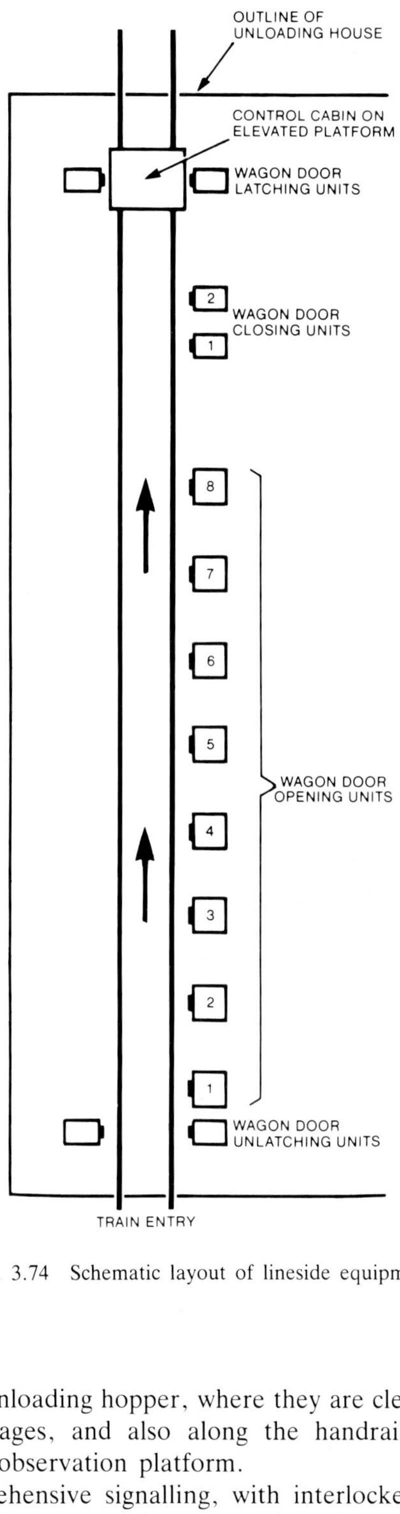

FIG. 3.74 Schematic layout of lineside equipment

of each unloading hopper, where they are clear of any coal spillages, and also along the handrails of the elevated observation platform.

Comprehensive signalling, with interlocked operation of railway points, is necessary for safe control of all rail traffic within the site boundary. The signalling scheme is also interlocked with signals controlling access to the railway company system. Figure 3.75 shows the rail traffic control mimic diagram for an open-loop track arrangement. Each adjacent section of track capable of holding a train is shown in a different colour. The mimic diagram indications show the status of all signals and points, and also which sections of track are occupied.

13.7.2 Commissioning

It is simpler to consider rail-borne coal handling as several independent areas for commissioning purposes, namely:

- Wagon unloading equipment
- Wagon weighing equipment
- Railway tracks, signalling and traffic control

Railway traffic control

Checking and testing of all equipment associated with running main-line railway goods traffic on site tracks must be carried out to the satisfaction of the railway company representative.

All signals, points and associated interlocking must be proved to give the correct safeguards under all normal and abnormal conditions of operation before any rail traffic is allowed on to the tracks. By simulating the presence of rail vehicles, all combinations must be thoroughly tested, along with indicators and controls located on the rail traffic control mimic diagram. The functions will be proved again when a locomotive and several wagons are on site for testing of the unloading hopper lineside equipment.

Wagon unloading equipment

All control and indicator cabling must be checked for correct installation, in accordance with latest wiring diagrams. The small compressed-air plant has to be commissioned and the quality of the air supply checked.

Satisfactory manual operation of the rail wagon door-operating gear is proved first. The automatic sequence is proved next, by simulation of the wagon proximity detectors. Once confidence has been established that the equipment is functioning correctly, a locomotive and group of empty wagons with long and short wheelbases are brought on site for several days of live testing and adjustment.

The rolling stock is moved slowly through the unloading shed while the following operations are conducted along each of the unloading tracks:

- Check and adjust wagon safety catch unlatching gear.
- Check and adjust each set of wagon door opening equipment.
- Repeat for door closing and safety catch relatching equipment.
- Prove correct operation of wagon detecting sensors and then prove fully automatic operation of the lineside equipment from the local control panel.

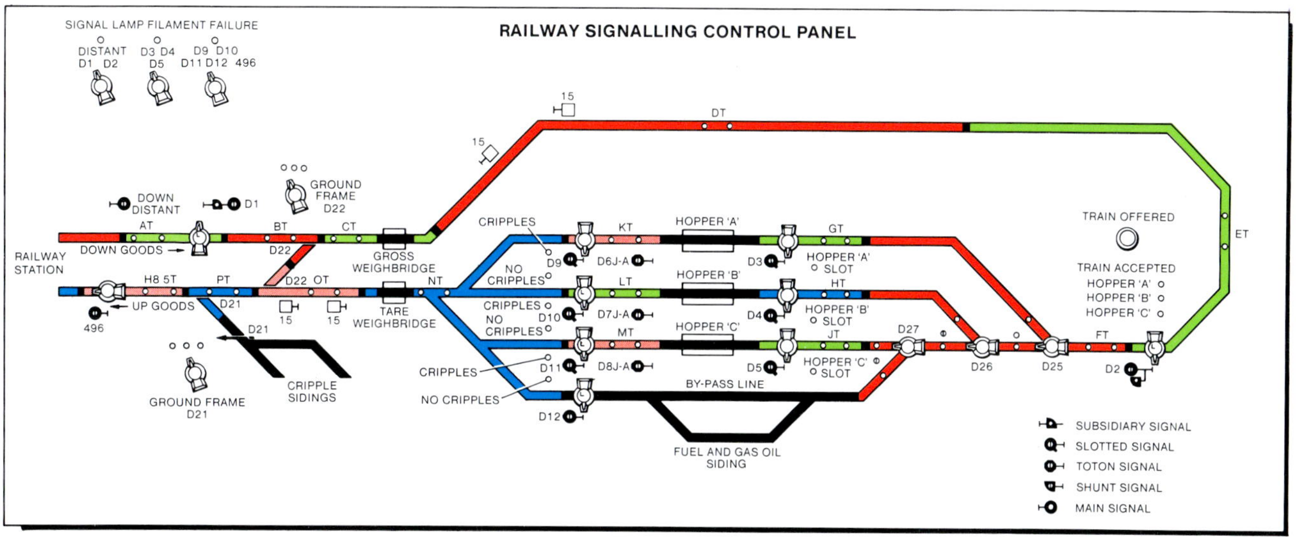

FIG. 3.75 Mimic diagram of rail traffic control system

- Simulate fault conditions and ensure that individual items of the plant can be operated manually without interfering with the overall sequence. Typical fault conditions are failure of lineside units to advance or retract when required, or failure of the wagon safety catch or door-operating actions. It is important to check that all lineside equipment remains retracted until the locomotive has passed each unit. If the train has a brake van, then the equipment must be in the retracted position while the van passes through the unloading house.

Correct operation of the unloading-hopper coal level indicator probes, high level alarms, and also the trackside safety tripwire actuating the train 'stop' signal, must be proved. The wagon-shaking equipment should be commissioned whilst wagons are on site.

When the remainder of the coal handling plant is ready to accept coal, then live operational tests can be carried out. Initially, one or two trainloads of coal would be ordered for proving the system.

Wagon weighing equipment

This is normally set up by the manufacturer after all cabling and indication equipment has been proved. Simulation of the wagon-detection scheme is necessary to check static operation. In conjunction with the lineside equipment tests, the locomotive and wagons are traversed over the weighers to prove correct functioning of the wagon detection scheme with long and short wheelbase wagons. The reception (gross) weight and the departure (tare) weight should, of course, be identical with empty wagons. This gives confidence in the initial adjustment of the equipment.

Later, when trainloads of coal are being accepted, printed totals from the weighing equipment are compared with the declared net coal tonnage of each train, so that any inconsistencies can be identified.

13.8 Paddle feeders

13.8.1 System description

Figure 3.76 shows the general arrangement of paddle feeders in the unloading house, while Fig 3.77 illustrates a typical twin-rotor paddle feeder. At Drax Power Station, a quadruple-rotor paddle feeder is used to remove coal from the base of each railtrack unloading hopper and feed it on to the first conveyor belt in the flow path to either stock or boiler bunkers. Each paddle feeder unit is mounted, along with all of its associated auxiliary equipment, on a rail-borne steel frame which is traversed along the base of the unloading hopper. As the paddle feeder is traversed, a specially shaped operating rail sequentially opens the unloading-hopper discharge flap-doors to allow the feeder to cut into the coal. The rotor of each paddle comprises a number of curved blades extending horizontally from a central hub.

Power and control supplies are connected via power-operated cable reeling drums, which automatically coil or uncoil the cable as the machine traverses the track. Twin limit switches are fitted at each end of the track. One provides signals for automatic reversal of the direction of travel, whilst the second trips the paddle feeder traverse motor should the auto-reverse fail.

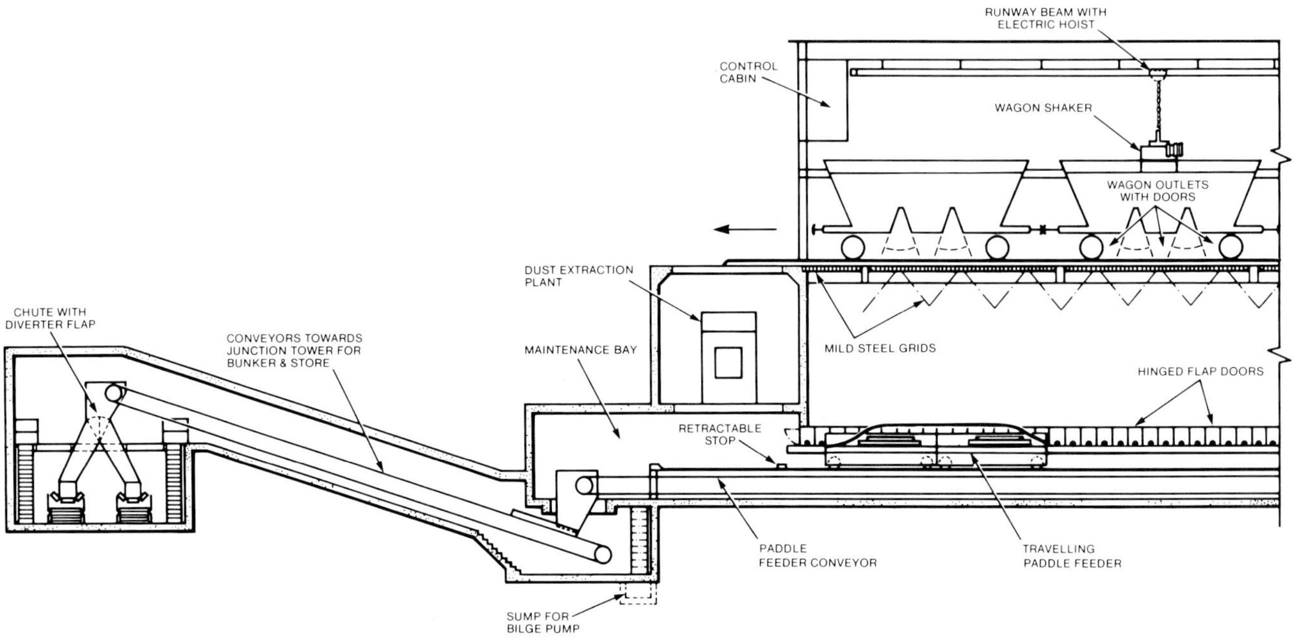

FIG. 3.76 Sectional arrangement of unloading house

Separate electric motors drive gearboxes via fluid couplings for the traverse motion and paddle rotation. The paddle rotor gearbox is usually of the infinitely variable speed type, positioned by an electric-motor-driven actuator. An electro-hydraulic pumpset is fitted to power hydraulic rams, which extend or retract to move the paddle rotor into or out of the unloading hopper base. Other designs of paddle feeders use hydraulic motors to drive the paddles. In this case there is a separate duplicated hydraulic pumpset installation. Control of the speed of rotation is by an actuator adjusting the swashplate angle of the hydraulic pumps.

Full manual control facilities are available at each paddle feeder, but they are normally under automatic control from either the unloading house control room or the master control panel.

Fixed dust suppression equipment may be fitted to prevent the accumulation of a volatile dust/air mixture in the paddle feeder bays. A duct, mounted along the bay, connects via a flexible adaptor with a hood mounted above the paddle feeder discharge-opening. A motor-driven exhauster draws the coal/air mixture over, or through, a water-sealed cyclone.

Coal dust slurry is returned to the conveyor, whilst the cleaned filtered air is returned to the paddle feeder bay.

13.8.2 Commissioning

First, it is necessary to ensure that the feeder is complete and can be manually traversed along the rail track, without binding. Traversing the feeder will check the special operating rail which opens and then recloses, the unloading hopper discharge-doors as the feeder moves along the track. This will be important later, since correct functioning of the flaps reduces coal spillage and dust nuisance.

A manual traverse also checks the power and control supply-cable reel operation, providing that the cable reel drive-motor is uncoupled. The cable should pay out and take up smoothly, without snagging, twisting or jerking. Operation with the cable reel motor-drive energised follows later.

After confirming that all wiring, terminations and earth-bonding is correct to drawing and that the insulation resistance is acceptable, electrical supplies are energised. Control and alarm function checks and motor rotation tests are carried out; then the hydraulic system is flushed until clean. When the drive motors have been coupled and all personnel safety guards are in place, paddle feeder rotation and traverse motions are proved. Paddle rotation at various speeds, as a check on dynamic balance and smooth speed control of the gearbox, is also carried out. Retraction and extension of the paddle feeder hydraulic-rams has to be proved. All interlocks, including end-of-traverse, reversing and any anti-collision limit switch operation, have to be proved by this stage. The paddle feeders are then given an extended (empty) run under automatic control to establish confidence in the reliability of the plant.

Having proved that each paddle feeder is an operational machine, it is necessary to commission the dust extraction plant associated with it before any coal is discharged into the hoppers.

When continuous deliveries of coal are being received, operation of the paddle feeders is proved, in conjunction with conveyors and other ancillary plant, whilst stocking-out or coaling to bunkers. Particular attention is paid to noise levels and evidence of dust nuisance or coal spillage.

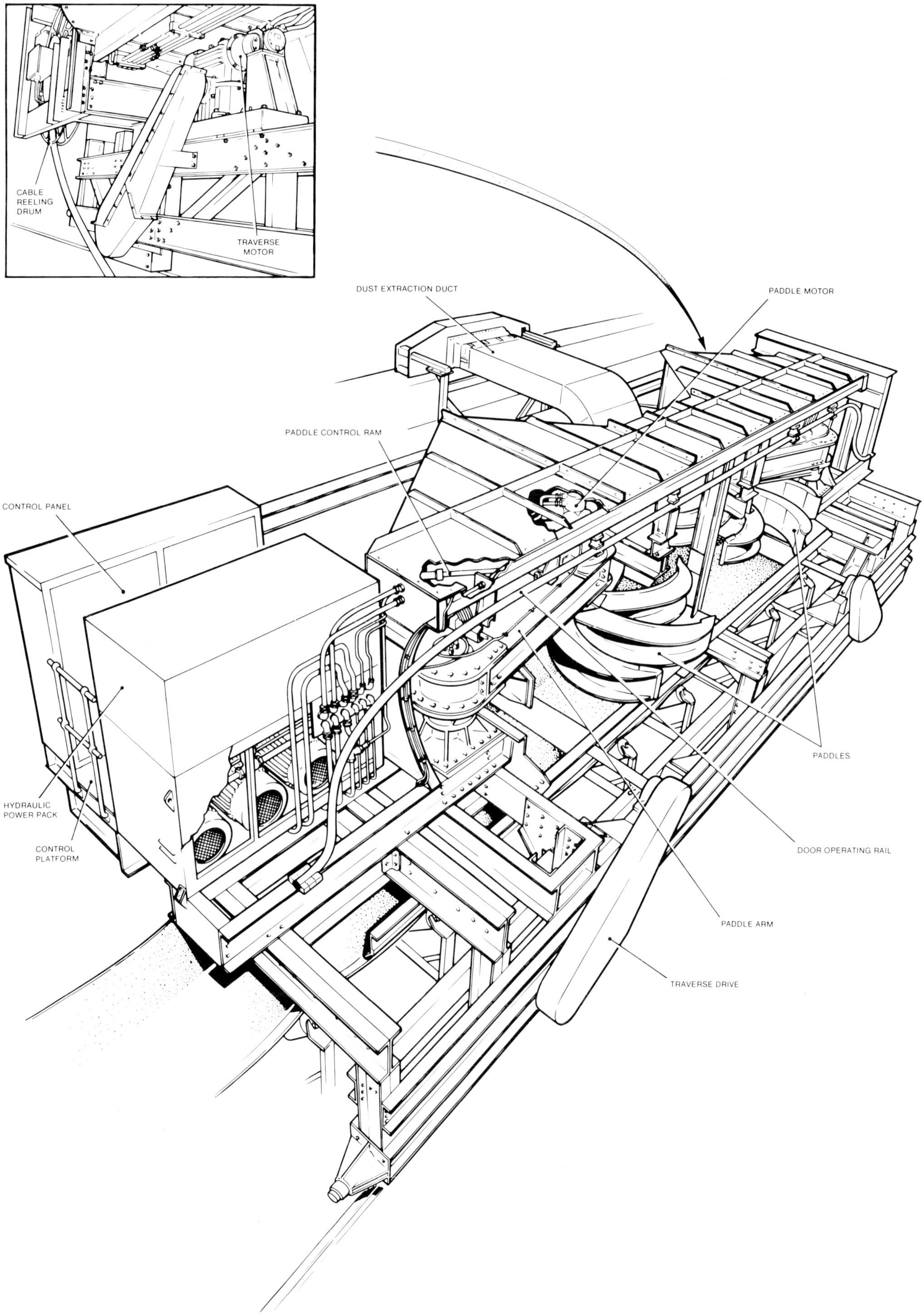

FIG. 3.77 Typical arrangement of paddle feeder

13.9 Conveyors

13.9.1 System description

Conveyor belting is manufactured from a specially developed fire-resistant rubber, reinforced with several textile plies in accordance with BS490. The longer runs of conveyor belting do, of necessity, contain several joins which must be assembled in compliance with BS6593. A typical conveyor belt would have a speed of between 2.5 and 3.0 m/s and would be constrained by idler rollers into a deep trough with the sides at an angle of 45°. A 1.83 m wide belt would be capable of handling 3000 t/h and a 1.37 m belt some 1200 t/h. All idler rollers now have sealed-for-life bearings, as do most other bearings employed throughout modern coal handling plants. Safety guards to BS2890 are fitted, where appropriate.

Figure 3.78 shows a diagrammatic arrangement of the conveyor layout at Drax Power Station. Referring to this figure, the coal flow path is as follows. Paddle feeders load coal on to conveyors, which transfer the coal to the top of junction house 2. From here, coal can flow to stock via conveyors 23, 24 and 25 and the reversible boom conveyor 26 of the bucket-wheel machine (refer to Fig 3.79). Alternatively, coal can be transferred to junction house 3, where it is screened and crushed before passing to the unit bunkers via conveyors 9 and 10. From junction house 3, coal can also be stocked-out via conveyors 28, 29, 30 and the reversible boom conveyor 31 of the second bucket-wheel machine.

Coal is reclaimed from stock either by one bucket-wheel machine which discharges on to its reversible boom conveyor 26 and thence via (reversed) conveyor 24 into junction house 4, or by the other bucket-wheel machine via conveyors 31 and 29 into junction house 5. In each junction house, coal discharges on to reclaim belts 27 and 32, respectively, and thence to the boiler bunkers via conveyors 9 and 10.

Each of the bucket-wheel machine main feed reversible conveyors (24 and 29) incorporates a collapsible tripper, which is mechanically coupled to the bucket-wheel machine. The collapsible tripper comprises a winch-actuated frame containing pulleys, which elevate and form the main feed conveyor into the shape of a 'Z' for stocking-out. A short conveyor transfers coal between the raised section of the main feed conveyor and the reversible boom conveyor along the luffing-boom during stocking-out. For reclaiming, the tripper is collapsed, allowing the main feed conveyor to extend and become level. Coal from the bucket-wheel boom conveyor then by-passes the short conveyor and discharges directly on to the (now reversed) main feed conveyor.

Every conveyor is fitted with some type of automatic belt-tensioning device. This often takes the form of a gravity take-up unit (GTU) in which weights suspended from a wire and pulley system ensure that gravity acting on the weights maintains the correct working tension for each belt. Usually the GTU is located near the tail pulley, where it tensions the return section of the conveyor belt. Another form of automatic tensioning device consists of a load-cell detector on the tail pulley shaft-bearings. The output from the load cell causes an actuator to adjust the position of the tensioning pulley to maintain the correct belt tension for all conditions of loading.

Motors used for conveyor and other coal plant ancillary drives are of the totally-enclosed squirrel-cage induction type. Conveyors are driven via speed-reduction worm-and-wheel gearboxes, some of which are coupled to their motors by fluid couplings. Fluid couplings ease the motor starting duty and enable the conveyor to be accelerated to its normal running speed smoothly, even when the conveyor is fully loaded with coal. Generally couplings used on coal plant drives are non-adjustable and are designed to stall at approximately 150% full-load torque. On the heavier-duty conveyors, where the motors are rated at more than 150 kW, the fluid couplings are automatic scoop-controlled units. An actuator ensures that the scoop is in the minimum speed position before the motor is started; then once the drive motor is up to speed, the actuator progressively withdraws the scoop to its full-speed position. The hydraulic oil is cooled by air circulation over external fins on the couplings. Some couplings are fitted with separate oil/air coolers, over which air is circulated by shaft-mounted fans.

Electromagnetic or hydraulic thrustor brakes, acting on drums or discs attached to the gearbox shaft, prevent rotation of the conveyor pulley in event of either a motor or an electrical supply failure. Brakes are held 'off' when they are energised, i.e., they fail to the 'on' position.

Emergency stopping facilities are catered for by the fitting of red PVC-covered pull wires running along the entire length of at least one side of each conveyor. Pulling the wire at any point operates manually-reset switches which trip the affected conveyor. Trip wire schemes are designed to comply with BS5304.

Every conveyor has a centrifugal speed switch, driven from an extension of the tail pulley shaft, to provide interlock and alarm signals to the sequence control equipment.

13.9.2 Commissioning

The basic assumption in this section is that the conveyor belting has been installed and jointed, and that the belt-tensioning units are adjusted to their design settings. Drive-motor rotation and electromagnetic brake operation is proved with the motor coupling disconnected from the gearbox shaft. Operation of the 'conveyor start' warning klaxon, the pull-trip wire and the tail pulley centrifugal switch are all proved to and from the control equipment. Where appropriate, fire detection trips are proved as well.

If practicable, a number of consecutive conveyors is treated similarly. When convenient, a full sequence

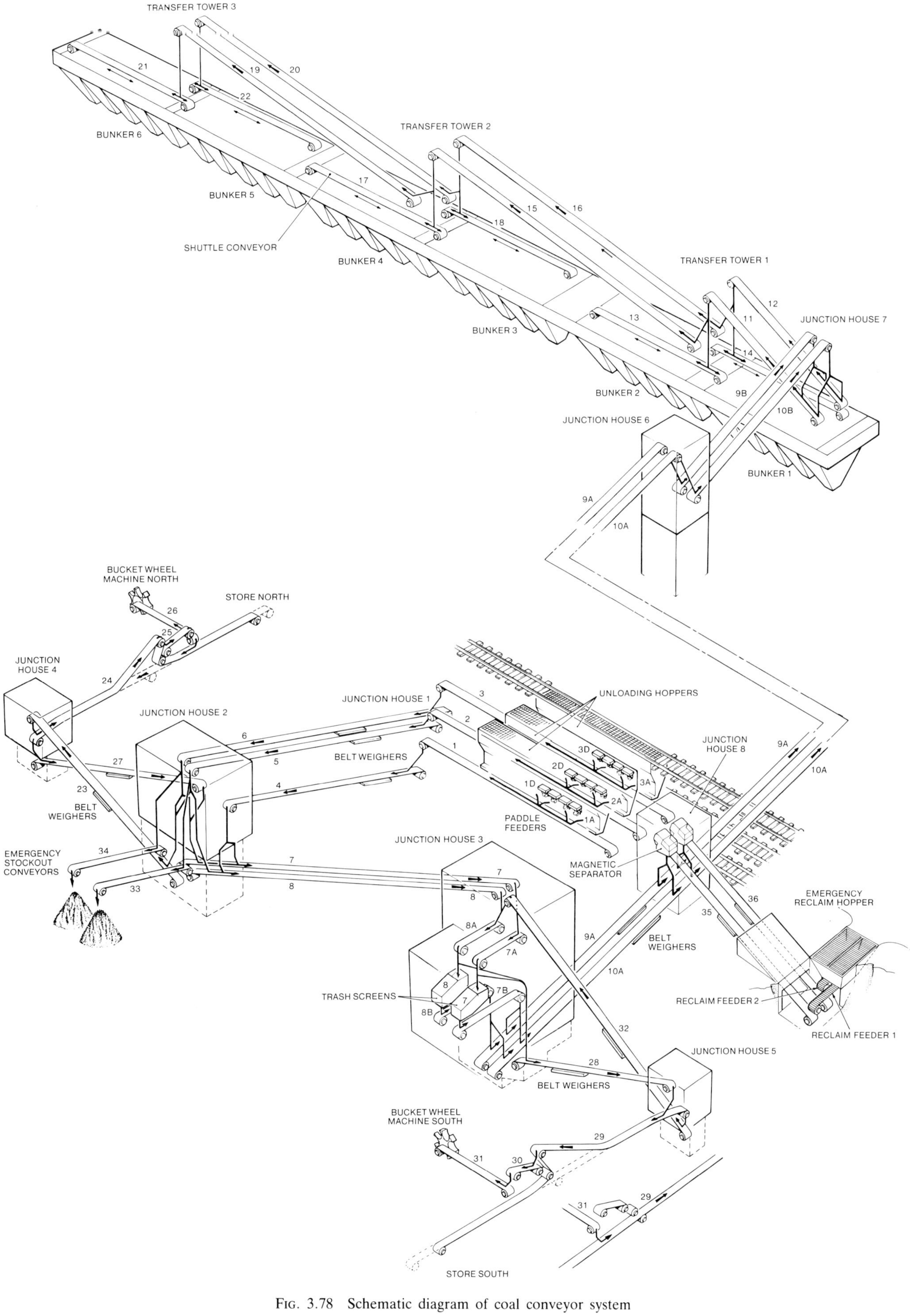

FIG. 3.78 Schematic diagram of coal conveyor system

operation is carried out to check the system, with the motors still disconnected from the conveyor gearboxes. Once the start-up, shutdown and fault condition operations have been proved satisfactorily, the motors are coupled and the fluid couplings charged with oil in readiness for conveyor operation. A check is carried out to ensure that all safety guards are securely fitted to each belt that is to be run.

Each conveyor is run independently at first, with personnel stationed within sight of each other along its length. Initially, a short run is carried out, followed by a longer continuous run for the following items to be observed and noted, where appropriate:

- Motor starting current and run-up time.
- Belt running speed and centrifugal switch operation.
- Adjust belt tension, if necessary.
- Operation of belt trough and return idlers, including a check for overheating of pulley bearings.
- Check that belt alignment remains true.
- Check and, where appropriate, adjust belt-weighing equipment.
- Observe operation of belt-cleaning equipment.
- Observe operating temperature of fluid couplings.

Before accepting the first delivery of coal, it is prudent to recheck thoroughly the full sequential operation of all plant along the selected coal route. Where appropriate, simulate various fault conditions to confirm correct response from the control and alarm schemes.

As coal supplies become available, load checks are carried out to ensure that belt alignment and tensioning remain correct. It is also important to check and adjust the seals between coal chute outlets and the conveyors to minimise spillages and dust emission. At some stage, it is advantageous to initiate a plant trip when the conveyors are fully loaded, especially with wet coal, to ensure that the electromagnetic brake prevents any run-back. The drive motor should be capable of restarting the loaded belt, without any slippage of either the belt or the coal on it, within the prescribed run-up time. The drive motor should not run in an overloaded condition during this test either.

13.10 Ancillary plant

13.10.1 System description

Chutes and flap valves

Steel-fabricated chutes, lined with either stainless steel plates or thick fireclay tiles, direct the downward flow of coal between conveyors. Motorised stainless-steel-faced deflector flap valves are installed in some chutes to direct material flow from one line of conveyors to another. Each chute incorporates special diaphragm-type switches to warn of a 'blocked chute' condition.

Belt-cleaning devices

Each conveyor is fitted with some means of removing any material which continues to adhere to the belt surface after the coal has been discharged. Head pulleys are now fitted with a motor-driven brush-type cleaner located beneath the pulley so that the debris falls into the coal chute. The head snub pulley surface is cleaned by a renewable rubber blade scraper. It is also practice to stretch a spring-loaded piano wire scraper across the inside of the belt between the head and associated snub pulleys. Reversible conveyors are sometimes fitted with a cleaner at both ends of the belt.

Belt-weighers

Weighers are installed on selected conveyors (see Fig 3.78) to monitor the weight of coal being transferred to boiler bunkers or for stocking-out. These indicate totalised weight and rate-of-flow both locally and in the coal plant control room. Modern belt-weighers comprise a freely suspended carriage, perhaps 4 metres long, complete with five or six sets of idler rollers. The force exerted on the carriage from the loaded belt is transmitted to an electronic module via a lever system. A pulsed speed signal, driven from the conveyor tail pulley, is fed into the module, where it integrates the force signal in relation to the length of belt passing in a fixed period of time to give the required output signal. It is common practice to locate belt weighers fairly close to the tail pulley, as this is where the belt tension is least.

Magnetic separators

These are fitted to remove any ferrous debris from the coal passing along the conveyors to the bunkers. Transverse over-belt separators are now being superseded by in-line separators. In both, a powerful DC electromagnet is used to magnetise spikes fitted to a short endless belt, which passes over the coal on the main conveyor. Recent advances in high power ferritic magnets has led to the adoption of permanent magnets in place of electromagnets for tramp iron removal. Ferrous material is thus attracted to the spikes which carry it around the belt and out of the magnetic field, to be dumped into a tramp iron chute.

Vibrating screens, crushers and vibrating feeders

Double-decked vibrating screens are fitted in the main junction house of the conveyor system, using an upper primary mesh to reject trash and a lower secondary one to separate oversized coal. Relatively large mesh sizes, e.g., 100 mm, are used so that only really large lumps are discarded as trash. Smaller lumps, i.e., between 40 and 100 mm, are diverted to a rotary crusher; the remainder flows on to the next conveyor. A typical crusher comprises a series of rotating hammers mating with anvils fixed to a short endless belt. Both items rotate in the same direction and are set a specific distance apart to achieve the required maximum coal size for returning crushed coal to the conveyors. There is a growing tendency to deliver treated coals, which do not need to be screened or crushed. For this reason, it is practice to provide a full-flow chute to by-pass the trash sizing screen and the crusher units whenever suitable coal is being delivered to the bunkers.

Vibrating feeders are often fitted to chutes in reclaim, transfer or junction houses to promote an even flow of coal on to the selected conveyor. Eccentric weights, shaft-driven from an electric motor, or electromagnetic vibrators are the two most common methods of inducing the necessary levels of vibration.

Coal sampling equipment

Automatic coal sampling equipment is necessary in order to obtain a truly representative sample of the coal being loaded into the bunkers. Duplicated equipment is arranged to sample from either of the conveyors leading to the bunkers. A bucket collects samples from the conveyor belt at predetermined intervals. The sample is crushed and separated into two; one portion is put into a container for moisture analysis, while the other is further crushed and divided into two. One part of this is put into a container for calorific value analysis, whilst the remainder is returned to the bunker conveyor stream.

Bunkers

Inverted, rectangular pyramid-shaped bunkers made of sheet steel are fitted for each coal milling group. The upper ends of the individual bunkers for any one boiler unit are usually joined together to form the one large bunker opening beneath the shuttle conveyor. Bunkers are lined with either HP-spray-applied, wear-resistant concrete or basalt-glazed tiles to promote a smooth flow of coal. Modern designs may also incorporate a specially profiled stainless steel lining nearer the bunker outlet to eliminate corners and also to permit slipping at the walls.

All modern coal bunkers are designed on the mass flow principle, where a generously proportioned outlet permits the movement of a large volume of coal. A typical 1000 t bunker may have a rectangular outlet, measuring some 2.5 m by 1.0 m, followed by a very short specially profiled stainless steel lined chute into the coal feeder unit. Bunker outlet isolating devices are usually of the sliding-plate design, where two opposing plates are driven across the coal flow by hydraulic rams. Electric-motor-driven lead-screw mechanisms have been prone to failure as a result of accumulation of coal dust on the screw, and are seldom used now. The multiple-poker type of isolating device is still quite common, but groups of pokers are now driven by hydraulic rams, rather than being hammered mechanically, through the coal.

Pneumatic coal flow stimulators are often fitted near bunker outlets to move sticking coal. They utilise a compressed-air jet which blasts air into the coal, either on a 'one-shot' basis or on a preprogrammed time sequence. Compressed air is usually derived from the General Services air system, but sometimes a dedicated compressor system is provided. The stimulators for each bunker are supplied from individual air receivers, normally sized to provide up to six air pulses. Control of the equipment is usually from a panel located at the bunker filling level, with provision for local manual control from the bunker outlet position.

Shuttle conveyors

Distribution of coal into the bunkers is by duplicated streams of shuttle conveyors, as shown earlier by Fig 3.78. Each shuttle conveyor spans the length of one bunker but is designed to shuttle across the top of a pair of adjacent bunkers. The shuttle conveyor comprises a reversible conveyor belt within a rail-mounted frame which has a reversible direction of traverse. Separate motors are fitted to drive the conveyor and to traverse the frame. In operation, the frame is arranged to traverse in the same running direction as the loaded conveyor, so that coal is distributed evenly along the length of the bunker. When the frame reaches the far end of one bunker, the conveyor is reversed and the frame is then traversed back again to distribute coal evenly into the other bunker of this pair. The whole sequence repeats continuously under manual control of the conveyor operator. Shuttle conveyors normally have manual belt-tensioning equipment, rather than the automatic tensioning fitted to the other conveyors. Power and control supplies are connected through cable reeling drums which may be power-operated.

13.10.2 Commissioning

Chutes and flap valves

Chute linings have to be checked to ensure that coal will flow smoothly over them, with no corners or ledges likely to promote blockages. Manual operation and indication of each flap valve is checked, then remote operation and indication is proved using the actuators. Correct receipt of the 'blocked chute' alarm is checked by simulation at the diaphragm-operated switch. In conjunction with proving sequential operation of the selected coal conveyor route, operation of the diverter flap valves is again demonstrated before any coal is admitted to the system.

Belt-cleaning devices

Most of the checks are aimed at ensuring that each device is correctly lined up to clean the conveyor belting or pulleys effectively, without causing undue wear. It is also necessary to confirm that all spoil falls into the coal chute, and not on to the floor. Brush-type cleaners must rotate in the correct direction and be adjusted to the distance from the belt that gives minimum brush wear. There must be safe access to adjust or replace the devices under service conditions within the safety rules, but without requiring the erection of an access scaffold.

Belt-weighers

Commissioning of belt-weighers is a specialised activity, usually carried out by the manufacturer to the satisfaction of the official Department of Metrology inspector, who also specifies the time intervals between recalibration. It is advantageous to compare belt weighing accuracy against a known weight of coal passed in a given period from the wagon weighing equipment. Typically, each weigher would be calibrated at four loadings: two at the extreme limits of flow, e.g., 20% and 100% flow, for which the specified accuracy may be guaranteed, and the other two at intermediate flows, say 50% and 80%. Three tests at each flow rate are needed to confirm repeatability of results.

Magnetic separators

It is safer to commission these when the associated conveyor belt is isolated. The drive motor and the rectifier unit for the magnet are energised, and correct operation proved. Selected pieces of ferrous material are placed on to the main conveyor belt in different positions within the influence of the magnetic field to prove that they will be uplifted and deposited into the tramp iron chute. When adjustments have given acceptable results, a mixture of coal and tramp iron is loaded on to the belt, and the separator shown to lift tramp metal out of various depths of coal.

As a final test, pieces of tramp iron should be dropped into the running conveyor belt loaded with coal to demonstrate that the separator is capable of removing the alien material from a moving belt.

Vibrating screens, feeders and crushers

Commissioning of screens involves checking the drive-motor rotation, followed by fitting and tensioning of the drive belts. The flexible supporting mountings are checked; then the unit is operated to ensure that the correct oscillation is produced. After a period of service it is important to check that the flexible bellows at the coal-flow entry and exit are adjusted properly to maintain an effective seal.

Vibrating-feeder commissioning follows a similar pattern, but the total vibrating mass is usually supported on specially designed rubber buffers, which must be checked to ensure that they have been fitted correctly. Drive-motor rotation is checked and then the drive belts fitted and tensioned. The flow regulation gate operation and position indication has to be proved.

The clearance between the hammers and anvils of coal crushers is set to suit the coal sizing desired. The hammer drive-motor and then the anvil belt-motor are commissioned, confirming that both directions of rotation are correct. An extended run then checks that the relative positioning of hammers and anvils is maintained and that the rotating mass is balanced.

When coal is flowing, all items of plant have to be checked to ensure that they are fulfilling their designed functions correctly. Where appropriate, adjustments are made and the results recorded for future setting-up. Drive-belt tensions and security of fastenings have to be checked and adjusted after some 50, 100 and 300 hours of service.

Coal sampling equipment

Each drive motor has to be individually checked for correct rotation, drive-belts fitted and correct operation of all integral conveyor belts and coal-proportioning flaps proved. Operation of the sampling scoop has to be checked, initially with the main conveyor belts isolated and later whilst coal is passing along the main belts. In some installations, it may be practicable to introduce a supply of coal directly into the scoop manually and to check on the ability of the whole sampling system to produce correctly divided samples.

Bunkers and shuttle conveyors

The lining has to be checked for completeness, absence of ridges or ledges, etc. All debris has to be cleared out and free operation of the bunker outlet isolating device confirmed. Where appropriate, pneumatic coal-flow stimulators and associated control equipment have to be commissioned. Bunker flow checks plus operation of the bunker outlet gates are carried out as part of main boiler commissioning in Chapter 6. Compressor plant is commissioned in accordance with the principles outlined in Section 21 of this chapter.

Commissioning of shuttle conveyors follows a similar pattern to that outlined earlier for coal conveyors, with the addition of proving the correct functioning of:

- Manual and automatic, forward and reversing controls for traverse and belt drive-motors
- Manual belt tensioning
- Cable reeling arrangements
- Rail alignment

13.11 Stocking-out/reclaiming plant

13.11.1 System description

Coal for stocking-out is rerouted at one of the junction houses on to a conveyor, which may feed either a simple boom-stacker or a stocking-out/reclaiming machine of the bucket-wheel type.

Boom-stackers

A boom-stacker comprises a conveyor running in a jib structure erected on a slewing platform, which is usually arranged to traverse along a rail track. Separate motors are used to luff the jib, usually from 10° below to 20° above the horizontal, and to slew the jib through 230° of arc. A further motor and gearbox mounted at the conveyor head rotates a bifurcated chute, which distributes coal into a wide pile. Since the boom-stacker only functions as a stocking-out device, separate facilities involving mobile plant have to be provided for reclaiming. These generally comprise a vibrating grid covering an underground hopper, which discharges on to a reclaiming conveyor belt, via either a vibrating or a paddle feeder. Mobile scraper units and bulldozers gather stocked-out coal and deliver it to the reclaim hopper.

Bucket-wheel machines

Bucket-wheel machines (Fig 3.79) are designed for both stocking-out and reclaiming from a relatively large stockpile. When engaged on stocking-out duties, the machine acts in a similar manner to the boom-conveyor, with coal being discharged over the outer end of the reversible conveyor within the boom. The bucket-wheel is inclined at an angle of some 20° to the vertical. In the reclaiming mode, coal is scooped up by the buckets and discharged onto the reversed boom-conveyor through an opening in the inner peripheral baffle around which the cell-less buckets rotate. The Drax bucket-wheel machine is capable of stocking-out at 3000 t/h and of reclaiming at 2500 t/h. It is capable of luffing and slewing within limits similar to the boom-stacker. The bucket-wheel is approximately 9 m in diameter and comprises nine cell-less buckets, each of some 2000 litre capacity. The machine traverses along a central rail track and can stack coal on both sides to a total quantity of some 17 000 t without assistance from mobile plant. Electro-hydraulic drives are used for bucket-wheel rotation, and for the slew drive and luffing system. Conventional motor drives are used for traversing, for driving the boom conveyor and for the short conveyor associated with the collapsible tripper mechanism, discussed earlier.

Mobile plant

The amount of the mobile plant at a site depends upon the capacity of the coal stock area and the plant arrangement adopted for stocking-out and reclaiming. Usually there will be a combination of the following:

- Large-capacity rubber-tyred scrapers
- Rubber-tyred articulated bulldozers
- Conventional heavy-duty tracked bulldozers
- Tracked or rubber-tyred loading shovel

Large power stations are able to store between 1.5 and 3.5 million tonnes of coal. Each bucket-wheel machine can only reclaim some 17 000 t without assistance, so it is obvious that at least two scrapers and bulldozers are needed to support each bucket-wheel machine if the rate of reclaiming coal is to be maintained. A greater number is required if only a boom-stacker has been fitted.

A comprehensive mobile plant maintenance workshop is often provided as part of the coal plant buildings.

13.11.2 Commissioning

In many respects the commissioning checks are similar to those for the ship unloaders, discussed earlier.

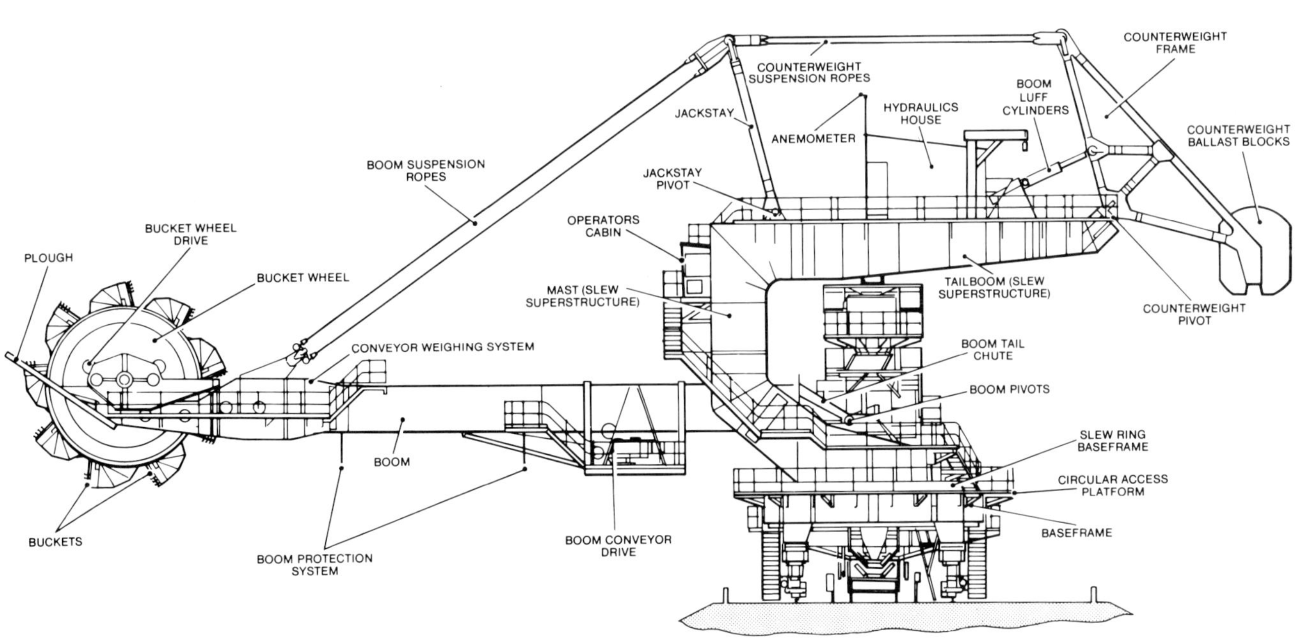

FIG. 3.79 General arrangement of bucket-wheel reclaiming machine

Conveyor belt commissioning has also been described earlier.

Boom-stackers

Generally, commissioning is carried out in a similar manner to that for cranes (see Section 5 of this chapter). Statutory certificates are needed for the jib or boom hoist ropes. Hinged safety-bridles are often fitted to safeguard the boom in event of rope failure. These must be tested before the machine is used for handling coal. As with bucket-wheel machines, it is important to ensure that coal is directed into the receiving chutes for all angles of boom inclination. Commissioning of the rotating discharge chute, with its drive motor, follows the practice outlined previously.

Bucket-wheel machines

Most of the initial checks are done with the machine mechanically secured to its running rails. The following items must be confirmed:

- Erection completed, electrical and hydraulic installation completed and checked to latest diagrams. Special attention is given to lightning protection, if appropriate, and to continuity of earth bonding.
- Works test and statutory certification in order.
- Hydraulic system temporary strainers fitted; also bypasses fitted around the hydraulic motors and the system filled with oil.
- Cable reels checked, power and control supplies energised to facilitate checking of controls and alarms. Motor rotation checks and energising of hydraulic tank heaters completed.
- Ensure that there is safe access to operator's cabin and for maintenance of equipment, even under adverse weather conditions. Confirm clear visibility from operating position. Prove correct operation of emergency escape equipment and ensure that communications channel to nearest manned control room is operational.
- Commission hydraulic pumping sets and flush pipework for each drive unit separately, until all debris has been removed. Remove temporary strainers; include the hydraulic motors into system and prove correct operation of each motion, i.e., slewing, luffing and bucket-wheel rotation.
- Confirm that the warning klaxon is clearly audible from all places where personnel may be at risk.
- Prove all safety features, interlocks and alarms, and then check the alignment over all degrees of motion up to extreme positions.
- Energise the traversing motors and operate the machine over the entire length of the rails. Check that horizontal and vertical alignment is maintained. Observe that the cable reels pay out and takes up without snagging, twisting or kinking the cable.
- Confirm the correct operation of the end-of-travel limit stops and ensure that the reclaiming chute remains aligned over the main feeder conveyor.
- Check that the main feeder conveyor runs through the machine satisfactorily during traversing in both directions.
- Confirm correct operation of the reversible-boom conveyor.
- Prove that the tripper conveyor operates correctly and check its collapsibility while transferring from stocking-out to reclaiming, and vice versa.
- Operate all control motions while machine is being traversed, and confirm that the equipment remains stable. Simulate plant failures to check that the control response is always to fail-safe.

It is important to check that any accumulation of storm water on the horizontal run of the main feeder conveyor can be safely disposed of when the conveyor is started and before coal loading commences.

- When coal is available, carry out load checks, rates of flow, etc., for both stocking-out and reclaiming operations.
- Demonstrate that the machine can create a coal stack to the designed dimensions and also reclaim most of it without using mobile plant.
- Confirm the stability of the machine in high winds, when handling coal at various rates of flow.

Reclaim hoppers

The hopper-mouth screen vibration equipment has to be checked and commissioned, including the interlocks which inhibit vibration when mobile scrapers approach the grid. One obvious check is to see that the hopper mouth is slightly above the level of the surrounding area to reduce ingress of ground water. The hopper has to be cleared of all debris before paddle or vibrating feeders and reclaiming-conveyors are commissioned, in accordance with principles detailed in earlier paragraphs.

13.12 Control systems

13.12.1 System description

Much of the plant is controlled from a mimic diagram in the coal plant control room. The coal conveying plant and the railway signalling controls are located on separate mimics. Overall handling of all rail traffic is conducted from the plant control room. The train unloading operation is carried out under the direct control of an operator stationed in the unloading house.

Sequence controls, interlocks and alarms, similar to those discussed in Sections 7 and 8 of this chapter, are all integrated to ensure that the plant cannot be operated incorrectly. Generally, the control system is arranged so that ship unloaders, paddle feeders, shuttle conveyors and stocking-out/reclaiming machines have to be started local to the plant. Thereafter, the conveyors may be started either individually or in a sequence from a single initiation from the control panel. In each case, all of the appropriate interlocks have to be satisfied before any plant can be started. Conveyor belt sequencing, for instance, requires a coal flow path to be set up by selecting the chute diverter flaps to the appropriate conveyors. Initiation of the 'start' sequence automatically starts the furthest downstream conveyor and then progressively works back against the direction of coal flow in the group selected. Prior to starting of the conveyors, a warning alarm is automatically sounded as part of the sequence. Should any conveyor trip, or be stopped out of sequence, all of the other upstream conveyors will also trip automatically.

Closed circuit television is fitted so that the coal plant control room staff can monitor safe operation of the railway sidings, unloading equipment, paddle feeders and some of the more remote coal conveyors.

13.12.2 Commissioning

Each of the control panels is commissioned generally in accordance with the principles outlined in Sections 7 and 8 of this chapter together with supporting sequence, interlocking and alarm equipment. Essentially, control supplies are made available: then, by simulation at the equipment terminals, all functions, indications and alarms are proved. When plant cabling is available, the exercise is repeated using plant-mounted devices as far as practicable.

Plant is individually operated from either the local-to-plant or the switchgear control location first, i.e., in an out of sequence mode. When satisfactory operation has been confirmed, individual items are proved from the control panel in a remote/manual mode. Controls for groups of plant are then selected to 'in sequence' to check that the plant availability interlocks are satisified. Having proved that all controls and interlocks are functioning correctly, a full sequential start is initiated. Various combinations of conveyors and flap valve selections are tested to confirm that the logic controls continue to provide the prescribed features. It is important also to conduct negative checks, i.e., to create fault conditions and to ensure that the control systems function correctly.

14 Ash and dust plant

This section outlines the principal activities for commissioning plant and equipment directly involved with the collection, handling and disposal of all ash and dust arising from a pulverised-fuel fired (PF) boiler plant.

Much of the information given in this section is based on the plant installed at Drax Power Station. Reference to other types of plant is also given, where appropriate.

14.1 General description

Modern ash and dust plant system designs follow a fairly well defined pattern, though the disposal arrangements have to be tailored to suit the location of the site.

A 4000 MW power plant may consume up to 39 000 t of coal per day and create some 7800 t of ash in the process. The generic term *ash* in this context covers three classes of product, roughly proportioned as shown below:

- Furnace-bottom ash: approximately 1250 t/day, or 9 t/h/boiler
- Boiler grits: approximately 300 t/day, or 2 t/h/boiler
- Precipitator dusts: approximately 6250 t/day, or 43 t/h/boiler

Modern plants seldom have mechanical cyclone-type grit arresters fitted upstream of the electrostatic precipitators. Most of the grits now accumulate in economiser and gas air-heater hoppers; the remainder are collected by the electrostatic precipitators.

The essential components of an ash and dust plant are illustrated in block diagram form by Fig 3.80. Furnace-bottom ash is abstracted and conveyed to ash pits by high pressure water jets. Water is recycled to a reservoir for reuse, while ash is grabbed out of the ash pits for either disposal or sale. Dusts and grits are withdrawn from the various collection hoppers and transferred to a common central dust bunker by compressed air. The contents of the bunker are then either sold or disposed of at an off-site dumping ground. Disposal of ash and dusts generally entails mixing it into a water-based slurry, which is pumped to large lagoons. After settlement, and decanting the water, the resultant bed forms an excellent foundation for land reclamation projects. Another method of disposal is to waterspray the dust into a 'just-damp' condition and then convey it, with the furnace-bottom ash, to a disposal ground using conventional belt conveyors. There is a commercial market for up to 30% of the total production of ash and dust, but the plant has to be designed for disposal of the entire quantity.

Control of the plant is exercised from panels, with the control switches and indicators located on a mimic diagram of the plant. Generally, separate panels are fitted, one for handling and the other for disposal of ash and dust. Dependent upon the plant layout, each panel may be in a separate building, or both in the same room. At Drax, both panels, plus the coal plant control panels, are all in the same room.

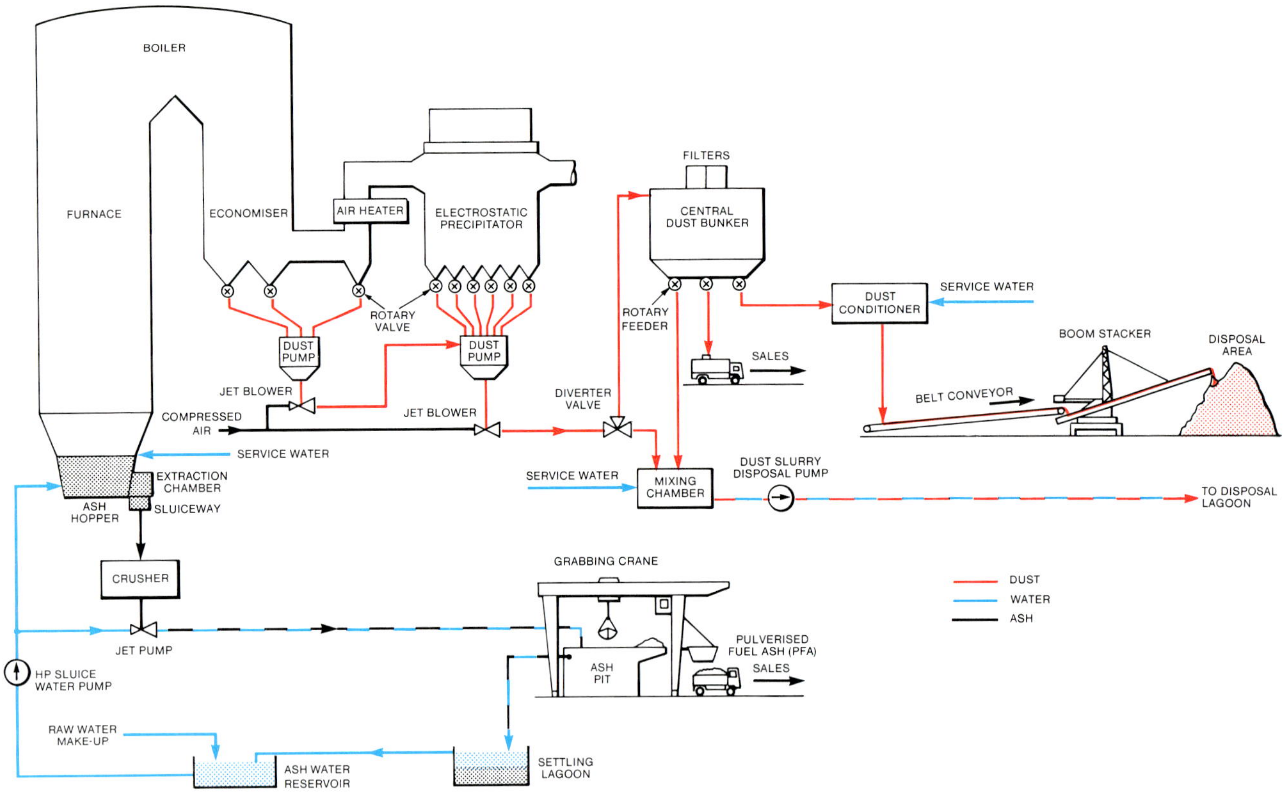

FIG. 3.80 Ash and dust plant — simplified block diagram

Duplicated main electrical supplies are usually provided, so that at least one half of the plant can be kept available to service all boiler units, in the event of an electrical system fault. The majority of the plant is driven by 415 V three-phase AC motors, but some of the higher rated drives need 3.3 kV motors.

In order to deal with spillages of dust, vacuum cleaning plant is installed, with pipes leading to service connection points strategically located about the ash and dust handling equipment. Dedicated compressed air plant is fitted to convey grits and dusts from the various collection hoppers to the central dust bunker.

Throughout the remainder of this section, the term *ash* refers to furnace-bottom ash, and *dust* refers to grits and dusts from the boiler flue-gas ducting collection hoppers.

14.2 Commissioning objectives

In addition to the general objectives listed at the beginning of this chapter, the following are specific to ash and dust plant:

- Ensure that sufficient plant is commissioned in time to provide ash and dust disposal facilities for main unit commissioning.
- Check and demonstrate that all plant is set to work in a safe manner, with all interlocks and safety circuits functioning correctly.
- Prove that air-compressor plant is capable of meeting the expected demands.
- Demonstrate that ash plant can adequately deal with the range of ash types and sizes from the furnace-bottom hoppers.
- Confirm that electrostatic precipitators can be set to their optimum collection efficiency.
- Demonstrate that the precipitator mechanical key interlocking scheme gives full protection to personnel, and that no keys are interchangeable.
- Prove that the dust plant can adequately handle the range of grits and dusts collected in the gas-pass hoppers.
- Show that emergency emptying of flue-gas-ducting dust hoppers can be carried out safely, even if the carbon-in-ash is relatively high.
- Check that the disposal system can handle the range of products at the specified rates. Grab-crane and belt-conveyor performance is of particular interest here.
- Prove that conveyor belts are capable of being restarted, when fully-loaded with damp ash and dust.
- Ascertain that hopper trace heating loadings are adequate to prevent formation of damp dust, particularly near the outlet points.

14.3 Safety considerations

The main risk to plant is from conveyor belt fires, and the same precautions that were taken with coal handling plant have to be adopted for ash conveyor belts (see Section 13.3 of this chapter).

If due care is not exercised, personnel can be exposed to the following risks during operation of ash and dust plant:

- Falling into unguarded sluiceways, ash pits, hoppers, etc.
- Being trapped by moving machinery, e.g., grab cranes.
- Getting caught up in moving conveyor belting.
- Suffocation in event of a major dust escape in underground chambers holding dust pumps, etc.
- Drowning as a result of falling into ash pits and water reservoirs or tanks.

The following items should be given proper consideration, in conjunction with those outlined at the beginning of this chapter, before commencing any commissioning activity:

- Safety circuits, including conveyor emergency trip wires, to be fully proved before any plant is run.
- Where appropriate, all safety guards must either be in place or temporary guarding provided.
- Conveyor belts must be inspected for damage and freedom from obstruction before initial running, then kept under direct supervision during the first few hours of operation.
- Adequate telephone and two-way radio communications to be available before starting any of the water, ash or dust systems.
- Work in hoppers, bunkers, etc., to be in accordance with the appropriate safety rules.
- Statutory test certificates must be available for all lifting equipment, including grab cranes and boom conveyors. Personnel-escape safety gear to be tested and available in control cabins.

Special care is needed during removal of ash and dusts arising from early operation of the boiler plant, when they are likely to contain a high proportion of combustible material. An explosive mixture is easily created, particularly should there be a direct discharge of very hot dust into the open air. The emergency disposal system should always be used.

14.4 Resources

14.4.1 Services

In addition to having the electrical power and control supplies systems commissioned and ready for use, the following will be needed:

- Two-way radio communications.
- PAX telephone system between various positions along the ash and dust routes.
- Fixed fire fighting system, including fire hydrants, where appropriate.
- Plentiful supply of portable fire extinguishers.
- Station drains system for storm and foul water.
- Domestic water supply for ablutions and fire hose reels.
- General services water system for ash plant services and compressor cooling.
- General services and/or control and instrumentation compressed-air systems for ash and dust plant pneumatic controls.
- Industrial-grade water supply for ash plant make-up and dust conditioning.
- Supply of sand or dry ash for initial protection of furnace-bottom hearths.
- Load of ash, and perhaps dust, to carry out initial proving of plant in advance of firing the main boiler with coal.
- HV AC test set suitable for 70 kV.
- Insulation testers 500 V and 5 kV DC.
- Digital micro-ohmmeter.

14.4.2 Staffing

Several teams of three will be needed to work concurrently on precipitators, ash and dust handling equipments and conveyors, including sequence and control systems and air compressor plant.

Personnel drawn from the station permanent staff complement will be needed to operate plant to the contractor's direction from the very early stages of commissioning activity. Progressive involvement of a supervisor and up to five men several months before the production of the first ash or dust would be advantageous.

The overall programme time for commissioning could occupy a three to nine month period, dependent upon the plant erection sequence for the particular station. Typical times for testing some of the individual plant items are:

• Control equipment	10 weeks
• Dust pump	4 to 6 days for each item
• Compressor	
• Conveyor	
• Grab crane	
• Ash water pumps	
• Ash crusher	

14.5 Commissioning

14.5.1 General

Figure 3.81 shows a typical basic commissioning network of the activities necessary to commission the plant for handling and disposal of ash and dust, arising from firing of the first boiler unit with coal.

Though the following notes are based on the arrangement at Drax Power Station, the reader should be able to identify the basic principles for testing similar systems.

For convenience, the subject matter has been subdivided, for both descriptive and commissioning purposes, into the following areas:

- Furnace-bottom ash handling
- Flue-gas dust handling
- Dust disposal system
- Electrostatic precipitators
- Vacuum cleaning and compressed-air plant

14.6 Furnace-bottom ash handling

14.6.1 System description

Figure 3.82 shows the general arrangement of a typical ash handling plant. Ash, formed during the coal combustion process, accumulates in hoppers at the bottom of the furnace. High pressure water jets propel the ash into external sluice ways, where further jets assist ash flow to ash crushers. Crushed ash falls into a large tank, from which water-powered jet pumps transfer the slurrified ash to reinforced concrete ash pits. Water continuously decants from the ash pits into an ash pit sump, whence it is pumped to settling lagoons.

After settlement of the fine particles of ash, clarified water flows into the return-water pit and is then pumped to the ash water reservoir. High pressure (HP) sluice pumps take their suction from the reservoir and supply water to the sluicing jets and ash-crusher jet pumps. The HP sluice pumps also supply water to valved outlets beneath the boiler gas-pass dust hoppers for use with the emergency dust jet pumps, which discharge into a separate sluiceway direct to the settling lagoons. The system uses fresh water with make-up into the ash-water reservoir from either the site towns water reservoir, a bore hole, or an adjacent river.

Accumulated ash is grabbed from the ash pits and deposited on to sloping aprons adjacent to each pit to allow surplus water to drain into the ash pit sump. Drained ash is later grabbed into a feeder hopper, on to a conveyor and thence to the combined ash and dust disposal plant. Alternatively, the ash-grabbing cranes may load drained ash into road vehicles for sale or for disposal.

A hopper and water ejector are usually provided to deal with coal rejected from the pulverising mills of each main boiler unit. This material is normally admitted to the sluiceway where it can pass through the ash crushers, in case there are any relatively large lumps.

Detailed description of some of the more important parts of the plant is given in the following paragraphs.

Furnace ash hopper

This usually is of the dry-bottom type, formed into one continuous hopper with a number of shallow hearths which slope downwards to the hopper extraction chambers at the rear of the boiler. Figure 3.83 shows the sectional arrangement of a typical dry-bottom ash hopper. The entire hopper is lined with fire-resistant bricks. It is capped by a continuous seal water trough, into which a steel dipper-plate fastened to the boiler waterwall tubes projects. This forms a flexible seal which prevents leakage of air into the furnace, yet will accommodate downward thermal expansion (some 250 mm) of the boiler. Refractory-covered heat shields are fitted on the furnace side of the dipper plate to protect the water seal from radiated heat. Each hopper is fitted with up to 30 quencher spray nozzles, to form a continuous water-curtain wall, which prevents combustion within ash in the hopper. Water supplies for the sealing trough and quencher sprays are taken from a service water system (see Section 18 of this chapter).

Every hearth is fitted with an ash extraction chamber, housing an inner semi-rotary air-sealing gate, an outer observation door and either one manually-operated oscillatory HP extraction nozzle or twin fixed extraction nozzles. Often a fixed nozzle is also fitted at the upper end of the hearth. Water supply to all of the fixed extraction nozzles is regulated by pneumatically-operated valves, controlled by manually-operated pilot valves grouped on top of each extraction chamber.

Sizing grids limit the lumps of clinker passing through into the sluiceway to approximately 250 mm. Ash discharged from all of the hearths flows into a covered sluiceway, fitted with a wear-resistant liner. Sections of the liner are actuator-operated to route the ash-flow into one or other of the two ash crushers. High pressure jet nozzles are set into the sluiceway liners to assist the flow of ash and clinker.

Ash crusher and ancillary plant

Both ash crushers discharge into a common basalt-tile-lined steel-fabricated crushed-ash tank (or jet pump loading box), fitted with high pressure agitation nozzles. Each ash crusher comprises a coupled pair of counter-rotating profiled wear-resistant rollers, electrically driven through a speed-reducing gearbox. The rollers are capable of reducing large lumps of clinker to 50 mm cubes. Either a fluid coupling or a belt drive is fitted, to reduce torsional vibrations being transmitted along the crusher shaft to the drive motor. A shear pin is fitted to prevent overloading the drive mechanism. Electrical supplies to the drive motor are interlocked

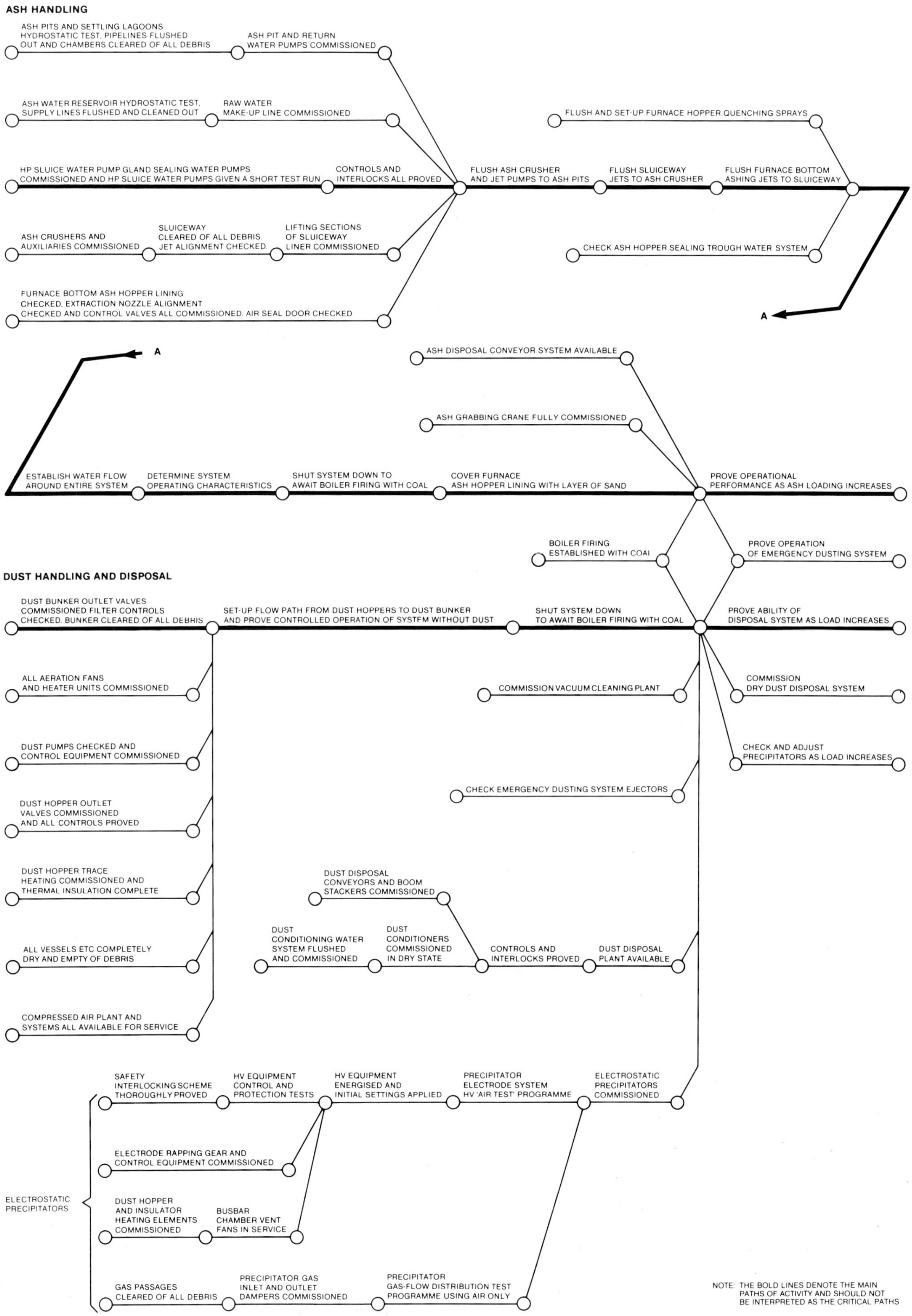

FIG. 3.81 Ash and dust plant commissioning network

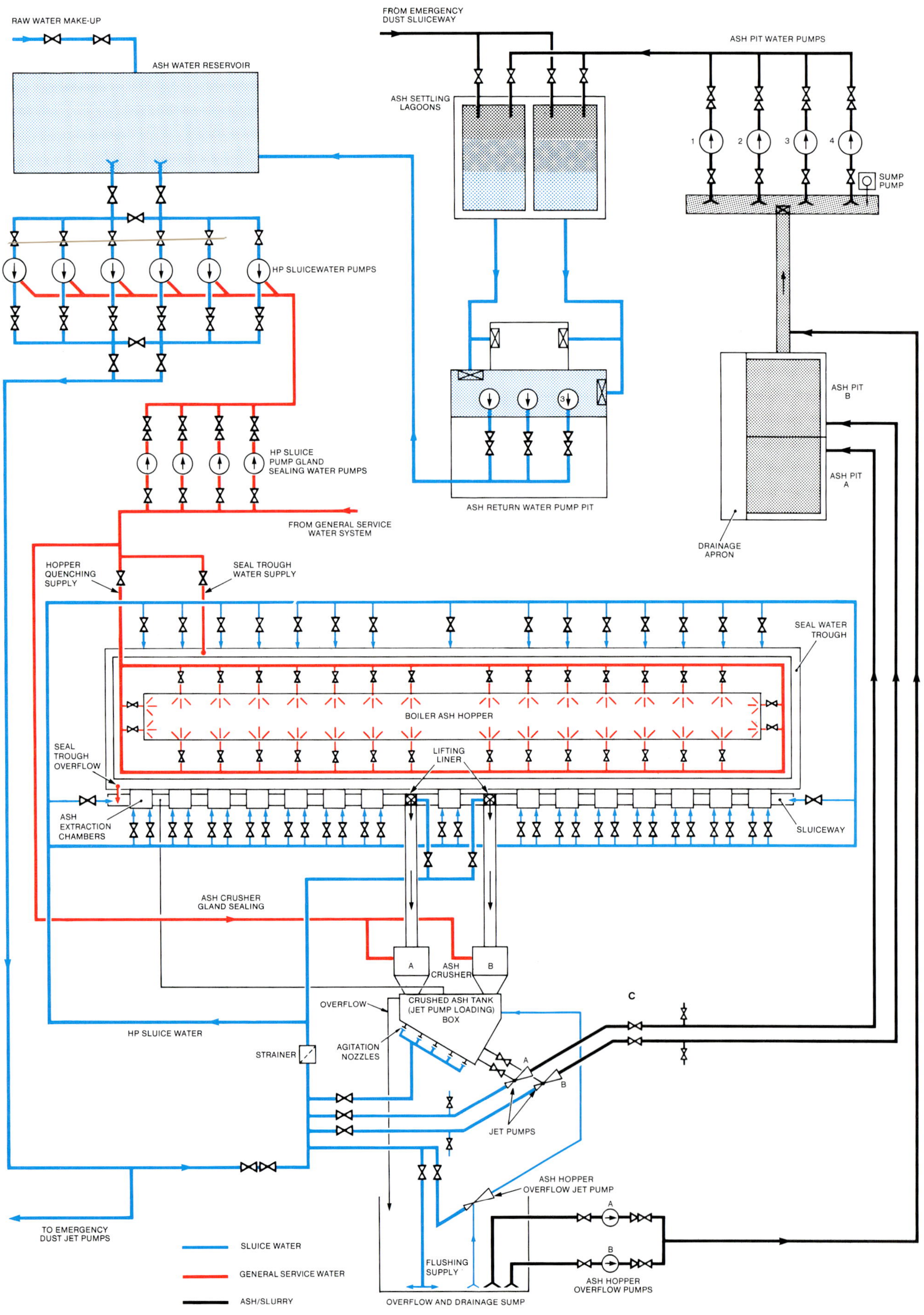

FIG. 3.82 General arrangement of ash handling plant

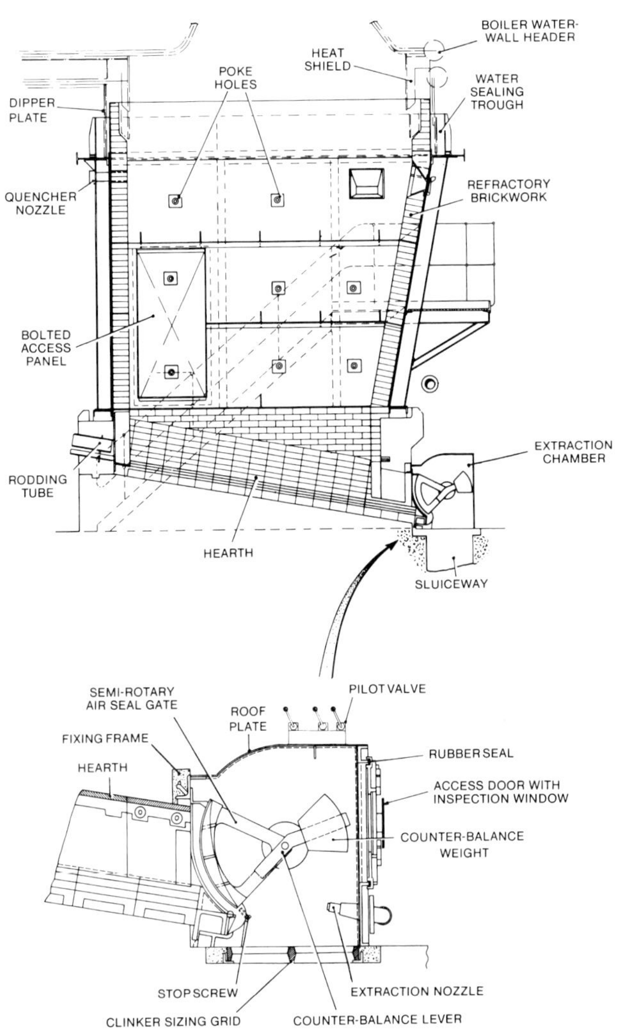

FIG. 3.83 Sectional arrangement of furnace-bottom ash hopper

with a hand winding facility in order to permit manual rotation of the crusher to free blockages.

Ash from the crushed-ash tank is pumped to the ash pits by one of two jet pumps, through pipes typically of 250 mm diameter with a 20 mm lining of basalt. These also derive their motive water supply from the high pressure sluicing system to maintain sufficient velocity to keep the ash in suspension.

Should the crushed-ash tank overflow into the concrete containment structure, the water is pumped to the ash pit drainage channels but ash spillage is returned to the crushed-ash tank.

Ash pits

Reinforced-concrete-divided ash pits, of which Fig 3.84 is a typical example, are provided for each main boiler. They are usually arranged in a line, so that all can be serviced by two or three 12 t capacity gantry-type grabbing cranes. The cranes travel along rails, so that ash can be grabbed out and deposited onto the ash pit drainage apron with the minimum of cross-traversing. For transfer of ash to road vehicles or into the ash conveyor feeder hopper, each gantry is fitted with an extension which overhangs the roadway.

While ashing is in progress, water decants through perforated plates into high level troughs. Drainage louvres and wire-mesh strainers are set into the walls of the ash pits: these are closed whilst ash is flowing into the pit, but are opened afterwards to permit drainage of lower-level water entrainment into the ash pit water sump. Water jets are fitted to backflush the louvres and associated mesh strainer material. Scum boards, fitted at water surface level, also serve to protect the screens from being damaged by the crane grab. Decanted water, plus drainage from the apron, is led to a common ash pit water sump.

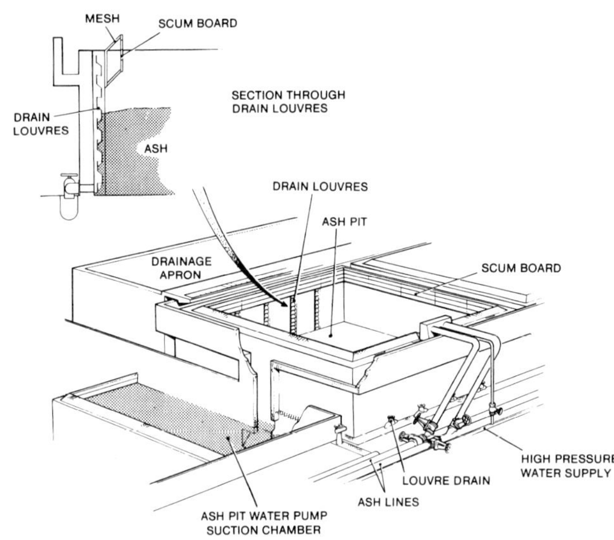

FIG. 3.84 Typical arrangement of ash pit

High pressure sluice pumps

These take their suction from the ash-water reservoir and supply high pressure water at 12 bar for the following functions:

- Furnace-hopper ashing jets
- Ash sluiceways jets
- Crushed-ash transfer jet pumps
- Crushed-ash tank overflow jet pump
- Dust hopper emergency sluicing jet pumps

The discharge branch of each pump is sometimes fitted with an orificed leak-off line to the ash-water reservoir to maintain a minimum recirculation flow through the pump at all times. Sealing of the pump glands is achieved from a separate system supplied with service water. Failure of sealing-water pressure trips the sluice pumps.

Relatively large-mesh strainers are fitted into the common supply lines to the water jets to reduce the chances of blockage. To give increased wear-resistance, the bore of each jet nozzle is fitted with a removable ceramic insert.

Ash disposal

Usually ash is grabbed from the ash pits, allowed to drain on the ash pit apron, then later transferred by grab into road vehicles, via a feeder hopper to reduce spillage. Road vehicles transport the ash to a dumping ground, or take it to a sales outlet.

Some of the larger power stations have a troughed belt conveyor alongside the ash pit drainage aprons. Partially drained ash is grabbed from the apron into a movable feeder hopper, which directs ash onto the conveyor. Thereafter, a system of conveyors transfers the ash to a remote dumping area, along with conditioned dust from the dust handling plant.

14.6.2 Commissioning

The basic aim is to commission each item of plant individually, using clean water. Then the entire system is operated with water only, until its operational stability has been proven. If practicable, controlled quantities of ash should be used to confirm the system capabilities before the first boiler is fired with coal (see Fig 3.81).

Before firing each boiler with oil, the furnace hopper should be covered with fine ash or sand to prevent damage to the hopper tiling and contamination of the sluiceways in the event of oil spillages.

Each of the reinforced-concrete structures forming ash pits, reservoirs or settling chambers has to be hydrostatically tested, by filling with water to the overflow level. The order in which this is done will depend upon local conditions, but an attempt should be made to transfer water from one structure to another to minimise waste. It is advantageous to be able to commission the electrically-driven pumps concurrent with this activity, as it shortens the overall programme by proving parts of the system at an early stage. Ultimately the ash-water reservoir should be left full for system flow trials.

Some of the more important aspects that need to be considered are as follows.

Furnace ash hopper

- Hopper sealing-trough watertight and overflow pipe large enough to deal with maximum filling rate.
- Furnace-bottom dipper plates set to give appropriate clearance for boiler downward expansion.
- Flame shield correctly fitted and tiling complete.
- Furnace-hopper hearth and wall tiling as smooth as possible, with all expansion joints correctly fitted. All construction debris removed.
- Swinging-nozzle linkage lubricated and checked to give uninterrupted travel over the prescribed arc.
- Alignment of all jets proved to be correct for their location.
- Air-door seals set to give best fit without binding, with due allowance for when the door is later exposed to hot ash in service.
- Sluiceway grids securely fastened, so that large clinker can be manually broken down on top of them.
- Quenching sprays to be as high up the hopper wall as practicable. Waterspray coverage to be checked for direction and rate of flow. Pressure to be recorded for future information.
- Removable hatches and safety grid fittings checked for ease of operation (if hopper is fitted with facilities for poking or rodding ash-bridges between furnace-bottom tube banks).

Sluiceways, crushers and ash pipework

- Access covers fitted properly to seal sluiceway, so that they do not cause any tripping hazards, yet can be easily removed if required.
- Liners correctly fitted with no ledges to promote blockages. All ashing jets checked for correct alignment.
- Movable liners for directing ash into duty crusher lubricated and set to work, including a check on the ability to seal-off the out-of-service crusher.
- Basalt-lined piping to have lining thickness checked before each length is fitted into position.
- Crusher tiling to be correctly fitted, with no ledges likely to impede flow.
- Crusher drive-motor rotation checked and all safety features proved. Fluid coupling or belt drive fitted and correctly adjusted, then drive proved with machine empty. Bearing sealing-water system checked out and seals adjusted.
- All debris cleared out of sluiceways, crusher and crushed-ash tank.

Ash pits, settling chambers and ash-water reservoir

- Ash and water-containing structures all cleared of construction debris and hydrostatically tested to confirm that they are water-retaining. Where appropriate, prove that overflow lines are adequately sized and form a clear route to a drainage point.
- Ash pit drainage louvres correctly adjusted. Back-flushing and draining facilities checked for correct operation.
- Ash pit inlet splash-pan adequately supported against thrust loadings.
- Ash pit straining grids and scum-boards correctly set and sealed to prevent leakage, where appropriate.

Electrically-driven pumping plant

- Motor insulation resistance, earth bonding and rotation checks completed.
- Control functions and interlocks from level controllers, etc., including auto/standby sequences, all proved with pump switchgear 'on jumpers' (see Section 8 of this chapter for details).
- Gland-sealing water for HP sluice pumps commissioned and seals adjusted for initial running tests.
- Drainage sump-pump automatic operation proved in advance of any other plant, to ensure that any accidental flooding of low level chambers can be dealt with expeditiously.
- Individual ash-water pumps primed with water and given a short run with the discharge valve closed, before being given an extended run pumping clean water from one structure to another. The extended run may be incorporated into the hydrostatic test programme or left until system testing is undertaken.
- It is important that the suction pipework to the HP sluice pumps is flushed clear before the pumps are started for the first time. Equally important is that the sealing-water supply and starting interlocks are checked, and that the leak-off line, if fitted, is proved to be clear.

System testing

Before commencing this stage, it is prudent to lift up sections of the sluiceway cover-plates to confirm that no debris is present and later to permit observation of the jet nozzles under flow conditions. Safety barriers must be erected around the openings. It may be advantageous to provide temporary pipework which would permit the settling lagoons to be by-passed for most of the tests, after the normal line has been flushed clear. Personnel should be positioned at strategic positions and be in two-way radio communication with each other before any water is pumped round the system. The make-up supply line to the ash-water reservoir has to be available, with preliminary settings applied to its control valve, if appropriate.

The system should be flushed in stages, with the ceramic nozzles removed from all jets. The first stage involves flushing the crushed-ash jet pumps and the ash discharge lines to the ash pits, with the sluiceway and ash-hopper jets isolated. The jet pump and ash discharge line nozzles are then refitted and the sluiceway jets flushed into the crushed-ash tank from which the jet pumps transfer the contents to the ash pits. After completion of this flush, the sluiceway nozzles are fitted, then each of the furnace-hopper lines flushed and the nozzles refitted. The HP water supply line strainers need to be inspected and cleaned, if necessary, after the flushing stage.

When the HP sluice pumps are restarted, the ashing-jet flow pattern and control of front and rear jets for each of the furnace-bottom ash hoppers is checked, as is the sluiceway jet flow distribution. As the entire system is gradually put into service, careful observation of the following items is necessary:

- Free flow of water through the whole system, recirculating eventually to the ash-water reservoir.
- Level control for each group of pumps confirmed to be operating correctly.
- Ash pits splash-pan distribution acceptable.
- Adjustment of make-up control to maintain level in the ash-water reservoir, as a working level is attained in all chambers.
- Check correct operation of ash crusher sluiceway diverter liner sections.
- Pump bearings and seals to be checked throughout an extended period of operation.
- Auto/standby functions checked, by creating simulated fault conditions, to confirm correct operation of all associated plant.

The plant should be run in this fashion for several days to establish confidence in its stability under various normal and abnormal modes of operation.

If practicable, several tonnes of ash should be obtained to test the plant in advance of the first boiler being fired with coal. This enables an assessment to be made of crusher operation and the handling of mill rejects. At some stage, it will be opportune to prove the water ejectors and emergency dust disposal arrangements, again in advance of coal firing.

When all tests have been satisfactorily concluded, covers are replaced, strainers cleaned out and the system left ready for use when required. The furnace-bottom ash hoppers should be filled with sand or dust to protect the tiling during initial firing of the boiler with fuel oil. The quenching sprays should be left shut off until coal firing commences.

Ash disposal

Much of the plant is also used elsewhere on the site and the commissioning activities are outlined in the appropriate sections of this chapter as indicated:

- Grabbing cranes commissioned in line with the principles discussed in Sections 5 and 13.
- Conveyor belt commissioning conducted in a similar manner to that outlined in Section 13.
- Movable feed hopper over ash conveyor confirmed to traverse, without fouling any structural steelwork or other plant. Its alignment with the associated conveyor to be correct at all positions. If more than one hopper is fitted per conveyor, then the anti-collision interlocks need to be proved.

Subsequent proving of the ash handling system is completed in conjunction with main boiler plant commissioning, as outlined in Chapter 6.

14.7 Flue-gas dust handling

14.7.1 System description

The dust plant collects grits and dust from the air heater, economiser and precipitator hoppers and transfers it to a common central bunker. Figures 3.85 and 3.88 show the general arrangement of typical dust handling systems. The transfer of dust from the various collecting hoppers to the central dust bunker is regulated by automatically sequenced pumps and valves, which simultaneously clear precipitator and economiser hoppers.

Hoppers and ancillary equipment

Up to 30 precipitator and 10 economiser dust collection hoppers may be fitted to a modern 500 MW or 660 MW boiler. The lower external surfaces of each hopper are electrically trace heated and thermally-insulated to prevent condensation within the hopper during off-load periods. Should the dust become damp it will not flow freely through the system. Thermostatic control is designed to maintain a temperature of about 35°C (under normal service conditions, some of the dusts could be at 200°C).

In an emergency, the dust may have to be emptied directly into separate ash sluiceways via portable jet pumps, rather than be conveyed to the surge hoppers. Every collection hopper is fitted with two outlets, each of which has a quadrant-type isolating valve, as shown on Fig 3.85. The in-line valve is either pneumatically- or electrically-operated for normal sequential dust disposal via the jet blower mounted directly beneath it. For emergency emptying purposes, the other manually-operated valve terminates at a flange, onto which a mobile emergency disposal device is coupled. The emergency disposal device incorporates a jet pump supplied with motive water from the HP sluicing system, via outlet points strategically located adjacent to the hoppers. The emergency dust jet pump discharges into a system of emergency dust sluiceways, which conduct the dust slurry to the ash plant settling lagoons. Periodically, each lagoon has to be taken out of service for mechanical removal of accumulated dusts.

In normal operation, the jet blowers transfer dust into two surge-hoppers: one for all precipitator dusts and another for economiser and air heater grits. Both hoppers are trace heated and lagged to prevent condensation during shutdown periods. It is usual to pump the economiser surge-hopper contents into the precipitator surge-hopper, then pump the combined dust to the central dust bunker. The precipitator surge-hopper is also fitted with aeration pads around the outlets to the dust pumps. Dry heated air to maintain the dust in an aerated state is provided by duplicated fan heater units, the air ducts of which are lagged to prevent heat loss. Each surge-hopper is vented to the boiler gas ducting to dissipate any accumulation of air pressure from the jet blower operation.

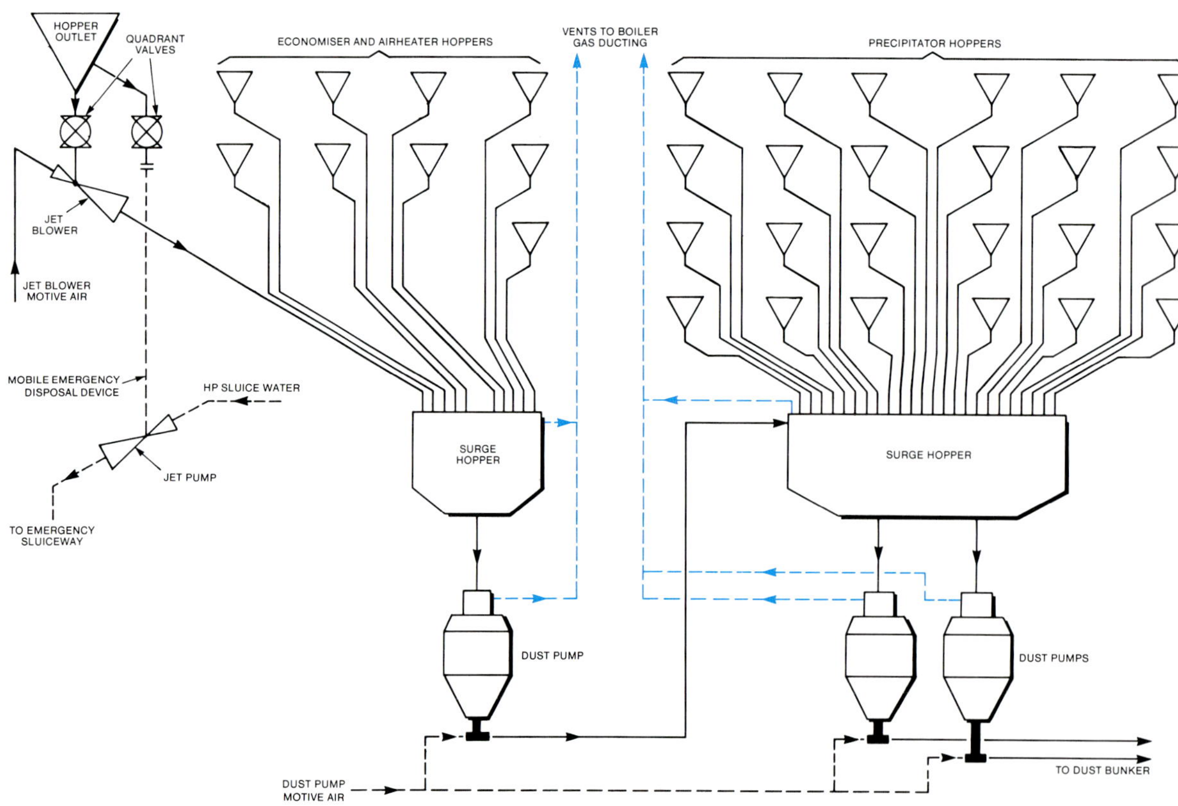

FIG. 3.85 Dust handling system

Another means of transferring dusts is by fluidising air conveyors. These are not generally adopted on modern plants, unless there is a particular advantage over the more effective jet blower system. Such conveyors comprise an inclined rectangular-section steel trunking divided along its entire length by a horizontal porous diaphragm. The space below the diaphragm is charged with low pressure air, at approximately 30 mbar, from small fans. Air percolating through the diaphragm forms a layer of fluidised dust which promotes a steady flow down the slight incline to the surge-hopper. Damp dust has a very deleterious effect on the porous diaphragm, so the aeration air must be kept warm and dry, if dust blockages are to be avoided.

Dust pumps and associated plant

Two forms of dust pump are commonly used. Though the means of regulating the flow is different, both employ compressed air for the actual dust-conveying operation.

The first, now less common method, comprises a motor-driven screw-impeller pump (see Fig 3.86). Dust from the surge-hopper enters the impeller, then advances toward a check valve disc and is compacted by the impeller screw. Compacted dust enters the check valve housing (or mixing chamber), where it is fluidised and transported to the central dust bunker by compressed air, typically at a pressure of 2.5 bar, introduced through a series of air jets.

Modern dust handling designs tend to use the pressure vessel pump, as shown by Fig 3.87. A cylindrical steel pressure vessel is supported on spring mountings. The inlet to the vessel contains a butterfly isolating valve above a mushroom valve at the top of the container. The vessel, which is vented into the surge hopper, fills with dust in a few seconds, and as it moves downwards on its spring mountings, the vessel actuates the control system to initiate the discharge cycle. The butterfly valve closes to stop the dust flow, then the mushroom valve closes to seal the vessel. Compressed air is applied automatically to pressurise the vessel and also to feed the aeration nozzles and discharge jet, which moves the fluidised dust into and along the transport line. When the vessel has emptied, the control equipment shuts off the compressed air and reopens the dust valves. The cycle repeats continuously as long as there is a dust load available.

Dust transport piping is manufactured from wear-resistant steels. The pipe run is designed to have straight runs, with the minimum number of bends. Bends have to be specially strengthened to compensate for the excessive wear encountered at these points.

Central dust bunker and associated equipment

The central dust bunker (refer to Fig 3.88) is usually either a reinforced-concrete or a fabricated steel plate structure, designed to contain the total dust from two days of operation. The bunker is usually divided into two equal parts for operational flexibility. At some power stations, dust from specific boilers can be selectively stored in separate smaller bunkers, should it be suitable for a particular purpose. Dust transport

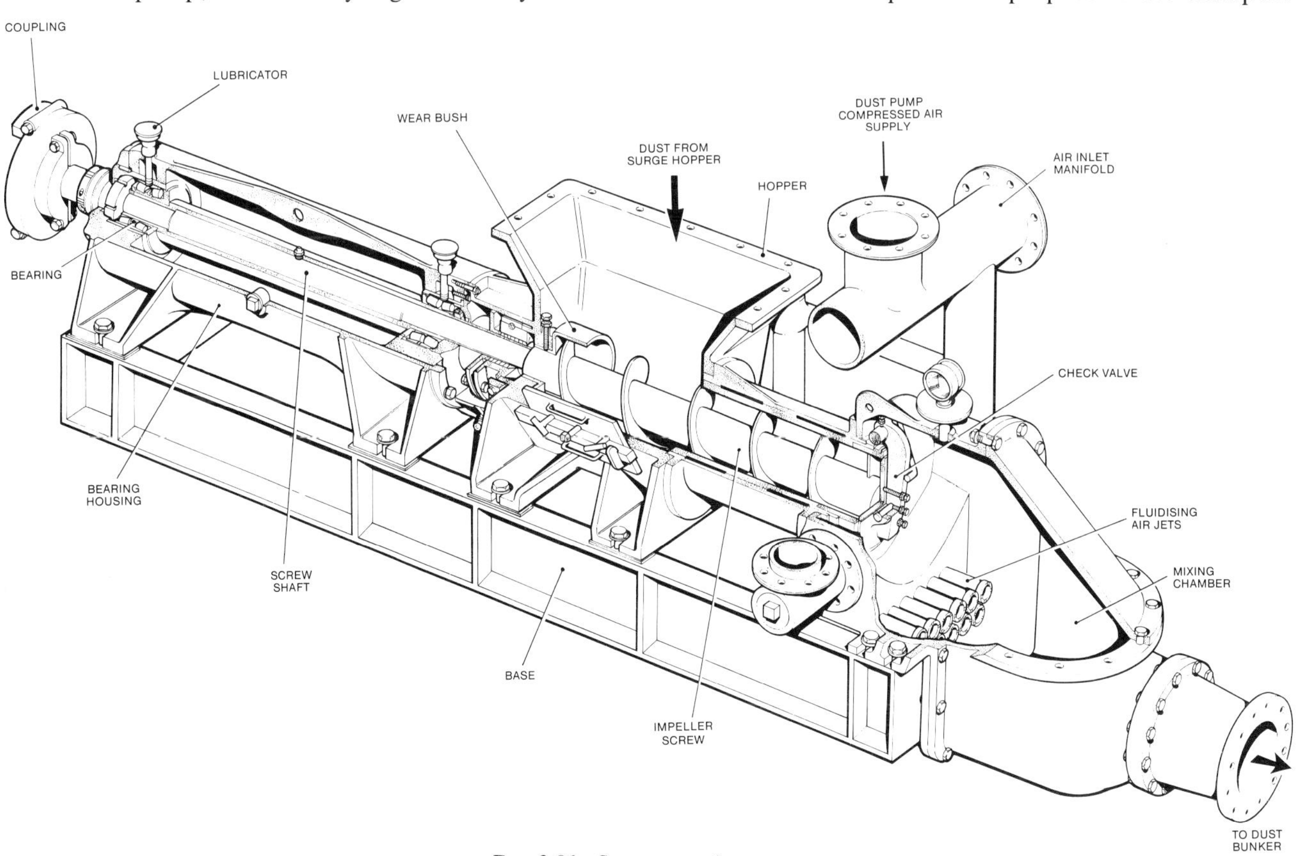

FIG. 3.86 Screw type dust pump

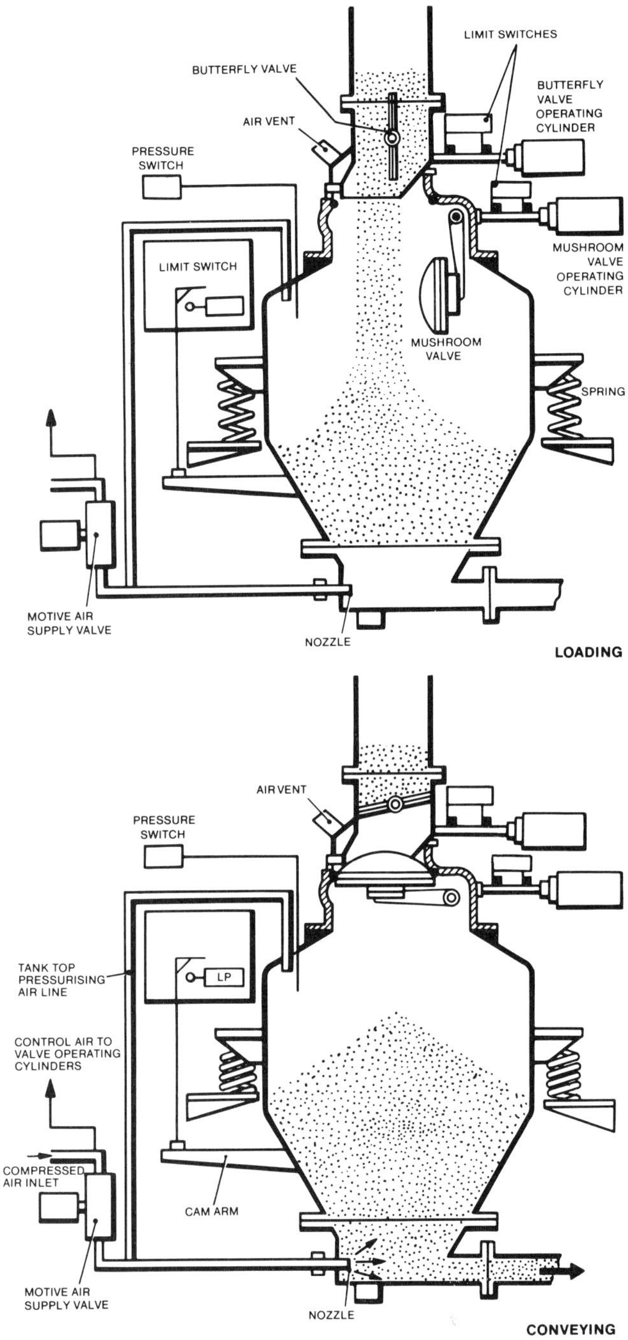

FIG. 3.87 Pressure vessel type dust pump

lines from the dust pumps for each main unit are led via inlet selector valves to either half of the dust bunker, as required by the plant operator.

Aeration pads set into the lower surfaces and base of the dust bunker fluidise the contents to promote flow of dust to the bunker outlets for disposal. Dry air for the aeration pads is supplied from a group of LP fan heater units, delivering heated air at a pressure of some 200 mbar.

Each dust outlet is fitted with a quadrant-type isolating valve, which is often followed by a rotary feeder to provide a constant rate-of-flow to the disposal plant. Both valve and feeder are usually electrically-driven and are interlocked to ensure that the valve is open only while the feeder is running.

On top of the bunker is a double filter unit, through which the conveying air bringing dust into the bunker, and also the fluidising air from the aeration pads, is filtered before being released to atmosphere. The filters can be selected to serve either of the bunker compartments. The filter comprises a number of open-ended cotton-sateen or terylene porous bags, arranged with the open ends on the clean air side. A travelling reverse-flow unit, fed with backflushing air from a pair of centrifugal clean-air fans, passes over the open ends of each row of bags in turn. The reversed flow of air dislodges accumulated dust on the bunker side of the bags.

Seals at the mouth of the machine ensure that each filter bag is backflushed individually in a timed sequence. Some installations are also provided with a low energy vibrator which shakes the dust bag supporting frame.

Commercial organisations have established their own dust handling plant at some sites. Generally this comprises a small bunker into which the dust from selected boilers is discharged, plus a conditioning plant similar to that mentioned later. The dust transport lines to the station dust bunkers have to be fitted with pneumatically-operated diverter valves to allow dust to be directed into the commercial bunker, when required.

14.7.2 Commissioning

The general aim is to commission individual items of equipment first, proving all sequencing and remote controls, then to run the entire system with no dust present. Dependent on the system design, it may be practicable to carry out further tests with a supply of dust brought to site for this purpose. Normally, however, proving of the plant is conducted in conjunction with the main boiler coal-firing commissioning activities outlined in Chapter 6.

The most important prerequisites for commissioning are:

- All surfaces in contact with dust must be clean, smooth and absolutely dry.
- There must be no construction debris left in any part of the system.
- Electrical trace heating of hopper and bunker outlets must be put into service at a very early stage.
- Concrete bunkers or hoppers need to be carefully dried out after erection to drive off excess moisture retained within the structure.
- All dust-containing hoppers, etc., to be checked for dust tightness, i.e., no air leaks.

Other main aspects to be given consideration are as follows:

- Quadrant-type valves and rotary feeders to be commissioned, with all remote controls and indications proved.

- Pneumatic dust pump sequence control system proved to operate correctly, by simulation of the 'vessel loaded' signal. Butterfly isolating valves indication checked as correctly set, and valve seating against dust flow checked. Vessel inlet mushroom valve proved to be pressure tight.
- Screw-type dust pump motor rotations checked, noting that the pump should not be run empty more than for a few minutes. An adequate, regulated flow of dust is needed to prevent shaft whip, excessive wear, or overloading of driving equipment.
- Jet blowers all checked to be free of any blockage.
- Surge-hopper vent lines proved to give a clear unrestricted path to the boiler gas ducting.
- Dust bunker pressure relieving features to be proved. All other system pressure relief valves to be shown to be at their correct setting.
- Dust bunker outlet isolating valves all checked and isolation proved.
- Dust pipeline diverter valves to be checked for correct operation, control and indication.
- Dust bunker filter cleaning system put into service and sequencing adjusted.

When the compressed-air service systems and all other low pressure air blower/heater units are available, system airflow tests can be carried out. The plant should be set up for operation from the control panel, with observers stationed at strategic locations to warn of any plant malfunction. Various operational modes have to be tested, to prove correct plant response, along with simulated fault conditions. During this stage, it is important to record all air pressures and to assess compressor performance, so that the air velocity at various parts of the conveying system can be set correctly. If this is not done, then there is no guarantee that dust will not settle out in the pipelines, eventually causing blockages when the plant is in normal service.

While the dust handling system air flow tests are in progress, the ability of the dust bunker filter units to cope with the compressed-air flow liberated into the bunker can be checked. Allowance must be made for future partial blockage of the filter elements, as this could happen in normal service.

When the plant has been satisfactorily adjusted and proved with air, it should be left available for service until main boiler coal-firing starts. The plant trace heating and aeration air should remain in service to maintain all equipment in a dry condition. It may be prudent to blow air through all lines weekly, if the plant is not to be used for several months. The emergency hopper-emptying equipment must be demonstrated at nominated outlets, preferably each outlet, over a period of time. Subsequent commissioning of the dust-handling plant is conducted in conjunction with main boiler commissioning, as outlined in Chapter 6.

14.8 Dust disposal system

14.8.1 System description

The method of disposal depends upon many factors, so only the most common will be outlined in these paragraphs. Figure 3.88 depicts a general arrangement of a typical system.

Dry dust is disposed of from the central dust bunker by some combination of the following methods:

- Dry-dust sales, using sealed road or rail tankers.
- Conditioned (i.e., damped) dust into road vehicles for sales or disposal.
- Conditioned dust onto belt conveyors for transfer to an off-site disposal area.

Dry-dust disposal

Dry dust disposal entails the use of special chutes, air slides or chain-type conveyors to interconnect the bunker outlet and the vehicle filling point. The vehicle filling connection takes the form of two concentric convoluted flexible tubes, where the inner tube conveys dust into the vehicle, whilst air displaced from the vehicle is exhausted via the annular space, then piped to small low pressure fans (reek fans) which discharge into the dust bunker filter units. Loading of dust is automatically controlled by means of a pneumatic probe which enters the vehicle with the chute. Low pressure fans supply air to the probe, which, as the dust level rises, senses an increase in pressure to shut off the dust flow control valve.

Conditioned dust disposal

Conditioned dust disposal involves feeding dry dust at a controlled rate into a dust conditioner, where it is mixed with regulated fine watersprays to produce a 'just damp' dust-free mixture. One form of dust conditioner comprises a horizontal U-shaped trough containing two counter-rotating paddles, shaped so that dust and water are mixed and kneaded as they are propelled toward the conditioner outlet. The conditioner drive motor and variable-speed gearbox also drives the rotary dust feeder through a slipping clutch.

Another form of conditioner, shown in Fig 3.89, is in the form of an inclined drum which rotates at 30 r/min. Dry dust, at a rate of some 150 t/h, is admitted to the upper end of the drum, where it passes through a curtain of fine watersprays. Mixing of dust and water is promoted by a fixed longitudinal scraper bar and a series of fixed transverse baffles, plus rotation of the drum. The dampened dust, which tumbles around and down the drum, is further conditioned by a line of secondary spray nozzles, so that it leaves the conditioner in a 'just damp' condition. Conditioning water is derived from the ash reservoir make-up line. The drum is kept at a slightly negative pressure, to prevent dust nuisance in the building, by means of a small fan and filter unit common to several conditioners.

Belt conveyor systems and boom-stackers for conditioned dust disposal are generally similar to those used for handling coal. Provision is made for emergency stocking-out of dust on site, should the main disposal conveyor system be out of service. This dust has to be reclaimed, in a similar manner to coal, for subsequent disposal. Tractor shovels, rather than bucket-wheels or grab cranes, are used to transfer the dust from the emergency stock on to the dust disposal conveyor system. Mobile scrapers and bulldozers are used to spread conditioned dust, discharged from the boom-stackers, into a landscaped area.

Where dust is disposed to distant lagoons, it has to be passed through a wetting ejector: the slurry is then pumped by means of centrifugal pumps. It is not uncommon for two-stage pumping to be employed to generate adequate pressure to maintain sufficent velocity of flow to prevent dust settling out of the water into the lines. Some system designs arrange for the dust handling transport lines to by-pass the central dust bunker, directly into a sump with a wetting ejector.

14.8.2 Commissioning

All items of plant should be set to work independently and then demonstrated to work together as a sequence controlled group, using the control panel. However, proving with ash or dust loadings has to await main boiler commissioning before there will be sufficient ash for disposal.

- Dust conditioner motor-rotation must be proved to be correct, including the auxiliary drive to the bunker outlet rotary-feeder, if appropriate.
- Spraywater lines to be flushed out and the spray nozzles calibrated for flow against control valve settings.
- Dry dust disposal LP fans to be commissioned, including checking for correct direction of air flow from the dust filter fans.
- Pumped-disposal plant to be proved, with water only first, to check for designed rates of flow.

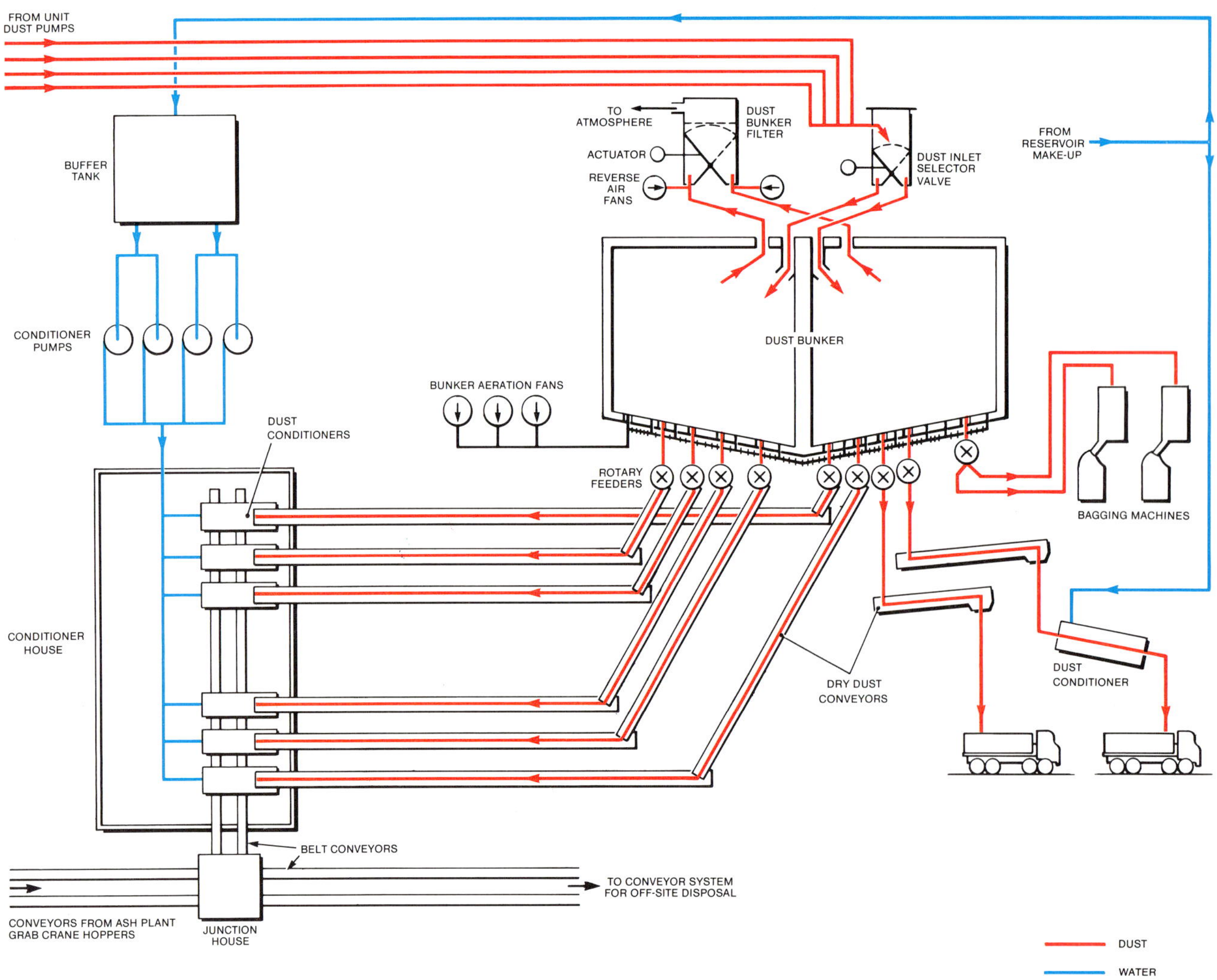

FIG. 3.88 Dust disposal system

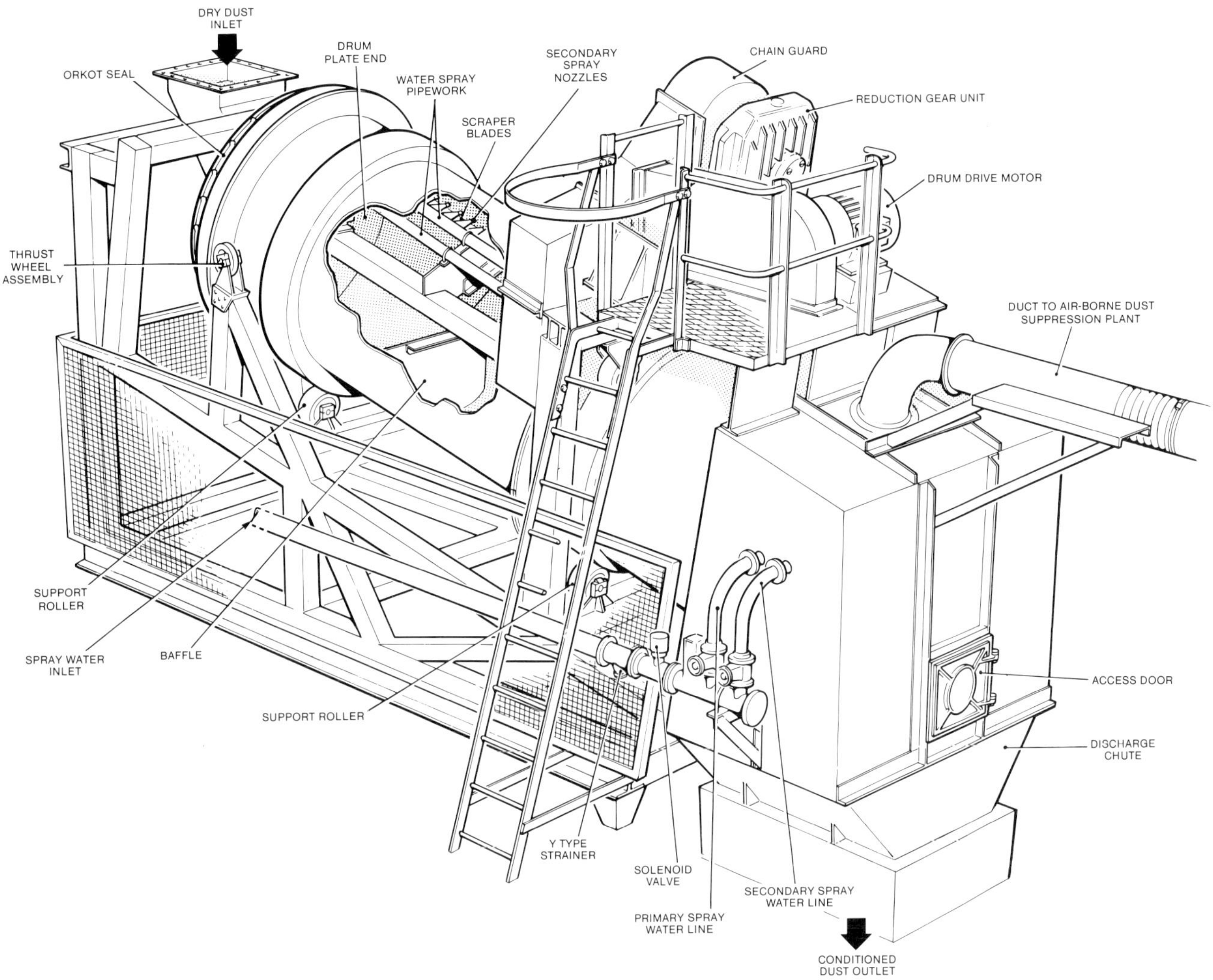

FIG. 3.89 Arrangement of drum type dust conditioner

As supplies of ash and dust become available, the following items have to be checked:

- Operation of conveyors when handling wet ash and/or conditioned dust, including ability to restart a fully-loaded conveyor without run-back or slippage of load.
- Ability of boom-stackers at emergency and normal stocking-out areas to create the designed heap. This should be followed by a demonstration of the reclaiming facilities.
- Dry dust disposal checks involve confirming that any associated pneumatic or mechanical conveyor functions correctly. The ability of the vessel venting scheme and overfilling protection must also be proved.
- Pumped-disposal systems performance to be assessed to ensure that the ash contents are maintained in suspension throughout the pipeline run.
- Dust conditioners adjusted to give the specified mixture of dust and water.

14.9 Electrostatic precipitators

14.9.1 System description

Figure 3.90 shows a side elevation of a typical electrostatic precipitator. Boiler flue-gases are passed through electrostatic precipitators to remove entrained dust before the gases enter the main chimney. Generally two, and sometimes three, parallel gas flow paths, each with inlet and outlet isolating dampers, are provided for every boiler unit. Modern precipitator shells are fabricated from welded mild-steel plates, which are thermally insulated and covered with profiled aluminium cladding for weather protection. The welded construction ensures that air does not leak into the structure.

Each precipitator structure contains a series of mild-steel-sheet collector plates, spaced some 200 mm apart, suspended in and parallel to the gas stream. Lines of 2.5 mm square-section alloy-steel discharge electrodes are interspersed with the collector plates. These wires are hooked onto a support frame, each wire being kept in tension by a weight which engages in another frame at the base of the precipitator to maintain relative

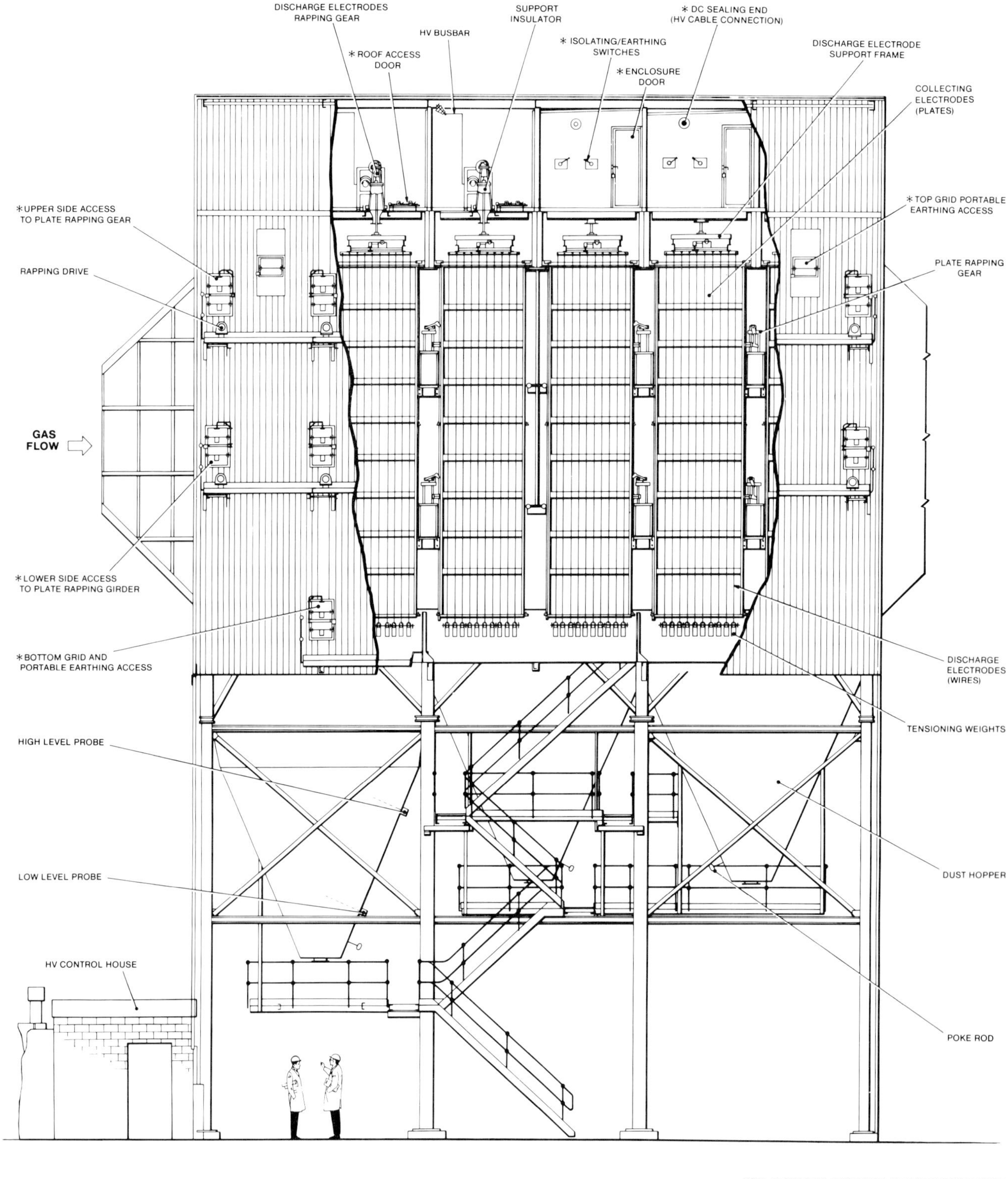

FIG. 3.90 Section through electrostatic precipitator

positioning of collector and discharge electrodes. The upper and lower frames are supported by electrical insulators which are trace heated to 65°C to prevent off-load condensation on the surfaces and also to minimise thermal shock, when hot flue-gases start to flow.

The collector plates are electrically bonded together and to the station main earthing system. They are also connected to the positive terminal of the HV DC busbars. The negative pole of the busbars is connected to the discharge electrodes. It is practice to divide the electrode system of each gas flow into several (possibly six) zones, each of which is equipped with its own HV DC power pack and electrode-rapping equipment.

Most of the dust accumulates on the collector electrodes, with the greater proportion in the gas inlet zones. A typical six-zone precipitator would have twelve collector-rappers and only six discharge-rappers, with the former operating much more frequently. Electric-motor-driven rapping equipment, controlled by a combination of motor-driven cam timers and process timers, impact mechanical shocks to the collector plates and discharge electrodes. Dust dislodged by the rapping action drops into the hoppers beneath, whence it is removed by the dust handling plant.

A negative discharge is more stable than a positive one, which allows a higher DC potential to be applied to ionise the dust particles. For practical purposes, dust collection efficiency is about 99% with an HV DC potential of about 34 kV. Re-entrainment of dust into the gas stream, particularly as a consequence of rapping gear operation, but also because of the dust extraction disturbances, tends to limit the collection efficiency; with no rapping and no dust extraction, it should be possible to achieve virtually 100% collection. However, dust build-up on plates and hoppers has to be cleared to maintain maximum collection efficiency. This reintroduces dust back into the gas stream, some of which escapes to the chimney. The optimum of all factors gives about 99% as the best practicable collection efficiency.

Each dust hopper is fitted with a detector to warn of high dust levels, which may cause the electrodes to short-circuit and render that zone ineffective. The lower ends of the hopper are externally trace heated and thermally-insulated to maintain an inner surface temperature of about 25°C. Electrical supplies for each six-zone precipitator would originate from a dedicated 415 V switchboard, with outlets to the following:

- Insulator heating (48 kW) and hopper heating (18 kW)
- Collector and discharge electrode-rapping gear (18 motors)
- HV busbar room ventilation fans
- Transformer/rectifier units (six units)
- Control and indicator supplies

Generation of the high voltage DC potential is by means of single-phase, double-wound transformer/full-wave HV silicon rectifier units. The whole assembly, including a primary winding series-connected inductor to increase the reactance to 40%, is mounted in a conventional oil-filled tank. The transformer/rectifier unit is fitted with conservator and breather plus Buchholz relay, pressure relief device and oil temperature trip protection (see Section 9 of this chapter). The units are ground-mounted outdoors, where the ONAN cooling classification is quite adequate, and fixed fire fighting can be applied.

Special high voltage cables connect the transformer/rectifier output to open-type air-insulated HV busbars located in rooms at the upper levels of the precipitators. Manually-operated isolation and earthing switches are fitted, to facilitate isolation and earthing of individual half zones and also of the rectifier output. Where appropriate, viewing windows are fitted, so that personnel can verify that the appropriate HV equipment is 'isolated' with the 'earth on' before maintenance work is commenced. Portable earthing leads are connected to the station main earth, and arranged so that they have to be led through opened access doors to be fixed to ball-shaped attachments, specially fitted for the application of additional earths to previously isolated and earthed HV equipment.

An extremely complex system of mechanical key interlocks is fitted to prevent personnel access to any HV equipment, including precipitator gas-pass access hatches, unless the plant is isolated and effectively connected to earth.

As mentioned earlier, the efficiency of electrostatic precipitation depends primarily on the voltage between the discharge and collecting electrodes. The optimum voltage is obtained when an acceptable level of sparking is achieved without the development of a continuous arc. For practical purposes, a spark in this context is defined as a self-extinguishing disturbance which depresses the output voltage for less than 10 ms. The thyristor control system varies the transformer primary voltage, such that the output voltage just allows sparking to take place but then takes corrective action to ensure that arcing does not follow. It should be appreciated that sparking and arcing voltages vary with the operating conditions of each zone of the precipitator. The control scheme continuously adjusts output voltage and current, subject to preset maximum and minimum values, to maintain optimum collection efficiency. Conditions which limit the auto-control system are maximum output current and voltage; but overcurrent and undervoltage protection trip the transformer unit.

14.9.2 Commissioning

Individual plant items are tested, first under local control, then under remote and/or automatic control, where appropriate. When all of the safety interlocks have been proved, the HV DC equipment is energised on manual control before the automatic controls are tested and adjusted. The final test, to calibrate and adjust the air flow distribution across the precipitators before each boiler is first fired, is described in Chapter 4.

Some of the more important details to be considered are as follows:

- Test that the precipitator casings are dust-tight, i.e., that there is no in-leakage of air which may dilute and cool the flue gas stream.
- Commission dust-hopper trace heating, checking and recording the temperature of each hopper after 4 hours and 24 hours. The temperature can be deduced by having the control thermostats set to their maximum value; then the actual operating temperature can be ascertained by downward adjustment of the setting knob.

- Commission discharge electrode support-insulator heater elements, again checking the temperature after 4 hours and 24 hours. The control thermostat setting must not be increased above the design point, to prevent damage to the insulator.
- Check that dust-hoppers and gas-passes are clear of all debris. Confirm that hopper outlet valves are operable. Check that all hopper dust poke-rods are free.
- Commission inlet and outlet dampers, as outlined in Section 1 of this chapter.
- Check correct installation of discharge electrodes, including support tensioning and spacing structures. Confirm that support insulators are clean and free from damage.
- Ensure personnel safety, then commission each set of rapping equipment in turn, confirming that motor rotation is correct and that alignment of hammers and anvils is maintained. Check drive-shaft bearing lubrication, casing dust-seals and safety shear-pins. On discharge electrode rappers check that the drive-shaft insulators are not cracked or damaged and that the alignment is not inducing any strain. Apply preliminary settings to, and check the timing sequence of, all the rapping gear.
- Recheck discharge electrode system after a lengthy period of rapping gear operation to ensure that relative spacing between discharge and collector electrodes is still maintained.
- Ensure that all metal parts, including portable earthing points and the collecting electrode assemblies, are securely bonded to the station main earth.

The safety interlocking scheme has to be checked thoroughly to ensure that it is impossible for personnel to gain access to any HV enclosure until the equipment has been isolated and connected to earth. The essential points to check are:

- Verify that each key and lock is correctly located and engraved according to its function. Check also that the appropriate key operates its associated lock.
- Check the key interlocking scheme associated with each electrical isolator, panel door and earth switch, then check scheme for all key exchange boxes, followed by each access door.
- Finally, ensure non-interchangeability of keys by random key checks at each lock.

Commissioning of the HV DC equipment usually follows checking of the safety interlocking scheme, so that personnel safety can be assured. This generally entails:

- Transformer/rectifier unit checks, including protection trip tests as outlined in Section 9 of this chapter.
- Auto control system tested by simulation, as far as practicable.
- Visual check of operation of all earth switches; contact resistance checked and recorded using a digital micro-ohmmeter.
- Insulation resistance of each zone of discharge electrode checked to be well clear of earth at 5 kV DC.
- Cables from rectifier unit to discharge electrodes tested, whilst disconnected at both ends. Insulation test, before and after an HV pressure test, using 5 kV DC insulation tester. Cable HV pressure test conducted at 70 kV DC for 15 minutes, recording voltage and leakage current.
- Ease of application of all portable earthing leads has to be proved.

After a final check that the precipitator gas-passes and hoppers are clean, personnel are withdrawn and all access doors secured; then an 'air test' is carried out. This involves energising each zone of discharge electrodes in turn from its associated transformer/rectifier unit, slowly increasing the system voltage under manual control until a stable leakage current is achieved. When that is satisfactory, the automatic current control is adjusted to give the optimum values.

Initial operation of the HV DC equipment is conducted with rapping gear isolated. The test is repeated, first with discharge electrode rappers in service, then with the collector-plate rappers in service. Finally, the test is repeated with all rapping gear operating in sequence, with adjustment to timers where appropriate. Discharge voltage and leakage current are noted at each stage for plant history purposes.

When these checks have been completed satisfactorily, the precipitators are left with all HV equipment isolated and discharged to earth. However, all electrical heating is kept energised until the precipitator is required for service.

Before the boiler plant is first fired, it is vitally important that airflow distribution tests are carried out to ensure that there will be a relatively uniform gas velocity across the collectors. Failure to do so may give rise to ineffective dust collection. This test is dealt with in Chapter 4.

On-load operation and checking of the precipitators will be covered in Chapter 6. Basically, this involves monitoring the automatic current control of the transformer/rectifier units and adjusting the timer sequencing of the different zones of discharge and collecting electrode-rapping equipments. At a later stage, effectiveness of the precipitator inlet and outlet isolating dampers has to be rechecked, whilst the boiler is on load.

14.10 Vacuum cleaning and compressor plant

14.10.1 System description

Vacuum cleaning system

A fixed vacuum cleaning installation is provided to collect accumulations of pulverised coal and flue-gas dusts from boiler surfaces, surrounding galleries, stairways, building support steelwork and the PF milling area. Generally one system serves the coal bunker bay and there is one associated with each main boiler unit.

Figure 3.91 shows a typical vacuum cleaning system, which comprises a belt-driven air exhauster, filter receiver and two primary receivers, connected to several sections of fixed pipework. Each section of pipework is equipped with 50 mm self-sealing hose connections at strategic locations around the main plant. Dust is picked up by operator-held portable tools, coupled to the fixed pipework by flexible hoses.

The air exhauster may be capable of maintaining a vacuum of 320 mbar, whilst handling air at a rate of 0.1 m^3/s. Each exhauster is fitted with a discharge silencer to limit the noise level from air being released into the buildings.

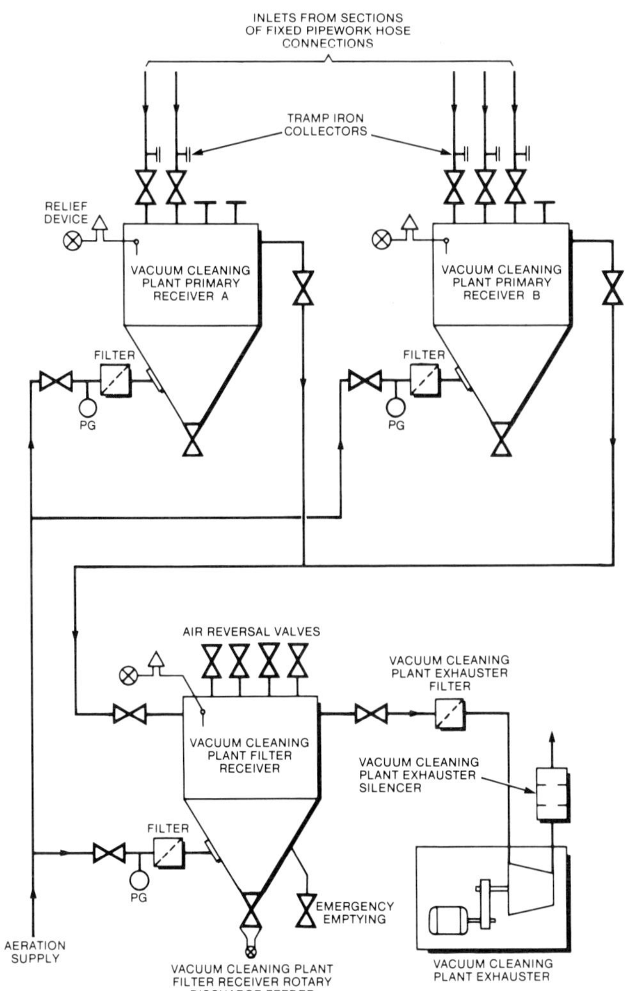

FIG. 3.91 General arrangement of vacuum cleaning plant

Each receiver, which typically has a capacity of 8 t, is provided with several ceramic relief panels and a self-resetting pressure relief flap valve. These protect the receiver and fittings from rupture in the event of either a coal dust explosion or accidental overpressurisation with aeration air. Electrical trace heating and thermal insulation is applied to the receiver bottom cones to guard against internal condensation. Should the dust level become too high, a detector trips the air exhauster.

A tramp iron collector is fitted at each dust inlet to the primary receiver to remove large debris and foreign bodies. The primary receiver collects heavy particles of dust as the suction (conveying) air expands into the enlarged space of the receiver body. Removal of fine dust takes place in the filter receiver, which is fitted with an automatic control system to clean each of the filter elements sequentially by air reversal.

Collected dust is discharged from the bottom of each receiver, with assistance from air fed through aeration pads. The low pressure regulated supply of compressed air is derived from the general services compressed-air system. Dust from the hoppers passes through a quadrant valve, often augmented by a rotary feeder, to either a jet pump or a jet blower for disposal via the ash or dust handling system, respectively. Coal dust collected in the bunker bay receiver-hoppers is discharged into one of the coal bunkers.

Compressed air plant

Compressed air is used for the following functions within the ash and dust plant:

- Operation of all pneumatic controls for dust pumps, changeover valves, etc.
- Motive air for jet blowers.
- Motive air for dust pumps.
- Aeration of dust bunkers, hoppers and air slides.

Control air pressure is generally at approximately 7 bar and is derived from either the general service (GS) or the control and instrumentation (C and I) compressed air systems, described in Section 21 of this chapter. Motive air for the jet blowers and dust pumps is provided by separate dedicated compressor installations for each service. Typically, four two-stage centrifugal compressors are fitted such that three can be selected for first, second or third duty, with the fourth as standby. Dependent upon the system design, compressors for the jet blowers produce compressed air at a pressure of either 2.9 bar or 1.7 bar, whilst those for the dust pumps discharge at 7 bar. The dust pump compressors are often driven by 3.3 kV motors and are usually water cooled. In some cases, the dust-pump services compressor plant is interconnected with the GS compressed air system.

Single-stage centrifugal fans supply low pressure air through electrically-powered heat exchangers for dust aeration, and for fluidised conveyors. Discharge pressures range from approximately 20 to 40 mbar dependent upon the system design.

14.10.2 Commissioning

Vacuum cleaning plants

Individual items should be checked to approved standard check lists, for example:

- Aeration equipment.
- Trace heating system.
- Filter cleaning equipment.
- Air-exhauster motor rotation, followed by drive-belt fitting and tensioning, if appropriate.
- Cleanliness of all vessels, receivers, etc.
- Safety devices all proved.

Running tests may comprise the following:

- Checking alarms, controls and interlocks, with air-exhauster motor switchgear 'on jumpers' (see Section 8 of this chapter).
- Proving run of air-exhausters to check general operational behaviour, vibration, noise, etc.
- Overall control of plant, involving filter air reversal valve operation, control sequences and correct functioning of all valves and trace heating.
- Pipework leakage tests involving checking each section, first by a pressure drop test over 24 hours, and then by subjecting it to the maximum vacuum obtainable by the air-exhausters for a further 24 hours.
- Reliability runs to prove satisfactory operation of the plant under normal service conditions, including demonstration of plant cleaning in specified areas.
- Confirm correct operation of the tramp iron interceptors.
- Confirm correct operation of vessel pressure relief devices.
- Check that noise level within 1 metre of the air-exhauster is less than 90 dB.

Compressed air plant

Commissioning of compressed air plant dedicated to dust handling duties follows the general principles outlined for other compressed-air systems in Section 21 of this chapter. The details are not repeated here; the reader should refer to that section.

Low pressure fans and blowers are commissioned in a manner similar to that outlined in Section 4 of this chapter. Safety cut-outs and temperature interlocks associated with the heat exchangers have to be proved to protect the plant satisfactorily and to prevent damp air contaminating aeration pads or fluidising conveyors.

15 Fuel oil handling

This section covers commissioning of the plant concerned with the reception, storage and transfer of boiler fuel oil, up to the heating and pumping plant associated with coal- and oil-fired main unit boilers.

Commissioning of the heating and pumping plant, oil burners and associated oil pipeline systems will be found in Chapters 4 and 6.

15.1 System description

Within this section, the term *boiler fuel oil* includes all black residual fuel oils with kinematic viscosities ranging between 75 mm^2/s and 160 mm^2/s at 80°C (3500 to 10 000 s Redwood No. 1 at 37.8°C), having a specific gravity of almost unity. The British Standard covering this range of boiler fuel oils is BS2869.

Figure 3.92 has been developed purely to illustrate different arrangements for the reception of boiler fuel oil: it also shows a typical storage and transfer system. Fuel oil is delivered to site either by some combination of road, rail or marine tanker, or by a direct pipeline connection from a nearby oil refinery. Dependent upon their geographical location, coal-fired power stations are usually able to accept boiler light-up oil deliveries by coastal tanker or rail. Oil-fired power stations must have a direct pipeline from one or more oil refineries, or else be able to accept frequent deliveries from large marine tankers.

Oil deliveries are stored in several large cylindrical tanks, from which oil flows by gravity through strainers to transfer pumps. These pumps discharge to each of the main boiler fuel oil heating and pumping plants, where the temperature and pressure of the oil is increased to ensure satisfactory firing at the oil burners. The transfer pumps also circulate fuel oil around the unit pipework when the boiler plant is shut down, so that it is at an even temperature ready for the next start-up. Duplicated pressure control valves ensure that the fuel oil pressure at the transfer pump discharge remains constant, irrespective of fluctuation in flow rate. The control valves recirculate (or spill) excess fuel oil into the recirculation line for return to the main oil tank that is in service. Steam from the auxiliary steam system is used for heating all equipment containing boiler fuel oil.

15.1.1 General

Fuel oil transfer system design is usually based on deliveries at the following rates:

- Large marine tankers: up to 1000 t/h.
- Rail or coastal tankers: either 250 t/h or 500 t/h.

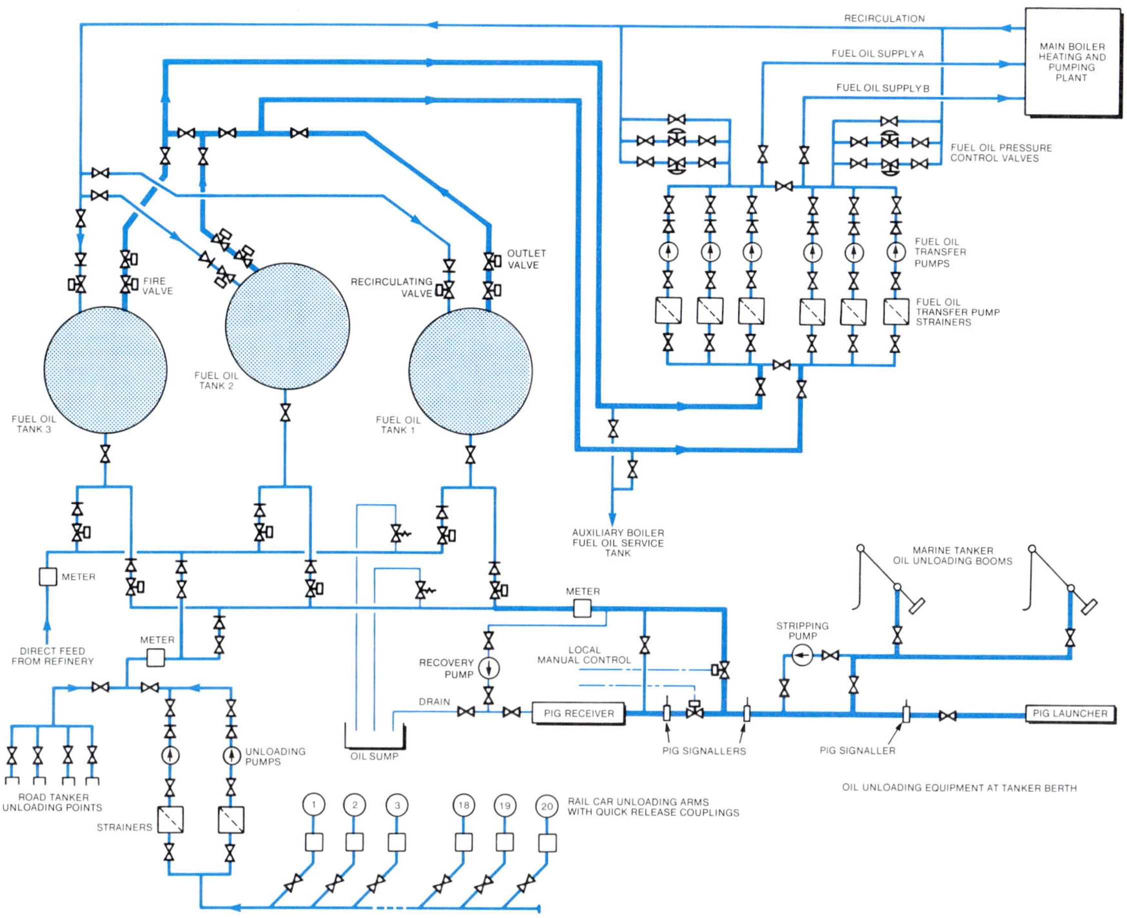

FIG. 3.92 Fuel oil handling system

- Refinery pipeline: either 250 t/h or 500 t/h.
- Road tankers: at 100 t/h from four 18 t vehicles.

With the exception of rail tankers, all other carriers are capable of discharging their own cargo by pumping. A separate pumping plant has to be provided to unload oil from rail tankers.

Pipelines, tanks, rail cars, pumps, strainers, etc., are all steam heated from the auxiliary steam system 7 bar manifold. Condensate is collected and it is either pumped, or returned by gravity, to the auxiliary boilers, via oil-in-condensate detectors (refer to Section 19 of this chapter). In most cases, fuel oil supply lines to the auxiliary boilers are fitted with electrical tracing elements to facilitate firing of an auxiliary boiler, should there be no heating steam available.

It is not normal practice to apply trace heating to long lengths of fuel oil supply pipeline, unless it is really necessary for operational convenience. Usually a thick layer of thermal insulation is applied instead, and the pipeline provided with some means of replacing the boiler fuel oil with either compressed air or a lighter grade of oil, as described later.

Thermal relief valves are fitted to prevent the heat input from tracing steam causing excessive pressure rise in sections of fuel oil lines which may have been isolated when full of oil. These are set to lift at a higher pressure than the normal pipeline overpressure-protection relief valves. Oil vented from relief valves, plus any drained off in preparation for maintenance work, is usually collected in portable tanks before being pumped into the recirculation line for reuse. Some of the larger power stations have a comprehensive fuel oil drainage installation, where oil in sections of pipeline can be drained direct into an intermediate drains collection tank from which fixed pumps return the drained oil into the recirculation line.

Oil spillages from fuel oil handling areas are led to underground oil holding tanks for subsequent removal and disposal using road vehicles. Surface drains from oil handling areas are directed into the station dirty drains system, and subsequently passed through an oil separator. Special arrangements have to be provided for the collection of oil drainage, surface water and any oil spillage at marine tanker berths to prevent pollution of the waterway.

Boiler fuel oil is not classed as a fire risk unless it is hot and under pressure, so fire protection for fuel oil tanks is not necessary. However, it is practice to provide fire protection in transfer fuel oil pumphouses. Generally, this is some form of sprinkler system, with a separate compressed-air fire detection scheme (see Section 12 of this chapter).

15.1.2 Rail-borne reception

Rail-borne boiler fuel oil is usually delivered as 760 t loads in ten thermally-insulated rail cars. The railway siding is often also used for gas-oil deliveries. Some sites are designed to accept 20-car train loads of boiler oil, though the facilities for gas-oil deliveries would continue to be based on 10-car trains. Special precautions are necessary to prevent deliveries of one class of fuel being transferred into the fuel system for the other class. Unloading equipment for rail cars takes the form of horizontal waist-height articulated pipes, fitted with flow indicators, and terminating in quick-release couplings to mate with those of the appropriate rail cars. Provision is made for application of steam heating (and condensate recovery) to rail cars should this become necessary. Though supply points are located beside each unloading arm, the system is normally sized to cater for simultaneous application to four cars only.

15.1.3 Marine tanker and refinery reception

Marine unloading booms, because of the degree of flexibility required to cope with rise and fall of the ship, are of the counterbalanced articulated-arm type. The apex joint has to be provided with a vacuum-breaking vent, operated by ropes from ground level, to facilitate priming and draining of the primary and secondary arms of the boom.

If a marine tanker berth or a direct refinery pipeline is to be out of service for perhaps 12 h or more, then the boiler fuel oil must be cleared out. Often this is achieved by 'plugging', that is replacing the boiler fuel oil with a lighter grade (e.g., 220 s) of oil. Typically, a 5 km line may need some 150 t of light oil for this exercise. When boiler fuel oil flow is resumed, the light oil may either be displaced into a separate holding tank for subsequent re-use, or transferred into the boiler fuel oil storage tank and fired in the boiler. Should the pipeline have been out of service for a lengthy period, then the boiler fuel oil will need to be heated to some 50°C before it is pumped into the relatively cold line.

Another method is to use compressed air to propel a purpose designed cylindrical plug or 'pig' (Fig 3.93) along the line to push the remaining fuel oil into one of the storage tanks. The figure shows a 'separation' pig which is used for purging and, in some instances, for separating different grades of oil. Special handling equipment is needed at each end of the line to send (or launch) and receive the pig (as indicated in Fig. 3.92). Typically, a 460 mm diameter pig may be about 750 mm long and would be capable of negotiating bends of some 700 mm radius at a speed of 3 to 5 km/h. A battery-operated radio transmitter is often fastened to the pig before it is placed into the pig launcher. If the pig becomes stuck during transit, it can be located by means of a radio receiver carried by an operator walking over the pipeline route. When a separation type of pig is being used for pipeline cleaning, by-pass plugs in the nose are removed and the leading neoprene cup is replaced with a purpose designed metal scraper disc. The propelling fluid, which is either a light oil or water in this case, leaks through the plug holes to lubricate the scraper and also to help in washing the scrapings along the line. As one of the pipeline construction checks, it is customary to replace one of the neoprene cups with a thin sheet-metal disc as a means of gauging the extent of any peripheral obstructions caused by welding protrusions, etc., as the pig traverses the line.

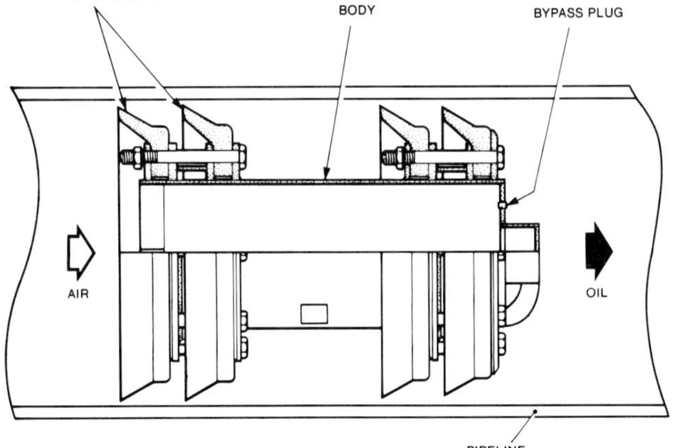

FIG. 3.93 General arrangement of a typical fuel oil separation pig

A combination of plugging and pigging may be adopted for preheating longer lengths of pipeline prior to reception of boiler fuel oil. Sufficient light oil to fill the pipeline is held in a separate tank at one end of the line. Pumps recirculate the oil through the tank and a bank of heaters supplied with auxiliary steam until the temperature approaches 50°C. The heated oil is then pumped through the heaters to fill the line with light oil at a temperature of some 60°C. After a period of time, the light oil is pigged back into the tank and the line charged with boiler fuel oil. Upon completion of the oil transfer, the boiler fuel oil is cleared out of the line by pumping in the light oil, i.e., the pipeline is plugged with light oil until the next delivery is due. Prior to the next boiler fuel delivery, the light oil is pigged back into the holding tank for recirculation and heating as before. The holding tank may take about 4 hours to have the oil temperature raised to 50°C.

15.1.4 Fuel oil storage tanks

Atmospherically-vented storage tanks range in capacity from several hundred tonnes up to 75 000 t or more. They are thermally-insulated and usually steam heated to a temperature of between 50°C and 60°C. Axial-flow low level belt-driven mixers are often provided to assist natural circulation of heated oil, to eliminate excessive temperature gradients within the oil. Marine oil cargoes often have a relatively high water content. Deliveries have to be allowed to settle for a few days before use, so that water can separate from the oil and then be drained off.

Fuel oil tanks are usually provided with three means of indicating the oil quantity in store:

(a) A manual direct-reading device whereby a weight, supported on a calibrated tape, is lowered through an ullage hatch until it touches the oil surface.

(b) A simple float-operated suspension-wire indicator to give local indication only.

(c) Either a pressure-sensitive or a float-operated transmitter, which provides local indication as well as remote indication in the central control room (CCR).

This last device is usually arranged to work in conjunction with the tank contents averaging thermometer, and also to provide warning of low or high level: it may also cause the inlet valve to close when a very high level is reached. Float-operated switches are also fitted to supplement these alarms and to trip the mixers when the oil level gets too low.

The averaging thermometer often takes the form of a series of resistance thermometers, suspended in the oil within a common sheath, with the sensing elements placed at perhaps 1 metre increments of tank height. The outputs are transmitted to the CCR for averaging in conjunction with the tank level signals.

Fuel oil inlet, outlet and recirculation valves are usually electrically-operated from the CCR. It is common practice for the outlet and recirculation valves to be interlocked and operated in a prescribed sequence. The outlet line is often also fitted with a quick-closing isolating valve, which may be either air or nitrogen actuated, to close in the event of fire in the transfer pumphouse.

Lightning protection is ensured by bonding the tank shell and all associated metalwork to a ring of earth electrodes spaced around the base. Though there is no statutory duty to contain boiler fuel oil tanks within a bund, a common retaining wall about 1 metre high is generally erected to contain storm water and any oil spillages from all of the fuel oil tanks.

15.1.5 Fuel oil transfer system

Transfer pumps are either positive displacement screw-type or centrifugal units fitted with a large flywheel to maintain output, in event of failure of a duty pump, until an automatically-started standby pump takes over. Constant discharge pressure is regulated by duplicate pressure control valves, which recirculate excess oil to the fuel oil tank in service. Minor pressure perturbations are smoothed by a nitrogen-charged accumulator fitted to the discharge main. Though the fuel oil transfer system provides a station service, the pumps are often interconnected, so that they can be unitised if desired.

15.2 Commissioning objectives

The following should be considered in conjunction with the general objectives listed at the beginning of this chapter:

- Carry out hydrostatic testing of each fuel-oil tank to BS2654.
- Conduct air pressure test of each fuel oil tank.
- Monitor fuel oil tank settlement and ensure that specified limits are not exceeded.
- Pressure test all pipework to BS806.
- Accept first delivery of oil into each tank and calibrate contents gauges.
- Prove that fuel oil tanks can be put on line and taken off line without jeopardising main boiler operation.
- Demonstrate that fuel oil transfer pumps operate successfully in parallel.
- Prove operation of transfer pump discharge-main pressure control under varying flows and with different combinations of pumps in service.
- Confirm that fuel oil tank and pipeline steam heating system performance is within specification.
- Accept rail, sea or refinery delivery and prove ability of system to cope with designed flow rates.
- Demonstrate that a fuel oil line can be safely pigged clear.
- Ensure safe containment of any oil that may be vented as a result of either normal service procedures or by operation of system relief valves.
- Demonstrate that boiler fuel and gas-oil rail-borne delivery couplings are not interchangeable.

15.3 Safety considerations

The following activities need special consideration in addition to those highlighted at the beginning of this chapter. It is prudent to enforce a NO SMOKING/NO NAKED LIGHTS rule when boiler fuels are being handled:

- Filling and draining fuel oil tanks for hydrostatic test.
- Air testing of fuel oil tanks.
- Initial fill of fuel oil.
- Handling of railway traffic.
- Handling of marine tankers.
- Oil spillages, and any attendant pollution and fire risks.
- Pigging of fuel oil lines.
- Adequate thermal insulation to safeguard personnel from burns.

- Creation of high temperatures in fuel oil storage tanks, resulting from extended periods of fuel oil recirculation from heating and pumping plant during main boiler commissioning.

15.4 Resources

15.4.1 Services

Apart from electrical supplies, the following will also be required:

- A plentiful supply of clean fresh water for hydraulic tests.
- Temporary pumps for handling test water.
- Portable oil flashpoint test unit, e.g., SETAFLASH.
- Temporary low pressure air-blower for air test of fuel oil tanks.
- Station drains and oil separator systems available in time for hydraulic testing of oil tanks.
- Portable oil-draining and steam-cleaning equipment, plus oil-absorbent material for coping with oil spillages.
- Fire protection equipment commissioned before commissioning of fuel oil transfer pumps.
- Chemicals and appropriate containment and collecting equipment for dealing with a major oil spillage at marine tanker berths.
- A suitable grade of fuel oil, if the service grade of boiler fuel oil is not to be used for initial commissioning.
- Portable nitrogen cylinders for charging accumulators.
- Auxiliary steam supply and condensate recovery system for pipeline and tank heaters.
- Compressed air for control valve operation.

15.4.2 Staffing

The commissioning team comprises several groups of two or three people engaged on different activities. For the more extensive tests, the teams need to be armed with effective two-way radio communications. Testing of each fuel oil tank could take about three weeks, whilst testing and commissioning the rest of the system with fuel oil may take a further eight to twelve weeks.

15.5 Commissioning

Figure 3.94 shows a simplified network for commissioning a fuel oil handling system. Each of the main aspects is covered in turn on the following pages, assuming that all relevant safety measures have been implemented.

Initially, the following basic conditions must apply:

- Oil, steam and condensate pipework hydraulically tested to 1.5 times normal working pressure before thermal insulation is applied.
- Fuel oil tank heater-banks hydraulically tested to twice normal working pressure.
- Test certificates available for all oil system relief valves, since it is undesirable to lift them purposely during commissioning.
- Full use of standard check lists (SCL) or test procedures for individual items of plant or equipment, including checking of controls, indicators, etc.

Generally, the construction programme ensures that the fuel oil tanks are erected and tested well in advance of the rest of the plant, so that application of the thermal insulation will be completed in time for fuel oil system commissioning. As far as practicable, the tanks are hydrostatically tested consecutively to minimise wastage of fresh water.

15.5.1 General

In parallel with finishing off erection work on the fuel oil storage tanks, preparation of the remainder of the system is progressing thus:

- Oil and trace heating pipework hydraulically tested, drained, and then thermal insulation applied.
- Where appropriate, electrical trace heating elements checked for electrical continuity and insulation resistance before thermal insulation is applied.
- Electric motors spun for rotation and coupled to their pumps, etc.
- Controls, alarms, interlocks, valves, etc., satisfactorily proved, as far as practicable.
- Steam applied to all tracers, and the lines thoroughly blown to waste, to clear construction debris. Operation of condensate drain traps, steam pumps and recovery system checked.
- Fire protection commissioned, where necessary.
- Railway-siding equipment, including unloading arms, all lubricated and checked for service.
- Pigging air-compressors commissioned and used to propel a gauging pig along the pipeline carefully. This confirms that the pipe is clear of any obstructions and also displaces any hydraulic test water remaining in the line.
- Tanker berth unloading-arms lubricated and proved to move freely in response to operation of the handling ropes. The mooring chains must be checked to be adequate, and all draining and venting arrangements proved. All facilities for dealing with fuel oil drains checked to be operational. This includes ensuring that adequate precautions have been taken to prevent oil spillage during normal operation.

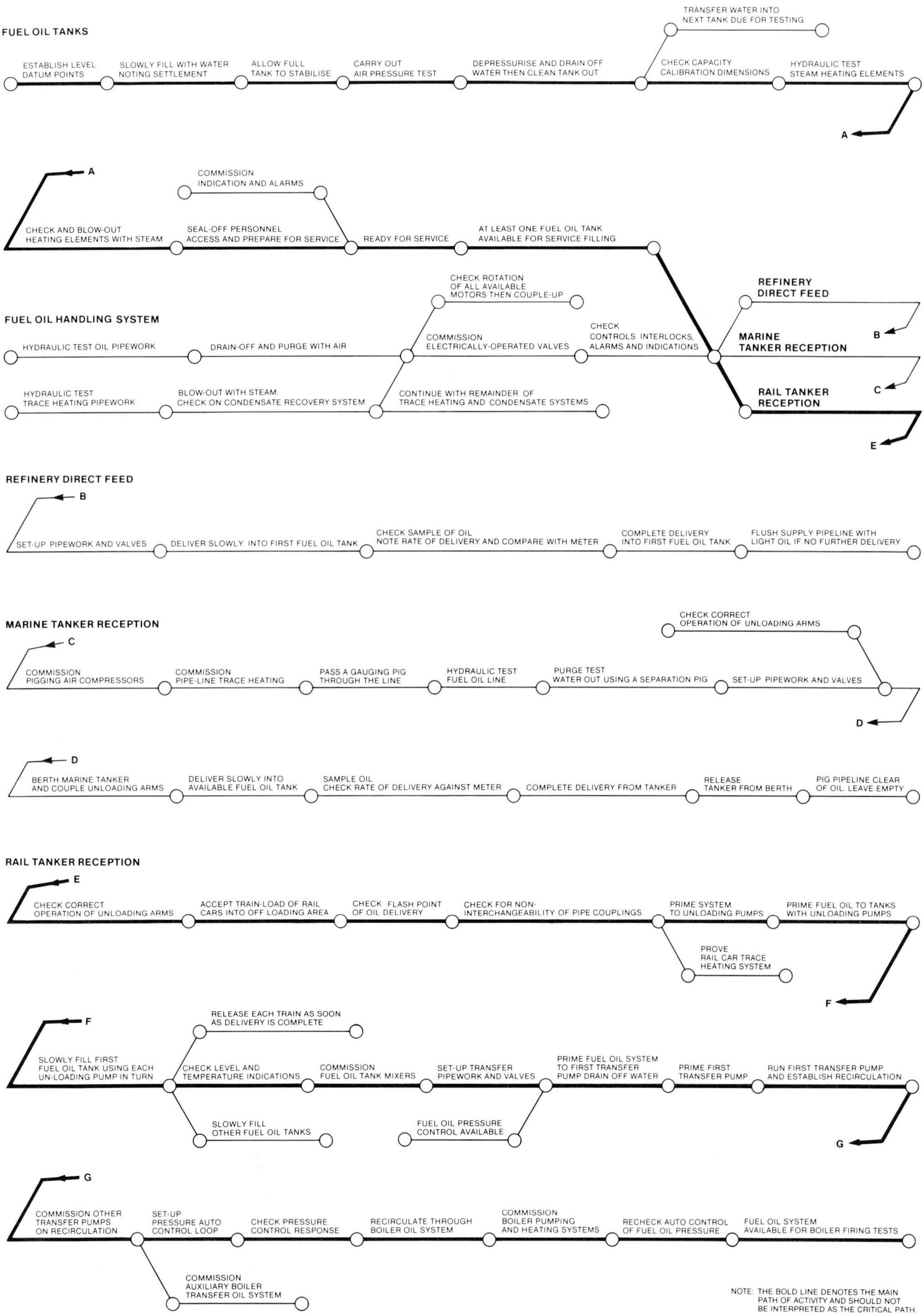

FIG. 3.94 Fuel oil handling system commissioning network

- Fuel oil tanks and all metallic attachments checked to be securely bonded to the station main earth (see Section 9 of this chapter).

15.5.2 Rail-borne reception

The following is written on the basis that the first delivery of oil will be by train, and that it will be the service grade of fuel oil, rather than a lighter-grade (expensive) oil.

Set up a flow path from rail sidings to a selected fuel oil tank. Ensure that steam tracing is operational. Accept the train into the sidings and check the flash-point of oil in each rail car. Refer to Section 10.5.1 for testing of the flashpoint. Though BS2869 specifies a minimum flashpoint of 66°C, for boiler fuel oils it is usually acceptable to use the same 60°C 'go/no-go' datum adopted for gas-oils. Confirm that the rail car quick-release couplings are not interchangeable with those on gas-oil unloading arms. Reposition and train-connect all cars to the boiler fuel oil unloading arms.

Open the outlet valve from each car in turn to charge the fuel oil main and all pipework, up to and including the unloading pumps. Carefully prime both unloading pumps. It is imperative that this type of pump be fully primed before it is run. Confirm that each pump is free to rotate by hand. Start and run one pump until fuel oil has charged the line into the selected fuel oil tank. Repeat this with the second pump, and then complete the discharge of all the rail cars. Check the contents indication of fuel oil tank against delivery notes.

Concurrent with this activity, couple the rail car heating system to each car in turn and prove that it will function correctly.

Prove the operation of all combinations of unloading arms and rail car heating connections with each successive trainload of fuel oil. It may be necessary to clean the unloading pump suction-filters frequently until the entire unloading system has been flushed.

15.5.3 Marine tanker and refinery reception

Commissioning of a refinery direct-feed or a marine tanker supply line is carried out in a similar manner. More staff are needed, equipped with excellent communications equipment, due to the distance between both ends of the line. Should there be facilities for pigging one or both lines, then it is prudent to practise the procedures for passing the separation pig along the line first, before any oil is accepted. Correct operation of all valves must be demonstrated before any oil is received into the system.

The compressor plant is commissioned as outlined in Section 21 of this chapter. The pig is placed into the launching trap after fitting of the locating transmitter, ensuring that the enclosure is correctly secured and sealed, then the pipeline is set up to release air (and any water present) from the receiving end. The pig location equipment is checked to be operating correctly before any further action takes place. With the compressor running, the launching trap is pressurised carefully (typically to 2 bar) until the pig signaller indicates that the pig has moved off; then the pig is allowed to traverse the line. The rate of air admission is manually regulated to ensure that the pressure necessary to set the pig in motion is not exceeded, otherwise the pig may travel at an excessive speed and be damaged. The compressor discharge pressure should always be greater than, say, 5 bar to safeguard the compressor. At the receiving end, when there is preliminary warning of the approaching pig, the valves are operated to bring the receiving trap into service until the pig has been safely collected. The compressors are shut down, then the line and receiving trap are carefully depressurised and drained before the pig is removed. Having established that the normal service pigging procedure is practicable, delivery of oil can be accepted.

The refinery operator may wish to use a lighter grade of oil initially in order to commission his pumping plant. In that event, a flow path is set up from the refinery to the selected fuel oil storage tank. Delivery is commenced slowly until the pipeline primes and all visible joints, seals, glands, etc., have been checked to be oiltight. The flow rate is steadily increased to its maximum and maintained for several hours. The oil tank contents indicator is checked against ullage tape readings and the refinery operator's declared rate-of-flow to ensure that the specified rate is being achieved. For effective fuel oil system management, it is important that the oil temperature measurement and indication equipment is proved to be correct during receipt of any oil. It is practice to commission oil flowmeters after the oil pipeline has been put into service. Samples of the oil have to be taken for calorific value analysis and also when the initial flow of a lighter grade of oil is being changed to delivery of the normal service grade of oil.

Before accepting any oil by sea, all procedures and equipment needed to deal with spillage into the waterway must be to hand. A flow path, using one unloading boom, has to be set up to the selected fuel oil storage tank. The selected tank has to be ullaged before, and again after, each delivery for Customs and Excise purposes. If the pipeline is fitted with trace heating, then this should have been in service for at least 24 hours in advance of the arrival of the tanker since marine deliveries of oil are normally at a temperature which is only just high enough to allow pumping.

When the tanker has been securely berthed, the scuppers are sealed, the unloading booms coupled up and the ship's tanks ullaged. Oil discharge is commenced slowly, using one boom, until the pipeline has primed. The discharging rate is increased to the maximum for one boom. The second unloading boom is brought into use, then the rate is doubled and maintained until the tanker is nearly empty, when it is slowly reduced. During this operation, the oil storage tank contents indicator is checked against tank ullage tape readings, and also with those from the ship's tanks. After the discharge has been completed, the secondary arm of the unloading boom is drained back into the tanker. The primary arm is drained into the fuel oil line by means of a small stripping pump. The stripping pump can also be used to pump the contents of the oil drains collecting sump into the boiler fuel oil line. The coupling is disconnected, blanked and the boom moored in its 'rest' position. If no further deliveries are expected within the day, then all oil in the line should be purged into the selected fuel oil storage tank. Boiler oil trapped in the pig receiving trap is recovered and pumped into the storage tank supply line. It is customary to arrange for the recovery pump to empty the pig receiving station drain sump into the fuel oil line.

15.5.4 Fuel oil storage tanks

The procedure for hydrostatic and air testing of large conical-roof fuel oil tanks is so important and time consuming that it is worth outlining the salient points here. For the air testing phase, it is customary to fit several large capacity quick-opening valved connections to prevent the tank being subjected to sub-atmospheric pressures at any time during the test programme:

- Establish datum level measurements at selected points around the base.
- Slowly fill the tank with clean fresh water, minimum temperature 7°C, at a rate equivalent to approximately 400 mm of tank height per hour. Filling must be stopped if the rate of settlement increases above the design values.
- Record settlement at 1 metre increments of water level until the tank is filled to the top of the cylindrical section. Stop filling when this level is reached.
- Allow ground settlement to stabilise over about five days, with the cylindrical section of the tank full of water.
- Carry out air test with all vents closed. This is a test which, once started, must be continued without interruption. Constant surveillance has to be maintained throughout. It is important to appreciate that a 2.5 mbar pressure differential can produce loads of up to 50 t on the roof of a 50 m diameter tank. Pressurise the roof space to a maximum pressure of 5 mbar and soap test all joints for leakage.
- Reduce pressure toward atmospheric, opening manhole doors and quick-opening valves to ensure that the tank will not go under vacuum as it empties.
- Drain water at the same rate as for filling, and record any change in settlement at 1 metre increments of water level.

Each tank is then cleaned out and a comprehensive dimensional check carried out for subsequent calibration of the contents indicator. The heater elements are hydraulically tested, then pressurised with steam. This is to check that the support and expansion arrangments are satisfactory with the elements hot. The tank access manholes are then bolted down, ready for receipt of oil. Finally the tank level alarms are proved, as is the correct operation of all valves directly connected to the tank, since it will be difficult to rectify any problem once the tank contains oil.

As each tank is being filled with oil, the foundation settlement is noted and compared with results from the hydrostatic test. Indication of oil level and temperature, along with appropriate alarm signals, is checked as the tank fills. When sufficient oil has been received, the mixers are commissioned, noting that drive-belt tension has to be checked and adjusted at service intervals of 50, 100 and then 300 hours from initial operation. The tank heating is put into service, when boiler fuel oil is being received, after the elements are below the oil level. Initially, the temperature control may have to be carried out by manual adjustment for some time before full automatic control can be commissioned. One important check is to confirm that there is no trace of oil in the condensate being returned from the heating elements.

15.5.5 Fuel oil transfer system

The following sequence is based on an oil-fired power station, but most of the procedure is equally applicable to the oil lighting-up (OLU) burner installation for a coal-fired plant.

Referring to Fig 3.92, a flow path is set up from a selected fuel oil tank through the transfer pumps up to the isolating valve for the first main-unit fuel oil heaters. It is important that correct operation of all valves is fully proved before any oil is admitted to the pipelines. Ensure also that a flow path exists through the transfer pump discharge pressure control valves, recirculation valves and recirculation line back to the selected fuel oil tank. Charge the accumulator to the specified pressure, but leave it isolated from the system. By this time, the fuel oil pressure control loop should be commissioned and the pipeline trace heating available for service. The contents of the selected fuel-oil tank should be at a temperatue of some 50–60°C before commencing to prime any part of the transfer system.

Slowly prime sections of the pipework up to the transfer pumps to drive out all air pockets plus any water remaining from the hydraulic tests. Carefully prime the transfer pumps and check that each is free to rotate by hand. Start one pump and prime the recirculation system through to the fuel oil tank, then repeat this with each of the available transfer pumps in turn. If there are no suitable air vents adjacent to the inlet isolating valves of the main unit fuel oil heaters, the fuel oil line may have to be vented, in a carefully controlled manner, through the isolating valves with the downstream pipeline disconnected. Throughout the whole exercise, extreme vigilance is needed to confirm that there are no oil leakages.

Bring each pressure control valve into service in turn. Prove all control features, including transfer pump auto-standby facilities. Check that the transfer pump non-return valves seat correctly against a reversal of flow.

At some convenient point in the programme, the fuel oil supply lines to the auxiliary boiler fuel oil service tank have to be primed and put into service. This is carried out in conjunction with the auxiliary boiler test team, generally as outlined in Section 19 of this chapter.

15.5.6 Fuel oil circulation tests

The next stage requires fuel oil to be circulated through the main unit heating and pumping plant and the oil burner ring main back to the fuel oil tank. The transfer system is made available to the appropriate main boiler test team for this activity to be carried out as a joint exercise. Refer to Chapter 4 and also to Chapter 6 for more detailed procedures.

With fuel oil recirculating around the whole system, the pressure control valves are tuned to give the best response. The response should be confirmed by testing various combinations of transfer pump failure modes. Testing has to be repeated with the accumulator introduced into service, to ascertain the optimum setting for its isolating valve.

Final optimisation of the fuel oil transfer system cannot be completed until the main boiler HP pumps are commissioned and the main unit operated at significant loads.

Commissioning of the remaining fuel oil tanks, pipework and transfer pumps is carried out as soon as practicable, in accordance with the basic philosophy outlined in this section.

16 Water treatment plant

The main purpose of the water treatment plant (WTP) is to supply high quality make-up water to the main and auxiliary boilers. The plant may also polish slightly contaminated unit condensate during start-up and up to low-load (10% MCR) operation, if a full-flow condensate polishing plant has not been provided. During the initial stages of a unit start-up, condensate from the turbine condenser is dumped to waste until the suspended solids are < 1 PPM. At this stage, the condensate is rerouted to the WTP for cleaning up (or polishing) to achieve the ultra purity required for feedwater to modern high pressure boiler plants. It is customary to make provision for recycling the water stored in the reserve feedwater tanks to improve its quality.

This section outlines the activities necessary to commission all of the plant, from the raw water inlet up to, and including, the unit reserve feedwater (RFW) tanks. The WTP is usually programmed for commissioning well in advance of the main units. Good quality water is needed for the hydraulic test and chemical clean of the first unit boiler, as well as for auxiliary boiler commissioning.

16.1 System description

WTP processes and their integration into the boiler feedwater system vary considerably. Figure 3.95 shows a single-line diagram of one design of a fully automatic twin-stream plant, which treats industrial (non-potable) quality water by three sequential processes:

- Pretreatment
- Demineralisation
- Polishing

For example, the plant on which this section is based has primary (make-up) and secondary (polishing) mixed-bed units. Other plants have the mixed beds of the make-up plant sufficiently rated to give a polishing facility as well. At some stations, the make-up plant is augmented by a full-flow, i.e., one unit MCR feed flow, polishing plant. Condensate polishing plant is of greater importance during commissioning of once-through boilers. Some modern power stations also have unit evaporators, whose distillate would be returned to the break pressure tank for polishing before use. Main boiler blowdown or dirty drains are dumped, for example, to the CW forebay, whereas clean drains are collected, sometimes cooled, then returned to the RFW tanks.

As a general guide, modern water treatment plants are based on the following:

- 100% make-up, being equivalent to 4.5% MCR evaporation for all main units.
- Polishing, equivalent to approximately 2.5 to 3 times normal make-up, assuming that 100%-flow polishing has not been provided.
- RFW storage equivalent to one hour of MCR evaporation.

The arrangement at a 1000 MW (Ince *B*) power station is shown in Fig 3.95. The demineralisation and polishing section capacities are based on 50% flow through each stream for normal operation.

Whilst one stream is being regenerated, the other operates at 100% flow. During unit start-up, the maximum unit make-up demand exceeds the demineralisation plant capacity and the balance has to be met from water stored in the RFW tanks and recycled through the polishing, or secondary, mixed-bed units.

16.1.1 Pretreatment

Not all WTP requires a pretreatment section, as it depends upon the chemical nature of the water supply.

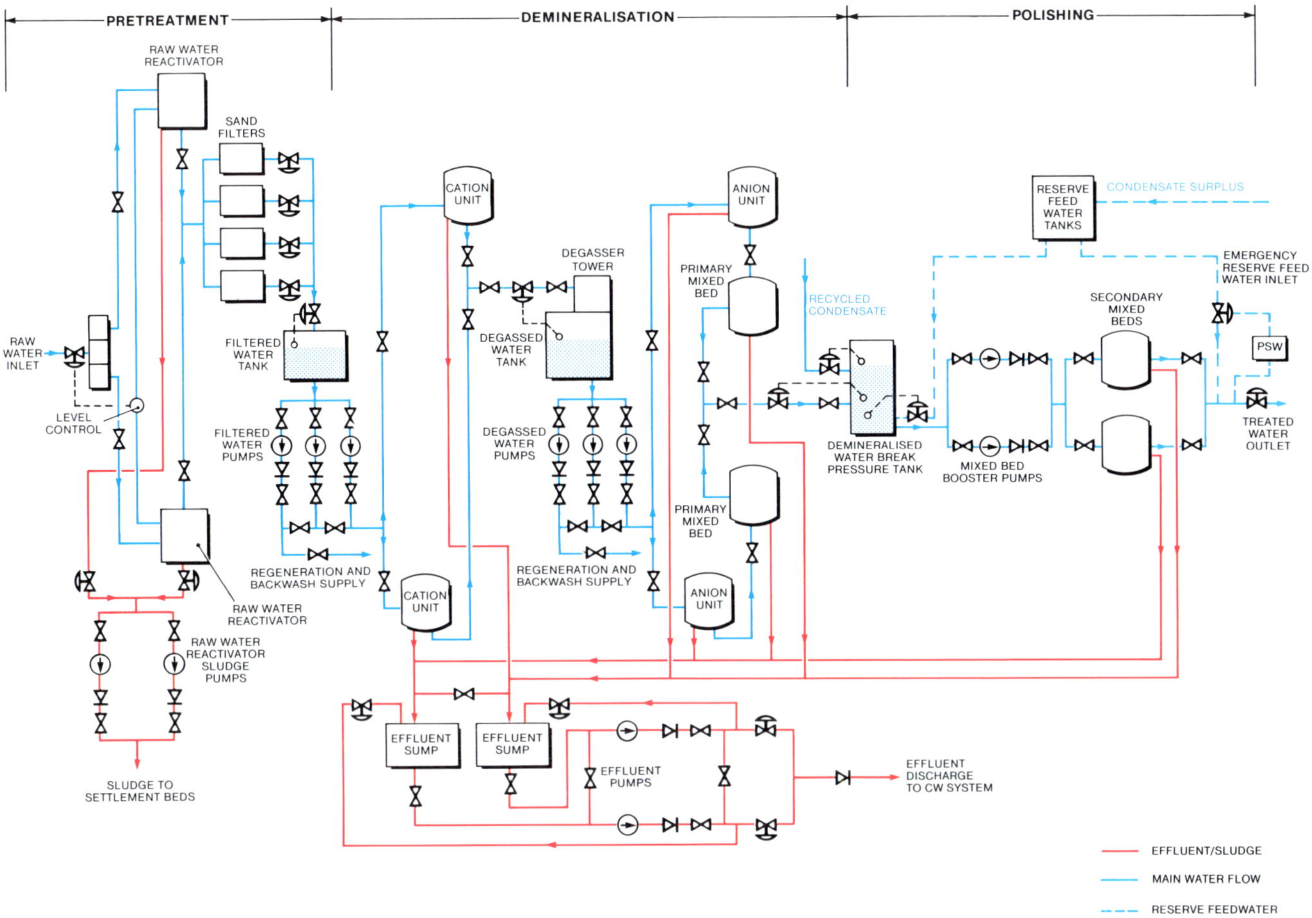

FIG. 3.95 Water treatment plant

The selection of a source of supply depends upon the economic availability of water of acceptable quality. Different locations may resort to any of the following:

- Industrial-grade fresh water from a site towns-water reservoir (Section 18). Potable water is used in some plants.
- Clarified river water from the CW make-up system.
- Site borehole supply into a separate WTP raw water tank.

Raw water is pumped into one or both of the coagulation and sludge settlement tanks, or reactivators, at a rate which is regulated to maintain a level in the reactivator outlet boxes. Chemicals are injected into the raw water as follows:

- Aluminium sulphate to promote a flocculant precipitate
- Sulphuric acid to give an optimum pH value for coagulation
- Polyelectrolyte to improve coagulation and flocculation

Each reactivator has two motor-driven units. One is an impeller to promote recirculation and formation of a flocculated sludge blanket within the tank. The other is a combined scraper and paddle. Settled sludge is scraped into a central well, where the paddle keeps it fluid for subsequent gravitation into an external sump. Sludge is pumped from the sump to a settlement bed, where water decants into the station surface water drains and the solids are removed manually every few years.

After precipitation of suspended matter in the reactivator, clarified water flows by gravity through a bank of four sand filters into a filtered water tank. The rate of flow is controlled by the water level in the filtered water tank. The filtered water is then pumped to the demineralisation section, and also used for regeneration and backwashing. Periodically each sand filter has to be scoured with compressed air and backwashed with filtered water to clean the filtration bed.

16.1.2 Demineralisation section

Filtered water is pumped through one or both cation ion-exchange resin units, then to a common degasser tower for removal of CO_2. Water flows downwards through polypropylene packing rings against an upward current of air from one of a pair of degasser fans. Degassed water gathers in the degasser tank, the water level of which regulates the rate of water flow into the tower. Pumps transfer water from the tank for treatment in the anion and primary mixed-bed units.

Degassed water is also used for regeneration and backwashing of the anion and mixed-bed units. The outlet from the mixed-bed units is led into the common break pressure tank via a control valve, which is manually set to the desired output from the plant. It has the facility of being overridden by the level in the break pressure tank, as described later.

After a certain quantity of water has been treated, the ion-exchange resins have to be regenerated as follows:

Cation units with sulphuric acid by means of a split-flow system. The top one-third of resin is regenerated with 1.5% acid downflow and the bottom two-thirds with 3% acid upflow.

Anion units with caustic soda solution at 5% co-current, i.e., downflow. In some plants the anion units are regenerated counter current, i.e., upflow.

Mixed-bed units by 5% acid downflow and 5% caustic soda upflow after resin separation by backwash. Resins are remixed by means of air blowers, after regeneration.

Facilities are also provided for periodic alkaline brine treatment of anion and mixed-bed units.

Regeneration effluents are collected in one of two underground effluent tanks for neutralisation before controlled disposal to the main CW system outfall.

16.1.3 Polishing section

The break pressure tank accepts water from three sources, each of which is at a different pressure and perhaps a different temperature. Since the tank is vented to atmosphere, the inlet pressures equalise and present a relatively constant pressure at the inlet to the mixed-bed booster pumps. The break pressure tank water level control scheme has overall control over the valves which allow water to flow in from the appropriate pipeline in the following order of priority as the water level falls:

- Recycled condensate from main turbine condenser.
- Treated water from demineralisation section of the WTP.
- Reserve feedwater, recycled from RFW tanks.

The tank level controls shut each inlet in reverse order as the level rises. Mixed-bed booster pumps withdraw water from the break pressure tank and pass it through the secondary or polishing mixed-bed units. The output from the polishing section is set by manual adjustment of the secondary mixed-bed outlet valve opening.

Regeneration of the secondary mixed-bed units follows the same procedure as for the primary mixed-bed units.

Treated water from the polishing section is directed to the following, dependent upon the operational state of the main units (see Fig 3.96):

- Condenser condensate make-up.

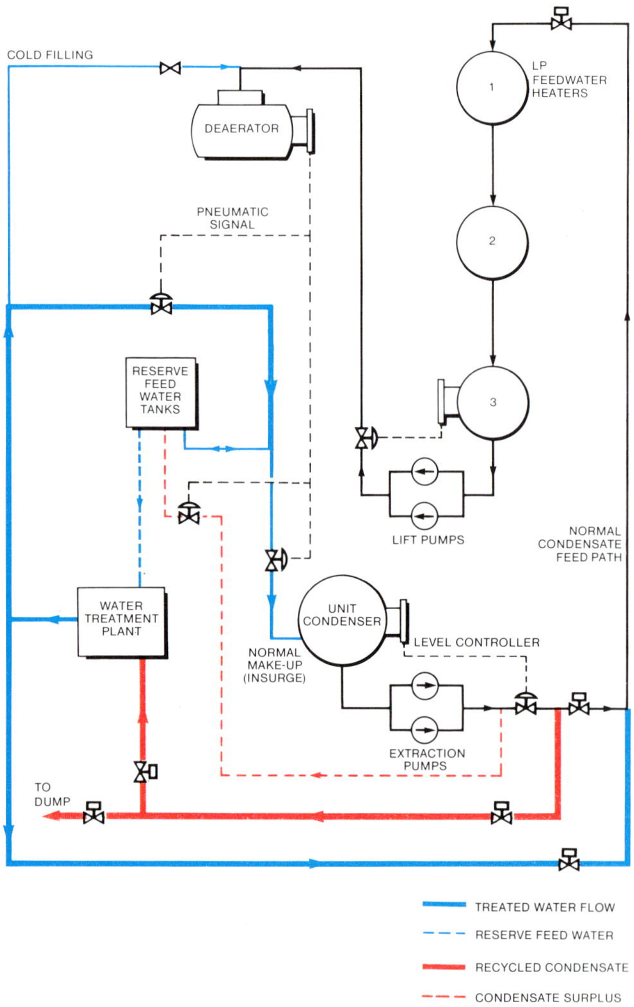

FIG. 3.96 Single line diagram of treated water system

- LP heater direct feed (during condensate recycle).
- RFW tanks (surplus).

The plant outlet is provided with an emergency RFW inlet branch to safeguard the polishing mixed-bed units from being subjected to unit condenser vacuum, in event of a failure of plant flow. The valve opens automatically as the plant discharge pressure falls toward atmospheric.

Branches from the make-up main supply water are directed to, for example:

- Main generator stator winding and hydrogen cooling systems.
- Off-load filling of the main unit de-aerators.
- Electric feed pump speed regulator make-up.
- Air extraction pump sealing water system.
- Hydrogen production plant (see Section 20 of this chapter).

16.1.4 Chemical storage and handling

Associated with the plant is a complex system of bulk chemical storage vessels, transfer pumps, water-powered ejectors, measuring and dilution tanks. Much of this section of the plant is automatically controlled to ensure that the correct strength of solution is available at the appropriate point when required. Typically, the bulk storage quantities would be:

- 8% (Al_2O_3) aluminium sulphate 70 t
- 16% caustic soda 300 t
- 96% sulphuric acid 200 t
- Industrial salt 18 t

Caustic soda is contained in electrically-heated tanks. Chemical suppliers usually wish to inspect the installation prior to making the first delivery. Safety equipment comprises at least two personnel emergency showers.

16.1.5 Effluent disposal

Effluent derived from the backwash, chemical injection and rinse stages of the regeneration processes is collected in two underground tanks. Should the chemical condition be outside the range of pH 6 to pH 9, then the effluent is recirculated and dosed with either sulphuric acid or caustic soda. When the pH is acceptable, effluent is discharged into the CW outfall or CW purge line, with pH control being exercised at all times.

16.1.6 Controls

The whole plant is designed for fully automatic control, with the minimum of manual attention, from a purpose built control panel. Modern equipment makes full use of microprocessors or solid state logic equipment for operation of relays and solenoid-operated air valves to control the action of pneumatically-actuated diaphragm valves. Control air for the pneumatically-operated valves, controllers and instrumentation is supplied by duplicated compressor/drier/receiver units within the plant building. These are similar to, but much smaller than, the unit control air compressors.

16.1.7 RFW tanks

General practice is to have two RFW tanks per main unit, normally mounted in the upper levels of the boilerhouse steelwork. Tank overflows and drains are usually led to the CW system. Tanks are usually externally braced to provide an unbroken interior surface, suitable for the application of an epoxy-based liner. Such lining materials normally have an upper temperature limit of 60°C. It is accepted that on occasions water at a higher temperature, for example from boiler clean drains, may be added to the RFW tanks. It is considered that water subsequently recycled to the break pressure tank is extremely unlikely to exceed the temperature limitation of the mixed-bed resins. This temperature limit is a function of the type of resin used within the polishing plant and is typically in the range of 30°C to 50°C.

16.2 Commissioning objectives

The following are specific to water treatment plants and should be considered in conjunction with general objectives outlined at the beginning of this chapter:

- Show that the plant has been constructed to specification and is free from hazards to operating staff.
- Flush pipework and vessels until each is clear of construction debris.
- Ensure that the plant is available to produce water for the commissioning of other plant, when required.
- Demonstrate that the plant can be operated safely under manual and automatic control with minimum operator involvement.
- Prove the adequacy of the effluent neutralisation and disposal facilities.
- Confirm that the flow rates, treated water quality and exchange-resin capacity are in accordance with the design intent.
- Check that, after using the specified quantity of regenerants, there has been no significant deterioration of the resins.

16.3 Safety considerations

Commissioning of this plant could produce a number of hazards, many of which may be unfamiliar to construction personnel. Each of the following should be considered in conjunction with those highlighted at the beginning of this chapter:

- Flooding as a result of ineffective isolation, pipe or tank failure.
- Adequate ventilation of enclosed tanks and vessels including testing of the internal atmosphere, before personnel entry.
- Secure temporary barriers, for areas containing hazardous chemicals.
- Clear emergency instructions posted at several locations.
- Personnel emergency showers to be always available.
- Personnel to be fully familiar with inherent dangers of working in an environment where caustic soda and sulphuric acid are being used.
- Appropriate protective clothing to be worn.
- Approved procedures prepared for safe receipt, storage and handling of all chemicals being used in the plant.
- Violent reaction with liberation of substantial heat, when sulphuric acid and water are mixed.

- When sulphuric acid comes into contact with most metals hydrogen is evolved. NO SMOKING/NO NAKED LIGHTS to be enforced in the vicinity of acid-containing vessels.

16.4 Resources

16.4.1 Services

Apart from availability of electrical supplies from dedicated 415 V switchboards, the following will be necessary:

- At least 1 tonne of sodium carbonate for acid neutralisation purposes *before* delivery of acid is accepted.
- Initial supplies of the requisite chemicals, namely:

96% sulphuric acid	18/20 t
46% caustic soda	18/20 t
8% (Al_2O_3) aluminium sulphate	15 t
Salt	12 t
Polyelectrolyte	25 kg bags

- Domestic water supply for personnel emergency showers, washdown hoses, salt saturator mixing, etc.
- Surface water drainage system.
- Protective clothing, safety barriers, first aid notices.
- Washing down hose adjacent to regeneration and bulk storage area.
- Adequate supply of sealed eyewash bottles mounted in clearly identified cabinets located at strategic areas. Clean water is now regarded as an adequate eyewash solution.
- Temporary submersible pump, perhaps rated for 15 litres/s at 10 m head, for emptying tanks and sumps, where necessary.
- Hot water heating supply at 82°C for caustic regeneration heat-exchanger.
- Supply of sand, pebbles and resins for loading into the appropriate vessels.
- Two-way radio communications.

16.4.2 Staffing

Specialist staff from each of the chemical suppliers will be required to conduct an inspection of the plant before the first delivery of chemical is authorised. Three teams of three people, supplemented by station operating staff when necessary, should be sufficient to commission the plant covered by this section. The commissioning period, assuming five-day working, may well occupy the whole team for some 25 weeks.

16.5 Commissioning

A detailed commissioning network should be prepared to highlight the interaction of different sections of plant. A simplified version of a typical diagram is given by Fig 3.97.

The following basic conditions should be applied, assuming that all relevant safety considerations have been taken into account.

- Full use of standard check lists (SCL) for individual items of equipment, including proving of controls, indicators etc.
- Pipework to be hydraulically-tested to 1.5 times normal working pressure and then fully-flushed clear of debris.
- Compressed air pipework to be purged to atmosphere after pressure drop testing at normal working pressures.
- All tanks to be hydrostatically tested, and then the protective lining applied.

Commissioning of each section of the plant is now considered in turn.

16.5.1 Prerequisites

- Ensure that all services will be available when required.
- Mechanical erection complete.
- Cabling terminated and checked.
- All motors checked for correct rotation (may be up to 40 items).
- Control panel energised. Controls and valve operation proved, including 'dry' runs of the programme sequence controllers, in advance of them being required for plant commissioning. Alarms and control checks programmed to be available, as each stage of plant commissioning dictates.
- Electrical trace heating proved and plant released for thermal insulation as soon as practicable.
- Chemical measuring and dilution equipment checked; test certificates available; plant left completely clean and dry until wanted.
- Personnel emergency showers proved to maintain a flow of water at 30–35°C for up to 10 minutes, typically at a rate of 0.9 litre/s.
- Control compressed-air system commissioned and all control functions proved, generally as given in Section 21 of this chapter.
- Degasser tower fans, sand filter and mixed-bed air compressors commissioned.
- Water and effluent tanks hydrostatically tested to prove that they are free of leaks. Anion, cation and

mixed-bed units all hydraulically tested as well. All linings properly applied and tested to be of adequate thickness. Similar checks are necessary on sections of lined piping.

- Pneumatic and electrically-operated valves commissioned.

16.5.2 Bulk storage

- Chemical tanks checked to be clean, dry, and, where appropriate, that the lining is intact and has been extended into all pipework branches. Test certificates to be available.
- Filling lines clearly identified, inlet valves fitted with locking facilities, and the tank inlet pipe extended down to within 150 mm of tank bottom.
- Emergency outlet plug valve checked for operation, then tank manhole secured.
- Tank level indicators and alarms all proved to work satisfactorily.
- Caustic tank heaters commissioned. Safety thermostat proved to operate at 25°C, i.e., 10°C above control thermostat setting.
- Bund chemical-resistant lining complete. Outlet drain valves proved to operate and drain lines to effluent tanks checked to be clear.
- Salt tank asphalt lining complete, then gravel bed installed. Towns water inlet ball valve proved. Tank and filling connection clearly identified, with locking facility fitted.
- Road drains at all filling points proved to be clear by checking for a flow at the nearest manhole.
- Bulk delivery of each chemical received to the approved procedure in the following order on separate days; sulphuric acid, caustic soda and aluminium sulphate. Delivery of salt can be deferred until later, if preferred.

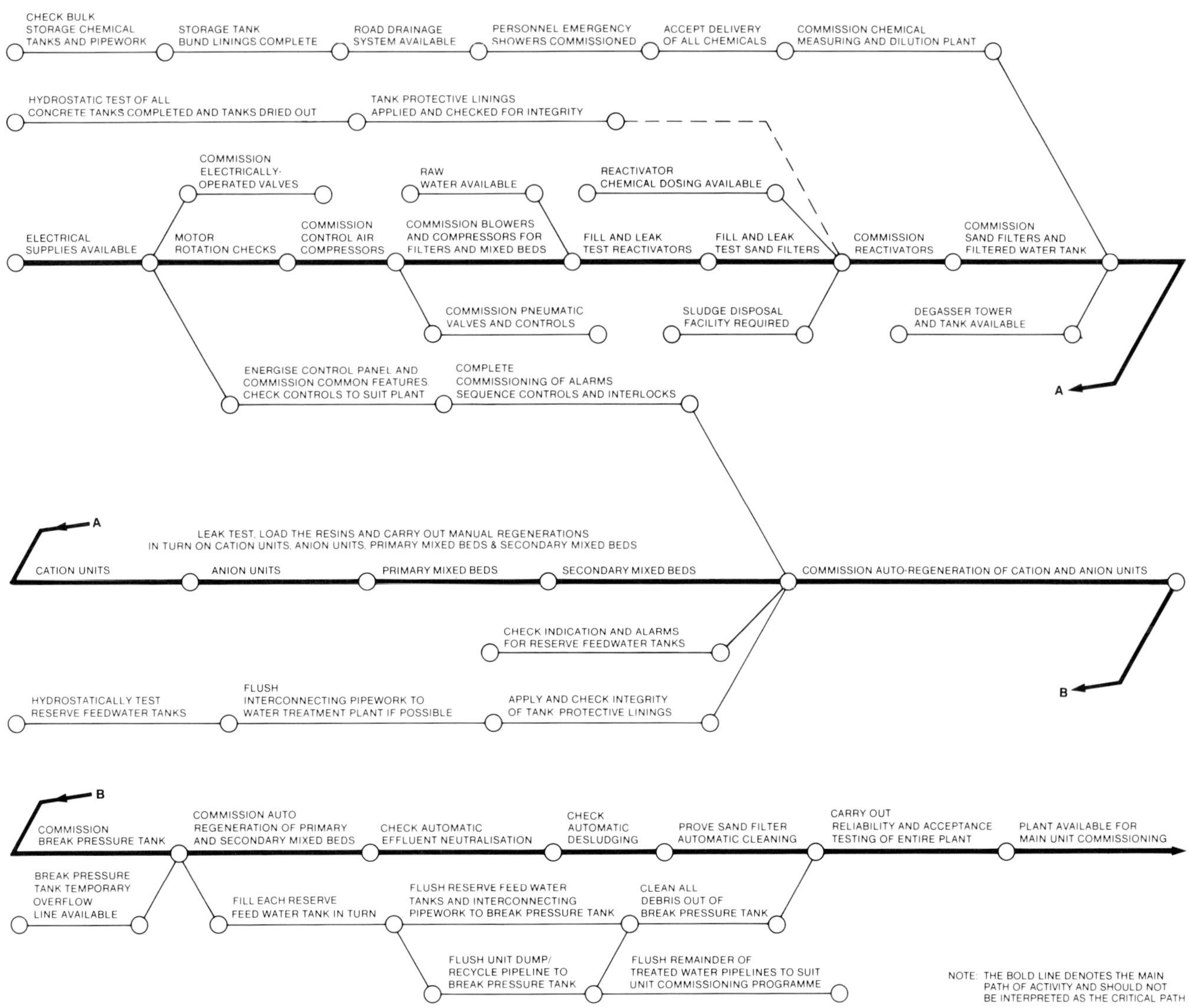

FIG. 3.97 Water treatment plant commissioning network

16.5.3 Pretreatment plant

- Sand-filter internal distribution system is checked for integrity, then graded gravel and sand loaded, ensuring correct layering of each material. The final level of the filtering material in the vessel should be recorded.
- The reactivators are filled with water, the dosing equipment set up and the impeller and scraper speeds adjusted.
- Sedimentation and coagulation checks are carried out on the reactivators while passing water across the face of the sand filters to drain, thus simulating the normal reactivator throughput.
- Water, equivalent to normal flow rate, is then passed through the sand filters, again to drain.
- Sand filter backwash flow and the automatic cleaning sequence is set up.
- Pretreatment plant is left on 'auto', maintaining the filtered water tank full, to provide water for all subsequent operations in the demineralisation and polishing sections of the plant.
- Chemical dilution ejectors tested in preparation for use.

16.5.4 Cation units

- Water is then fed to the cation units and the distribution pattern of the internal spreader systems is checked prior to loading of the resin.
- Take sample of resin during loading for quality control checks.
- Fines in the resin are removed by backwashing the resin bed and then by skimming off the settled surface.
- The resin bed is then adjusted to the correct level and this level recorded in relation to a fixed datum.
- Regeneration flow rates are set up, using the installed instrumentation and weir boards.
- Three manual regenerations are carried out, checking the acid injection strength, to regenerate the new resin fully; then the degassed water tank is pumped to waste. Prior to this, the tank would have been full of poor quality water resulting from the passage of filtered water over partially exhausted cation resins in order to provide water for cation regeneration.
- The degassed water tank is now refilled with good quality cation-treated water, ready for the next stage in the programme.
- An effluent neutralisation and discharge may be carried out at this point to prove correct functioning of the controls. However, this would probably require a large quantity of caustic. If sufficient sump capacity is available, it is advantageous to defer this activity until after the anion regeneration phase.

16.5.5 Anion units

- The anion units are treated in a similar manner to the cation units, by admitting water for checking the internal spreader distribution patterns.
- The resins are loaded, fines removed and the resin level corrected. Take sample of resin during loading for quality control checks. The final level should be recorded in relation to a fixed datum.
- Good quality degassed water is used to set up regeneration flow rates and to perform three manual regenerations. During the third regeneration, the strength of the caustic soda injected onto the resin is checked.
- The cation/degasser/anion combination is then run to rinse/recycle back to the filtered water tank. When the quality, i.e., conductivity, of the rinse outlet is better than that of the filtered water being used for the rinse procedure, it is redirected from waste to then recycle to the filtered water tank. There is a significant economic advantage with this approach, particularly in mixed-bed unit rinsing.
- Effluent neutralisation and discharge to waste is carried out, if required.

16.5.6 Primary and secondary mixed-bed units

- Water is next fed to the primary mixed-bed units and the spreaders and collectors are checked. Each unit is charged with cation resin, the fines removed and careful attention paid to the resin level with respect to the central distribution laterals.
- The anion resin is placed in the unit and fines again removed, then the unit is backwashed and the resin interface and level checked and recorded, as for other resin-filled vessels. The interface should be clearly visible in the viewing port provided for this purpose.
- Take samples of all resins as they are being loaded for quality control checks.
- Regeneration flows are set up, followed by three manual regenerations. The demineralisation section of the plant is then run on recycle and finally the system is run to waste to check the chemistry of this section of plant.
- When water of adequate quality is obtained from the demineralisation section of the plant, water is fed forwards to the break pressure tank and then to the secondary mixed-beds, which are treated in an identical manner to the primary mixed-bed units.

- The whole plant is then run on manual control, passing water to the RFW tanks while control and quality checks are carried out.

16.5.7 Reserve feedwater system

- Available tanks hydrostatically tested, cleaned out and lining applied, and lining integrity checks completed.
- Pipework to and from WTP all checked and tested.
- Drains and overflow lines all proved to be clear.
- Level indication and alarms commissioned.
- Pipework up to the condenser make-up, LP heater and de-aerator direct feedlines isolating valves all checked. Temporary drain piping may need to be fitted to convenient points, to facilitate flushing each line as far as possible.
- With RFW tank outlet valves shut, fill each tank in turn to prove the capacity of the overflow line, using water from the secondary mixed-bed outlet.
- If necessary, fit a temporary overflow line from the break pressure tank into an adjacent surface water drain. The normal overflow pipework may not be of adequate capacity to cope with the abnormal mode of operation associated with the flushing procedure.
- Flush the RFW return main into the break-pressure tank from each RFW tank in turn.
- Flush the emergency RFW inlet line to waste via a temporary pipework branch.
- Drain and clean out break-pressure tank after completion of the RFW tank flush.
- Flush condenser make-up and de-aerator direct feedlines to waste from WTP outlet.
- Flush recycle main through the LP heater direct feedlines to the CW forebay and, when clear, flush the short length into the break pressure tank.
- Clean out the break pressure tank.

Note Flushing of extensions to the common make-up and recycle pipework for each unit has to be considered individually, as each is released for commissioning.

16.5.8 Plant proving

Proving tests are all carried out with the plant recycling treated water between the break pressure and RFW tanks. For some tests, the RFW tanks are allowed to overflow in order to create the desired flow from the demineralisation plant. Commissioning of the anion resin brine/caustic wash part of the regeneration cycle can be proved after the bulk delivery of salt has been received.

Each section of the plant has to be shown to operate successfully under manual control before automatic sequence controls can be commissioned. Automatic control of various sections is integrated progressively, until the entire plant is operating successfully. Operational tests have to be carried out to prove that the plant can be easily started up and shut down, and that the output can be regulated to suit the demand.

Selected chemical parameter faults have to be simulated to ensure that the appropriate sections of the plant react in a safe and consistent manner. Similarly the effects of motor-driven auxiliary plant failure have to be demonstrated. Electrical power and compressed-air failure tests should be conducted and the effects assessed.

Capacity, or throughput, tests must be carried out to prove that each type of ion-exchange resin is capable of treating the specified quantity of water before it becomes exhausted. Subsequent regeneration cycles should be carried out within the prescribed time periods, using the specified quantities of regenerant chemicals. Performance of the effluent neutralisation and the discharge pH-control loops has to be demonstrated and proved to be reliable.

Finally, the plant is subjected to a four-week reliability run prior to conducting the contractual acceptance test programme. At this stage, the plant is generally released for operation by station staff to meet the water demands imposed by commissioning in other plant areas.

As the main unit commissioning progresses, the ability of the plant to meet the make-up and polishing demands under normal service conditions, whilst maintaining the prescribed quality of output, must be demonstrated. This will be carried out in conjunction with the tests described in Chapter 6. However, as this may not be able to be completed until more than twelve months after the official takeover date, special arrangements may have to be reachèd to extend the contractual guarantee.

17 Cooling water systems

Within the scope of this section, commissioning of two types of cooling water (CW) system is considered:

Direct One that uses natural heat sinks, such as the sea, estuaries or wide rivers, where the water flows only once through the turbine condensers.

Indirect Where river water is recirculated through the turbine condensers and a number of cooling towers. This type needs a considerable source of make-up water from a river or canal to replace the inherent evaporation losses, plus water that has to be purged from the system for regulation of suspended solids.

Description and commissioning of typical CW system anti-fouling systems is given toward the end of this section.

17.1 CW systems

17.1.1 Systems description

General

The CW system provides coolant for the main turbine condensers. An auxiliary cooling water (ACW) system, tapped into the CW system, provides coolant for many of the heat exchangers associated with unit auxiliary plant. Commissioning of the turbine condensers and all of the ACW-cooled heat exchangers is dealt with in the appropriate sections of Chapters 5 and 6.

Other auxiliary plant heat exchangers use coolant from separate independent service water systems which may, but frequently do not, utilise the same water as does the CW system. Commissioning of these systems is covered in Section 18 of this chapter.

If the conditions attached to the abstraction licence issued by the local water authority can be met, it is customary to use the CW system for disposal of the flushings from the main boiler chemical cleaning process, as well as the neutralised effluents from the water treatment plant. Because of this, the commissioning programme for the CW system has to be integrated with that of the first main boiler.

CW systems usually are provided with some means of controlling the growth of bio-fouling organisms, which reduce the effectiveness of heat transfer in coolers. The methods employed depend very much on local conditions, but most generally involve some form of chlorine injection. Some power station condensers are provided with condenser tube on-load cleaning systems to prevent fouling of the condenser tube bores.

Direct CW systems

Figure 3.98 shows an arrangement of one direct CW system where sea water is abstracted from an estuary through a pair of intake tunnels whose inlet is some distance off shore. After passing through coarse screens, CW is pumped through fine strainers into a system of large cross-section interconnected culverts to the turbine condensers. Similar culverts convey CW discharged from the condensers to an outfall structure located on the banks of the estuary. The location of the outfall in relation to the intake has to be carefully considered with respect to the prevailing tidal flows to minimise the proportion of heated water likely to be recirculated to the intake. Supply and return branches for the unit ACW system are tapped into the appropriate unit culverts adjacent to the condenser branch pipes. Chlorination of the sea water to limit bio-fouling and mussel growth is usually achieved by injection at the off-shore end of the intake tunnels and either at the CW pump discharge branch or at the condenser inlets.

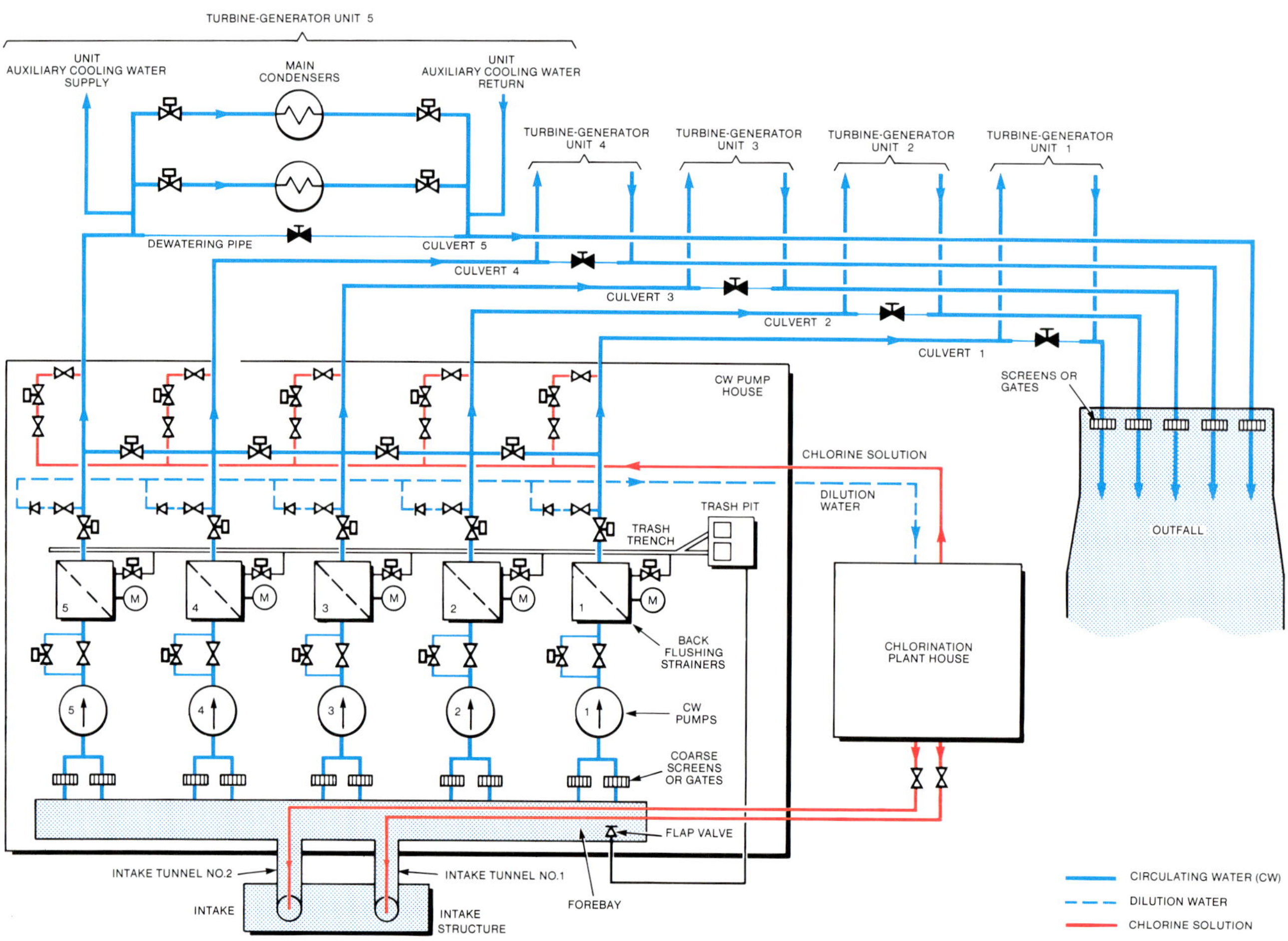

Fig. 3.98 Direct CW system

Where the intake works are located on the shore line, injection may be carried out at the forebay.

Stop gates, manufactured from steel or a light alloy plate, and fitted with peripheral hard rubber sealing strips, are provided for isolation of the forebay, outlet culvert sections and CW pump inlets. The stop gates at the inlet to the forebay and at the outfall are replaced with coarse-pitch bar screens for normal operational service. Isolation of intake tunnels is usually achieved by means of large dome-shaped covers lowered on to the off-shore entrance under direct control of a diver.

Some system designs, like that shown by Fig 3.98, have strainers fitted into the CW pump discharge branch, or high pressure side. Other designs may incorporate fine screens located at the CW pump suction branch, or low pressure side.

Low pressure screens may be either drum- or band-type, driven by an electric motor via a speed-reducing gearbox equipped with variable-speed control. The screening material typically has 10 mm apertures, and is usually formed from either stainless steel wire mesh or plate. A pressure jet screen cleaning system, utilising a pumped supply of screened water as a medium, is usually provided for removal of debris, fish, etc. Trash accumulates in baskets from which it has to be removed manually and disposed of. Water drains back into the screen inlet chamber.

Discharge strainers usually are of the automatic backflushing type, where the screening material, formed into either a tube or a reel, is contained in a steel body. Cleaning of debris is effected by means of a semi-enclosed duct, the open side of which fits relatively closely to the 'dirty' side of the screen. When the outlet pipe from the duct is opened as part of the automatic back-flushing sequence, internal CW pressure induces a reverse flow through the screen to wash debris into the trash pit. An electric motor is used to move the duct over the surface of the screening material to remove all of the debris. Automatic operation of the backflushing sequence is initiated from a measurement of differential pressure across the strainer.

It is customary to provide one CW pump per unit, subject to there being a minimum of four pumps for operational security. The drive motors are connected to an 11 kV unit board because of the power requirement, typically 4 MW for a 500 MW unit size station. The pumping units may be either horizontally- or vertically-mounted with the motor drive usually via a speed-reducing gearbox. A separate pressure-fed lubrication system is required for the gearbox, and it may also include the motor and pump bearings. Electrically-operated butterfly type suction valves may be fitted to the CW pumps, but it is becoming more common only to provide stop gates for isolation purposes. CW pump discharge valves are of the hydraulically-opened self-closing butterfly type (see Section 2 of this chapter). Most of the other valves in the CW system are electrically-operated butterfly style.

The CW outfall structure incorporates a low tide seal-weir, which ensures that the system operates under syphonic conditions at all predicted tide levels, thus utilising syphonic recovery to reduce the pumping power required. At the lowest tides, the CW pumps may be unable to achieve complete priming of the condenser water boxes and some of the ACW-supplied heat exchangers. A small centralised vacuum exhauster plant with interconnecting pipework to the highest point of each water box may be provided to overcome this problem.

Frequently a comprehensive interconnected (sometimes pumped) dewatering system is provided for emptying sections of culvert that have been isolated for internal inspection. Often the system can be temporarily adapted for initial priming of the culverts, to save having to use the main CW pumps for this purpose.

Indirect CW systems

An arrangement of an indirect CW system is shown by Fig 3.99, where water is pumped from a forebay through large cross-section interconnected culverts to the turbine condensers. Water discharged from the condensers is conducted through similar culverts to a group of cooling towers. Cooled water then flows back to the forebay for recirculation to the turbine condensers. Unit ACW supplies are usually tapped off the main CW culverts to each unit, but the returns are often commoned and led direct to the forebay, thus bypassing the cooling towers. The system depicted by Fig 3.99 also uses the forebay as a source of water and as a heat sink for the general service water system (see Section 18 of this chapter for further details).

CW make-up, equivalent to approximately 3% of the CW system throughput, is abstracted from a nearby river or canal. This is to replace evaporation losses from the cooling tower (typically 1% of throughput) and also to compensate for water purged from the CW system to control suspended solids as a result of evaporation. The purge flow rate is equivalent to approximately 2% of the CW system throughput and is returned to the river or canal.

Chlorine injection is applied, for control of sliming and scaling organisms, usually at the condenser inlets. At some sites, this is supplemented by mechanical on-load cleaning of the condenser tube bores by recirculating sponge-rubber balls between the condenser inlet and outlet waterheads.

One CW pump per unit, subject to a minimum of four pumps, is usually installed and current practice is to use vertical-spindle concrete-volute pumps, either directly- or gearbox-driven by motors connected to 11 kV unit boards. Power ratings are similar to those for a direct CW system, i.e., 4 MW each, giving a flow rate of some 14 to 16 m^3/s. Stop gates are provided for isolating purposes at the inlet to each pump. Hydraulically-opened self-closing butterfly valves are fitted at the pump discharge. Most of the other valves in the CW system are of the electrically-operated butterfly type.

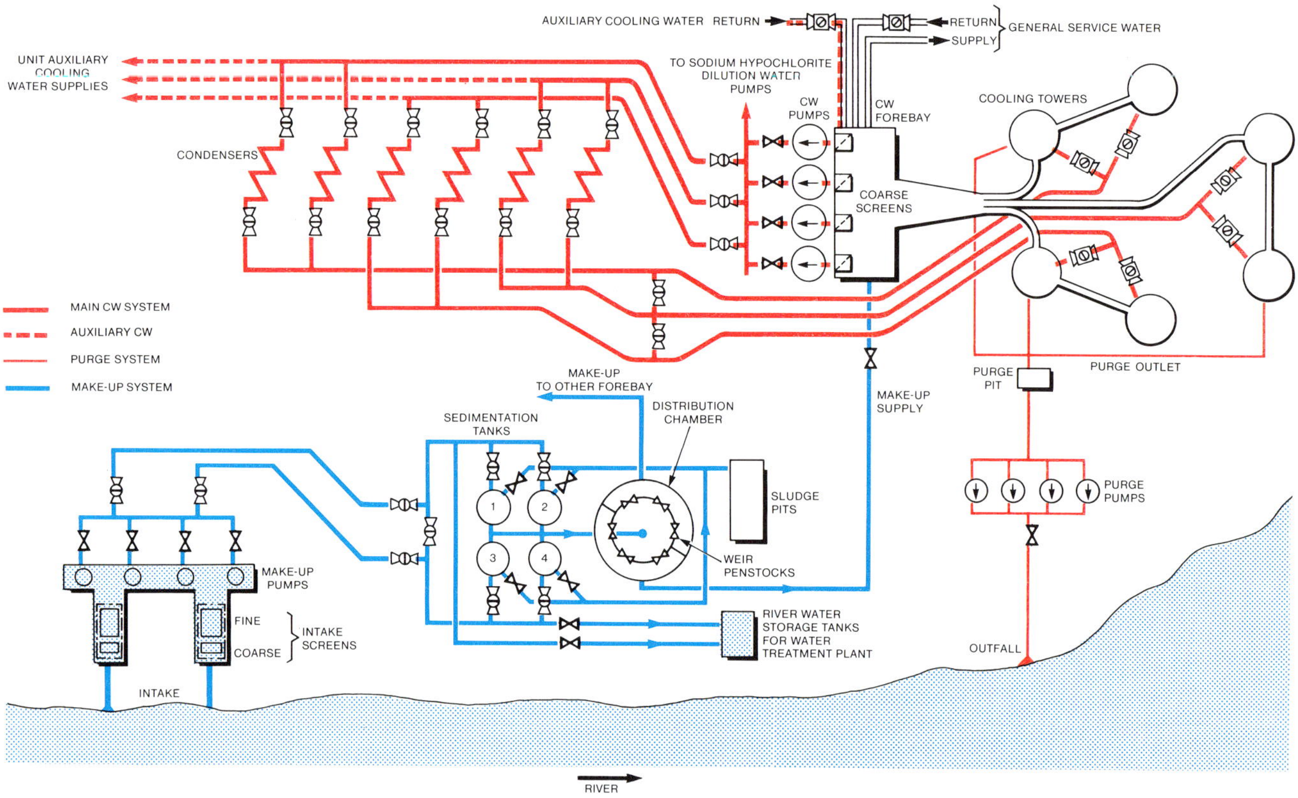

FIG. 3.99 Indirect CW system

The cooling water make-up (CWM/U) is usually added to the forebay, along with the common return from the unit ACW systems and also oil-free water drains and overflows from various other items of plant. Three or four make-up water pumps operate under automatic control in response to water level changes in the CW forebay. At some power stations variable-speed drives are fitted, though most stations utilise simple stop/start sequences. Typically make-up pumps are rated at 150 kW, are connected to a 415 V Station Services Board, and have a flow rate of some 0.8 m^3/s. Hydraulically-opened self-closing butterfly discharge valves are usually fitted to each pump.

Make-up water is normally drawn through coarse-pitch bar screens, then pumped through either automatic backflushing strainers or low pressure screens, similar to those described earlier. If the make-up water contains a high concentration of silt, it may be necessary to pass the water through sedimentation tanks, so that the silt can settle out and then be desludged from the system. Make-up water from the sedimentation tanks flows over peripheral V-notched weirs into a high level sluiceway feeding a common distribution chamber. The distribution chamber may be a concentric structure with the outer annulus divided in two along a diameter. Flow from the inner chamber to the outer annulus is regulated by a series of weir penstocks, as shown in Fig 3.99, before passing to the CW forebay by gravity.

Natural draught cooling towers are usually arranged on a two per unit basis, with stop gates provided to isolate the ponds of individual towers. Each cooling tower has a water purging connection led to a common purge pit, whence silt-laden water is usually pumped back into the river or canal. Generally three or four fixed-speed purge pumps of capacity similar to those of the make-up system are provided. These pumps operate on start/stop control from the purge pit water level. Another design of purge system taps the purge off the CW pump discharge culverts and pipes it direct to the river or canal, thus dispensing with the need for purge pumps.

One prototype assisted-draught cooling tower (ADCT) has been constructed at Ince *B* Power Station (1981). This tower has a thermal performance equivalent to three similar-sized conventional natural draught cooling towers. Air is induced horizontally, through the splash packings located around the outer periphery of the tower base, by thirty-five 212 kW 3.3 kV slow-speed axial fans on a 10° pitch. Figure 3.100 shows a cross-section through the base of this special tower design.

The abstraction licence granted by the local water authority normally lays down strict conditions in respect of the rate of make-up and purge flows, and also the abstraction flow in a given period. The permissible temperature and chemical condition of CW purge returned to the waterway is also subject to strict limitations.

17.1.2 Commissioning objectives

Irrespective of the type of CW system, most of the following objectives are common to each. They should

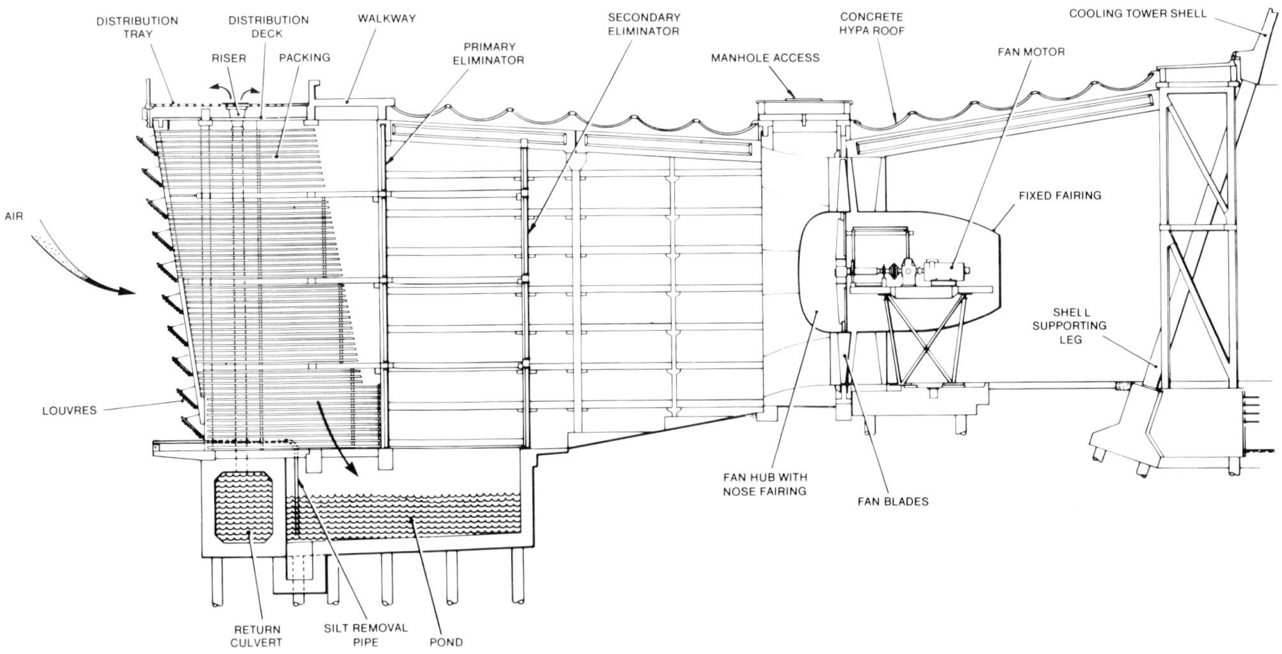

FIG. 3.100 Section through assisted draught cooling tower cell

be considered in addition to those outlined in the introduction to this chapter:

- Hydrostatically test all culverts and prove that all isolation valves are leak free.
- Ensure that all culverts, screen chambers, etc., are clean and free from debris, and check that any protective coatings are free from damage.
- Demonstrate that all stop gates provide effective means of isolation.
- Prove correct functioning of all control and protective devices.
- Energise and prove operation of all butterfly valves and carry out initial timing tests of hydraulically-operated valves.
- Rotation check main CW pump motors and all auxiliary drives, then flush the CW pump sealing water and lubrication systems.
- Carefully flood the CW system as far as practicable, using gravity flow.
- Carry out coupled runs of main CW pumps and prove correct operation of all auxiliaries.
- Commission chlorine injection system (water only, initially).
- In conjunction with condenser test team, prime and circulate CW through unit condensers and ACW systems.
- Carry out programme of system surge tests, using different combinations of CW pumps, flow rates and pump discharge valve operating times, to simulate all foreseeable normal and abnormal operational conditions.

In the case of indirect CW systems, the following objectives have to be incorporated into the appropriate part of the programme:

- Commission and prove the CWM/U system, which may also involve a system surge test programme for hydraulically-operated valve time settings.
- Commission the purge system.
- Commission the cooling towers (particularly if they are fan-assisted).

17.1.3 Safety considerations

The following should be considered, in addition to those listed at the beginning of this chapter:

- The main dangers are of personnel becoming trapped inside sections of enclosed culverts or being drowned as a result of falling into deep open culverts.
- Asphyxiation or explosion are slight possibilities in drained culverts, which have retained still water for extended periods of time. No personnel should be allowed into any enclosed culvert until the atmosphere has been certified as being 'safe'.
- It is prudent that personnel be temporarily withdrawn from adjacent sections of culvert during initial flooding until the isolation has been proved. Similarly, great care must be taken to ensure that all personnel are accounted for before access manholes on any section of culvert are bolted down.
- Lifebelts must be strategically located around open water-culverts and, if necessary, temporary handrails provided.

- Additional safety precautions should be drawn up for flow measurement tests which require the use of radioactive isotopes.

During system surge testing, it is not unknown for peak pressures of the order of 6 bar to be attained, where the normal working pressure is between 2 and 3 bar. Though the majority of modern CW pumps are designed to be safe when running at full speed in the reverse direction, emergency procedures should be fully understood in case the safe speed is exceeded during testing.

Commissioning of large diameter slow-speed fans for an ADCT requires stringent restrictions governing access to the fan unit during initial operation, particularly until confidence in the design has been established.

17.1.4 Resources

Services

Typically some 800 to 1000 m^3 of fresh water will be required for hydrostatic testing of the culvert system and water-retaining concrete sections of cooling towers. It is customary to test the system in definable sections, transferring water from section to section as testing proceeds before finally discharging it into the CW forebay or intake chamber, or to waste.

Unless a permanent dewatering scheme has been provided, portable pumps will be necessary for transferring test water from culvert to culvert.

Staffing

Due to the geographic location of the components of a CW and/or CWM/U system, the main resource required is a relatively large number of personnel (say seven to ten) equipped with reliable two-way radio communication. The commissioning activity could well be spread over a period of six to twelve weeks.

17.1.5 Commissioning

General

The principles described apply to direct or indirect CW systems, though any differences will be highlighted, where applicable. Commissioning is carried out in defined stages as the various culverts, pumps and associated unit CW systems are completed. The construction programme should be such that two CW pumps and associated auxiliaries are available at the same time as the culvert system for the first turbine-generator unit. Figure 3.101 shows a simplified commissioning network for CW systems. During commissioning of the first sections of culvert, effective isolation must be maintained between finished and unfinished sections to safeguard personnel and plant.

Culverts

Following completion of erection, the culverts with their isolating and crossover valves are checked to standard check lists (SCLs), which include mechanical operation of the valves and checks on cleanliness, plus the integrity of any culvert lining. The culvert system has to be subjected to a hydraulic test for at least a 12-hour period at a pressure of between 3 and 5 bar, dependent upon the system design pressure.

A loss of up to 0.3 bar over 12 hours is generally considered as acceptable. Note that the age of new concrete culverts determines the maximum test pressure, typically 2.4 bar at one month, rising to 4.5 bar after six months. Furthermore, the test water temperature must not be more than 3°C below the ambient temperature.

Immediately prior to flooding of culverts with CW for system commissioning, there should be a further inspection for cleanliness and, where appropriate, lining integrity. Securing of access manholes must be conducted in an organised manner, such that the last two, preferably one at each end, are continuously manned while a final check is carried out to ensure that no personnel or debris has been left inside. These two covers are then immediately bolted down. Electrical operation of isolating and crossover valves should be proved, along with remote controls and indications, where practicable, before flooding is carried out.

Direct CW systems

The CW pump inlets, screen chambers and forebay are cleaned out and all stop gates with associated lifting tackle checked, prior to initial flooding up to the CW pump intakes.

In the case of direct systems, water is often pumped over the construction cofferdam to prove the intake gates before the cofferdam is removed and the area dredged to the correct contour. Penstock valves are then opened to flood the intake structure and balance water levels before the stop gates are removed.

The low pressure screens and their wash water system can now be commissioned. One of the important checks is to ensure that the edge-seal gap between clean and dirty sides of the screening material is less than 25% of the mesh size. The backflushing control sequence and operation of the duct drive mechanism of pressure strainers is checked with the system dry. Completion of commissioning has to await operation of the appropriate CW pump to make water available for proving the backflushing action. Part of the chlorine injection plant has to be commissioned in order to control mussel growth, particularly during the summer months, if there is likely to be intermittent or no CW flow.

CW pump hydraulically-operated discharge valves are commissioned, including proving of all controls, alarms and indications, and the closing times set to the longest available. Typically these may allow for the

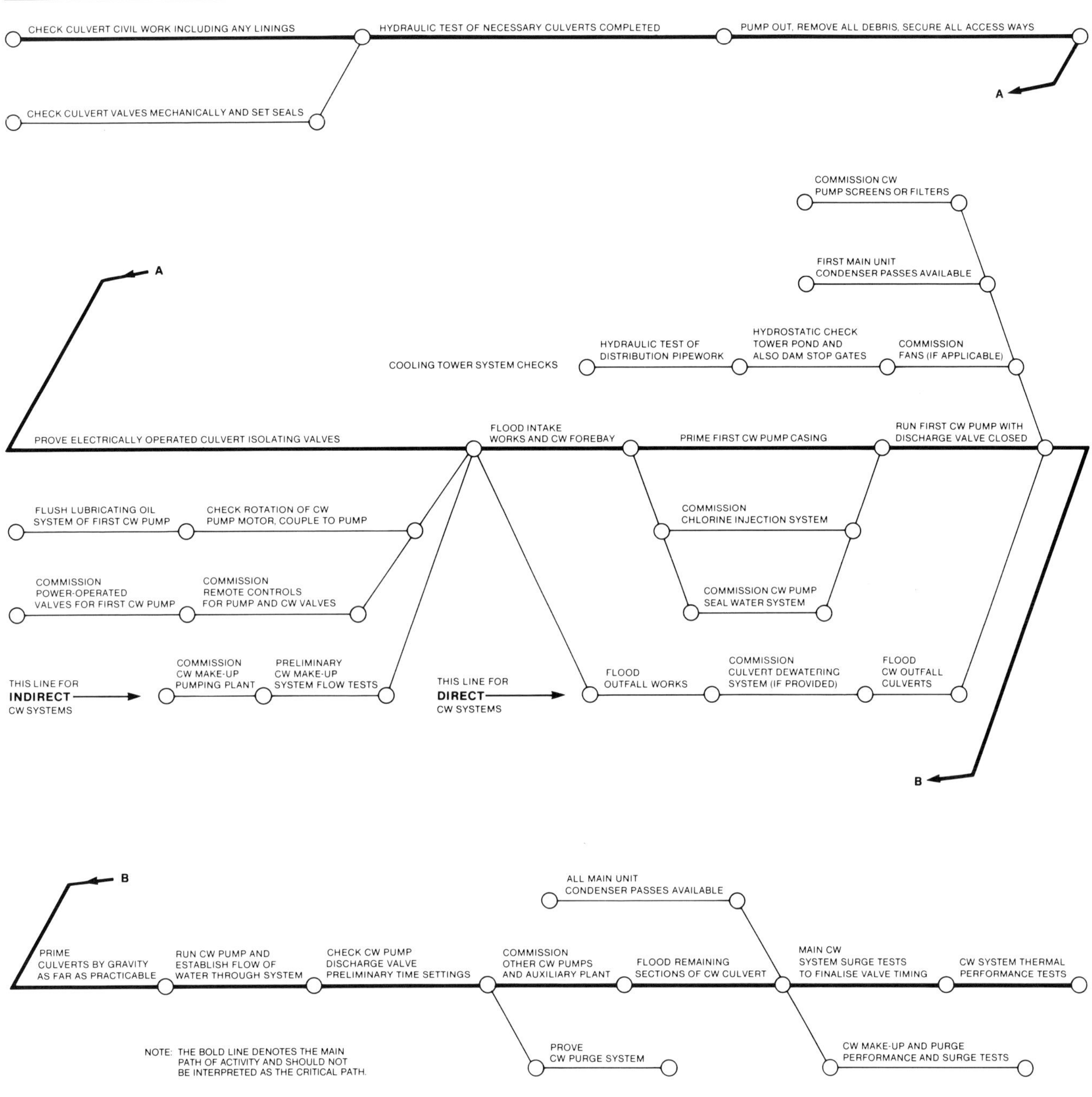

FIG. 3.101 CW systems commissioning network

valve to close over the first 80% of travel in 30 s and completely close in a further 30 s.

Indirect CW systems

Cooling tower ponds are filled with water to the top of the wall to prove their ability as a water-retaining structure, during which the pond stop-gates can be checked for tightness. The gates are normally expected to seal against flow into the pond area, and this is checked at a later stage. Prior to flooding of the CW system, the ponds are cleaned out and all stop gates and lifting tackle checked.

Indirect CW systems require the CWM/U system to be commissioned first, before flooding the forebay and CW pump inlets to prove sealing of the stop gates at the CW pumps and the cooling tower ponds. Commissioning of CWM/U system follows the basic principles laid down for the main CW system. If the CWM/U pumps have hydraulically-operated discharge valves, typical times may be 5 s for the first 80% of travel and 15 s for the final 20% of travel.

CW pumps

Each CW pump is prepared for running by flushing and setting up the lubricating oil and sealing water systems, including initial adjustment of the pump shaft running-seal.

The drive motor is given a one-hour uncoupled run, during which all associated alarms, protective devices, controls and indications are checked. The pumps are then coupled to their motors, the pump casings flooded and the rotor hand-barred to ensure that it is free, then the pump unit given a five-minute run against a closed discharge valve. It is most important that the design limitations for operation against either a closed or a cracked-open valve are adhered to, as damage to the pump and/or discharge valve may occur. Similarly, with large HV motors, the number of starts is limited to three evenly-spaced or two consecutive starts in any one-hour period. A start is defined as a normally loaded run-up to service speed.

CW system priming

The inlet and outlet culvert systems and cooling towers are rechecked to ensure that the system is clear of debris, then all access manholes are secured. All valves should be checked to be in the correct state and locked, where appropriate, to maintain isolation. The inlet culvert system is usually gravity primed, to the level of water at the CW pump intakes, by cracking open the pump discharge valve by hand. Arrangements have to be made to release the air trapped beneath the condenser CW inlet valves as the system primes.

Priming of the outlet culverts of a direct system is often accomplished by backfilling from the outfall. Water is pumped into the outlet side of the stop gates to prove the isolation before the construction cofferdam is removed and the area dredged clear. Water is then pumped over the stop gates until the levels balance, after which the gates are withdrawn by a crane or fixed hoist. If a culvert dewatering system is fitted, it may be convenient to use this to prime the outlet culverts from the inlet culverts. Where appropriate, the vacuum exhauster system has to be commissioned at an early stage to assist with priming to the highest points of the condensers. It is also prudent to carry out much of the initial flooding and priming when the tide levels are highest.

CW system circulation

Since the next stage involves pumping water through the main turbine condensers, it has to be carried out in conjunction with the condenser test team. Personnel, with two-way radio communication, should be stationed at vantage points around the system. The condenser CW valves must be opened, then one CW pump started and its discharge valve cracked open to complete the priming of the system, including the condensers. During the priming process, a close watch must be kept on all vents, drains, blanks, manhole access covers, valve pits and chambers for leakage. There may be a limitation on the length of time that the CW pump can be kept running while the discharge valve is only cracked open.

Where cooling towers are involved, providing the initial checks for cleanliness and integrity of the water-retaining structures and stop gates have been completed, then it is a matter of observing water distribution within the tower galleries and ponds, with the CW system running.

In the case of assisted-draught cooling towers, all fans should be commissioned before the first main turbine-generator is commissioned. When the tower is on-load, atmospheric conditions may cause visibility so poor that it inhibits safe access to the fans. Commissioning of each fan normally entails an initial one-hour uncoupled run of the drive motor, followed by a four-hour coupled run to allow bearing temperatures and vibration levels to be monitored as conditions stabilise. Unless the design is well proven, the first fan may be given an extended period of running over several weeks to establish confidence in the plant.

Operation of the CWM/U pumps has to be monitored during priming of the CW system and also later, when commissioning of the purge system is being done. Initially, the purge control valves and instrumentation are commissioned, then the pipeline is checked to ensure that all drains and vents are set correctly. During an extended run of the CW system, each of the purge valves is operated in turn and the rate of make-up and purge flow measured to confirm correct indication from the installed equipment. If applicable, the condenser tube cleaning system should be commissioned at this stage.

Further CW pumps are commissioned in a similar manner to the first, then operation of the various isolating and crossover valves proved under system flow conditions. When at least two pumps are available, a detailed programme of preliminary surge and flow tests has to be carried out, using different combinations of CW pumps and condenser valve openings. Tripping CW pumps under different operational conditions will determine whether further adjustments to the pump discharge valve closing times are necessary. The aim is to achieve the fastest closing time, consistent with minimum hydraulic shock, whilst ensuring that any reverse rotation of a tripped pump is within the safe design speed.

After all pumps (and cooling towers, if appropriate) have been commissioned, a further series of tests is carried out to verify or modify the preliminary valve timings established from the earlier tests.

It will be necessary to prove handling and sealing of all stop gates after the CW system has been in service for some months. At a later stage, thermal performance tests will be carried out, particularly for indirect systems. For details of these tests, reference should be made to the appropriate volume in this series.

17.2 CW system anti-fouling systems

17.2.1 Systems description

General

In order to eliminate mussel and marine growth and to regulate slime formation on heat exchanger surfaces, most CW systems have to be dosed with an aqueous solution of chlorine. At seawater-cooled plants mussel growth is extremely active during the summer period; however, when the water temperature is below 10°C, the treatment is usually applied for slime control only.

Chlorination plants incorporate a very comprehensive electrical interlocking system to ensure safety of operation under all modes of control. The specified treatment regime may require a plant which is capable either of continuous chlorination of all or a number of selected points, or alternatively intermittent dosing at each point. Dosing of each point is carried out automatically, with adjustment of the dosing rate to suit seasonal demand under manual control.

Water, usually from the CW pump discharge manifold, is boost-pumped and strained for use as the operating or diluting water with which gaseous chlorine or liquid sodium hypochlorite is mixed before injection into the CW, ACW or service water systems at various points of application (POA). Figures 3.102, 3.103, and 3.104 show typical arrangements of chlorine and sodium hypochlorite systems. Generally, there is an injection point at the CW inlet to each main condenser pass and ACW strainer for indirectly cooled stations. Directly cooled plants normally have injection points at the CW intakes, and also at the CW inlet to each main condenser pass or at the discharge branch from each of the CW pumps.

The intermittent dosing control sequence for each POA usually provides for five minutes priming, then up to 15 minutes treatment, followed by five minutes of flushing. The electrically-operated isolating valve at each POA is interlocked to ensure that water is flowing through the appropriate section of the CW or ACW system before any chlorine solution is admitted to it.

Most plants are designed to give a maximum injection rate of up to 20 PPM of chlorine, though the general aim is to create a chlorine concentration which ensures that there is a residual of 0.2 PPM of chlorine at the CW system outlet. If seawater stations are to be shut down for more than a few days during the summer period, then the CW system should be dosed to 20 PPM first to prevent mussel growth in still water.

Bulk storage of chlorine is being phased out on grounds of environmental safety. Modern plants now use liquid sodium hypochlorite, which is much safer in that no chlorine gas is ever present at any stage of the process. Commercially available sodium hypochlorite contains approximately 15% by weight of chlorine. Because a much larger quantity of sodium hypochlorite is required for treatment at seawater- or estuarine-cooled power stations, it is more economic to produce it on-site by some form of electrochlorination plant. Dependent upon the design selected for a particular site, the electrical power required could range from 3 to 5 MW.

Many power stations supplement treatment of CW with a mechanical on-load condenser tube cleaning system, e.g., Taprogge, for control of fouling and scaling. Figure 3.105 depicts a diagrammatic arrangement of this system.

Bulk chlorine system

Some power stations store liquid chlorine in duplicate 50 t capacity tanks, which are designed for a maximum working pressure of 10 bar, but tested to 24 bar (refer to Fig 3.102). A 10% capacity relief tank is provided to contain any excess chlorine gas pressure, which may be relieved from the main tanks should the bursting-disc protective devices operate. A small compressor/drier plant is installed to supply dry compressed-air for creating the pressure differential necessary to transfer liquid chlorine from a road tanker into the bulk tanks. Liquid chlorine then flows by gravity to electrically-heated water-jacketed evaporators, temperature controlled to 75°C. The rate of flow of evaporated chlorine is regulated by a chlorinator: an injector then induces the gas into the operating water, which conveys the aqueous solution to the selected POA.

Sodium hypochlorite system

Liquid sodium hypochlorite is delivered by road tanker and stored in duplicate 40 t to 50 t steel tanks at atmospheric pressure (refer to Fig 3.103). The liquid flows by gravity to positive displacement adjustable-stroke metering pumps which discharge a regulated quantity into the diluting water system, which then conveys the aqueous solution to the POA, as before. A small relief tank is installed to collect any excess fluid which has been pressure-relieved from the metering pump discharge line relief valves.

Electrochlorination system

Some plants have the facility for on-site production of sodium hypochlorite by electrolysing brine. This may be obtained either from the CW system, if its salinity lies between 18 900 and 11 300 PPM of chloride ions, or from a specially prepared brine solution for river-water-cooled power plants. Sodium hypochlorite generated from seawater is inherently unstable and the rate of chlorine decay is such that long-term storage is rarely practicable.

Four designs of electrolysis plants are installed at different power stations in the UK, and Fig 3.104 shows the general arrangement of one design. The quantity of sodium hypochlorite generated is directly proportional to the DC current applied to the electrodes. Thus, with a number of generators, each capable of operating over a range of 10 to 100% of its rated output, it is very easy to regulate the rate of production to suit various operating conditions.

Seawater pumps discharge through strainers into the generators, where the sea water is electrolysed to produce sodium hypochlorite. Hypochlorite solution then passes into large open-topped hydrogen removal

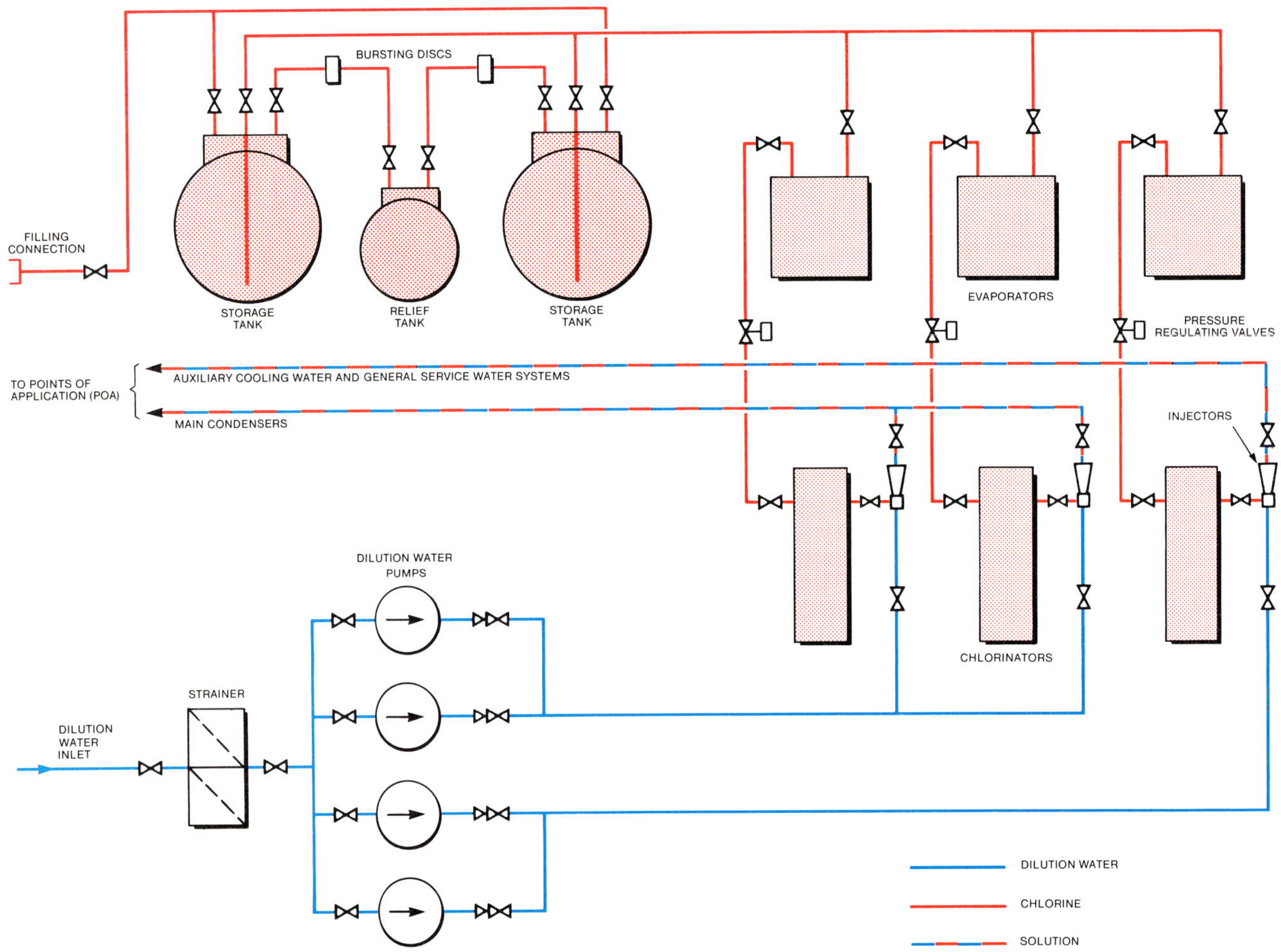

FIG. 3.102 Bulk chlorine system

tanks, so that hydrogen can be detrained and vented safely to atmosphere without approaching the lower explosive limit (LEL). Each hydrogen removal tank is provided with a set of manually-controlled sprays, which can be used to reduce the formation of foam on the liquid surface when required. Dosing pumps then transfer the solution for injection at the points of application as before. The DC current to the generators is automatically regulated to maintain the specified chlorine residual in the CW system.

The seawater strainers are of the automatic back-flushing type formed from stainless steel wire wound on a drum to give filtration of particles over 0.8 mm.

Typically, each hypochlorite generator may comprise 192 electrolysing cells arranged in 12 parallel banks of 16 series-connected cells. Each cell is formed from a pair of concentric tubes, where seawater flows through the annular space between the tubes. The inner tube is a continuous bipolar electrode, whereas the outer tube is in two halves which are insulated from each other and connected to the poles of the DC supply.

DC power supplies for each generator are derived from a special overwound delta/double-star 3.3 kV 350 kVA transformer/rectifier unit arranged to give 12 positive outputs each of 265 A at 78 V. The transformer primary winding incorporates phase shifting links, which can be selected to give either plus or minus 15° of shift. When alternate transformers are set to +15° and −15°, the whole installation operates as a 12-pulse system, to reduce harmonics in the 3.3 kV electrical distribution. The positive outputs are switched by individual circuit breakers, whilst the common negative pole, which is earthed via a link, is equipped with an isolating switch. One generator at full output may require up to 60 m^3 of seawater and 300 kW of electricity to produce 60 kg of available chlorine plus up to 20 m^3 of hydrogen. A towns water supply is used to flush brine from the generator cells before any maintenance work is undertaken.

A microprocessor-based programmable logic controller (PLC) ensures that the plant is operated automatically to meet the demand. The PLC starts seawater pumps and strainers, and opens seawater outlets to the required number of hypochlorite generators, as a function of the number of main CW pumps running. Once seawater flow has been established through all banks of each selected generator its DC power supplies are automatically energised by the PLC. As a predetermined level is reached in the hydrogen removal tanks, the PLC starts a number of dosing pumps dependent upon the number of generators in service. It also ensures that the selected POA isolating valves are opened to all condenser passes that are in service. When the system is running, the hydrogen removal

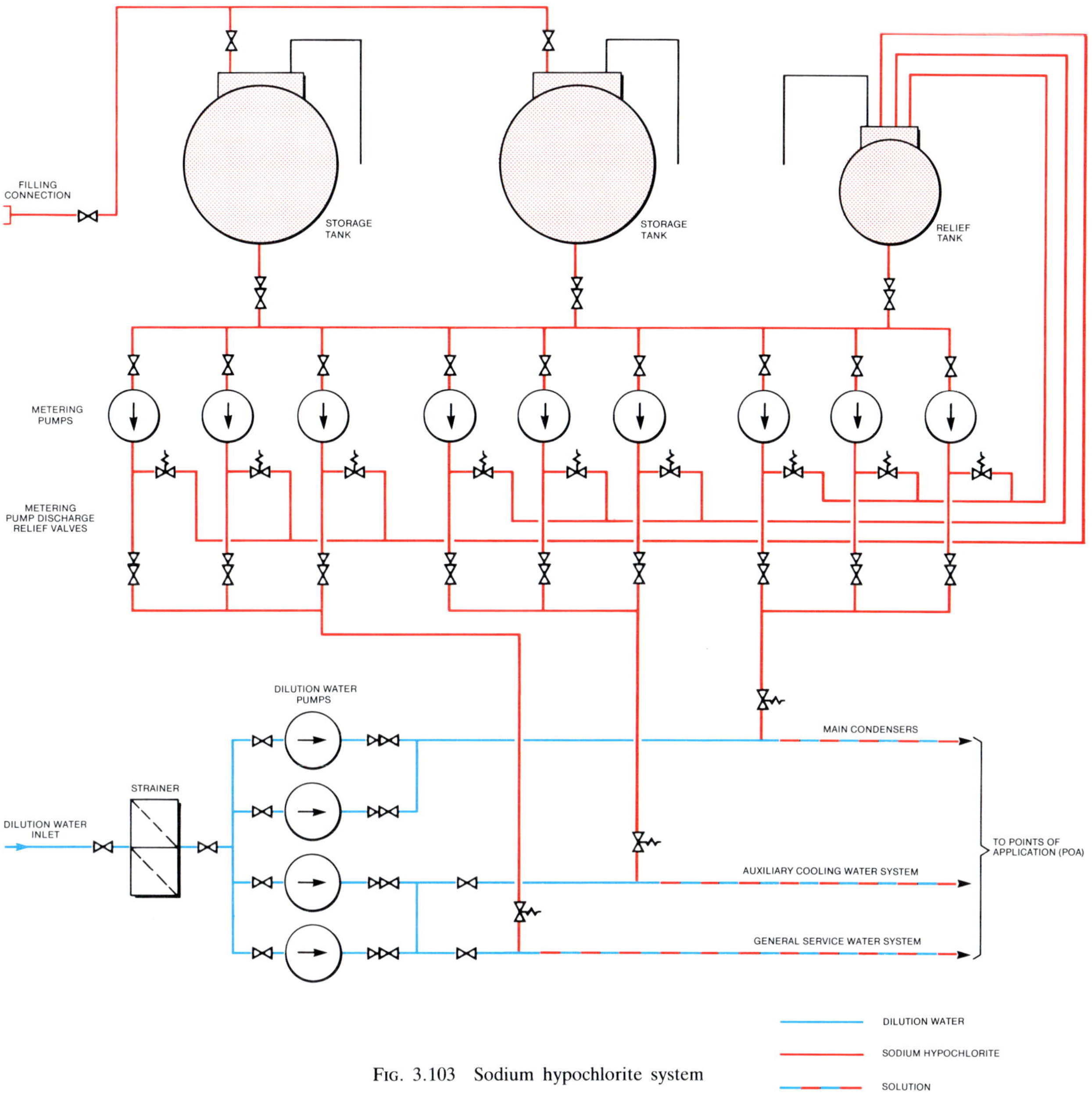

FIG. 3.103 Sodium hypochlorite system

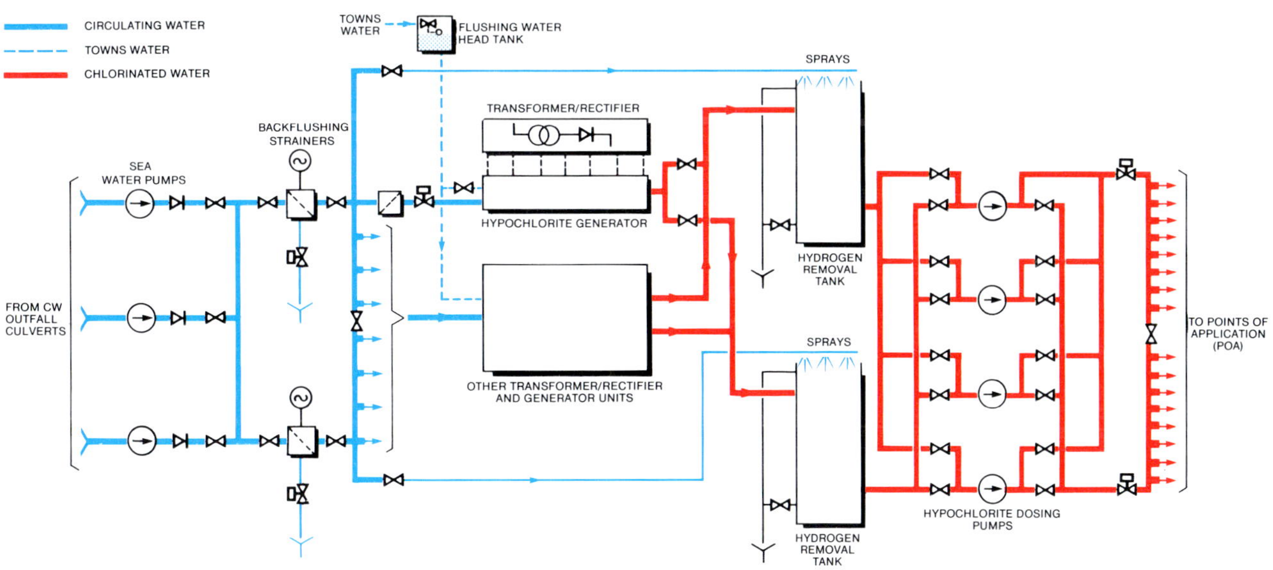

FIG. 3.104 Electrochlorination system

tank level-controller regulates the discharge of the dosing pumps to maintain the prescribed level.

Condenser tube cleaning system

The system shown in Fig 3.105 is designed to inject a number of sponge-rubber balls into each CW inlet and to recover these balls from each CW outlet, after they

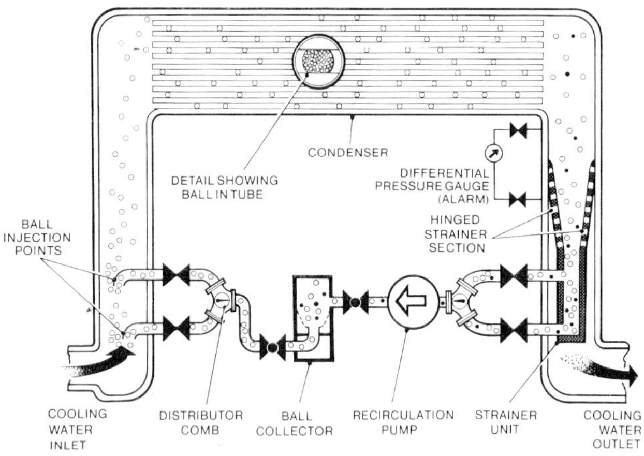

FIG. 3.105 Condenser tube cleaning system schematic diagram

have passed through the condenser tubes. A separate system is provided for each CW pass of the turbine condenser. The balls are some 10% to 15% larger in diameter than the tube bore and thus remove fouling by a gentle scraping action as the CW flow forces them through the tubes. In the unlikely event of severe fouling, special abrasive-coated balls can be used to speed the process, but they must only be allowed to circulate for a short time.

The balls, when fully wetted, have a similar density to the CW and they are injected through several distribution pipes against the CW flow at the condenser CW inlet. A specially designed strainer section incorporated into the condenser CW outlet pipe collects the balls, which are then drawn off by the suction of the recirculation pump, and transported back to the CW inlet. The strainer comprises two sets of screens whose surfaces are inclined against the CW flow to direct the balls into the recirculation pump suction funnel. The screens are pivoted and are operated by means of external linkages so that, when no balls are in circulation, the screens can be partially rotated to enable the CW flow to flush any accumulated debris back into the CW system. Interlocks ensure that the screens are returned to the operating position before circulation can be re-established. A differential pressure indicator warns of fouling of the screening section.

Typically, some 500 balls are kept in continuous circulation whilst each condenser pass is in service. The number of balls in circulation is calculated to ensure that each tube is cleaned at least once every five minutes. At intervals, the ball collector unit is selected from 'recirculation' to 'catch', so that all of the balls can be retrieved for counting and sizing. Undersized balls are replaced with new ones to re-establish the desired total, and the system returned to normal recirculation.

17.2.2 Commissioning objectives

The following are based on a plant utilising bulk storage of sodium hypochlorite, unless otherwise noted. In each case, the general objectives outlined at the beginning of this chapter also have to be adopted:

- To check that personnel emergency showers are safely accessible from the road tanker unloading point (not applicable for liquid chlorine plants).
- To prove that the showers function effectively, particularly in respect of quantity and temperature of water. Typically a flow rate of 0.9 litre/s at 30 to 35°C for up to 10 minutes may be specified.
- To pressure test the storage tanks and all pipework up to each POA.
- To check correct functioning of the storage tank filling connections and to ensure that they are secured to prevent inadvertent contact with any other bulk chemical deliveries.
- To calibrate the contents indicators, plus high and low level alarms, and to prove correct functioning of the storage tanks during filling.
- To prove that all controls, alarms, interlocks and protective systems are adequate and effective.
- To verify that the water flow rate/pipe sizes give a complete purge of the distribution system to the furthest POA in less than 5 minutes.
- To check the metering pump injection rates with variations of pump stroke length and to prove operation of the discharge relief valves.
- To establish that the dilution water injection pumps provide at least 200 times dilution of sodium hypochlorite solution, to minimise risk of precipitation.
- To confirm that the sequence control facilities give automatic injection of sodium hypochlorite over the specified periods and that the optimum chlorine residual is well within the plant capacity.

If on-site production of sodium hypochlorite is involved, then the following are appropriate:

- To prove that each transformer/rectifier unit is capable of smooth load control over the specified range of output.
- To demonstrate that each hypochlorite generator is capable of delivering the rated output at different demand levels.
- To ensure that all hydrogen is detrained effectively and vented safely.
- To prove that the control system operates correctly and fails to a safe condition in event of any type of fault.

- To demonstrate that the specified residual chlorine can be maintained at full CW system throughput with the prescribed plant items in service.

In the case of liquid chlorine bulk storage installations, the following objectives also require consideration:

- Prove correct operation of the compressor/drier plant and confirm that the quality of the compressed air is satisfactory (refer to Section 21 of this chapter).
- Confirm correct functioning of the chlorine gas leak detectors and prove automatic operation and effectiveness of the plant room heating and ventilation equipment.

Where condenser on-load tube cleaning systems are fitted, it is necessary to demonstrate that the plant is capable of recirculating the balls without losing any into the CW system. At a later stage, the condenser has to be inspected to confirm that effective cleaning is taking place.

17.2.3 Safety considerations

The basic considerations outlined at the beginning of this chapter should be taken into account, in addition to the following:

- Before any delivery of chemical is received, comprehensive safety procedures for its receipt, storage and use must be approved. All equipment required to implement the procedures must be fully available.
- All personnel involved in commissioning the plant must be completely familiar with the dangers, effects and appropriate emergency action necessary in event of a chemical spillage.

  Note Should sodium hypochlorite and acid come into contact, large volumes of chlorine gas are given off.
- Appropriate protective clothing must be worn at all times.
- Since sections of the main CW and ACW systems may be incomplete when chlorination plant is being commissioned, special attention must be paid to ensuring safety of construction personnel in a similar manner to that involved with main CW system commissioning.

17.2.4 Resources

Services

Two standard-sized drums of chlorine will be needed for initial testing of the chlorine plant. The minimum delivery of 20 t of liquid sodium hypochlorite will be required for the first fill of half of the load into each storage tank.

An initial charge of sponge-rubber balls for the condenser tube cleaning system is needed. Typically, some 5000 may be required per unit to cater for initial losses, and damage sustained during commissioning.

Staffing

Personnel involved could total up to six people, equipped with good two-way radio communication for the more extensive activities. The commissioning period could occupy approximately six to eight weeks.

17.2.5 Commissioning

Sodium hypochlorite system

The sodium hypochlorite storage and relief tanks are open vented and, as such, they and the associated filling pipework are hydrostatically tested, using either fresh water or demineralised water at atmospheric pressure. The tank outlet pipework, up to the outlet isolating valves, is also included in this test. No leakage of water can be accepted during the test which may be over any period from six to 24 hours.

The dilution water and sodium hypochlorite pipework are hydraulically tested, using water from the boiler make-up water main, via a temporary connection, directly into the suction of the metering pumps. One or more of the metering pumps is used to achieve a test pressure of 14 bar in each section of the pipework system. The total extent of the pipework covered is from the storage and relief tank isolating valves through to the POA valves at the unit condensers, to the ACW strainers and GSW strainers including the dilution-water pipework downstream of the dilution-water pump discharge valves. For this part of the system, negligible pressure loss is required over a six-hour period.

Sections of pipework between the dilution-water pump inlet valves and the supply point isolating valves at the main CW pump discharge pipework are hydraulically tested at 4 bar, with no pressure loss over six hours.

Once pressure testing has been completed, commissioning of the metering pumps with their relief valves, protective devices and controls can be progressed using the temporary water supply connection flushing the pipework to waste via system drains. After this stage, the temporary connection is removed and the metering pump suction pipework reassembled.

With one main CW pump running, each dilution-water pump is commissioned in turn, including proving of all interlocks and controls.

The storage tank bunded area is cleared of all debris, filled with water to prove that it is capable of retaining water, then drained dry. Contents indicators for each tank are calibrated and the personnel emergency showers tested. The loading points are checked to ensure that the road tanker connection cannot be used to discharge incorrect chemicals.

When all equipment has been satisfactorily proved, the first road tanker of sodium hypochlorite is discharged into one storage tank at a time, preferably dividing the load between the tanks. During this exercise, the tank contents gauges should be compared with the tanker delivery rate.

To commission the dosing system, one CW pump has to be in service along with at least one dilution-water pump. The distribution pipework should be available to inject into the points of application associated with the appropriate cooling system, e.g., CW, ACW or GSW. One sodium hypochlorite storage tank outlet is opened to prime the pipework through to the metering pumps, then each metering pump is run in turn to set the injection rates for each cooling system.

It is now opportune to set up the sequence control system to give automatic injection of sodium hypochlorite over the required periods, whilst confirming that the mimic diagram correctly indicates the state of the plant under all operational modes. Finally, checks are carried out to establish that a complete purge of the distribution system to the furthest POA can be achieved within 5 minutes, and that the dilution-water pumps provide adequate dilution of the chemical.

Bulk chlorine system

For plants based on bulk storage of liquid chlorine, it is most important that all parts of the system which will contain chlorine gas or liquid are kept absolutely dry. For this reason, pressure testing of pipework, storage tanks and relief tanks is conducted using nitrogen or dry compressed air. Compressed air should have a dewpoint of better than −30°C at atmospheric pressure.

The storage tanks and relief tank are all pressure tested with nitrogen at a pressure of 24 bar, whereas all pipework downstream of the tank outlet valves is tested at 10 bar. In each case, no leakage should be accepted. During testing, the pressure should be increased by stages over several hours, allowing conditions to stabilise at each stage whilst a check for obvious leaks is carried out, using a soapy-water solution.

All operational modes, alarms and indications are proved, with the plant operating on water before any chlorine is admitted to the system. Initially one or two cylinders of liquid chlorine are discharged either into the storage vessels or direct to the chlorinators via temporary pipework to prove operation of the plant with chlorine before any bulk delivery is accepted. Bulk liquid delivery is accepted almost as the last part of the procedure.

Electrochlorination system

Essentially the commissioning procedures are similar to those outlined earlier for sodium hypochlorite systems. The main activities are as follows:

- Pressure test all sections of pipework, using clean water.
- Hydrostatic test of hydrogen removal tanks.
- Check motor rotation and couple to pumps.
- Commission electrically-operated valves.
- Calibrate all instrumentation.
- Commission backflushing sequence operation of the strainers.
- Prove interlocks between condenser CW inlet valves and POA isolating valves.
- Confirm that hydrogen can be vented safely without any risk of fire or explosion.
- Prove that towns water flushing system is available.
- Load the PLC program and check the operation of the control logic as far as practicable.
- Commission transformer/rectifier units on open circuit.

Now set up a flow path from the seawater source into the hydrogen removal tanks and thence to drain:

- Commission the seawater pumps and use them to pump water through each generator in turn into the hydrogen removal tanks.
- Drain water from the tanks continuously to control the level of water. Prove level indication, controls and alarms.
- Cause each tank to overflow to prove adequacy of overflow line.
- Set up correct flows through each bank of every generator unit.
- Commission dosing pumps and flush seawater through all legs and POA valves.

Once it has been demonstrated that the plant is operational, that it is free from leaks and that all controls function correctly while seawater is passed through the plant, generation of hypochlorite can commence:

- Confirm that hydrogen leak detectors are operational.
- Set up the plant to pass water through one generator. With controls selected to 'manual', energise the generator DC supplies and slowly raise the current by increments to its full-load value.
- Check the quality of hypochlorite output at each current increment.
- Commission all generators in a similar manner.
- Check the 'auto' operation of the entire plant and confirm that the desired chlorine residual can be maintained, using different numbers of CW pumps and condenser passes.
- Create fault conditions and prove that the plant reacts safely when selected for 'manual' or 'auto' operation.

At a later stage, it will be necessary to prove that the generators produce hypochlorite with the most economic consumption of seawater and electrical energy.

Condenser tube-cleaning system

First, carry out a visual check that the external operating levers give the required free movement of all screens and flap valves. This activity also involves checking that the local mechanical position indicators, including the remote (i.e., at the local control panel) indicators are correct, as well as proving the interlocks associated with the CW outlet-leg strainer. The recirculation pump motor is checked for rotation, then coupled. Next, the condenser has to be primed, though a flow is not really necessary at this stage, so that the tube cleaning equipment can be filled and checked for leaks. Operation of the strainer screens and associated interlocks and flap valves is rechecked, then the recirculation pump given a short run, with the discharge valve closed. This is followed by another short run, pumping water round the system.

The ball charge is loaded into the collector and the strainer confirmed to be in the 'operating' position. With the CW system in service, the recirculation pump is restarted and the balls gradually released into the condenser. It may be necessary to adjust the ball distributor to obtain an even distribution of balls to the inlet pipes. Once recirculation has been established and the system run for several hours, the ball-catching facility is checked, then the balls counted. While the CW is flowing, each section of the strainer is selected to the 'backwash' position to confirm freedom of movement and correct indication of differential pressure before being returned to the 'operating' position.

If practicable, the condenser should be drained, so that a check can be carried out to ascertain if any groups of tubes are not being cleaned. Any stuck balls can be identified and recovered, then the affected tubes sighted in case they are blocked by debris.

Afterwards, the system should be left in service for an extended period, then a further inspection carried out.

18 Service water and drains systems

This section deals with commissioning of the following systems:

- General service water (GSW) system
- Towns water (TW) system
- Station drainage systems

The two service water systems provide fresh water cooling services to particular items of station and unit auxiliary plant. Station drainage covers all surface water and oil-contaminated drains, and also includes sewage treatment.

18.1 Systems description

18.1.1 General service water system

It is neither economic nor operationally desirable to provide cooling services to all items of auxiliary plant from the auxiliary cooling water (ACW) system. This is particularly so in power stations where salt water is the primary coolant. GSW systems are designed so that the water flows in a closed loop or recirculating arrangement. The system uses fresh water, derived either from an industrial towns water supply or from a CW system, where fresh water is the primary coolant. General service water systems fed from the towns water supply usually incorporate either an auxiliary cooling tower or heat exchanger to dissipate heat rejected from the auxiliary plant coolers. Depending upon the distribution of heat exchangers between ACW and GSW systems, up to 10 MW of heat per main unit may have to be rejected by the GSW system.

Figure 3.106 shows a typical semi-closed loop GSW system, where the CW system acts both as a water source and as a heat sink. The figure also shows interconnections with a typical TW system and other plant systems. Water is drawn from the CW forebay through duplicate double-reel strainers by four pumps, which discharge into several 320 m^3 storage tanks located at a high level in the boilerhouse. Water from the tanks is distributed to heat exchangers, pump bearings and glands, valve glands, etc. Where GSW is used for cooling of the generator and unit transformer oil coolers, the supply line incorporates a break pressure tank, which is often arranged with a weir which overflows continuously. The mounting height for the tank is selected so that the pressure exerted by oil in the transformer conservator tank is always greater than that of the cooling water in the coolers. This ensures that water cannot leak into the transformer oil circuit.

GSW discharged from these services is collected into a common return main and flows by gravity direct into the CW forebay, along with drains and overflows from the storage tanks. The GSW system is designed to operate independently of the CW system, as long as the CW forebay remains filled. The relatively small amount of heat rejected to the GSW system when all units are shut down (i.e., no CW pumps in service) can be dissipated naturally from the static reserve of water in the forebay.

In the case of a TW-filled GSW system this is usually arranged as a fully-closed loop which includes either an auxiliary cooling tower or a water/water heat exchanger for heat rejection. The auxiliary tower is usually of the direct contact type and is of conventional wooden construction, with several electrically-driven fans mounted in either the sides or the top. The fans induce a counter-flow of air against the downward fall of the water spraying over the slatted wooden packing. Towns water make-up is supplied to the cooling tower pond. The wooden packing must be kept in a damp condition at all times, after the initial wetting, should the tower or part of it be out of service for any length of time. If, instead of a cooling tower, a water/water heat exchanger is fitted, then the heat is rejected from the GSW into the ACW system. Make-up may be added to the storage tanks to compensate for system losses.

The double-reel strainer may have a mesh of 5 mm and be capable of dealing with water at a flow rate of 1.5 m^3/s. Water from the strainer outlet is often also connected to the suction main for the fire protection system pumps. Another pipework branch conducts water to the dust plant water service pumps, if the station disposes of its dust as a pumped slurry. The GSW pumps typically are capable of delivering 0.4 m^3/s against a head of 5.5 bar. Two pumps may be selected for 'duty', with a third as 'second duty' and the fourth as 'standby'; all would be controlled from float switches sensing water level in the storage tanks. The water level in individual tanks is regulated by ball-float-operated inlet valves.

In an emergency, the GSW pumps can supply the cooling services directly, by-passing the storage tanks, when the state of the interlocked valves is interchanged. To prevent over-pressurisation of the heat exchangers under this mode of operation, an atmospherically-vented loop is fitted. The interlocked valves and over-pressurisation loop are shown in Fig 3.106.

Non-return valves in the storage tank outlet branches prevent the tanks backfilling.

18.1.2 Towns water system

Frequently, two separate metered supplies are provided from the local water authority distribution mains. One is of industrial-quality raw water, which is stored on site in a large (up to 15 000 m^3) towns water reservoir, which also serves to ease the effects on the water authority supply system of large fluctuations in demand. The other is a direct feed domestic supply used for personal hygiene and food preparation purposes. Hot water calorifiers, fire hose reels, etc., may be connected to the direct supply, but are usually supplied from a pressurised distribution system fed from the towns water reservoir (Fig 3.106).

The figure also shows a separate closed-circuit cooling system for a centralised compressed air facility. At some sites, the water used for the GSW system has a scaling propensity which precludes its use for compressor cooling, mainly due to the large temperature difference between compressed air and cooling water. Again, because of the relatively large rate-of-flow required, a recirculating system using industrial-grade towns water has distinct advantages. The circuit shown

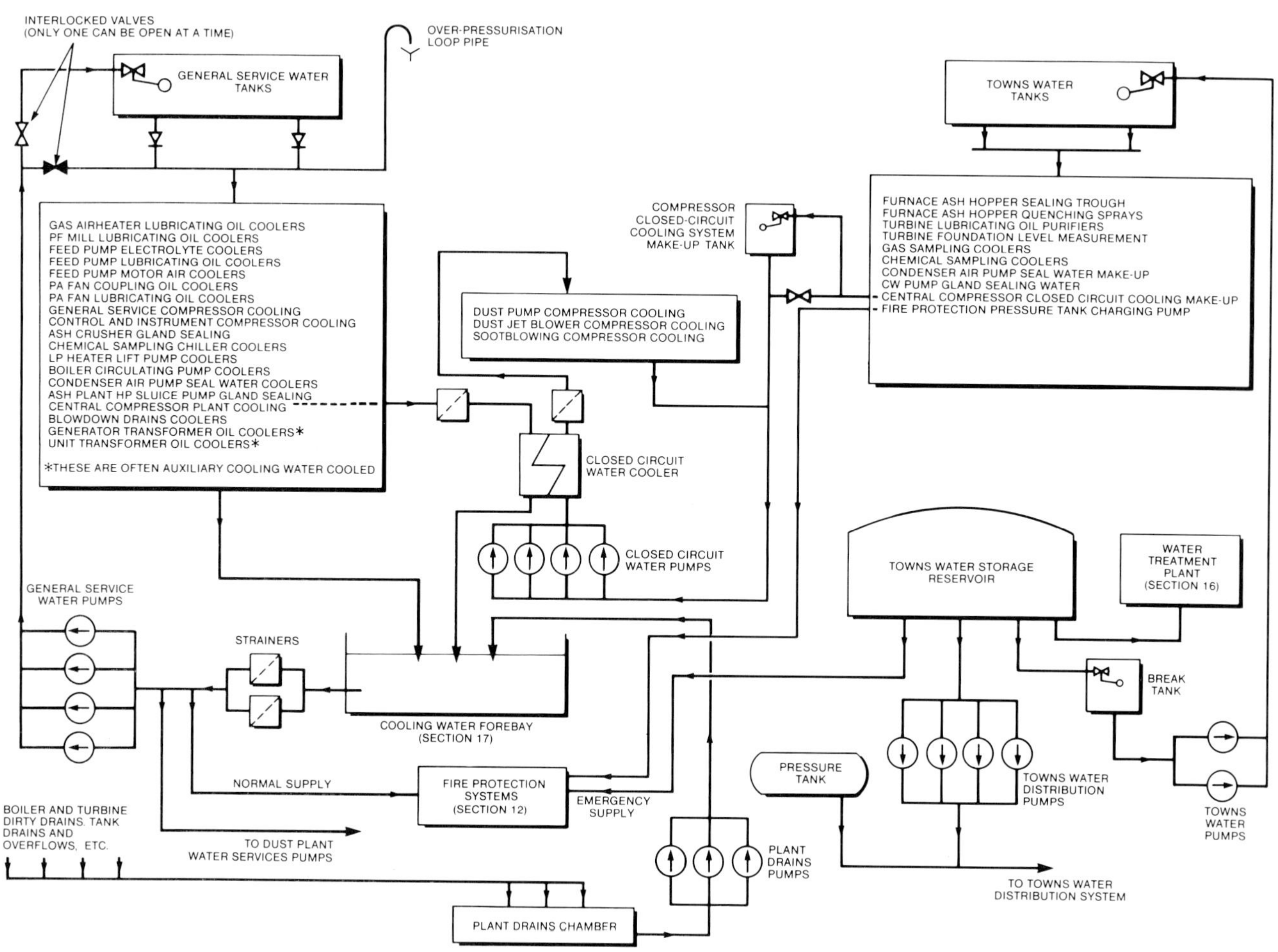

FIG. 3.106 General arrangement of service water systems

uses a water/water heat exchanger to reject the heat, but it is also relatively common to employ either a small auxiliary cooling tower, similar to that mentioned earlier in the GSW system, or an air/water non-contact heat exchanger equipped with four or six fans.

Water from the site towns water reservoir is piped to the water treatment plant and pumped to storage tanks to supply various unit and station services which do not justify a domestic-quality water supply. Services connected to the TW system normally have either an intermittent or a low rate-of-flow water requirement. For this reason, no attempt is made to collect and recirculate water discharged from the services. Instead, each service drains into the CW, ashing or site drainage systems, as appropriate. Other services, not within the main plant buildings, are often connected to a pressurised TW distribution system, e.g., those located in coal and ash plants, CW make-up plant, etc. Construction site temporary fire protection pumps often take their suction from the towns water reservoir, which also later doubles as an emergency source for the station fire-pumps (see Section 12 of this chapter).

It is becoming the practice to provide several deep-well borehole pumps close to the site. These pumps usually discharge into the TW reservoir to augment the industrial-quality supply. In some cases, the borehole water is directed to the water treatment plant raw water tank only.

18.1.3 Station drainage systems

Four separate systems are involved, as follows:

- Surface (or storm) water
- Dirty water
- Oil and contaminated liquid drains
- Foul (or sewage) water

Almost all comprise extensive buried pipelines, equipped with strategically placed catch pits and inspection manholes unique to each system. Figure 3.107 shows a general arrangement of the various drainage systems at Ince *B* Power Station.

Surface water drains

This system collects all of the roadway and building drains that are not susceptible to oil contamination. Water gravitates into low level sumps, from which submersible pumps transfer it into the main drainage system, along with other gravity-fed drains. Most power stations are provided with a relatively large capacity open-drainage ditch around most of the site perimeter. All site surface water drains usually end up in this ditch, from which water is transferred either into the land drainage system of the local water authority or to the CW purge system. The coal plant stockyard is provided with a comprehensive network of land drains, equipped with strategically located catchpits, to contain fine coal dust washed along with surface water. All of the submersible pumps employed are fully automatic in operation, and most are duplicated in a duty/standby configuration. At many locations, each outfall to the local waterway is equipped with an oil-in-water detector/alarm unit, as a final safeguard in the unlikely event of oil contaminating the perimeter drainage ditch.

Dirty water drains

These drains deal with plant and floor washings, turbine and boilerhouse drains and any road drains susceptible to oil contamination. Drains from transformer bays and other oil handling plant areas are also led to this system, via isolating valves which are normally only opened to drain off accumulations of water (refer to the section on oil and contaminated liquid drains). Drainage either flows by gravity or is pumped to one or more oil separators, whence clean water is discharged to join other surface water drains. Oil and sludge which collects in the oil separator has to be removed periodically for disposal, using road vehicles. Oil separators that are likely to have to deal with boiler fuel oil usually need an auxiliary steam supply to heat the oil separation chambers and to provide steam cleaning facilities.

Two forms of oil separator are in general use; both are designed to comply with American Petroleum Institute (API) standards and treat oily water at a rate of up to 11 litres/s. Peak flows of twice that rate can be handled, but the efficiency of separation depends to a large extent upon the specific gravity (SG) of the contaminant. In the case of transformer and lubricating oils, with an SG of 0.9, the design flow rate can be maintained continuously and the outlet particle size will be limited to about 120 μm. However, with boiler fuel oils having an SG of 0.99 or so, the performance decreases so rapidly that the flow rate may have be reduced to 25% to achieve the same quality of clean water discharged from the separator.

Both forms of separator are designed with such a slow transit velocity, typically 0.2 m/s, that oil and water separate by gravity. Debris is screened off at the inlet to the separator chamber; then, as the oil/water mixture progresses through the chamber, sludge settles out into a sump. At the outlet end, oil is removed from the surface by a manually-adjustable skimmer pipe and deposited into a heated collection sump. The cleaned water flows under a baffle into the outlet chamber, to be returned into the surface water system.

Figure 3.108 shows a sectional elevation of one design of separator, which incorporates a pack of corrugated glass-reinforced plastic (GRP) sheets, tilted at an angle of 45°. An evenly distributed flow into the separator pack is achieved by means of the distribution baffle. The upper part of the baffle is not slotted and thus allows oil, separated before the baffle, to flow directly to the skimmer pipe. Contaminated water flows down through the pack and, as it does so, the oil globules separate and rise the short distance to the smooth underside of the plates. Oil then flows upwards

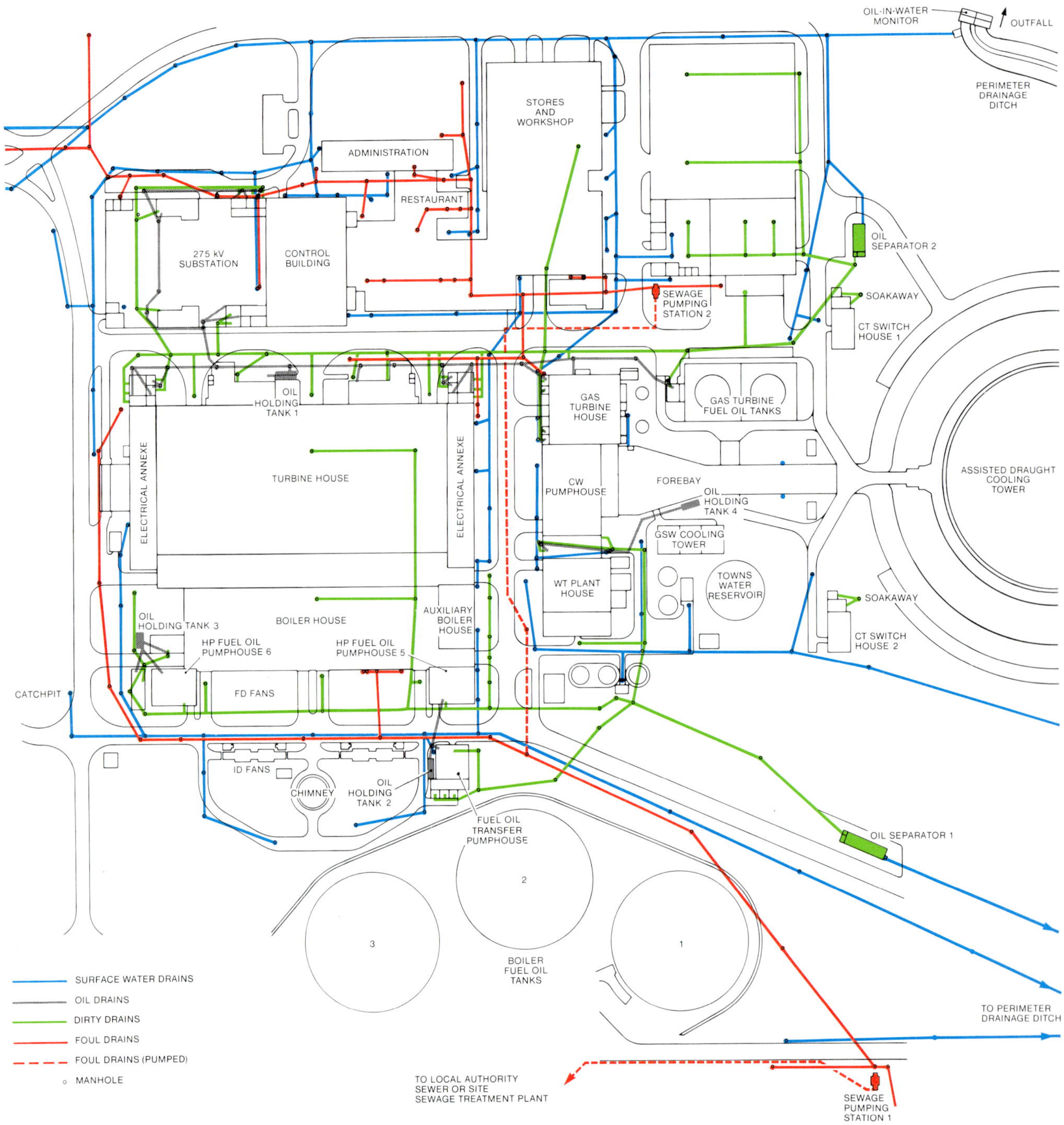

FIG. 3.107 Typical arrangement of site drainage systems

along the top of the corrugations and rises toward the skimmer pipe. Any sludge remaining in the effluent settles and collects in the troughs of the corrugations, then slides down into the sludge compartment.

The overflow edge of the slot in the skimmer pipe is usually set above the highest water level for the separator, so that there should always be a relatively even layer of oil on the surface of the separator chamber. The inlet and outlet weir heights are adjusted to suit the designed flow rate capacity of the separator. Typically, the inlet weir may be set 70 mm higher than that at the outlet, with the skimmer pipe overflow edge perhaps 50 mm above the outlet weir.

The other design of separator is similar in principle, but has a relatively long length of plain separator chamber. Distribution baffles streamline the inlet flow across the width of the chamber. The length permits a slow passage of contaminated liquid toward the skimmer pipe, thus allowing oil globules to conglomerate and rise to the liquid surface. The outlet flow is taken under a baffle and over a weir, similar to the tilted-plate design, before discharge into the surface water drains.

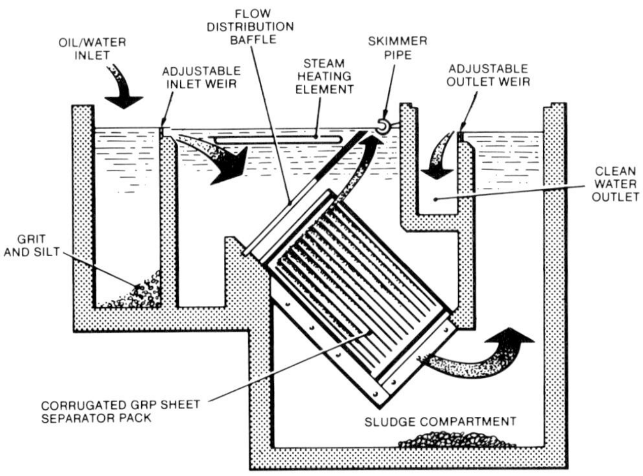

FIG. 3.108 Section through tilted-plate oil separator

Oil and contaminated-liquid drains

Special arrangements have to be made for drainage of areas containing transformers and oil handling plant, particularly where operation of fire protection would give rise to a 'flash flood' of oily water into the drains system. Fire system water flows impose too great a load on oil separators, so oil holding tanks are provided for reception of these discharges. Each of the above areas is normally provided with a dual-drainage system, the inlets to which are usually separated vertically by an 80 to 100 mm upstand. The lower, normally isolated, outlet is led to the surface water drains system, so that accumulations of rain water can be released manually, when required. The higher, emergency, open outlet is led directly into an oil holding tank or contaminated-liquid chamber (CLC), which has the capacity to cope with a lengthy fire protection discharge. Periodically, the tanks have to be pumped out to remove any slow accumulation of surface water. In event of either an oil spillage or fire deluge, portable submersible pumps may be used to pump the contents into road vehicles for disposal. At some power stations, the oil-holding tanks are provided with a pair of submersible pumps which discharge the contents to the oil separator, at a rate which can be effectively handled by the separator. Typical oil-holding tanks may have capacities ranging between 50 and 250 m^3 and are provided with an access hatch and an atmospheric loop-pipe vent.

Foul water drains

The foul water system collects the sewage and wash-water arising from all personnel on site. Foul water flows by gravity into several underground collecting sumps to be pumped to a main storage and transfer station for subsequent discharge, either to the public sewer or to a site sewage treatment plant. From time to time, the foul water flow rate may exceed the maximum rated throughput of the site sewage treatment plant. It then becomes necessary to hold back a relatively large volume of sewage, and so short-term storage capacity is provided at the main transfer station.

Sewage pumps are usually of the electrically-driven submersible type, normally duplicated on a duty/standby basis. Alternatively, compressed-air sewage ejector units may be used to deliver into rising mains, which in turn deliver into the foul drains system to gravitate to the nearest pumped collection chamber. Each ejector system comprises a small compressor/receiver and an ejector chamber capable of operating at a pressure of some 10 to 14 bar. Sewage flows into the chamber by gravity and displaces air through a vent. At a predetermined level, the vent valve closes and the vessel is pressurised with compressed air, which discharges the contents into the rising main. When the fluid level falls, the air inlet closes and the vent valve opens to allow the sequence to repeat.

Each collecting sump is provided with a high-fluid-level alarm, which is often grouped at a common point with a single alarm relayed to the central control room. Where the number of collecting sumps justifies it, a diagnostic switch panel is provided, so that the operation of each sump can be monitored from the common location.

Figure 3.109 shows a simplified flow diagram for a typical power station sewage treatment plant, which may be designed for a peak flow of 400 m^3/day. Raw sewage is first passed through a grit extractor, where a motor-driven slow-speed paddle causes grit and other dense material to fall to the bottom. The sewage flows to a comminutor pump, in which it is reduced to small pieces of maximum size 6 mm, before being discharged to a pre-aeration tank. Air from low pressure blowers bubbles into this tank so that the sewage, now in the form of a liquor, starts decomposing into a jelly-like sludge before flowing by gravity to a primary settling tank.

Clarified effluent passes over a weir into the main aeration basin to be brought into contact with activated sludge and be subjected to aeration to break down organic contamination. The mixture of treated sewage and activated sludge flows by gravity to the final settling tank, where it passes over a further weir and into the surface water drainage ditch. Clean water flow is recorded and regular samples are taken for determination of biological oxygen demand (BOD), total dissolved solids (TDS) and pH.

Successful operation of the plant is completely dependent upon maintaining optimum quantities of sludge in each vessel. Most of the sludge transfer operations are carried out automatically at prescribed time intervals, which can be manually adjusted to suit the demand on the plant and the season of the year. Bacteria multiply rapidly under normal operating conditions and very high concentrations of sludge could result. A proportion of the activated sludge carried into the final settling tank is recirculated to the main aeration basin. The excess sludge is routed into a sludge sump, to be periodically pumped to the pre-aeration

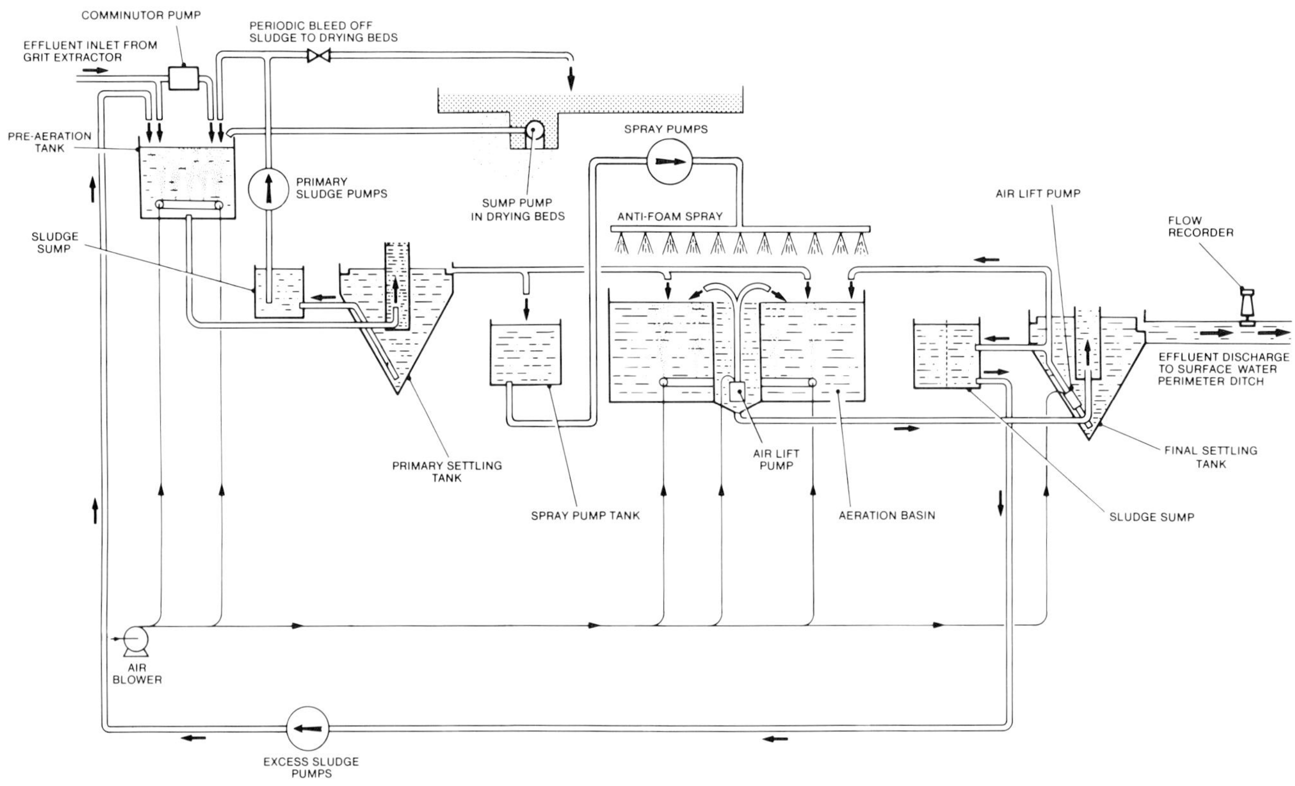

FIG. 3.109 Sewage treatment flow diagram

tank, along with sludge recirculated from the primary settling tank. Providing that the contents of the pre-aeration tank are an inky-black colour, then sludge recirculation rates are satisfactory. Periodically, excess sludge in the system has to be dumped to coke-based sludge beds where, after drying, it has to be manually removed. Liquor decanted from the sludge beds must be returned to the plant for treatment.

Air blowers supply low pressure air for aeration purposes, and to act as the motive force in the sludge transfer air-lift pumps.

18.2 Commissioning objectives

Apart from the general objectives outlined at the beginning of this chapter, these are specific to the systems encompassed within this section:

- To ensure that all pipework is free from leaks, is pressure tested (where required), then flushed to waste to remove all construction debris.
- To commission the strainers, pumps, tanks, calorifiers, etc., and prove that all controls and indications function correctly.
- To establish flow through the GSW and TW distribution systems and ensure that the flow to individual services is balanced to suit the design requirements.
- To confirm that the design flow distribution is correct and, if not, to modify it and obtain the best balance in respect of heat rejected from the various services under full-load conditions.
- To ensure that there is an adequate supply of hot and cold water from all TW outlets.
- To confirm that all drinking and food-preparation water is direct from the supply main and not from storage tanks on site.
- To prove that one drainage pipeline system does not leak into another and also to check that all access manholes are correctly identified for the system function.
- To demonstrate that each oil holding tank or contaminated-liquid chamber (CLC) is a water-retaining structure.
- To ensure that each oil separator effectively prevents carryover of oil into the drainage outfall. In parallel with this, correct operation of the oil-in-water monitoring equipment has to be proved.
- To confirm correct operation of the sewerage system, and prove that it can effectively treat the maximum designed flow of raw sewage.
- To ensure that the *as constructed* drawings reflect the actual layout of each system, particularly those associated with TW distribution and drainage underground services.

18.3 Safety considerations

The following aspects should be considered, in addition to those listed at the beginning of this chapter:

- Personnel may be at risk of getting wet from leaks, or from water entering incomplete parts of the system from ineffective isolation.
- Accidental flooding is conceivable, so care must be taken when filling water stòrage tanks to ensure that they do not exhibit any structural weakness likely to lead to failure.
- Special care is required to prevent personnel falling into underground chambers. Access covers must be fitted securely or the openings safely protected by barriers.
- All points of isolation for underground water supply or drainage services must be clearly identified above ground. The *as constructed* drawings must be kept up-to-date as erection/commissioning progresses.
- Entry into underground chambers that have contained water or other fluid must be prevented, until the atmosphere has been declared free from dangerous gases (e.g., methane).

18.4 Resources

18.4.1 Services

The following will be needed to help with commissioning each area at the appropriate time:

- Supplies of fresh water for hydraulic testing of pipework and for hydrostatic testing of tanks and drainage systems.
- Supply of auxiliary steam for the heating of domestic hot water calorifiers and waste-oil sumps.
- Regulated quantity of oil for testing effectiveness of oil/water separators.
- Fire protection deluge-test water (see Section 12 of this chapter) for testing adequacy of drainage system to cope with excessive water flows.
- A load of bacterially-active sludge (typically, 5 to 8 m^3) to 'seed' the sewage treatment plant.
- Portable submersible pump for emptying underground chambers (typically capable of pumping water at 15 litres/s against a discharge head of 10 m).

18.4.2 Staffing

A small team of three, sometimes four, people will be required to work in each area concurrently. In all cases, the activities will continue over most of the construction phase, since extension of each system has to be progressed concurrently with erection of successive main units and other plant buildings. The same teams may be involved with alterations, particularly to the TW and drainage systems, as contractors' temporary accommodation is removed or repositioned.

18.5 Commissioning

18.5.1 General service water systems

Storage tanks should be individually isolated, then slowly filled up to the maximum water level with water from a temporary source to prove the integrity of the welds, and to check tank settlement. Following draining, shot-blasting and the application of the internal lining, each tank has to be thoroughly inspected to check for consistency of thickness and coverage of the lining material.

All instrumentation, controls, etc., have to be proved and the GSW pump motors given a rotation check before being coupled. Sections of the pipework system have to be filled with water and checked for leaks under the normal working pressure. Typically, the pump discharge pipework would be subjected to a pressure test at 12.4 bar for four hours.

Water from the CW system is admitted to the strainers and the pumps, then operation of each strainer and its cleaning facilities proved, followed by running each pump for up to five minutes with the discharge valve closed. Next, one pump is run carefully to prime the discharge pipework up to the tanks, controlling the rate-of-flow to avoid overloading the pump, whilst taking care to vent any trapped air pockets. Each tank should be filled, preferably using a different pump, and then overfilled to prove the adequacy of the overflow pipework. Correct operation of the level indication and alarm equipment should be observed during this exercise.

Various combinations of pumps should be tested to check the hydraulic performance under different rates-of-flow. This will also involve:

- Proving that the discharge non-return valve of each pump closes effectively when it is stopped.
- Assessing the run-up time for a standby pump, and the consequent effect on flow, when a running pump trips.
- Checking for hydraulic shock in event of 'duty' pump failure, and also during routine stopping and starting of the second 'duty' pump.

The adequacy of the over-pressurisation loop and its drain line should be demonstrated: it is also prudent to prove the effectiveness of the storage tank outlet branch non-return valves at this stage. It is advisable to drain, inspect and, if necessary, remove any debris from the tank into which the pump discharge line was flushed.

Commissioning of the distribution system is carried out in clearly defined stages. Sections of pipework are charged, then flushed to waste from terminal point valves, some of which may have been fitted temporarily for this exercise. The static pressure from the storage tanks is normally adequate to identify any leaks. In conjunction with members of other plant test teams, a flow of GSW is established through various auxiliaries into the return main, and thence to the forebay. Later, as more plant is commissioned, the ability of the system to reproduce the designed flow distribution should be proved. Once significant quantities of heat are being rejected into the GSW system, it will be necessary to re-examine, and perhaps rebalance, the flow distribution to obtain optimum heat exchanger performance for each of the auxiliaries.

Auxiliary cooling tower commissioning involves the following activities, most of which are similar to those outlined earlier in Section 17 of this chapter:

- Check that cooling tower pond is a water-retaining structure.
- Check operation of TW make-up and purge valves and also isolating penstocks, proving that the latter do seal against flow into the pond from the rest of the system.
- Check rotation of fan motors, then carry out an extended run of each fan and its associated gearbox, checking for vibration, overheating bearings, etc. Prove that the fan vibration detectors trip the drive motor.

When water is flowing round the system (through the tower), these checks have to be carried out, initially with cold water and later when heat is being rejected:

- Confirm that water distribution within the tower is correct according to design.
- Check that spray eliminators are effective, even with strong sidewinds.
- Ensure that water in the pond neither overflows, nor runs at too low a level.
- Test the thermal performance under various loads and atmospheric conditions.

Any chlorinated-solution injection scheme should be commissioned as soon as practicable (see Section 17 of this chapter).

18.5.2 Towns water systems

The towns water reservoir, usually the first major source of water on site, is cleared of construction debris, and the internal surface treated or lined, if required. It is then gradually filled over a period of several days, whilst checks for structural integrity and settlement are carried out, similar to those for fuel oil tanks described in Section 15 of this chapter. Thereafter, the procedure for flushing and commissioning of the plant supplied from it follows the principles outlined for the GSW system.

Borehole pumps are commissioned, initially discharging to waste near the well-head, until the draw-down water levels and sand content stabilise. After the discharge pipeline has been pressure tested, water is pumped through the line and flushed out at a point close to the appropriate reservoir or water tank. Once all remote controls and indications have been proved, water is then pumped into the appropriate reservoir or tank.

Charging of the domestic-water system is carried out in distinct stages, as for the GSW system, following similar principles, but testing is carried out in accordance with BS CP310 'Water supply'. Ball-float-operated inlet valves and overflows for head tanks must be checked for correct operation and adequacy of flow rate, respectively.

Cold water flow checks are carried out at all hot and cold water outlets. The steam and/or electrically-heated hot water calorifiers are commissioned, then availability of hot water, within the prescribed temperature band, is checked at each outlet. Tests are conducted to prove that drinking water and food preparation hot- and cold-water supplies are connected directly to the supply mains.

Particular attention must be paid to ensuring that all temporary branches used for construction services are properly sealed off as each becomes redundant. Drawings must correctly reflect all changes.

18.5.3 Station drainage systems

All oil holding tanks, contaminated-liquid chambers, collection sumps, and sewage transfer and treatment stations must be hydrostatically tested to prove that they are water-retaining structures. They must then be pumped out and any remaining construction debris removed. As the various drainage systems are flushed and put into service, collection sumps should be emptied periodically to clean out any accumulations of debris. It is also important to ensure that atmospheric vent lines from enclosed chambers do not become obstructed by adjacent construction activities. Permanent safety barriers should be fitted around access manholes and periodically checked that they remain in a safe condition.

All drains pipework and manholes must be tested with a water flow, preferably from manhole to manhole along the route to prove the correct flow path. Every manhole should be checked to ensure that it is correctly identified for its drainage service be it *surface, dirty, oil* or *foul* drains. Each section of pipework below 300 mm diameter should be hydrostatically tested, before back-filling, at a pressure of 1.2 m above the crown of the pipe at the high end, providing that the resultant static pressure is not more than 6 m at the lower end. It is not normal practice to hydrostatically test drains pipelines of diameters greater than about 300 mm.

Oil separators should be proved with clean water first, to establish that fluid levels will be correct for various rates of flow. The ability to separate oil from the drains is proved by adding metered quantities of an oil having a specific gravity of say 0.9 and 0.99 at 10°C, at least 2 m from the drains inlet to the separator, then analysing the quality of the outlet water over a one-hour period. Typical oil dosage rates would be 0.5, 5 and 10 g/litre of drainage flow. The auxiliary steam supply to, and the operation of, the oil heating elements should be proved in advance of any oil separation tests being carried out. It may also be prudent to dump a small quantity of boiler fuel oil direct into the skimmed-oil sump, then prove that the heater effectively raises the temperature, so that oil can be pumped out again without difficulty.

Concurrent with proving of the oil separator, commissioning of the surface water oil-in-water monitors should be progressed. At this stage it is likely that temporary visual and audible alarms will have to be provided at the monitors, since the cabling to the station alarms system is unlikely to be available.

Sewage and other drains pumps are usually first energised from the construction site supplies network (see Section 9 of this chapter). They would be commissioned, using clean water initially, to prove correct operation of the automatic controls and as a check on their pumping capacity. Commissioning of the sewage treatment plant electrical auxiliaries will also be carried out, using construction site supplies.

Once all of the electrical auxiliaries and their controls have been proved, using clean water, the main aeration basin has to be loaded with clean water, followed by 'seeding' with an initial charge of activated sludge obtained from the local sewage authority. The plant must then accept sewage within a few days to realise the full potential of the activated sludge. Again, all alarms associated with the pumping and treatment plant will have to be provided by means of temporary audible and visual means local to the plant, until the permanent cabling is installed. These too will need to be recommissioned at a later date.

During the construction period, the foul drains system may be subjected to periodic overloading by sewage arising from the contractual labour force. It is normal practice for the civil construction contractor to carry out routine servicing of the sewage treatment plant on behalf of the CEGB until the contractual labour force reduces in numbers. At a later date, the sewage plant, along with surface water drainage sump pumps, must be refurbished to the *as new* condition and re-energised from their permanent station switchboards.

Particular attention must be paid to ensuring that redundant foul water lines from contractor's temporary facilities are properly sealed off so that surface water does not percolate into them at a later date and give rise to unnecessary overloading of the sewage plant.

19 Auxiliary steam supply systems

Modern fossil-fuel-fired power plants need a supply of relatively low grade steam, primarily for fuel oil heating, and also for various unit off-load heating services. It is both economic and practical to meet this demand by oil-fired auxiliary boilers, rather than use degraded high energy steam from main unit boilers.

This section deals with commissioning of the constituent parts of the auxiliary steam supply system as follows:

- Auxiliary boiler condensate and feedheating system
- Auxiliary boiler fuel oil and propane gas systems
- Auxiliary boilers, including automatic control schemes
- Auxiliary steam distribution system

19.1 System description

Figure 3.110 shows a simplified line diagram of the auxiliary steam system at a 1000 MW oil-fired power station, where the auxiliary steam demand is broadly similar to that of a 4000 MW coal-fired plant.

Three self-contained oil-fired package boilers are connected to a common high pressure (HP) steam ring main, which encircles the unit boiler house for distribution of auxiliary steam. HP steam is also reduced in pressure to feed a low pressure (LP) steam manifold for distribution to other auxiliaries. Condensate is recovered and either flows by gravity or is pumped back to the auxiliary boiler house.

Ancillary systems, common to all three auxiliary boilers, handle fuel oil, condensate, feedwater, make-up, blowdown drains and chemical dosing. Most auxiliary boiler plants are designed for fully-automatic operation, requiring only minimal manual attention.

At some power stations, a pressure-reduced and desuperheated steam supply from the main boiler primary superheater outlet-header is connected into the HP ring main. Usually this is intended for use in an emergency, should there be a failure which affects auxiliary boiler output.

19.1.1 Auxiliary boiler ancillary systems

Condensate and feedwater system

Condensate recovered from plant served by the auxiliary steam system is returned to the condensate flash vessel. Figure 3.111 shows that it is pumped into the 14 m^3 monitoring section, then transferred into the 55 m^3 reserve feedwater (RFW) section of a combined monitoring and RFW tank. Feedwater is pumped from the RFW section into a steam heated direct-contact de-aerator from whence the boiler feed pumps pump it into each of the three boilers. The de-aerator lift pumps and boiler feed pumps are fitted with continuous leak-off (or bleed) lines tapped off the discharge branch of each pump. These recirculate water to the RFW section of the combined feed tank to maintain a specified minimum flow of water through each running pump at

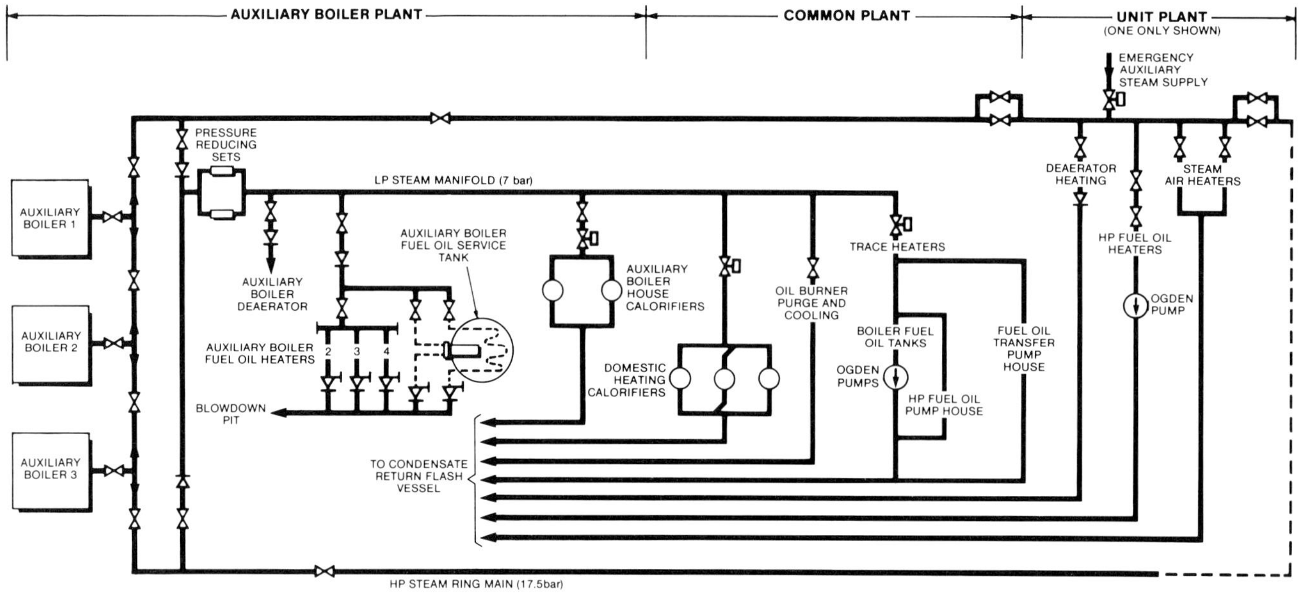

FIG. 3.110 Auxiliary steam system single-line diagram

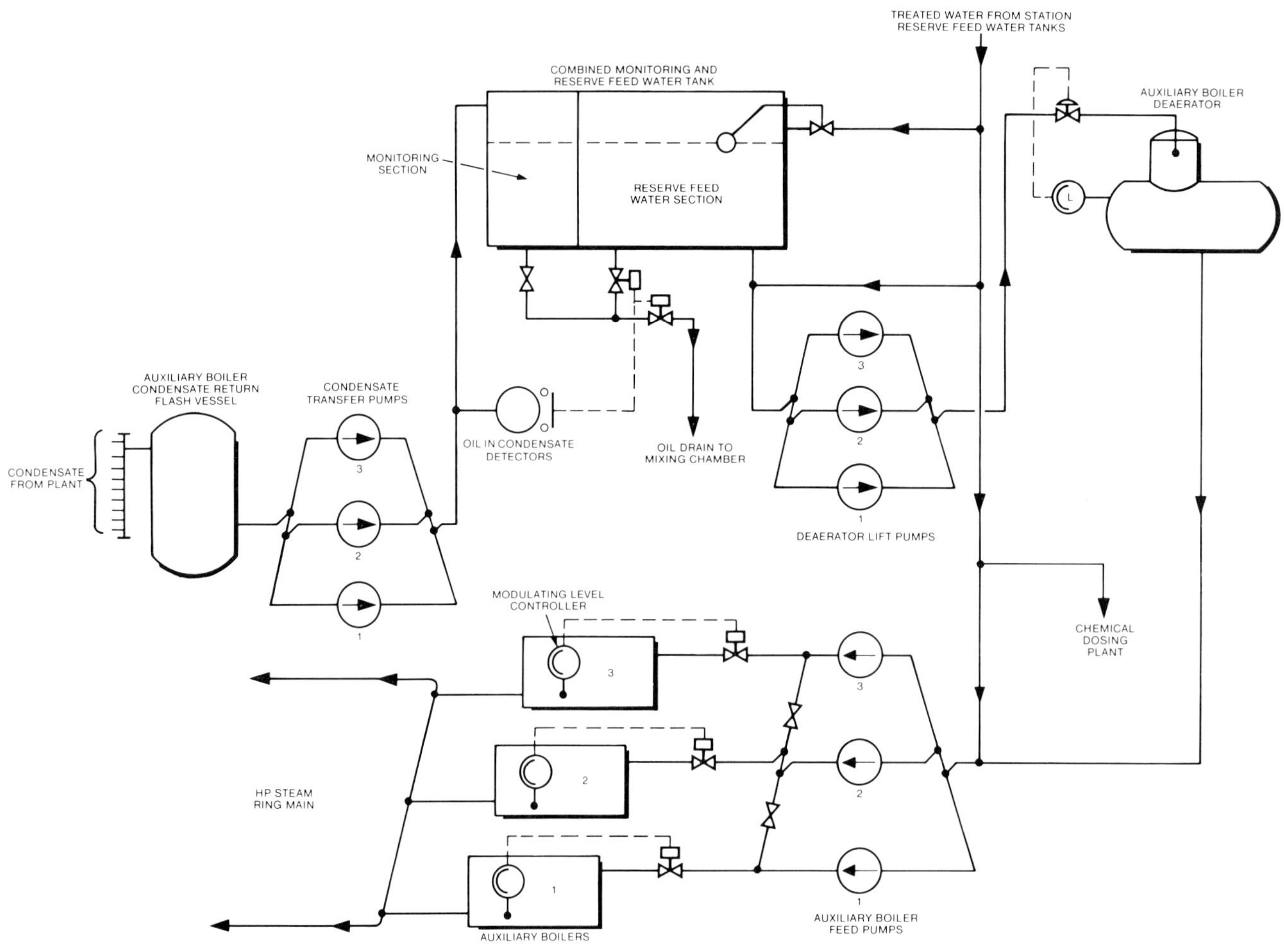

FIG. 3.111 Auxiliary boiler feedwater system block diagram

all times. The boiler feed pump leak-off flow is redirected to the de-aerator storage vessel when at least one boiler is on steady load.

In event of traces of oil being detected in condensate returning from the plant, electrically-operated valves dump the contents of the monitoring section into a mixing chamber, to be cooled by water from the towns water (TW) system before flowing into the station dirty drainage system. Boiler water drains are led to a blowdown pit, where they too are cooled by water from the TW system before flowing to the CW forebay.

To ensure a supply of feedwater under all conditions of operation, treated water from the main unit RFW tanks can be injected:

- At the RFW section of the combined monitoring tank, via a ball float control valve.
- Direct into the de-aerator lift pump suction manifold.
- Direct into the feed pump suction manifold.

Three chemical dosing pumps, each with chemical tank and stirrer, are interconnected to permit simultaneous dosing of appropriate chemicals into either the drum or the feedwater inlet of different boilers. Water for mixing the chemicals is also derived from the main unit RFW tanks.

Fuel oil and propane systems

Fuel oil is stored in a steam-heated 30 t fuel oil service tank, which is automatically replenished from the main boiler fuel oil tanks. It flows from the service tank, through a combined steam/electric outflow heater, followed by strainers, before being pumped to the oil burner ring-main via steam and electric heaters, as shown in Fig 3.112.

Tappings on the oil ring main provide individual flow and return lines to each horizontal rotary-cup oil burner assembly. A proportion of the oil is continuously recirculated to the strainer inlet manifold, to maintain the system at firing temperature when burners are shut down. Each burner can use fuel oils with a wide range of viscosity, providing that heating keeps the firing viscosity equivalent to about 130 mm^2/s (130 cSt) at 80°C. This equates to a firing temperature of approximately 90°C to achieve a viscosity of 85 mm^2/s (85 cSt) at the burner tip.

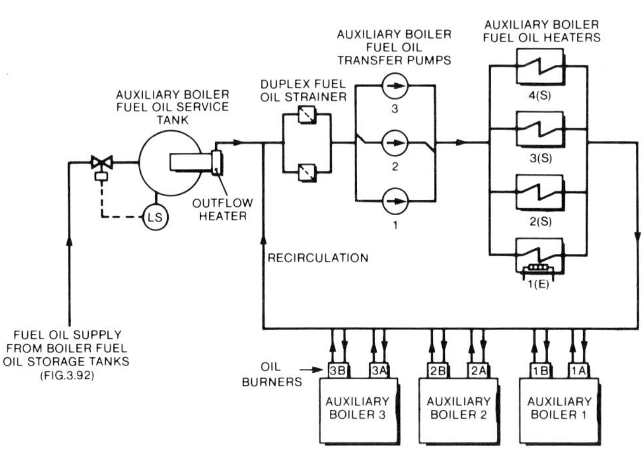

FIG. 3.112 Auxiliary boiler fuel oil system block diagram

A separate ignition system is provided for each burner, comprising a high-tension spark-ignition unit, which causes a propane gas burner to light. Propane gas is derived from the propane distribution system, which supplies all of the boiler plant on site (see Section 20 of this chapter).

It is customary for all of the fuel oil lines between the main storage tanks and the auxiliary boiler burners to be electrically trace heated. This, in conjunction with electric fuel oil heaters, facilitates cold start-up when steam is not available.

Waterspray fire protection systems are installed around the fuel oil service tank, transfer pumps, heaters, etc., as well as the oil burner ring main and burner assemblies. Fire at any one burner assembly automatically isolates the fuel supply to that burner.

19.1.2 Auxiliary boilers

Figure 3.113 shows a general arrangement of a typical auxiliary boiler. Each boiler consists of twin furnace tubes, two wet-back combustion chambers and two passes of small-bore smoke tubes, all contained within the shell and end plates of the boiler. A rotating-cup automatic oil burner assembly and a forced draught (FD) fan is associated with each furnace tube.

Twin outlet flues are at the rear of the boiler and a small superheater section is located in the front smoke box. Four fully-automatic sootblowers, using saturated steam direct from the drum, are located in the superheaters. Unlike a dry-back boiler, the wet-back type has virtually no high-grade heat to be absorbed when the boiler shuts down.

Typically, feedwater inlet temperatures are between 60°C and 120°C and the boiler evaporates 6.5 kg/s of slightly superheated steam at 17.5 bar and 220°C.

19.1.3 Controls and sequencing

A comprehensive system of controls and sequence logic equipment is provided for automatic operation of the boilers and the ancillary plant. Each boiler is provided with four manually-selected modes of automatic control:

Modulating Where the rate of firing is varied in response to an output from the master steam pressure control loop.

First duty Firing is fixed at the 100% rate, providing another boiler is selected for modulation.

Standby The boiler will automatically transfer from a banked condition to a preselected fixed firing-rate of either 100% or 50%, upon failure of another boiler.

Banked While the boiler is off-line, it is fired intermittently at a low rate to maintain a preset superheater steam pressure.

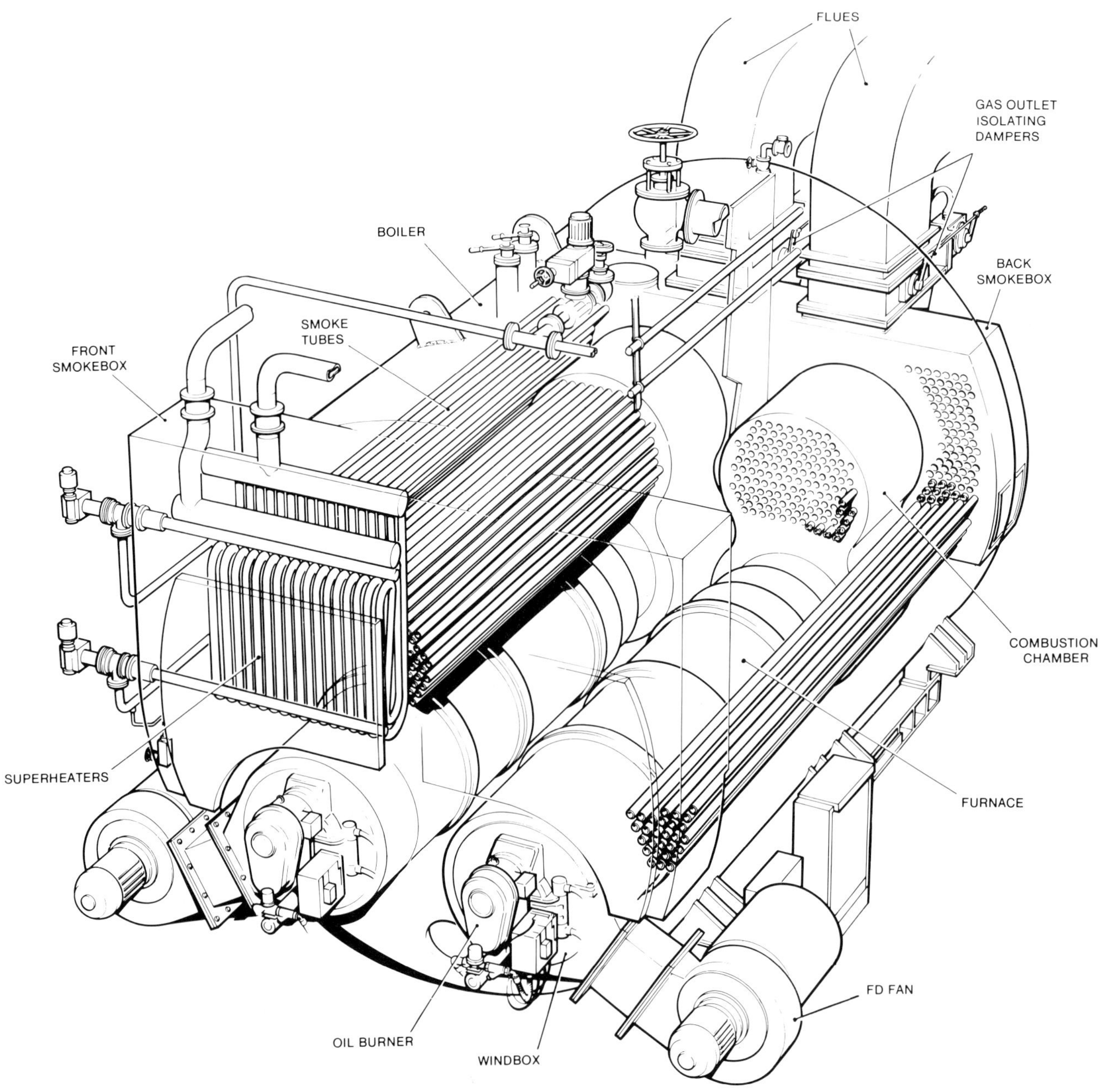

FIG. 3.113 Auxiliary boiler general arrangement

Ancillary plant control logic generally provides for automatic starting of a standby pump, in the event of failure of a pump selected to 'duty'. This feature covers the condensate, feed and fuel oil pumps only; all other small pumps are usually manually started, when required, and are not provided with automatic starting facilities.

Compressed air for pneumatically-operated control valves is usually derived from either a unit or a station-services control and instrumentation compressed-air system (see Section 21 of this chapter). Sometimes a small-capacity compressor/receiver/drier installation may be provided, purely for auxiliary boiler controls and instruments.

19.1.4 Auxiliary steam system

The three boilers are connected to a common HP ring main, which encircles the main unit boilerhouse. Duplicated steam pressure-reducing sets, tapped into the HP ring main, reduce steam from 17.5 bar at 220°C to 7 bar at 195°C for distribution from an LP manifold located in the auxiliary boiler house.

As shown earlier by Fig 3.110, the following services are tapped off each section:

HP ring main

- Unit de-aerator off-load and low-load heating up to 20% MCR.

- Unit fuel oil heaters.
- Unit steam air-heaters up to 40% MCR (oil-fired boilers only).

LP manifold

- Auxiliary boiler de-aerator.
- Auxiliary boiler fuel oil heating.
- Main boiler fuel oil tank heating.
- Fuel oil line trace heating.
- Calorifiers for space heating, ventilation and domestic hot water systems.
- Miscellaneous small plant items for other heating or steam purging purposes.

Various types of steam trap are used extensively throughout the auxiliary steam supply system for control of drainage from steam condensed in pipelines and heat exchangers. Each design automatically prevents the passage of live steam, yet allows free flow of accumulated condensate. Most contain an integral, removable, stainless steel strainer to protect the internal mechanism from damage by swarf, etc.

Condensate from pipeline tracers and fuel-oil tank-heating coils generally does not have sufficient energy for gravity drainage over long lengths of pipework to the auxiliary boiler house. To overcome this problem the condensate recovery system is divided into separate sections, each of which serves a geographical area. The condensate recovered from each section is then pumped to the boiler house.

A typical fully-automatic pump for this duty is the Ogden pump, which is shown in Fig 3.114. This pump requires an operating steam supply from the LP distribution system.

19.2 Commissioning objectives

The specific objectives for commissioning an auxiliary steam supply system are given below. They should be read in conjunction with the general objectives listed at the beginning of this chapter:

- Prove satisfactory operation of each ancillary plant system.
- Complete an alkaline boil-out of each boiler.
- Float the safety valves and obtain statutory certification for each boiler.
- Carry out a pressure-accumulation test with each boiler to prove the capability of its safety valves to discharge MCR steam flow.
- Establish safe manual, and then automatic, control of each boiler.

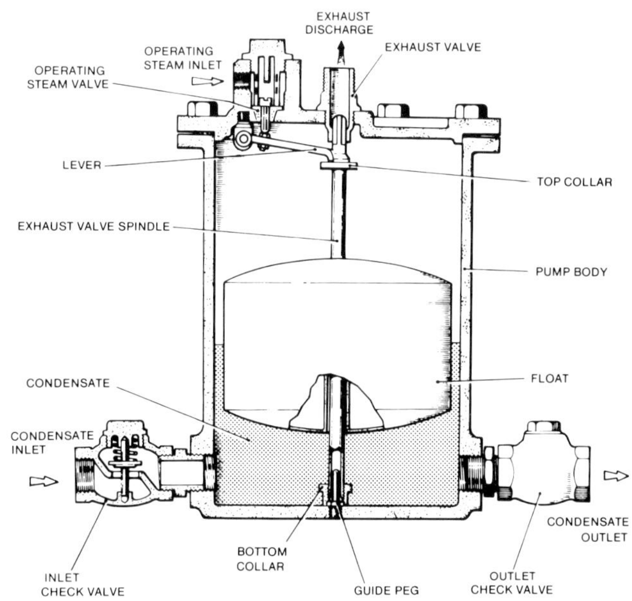

FIG. 3.114 Ogden condensate pump

Condensate flows by gravity into the body of the pump and makes the float move progressively upwards along the exhaust valve spindle. The operating steam inlet valve is kept closed and the exhaust valve kept open by the weight of the spindle. When the float presses against the top collar, it opens the operating steam valve which then causes the exhaust valve to snap closed and also pressurises the body. Discharge of condensate continues until the float descends to engage with the bottom collar on the exhaust valve spindle. The weight of the float simultaneously closes the operating steam valve and vents the body through the exhaust valve.

- Confirm the capacity and efficiency of each boiler under different conditions of operation.
- Achieve reliable indication from each oil-in-condensate monitor.
- Check that all steam trap drainage-sets function correctly.
- Satisfactorily purge the HP and LP auxiliary steam mains, then purge each of the distribution branches in turn.
- Commission and adjust the auxiliary steam LP pressure-regulator stations, with varying steam flows.
- Prove that the steam supply and condensate return systems are adequate for handling the specified steam loads.
- Commission and prove the adequacy of the emergency auxiliary steam supply connection (if fitted).

19.3 Safety considerations

This is the first steam-raising activity on-site and it generally takes place in a very congested partially completed construction area. Usually only the absolute minimum of plant is available for firing the first boiler.

Because of the hazards of commissioning in such a congested area, the following points need to be considered in addition to those outlined at the start of this chapter:

- Fuel oil, steam and condensate pipework and flue-gas ductwork must have adequate thermal insulation to safeguard personnel from contact with hot metal surfaces.
- Many of the operations give rise to unacceptably high noise levels. Clearly understood audible warnings must be given before any noisy activity is undertaken. Hearing protection must be worn, where necessary. In some cases, the testing may have to be done outside normal working hours, so that the erection programme is not delayed.
- All plant safety constraints must be observed during operation of the boilers.

19.4 Resources

19.4.1 Services

Apart from electricity supplies, the following main services will be needed, for which cross-reference should be made to other sections of this chapter as indicated:

- Compressed air for control valves and instrumentation (Section 21).
- Propane gas for burner ignition (Section 20).
- Treated water for boiler filling and pipework flushing (Section 16).
- Suitable fuel oil (Section 15).
- Fire protection systems (Section 12).
- Main cooling water (CW) system for disposal of drains and effluent (Section 17).
- Towns water (TW) system for cooling of drains, etc.
- Boiler chemicals and temporary disposal tanks.
- Temporary pipework for system flushing and steam purging.
- Safety barriers, protective clothing, hearing protection, etc.
- First main unit drains flash vessel for accepting discharge of steam during auxiliary boiler performance tests.

19.4.2 Staffing

Commissioning of each auxiliary boiler system with its controls, plus load testing, may occupy three teams, each of two or three people, for at least ten weeks. The commissioning period for the auxiliary steam system is dependent upon main plant commissioning; completion requires all main units to be in service to create the maximum steam demands necessary to test system capacity. However, the bulk of the work associated with proving of the HP and LP main branches should be possible within three weeks. A team of up to five people may be involved in these activities.

Members of the station operating staff will need to be involved, both for manipulating the plant and for training.

19.5 Commissioning

Figure 3.115 is a simplified commissioning network which highlights the interaction of the main activities. The basic intention is to commission sufficient ancillary plant to carry out the alkaline boil-out and float the safety valves of the first boiler safely. It is not customary to conduct a boiler hydraulic test on site, since the test is carried out by the manufacturer at works, who then issues a test certificate as proof. Steam from the first boiler is then used to achieve de-aeration of its own feedwater, fuel oil heating and sootblowing. Commissioning of the remainder of the ancillary plant, auxiliary steam systems, other boilers and all plant control facilities then continues in parallel.

In practice, commissioning of the auxiliary boilers and the auxiliary steam system is usually carried out as two separate yet interrelated exercises, often by different teams.

Before commencing, the following questions need to be resolved in relation to availability of other station services:

- Is normal service fuel oil to be used, or will the fuel oil service tank have to be filled with a special delivery of expensive gas-oil?
- If normal boiler fuel oil is to be used will the main boiler fuel oil storage system be available or will initial fuel oil supplies have to be delivered by road into the service tank?
- Is the bulk storage plant going to be available? If not, portable cylinders of propane ignition-gas will have to be fitted on a temporary basis.
- Will the water treatment plant be available for service? If not, then numerous road tankers of demineralised water will be needed.
- Is the CW forebay available for receipt of boiler drains and chemical clean rinse water?
- Will temporary water supplies be required for cooling the oily and blowdown drains?
- Will a temporary control compressed-air supply be needed?

For simplicity, the commissioning activities have been divided into three distinct areas:

- Auxiliary boiler ancillary systems comprising condensate and feedwater (including make-up and chemical dosing), fuel oil and propane.

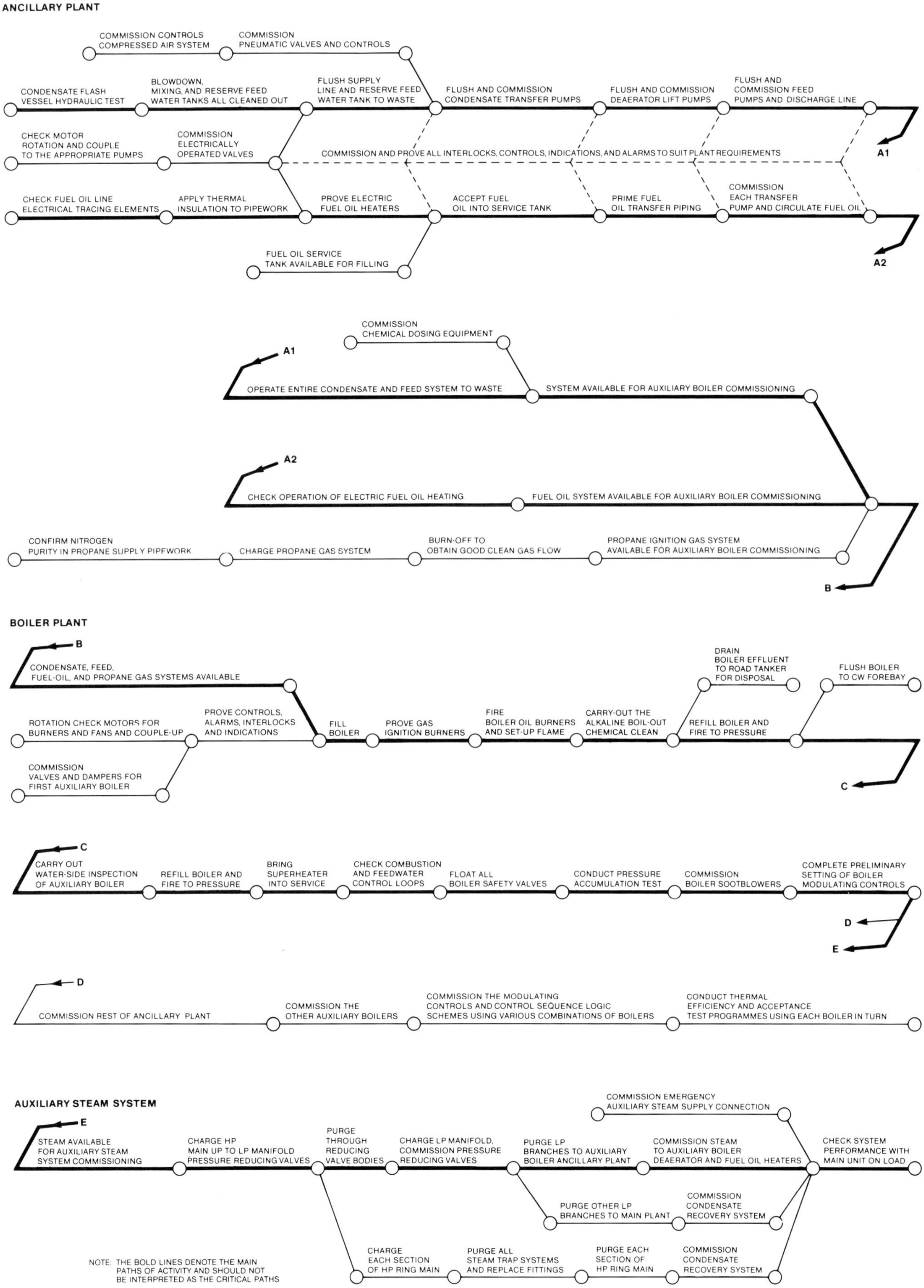

FIG. 3.115 Auxiliary steam supply system commissioning network

- Auxiliary boilers and their controls.
- Auxiliary steam systems.

19.5.1 Auxiliary boiler ancillary systems

Condensate and feedwater system commissioning (Fig 3.111)

After confirming that plant erection is complete, all individual items of equipment are checked to the appropriate standard check list (SCL), including remote indications and controls on the ancillary plant control panel.

All pipeline blanks should have been removed from the condensate flash vessel following its hydraulic test, except those blanking-off condensate returns from the plant. The combined monitoring and RFW tank, mixing chamber and blowdown pit should be cleared of all debris and the drainage outlets proved to be clear.

All motors are checked for correct direction of rotation, then coupled to their pumps, fans, etc. Where practicable, the remote control features are proved as well. Control air compressor plant has to be commissioned and available for setting up pneumatic control valves. Commissioning of pneumatic and electrically-operated valves should be completed, as far as possible.

The treated water make-up line from the unit RFW tanks is flushed into the RFW section and then into the monitoring section. Both sections are then filled, and flushed to waste, via the mixing chamber and station dirty drains. The RFW section inlet ball-valve and all of the RFW tank alarms are proved, along with demonstrating the ability of the overflow to cope with full inflow of make-up.

A temporary connection from the treated water make-up line is used to fill the condensate flash vessel to check alarms, followed by proving correct operation of the drain to the blowdown pit, and thence to the CW forebay. This temporary line enables each condensate transfer pump to be run for 30 minutes, discharging into the monitoring section of the combined feed tank, which is allowed to drain to waste. After proving that the condensate transfer pump discharge non-return valves operate correctly, the auto/standby control facilities are also proved.

Treated water from the RFW section of the combined feed tank is flushed to waste from the suction branches of the de-aerator lift pumps. Then each pump is run for 30 minutes with its leak-off line discharging into the RFW section. Next each pump is run to flush water through the discharge pipework into the de-aerator. Correct operation of the pump discharge line non-return valves is proved and, if practicable, the auto/standby facilities are tested. Water from the de-aerator is flushed through the feed pump suction line strainer bodies and out to waste via temporary pipework. Each feed pump is run for 30 minutes with its leak-off line selected to discharge firstly into the de-aerator and then to the RFW tank. Correct operation of the pump discharge line non-return valves and auto/standby facilities are tested.

Temporary pipework is fitted upstream of the feed inlet isolating valve for each boiler. Each feed pump is run in turn to flush the discharge pipework to waste via this temporary piping, ensuring that the feed pump discharge pressure is maintained at approximately 18 bar to safeguard the feed pump.

Finally, the feed system is set up so that the de-aerator lift pumps and feed pumps are run continuously. During this exercise feedwater is being replenished via the make-up into the RFW tank to replace water being drained to waste at a controlled rate from the de-aerator.

The branch from the treated water make-up main is flushed into the chemical-dosing mixing tanks, which are drained, cleaned out and refilled. The dosing pumps are then run for half an hour each, flushing water through the distribution pipework to waste via temporary disconnections at the boiler drum and boiler feedwater inlet. Later, suitable chemicals are injected into the feed system in preparation for boiler filling. At this stage, the condensate and feedwater system is available for auxiliary boiler commissioning.

Fuel oil and propane system commissioning (Fig 3.112)

After checking correct erection of plant and equipment, and the pipeline electrical trace heating installation in particular, the thermal insulation is applied. Individual items of equipment, including remote indicators and controls on the ancillary plant control panel, are checked to the appropriate SCL.

Fuel oil transfer pump motors are checked for correct direction of rotation, then coupled to their pumps. Electrical heaters, including the pipeline tracing, are tested, then left de-energised until required.

Providing that the fuel oil service tank is free from debris, that the float switches operate correctly, and that the fire protection is available, the first delivery of fuel oil can be accepted. The initial deliveries of oil may be by road tanker direct into the service tank, because the boiler fuel oil storage and transfer system is not available. During delivery, correct functioning of the contents gauge and level alarms is proved.

If the first fill is boiler fuel oil (rather than gas-oil), then temporary heating of the service tank to 50°C may be necessary until fuel oil circulation, with electric heating in service, has been established.

Fuel oil is then allowed to flow from the service tank through the system into the transfer pumps, whilst all trapped air is vented. One fuel oil heater and the rest of the circulatory system is set up for service, then a transfer pump is run to charge the pipework carefully. Each available pump is run for half an hour in turn, as flow is established through the remaining oil heaters.

The suction strainers are exercised frequently over the next few hours, whilst two and three pump parallel operation is attempted. Correct operation of the transfer pump discharge non-return valves and auto/standby facilities is checked at this stage.

If commissioning has been carried out using gas-oil, then the electric heating may be proved at this stage, though it should be left isolated until the reception of boiler fuel oil.

The propane supply pipework (see Section 20 of this chapter) is checked to ensure that it has been purged with nitrogen, such that oxygen in the nitrogen atmosphere is less than 5%. A burning-off connection is attached to the remote end of the auxiliary boiler gas main. The system is then charged with propane gas, either from the bulk store or from temporary cylinders, and the gas flared off until a clean steady flame is established. The small volume of nitrogen/propane gas present in the individual burner supply lines can be purged into the furnace as part of the boiler firing exercise.

At this point, the fuel oil and gas systems are available for boiler commissioning.

19.5.2 Boiler commissioning

All pneumatic and electrically-operated valves and dampers are commissioned. Rotation checking of the FD fans and oil burner motors is completed, along with preliminary adjustment of dampers and burner controls. At this stage, fuel oil, propane gas, feed, condensate, blowdown and chemical-dosing plant commissioning must be substantially complete. The boiler is filled to working level then, with the fuel oil supply isolated, the propane flame-ignition sequence is established. Both oil burners are commissioned and the combustion controls trimmed to achieve a good clean steady oil-flame as soon as practicable.

Before commencing the alkaline boil-out, the boiler water temperature must be cooled to below 30°C. Chemicals, typically trisodium phosphate (TSP) and a wetting agent, are manually added direct into the boiler drum through an open manhole, which is then secured in place. A temporary blank has to be fitted into the outlet from the blowdown pit before the boiler is fired.

Observing the manufacturer's operating restraints, the boiler pressure is gradually raised to 13.5 bar. Note that the superheater remains isolated for this part of the process. Firing continues intermittently over the next 24 hours to maintain the boiler pressure at approximately 13.5 bar, then the pressure is allowed to decay to 3.5 bar. During the boil-out, chemicals are added to maintain the specified minimum level of TSP. Also, all boiler blowdown drains are operated regularly to a set pattern. When the boiler is drained empty at the end of the boil-out stage, from a pressure of 3.5 bar, the effluent plus cooling water has to be pumped from the blowdown pit into tankers for disposal off-site. The temporary blank is then removed from the blowdown pit outlet.

The boiler is refilled, fired to a pressure of 3.5 bar, then drained into the blowdown pit, where the drains are cooled and dumped to the CW forebay. Providing that an internal inspection of the boiler pressure parts is satisfactory, the boiler is refilled and slowly fired to normal working pressure. During this exercise all blowdown drains are operated in turn, the superheater is put into service, the combustion controls trimmed and the feedwater auto control loop checked.

Now the boiler pressure is carefully raised to check and (if necessary) adjust the setting of each safety valve in turn. When floating of the safety valves has been completed, a pressure accumulation test is carried out as follows.

Initially, the boiler is fired at a rate which maintains normal superheater outlet steam pressure. The pressure accumulation test then involves gradually increasing the boiler firing rate to its maximum value, whilst maintaining normal drum water level. As the boiler pressure rises, confirm that each of the safety valves lifts in turn and that the boiler pressure does not continue to increase beyond 10% above the normal working pressure. When the test has been completed, firing is reduced until all safety valves reseat. Providing the boiler insurance inspector is satisfied with the result, the statutory certificate is issued.

Finally, automatic feedwater control is established, sootblowers commissioned and oil burner safety trips re-proved. This boiler is now available with full control from its local control panel and can be used for auxiliary steam system commissioning.

Each boiler is commissioned to the same status. When at least two boilers are available, automatic modulating control and sequence logic proving tests are conducted. Boiler loading is varied over the full range by a combination of meeting system demand and by dumping to waste via a large-bore temporary branch led from the auxiliary steam HP ring main into the drains flash vessel of the first unit boiler.

The principles governing modulating control/sequence logic commissioning have been outlined earlier in Sections 3 and 8 of this chapter. When conditions are acceptable, thermal efficiency and boiler acceptance tests are conducted, generally in accordance with BS845 and BS2885.

19.5.3 Auxiliary steam system commissioning

Commissioning of the auxiliary steam system is usually carried out in the following sequence:

- Charge the HP ring main from boiler outlet up to the LP pressure-reducing sets.
- Purge the HP pipework out to atmosphere through the pressure-reducing valve bodies with the internals removed.
- Charge the LP manifold, commission pressure-reducing sets and test relief valve operation.

- Purge each of the branches from the LP manifold at the plant terminal point isolation valves.
- Commission auxiliary boiler fuel oil heating with the condensate running to waste for several days to clean out the pipework and gain confidence with the oil-in-condensate monitors.
- Commission the auxiliary boiler de-aerator steam supply and prove its automatic controls.
- Charge and purge the HP ring main in sections.
- Purge each of the distribution branches from the HP ring main at the plant terminal point isolation valves.
- Commission condensate return pipework, with each section run to waste for several hours to flush all debris before the connection at the condensate return flash vessel is made off.
- Commission the oil-in-condensate monitoring equipment associated with the main boiler fuel oil heating condensate return system.
- Prove that all condensate drain trap sets and condensate pump units are functioning correctly.
- Commission the emergency auxiliary steam supply from the main boilers.

Purging of steam lines

The internals and associated strainers have to be removed from all condensate drain trap sets. Each section of main and/or distribution pipework is charged gradually until all debris has been blown clear from vents and drain trap bodies, and then the drain traps are reinstated. The supply main is disconnected at, and separated from, the upstream face of a convenient terminal point valve, and a temporary deflector plate is fitted to protect the valve discs. This section of pipework is then purged through the disconnection until all debris has been ejected. Other sections of HP and LP steam supply pipework are treated in a similar manner, though it may not always be necessary to fit deflector plates if the open end of the pipe can be directed into an unobstructed area.

Throughout this exercise extreme vigilance is necessary, with particular attention to the following points:

- Temporary purging outlets must be adequately anchored to cope with thrusts which may have a greater magnitude and a different direction from those imposed by normal service operation.
- The purge outlet area must be barriered off and precautions taken to protect plant and personnel from steam jets and ejected debris.
- Ear protection must be worn and noise levels recorded in case of complaints.

In conjunction with the auxiliary boiler test team, part of the HP ring main is used to conduct the steam flow of two boilers to waste via temporary pipework led to the drains flash vessel for one of the main boilers. This proves control of the auxiliary boilers with one at 100% fixed-firing and the other modulating to meet a steam demand which is being manually regulated by means of a valve in the temporary pipework.

Commissioning of the emergency auxiliary steam connection is usually conducted by one of the main boiler test teams at some convenient point in the unit commissioning programme. The procedures will be detailed in Chapter 6, but the essential points are as follows:

- Flush desuperheating water supply pipework.
- Steam purge HP steam supply pipework.
- Commission steam pressure-reducing and desuperheating temperature control loops and associated valves.
- Steam purge the supply lines to the auxiliary steam HP ring main.
- Check dynamic response with variations of steam demand.
- Demonstrate smooth transition from 'live' steam to auxiliary boiler steam, and vice versa.

20 Bulk gas systems

This section deals with commissioning of each of the following bulk gas storage systems. Most of the gases are actually stored as liquids under pressure before being vaporised for distribution and use.

Nitrogen (N_2) Used as a low pressure gas for the exclusion of atmospheric oxygen in steam spaces of off-load hot boilers, turbine condensers and feedwater heaters.

Ammonia (NH_3) Gaseous ammonia is injected into the flue gases of oil-fired boilers to neutralise acidity arising from the combustion of residual fuel oil, thereby eliminating the emission of acid smuts.

Propane (C_3H_8) Used for primary ignition of oil burners in oil- and coal-fired boiler plants.

Carbon dioxide (CO_2) Used as an interface between atmospheric air and hydrogen during filling or purging of hydrogen-cooled generators and associated auxiliary plant to prevent the formation of an explosive hydrogen/air mixture.

Hydrogen (H_2) Coolant gas in all large modern electrical generators. It is now frequently produced on site from a methanol/water mixture, then compressed to a very high pressure for storage.

Each bulk storage plant is dealt with independently within this section. The commissioning objectives and the safety considerations specific to each plant must be read in conjunction with those given in the introduction to this chapter.

20.1 Nitrogen bulk storage

20.1.1 System description

The use of a nitrogen-rich atmosphere as an oxygen-excluder in turbine and condenser steam spaces of modern power plants is becoming less common. However, there is a growing usage for sealing, or capping, the steam spaces of hot depressurised unit boilers, feedwater heaters, and other similar tanks or vessels. This has resulted in a change of emphasis from large volumes of nitrogen at high flow rates to smaller volumes with lower flow rates.

Nitrogen generator module equipments, having outputs of up to 500 m^3/h when using ordinary compressed air, are now available. Whether or not on-site generation is economic compared with the more established cryogenic bulk storage system depends upon the likely usage pattern at each individual location.

Bulk storage plant

A typical plant for a 2000 MW power station (see Fig 3.116) may comprise a vacuum-insulated (i.e., double-shell) cylindrical storage tank, containing perhaps 30 t of liquid nitrogen (approximately 23 000 m^3 of gas at NTP) at a pressure of approximately 17 bar. Delivery is by vacuum-insulated road tankers with a capacity of 24 t and equipped with an engine-driven transfer pump.

Four 45 kW electrically-heated vaporisers, each capable of producing up to 1800 m^3/h of gas at 10°C, are installed. The plant is safeguarded against vaporiser failure by a low temperature alarm and an automatic shut-off valve, which operate at −5°C and −10°C respectively. A pressure regulating valve ensures that the distribution system pressure is maintained at approximately 5 bar. Additional pressure control valves are fitted close to the points of injection to ensure that the nitrogen pressure is, typically, between 10 and 50 mbar above atmospheric pressure. For effective exclusion of oxygen, the nitrogen concentration must be approximately 97%.

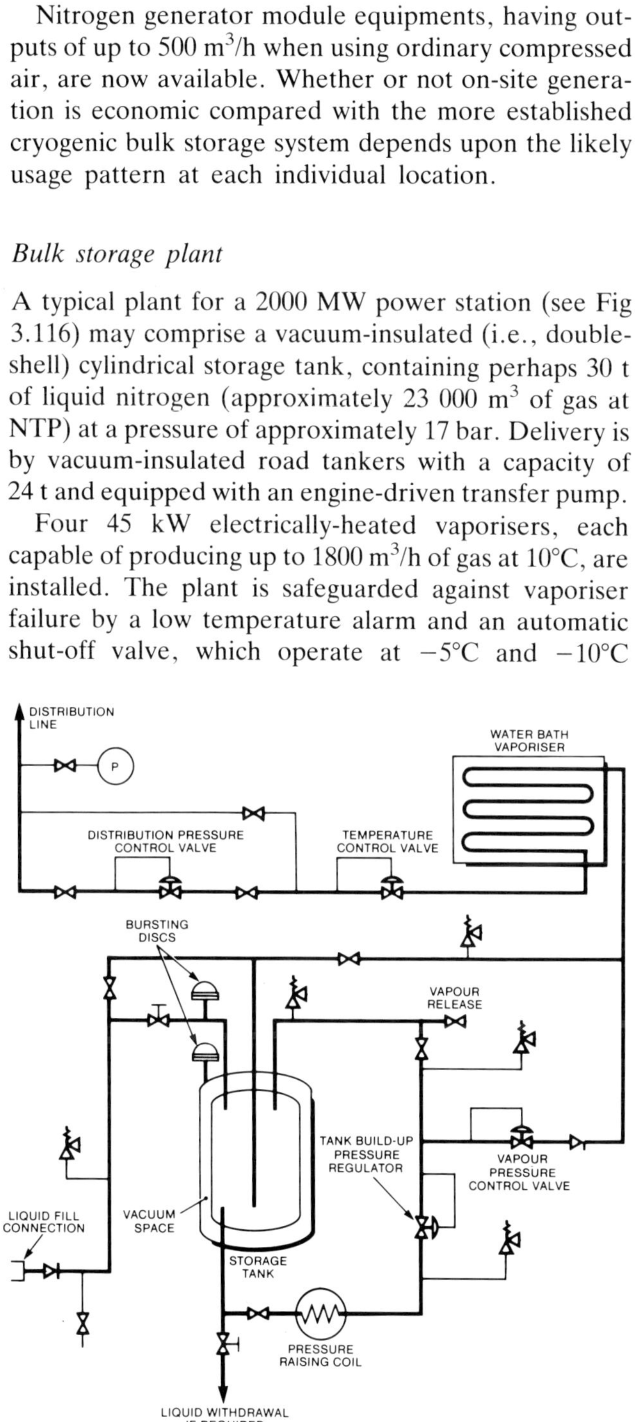

FIG. 3.116 Liquid nitrogen bulk storage system

Nitrogen generator systems

A typical system may consist of two ambient-temperature generator modules, each rated at 30 m^3/h, producing pure nitrogen from a supply of general service compressed air. The nitrogen produced is stored in a cylindrical pressure vessel. Each generator comprises two vessels containing a bed of special carbon molecular-sieve material which has the ability to remove oxygen from air preferentially. One vessel is in service producing nitrogen, while a small quantity of nitrogen is purged through the other for regeneration. At fixed time intervals, the vessels change duty in a cyclic process known as 'pressure swing adsorption'. The generator module also incorporates a prefiltration system, which can cope with wet oily compressed air.

When supplied with compressed air at 7 bar, the generator module is capable of producing nitrogen at a pressure of up to 5 bar. The purity of the nitrogen product is a function of the demanded gas flow. Typically, the O_2 in nitrogen could be 5% at maximum flow, but at only 20% flow it would improve to less than 0.05% with a dewpoint of better than −60°C. A control unit continuously monitors O_2 in nitrogen and, if a preset level is exceeded, the output is automatically dumped to waste to avoid contaminating the storage vessel.

As an example, some 10 m^3/h of nitrogen would be needed to establish a nitrogen atmosphere in a boiler which had been depressurised but was still full of hot water. After a few days the required rate of flow would be reduced as conditions stabilised, until the normal make-up rate may be only 1 m^3/h. With this type of generator module, a large volume of relatively impure nitrogen can be used to establish the nitrogen concentration. This is followed by a low flow rate of very pure gas for normal make-up, to maintain the nitrogen pressure in the boiler between 10 and 50 mbar above atmospheric pressure, with a concentration of better than 97%.

20.1.2 Commissioning objectives

- To pressure test the storage tank and pipework to prove that they are acceptable for the initial fill of liquid nitrogen.
- To commission the vaporisers and prove correct operation of the protective devices.
- To commission each nitrogen generator and prove that its controls function correctly.

- To prove that the plant is capable of giving the specified rate of flow.

20.1.3 Safety considerations

Though nitrogen may be inert, it is not without hazard and the main dangers to personnel are:

- Asphyxiation by excessive leakage of the odourless, colourless gas into confined spaces. Personnel experience no distress symptoms and they can lose consciousness in as little as 20 seconds.
- Contact with liquid nitrogen or bare metal pipes containing liquid could cause *cold burns*.
- Liquid nitrogen is usually stored at a temperature of −200°C and it can cause brittle fracture of materials, with disastrous results if it gets into or onto parts not designed for this duty.
- The liquid boils at −196°C and can have a vapour pressure of 30 bar at −148°C. Thus, if vents or relief valves become blocked, the pressure generated may be sufficient to rupture storage tanks.

20.1.4 Resources

Providing that a tanker of liquid nitrogen is available, a team of three should be able to commission a bulk storage plant within three days. A supply of compressed air will be necessary for on-site generation plant, and it should be possible to commission this system within two days.

20.1.5 Commissioning

Bulk storage plant

Current test certificates must be available for all presure vessels before any testing is commenced. The vacuum insulation interspace between the two shells should be evacuated to the best possible vacuum and an air in-leakage test conducted over a 24-hour period. During this time, there should be no detectable loss of vacuum. The storage tank, filling lines and pipework up to the first-stage pressure regulating valve should be charged with nitrogen gas from a road tanker, for a 24-hour pressure test at 24 bar, then operation of the relief valves should be checked and adjusted if necessary. The distribution system should be tested at 10 bar over a 24-hour period.

The vaporisers should be commissioned, as far as practicable, without liquid nitrogen. Then all instrumentation and protective devices should be calibrated and proved to operate correctly. When this is complete, delivery of liquid can be accepted and the storage plant allowed to stabilise for several days. Now, a path can be set up for expelling nitrogen gas from a terminal point of the distribution pipework (perhaps into a turbine condenser) and the vaporisers put into service. The low temperature alarm and shut-off valve can be proved and then a plant flow rate test carried out.

Nitrogen generating systems

Commissioning of a nitrogen generator is an activity similar in many respects to that for a compressed-air drier unit. After verifying that the works test certificate for each pressure vessel is in order, the entire system has to be subjected to a compressed-air pressure test to confirm that there are no leaks. Prior to that, the compressed air supply line has to be purged to clear out any construction debris. The control power supplies are energised and the vessel changeover and nitrogen dump sequences proved. The purity monitoring and alarm unit should be calibrated too.

Compressed air flow is applied to each module with the output being dumped to waste at various rates of flow to check that the specified purity and gas dewpoint can be attained. When the output and performance is acceptable, the storage vessel should be brought into service and fully charged. The performance of the system should again be monitored over several nitrogen sealing exercises to ensure that the designed flow rate/purity relationship is being maintained.

20.2 Ammonia bulk storage

20.2.1 System description

Gaseous ammonia is injected into the flue gas to neutralise acidity arising from the combustion of residual fuel oil, thereby minimising corrosion within the induced draught fan ducts and station chimney and eliminating the emission of acid smuts. Sulphur trioxide is the principal acidic component of the flue gas, reacting with water vapour to form sulphuric acid which then condenses out on cooled surfaces in the ducts and chimney. The ammonia reacts directly with the sulphur trioxide to form the neutral salt, ammonium sulphate, thus preventing the formation of sulphuric acid. Ammonia must always be present in excess of the sulphur trioxide otherwise ammonium bisulphate is formed; the latter, at the temperatures existing within the ducts, is a sticky liquid causing severe fouling of the ductwork.

Liquid anhydrous ammonia is stored in an unlagged mild-steel cylindrical storage tank, typically of 75 t capacity, but having a working load of 65 t. The pressure in the tank may vary between 3 and 9 bar as a result of normal variation of ambient temperature. Two sets of duplicate pressure relief valves, set at 13.5 bar and 15 bar respectively, are provided as a safeguard. The liquid filling line and a vapour balance line are led to a road vehicle filling point, where they terminate with non-interchangeable couplings. The combined liquid filling and plant draw-off line is fitted with an emergency shut-off valve, which is pneumatically-actuated, using air from adjacent standard-sized compressed-air cylinders. Figure 3.117 shows the general arrangement of a typical system.

Steam heated self-regulating evaporators produce gas at a rate of up to 280 kg/h, the pressure of which is reduced to 2 bar before injection into the boiler flue-gas stream through a system of distribution grids. In event of an excessive gas flow being demanded, liquid may pass through the evaporator, and a detector will initiate a 'low temperature' alarm at −5°C. All equipment containing liquid ammonia is located within the confines of the storage tank bunded area. Gas, either automatically-relieved or manually-vented, is led to a tall cylindrical ventpot, the open end of which is approximately 9 m above ground level to allow gaseous ammonia to disperse safely into atmosphere.

Auxiliary steam, at 7 bar, is used for evaporator heating, for distribution pipework purging and for trace heating the evaporator gas-outlet pipework. General service compressed air, also at 7 bar, is used for purging of the gas distribution pipework. The pipework is first purged with air to remove most of the gas; the small quantity of ammonia remaining is then removed by steam purging. The distribution pipework also must be purged with compressed air, before being recharged with ammonia gas. Care has to be taken with steam purging of ammonia lines, since ammonia gas is extremely soluble in condensed steam and the resultant collapse of pressure can cause severe pipework vibration.

Personnel safety equipment comprises an emergency shower and several sets of self-contained breathing apparatus. It is practice to fit at least one windsock to give a clear indication of wind direction close to the ventpot outlet.

20.2.2 Commissioning objectives

- To prove satisfactory operation of all controls, alarms, indication and protective devices.
- To pressure test the storage tank and all ammonia pipework.
- To prove that the personnel emergency shower is functioning correctly and meets the specified flow rate and water temperature of 0.9 litre/s at 30 to 35°C, for up to 10 minutes.
- To purge the storage tank and receive the first fill of liquid ammonia.
- To commission the evaporators and ensure that each can meet the specified flow rate.
- To prove that the steam and air purging facilities are effective.

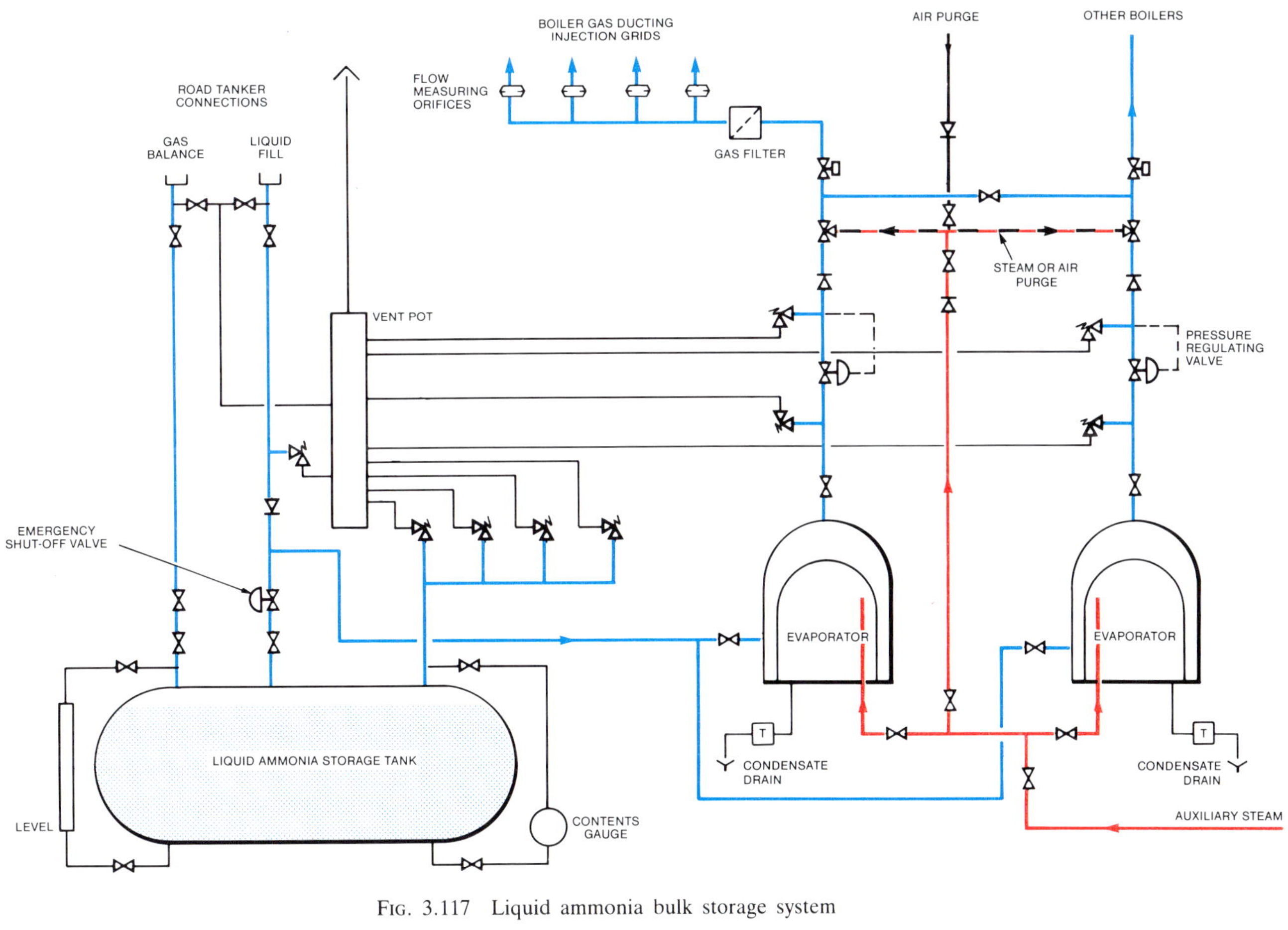

FIG. 3.117 Liquid ammonia bulk storage system

20.2.3 Safety considerations

Ammonia as a liquid boils freely in air at a temperature of −33°C and splashes on the skin produce severe cold burns. Gaseous ammonia is flammable in air at between 16% and 27% by volume and is extremely soluble in water. Even minute concentrations of 25 PPM can be detected by smell. Exposure to increasing concentrations results in greater irritation, and breathing becomes progressively more difficult. Rapid asphyxia may ensue in the event of extreme concentrations. The gas is lighter than air.

Before any chemical is delivered, comprehensive safety procedures for its receipt, storage and use must be approved. All equipment required to implement the procedures must be available. This involves fire hydrant supplies for hosing down spillages, breathing apparatus and protective clothing, an operational personnel emergency shower and several approved eyewash bottles.

20.2.4 Resources

Services

- Availability of auxiliary steam and general service compressed air must be guaranteed.
- A supply of C and I compressed air may be required to operate the road tanker outlet valves, but this will depend upon the supplier who has the contract for bulk supply of ammonia.
- Two 50 kg cylinders of gaseous ammonia.
- A bottle of sulphur dioxide (SO_2) will also be required for detection of ammonia leaks.
- Later, a road tanker containing 20 t of liquid anhydrous ammonia will be necessary.

Staffing

The commissioning period can extend to 10 or 12 days and involve up to four people. The attendance of a specialist from the ammonia supplier is necessary to inspect the installation, prior to receipt of the initial fill.

20.2.5 Commissioning

Current test certificates for the storage tank, evaporators, ventpot and all relief valves must be available. The bunded area should be filled with water, to ensure that the wall will retain a liquid spillage, and then it should be pumped out again. The compressed air system for the emergency shut-off valve is pressure tested, with no measurable loss, at some 7 bar for 2 hours.

The storage tank is isolated from the rest of the system and is then charged with general service compressed air to a pressure of 2 bar. If there are no leaks, then both 50 kg cylinders of ammonia are discharged into the tank and a further leak search carried out, using the bottle of SO_2 as a detector. In the presence of ammonia, SO_2 forms a white smoke. The air pressure is increased to 4 bar, and then to 6.5 bar, with an SO_2 leak search at each stage. When the leak search is acceptable, gas in the storage tank is used for charging (and leak testing) successive sections of pipework, extending from the tank up to the last valve on the distribution system. In order to check the final section of the distribution pipework for leaks, gas in the entire system is discharged through the boiler gas duct injection grids.

Subject to all controls, protective devices, etc., and the personnel emergency shower having been satisfactorily proven, the next stage involves bringing a road tanker of liquid ammonia to site for purging the storage tank.

With the gas balance line coupled, gas from the tanker is carefully passed into the storage tank, so that its pressure remains at approximately atmospheric. The storage tank is continuously vented to the ventpot, until the vent sample indicates a concentration of between 90% and 95% ammonia in air. At this stage the vent is shut, the storage tank pressure is balanced with that of the road tanker, and then delivery of liquid is carried out. The storage tank contents indicators are checked at 2 t increments during the filling operation.

The steam supply to both evaporators is commissioned and, when each is at working temperature, liquid ammonia is admitted. The ammonia pressure control valve is set to control at 2 bar, then gas is passed through to the boiler flue-gas ducting to prove operation of the flow indicators and the gas flow regulating valves for each duct. After activating as many of the protective devices as possible, the plant is shut down and a purging exercise carried out to confirm that a satisfactory purge can be achieved without excessive use of steam and compressed air.

20.3 Propane bulk storage

20.3.1 System description

The general arrangement of a typical propane bulk storage plant is shown in Fig 3.118. Propane is stored in unlagged mild-steel cylindrical storage tanks, typically of 10 t capacity each. The tanks are mounted on an elevated shallow concrete bund with a sloping floor, in an open area enclosed by a steel palisade fence. The pressure in the tanks may vary between 2.5 bar and 7.5 bar, in response to the normal ambient temperature variation.

Duplicate pressure relief valves, set at 13.8 bar, are arranged such that only one can be isolated at a time. Evaporation of propane within the tanks is by absorption of heat from the surrounding atmosphere. The gas distribution pressure is regulated to perhaps 2 bar, before being piped to the oil burner ignition equipment.

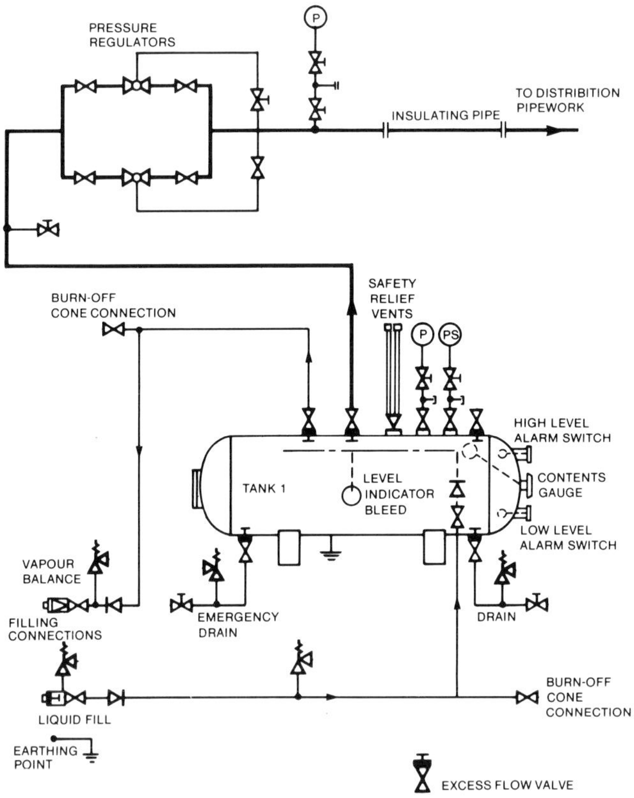

FIG. 3.118 Propane bulk storage system

All filling, drain and vapour take-off lines are fitted with special spring-loaded valves, which close in the event of excessive flow through them in the normal direction of flow.

The road tanker vapour balance line has a connection whereby a 16 m length of flexible hose with a burning-off cone can be fitted to flare off propane at a safe distance in an emergency, or for tank emptying. The contents indicator is backed up by a valved bleed-hole which emits liquid, rather than vapour, when the storage tank is full.

Electrical equipment must be either flameproof or intrinsically safe. All metalwork, including the compound palisade fence, must be effectively bonded together and bonded to earth electrodes local to the plant. The distribution pipework is bonded to earth but is electrically separated from the storage plant by either a section of insulated pipe or an insulated flange joint to prevent circulation of earth currents.

The storage area is protected by an automatic waterspray fire protection system, which is designed to keep the plant cool rather than to extinguish a propane fire.

20.3.2 Commissioning objectives

- To prove that the storage tank resistance-to-earth is less than 10 Ω.
- To purge and pressure test the entire system with nitrogen.
- To clear all moisture from the storage tanks with methanol.
- To confirm that the residual oxygen-in-nitrogen in the storage tanks is less than 5% and then fill them with propane.
- To prove the plant and ensure that it is capable of meeting the specified gas draw-off rate.

20.3.3 Safety considerations

Propane is 1.5 times heavier than air and disperses very slowly. It is flammable between 2.2% and 9.5% by volume and thus presents an explosion risk. Personal contact with liquid spillage may give rise to cold burns. Part of the commissioning procedure involves the use of the toxic and highly flammable chemical methanol, which must be handled with extreme care.

In the event of fire, the area must be cooled with watersprays and any propane fire allowed to burn until the gas can be shut off. On no account must foam be used to extinguish propane fires, as it is ineffective on liquid and may trap pockets of explosive gas.

The road tanker must be bonded to the storage area metalwork before any pipework connections are made.

20.3.4 Resources

Services

- Waterspray fire protection system commissioned and fire hydrants available.
- At least 30 m^3 of nitrogen may be required for leak testing.
- Approximately 100 litres of methanol within the storage tanks to absorb moisture, with a peristaltic pump for injection.
- Burning-off pipework and cone, plus propane gas torch with portable gas cylinders.

Staffing

Commissioning may require up to four people over an eight-day period. This will include the attendance of a specialist engineer from the gas supplier to inspect the installation prior to delivery of the initial fill.

20.3.5 Commissioning

The first stage is to confirm that there is safe access to the plant and that all warning notices and barriers are in position. Current test certificates for storage tanks and relief valves must be available. Instrumentation calibration and all earth bonding must be completed and the latter tested to give less than 10 Ω to earth. The insulation resistance of the distribution pipework insulated-flanges should be greater than 2 MΩ at 500 V DC.

The storage tanks and all pipework up to the outlet from the pressure regulating valves are charged with nitrogen to check for leaks and the pressure increased to 13.8 bar to confirm that all relief valves are correctly set. Next, the nitrogen pressure is reduced to 6 bar by purging from each vent in turn, and then a pressure test is carried out, with no measurable loss over a 2-hour period. The pressure regulating valves have to be reset temporarily to 3.0 bar, then the entire distribution system up to terminal points at the auxiliary and main boilers is charged to 3 bar, using nitrogen from the storage tanks. A pressure test at 3 bar, with no measurable loss over a 2-hour period, is then carried out.

Correct operation of the excess flow valves is checked by opening a valve downstream of each to atmosphere. The system is recharged with nitrogen, if necessary, and all the vents purged to confirm that there is less than 5% oxygen-in-nitrogen at each point, before the pressure regulating valves are reset to their normal working pressure, and the entire system vented to a pressure of 0.5 bar.

When delivery of propane is imminent, the water-spray fire protection system must be operational. Each storage tank is dried out by injecting the correct volume of methanol and the oxygen concentration is confirmed to be still less than 5%. Because moist mild steel reacts with propane, it is necessary to add a measured quantity of methanol to absorb any moisture within the tank, using 4.5 litres for every tonne of storage capacity. Methanol and propane are completely miscible, so there are no disposal problems after the propane tanks have been filled. The burning-off connection is installed, then the road tanker of propane is accepted and earthed. Each storage tank is isolated and charged in turn as follows:

- Gas from the road tanker is allowed to flow into the storage tank via the liquid filling line until the pressures equalise, then the line is disconnected.
- Liquid from the tanker is transferred into the storage tank until the pressure is approximately 2 bar greater than the propane-vapour balance pressure, corresponding to the ambient temperature (say 2 + 5 = 7 bar).
- After isolating the tanker, the gas in the storage tank is flared off until a good steady flame is achieved, then the burning-off branch is isolated and the gas balance pipework reconnected.
- Liquid is now transferred to fill the storage tank, and it is confirmed that the contents indicator agrees with the tanker delivery meter. The actual contents, when between 90% and 100% full, are verified by checking for liquid being emitted from the valved bleed-hole level indicator.

The whole exercise is repeated for the second tank. The distribution system, from the storage tank outlet isolating valves, remains charged with nitrogen until either the auxiliary or the main boiler oil burner test team requires a supply of propane gas.

At that stage, the oxygen concentration in nitrogen is again confirmed to be less than 5%, with further purging if required. Propane is then passed through the system until a good steady flame is established at each burner.

Finally, when sufficient burners are available, the ability of the system to cope with the designed draw-off rate has to be demonstrated.

20.4 Carbon dioxide bulk storage

20.4.1 System description

A typical plant is shown in Fig 3.119, and may comprise one or two thermally-insulated steel cylindrical storage tanks, each of 6 t capacity. Independent self-contained refrigeration units are provided for control of CO_2 pressure by circulating refrigerant through

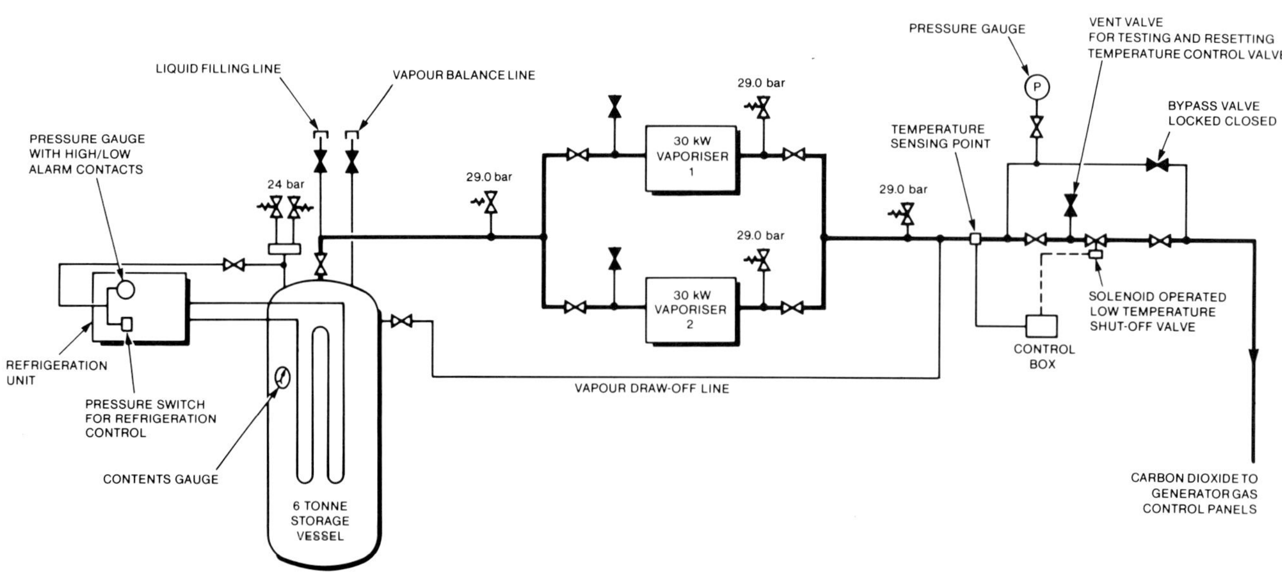

FIG. 3.119 Carbon dioxide bulk storage system

an evaporator coil mounted in the gas space of each tank. Duplicate pressure relief valves, set at 24 bar, are fitted to each tank. CO_2 is stored at its boiling point of −17°C at 21 bar and the refrigeration unit is set to maintain the pressure between 20.3 bar and 21 bar. Alarms are raised if the pressure goes outside the range of 12.5 to 22 bar. Should the tank pressure drop to 10 bar, the liquid may solidify. Equally if the tank temperature should rise, due to fire, there will be a rapid increase in pressure.

Two 30 kW electrically-heated vaporisers produce gaseous CO_2 at 19 bar and 55°C for distribution to the electrical generators and hydrogen driers for purging purposes. The common outlet from the vaporisers is protected by a low temperature shut-off valve, designed to close automatically if the CO_2 gas temperature falls to −10°C.

20.4.2 Commissioning objectives

- To prove the integrity of the plant by checking correct operation of all controls, alarms and protective devices.
- To purge and prove the plant with gaseous CO_2, confirming that it is leak-free.
- To carry out the first fill of liquid, and then prove that the refrigeration units function correctly, before commissioning the vaporisers.
- To prove the ability of the plant to purge one electrical generator within the specified time, using the appropriate quantity of liquid (e.g., 90 minutes at 270 kg/h).

20.4.3 Safety considerations

- CO_2 gas is 1.53 times heavier than air. It is non-toxic, but can cause asphyxiation if inhaled in large quantities. Inhalation of small concentrations stimulates breathing and may produce slight narcotic effects.
- The main hazard to personnel is in handling bare metal pipes which contain liquid carbon dioxide as this will cause cold burns.
- Filling hoses should be restrained in a harness to prevent them flailing about, in the unlikely event of a coupling or hose failure.

20.4.4 Resources

Services

No specialised equipment or services are needed other than a CO_2 relief valve test rig and a tanker of liquid CO_2 for the initial fill.

Staffing

Commissioning may require the services of three people over a five-day period. The attendance of the CO_2 supplier's specialist engineer will be needed to inspect the installation prior to receipt of the initial fill.

20.4.5 Commissioning

Current test certificates for storage tanks, vaporisers and all relief valves must be available. After all relief valves have been set using the special test rig, and instrumentation calibrated, the refrigeration units, vaporisers and low temperature shut-off valve are proved to operate correctly without any CO_2 in the system.

The CO_2 tanker is received on site and the gas balance connection coupled up; each storage tank is charged to 7 bar and purged of all air, using gaseous CO_2 only. The gas pressure is steadily increased to 21 bar and any leaks are identified and cured; then a pressure drop test is carried out to confirm no measurable loss over a 2-hour period.

Operation of the storage tank alarms and refrigeration unit controls is checked, and then the storage tank is filled with liquid. The contents gauge is checked against the tanker delivery rate.

The liquid CO_2 pipework within the storage area is charged up to the vaporisers; then operation of the vaporisers and the low temperature shut-off valve is proved by venting CO_2 gas to waste. During this exercise the distribution pipework is checked for leaks, but the pressure test is usually deferred until the distribution mains to the electrical generators can be commissioned.

Arrangements have to be made for purging CO_2 out of the remote end of the distribution system to displace all air and any debris in the lines. Having completed that stage, the entire distribution system is pressurised up to the available system pressure for a pressure drop test. Again, there should be no measurable drop in pressure over a 2-hour period.

Performance of each refrigeration unit should be kept under observation for the first week to ensure that the compressor is not running for an excessive length of time. In conjunction with the turbine-generator test team, the capability of the plant to purge one electrical generator casing, typically within 90 minutes at an average CO_2 flow rate of 270 kg/h, using only one vaporiser, has to be proved.

20.5 Hydrogen production and storage

20.5.1 System description

General

As an example of the principles involved, a plant sized for a 1000 MW power station (see Fig 3.120) will be considered. Hydrogen gas, produced by reacting a 64/36% mixture of methanol and demineralised water in hydrogen generator modules, is stored at pressure of up to 10 bar in a cylindrical LP storage vessel.

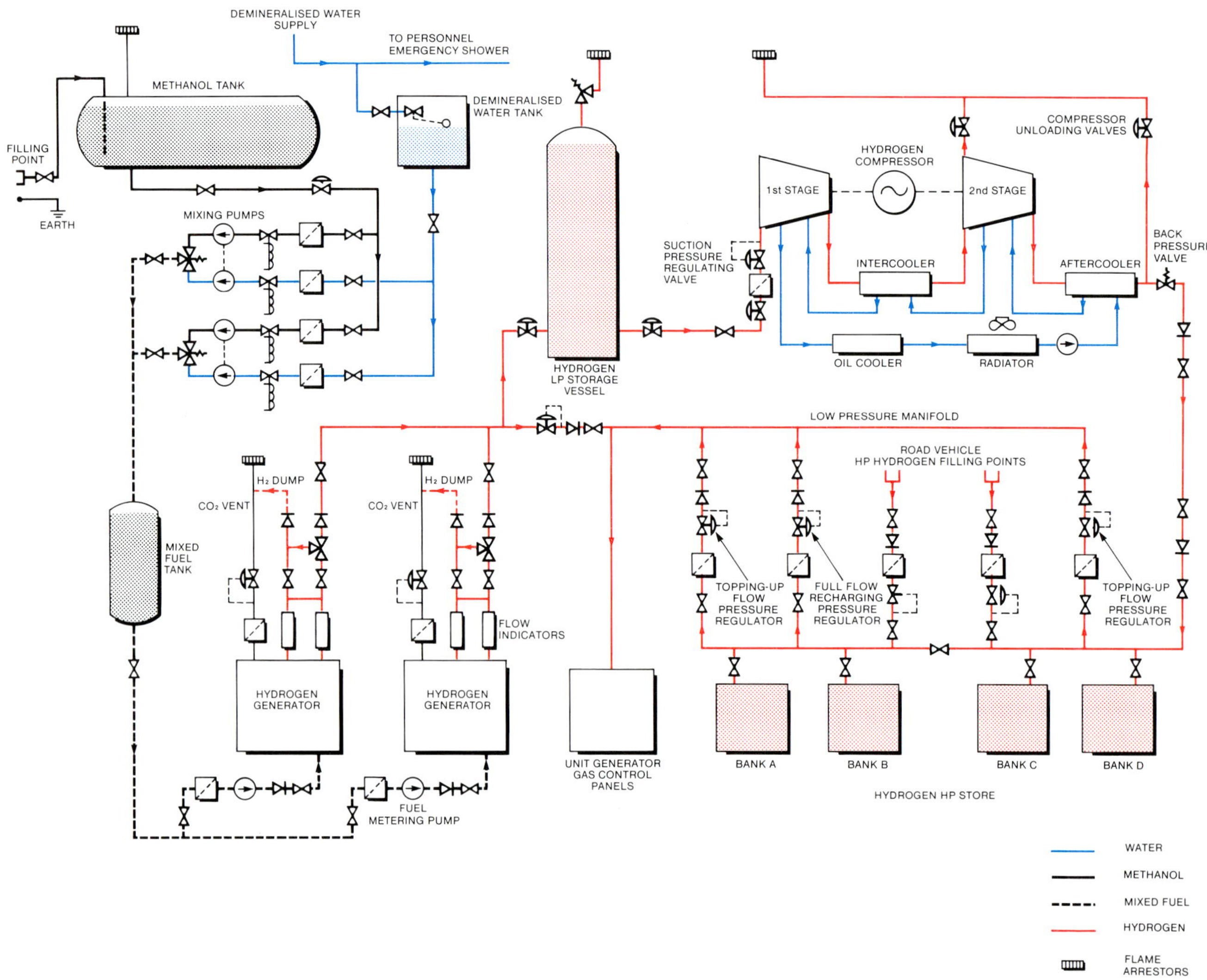

FIG. 3.120 Schematic diagram of hydrogen plant

Gas is withdrawn from the LP storage vessel, either as a direct feed to supply the on-load hydrogen requirements of the main unit generators, or to be compressed up to 172 bar for addition to the HP storage cylinders.

The HP store has facilities for accepting road tanker deliveries of gas at a pressure of 228 bar and reducing it to a pressure of 172 bar for storage. Other pressure regulating valves reduce the gas to a pressure of 10 bar for distribution to the electrical-generator gas control panels.

Automatic waterspray fire protection is provided for the methanol tank, LP storage vessel and the HP store, mainly as a coolant in the event of fire until the methanol or hydrogen can be isolated. Portable appliances, containing a special alcohol-resistant fluoroprotein-foam compound, are deployed to deal with fires involving methanol.

All metalwork associated with the plant must be effectively bonded together and to earth. Furthermore, methanol and hydrogen road tankers must be bonded to the plant before any hoses are coupled, to discharge any accumulation of static electricity from the road vehicle. If the plant is not naturally protected against lightning discharge by being within the shadow of an adjacent building, then special provisions will be required, as outlined in Section 9 of this chapter.

Hydrogen production plant

The 40 t methanol tank and 0.5 t demineralised water tank are both atmospherically-vented vessels. Duplicate positive displacement mixing pumps are provided to mix methanol and water in the desired proportions and then to discharge it into a stainless steel mixed-fuel tank.

Each hydrogen generator module, rated at 10 m^3/h, contains a positive displacement metering pump to pass fuel through a preheater into the electrically-heated vaporiser section. The resultant vapour is passed over a heated catalyst for conversion into CO_2, water vapour and H_2. Pure hydrogen is separated by diffusion and led to the LP storage vessel, whilst the CO_2 and water vapour are vented off to atmosphere, after giving up some of their heat to the incoming fuel in the preheater. The CO_2 vent-line and its pressure-regulating valve are electrically trace heated and thermally-insulated to prevent the formation of ice blockages in low ambient temperatures. The temperature of the vent gas has to be kept above 25°C, otherwise the pressure regulating valve will freeze due to the endothermic reaction when CO_2 expands to atmospheric pressure.

The two-stage diaphragm-type electrically-driven hydrogen compressor has its own pressure-fed lubrication system and a closed circuit cooling water system with an air-blast heat exchanger. The compressor takes in gas, via a 3 μm filter, followed by a suction pressure-limiter, which is set to a maximum pressure of 3 bar to avoid overloading of the motor. Three pneumatically-actuated valves are fitted to isolate the compressor suction branch and to vent both stages to atmosphere, when the compressor shuts down. The valves are time-delayed to operate in the reverse direction when the plant starts up, thus allowing the motor to reach full speed before any pumping load is applied. The discharge line from the compressor incorporates a back pressure valve, set at 150 bar, to maintain a fixed minimum pressure at the compressor outlet at all times.

The whole plant is designed for fully automatic operation and comprehensive instrumentation, controls and protective devices are provided. Control equipment is mounted on a 415 V motor-starter board and on a control panel, both of which are located in a non-hazardous (i.e., hydrogen-free) area close to the plant.

Compressed air for the pneumatic valves is derived from the unit controls and instrumentation compressed air system, though at some power stations a dedicated compressor/receiver/drier is provided for this service. In addition to the three valves associated with the H_2 compressor mentioned earlier, pneumatic isolating valves are also fitted to the methanol tank outlet and to the LP store inlet and outlet. These all close in the event of a plant trip arising from detection of fire anywhere on the hydrogen plant.

A personnel emergency shower is provided adjacent to the methanol storage and handling area should there be any spillage or splashes which may have caused personnel injury.

HP store

The HP store has a capacity of almost 10 000 m^3 of gas at STP (i.e., 0°C and 1 bar absolute) and comprises 900 interconnected standard-sized hydrogen cylinders. A control panel within the HP storage enclosure houses the 228/172 bar and 172/10 bar pressure regulating valves, along with pressure gauges and a 'HP storage pressure low' alarm pressure switch.

LP distribution system

Distribution mains to the electrical generator gas-control panels are led from the HP storage enclosure and are provided with an insulated flange joint adjacent to the coolant control panel for each electrical generator to prevent circulation of earth currents.

20.5.2 Commissioning objectives

- To ensure that valid test certificates are available for all vessels, tanks and relief valves.
- To prove that all controls and protective devices function correctly.
- To purge and pressure test all vessels and pipework with nitrogen.
- To commission all plant items, as far as practicable, before delivery of any methanol and/or hydrogen.
- To accept first delivery of methanol, then to prove operation of the fuel-mixing pumps.
- To commission each hydrogen generator module and to charge the LP storage vessel to working pressure.
- To charge part of the HP store from a road tanker and to prove the HP pressure regulating valves and LP distribution system.
- To commission the hydrogen compressor plant and to use it to complete the charging of the HP store.
- To prove that the production plant can meet the specified output under differing operational conditions.
- To prove that the entire plant is flexible in operation and can supply the hydrogen requirements for initial charging and for on-load make-up to the electrical generators.

20.5.3 Safety considerations

The main dangers are those concerned with the fire and explosion hazards associated with hydrogen, as well as those arising from handling methanol, which is toxic and highly flammable.

Hydrogen is the lightest known gas and forms an explosive mixture at concentrations between 4% and 74% hydrogen in air. Almost all areas of the plant are classified as either Zone 1 or Zone 2 dependent upon their location, and no form of spark-producing equipment must be permitted once methanol or hydrogen is on site. All electrical equipment must be suitable for the particular zone and gas, as specified in BS5345.

In common with all chemicals, approved procedures for the receipt, storage and use must be available and all necessary safety equipment to hand when required.

Personnel access should be carefully controlled and all involved should be fully aware of the dangers and of the appropriate action to be taken in emergency.

20.5.4 Resources

Services

- The waterspray fire protection, fire hydrants and special fire protection foam-making equipment must be available when required.

- Special HP test rig for setting pressure switches and relief valves will be required. This will involve a nitrogen booster pump to produce the high pressures necessary when using a standard nitrogen cylinder at a pressure of 160 bar.
- Test gases for checking H_2 purity and leak detection equipment.
- Burning-off connection and cone for purging HP store, or for use in an emergency to flare-off hydrogen.
- Plentiful supply of nitrogen for leak testing and vessel proving. The quantity lies between 500 and 800 m^3 dependent upon the success with which leaks are found and cured.
- Supply of treated water from the water treatment plant (see Section 16 of this chapter) for commissioning and hosing down purposes, and also for the personnel emergency shower.
- Supply of control and instrumentation compressed air for controls.
- Initial fill of 20 t of methanol. Road tanker of H_2, if applicable.

Staffing

Commissioning of an extensive plant like this could take from eight to twelve weeks of intensive effort by two teams, each of three or four people. A specialist from the hydrogen supplier will need to inspect the HP storage plant before the initial fill is received, and similarly before the first delivery of methanol is accepted.

20.5.5 Commissioning

Initially, all erection should be complete and non-destructive tests carried out on all joints of the high pressure pipework. Calibration of all instrumentation and protective devices, together with bonding of all metalwork to earth, should be complete. Figure 3.121 shows a simple network linking the main commissioning activities.

Pressure testing

For pressure testing purposes, the plant is divided into several sections. Each section is first purged with nitrogen to achieve a residual oxygen of less than 1%, and then is subjected to a pressure drop test as follows:

- HP store tanker delivery pipework up to the 228/172 bar pressure regulating valves tested at 342 bar for 2 h, with no loss of pressure.
- All H_2 high pressure pipework and storage cylinders tested at 186 bar for 24 h, with a maximum permitted loss of 1 bar.
- All H_2 low pressure pipework, including sections up to the main electrical generator gas control panels, tested at 15 bar for 24 h, with an acceptable loss of less than 350 mbar.
- Methanol tank (with vent temporarily blanked-off), and all methanol pipework up to the mixed-fuel tank, tested at 0.7 bar, with no measurable loss over 2 h. (The prescribed test pressure will depend upon which British Standard was adopted for the construction of the tank.)

All H_2 pipework and vessels are left charged with nitrogen at a pressure of at least 5 bar, until required for use. It is not customary to pressure test the H_2 compressor and generator modules as they have been tested at works; valid test certificates, including those for all relief valves and storage cylinders, should be available. The methanol tank is left sealed with the nitrogen pressure at just about atmospheric pressure until it is required. Care will be necessary to ensure that ambient temperature variations do not cause the methanol tank to become pressurised, or to be subjected to sub-atmospheric pressures. The precise restriction will depend upon which British Standard was specified for the construction of the tank.

General preparation

Electrical checks of cabling, controls and protective devices (including rotation checks of motors) are carried out, as well as proving the correct functioning of all control and protection systems. It may not be practicable to disconnect some of the motors from their pumps for a rotation check. However, they can be given a short spin, providing that the pumps have been charged with water. Proving of the interlocks between the compressor drive-motor and its auxiliaries can be carried out by allowing the water pump and the air-blast fan to run, but with the main motor drive-belts disconnected.

Compressed air and treated, or demineralised, water supplies are made available; then the pneumatic controls and anti-frost protection are commissioned. Operation of the personnel emergency shower (water pressure, temperature and adequacy of flow) is proved. Typically this may be a flow rate of 0.9 l/s at 30 to 35°C for up to 10 minutes. The demineralised water tank is filled, the alarm commissioned and the supply line to the mixing pumps flushed to waste. A temporary pipe is used to connect the mixing-pump methanol inlet-branch to the demineralised inlet branch; then the pumps are run to fill the mixed-fuel tank with demineralised water. This arrangement proves that the pumps, pipework and mixed-fuel tank are clear and leak-free. The system is preferably drained off afterwards, before any attempt is made to pump a methanol/water mixture, as this will prevent unnecessary waste of fuel before the requisite specific gravity can be achieved.

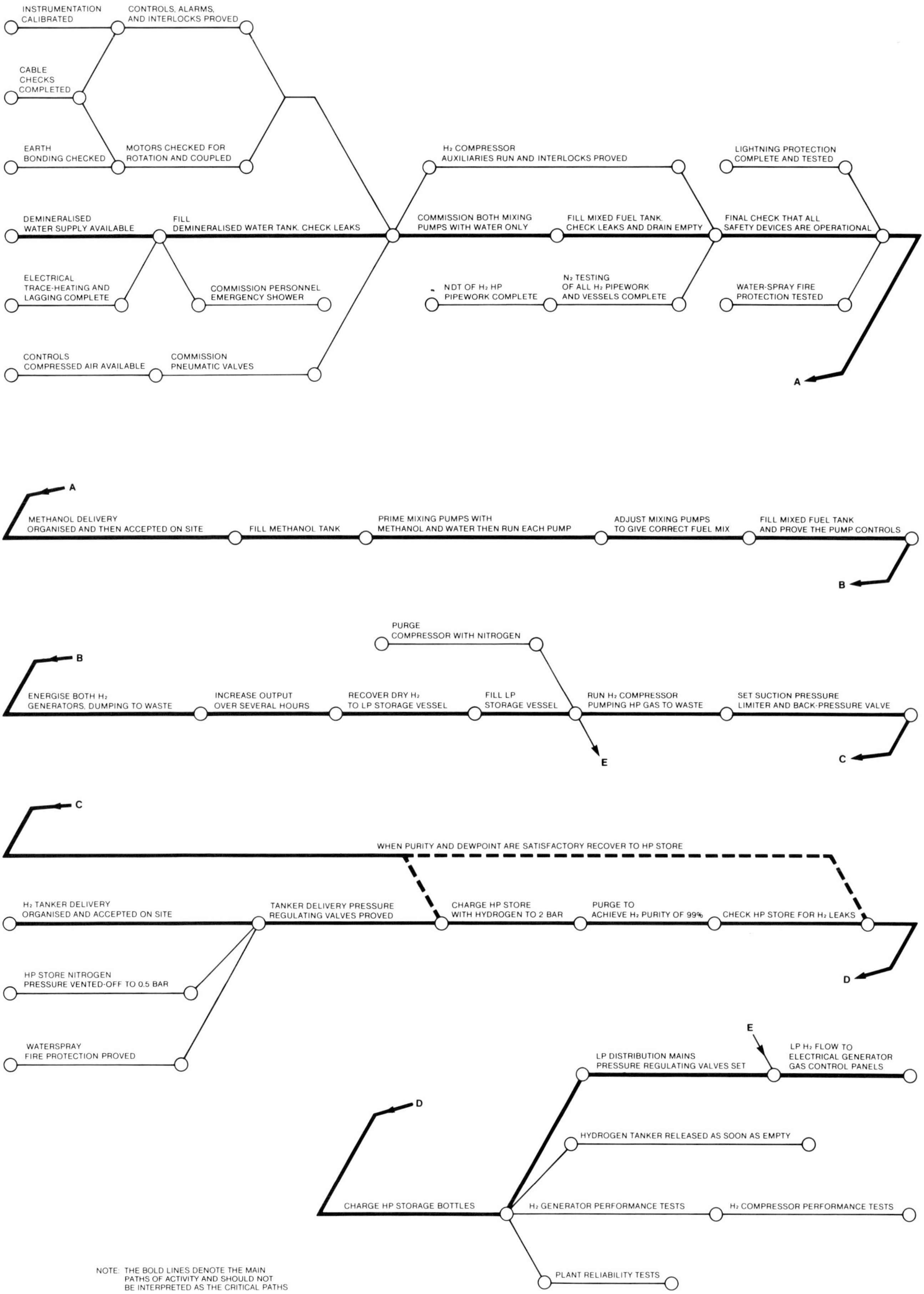

NOTE: THE BOLD LINES DENOTE THE MAIN PATHS OF ACTIVITY AND SHOULD NOT BE INTERPRETED AS THE CRITICAL PATHS

FIG. 3.121 Hydrogen plant commissioning network

Waterspray fire protection systems are proved (refer to Section 12 of this chapter) for coverage and adequacy of drainage from the area. This will also recheck the water-retaining capability of the concrete bund around the methanol tank.

After confirmation that the plant has attained an acceptable state of commissioning, deliveries of methanol and, if appropriate, hydrogen can be organised.

HP store

The HP store is isolated from the rest of the system, and then the nitrogen pressure is released down to about 0.5 bar. The tanker is accepted, earthed, coupled up and used to charge each bank of storage cylinders to approximately 2 bar, after which all vents are purged to atmosphere until a hydrogen purity of 99% hydrogen-in-air is achieved at each vent. At this stage, a preliminary leak test is carried out to ensure that no joint is leaking hydrogen. Gas remaining in the tanker is discharged into the storage cylinders to prove correct operation of the 228/172 bar pressure regulating valves. A further check has to be undertaken at several increments of pressure, as the storage pressure is increased up to 172 bar. Should it be uneconomic to use a tanker to commission the HP store, then it will need to be carried out over a much longer period of time, using gas from the on-site production plant. However, most of the checks mentioned are applicable, irrespective of which method is selected.

Hydrogen production plant

As a final preparatory check, overall protection tripping tests are repeated to prove that all protective devices function correctly. The temporary blank is removed from the methanol tank vent line, and then the methanol tanker is received, earthed and used to fill the tank with 20 t of methanol.

Methanol and demineralised water are supplied to the mixing pumps, both of which are run independently to fill the mixed-fuel tank. Samples are taken from the pump discharge branch and their specific gravity checked to determine the adjustments of the pump-stroke needed to achieve the correct fuel mix. Each H_2 generator module is put into service, and the fuel input increased in increments of 10% of metering pump stroke. Hydrogen gas is vented to waste until a gas dewpoint of better than −60°C (at atmospheric pressure) is achieved.

During early operation of the generator modules, it is important to check that the H_2 flow from the twin outlets is balanced and also that the heating elements are maintaining the desired temperature (350°C) with intermittent (rather than continuous) operation. The temperature of the vent gas outlet should be monitored. Samples collected from the vent gas drain-traps need to be checked for any methanol carryover. The LP storage vessel is then charged to its nominal pressure and then to overpressure, to test the vessel relief valve. Checks for leaks are carried out on all LP pipework and the LP storage vessel, as the hydrogen pressure increases.

The hydrogen compressor has to be purged with nitrogen until the residual oxygen is less than 1%; then the compressor and its auxiliaries can be run. Hydrogen from the compressor is discharged to waste until a hydrogen purity of 99% with an atmospheric dewpoint of better than −60°C is attained. During early operation of the compressor, a close watch is kept on motor current, lubricating oil pressure and cooling water temperature. The suction pressure limiter is checked to be regulating the pressure at 3 bar, while the discharge back pressure control valve is adjusted, if necessary, to maintain 150 bar.

High pressure hydrogen is then redirected into each bank of the HP store in turn, to increase the storage pressure to 172 bar. During this exercise, automatic operation of the plant at differing generator outputs can be checked, along with a preliminary assessment of generator and compressor capacity.

LP distribution system

The LP distribution pipework to the electrical generator gas-control panels should be rechecked to confirm that the residual oxygen-in-nitrogen is still less than 1%. If not, the pipelines have to be purged with nitrogen to achieve the desired standard: then each line can be charged with hydrogen from the HP store. During this exercise, satisfactory operation of the 172/10 bar pressure regulators is checked, and all pipeline vents are purged until a hydrogen purity of at least 99% in air is achieved. Samples of the gas at the outlet end of the LP distribution mains should be checked to ascertain the dewpoint of the gas, before any is admitted to the electrical generator casings.

System proving tests

In conjunction with the electrical generator test team, the LP hydrogen direct feed from the hydrogen production plant will be proved, as will the capability of the plant to charge an electrical-generator casing with H_2 (from CO_2 purity) within the specified time.

Finally, the production plant is subjected to a series of reliability and performance tests at various outputs to prove that it can meet the specified output and operational flexibility.

21 Compressed air systems

Modern power stations continue to utilise large volumes of compressed air, one of the most costly forms of energy, for a variety of applications. Several separate fixed-compressor systems are strategically located to meet the total demand for compressed air.

Typical usage of compressed air is for:

- Control and instrumentation (C and I): e.g., pneumatic valves, oil burner controls, etc.
- General services (GS): e.g., tools and general operational purposes.
- Boiler sootblowing: primarily main boilers, but sometimes auxiliary boilers too.
- Turbine forced cooling: for cooling the HP and IP cylinders, to reduce the time that the shaft has to be kept rotating before the turbine is cool enough for maintenance work to commence.
- Ash handling and disposal plants: for conveying and disposal of pulverised fuel ash (PFA).
- Fire detection.

This section deals with the principal commissioning aspects of the compressors and pipework systems for only the first four of the above services. Other compressed air systems are described in the appropriate plant section of this chapter, but they are commissioned in accordance with the principles outlined in this section.

21.1 System description

The pressures employed range from 6 to 14 bar, with individual compressor outputs of up to 5 m^3/s, dependent upon station design, though outputs of up to 8 m^3/s are not uncommon. Reciprocating compressors are almost always used for lower outputs, screw compressors for the intermediate, and multi-stage centrifugal compressors for the higher outputs. Vane compressors are generally adopted for portable equipments; though, by improvements in design, they are challenging the established role of reciprocating compressors in modern power stations.

For specific applications, very clean dry compressed air is required. Though the bulk of the drier plant at present is of the heatless automatic-regeneration dessicant type, operating on a fixed time cycle, driers of the refrigeration/condensing type are becoming common.

21.1.1 C and I compressed air systems

The standards for the quality of compressed air for pneumatic control equipment are much more stringent than those for most other applications. Typically the specification calls for compliance with the following parameters:

- Moisture content to be so low that the air has a dewpoint of at least 10°C lower than the lowest ambient temperature expected, when measured at the nominal system pressure.
- Oil content to be less than 1 PPM (wt) of oil at 20°C, again at the nominal system pressure.
- Dust particles must not be larger than 1 μm.

A typical system, as shown in Fig 3.122, comprises two or three 100% duty two-stage reciprocating oil-free compressors, complete with suction filters, two 100% duty air driers and two storage receivers for every boiler/turbine unit. Each compressor may have a rating of 0.2 m^3/s at 8.5 bar and is driven by a motor connected to a 415 V unit essential services switchboard. Unit compressed air systems often can be interconnected to that of an adjacent unit in an emergency only, since the failure of one unit system should not affect that for an adjacent unit. In addition, there has to be sufficient receiver capacity to meet the compressed air requirements of each main unit for up to 10 minutes should there be a complete failure of compressors.

Branches may be fitted to the distribution pipework to supply the compressed air requirements of station plant control equipment, if separate compressor systems have not been installed for this duty. Large coal-fired power stations may have one (sometimes two) similar compressor installations to meet the compressed air requirements for control of station service equipment, such as water treatment, ash and dust and auxiliary boiler plants.

Coolant from a service water system is required to cool the compressed air in the compressor intercooler and aftercoolers, and perhaps the compressor lubricating oil as well.

Compressor inlet valves are fitted with time-delayed pneumatic actuators, which hold the inlet valves open during start-up, then allow the valves to operate normally when the compressor is up to speed. The control equipment usually provides for each compressor selected to 'duty' to be running continuously, but to be unloaded and loaded within a specified band of system pressure by automatic operation of the inlet-valve actuators. Generally, the three compressors are arranged for a first, second and third 'duty' mode of control, such that the second and third 'duty' compressors will start in turn should the system pressure progressively decay below the normal operating band.

Dessicant-type driers usually have two enclosed vessels or columns charged with activated alumina beads, which adsorb moisture from wet air passed over the beads. While one column is in service on air drying duty, a small proportion of the dried air is passed back through the out-of-service column to regenerate the dessicant before exhausting to atmosphere.

Some driers incorporate electric heating elements to assist the regeneration process. Most, however, are heatless and work on the principle that expansion of dry compressed air to atmospheric pressure makes the air extremely dry indeed. When this air is passed over moist dessicant, it rapidly reabsorbs moisture and carries it off to atmosphere. At predetermined time intervals, pneumatic valves operate so that the columns change over duty, and the cycle repeats.

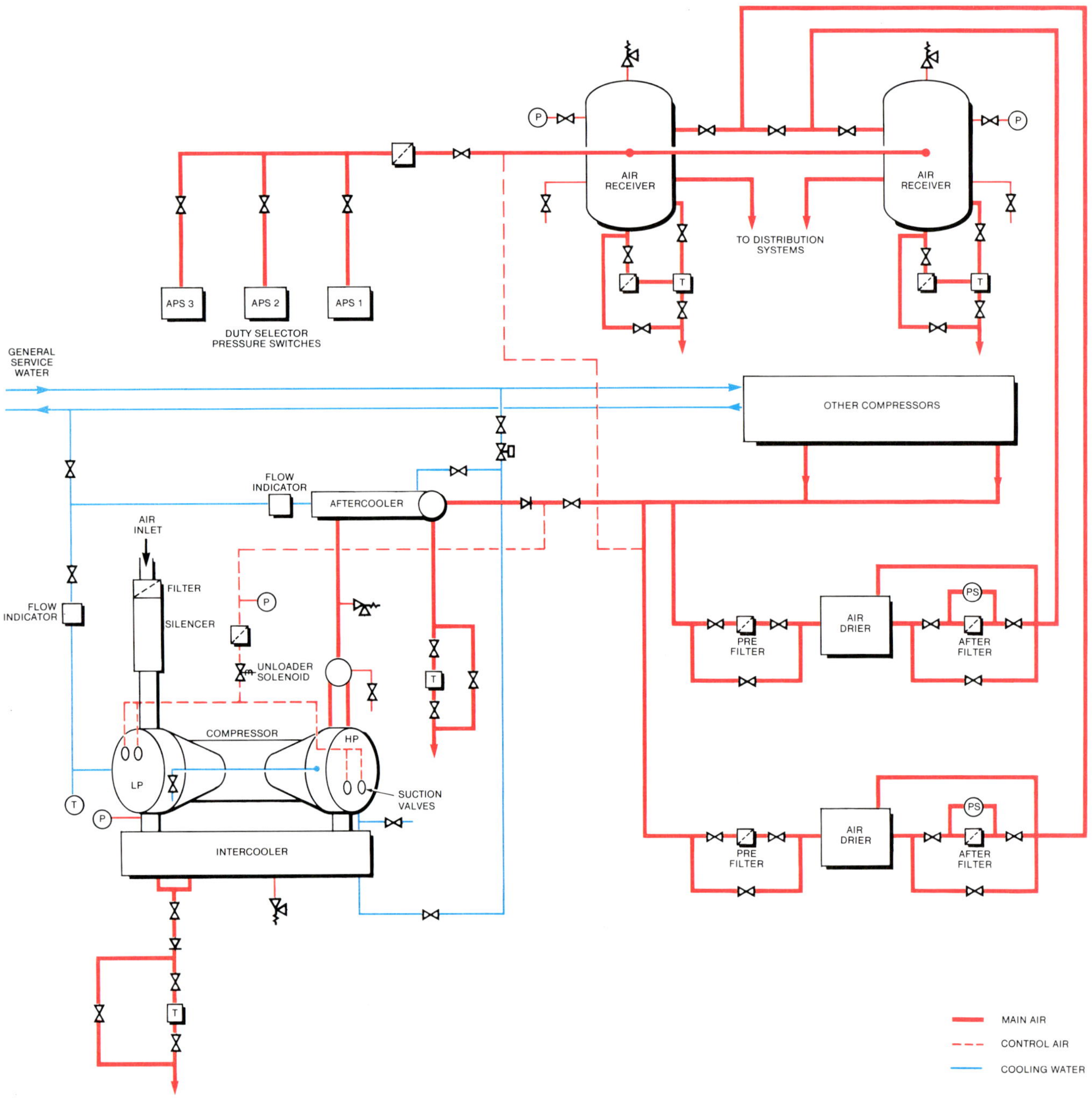

FIG. 3.122 Control and instrumentation compressed air system

One type of dessicant drier incorporates a microprocessor-based control scheme, which matches the regeneration cycle to the actual moisture load and ensures that the purge airflow is shut off when regeneration has been completed.

The refrigeration/condensing type of drier employs a refrigerant in a closed circuit to cool the compressed air to its dewpoint and then removes the moisture by centrifugal action in a vortex separator, before the dried compressed air is passed to the air receivers.

Fine filters are usually fitted before and after the driers. Typically, the inlet filters limit the particle size to less than 25 μm, whilst the outlet filter allows only particles of less than 1 μm to pass. It is an established fact that when activated alumina becomes saturated, as a result of the drier controls failing, the beads disintegrate into a fine abrasive powder which rapidly damages pneumatic equipment. The 1 μm outlet filter effectively traps the dust but, as a consequence, it blocks very quickly and could cause air failure. Some designs incorporate an automatic compressed air reverse-flow feature to purge the outlet filter and then collect the dust in a separate vessel for disposal.

21.1.2 GS compressed air systems

Air is used to power pneumatic maintenance tools and for selected operational functions, where use of high quality C and I air cannot be justified. Examples of operational uses are:

- Purging of some types of oil burner.

- Blowing out blocked boiler draught-gauge impulse lines.
- Air cooling for boiler furnace flame-viewing television cameras.

Outlets for tools normally terminate in self-sealing hose couplings. These are strategically located all over the unit and station services plant areas, including the maintenance workshops.

Since the quality of the compressed air is not so demanding, oil-free compressors are not normally necessary. In fact, the presence of a small quantity of oil in the air stream may be beneficial. Filtration for dust is usually limited to an inlet filter for each compressor, similar to that provided for the C and I compressors.

A typical system as shown in Fig 3.123 comprises two, three or four two-stage double-acting reciprocating compressors, supplying a pipework ring main via at least two air receivers. Each compressor may have a rating of 0.4 m^3/s at 7 bar and is driven by an electric motor connected to a 415 V station services switchboard. Some installations, where the steady air demand is higher, are provided with screw compressors, perhaps of 1.0 m^3/s capacity and driven by 3.3 kV motors.

Another less common type is a multi-stage single-shaft reciprocating compressor. This is designed to compress air successively to several distinct pressures, which are individually bled-off for different services. Each stage is provided with an automatic off-loading or 'surplussing' control system to maintain a minimum airflow through each stage of the machine at all times. A typical example of this type of compressor is rated at 8 m^3/s, where 1.2 m^3/s at approximately 2 bar would be used for dust plant conveying air, 2.3 m^3/s at 7 bar for GS purposes and other dust plant functions, and 4.5 m^3/s at perhaps 14 bar for boiler sootblowing. A compressor with this rating would have its drive motor connected to an 11 kV station board.

GS compressor plant is normally situated at one, sometimes two, locations on the ground floor, adjacent to the boiler/turbine house. The compressor control scheme is usually arranged to cater for automatic starting of each compressor in cascade, as a function of air system pressure, with automatic loading and unloading within a band of 6.5 to 7 bar. In some cases, the first duty compressor may stop automatically, and shut off its cooling water, if there is no air demand within a specified period of time.

Coolant from a service water system cools the compressed air in the compressor intercooler and aftercoolers, and also the compressor lubricating oil and drive-motor air cooler.

It is not usually necessary to commission this plant at an early stage for plant construction purposes, but it must be available to commission the first boiler/turbine unit.

21.1.3 Sootblowing and/or turbine-cooling compressed air systems

This plant provides compressed air for boiler sootblowing and for turbine forced-cooling purposes. A typical system (see Fig 3.124) may have a two-stage oil-free screw compressor, feeding its own air receiver for each boiler/turbine unit. Facilities for interconnecting with adjacent similar systems are usually provided.

Typically, the compressor is rated at 2.2 m^3/s and is driven by an electric motor connected to a 3.3 kV station board. The compressor unit has a separate forced-lubrication system, and each stage is driven via its own gearbox from a common double-ended drive motor. The compressed air pressure is 14 or 17 bar for coal-fired boiler sootblowing and perhaps 12 bar for an oil-fired boiler. This pressure subsequently would be reduced to about 7 bar for turbine cylinder forced-cooling purposes.

Some coal-fired power stations may take the turbine forced-cooling air requirements from an enlarged capacity GS compressed air system. In this case, duplicate 11 kV motor-driven five-stage centrifugal compressors rated for 5 m^3/s at 17 bar discharge pressure are fitted for sootblowing only. Each stage of this compressor has its own intercooler.

Since screw compressors are very noisy, several silencers are provided in the compressed air circuit. Frequently, the whole unit is also enclosed in a sound-absorbent housing, which is fitted after the plant has been commissioned so as not to impede access during testing.

Coolant from a service water system cools the compressor casings, intercoolers, aftercoolers and the forced lubricating oil systems. The drive-motor air cooler for the larger-rated compressors is water cooled.

The compressor air discharge line is usually provided with a by-pass, incorporating a pneumatically-operated modulating surplussing valve, which discharges excess air to atmosphere to maintain a preset air pressure at the receiver. The surplussing valve opens when the air demand is low, so that there will always be a minimum flow of air through the machine. Compressed air from the C and I system is used to control this pneumatic valve.

21.2 Commissioning objectives

The following are specific for compressed air systems and should be read in conjunction with those outlined at the start of this chapter:

- To ensure that each compressed air system is available for reliable operation before the main plant needing the service is due for commissioning.
- To confirm that the pipework is correctly installed, with particular regard to drainage of moisture, and that it is leak-free.

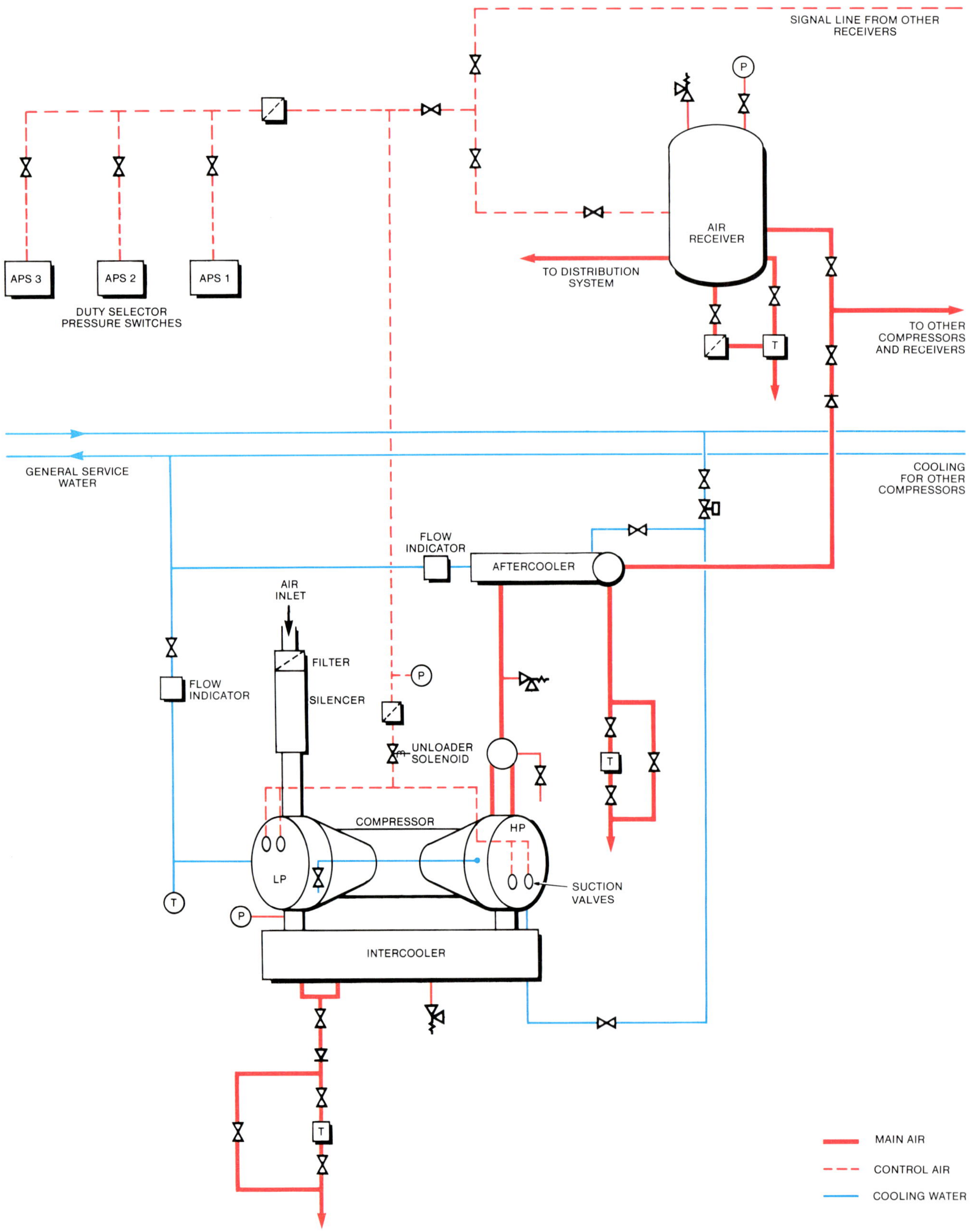

FIG. 3.123 General service compressed air system diagram

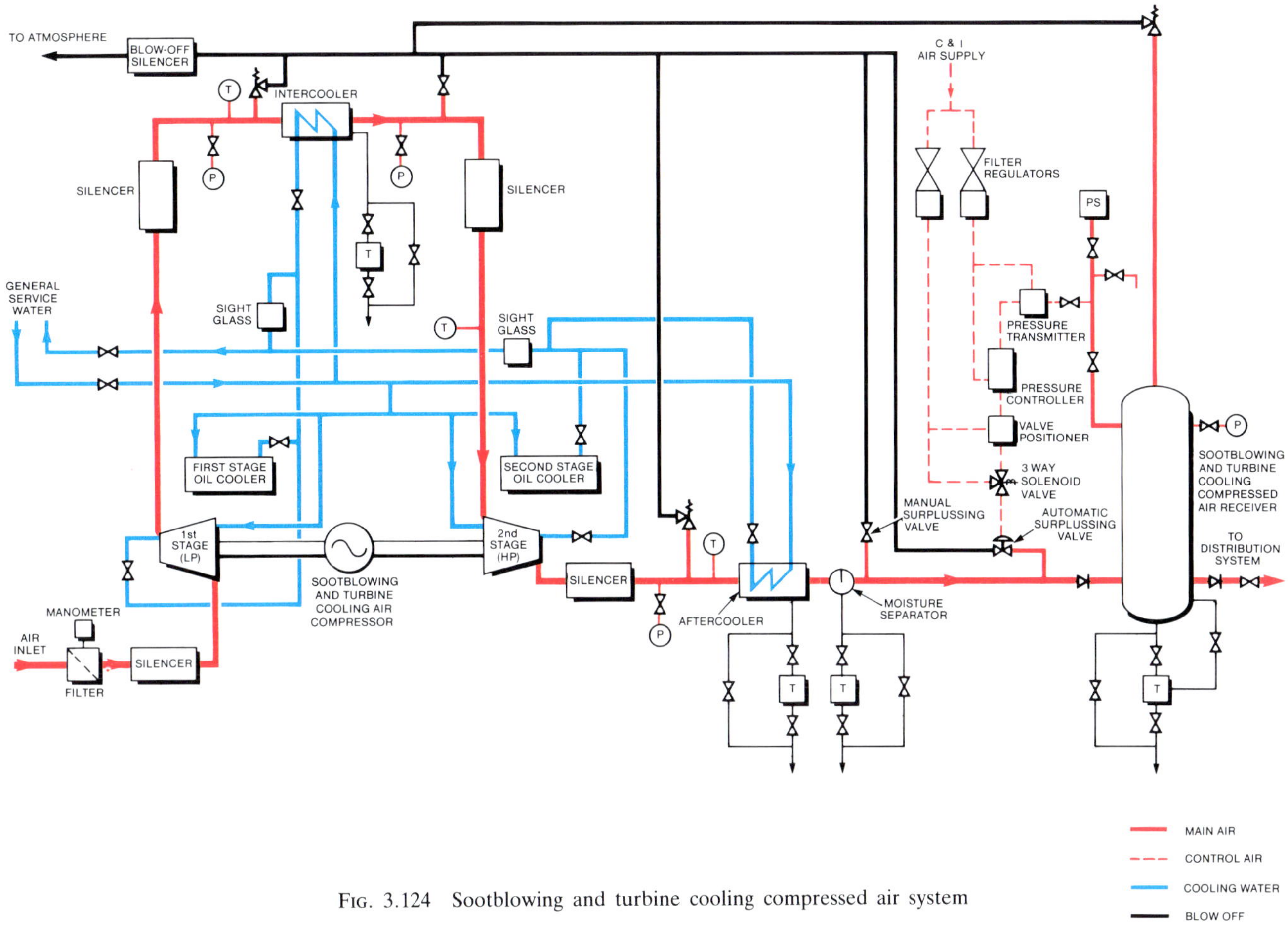

FIG. 3.124 Sootblowing and turbine cooling compressed air system

- To confirm that all controls, alarms and protective devices function correctly.
- To check that the plant attains its specified output.
- To prove that, where appropriate, the air quality meets the specification.
- To ensure that the operational noise levels are within the specified tolerances.

21.3 Safety considerations

The following points should be taken into account, in addition to those highlighted at the beginning of this chapter:

- The stored energy of compressed air should be recognised as a safety hazard and suitable precautions taken to prevent danger to personnel and plant.
- Compressors generate relatively high noise levels, particularly when lines are being purged or vented, so hearing protection must be used by all personnel working in the vicinity of the plant.
- Special care must be taken whilst blowing-out compressed air lines to prevent injury to personnel or plant by debris ejected from the pipeline.
- Pneumatic testing of pipelines must be carried out by gradually increasing the pressure in stages up to the desired test value, whilst checking for leaks at all stages.
- All relief valves and pressure vessels must have valid test certificates acceptable to the insurers before the plant is pressurised.

21.4 Resources

Services

Apart from electrical supplies, the other essential resource is cooling water from the appropriate service water system before commissioning can be progressed.

Staffing

Each plant area will require a team of two or three people to carry out the necessary checks. The overall programme time will be a function of release of plant for commissioning, but individual compressors may take from three to seven days to prove fully.

21.5 Commissioning

The principles outlined are based on commissioning of a C and I compressed air system, but they are applicable to any compressor system unless otherwise noted. Interaction of the main activities is shown on the simplified commissioning network given by Fig 3.125. Initially, the commissioning team has to be satisfied that certain erection activities have been carried out, for example:

- All plant and pipework installed to drawing and completed up to defined lockable terminal points.
- Works test certificates available for all items having statutory requirements.
- All power and control cabling installed to drawing and all electrical equipment adequately bonded to earth.
- All necessary instruments and protective devices calibrated and set at their correct operating values.
- Compressor/drive motor shafts correctly aligned, but uncoupled.

21.5.1 C and I compressed air systems

Control scheme checks

The compressor electrical control supplies are energised and all modes of control proved with the main drive-motor isolated. Most of the protective devices and controlling functions may have to be simulated at this stage, though it is better to operate each device where practicable.

Correct setting of the compressor drive-motor electrical protection in relation to the motor rating has to be confirmed (see Section 4 of this chapter). After the drive-motor insulation resistance and phase balance have been checked, the motor is energised for a short period to check for correct direction of rotation and, where appropriate, location of the magnetic centre. The compressor is then coupled, the lubricating oil system primed and the compressor unit checked for freedom of movement by manual rotation of the shaft.

By this stage, all remote controls, protective devices, indicators and alarms should have been proved in readiness for plant running tests. Air drier controls should be proved, as far as is practicable without any compressed air being available. Service water cooling is made available: all components are checked for absence of leaks and for an adequate flow through all parts of the cooling circuit.

Initial running checks

One compressor, a drier and an air receiver is set up for testing first, with the rest of the plant and the pipework

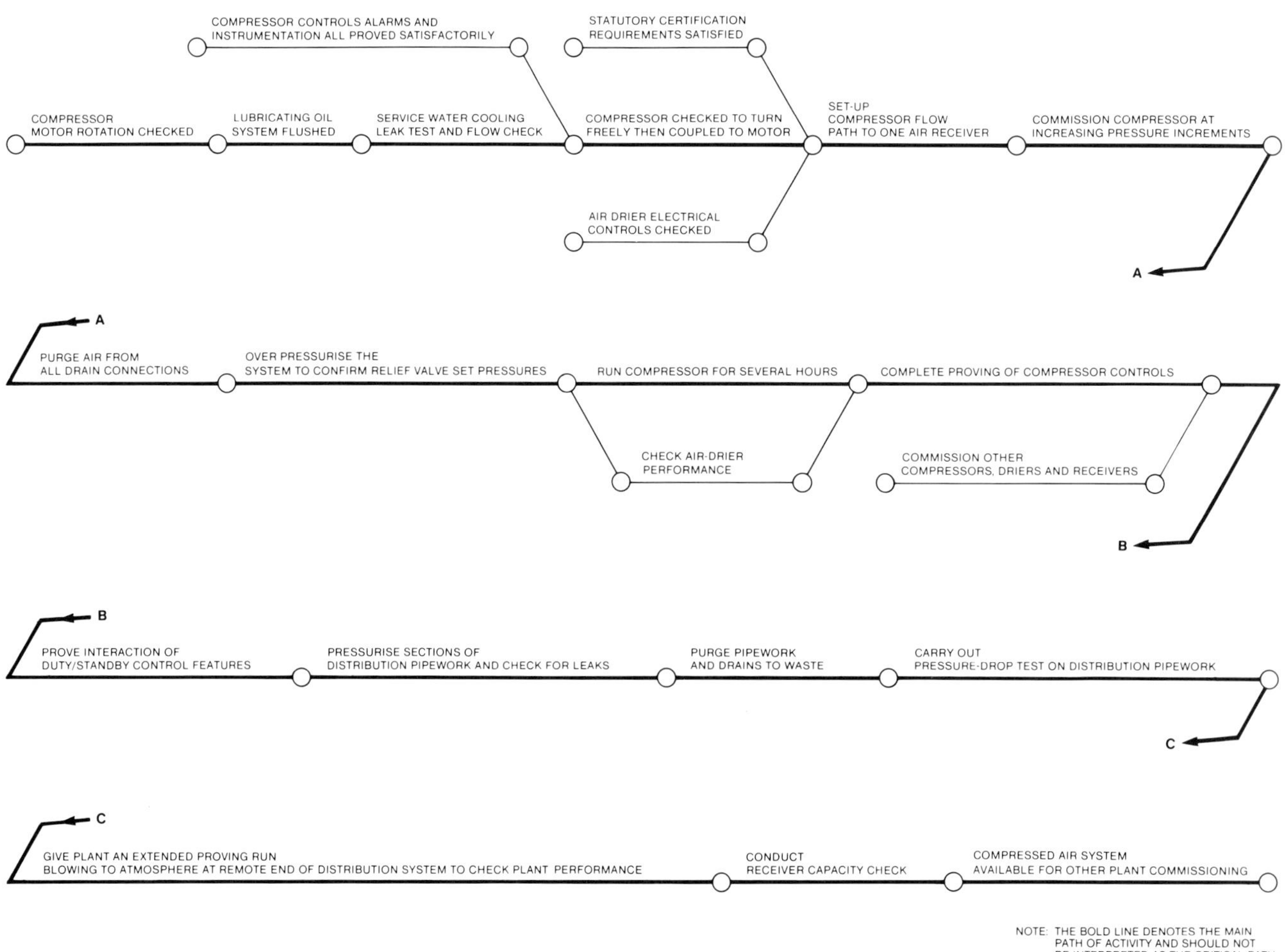

FIG. 3.125 Compressed air system typical commissioning network

system isolated. The compressor automatic loading control pressure switches, which typically operate in a band of 8.1 to 8.5 bar, are temporarily adjusted to approximately 2.6 and 3.0 bar.

The compressor is then started and allowed to charge the drier and receiver, control of pressure being exercised by manual operation of the air receiver drain valve. The system is allowed to operate for some 30 minutes at this low pressure, and the plant is kept under close observation. Any pipework leaks are identified and air drier operation is checked, if practicable.

Over the next hour, the control pressure switch settings are increased in increments up to their nominal values. After blowing compressed air through all available drains to prove that they are clear of dirt, swarf and any accumulated moisture, the system is carefully overpressurised to check the design settings of the various relief valves. When the system pressure has been returned to its nominal value, each of the compressor protective devices is operated to prove that the compressor will trip in event of a fault condition arising.

After a further period of operation, the compressed air dewpoint at the inlet and outlet of the air drier is measured, then auto-changeover of the drier unit observed before rechecking the dewpoint. In each case, the outlet dewpoint must be at least −25°C (at system nominal pressure) even at the point where the drier unit is due to change over. All of these tests are carried out using every compressor, drier and receiver in turn. The interaction between compressor controls should be proved, e.g., first duty/second duty/third duty, in accordance with the overall control philosophy for the installation. When all compressors have been commissioned, it is prudent to confirm that the loaded compressor does not recirculate its air through the non-return valves of off-load machines.

Distribution system checks

With a compressor in service and an air receiver charged up to working pressure, a manageable length of pipework downstream of the receiver is slowly pressurised up to its nominal pressure whilst a regular search for leaks is undertaken. Providing no serious leaks are identified, all available drains are blown to waste to clear any accumulation of condensation, swarf or debris. Other manageable sections of the pipework system are treated in a similar manner. Where practicable, each terminal point valve is blown through in turn, working outward from the air receivers. Whilst blowing out the lines, care should be taken to ensure that the system pressure is kept between 60% and 100% of its nominal value to achieve effective movement of debris. Arrangements should be made to collect a sample of the debris to measure the effectiveness of the blow. A pressure drop test is conducted over a 24-hour period, with the pipework system charged to nominal pressure then isolated. After compensation for temperature variations, there should be a negligible loss of pressure. Typically, a loss of not more than 15 mbar over 24 hours should be achievable.

System proving checks

Finally, the plant is given an extended proving run of at least six hours, with compressed air blowing off from a selected outlet at the remote end of the pipework system. The system pressure is regulated to maintain the compressor (or compressors) within the normal automatic pressure control band. During the extended run, the outlet air is sampled at intervals to check that the plant meets the specification in respect of dryness, oil content and dust entrainment. This is followed by a receiver capacity check, with the compressors isolated from the receivers. The time taken for the receiver pressure to drop to 50% of its nominal value is checked, with a system airflow equivalent to 100% air demand. This time should be in excess of ten minutes.

Upon completion of this stage, the system can be released to other commissioning test teams engaged in various plant areas needing supplies of compressed air for control equipment commissioning. The test teams will be responsible for blowing out their own sections of pipework, up to the controlled equipment.

21.5.2 GS and sootblowing compressed air systems

The test procedures follow those outlined above, with the deletion of air drier commissioning. In addition, it is most important that automatic oil drain traps fitted to the compressor intercoolers and aftercoolers are proved to operate correctly, lest accumulation of oil vapour in the hot airstream creates an explosive mixture. Even greater attention has to be given to ensuring adequate safety for plant and personnel, since commissioning is often taking place in areas which are at a much less advanced stage of construction.

Lubricating oil systems are flushed, using the normal service grade of oil heated by means of a temporary electric heater unit. The oil is circulated either by temporary or permanent oil pumps, until the normal service filters are observed to be clean. Initially, this circulation is established for up to six hours, with the supply lines diverted to by-pass the bearings, gearbox, etc., and connected into the return lines. The supply and return lines are then restored to normal and the whole system circulated again for up to six hours.

Fine-mesh gauze filters are fitted between flanges in the air suction to the appropriate compressor stages. After completion of its electrical tests, the motor is energised for a rotation check, confirmation of its magnetic centre position and then coupled. The service water cooling circuits are charged to check for leaks, followed by confirmation that there is an adequate flow through each circuit.

The compressor is then started, with surplussing or unloading valves kept open and with the air receiver in circuit up to its outlet isolating valve. Initially, the plant is run for five or six hours, then the gauze filters are inspected, cleaned and refitted. The purging runs continue until the temporary filters remain clean, after which they can be removed.

The plant is restarted to prove the protective devices and then, by regulation of the manual-unloading valves, control of the surplussing valve is checked and adjusted to give the best response. Correct operation of the compressor discharge non-return valve should be proved by reversing the flow from the pressurised air receiver, with the compressor shut down.

The compressed air distribution main is pressurised and checked for leaks, then the system subjected to a pressure loss test, as for any other compressed air system.

The air mains are blown out at all remote terminal points and the air quality sampled to provide datum values for dust, oil and moisture content.

Dynamic performance of the compressors and the controls must be proved in conjunction with the test teams commissioning the various plant systems requiring the compressed air, e.g., boiler sootblowers. The majority of this work is covered in Chapter 6, though some preliminary testing will be dealt with in Chapter 4.

C H A P T E R 4

Boiler pre-steam-to-set commissioning

Introduction

This chapter deals with the approach to commissioning of all boiler plant from completion of erection up to steam-to-set conditions: this will include all activities from the boiler hydraulic test onwards. Commissioning of the boiler from steam-to-set is covered in Chapters 5 and 6.

1 Scope

This chapter will cover the following main boiler areas:

- Boiler hydraulic test.
- The pressure parts, including high pressure (HP) pipework, blowdown drains, chemical sampling and chemical dosing.
- Boiler air and gas systems, from forced draught (FD) fan inlets to the chimney exit.
- Boiler firing equipment.
- Fuel oil plant for oil- and coal-fired stations.
- Sootblowing.
- Control and instrumentation, including interlocks, protection and sequence control systems.
- Initial boiler operation.
- Boiler chemical cleaning.
- Boiler and high pressure (HP) pipework steam purging.
- Floating of boiler safety valves.

1.1 Content

As a general rule, each section will cover the following:

System description This gives the reader sufficient detail to understand the commissioning activities. More detailed information will be found in other volumes of this publication or manufacturers' literature.

Safety considerations Where particular hazards associated with the system exist, the safety considerations are highlighted.

Resources If additional resources are required to progress the commissioning, these are highlighted.

Commissioning The basic principles of the activities required to set the plant to work are outlined to meet the specific objectives applicable to the system involved.

General commissioning objectives are:

- To make plant available for commercial operation as soon as possible, consistent with contractual requirements and statutory obligations.
- To establish that the plant meets the Station Development Particulars.
- To demonstrate that the plant is capable of safe and reliable operation.
- To identify any limitations of performance or flexibility of operation.

A simplified overall commissioning programme for a coal-fired boiler is shown in Fig 4.1. An oil-fired boiler commissioning programme is similar, but without the milling plant.

2 Post-erection checks

On completion of erection in a particular plant area, the GDCD engineers in charge of the contract carry out an inspection of the plant to check that it has been built to design requirements and that it is sufficiently complete for plant completion and station commissioning activities to take place.

The plant is inspected for safety purposes, to ensure that all construction debris is removed from the commissioning areas and that safe access and good lighting are provided. The plant should be suitably identified, whenever possible, with its permanent identification labels, but adequate temporary labels must be used if this is not possible.

Any terminal points, where the plant interfaces with other plant areas not yet ready for commissioning, are locked closed with locks and chains, and suitably identified as 'commissioning terminal point isolations', in accordance with the procedures discussed in Chapter 2.

A Safety Rules Clearance Certificate (SRCC) is raised, when necessary, to notify all contractors and CEGB personnel that the plant is to be pressurised or charged, and that operational services such as steam, water, fuel oil, electrical supplies, etc., are to be made available. The date and time that these operations are to commence must be clearly stated.

Before any commissioning starts, the working party responsible for commissioning the plant area will ensure that any services required are available and that any relevant safety systems are in service. For example, fire protection equipment should be in service before filling pipework with fuel oil and adequate fire extinguishers of the correct type should be available and placed in predetermined areas.

The members of the commissioning working party and test teams must be made aware of any safety considerations which are of particular importance, such as the risk of fire when commissioning fuel oil plant or the risk from injury due to handling corrosive chemicals during the chemical clean programme.

3 Hydraulic testing

3.1 Introduction

The boiler hydraulic test is generally carried out before lagging is finished and while reasonable access is available for inspection and repair. The pressure circuit from the economiser inlet to the superheater outlet, including all welded attachments, drain pipework and blowdown vessels must be completed. The procedure laid down for hydraulic testing must be strictly observed, as three drum type vessels have failed in recent years whilst undergoing hydraulic tests. Because of these failures, it is now recommended that increased quality control should be applied during boiler drum manufacture and that the number of full overpressure tests before operation should be limited to one at the works and one at site, wherever possible. Increased safety precautions should be taken during the actual testing; these are described in Section 3.2 of this chapter.

Pre-operational hydraulic testing should be carried out as near to the chemical clean stage as possible, so as to avoid the boiler remaining with wet surfaces for long periods, which would give rise to corrosion problems; it is also essential to protect the boiler from frost if ambient temperatures fall below freezing.

3.2 Limiting requirements

3.2.1 Pressure

Test pressure The hydraulic test pressure depends on the works testing. If this was at 1.5 times the design

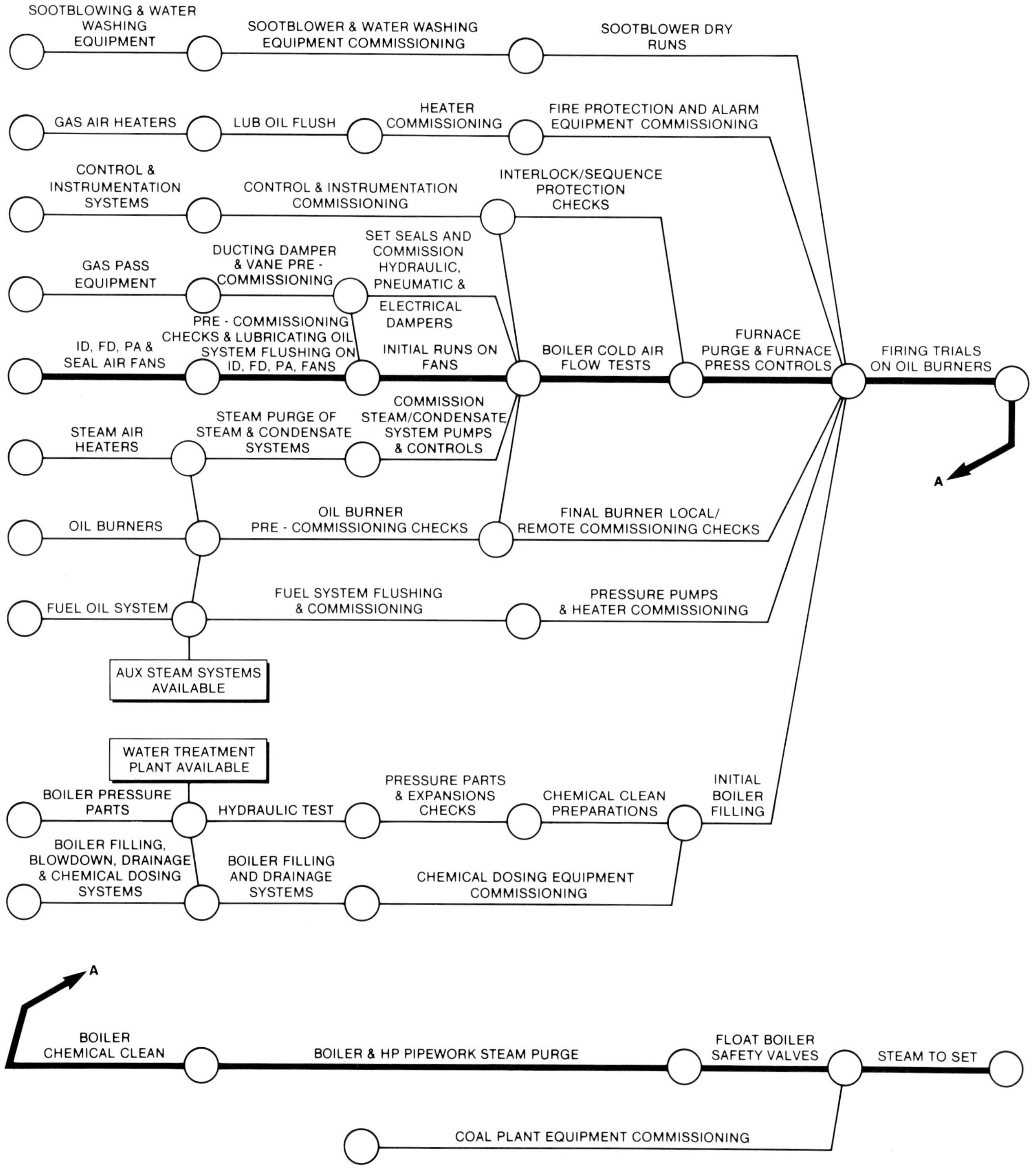

FIG. 4.1 Coal-fired boiler — simplified commissioning network

pressure, then the site test will be at 1.25 × design pressure. If works testing has not been carried out at 1.5 × design pressure, then the site pressure test will be at 1.5 × design pressure.

Note Design pressure for natural and assisted circulation boilers is the maximum permissible working pressure in the steam drum of the boiler. This pressure is not less than the highest pressure at which any of the safety valves is set to lift.

Pressure raising The best possible means of removing air from the pressure circuit must be used and any sudden pressure surges or rapid pressure increases which would give rise to high transient pressures must be avoided. The pressure must be applied gradually until the design pressure is reached: this is then maintained whilst a thorough examination of the drum and all pressure parts is made. All leaks must be repaired before pressure is raised above the design pressure.

The pressure can then be raised to the full test pressure and maintained for not less than 30 minutes. The pressure should then be reduced by 10% and maintained at this level whilst the drum and pressure parts are thoroughly examined. It should not be necessary to raise the pressure above design for any further testing.

3.2.2 Temperature

- The temperature of the water in the boiler must not be less than 7°C at any time in order to eliminate the possibility of freezing.
- The temperature of the water shall not exceed that of the outside surface of the drum by more than 15°C.
- The temperature measured on the outside surface of the drum shall not exceed 20°C or the prevailing ambient temperature, whichever is the greater.

3.2.3 Water quality

The water for hydraulic testing must be of the highest quality and where austenitic steels are present in the boiler pressure parts the water used for hydraulic testing must be free from significant quantities of chloride. This requirement is satisfied by ensuring that the quality of water leaving the water treatment plant has a conductivity not exceeding 10 μS/m at 25°C.

3.2.4 Terminal point isolations

Before filling the boiler for the hydraulic test, terminal point isolations must be applied, as follows:

- Boiler stop valves and by-passes.
- Economiser inlet isolating valve.
- Any plant not included in the test, particularly transmitters, gauges, etc., which will not withstand the overpressure test.
- If the reheater is to be included, then the isolating devices used to isolate the boiler from the turbine must be fitted. This can be achieved by fitting temporary blanks in the reheater inlet and outlet pipework adjacent to the boiler. To prevent problems of cutting and disconnecting pipework, the CEGB fits pressure lock devices to all its 500 MW and 660 MW units: a typical example is shown in Fig 4.2. Pressure lock devices are preferred to isolating valves due to their lower cost. The positions where they are fitted can be seen in Fig 4.13 (shown as 'blanking bodies').
- Special hydrostatic plugs are fitted to the boiler safety valves to obviate damage due to excessive pressure.

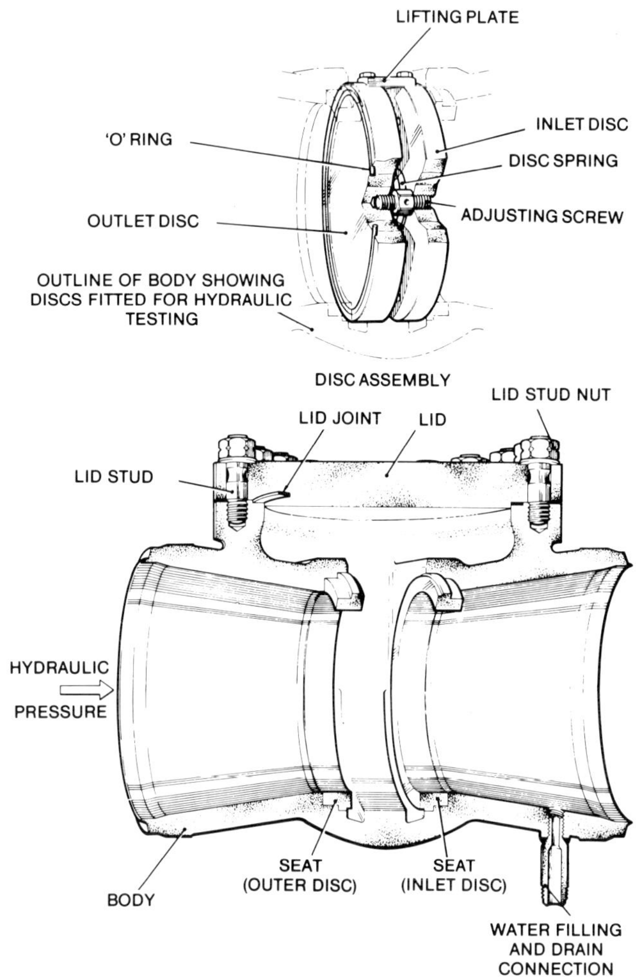

Fig. 4.2 Typical pressure lock device

3.2.5 Safety and resources

Safety

- Safe access to the boiler must be ensured when inspecting for leaks.
- Safe means of handling large amounts of water require that the station drainage systems be available.
- The limitations outlined in Sections 3.2.1 to 3.2.4 above must be strictly observed.

Services

For a boiler hydraulic test, large quantities of high quality water are required and it will therefore be necessary for the station water treatment plant and the reserve feedwater tanks (RFWT) to be available. Typical quantities of water to fill a 500 MW oil-fired boiler are given below:

Boiler	183 m^3
Economiser	79 m^3
Superheater	190 m^3

Reheater	132 m^3
Total	584 m^3

Calibrated test pressure gauges will be needed.

Staffing

A team of approximately six people will be needed for the boiler hydraulic test and will require about one week to complete the task, depending upon the number of leaks found.

3.3 Boiler filling procedure

When the boiler filling pump is not available, a fire pump or other suitable portable pump may be used for initial filling of the boiler with deionised water, to which a suitable corrosion inhibitor could be added. The water must be of the highest quality as stated in Section 3.2.3 of this chapter.

Before water is introduced to the boiler, the filling lines must be flushed and the water quality checked. At intermediate stages of the filling, the drains are opened to prevent debris being flushed into the boiler. Figure 4.3 shows a suitable filling and pressurising arrangement. If the boiler filling equipment is available, then refer to Section 4.4.1 of this chapter for the boiler filling procedures.

It is essential to eliminate all air from the pressure circuit and consequently most boilers are part-filled with the drum access doors removed. Water is introduced through the economiser, keeping the economiser recirculation valve closed at all times. When the economiser and drum interconnecting pipework are full, water will spill over into the downcomers from the drum to the bottom headers. The downcomer and bottom header drains are then opened to allow the flow to displace any loose debris. As the water level rises in the furnace wall tubes, air will be expelled through the drum access doors. When the water level in the drum approaches the bottom of the manhole doors, filling is stopped and the drum doors are fitted. Drum level gauge glass drains are operated at this stage to remove any debris from the drain lines and to ensure that the glass is free to fill.

Following fitting of the drum doors, all drains and air cocks are closed and a vacuum pump is used to produce a vacuum of 840 mbar in the remaining steam spaces. The vacuum pump can be connected to one of the drum air cocks. At the specified vacuum, filling is recommended from a suitable connection on the superheater outlet manifold, using a portable pump. When the superheater is filled, the vacuum pump must be isolated to prevent it becoming waterlogged. Alternatively, the superheater could be filled by overflow from the drum.

Once the superheater is filled, pressure should be raised to 6.9 bar, using the filling or intermediate pressurising pump, and then held for boiler expansion and clearance checks.

3.4 Hydraulic testing procedure

Before the hydraulic test commences, the following actions are required:

- All major boiler expansions and clearances checked.
- Communications system to be set up between the test gauge, the boiler test unit pump, drain and air release valve positions.
- Access to the pressure parts area restricted to personnel involved in the test, and warning notices displayed to exclude unauthorised persons.
- Calibrated test pressure gauges are installed at the drum, superheater and reheater outlets. The instruments and connecting pipework must be checked to ensure they are mechanically secure and the pipework should be purged with compressed air before fitting.
- Finally, check that the conditions specified in Sections 3.2.2, 3.2.3 and 3.2.4 of this chapter are being complied with.

Pressure can now be applied gradually, using the boiler test unit pump shown in Fig 4.3, at about 6.9 bar/h, until the design pressure is reached. At this pressure (187 bar for 500 MW and 660 MW units), all sections of the boiler must be examined thoroughly and sufficient time allowed in the pressurised condition to permit very small leaks to show. Sufficient boiler casings should be left off to facilitate the inspection of all welds.

When all detected leaks have been repaired, the boiler should be retested and proved watertight at the design pressure. The pressure can be gradually increased to the test pressure and maintained for not less than 30 minutes. At the end of this period, the pressure is reduced by 10% and maintained at this level for leakage, drum deflection and boiler expansion checks.

When the hydraulic test is completed, ultrasonic testing can be used to check strategic areas of tubes for debris blockages or weld intrusion. The external tube surface is wire-brushed, and grease or glycerine is used to couple the ultrasonic beam into the tube.

Figure 4.4 shows typical sound echoes obtained on an oscillograph. Figure 4.4 (a) shows a trace for a normal tube; the first series of pulses are those reflected from the adjacent steel/water interface and the right-hand series represent echoes obtained after the ultrasonic energy has passed through the water and has been reflected by the steel/water interface at the rear tube wall. Figure 4.4 (b) shows the trace obtained when objects, such as nuts, bolts, or welding rods, are blocking a tube. Figure 4.4 (c) shows the trace from a normal dry tube. If blockages are found, the tubes must be cut and the obstructions removed.

On completion of the test, the boiler should be depressurised and drained. Precautions must be taken to protect any part of the pressure circuits that cannot be drained, for example, superheater pendants, if the

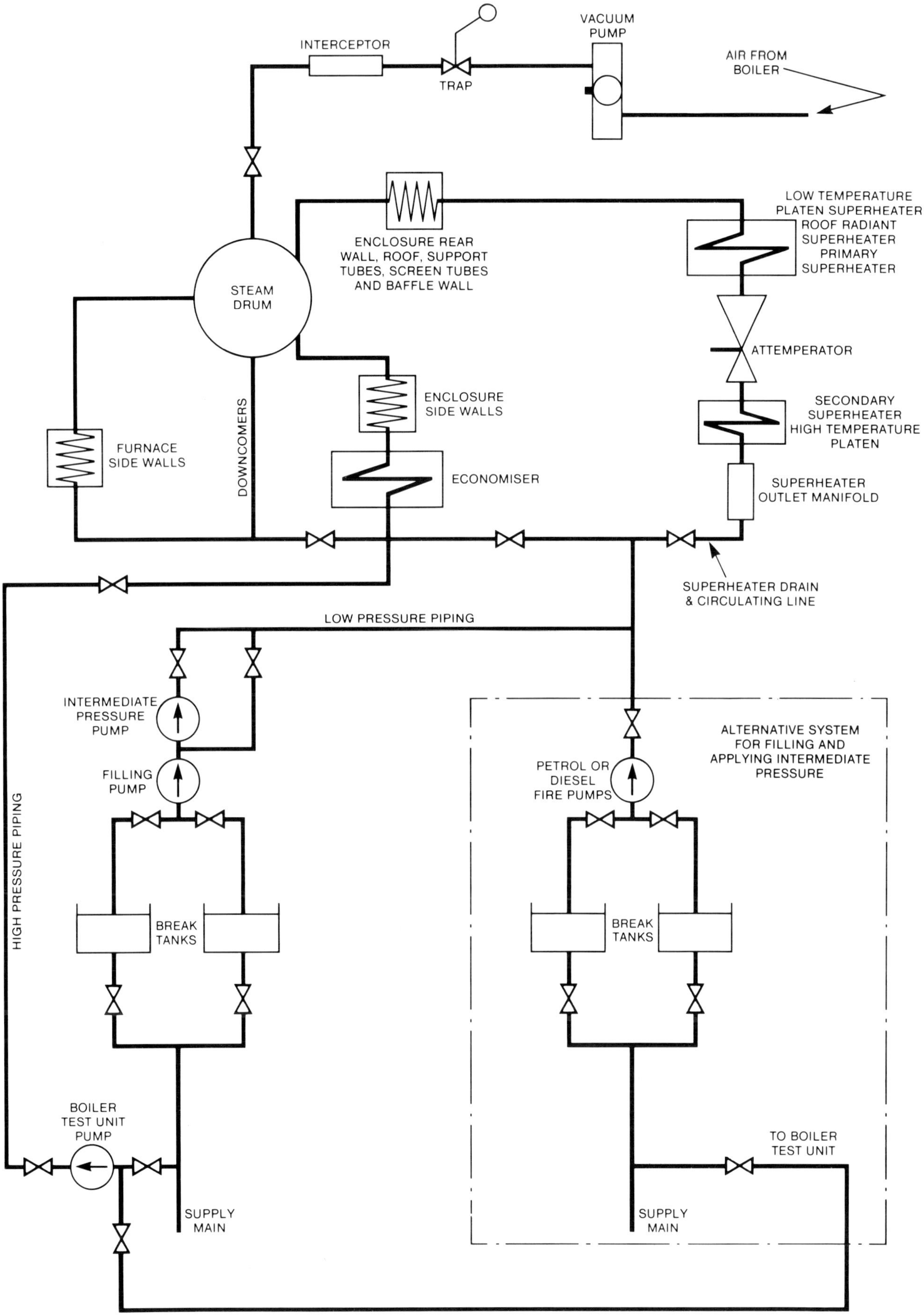

FIG. 4.3 Boiler filling circuit

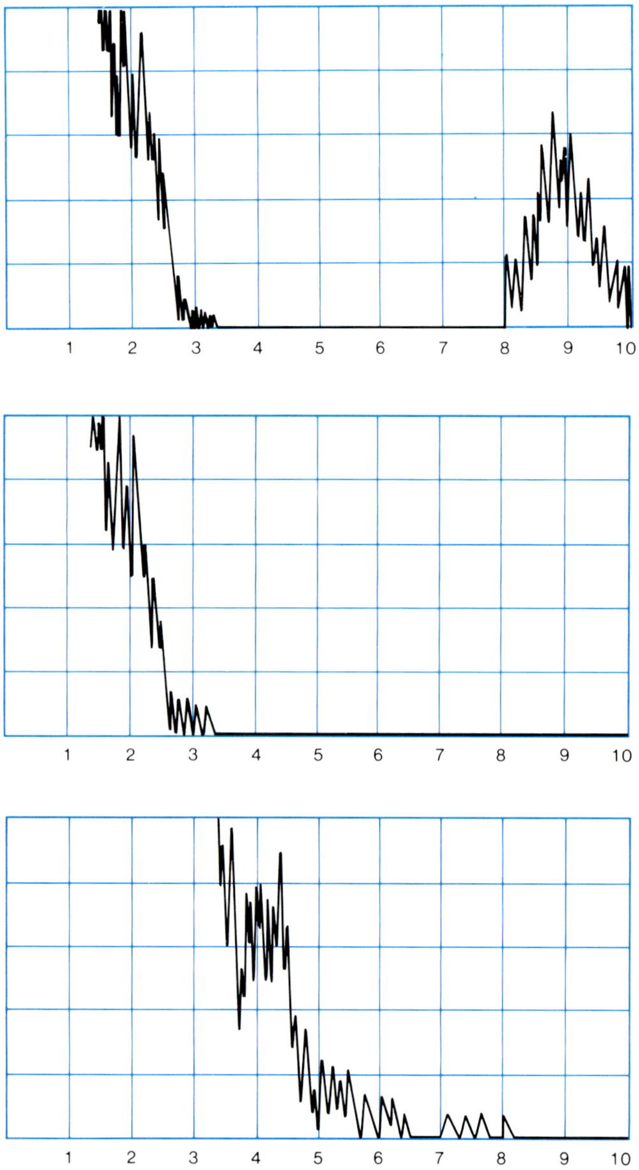

FIG. 4.4 Ultrasonic testing of boiler tubes — sound echoes obtained on an oscillograph

ambient temperatures could fall below freezing before the boiler is fired.

If long-term storage is necessary, then the boiler can be stored using dry air, or nitrogen, to protect the drum and tubes from corrosion.

4 Pressure parts and expansions

4.1 Introduction

This section covers the pressure parts from the economiser through to the HP steam pipework to the turbine.

The following systems are covered:

- Boiler water systems, including economiser, drum, natural- or assisted-circulation furnace, boiler circulating pumps, drum level indication equipment and boiler feedwater control.
- Superheater and reheater, with the differing methods of temperature control.
- Boiler filling, blowdown and drainage systems.
- Boiler water sampling equipment.
- Boiler supports and expansions.
- HP and LP chemical dosing equipment.
- High pressure piping systems.

A simplified commissioning network diagram is shown in Fig 4.5, which covers boiler commissioning from the completion of erection up to the boiler chemical clean stage. It will be seen that the boiler circulating pumps A and D are commissioned before the chemical clean. Pumps B and C are commissioned after the chemical-clean, but before the steam purge.

4.2 Boiler water systems

4.2.1 Boiler feedwater and circulating systems

A typical 500 MW oil-fired boiler feedwater and circulating system is shown in Fig 4.6.

Feedwater enters the boiler via the feed control valves to the economiser and into the boiler drum, then to the water-wall bottom manifolds. The flow circuit is arranged such that, for a boiler operating at 164 bar, the water/steam mixture will enter the bottom of the water-wall tubes at approximately 1.5 m/s.

By the time the mixture has reached the top of the tubes, about one-third to one-quarter of the weight of the water will have evaporated, giving a *circulation ratio* (defined as weight of water entering the circuit/weight of steam leaving) of three or four. These are approximate values only and vary for differing boiler designs.

In the drum, water which has not been evaporated is mixed with the additional feedwater from the economiser, and returns to the downcomers to continue the process.

Modern 500/660 MW HP boilers are all designed with reheat, and are either natural- or assisted-circulation boilers. The steam-to-water capacity in these boilers is relatively high, so the maintenance of a continuous supply of feedwater is critical in preventing the furnace tubes from overheating.

At high boiler operating pressures, the density of water approaches that of saturated steam, so some boiler designs incorporate circulating pumps to assist the natural circulation of water in the furnace tubes to maintain them below the permitted maximum metal temperatures. As the furnace gas temperature exceeds 1250°C and the maximum working metal temperature of the furnace tubes is generally below 420°C, even slight disturbances or interruptions in the water supply can result in the breakdown of the protective magnetite layer on the internal surface of the tubes, which leads to

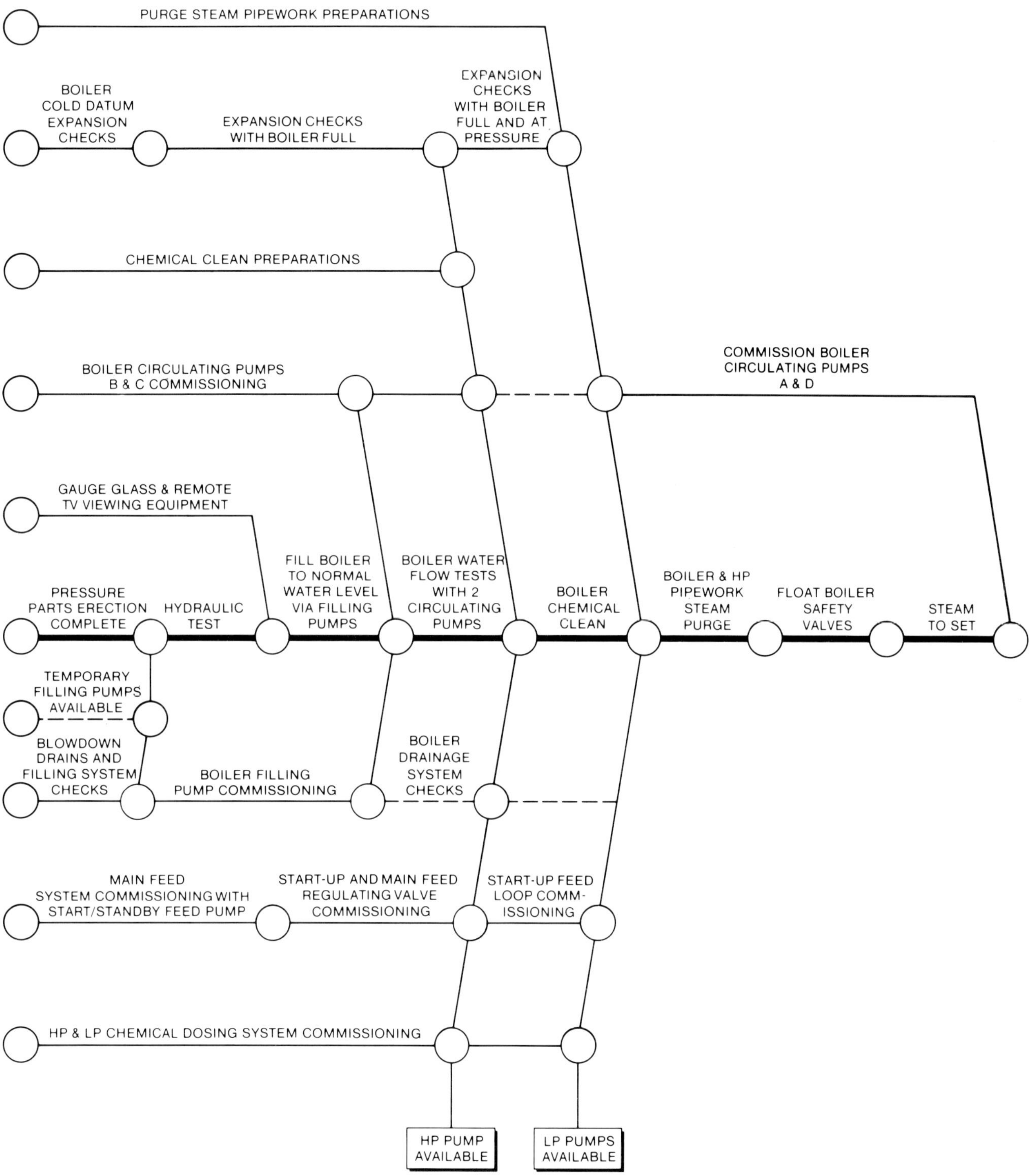

FIG. 4.5 Boiler pressure parts — simplified commissioning network

tube metal loss. This may eventually lead to tube failure due to 'on-load corrosion' or, if there is sufficient reduction of water flow due to deposits or other causes, failure will eventually occur due to overheating. Hence, it is vital to maintain the correct drum water level at all times and to monitor the performance of circulating pumps.

With an assisted-circulation boiler, smaller diameter tubes can be used, compared with natural-circulation boilers, and with parasitic orifices at the inlets of the feed headers, control of flow to individual tubes can be achieved.

Oil-fired boilers are normally fitted with assisted-circulation furnaces to ensure that adequate circulation is maintained for heat absorption, due to the high radiant heat fluxes generated by oil firing. The quantity of water which has to be circulated by the pumps in an assisted-circulation boiler is the feedwater quantity

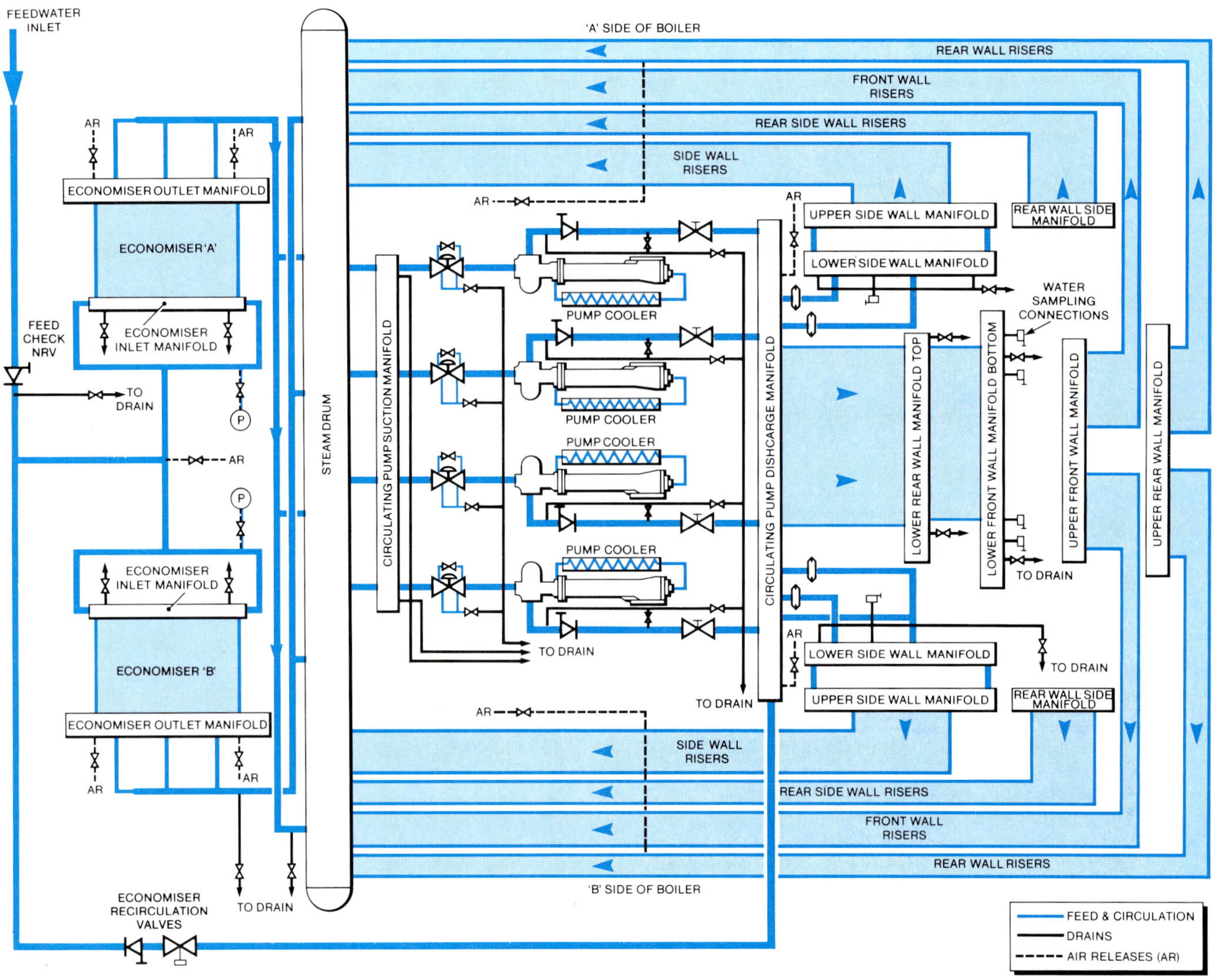

FIG. 4.6 Boiler feedwater and circulation system

multiplied by the circulation ratio, and the duty is met by having a number of pumps working in parallel. A 500 MW boiler may have four pumps, of which one is normally standby, each with a capacity of about 2400 l/min.

A typical pump consists of a casing (usually of the double-volute type, to reduce radial thrust), impellers of the medium specific-speed mixed-flow type, and a heat barrier to separate the 'hot' pump from the 'cold' motor. This heat barrier consists of a series of baffles to ensure a uniform temperature gradient and also to break up any streams of hot water that may attempt to enter the motor from the pump.

The squirrel-cage induction motor, which is enclosed in a blind barrel bolted to the pump casing, is suitable for operating 'wet', with water at 93°C and at full boiler pressure.

The internal circulating fluid, normally demineralised water, is charged into the motor and pump unit from an external source situated near the pumps, through a combined priming/drain connection. This water serves as a lubricating and cooling medium for the motor and bearings. With a pump operating at 316°C in the same casing as a motor suitable for only 93°C, the cooling circuit requires careful attention. Figure 4.7 shows a typical pump cooling system. The cooling system generally adopted has HP and LP circuits as follows. The water in the motor is circulated by an impeller, usually mounted adjacent to the thrust bearing. This impeller pumps the water through the motor to a position near the heat baffle, where it is piped to a heat exchanger and then back to the bottom of the motor. The heat exchanger (low pressure circuit) is supplied with water from some convenient source in the unit auxiliaries cooling system (UACS) and flows by gravity through the heat exchanger and then back into the cooling system return main.

Most items of plant can be considered safe when shut down, but this pump is only safe when cooling water is circulating, if the boiler is hot. The supply source for cooling water must be extremely reliable, so it is taken from the UACS, with an emergency feed taken from elevated storage tanks and gravity fed to the heat exchanger on loss of UACS flow; therefore, even if all

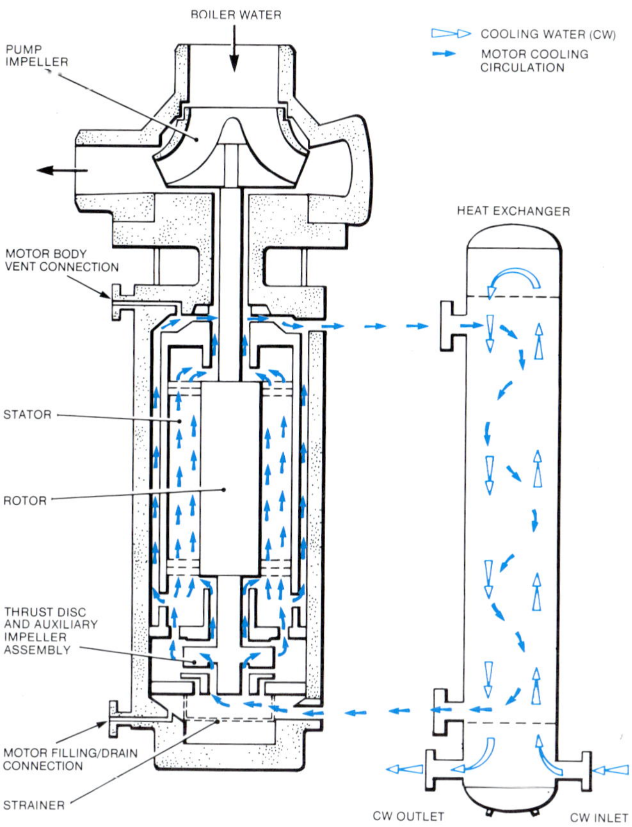

FIG. 4.7 Pump cooling system

auxiliary supplies are lost, there is no risk of overheating the circulating pump motor. Interlocks and alarms are provided, in case the pump cooling system fails.

Electrical connections to the motor windings, which are made from pressure-tight submersible cable, are led through the motor casing and sealed by high pressure glands, each of the three-phase cables being led into its own terminal box. Terminal box sealing to the motor casing and atmosphere is effected by Neoprene gaskets; bursting discs and dessicators are also incorporated in the terminal box casings.

4.2.2 Drum level indication equipment

Gauge glass and Hydrastep level indication

On all modern high pressure boilers, the CEGB uses the Hydrastep drum level indication system for normal operational use, the Hydrastep now being accepted by the Factory Inspectorate as a means of drum level indication to replace the conventional gauge glass. Conventional gauge glasses are still used for drum level monitoring during chemical cleaning and steam purging operations, when the Hydrastep unit cannot be used.

For chemical clean and steam purge purposes, a double glass is normally used to give the operator a better indication of level during periods of maximum disturbance in the drum. A typical conventional level indication system, consisting of twin bull's-eye gauge glasses connected in series with the bottom long-range drum tappings and the drum-top air vent connection, is shown in Fig 4.8 (a). Closed circuit TV is used to transmit the indicated boiler water level back to the unit control desk.

Hydrastep drum level indication equipment consists of a system of electrodes positioned at discrete levels in a measuring vessel, the measuring vessel being connected to the boiler drum in the same way as the gauge glass. The water level is determined by detecting whether each electrode is immersed in water or steam/air by measuring the electrical resistance between the electrode and the measuring vessel immediately surrounding. Resistance values below a preset value indicate 'not water' (i.e., steam/air). A detector circuit connected to each electrode produces a high voltage output, if the electrode is in steam or air, and a low voltage output, if the electrode is in water. This voltage output is transmitted, via a set of relays and logic units, to a level indicator on the unit desk which has red/green lights to indicate water/steam. Alarms and trips can also be operated from these electrical signals.

To minimise the effects of density errors in Hydrastep gauges, the waterside arm connections and valves are lagged, so that the temperature drop in the water between drum and vessel is minimised. The steamside connections, together with the vessel, are left unlagged so that a high condensate flow is maintained. A typical Hydrastep is shown in Fig 4.8 (b).

Remote drum level indication

In addition to the gauge glass and Hydrastep drum level indication equipment, remote drum indication is provided on the unit control desk in the form of a meter or a recorder chart. Short-range transmitter columns are connected at both ends of the drum and, from these columns, the water level in the drum is transmitted, via level transmitters, to the control panel. Density compensation is provided for each level column. A typical system is shown in Fig 4.8 (c), which gives an analogue output proportional to drum water level, corrected for density changes over the whole range of boiler operation from zero to full pressure.

The inputs come from a pressure transmitter which monitors drum pressure, from which is obtained average saturation temperature and a differential pressure transmitter which compares the head of water above the bottom tapping point of the drum with that of an associated constant head reference column.

The output signals from these two transmitters are fed into the density compensator, which carries out a continuous computation and produces a signal analogous to true drum water level, corrected for the change in water density. The signal is fed to the remote level indicator in the control room and is also used as an input for the feedwater controls.

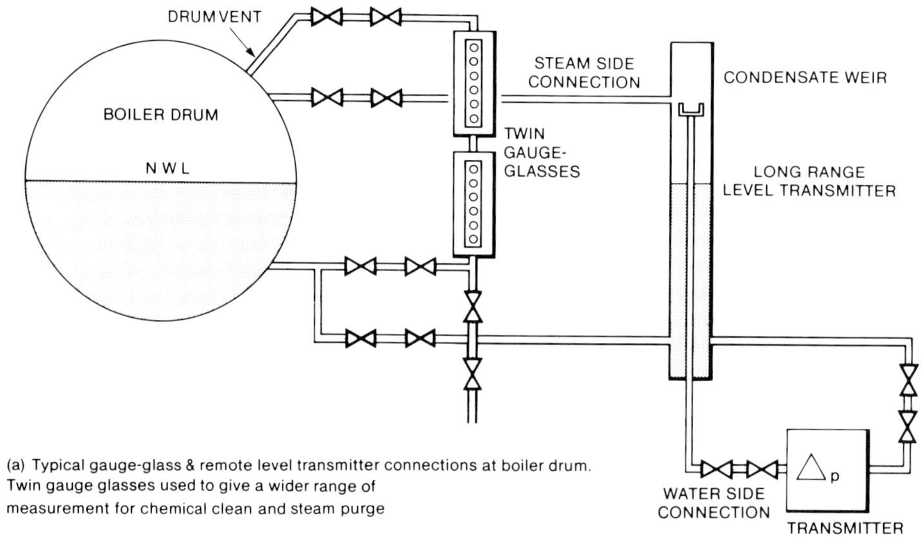

(a) Typical gauge-glass & remote level transmitter connections at boiler drum. Twin gauge glasses used to give a wider range of measurement for chemical clean and steam purge

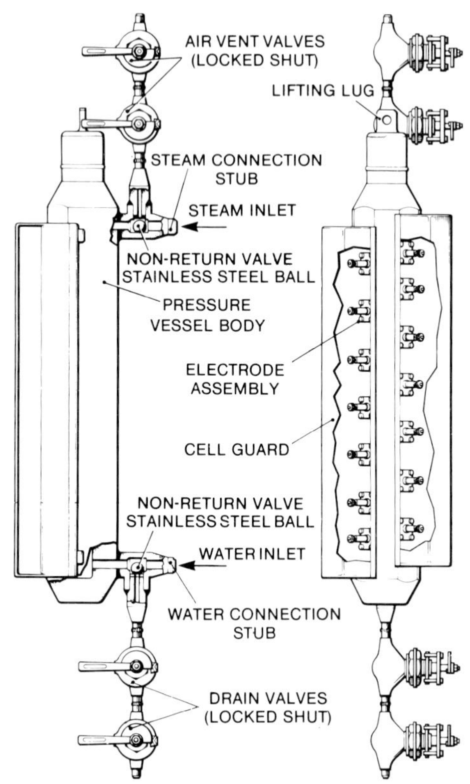

(b) Hydrastep system pressure vessel

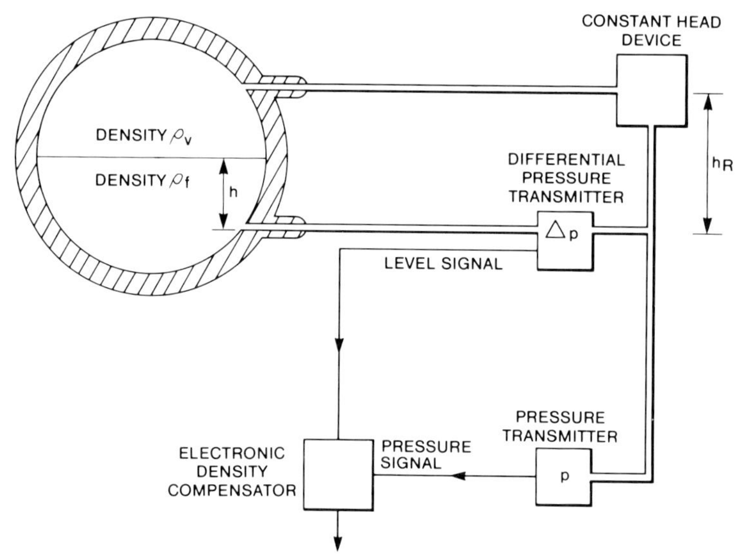

(c) Density compensation system for boiler water drum level

FIG. 4.8 Gauge glass, Hydrastep and remote drum level indication systems

4.2.3 Boiler feedwater control

Boiler feedwater control equipment regulates the supply of feedwater to the boiler drum to maintain the correct level, normally by modulation of the feedwater control valves, or by boiler feed pump speed control. A typical three-element feedwater control scheme is described in Chapter 3.

The different methods of effecting feedwater control are initiated at different load levels, and this is covered in Chapter 6.

For the initial boiler light-up, for chemical clean and steam purge it is essential to have the feed regulating valves and the starting and standby boiler feed pump speed controls available on manual control from the unit control desk. Where possible, the single-element feed control should be available on automatic; this will be particularly advantageous during the steam purge and float safety valve programmes.

4.2.4 Safety

The following safety points should be considered:

- Safe access to the boiler for metal checking.
- Adequate thermal insulation to ensure that personnel cannot be burned when pressure raising is in progress.
- Safe means of handling large amounts of drainage, which will require the station drainage systems to be commissioned prior to boiler firing.
- Extreme care to be taken when operating drainage valves with the boiler at pressure, to prevent damage to valve seats.
- When the boiler filling system is connected to the boiler via the movable bend, a warning notice must be fixed to the unit control panel.
- Facilities to deal with large amounts of leakage, both internal and external to the boiler, should be available.

4.2.5 Resources

Services

- Large quantities of pure water are needed to fill the boiler. See Section 3.2.5 of this chapter.
- Cooling water is needed for boiler circulating pumps, so the station auxiliaries cooling system must have been commissioned.
- Hoses for filling boiler circulating pump motors.
- Hydrastep test box.
- Electrical supplies to valves, dampers, motors, etc.
- Hydraulic/pneumatic supplies to valves.
- Metascope instrument for metal checking.

Staffing

The commissioning team will range between four and eight people, and commissioning will take several weeks.

4.2.6 Boiler water systems commissioning

Boiler water and circulating systems

It is essential to ensure that the headers, tube welds, supports and attachments are of the correct materials. The metascope instrument has been widely used for the initial checking of the differences between carbon steel and ferritic alloy materials. Any inadvertent use of incorrect materials clearly leads to failure at high temperatures. Figure 4.9 shows the location of various alloys in a typical 500 MW oil-fired boiler.

Any thermocouples associated with the waterside and metalside temperatures should be checked for correct fitting and then checked for response to changes in temperature by the application of heat from a blow torch. The cabling, cold junctions and recorders should be checked in accordance with the procedures in Chapter 3 and the appropriate standard check list (SCL).

Any mechanical and electrical valves on the water systems should be checked for correct installation; commissioning checks should be completed, following the procedures in Chapter 3. The valves should then be set up in the correct positions for boiler filling (see Section 4.6 of this chapter).

Boiler drums for modern high pressure steam plant have a shell thickness of about 100 to 150 mm, and the specified differential metal temperatures across the shell must not be exceeded. The permitted temperature differentials are shown on charts produced by each manufacturer, for both the heating and cooling cycles at various pressures. Figure 4.10 is a typical chart for a 500 MW boiler drum; the curves show the maximum permissible temperature differential through the shell for the specified temperature differentials between the top and bottom of the drum in the same plane. The outer and inner drum metal temperatures are measured by thermocouples fitted to the drum exterior and to the furnace steam-riser tubes, respectively. If the specified limits are exceeded, firing should be stopped until the permitted differentials are restored. This process can be accelerated by filling the drum to equalise temperatures, but care must be taken to prevent water passing over to the primary superheater. In addition to the drum metal differential temperatures, the boiler manufacturers also stipulate the maximum temperature difference allowed between the feedwater and the drum metal temperatures; a typical figure is 70°C.

Boiler circulating pumps

A typical commissioning sequence diagram is shown in Fig 4.11; here, B and C pumps are commissioned before the chemical clean, followed by the A and D pumps.

The motor and cables are commissioned in accordance with the procedure in Chapter 3.

The cooling water supplies to the motor coolers are flushed, to remove any debris from the system, by breaking flange joints on the supply and return water manifolds as near to the pump cooler as possible, and flushing until clean. The boiler flange joints are then remade. The instrumentation must be checked to ensure alarms and trips function correctly for the following conditions:

Starting interlocks

- Cooling water flow satisfactory

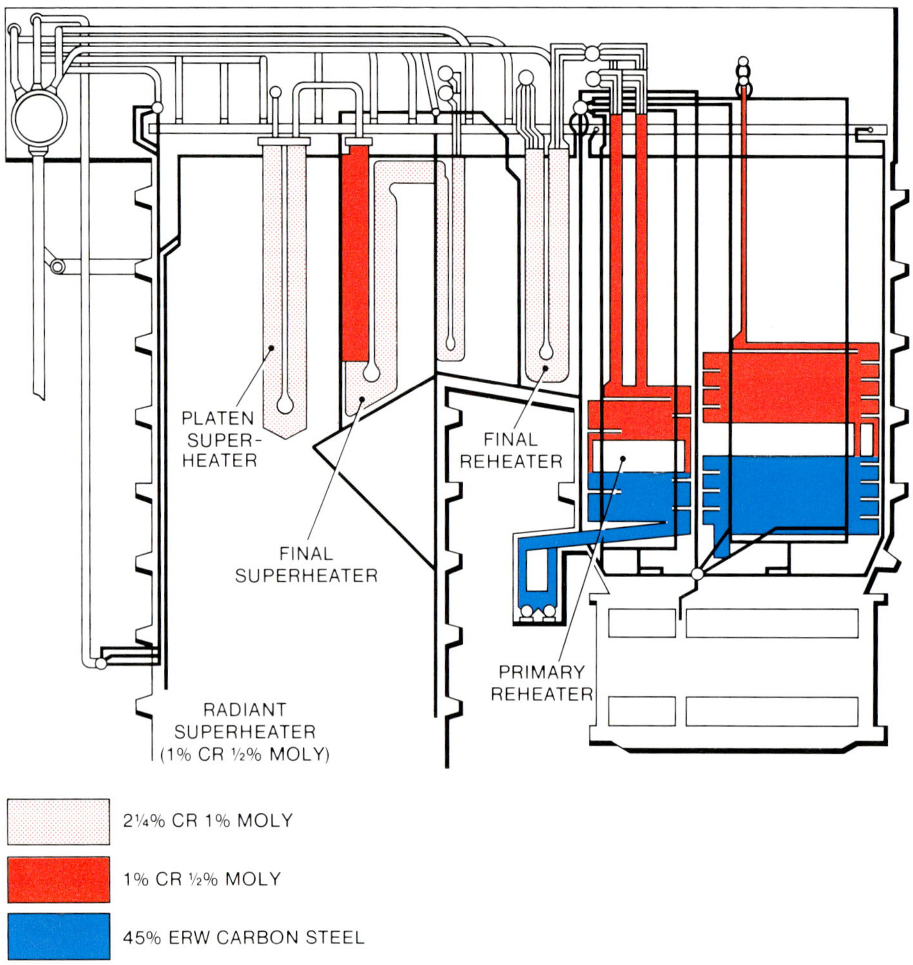

FIG. 4.9 Layout of materials for a 500 MW oil-fired boiler

- Suction and discharge valves open — monitored by limit switches

Running interlocks

- With three pumps running, the trip of one pump starts the selected standby pump. Note that there is normally a limit on the load that can be carried with fewer than three pumps in service so, if the standby pumps fail to start, the load must be decreased to the maximum permitted with two pumps in service (350 MW for a 500 MW boiler). This action means that heat flux is reduced in the furnace to acceptable limits. The action is manual and carried out by the operator.
- If the pump suction/discharge common-manifold differential pressure falls, an alarm is initiated and the standby pump is started (independent of the trip of one pump, above). If the standby pump fails, action is the same as for the tripping of a pump.
- If three boiler circulating pumps trip, a unit trip is initiated.
- An alarm is initiated for loss of pump cooling water flow and the pump is tripped.
- Pumps are tripped if the motor water temperature reaches 60°C.
- An alarm is initiated for 'high motor water temperature' at 50°C.
- Alarms for 'pump tripped' and 'differential pressure low' should also be checked.

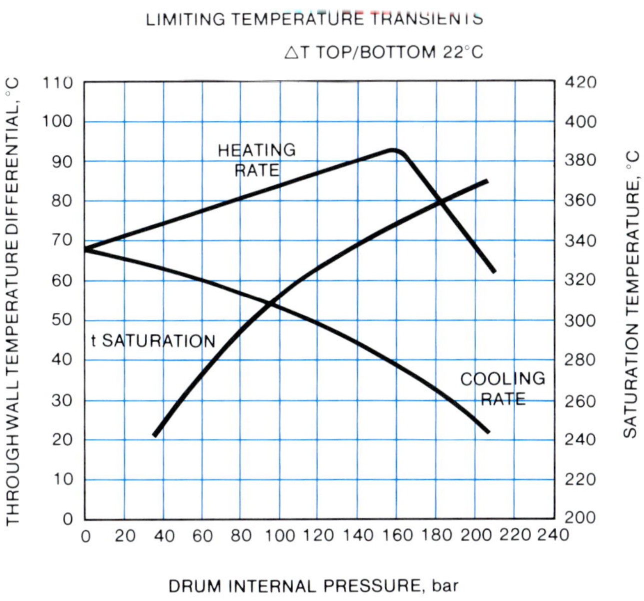

FIG. 4.10 Maximum allowable drum temperature differentials

All valves are checked and commissioned in accordance with the procedure in Chapter 3.

The motor is filled with demineralised water of conductivity 2 μS/m, and fully vented to ensure that any air is expelled. The maximum conductivity allowed is 10 μS/m. The motor and cable insulation should now be rechecked.

It would be possible to 'flick' the motor for rotation tests at this stage, providing the pump volute has been filled with water, but it is better to carry out the initial test with the boiler full of water. With the boiler full of water, the pump should be vented, then run under local control for a few minutes. The pump should then be stopped and the motor-prime checked. The pump can then be restarted and carefully monitored for any signs of vibration. The starting and running current are noted, a drop in boiler water level should be seen and a steady pump differential pressure reading should be observed.

The other boiler circulating pumps are commissioned in a similar manner. With forced-circulation boilers, it is now normal practice to have two boiler circulating pumps available for the chemical clean, with the other two pumps isolated. Following completion of the chemical clean, the pumps are inspected to ensure that they are not damaged. The other boiler circulation pumps are then introduced into the system during or after the steam purge operations. The boiler cold water flow tests should then be carried out with all four pumps in service. If any changes to parasitic orificing at the water headers are needed, the pumps will have to be isolated, the boiler drained and a Permit for Work (PFW) issued to provide access for cutting the pipework, readjusting the orifices and rewelding the pipework. The flow tests are then repeated.

Drum level indication equipment

(a) *Drum gauge glass commissioning* Check for correct installation of all valves in accordance with the SCL. For chemical clean and steam purge purposes, a double gauge glass is normally used to give the operator a better indication of level during periods of maximum disturbance in the drum. The closed circuit TV equipment should also be checked in accordance with the manufacturer's instructions; it is essential to align the camera correctly for maximum effect and to ensure that the picture is not affected by lighting conditions, such as bright sunlight. The gauge glass illumination system should also be checked and energised.

(b) *Hydrastep commissioning* The Hydrastep unit is not used during the chemical clean and steam purge operations and must be effectively isolated from the drum by locking the isolating valves closed. All the Hydrastep equipment can be commissioned before the electrodes are fitted with the exception of the indication and trip vessels, an electrode-simulation test box being used to simulate the water levels. The use of this test box allows most of the commissioning to be carried out independently of the state of the boiler. When the steam purge has been completed, it is then possible to fit the Hydrastep electrodes, commission the trip and indication vessels, fill the boiler and complete the final commissioning with the whole system in service and the boiler drum at normal working level (NWL). The Hydrastep equipment is put into service and the indications checked against the bull's-eye gauge glass. The 'high' and 'low' alarms are checked by variation of the drum water level. The trip initiation is checked (with the trip inhibited) in the same way. These tests are carried out with the boiler either hot or cold, but not firing. Radio interference susceptibility tests should be carried out in all the areas of the Hydrastep equipment.

When putting a Hydrastep or conventional gauge glass into service with the boiler hot and at pressure, the following procedure should be adopted to ensure that bull's-eyes and/or electrodes do not fail:

- Water-side isolating valves left *closed*.
- Vent valves *open*.
- *Crack open* the steamside isolation valves and gradually warm through the vessels by allowing them to fill slowly with water from the condensing steam.

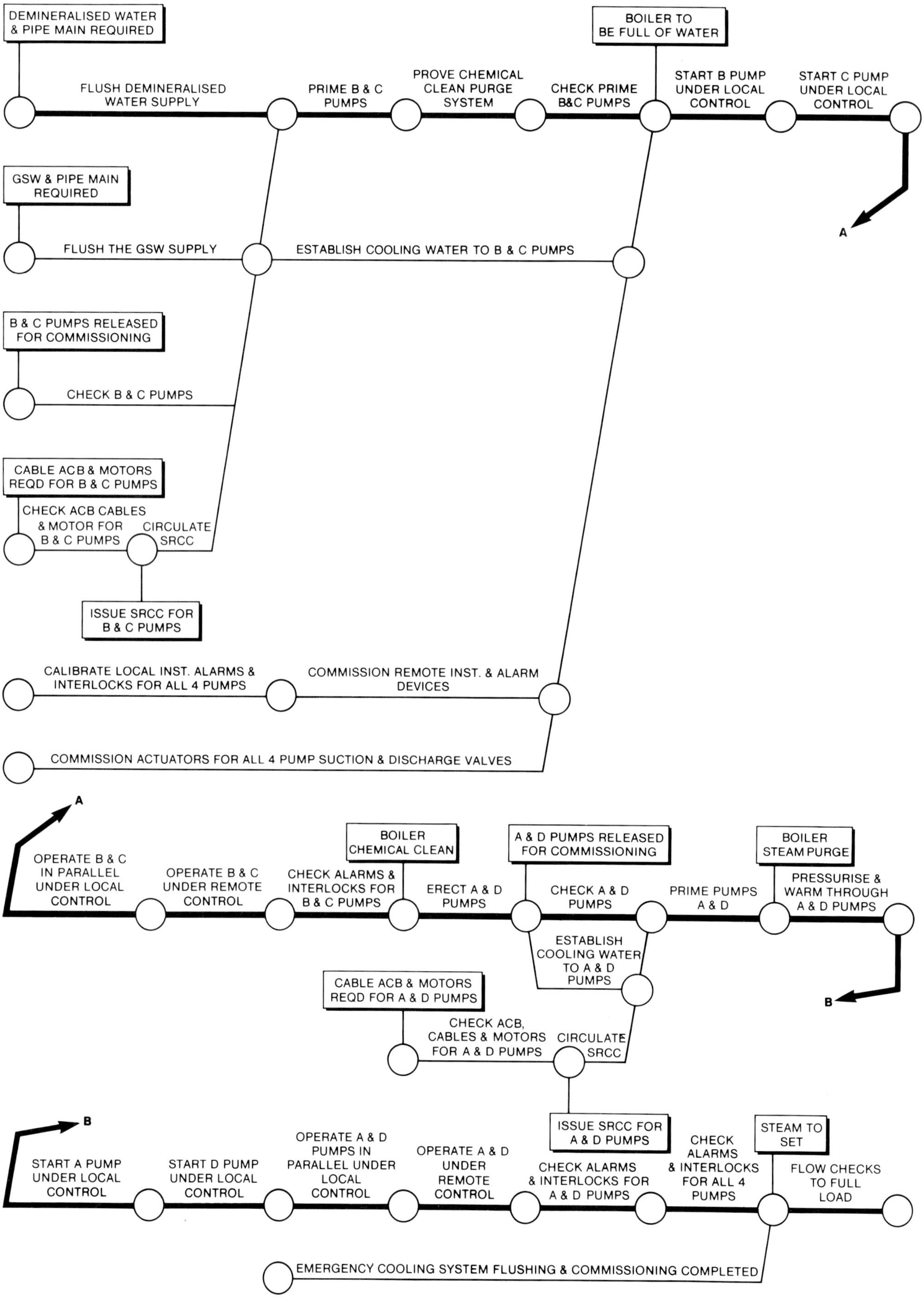

FIG. 4.11 Boiler circulating pumps commissioning network

- When the vessels are full and completely heat-saturated, the waterside isolating valves should be *slowly opened*.
- The vent valves should then be shut and the steam-side isolating valves opened fully to put the vessels totally into service.

(c) *Remote drum level indication equipment* Commissioning consists of checking all level column isolating valves in accordance with the procedure in Chapter 3, Section 1 and SCLs.

The transmitters and cables should be checked according to the procedures in Chapter 3, Section 3 and the appropriate SCL. By injecting 0–10 mA or 4–20 mA signals into the transmitter, a check can be made that the control desk indicator responds accordingly. When the boiler is filled and a water level obtained in the drum, a check should be made that the level indication corresponds to the correct water level, as shown on the drum level gauge glass. Finally, when pressure is raised on the boiler, the system is checked for pressure and density compensation, and final calibrations are completed.

Boiler feedwater controls

The valves, electrical and hydraulic actuators will be checked according to the procedure in Chapter 3, Sections 1 and 2. A commissioning procedure for the three-element control system is described in Chapter 3, Section 8. For boiler commissioning up to steam-to-set, the feedwater controls will only be available on single-element, the commissioning procedure being similar to the three-element system.

4.3 Superheater and reheater

This section describes a typical 500 MW oil-fired boiler system, as shown in Figs 4.12 and 4.13. It has four stages of superheat, with temperature control by spraywater desuperheaters. Steam from the boiler drum passes through radiant superheater elements, incorporated in the combustion chamber front wall and roof, to the horizontal-pass primary superheater elements located above the economiser in the rear section of the convection chamber. From this stage, the steam passes through desuperheaters to a vertical-pass platen and final pendant elements suspended in the top of the combustion chamber. Connected to the underside of the platen superheater inlet manifold are 53 distributor/collector manifolds. These are divided internally into front (distributor) and rear (collector) manifolds, from which the element tubes hang downwards in U-form. Outlet pipes from the top of each collector section transfer the steam into the final superheater inlet manifold. The pendant elements make four vertical passes, with the final pass extending through the roof tubes to the final superheater outlet manifold assembly, which consists of a central long section and two outer sections. The four discharge pipes from this assembly are connected to the main boiler stop valves.

For control of steam and metal temperatures during start-up, steam dumping branches are provided at the final superheater outlet. Back-up steam supply to the auxiliary steam system may be incorporated by means of a tapping from the primary superheater outlet, upstream of the desuperheaters. Steam flow from the superheater is controlled by four main boiler stop valves.

Steam discharged by the HP turbine is reheated in two stages before passing to the IP turbine. From the primary reheater, which is in the form of horizontal-pass elements located above the economiser in the front section of the convection chamber, the steam flows to a final vertical-pass pendant reheater suspended in the combustion chamber rear-pass, and then to the IP turbine.

Reheat steam temperature is controlled by gas biasing dampers located below the economiser. The dampers control the gas flow through the primary reheater and superheater elements in the convection chamber, which is vertically divided into a front and rear chamber by a wall formed by refractory-covered primary superheater feeder tubes. Steam flow to the reheater is controlled by four butterfly valves.

Equipment is fitted for saturated and superheated steam sampling, and also branches for chemical cleaning on the primary superheater outlets upstream of the desuperheaters.

4.3.1 Methods of temperature control

It is a specified requirement that CEGB boilers should deliver steam at a controlled temperature from both the superheater and the reheater at loads from 70% to 100% of the maximum continuous rating (MCR) of the boiler. The natural characteristic of a superheater is to provide a steam temperature which varies with load, so that a means of control has to be introduced.

In a radiant superheater section, the steam temperature falls as load increases, whereas the steam temperature rises with load in a convection superheater section. The combination of these two sections in a boiler forms an overall characteristic such that the steam temperature does not change greatly with load, but nevertheless some control is necessary. This enables the designer to make the superheater large enough to give the required temperature when it is dirty with ash dust and yet not produce too high an outlet temperature when quite clean (as, for example, just after sootblowing). There are six basic methods of control in use on present designs:

Tilting burners Generally used for corner firing. The burner tilt affects the furnace-exit gas temperature and the proportion of heat absorbed by the superheater and reheater. An upward tilt increases the temperature: a downward tilt decreases the temperature. This

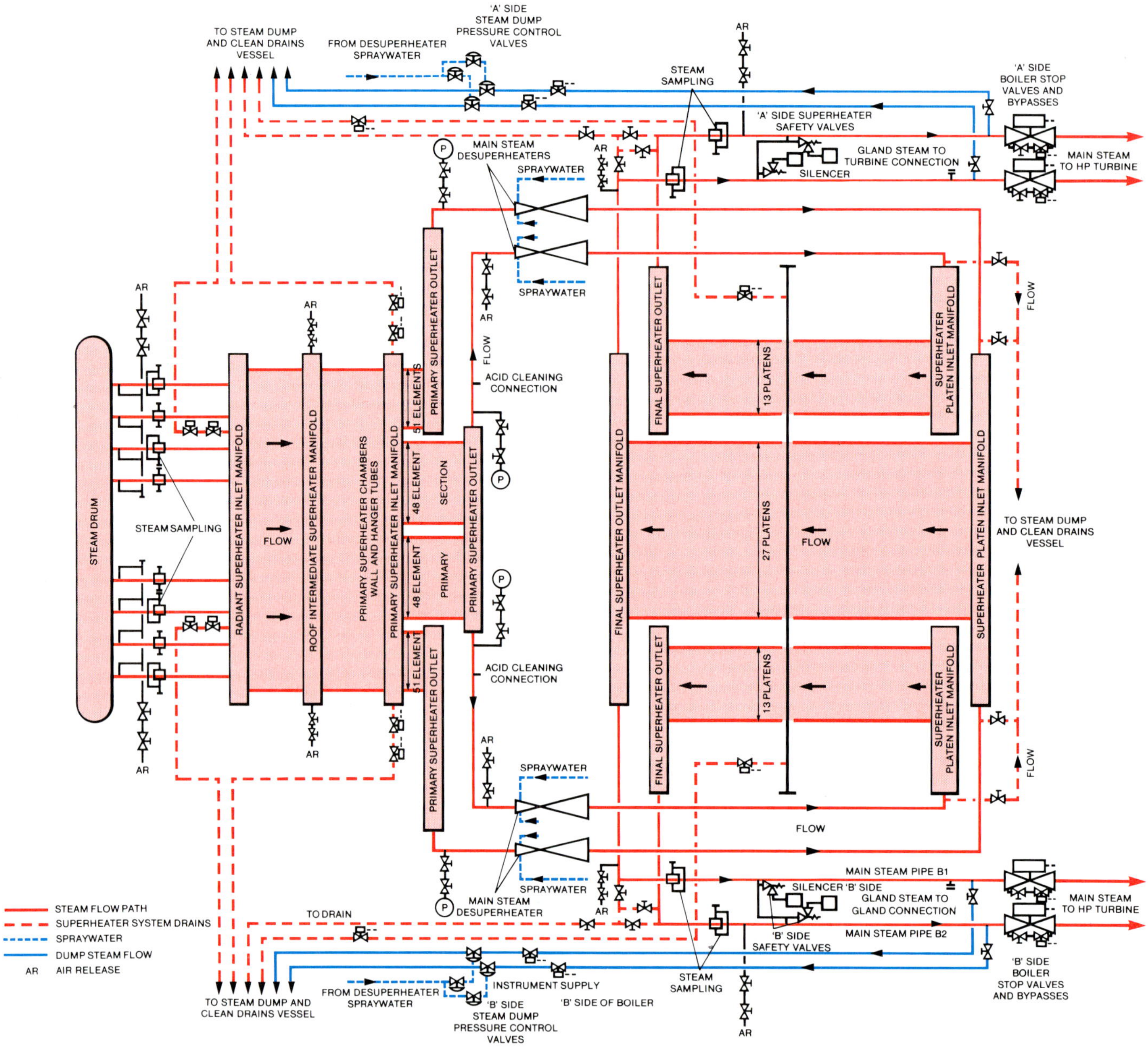

FIG. 4.12 Superheater system; showing steam flow, steam dump, drains, steam sampling and valve locations

type of control does not give an effective simultaneous control of both superheater and reheat temperatures because of the different superheater and reheater characteristics, so the normal practice is to control the reheater temperature by this means, with spray attemperation for final control of the superheater temperature.

Gas by-pass Control dampers are used to regulate the flow of gas over parts of the superheater and reheater. This method is also normally used to control reheat temperatures, with spray attemperation for superheater control.

Gas recirculation Gas recirculation fans are also used as a means of superheat/reheat control. This system is regarded as a means of increasing or boosting superheat temperature rather than a precise form of control. Approximately 10% of gas flow is recirculated from the economiser outlet to the bottom of the furnace by the GR fans. This reduces the furnace heat absorption by decreasing the residence time of the hot gas stream in the furnace and therefore boosts the superheat temperatures. It is only used up to approximately 50% of full-load.

Spray desuperheaters Also known as *attemperators*. These provide direct-contact heat exchange, whereby feedwater which has by-passed the economiser is sprayed into the steam flow and cools the steam by absorbing latent heat. The water joins the main steam flow and evaporates. This method is used by all boiler manufacturers for final superheater temperature control, the spraywater being supplied either directly from the feed system via the boiler feed pumps, or by

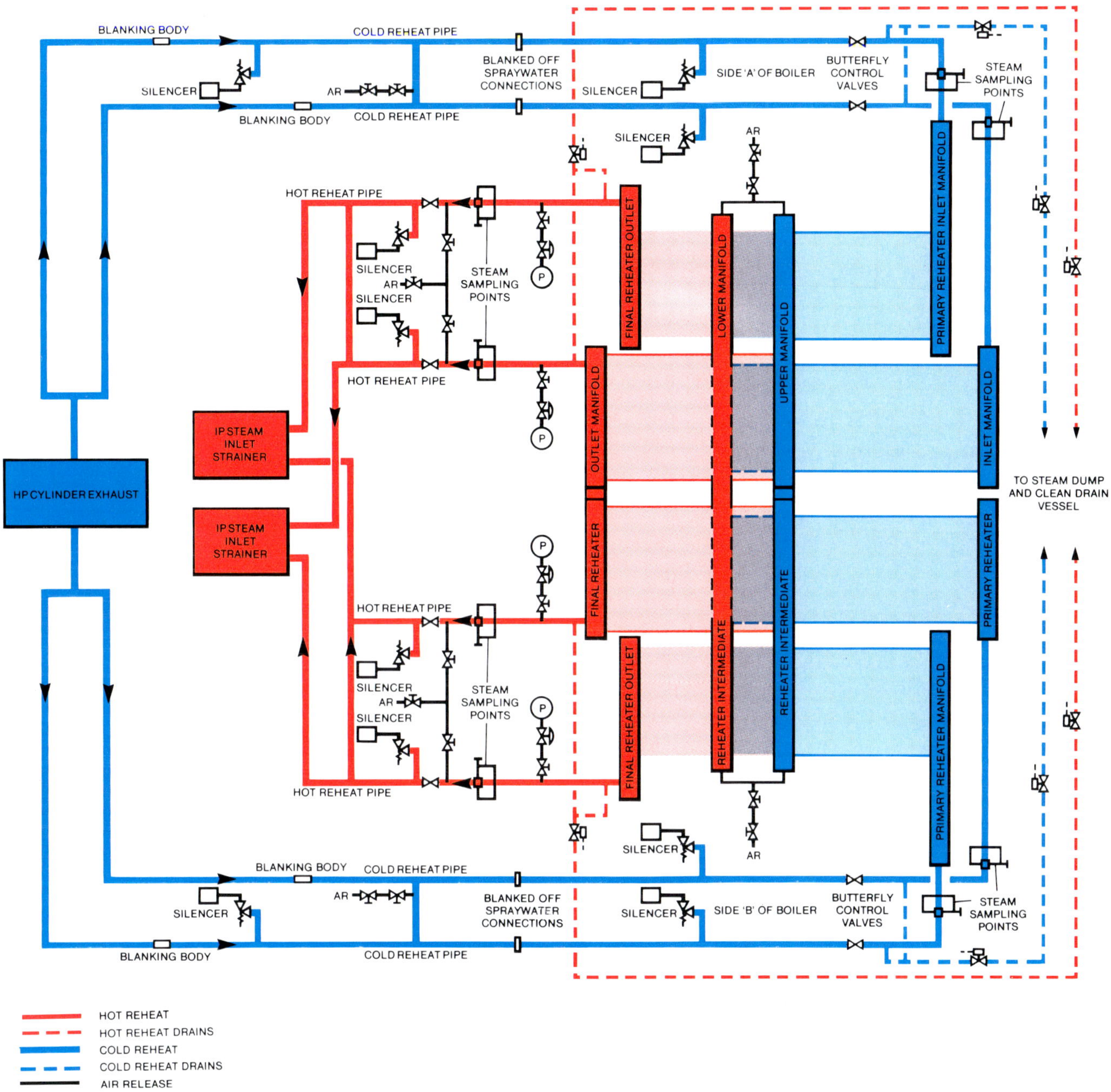

FIG. 4.13 Hot and cold reheat steam flow with valve positions shown

spraywater pumps which take their supply from the feed system.

Non-contact attemperators (or desuperheaters) These are a type of shell and tube heat exchanger. The water on its way from the economiser to the circulation system is passed through finned U-tubes (forced flow). Part of the steam is passed over these tubes, the rest of the steam by-passing the exchanger; control is effected by a proportioning valve on the steam side.

The non-contact attemperator finds favour for reheat control, since sprays are not permitted because of the adverse effect on overall cycle efficiency. It has been shown that a spray quantity of 1% of turbine steam flow would increase the heat rate by approximately 0.2%.

Steam temperature and pressure control on start-up and shutdown During boiler start-up, high superheater metal temperatures are usually a problem, due to the steam flow being relatively low compared with furnace gas flow; on modern boiler designs, steam dumping equipment has been incorporated to overcome this problem. The steam dumping facilities are provided at the superheater outlets, but before the main boiler stop valves, the steam being discharged to steam dump and clean drains vessels. The boiler can therefore be 'overfired' to raise steam temperatures quickly, without exceeding the superheater temperature limitations, and without overpressurising the boiler.

Metal temperatures are monitored by thermocouples which indicate when the steam dumping valves should be opened to give a cooling flow of steam through the associated section of the superheater. The system can be set to control pressure instead of temperature, if required. To regulate the steam temperature entering the steam dump vessels, spraywater from the boiler superheater spraywater system is supplied to the dump valves. The spraywater quantity is controlled by pneumatically-operated control valves.

Steam dumping facilities are also provided by motorised drain valves at various superheater intermediate stages. These can be clearly seen in Fig 4.12.

4.3.2 Safety and resource considerations

The safety and resource considerations are the same as those described in Sections 4.2.4 and 4.2.5 of this chapter.

4.3.3 Superheater and reheater systems commissioning

Superheater and reheater

As with the waterside, it is essential to ensure that the headers, tube welds, supports and attachments are of the correct materials. The metascope instrument can again be used for the purpose. Figure 4.9 should again be referred to for typical 500 MW oil-fired boiler materials.

Any thermocouples associated with the superheater and reheater steam and metal temperature indicators should be checked for correct fitting and then checked for response to changes in temperatures by application of heat from a blowtorch. The cabling, cold junctions and recorders should be checked in accordance with the procedures in Chapter 3, Section 1 and any relevant SCLs.

The boiler manufacturer usually indicates limitations on various metal temperatures in the operating instructions, but as a general guide the following apply:

Tube material	*Maximum working metal temperature*
Carbon steel (ferritic)	482°C
1% chromium 0.5% molybdenum (ferritic)	565°C
2.1/4% chromium 1% molybdenum (ferritic)	582°C
18% chromium 11% nickel (austenitic)	676°C

For convenience and to increase thermocouple life, metal temperature measurements are sometimes made in the dead space above the furnace roof. The readings therefore represent a temperature which normally indicates steam rather than tube metal and allowance must be made for the actual metal temperature in the hot gas stream. This is usually derived from test results taken during early boiler operation.

Any valves on the system should be checked in accordance with procedures in Chapter 3, Section 1 and relevant SCLs.

Steam temperature control system

Tilting burners Commissioning of tilting burners is covered in Section 6 of this chapter.

Gas by-pass dampers Damper commissioning is covered in Chapter 3, Section 1.2, and in Section 5 of this chapter.

Once the damper and actuators are commissioned, the operation of the dampers from the remote control station on the unit control desk should be proved in the manual mode. For the initial commissioning of the boiler, up to steam-to-set, the dampers will be used in the manual mode. The control-loop commissioning is covered in Chapter 6.

Gas recirculation The gas recirculation fans are commissioned using the procedures in Section 5 and the dampers and control loops as above.

Spray desuperheaters All valves and actuators are checked, using the procedures of Chapter 3, Sections 1 and 2, and the SCL. When the boiler is filled from the boiler feed pump for the first time, the spraywater lines are flushed through and the system checked for leaks. The control loops are checked, using the procedures in Chapter 3, Section 8; the final commissioning of the controls will be covered in Chapter 6. The spraywater system will only be used during the boiler commissioning up to steam-to-set to backfill the superheater during the chemical clean and provide continuous flushing to prevent chemicals being injected into the superheater section of the boiler.

Non-contact attemperators All valves and actuators are checked, using the procedure in Chapter 3, Sections 1 and 2, and the SCL. As for the spray desuperheaters, the spraywater line will be flushed and the system checked for leaks.

The spraywater vessels will be pressure tested. Any associated relief valves are tested to lift at the correct setting.

The control loops are checked, using the procedures in Chapter 3. Final commissioning of the control loops is also covered in Chapter 6.

Steam-dumping on start-up and shutdown Once again, the valves and actuators for the steam dump and spraywater control valves are checked in accordance with the procedures in Chapter 3, Sections 1 and 2, and the SCL. All discharge pipework to the steam dump and clean drains vessels, and the spraywater supply pipework, should be checked for correct installation and supports.

During the initial filling of the boiler, the spraywater lines must be flushed until clear. This is done after the steam dump and clean drains vessels have been commissioned. The controls from the unit control panel to the dump steam pressure and spraywater valves are checked to ensure that they work in the manual mode for the initial boiler pressure raising. The automatic control will not be available until after steam-to-set; this will be covered in Chapter 6.

4.4 Boiler filling, blowdown and drainage systems

4.4.1 Boiler filling

On most 500 MW and 660 MW units, the boiler can be filled from the main boiler feed pumps via the feed system to the economiser and drum, or by the boiler filling system, which uses low pressure boiler filling pumps connected by a removable section of pipework, known as the 'movable bend', to the boiler blowdown and drains system.

The boiler filling pumps usually take their suction from the reserve feedwater tanks (RFT) or unit make-up main and pump on a common discharge main to any boiler. The connection to the boiler is made through the movable bend on the drainage system, the connection being done this way to prevent the HP boiler system being directly connected to the LP filling system. A typical system is shown in Fig 4.14.

When the movable bend is connected to the filling system, a note must be put on the unit control panel as a precaution to prevent the boiler from being fired with the LP and HP systems connected.

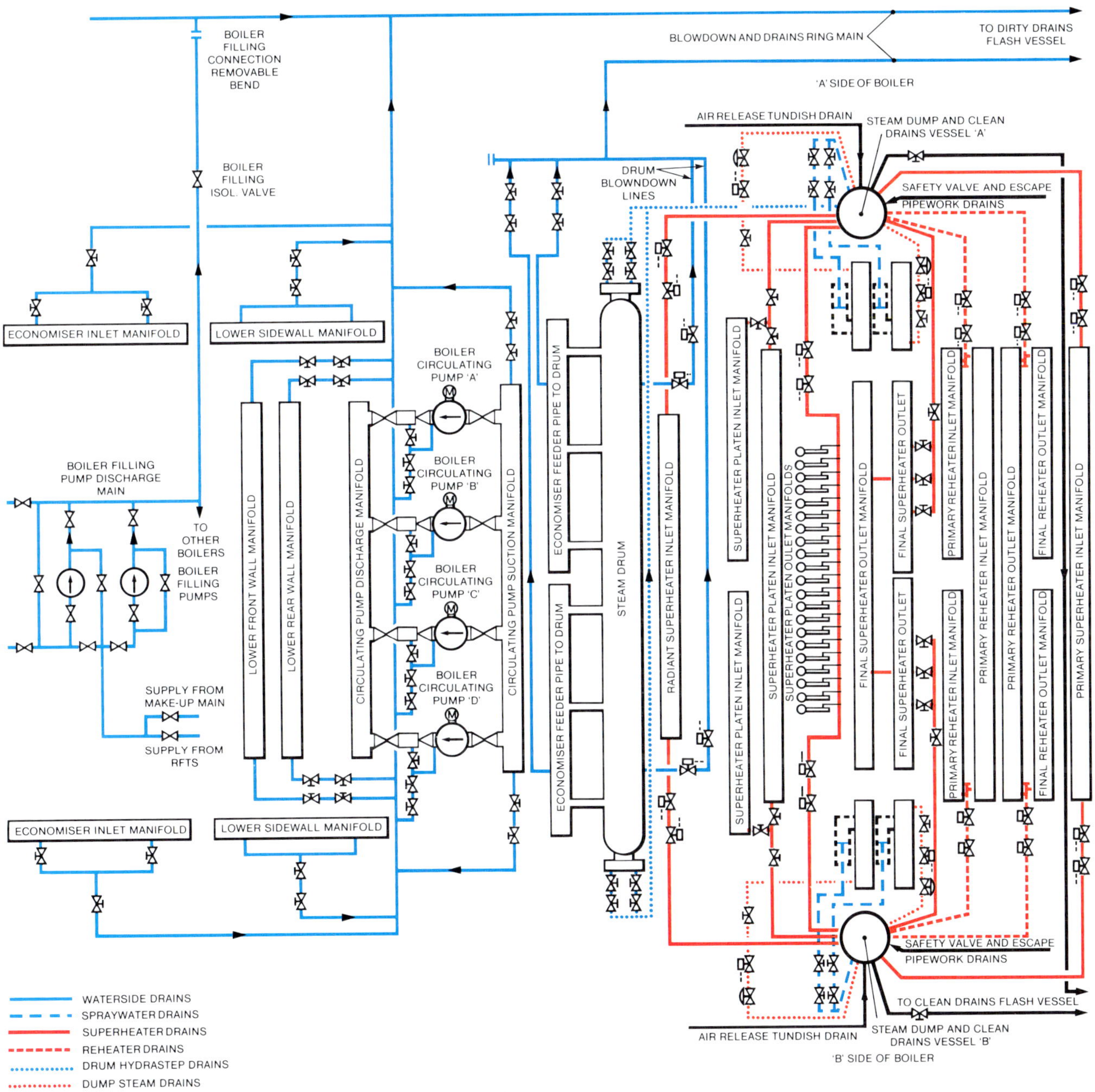

FIG. 4.14 Boiler filling, blowdown and drains system

Effectively, the boiler is therefore filled through the boiler drain connections on the following headers:

- Economiser inlet headers
- Boiler circulation pump suction header
- Lower front-wall header
- Lower rear-wall header
- Lower side-wall headers

Filling from the drainage system, i.e., from the bottom of the boiler, effectively removes all entrained air, as the boiler fills through the headers to the drum.

4.4.2 Boiler blowdown and drainage systems

A typical boiler blowdown and drainage system is also shown in Fig 4.14 and utilises two steam dump vessels, positioned one on each side of the boiler. They receive steam from the final superheater outlet pipes and various superheater and reheater manifolds and clean drains from the desuperheater spraywater, safety valves, air release valves and Hydrastep water level gauges. Any condensate collected in the vessels is led away to a clean drains flash vessel, where the condensate can be reclaimed. It is then pumped to the reserve feedwater tanks (RFT), or dumped to waste if not suitable for recovery. During early boiler operation, when systems are dirty, the clean drains are normally dumped to waste, the condensate only being recovered when the water quality is chemically satisfactory.

Drains from the lower water walls, economiser inlet manifolds, circulating water pump suction and discharge manifolds and the steam drum blowdowns are connected into the boiler blowdown and drains ring main, which discharges to the dirty drains flash vessel (DDFV). The DDFV normally discharges the dirty condensate to waste (CW forebay or station drainage system), via anti-syphon drain pipes.

4.4.3 Boiler filling, blowdown and drainage system commissioning

Boiler filling

All valves and actuators on the system are checked in accordance with the procedures in Chapter 3, Sections 1 and 2, and all pipework checked for correct installation and support/expansion arrangements.

The boiler filling pumps suction pipework from the RFTs are flushed to a point as near to the pumps suction as possible, flushing continuing until the water is seen to be clean and free of debris.

The filling pump motors are checked in accordance with the procedures in Chapter 3, Section 4 and the alignment between the pumps and motors is also checked. The pumps should then be primed and spun for rotation checks, with the discharge valves closed. The pump discharge pipework to the movable bends should then be flushed, the pump discharge pressure being carefully controlled by regulation of the pump discharge valves. During this exercise, the pump/motor should be checked for signs of vibration and the running current monitored. Once the flushing exercise is complete, the movable bend on the boiler that is to be filled should be connected to the boiler filling system.

At this stage, it may be the first time the boiler is filled, i.e., for the hydraulic test, unless the boiler filling system was not complete at the hydraulic stage and a temporary system, as described in Section 3.3 of this chapter, was used instead. The necessary drain valves should be set open to allow the boiler to be filled through the drains system and the boiler filling pump started, the discharge pressure being regulated by the degree of opening of the discharge valve.

As the boiler fills, a thorough check must be made of the boiler supports and expansions, see Section 4.6.2 of this chapter. As the boiler water level rises, the various air release valves should be closed when water issues freely from them. This method of boiler filling ensures that water levels in the evaporator and economiser circuits rise at the same rate; thus it is relatively simple to anticipate when the water is about to reach the drum, by observation of the discharge head on the pressure gauges of the filling pumps. Once water is in the drum, filling should be stopped for the drum manhole door to be fitted, if filling for the hydraulic test.

Note It may be necessary to drain the boiler to remove any entrained debris after the first fill. If so, the boiler should be drained before the manhole door is fitted and the above filling procedure is then repeated.

Subsequent to completion of the boiler hydraulic test, the next time the boiler will be filled is in preparation for boiler circulating pump commissioning: this should present the most suitable opportunity for assessing optimum filling time, using the boiler filling pumps. With the boiler at NWL, the drain valves used for boiler filling should be closed before commissioning the boiler circulating pumps. During this commissioning, the boiler level will drop slightly and the boiler will then require topping up by use of the filling pumps. On completion of the boiler filling and before the boiler is fired, the movable bend must be changed over to the drain position.

When filling a hot boiler with either the main boiler feed pump or the boiler filling system, the boiler drum through-wall differential temperatures must be monitored (see Fig 4.8 for typical drum differential limits). The feedwater temperature with respect to the drum metal temperatures must also be carefully controlled.

Blowdown and drainage system commissioning

All valves and actuators on the system are checked in accordance with the procedures in Chapter 3, Sections

1 and 2, and all pipework checked for correct installation and support/expansion arrangements. All valve controls and position indicators should also be checked and proved from the control desk.

The steam dump vessels should be flushed at some convenient time prior to boiler filling and checked that there is no restriction of flow to the clean drains recovery tank. This is done by opening manhole door covers and filling them with clean water from a hosepipe, to a depth where the vacuum breaker pipes are filled. The dirty drains flash vessel is similarly flushed. On completion of the flushing exercises, the manhole doors are replaced. The cleanliness of the drain lines will have been checked during the boiler draining after the first fill: the drainage system should next be proven and all drain lines checked to be clear during the pressure raising exercise for the initial boil-out in the chemical clean programme. All drain valves should be operated and a check made that the drain lines become hot, thus proving that they are clear and passing water to the steam dump and/or dirty drains vessels.

During blowdown exercises, there is a risk of scoring valve discs/seats, due to the relatively high debris content in the boiler; it is therefore essential to open the isolating valve first, followed by the drain regulating valve.

4.5 Boiler sampling systems

4.5.1 Description

To ensure that the boiler water and steam systems are maintained to the required purity, comprehensive sampling systems are installed. The relevant sample points within the water and steam systems can be clearly seen by reference to Figs 4.6, 4.12 and 4.13.

The water and steam has to be cooled before it goes to the sample laboratory, using sample water coolers which are provided with cooling water from the general service water (GSW) system. To further cool the sample water, it is put through refrigerated chiller units. The boiler feedwater sampling laboratory equipment is described in Volumes E and F.

4.5.2 Boiler sample system commissioning

The system is set to work by flushing all the pipework to waste during the boiler filling and drainage exercises, until the samples show that the system is clean. The GSW system is flushed during the GSW commissioning phase; then, when the boiler is fired after the safety valve floating exercise, the sample water is fed to the relevant chemical instrumentation in the sample laboratory, via the coolers and chillers.

Details on commissioning of analytical instrumentation is covered in Chapter 3, Section 3. The on-line commissioning will be covered in Chapter 6.

4.6 Boiler supports and expansions

4.6.1 Description

All modern boilers are suspended from the boilerhouse steelwork and, depending upon the weights and positions of the headers/pipe/tube assemblies, are suspended either directly from the steelwork, or from rafts, which are themselves suspended from the main steelwork.

Ducts and pipework outside the main boiler support envelope are hung from the main steelwork or ductwork using constant-load supports, which may be either combinations of spring/rod or rod alone. Full details of boiler construction will be found in Volumes A and B. A typical constant-load pipe support is shown in Fig 4.15.

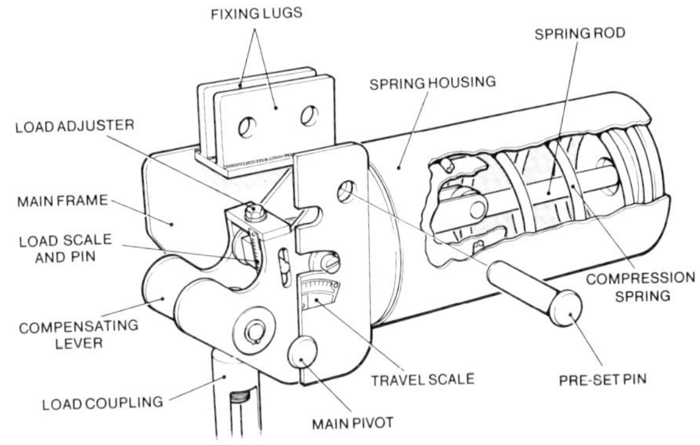

FIG. 4.15 Typical constant-load pipe support

4.6.2 Boiler support and expansion system commissioning

The boiler expansion and support arrangements must be checked for completeness and absence of restriction before the boiler has heat applied.

During the erection phase and boiler hydraulic test, some spring supports may be 'pinned' to prevent their overloading; these restraints must be removed from the pipework supports of the final superheater, reheater and HP steam pipework. All constant-load supports must have their 'loading pins' removed and movement indicators set, following the hydraulic test. Records of all spring settings and constant-load support settings must be made with the boiler cold and empty.

Expansion indicators should be fixed at points on the system where large movements are expected and a note made of their settings with the boiler cold and empty. These measure the relative movement between the boiler steelwork and the building steelwork. Recommendations of the type number and locations of thermal expansion indicators are given in Generation Design Memorandum 068/27 'Boiler thermal expansion indications and their use'.

Expansion checks should be made at the following stages in commissioning of the boiler:

- When the boiler is filled to NWL.

- When the boiler is filled completely for hydraulic testing.
- When the boiler is filled to NWL and the reheater has been filled for hydraulic test.
- When the boiler is fired for the first time, and boiling-out and steam purging take place.
- When steam pressure is being increased, at intervals up to the rated temperatures and pressure.
- When the boiler has reached steady temperature conditions after a continuous period of operation at MCR conditions.

At any time, if constant-load supports, or spring supports, exceed their maximum settings, firing must be stopped and adjustments made before there is any further increase in pressure/temperature. Closing plates on the boiler casings, through which boiler ancillary pipework passes, must be checked to have the necessary freedom and clearance.

Gas seals should be inspected to ensure that they are not binding. The flexible seals connecting the hot air ducting to the windbox and to the economiser/primary superheater enclosures must accommodate a large relative movement of the order of 200 mm; these should be inspected for possible tearing and leakage.

4.7 Chemical-dosing equipment

The maintenance of strict chemical control of the boiler water is required for all high pressure and temperature boilers and the limits on impurities are given in Generation Operation Memorandum 72. In order to maintain these chemical conditions, HP and LP chemical-dosing equipment is provided and is operated to strict limits, specified by the Station Chemist.

The chemical control limits for the boiler and feed system from steam-to-set are covered in Chapter 6.

4.7.1 LP chemical-dosing equipment

The LP chemical-dosing equipment is designed to inject hydrazine and/or ammonia into the boiler feed-water to control the feed system pH and dissolved oxygen (O_2) levels prior to boiler entry. This maintains the magnetite film created in the boiler water and feed circuit during the chemical clean.

A typical system consists of micrometer-adjustment variable-stroke dosing pumps, complete with mixing tanks, vent fans, and pipework. It is connected through valves to the de-aerator downcomer and condensate extraction pump discharge. A treated water make-up line supplies the mixing tanks and an electrically-driven tank mixer ensures adequate mixing of the chemicals. An ejector system, using water as the power source, transfers chemicals from the storage carboys to the mixing tanks.

LP dose is pumped to the de-aerator downcomer during unit start-up and is then transferred to the condenser extraction pump discharge, when the unit is on-load and condensate is being pumped through the feed system by the condenser extraction pumps. Dosing is continuous: the amount injected depends upon the feed system chemical conditions.

4.7.2 HP chemical-dosing equipment

The HP chemical-dosing equipment injects a controlled dose of chemical solution (caustic soda or disodium hydrogen phosphate) into the boiler steam drum. The equipment is similar to the LP dosing, except that the pumps are of the positive displacement type, designed to operate at pressures up to 200 bar. The HP dose is piped through high pressure pipework to the injection point on the boiler drum. HP dosing maintains the correct alkalinity, controls chloride levels and preserves the oxide film on boiler water surfaces.

HP dosing is intermittent, a controlled amount being pumped on-load to the boiler drum, as boiler water conditions dictate.

4.7.3 Safety

Before commencing commissioning, the following safety requirements must be satisfied:

- A correctly identified escape door must be available in the chemical-dosing house.
- Approved eyewash facilities must be available.
- Safety notices concerning the use of eyewash bottles must be displayed.
- Clean water hoses for flushing of spilt chemicals must be available.
- Breathing apparatus must be available outside the dosing house.
- Chemical handling precautions and first aid instructions must be available.
- PVC gloves, goggles and rubber aprons must be worn when handling chemicals.
- Instructions on what to do in the event of chemical spillage must be readily available.
- The hazards associated with ammonia, hydrazine, caustic soda and disodium hydrogen phosphate must be made known to all members of the test team involved in commissioning of the chemical-dosing plant.

4.7.4 Resources

Services

The following are required:

- Supplies of demineralised water to the mixing tanks.
- Supplies of ammonia, hydrazine and caustic soda.

- Protective clothing.
- Electrical supplies to the pumps/mixers and valves.

Staffing

A team of approximately five engineers will be required. The length of time taken will depend upon the boiler and feed system commissioning programme.

4.7.5 Commissioning of LP and HP chemical-dosing equipment

A typical commissioning network is shown in Fig 4.16.

All mixing tanks, fittings and pipework should be inspected for erection completion and cleanliness. The HP pipework should have been hydraulically tested to 1.25 × normal working pressure during the erection phase.

The tanks are filled with clean demineralised water. The high-speed mixers are commissioned by checking their drive motors in accordance with the procedure in Chapter 3, Section 4 and the SCL; they are then checked for direction of rotation, followed by a two-hour test run.

The dosing pumps are checked for correct installation. Initial proving runs will be carried out discharging the water to drain; the motors are again checked as above, and the gearboxes are checked for correct lubrication.

Once the pumps have been proven, the pipework is flushed. In the LP system, flushing is from the LP dosing pumps to the de-aerator downcomer and the condenser extraction pump discharge pipework, this operation being done when the feed system has been commissioned and is full of water. Similarly, the HP dosing lines to the drum will be flushed, a suitable time being when the boiler drum is filled for the chemical clean or steam purge activity. In each case, a check should be made that the dosing tank levels are falling, to prove flow. The pressure relief valves on the system should then be set to the correct settings, the LP pumps being set at 19 bar and the HP pumps at 185 bar for units operating at 164 bar. This can be done by throttling the pump discharge valves gradually until the correct discharge pressure is reached, the relief valve then being carefully adjusted to relieve the system pressure.

On completion of these procedures the system is ready for use with chemicals and the safety precautions for handling hydrazine, ammonia and caustic soda must be strictly maintained. The concentrations of the chemical mixtures are advised by the Station Chemist. Final tuning of the system is done when the unit is on-load and operating at design MCR conditions.

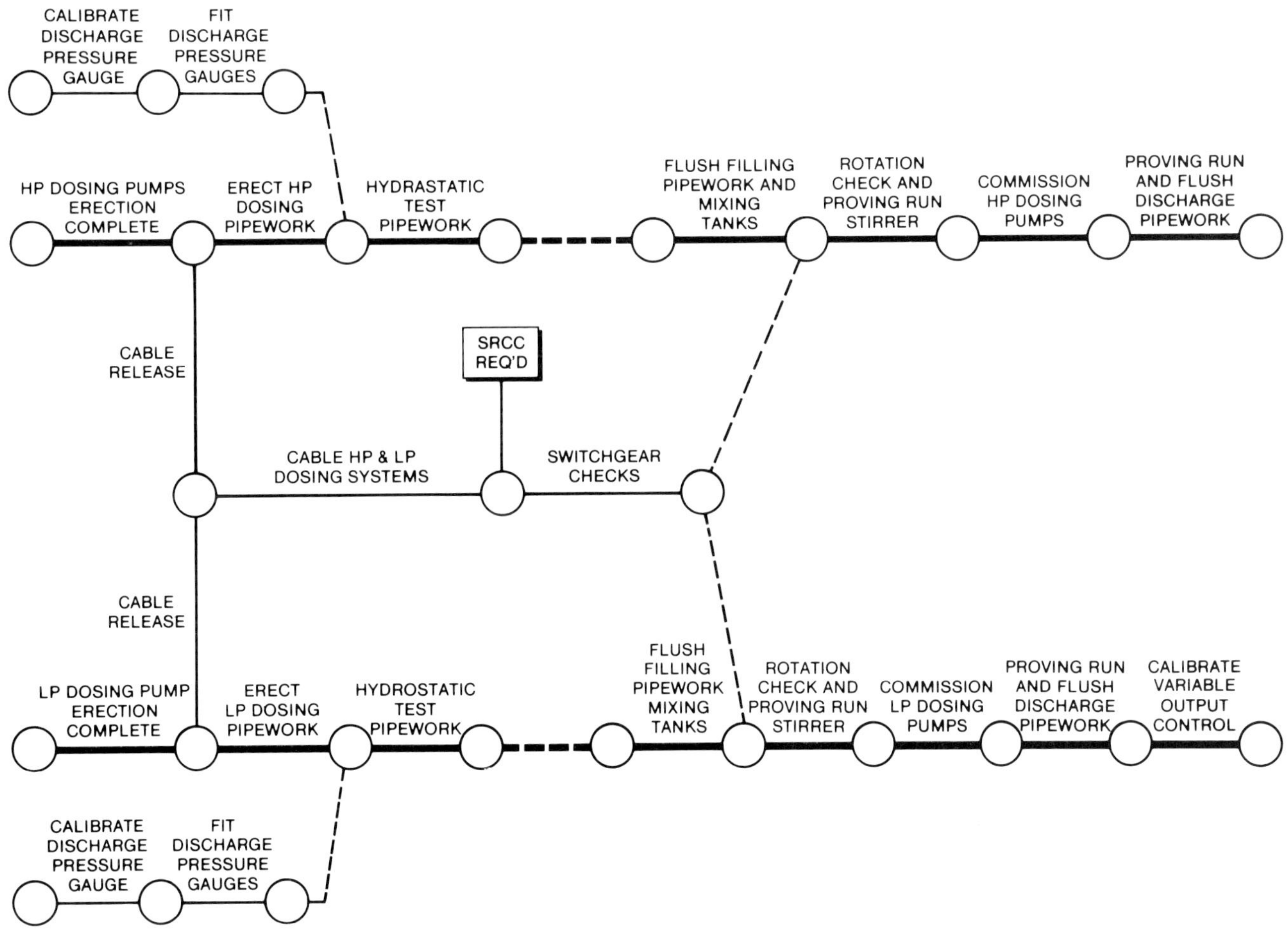

FIG. 4.16 HP and LP chemical-dosing commissioning network

4.8 HP piping systems

4.8.1 Description

The systems included in the generic title 'HP piping systems' are:

- The boiler feedwater and spraywater pressure lines, from the boiler feed/spraywater pump discharges to the boiler.
- The main steam supply pipework from the boiler superheater outlet manifolds to the turbine steam chests. On 500 and 660 MW units, there are four pipes, each carrying 25% of the total steam required.
- Low temperature (LT) reheat return pipes (four), carrying exhaust steam from the turbine HP cylinder back to the boiler reheater, and passing a proportion of the steam to the main boiler feed pump turbine on the way.
- High temperature (HT) reheat pipes, carrying reheated steam back to the intermediate pressure (IP) cylinder(s).

Detailed design and layout of HP piping is dealt with in Volumes A and C, but certain aspects require special care during erection and commissioning.

Each pipe service is carefully designed as a flexible system, i.e., in expanding from the cold condition, or contracting after shutdown, no excessive loads or turning moments should act upon the principal terminal points at the turbine, boiler or boiler feed pumps. The maximum loads and moments which the plant manufacturers have agreed to accept are shown on the flexibility diagram, which is always supplied under the terms of the contract.

There must always be some practical tolerance, but a first essential is to ensure that the pipes are erected as precisely as possible to the dimensions shown on the detail drawings, prior to applying the 'cold pull-up', which prestresses the pipe in the cold condition. The accuracy of the pipe position can be readily checked using a surveyor's theodolite.

HP piping ranges must be adequately supported using the specially designed spring-loaded supports described in Section 4.6 of this chapter. Because of their geometry, they provide near constant tension in the support rods, however the pipe moves when expansion takes place.

The pipework supports must be regularly checked during commissioning and should be adjacent to, or provided with, safe and satisfactory access prior to filling the boiler with water.

Safe access to the creep measuring bosses on the high temperature steam mains must also be provided. The datum 'as made' measurement across these is recorded by the pipe contractor and thereafter measured and recorded annually by the operator against operating hours.

4.8.2 HP piping system commissioning

Pre-commissioning checks

The HP piping butt-welds are either radiographed or ultrasonically tested. There is no question of employing hydraulic testing, except for the portion of feed piping which, depending upon design, must be included in the boiler hydraulic test.

At this stage, the HP systems form an important part of the chemical cleaning and steam purging processes, discussed in Sections 11 and 12 of this chapter. Spring-loaded supports, designed for steam-filled pipe loads, must be 'gagged' or fixed in some way in order to support the same pipes when filled with water.

The latest designs now being built are provided with built-in branches to suit chemical cleaning. Nevertheless, both for chemical cleaning and steam purging, much temporary piping is needed: the steam and feed systems must be made off permanently afterwards.

All piping drains and air release pipework must be properly installed in accordance with drawings and specification. It should not be possible to 'lock' feed-water between two main valves on the feedwater system without the facility to vent and drain.

The steam drain-point thermocouples should be installed and operable and the bulk of the thermal insulation applied, except where welded joints are left exposed as a precaution, to permit detection of possible leaks.

Commissioning

The feed system is required for commissioning the boiler during the chemical clean, steam purge and floating safety valves programmes, but the steam supplies must essentially be seen as part of turbine commissioning on main boiler steam, irrespective of any previous turbine running on temporary steam supplies.

A comprehensive check list, agreed between the piping contractor and the CEGB commissioning engineer, should be available and all concerned should know precisely what is involved and their individual duties.

Broadly, with piping, a *search* and a *watch* are required:

- Before commissioning, it is necessary to *search* the piping systems throughout because at some point they will change their position by as much as 460 mm when hot. Nothing must be permitted to obstruct this movement; and particular attention must be paid to places where a pipe passes through a floor or is adjacent to a structure. Any scaffolding remaining for construction purposes must be carefully inspected to ensure that it will not restrict this movement.

- All gags or supplementary supports put in for chemical cleaning must be removed and the systems made ready in all respects for warming through.
- The performance of the piping and its drains must be *watched* in all respects as it receives warming steam and begins to expand. Critical positions should have been identified and special measures taken.

Progress will be slow and some considerable time, often several hours, will be spent in raising metal temperatures with the boiler operating in the difficult areas of low pressure and low steam output relative to MCR. These conditions call for some superheater circulation to blowdown (to keep superheater metal temperatures within acceptable limits), and possibly some steam drainage from the main steam piping immediately adjacent to the turbine steam chests. There is usually a high level of noise associated with this operation. If the noise output is greater than 80 dB, it will be necessary to issue staff in the vicinity with ear defenders or plugs. The effect on people living in the surrounding district should also be considered.

Once the steam piping is up to temperature, search once more for any obstructions to expansion and thoroughly examine the operating positions of indicators on constant-tension spring supports and the attitude of supporting rods, which are generally designed to be vertical (or plumb) in the operating position.

It is the contractor's responsibility to provide final records of support settings and pipe movement from cold to hot conditions. It is essential to have this information in the station records before the contract is cleared.

5 Gas and air system

5.1 Introduction

The gas and air system comprises:

- Induced draught (ID) fans
- Forced draught (FD) fans
- Primary air (PA) fans
- Gas recirculation (GR) fans
- Seal-air (SA) fans
- Steam air heaters
- Gas air heaters
- Precipitators
- All interlocking ductwork, crossover ducts, dampers and expansion joints through to the chimney.

A typical gas/air system for a 500 MW oil-fired boiler is shown in Fig 4.17, which shows the location of the plant listed above.

The ducting is normally formed from 4.76 mm (3/16 in) mild steel plate, suitably stiffened externally with

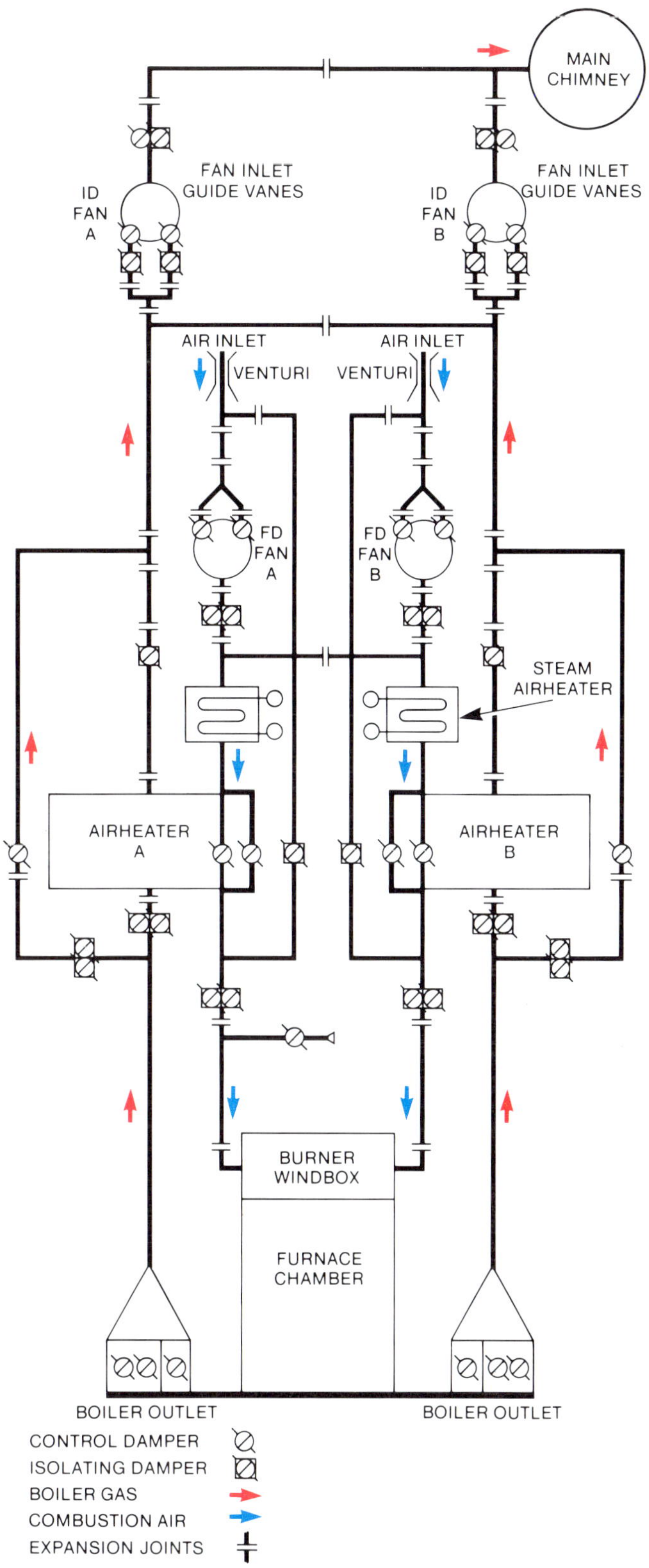

FIG. 4.17 Location of ducting expansion joints and dampers for a 500 MW boiler

RSCs, RSAs, brackets, etc., to form a rigid structure which can be supported or hung from the main boiler steelwork. Additional internal bracing is used on the large-section ducts. Internal handrailing is used at entry into vertical, or near vertical, sections of ducting.

Access doors are either bolted or of the cast iron quick-release type, fitted with glass-fibre joints ensuring air/gas-tight sealing.

At all 90° changes in direction of flow, guide vanes are fitted to reduce turbulence and draught losses. Expansion joints cater for relative duct movements; flow plates covering each joint reduce gas/air turbulence, whilst protecting the seal from accidental damage.

Dampers for control purposes are operated by either hydraulically or pneumatically operated actuators, whilst isolating dampers are operated by electrical actuators.

5.2 Draught plant

5.2.1 Fan types and drives

Most CEGB large unit fans have an impeller with runner blades of backward-inclined aerofoil section welded to a runner sheet and cone sheet, which also incorporates the inlet sealing rings. Figure 4.18 shows a typical fan layout, with fixed-speed drive and inlet guide vane (IGV) throughput control.

Various types of fan drives and arrangements are available to provide good combustion control. The main types used on the CEGB 500/660 MW units are:

- Variable-speed motors using induction regulators
- Fixed-speed motors with inlet guide vane control
- Two-speed motors with inlet guide vane control
- Fixed-speed motors with hydraulic couplings

The most economic arrangement for large coal-fired boilers is the two-speed motor with IGV control for the ID fans. These fans are designed to operate just below the speed change point under MCR conditions and the vanes control the fan output. The FD fans usually have fixed-speed motors with IGV control.

Large oil-fired units use either the fixed-speed motor with IGV control, or variable-speed motors with induction regulator speed control.

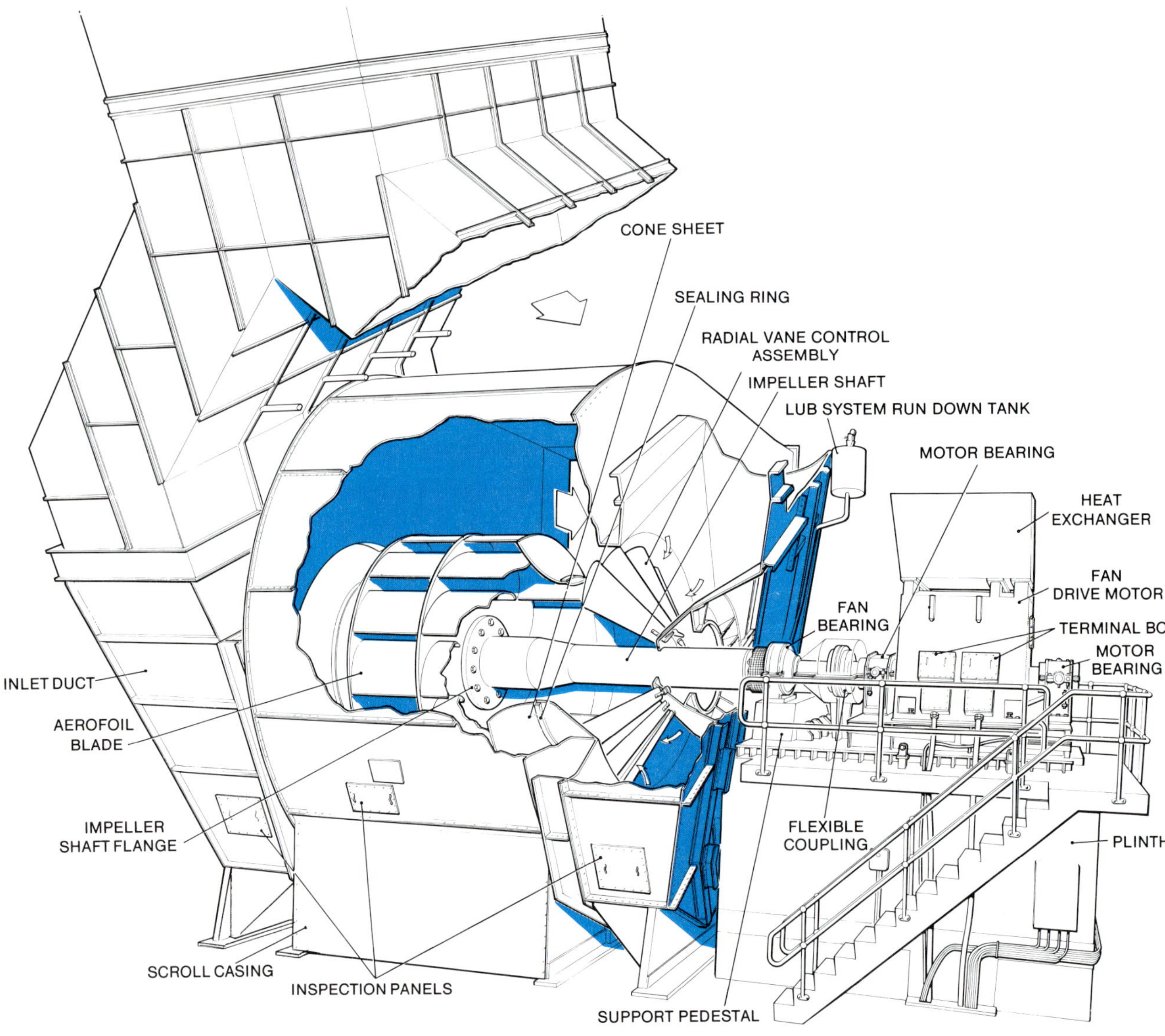

FIG. 4.18 FD fan and drive

A typical radial inlet guide vane is shown in Fig 4.19, a radial vane assembly being fitted to each suction inlet. The vanes are fully closed to reduce load on the motor when starting and are then progressively opened to control the fan output, as required. Inlet guide vanes are normally moved by a hydraulic actuator or an electrical actuator with thyristor control.

The induction regulator is normally used in conjunction with an AC commutator motor and controls motor speed by altering the strength of the magnetic link between the rotating field of the motor stator and the motor rotor, so changing the motor slip. The induction regulator is usually controlled by a hydraulic or an electrical actuator.

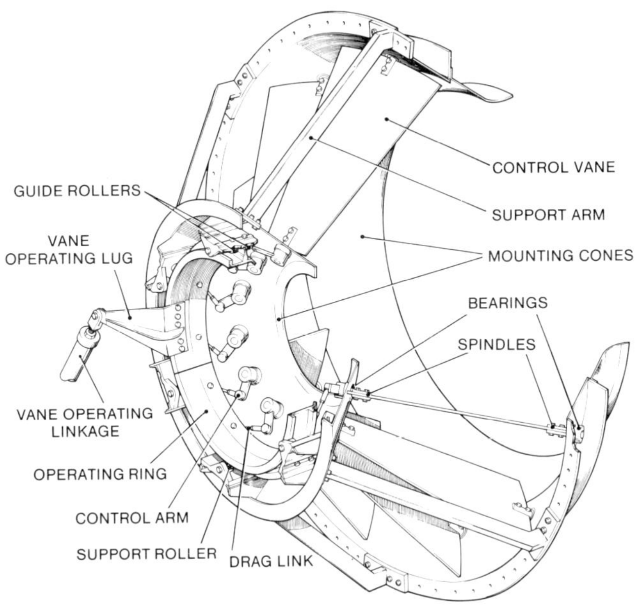

FIG. 4.19 Radial inlet vane control gear

On some 500 MW oil-fired units the speed control of the ID, FD and GR fans is by hydraulic fluid couplings. The primary air fans on the 660 MW second stage at Drax Power Station also have this type of speed control. A typical fluid coupling is shown in Fig 4.20 and is designed to produce a stepless speed control at the fan from a constant-speed motor. Power is transmitted by viscous drag and the kinetic energy flowing between a set of opposed rotating cups. The output speed depends upon the amount of oil carried between the two sets of cups. If the cups are full, then maximum torque is transmitted; as the quantity of oil is reduced, so also are the torque and speed. The quantity of oil between the cups is varied by the vertical position of a scoop, which can be lowered or raised into an oil reservoir. The reservoir is formed by the outer casing of the coupling, rotating at the same speed as the runner. The slip in the coupling heats the oil, so a water cooler is incorporated in the scoop circuit to cool the oil. This oil circuit is independent of the main lubricating oil system.

5.2.2 Fan lubricating oil systems

Large fans require an auxiliary lubricating oil system, which normally falls into one of the following categories:

- The main lubricating oil pump is driven from the fan shaft with an auxiliary AC pump for start-up and shutdown and a DC pump for emergencies, if AC supplies were lost; or
- The main lubricating oil pump is driven from an AC motor, with a DC motor-driven pump for emergencies; or
- A motor-driven pump transfers oil to a high level header tank, from which it is gravity-fed to the bearings.

Each system will contain oil coolers, filters, and alarm equipment for temperature and pressure. The standby pump is automatically started when the pressure falls. The temperature is maintained by a thermostatically controlled fan, which cools an air radiator. The oil in the lubricating oil tank is maintained above 20°C by a thermostatically controlled heater. Figure 4.21 shows a typical system layout.

5.2.3 ID fan brakes

Each ID fan is fitted with a brake to prevent rotation due to natural draught caused by the chimney. Figure 4.22 shows a typical brake assembly and is basically a set of shoes which clamp to the fan drive shaft. The brake can be applied manually or remotely-operated from the fan sequence shutdown equipment once the fan has come to rest, via a solenoid-operated mechanism.

5.2.4 Seal-air fans

These are installed on large 500/660 MW coal-fired units to provide sealing air to the mills. A typical seal-air fan is shown in Fig 4.23. They are normally direct-driven by an electric motor through a 'Bibby' flexible coupling. Air is discharged into a common main, through a non-return flap valve, and is then fed through isolating valves to the mills. An alarm, fitted on the common main, operates if the seal-air pressure drops below 135 mbar. The inlet air is drawn through a filter system, which removes dust down to 5 μm. Pressure drop is measured across the filters, which are changed when it reaches 62 mbar.

5.2.5 Safety

- When commissioning dampers/vanes, care should be exercised to ensure that people do not become trapped in the vanes or ducts.
- Facilities should be available for dealing with lubricating oil spillages.

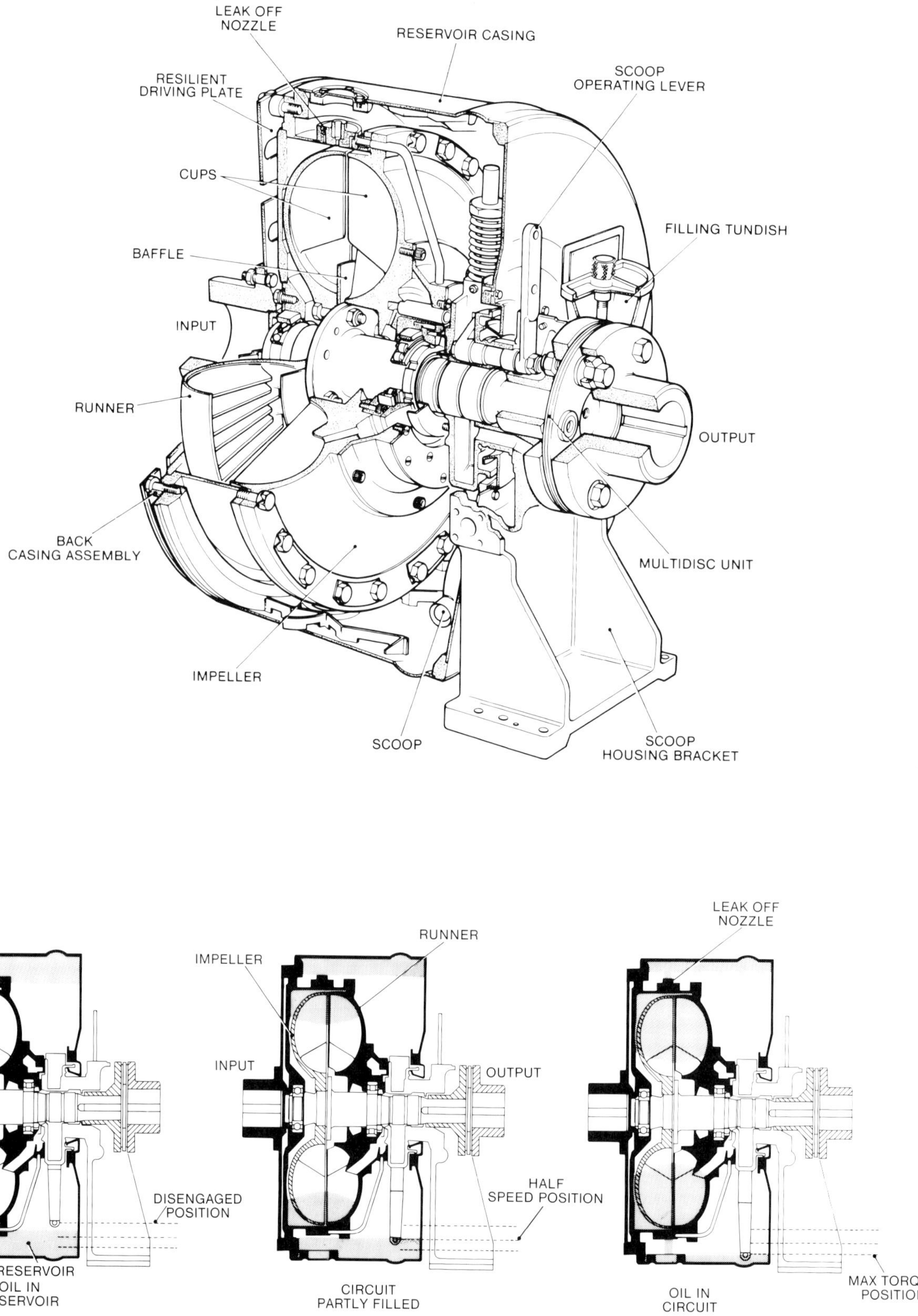

FIG. 4.20 'Fluidrive' coupling

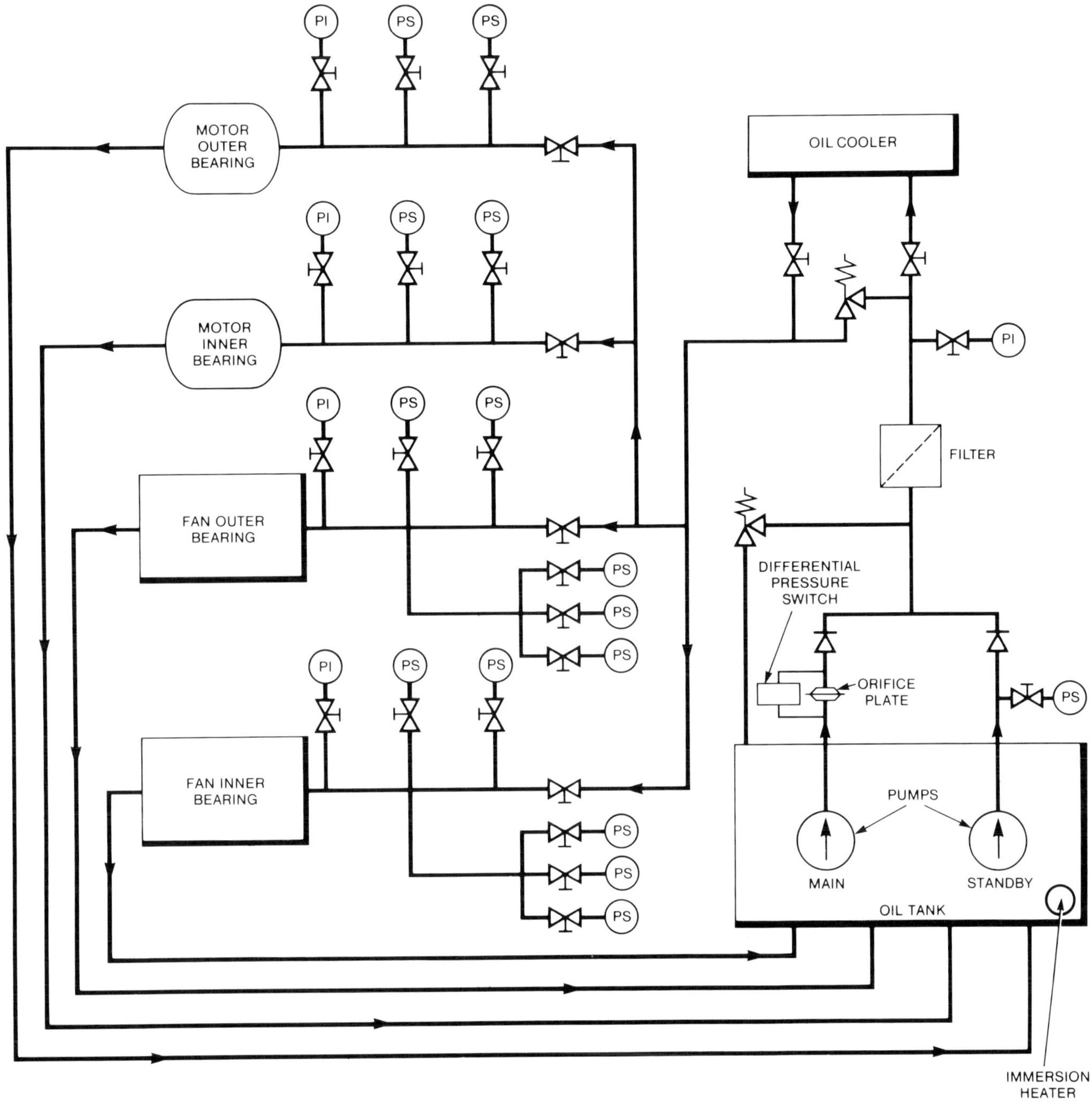

FIG. 4.21 Typical fan lubricating oil system

- High voltage switchgear operation must only be carried out by authorised persons.
- When running motors uncoupled, the coupling halves must be adequately secured and the fan motor plinth restricted to personnel involved in the tests. Coupling guards must be fitted prior to running a motor/fan unit.
- A thorough check must be made of all gas/air ducting prior to fitting access doors and running fans. Personnel access should be carefully controlled by the use of Permits for Work.

5.2.6 Resources

Services

- Electrical supplies to motors/dampers.
- Flushing and lubricating oil.
- Fine-gauze filters for the oil flush.
- Jumper pipes to by-pass motor bearings during the oil flush.
- Oil-absorbent material to contain oil spillages.
- 2.5 kV Megger.

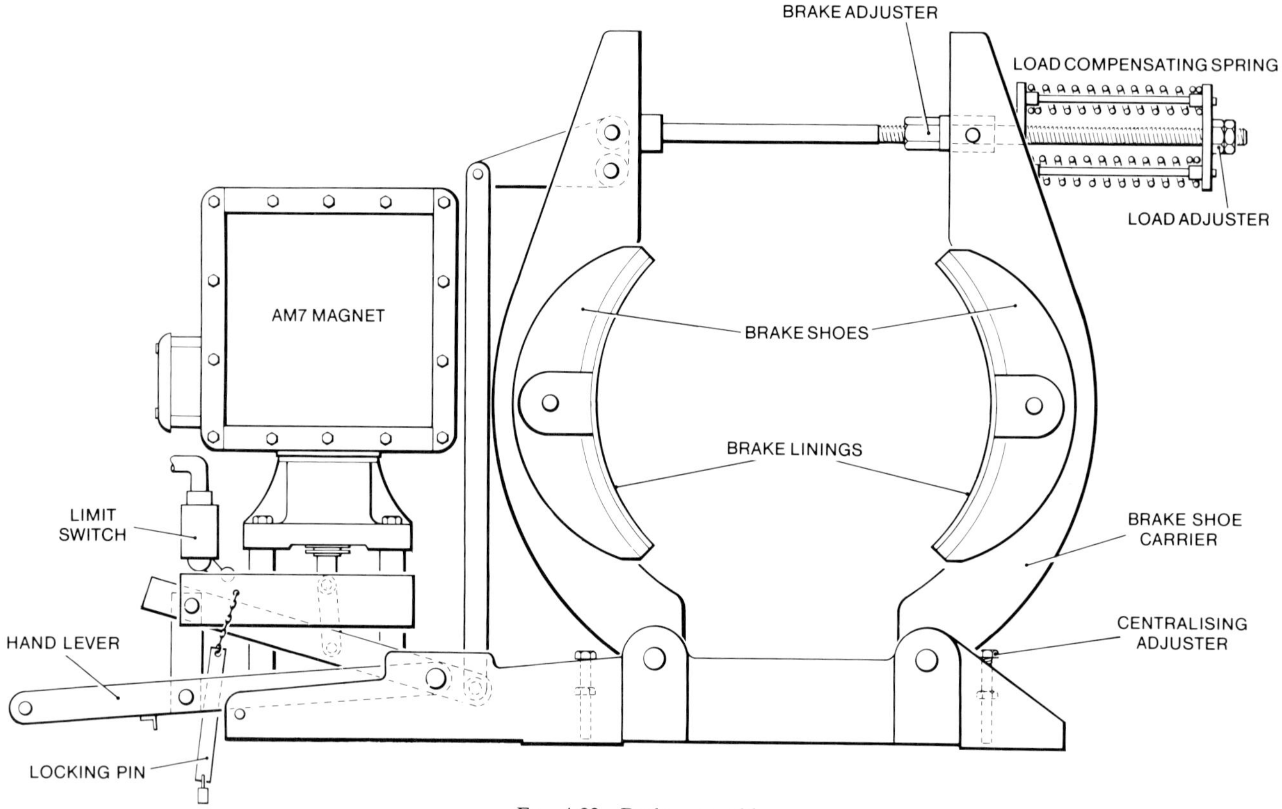

FIG. 4.22 Brake assembly

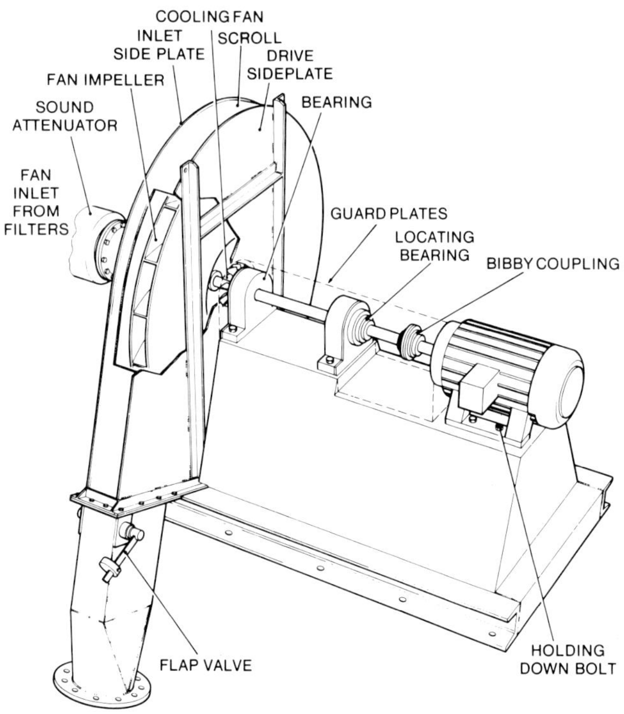

FIG. 4.23 Seal-air fan

- Tachometer.
- Vibration monitoring gear.

Staffing

A team of six to eight people will be needed and four to six weeks will be required to commission each fan.

5.2.7 Draught plant commissioning

The general commissioning sequence for an ID, FD, GR or a PA fan is as follows:

- To commission local and remote instruments, interlock devices and alarm inputs.
- Commission and flush the lubricating oil system.
- Commission inlet guide vanes, dampers and ductwork.
- Commission the driving motor/induction regulator (see Chapter 3, Section 4).
- Commission the hydraulic coupling on this type of fan.
- Initial run and fan balancing checks.
- ID fan brake commissioning.
- Carry out full boiler airflow tests.

A typical commissioning sequence diagram is shown in Fig 4.24 and can be adopted to suit most types of fan installations.

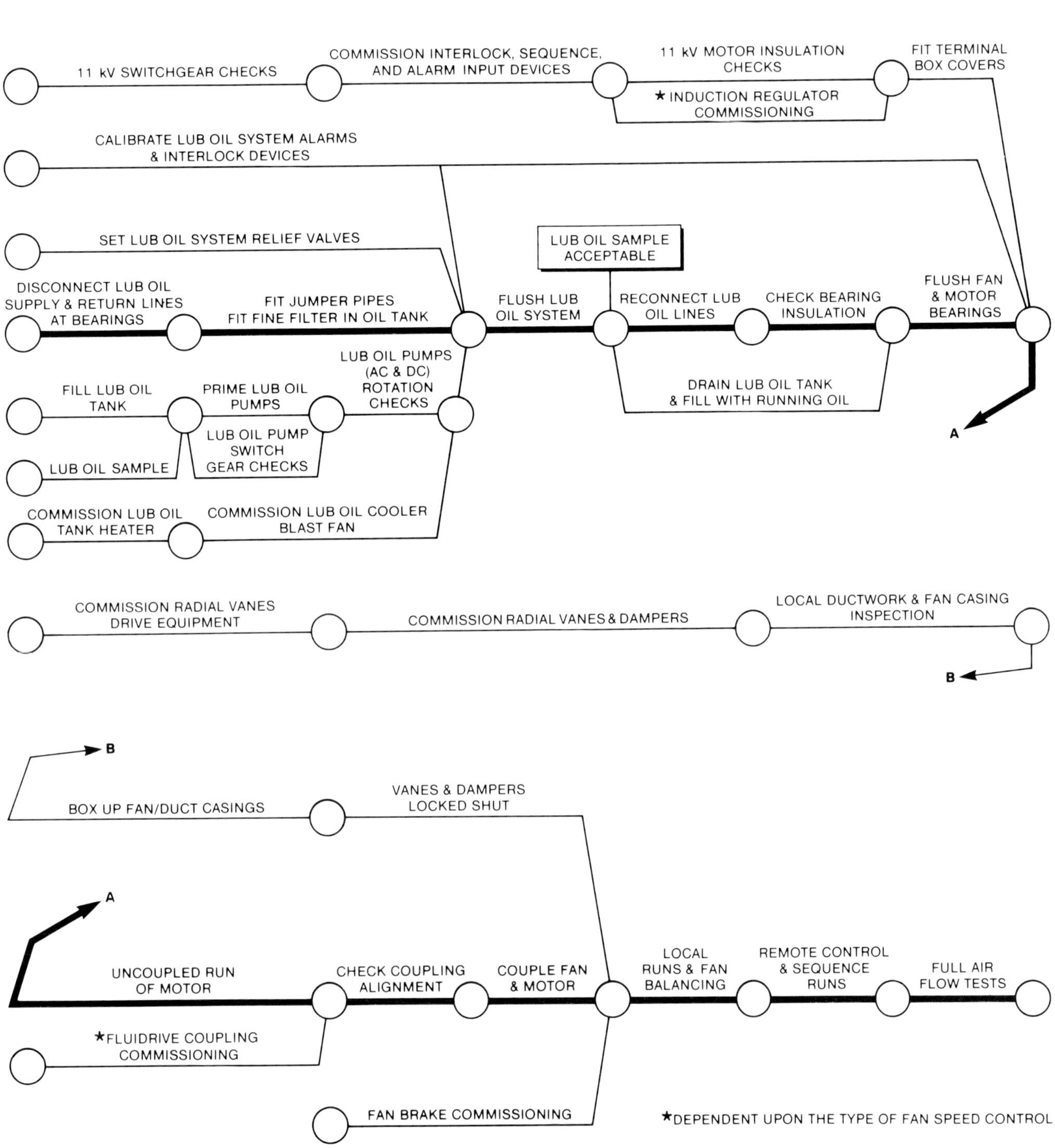

FIG. 4.24 Typical fan commissioning network

Local and remote instrumentation

All local and remote instrumentation should be checked, using the procedure in Chapter 3, Section 3 and relevant SCLs. This will include pressure gauges, pressure switches, temperature gauges, limit switches, etc.

Lubricating oil system commissioning

The objective is to achieve a satisfactory clean lubricating oil supply-and-return system for the fan and its drive motor bearings. Flushing depends on the design of system and the manufacturer's requirements; basically, the fan and motor bearings are filled with their correct grade of oil, which is retained by fitting blanks on the inlet and outlet connections. Jumper pipes are fitted to connect the oil supply-and-return pipes to provide a flushing path for the oil system pipework.

The lubricating oil tank is filled with the running grade of oil sufficient to cover the oil pump suctions. The AC and DC lubricating oil pumps are checked for direction of rotation. All switchgear, instrumentation, tank heaters, oil cooler fans and lubricating oil system interlocks are checked for completeness. The AC lubricating oil pump will then be primed and the lubricating oil system flush commenced. Before starting the flush, a fine-gauze filter should be fitted to the oil return inlets in the storage tank to catch debris from the system. The flush will continue, using both the AC and DC lubricating oil pumps, until the system is declared clean. During the flush, oil system relief valves should be set to their correct settings. The lubricating oil tank should be drained and thoroughly cleaned, then filled with the running charge of oil. The oil pipework jumper-pipes at the bearings and the bearing blanks should be removed and the permanent pipework connected. During this operation, it is particularly important to check that the bearing electrical insulation is satisfactory. A further flush of the fan and motor bearings should now be completed and the correct oil flow to each bearing set, using the bearing oil supply regulating valve.

During the oil flush, the following instrumentation checks should be completed:

- Bearing 'low oil pressure' alarms commissioned and correctly set (0.7 bar).
- Oil tank 'high' and 'low' level alarms tested.
- Tank heater commissioned to control at the recommended setting (27–34°C).
- Bearing temperature high alarms commissioned at the recommended setting (50°C).
- Oil blast-cooler fan adjusted to cut in/out at the correct settings.
- AC and DC lubricating oil pumps set to cut in at their recommended pressures (AC 1.5 bar; DC 1.3 bar).
- Bearing oil pressure starting interlock set at the correct pressure (0.6 bar).
- Oil pressure relief valves correctly set.

Radial guide vanes, dampers and ductwork commissioning

A procedure for commissioning dampers/vanes is covered in Chapter 3, Section 1.2, but the main points to be observed are highlighted as follows:

- Check correct assembly of damper blades and linkage, and ensure free movement over full travel.
- Confirm that indication marked on damper shaft truly agrees with the actual blade alignment.
- Check that the blades are parallel with gas flow, in the fully-open position.
- Check that damper blades seal effectively in the closed position.
- Confirm that the environmental location is unlikely to cause operational problems, e.g., bearing seizure, blades fouling each other or the ducting, dust build-up, etc.
- Ensure that each damper can be secured (and locked) at the fully-open and fully-closed positions. This is important where access to certain sections of ducting may be achieved by climbing through the open damper. Usually, however, dampers are used for safety isolation when in the shut position.
- Check that limit switches operate some 3–5° before the damper reaches its extreme limits of travel. If an actuator is fitted, then the actuator limits must operate just before the damper reaches the mechanical travel stops, with due allowance for any over-travel of the linkage.
- The vane/damper drive actuators should be checked in accordance with the procedures in Chapter 3, Section 2.
- Once the above checks are completed, the damper/vane and actuator drive can be coupled.
- The dampers/vanes can then be driven by their actuators in the local control and all limit switches checked for correct operation.
- Remote/auto control of the dampers/vanes can now be checked.

Once the damper commissioning is completed and before the ducting access doors are boxed in, the ductwork must be thoroughly checked and inspected for the following:

- Correct installation of seals and expansion joints
- All bracings and fixed guides correctly installed
- All debris removed from the ductwork and fan casings

The access doors should then be closed and bolted ensuring that the correct sealing gaskets have been applied.

Note When ducting inspections are being carried out, all dampers and fans must be isolated and rendered inoperative.

Main drive motor/induction regulator commissioning

The 11 kV main drive motors/induction regulators and associated switchgear are described in Volume D, and commissioning should be carried out as described in Chapter 3, Section 4. When the switchgear and cabling checks are completed and power supplies are available to the motor, a complete physical check of the motor/induction regulator should be made to satisfy commissioning personnel that the installation is complete and satisfactory. Particular attention must be paid to cleanliness, and clean and safe access must be available.

The motor/fan coupling should be broken and checks made to ensure that both coupling halves are kept clear for the motor uncoupled runs. Motor/cable insulation checks must be made just prior to the uncoupled runs to ensure that they are satisfactory. Motor pedestal insulation checks must also be made. For the uncoupled runs the motor should be operated from a local-to-plant pushbutton and communication established between the motor and control point. For the initial run, the lubricating oil system should be in service local-to-plant and the drive motor allowed to run up to speed to check the direction of rotation. If the direction of rotation is incorrect, then the motor should be stopped and the motor terminal connections changed over. A four-hour uncoupled run should be carried out to check the performance of the motor bearings.

When the uncoupled runs are completed and the fluid couplings have been commissioned, the motor/fan should be isolated and coupling alignment checks carried out, followed by the coupling of the fan and motor.

Fan balancing is usually done in the works prior to delivery, but some balancing may be necessary on the initial low-load runs.

If the fan has induction regulator speed control, then the induction regulator must be checked for correct installation and electrical insulation, and the following interlocks proven prior to starting:

- The fan is prevented from starting, unless the induction regulator is in the starting position. The starting position is checked by a limit switch in the induction regulator.
- The fan can be stopped or prevented from starting, if the supply voltage falls below 70% of nominal for more than 3 seconds by means of a low volt relay and timer.
- Speed must be below 500 r/min for starting: this is detected by the fan speed pulse counter.
- Inlet guide vane or damper shut (limit switch in drive actuator).
- Operation of the start switch for the fan starts the AC lubricating oil pump, motor cooling fans and induction regulator cooling fans and releases the fan brake.
- The induction regulator drive system actuator is commissioned in accordance with the procedures in Chapter 3, Section 2.

Fluid coupling commissioning

If the fan speed-control is by fluid coupling, then this must be checked prior to connecting the fan to the motor coupling:

- Check that all driving plate lengths and coupling alignments are as recommended by the manufacturer.
- The concentricity of the scoop housing bracket and the half-coupling must not exceed 0.13 mm.
- The fluid coupling should be filled with the correct grade of oil (recommended by the manufacturer).
- The scoop operating lever must be moved to the minimum position.

The fan lubricating oil system should now be put in service local-to-plant and the fan motor started locally from the switchgear. The scoop operating lever is now slowly moved to the maximum speed position to circulate oil through the pipes and coolers. After the motor and coupling have been run for a short period, the scoop should be returned to the minimum position and the filling tundish cover cautiously removed. More oil should then be poured into the tundish until a definite 'welling up' of oil is seen there. The coupling is then fully primed. The tundish cover can now be replaced and the fan motor shut down. The fan should be coupled to the motor and fluidrive coupling unit: fan commissioning can now proceed.

Initial run and fan balancing

The flexible or fluid coupling should now have been made up; during this period, the final fan inspection should be carried out, together with any final clearing of ductwork.

Isolating dampers should be locked off in the 'shut' position. Inspection doors should be left off the fan suction-boxes and the discharge volute, with the access restricted by positioning a crossbrace of scaffolding poles across the entry.

Note Upon completion of making up the flexible or fluid coupling, the anti-windmilling brake will have been fitted to the ID fan. Manual operations of the brake must be checked prior to commencement of the fan initial run. Obviously it will be preferable if the brake power supplies can be made available for the initial run; however, it is not entirely necessary and hand operation would be quite adequate.

Once all inspections and work on couplings are completed, the safety guards must be fitted.

The radial inlet control vanes should be set in their minimum positions, or, in the case of a hydraulic coupling, the fluidrive coupling should be set with its scoop operating lever in the minimum position.

Again, the area around the fan and motor unit should be checked for safety of access and cleanliness, prior to allowing the initial run to commence.

Once satisfactory, power supplies to the lubricating oil system should be made available, and the lubricating oil system put into service, together with a check on the availability of the emergency lubricating oil pump.

With communications established between the switchgear room and fan unit, the breaker should be checked for closing and opening action from the local-to-switchgear controls and emergency pushbutton, with the breaker in the 'test' position.

Once established as correct, the test controller should authorise initiation of the main motor. Personnel should be stationed:

- At the 'Emergency Stop' pushbutton in case an immediate trip is necessary
- At the switchgear to note the maximum starting current
- At the fan to monitor time of run-up to full speed

It should not be necessary to subject the fan to any type of rotation test; thus, once started, it should continue for a six-hour proving run. During this time the running current should be recorded. Again, throughout the proving run, the fan and motor bearing temperatures should be logged and studied for any undue trend away from normal running temperature. Also close observation for vibration of the fan casing, bearings, joints, vane gear, fluid couplings and foundations should be maintained.

Fan balance will probably have been carried out in the works and therefore balancing of individual fan units should not be required. However, when a unit set of fans is available (all FD, PA, GR and ID fans), the manufacturer's test engineer should attend to carry out balance checks.

Upon completion of the six-hour proving run, the fan unit should be tripped and the fan 'run down' time noted, with particular attention to the period over which the last 100 r/min decays. This information will be of value in establishing when the anti-windmilling brake on ID fans will be applied by the sequence system. Once the fan has stopped, the oil system should remain running for a further 30 minutes, and the main motor heaters should be energised.

In order to complete full-sequence load runs, it will be necessary for all other fan commissioning to be completed and the boiler available for full airflow from FD fans to the chimney. This will be dealt with later.

ID fan brake commissioning

Fan brake commissioning under power takes place after the initial fan run: the brake cover must be removed to allow clear observation of the brake action.

On energising from the local control position, the brake shoes should 'nip' the brake drum all round. If this is not so, the brake solenoid should be de-energised and the clearances rechecked: any necessary adjustments are made, using the shoe-positioning screws and tensioning springs. On completion of adjustments, the brake should be re-energised and clearances again checked, until the brake operation is satisfactory.

Once satisfactory operation of the brake is proved, the proximity limit switches should be set to indicate 'brake on' and 'brake off' positions, for alarm and sequence logic purposes.

Seal-air fan commissioning

A check is made that erection is complete on both the fan and the discharge pipework/ducting to the mills. Check that all bearings have been greased and that the flexible coupling has been aligned and coupled, with coupling guard fitted. The inlet V-bag filters should be in place and the non-return valve flap set to open with discharge from the fan.

The seal-air valves to the mills are then locked closed. The fan motor should have been checked in accordance with the procedure in Chapter 3, Section 4 and the SCL, and a flick test for correct rotation carried out. If rotation is correct, a four-hour run will be carried out to check motor current, bearing temperatures and fan vibration.

When the mills are available to receive sealing air, the fan will be run to ensure that it will maintain sufficient seal-air pressure to satisfy the mill interlocks.

5.3 Gas air heaters

On CEGB 500/660 MW units, two types of regenerative gas air heaters are used:

- Rotating-hood (Davidson type)
- Rotary heat exchanger (Howden type)

5.3.1 Rotating-hood air heater (Davidson type)

These are fitted on the 660 MW oil-fired units at Grain Power Station and the 500 MW coal units at Ironbridge Power Station. Figure 4.25 shows a typical rotating-hood air heater.

Each air heater comprises a number of heating elements contained within a stator. Two air-hoods mounted on a common shaft are positioned above and below the stator surfaces and are rotated slowly in synchronism (0.75 r/min) by an electric motor/gearbox unit driving a rack and pinion attached to the cover air-hood.

In operation, hot flue gas from the economiser outlet enters the air heater via the gas inlet flue and passes downwards through the stator elements, giving up a large proportion of its heat *en route:* the gas is then discharged via the gas outlet flue at the bottom to the ID fan inlets.

Cold air from the FD fan is preheated by the steam air heater before entering the gas air heater. It then passes upwards via the lower air-hood to the stator, where it absorbs the heat previously imparted to the stator elements by the flue gas. The air-hoods rotate in synchronism and the heated air is collected by the upper air-hood, passing via the air heater outlet duct to the burner windboxes. A system of sliding seals is incorporated in each air-hood to prevent leakage between the air and gas sides of the air heater.

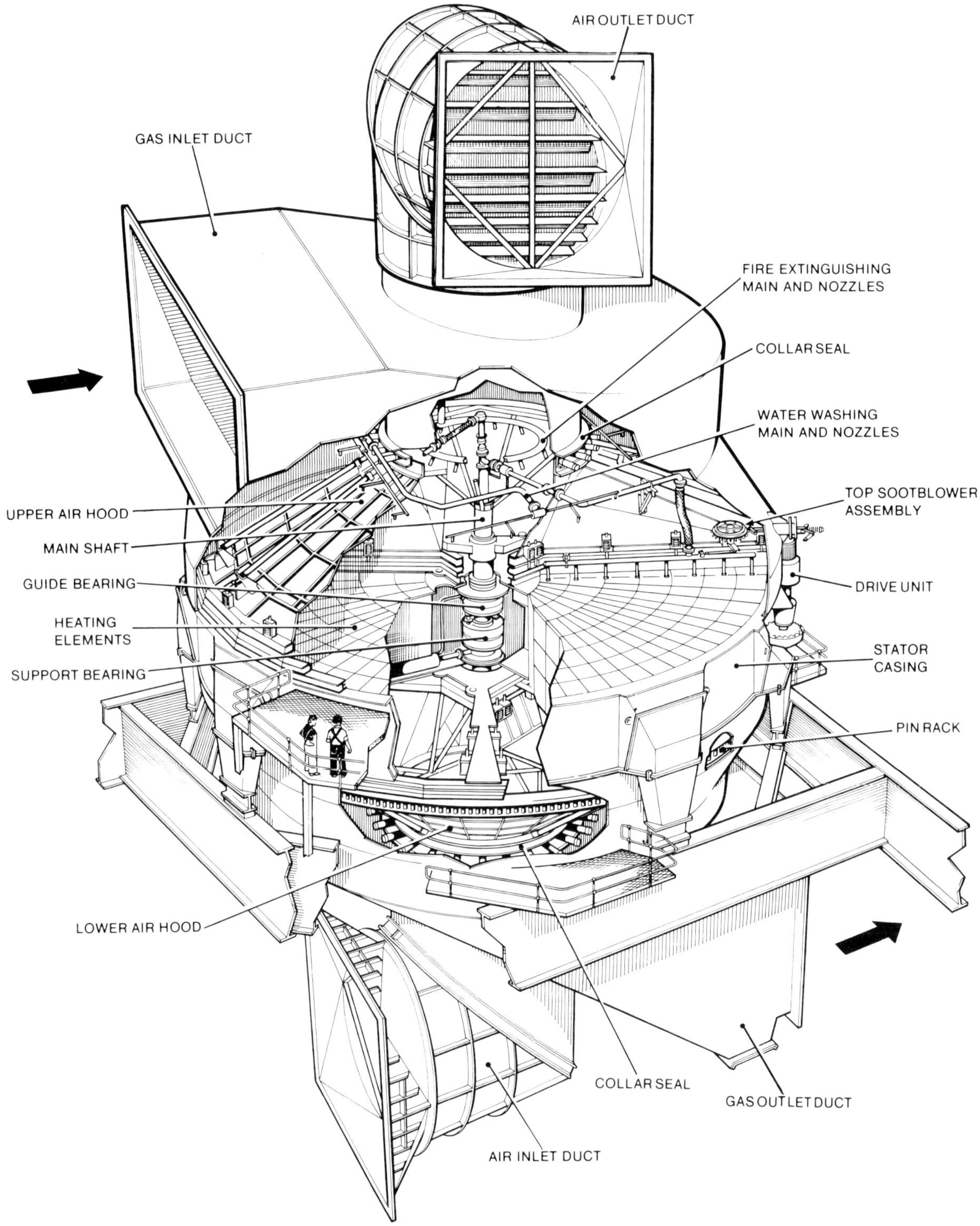

FIG. 4.25 Gas air heater — rotating-hood type

Provisions for on-load sootblowing and off-load water washing are provided.

The gas air heaters are each supported in a steel girder framework by means of eight feet attached to the stator periphery, each foot incorporating a sliding bearing to allow limited movement for thermal expansion.

The main shaft, which carries the two air-hoods, is positioned vertically within the stator and is located by three roller bearings of the split type carried in two housings to the stator hub. The top and bottom bearings are radial thrust, single-row units and the centre bearing is a vertical thrust unit.

The bearings are oil-cooled and lubricated from the external lubrication unit, oil passing through the top bearing housing to the lower bearing housing and thence back to the tank. Both bearing housings are fitted with labyrinth glands to prevent oil leakage into the air circuit. The main components of the lubricating oil system are an oil sump tank, two lubricating oil pumps (one for normal running and the other standby), a duplex oil filter and two air-blast oil coolers. The system is protected by a series of pressure switches, temperature indicators and transmitters for control and interlock purposes. A typical lubricating oil system is shown in Fig 4.26.

An interesting feature is the barometric loop fitted in the oil-return line from the bearings. The height of the U-bend on each barometric loop is such that the static head balances the positive pressure inside the air heater at the bearings. Any air that is entrained in the oil returning from the bearings is expelled to atmosphere through a small vent pipe at the top of the barometric loop, before the oil returns to the sump tank for recirculation.

The air heaters are provided with full sootblowing and water washing facilities. These are covered in Section 8 of this chapter.

The air heater is provided with gas inlet and outlet, air inlet and outlet, and gas by-pass dampers.

Each air heater is provided with fire detection and extinguishing equipment. This is covered in Section 5.3.3 of this chapter.

5.3.2 Rotary heat exchanger (Howden type)

These air heaters are more common in the CEGB: they consist of a large-diameter heat transfer matrix of metal plates, which is heated by rotation through the hot gas system. The matrix then rotates through the combustion airstream, thus heating the air and cooling the matrix. To ensure maximum heat transfer, contraflow is used, the air flowing upwards and the hot gas downwards.

The rotor is driven at 1.5 r/min by an electric motor, through a flexible coupling and associated reduction-gearing arrangements attached to the top frame, which spans the rotor. An alternative drive is provided from an auxiliary motor, which turns the rotor at 0.25 r/min through a clutch coupling, when water washing the heater elements. A shaft-drive centrifugal fan is fitted to blow cooling air over the gear casings. The sealing system is designed to permit minimum operational clearances to be maintained over the wide range of temperatures associated with variable-load conditions. This reduces the leakage of combustion air into the flue gas when the rotor bends, due to the temperature gradient from the top to bottom faces. The rotor should not come into rubbing or rolling contact with the sealing surfaces: this is achieved by adjustment of the seal positions to maintain minimum clearances. Figure 4.27 shows a typical Howden-type regenerative rotating-matrix gas air heater.

The rotor is supported by a roller thrust bearing, which requires constant lubrication. The bottom support bearing and the top guide bearing are continuously lubricated by a self-contained recirculation system.

The oil storage tank is fitted with a level switch and a thermostatically controlled immersion heater. Two electrically-driven oil pumps are installed, either one of which can be selected for standby duty. Figure 4.28 shows a typical lubricating oil system. The pump

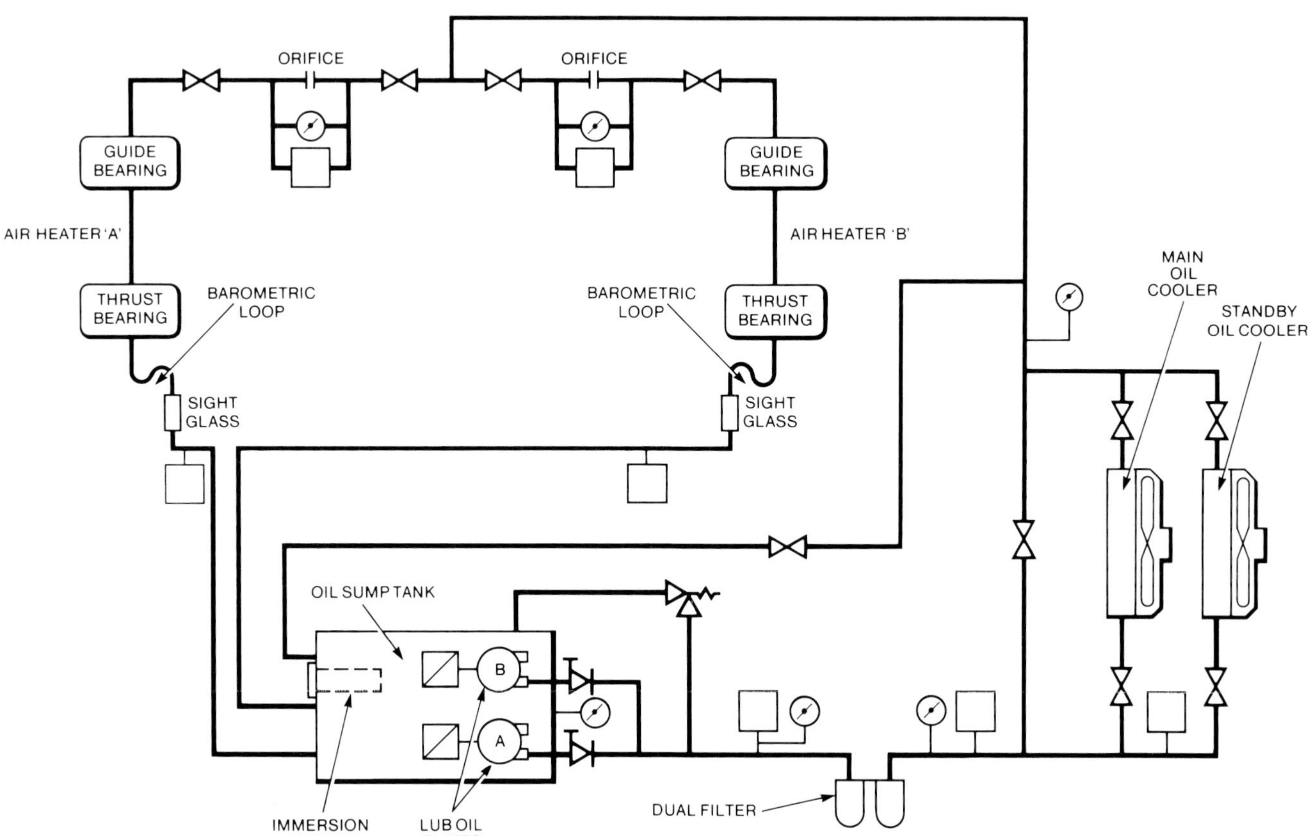

FIG. 4.26 Rotary-hood gas air heater lubricating oil system

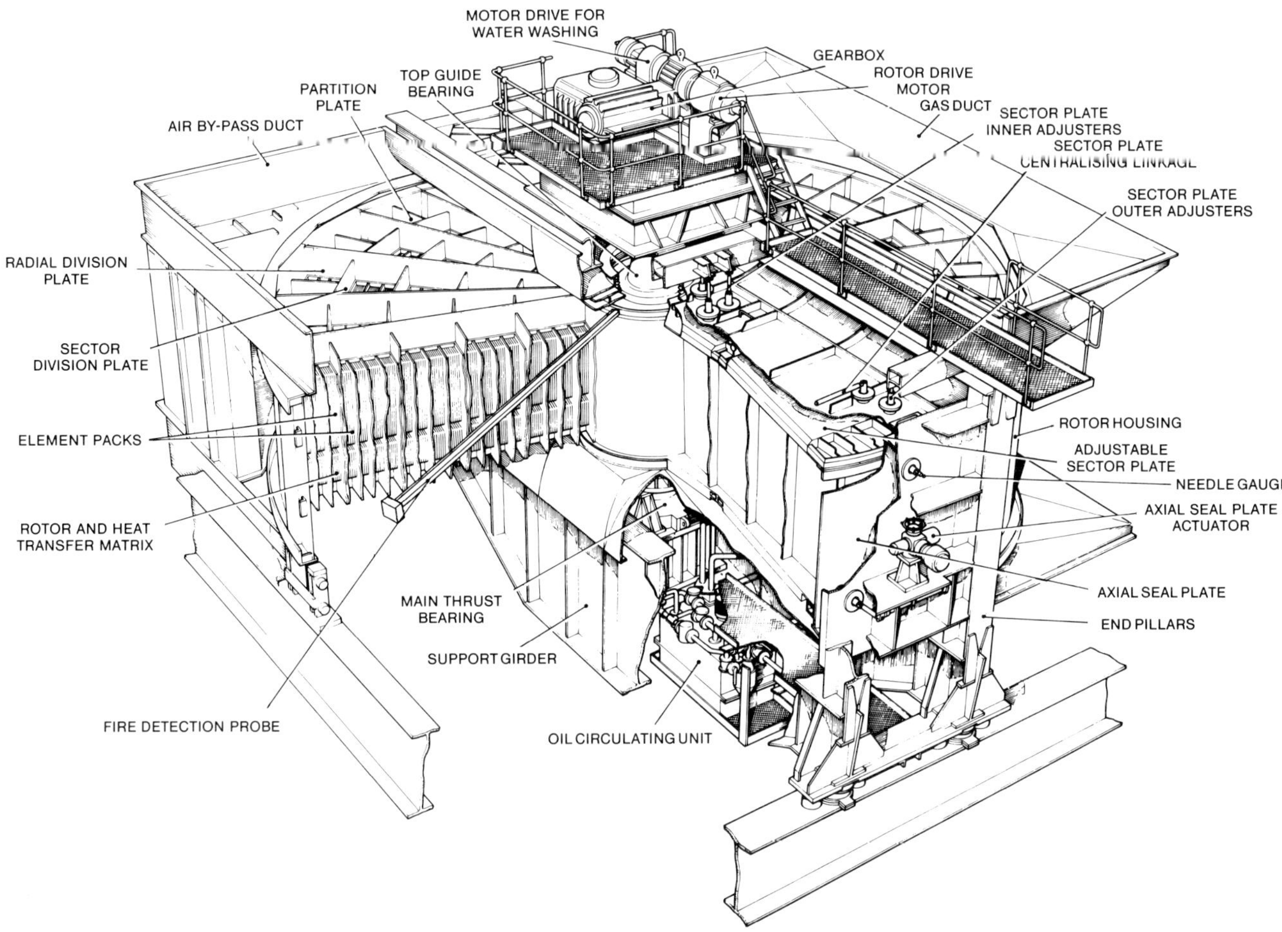

FIG. 4.27 Regenerative gas air heater — rotating matrix type

supplies oil at a rate of 11.5 l/min. This is discharged through externally-mounted non-return valves to a duplex oil filter and a Howden U-type oil cooler, which is fed with cold water from the general services water system at a rate of 700 l/h. A pressure relief valve is incorporated in the oil supply line from the pump to the oil filter and is set at 7 bar. A pressure switch and a pressure gauge are incorporated in the common discharge line from the pumps.

From the oil cooler, the oil is directed through a four-way distributor piece to the top steady-bearing, to the bottom thrust-bearing supply lines, and to the oil reservoir tank drain, which incorporates an isolating valve. The supply line to the bottom support bearing incorporates a non-return valve, a flow switch with by-pass, an orifice plate connection and a pressure switch. A drain line incorporating an isolating valve is fitted after the non-return valve to allow the system to be drained.

The supply line to the top bearing inlet incorporates a non-return valve, a flow switch with by-pass, an orifice plate connection and a pressure switch. A three-way distributor at the top bearing-housing assembly divides the oil supply into two lines, one to the bottom of the housing and the other to the top of the housing, where the oil supply is sprayed onto the spur gear. A drain connection, incorporating an isolating valve, is located after the non-return valve to enable the system to be drained.

Oil is returned to the reservoir tanks from the bearing housings by means of a drain return pipe from each housing. Each return pipe terminates in a sight glass mounted on top of the tank to allow visual checking of the oil flow in the lines.

The air heaters are supplied with full sootblowing and water washing equipment, which are covered in Section 8 of this chapter.

The air heater is provided with gas inlet and outlet, air inlet and outlet, and gas by-pass dampers.

5.3.3 Gas air heater fire detection and extinguishing systems

These are similar for both types of gas air heater.

The main problem in air heater operation is the prevention of deposits accumulating on and between the plate elements; these impair performance and, in the case of combustible deposits, create a serious fire hazard.

Air heater fires in pulverised-fuel-firing systems are generally associated with frequent and prolonged use of lighting-up oil burners for pre-commissioning steam raising, low-load operation and the lighting-up and shutting down of a boiler. On oil-fired boilers, the risk

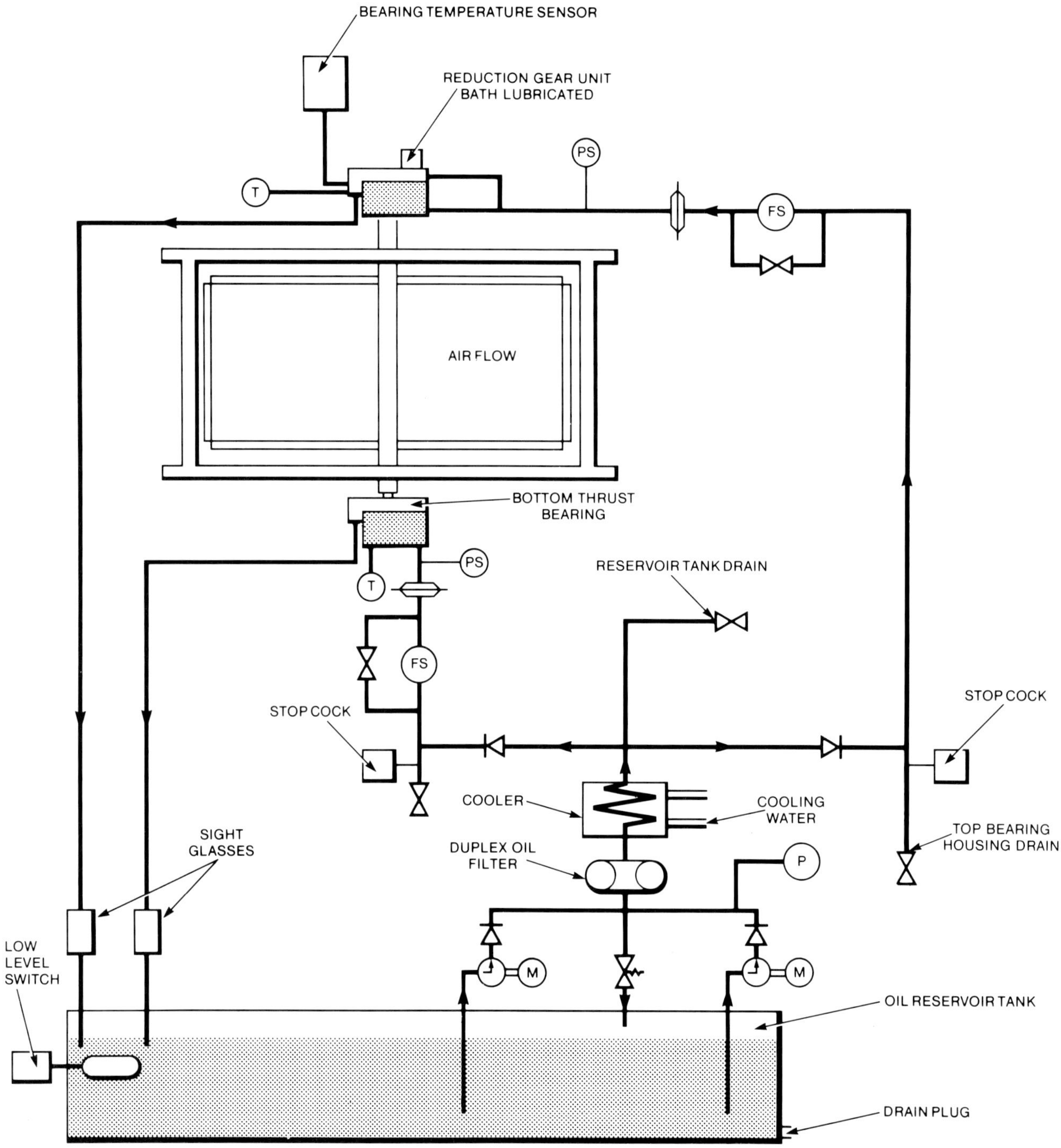

FIG. 4.28 Rotary gas air heater lubrication system

is always present, particularly during boiler start-up and shutdown. The main causes of air heater fires are usually:

- Carryover of unburnt fuel oil, due to delayed ignition or leakage, depositing on the air heater elements.

- Inefficient combustion, due to poor oil burner performance, again causing carbon to deposit upon the air heater elements.

Good operating practices and regular sootblowing reduce these risks, but should a fire start it is essential to detect its presence as soon as possible. Each heater is therefore provided with a fire detection system arranged in two equal sections, one of which monitors 'gas inlet/air outlet' temperatures over the range 0–450°C, while the other monitors the 'gas outlet/air inlet' temperatures over the range 0–250°C.

Each section comprises eight duplex thermocouples, six 'gas in' or 'gas out' and two 'air in' or 'air out', all feeding into a motorised switch that rotates continuously and connects each thermocouple in turn to the alarm equipment housed in a local panel. The thermocouple amplifier outputs in each section are averaged, and the average temperature thus obtained has a bias factor added to it — equivalent to a temperature of between 45°C and 55°C. The 'average + bias' total is then compared in turn with each individual thermocouple temperature and, should any one of these temperatures exceed the 'average + bias' value, an alarm is activated on the local panel and in the main control room. The system is based on the principle that, if a fire were to occur in the gas air heater, the temperature rise would be local at first — causing only one thermocouple temperature to increase.

As well as the alarm being initiated in the central control room (CCR), a recorder is started to record temperatures in that band or section of the air heater causing the alarm. Local temperature indicators and zone alarm lamps also pinpoint the zone, initiating the alarm.

A 'monitor failed' alarm is also incorporated in the system, to operate if the motorised switch stops, or an electrical failure occurs, or in the event of a thermocouple burn-out. The scanning system will compensate for one thermocouple going open-circuit, but is then obviously not so effective in detecting a fire. Failed thermocouples must be replaced at the earliest opportunity.

When an air heater fire alarm is initiated, it is the responsibility of the Shift Charge Engineer to assess the charts and information to decide whether the fire extinguishing system should be put in operation, the system being manual and not automatic. This is because it is not good practice to apply cold water to a hot metal matrix and therefore it is left to the engineer to judge the need.

Figure 4.29 shows three typical chart traces when a fire exists. Type I is a localised fire which would burn itself out, Type II is an extensive surface-deposit fire, spreading but then burning itself out, and Type III is a dangerous-deposit fire, which would require the air heater to be taken out-of-service and the fire extinguishing equipment activated.

The fire extinguishing equipment consists of a series of nozzles situated in the air heaters, so that water spray can be directed onto the elements. The nozzles are supplied with water from the main station fire pumps and the water spray is controlled by a hand-operated isolating valve, a separate valve being supplied for each air heater. The nozzles are protected from dirt and dust ingress by bursting discs, which rupture when fire pump pressure is applied. The water drains from the air heaters via the gas/air duct counter-balanced flap drains.

5.3.4 Safety

- Current operating practices for the prevention of gas air heater fires must be strictly observed.
- Access to gas air heaters must be carefully controlled.

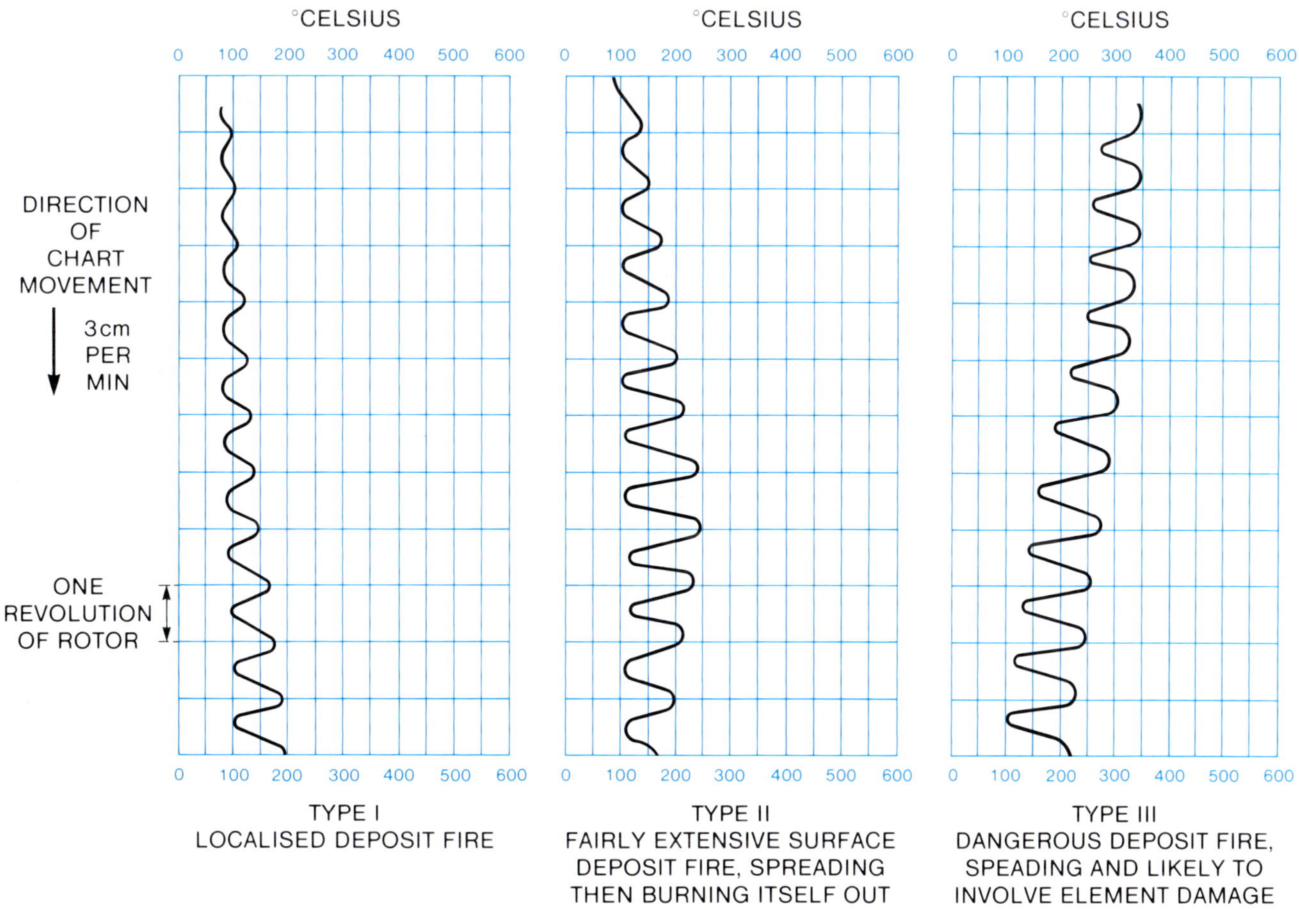

FIG. 4.29 Interpretation of recorder trace

- Fire detection systems must be commissioned before the boiler is fired.
- Sootblowing equipment must be available before the boiler is fired.
- Procedures for dealing with lubricating oil spillages must be available.
- Gas air heaters to be by-passed during off-load and low-load boiler operation.
- Air heater blanking plates to be left in position until after the float safety valve programme.

5.3.5 Resources

Services

- Electrical supplies to motors, fire protection equipment, dampers, sootblowers.
- Charge of flushing oil required (1300 litres).
- Absorbent material for containing oil spillages.
- Pumps/motors/containers for filling/draining lubricating system.
- Cooling services from station cooling system.
- Orifice plates for fitting in lubricating oil systems to balance oil flows.

Staffing

A team of six people will need up to six weeks to commission each air heater.

5.3.6 Gas air heater commissioning

A typical commissioning sequence diagram is shown in Fig 4.30. This diagram can be adapted to commission both types of gas air heater.

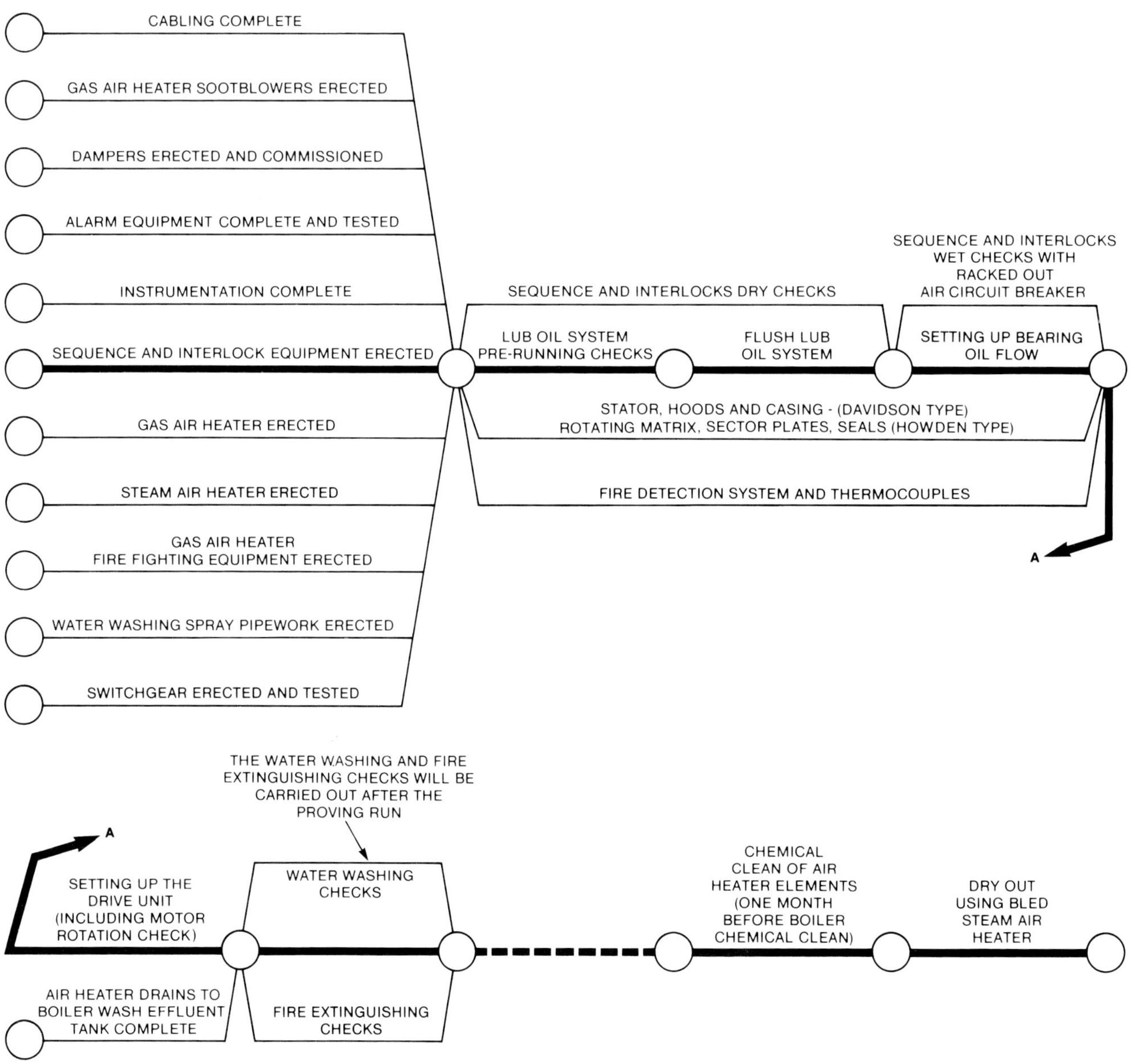

FIG. 4.30 Gas air heater commissioning network

Rotating-hood air-heater (Davidson type)

(a) *Pre-commissioning checks*

- Ensure that all mechanical erection is complete and that the stator elements are fitted.
- Check that blanking plates are left in place to protect the elements during early boiler operation.
- The upper and lower hoods are inspected to check that the sealing shoes have been fitted, are free to move and that the seal-frame settings are correct, in accordance with the manufacturer's recommendations.
- All sootblowing, water washing and fire fighting pipework and spray nozzles should be completed. Bursting discs are to be in place on spray nozzles.
- Oil tank level indication is checked for correct fitting and all tappings are checked to be clear of obstruction.
- Temperature indication and thermostat operation are proved.
- The position and calibration of the fire detection thermocouples, along with alarm operation and initiating values, are checked.
- All local/remote instrumentation is checked, in accordance with the procedures in Chapter 3, Section 3 and the SCL.
- Ensure that labyrinth seals on support and guide bearings have been greased.

(b) *Lubricating oil system flush*

The lubricating oil tank is checked for cleanliness, with gauze strainers fitted to the pump suction pipes. The tank is then filled with flushing oil (600 litres) and flexible by-pass connections made to the air heater guide and support bearings to facilitate flushing.

The air-blast cooler by-pass is opened and a lubricating oil pump started on local-to-plant pushbutton. Checks will be made for motor current, for system leaks and for a positive return flow at the tank. The oil temperature should be maintained at 65°C by use of the oil tank thermostatically-controlled heater. After 30 minutes of flushing, the oil cooler by-pass valve should be closed and the main oil cooler put into circuit, checks again being made of motor current and oil flows. Flushing should then continue until the system is proved to be clean by continual checking and draining of the duplex oil filters. The Station Chemist will determine when the system is satisfactory by sample analysis: once a clean system is obtained, the pump is stopped and the flushing oil drained from the tank and lubricating system, the fine-gauze strainers removed from the pump suctions and the lubricating oil tank is refilled with clean lubricating oil (1600 litres). The by-pass connections are removed from the bearings and the permanent pipework remade.

After this work, the oil system is run to set up the correct oil flow to each bearing. During this operation, the sequence operation and standby cut-in facility of the lubricating oil pumps is checked. All alarms associated with the system should also be checked.

The air-blast oil cooler should also be checked for direction of rotation and that the fan starts/stops at the correct temperature settings. This can be done by withdrawing the lubricating oil temperature bulb from its pocket and inserting it into a container of water. The water is heated to 60°C to check the fan cut-in setting and then allowed to cool to 50°C to check the cut-out setting. The lubricating oil temperature 'high' alarm at 80°C should also be checked.

(c) *Damper commissioning*

The air inlet and outlet, gas inlet and outlet, and by-pass dampers should be commissioned in accordance with the procedure in Section 5.2.7 of this chapter and Chapter 3, Sections 1 and 2.

(d) *Drive unit commissioning*

Prior to operating the drive unit, the following checks will be carried out:

- The main drive motor should be checked in accordance with the procedure in Chapter 3, Section 4, together with its direction of rotation: the drive coupling can then be made, coupling alignment being checked at this time.
- Gearbox oil level must be checked, the pinion drive-shaft bearings greased and, with the 415 V drive supplies isolated, the upper and lower hoods checked for freedom of rotation by use of the hand-barring equipment. Inspection of the hoods should be made at this time, by removal of the top and bottom inspection doors.
- An SRCC should be available for the air heater, and all terminal-point isolating dampers locked closed and access doors refitted.

(e) *Proving run and sequence operation*

The 415 V supplies are then reinstated and the lubricating oil system pumps and drive selected to 'remote control'. The lubricating oil pumps will be selected, one to duty and one to standby.

The air heater can then be started and checks made that the duty lubricating oil pump starts and, with satisfactory oil pressure, the air heater will start. Checks must be made that noise levels, drive-motor current, lubricating oil system pressure, flows and temperature are all satisfactory.

The air heater should now be left to rotate for 24 hours with continual checks being made that lubricating oil pressure is 0.14 bar, that oil temperature is controlled between 50°C and 60°C and that the motor drive current is satisfactory.

Full remote sequence operation should then be checked.

Following the satisfactory proving run, the water washing, sootblowing and fire detection/extinguishing systems are checked for satisfactory operation. Sootblowing and water washing equipment commissioning is covered in Section 8 of this chapter.

The air heater elements may have been protected during the construction phase by coating to prevent corrosion and this will have to be removed by chemical cleaning before the boiler is fired, using the water washing facility. The air heater is dried out after the clean, using the steam heater and fans to blow warm air through the stator matrix.

Rotary heat exchanger (Howden type)

(a) *Pre-commissioning checks*

- The local and remote instrumentation, alarms and interlock devices are checked in accordance with the procedure in Chapter 3, Section 3 and the SCL.
- The oil tank level indicator is checked to ensure correct fitting and that tappings are clear of obstruction.
- The temperature indicator and thermostat operation/setting are proved.
- The local oil pressure gauge is deadweight tested and all clear-glass indicators checked for cleanliness and correct fitting.
- The pressure switches are then proved, by manual operation.
- The position and calibration of the fire detecting thermocouples, along with the alarm operation and initiating values, are checked.
- Ensure that all mechanical erection is complete and that stator matrix elements are fitted.
- Check that blanking plates have been left in place to protect the elements during early boiler operation.
- Verify that all seals are correctly fitted and checked for freedom of movement.
- Check that all sootblowing, water washing and fire fighting pipework and spray nozzles are completed. Bursting discs are to be in place on fire nozzles.

(b) *Lubricating oil system*

Flushing of the system will proceed to establish a clean oil supply and return circuit via the top and bottom bearings, free from leaks and satisfying the flow requirements dictated by the works-calibrated flow switches.

The reduction gearbox should be standing fully-charged with the correct grade of running oil since rotor machining during erection phases, but it is necessary to check the oil level in the reduction gearbox, as it is independent of the main lubricating oil circuit.

The reservoir tank, factory cleaned and sealed during erection against ingress of dirt, etc., should be filled through the combined dipstick/filler hole with the correct grade of flushing oil up to mark *A* on the dipstick (approximately 120 litres). As the tank is filled, the operating point of the Mobrey low level alarm can be noted, either by checking for continuity across the contacts, or as the alarm is initiated in the CCR.

The lubricating oil pump motors will be checked using the procedure in Chapter 3, Section 4, and then checked for direction of rotation. Once this has been completed, the oil flush can commence.

The top bearing, bottom bearing and reservoir tank drain valves should be shut and the cooling water drain-plug on the oil cooler removed (cooling water isolating valves shut) to check for possible leaks existing in the cooler circuit between oil and water sides.

One pump should then be selected for priming the lubricating oil circuit and filling the bearings, which amounts to 360 litres in the top reduction-drive unit and 80 litres in the bottom bearing, each pump supplying approximately 11.5 litres/min.

The selected duty pump will be started local-to-switchgear, ensuring that the pump picks up suction and creates a flow; this can be checked by cracking one of the drain lines. Careful observation of the tank level is maintained as the lubricating oil system floods, stopping the pump to recharge the tank as required. The bottom bearing housing will fill first and, once oil returns are flowing as observed at the return sight-glasses, retopping of the reservoir tank will be at longer intervals.

The correct quantity of oil for topping up the reduction-drive unit should be ascertained in advance of topping-up activities, as it will help in estimating at what point the top drive unit will be filled. If random filling continues during pumping activities, there is a likelihood that the slug of slow-moving oil in the 76 mm top bearing return line will cause the tank to overflow.

If, during filling, the oil level in the reservoir tank remains unchanged without oil returning from the top bearing, it will be necessary to remove the drive unit inspection cover and confirm that oil is not reaching the top bearing assembly. Then stop the duty oil pump whilst an orifice plate is fitted into the oil supply line to the bottom bearing, before the discharge NRV, thereby rebalancing the oil flow.

Once the system is completely flooded, with steady lubricating oil returns observed from both bearing assemblies and with tank level maintained at mark *A* on the dipstick, the pump can be left in operation to flush the system. A note is taken of the running current of the duty pump; after four hours, it can be tripped and flushing continued on the standby pump, noting starting and running currents.

Note Do not attempt to run both pumps in parallel, as the increased flow rates could overfill bearing housings.

As stated earlier, the lubricating oil supplies to each bearing may require rebalancing to prime the full circuit: it must be ascertained that both bearings are receiving approximately the same quantities and that lubricating oil is spraying on to the main gear-wheel meshing point. Direct measurement of supply quantities is not practicable and flow can only be checked by observation at the return sight glasses and the drive unit inspection cover, flow switches giving only the minimum flow rates required. If redistribution of oil flows is required, various-sized orifice plates are provided for this purpose: it can be carried out at any convenient time after the start of lubricating oil system flushing operations.

Flushing operations, alternating the duty pumps, should continue for two full working days. It is also necessary that the filter be checked and cleaned regularly throughout the flushing exercise. Shortly after starting the flush, it will be obvious if there is any leakage within the oil cooler, as oil will issue from the water drain, the water drain-plug having previously been removed. If not, the oil cooler water supply line can be cracked open and the source checked for cleanliness. The drain-plug will then be replaced and the water supply and return lines isolated.

Note Do not introduce the oil cooler until after flushing.

After two days flushing, a sample taken from any one of the three drain valves should be passed to the Station Chemist for analysis. Upon verification that the lubricating oil sample is satisfactory, the pump motors are isolated and the reservoir tank and both bearings are drained down. A separate drain-plug is provided in the bottom of the reduction gearbox, and if this and the driven-gear bottom coverplate are removed, following flushing, the gearbox will be completely drained of oil.

An inspection cover is provided in the side of the reduction gearbox to inspect the oil feed to the gear meshing point and this will be opened for inspection following flushing and draining.

The system is then charged with the running oil. As with filling for flushing, a careful watch must be kept on the oil tank reservoir level, maintaining the level between marks *A* and *B* on the dipstick.

A check can now be made on the motor running current, discharge pressure increase, flow reductions, etc., when the lubricating oil cooler is introduced into the circuit, as this effectively increases the viscosity to the maximum possible under the most adverse conditions the pump will encounter under normal start-up conditions.

(c) *Damper commissioning*

The air inlet and outlet, gas inlet and outlet, and by-pass dampers should be commissioned, following the procedures in Section 5.2.7 of this chapter and Chapter 3, Sections 1 and 2.

(d) *Main drive and washing motor drive commissioning*

Prior to operating the air heater matrix, the following checks must be carried out:

- With the main drive coupling removed and the washing motor drive claw-clutch mechanism disconnected, both motors are checked in accordance with the procedure in Chapter 3, Section 4 and then checked for direction of rotation.
- The main and washing drive-couplings are then reconnected, coupling alignment checks also being made.
- All matrix elements are fitted.
- The gas and air by-pass dampers are satisfactory, and locked closed.
- The actuator-controlled axial seal-plates operate satisfactorily and operating limit switches are set up.
- The sector seal-plates are set back to their fullest extent, and locked in that position.
- The observation windows are cleaned, top and bottom gas and air pass access doors can be left off and the rotor turned by hand to check for freedom of rotation. The access doors must then be replaced.
- An SRCC is issued for the gas air heater and the terminal point dampers are locked closed.
- Adequate GSW supplies are available to the oil cooler.

(e) *Proving run and sequence operation*

The following interlocks and sequence operations can now be checked:

- When one lubricating oil pump is started, the other is set for standby duty.
- The air heater driving motor may only be started when the lubricating oil flow has been established by pressure switch operation. Failure of the pump to start, or of pressure switches to operate, prevents the rotor driving motor starting, until the standby pump has started and satisfied the interlocks. Failure of the first oil pump initiates an alarm, and starts the standby pump.
- The air heater driving motor is interlocked with the air heater inlet and gas outlet dampers, so that dampers cannot be opened until the air heater rotor is turning.
- The simultaneous operation of air heater main drive and washing drive motors is prevented by interlocks.

There are no unit tripping interlocks and the following faults give alarms only:

- Air heater trip (alarm for each heater).
- Running lubricating oil pump failure.

- Pressure loss at one or both oil switches (common alarm).
- High oil temperatures.

Note On coal-fired stations, loss of one gas air-heater will trip selected mills.

These alarms are checked by simulating the operating conditions. Effective remote operation is then checked and the air heater should be left rotating for a period of approximately 8 hours. During this period, checks on motor current, lubricating oil flows, pressures and temperatures, bearing temperatures and any signs of rotor rubbing must be made frequently. A check of the rotor speed (1.5 r/min) should be made.

Following this successful running, the auxiliary drive can then be checked, demonstrating that the drive cannot be started unless the lubricating oil system is in service, and also that the rotational speed is correct at 0.25 r/min.

Following the proving run, the air heater should be isolated for internal access so that the axial and radial seals can be finally set.

The axial seal-plates can now be set up, the actuators being commissioned following the procedure in Chapter 3, Section 2 and the SCL, and the limits and torque settings set according to the manufacturer's limits for cold seal-settings. The seals should then be fully retracted, and must not be advanced during boiler operation until the boiler is at MCR.

The radial seals should be fitted to their optimum settings; final adjustments will then be done with the unit at MCR.

During the boiler pre-commissioning stages, up to floating safety valves, the air heaters will be rotating and by-passed on the gas side by having the gas inlet and gas outlet dampers shut. The gas by-pass dampers will be open, to control furnace suction. Blanking plates will also be left tacked in place covering the elements to protect the basket matrix from any debris that may be blown through the ductwork by the fan operation.

The blanking plates will be removed after the float safety valve programme is completed. The air heater elements may have been protected from atmospheric corrosion during the erection phase by coating with a protective layer of impregnated paper; this and all putty and tape should be removed prior to boiler operation, to avoid possible fire risk.

The sootblowing and water washing equipment should also be commissioned at this stage (see Section 8 of this chapter).

Gas air heater fire detection and extinguishing system commissioning

(a) *Fire extinguishing equipment*

This equipment must be commissioned prior to operating the boiler for the chemical clean. The fire protection equipment will be supplied from the main station fire pumps and checks should be made that all pipework is complete and that each water supply nozzle is securely fitted to its terminal and that the bursting disc is complete. Flushing of the fire protection pipework should have been completed by the fire protection working party at an earlier date. It is not usual to spray water into the air heater baskets prior to boiler operation, subject to the above checks being completed.

(b) *Fire detection equipment*

Prior to boxing up the air heater, a thorough check of all thermocouple fixings and connections will be made and a continuity check made that all thermocouples are satisfactory. The thermocouple connections to the control panel will be checked from the wiring diagram.

The control panel will be checked for correct installation and that 110 V AC and 50 V DC supplies are available.

The fire protection panel will then be checked out for the following:

- The equipment fault lamp is illuminated.
- The auto/manual switch is in the 'auto' position, and the on/off switch is in the 'on' position.
- The 110 V AC isolating switch is closed.
- The gas/air indicator should indicate 'gas'.
- The numerical indicator should show 01.
- The three lights on the power supply unit should be illuminated.

Then the flowing operation should be carried out:

- Set the auto/manual switch to 'manual'.
- Set the 'manual search' switch to 'high temp' and press the 'fast scan' pushbutton (this should be illuminated while depressed).
- Check that the numerical indicator numbers follow in rotation.
- Check that the 'gas/air' indicator changes over at appropriate points.
- Switch the recorder on/off switch to 'on'.

The alarm test facility should then be checked by depressing the 'Test' pushbutton. The following should occur:

- 'High temp' alarm lamp starts flashing.
- Recorder drive starts.
- Recorder starts to indicate probe temperature.
- Remote alarm operates.

The manual search facility is then checked by operation of the 'manual search' switch:

- In the 'open circuits thermocouple' position

• In the 'high temp' position.

The open-circuit thermocouple circuit test is then checked by removing one of the input connections from a point well in front of the scan and letting the equipment scan round to this point.

When this point is reached the following should occur:

- 'Equipment fault' lamp indicates.
- 'O/circuit T/couple' lamp indicates.
- Scan continues with 'Equipment Fault' lamp on and 'O/circuit T/couple' lamp off.

When the scan has passed the point, and the above functions have been checked, reset the equipment by depressing the 'Equipment fault' pushbutton, and replace the broken input connection.

The high temperature alarm circuit test is then checked by setting the equipment to 'manual' control and switching off the recorder to avoid driving the instrument to full scale by the test signal. Remove one of the thermocouple circuits from the control cabinet and inject a dummy input signal to simulate a high temperature, using a suitable potential source.

The equipment is then switched to 'auto' control and allowed to scan to the selected high temperature point:

- Check operation of local alarms.
- Check 'alarm cancel' action.
- Reconnect the thermocouple circuit in its normal position.

The probe sensor flame test is then checked as follows: With the probes removed from the air heater and the Plessey plugs reconnected, switch the scanning unit to automatic operation. Apply a flame to the fast-response thermocouple of each sensor in turn and check that the alarm operates, with correct point indication in each case.

5.4 Steam air heaters

5.4.1 Description

Steam air heaters are installed on oil-fired boilers, and in some instances on coal-fired boilers, in order to maintain the boiler 'cold end' metal temperature sufficiently high to avoid corrosion during start-up on a cold boiler and during normal operation of the unit. The arithmetical mean of the gas outlet and air inlet temperatures should not be less than 110°C for oil-fired and 78°C for coal-fired boilers. For example, on an oil-fired boiler the gas outlet temperature is 150°C and the air inlet temperature is 70°C, so the mean temperature is (150 + 70)/2 = 110°C.

Steam air heaters are located in the FD fan discharge ducting at the inlets to the regenerative air preheaters. Figure 4.31 shows one type of preheater and Fig 4.32 outlines the steam and condensate system.

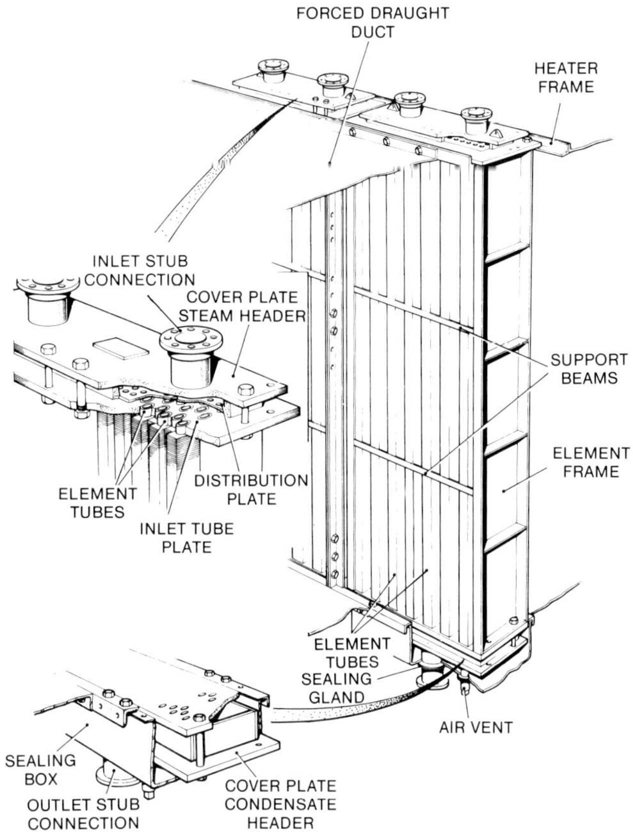

FIG. 4.31 Steam air heater

Steam enters the preheater vertical heating elements from the top header and condensate is collected in the bottom header. The heating elements consist of thin tubes installed in panels which are free to expand downwards. The ancillary plant consists of a condensate tank complete with level controller, and two condensate extraction pumps, one duty and one standby, for each heater. For start-up purposes, a steam trap is used, when on auxiliary steam supply.

At start-up and during low-load running, the heater takes its steam supply from the auxiliary boilers at 17 bar and 238°C, until suitable quality bled-steam is available from the turbine IP cylinder at 5.5 bar and 155°C (i.e., at approximately 45% MCR), at which point a changeover to the bled-steam supply takes place by operator action. When steam is supplied from the auxiliary boilers, the condensate is discharged to the auxiliary boiler flash vessel, and when it is supplied from the turbine-generators, the condensate is collected in the condensate tank and pumped back into the unit condensate system, provided the quality of the condensate is acceptable. If the quality is not good enough, the condensate can be pumped to the auxiliary boiler flash vessel, or to the boiler dirty drains flash vessel.

The auxiliary and bled-steam control isolating valves are interlocked so that they cannot both be used at the same time, and the bled-steam isolating valve cannot be opened if the steam pressure is <350 mbar (5 psig). Pneumatically-controlled butterfly valves, situated in

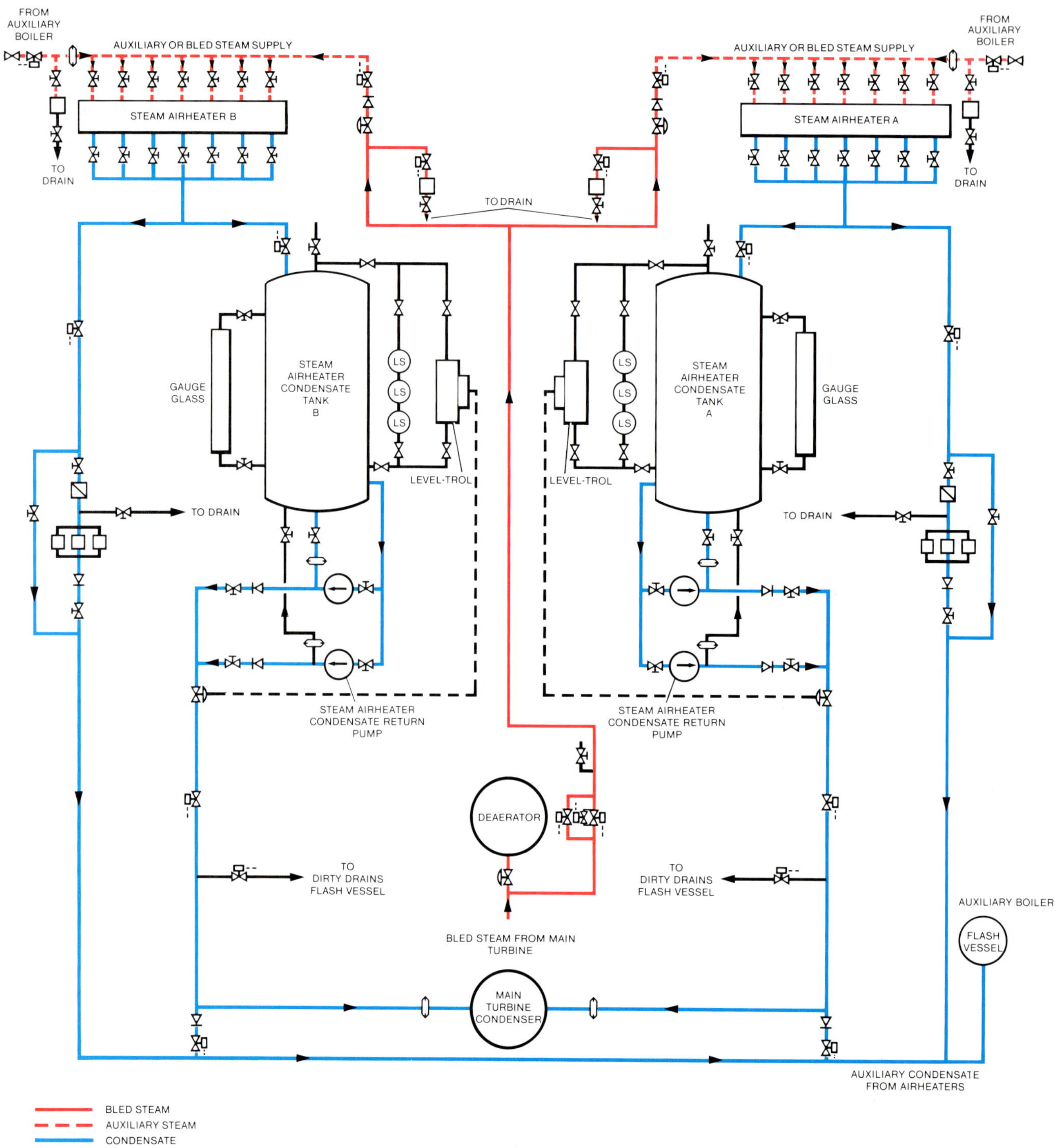

FIG. 4.32 Steam air heater — steam and condensate system

each bled-steam supply line, control the heater temperature, the controller receiving signals from the air inlet and gas outlet temperatures associated with the gas air heater, which is set to give the desired combined temperatures.

The condensate tank is fitted with level switches. When the tank level is high, the duty and standby pumps are switched in; when the tank level is within its normal operating range, the duty pump only is used. When the tank level is low, both pumps are stopped.

A recirculation valve is provided to recirculate condensate back to the condensate tank, so that in the event of the tank outlet control-valve closing, overheating of the pumps is prevented. The condensate tank outlet control-valve controls the condensate discharge, so that a constant level is maintained in the condensate tank.

5.4.2 Safety

- Adequate thermal insulation to safeguard personnel from burns.
- Station drainage systems available to deal with large amounts of flushing water.

- Careful control of personnel access into ductwork, particularly when steam is being supplied to the air heaters.
- Barrier off areas of steam escape during steam purges.
- Any activities that require access to pressure parts or gas-passes must be controlled in accordance with the requirements of the Safety Rules. The issue of Permits for Work should be used for this purpose.

5.4.3 Resources

Services

- Temporary hoses and pipework available for flushing purposes.
- Station auxiliary boilers commissioned to provide steam.
- Station drainage system available.
- Auxiliaries cooling water system available.
- Electrical supplies to motor/valves.
- Pneumatic air supplies to control valves.
- Barriers to control the commissioning area.
- Supplies of demineralised water for flushing condensate systems.

Staffing

A team of six people will require several weeks to complete commissioning of the air heaters. Commissioning will not be completed until control loop checks are satisfactory, with the unit at full-load.

5.4.4 Steam air heater commissioning

A typical steam air heater commissioning sequence diagram is shown in Fig 4.33.

All instrumentation, valves and controllers will be checked in accordance with the procedures in Chapter 3 and SCLs: once the erection completion checks are satisfactory, a steam purge of the elements will take place. The procedure is to blow through each element separately using steam from the auxiliary boiler system. The ejected steam/condensate can be released to atmosphere by removing the steam-strainer top cover and internals. The area around the strainer must be barriered off and only personnel involved in the steam blow must enter the area. The degree of valve opening during the purge will depend upon the amount of steam escaping from the strainer and this will have to be regulated to suit local conditions. If necessary, temporary pipework could be erected and routed outside the building.

The auxiliary steam mains will have been steam purged by the auxiliary steam working party and only the pipework involved within the steam air heater isolating valve will be subject to steam clean.

Each element will be steam blown in turn and during the blow checks should be made for leaks, both inside and outside the ducting. Personnel access inside the ducting must be subject to the Safety Rules.

Finally, after each element has been purged in turn, the whole system should be purged; this will give minimum steam resistance, and help further to clean up the auxiliary steam supply pipework.

The bled-steam supply pipework to steam air heaters should also be given a steam blow to remove any erection debris and this can be done by steam blowing the line in reverse from the auxiliary steam system. The bled-steam pipework will have to be broken at a suitable position, as near to the supply point as possible, and temporary pipework fitted to enable the escaping steam to be safely blown to atmosphere. Again, only personnel involved in the steam blow activities should be allowed in the area.

Following the purge of the heater elements, the condensate system should be flushed, this being done by filling the condensate tank from a demineralised water supply with a temporary hose and flushing to waste via the pump suction pipework. The condensate tank should then be drained, the manhole door removed and the internals inspected for cleanliness. The tank manhole door should then be replaced. Once the cleanliness of the tanks and suction pipework is established, the condensate pumps can be commissioned.

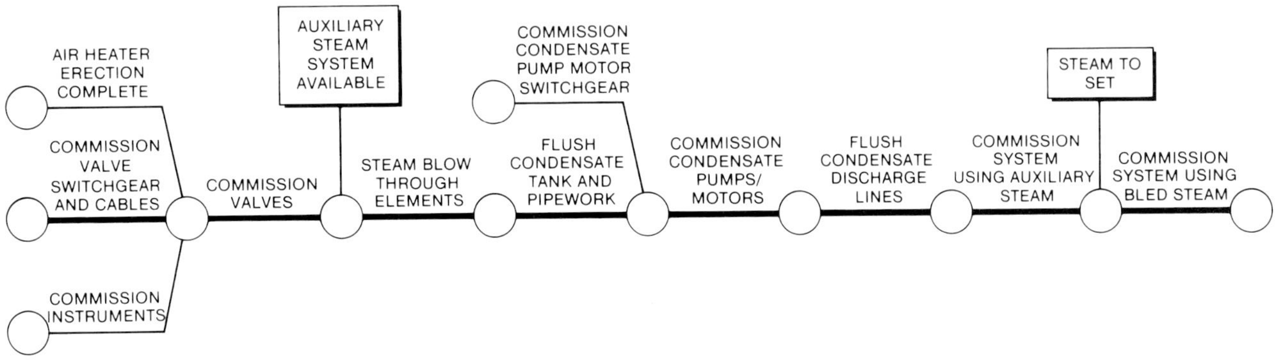

FIG. 4.33 Steam air heater commissioning network

It is imperative to ensure that the pump gland cooling water supply is available and in service, and that the pump is properly primed and vented, prior to running. The seal chamber should be checked to be full of water prior to running by checking that there is a good flow of water at the gland vent.

The tank will be filled to NWL with demineralised water from the temporary demineralised water supply and the pumps set up to run on recirculation back to the tank with the pump discharge valves shut.

Checks should be made that both the switchgear and motors have been commissioned in accordance with the procedure in Chapter 3 and SCLs, and the pumps should then be run to check for direction of rotation. If rotation is correct, the pump should be checked for motor current, discharge pressure and any leaks/ vibrations, over a period of approximately two hours. Once the condensate pumps have been satisfactorily run, the three condensate discharge lines should be flushed, using both pumps in parallel to ensure maximum velocities. Flanges should be broken at suitable points on the discharge lines, nearest to the condensate discharge point in turn, and temporary pipework fitted to discharge the condensate to a suitable drain. Adequate supplies of demineralised water will be required for this purpose.

During the flush, the pumps must be kept under constant supervision and checked for overheating and vibration. A constant check on the condensate tank level should be made to ensure that the pumps do not lose suction. Following satisfactory flushing of the three condensate discharge lines, the temporary pipework should be removed and the permanent pipework reinstated.

Automatic control of the condensate pumps can now be checked, and the alarms and interlocks proved. This can be done by raising and lowering the water level in the condensate tank to check the high level cut-in of the standby pump. The tank high-level alarm, then the standby pump cut-out, the duty pump stop/start and finally the tank low level alarm are checked. The tank level controller can also be set up to control tank level, the control valve and positioner first being checked in accordance with the SCL.

The steam air heaters will first be used on auxiliary steam during early boiler operation for fan tests and the chemical clean.

With the fans in service, auxiliary steam is applied by manual operation of the steam valves and the whole system allowed to warm through slowly. The condensate will first be drained to waste through the drain valves to the dirty drains flash vessel and, when the condensate is clean enough, the traps will be put in service; checks are made that they operate correctly and that the condensate is effectively removed from the steam air heater elements. It is essential to check that waterlogging of the steam air heater elements does not occur. Once the system has been proved, using the steam traps, the condensate should then be diverted to the condensate tank and the trap inlet valves shut. The condensate pumps can then be checked for correct operation, as the condensate level rises in the tank. The condensate system level controller is tuned for correct operation at this stage.

During the above operations, the steam traps in the auxiliary steam supply lines should be checked for correct operation.

Once manual operation of the system is satisfactory and it has been proven that the system operates correctly with no waterlogging of the elements, the remote sequence operation of the valves should be checked.

Final commissioning of the system will be completed during the early running of the unit after steam-to-set and the bled-steam controls, interlocks and control loops will then be proven on-line. This will be covered in Chapter 6.

5.5 Full boiler airflow tests

5.5.1 Introduction

Once the ID, FD, GR and PA fans have been commissioned, it will be necessary to conduct airflow tests on the full boiler to ensure that the fans are capable of supplying the boiler air demand in a controlled manner.

The tests are as follows:

- FD fan inlet venturi calibrations.
- Fan group performance characteristic tests.
- Ducting noise and vibration level checks.
- Primary air to PF burners to establish the correct air/fuel ratios.
- Furnace purge trial to ensure that the combustion gases can be adequately purged from the furnace.
- Oil burner airflow tests using anemometers within the furnace (oil-fired boilers).
- Precipitator airflow tests.

Before running fans for the above tests, a thorough check must be made that dampers are correctly set, to enable adequate airflow through the boiler. The dampers should be physically examined to ensure that they are in the desired position. The gas air heater should be put in service and the steam air heaters used to test their performance, and also to assist in drying out ductwork and gas passes. Sufficient oil burner airdoors or secondary air dampers should be open to allow for the air mass flow conditions.

5.5.2 Boiler FD fan inlet venturi calibrations

The ID and FD fans should all be commissioned, complete with speed control (if used), all mechanical and remote control checking of dampers and control vanes should be complete and furnace pressure control loops available to be put in service. All boiler internal pressure parts and gas-pass work should be complete, and all boiler access doors fitted, the boiler ducting being thoroughly inspected for cleanliness prior to the doors being fitted. Any personnel working external to the boiler should be warned that the airflow tests are to proceed.

The test procedure for venturi calibrations will depend upon the type of installation, but should be based on BS848, Part 1, 1963 and BS1042, Part 1, 1964. These basically consist of flow tests to determine the coefficient of discharge of the venturi section. The method consists of the correlation of measurements of mean velocity head by Pitot transverse across the throat with the differential pressure generated by the venturi section, to establish the relationship between the actual and theoretical quantities of air flowing for fan outputs of 40, 60, 80, 90 and 100%. Two ID fans will be needed in service to meet the airflow demand, because of the limitations imposed by the gas air heater blanking-plates (unless these have been removed prior to the test).

Special measuring equipment will be required, as follows:

- A barometer to measure atmospheric pressure.
- Inclined manometers, suitable for obtaining velocity-head pressure readings.
- U-tube manometers, to measure static-head pressure readings.
- Pitot tubes suitable for insertion in the sampling points, to measure velocity head.

The sample holes in the duct must permit an adequate traverse to be made with the Pitot tubes, so that a representative average velocity can be measured.

Evaluation of the venturi coefficient enables the measured differential pressure/airflow relationship to be completed over the range of the boiler airflow controller and meter.

The flow coefficient can be defined as the actual mass flow of air through the duct as a function of the measured square root of the differential pressure across the venturi section. It is obtained as follows. Let

V = velocity, m/s
g = acceleration due to gravity, m/s^2 (= 9.81 m/s^2)
H = pressure measured as head, metres of air
Δp = pressure difference, mbar
q = mass flow of air, kg/s
A = duct (venturi) area, m^2
ρ_A = air density, kg/m^3
ρ_W = water density, kg/m^3
K = constant associated with each venturi

$$V = \sqrt{(2gH)} \quad \text{m/s}$$
$$= \sqrt{(2 \times 9.81 \times H)}$$
$$= 4.43 \sqrt{H} \tag{4.1}$$

Since $H_A \times \rho_A = H_W \times \rho_W$ and the density of water is 1000 kg/m^3, then

$$H = \frac{\Delta p}{\rho}$$

Substituting in Equation (4.1),

$$V = 4.43 \sqrt{(\Delta P/\rho)} \tag{4.2}$$
$$q = \rho V A \quad \text{kg/s} \tag{4.3}$$

Combining Equations (4.2) and (4.3),

$$q \times \rho_A \sqrt{(\Delta P/\rho)}$$
$$\text{Therefore } q = K \sqrt{\rho} \sqrt{\Delta p} \tag{4.4}$$

A series of tests are carried out measuring p and velocity head, using the Pitot tubes for various IGV openings (or fan speed), and a value for K can then be determined. Using suffix 1, 2, etc. for each test,

$$q_1 = K \sqrt{\rho_1} \sqrt{\Delta p_1} = \bar{K}_1 \sqrt{\Delta p_1} \quad \text{Test 1}$$
$$q_2 = K \sqrt{\rho_2} \sqrt{\Delta p_2} = \bar{K}_2 \sqrt{\Delta p_2} \quad \text{Test 2}$$
$$q_3 = K \sqrt{\rho_3} \sqrt{\Delta p_3} = \bar{K}_3 \sqrt{\Delta p_3} \quad \text{Test 3}$$
$$q_n = K \sqrt{\rho_n} \sqrt{\Delta p_n} = \bar{K}_n \sqrt{\Delta p_n} \quad \text{Test n}$$

$$\text{Then} \quad K = \bar{K}_n/\sqrt{\rho_n}$$

The value of K is then calculated for each of the tests and this value is fixed for that venturi and independent of density. This value of K is then substituted in Equation (4.4) to give:

$$q_A = K_A \sqrt{(\Delta p_A \rho_A)}$$

Subscript A refers to dry air, and therefore a factor M is introduced to allow for moisture in the air:

$$\text{i.e., } q_A = K_A \sqrt{(\Delta p_A \rho_A M)} \tag{4.5}$$

For moist air at approximately 1 bar, Equation 25 of BS1042: Part 1 can be expressed in metric units as follows:

$$\rho_A = 0.3485\,[(p_2 - p_v/T_2) + 0.62(p_v/T_2)]$$

where

p_2 = air inlet pressure, mbar
T_2 = air inlet temperature, K
ρ_{sv} = saturation vapour pressure of water vapour at T_2, mbar
p_v = partial pressure of water vapour, mbar: $p_v = \phi p_{sv}$
ϕ = relative humidity
ρ_A = air density, kg/m^3

Hence ρ_A can be calculated.

In addition, the mass fraction of dry air within the moist air steam is given by:

$$M = \frac{1-(\phi p_{sv}/p_2)}{1-0.38(\phi p_{sv}/p_2)}$$

Substituting the values of M and ρ_A into Equation (4.5), the mass flow q_A can be calculated and the relationship between Δp and mass flow determined for the operating range of the fans.

The transmitters and control equipment are then set up to this relationship by normal calibration procedures.

5.5.3 Fan group performance characteristic tests

This procedure is conducted to demonstrate by means of a cold air flow test that one ID fan is capable of extracting the combined weight of the air supplied by one FD fan plus half the weight of the flue gas generated by the fuel input at MCR conditions.

The test should follow on from the FD fan venturi tests and all pressure recorders for the furnace and gas passes should be fully calibrated and in service. The furnace pressure automatic control loop should be available and in service. The procedure involves running the fans and establishing furnace pressure conditions for design full-load airflow (MCR rating). Under these conditions, it must be established that no pressure recorder has exceeded design criteria and that the ID fan vane (or speed) control has not reached 100% conditions.

In order to simulate the MCR conditions for the above tests, it is necessary to calculate the design flow rate required for the FD fan. This can be done as follows.

From the boiler design parameters at anticipated MCR conditions establish:

(a) Fuel consumption = x kg/s

(b) Output of FD fans = y kg/s at design inlet temperatures

(this can be determined from the venturi tests)

Therefore total input mass flow to furnace is given by

$$q = x + y \text{ kg/s}$$

The input mass flow per ID fan = q/2 kg/s. Thus, using the formula:

$$q = Q \times \rho \text{ kg/s}$$

where Q = volume flow, m^3/s

ρ = density of air, kg/m^3 (determined from tables for the corresponding air inlet temperature)

then $Q = q/\rho$ m^3/s can be determined.

One ID fan and two FD fans are now put into service in turn and the FD fan output set to the design volume flow rate Q. The ID fan should be monitored to check that it has not exceeded the design criteria and that the ID vane or speed control has not reached 100%. This will demonstrate that one ID fan will extract the combined weight of air supplied by one FD fan and half the weight of the flue gas generated by the fuel at MCR conditions.

5.5.4 Noise and vibration level checks

The purpose of these tests is to ensure that the draught plant conforms to the CEGB Standard 989907 and that the FD and ID fans do not exceed acceptable noise and vibration levels.

A portable sound level meter is needed and the fans must operate through the full speed/IGV range, from start-up to MCR conditions. Random spot readings of the noise levels are made at each selected load level and sufficient information noted so that remedial action can be taken should the noise levels exceed the criteria in the appropriate standard. A check for off-site noise levels should also be made. During the above tests, the ductwork should also be carefully examined for any excessive vibrations.

5.5.5 Furnace purge trial

This test checks the effectiveness of the furnace purge sequence and ensures that the furnace is adequately purged at the minimum purge airflow setting. For these tests, smoke-generating equipment (smoke bombs) are installed internally in the boiler and sufficient internal lighting is provided to enable the boiler furnace to be examined by external observation. The boiler is boxed-in by closing dampers and the smoke bombs set off, to allow the boiler to fill with smoke until it obscures the internally-installed lighting. The boiler should then be set to allow natural ventilation, by damper adjustment, for the minimum recommended period of 20 minutes and then the fans run up and a furnace-purge sequence operation carried out. This is usually for 5 minutes duration at 25% MCR airflow. Finally, inspection of the furnace to be clear of smoke will demonstrate an effective furnace purge.

5.5.6 Primary air to mills/PF burners — establishing the calibration constant (K) relating primary airflow to primary air differential pressure

The amount of fuel supplied to a mill group is controlled by relating it to the quantity of primary air (PA) supplied; the object of these tests is to find a calibration constant K, relating primary airflow to primary air differential pressure, so that mill instrumentation and control can be set up, prior to coal being supplied to the mills.

The procedure requires accurate measurement of the airflow at various temperatures. Calibration consists of duplicate measuring and comparison of the installed instruments and test equipment of the air flowing through the pulverised fuel (PF) burner pipes. This test is usually carried out with the boiler off-load, using air at ambient temperature, but circumstances may dictate, on certain boilers, that calibration can only be carried out using hot air with the boiler on-load.

Before the tests can proceed, the mill supplying the burners to be tested must be available, with all its instrumentation installed and calibrated.

Before calibration commences, the coal inlet to the mill must be isolated by sealing the bunker outlets with coal or, if provided, closing a damper in the coal chute.

The mill to be tested, along with its associated pipework and draught transmitter lines, must be pressurised and checked for air leakage.

Normally a number of burners are supplied from one mill and, before airflow measurements are taken on any burner supply line, the remaining pipes must be effectively isolated from the mill.

In addition to the commercial instruments provided on the plant, the following test instruments are required: three manometers (two U-tubes and one inclined draught gauge), one Pitot tube and two temperature measuring instruments.

The test instruments are positioned as follows.

The Pitot tube is used for taking a traverse of velocity head in a position in the PF pipe, such that at least ten diameters of straight pipe exist upstream and five diameters downstream of the measuring position. Sampling holes must allow for adequate traverses to be made with the Pitot tube, so that a representative average velocity can be measured. The manometers are used in parallel with the commercial indicators of primary air (PA) and mill differential pressure. Air temperature measurements are taken at the positions used for the Pitot traverse and the normal PA differential measurement. The discharge static head at the PA fans is also measured.

When the system has been proved airtight and air is being supplied to one burner only, with all the test instruments in position, readings of the PA flow and PA differential pressures are taken. The results are plotted on graph: an example is shown in Fig 4.34. A set of readings should be taken for airflows between 20% and 100% of the maximum PA differential or the maximum PA fan motor current.

Whenever the PA fan is started, the isolating damper must be closed to limit the starting current. The amount of air leakage past the closed dampers, ducting and any inspection doors is measured.

Determination of the formulae for velocity and flow are as described in Section 5.5.2 of this chapter, i.e.,

$$V = 4.43\ \surd(\Delta p/\rho) \quad q = K\ \surd(\Delta p\rho)$$

In the following, subscript o refers to the Pitot position and subscript 1 to other measuring positions. From the above, the velocity of primary air at the Pitot position can be determined from:

$$V_o = 4.43\ \surd(\Delta p_o/\rho_o)$$

and the corresponding weight of primary air is

$$q_o = V_o \rho_o A_o$$

where A = primary air pipe area (m^2). Since the mass flow of air along the primary air pipe is constant for any given boiler load,

$$q_o = q_1 = q$$

and

$$q = K_1\ \surd(\Delta p_1 \rho_1)$$

where K_1 depends on the type of primary air measuring device. Values of K can be calculated for each test (i.e., airflow), and these should not deviate more than 5% for a satisfactory primary air measurement. An average value of K for each mill group should be found and these should not deviate by more than 5% on any boiler. By substituting the average K value in the formula $q = K\surd(\Delta p\rho)$, the mass of air flowing for various primary air fan outlet temperatures can be found.

To find the volume of air per second for various mill outlet temperatures, the mass flow of air must be divided by the density corresponding to the temperature.

A number of curves can be drawn from the data obtained as shown in Fig 4.34. The relationship between PA differential pressure and mass flow of air at various PA fan outlet temperatures (mill inlet) is shown.

Another useful set of graphs can be drawn to show the relationship between the weight and volume of air flowing at various air temperatures. These are again shown in Fig 4.34 for various mill outlet temperatures.

The primary air differential pressure at any mill inlet temperature for a designed volume of air to the furnace at a specified temperature can easily be found by following the examples of Fig 4.34.

Once established for a particular mill group, the 'K' factor should remain unchanged. Any interference with the primary air differential measuring device should be followed by a calibration check that the 'K' value has not been affected.

Repetitive full stroke operation is also carried out on the vanes or dampers controlling the supply of primary and attemperating air to the mill. All other dampers must be checked for 'tightness' in the closed position and operated over the full range.

5.5.7 Burner airflow tests (oil-fired boiler)

All oil-fired boiler air registers are designed to give the correct air throughput at MCR windbox conditions and are also tested at the manufacturer's works on specially constructed rigs. After construction is complete, it is vital that the windbox installation is fully inspected for correct installation and burner quarls must be accurately made to the design dimensions. The burner quarl is made from refractory material and fits into the furnace wall at each burner position; it has a nozzle protecting recess at the back and a tunnel on the firing side to give the correct airflow shape and throughput.

When the boiler is complete and the airflow tests are being carried out, it is normally necessary to measure the airflow through each burner register in a multi-windbox installation. This can be done using Pitot tubes sampling at the burner throat, or with an anemometer.

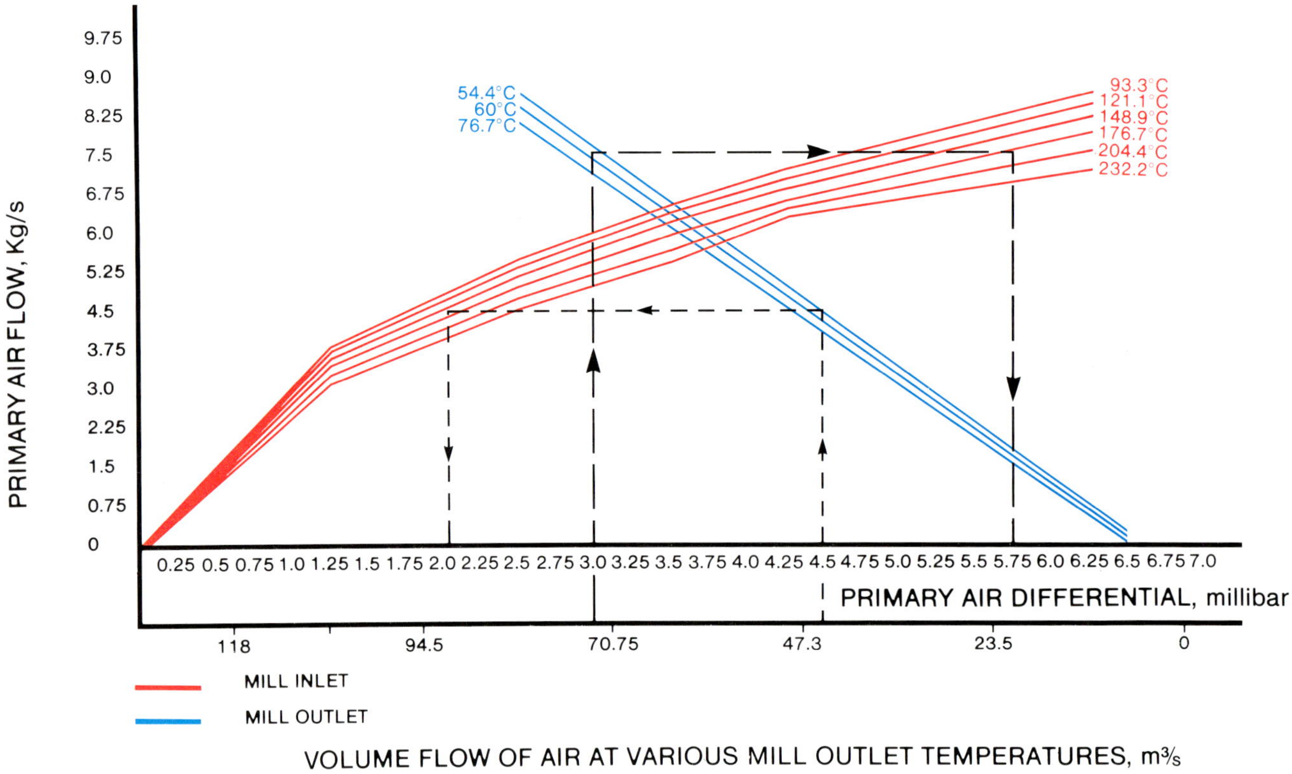

FIG. 4.34 Primary airflows plotted for various mill inlet/outlet temperatures

To use an anemometer, a scaffold has to be erected in the furnace to give access to each burner outlet, and it is essential that the scaffold is adequately secured to withstand the air velocities involved at MCR conditions. Personnel should take every precaution to ensure that they are safely secured at each burner level with a safety harness before the tests proceed, and communication must be maintained with personnel outside the boiler at all times by use of radios.

The tests are carried out by running the ID and FD fans and setting the windbox and furnace suction to the design conditions. Measurements are then taken of the airflow through each burner. Following the tests, the air doors should be shut on each burner in turn to check for a tight shut-off. Leakage through the damper in the closed position must not exceed 1% of the burner full-load airflow.

Following the above tests, the boiler air system is now ready for use. Further tests and tuning of control loops will be made during the initial boiler operations for chemical clean and steam purge. When the unit is available for load, further adjustments will be made during the load-raising programme, from synchronising to full-load. Final adjustments will be made at MCR conditions.

5.5.8 Precipitator airflow tests

Precipitator commissioning is covered in Chapter 3, Section 14, but gas distribution tests will also have to be carried out. The tests may be carried out before the full boiler airflow tests, soon after the ID and FD fan commissioning, if required, but a more suitable time is during the boiler airflow tests.

Special arrangements have to be made for these tests and it must be ensured that all safety precautions for access into the precipitators are taken. The object of the tests is to verify that the gas velocity distribution throughout each precipitator casing is of a satisfactorily high standard to enable contractual guarantees to be met. Model tests will have been carried out and correction devices built into the plant during erection. This saves considerable time in plant testing, but it is still necessary to measure actual flow conditions within each precipitator and correct as necessary to overcome scale effects.

Method of testing

A typical test method is described, based on the Drax Completion tests procedure.

The flow rate is measured at test positions located in the approach ducts to each precipitator. Corrections to obtain the required total airflow may be achieved by alterations to the fan vane controls.

Flow balance between precipitators is corrected by adjustments to the baffles fitted on the primary outlet dampers of the outer (wing) precipitator only.

When the correct total flow and balance between the precipitators is achieved, fan vane control settings and motor current should be recorded.

Access into the precipitator casing is needed for these tests and adequate safety precautions must be taken prior to entry: all high voltage supplies must be isolated and earthed and the rapping-gear supplies isolated.

As the high suction within the precipitator casings will prevent the opening of access doors, fan vane controls should be set to the closed position, necessary personnel installed in the precipitator casings and the fan vane settings returned to the previously recorded positions.

A further check should then be carried out at the approach duct test positions to confirm correct airflow has been re-established. Contact should then be made with the personnel within the precipitator casing to start measuring air velocities at the preselected positions between the collecting electrodes.

A series of air velocity measurements is taken across the full face of the precipitator inlet zone. The anemometer is lowered in turn down the centre of alternate pairs of collecting electrodes and readings taken at approximately one metre intervals; from these, a graph of velocity profile across the measurement plane is plotted.

The distribution pattern in the precipitator inlet plenum may be determined by analysis of the graph. Should adjustment be necessary, this can be achieved by selective blanking of holes in either or both baffles. It will then be necessary to repeat the previous step.

The tests should be repeated until a satisfactory air velocity profile is obtained across this plane.

Velocity profile measurements are then taken at the rear of each collector zone to ensure that a satisfactory pattern of airflow is maintained throughout the precipitator.

Should the velocity profile in the last zone be unacceptable due to the effect on the airflow pattern brought about by the precipitator outlet plenum, adjustments should be made to the baffles located at the entry to this plenum.

Additional velocity measurements should be made immediately below the collector electrodes and within the hoppers to ensure that air is not by-passing the electrodes and that circulation within the hoppers is kept to a minimum.

Finally, a check should be carried out to determine that sufficient purge air is flowing down the flowerpot support insulators.

6 Boiler combustion equipment

6.1 Introduction

The CEGB uses many different types of burner in its oil- and coal-fired boilers and numerous types of pulverising mills, so it would be impossible in this book to include information on commissioning procedures for all types. A description and commissioning procedure for one of the large oil- and coal-fired boiler installations will be covered.

A full description of boiler-combustion equipment is covered in Volume B. Oil burner design requirements are covered in the appropriate CEGB Standards.

In this section, the commissioning of burners on oil only is covered, as the coal-firing equipment is not normally used until after steam-to-set. The milling plant commissioning is covered up to the point where the mill is required to receive coal. Operation of the mills and coal burners using coal is covered in Chapter 6.

6.2 Burner description

6.2.1 Atomisers

There are three basic types of burner in common use on both oil- and coal-fired boilers:

- Mechanical (or pressure jet) atomisation
- Steam atomisation
- Air atomisation

Only the pressure jet and steam types are used on the CEGB 500/660 MW units.

In *mechanical or pressure jet atomised burners*, the oil is raised to the pressure required for atomising by means of a high pressure pump (described in Sections 7.2 and 7.3 of this chapter) and the oil flow is controlled by a common burner pressure control valve. Turndown ratios of approximately 4:1 can be obtained with this type of burner; further turndown is avoided, as the minimum pressure energy for satisfactory atomisation is then reached. On a 500 MW oil-fired boiler with 32 oil burners in service, the pressure at the burner atomiser for MCR conditions is 56 bar. Figure 4.35 shows a typical pressure jet sprayer assembly. A development of this type of burner uses a spring-loaded valve to spill surplus oil, thereby maintaining maximum pressure for atomisation at low loads. At higher loads, this valve closes and allows all the oil to flow through the sprayer plate. This development has only been used on small burners.

Mechanical or pressure jet atomisers with a tip recirculating system have been developed for use in coal burner lighting-up boilers. The operating principles are described in Section 6.2.3 of this chapter.

Atomisation in *steam atomised burners* is achieved by steam pressure energy and the burner produces a particularly fine spray, which gives it a better turndown ratio than the mechanical burner, ratios of 10:1 being achieved. Steam is fed down a central burner tube at a pressure of 2 to 8 bar (depending on the design) to a perforated plate where it meets the oil which has passed along an annular space between the central steam-tube and the concentric outer tube. The steam is at a pressure of 1 bar, or so, above that of the oil. In this type of burner, the oil temperature before entering the central tube need not be as high as in the pressure-atomising type, since additional heat is imparted to the oil in its passage down the central tube which raises it to the required final temperature.

The steam-atomising burner produces a particularly fine spray and for this purpose is accepted as a lighting-

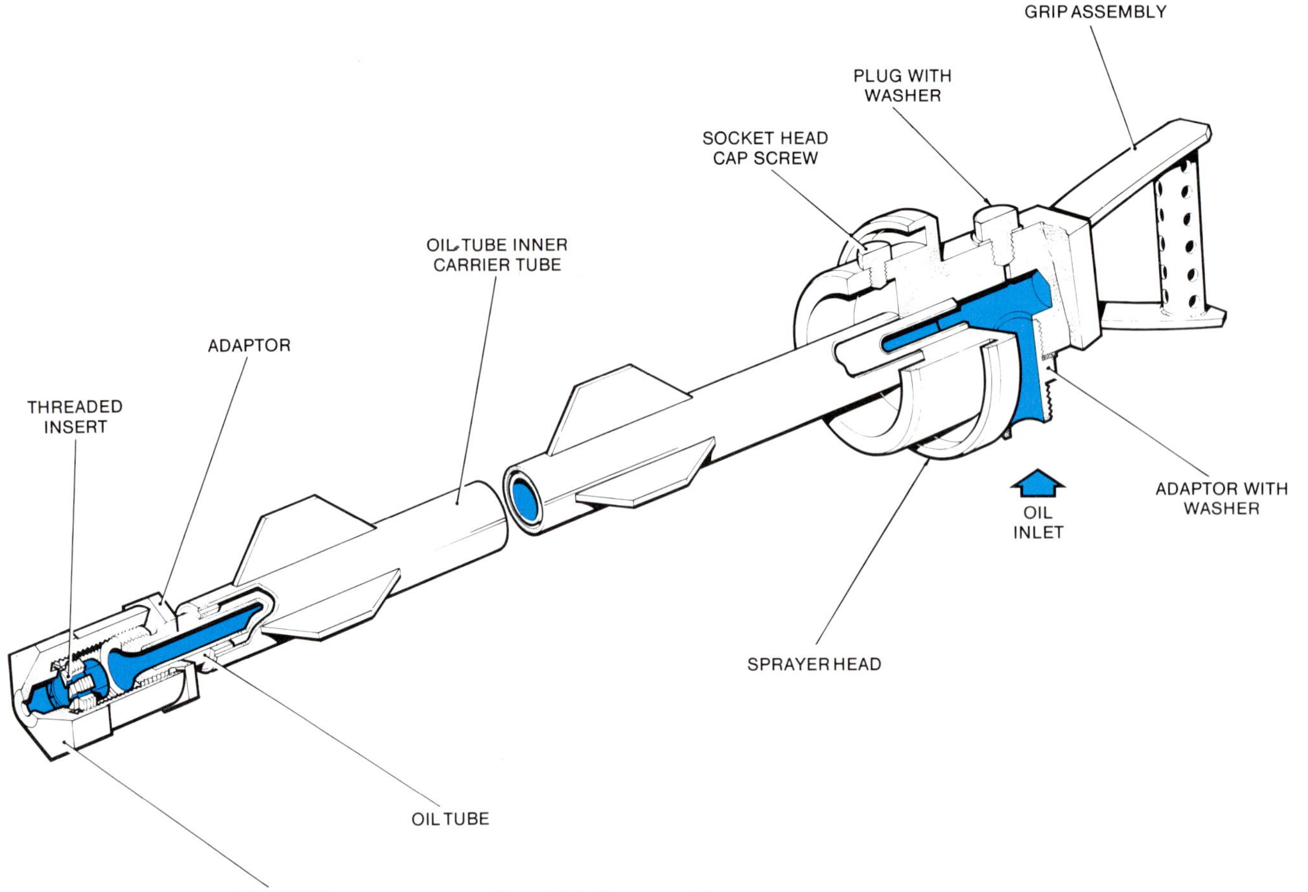

FIG. 4.35 Pressure jet sprayer

up burner in a cold furnace. This type of burner was not used on the early CEGB 500 MW oil-fired boilers because of its high steam consumption (at least 0.5% of the total evaporation); however, with the present requirement for the oil-fired boilers to operate on a regular two-shifting regime, it is used on the later 660 MW oil-fired boilers.

A typical steam-atomised burner is shown in Fig 4.36. In both the pressure jet and steam-atomised burners the oil ways are purged with steam to clear any remaining oil from the lance tubes during the shutdown sequence of the burner, and the atomiser is withdrawn into the air register by means of pneumatic rams to remove it from the high radiant-heat zone.

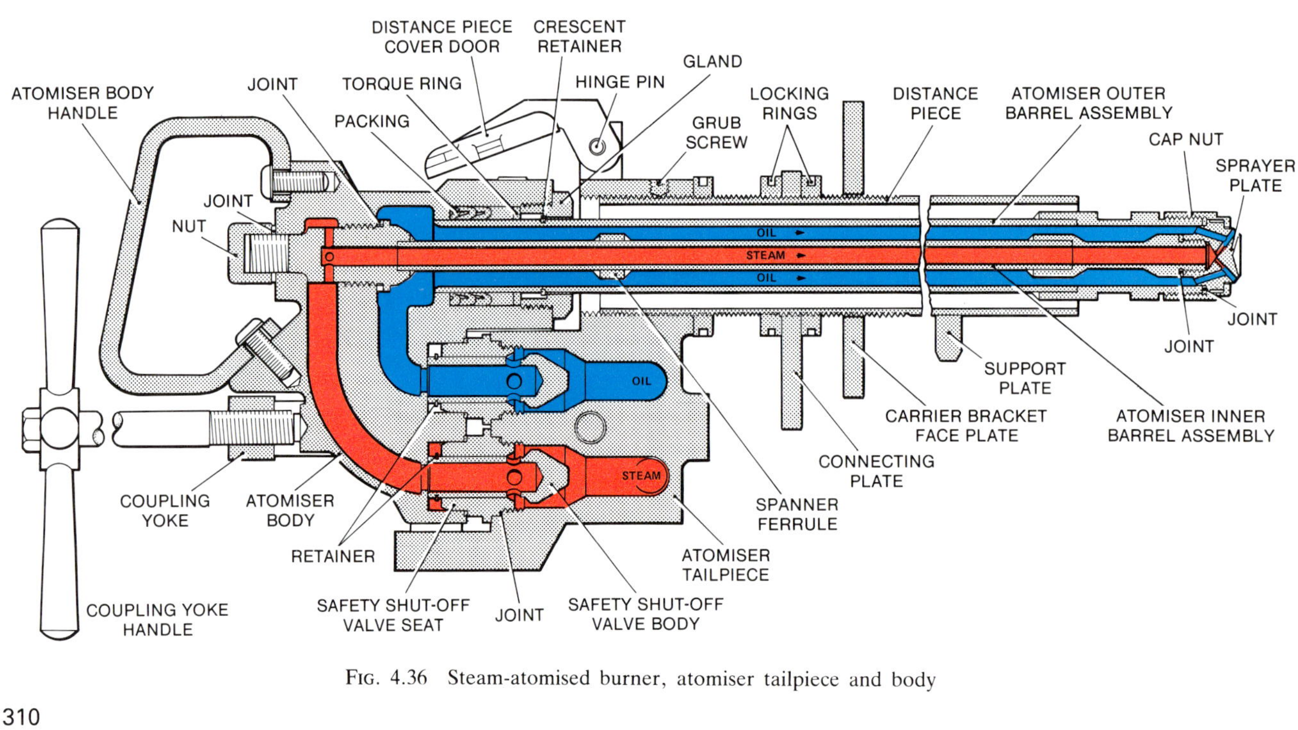

FIG. 4.36 Steam-atomised burner, atomiser tailpiece and body

6.2.2 Air register and burner layout (oil-fired boiler)

The layout for both steam- and pressure-jet atomised burners is similar, the essential features being:

- The atomiser and igniter
- The air register
- The shape of the burner throat
- The position of the burner tube and its diffuser in relation to the burner throat

A typical steam-atomised burner assembly for a 660 MW oil-fired boiler is shown in Fig 4.37 and comprises:

- The atomiser and igniter carrier
- The air register
- The venturi throat

The atomiser and igniter assembly is that part of the burner in which the preheated oil is atomised under steam pressure and ignited by a gas/electric torch during the start-up procedure; a double-acting air cylinder drives the atomiser both into and out of the furnace. The torch igniter is supported alongside the atomiser in a fabricated carrier which is mounted through an opening in the centre of the air register front plate. Fuel oil and steam for atomising purposes are supplied to opposite sides of the atomiser through a pair of swivel-jointed pipe assemblies, which allow for the forward and backward travel of the atomiser and for the expansion movement of the boiler. Limit switches at each end of the atomiser travel provide a position signal to the control system in the burner control cubicle.

The burner air register, which regulates the airflow from the windbox into the venturi throat, is mounted through an opening in the front of the windbox, the front plate of the register being bolted to the windbox. The register air-sleeve is also operated by a double-acting air cylinder, which is anchored to a bracket attached to the register front plate. The air cylinder is fitted with two limit switches to signal the register position — open or closed — to the control system in the burner cubicle.

Four ports are provided in the air register front plate; the two ports on the horizontal centre line are fitted with observation windows for viewing flame conditions, while the two lower ports accommodate the flame-sighting heads for the dual-signal flame monitoring equipment, which is housed in the adjacent burner control cubicle.

The venturi throat is designed to form a smooth passage for the combustion air admitted from the windbox into the furnace when the air register is open; it provides a pressure tapping point for use during

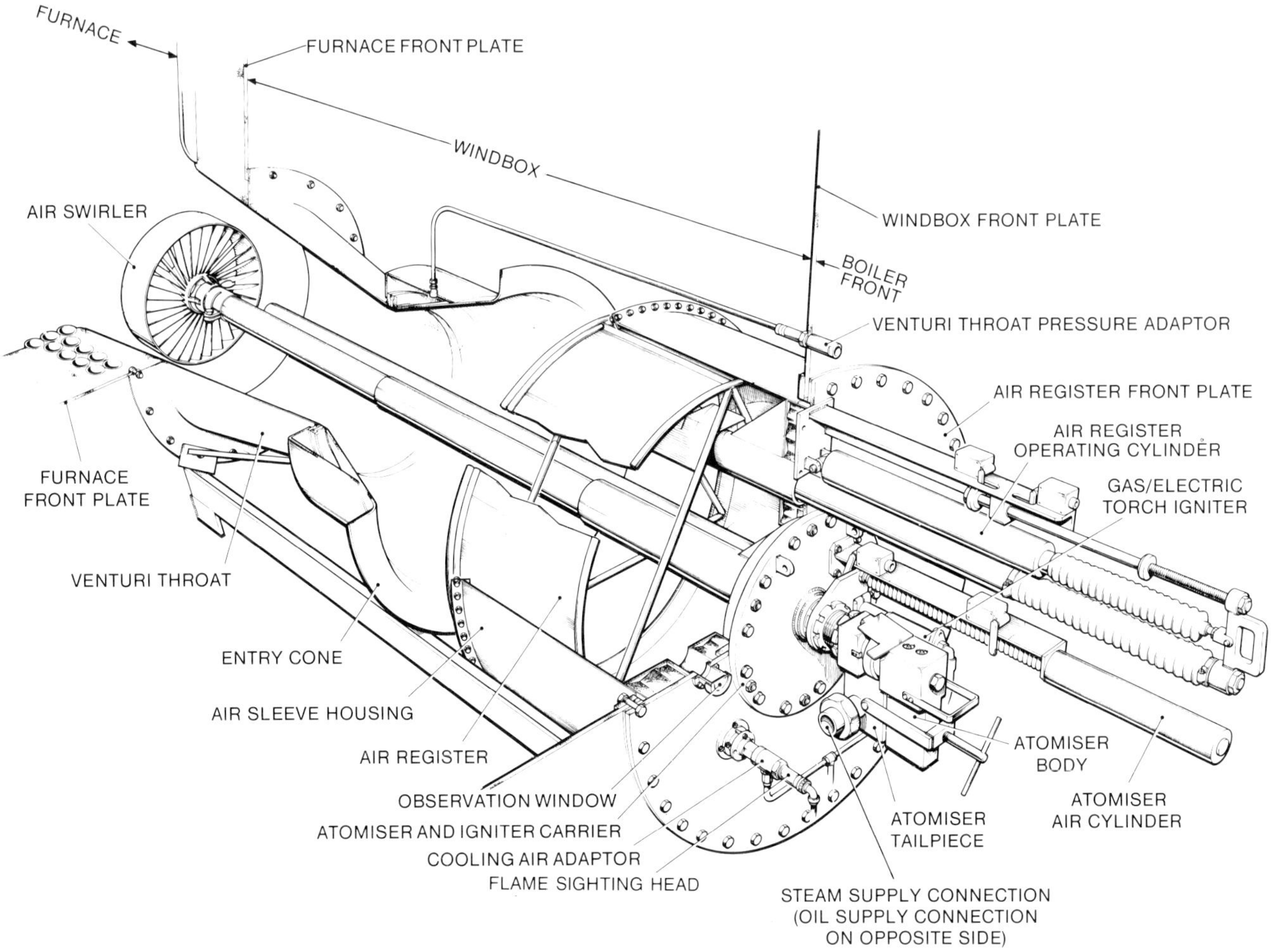

FIG. 4.37 Steam-atomised oil burner assembly

commissioning to balance the airflow through the registers. The throat assembly is bolted to a front plate attached to the furnace wall, and extends into the windbox on a common axis with the air register. The bell mouth of the venturi entry-cone forms a sealing surface for the air sleeve when the register is in the closed position; it is essential that this is accurately adjusted to prevent tramp-air ingress into the boiler with the burner out-of-service.

An air swirler is provided as an integral part of the atomiser, for flame stabilising purposes, and it is positioned centrally in the venturi throat. The combustion air supply from the windbox, admitted via the air register, passes through the throat and the front exit-cone in parallel flow and, consequently, no swirl is produced by the air register, so unrecovered pressure losses are reduced to a minimum.

The swirler is mounted at the extreme inboard end of the atomiser distance piece, in the centre of the venturi exit, and is designed to produce a zone of air that is swirling within an outer envelope of air that is flowing in a parallel direction, thereby stabilising the flame created at the burner tip.

In a 660 MW boiler, each burner will have an oil throughput of 6.6 t/h; and 24 of them are installed in windboxes in two rows of six at the front and two directly-opposite rows of six at the rear of the furnace.

A typical CEGB 500 MW oil-fired boiler burner layout, using the pressure jet principle, is shown in Fig 4.38 and, as can be seen, the operating principles and applications are very similar to those described for the steam-atomised burners.

Burner control and operation (steam-atomised and pressure jet)

The control scheme for both pressure jet and steam-atomised burners uses an electropneumatic system and each burner is provided with its own control cubicle, situated adjacent to the burner on the same level, through which the control air and electrical services are supplied; the propane gas for lighting up the burner torch igniter is also supplied via a premix unit in the control cubicle. The fuel oil and steam supplies are fed through branch pipes from the ring main that runs from one side of the boiler to the other underneath the burner galleries. Pneumatically-operated shut-off valves are provided in the oil and steam supply lines, adjacent to each control cubicle, and an oil safety shut-off valve and a purge steam valve are positioned at the burner front.

On the steam-atomised burner, high and low pressure steam supplies are provided for atomisation and purging purposes.

The burner control cubicle contains all the electronic and electropneumatic control equipment required to operate the burner automatically during the start-up and shutdown procedures. The burner may be started or stopped by pushbutton operation from the control panel of this cubicle, or may be selected for remote operation and operated from a module on the unit control desk in the central control room.

The main flame of each burner is monitored by ultraviolet (UV) flame monitors, which are housed in the control cubicle, the flame sighting UV heads being

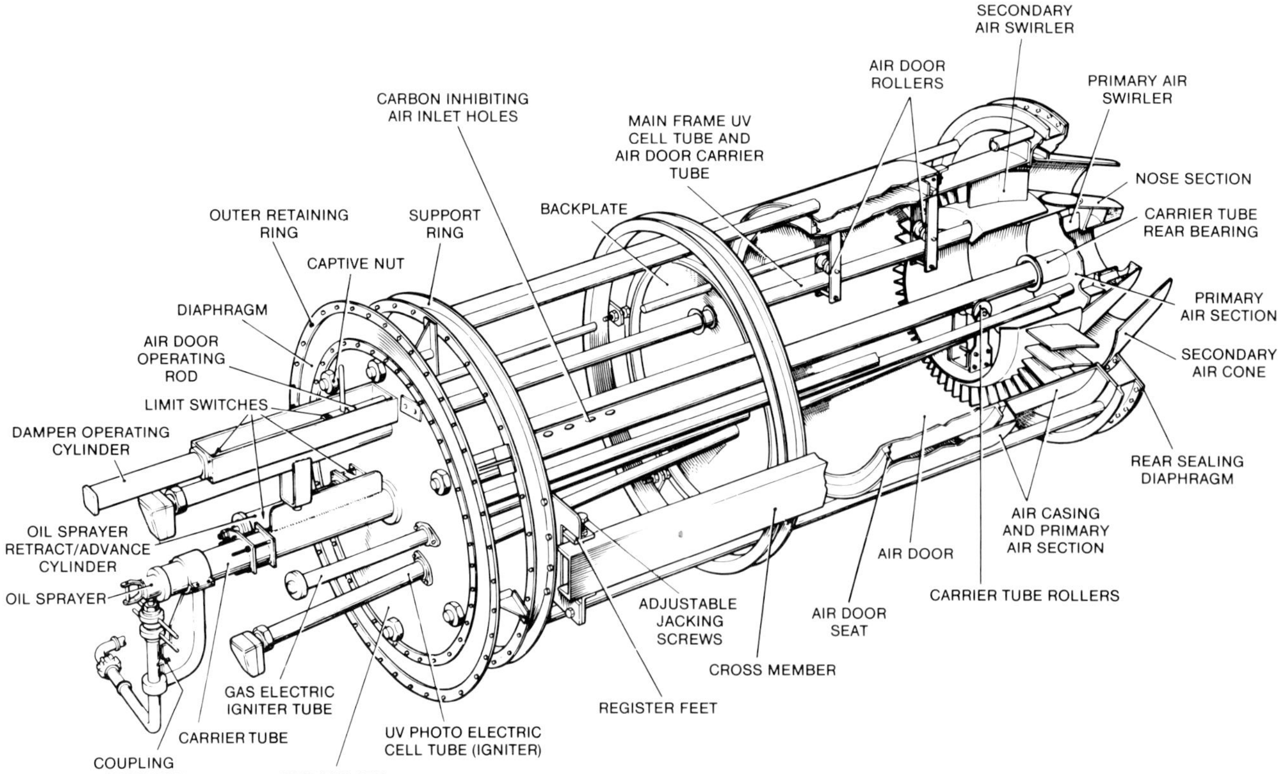

FIG. 4.38 500 MW oil-fired boiler pressure jet fuel oil burner assembly

installed in ports on the air register front plate; a gas flame UV monitor is also provided in the control cubicle, the sighting head of which is also installed on the register front plate. Compressed air is taken from the cubicle to each of the main UV flame-sighting heads for cooling, and to prevent the accumulation of dust or dirt on the lens.

Before an oil burner can be fired, the following starting interlocks must be satisfied:

- Oil pressure and temperature above the required minimum value.
- Boiler circulation pumps differential pressure satisfactory (on assisted-circulation boilers).
- Ignition gas pressure satisfactory.
- Control air pressure satisfactory.
- HP and LP steam pressures to burners satisfactory (steam-atomised).
- Purge steam pressure satisfactory (pressure jet).
- Furnace purge sequence completed and windbox air pressure satisfactory (combustion air).
- UV head cooling air pressure satisfactory.

To start a burner, the 'start' pushbutton is pressed — either at the local control cubicle or at the unit control desk. The air register opens and the atomiser moves forward automatically into the firing position; steam is then supplied to the burner pipework and to the atomiser to warm them through prior to burner light-up.

On the pressure jet burner, a warming bleed-steam supply is maintained on shutdown burners when in the 'ready' state. This is provided from the purge steam system, via an orifice valve operated pneumatically by the burner sequence. In the start-up sequence, an initial purge is also carried out as with the steam-atomised burner.

When the initial purge is complete on steam-atomised burners, the torch igniter is lit automatically and, when the gas flame is established, the oil and steam valves open to allow steam and oil to flow to the sprayer plate at the burner tip, where the oil is entrained with the steam to be ignited by the gas-torch flame.

On the pressure jet burner, with the ignition flame established, the purge steam valve closes and the main oil valve opens to establish an oil flow. In both burners the burner will trip out into the purge sequence, if the main flame UV signal is not established within 3 seconds.

The burner will fire on satisfactory completion of the start/interlock sequence:

- The 'burner ready' lamp is extinguished and the 'burner run' lamp is illuminated, indicating that the start interlocks are made.
- Operation of the 'start' pushbutton initiates the start monitor and lockout timer.
- The burner register opens and makes 'open' limit switch.
- The atomiser is then inserted and makes the 'inserted' limit switch.
- Atomiser steam valve opens and satisfies the steam-pressure switch (steam-atomised burner).
- The burner then pre-purges by the purge steam valve opening for 45 seconds.
- Pre-purge operation takes place on pressure jet burner.
- On completion of the pre-purge, the purge valves close (both types of burner).
- The igniter spark is initiated and the propane gas valve opens, subject to a signal from a pressure switch in the oil line that the purge steam pressure has fallen (steam-atomised burner).
- 'Gas flame on' signal received.
- Oil valves then open and the steam valves close.
- The oil valve will close, if an oil flame is not established within 3 seconds.
- The burner 'main flame on' indication is initiated.

The burner shutdown sequence is:

- Operate the 'stop' pushbutton.
- Igniter light-up sequence initiates.
- The purge steam shut-off valve opens.
- After 15 seconds, the atomising steam shut-off valve closes and the HP purge shut-off valve opens (steam-atomised burner).

After the purge period of 60 seconds:

- The igniter shuts down.
- The purge steam shut-off and HP purge valves close.
- The burner is retracted.
- The air register closes.

Running interlocks are provided to trip all burners, leaving their air register open. These are:

- Unit trip.
- Tripping of any ID or FD fan.
- Low fuel oil pressure.
- Low control air pressure.
- Loss of electrical supplies.

If the ID or FD fans trip, they have to be restarted and the furnace purge sequence carried out; the burners can then be purged and reset to operational state.

Oil burners are tripped individually on loss of a satisfactory flame monitor signal.

6.2.3 Air register and burner layout (coal-fired boilers)

Burner air register assembly

A typical burner air register layout is shown in Fig 4.39, which is used on the 660 MW coal-fired boiler at Drax Power Station; it incorporates a recirculating-tip shut-off lighting-up oil burner, with a gas/electric igniter.

The oil burner unit comprises a local actuator assembly and an airbox with a centre carriage assembly into which are fitted the igniter, two UV detector heads and the fuel oil injection. A combustion air fan provides the combustion air for the burner and the airflow is closely matched with the firing flow of fuel oil to maintain the correct air/fuel ratio.

The oil burner is positioned in the burner assembly to project through the air register and windbox sufficiently for the fuel oil injection-tip just to enter the burner throat. When not in the firing mode the oil burner tip is kept cool by the recirculating oil flow through the burner.

The pulverised-fuel (PF) burner tube is supplied with a mixture of primary air and pulverised fuel. As the mixture passes from the burner tube into the furnace, turbulence is imparted to it by the impeller and the mixture joins the secondary air from the windbox, which ensures the correct air/fuel ratio for correct combustion.

The burners are situated in groups of six horizontally at five levels on the boiler, front and rear. An oil burner is provided with each PF burner.

Control and operation

The principle of operation of the recirculating-tip oil burner is shown in Fig 4.40; it depends on a continuous supply of recirculating fuel oil, which operates the injector when in use, and cools the tip-valve assembly to prevent damage due to overheating, when not in use.

Under normal operating conditions, the injector is set to one of two modes, 'firing' or 'recirculating', by the changeover valve. Within the changeover valve, oil ports in the inlet and outlet sleeve enable the oil flow to the tip-valve to be reversed, this reversal changing the position of the bobbin valve in the tip assembly. The bobbin valve is held shut in the 'recirculating' mode by both a spring and the oil supply pressure to give a leak-tight seal, the recirculating oil also cooling the atomiser components. In the 'firing' mode, the oil flow is reversed, opening the bobbin-valve and allowing oil to flow to the atomiser nozzle. Normal differential pressure between the supply and return valves is 9 bar, and a minimum of 6–8 bar is required when initiating an injector light-up to allow the bobbin-valve to open against spring pressure.

Firing mode The changeover valve spindle is positioned forward by pneumatic control from the local actuator unit, i.e., toward the injector tip-valve assembly. Fuel oil then passes from the supply port of the changeover valve, around the valve spindle and inlet chamber, into the annular duct between the inner and outlet tubes, where it then passes to the tip-valve assembly.

At the tip-valve, a drilling in the distribution body allows fuel oil pressure to act on the front face of the bobbin, causing it to move against the spring and setting it in the open position to allow fuel-oil through the atomiser nozzle. As the oil flows through the atomiser, it forms a finely-divided spray, necessary for good combustion; this is called the 'firing flow'.

A return oil path is provided through an orifice in the bobbin-valve, the central inner tube and the centre of the changeover valve spindle to the return port of the changeover valve. The bobbin is maintained in the 'firing' mode due to the return oil pressure acting on the rear face of the bobbin being less than the supply oil pressure acting on the front face.

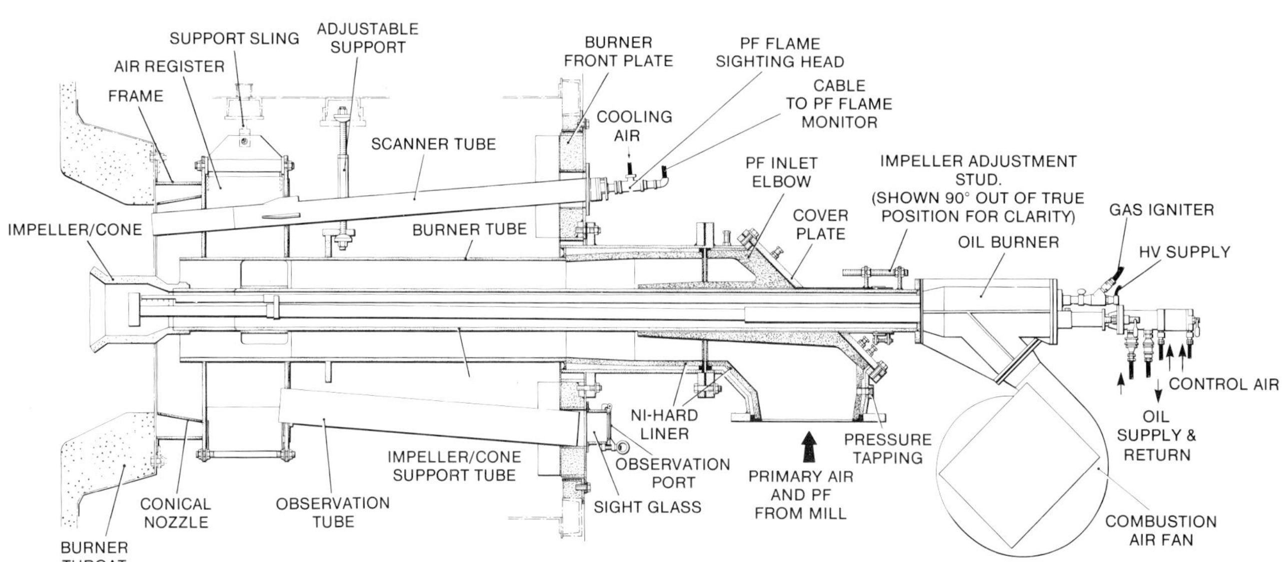

FIG. 4.39 Pulverised-fuel (PF) burner

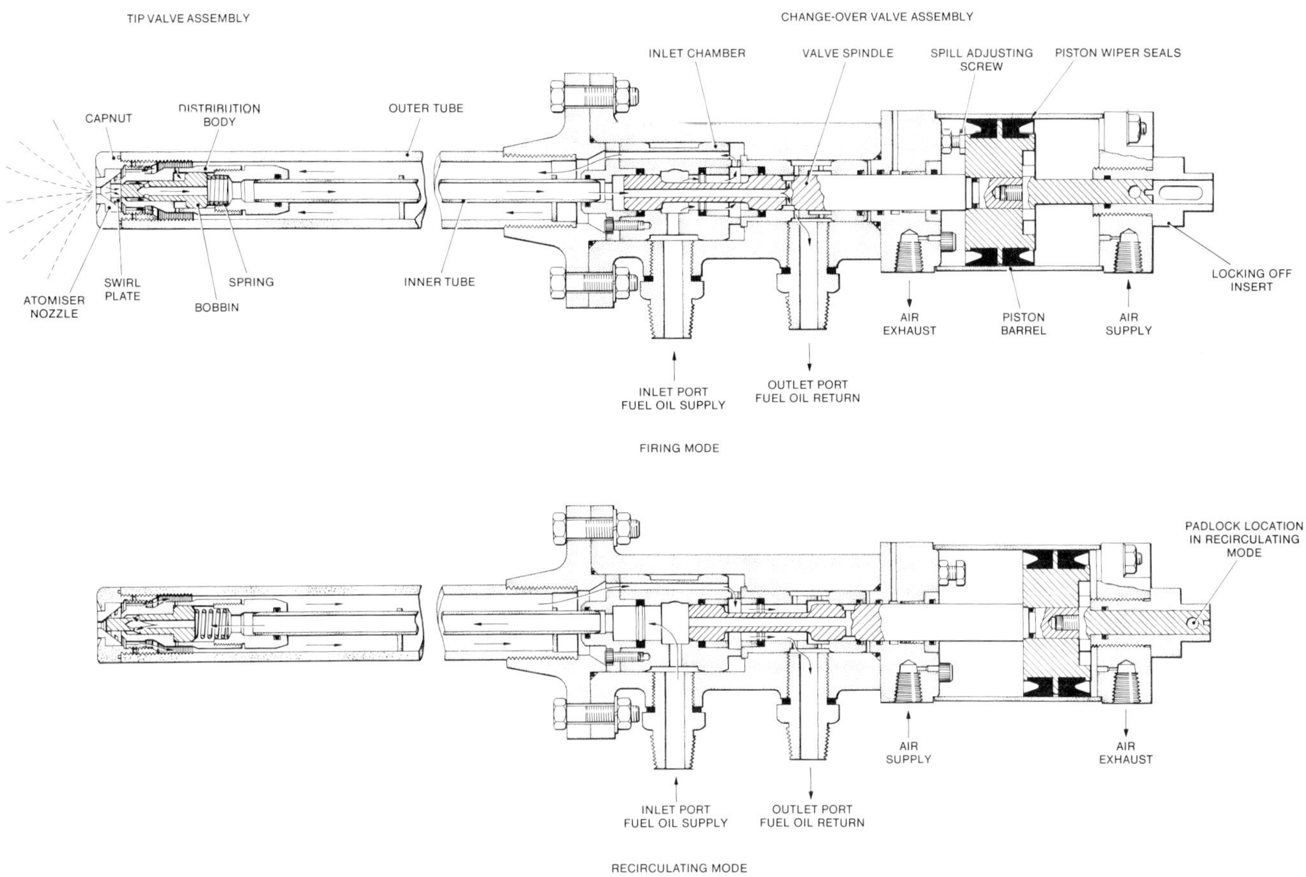

FIG. 4.40 Fuel oil injector — principle of operation

Recirculating mode Under control of the local actuator unit, the changeover valve spindle is positioned to the rear, i.e., away from the injector tip-valve assembly.

At the tip-valve, fuel oil pressure, acting on the rear face of the bobbin-valve, causes it to move forward and seat in the atomiser nozzle, shutting off the flow of oil being sprayed through the atomiser.

A return oil path is provided through an orifice in the annular bobbin-valve, around the atomiser, back through the duct to the inlet chamber, around the changeover valve spindle and through drillings in the outlet sleeve to the return port of the changeover valve. The fuel oil now flowing through the injector is termed the 'recirculating flow'.

Oil supply to the burners is controlled by pneumatically-operated shut-off valves. The valves are arranged to allow a 'parking' facility which enables the supply and return pipes to be parked whilst not affecting the total fuel oil flow. In the parked position, fuel oil flow is diverted to the recirculation main, via a restrictor-block, which simulates flow-loading through the injector.

As with the oil-fired boiler, the burners are controlled by an electropneumatic system; flame monitoring is by UV flame detector equipment. The starting interlocks are such that the gas air heaters and one group of ID and FD fans must be in service before the oil-fired lighting-up burner start is permitted. Control air, fuel oil and ignition gas pressure interlocks must also be healthy.

The oil lighting-up burner start-up sequence is as follows, subject to the starting interlocks being satisfied:

- Combustion air fan starts and satisfies the air pressure switch
- Ignition gas valve opens and gas flame is established for 5 s, monitored by UV signal
- The solenoid-actuated valve then operates and changes the fuel-injector from 'recirculation' to the 'firing' position
- An oil flame is established and monitored by UV signal; if satisfactory, the 'flame on' signal lamp is lit
- Loss of the 'flame on' signal will trip the fuel injectors within 5 s.

If, during the light-up sequence, the sequence equipment detects a 'flame circuit fail', or the 'ignition check' detects no flame, the combustion air fan continues to run and the sequence goes to the locked-out state, a 'fail' indication lamp being illuminated.

If the 'fail' indication is illuminated, it can only be cancelled with the reset button in the burner control cubicle. When the local reset button has been pressed, an oil valve check is made. If the oil valve is open, a check is made to ensure that the burner is still locked out.

If the oil valve is closed, the combustion air fan is stopped and a 'combustion air off' check is made. This signal is fed to the fail monitor which will detect 'tripped' or 'satisfactory'.

With the selector switch in the 'air' mode, pressing the burner start button will start the combustion air fan and establish a flow of combustion air. An air check will give either pass or fail. If it is fail, the fault will be either in the fan or in the fan discharge damper. If it is pass, the 'air on' indication (blue) will be illuminated.

If the start button is held down for short periods (less than 5 s) only, the ignition transformer and gas solenoid-valve are energised. This enables the igniter operation to be checked. Releasing the start button de-energises the ignition transformer and gas solenoid valve.

If the 'start' button is held down for at least 10 s, fuel oil will be injected after the first 5 s and for a maximum period of a further 5 s. This enables the complete ignition sequence to be checked, up to the point of oil ignition.

Tilting burners

As explained in Section 4.3 of this chapter, some coal-fired boilers use tilting burners in conjunction with superheater and reheater sprays to control temperatures.

A typical tilting burner used in a 500 MW coal-fired boiler is shown in Fig 4.41. The boiler uses the corner-firing technique and the burner tilt mechanism is moved by a hydraulic actuator, each corner group of burners having a single actuator unit. The burner tilts have mechanical stops to prevent them operating beyond ±20° to the horizontal.

The burner control works in conjunction with the reheater spraywater control valves, and consists of control loops which measure the reheat outlet temperature and vary the tilt of the burner to control to the desired temperature. When the burner reaches the mechanical limit of tilt, −20° from horizontal, the reheater sprays are brought into operation. The burner tilt can also be controlled manually from 'raise/lower' pushbuttons on the control desk. In normal operation, with little boiler fouling, the boiler should be able to operate with the tilting burners controlling the reheat temperature and the reheater spraywater control valves staying closed.

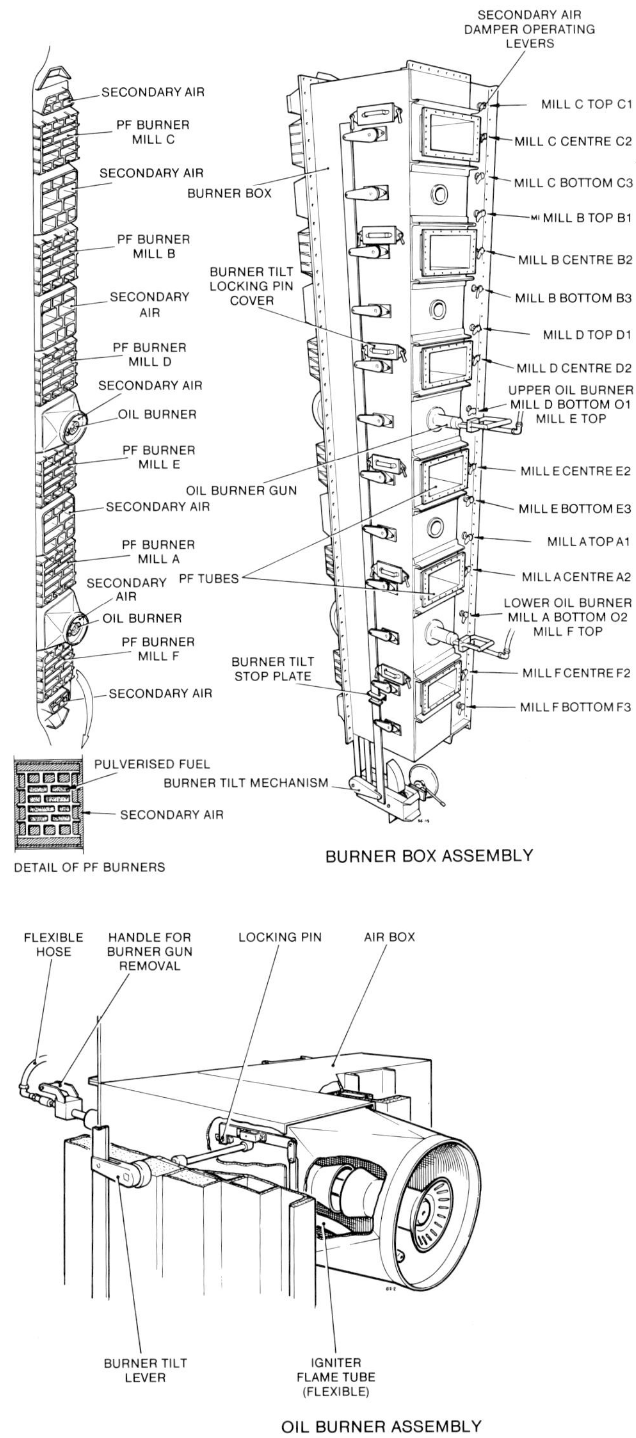

FIG. 4.41 PF boiler corner-firing tilting burner assembly.

6.2.4 Safety and resources

Safety

The following safety points should be considered:

- Safe access must be provided to all burner registers within the windbox and furnace. On oil-fired stations, a scaffold will have to be erected in the furnace which will withstand full burner airflows.
- Good lighting required in furnace, windbox and burner galleries.
- Access must be restricted to personnel involved in the tests to furnace, windbox, burner galleries and any adjacent air and gas ducting, particularly during burner purging and igniter test firing exercises.

- Fixed and portable fire extinguishing equipment to be commissioned and available.
- All combustible materials to be removed from the vicinity of the boiler.
- The risk of fires and explosions from propane and fuel-oil to be fully understood by all test team personnel. Propane is heavier than air and very explosive when mixed with air in the range 2.5% to 9.5% by volume.
- All propane pipework to be purged with nitrogen before propane is admitted.
- Smoking and naked lights to be prohibited in the burner vicinity when propane is being used.
- Access to pressure parts, gas passes, windbox, and any work involved in breaking flange joints to be in accordance with the Safety Rules and controlled, using Permits for Work.

Resources

The following resources will be required during oil burner commissioning:

- Nitrogen bottles, complete with pressure regulators and pipework, for purging propane pipework.
- Manufacturer's burner control test box.
- Megger.
- Multimeter.
- Lubricating oil for the mist lubricators.
- Electrical supplies to the burners (440 V and 110 V AC; 110 V and 50 V DC).
- Boiler fans to be commissioned and available.
- Fuel oil system commissioned and oil available.
- Propane system commissioned and propane available.
- Auxiliary steam system available.
- Control air system available.

Staffing

A team of approximately eight people will take up to fifteen weeks to commission the burners on oil firing.

6.2.5 Oil burner commissioning

Steam-atomised and pressure jet burners in oil-fired boilers

A typical commissioning network diagram is shown in Fig 4.42, which can be adapted to suit either type of burner installation.

Checks will be made that all erection work is complete within the windbox and that control air, ignition gas and steam pipework, and cabling are complete.

Each burner register will be checked from within the windbox, ensuring that venturis are correctly installed, that the register seals are complete and that the register sleeve makes a good seal with the venturi inlet-flange by manual operation of the register. Extreme care must be exercised when carrying out burner register movements, with personnel inside the windbox, to prevent injury.

All dimensions and alignments must be checked to ensure correct installation of igniters, burner actuators and flame viewing equipment. From within the furnace, the concentricity of the burner within the quarl and all swirler, venturi and quarl dimensions must be checked. The setting of the burner tip with respect to the swirler should also be checked.

When all the above mechanical checks are complete, the boiler airflow tests, described in Section 5.5 of this chapter, are carried out at a convenient time during the burner commissioning programme.

On completion of mechanical checks, the following steps are taken:

(a) An air purge of control air and ignition gas pipework is carried out. This will be done by breaking flange joints at suitable points in the system and blowing through the pipework until all scale and debris is removed. Temporary connections from the air to the ignition gas pipework will have to be made. For the above activities, the necessary safety instructions should be issued and only personnel involved in the test should be allowed in the commissioning area. On completion of purging, the ignition gas pipework should then be filled with nitrogen to displace all the air in preparation for pressurising with propane.

(b) With the control air systems pressurised, all pneumatic checks of the oil-firing equipment can then take place, all valves and burner moving parts checked for correct and free movement. All pressure switches should be set to operate at their correct settings.

(c) Following the air purge and pressurising of the propane gas pipework with nitrogen, the system should then be pressurised with propane gas to the burner fronts. Before this activity, the following safety features must be observed:

   - Portable fire extinguishers available
   - All combustible equipment moved from the burner area
   - Portable ventilation equipment set up for igniter commissioning external to the burner register
   - Permanent fixed fire-fighting system available
   - All personnel not involved in the tests removed from the area

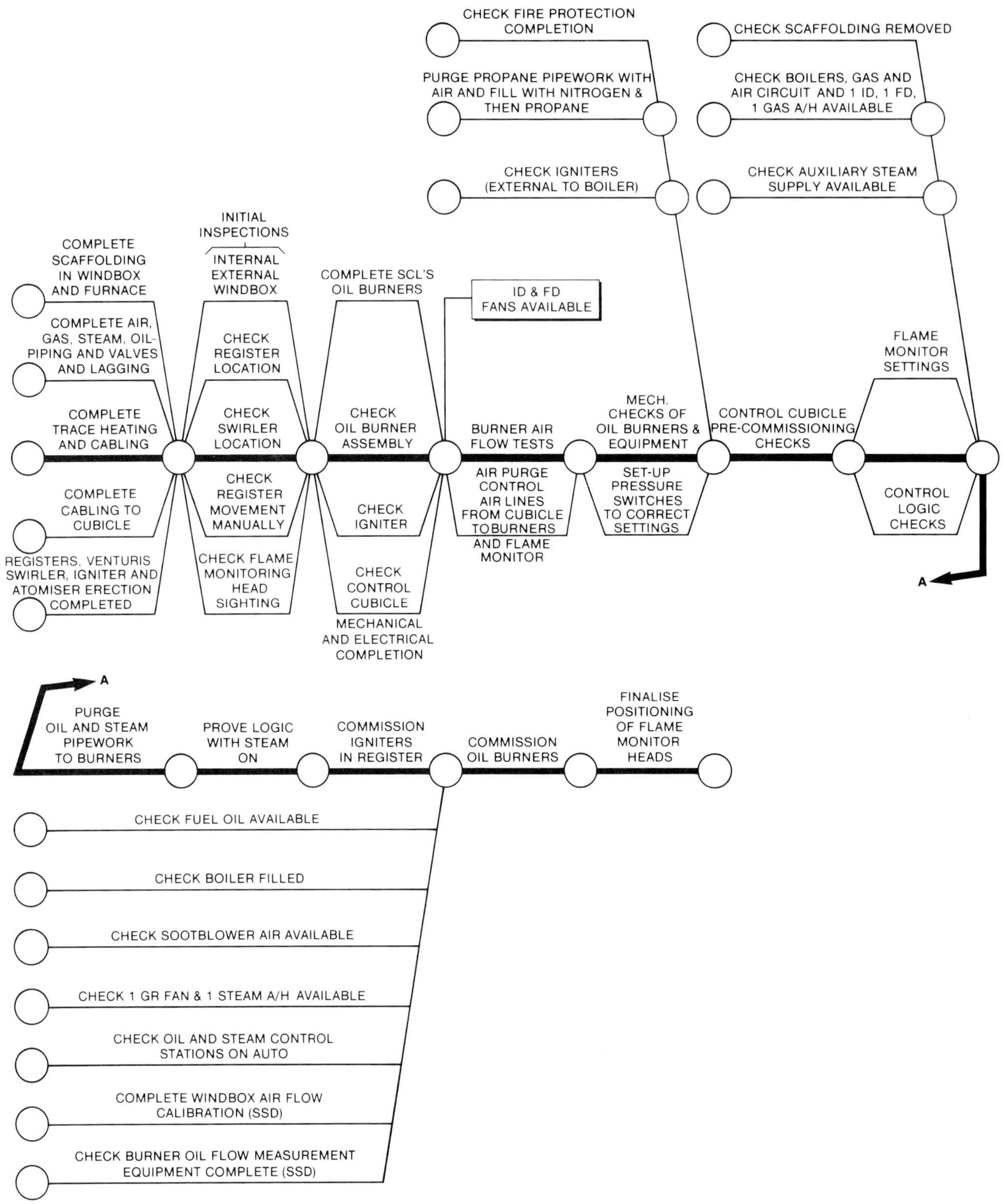

FIG. 4.42 Oil burner commissioning network

The gas/electric igniters can then be fully checked and, using the burner test box, can be fired to check the spark, ignition and flame length. The igniters should then be isolated and refitted to the burner register. *Note* The manufacturer's burner test box is a device provided for burner testing. It enables all the individual burner operations to be carried out without recourse to having all the starting and sequence interlocks made. For example, the burner can be put through a 'dry run' sequence without the starting interlocks being made.

(d) At this stage, steam lines to the boiler front should be purged with steam from the auxiliary boilers until all debris is removed; again, extreme care must be exercised when flanges, etc., are broken at suitable points on the system.

(e) The fuel oil lines should also be flushed; these activities are normally carried out during the fuel oil system flushing, as described in Section 7 of this chapter.

(f) In parallel with items (a) to (e), the full burner-sequence logic and interlock system, described in Section 6.2.2 of this chapter, must be commissioned for each burner, on both the local and remote controls.

With the fuel oil and steam isolating valves closed, the burners should then all be tested on a 'dry run' sequence to operate correctly. It is normal practice to use the manufacturer's test box, from which each individual burner operation can be tested. The test box will be widely used in the above operations.

(g) When all the pre-commissioning activities in (a) to (f) are complete, the boiler should be prepared for test firing the gas/electric igniters within the register and for proving the logic and full sequence controls. The oil and steam pipework from the burner front to the burner are also purged at this stage.

For this purpose, the boiler will be set up with one group of steam and gas air heaters and fans in service with the burner steam, air, propane gas and electrical supplies all available. The fuel oil to the burners will be isolated and locked off. For this exercise, it is not necessary to fill the boiler with water.

Each atomiser should be removed from the burner and its sprayer plate removed; the burner atomiser is then replaced in the register and, using the test box, the purge steam supplies are put onto the burner, the isolating valves opened, and the pipework and lance purged of entrained debris. The steam supplies are then isolated, the sprayer plate reinserted and the atomiser replaced in the register. Each burner is purged in turn, as above. For test firing the gas/electric igniters, the boiler should again be set up with steam on the steam air heaters, a gas air heater in service and one group of fans running. A furnace purge should be completed, the full furnace purge sequence being tested and the windbox air pressure set up to design conditions. Using the test box, each gas/electric igniter should then be fired in turn and the flame length and shape set up. After each igniter test firing, a furnace purge must be completed before firing the next one.

A full simulated 'dry run' test of each burner should now be completed and all interlocks and sequence controls fully proved. The burner is now available for firing on oil, which will require the boiler to be filled with water; this activity is normally carried out for the first time during the boiler chemical clean programme. During these early firing trials and then later when the boiler is operating at MCR conditions, the UV monitoring system of the burner flames will be finally positioned for optimum performance.

Before firing on oil for the first time, the starting interlocks described in Section 6.2.2 of this chapter must be fully operational and proven, and the running interlocks should also be commissioned and proven; final testing will be covered in Chapter 6.

Coal-fired boiler recirculating-tip oil burners

The commissioning checks described in Section 6.3.5 (a) to (c) and (e) to (g) will be similar for this type of burner. Each of the six burners for each mill group must be checked; this is then repeated for all other mill groups. Obviously there will be no purge steam systems to commission and the burner combustion-air fan will have to be checked by commissioning the motor in accordance with procedures described in Chapter 3, Section 4, direction of rotation then being checked and the fans run to blow out any dirt or debris left in the burner air pipework.

The full sequence control/interlock scheme described in Section 6.2.3 of this chapter should be commissioned. The burners will be used initially firing on oil only for the chemical clean, steam purge and floating safety valves programmes, coal-firing only being introduced during the unit loading programme. This is covered in Chapter 6.

The PF burner airflow tests should be completed at this stage: this has been covered in Section 5.5 of this chapter.

Tilting burner commissioning

The commissioning checks described in Section 6.3.5 will be similar for this type of burner; in addition, the burner tilting mechanism must be checked.

The burner box assembly is checked for correct installation, and each PF oil burner and secondary-air register checked. The burner tilting mechanism is operated manually to the +20° and −20° positions, and the maximum position stops set. Any limit switches associated with the operating positions should be correctly set.

Secondary-air damper operation should be checked and the operating linkages set and locked in the correct positions.

The hydraulic actuator and pump for the burner tilt mechanism are checked, following the procedure in Chapter 3, Section 2.3. The burner tilting mechanism should then be checked by operation from the unit control panel, and then in the automatic control mode.

6.3 Pulverised fuel (PF) milling plant

6.3.1 Introduction

Milling plant may be divided into three main types — low, medium and high speed; each has its own advantages and disadvantages.

The CEGB uses all three types in its power stations and it would be impossible in this book to cover them all. Thus one type will be covered, namely, medium-speed pressure mills used with the latest CEGB 660 MW coal-fired boilers.

This section covers the description and commissioning of the coal feeders and the mills up to the stage where they are ready to receive coal. Operation of the mills with coal is covered in Chapter 6.

Further information on other types of mill can be found in Volume B.

6.3.2 PF milling plant description

PF milling plant layout

A simplified diagram of a PF mill and associated equipment is shown in Fig 4.43 and a diagram of the layout of the air pipework, dampers and fans is shown in Fig 4.44.

On a 660 MW boiler there are ten mills supplying 60 PF burners; each mill supplies six PF burners with the coal/air mixture.

Each mill is driven by its own electric motor via a reduction gear unit equipped with its own lubrication system. The mills are supplied with coal from their own coal feeders and air from their primary air fan. Sealing air is also supplied from a separate seal-air fan.

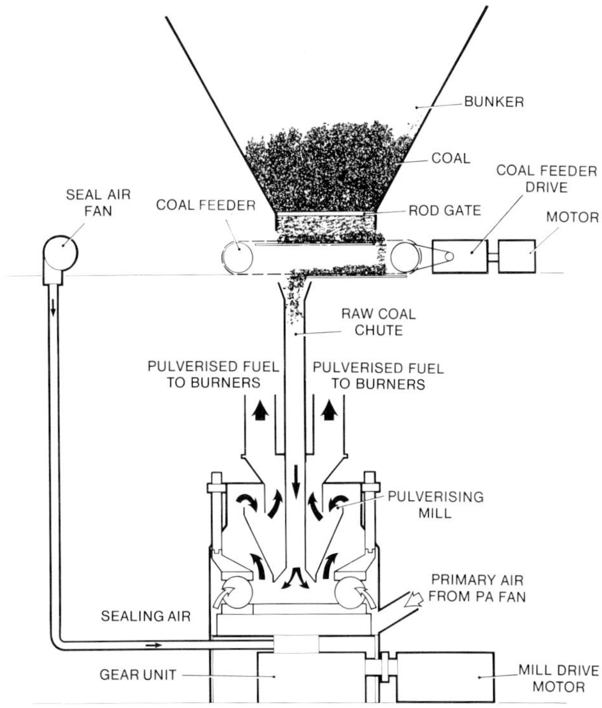

FIG. 4.43 Simplified diagram of PF mill and associated equipment

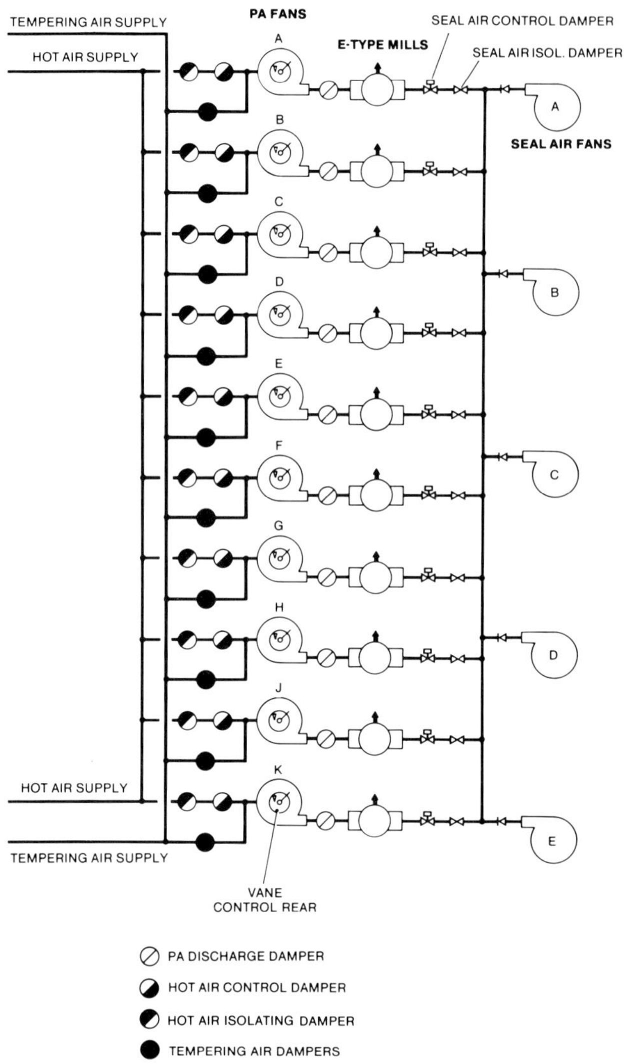

FIG. 4.44 Layout of PF mills, air pipework and dampers for a 660 MW boiler

Coal feeders

Each pulverising mill is served by an associated drag-link coal feeder which provides a controlled flow of coal from the storage bunker to the inlet chute of the mill. The coal feeder is driven by a variable-speed compound gear unit powered by an 11.2 kW flange-mounted electric motor and has a capacity of 37.0 t/h.

Figure 4.45 shows a typical drag-link coal feeder and Fig 4.46 shows a schematic diagram of the coal feeder drive.

The coal feeder comprises an endless conveyor drag chain, enclosed in a casing fabricated from mild-steel plating, and driven by an electric motor through a positive infinitely variable (PIV) gear unit, a reduction unit and a chain drive. The casing forms an enclosed trough in which the conveyor chain runs between a drive shaft and a tailshaft, each carrying two chain sprockets. The chain is made up of a number of bushed links joined by flight bars; every other link is connected with these flight bars.

The feeder is sited below the bunker outlet on steel supports. The upper part of the casing is connected via

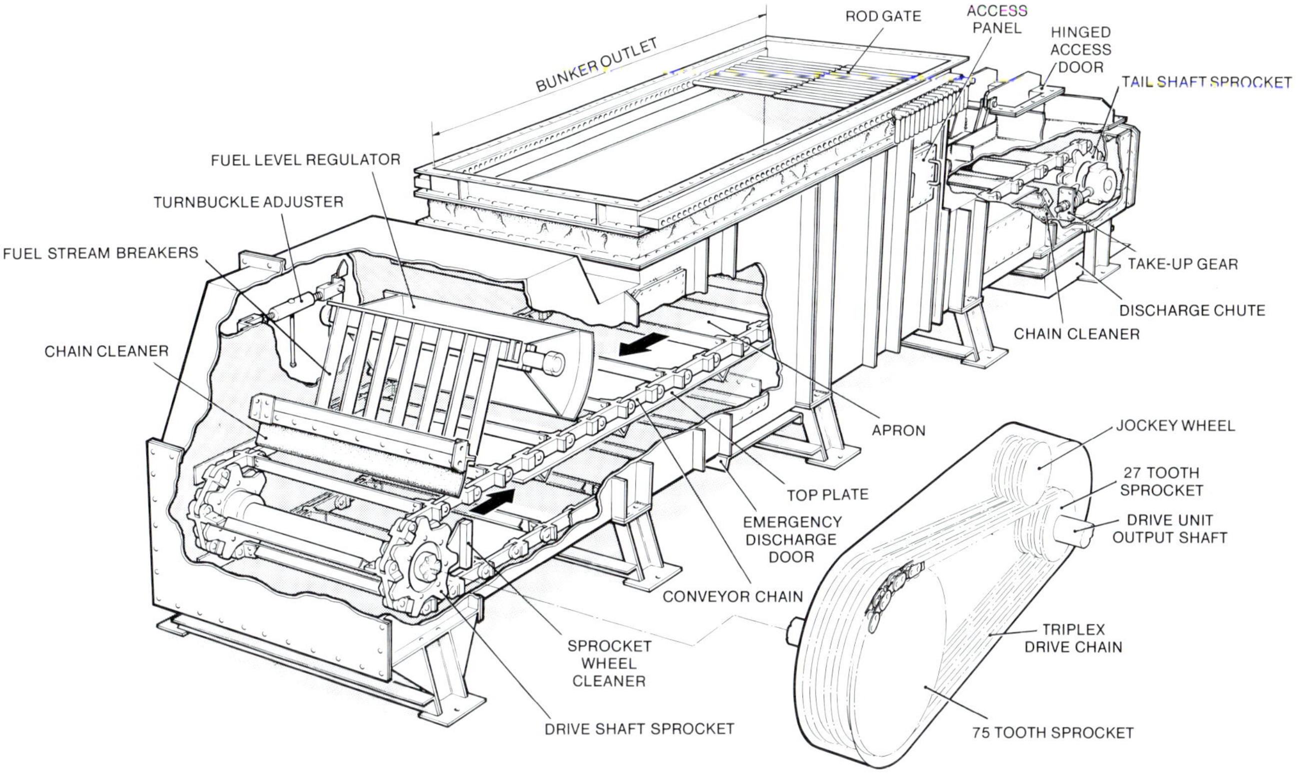

FIG. 4.45 Coal feeder

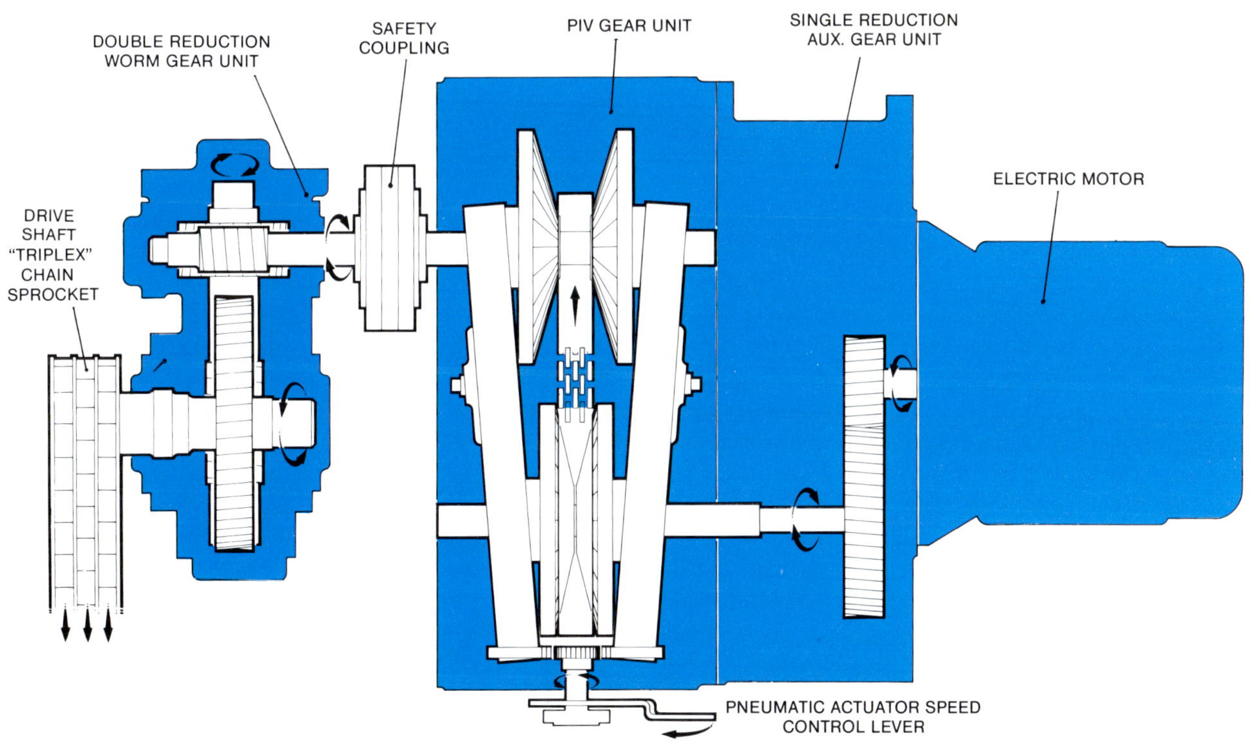

FIG. 4.46 Coal feeder drive — schematic diagram

a rod-type cut-off gate to the bunker outlet. The cut-off gate is used to control the flow of coal into the feeder. A discharge chute is located on the underside of the feeder, at the non-drive end.

Coal falls under gravity from the bunker outlet through the rod gate and onto the feeder apron, from where it is dragged along by the upper strand of the conveyor chain which moves over the top plate. When the coal reaches the end of the top plate, it drops through the upper strand of the conveyor chain to the bottom of the casing where it is dragged along by the lower strand of the conveyor chain to the discharge chute.

The depth of the coal bed is determined by the setting of a fuel level regulator which is adjusted by means of a turnbuckle adjuster, the setting being indicated on a scale on the feeder casing. A 'loss of coal flow' alarm is also incorporated.

Fuel stream breakers, fitted above the upper strand of the conveyor chain, force the conveyed fuel to collapse if it does not fall freely after leaving the top plate. Sprocket wheel cleaners are located at the driving end and there are chain cleaners at both ends of the feeder.

The tail shaft is provided with spring-loaded take-up gear which automatically takes up any extension in the conveyor chain due to wear or stretch. A limited amount of adjustment may be undertaken on the take-up gear, scales being fitted to indicate when adjustment is necessary.

Feed control is achieved by setting the fuel level regulator to give a satisfactory depth of coal bed, and thereafter adjusting the variable-speed drive to give the required feed rate.

The drive unit is a compound gear unit comprising a single-reduction auxiliary gear unit, a horizontally-mounted PIV gear unit and a double-reduction worm-gear unit. The auxiliary gear unit is bolted to the side of the PIV gear unit and is driven by a flange-mounted electric motor.

The PIV gear unit incorporates the mechanism whereby the speed of the coal feeder is varied and comprises two pairs of opposed shallow conical wheels, one pair mounted on the constant-speed input shaft and the other pair mounted on the variable-speed output shaft. These pairs of opposed wheels are attached to their respective shafts by feather keys which enable them to slide axially on their shafts. Axial movement is imparted to the wheels by angular movement of a speed control lever which is located on the casing of the PIV gear and is controlled by a pneumatic actuator.

Power is transmitted from the input shaft to the output shaft by a specially constructed all-metal chain which locates in the 'Vee' formed between each pair of opposed wheels, thereby taking up a diameter depending on the axial distance separating the wheels. Operation of the speed control lever causes one pair of wheels to move towards each other and the other pair to move away from each other, thereby causing the chain to rise in one pair of wheels by the amount that it descends in the other pair. Two compensating fulcrums (compensators) ensure that the chain tension is kept uniform throughout the whole range of speed variation.

Speed control is effected by a pneumatic actuator connected to the speed-control lever on the PIV gearbox. This lever operates through a slipping clutch and a rack and pinion to move the two operating levers which are pivoted on compensating fulcrums inside the casing. Movement of the PIV control shaft causes the operating lever arms to move in opposition and vary the gaps in the wheel faces. This causes the wheels to slide axially along their shafts, closing one pair of wheels and opening the other pair, dependent on the direction of movement of the speed control lever, thereby changing the effective meshing diameters of the chain and altering the speed of the output shaft.

The mechanism is confined to angular movement of the pinion shaft of 87° of arc. Any movement outside this arc or any mechanical seizure in the speed control mechanism will cause the clutch to disengage and a microswitch to operate an alarm. The clutch will also disengage if the lever is operated while the PIV gear is stopped.

A safety coupling connects the output of the PIV gear to the double-reduction wormgear and affords protection to the drive unit in the event of a feeder jam. The final drive from the wormgear unit is by 'Triplex' chain to the feeder drive shaft.

Should the safety coupling disengage due to overload or a jammed feeder, the action of the coupling as it disengages operates two microswitches which initiate an alarm to warn that the feeder has stopped and also send a signal to stop the drive motor.

Pulverising mills

A typical vertical spindle medium-speed 10E mill of the Babcock design is shown in Fig 4.47.

The main structure of the pulverising mill is fabricated from mild-steel in three cylindrical sections: a bottom section which supports the entire unit and encloses the mill drive gear unit, a centre section that contains the rotary grinding elements and an upper section comprising an outlet turret and classifier assembly. The classifier assembly also accommodates the hydropneumatic units of the mill loading gear. A platform around the upper section provides access to an inspection door and to the top of the mill for routine maintenance and is served by a detachable ladder.

The grinding elements comprise ten hollow cast-steel balls which run between two troughed grinding rings. The lower grinding ring is keyed to a yoke which is driven by the mill drive motor via a gear unit.

The upper grinding ring is keyed to a cast-iron spider which is located by guides in the mill centre section. The spider and upper grinding ring are free to move vertically, but are prevented from rotating. The grinding rings and balls are loaded by means of eight

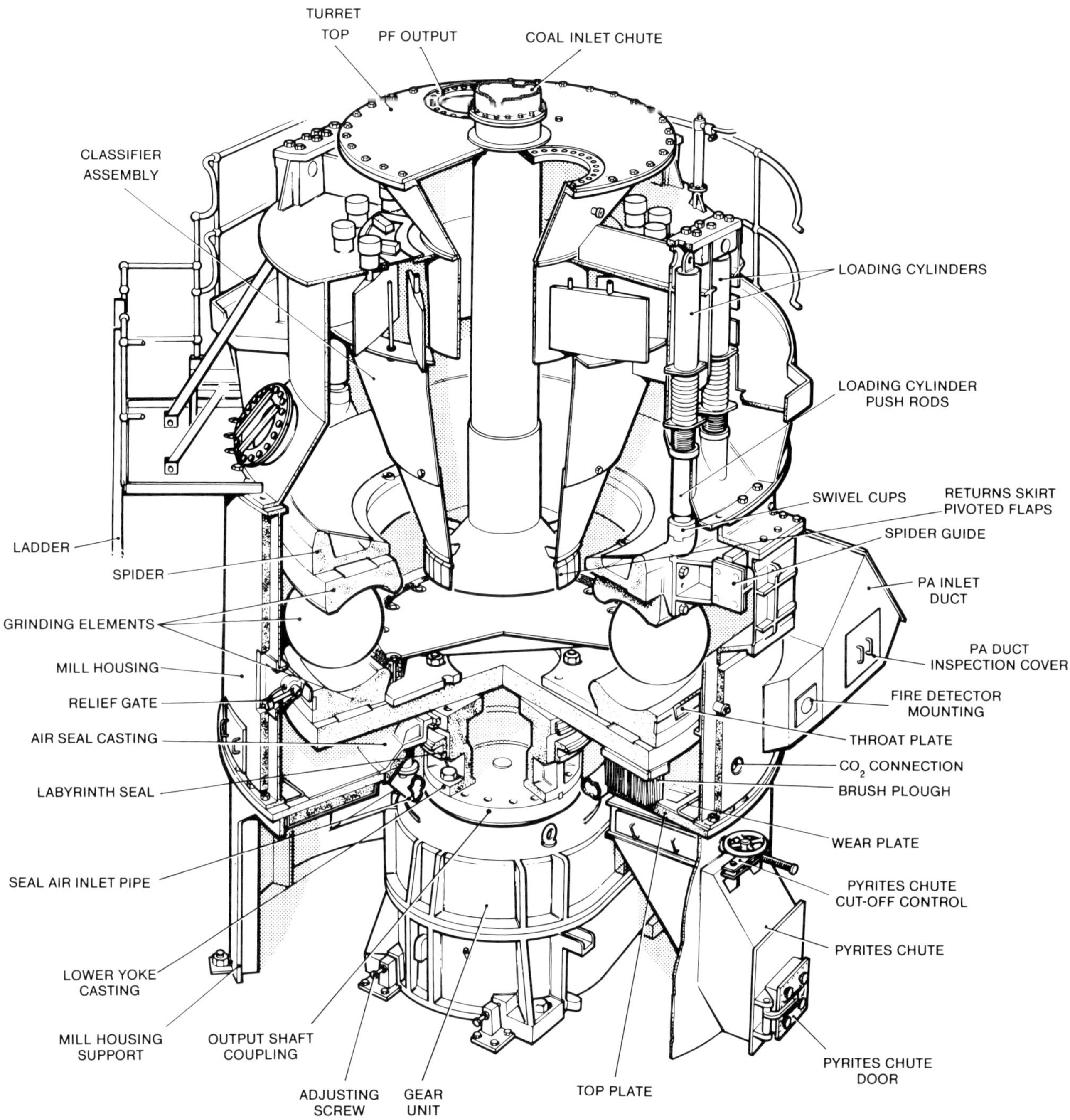

Fig. 4.47 Pulverising mill

hydropneumatic units which apply downward thrust onto the spider. A diagram showing the hydropneumatic mill loading equipment is shown in Fig 4.48.

The hydropneumatic units are gas operated with oil pressure acting between two sets of seals in each cylinder to contain the gas and lubricate the ram. The oil pressure is applied by a manually operated hydraulic pump which draws oil from a supply tank via a filter unit.

The oil pressure is maintained at a higher value than the gas pressure and any leakage past the oil seals in the loading units will be from the oil side to the gas side. The oil will drain down to the lowest point in the pipework, where a sight gauge is fitted. Isolating and drain valves are provided so that the oil may be drained off while the mill is running. A drain connection at the flange end of each loading-unit cylinder returns any oil leakage from this end to the supply tank, via a filter in the drain return line.

Loss of oil for any reason will cause the accumulator piston to move, the fluid content of the system being indicated by the projecting end of the ram against a linear scale adjacent to the accumulator. In the event of 75% of the oil being lost from the system, a limit switch is operated by the ram to initiate an alarm 'mill ball loading fault' on the unit control panel in the central control room.

Gas is supplied from a nitrogen bottle through an external connection, the bottle pressure being indicated on the pre-charge pressure gauge. A regulating valve

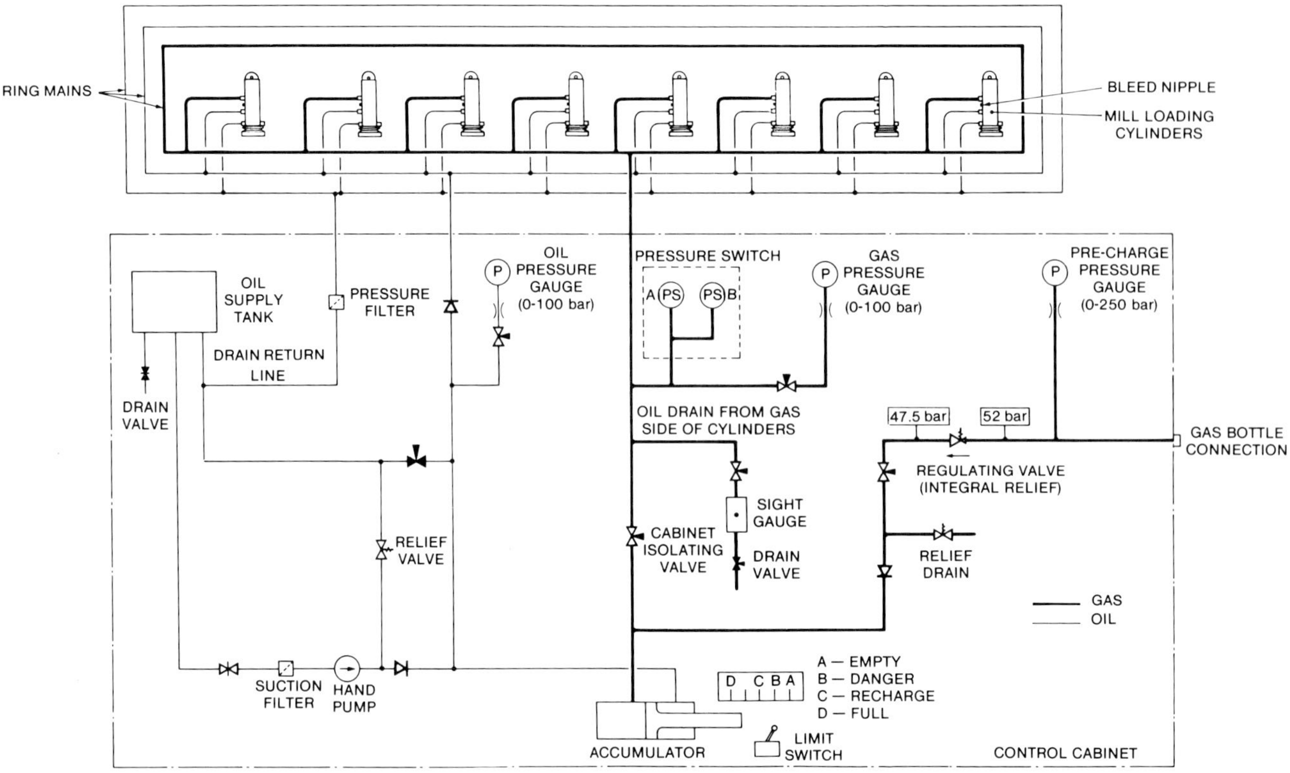

FIG. 4.48 Mill loading equipment

maintains the system at a constant pressure, provided that the bottle pressure exceeds the system pressure, and a gas relief valve protects the system against excessive pressure. Normal system pressure is 48 bar and is indicated on a pressure gauge fitted in the main supply line to the loading units.

A dual-pressure switch, comprising switches A and B, is also fitted in the gas line to the loading units. On charging the system, the pressure rising above 24 bar opens switch A, which is one of the permissives, thereby allowing the mill motor to be available for start-up. In operation, if the gas system pressure falls to 45 bar switch B closes and initiates 'mill ball loading fault' alarm on the unit panel in the central control room. Should the pressure continue to fall to below 24 bar, switch A closes and trips the mill motor.

Gas is also taken to an accumulator, which consists of a cylinder containing a piston and ram assembly. The gas acts on the full area side of the piston and oil acts on the ram side, which is of smaller area. The piston is normally held in the equilibrium by the oil pressure being maintained higher than the gas pressure. The accumulator is used to contain sudden large demands, for sustained efforts over long periods and for smoothing out pulsations or shock loads. A limit switch with four positions provides indications/alarms to show when the accumulator is full, requires charging, is reaching danger level or becomes empty.

The classifier assembly is of the multi-vaned cyclone type, comprising an inverted open cone and incorporating 16 pivoted vanes around the top. The lower part of the cone is closed by a 'returns' skirt formed by a number of pivoted flaps which hang vertically downwards against the belled end of the coal inlet chute which passes through the classifier cone.

Raw coal supplied by the feeder enters the mill through the inlet chute and falls into the grinding zone. Primary air is fed to the mill through two inlet ducts in the centre section of the mill, then passes into the throat formed by the annular space around the lower grinding ring and the mill housing, and flows upwards through slots machined in the throat plate.

Ground fuel particles are picked up by the primary air stream and are carried upwards towards the classifier. The larger particles are initially carried upwards by the air stream and circulate over the upper grinding ring before falling back into the grinding zone by virtue of their weight. The coal/air mixture then passes into the classifier, where any remaining oversize particles are separated out and fall down to the 'returns' skirt until their cumulative weight is sufficient to deflect the flaps and return them into the grinding zone. The setting of the classifier vane angle controls the fineness of the ground product: this is set during commissioning.

Heavy material such as pyrites and tramp iron, which has passed through the grinding zone without being pulverised, is carried around the throat plate and discharged through a counterbalanced relief gate into the space below the yoke. The counterbalanced arm of the relief gate extends outside the mill housing and may be used as a handle to test for unrestricted movement of the gate.

Wire brush ploughs, bolted to the underside of the yoke, sweep the rejected material into one of the two pyrites chutes in the mill base. A manually-operated sliding cut-off gate is sited at the inlet to each pyrites chute so that the chute may be isolated and emptied while the mill is in service.

The brush ploughs and spider guides are accessible for renewal and routine maintenance through doors and cover plates fitted to the centre section of the mill. Similar inspection doors and plates are provided on the primary air ducts and the classifier housing.

Primary air is supplied by a PA fan which has been discussed earlier in Section 5.2 of this chapter and the air ducting and valve layout is shown in Fig 4.44.

Sealing air, at a pressure higher than that inside the mill, is piped to a labyrinth seal around the lower yoke casting to prevent coal dust escaping to the atmosphere. The seal-air inlet pipe to each mill is fitted with two isolating valves which control the seal-air pressure; one is manually operated and the other pneumatically operated. A tapping on the mill side of these valves is taken to a pressure switch to operate an alarm if the sealing air pressure should fall below 135 mbar.

Two mountings for fitting CO_2 fire extinguishers are located diametrically opposite on the centre section of the mill. Provision is also made on the mill for fitting infra-red fire detection equipment and for monitoring temperature and differential pressure.

The mill drive gearbox is a double-reduction unit which transmits the drive from the 3.3 kV drive motor to the lower grinding ring of the mill. The gear unit is pressure lubricated from an external lubrication unit comprising a gear pump, driven by a 415 V motor via a flexible gear-type coupling, a dual-basket gauze filter and a water cooled oil cooler. The pump draws oil from the reservoir in the sump of the gearbox and discharges it, under pressure, via the filter and cooler back to an internal distribution system within the gearbox. Figure 4.49 shows the mill lubrication system.

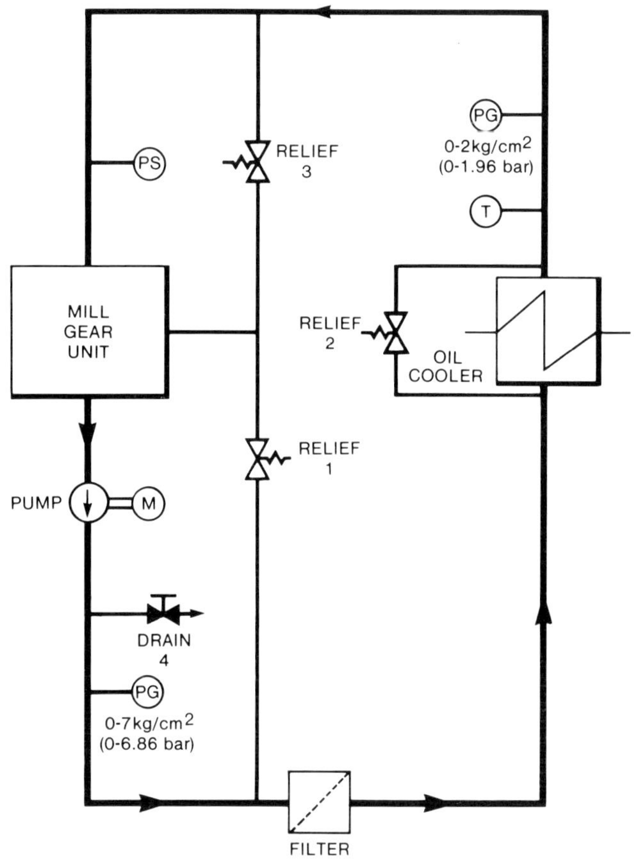

FIG. 4.49 Mill gearbox unit lubrication system

Mill operation and control

Coal is fed at a controlled rate from the coal feeder into the mill where it is pulverised into a very fine powder; this fine powder is conveyed to the burners by high velocity air supplied by the mill PA fan.

The PA fan draws air from two sources, one from the FD fan discharge before the air heaters (tempering air) and the second after the air heaters (hot air). The air from both sources is mixed automatically and delivered to the mills at a controlled temperature sufficient to dry the fuel and prevent cohesion of the coal particles.

The mill operational temperature limits are as follows:

- Normal operating set point 75°C
- High temperature alarm set point 85°C
- Mill classifier temperature high alarm 110°C

The extreme mill operating temperature limits are:

- Each mill can be operated above 100°C but below 110° for ten minutes only; the mill must then be tripped on the emergency trip button.
- The minimum temperature at which the mill can be operated is 60°C.

The primary air that carries the pulverised fuel to the burners is approximately 25% of the total air required for complete combustion of the fuel; the balance is made up from secondary air taken from the air heater outlets to the burner windboxes. The output of each mill is regulated by varying the quality of primary air through the mill. The variation is achieved by regulation of the PA fan speed via its fluid-coupling speed control scoop.

The desired value of primary air is compared with the temperature-compensated primary air differential pressure in a PI (proportional plus integral) controller, the output of which drives the PA fan fluid-coupling scoop actuator. The controller desired value is derived from a total fuel-demand circuit, which is itself derived from the unit load control system.

The total fuel throughput of each mill is deduced from feeder speed. The raw coal input to each mill is adjusted by varying the speed of the associated coal feeder. The measured value of the feeder controller is feeder speed, with an engineer-set gain adjustment to compensate for variation from the normally constant coal-bed depth following feeder maintenance. The desired value is the same temperature-compensated primary air differential pressure signal as utilised for the primary air controller measured value, but subject to engineer-set gain and bias adjustments. The measured and desired values are compared in a PI controller and the output from the controller drives the coal feeder speed-regulation actuator.

The mill exit temperature is controlled by varying the proportion of hot and cold air in the constitution of the primary air. The preset desired value is compared with the measured mill exit temperature in a PI controller, the output of which drives the hot air and cold air dampers in parallel opposition.

A comparison of the measured and desired values of the mill differential pressure is used to initiate high and low differential alarms to indicate mill 'full' or 'empty'.

The mill start-up sequence control is in three distinct parts and can be preselected to be held in the following conditions:

- At Hold 1 — PA fan running and dampers open
- At Hold 2 — PA fan and mill motor running
- At Hold 3 — PA fan, mill and feeder motors running

Before a sequence start is permitted, the following prestart checks have to be satisfied:

- Mill hydropneumatic pressure in the mill load system >24 bar
- Mill hot air control damper closed
- Mill hot air isolating damper open
- Mill isolating valves open
- One FD fan running
- Seal-air fan running

The sequence start will then proceed to start the PA fan lubricating oil pump, then the fan if the lubricating oil pressure is satisfactory.

A signal is also sent to open the seal-air isolating valve. Following the start of the PA fan, its discharge damper is opened and a permit 'auto' signal is applied to the mill temperature controls and PA fan speed control. Hold 1 is now reached. When the hold is released, checks are made as follows:

- Sufficient oil burners lit in mill group
- Mill temperature correct (>55°C but <105°C)
- PA fan differential pressure satisfactory (3.5 mbar)
- Seal-air pressure 135 mbar

The sequence proceeds as follows. The mill lubricating oil pump is started and a check made for satisfactory oil pressure (0.78 bar) and a further check that sufficient oil burners are in service. The mill motor is then started. Hold 2 is now reached and, when released, will proceed to the next point subject to the following checks:

- Oil flame check
- Mill permitted to run from selection circuit
- Mill temperature satisfactory
- PA fan control in manual or auto
- Coal feeder control on auto
- Mill temperature control on auto
- Secondary-air temperature control on auto.

If the checks are satisfactory, the coal feeder then starts and, after a time delay, the 'permit auto-feed' is released and coal is fed to the mill.

To shut down a mill on sequence control, the following takes place. There are no hold points in the shutdown sequence but it is possible for the mill and feeder motors to be stopped locally at the switchgear or pushbuttons on the unit control desk.

The sequence is not permitted to start unless sufficient oil burners in the group are lit.

The starting of the sequence is followed by the immediate switching of the mill temperature control into a mode ready to receive open and close signals which, if satisfactory, set the manual control monitor and starts a mill cooling time delay. After this delay, the feeder control is taken from 'auto' and the feeder run to a minimum position at which it is checked for at least 20 seconds before removing the feeder run-down signal. The feeder is then stopped.

The feeder run-down timer is inserted to permit the mill to empty before the main mill motor is stopped and the mill lubricating oil pump stopped, also after a cooling time delay. At the same time, the PA fan control is removed from 'auto' and, as the PA fan is stopped, its discharge damper is closed via the interlock equipment. The check that the PA fan ACB is open is followed by the closing of the mill seal-air isolating valve.

Also at this time, the PA fan coupling scoop is run to the start (out) position and, when reached, the lowering signal is removed. This is followed by the setting of the manual control monitor which leads to 'sequence complete'.

The post-sequence then permits the run-down of the PA fan before stopping its duty and standby lubricating oil pumps, with an alarm feature if they fail to do so.

6.3.3 Safety

- The internal access to any mill or feeder equipment must be carefully controlled and the conditions laid down in the CEGB Safety Rules observed.
- Procedures for dealing with fires must be available and understood by test team members.
- Operation of pulverised-fuel plant must be in strict accordance with an approved PF Code of Practice, which must be approved by the Station Manager as applicable to the type of mill installation.
- Fire detection systems must be commissioned before hot air or coal is allowed into a mill.
- Procedures for dealing with lubricating and hydraulic oil spillages must be available.

- High voltage switchgear must only be operated by authorised persons.
- When handling hydraulic fluids, care should be taken not to get them on the skin. Protective gloves should be worn.

6.3.4 Resources and staffing

Resources

The following resources will be required:

- Electrical supplies to motors, dampers, feeder drives, etc.
- Control air supplies to pneumatic actuators and valves
- GSW supplies to oil coolers
- Nitrogen supplies to mill loading equipment
- Portable radios
- Fire extinguishers (two 9 kg dry powder, two foam)
- Flushing oil for hydropneumatic loading system
- Coal in bunkers to commission coal feeders
- Lorry available to discharge coal from feeder outlets
- Primary air and seal-air fans to be available

Staffing

A test team of approximately six people will be required and will take up to one week to commission each mill and coal feeder.

6.3.5 Pulverised fuel (PF) milling plant commissioning

A simplified commissioning network for a coal feeder and mill is shown in Fig 4.50.

Coal feeders

(a) *Pre-commissioning checks*

- Inspect feeder for completion of erection and check that holding-down bolts are tight
- Check that the feeder is connected to the coal bunker
- Check that the cut-off rods are fitted
- Check that the tail shaft adjusters are between the marks on the scales and that adjustment on both sides are equal
- Check that drive chain tension is correct
- Check that coal failure alarm switch is cabled

Remove all access doors and check the following:

- Feeder is free from all debris
- Conveyor chain is correctly located on headshaft sprocket
- Fuel stream breaker is correctly fitted
- Sprocket cleaners are correctly fitted
- Fuel depth regulator is set to a coal bed depth of 254 mm
- The fuel level indicator is free to move and the alarm 'feeder loss of coal' is initiated when the flap is down

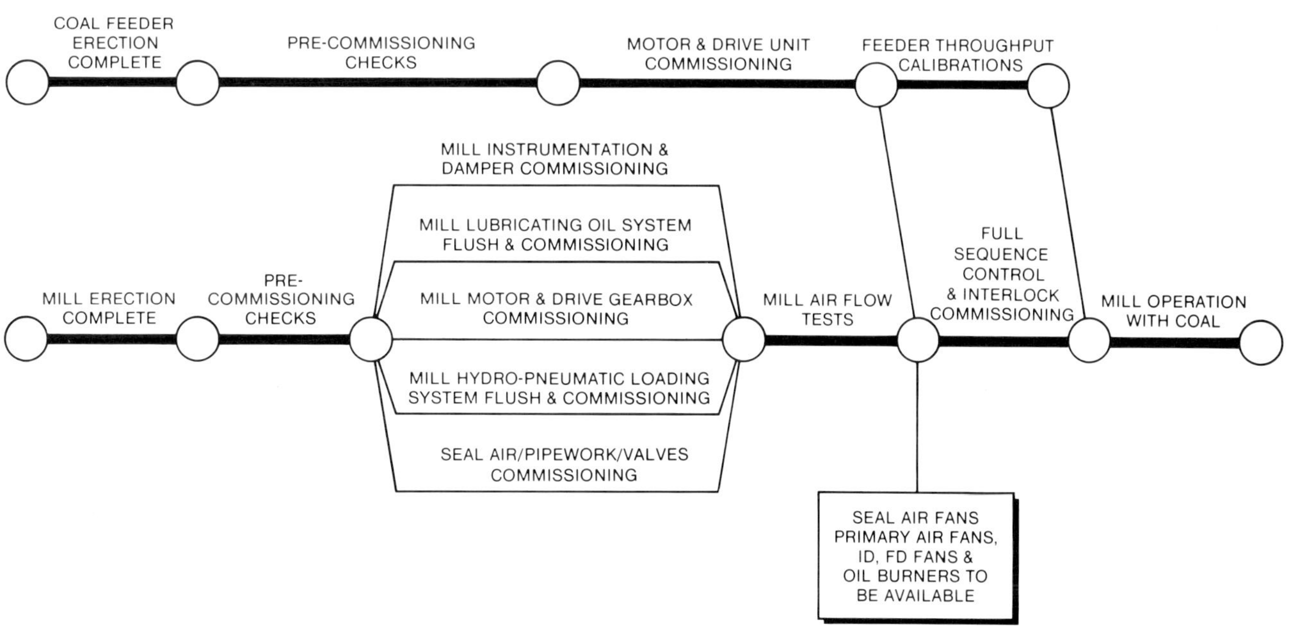

FIG. 4.50 Coal feeder and PF mill — simplified commissioning network

(b) *Motor and drive unit commissioning*

The motor and switchgear will be commissioned in accordance with the procedure in Chapter 3, Section 4 and, with the drive coupling broken, the direction of rotation checked. It is imperative to ensure the correct direction as the coal feeder is only designed to operate in one direction.

The oil levels should be checked to be correct in the single-reduction gear unit, the PIV gearbox and the double-reduction wormwheel gear unit, and checks made that the gearbox breathers are not obstructed. The tension on the PIV gearbox drive chain should then be checked to have a total up and down movement of 51 mm. When tension is correct, the inspection cover should be replaced. The safety coupling between the PIV gearbox and double wormgear unit should then be set up. With the coupling engaged, the double-reduction gears are prevented from rotating. The PIV gearbox output shaft is rotated in a clockwise direction until the coupling disengages. The microswitch should then be checked to operate and initiate the alarm. The preventive gear can then be removed from the double-reduction gear, the safety coupling reset, and all guards and covers replaced.

The speed control pneumatic actuator should be checked in accordance with the procedures in Chapter 3, Section 2.2, and all control air pipework blown through to remove debris. The actuator is then operated to disengage the PIV speed control slipping clutch and checks made that the 'slipping clutch alarm' operates. The speed control operating lever can then be moved back to its original position and a check made that the slipping clutch re-engages and that the alarm clears.

The coal feeder can now be started from the local-to-switchgear position (this will by-pass milling plant interlocks) and the speed actuator operated to the maximum position to check the feeder speed which should be 2.6 r/min to give 37 t/h of coal feed with a coal-bed depth of 254 mm. If the coal feeder speed is not correct, the feeder should be stopped and the linkage between the speed control actuator and the PIV speed control lever adjusted. This process should be repeated until the speed is correct.

The speed control is then proved by operation of the speed controller at the local and remote station in the control room.

(c) *Calibration checks on the coal feeder*

Once the above checks have been completed, the feeder is ready to receive coal to check its output. The outlet chute from the feeder is disconnected from the mill and redirected to a lorry or dumper truck positioned beneath the feeder operating floor. Coal should be delivered to the bunker which supplies the coal feeder. In this instance, the coal feeder would be supplying a Babcock 10E mill and the design operating range of the mill is 50% to 100% MCR, corresponding to 5.2 kg/s to 10.44 kg/s (37 t/h) of coal. The coal feeder speed for this range is 1.3 to 2.6 r/min, with a coal depth of 254 mm.

The feeder speed is set to 1.4 r/min and coal allowed into the feeder and discharged into the lorry. A speed of 1.4 r/min gives 20 t/h, so 15 minutes of operation represents a throughput of 5 t. The lorry is weighed before and after the test and the throughput thus measured.

The above tests should be completed through the full speed range of the feeder and the corresponding weights of coal delivered checked to be correct. If the throughputs are not correct, the fuel regulator should be adjusted to the correct settings. On completion, the feeder is isolated from the bunker by shutting the bunker-outlet rod gates, the feeder emptied of coal and the temporary coal chute removed. The feeder outlet chute is then reconnected.

PF mill commissioning

(a) *Pre-commissioning checks*

- Check that mill erection is complete
- Check that all debris is removed from inside of the mill
- Check that mill holding-down bolts are secure
- Check spider guides for correct installation
- Check that chamber below revolving yoke is clear of debris
- Check that the plough brushes are correctly fitted
- Replace all access doors
- Check that all mill loading hydraulic cylinders are correctly installed
- Check that mill classifier blades are correctly set and locknuts secured (blade setting 40° to the mill radial)
- Check that the classifier flaps in the skirt are free to be opened and closed.
- Check that the grinding rings are correctly installed
- Close and secure mill access doors

(b) *Instrumentation and damper commissioning*

- Check all local instruments (pressure gauges, transmitters, pressure switches, etc.) against the standard check lists and procedures in Chapter 3.
- Check that the mill outlet temperature transmitter is fitted, cabled and commissioned in accordance with procedures in Chapter 3, Section 3.
- Check that the mill outlet pressure standpipe is fitted and connected to the differential pressure transmitter. Transmitter to be commissioned in accordance with procedures in Chapter 3, Section 3.

- Check all dampers to be operable and commission the actuators and dampers using the procedures in Chapter 3, Sections 1 and 2, and Section 5.2.7 of this chapter.

(c) *Seal-air pipework and valves*

The seal-air isolating and control valves will be commissioned, using the procedures of Chapter 3. The control air pipework to the pneumatic control valves is blown through to clear debris from the pipework. The main seal-air pipework to the mill is blown through, using the seal-air fan, to clear debris from the pipework.

(d) *Hydropneumatic loading system commissioning*

The oil tank should be filled with a suitable flushing oil and all valves set to the operating positions. A nitrogen bottle is connected to the gas bottle and the gas pressure on the system adjusted to 14 bar. The flexible oil hoses at each of the loading cylinders are now disconnected and the oil system circulated to flush out the pipework by use of the hand pump. Oil catchment pots must be available to contain any spillages and the system flushed until all pipework is clean of swarf, debris, etc. Once flushing is complete, the oil tank is drained and then refilled with the correct grade of hydraulic fluid. The system should then be recirculated with the hand pump until *all* residual flushing oil has been expelled from the oil pipes. Flexible hoses are then reconnected, ensuring that all air is expelled before final tightening, and the oil tank topped up to normal working level. The gas pressure is now increased to 48 bar and the system left for 24 hours for gas and oil leak checks. Following a satisfactory leak check, the gas pressure can be reduced by shutting the bottle valve and gradually 'bleeding off' until the pressure falls to 45 bar. At this pressure, the contacts should close in the *B* pressure switch and the alarm 'mill ball loading fault' initiated. The 3.3 kV ACB is then put on 'jumpers' at the switchgear and the ACB closed local-to-plant. The gas system pressure should then be allowed to fall to 24 bar, at which the 3.3 kV ACB should trip. On completion the gas pipework should be secured and the pressure reinstated to 52 bar.

(e) *Mill lubricating oil system*

The following pre-commissioning checks must be made:

- Lubricating system erection complete
- Lubricating oil system pressure relief valves set to operate at the correct settings — valve 1 at 3.9 bar, valve 2 at 1.4 bar and valve 3 at 1.0 bar (see Fig 4.49)
- Gearbox filled to NWL with the correct grade of lubricating oil
- Cooling water supplies available to oil cooler
- Pressure switch calibrated to 'make' on rising pressure of 0.75 bar and 'break' on falling pressure of 0.50 bar

The lubricating oil pump motor and system switchgear are checked in accordance with the procedure in Chapter 3, Section 4 and the motor then checked for direction of rotation. The pump/motor unit can then be coupled and the pump put in service to circulate oil and flush the system. The filter should be checked frequently and, once it is assured that the system is clean, the instrumentation should be commissioned and a check made that the oil pressure switch satisfies the mill permissive-start interlocks. The systems should be checked thoroughly for any leaks. GSW should be put through the cooler and checks made for satisfactory flow. The mill 3.3 kV ACB is put on 'jumpers' at the switchgear and the ACB closed from the local position to simulate the mill running. The lubricating oil pump is then tripped and, when the oil pressure falls to 0.50 bar, the 3.3 kV ACB should trip on 'low oil pressure'.

(f) *Mill drive motor and gearbox*

The mill 11 kV drive motors and switchgear must be commissioned, using the procedure in Chapter 3, Section 4. The coupling between the motor and the gearbox is broken and the coupling halves safely secured. The motor bearings are checked to see that they contain the correct lubricating oil, and the bearing high temperature alarms are proved to operate at 80°C. The mill motor 'air temperature high' alarm should be proved to operate at 90°C. The motor can then be test run from the local-to-plant pushbutton and checked for correct direction of rotation. If correct, the motor is given a four-hour proving run, checking bearing and motor air temperatures frequently during this time. The mill gearbox can now be filled to NWL with the correct grade of lubricating oil, the motor/gearbox drive coupling checked for alignment and then the coupling bolts fitted.

(g) *Mill airflow tests*

The airflow tests, as described in Section 5 of this chapter, can now be completed.

(h) *Sequence, controls and interlock commissioning*

The full mill start-up and shutdown sequences, controls and interlocks can now be proved. The sequence and controls are described in Section 6.3.2 of this chapter, and these should be commissioned using the procedures described in Chapter 3, Section 8. This may initially be done by simulating boiler conditions to allow mill start-up, but to check the sequences/interlocks completely the boiler will have to be ready to fire on coal. This is covered in Chapter 6. Once the mill has been commissioned to this stage, it is ready to be operated with coal. This is covered in Chapter 6 and the complete sequence/interlock checks will be carried out with operating conditions. With the mill on-load, grinding performance tests should also be completed.

7 Fuel oil systems

7.1 Introduction

This section covers the fuel oil systems from the high pressure pumps to the oil burners. The fuel oil storage, delivery and transfer systems are covered in Chapter 3, Section 15. HP oil systems vary, depending on the type of burner installation that is being supplied with residual fuel oil, and therefore typical systems will be described for a large oil-fired power station and for lighting-up oil burners on a coal-fired power station.

The description of the HP oil system for the oil-fired station is based on the installation for a 500 MW oil-fired boiler with pressure jet atomisers, and for the coal-fired boiler system with a recirculating-tip type burner on a 660 MW boiler.

7.2 HP fuel oil system for 500 MW oil-fired boiler with pressure jet atomiser

7.2.1 Oil/steam pipework system

The HP fuel oil systems are generally provided on a unit basis and a typical layout is shown in Fig 4.51.

Fuel is supplied by the transfer fuel oil pumps at 14 bar through the common inlet manifold to a system of 5 × 25% duty parallel-linked fuel oil heaters, then through a common outlet header to the suction manifold of 3 × 50% duty HP fuel oil pumps. Each pump has a simplex suction filter.

The pumps discharge the oil at a pressure of 117 bar through a common discharge header, via three parallel-linked discharge filters and then one of the 2 × 50% hydraulically-operated pressure control valves, to feed the oil burners by a ring main system. The pressure control valves are used to control the oil pressure to the burners at pressures between 23 and 78 bar, depending upon the unit load and the number of burners in service. On a 500 MW boiler with 32 oil burners in service, the burner rail pressure required for MCR conditions is 56 bar.

The ring main system is then extended to provide off-load recirculation by the transfer fuel oil pumps and during periods of low-load operation by the HP fuel oil pumps back to the main oil storage tanks. Each high pressure fuel oil pump has a by-pass line and the off-load recirculation line is sized by an orifice plate for a recirculation flow rate equal to three times the by-pass line flow, with the transfer pumps in service. On a 500 MW oil-fired boiler, this is equivalent to 13.6 t/h. The low-load recirculation line is sized, again by an orifice plate, to accept a flow rate from one HP pump which ensures that there is no possibility of overheating the pump casing and rotating element, yet limiting the pressure of the oil in the burner ring main to within normal operating levels. For a 500 MW boiler, this is equivalent to 18 t/h of oil flow. The recirculation system from the low-load and off-load recirculation valves, including the viscometer and density meter line back to the storage tanks, operates at a pressure of less than 17 bar. If the pressure in the line exceeds 17 bar, the HP pumps are tripped.

During off-load recirculation, the pump by-pass valves are open and the transfer pumps circulate oil via the HP fuel oil heaters and recirculation main. This maintains the oil at the correct temperature for start-up purposes. The HP pumps also have warming bleed-valves which are sequence operated with the by-pass valves. On start-up of the first HP pump, the by-pass valves close and the warming bleed-valves open on all three pumps.

When the pump has started, its warming bleed-valve closes, but the warming bleed-valves on the other pumps remain open to provide a warming flow of oil through the pumps, keeping them at a temperature above the low-temperature starting-interlock of 110°C. The valves close in sequence on start-up of the pumps.

The whole of the oil pipework is trace heated with steam at 7 bar. Thermal relief valves are provided throughout the system to ensure that overpressurisation does not occur should isolating valves be left shut with trace heating steam remaining in service.

Flowmeters are provided in the main oil line between the heaters and HP pumps and in the recirculation line back to the main oil storage tanks, and a separate line containing a viscometer and density meter is provided after the HP discharge filters feeding back into the recirculation main, via an orifice plate.

7.2.2 Fuel oil heaters

Five fuel oil heater units are fitted in the fuel system for each boiler and each unit is capable of passing 25% of the total demand. They can be isolated both on the fuel oil and on the steam/condensate side.

Each unit consists of four fuel oil heater elements mounted above an automatic condensate pump unit. Figure 4.52 shows a typical heater.

Each heater comprises a mild steel cylinder in which is fitted a tube nest. Steam from the auxiliary boilers at 17 bar, using air-operated control valves, passes up a metering tube inside each of the main tubes in the nest. The steam then returns down the outside of the metering tube giving its heat to fins fitted on the outside of the main tube and passes out of the cylinder to a steam trap. Providing the condensate pressure is above a nominal 1.7 bar, it will flow back to the auxiliary boilers. If the condensate pressure is below the nominal figure, it is directed from the steam trap to the automatic condensate pump, which then forces it back to the auxiliary boiler flash vessel, using 7 bar auxiliary steam.

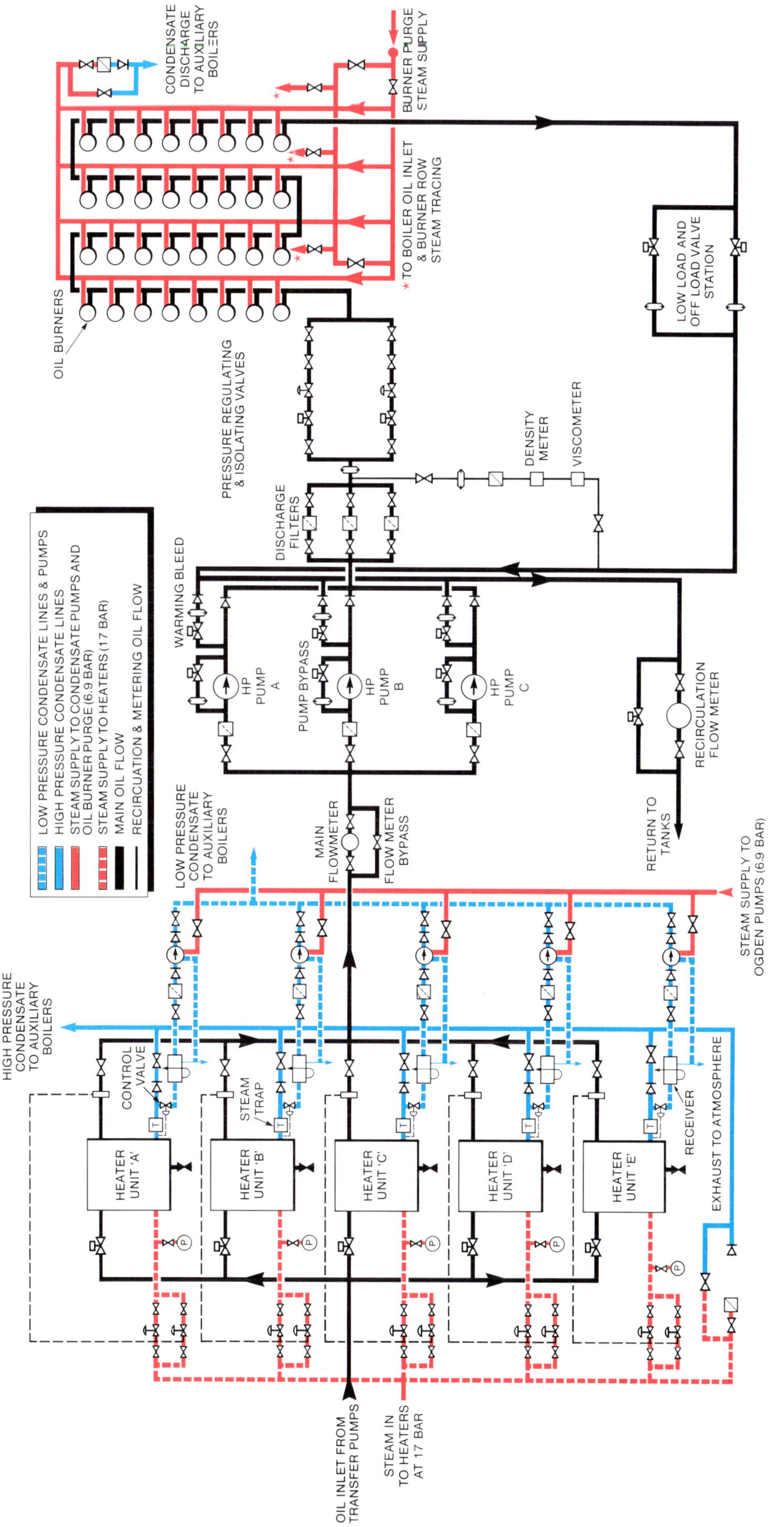

FIG. 4.51 HP fuel oil system for a 500 MW oil-fired station

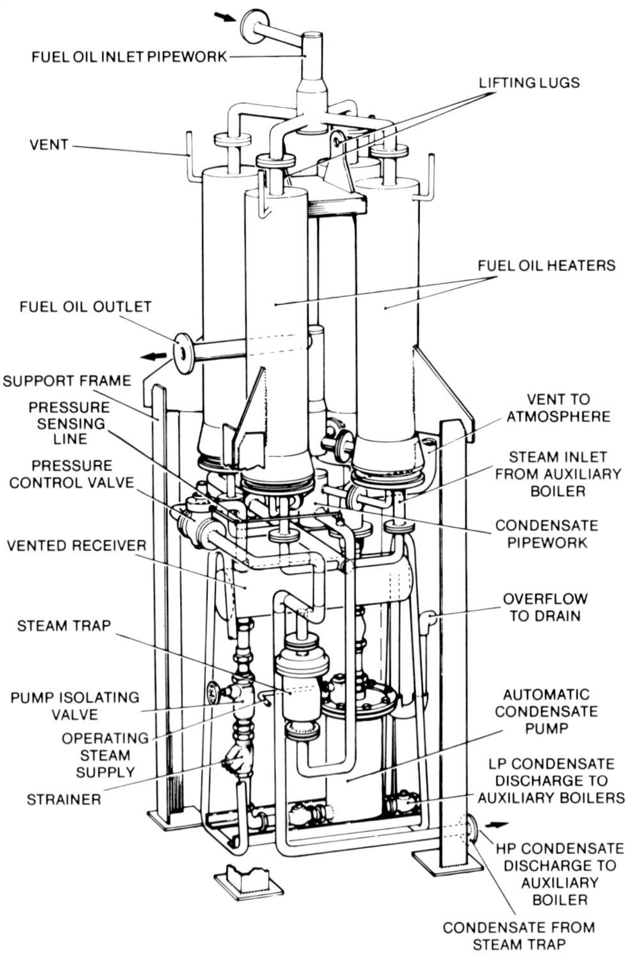

FIG. 4.52 Fuel oil heating set

Fuel oil is supplied to the top of the cylinder and is constrained to flow at right angles to the finned main tubes by baffles, alternately large and small. The large baffles are a close fit inside the cylinder and have a hole in the centre; the small baffles are solid but leave an annular space between their circumference and the inner surface of cylinder. After passing down the cylinder, changing direction 52 times, the heated oil passes out of the bottom of the cylinder to a common discharge outlet, where it passes to the HP oil pump suction manifold.

7.2.3 HP fuel oil pumps

The HP pump is a ten-stage centrifugal type, driven through a speed-increasing single-stage gearbox at 4000 r/min by a 3.3 kV three-phase, squirrel-cage induction motor, operating at 1450 r/min. The gearbox and bearings are supplied with lubricating oil from a lubricating oil storage tank, using an electrically-driven pump for start-up and a shaft-driven pump, once the pump is running. The lubricating oil is cooled by an oil cooler supplied from the station auxiliaries cooling system. (It should be noted that residual fuel oil is not suitable for lubrication of pumps of this type.) Figure 4.53 shows a typical HPFO pump.

Each pump is protected by a differential temperature unit, known as the 'Thermaster' unit. The Thermaster unit measures the differential temperature across the pump, the temperatures being measured by two resistance thermometers positioned at the suction and discharge ends of the pump. The differential temperature generated is proportional to the oil flow through the pump, and a rise in differential temperature to +15°C is almost certain to indicate that flow has dropped below the safe level to contain the heat generated by the pumping action within safe limits. The Thermaster unit has a meter graduated from −10°C to +25°C, and a set-point switch is provided by which a trip level can be set in any one of the six positions. The trip level is set at +15°C differential.

The pumps operate at a discharge pressure of 117 bar.

7.2.4 System operation

The specified range of fuel oils to be handled by the pumping and heating plant varies between 145 $\mu m^2/s$ (145 cSt, 600 Redwood No. 1 seconds) and 1500 $\mu m^2/s$

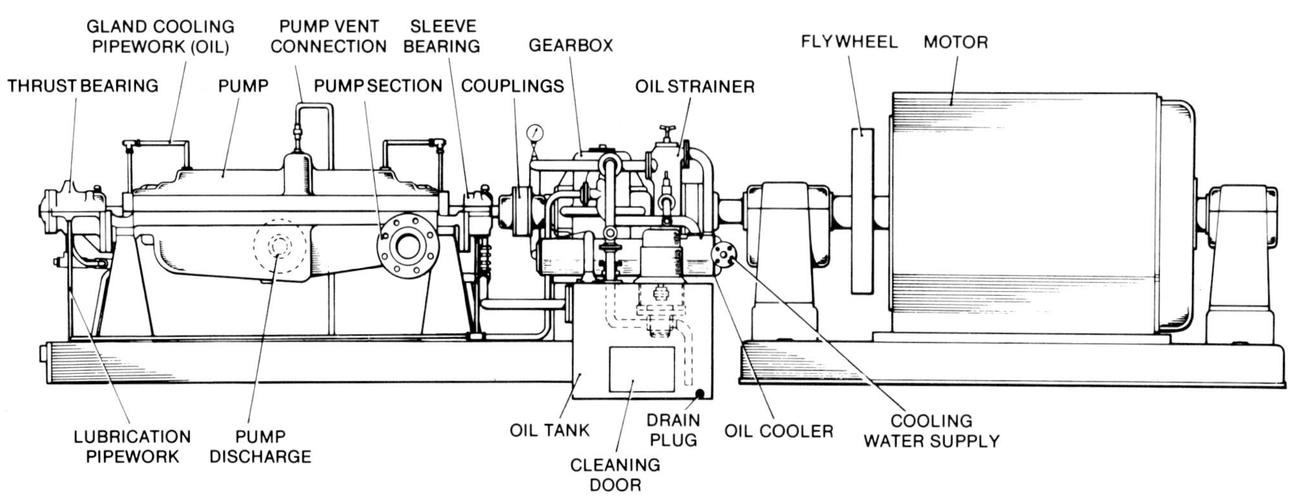

FIG. 4.53 Fuel oil pump and drive

(1500 cSt, 6500 Redwood No. 1 seconds) at 38°C with supply temperatures between 50°C and 80°C. Consequently the mode of operation can require up to a maximum of five heaters in service and two HP fuel oil pumps for MCR conditions on the boiler, but normally operation will be with four heaters in service, with the fifth heater as standby.

Residual fuel oils with viscosities up to 3000 $\mu m^2/s$ (12 000 Redwood No. 1 seconds, 3000 cSt) have been handled by the plant.

Prior to start-up on the HP fuel oil pumps, oil is circulated by the transfer pumps through the oil heaters (where the oil temperature is progressively raised), through the HP fuel oil pump by-pass lines and around the boiler front, and is recirculated via the off-load recirculation line back to the storage tanks.

Once the temperature of the circulating oil is raised to the level required to satisfy the pump starting interlock (110°C), the first HP oil pump can be started. The fuel oil pressure control valve is set to approximately 25% open to provide adequate oil flow through the pump and to ensure that the pump differential temperature limit is not exceeded.

On initiating a start-up for the first HP fuel oil pump, the sequence logic/interlock system carries out the following:

- Checks suction pressure satisfactory (7 bar)
- Checks fuel supply temperature is satisfactory (110°C)
- Checks cooling water flow to lubricating oil coolers satisfactory
- Starts electrically-driven lubricating oil pump and establishes pressure to the gearbox/bearings of 0.8 bar
- Checks if any other pump is running and, if not
- Signals for the low-load recirculating control valve to open and the off-load recirculating valve to close in sequence
- Opens selected pump warming bleed-valve, together with the bleed-valves of the other two pumps, and checks that these are accomplished
- Shuts selected pump by-pass valve, together with the by-pass valves of the other two pumps, and checks that these are accomplished
- Checks that no high differential-temperature exists between pump inlet and outlet (monitored by Thermaster unit) and, if so
- Checks selected pump to establish if it has been selected to standby duty and, if not
- Starts selected pump, followed by a check to establish that the ACB has closed
- Shuts the running pump hot bleed-valve, and checks that this is accomplished
- Sets the duty interlocks and checks that this is accomplished. The lubricating oil pump monitor stops the electric lubricating oil pump, providing the lubricating oil pressure is greater than 1.25 bar. The sequence is then complete

Once the first HP fuel oil pump is established in service, the fuel oil pressure control valve is adjusted to maintain a pressure of approximately 35 bar at the burner supply main and an oil flow in excess of 18 t/h.

Insufficient flow would manifest itself in a pump trip, initiated by a differential temperature across the pump in excess of 15°C.

The two remaining HP pumps continue warming-through, via the hot bleed valves.

With the obvious increase in oil flow once the first pump has started, the temperature of the oil being supplied to the pump has to be maintained by increasing the steam flow to the fuel oil heater.

The second pump, which has been warming-through via its hot bleed valve, is started by the operator under sequence control; the logic being to all intents and purposes similar to that when starting the first pump, when the unit load is approximately 200 MW. The fuel oil pressure and temperature are controlled to ensure that the minimum combined oil flow of 36 t/h is being maintained.

With a total oil flow of 36 t/h established, control of fuel oil pressure and temperature are transferred to automatic control and set points adjusted to the required burner light-off conditions. These are 35 bar oil pressure and an atomising temperature for the viscosity of fuel oil being utilised, to ensure a viscosity of 16 $\mu m^2/s$ (70 Redwood No. 1 seconds) at the burner. The low-load recirculating valve is shut by the operator once total oil flow increases to 68 t/h, i.e., 34 t/h to burners plus 34 t/h recirculation. This corresponds to approximately 200 MW.

Under conditions of recirculation and low-load operation, condensate from the fuel oil heaters is required to be removed by the automatic condensate pump system provided with each heater, condensate being delivered by the pumps into the LP condensate header and back to the auxiliary boiler condensate recovery system.

Once pressure at the steam trap discharge rises above 1.7 bar, a pressure-initiated control valve will close and isolate the LP condensate main, the steam traps discharging directly back into the condensate recovery system.

7.3 Pumping and heating system for a coal-fired boiler lighting-up system

7.3.1 Introduction

The system is installed on a station basis and uses the recirculating-tip system. The recirculating-tip burner has two operating modes, as described in Section 6 of this chapter:

- When firing
- When recirculating

Figure 4.54 shows the fuel oil flow system for this type of installation.

Transfer pumps of the positive-displacement type discharge fuel oil through U-tubed oil heaters, to maintain the firing viscosity of 16 $\mu m^2/s$ (16 cSt, 70 Redwood No. 1 seconds) from a storage system which normally contains oil at 520 $\mu m^2/s$ (520 cSt, 2100 Redwood No. 1 seconds) at ambient temperatures. Excess capacity is recirculated back to the oil storage tanks though pressure control regulating valves.

From the oil heaters, the oil is fed to the common suction of the pressure pumps via the hot oil filters.

The pressure pumps are seven-stage centrifugal type, driven by a 415 V, three-phase motor, through a flywheel and flexible coupling.

The pumps discharge at 44 bar and the oil pressure is controlled to 35 bar by pressure control regulating valves, the oil being fed, via hot oil filters and a flowmeter, to the recirculating-tip oil burners.

To prevent overheating at light loads, an automatic spring-loaded leak-off non-return valve is fitted at the discharge of each pump, which permits a minimum flow under all conditions.

Recirculated oil is then fed back, through pressure control regulating valves and a flowmeter at 27 bar, to the common recirculation line, whence it goes either to the fuel oil heater suction or to the boiler fuel oil storage tanks.

Drainage is provided to empty the system for maintenance purposes, the drained oil being pumped back to the oil storage tanks, via drains tanks and pumps.

7.3.2 System operation

In the firing mode, the total flow to each burner is 640 kg/h, of which 476 kg/h is fired and 164 kg/h is recirculated. In the recirculating mode, the total flow to each burner is 640 kg/h. To maintain these flows, the supply pressure from the HP pumps has to be 35 bar and the recirculation pressure at the burner outlet is 27 bar.

When burners are in the recirculating mode, all the oil pumped by the pressure pumps is returned to the pressure pump suction and no oil is supplied to the system by the transfer pumps, although a transfer pump must be operating to maintain the supply pressure to the pressure pumps. When burners are in the firing mode, the quantity of oil fired is made up by the transfer pumps.

Each transfer pump and each pressure pump is sized to supply the firing requirements of one boiler, with sufficient surplus to supply adequate spill-flow for pressure control purposes.

The pressure-sensing point for transfer pump delivery pressure control is located at the pressure pump suction manifold, and the pressure-sensing point for the pressure pump delivery pressure control is located at the junction point where the flow to the three boilers divides.

The pressure pump delivery pressure is controlled by spilling the surplus capacity of the operating pumps back to the delivery header of the fuel oil heaters. The control valve is again sized such that it is 10% open when one pump is operating and providing maximum demand for one boiler, i.e., the recirculating mode, and 60% open when passing the full capacity of one pump at normal output pressure. Pumps are started and stopped, as necessary, according to station demand.

The maximum operating temperature of the oil is 127°C. The centrifugal-type pressure pumps are suitable for operating temperatures up to 150°C. The oil temperature to the pumps is monitored and, if the temperature exceeds 135°C, an alarm is initiated and, if the temperature exceeds 140°C, the pump is tripped.

A viscosity meter is fitted at a point local to the junction point, where the flow to the three boilers divides.

The return pressure at each boiler is controlled to give the desired differential between the burner supply and the return manifolds. The differential controller modulates the control valve to maintain its upstream pressure at the correct value to maintain the desired differential.

Flowmeters are fitted in the supply and return pipelines for each boiler.

All recirculated oil is normally returned to the fuel-oil heaters inlet-header, but if the temperature of the returned oil exceeds the desired value (due to heat pick-up from non-firing burner recirculation), a small quantity of oil is returned to the day tanks. This is initiated by a temperature switch with a sensing point downstream of the leak-off point, which operates a motorised isolating valve in the leak-off line; an orifice limits the quantity returned.

7.3.3 Fuel oil heaters

The fuel oil heaters are positioned between the transfer pumps and the pressure pumps. The heating medium is saturated steam at 17 bar (auxiliary boiler supply pressure). Steam is supplied to the heater at full pressure, with no restriction to flow. Condensate is removed as it is formed by conventional steam traps, thus the heating surface is at constant temperature.

Oil temperature is maintained at the desired value by mixing unheated oil with overheated oil in the correct proportion. The oil enters the shell of the heater at 8.5 bar pressure up to a maximum flow of 57 t/h. The oil follows an undulating path across the steam tubes and passes out at the oil-firing temperature of 127°C.

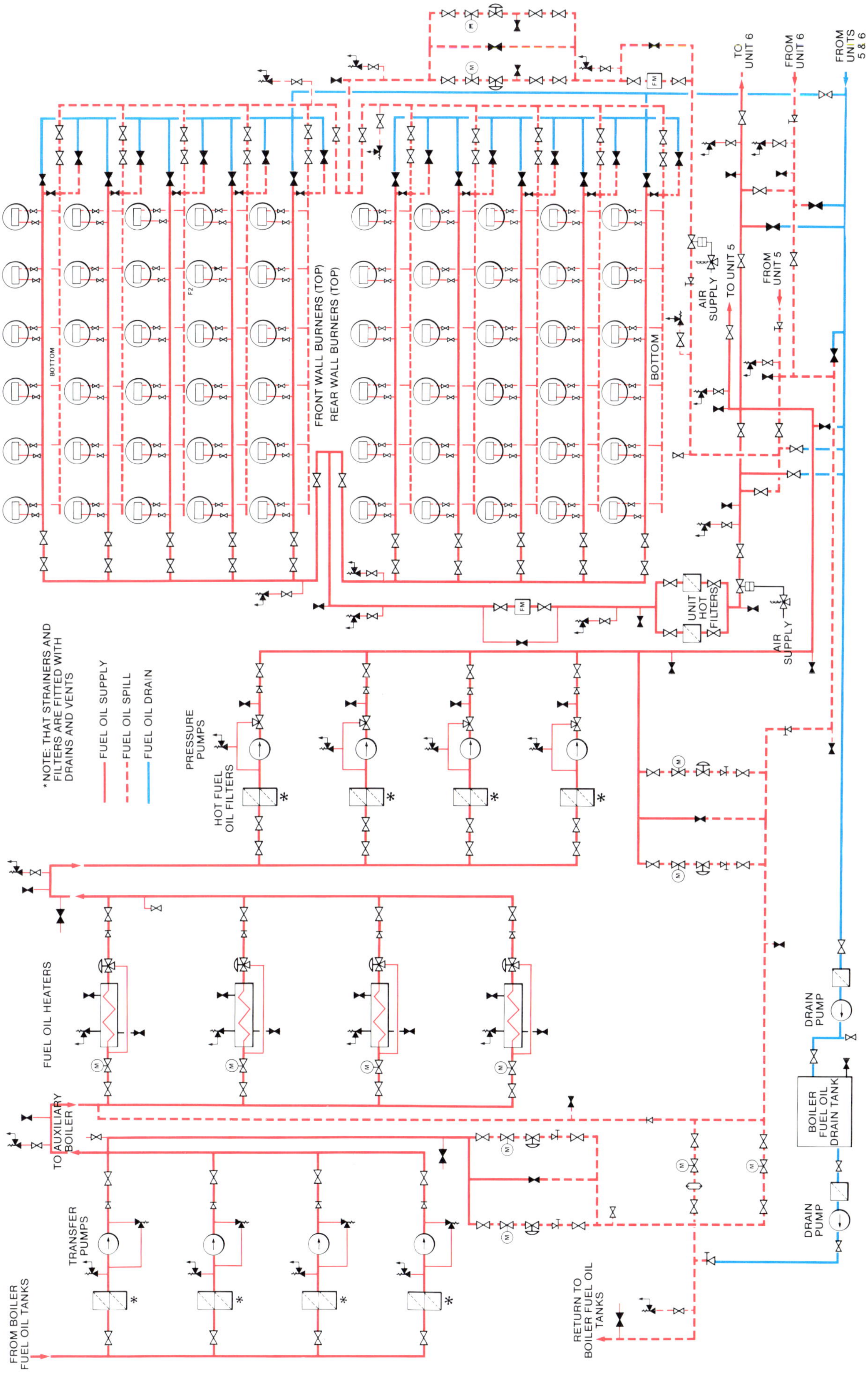

Fig. 4.54 Recirculating-tip oil burner, boiler fuel oil firing arrangement

Control of oil temperature is by means of a three-way valve; this modulates oil flow through the heater and/or the by-pass to maintain the desired temperature.

Each heater is sized to heat one boiler fired quantity of the heaviest grade of fuel oil from 520 $\mu m^2/s$ (520 cSt, 2100 Redwood No. 1 seconds) to 16 $\mu m^2/s$ (16 cSt, 70 Redwood No. 1 seconds).

Recirculated oil is also returned to the day tanks during cold-start recirculation, when the oil is circulated around the system by the transfer pumps, with the pressure pump in circuit but not runnning. When the oil circulating around the system is up to temperature, the centrifugal pump is started and the leak-off valve closed.

All pipework on the system is trace heated, using electrical trace heating.

7.4 Safety and resources

7.4.1 Safety

- Operation of HV switchgear should only be carried out by authorised personnel
- Fixed and portable fire fighting systems should be commissioned and available before filling fuel oil systems
- Adequate procedures should be available for dealing with oil spillages
- Areas being steam-blown should be barriered off and personnel access restricted
- Sufficient thermal insulation to prevent burns to personnel should be available
- Commissioning areas should have good clear access
- Good lighting should be available in commissioning areas
- Any work involved on the steam and oil systems, involving access to pressure parts or rotating plant, should be controlled by the use of Permits for Work

7.4.2 Resources

Services

The following equipment and supplies will be required:

- Electrical supplies to motors, valves, instruments, etc.
- Hydraulic and pneumatic supplies to valves
- Gearbox oil supplies available
- Oil collecting pots/trays available at vulnerable leak points, for example, thermal relief valves
- Decade resistance boxes
- Megger (500 V)
- Multimeter
- Thermometers
- Vibration monitoring equipment
- Fire fighting equipment (extinguishers, foam-making equipment and fixed installations)
- Control air supplies
- Auxiliary steam supplies
- Station auxiliaries cooling system
- Transfer fuel oil system commissioned, with the correct grade of fuel available
- Oil burners isolated and terminal point valves locked shut
- Barriers for use when steam purging pipework
- Blanks to blank off the HP pumps during the pipework hydraulic test
- Temporary pipework for steam purging purposes

Staffing

A team of eight people need four to six weeks to commission the HP fuel oil systems.

7.5 Commissioning of oil-fired station HP fuel oil systems

7.5.1 Hydraulic test of pipework

A typical commissioning network for HP fuel oil systems is shown in Fig 4.55. Once the pipework erection is complete, a hydraulic test will be carried out to 1.5 × design pressure. The hydraulic test should be left as late as possible in the programme to avoid corrosion attack after draining the water and before the first fill of oil. Chemical dosing of the water should be carried out to the recommendations of the Station Chemist. The procedure will be as follows:

- The high pressure pumps are not included in the site hydraulic tests; blanks are fitted to the suction and discharge lines before testing.
- The high pressure heaters and all pipework are included and, during the hydraulic test, a check should be made for leakage to the steam side of the heater.
- The hydraulic test will be completed in two parts:
  - (a) High pressure system
  - (b) Low pressure recirculation system
- During water filling, all main circuit valves are opened and the relief valves gagged or blanked.
- As the system is filled, air is released through the vents.
- Before testing the HP circuit, the off-load pump by-pass valves on all pumps, the isolating valves on

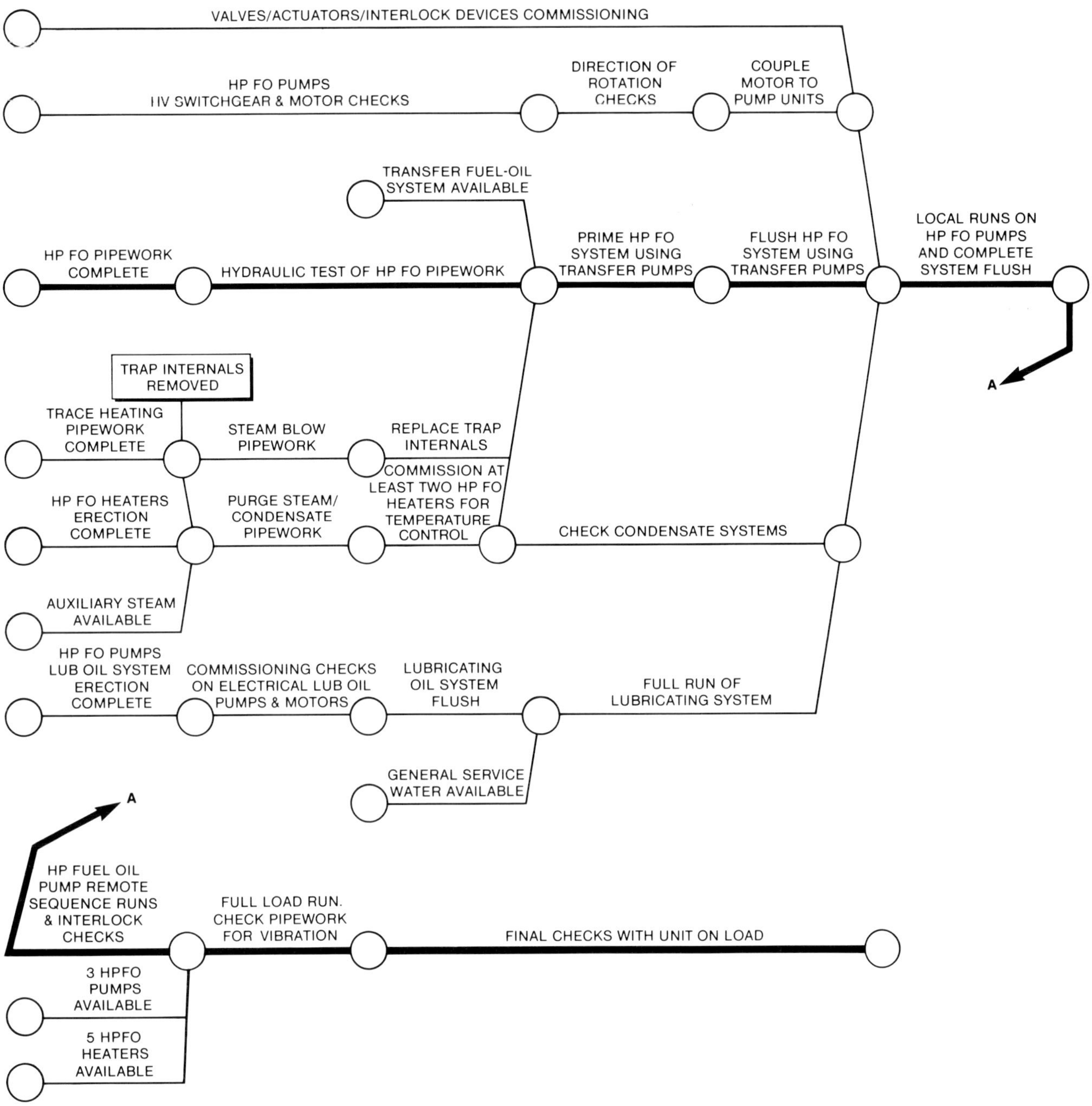

FIG. 4.55 HP fuel oil system commissioning network

the viscometer bleed-line and the main recirculating line isolator are closed to isolate the low pressure system.

- The HP system is pressurised, using a ram-type pump connected to the system at a suitable point, such as a relief valve connection.
- The pressure is then released from the HP system and the valves in the LP system opened to permit the LP system test. This includes all recirculation lines from the off- and low-load recirculation lines from the HP pump by-pass valves and the viscometer and density meter line.
- The LP system is then similarly tested.
- The steam heating and trace heating systems are not normally hydraulically tested, the welded joints being radiographed during erection.

Once the hydraulic tests are completed and the trace heating pipework fully installed, the cladding should then be put on the pipework, pumps and heaters.

7.5.2 Steam-purge of the steam heaters, pipework and trace heating pipework

The steam lines from the auxiliary boilers to the pumping and heating systems will have been blown clean by the auxiliary boiler working party to a terminal point on the HP fuel oil system, but some pipework within the HP pumphouse may have to be blown through.

The trace heating pipework steam-trap internals are removed for this exercise and suitable temporary pipework connected to ensure safe steam escape, or the area of steam escape must be adequately barriered off. It may be necessary to remove the steam control valves on the inlets to the heaters to prevent damage to the valves. Temporary pipework is fitted in place of the valves.

With the above arrangements completed and the areas of steam escape adequately barriered off, the steam pipework is gradually warmed through and all drains and vents blown, to remove any accumulated water. The steam blows should be carried out in small sections at a time to ensure that debris is not lodged in other systems: to this end, each heater with its individual pipework and each tracer section should be steam-blown separately.

Following the steam-blow, all temporary pipework should be removed and all permanent valves, traps and pipework reinstated.

7.5.3 HP fuel oil pump lubricating oil system commissioning

This system is flushed using the electrically-driven auxiliary oil-pump. The switchgear and motor are checked, in accordance with the procedure in Chapter 3, Section 4, and the lubricating oil tank is filled with the correct grade of oil after thoroughly checking for tank cleanliness. The pump bearing caps are then lifted and the top-half shell bearings removed. The bearing caps are then replaced in order to allow the supply and return pipework to be flushed. The lubricating system is flushed in two stages — first the pipework and bearings, then the gearbox and shaft-driven oil pump. The gearbox is thus blanked off for the initial stage of the flush.

The auto-clean filters must be in circuit for both stages of the flush.

The auxiliary oil pump will then be run from local-to-plant, ensuring that direction of rotation is correct and that positive discharge pressure is achieved. The discharge pressure should be set to 3.5 bar by adjustment of the discharge relief valve. Flushing should continue for approximately 8 hours, the filters being checked and cleaned regularly. Once the system is declared clean by chemical analysis of the oil, the second stage of the flush will be carried out.

The blank in the supply to the gearbox should be removed and the spade then inserted to blank off the supply to the pump bearings; the flushing exercise is then commenced on the gearbox, using a procedure similar to that adopted on the bearings. Whilst the gearbox is flushing, it should be possible to clean the pump bearing-housings thoroughly and reinstate the top-half shell bearings.

Once a satisfactory flush of the two circuits has been achieved, all temporary connections must be removed, care being taken to ensure that all joints are made up correctly.

With all circuits prepared, a flush of the complete circuit is now undertaken, with the bearings in the circuit for the first time, observation being maintained for any signs of leakage on the circuit, or at the pump bearings. Once again, the Autokleen strainer is changed over and checked regularly throughout this further 8-hour flushing period.

At completion of the flushing exercise, the Station Chemist will determine whether or not it is necessary to renew the oil charge.

The final operation in commissioning the lubricating oil system is a flush of the cooling water supply to each lubricating oil cooler. In order to facilitate the flush, temporary connections are made to the cooling water inlet to the flow switch, a flexible pipe being run from this temporary connection to a suitable floor drain.

Each supply line should be flushed in turn for approximately 20 minutes duration to clear any construction debris from the lines. Following completion of the flush, the flow switches should be reinstated ready for service.

7.5.4 HP fuel oil pumps, gearbox and motor commissioning

The motors and switchgear are checked in accordance with the procedure in Chapter 3, Section 4 and the motor given an uncoupled run to establish correct direction of rotation and satisfactory motor conditions. Prior to the run, the motor bearings must be checked for the correct grade and cleanliness of lubricating oil and the motor/gearbox coupling split. Checks should be made to ensure that the coupling halves are adequately secured. The motor should be run from local-to-plant and, if satisfactory on the initial run to speed, is given a four-hour proving run. Upon completion of the motor test run, the motor/gearbox coupling is remade. The electrically-driven oil pump is now put into service local-to-plant and the lubricating oil flows and pressures checked; the motor/gearbox unit can now be run, again, local-to-plant. Starting and running motor currents should be noted. With the gearbox running at full speed, the mechanical oil pump will have picked up suction and commenced discharging, as indicated by a rise in the lubricating oil pressure. The electrical lubricating oil pump should then be taken out of service and the lubricating oil pressures checked to be satisfactory. Any changes to system oil pressures should now be made by adjustment of the pressure relief valves.

A four-hour proving run should now be carried out, in which gearbox temperatures and oil pressures are carefully monitored. Following the test run, the unit should be stopped for coupling of the gearbox to the pump and replacement of all coupling guards.

7.5.5 Valves, instruments and controls

All thermal relief valves should be checked by removing them to a workshop and bench testing to ensure that the actual setting agrees with that stamped on the valve body. They should be fitted after the hydraulic test. Due to the nature of fuel oil and the mess it makes when spilt, on-line testing of the relief valves is not carried out.

All mechanical, electrical, pneumatic and hydraulic valves should be checked in accordance with procedures in Chapter 3, Section 1 and instrumentation as Chapter 3, Section 3. Detailed sequence and interlock commissioning procedures are covered in Chapter 3, Section 8 and the sequence and interlock controls should be commissioned and set up using these procedures. These controls are not essential for the initial pump running and may be left until after the commissioning procedures described in the following section. The Thermaster units must be set up using the manufacturer's instructions for the type of fuel to be used.

7.5.6 Initial charging and flushing of the system with fuel oil

During commissioning of the early 500 MW oil-fired units at Fawley/Pembroke, a residual fuel oil of 150 $\mu m^2/s$ (150 cSt, 600 Redwood No. 1 seconds) which did not require preheating was used for this operation. However, due to the high prices of fuel oil this is now done using the higher-viscosity grades of fuel oil normally burnt by the oil-fired stations: this has proved to be acceptable, with few problems. The procedure given here is for flushing and priming, using the heavy grades of residual fuel oil.

Prior to priming the HP fuel oil system with residual fuel oil, the fixed spray fire protection system must have been commissioned, fully proved and left in service.

The HP and LP system hydraulic tests must be completed.

The transfer fuel oil system will have been commissioned, as described in Chapter 3, Section 15, and one transfer fuel oil pump must be in service recirculating oil back to the main oil storage tanks. The trace heating system must also be in service on the pipework and pumps prior to the priming exercise. The HP fuel oil pumps lubricating-oil systems should be put in service, and the cooling water supply established to the lubricating oil coolers.

All pressure control valves should be available on manual control.

Priming of the fuel oil system can then commence as follows. Oil is admitted to the heaters through the inlet pipe and the discharge valves left shut until air is completely expelled through the heater vent valves. The heater vents are closed and the discharge valves opened to feed oil to the HP pump suction valves. Each HP pump is then primed in turn, oil being fed up to the pump discharge valves, venting to atmosphere as before. The rest of the system is slowly filled and vented in small sequenced sections in a similar manner up to the burners and back along the recirculating line to the storage tanks, both the off-load and low-load recirculating valves being open. Continual checks for leakages must be made throughout the priming and circulating exercises. The flowmeters are blanked off and by-passed to prevent any debris getting into the meters.

When the system is fully primed, the oil is circulated around the system using the transfer pumps. The system should be carefully checked for any signs of leakage and vent valves operated intermittently to expel any entrained air. Circulation should then continue for approximately 24 hours, during which time the HP pump discharge strainers should be operated regularly and the temperature of the circulating oil progressively raised to 100°C, using one or more of the HP heaters.

If necessary, the strainers should be cleaned after the exercise.

7.5.7 Initial proving runs of the HP fuel oil pumps and increased velocity flushing

After the 24 hour circulation with the transfer pumps, and with the fuel oil temperatures at 110°C to satisfy the HP fuel oil pump starting interlock, an HP fuel oil pump should be started to increase the flow velocity of the system and carry out initial proving runs of the HP fuel oil pump.

Prior to running the HP pump, its suction filter should be checked for cleanliness. The pump by-pass valves should then be closed, with the remaining pump warming via its hot bleed line.

A final check of the complete system should then be undertaken to see that all necessary isolating valves in the system are open and that the oil burner isolating valves are shut. The low-load recirculating valve should be open, and the off-load recirculating valve shut. As previously stated, the main and return flowmeters should be isolated and by-passed.

Performance of the second heater introduced into service should be observed and assessed as satisfactory, prior to commencement of the HP pump proving run.

Local pressure gauges, which have previously been charged with glycol between tapping point valve and gauge, are put into service and their operation checked under the circulating transfer pump pressure.

Personnel should be stationed around the plant at suitable positions to observe vulnerable points about to be subjected to high pressure oil for the first time, i.e., bolted flange joints and thermal relief valves, catch pots being positioned underneath vent valves, thermal relief valves and drip trays being positioned underneath the burner flexible connections.

Portable radio communications should be established between the transfer pump house, pumping and heating house, switchgear room and boiler front area.

The selected fuel oil pressure control valve can now be closed in to approximately 20% open in preparation for pump start-up.

With the complete communication link established, the lubricating oil system operating satisfactorily, the suction pressure at 10 bar and the temperature of fuel oil at 110°C, the HP fuel pump is started from the local manual position at the switchgear. Once the pump has started, the hot bleed valve should be shut. The maximum starting current, the time for peak current to fall off to running current, and the time for the pump to run up to full speed must be recorded.

Once up to full speed, the gearbox-driven lubricating oil pump should take over from the electrically driven pump and it should be checked that the electric pump automatically cuts out; however, it may be prudent to leave the electric pump running under local control for the initial runs on the HP pumps.

Close observation must be maintained at the Thermaster unit local to the pump motor for any significant rise in temperature differential across the HP pump, which would indicate insufficient flow through the pump. In the event of indicated temperature differential rising to say 10°C, immediate instruction should be given to ease open the fuel oil control valve to increase pump throughput; almost immediate reduction in differential temperature should be observed if the control valve has been opened enough.

Once the pump has settled down at adequate throughput, the fuel oil temperature must be carefully controlled, together with the performance of the fuel oil heater inlet pressure control valve.

Throughout the proving run, the performance of the pump and motor must be monitored and records of bearing temperature maintained at, say, 15-minute intervals, particular attention being given to the start of any unusual noise or vibration on the pump or motor, excessive rate of rise of bearing temperature or differential temperature rise across pump, deterioration in lubricating oil pressure, or onset of leakage at pump mechanical shaft seals. The Autokleen strainers on the lubricating oil system and fuel oil pump discharge line should be operated regularly throughout the proving and flushing operation to ensure maximum circuit cleaning.

Upon completion of the flushing and proving run, preparation should be made to trip the pump motor from the switchgear; with communications established, the electrically-driven lubricating oil pump should be started by local manual action at the switchgear and, once the lubricating oil pressure has risen, the main motor is tripped from the switchgear, and the time for the pump to come to rest noted for future reference. To keep the pump warm, the hot bleed valve should be opened. If, however, no further pump operation is envisaged, the by-pass valves of each of the pumps can be opened up to maintain the boiler front-pipework recirculation and the hot bleed valves are then closed. The off-load recirculating valve should then be opened and the low-load recirculating valve closed.

Once the fuel pump is tripped, the fuel oil heater steam control valve must be closed sufficient to keep the temperature of the oil at approximately 80°C, and the fuel oil pressure control valve opened up to approximately 50%.

Following the proving run, the fuel oil pump suction filter and the HP discharge strainers should be inspected and cleaned.

The other HP fuel oil pumps should be commissioned similarly; if flushing is complete, then a 4-hour proving run will be sufficient.

During the flushing exercise and proving runs on the HP fuel oil pumps, the performance of the fuel oil heaters should be checked for operation of the temperature control valves and that the condensate system pumps and traps function correctly.

7.5.8 Interlock checks and remote sequence running of the HPFO pumps

Once the HP fuel-oil pumps have been commissioned, under local manual control, preparation should be made to commission the plant under sequence logic control from the unit control desk; the sequence interlocks are described in Section 7.2.4 of this chapter. The sequence logic inputs will be checked and then tested, where possible, with the HP motor 3.3 kV ACBs in the 'test' position, as follows.

Starting interlocks

- Shut down the transfer fuel oil pumps so that the suction pressure is less than 7 bar and prove that the pump will not start.
- Reduce the fuel oil temperature to less than 110°C and prove that the pump will not start.
- Switch off the electrically-driven oil pump and prove that the pump will not start with the lubricating oil pressure less than 0.8 bar.
- Shut off cooling water to the lubricating oil cooler and prove that the pump will not start.
- Shut the pump by-pass valve and switch off the electrical supplies and prove that the pump will not start without the valve being open.

Pump trips

- With the HP pump 3.3 kV ACB closed in the 'test' position, trip a transfer fuel oil pump and check that the ACB trips when suction pressure falls to less than 7 bar.
- With the HP pump 3.3 kV ACB closed, operate the Thermaster unit so that the differential temperature is greater than +15°C: the ACB should trip.
- With the HP pump 3.3 kV ACB closed (electrical oil pump in service), trip the oil pump and check that ACB trips when oil pressure is less than 0.35 bar.

- With the 3.3 kV ACB closed, simulate high pressure in the recirculation main (greater than 17.5 bar) and check that the ACB trips.

Standby starting interlocks

- With the pump running and one selected as standby, with its standby monitor set, trip the running pump and check that the standby starts.
- With one pump running and one selected to standby, trip the running pump and check that the standby pump starts when the fuel oil discharge pressure falls to less than 85 bar.
- Check that, with a pump selected to standby with its monitor set, the electrical lubricating oil pump is running.

Final tuning and setting up of the HP fuel oil system will be completed when the unit is at sufficient load for the combustion control loops to be introduced. This will be covered in Chapter 6, and will involve the final settings of the HP fuel oil heater temperature controls, the fuel oil pressure control and the operation of the off-load and low-load recirculation valves.

7.6 Commissioning of the HP fuel oil system (recirculating-tip) for coal-fired boiler lighting-up

7.6.1 Hydraulic test

As with the oil-fired power station systems, the lighting-up oil systems for coal-fired boilers are subject to a hydraulic test of 1.5 times the design operating pressure. The procedure is similar to that described in Section 7.5.1 of this chapter. The transfer and pressure pumps are not included and will be blanked off prior to the hydraulic test. The hydraulic test will be completed in three stages:

- Pipework between transfer pumps and oil heaters, including the recirculation system back to the oil storage tanks (test pressure 16.5 bar).
- Pipework between the pressure pumps and the boiler front and rear burner galleries, and all recirculation pipework down to the differential pressure controllers (test pressure 70.5 bar).
- Recirculation pipework from differential pressure controllers back to the suction side of the heaters (test pressure 16.5 bar).

7.6.2 Steam purge of the HP fuel oil heaters

Again the steam supply manifolds and heaters are subject to a steam-blow exercise, the procedure being similar to that described in Section 7.5.2 of this chapter. On completion of a satisfactory purge, all steam and condensate lines should be reconnected and the steam-trap internals replaced. The heater oil-outlet control valves will be set up when a flow of oil is established through the heaters.

7.6.3 Electrical-tape trace heating commissioning

Electrical-tape trace heating should be commissioned before the thermal insulation is fitted. This is covered in Chapter 3, Section 15.

7.6.4 Instrument and control valve commissioning

Before running the pressure pumps, it is necessary to commission all valves, instruments and controls. All mechanical, electrical and pneumatic valves should be checked in accordance with the procedures in Chapter 3, Section 1. Thermal relief valves should be checked by removing them to a workshop and bench testing to ensure the correct lift pressure.

The pressure pump differential-pressure control valves should be set up to control the fuel oil to the correct pressures, in accordance with the manufacturer's test procedures.

The pressure pump high-temperature trips should be set to trip the pressure pumps at 140°C and the high-temperature alarms set at 135°C.

7.6.5 Initial runs of the HP fuel oil pumps during the priming and flushing exercise

The HP fuel oil system will again be flushed using the transfer fuel oil pumps, using a procedure similar to that described in Section 7.5.6 of this chapter, and the HP pumps will then be introduced to increase the flow velocity and to carry out proving runs on the pumps.

The HP pumps are primed by breaking the motor/pump coupling, opening the oil supply from the hot-oil filters, rotating the HP pump by hand and venting until all air is expelled. If necessary, a transfer pump can be run to provide sufficient oil pressure for pump priming.

The motors are then checked using the procedure in Chapter 3, Section 4, and should be given a proving run coupled to the flywheel. Following this run, the motor/pump unit is coupled in readiness for the pump running and flushing tests. The electrical trace heating must be in service at this stage. Before running the pumps, the fuel oil must be up to the system operating viscosity, i.e., 16 $\mu m^2/s$ (16 cSt, 70 Redwood No. 1 seconds). To achieve this, the oil heater temperature controls are set up by pumping oil through the heaters using the transfer pumps and by setting the three-way control valve to give the correct temperature. After priming the pressure pumps, they are commissioned individually on leak-off. Following this, oil is next pumped up the discharge pipework to the boiler front and rear supply mains and the whole system checked at maximum pump-generated head. Careful attention must be paid to venting the pipework at strategic points and watching carefully for any leaks.

The system will then undergo a flushing operation, using the pressure pumps via the differential pressure control valve by-pass line, then back to the storage tanks via the heater recirculation line.

Each of the burner supply and return mains is flushed in turn via the recirculation line to the main oil storage tanks.

Flushing is continued until the pressure pump hot-oil filters remain clean. The transfer pump cold-oil strainers should also remain clean during this operation.

During this exercise, the pressure pumps are checked to establish that flow capacity and pressure control are satisfactory and that they are capable of operation in parallel with the other pumps.

The fuel oil drainage system, pumps and tank are commissioned during system flushing. The motors are checked in accordance with the procedures in Chapter 3, Section 4 and given a four-hour proving run. The pump/motor units are then coupled and the coupling guards fitted. During the flushing drains from the boiler system are used in turn to discharge oil. The drains recovery pumps are operated until the various sections are clean, as indicated by the drain pump strainers.

The drainage system is then tested to establish its ability to remove system contents in a satisfactory manner. In addition, checks on the tank immersion-heater and electrical trace heating are made to ensure that the oil remains in a condition suitable for pumping.

Following satisfactory flushing operations and proving runs on the pressure pumps, the oil burner supply pressure, temperature and recirculation pressure control systems will be finely tuned to control at the correct pressures and temperatures. This involves circulating through all the front and rear burners to give fuel oil flows through the system.

Full operational tests on the pressure control valves will not be possible until all oil burners are in service and the design maximum demand is on the system.

8 Sootblowing and water washing systems

8.1 Introduction

Sootblowers are provided to remove ash and slag build-up from the boiler tubes and combustion deposits from gas air heaters during normal boiler operation. Sootblowers must be available for operation immediately the boiler burner operations commence. The first priority is for the sootblowers of the gas air heaters to be available, to minimise the risk of fires due to the deposition of combustible deposits and condensed oil vapours. The furnace, superheater, reheater and economiser section blowers cannot remove hard deposits formed over long periods and therefore must be available before commercial operation of the unit commences. Regular sootblowing of the boiler and air heaters must be carried out in order to avoid the deposit build-up, the frequency being at least once per shift. In many cases, where high-fusion coal or high-asphaltene oil is being burned, a higher frequency may be required. On all early 500 and 660 MW boilers steam was used as the blowing fluid, the steam being supplied from auxiliary boilers and/or bled from the boiler cold reheat section. On later designs, the CEGB has gone over to compressed-air type sootblowers, the air being supplied from specially installed large capacity compressors. The reasons for using air instead of steam for sootblowing are:

- Air is more economical than steam.
- Boilers can be cleaned when steam is not available (on coal-fired boilers without auxiliary boilers).
- Less maintenance on valves. With steam blowing, the condensate cuts the valve seats.
- Less erosion of superheater tubes.

Commissioning of the compressors and receivers is covered in Chapter 3, Section 21.

Even with regular sootblowing, air heaters still have a tendency to clog, due to the fine clearances in the air heater baskets and the fact that they are operating at the 'cold end' of the boiler. Burning of high-asphaltene oils or poor coals often leads to imperfect combustion and consequently some air heater fouling does take place. Starting up and shutting down a unit are other situations where unburnt combustible material can carry over to the gas air heaters and cause fouling which regular sootblowing does not remove. Gas air heater water washing facilities are therefore provided on all large units, especially oil-fired boilers. Water washing is carried out with the air heater isolated from the gas and air streams and the heater rotating on a slow-speed drive.

8.2 Sootblowing systems

8.2.1 Description

A typical sootblowing installation for a 500 MW oil-fired boiler *using steam as the blowing fluid* is shown in Figs 4.56 and 4.57.

A coal-fired boiler installation will be similar but more blowers will be installed, as fouling is often more extensive on coal than oil; also the coal-fired boiler physical size is larger.

The blowers are arranged in three groups, each group having automatic steam, condensate, drain and manual isolating valves. Because of this grouping, the air heaters can be sootblown without the other group steam-lines needing warming through.

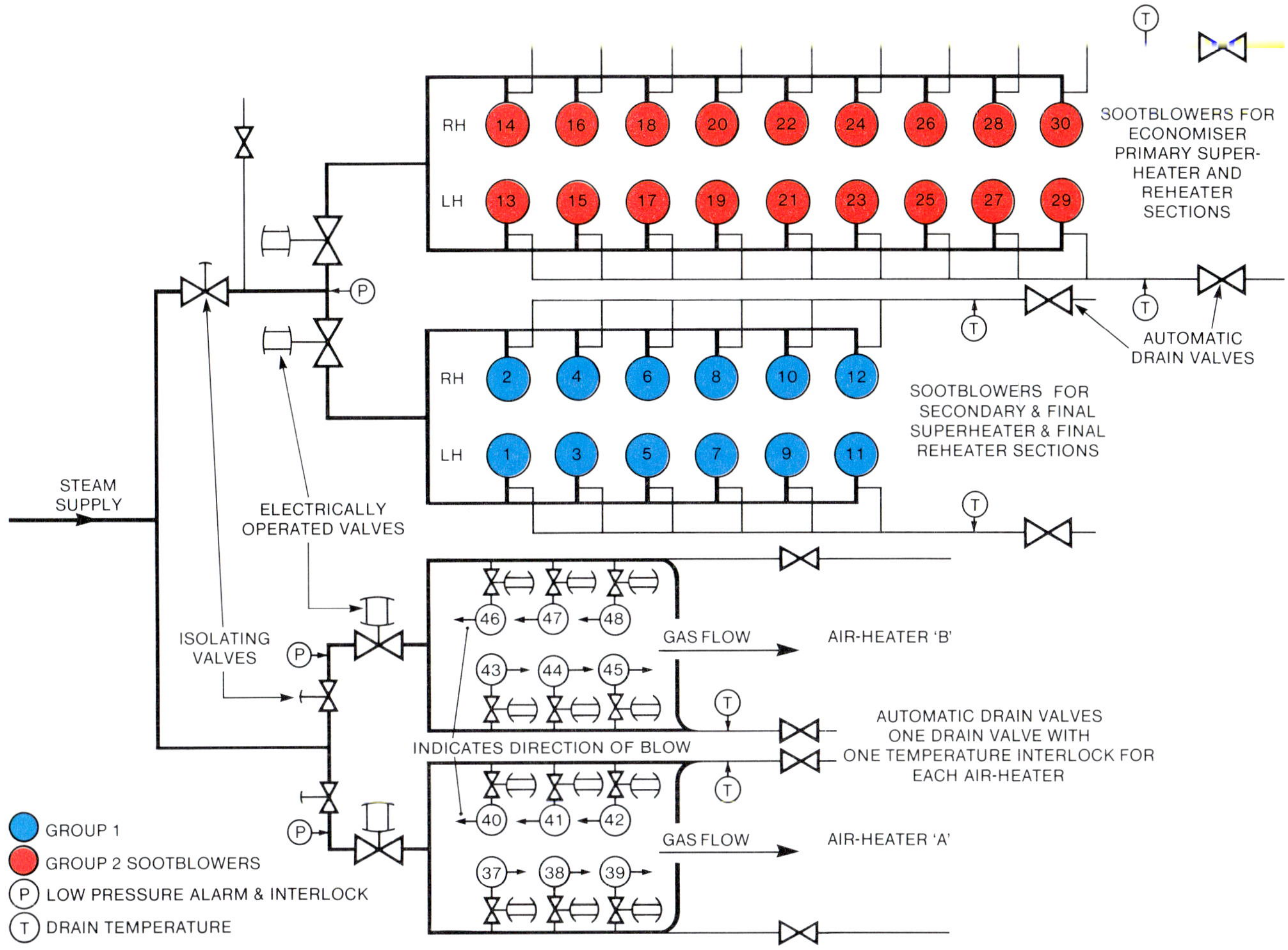

FIG. 4.56 Sootblower steam flow diagram for a 500 MW oil-fired boiler

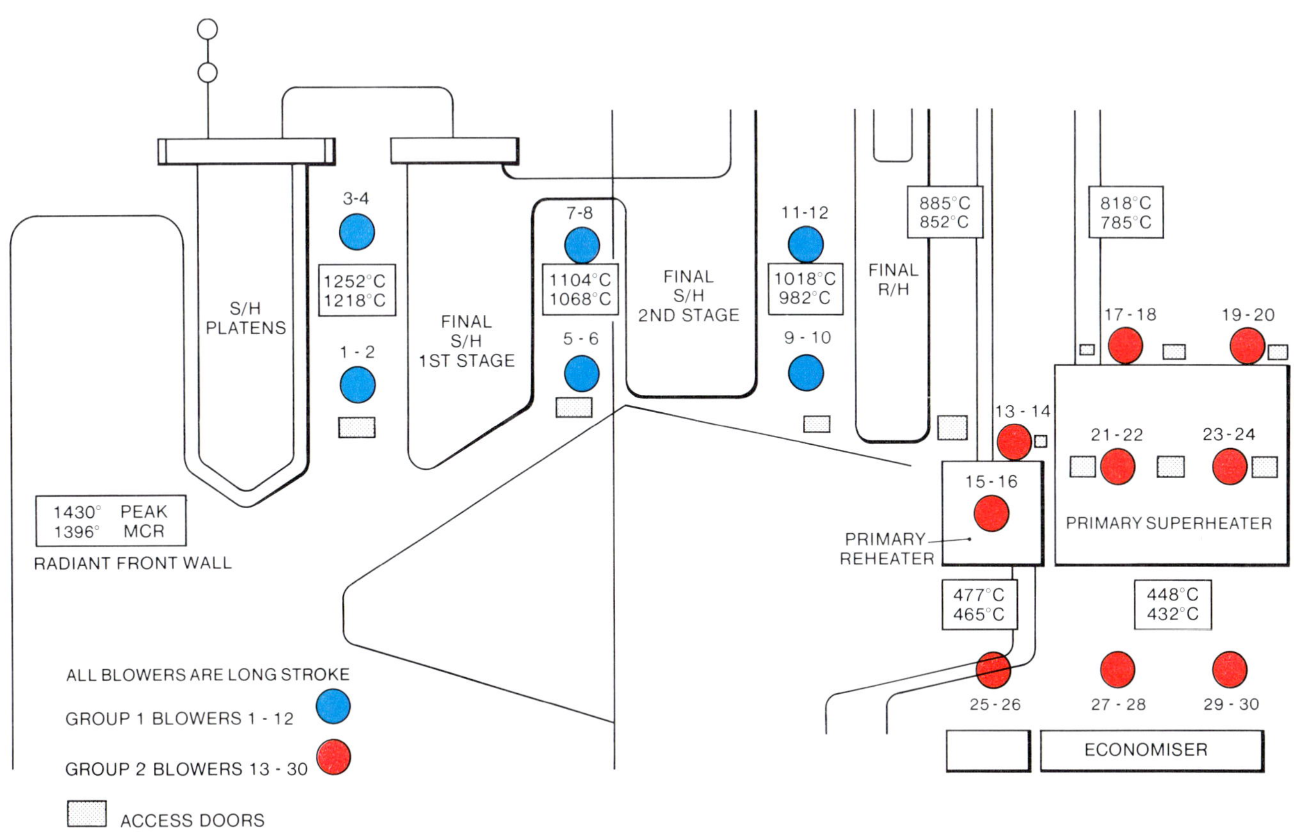

FIG. 4.57 Location of groups 1 and 2 sootblowers for an oil-fired boiler

The three groups comprise:

1. The secondary and final superheaters, and the final reheater sections (hot zones) (operating temperatures shown in Fig 4.57).
2. The economiser, primary superheater and reheater sections (cooler zones) (operating temperatures shown in Fig 4.57).
3. The regenerative air heaters.

The blowers in groups 1 and 2 are of the long stroke type, each operated by two reversing electric motors; one motor traverses and retracts the steam lance and the other rotates the lance. Forward movement of the blower carriage operates a cam which opens the steam/air inlet poppet valve. This cam also closes the poppet valve on the reverse travel.

The sootblowers in group 3 are of the multi-jet type, three blowing with the gas stream and three against in each air heater. The system is controlled by sequential control units.

Group 1 blowers are fitted with air-cooled wall boxes having heat-resisting tips to house and protect the heat-resisting double-ported nozzle. Air is supplied to the wall boxes from two fan units.

Group 2 blowers are similar to group 1, with the exception that the wall boxes are not fitted with heat-resisting tips.

Group 3 air heater blowers are of the two-way closed/open valve type, with timed control of the open/blowing period.

Each electrical steam valve and blower unit is equipped with an isolator, local control pushbutton, limit switches and a handwheel for manual operation.

The sootblower control panel typically includes the following facilities:

- Sequence operation is by blower selector switches. There is one switch fo each pair of blowers, with a 'blow' position for operation in numerical sequence and a 'skip' position for non-operation, with by-pass of the sequential interlock.
- The sequence programme is initiated by operating the sequential start pushbutton, group 1 blowers operating first, followed by groups 2 and 3 in turn. The blowers operate in pairs, one right-hand and one left-hand. This is shown in Fig 4.57.
- Faults due to motor overload, steam valve failures and low steam pressure are alarmed. Electrical steam valve operation is indicated by 'closed', 'cracked' and 'open' illuminated monitors.
- A reset pushbutton cancels all alarms and allows motor restarts when an alarm indicates failure to retract.
- The blower groups may be operated independently, once the control selector switch has been set to the particular group.
- Control may be transferred local to the individual blower units by selecting the control selector switch to the 'local' position. The units can then be controlled by the local pushbuttons, the steam supply valves operating independently of the drain temperature control.
- During local operation, 'valve and blowers' indication remains, together with all the protective and alarm features, the sequence control being inoperative.

The following alarm and protective features are incorporated:

Alarm indication By alarm lights on the control facia, and by audible alarms. Fault indication by semaphore of both groups on group 3 control facia.

Low steam pressure Initiated by a pressure switch, common to groups 1 and 2, in the steam supply to the steam supply valves. Alarm indication is given at the control panel. Operation of the switch prevents the sequence commencing and retracts any blower in operation, preventing further sequence operation. Blowers using compressed air have a similar pressure switch monitoring air pressure.

Steam supply valve motor overload or excessive torque When the motor stops, the 'steam supply fault' indication shows at the control panel.

Blower motor overload The traverse and rotate motors have separate thermal overload relays, fault indication being given by 'traverse' or 'rotate' alarm lights. The appropriate blower master contactor is tripped and the blower indication remains to identify the faulty blower.

Starting steam flow In the start sequence, the main steam supply valve goes to the cracked position, allowing only a small steam flow to warm up the steam pipework. The automatic drain valve solenoids are energised and held open until the temperature reaches 320°C, when they de-energise and close under steam pressure: this action energises the operating relays for the steam supply valves, allowing full steam to flow.

On compressed-air type sootblowers, the equipment is similar to that described above, but there is no requirement to warm up the air pipework using automatic drain valves, the system being ready when air at the correct pressure is available from the compressor.

Traps with inlet filters are provided to remove the water vapour that is produced when compressed air is cooled to ambient temperature, the traps being positioned at appropriate points in the system.

8.2.2 Safety and resources

Safety

- Access within the gas passes to set blower travel should be carefully controlled. When operating blowers with personnel inside the boiler, a special communication system should be used.

- Permits for Work (PFW) must be used to control boiler access during sootblower commissioning. The use of PFW with a restoration of motive power (ROMP) should be used when electrical supplies are used to drive blowers with men inside the boiler.
- When steam/air purging pipework, areas of steam/air escape to be barriered off and only personnel involved in the test allowed in the area.
- Ear protection to be worn.
- Steam trap internals to be removed prior to steam blowing.

Resources

The following resources will be required:

- Auxiliary boilers or sootblowing compressors commissioned to provide blowing fluid.
- Temporary pipework for use when steam purging pipework.
- Ear defenders.
- Electrical supplies to blowers, valves and instrumentation.

Staffing

A team of four persons will require approximately one week to commission the equipment. Final commissioning, with the unit on full-load, will take a further week.

8.2.3 Sootblower commissioning

A commissioning sequence diagram is shown in Fig 4.58.

General mechanical checks and pipework steam-blowing

Sootblowers should first be checked to see that they are correctly positioned in relation to the areas to be cleaned. The steam or air supply pipework will require blowing through to remove any entrained erection/atmospheric corrosion debris. Steam supplies must be available from the auxiliary boilers, or air from the compressors. Flanges must be broken at strategic points in the system and drain trap internals removed,

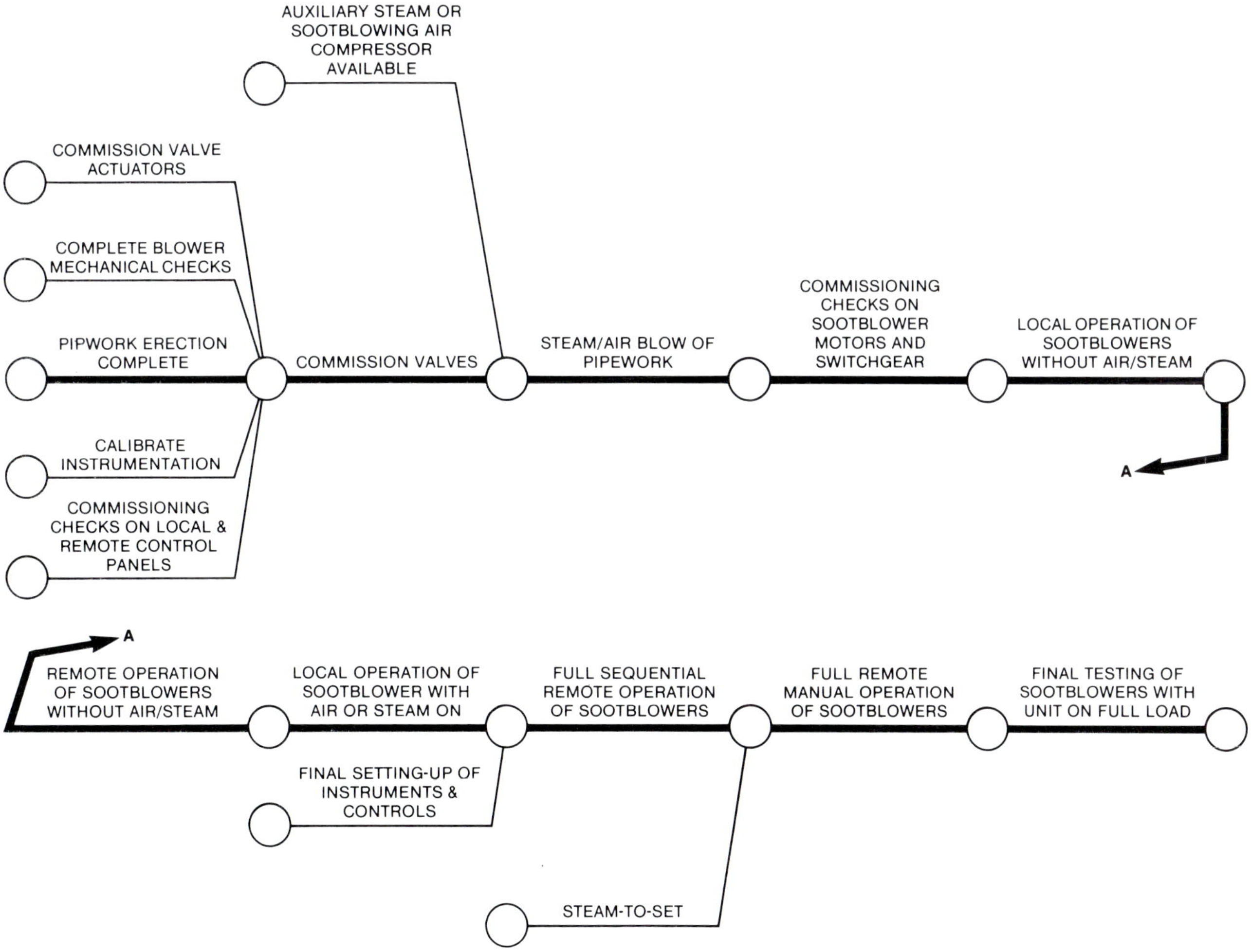

FIG. 4.58 Sootblower commissioning network

prior to the blowing operations. When the preparatory work is complete, the areas of steam/air escape should be adequately barriered off. Only personnel involved in the tests must be allowed in the area and effective communications must be established between the test team personnel.

Following satisfactory blowing of the pipework, all broken flanges should be remade and drain trap internals replaced.

The fans which supply the cooling air to the wall boxes should be commissioned against closed discharge dampers, the motors being checked using the procedure in Chapter 3. The ductwork and wall boxes should then be blown through, using the fans to clear out any entrained debris.

Pre-commissioning dry runs

All manual and electrically-operated valves should be checked, using the procedures in Chapter 3, Sections 1 and 2 and the SCL. Control equipment should be commissioned to the manufacturer's instructions in accordance with the procedures in Chapter 3, Section 8.

The following pre-commissioning dry runs should then be carried out on the blowers but, before proceeding, access to the gas passes in the area of the sootblowers being tested must be restricted to the test personnel and the steam or air supplies to the blowers must be locked off. It is also necessary to inhibit the low pressure and temperature interlocks to enable the blowers to operate.

(a) *Group 1 and 2 blowers (long-stroke blowers)*

- Check that the blowers are fitted with heat-resistant shrouds and also check air cooling arrangements, where provided.
- Check that the blowers are in the fully-retracted position, i.e., the nozzle should be 5 mm inside the shrouds, aligned with the centre line of the blowing holes in the horizontal plane.
- Check that each blower gearbox oil level is correct and its oil reservoir filled.
- Manually operate each individual sootblower to ensure freedom of movement and, whilst this is being carried out, check the blowing positions. After 250 mm forward traverse, the steam or air poppet-valve should be fully open (see Fig 4.59). The steam or air poppet-valve should reclose at this position on the reversing cycle. With the blowers in the fully-forward position, the maximum deflection of the steam lance should not exceed 90 mm. The beams which support the lance should be adjusted so the blower lance slopes downwards towards the boiler, to ensure that any condensate that collects in the lance can drain out through the holes provided in the nozzle.

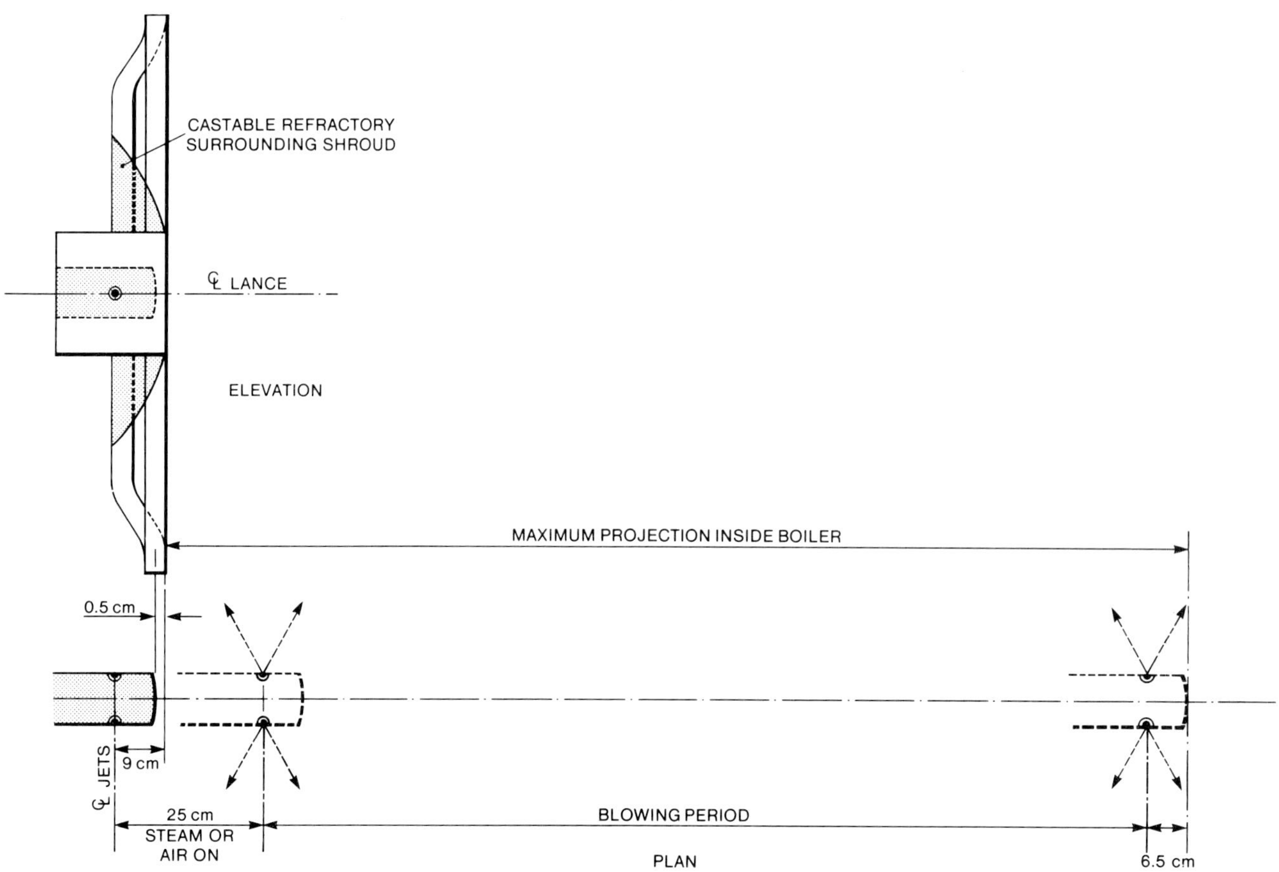

FIG. 4.59 Setting details for long-stroke sootblowers

- Check the rotation of the blower and valve motors. The motors should be commissioned using the procedure in Chapter 3, Section 4.
- After checking that all sootblowers are in the fully-retracted position with their isolators closed, select the control to 'local' sequence and close the main isolator at the contactor cubicle.
- Open the steam supply or air valve from the 'open' control pushbutton.
- Start each sootblower, using the 'local' control button. Check the performance of the traverse and rotate motors and observe that the blower steam or air poppet-valve opens after 250 mm forward traverse. At the full-forward position of the blower, the motors should stop and, after a short pause, reverse and retract the blower. As the blower retracts, the steam or air poppet-valve should close at the position 250 mm from the retract position and, in the fully-retracted position, the motors should stop. All blowers within the group should be checked in this manner.
- Following the above checks, the full sequence operation of the blowers should be checked, without steam or air on. The necessary interlocks should be defeated and the sequence started by operation of the 'start' pushbutton. The operation of each blower should be carefully observed. On completion, reinstate any defeated interlocks.

(b) *Group 3 blowers*

- Check the open and shut movement of the manually-operated isolating valves and of the blower steam (or air) poppet supply valves.
- Operate the steam (or air) supply electrical valves electrically, from the local position.
- Manually operate each blower to the forward traverse position and set the retract limit switches. Rewind the blower to the retract position and again set the limit switches.
- Defeat the 'low pressure' (and 'temperature' on steam blowers) interlocks and select the controls to 'local'.
- Commission the blower drive motors, in accordance with the procedure in Chapter 3, Section 4.
- Start the sootblowers from the 'local' pushbutton and check the indication on the control facia.
- Open the supply steam (or air) valves from the control panel 'open' valve pushbutton.
- Start the blowers by the 'start blower sequence' pushbutton and check that the blowers operate in sequence.
- When the blower sequence is complete, the steam or air supply valves should automatically close and indicate 'sequence complete'.
- Start the sequence by the full sequence 'start' pushbutton and check that the steam (or air) supply valves open.

When all the blowers have operated, the steam (or air) supply valve should reclose.

Full commissioning tests with air or steam

Following successful dry-run commissioning tests, the blowers should then be operated on steam or air, with all interlocks in service. In order to do this without the boiler firing, it will be necessary to withdraw all personnel from the boiler.

An airflow must be provided through the boiler to the chimney, but it is not necessary to run fans. The gas air heaters should be put into service and must be rotating before the air heater blowers are operated. It is more usual to operate the blowers with the boiler operational and an opportune time for the above to be carried out could be during the boiler chemical clean or steam purge operations.

Prior to these tests, it will also have been necessary to have checked all pressure and temperature instrumentation, in accordance with the procedure in Chapter 3, Section 3 and SCLs, and to have steam available from the auxiliary boilers, or air from the sootblowing compressors.

Having completed all the preliminary tests and established that the boiler is ready for sootblower operation, the following sequence runs on groups 1 and 2 blowers can be carried out.

The sootblowers are prestart checked to ensure that all manual isolating valves are open and that the electrical steam/air supply valves are shut and selected to remote. All blowers must be fully retracted. Steam or air services will be made available and electrical supplies switched on at the blower controls. The sequence will then commence as follows for steam sootblowers:

- The control selector is switched to groups 1 and 2 blowers.
- The blower selector switch is set to the 'blow' position.
- Depress the sequence 'start' button and check that the following sequence proceeds.

  (a) The steam supply valve motors start to open the valves of both groups 1 and 2, stopping when the valve reaches the 'crack' position. (Steam supply valve indicators show 'cracked' position.)

  (b) Steam then flows through the pipework via the open drain valves. The pipework should be warmed and any condensate lying in the pipework discharged. The drain valves are held open pneumatically by air solenoids, which remain energised from the drain temperature switches.

(c) When the pipework temperature reaches 320°C the temperature switches should operate, de-energising the drain valve solenoids. The valves should then close under steam pressure, energising the operating relays for 16 steam supply valves.

(d) The Group 1 steam supply valve then opens and the indicating lights show steam supply valve 'open' on Group 1 and 'crack' for Group 2 valve.

(e) When the steam supply valve is open the sequence commences, the flowers operating in pairs, one left hand, one right hand, individual blower indication being shown on the control panel.

(f) After 250 mm forward traverse, the blower steam valve opens fully.

(g) At a point just before completion of the reverse traverse, the blower steam valve mechanism operates in the reverse direction, closing the valve.

(h) The blowers return to the fully-retracted position and the traverse and rotate motors trip automatically, individual blower indication being extinguished.

- When the last blower in a pair completes its cycle, the next pair of blowers in the sequence is started and the cycle is repeated.
- The sequence progresses to operate all blowers in the group, except any selected to 'skip'. The sequence order is 1 and 2, 3 and 4, etc., together through to the last pair.
- When the last blower-pair completes its cycle, the Group 1 steam supply valve closes and the pneumatic drain valves reopen, 'sequence complete' indication being shown.
- The steam valve of Group 2 blowers then opens fully and the sequence for them progresses as given above for Group 1 blowers.

The sequence for air-operated blowers is similar, but all that is required to supply air to the blowers is satisfactory air pressure of 12 bar from the compressors. If the pressure is satisfactory, the air isolating valves to the Group 1 and Group 2 blowers open, the sequence then proceeding as for the steam blowers.

Before operating the air heater blowers, any protective paper, putty and tape should be manually removed from the air heater baskets; the sootblowing process should then remove any residue of grease left behind.

Final commissioning of the sootblowing system takes place with the unit on load and it is advisable to sootblow the gas air heaters at high loads to avoid the risk of condensation; as the element plates will be operating at design surface temperatures, the risk is greatly reduced. Sootblowing should be carried out just prior to shutting down a unit and also after start-up, when steady load is achieved; it is also advisable to sootblow following a situation which produced poor combustion conditions.

8.3 Air heater and water washing systems

8.3.1 Description

On Howden-type regenerative air heaters, the water washing equipment basically consists of two wash-pipes in the gas side of the air heater. One pipe is positioned above the rotor and the other below. Each pipe or lance is connected to the sootblowing lance to form dual lance units. The water supply is provided by flexible pipework connected to the lance water washing flanges prior to the wash commencing.

Each washing lance is fitted with sufficient nozzles to cover the rotor surface adequately. The function of the lower lance is to remove the cold-end deposits on the heating surface elements.

The upper lance is installed primarily to clean the deposits from the hot-end elements and to provide a flushing action to eliminate the possibility of any loose deposits remaining in the air heater. When washing the air heater, the bottom lance is operated first, followed by the upper lance.

The upper, or hot-end, washing lance is offset in the direction of rotation from the lower washing lance so that the washing action of the lower nozzles precedes the flushing action of the upper nozzles.

Whilst water washing, the air heater rotor is driven slowly at 0.25 r/min by a washing drive motor, and the combined sootblower and washing lances are driven into the heater in stages; a timer controls the sequence programme. In a typical installation, the lance moves forward a distance of 38 mm in 48 seconds, pauses for 3 minutes 12 seconds, then moves forward in steps until it reaches the extended position; the forward limit then operates and the lance is withdrawn normally.

Hot water for the washing lances is provided from a calorifier, which uses auxiliary steam to heat it to approximately 60°C; this temperature should be maintained by regulating the flow of water through the calorifier. The water drains from the air heater into the gas outlet ducts and from the ducting drains to a catchment tank or pit, where the washings are contained.

The effluent from the air heater wash is highly acidic and great care must be exercised to ensure that the effluent is carefully contained, until neutralised before disposal. In some stations, the water used for air heater washing is contained in storage tanks which are dosed with lime to a pH of 12 prior to the wash; water is then pumped from the tanks through the calorifier to the air heater lances, and then returned to the storage tanks to await neutralisation before disposal. A water pressure interlock which prevents the lances advancing is provided to ensure that sufficient water pressure is available to provide an effective cleaning action. This pressure is set at 4 bar.

8.3.2 Safety and resources

Safety

- Protective clothing is to be worn when handling lime for preparing and neutralising the wash water.
- The washing from the air heaters will be strongly acid and care must be taken to contain the water within the wash water system.
- The ducting drains should be checked to be free to operate before putting water on the lances.
- Thermocouple probes and gas and air sampling probes to be removed from the air heaters prior to washing, to prevent damage.
- Effluent disposal to be carefully monitored and the conditions specified in the Control of Pollution Act strictly observed.
- Any access to the air heaters or washing equipment must be carefully controlled, using Permits for Work.

Resources

- Flexible hoses for connecting the wash water pipework to the lances.
- Lime for preparing and neutralising the wash water.
- Auxiliary steam available to the wash water calorifier.
- Protective clothing of PVC gloves, apron, goggles and disposable coverall.
- Sootblower lances commissioned for sootblowing purposes.

Staffing

A team of four people will require approximately two weeks to commission the equipment.

8.3.3 Air heater washing systems commissioning

The air heater washing system may not be fully commissioned until after the boiler has operated for some time: it it at the discretion of the operating staff when it is deemed that the air heater surfaces are restricting gas flow, due to the build-up of deposits on the heating surfaces.

The air heater washing system will have been checked with all tanks, pumps and pipework proven to be leak-tight and auxiliary steam will be available to the heating calorifier. Any flushing of steam steam or water lines should be completed prior to carrying out an air heater water wash.

The air heater washing drive-motor and the sootblower/water washing control panel will have been fully checked during the air heater and sootblower commissioning activities.

Water washing is carried out with the air heater off-load, but can be done with the boiler firing. The unit load is reduced to a level where it can be carried on one in-service air heater, and the air heater to be washed is isolated on the gas inlet and air sides, by closing the respective dampers. The gas outlet damper is left open to allow a free flow of water to the ducting drains. A check should be made that the gas outlet and air-side ducting drains are fully operable and free to move, prior to putting water on the lances.

If washing of the air heater takes place with the unit off-load, the wash must be started immediately the boiler is shut down, so as to ensure that the metal temperatures of the element surfaces will still be warm for deposit removal. Water washing must commence well before the elements have completely cooled. Once water washing has commenced, complete cleanliness of the heating surfaces of the elements should be aimed at, because partial washing leaves a deposit on the heating surface which cakes hard when the air heater returns on-load. Air heaters are dried out after washing by returning to service, or by using hot air from the steam air heaters.

The following water washing procedure will then take place:

- Electrically isolate the air heater main drive motor.
- Check that the steam or air supplies to the sootblowers are isolated.
- Rotate the lower lance through 90°, so that the nozzles are facing upwards towards the rotor.
- Connect a water supply flexible hose to the top and bottom sootblower water washing flanges.
- Remove the fire detection probes to prevent damage.
- Remove any gas on air sampling probes to prevent damage.
- Check that the drain valves in the ducting below the air heater are operable and that the effluent drainage is routed in a way to ensure that other items of plant cannot be damaged.
- Engage the air heater low-speed drive.
- Start the air heater in low speed at 0.25 r/min.
- Commission the water washing system by warming through the calorifier and then apply water to satisfy the pressure interlock.
- Open the water-wash inlet isolating valve to the bottom lance.

- Push the 'start' button for the bottom lance on the local sootblower/water wash control panel.
- The lance will now advance 38 mm and then pause for 3 minutes 12 seconds, this sequence continuing until the lance is fully inserted.
- During the washing operation, regular checks must be made that the water is draining effectively from the ducting drains to the effluent storage system.
- When the bottom lance has completed its traverse, open up the water supply to the top lance, shut the water off to the bottom lance and push the sequence 'start' button for the top lance.
- The top lance will now progress, as for the bottom lance.
- When the top lance has completed its sequence, the water washing system should be shut down and the heater-element matrix inspected for cleanliness through the inspection windows, using powerful lights. (If the boiler is off-load, access to the heater is possible, but adequate protective clothing must be worn, because of the acidic nature of the water.)
- If the heater is clean, then it can be returned to service, but if further cleaning is required, the above operation should be repeated.

On completion of the air heater wash, the slow-speed drive should be disengaged and the heater normal drive re-energised, the flexible hoses disconnected and the air heater then returned to load.

If the air heater has been washed with the boiler on-load, the heater should be returned to service slowly over a period of approximately 2–3 hours to avoid uneven heating and expansion of the matrix. The air heater seals should not be applied until the heater matrix is thoroughly heat-soaked. If the air heater wash was completed with boiler off-load, then the boiler should be returned to load as soon as possible to ensure effective drying out of the air heater matrix. If the unit is not to be returned to load, the air heater can be dried out by running fans and putting hot air from the steam air heater through the matrix. The fire detection and sample probes must also be replaced on completion of the clean.

The effluent residue contained in the holding tanks will be acidic and this should now be neutralised to a pH of 7, using lime, to the satisfaction of the Station Chemist, before it is discharged to the disposal system.

9 Control and instrumentation

9.1 Introduction

Setting to work of control and instrumentation systems is normally one of the most critical areas of power station commissioning.

There is such a wide range of commercially available instrumentation that it would be impossible to describe it in detail and how to commission it in these pages. In the previous sections on boilers, a procedure for commissioning the necessary instrumentation, alarms, sequence and control equipment has been covered in sufficient detail to bring the plant into commission for first firing of the boiler for chemical clean and steam purge purposes.

In Chapter 3, Section 3, commissioning procedures have covered the broad principles involved in testing the different types of measuring and indication equipment. The areas covered are:

- Thermocouples
- Resistance thermometers
- Pressure gauges
- Pressure switches and transmitters
- Flow and level switches/transmitters
- Recorders
- Analytical instrumentation
- Electrical transducers

Sequence control and interlock cubicles are covered in Chapter 3, Section 8 and commissioning of alarm equipment in Chapter 3, Section 7.

A full description of control and instrumentation systems can be found in Volume F.

This section, therefore, is a brief summary of the control and instrumentation requirements for the first firing of the boiler.

The CEGB philosophy for the commissioning of control and instrumentation system is for the Control and Instrumentation Working Party to check all instrumentation, alarm inputs, alarm facias, VDUs, equipment cubicles, controls, interlocks and cabling, and to test them to the nearest junction box from the plant-initiating devices; also, the initiating device is calibrated. The Control and Instrumentation Working Party then hands the systems over to the Plant Area Working Parties for dynamic checking.

9.2 Control and instrumentation commissioning

9.2.1 Summary of the control and instrumentation commissioning for first firing of the boiler

The following is a list of control and instrumentation which is required before firing the boiler:

- Boiler temperature, pressure and flow indications for the respective parts of the water and steam circuits.
- Drum level gauge glass, complete with remote TV viewing to provide level indication on the unit control panel (UCP).

- Drum high and low level alarms (when these are provided from Hydrastep equipment, they may not be available until the steam purge operations).
- Remote drum level indications from long- and short-range transmitters.
- Single-element feedwater control loops.
- Boiler circulating pump controls and indicators.
- Oil burner controls and indicators.
- Gas and steam air heater controls and indicators.
- Fan controls and indicators, complete with the start-up furnace pressure control loop.
- Combustion air pressure and temperature controls and indicators.
- Fuel oil pressure and temperature controls and indicators.
- Boiler valve controls and indicators for water, steam and oil systems.
- Gas and air damper controls and indicators.
- All alarms associated with the above, on both conventional facias and computer VDUs.

9.2.2 In-service checks

As each section of plant is put into service, instrumentation loops should be put on-line, in accordance with an approved test procedure. An example of putting a transmitter into service measuring steam or water flow is given in Chapter 3, Section 3.

Because the boiler is to have heat applied to it for the first time, it is essential to recognise its limitations with respect to temperature, expansions, supports, etc., and it is therefore necessary to ensure that instruments providing data for the required checks are functioning correctly.

All alarms associated with the boiler will have been checked during the individual plant area commissioning programme: it is essential to check that they have been fully tested and that all records are signed as complete.

Each alarm facia should have an identification 'spot' attached to each alarm way, when it has been dynamically checked.

Commissioning of drum water level indicators was covered in Section 4.4 of this chapter, but it is essential to check before firing the boiler that the gauge glass and connecting pipework to the boiler drains vessel is adequately supported and correctly connected. Pipes should be checked for possible leakages or restrictions. The gauge glass should be aligned with the drum so that the water level is in the centre of the glass at the normal drum working level. Correct illumination and alignment with the TV camera must be established to provide a clear visual position of the water level, both directly and in the control room.

A calibration check, in the cold condition, can be carried out on the long- and short-range drum water level transmitters, and the drum-to-transmitter interconnecting pipework blown out with the compressed air. The connections to the high and low level alarms and the correct operation of these alarms should be checked by raising and lowering the drum water level. This may not be possible if they are from the Hydrastep, which will be left isolated until after the steam purge.

The instrumentation, cabling and cubicles associated with the sequence and interlocks are normally calibrated and set up with the correct operating parameters, using the procedures outlined in Chapter 3, Section 8, and handed over by the Control and Instrumentation Working Party for on-line (dynamic) testing on the plant. The working party responsible for the plant area being commissioned will then function test the pressure switches, transmitters, etc., by changing the operating conditions in a controlled manner to test dynamically for the correct indications and operation of the sequence, interlocks and alarms.

As an example, commissioning the control and instrumentation on an ID fan with inlet vane control is carried out as follows. The fan 11 kV ACB will be put on 'jumpers' (i.e., the fan ACB is racked into the out-of-service position, but with its carriage interlock switch auxiliary contacts made, to 'show' it in the service position) and all other auxiliaries made available for normal operation. Attempts are made to close the ACB with the lubricating oil pump out-of-service or discharge damper open or IGVs open, etc., to prove the starting interlocks. After proving all the starting interlocks, the fan lubricating oil system, dampers and brake are set to operate in the correct starting mode and the ACB is then closed via the sequence equipment. The ACB is then proved to trip by, say, taking the lubricating oil pump out-of-service to reduce the lubricating oil pressure to the trip point. When all the starting and running interlocks have been proven in this way, the fan ACB is taken off 'jumpers', racked into the service position and then test run, subject to the boiler conditions. During these tests, any alarms associated with the plant are function checked and accurate records are kept of the tests. Alarms to both conventional facias and to the computer should be checked. This type of procedure is adopted for all items of boiler plant prior to running.

Before firing the boiler for the first time, it is essential to have proven all the necessary starting interlocks and controls for the fans, air heaters, fuel oil pumps, oil burners, start-up furnace pressure controls and boiler circulating pumps to ensure correct plant operation. Figure 4.60 shows a simplified coal- or oil-fired boiler starting sequence interlock scheme.

Trips associated with boiler firing equipment that will ensure the oil burner trips under fault conditions must

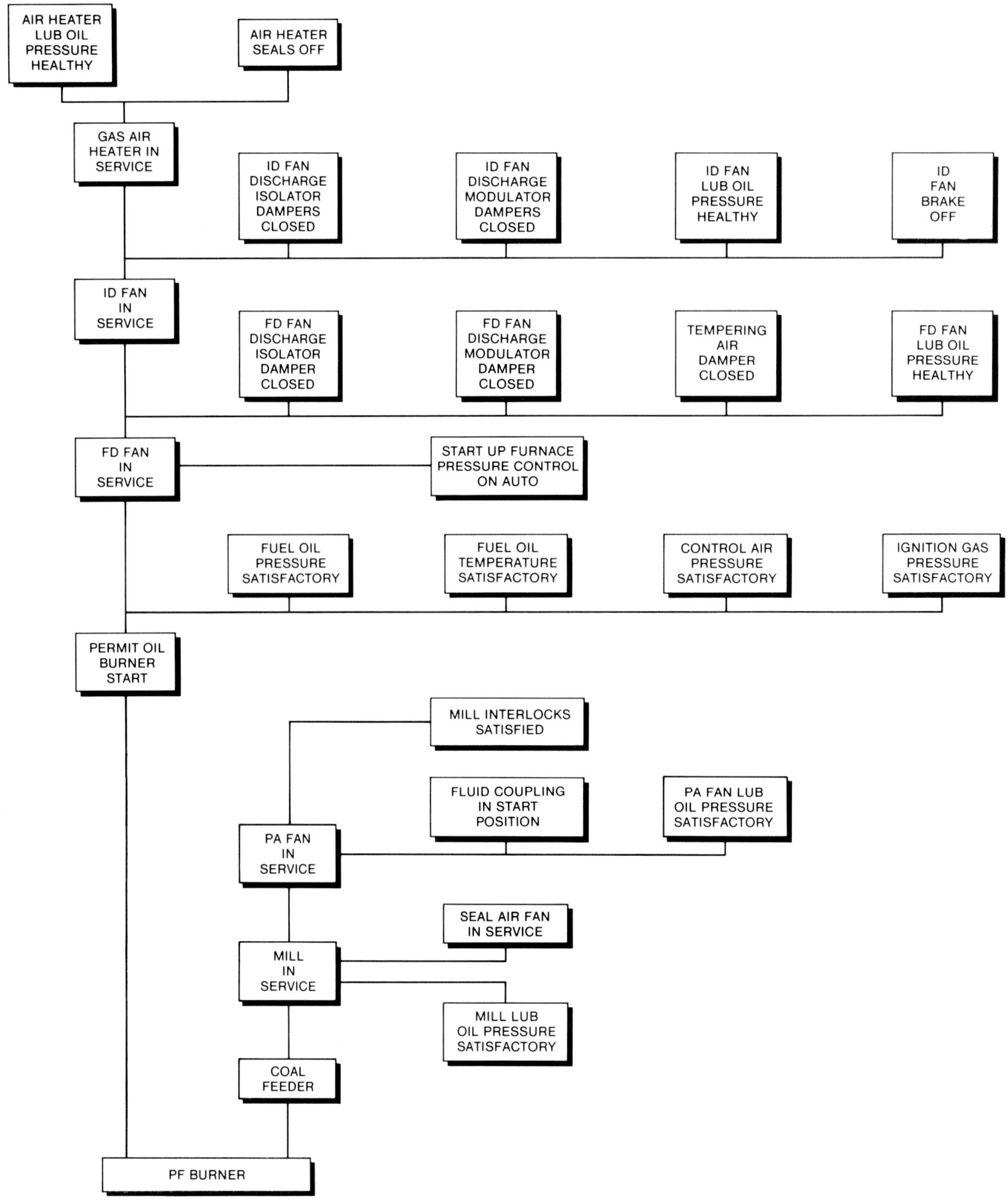

FIG. 4.60 Simplified starting sequence interlock scheme for a coal- or oil-fired boiler

also be proven. Typically, for a coal-fired boiler tripping interlocks will be as follows:

- Loss of one air heater trips selected coal feeders through mill group selection. The air heater inlet and outlet dampers will close.
- Loss of both air heaters causes a cascade trip of the ID fans and FD fans, with all milling plant and oil burners.

In addition to being tripped as above, an ID fan can be tripped by:

- Lubricating oil pressure failure.
- Furnace under-pressure condition.

When an ID fan trips, the following actions are automatically initiated:

- Fan control trips to manual

- Closes and locks the fan discharge modulating and isolating dampers, if the other fan group is running
- Starts the jacking/lubricating oil pump
- Applies the brake after 20 minutes
- Trips selected coal feeders through a selection circuit
- Trips any FD fan through a selection circuit

In the event of both fans tripping, all the dampers remain open.

An FD fan can be tripped on low lubricating oil pressure and the actions initiated are as follows:

- Fan control trips to manual
- The tempering air, discharge modulating and isolating dampers close and lock, if the other fan group is running
- The lubricating oil pump starts
- Selected coal feeders are tripped
- Loss of both FD fans trips all mill groups and oil burners
- Loss of one FD fan and/or one air heater will reduce the number of running mills in the selected tripping order

A primary air fan (PA fan) can be tripped on low lubricating oil pressure and mill group trip: this closes and locks the PA fan discharge damper.

Before operating the boiler on oil burners, the above fan trips and interlocks must be proven.

The above is a brief outline of the control and instrumentation (C and I) requirements. C and I is one of the most difficult areas of plant to co-ordinate commissioning activities and therefore a close working relationship between the C and I Working Party and Plant Area Working Parties is required.

10 Initial boiler operations

10.1 Introduction

Sections 1 to 9 of this chapter have described the commissioning requirements to prepare a boiler for the initial light-up, which normally takes place just prior to the boiler chemical clean.

Before the boiler is fired for the first time the commissioning tests referred to must have been completed. Milling plant is not normally introduced until after steam-to-set on coal-fired boilers, as sufficient heat is available from the oil burners for chemical cleaning, steam purge and floating safety valve operations. Operations with milling plant and coal burners are therefore covered in Chapter 6.

The objective of the initial boiler light-up is to prove sufficient burner availability for the chemical clean and steam purge operations and to prove that the rest of the boiler plant will operate in a satisfactory manner, according to the design parameters. The boiler will also be tested for leaks prior to adding chemicals.

The boiler pressure will not be allowed to go above approximately 40 bar, which is the pressure required for the alkaline boil-out for stage 2 of the chemical clean, but it is unlikely to reach this level, as only a limited number of burners will be used and they will be fired for short periods only.

10.2 Safety and resource considerations

Safety

- All boiler safety interlocks to be commissioned and in service.
- Furnace to be purged prior to lighting the first burners and for any subsequent light-ups should all firing burners trip.
- Attempts to light oil burners from adjacent oil burners must *not* be made.
- Oil burner safety interlocks to be fully in service. No attempt should be made to light oil burners with safety interlocks 'defeated'.
- Furnace pressure trips to be in service prior to attempting to light the first burner.
- Furnace pressure explosions have been caused by restoring the fuel supply to a furnace too frequently, when ignition has been lost. If problems with light-up are experienced, the furnace should be inspected for oil deposits before further light-ups are attempted.
- All scaffolds that restrict boiler expansion to be removed.
- Boiler cladding to be complete, in all areas where personnel may come into contact with hot surfaces.
- All boiler fixed fire protection systems to be in service and portable fire extinguishers to be available in strategic areas.
- Movable bend to be removed from the LP boiler filling system and connected to the boiler drains system.

Resources

The following resources will be required, prior to attempting to light-up the first oil burner:

- Auxiliary boiler available to provide steam for fuel oil heating and steam air heaters.
- Portable combusters to be available for use in the FD fan inlets on boilers without steam air heaters, particularly if ambient air temperature is low.

- Water treatment plant to be available to provide demineralised water for the boiler.
- Boiler to be full of water to the normal working level.
- Electric start-up feed pumps and start-up feedwater control valves to be available.
- Propane gas to be available to the burner igniters.
- Control air to be available to oil burners, dampers and air-operated valves.
- All boiler dampers to be commissioned and available for operation.
- Fuel oil to be available, with transfer fuel oil pumps in service.
- General service water cooling systems to be in service to provide cooling water to HP fuel oil pumps, boiler circulating pumps, etc.
- Boiler drainage systems to be available and the station drainage system should be ready to receive large quantities of water.
- Qualified unit operating staff to be available and trained for their duties.

10.3 Initial light-up of the boiler

A typical boiler initial light-up network diagram is shown in Fig 4.61.

The boiler should be fully inspected to ensure that it is free of all debris in the gas passes and that all unwanted scaffolding, or scaffolding that will restrict boiler expansions, has been removed. Boiler accessways must be clear and adequate lighting provided. All access doors and inspection ports should be checked to be shut.

A complete check must be made to ensure that the auxiliary services stated above are available.

A complete prestart check on the boiler pressure parts and gas passes is made to set the boiler up for filling and pressure raising; the boiler is then filled to NWL by the boiler filling pumps or main boiler feed pumps and initial boiler casing expansion readings taken. Transfer fuel oil pumps are put into service to circulate fuel around the boiler fuel oil systems, and steam is applied to the HP fuel oil heaters to warm the HP fuel pumps to their operating temperatures. Steam air heaters are put into service and gas air heaters put on rotation. The steam air heaters must be in service at all times to ensure that the combustion air temperature is satisfactory. On coal-fired boilers, where steam air heaters are not provided, the use of portable combustors in the FD fan inlets, fired by propane gas or lighting-up fuel oil, should be used to raise the combustion air inlet temperatures. The gas air heater inlet and outlet isolating dampers are left closed and the by-pass dampers opened. One ID and FD fan is then put into service and, when the windbox air inlet temperatures are satisfactory, a furnace purge operation will be carried out. Start-up furnace pressure control should then be put in service, if it is available at this stage of commissioning, and windbox and furnace air pressures set to their correct design settings for burner light-up.

The HP fuel oil pumps are then started and, when the required oil temperature is achieved and the oil burner starting interlocks are all satisfied, a burner light-up should be attempted. If the burner fails to light, a furnace purge will have to be carried out for approximately 5 minutes; then furnace and windbox air pressures are again set up and further attempts made to light-up the first burner.

If problems are experienced with burner light-ups and where the oil viscosity is correct and the equipment functions satisfactorily, a frequent cause of failure is an excessive supply of combustion air, or the oil pressure to the burner is too high.

On coal-fired boilers, when oil lighting-up burners are used, it may be necessary to run the primary air (PA) fans associated with the oil burners to prevent the PF impellers (swirl vanes) overheating. Air supplied to the boiler should however be limited, otherwise serious condensation and corrosion conditions may be produced in the gas passes.

When the first oil burner is in service, checks should be made that the combustion is good (0.5% O_2 should be the aim) that the smoke density is satisfactory, and that the burner flame shape and length are correct by viewing through an inspection port with an inspection safety glass.

The furnace must be kept under slight suction conditions for this operation. The CO, CO_2 and dust monitoring systems will only be reliable when a number of burners are in service after steam-to-set. This equipment will be fully tested during load raising and is covered in Chapter 6.

With one burner firing with a stable flame, other burners can then be test fired to check correct operating sequences on light-up or shutdown and to prove that the start-up furnace pressure control keeps the furnace at the design conditions. With increased oil flow on firing burners, checks should be made that the fuel oil temperature, pressure and viscosity are being maintained at their correct values and the boiler should be monitored carefully for expansion as it warms up. Burners should only be fired for limited periods at this time to prevent the boiler pressure going too high. When it is required to raise the boiler pressure the firing rate should be carefully controlled to ensure that the rate of rise of saturated steam temperature is not greater than 1°C per minute. The gas temperature entering the primary superheater must not exceed 540°C: the gas temperature entering the reheater should not exceed 480°C.

In general, high superheater metal temperatures are most likely to be encountered during the starting period when steam flow is relatively low in relation to the

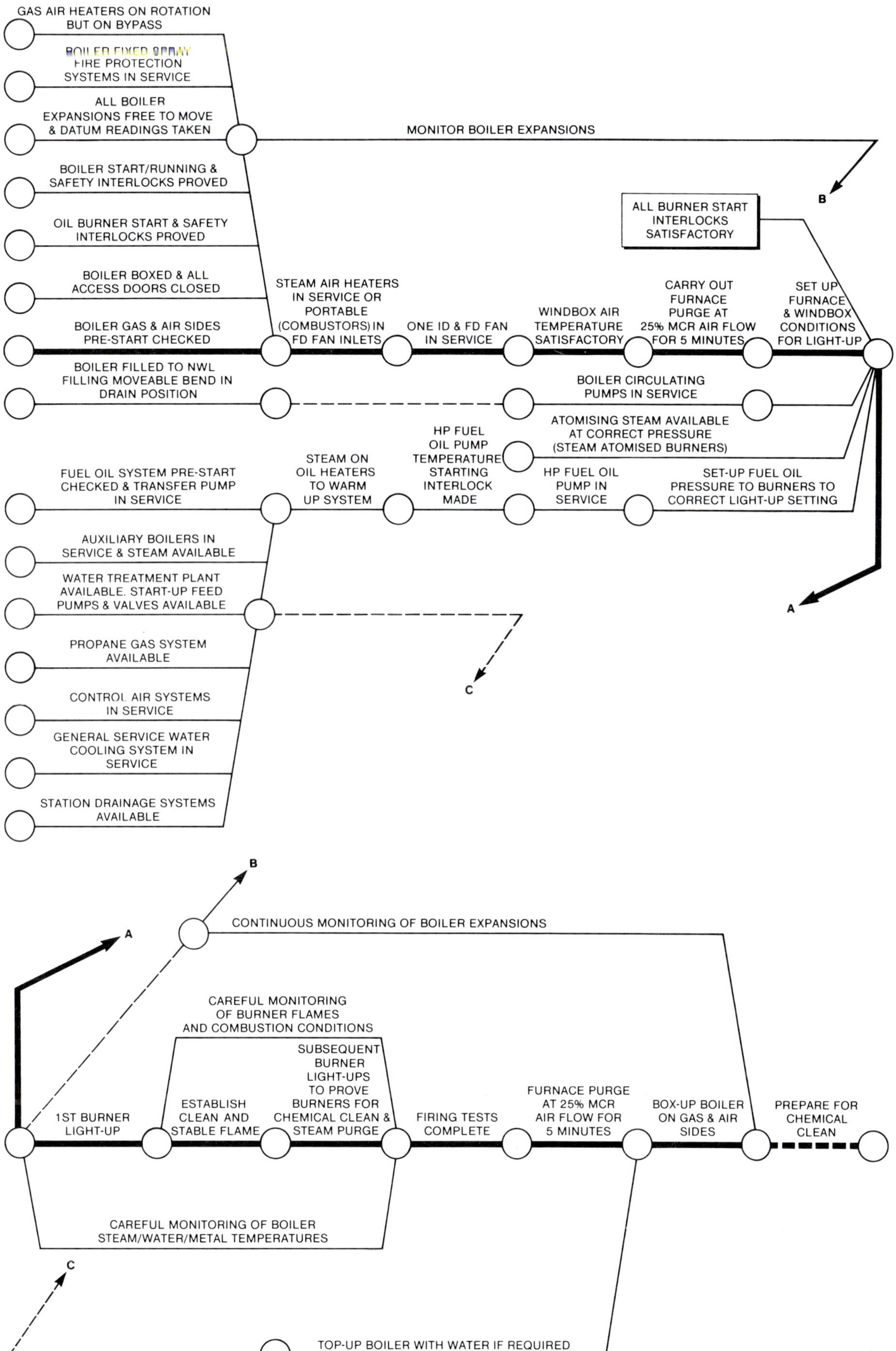

FIG. 4.61 First boiler light-up commissioning network

furnace gas flow and temperature. When spray-type attemperators are employed, their use at low load is restricted, so superheater metal temperatures must be controlled by other means. This can be done by reducing or limiting the firing rate, by using PF burners tilted downwards or situated low in the furnace, or by introducing cooling air above the flames and by reducing combustion air to the minimum necessary. Boilers fitted with superheater steam dumping valves can overcome the problem by regulating the steam flow through the superheater with these valves.

As no flow of steam through the reheater occurs until steam is admitted to the turbine, care must be taken to prevent the reheater metal temperature limits from being exceeded during initial firing.

The gas temperature entering the reheater may be limited by gas by-passing or recirculation where this facility is provided, by control of the firing rate or by tilting PF burners. It is usually advisable to limit the reheater metal temperature during the initial firing of the boiler so that, when steam is admitted to the turbine, there is no excessive rise in the reheater steam temperature due to heat stored in the reheater metal. If this is not done, the turbine loading rate may be restricted by a high differential expansion of the IP turbine rotor.

The boiler drum heating temperature-differentials must not exceed those given on the drum heating cycle curves.

It is essential that very careful attention is paid to the boiler and steam pipework expansions during all the above operations.

During this early firing and pressure-raising of the boiler, the opportunity should be taken to check the boiler blowdown and drainage systems, by operating drain valves for short periods to check that pipework is clear and that no undue expansion or vibration problems occur on the drainage pipework and dump vessels.

The possibility of furnace explosions, corrosion fouling and fire damage to air heaters, etc., during early commissioning of a new boiler requires that extra caution is maintained.

Whilst these problems may always be present, the commissioning period is particularly hazardous, in that the boiler may be operated for prolonged periods at low loads and with relatively low combustion air temperatures. During this early commissioning phase, prior to the chemical clean and steam purge programmes, gas air heaters should be left with their blanking plates in service and by-passed, where possible.

To ensure minimal operation problems, the following should be observed:

- Oil should be fired with the maximum efficiency obtainable under the prevailing circumstances.
- Cleaning and adjustment of burners to minimise the quantity of unburnt particles should be carried out at regular intervals.
- The elimination of infiltrating air and the consequent chilling of the flame should be avoided by ensuring that furnaces are gas- and air-tight and, if necessary, by the use of low viscosity oil during the early commissioning periods.
- Flue dampers should be tightly closed when shutting down after purging the furnace: effective means must be provided to ensure that the plant cannot be started or operated with a closed flue.
- Air heater temperatures should be recorded at frequent intervals during shutdown periods to monitor for fires, which can be caused by the carryover of unburnt particles of fuel. It is recommended that air heater temperatures are logged every hour for approximately 12 hours after the boiler has been shut down.
- The boiler water chemical conditions during the early boiler operations should be controlled by the Station Chemist.
- The conditions required during the chemical clean, steam purge and safety valve floating programmes are covered in Sections 11, 12 and 13 of this chapter.
- Chemical control from steam-to-set is covered in Chapter 6.

11 Boiler chemical cleaning

11.1 Introduction

Chemical cleaning is carried out in accordance with the instructions laid down in Generation Operation Memorandum 75: The Operational and Safety Aspects of the Chemical Cleaning of Power Station Plant Internals (Excluding AGR Boilers) (GOM 75). The choice of chemicals to be used will be based upon the type of chemical clean to be carried out. Citric acid is used more widely today in boiler chemical cleaning than hydrochloric acid: although more costly than HCI, it is preferred because of the problems of stress corrosion caused by HCl in austenitic regions of the boiler.

In recent years, oxide fouling rates of large boilers have been increasing because of the more frequent two-shift operations. When chemical cleaning to remove the fouling, it was found that a single application of 3–4% citric acid could no longer guarantee to completely clean the boilers, and hydrochloric acid, which is used in stronger solution, could yield unacceptable corrosion rates, so a new method using hydrofluoric acid has been used.

Hydrofluoric acid has been used in Europe for many years for boiler chemical cleaning, its main advantage being that it is a rapid and high capacity solvent, exhibiting low corrosion rates. It appears to offer a good solution for cleaning two-shifting large boilers. The CEGB use of the hydrofluoric cleaning technique has been fairly limited at the time of writing this book, but the Scientific Services Department has completed trials in some power station boilers and the process has now been approved for use in CEGB coal-fired boilers; GOM 75 is at present being updated to include the hydrofluoric acid chemical clean process.

Because the hydrofluoric acid process has had limited use in the CEGB and is restricted at present to coal-fired boilers (some additional work on the corrosion of low chromium steels by hydrofluoric acid is needed to clear its use for oil-fired plant), this section will describe the more widely used citric acid process for chemical cleaning a *pre-service boiler.*

Further information on the inspection and cleaning of plant can be found in Volume E.

11.2 Objectives

The objectives of chemical cleaning of the boiler circuits are:

- To commission and operate plant with the internal surfaces of the water/steam circuit as free as practicable from all foreign matter, such as adventitious metal, oxides, grease, oil and dirt.
- To establish a satisfactory basis for the protective magnetite film to grow on the plant internal surfaces.
- To remove debris and thick oxide accumulated during construction, and operation on older plants.
- To ensure that the chemical control requirements for subsequent operation are obtained with the minimum of delay and restriction of load.

As a result of increased standards of cleanliness during erection, only the furnaces, steam drum and boiler circulating pumps, along with their suction and discharge headers and economisers, are required to be chemically cleaned on the CEGB 500/660 MW units. The superheaters and reheaters are water-flushed and steam purged.

A typical layout of the chemical-cleaning circuit is shown in Fig 4.62. As can be seen on the diagram, all of the superheater section is water-wedged (i.e., kept full of water to prevent chemical ingress) by the spraywater system to prevent any access by chemicals. The chemical clean is effected by a system of temporary pipework and pumps. It should be observed that boiler circulating pumps A and D have their motor impeller units removed for this exercise to prevent acid damage to the impellers and that B and C pump bodies (which are used when the boiler is fired) are continuously purged with demineralised water to prevent any chemical access.

On natural circulation boilers, the chemical clean contractor will provide temporary pumps or use a gas-induced circulation system when the boiler is fired. In a gas-induced circulation system, a gas such as nitrogen is injected into the furnace bottom water-wall headers; this rapidly changes the water/acid solution density and promotes rapid circulation, which is not possible with natural circulation boilers without pumps. It is sometimes cheaper to use gas than temporary pumps.

11.3 Principles of citric acid chemical clean

The seven chemical clean stages are described as follows.

11.3.1 Stage 1 — initial flush

The system is flushed:

- To remove any contamination which has entered the plant during erection, including loose corrosion products.
- To remove any volatile corrosion inhibitors that have been used to protect the plant during storage and erection.

This is effected by filling and flushing the superheater sections into the boiler drum, and thence to waste, using the boiler feed pump and desuperheater spray lines. The boiler and economiser is initially filled to normal working level, using a boiler feed pump or the boiler filling pump, and then drained as quickly as possible through temporary connections on A and D circulating pumps and the blowdown ring main. This is followed by flushing the boiler and economiser to waste, using the boiler feed pump as before, until the effluent is free of suspended solids.

11.3.2 Stage 2 — alkaline boil-out

The purpose of the alkaline boil-out is to remove any grease and oil from the surfaces: these not only could lead to difficulties in operation but would certainly reduce the efficiency of the chemical cleaning process by preventing the acid solution from making contact with the surfaces.

In addition, the alkaline boil-out is intended to loosen and remove some of the metal oxides which will be present and thus enable the acid stage to be carried out more efficiently.

The two chemicals used in the boil-out are:

- Trisodium phosphate at 0.20% concentration
- Berol EMU 048 at 0.025% concentration

The first of these provides the necessary alkaline detergent action for detaching loose scale and emulsifying any oils and greases, thus making them dispersible in the solution. The second is described as a 'wetting agent' which reduces the surface tension of the liquid and thus allows better penetration into the material to be removed. The boil-out is effected by filling the boiler through the economiser, starting B and C boiler circulating pump or the temporary circulating pumps, and firing the boiler to a temperature of 50–60°C. The chemicals are then injected until the design concentra-

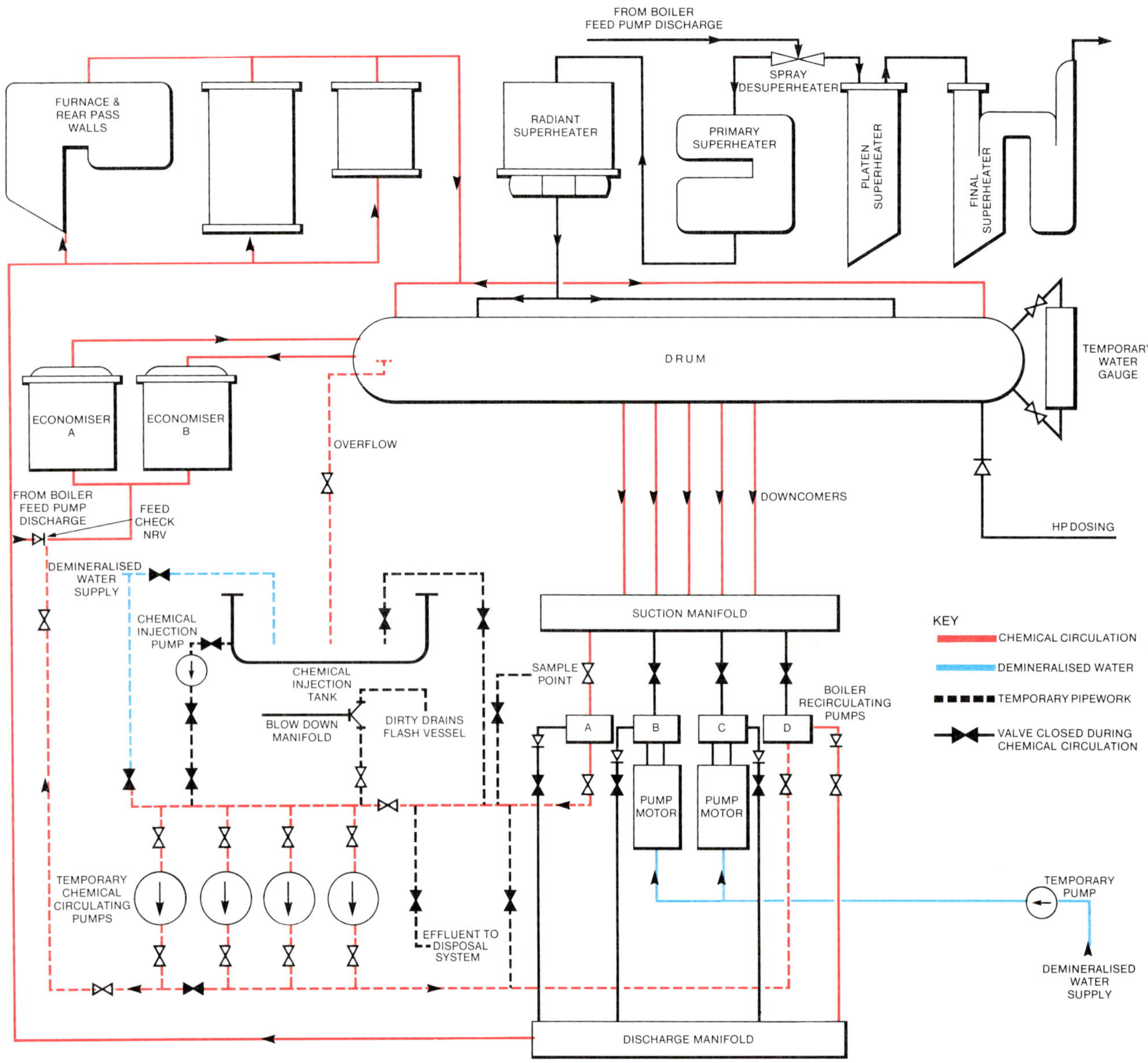

FIG. 4.62 Boiler chemical clean pipework layout

tions are reached: then the boiler is fired up to 40 bar and this pressure is maintained for 24 hours. Each blowdown valve is operated for 10 seconds in every hour for the 24-hour period. The pressure is then allowed to decay to 7 bar: the boiler is blown down from this pressure until empty.

11.3.3 Stage 3 — iron oxide removal

In order to remove the firmly attached scale and other debris which remain after the alkali boil-out, it is necessary to employ chemicals which are capable of dissolving the debris. The composition of the solution used is as follows:

- 3.0% citric acid
- 0.05% stannine LTP
- 0.5% ammonium bifluoride
- Ammonia, sufficient to bring the pH to 3.5–4.0

Note The solution overall must have a chloride-ion concentration of less than 2 PPM.

Some of the scale present is 'mill scale', which is a dense layer of magnetite produced on steel surfaces during rolling or fabrication at high temperatures. This mill scale is a complex mixture of iron oxides (mainly magnetite Fe_3O_4) which is subject to discontinuities. These discontinuities, being anodic to the surrounding mill scale, become corrosion sites in an aquatic medium. Therefore the acid solution is used to remove all trace of oxide from the tubes and leave a bare metal surface onto which can be 'grown' a controlled and continuous magnetite layer in the 'passivation' stage, which will protect the tubes from corrosion.

When citric acid was first used in chemical cleaning, it was employed by itself as a 3% solution of the free acid: it was observed that insoluble phases sometimes appeared which were identified as iron citrates. Investigation showed that the solubility of iron oxides in citric acid could be increased by the addition of ammonia, which led to the use of ammonia dihydrogen citrate which forms ferrous ammonium citrate, a complex salt of sufficiently high solubility to prevent precipitation in the cleaning solution. For this reason, and because the acid alone is very slow in dissolving mill scale, citric acid is now commonly ammoniated to a pH of 3.5 to 4.0 which corresponds to the formation of ammonium dihydrogen citrate.

The addition of fluoride (as ammonium bifluoride NH_4HF_2) to the solution has a pronounced accelerating effect on the rate of dissolution of mill scale by citric acid and, to a much lesser extent, on the rate of dissolution of siliceous material, thereby increasing the efficiency of the cleansing solution.

Stannine LTP is called an 'inhibitor' and is added to the acid cleaning solution to prevent as completely as possible the solvent action of the solution on the bare metal. This is achieved without significant reduction of the main function of dissolving the oxides of iron. These inhibitors are complex organic compounds whose protective mechanism is thought to involve an absorbed film on the surface of the bare metal, which thereby slows down the rate of attack by the acids on the metal.

The cleansing process is effected by filling the boiler through the economiser and water-wedging the superheater sections using the desuperheater sprays. The sealing water to B and C boiler circulating pumps is established. The boiler is then fired to 90–95°C with circulation from boiler circulating pumps B and C. When this uniform temperature is achieved, the boiler circulating pumps B and C are shut down and isolated, but the sealing water is left on. The above chemicals are then added and circulated, using the temporary pumps. Circulation is maintained through temporary pumps to give the solution a contact time of 8 hours. This should be sufficient time for the iron concentration to reach its plateau. The boiler is then drained to the effluent disposal system.

11.3.4 Stage 4 — citric acid rinse

The composition of the solution used in this stage is:

- 0.2% citric acid
- Ammonia to bring pH to 3.5–4.0

This stage is carried out by filling the boiler, as before, establishing circulation with the temporary circulation pumps and injecting the above chemicals. The solution is circulated for one hour and then drained to the effluent disposal system.

This stage serves merely to rinse out any residual iron by a solubilising effect, thus preventing the precipitation of iron hydroxides as the pH increases during the flushing process, and also to impart a small degree of rust-resistance to the steel surfaces. If the pH is >7 a precipitate of iron hydroxide may be formed. Consequently before the flushing stage which has pH >7 the system is rinsed with the above to remove any pockets or dregs of cleaning liquor remaining.

11.3.5 Stage 5 — flushing

The system is now flushed with demineralised water to waste until the conductivity of the water is 2000 μS/m, or less. This removes any residual acid and loose debris which remains undisolved.

11.3.6 Stages 6 and 7 — passivation

Passivation means simply the treatment applied to steel surfaces with the intention to produce a continuous coherent corrosion resistant layer of magnetite. This is done in two stages, interim passivation and final passivation. The purpose of the interim passivation is to establish a coating of magnetite to afford metal surfaces some protection during the period of boiler reinstatement before the final passivation stage.

The solutions used are:

- Stage 6 (interim): 100 PPM hydrazine, 50 PPM ammonia
- Stage 7 (final): 50 PPM hydrazine, 50 PPM ammonia

The interim passivation is carried out at 85–90°C for a period of 12 hours, with the temperature maintained by intermittent firing, if necessary.

If any firing is needed to maintain temperatures above 80°C, B and C boiler circulating pumps, or the temporary circulation pumps, must be started.

The final passivation is carried out at 40 bar for 24 hours with each blowdown valve being operated for 10 seconds every hour. After this time, the pressure is allowed to decay to 15 bar; then the boiler is blown down until empty. The boiler can then be inspected to ascertain the effectiveness of the clean.

11.4 Initial preparations

Before the chemical clean on the boiler can start, commissioning will have progressed as far as having oil burners available for lighting-up. This will require the pressure parts, including blowdown and drainage systems (some temporary drainage system for effluent will also be required), to be available. The gas and air systems and fuel oil systems will need to be completed to the stage where oil burner firing is available (this was completed in Section 10 of this chapter).

Auxiliary plant systems that are required are auxiliary boilers, effluent disposal, auxiliary cooling systems, boiler feed pumps and water treatment plant. Large quantities of demineralised water will be required and it will be essential to have the reserve feedwater tanks available *and full* prior to starting the chemical clean.

All temporary pipework required for the chemical clean will have been installed and pressure tested to 1.5 times the discharge pressure of the temporary acid circulating pumps. On forced-circulation boilers, at least two boiler circulating pumps will be available and the other boiler circulating pumps will have their motors removed and blanked off. On natural circulation boilers, temporary pumps or a gas-induced system will be installed by the contractor for circulation purposes.

The removal of the drum internals for cleaning is desirable to facilitate circulation of reagent, and to make inspection and the removal of debris easier, but their removal/replacement is time consuming. Programme considerations usually dictate that both pre- and post-commissioning cleaning is carried out with most of the drum internals in place.

A temporary overflow device and water level gauge is fitted to the drum. Figure 4.63 shows a typical overflow device: it is fitted because, in addition to water-plugging and backflushing the superheater, a constant-head overflow device must be used to ensure that chemical cleaning solutions do not enter the superheater. The overflow device may be fitted using a temporary drum door. It is not required to take full flow of the chemical circulation pumps and a 100 mm line should be adequate. Particular care is necessary not to exceed the working level when initially adding the acid to the water in the boiler. In this context, a temporary gauge glass extending the full height of the drum should be provided.

In the event of a stoppage of the acid circulation pump, the level in the drum may rise, due to water draining back from the tubes. This should be taken into consideration when determining the height of the overflow, which should be positioned 75–100 mm above the required working level.

The valve shown in Fig 4.63 should be open at all times, except when it is required to rinse the upper surfaces of the drum. Before such rinsing is carried out, it should be ensured that the solution contains less than 2 PPM chloride.

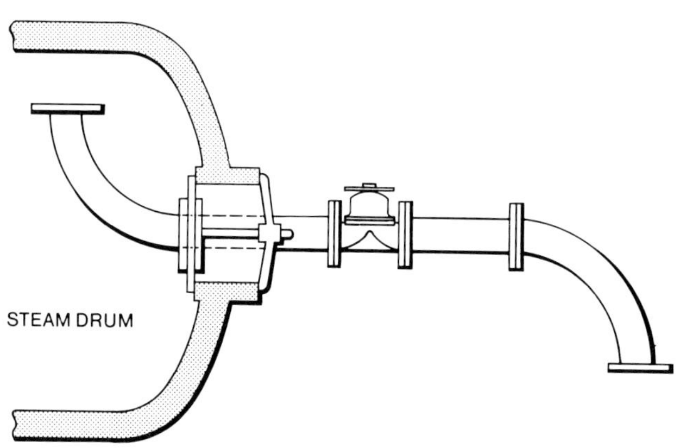

FIG. 4.63 Diagrammatic arrangement of constant-head overflow device fitted to boiler drum

Provision should be made for the overflow outlet from the boiler to be piped outside the building in a situation inaccessible to personnel and for any liquid overflowing to be led to a convenient sump, where it may be neutralised, if necessary. This may be achieved by leading the overflow outlet into a small tank from which the liquid can be drained and the vapour vented outside the buildings.

A temporary gauge glass is needed since the normal means of level indication cannot be used for several reasons:

- The temporary level during cleaning is higher than the maximum of the permanent indication.
- The cleaning chemicals cannot be rinsed out adequately from the permanent indicators without a risk of corrosion.
- The difficulty of reading the level on an otherwise out-of-service boiler by an operator stationed locally to the permanent glass.
- Increased use of Hydrastep water level gauges (i.e., no permanent gauge glasses).

Experience shows that the boiler drum becomes slightly pressurised during temperature raising and, with an open-ended temporary gauge glass, a false indication of drum level is obtained. Both links of the indicating column must be connected to the drum. It is, therefore, advised that permanent tapping points are provided for drum level indication during chemical cleaning. Each tapping point should be fitted with double-isolating valves and a blanked flange suitable for full boiler pressure. The lower tapping point should preferably be from a line other than the permanent gauge glass impulse line, for example, from the boiler dosing or sampling line. The top tapping point could be T-jointed into a drum air-release line before the normal air valve.

To check for the effectiveness of the chemical clean, it is often necessary to cut out sections of boiler tubing for inspection purposes. To avoid this and to reduce cost, a temporary rig containing sections of boiler tubes can be installed in the main circulation system by the chemical clean contractor when he is erecting his temporary pipework. If a temporary rig is to be used, it should be installed at this stage.

During the chemical clean, large quantities of effluent and rinse water have to be disposed of, and special arrangements have to be made for this purpose. For initial and final flushing, the normal boiler drainage systems can be used; these usually drain to the CW or station drainage systems and these areas will have to be ready to receive the drainage at the specified time. During the alkaline boil-out and citric acid stages, the effluent will not normally be allowed to be drained to CW or station drainage systems due to local river/water authority regulations, so arrangements will have to be made for the drainage of these chemicals and their removal. On the Ince *B* Power Station chemical clean, the boiler wash water tanks were used to contain the drained effluent and this was then removed from site by road tankers to licensed dumping facilities. Individual stations have to make their own arrangements for the effluent drainage disposal prior to the commencement of the chemical clean. Notification of the disposal of effluents will have to be made to the appropriate authorities under the Deposit of Poisonous Waste Act 1972.

During water-wedging of the superheater and reheater sections, it will be necessary to restrain the main steam pipe and superheater crossover-header pipework supports, in order to avoid the supports being overstressed due to the weight of water within the pipework. The *A* boiler circulating pump discharge non-return valve will be gagged open prior to the chemical clean to provide a positive flow circuit through the boiler circulating pump suction and discharge header with the temporary pumps.

A typical commissioning network diagram for the chemical clean of a 500 MW oil-fired boiler is shown in Fig 4.64.

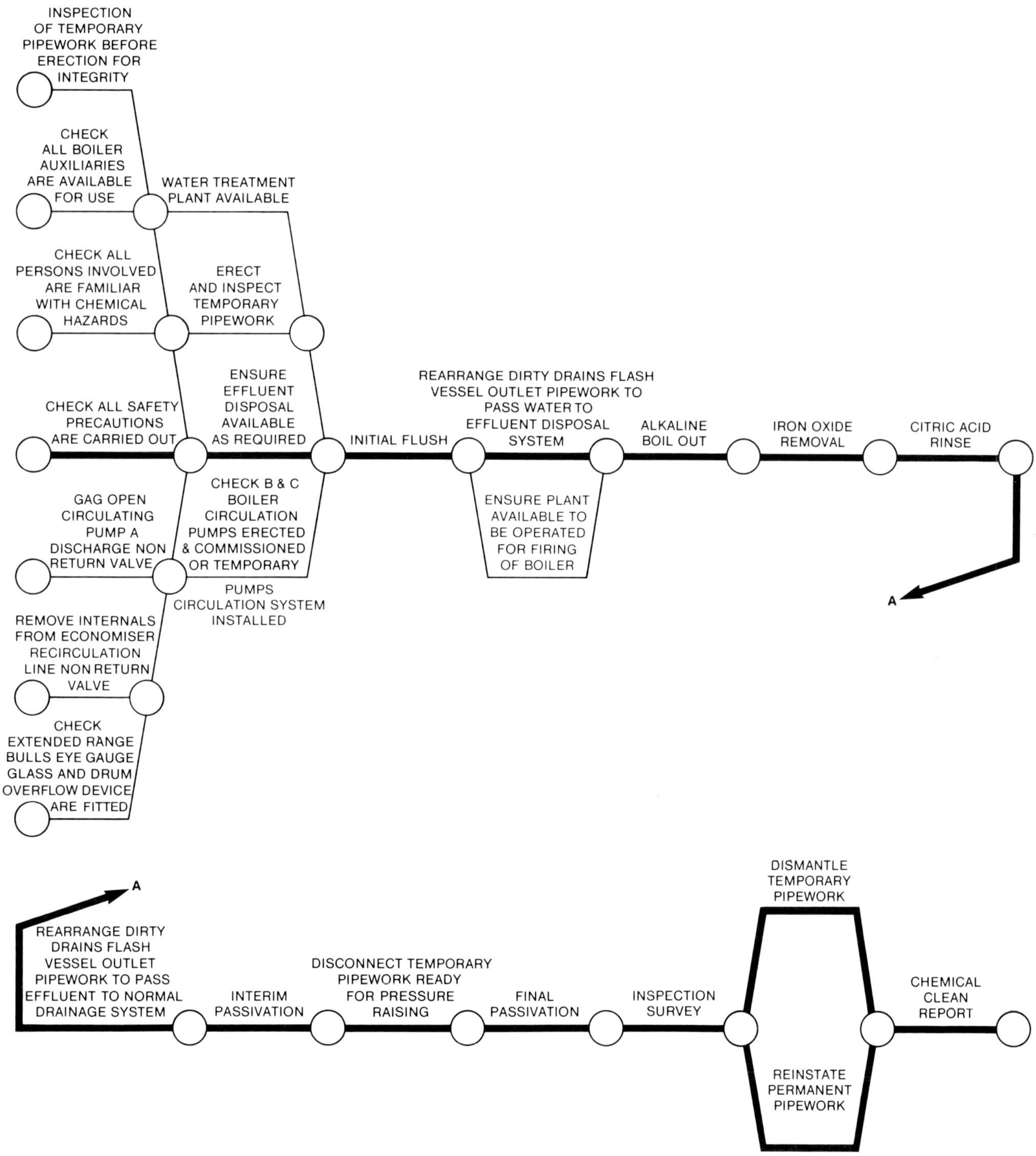

FIG. 4.64 500 MW boiler chemical clean commissioning network

11.5 Safety and resources

11.5.1 Safety

The following safety precautions and arrangements are required before commencing the chemical clean:

- The chemical storage and unloading areas must be cool and well ventilated, be roped off, have warning notices displayed and all chemicals clearly labelled. Facilities will be available for both bags and tankers.
- The chemical injection tank must be located outside the building, and protected from the weather.
- Chemical handling and injection areas must be clear of all unnecessary obstructions, roped off and have warning notices displayed.
- Notices detailing the actions to be taken in the event of spillage or emergency situations, and first aid treatment, must be prominently displayed in the chemical storage and mixing tank areas.
- Water shower, eye wash bottles and washing-down hoses must be provided in the chemical injection tank area.
- Adequate lighting must be available in the chemical injection tank area.
- Sufficient protective clothing, including goggles, gloves and resistant overalls, must be available.
- Medical Centre personnel must be informed of the nature of the chemicals to be used. The Medical Centre must be manned continuously for the duration of the acid stages.
- All valves, both permanent and temporary, must be clearly labelled. A valve position display board must be located in the CCR and be updated at the commencement of each stage of the clean procedure.
- One 9 kg dry powder portable fire extinguisher must be located in the area of the chemical circulating pump.
- 1000 kg of soda ash must be available in the injection area, together with shovels for spreading it, in case of acid spillage.
- Any electrical equipment likely to get splashed with acid must be covered with polythene sheeting.

The test team members must be familiar with the general precautions and standards of personal hygiene required. That is, whenever and wherever there is danger of contact with chemicals all personnel must:

- Be warned of the nature of the potential hazards and the necessary precautions and be instructed in the correct use of protective clothing and equipment by the supervising officer.
- Observe a high standard of personal hygiene, avoiding contact with chemicals, their vapours or contaminated residues.
- Not eat, drink or smoke.
- Avoid wiping nose, eyes or face, other than with clean paper tissues.
- Place all debris and chemical residues removed from the plant in clearly-labelled polythene bags for subsequent disposal.
- On leaving the plant, remove their protective clothing, which will be cleaned before future use.
- Clean themselves and their equipment using disposable materials.
- Wash thoroughly as soon as possible after leaving the operational area.

The following precautions must be taken during the chemical clean process:

- The chemical handling and injection areas must remain roped off with warning notices displayed. Additionally, whenever chemicals are being circulated, the working area around the boiler must be roped off and warning notices displayed at all levels up to and including the drum level. Personnel not normally involved with the cleaning process will be excluded from within the roped off areas. GDCD must make the necessary arrangements with the various erection contractors.
- All personnel within the barriers must have goggles and gloves for use in the more potentially hazardous zones.
- The citric acid injection and circulation stage should, if possible, take place outside normal working hours.
- The test team leader will ensure that a CEGB chemist is present in the injection tank area whenever chemicals are added to the cleaning circuit. He will also ensure that protective clothing is worn by all personnel involved in the addition of chemicals. The cleaning circuit must be hydraulically tested to 1.5 times the working pressure of the temporary circulating pump prior to the addition of chemicals, and any leaks repaired.
- The circuit must be checked whilst chemicals are being circulated and any leakage contained and, if possible, stopped. Areas contaminated with chemicals must be washed down thoroughly with water, taking care to protect electrical equipment, as necessary.
- Particular care must be exercised during the addition of inhibitor to ensure complete solution and mixing in the citric acid.
- Venting of the boiler during the process should, where practicable, be to the outside of the station

buildings, into a restricted area. Warning notices should be located at vent outlets.

Chemical hazards involved with the chemicals used in the cleaning process are as follows:

- Trisodium phosphate, supplied in powder form, can cause irritation of skin, nose and throat. Solutions are alkaline and can cause skin irritation.
- Berol EMU 048, supplied as liquid, can cause irritation of eyes and skin.
- Citric acid, supplied as a powder. The solid in contact with oxidising agents, such as hydrazine, may cause combustion, Solutions of citric acid can cause burns, particularly of the eyes.
- Stannine LTP inhibitor, supplied as a liquid, is an eye and skin irritant and is combustible.
- Ammonium bifluoride, supplied as a solid, can cause irritation of nose, eyes, mouth and lungs. Solutions cause burning of eyes and skin and are toxic.
- Ammonia, supplied in solution, can cause burns of eyes and skin. The vapour causes irritation of eyes, nose, throat and lungs. Particular care is required when opening drums of ammonia solution, since the contents may be under pressure.
- Hydrazine, supplied as a liquid, causes burning of eyes and skin. Vapour can cause irritation of eyes, nose and throat and internal damage.

Precautions to be taken for inspection of plant during or after the cleaning process:

- The test team leader will ensure that entry into plant for inspection is restricted to the minimum of personnel, and clean conditions working is observed. The inspection team will normally consist of representatives of the chemical cleaning contractor, GDCD, the Region and the boiler supply contractor. Permits for Work will be issued prior to entry.
- A penthouse must be erected at one end of the drum, with restricted access and an internal barrier to indicate the zone beyond which inspection clothing must be worn. Clean conditions clothing, consisting of head covering, cotton coverall and cotton overshoes must be available in the penthouse, together with a disposal container for dirty clothing. There must be telephone or radio contact between penthouse and CCR.
- There must be a minimum of three hours between opening up plant and inspection being carried out. If bodily entry is required, such as into the boiler drum, the appropriate plant will be ventilated at a rate of at least 10 air changes per hour. The temperature must be less than 30°C before bodily entry. If these conditions are not satisfied, then a full-face positive pressure hood will be worn by personnel entering.
- When personnel have gained access to the inside of plant, there will be standby personnel positioned at the point of entry.
- Personnel entering the plant must wear coveralls, gloves and goggles at all times.
- No possessions, other than those required to carry out inspection, will be taken inside the plant. Any implements such as torches, mirrors, spanners, etc., must be strapped or tied securely. All items taken in will be monitored and recorded.
- Removal of debris or chemical residues must be carried out so as to avoid contamination of personnel or external plant. The debris will be scooped into impervious containers and sealed.
- The conditions specified in the CEGB Safety Rules must be observed at all times and PFW used to control safety for all access to boiler pressure parts equipment.

The following first aid measures are applicable to all chemicals in use during the cleaning:

- Splashes in the eye — immediately flood the eye with water, keeping eyelids apart. Irrigate for 10 minutes. Refer to Medical Centre.
- Irritation of the skin — in the event of splashing, wash thoroughly with water, taking care to avoid spreading of contamination into eyes. Signs of skin irritation, even in the absence of splashing, should be referred to the Medical Centre.
- Gassing — remove casualty to fresh air, remove contaminated clothing. Refer to Medical Centre.

11.5.2 Resources

The following resources will be required.

Demineralised water

Large quantities of demineralised water will be necessary. Typical capacities of a 500 MW oil-fired boiler are:

- Capacity of the boiler, including steam drums, downcomers, circulating pumps/pipework, furnace walls and rear pass-wall tubes and headers to NWL — 180 tonnes
- Economiser — 77 tonnes
- Superheater section up to boiler stop valves — 187 tonnes

Chemicals

The chemicals required for a 500 MW oil-fired boiler chemical clean are:

- Stage 2: alkaline boil-out

0.2% w/w trisodium phosphate	10.5 × 50	kilograms
0.25% w/w Berol EMU 048	3 × 25	litres

- Stage 3: removal of iron oxides

3.0% w/w citric acid	157 × 50	kilograms
0.05% w/w stannine LTP	6 × 25	litres
0.5% w/w ammonium bifluoride	26 × 50	kilograms
Ammonia to pH 3.5–4.0	12.5 × 205	litres (32% solution)
Neutralise with ammonia	19.5 × 205	litres (32% solution)

- Stage 4: citric rinse

0.2% w/w citric acid	10.5 × 50	kilograms
Ammonia to pH 3.5–4.0	1 × 205	litres (32% solution)

- Stage 6: interim passivation

100 PPM hydrazine	8 × 25	litres (15% solution)
50 PPM ammonia	2 × 25	litres (32% solution)

- Stage 7: final passivation, magnetite formation

50 PPM hydrazine	4 × 25	litres (15% solution)
50 PPM ammonia	2 × 25	litres (32% solution)
Make-up water dosed to 50 PPM hydrazine	4 × 25	litres (15% solution)

Clothing

The protective clothing required for a 500 MW oil-fired boiler chemical clean is:

- Goggles
- Coveralls
- Acid-resistant overalls
- Gloves
- Breathing apparatus

11.5.3 Staffing

A team of approximately six people will require up to seven days to complete the chemical clean. Initial preparations will take approximately one week to complete.

11.6 Chemical clean procedure

11.6.1 Stage 1 — initial flush

The superheater and main steam pipework support-restraints will be fitted and the desuperheater spraywater lines and nozzles are flushed, using a main boiler feed pump (BFP).

The temporary connections to A and D boiler circulating pumps should then be put into circuit and the superheater sections filled and flushed into the boiler drum and thence to waste, using the BFP and desuperheater spray lines, until the effluent is free of suspended materials. The conductivity of water from the superheater should be less than 500 μS/m and the chloride ion concentration less than 1 PPM.

If flushing operations are carried out well in advance of first firing, the superheater should be left water-wedged, but if the boil-out is imminent, the radiant and primary superheater headers should be drained down in readiness and the restraints removed from the superheater crossover and main steam pipe supports. The superheater pendants will remain full, as they cannot be drained. The boiler and economiser is filled with demineralised water to normal working level, using the boiler feed pump, and a complete hydraulic test of the temporary pipework completed, first with a static head pressure on the temporary chemical cleaning pipework, and then at 1.5 times the pressure at which the process will be carried out.

The boiler and economiser, including downcomers, should then be drained as rapidly as possible to remove heavy debris, using the temporary connections on A and D boiler circulating pumps and the blowdown ring main.

Using the main BFP, the boiler is flushed through the economiser and simultaneously drained to waste through the temporary connections until the effluent is free of suspended material.

The temporary chemical cleaning pipework should then be isolated.

11.6.2 Stage 2 — alkaline boil-out of boiler and economiser

Fill the boiler with demineralised water through the economiser to NWL, using the main BFP(s), and water-wedge the superheater using the spraywater lines. Put sealing water on to B and C boiler circulating pumps and check the prime of the pumps at vents underneath the suction and discharge valves.

Start B and C boiler circulating pumps and then fire the boiler, raising the temperature to 50–60°C. Stop firing at 60°C. Observe carefully the boiler expansion as the temperature increases. The temporary injection circulating pump(s) should then be started and the injection system commissioned. The chemicals are then injected to give the following concentrations:

- 0.2% w/w trisodium phosphate (TSP)
- 0.25% w/w Berol EMU 048

The temporary circulating pump(s) are then stopped and the temporary low pressure pipework connection from A and D boiler circulating pumps is removed. Firing should then continue to raise the boiler pressure to 40 bar.

The pressure should be maintained at 27–40 bar and boil-out continued for 24 hours. After 2 hours at pressure, commence a regular blowdown routine so that each blowdown valve is operated for 10 seconds in every hour. Maintain chemical concentration above 1000 PPM TSP using HP dosing equipment.

After 24 hours, the firing is stopped and the pressure allowed to decay to 5–7 bar, when the boiler is blown down until empty.

Allow the boiler to cool, open up the steam drum and remove debris as necessary. Identify and photograph several areas inside the drum for comparison with the final inspection stage. Make final chemical cleaning connections to the boiler and economiser.

11.6.3 Stage 3 — mill scale and iron oxide removal

Before this stage commences, the economiser feed check non-return valve should be jacked shut to prevent chemicals entering the feed lines. The feed line is then water-wedged.

Fill and prove the temporary pipework with demineralised water. Fill the boiler with demineralised water through the economiser, using the temporary pump(s) to normal working level. Prove associated circuits.

Refit the restraints on the superheater crossover and main steam pipe supports.

Water-wedge the superheater section by backfilling through the desuperheater sprays, using a BFP, until a rise in drum level is seen.

Check that B and C boiler circulating pumps are primed using the temporary pump purging system, and leave the system flushing.

Establish circulation of boiler and economiser, using B and C boiler circulating pumps and temporary pump(s). Adjust the drum level as necessary and continue circulation. Fire the boiler and raise the temperature to 95°C, then stop firing and box-in the boiler. Continue circulation to stabilise temperature at 95°C. Isolate B and C boiler circulating pumps and ensure that the sealing water to these pumps is satisfactorily established.

Backfill superheater sections through the desuperheater sprays, using a BFP, until a rise in drum level is seen.

Adjust the drum level to accommodate chemicals by draining to waste. Inject chemicals to give the following concentrations:

- 3% w/w citric acid
- 0.05% w/w stannine LTP
- 0.5% w/w ammonium bifluoride
- Adjust pH to 3.5–4 with ammonia

Backfill the superheater sections through the desuperheater sprays until a rise in drum level is witnessed. Continue backfilling, and raise the drum level to the point of the temporary drum door fixed-overflow device.

Circulate the chemical solution until the iron concentration stabilises, or to give a contact time not greater than 8 hours. (The free citric acid concentration must be maintained at greater than 0.5% during this time.)

Continue circulation and raise the pH to 5 by controlled ammonia injection, then stop circulation and drain the boiler and economiser to waste.

11.6.4 Stage 4 — citric acid rinse

Fill the boiler through the economiser, using the temporary pump(s), with a solution containing 0.2% w/w citric acid. Establish circulation and inject ammonia to give a pH of 3.5–4 and circulate for one hour, then drain the boiler and economiser to waste. Backflush the superheater sections for 5–10 minutes.

11.6.5 Stage 5 — flushing

Fill and flush the economiser with demineralised water using temporary pump(s) into drum and thence to waste using the downcomers, until the conductivity of the effluent is 2000 μS/m or less. Leave the economiser full.

Flush the suction and delivery pipework of each boiler circulating pump, in turn, from the suction manifold through to the discharge manifold and thence to waste until the conductivity of effluent is 2000 μS/m, or less.

Fill and flush the furnace walls with demineralised water using the temporary pump(s), discharging up the walls and into the drum, and thence to waste using the downcomers, until the conductivity of the effluent is 2000 μS/m, or less. Leave the system full.

Fill the boiler and economiser to normal working level, then establish circulation of the boiler through the economiser, furnace rear and front walls, and drum (evaporative sections), using B and C boiler circulating pumps. Circulate and check that the conductivity is 2000 μS/m, or less.

If the conductivity is greater than 2000 μS/m, drain the boiler evaporative sections and refill to normal working level.

Re-establish circulation using temporary pumps, then backfill the superheater sections using the desuperheater sprays until a rise in drum level is seen. Leave the boiler full to normal working level.

11.6.6 Stage 6 — interim passivation

Continue circulation and inject chemicals, giving the following concentrations:

- 100 PPM hydrazine
- 50 PPM ammonia

Fire the boiler and raise the temperature to 85–90°C and then stop B and C boiler circulating pumps.

Note If any further firing is required, start B and C boiler circulating pumps.

Backfill the superheater sections using the desuperheater sprays until a rise in drum level is witnessed. Continue backfilling and raise the drum level to that of the temporary drum door overflow device.

Continue circulation and maintain the temperature at 85–90°C by low or intermittent firing, for 12 hours; then stop firing, stop the temporary pumps and drain the system to waste hot.

Stop the temporary purging pump and isolate the system. Prepare a PFW for the temporary connections from boiler and economiser to be removed. Open steam drum doors and allow sections to dry hot.

Remove the restraints from the superheater crossover and main steam pipe supports.

Reinstate pressure parts in preparation for final boil-out at 40 bar. Dose feedwater in the de-aerator with 50 PPM hydrazine and 50 PPM ammonia, using the LP dosing equipment to the feed pump suctions and circulating the de-aerator contents by the boiler feed pumps with the leak-off valves open and the BFP discharge valves closed.

11.6.7 Stage 7 — final magnetite formation boil-out

Fill the boiler through the economiser, using a main BFP, to normal working level with the demineralised water containing 50 PPM hydrazine and 50 PPM ammonia, then circulate the boiler using B and C boiler circulating pumps. Fire the boiler and raise the pressure to 40 bar. Maintain pressure between 27–40 bar and boil out for 24 hours; ensure the blowdown valves are operated for 10 seconds in every hour.

Note All water supplemented to the boiler during this period should contain 50 PPM hydrazine and 50 PPM ammonia. This concentration will be maintained as above, with the LP chemical-dosing system.

Experience has shown that, during the final passivation stage, it is extremely difficult to maintain the hydrazine concentration at 50 PPM for the 24 hour period. It is usual practice to accept a small positive residual amount of hydrazine in the water.

After 24 hours, stop firing and allow the boiler pressure to decay to 15 bar, then blow down the boiler until empty. Allow the boiler to cool and open up for inspection. When inspecting the boiler after the chemical clean, the following points should be observed to check the effectiveness of the chemical clean:

- The surface condition of headers/drums/pipework should be free of any 'pitting' from acid attack. (This proves that the inhibitor has been successful.)
- The coating left after the final passivation should be an even layer of black/grey iron oxide.
- Any debris left behind by the cleaning process in the drum and headers should be removed.
- If a rig with sample tubes has not been used to assess tube cleaning, then it may be necessary to cut a sample of tube from the various boiler sections to check for a suitable iron oxide coating with no acid 'pitting' attack.

11.6.8 Boiler storage following chemical clean

Following the completion of the chemical clean and satisfactory inspection, the temporary pipework will be removed and the boiler reinstated for normal operation. The storage of the boiler is of paramount importance and it must be ensured that the integrity of the protective iron oxide layer is maintained for future operational well-being. Past experience has shown that, following a chemical clean, some valves which have been in the circuit may have to be changed or overhauled.

Any test samples of boiler tube removed must be replaced and the welds tested by radiography.

12 Steam purge

12.1 Introduction

It is inevitable that, in spite of high standards of erection cleanliness and chemical cleaning, some debris will remain in the boiler tubes and HP pipework. The chemical clean will have disturbed and removed some debris: the steam purge completes the task of providing a completely clean system for passing steam to the main turbine and boiler feed pump turbine. Particulate matter, if left in the system, would give rise to increased turbine blade erosion and a relatively rapid deterioration in turbine efficiency.

Steam purging requires the use of extensive temporary pipework to allow the discharging steam to be blown safely to atmosphere through a silencer: this involves a complex theoretical treatment of the design parameters of the plant concerned.

Generation Design Memorandum No. 068/25 — Steam Purging — details the requirements for steam purging and provides a basis for the process design of a suitable system.

This section covers the objectives and actions needed to carry out a steam purge on a 500 MW reheat oil-fired boiler operating at 164 bar and 541°C.

12.2 Objectives

- Increasing the steam flow to the turbine increases the disturbance effect (or drag) on any particulate matter: this reaches a maximum under MCR/overload conditions. Steam purging (as part of the commissioning) aims to create a disturbance of the order of 60% greater than that likely to be achieved in normal service, so that any debris remaining in the system will not be disturbed by normal operation.

- Assessment of a satisfactory steam purge programme uses two indications, (a) from indentations made in alloy target plates inserted in the purge ventpipe, and (b) from calculations of the disturbance factor.
  - (a) Alloy target plates are used for the final purge assessment as the material content of the plates is similar to that of the turbine blades. Because of their cost, they are only used when the system is clean after completing a number of purges, mild steel target plates being used initially.
  - (b) *The disturbance factor* is the ratio of the drag force under purging conditions to that under normal conditions, where drag is a function of the density and velocity of the steam at the point of measurement. A disturbance factor of between 1.3 and 1.6 is generally accepted as indicating that a satisfactory purge has taken place. Disturbance factors achieved when a 'target plate' has been inserted are generally lower because of the added resistance to flow of the target plate. Disturbance factor determination is covered later.

12.3 Principles

Steam purging operations are normally carried out in three stages:

- *Superheater steam purge* The superheater and main steam pipes are blown through to the turbine main steam-chests, and from there to atmosphere using temporary pipework and a silencer. The turbine steam valve internals and headgear are not fitted and the valve covers are replaced by temporary blanks. A typical pipework arrangement is shown in Fig 4.65.
- *Superheater and reheater steam purge* After the superheater purge is complete, the temporary pipework is rearranged so that the turbine main steam-chest discharge pipework is replaced with temporary connections to the cold reheat steam pipes, discharging to atmosphere using the turbine reheat intercept steam-chests, temporary pipework and a silencer. Thus, in this stage, the steam flows in turn from the drum through the superheater, main steam pipes, cold reheat pipes, reheater, and hot reheat pipes. Figure 4.66 shows a typical arrangement. Again, turbine valve internals are removed and temporary blanks fitted.
- *Main boiler feed pump and turbine gland-steam system purge* The arrangements for the steam-driven feed pump turbine and turbine gland-steam system blows depend upon the system design, i.e., whether or not the feed pump turbine has a LIVE steam supply direct from the boiler. This steam purge has the same objectives as with the superheater and reheater — to remove particulate matter that would damage turbine blading. It may be carried out at a later date than the superheater and reheater purges, dependent upon the plant erection programme.

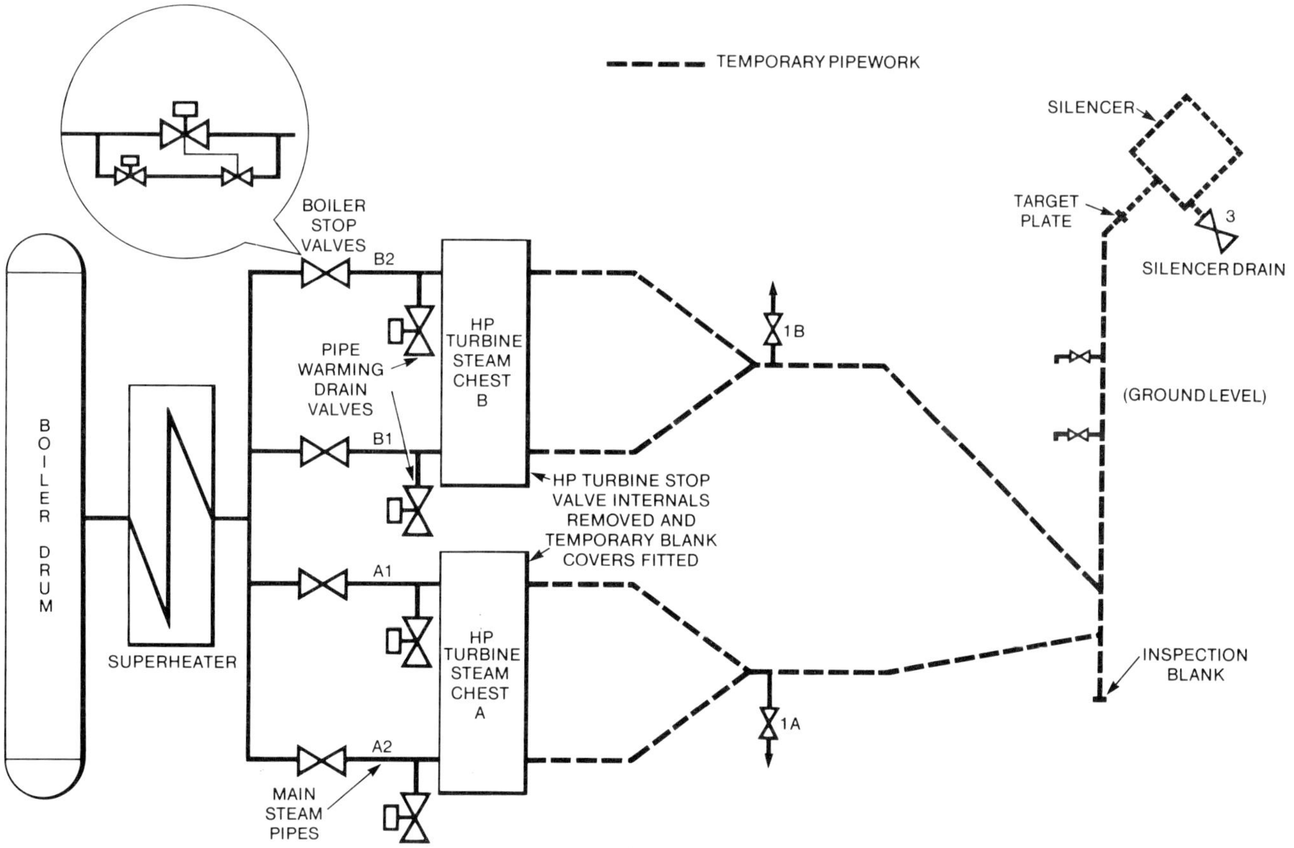

FIG. 4.65 Typical pipework layout for superheater steam purge

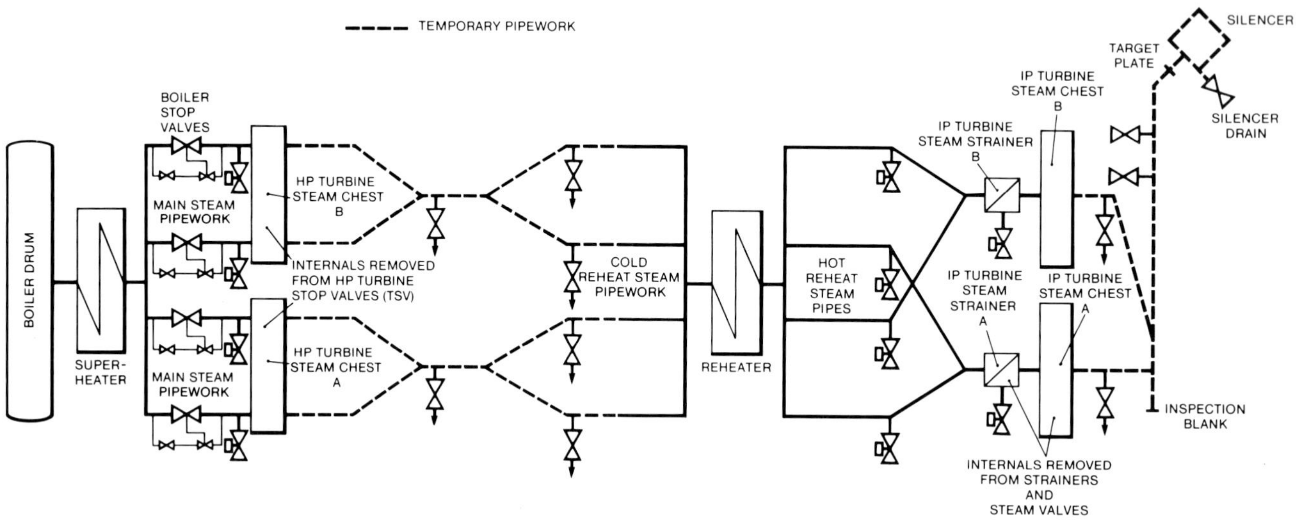

FIG. 4.66 Typical pipework layout for superheater and reheater steam purge

Temporary pipework will again be erected to purge the steam leads to atmosphere. The temporary pipework will depend upon the design of the system to be blown.

12.3.1 Disturbance factor determination

To enable the disturbance factor (DF) to be calculated at strategic points in the system, pressure and temperature measurements have to be made: suitable instrumentation is needed for this. Figure 4.67 shows a typical steam path with location references used for DF calculations. The diagram also shows a suitable position for fitting target plates.

The (DF) at any point in the system is calculated according to the following formula:

$$DF = \frac{q^2v \text{ during purge}}{q^2v \text{ MCR}}$$

where q = mass flow, kg/s

v = specific volume, dm^3/kg

The drag force q^2v at turbine MCR conditions is easily determined from design data for the various points under consideration. The drag force q^2v during the purge should be calculated at 10-second intervals. To do this, the use of minicomputers linked to a data

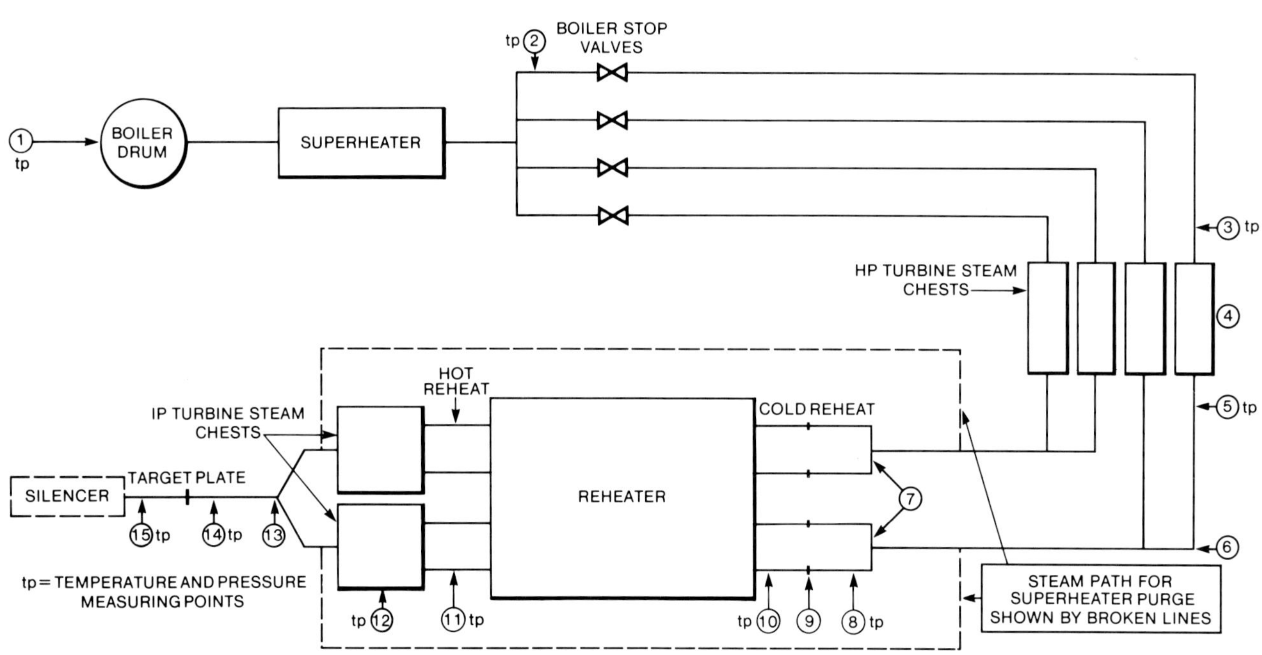

FIG. 4.67 Steam path and location references used in calculations

logger should be considered. The use of computers ensures instant results and saves time between purges, compared with manual calculations of purge flow drag forces.

The commercial and temporary instrumentation readings are fed into the computer and data logger: the computer is programmed to offset the response time of the thermocouple pockets by pairing temperature readings with pressure readings taken 10 seconds previously. All thermocouple pockets should be of the CEGB standard quick-response type.

The methods of calculating mass flow are given in GDM 068/25 but, due to errors that can take place in flow measurement, in the effects of resistance to flow, in pressure measurement errors due to high velocity heads and the possibility of the silencer causing sub-sonic flow at the exit pipe, more than one method of calculating flow should be considered.

Specific volume v is determined from steam tables, using the pressure and temperature at each point under consideration.

The method set out in GDM 068/25 is now available in the form of a computer programe, which can be run at the CEGB Computing and Information Systems Department to give the conditions necessary for a specified DF at any nominated stage in the system.

Typical areas where disturbance factors should be measured are shown in Fig 4.67 and identified as follows:

1. Boiler drum
2. Superheater outlet
3. Turbine inlet (before ESVs)
4. HP cylinder exhaust (after any bled-steam tapping)
5. Reheater inlet
6. Reheater outlet
7. IP cylinder inlet (before interceptor valves)
8. Temporary pipework before target plate
9. Temporary pipework after target plate

The graphs in Figs 4.68 and 4.69 show the predicted DFs at principal points on the system plotted against drum pressure for unit 6 superheater and reheater steam purges at Ince *B* Power Station.

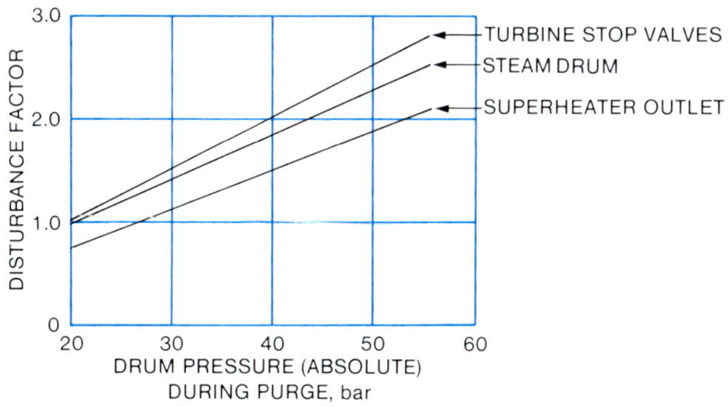

FIG. 4.68 Superheater steam purge. Ince *B* Power Station graph showing predicted disturbance factor at principal points, plotted against drum pressure

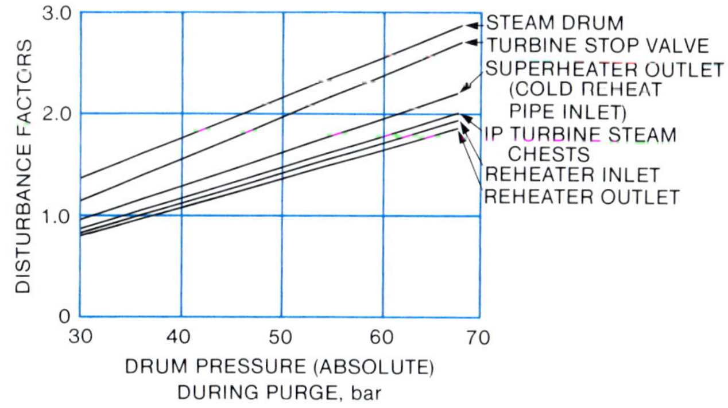

FIG. 4.69 Reheater steam purge. Ince *B* Power Station graph showing predicted disturbance factor at principal points, plotted against drum pressure

12.4 Initial preparations

The temporary pipework as described in Section 12.3 will have been designed and erected in accordance with Design Memorandum 068/25 — Steam Purging and will be the responsibility of the GDCD design department. The calculated conditions expected to be achieved will have been completed and a purging procedure developed.

In new stations, this procedure, which is the responsibility of GDCD, the Station Manager and the boiler manufacturer, should be developed in a series of trials starting at low operating conditions and working up to the calculated conditions. The procedure should be such that the boiler drum is not subjected to heavily sub-cooled water after each purge is completed. The time of opening of the boiler stop valves will have to be decided and it may be necessary to alter valve headgear to achieve the recommended times. Valve opening times in excess of one minute may be required for satisfactory steam purges. With boiler stop valves having short opening times and with flow rates approaching MCR, problems with control of boiler water levels will be experienced. It is essential to have good water level indication and, where possible, the maximum range of water level indication should be arranged. The double gauge glass, described previously, is normally acceptable. Temperature differentials of steam drums, boiler headers, permanent steam pipes, turbine steam-chests, and temporary pipework will apply and must not be exceeded. Where the purge path is through sections of steam-chests, special arrangements may have to be provided for heating steam or a special warming procedure adopted, the latter being more likely.

The necessary instrumentation to monitor the temperature differentials of the above equipment will have to be installed and commissioned prior to starting purging operations.

Mild steel and alloy target plates, suitably machined and identified, should be available for fitting.

To avoid environmental nuisance, an exhaust silencer will be necessary and this should be such that it does *not* alter the pressure at the end of the vent pipe.

The pressure and temperature instrumentation required for disturbance factor calculations shown in Fig 4.67 should be installed and the minicomputer programmed and connected, as required.

Where temperatures and pressures on the system are likely to exceed design limitations, permanent display instrumentation should be installed. Thermocouples should be installed in pockets rather than from pipework material, where possible, as response times are quicker.

A publicity campaign, warning the general public that a steam purge is to be carried out, should be made. This is particularly necessary for people living near to the site.

The steam purge generally takes from two to three weeks to complete and operations of this type raise the general noise level of the station. Purging operations should be restricted to 0800–1900 on each of six days, with no activities on Sundays. This should be made clear in the publicity campaign.

12.5 Safety and resources

12.5.1 Safety

The following safety points should be considered:

- All personnel not involved in the steam purge operations should be moved away from the area, particularly from areas of the temporary pipework and steam exhaust. Much of the temporary pipework will be unlagged and therefore barriers should be erected to prevent personnel access.
- An adequate communications system should be set up to warn personnel when each blow is about to commence, and its completion. It must be ensured that all personnel involved in the test can hear the warning siren from their observation positions.
- Personnel should be stationed at strategic points of the pipework flow path. They should be in a position to see and hear the signalling system and then prevent any encroachment from personnel not involved in the steam-blow operations.
- All personnel in the vicinity of the boiler and steam pipework must wear ear defenders. The Control of Pollution Act 1974 with respect to noise must be observed.
- When fitting target plates, the steam supply will be isolated by locking closed the boiler stop valves and by-passes. A Permit for Work will be issued.
- Before commencing the steam purge, all personnel involved must be familiar with the type of signalling system used and an emergency procedure must be set up and fully understood to deal with any plant emergencies that may occur.

Typical emergency procedures to be considered are:

- Procedures to deal with any failure of temporary or permanent pipework (reported by observers during each purge using the communications system).
- Boiler stop valves fail to shut.
- Boiler stop valves fail to open.
- Feed regulating valves fail to open.
- Feed regulating valves fail to close.
- Drum, header or temporary pipework temperature differentials going outside limits.
- Drum level high/low during purge.
- Reports of any dangerous plant conditions.
- Boiler or HP pipework going outside expansion limits.

12.5.2 Resources

The following resources are required:

- All emergency procedures to be available for use.
- Adequate sets of ear defenders available.
- Communication warning system commissioned and tested.
- Auxiliary boilers and steam systems available.
- Temporary barriers to prevent encroachment in areas of danger.
- Warning notices for use with the barriers.
- Scaffold platforms.
- Stopwatch to time the length of the purges.
- Target plates machined and ready for use.
- Machining facilities for cleaning up used target plates.
- Data logger and computer programmed for use.
- Boiler available for pressure raising, complete with chemical-dosing and feed systems.
- Hydrazine and ammonia for dosing the feed water.
- Adequate supplies of demineralised water. Large quantities are required. Typical consumption rates for the Ince *B* 500 MW oil-fired boiler steam purge were as follows.

  Each reheater purge needs approximately 40 t. Also, between purges the boiler stop valve by-passes normally remain open to keep the pipework and steam chests warm or, if this is not required, the boiler is kept firing with steam being dumped to drain, using the steam dump system. This amounts to 6 kg/s (22 t/h).

To maintain the boiler at pressure with one oil burner in service, the total boiler drainage including stop valve by-passes will be approximately 10 kg/s (36 t/h).

Therefore, assuming a 12 hour day with one purge per hour, total consumption would be:

$$12 \times 40 = 480 \text{ t/day}$$
$$12 \times 36 = 432 \text{ t/day}$$
$$\text{Total} = 912 \text{ t/day}$$

The water should be dosed with ammonia and hydrazine and the concentrations should be maintained to 50 PPM and 15 PPM, respectively, by the use of the HP and LP chemical-dosing systems.

The water should not contain more than 0.1 PPM of chloride ions or more than 2 PPM of sodium ions.

- Adequate supplies of fuel oil. Typical consumption rates for the Ince *B* 500 MW oil-fired boiler steam purge were:

Average of $1\frac{1}{2}$ burners, firing for 14 hours	= 50 t/day
Auxiliary boilers to provide steam services for fuel oil heating and steam air heating	= 30 t/day
Total	= 80 t/day

12.5.3 Staffing

The staffing required will depend upon the type of system, and the lengths of temporary pipework needing observers. Approximately ten people will be required, including an experienced unit operator and assistant. A fitter is needed to fit and remove target plates.

The time taken to complete the purging operations will again depend upon how clean the boiler is, which dictates the number of purges. Between two and three weeks should be allowed for the purging operations.

12.6 Steam purge procedure

12.6.1 Superheater steam purge

Prior to boiler pressure raising for the first purge, a number of checks is required to ensure that all aspects of the procedure have been taken into account. In particular, it will be necessary to ensure that the pre-purge public relations exercise has been completed, and that a procedure for dealing with public complaints during the purging has been made available. In terms of safety, the temporary pipework system must be inspected for mechanical integrity; safety of personnel and plant during purging will require thorough reviewing at this stage. Operationally, the signalling systems and boiler stop valves must be proved, the former every day.

An SRCC will be issued for the steam purge system.

The temporary pipework is arranged to allow the superheater and main steam-pipework to be purged to atmosphere; no target plate will be fitted at this stage. Discharge of debris over the site must be considered and steps taken to ensure that any buildings or equipment in the steam exit-area are protected.

The boiler pressure will be raised to 27 bar and the boiler stop valve by-passes opened to pre-warm the HP pipework and turbine steam-chests. If conditions are acceptable after pre-warming, then purging can commence.

All access into the restricted zones must be prevented by locked doors and manned barriers. Audible warning to clear the restricted zones will also be given.

The signalling system is used to verify that all conditions are satisfactory at the manned stations, the central control room, in turn, informing the manned stations that purging is about to commence.

An open signal will be initiated on all four boiler stop valves, in order that all four main steam pipes are purged simultaneously. The valves should be left open for approximately 20 seconds before all four valves are simultaneously shut. The boiler stop valves will not have gone fully-open in this time, but the intention should be to work up from partly- to fully-open valves, and purge periods of at least one minute.

Once all four boiler stop valves are verified shut, the signalling system is used to give the 'all clear'.

The first series of purges is intended to prove the temporary pipework and supports, to allow the purging technique to be developed, the disturbance factor instrumentation and safety procedures to be proved, and the 'system' to be brought up to temperature.

Each purge commences with the drum at a low water level in order to avoid carryover on opening the boiler stop valves. The drum level will rise as the boiler pressure falls and will collapse when the boiler stop valves are shut, so the intention should be to maintain the drum level in view, not higher than the top bull's-eye. An unacceptably low level on shutting off the stop valves can be avoided by commencing to feed the boiler as soon as the drum level indicates a maximum and maintaining the feed-rate to match the steam flow rate. Particular attention should be paid to drum differentials throughout.

To minimise the time between purges, oil burners can be kept in service continuously, the amount being dependent upon the type of plant.

Once the purging technique has been proved and satisfactory disturbance factors obtained, further purges are carried out with increasing drum pressure and increasing boiler stop valve opening-times until approximately four purges have been completed with an achieved disturbance factor of 1.6 and a maximum boiler stop valve opening time of approximately 1 minute.

The boiler pressure will now be held at the pressure giving the satisfactory disturbance factor, the boiler stop valves will be isolated and a PFW issued to fit a mild-steel target plate. The Permit for Work will then be cancelled and a further purge with a disturbance factor of 1.6 and maximum boiler stop valve opening-time (1 minute) carried out. The target plate is then removed and inspected. The condition of the target plate will determine whether further purges are required without target plates, or whether the system is clean enough for an alloy target plate to be fitted.

When a mild steel target plate is deemed to be acceptable by the target plate inspection party, then the senior commissioning engineer should be informed by the test team leader. Purging will then recommence, using an alloy target plate. The number of purges to get to this point will depend upon how clean the boiler and HP pipework was. At Ince *B* Power Station on unit 6, 25 purges were required (see Fig 4.70).

Once an alloy target plate has been accepted as satisfactory by the inspection party, which is normally the GDCD Site Manager and the Station Manager, then the boiler can be isolated and a PFW issued for the temporary pipework to be changed for the next stage of steam purging (reheater purge).

The final target plate should be photographed for plant history records, before being released for restoration to the 'as new' condition.

PURGE NUMBER	S3	S4	S5	S6	S7	S9	S10	S11	S12	S13	S14
DRUM PRESSURE, bar	29.0	29.3	41.8	40.9	52.7	58.1	59.3	65.3	65.2	64.4	62.6
S/H OUTLET PRESSURE, bar	15.0	15.7	23.1	22.4	29.5	31.6	32.6	37.8	37.6	37.5	36.3
S/H OUTLET TEMPERATURE, °C	278	289	325	334	406	344	346	356	377	341	303
TSV INLET PRESSURE, bar	9.1	9.5	14.0	13.8	17.5	19.4	20.1	23.3	23.2	23.4	22.8
TSV INLET TEMPERATURE, °C	239	251	283	298	350	308	306	326	341	318	292
10in PIPE ENTRY PRESSURE, bar	o/c	o/c	10.8	10.6	13.4	14.9	15.4	17.9	17.7	17.5	17.6
10in PIPE ENTRY TEMPERATURE, °C	224	239	273	287	332	290	292	308	325	305	262
24in PIPE ENTRY PRESSURE, bar	o/c	o/c	7.12	7.02	9.00	9.85	10.21	11.98	11.80	11.85	11.83
24in PIPE ENTRY TEMPERATURE, °C	211	225	257	273	318	o/c	280	302	316	293	265
24in PIPE EXIT PRESSURE, bar	2.40	2.60	3.80	3.74	4.71	5.32	5.53	6.34	6.20	6.25	6.29
24in PIPE EXIT TEMPERATURE, °C	173	189	216	239	270	238	o/c	o/c	o/c	254	234
TARGET PLATE FITTED	NO	NO	NO	NO	NO	NO	NO	NO	NO	NO	NO
PERIOD STOP VALVES FULLY OPEN, SECONDS	10	30	45	60	60	30	30	45	45	45	45
DRUM PRESSURE BEFORE PURGE, bar	33.0	33.5	47.5	48.9	60.0	70.8	70.6	75.4	75.4	73.9	75.0
STEAM FLOW, kg/s	141	148	208	202	249	289	292	336	327	335	343
DISTURBANCE FACTORS											
DRUM	1.18	1.29	1.78	1.72	2.03	2.48	2.48	2.98	2.83	3.00	3.24
SUPERHEATER OUTLET	0.80	0.86	1.22	1.21	1.57	1.76	1.73	2.00	1.99	1.94	1.93
TSV'S	1.19	1.28	1.83	1.81	2.34	2.63	2.56	3.03	2.98	2.96	3.00

NOTES:
1. RESULTS ARE FOR PURGES WITH BOILER STOP VALVES FULLY OPEN
2. READINGS ARE AT TIME OF MAXIMUM DISTURBANCE FACTOR UNLESS OTHERWISE STATED
3. ALL PRESSURES ARE ABSOLUTE

PURGE NUMBER	S15	S16	S17	S18	S19	S20	S21	S22	S23	S24	S25
DRUM PRESSURE, bar	64.4	61.4	63.2	62.8	50.2	66.1	65.5	63.9	62.4	59.4	54.5
S/H OUTLET PRESSURE, bar	37.5	36.0	37.0	37.0	27.5	37.0	37.9	36.9	35.9	34.4	30.7
S/H OUTLET TEMPERATURE, °C	355	346	372	284	355	318	366	376	384	384	306
TSV INLET PRESSURE, bar	23.4	22.4	22.9	23.0	17.5	22.8	23.5	23.0	22.5	21.8	19.7
TSV INLET TEMPERATURE, °C	318	316	330	345	331	280	319	335	352	360	301
10in PIPE ENTRY PRESSURE, bar	17.5	17.1	17.8	17.5	13.5	18.2	18.2	17.4	17.0	16.8	15.5
10in PIPE ENTRY TEMPERATURE, °C	297	292	308	321	314	270	302	311	328	338	295
24in PIPE ENTRY PRESSURE, bar	11.76	11.35	11.67	11.73	9.17	o/c	o/c	o/c	o/c	o/c	10.56
24in PIPE ENTRY TEMPERATURE, °C	287	288	301	317	309	252	289	304	324	334	284
24in PIPE EXIT PRESSURE, bar	6.30	5.98	6.19	6.15	4.62	6.17	6.42	6.19	6.10	5.97	5.28
24in PIPE EXIT TEMPERATURE, °C	223	243	245	274	260	212	240	257	282	302	255
TARGET PLATE FITTED	NO	NO	NO	NO	MS3A	NO	NO	NO	NO	NO	SS11B
PERIOD STOP VALVES FULLY OPEN, SECONDS	45	45	45	45	45	20	45	45	45	45	45
DRUM PRESSURE BEFORE PURGE, bar	75.2	72.1	74.5	74.0	60.7	74.0	75.7	73.8	75.4	76.2	62.2
STEAM FLOW, kg/s	334	322	327	325	244	345	349	331	319	306	287
DISTURBANCE FACTORS											
DRUM	2.98	2.91	2.92	2.89	2.04	3.10	3.20	2.96	2.45	2.72	2.61
SUPERHEATER OUTLET	1.99	1.90	2.01	2.03	1.48	1.99	2.21	2.08	1.77	1.95	1.65
TSV'S	2.94	2.85	2.96	2.98	2.18	2.90	3.20	3.03	2.61	2.90	2.51

NOTES:
1. RESULTS ARE FOR PURGES WITH BOILER STOP VALVES FULLY OPEN
2. READINGS ARE AT TIME OF MAXIMUM DISTURBANCE FACTOR UNLESS OTHERWISE STATED
3. ALL PRESSURES ARE ABSOLUTE

FIG. 4.70 Ince *B* Power Station, unit 6: superheater steam purge, summary of results

12.6.2 Reheater steam purge

As before, the purging will be with four parallel paths. The procedure employed in the first phase is repeated. The initial purges are again at lower-than-designed pressures, but here the aim is to assess the effect of the increased system resistance. Steam purging will then continue, as for the superheater purge, until a satisfactory clean system is obtained. Again, at Ince *B* Power Station unit 6 boiler required 42 purges (see Fig 4.71).

Once a satisfactory alloy target plate has been obtained and accepted by the inspection party, the decision on whether to blowdown the boiler will be determined by the unit requirements for further commissioning under steaming conditions.

Final target plates should be kept for historical purposes, the target plates also being photographed for record purposes.

12.6.3 Boiler feed pump and gland-steam system purging

This will be carried out in such a manner as determined by the programme. Drum level disturbance factors of approximately 1.6 should be aimed for. The procedure is covered in Chapter 5.

12.6.4 Assessment of results

The number of purges required to obtain a satisfactorily clean system will depend upon the type of plant and the condition of the boiler/pipework prior to steam purging. Typically, as previously stated, for a relatively clean system at Ince *B* on unit 6, 25 superheater and 42 reheater purges were needed to obtain an acceptable system.

The tables shown in Figs 4.70 and 4.71 are the summary of the results for this unit and the respective disturbance factors obtained; Fig 4.72 shows target plate conditions for the superheater and reheater for the Ince *B* unit 6.

As can be seen from the tablets in Figs 4.70 and 4.71, the boiler pressure was gradually raised from approximately 30 bar to 70 bar for the superheater system, and from 60 bar to approximately 90 bar for the reheater purge.

As stated earlier, the final assessment and acceptance is made by senior members of the Station or Plant Completion Committee.

PURGE NUMBER	R6	R7	R8	R9	R10	R11	R12	R13	R14	R15	R16	R17
DRUM PRESSURE, bar	66.0	66.5	71.0	73.4	76.9	77.6	78.2	72.9	75.8	80.4	79.4	71.8
S/H OUTLET PRESSURE, bar	45.0	38.8	43.2	43.8	47.2	49.3	48.4	45.7	44.4	47.6	47.3	43.7
S/H OUTLET TEMPERATURE, °C	324	303	366	336	384	358	358	372	392	434	436	287
TSV INLET PRESSURE, bar	38.1	28.8	31.6	32.8	34.4	36.7	35.5	33.9	30.4	34.5	35.1	33.1
TSV INLET TEMPERATURE, °C	318	292	334	321	348	341	357	353	365	397	409	298
10 in PIPE ENTRY PRESSURE, bar	35.7	25.2	27.2	28.4	29.5	31.5	30.2	29.4	28.1	29.2	29.3	28.6
10 in PIPE ENTRY TEMPERATURE, °C	299	273	314	300	315	304	327	320	343	370	383	290
COLD R/H PIPE ENTRY PRES. bar	34.5	23.0	24.7	26.2	26.9	28.7	27.7	26.9	25.6	26.3	27.1	26.2
COLD R/H PIPE ENTRY TEMP. °C	299	273	309	296	309	296	318	317	340	363	382	290
REHEATER OUTLET PRESSURE, bar	18.7	20.6	22.1	23.3	24.1	25.7	24.8	24.1	23.0	23.5	24.2	23.6
REHEATER OUTLET TEMP. °C	340	329	367	358	386	387	391	384	393	422	424	337
IP STEAM CHEST PRESSURE, bar	17.7	19.6	21.1	22.0	22.9	24.5	23.6	22.8	21.8	22.4	23.2	22.4
IP STEAM CHEST TEMP. °C	348	335	352	351	369	376	379	378	380	397	406	360
16 in PIPE ENTRY PRESSURE, bar	15.2	17.1	18.1	19.0	19.7	20.6	20.5	19.6	18.7	19.2	19.7	19.6
16 in PIPE ENTRY TEMP. °C	330	321	336	336	353	360	364	364	366	381	391	351
24 in PIPE ENTRY PRESSURE, bar	11.00	12.13	12.95	13.64	14.26	15.07	14.45	14.18	13.6	13.7	14.3	14.3
24 in PIPE ENTRY TEMP. °C	321	313	326	327	345	352	354	356	358	368	381	341
24 in PIPE EXIT PRESSURE, bar	5.82	6.52	6.89	7.13	7.50	7.78	7.69	7.58	7.29	7.25	7.61	7.16
24 in PIPE EXIT TEMPERATURE, °C	271	279	284	286	308	318	308	322	324	312	334	295
TARGET PLATE FITTED	NO	NO	NO	NO	NO	NO	NO	NO	NO	NO	NO	MS4A
PERIOD STOP VALVES FULLY OPEN, S	15	30	45	45	45	45	45	45	45	45	45	45
DRUM PRESSURE BEFORE PURGE, bar	77.5	79.9	85.8	90.3	94.0	96.1	95.3	95.2	98.5	98.5	98.4	87.1
STEAM FLOW, kg/s	303	337	355	374	386	405	388	380	363	364	376	370
DISTURBANCE FACTORS												
DRUM	2.40	2.94	3.07	3.29	3.33	3.64	3.32	3.42	3.00	3.84	3.07	3.29
SUPERHEATER OUTLET	1.25	1.73	1.98	2.02	2.20	2.19	2.17	2.16	2.12	2.13	2.30	1.73
TSV'S	1.43	2.26	2.50	2.58	2.76	2.79	2.75	2.75	2.66	2.69	2.88	2.36
COLD R/H PIPE INLET	0.76	1.38	1.55	1.56	1.66	1.66	1.66	1.64	1.66	1.70	1.81	1.51
REHEATER OUTLET	1.11	1.21	1.35	1.39	1.50	1.55	1.48	1.45	1.41	1.46	1.51	1.30
IP STEAM CHESTS	1.13	1.22	1.31	1.38	1.45	1.51	1.45	1.44	1.39	1.38	1.46	1.34

NOTES:
1. RESULTS ARE FOR PURGES WITH BOILER STOP VALVES FULLY OPEN
2. READINGS ARE AT TIME OF MAXIMUM DISTURBANCE FACTOR UNLESS OTHERWISE STATED
3. ALL PRESSURES ARE ABSOLUTE

FIG. 4.71 Ince *B* Power Station, unit 6: reheater steam purge, summary of results

PURGE NUMBER	R18	R19	R20	R21	R22	R24	R25	R26	R27	R28	R29	R30
DRUM PRESSURE, bar	80.0	74.2	80.0	80.0	75.6	76.1	82.8	85.3	85.5	84.5	68.7	74.8
S/H OUTLET PRESSURE, bar	48.7	45.9	49.3	49.8	46.1	44.6	49.7	51.9	52.6	51.9	42.5	46.3
S/H OUTLET TEMPERATURE, °C	390	389	416	425	421	399	425	384	374	377	345	266
TSV INLET PRESSURE, bar	36.2	34.3	36.2	36.3	34.4	32.4	35.4	38.2	38.5	38.1	31.8	35.7
TSV INLET TEMPERATURE, °C	359	368	386	396	400	360	389	370	359	356	332	268
10 in PIPE ENTRY PRESSURE, bar	30.9	29.1	30.2	30.3	29.4	27.7	29.6	31.8	31.7	32.0	26.9	31.2
10 in PIPE ENTRY TEMPERATURE, °C	336	344	356	365	372	345	368	354	345	337	315	260
COLD R/H PIPE ENTRY PRES. bar	28.1	26.9	27.6	27.2	26.5	25.3	26.7	29.1	29.2	29.3	24.9	29.0
COLD R/H PIPE ENTRY TEMP. °C	331	338	352	362	372	341	365	354	339	324	304	245
REHEATER OUTLET PRESSURE, bar	25.1	24.1	24.7	24.3	23.8	23.0	24.0	26.1	26.2	26.3	22.4	26.0
REHEATER OUTLET TEMP. °C	393	392	412	420	415	388	410	392	387	391	366	353
IP STEAM CHEST PRESSURE, bar	24.1	22.8	23.7	23.0	22.4	21.9	22.9	24.8	24.9	25.1	21.3	24.5
IP STEAM CHEST TEMP. °C	378	383	395	403	405	371	386	383	381	381	367	359
16 in PIPE ENTRY PRESSURE, bar	20.6	19.8	20.2	20.0	19.5	18.8	19.9	21.2	21.6	21.5	18.7	21.4
16 in PIPE ENTRY TEMP. °C	364	369	380	388	391	355	370	369	367	367	356	348
24 in PIPE ENTRY PRESSURE, bar	14.9	14.4	14.4	14.5	14.1	13.5	14.0	15.30	15.30	15.60	13.80	15.31
24 in PIPE ENTRY TEMP. °C	353	361	371	378	384	345	358	356	355	356	347	339
24 in PIPE EXIT PRESSURE, bar	7.89	7.60	7.72	7.67	7.45	7.17	7.43	8.29	8.15	8.08	6.81	8.20
24 in PIPE EXIT TEMPERATURE, °C	301	320	327	334	345	313	313	303	299	299	305	311
TARGET PLATE FITTED	NO	NO	NO	NO	NO	NO	NO	NO	NO	NO	MS4B	NO
PERIOD STOP VALVES FULLY OPEN, S	45	45	45	45	45	45	45	45	45	45	45	45
DRUM PRESSURE BEFORE PURGE, bar	98.6	98.0	96.2	97.2	98.3	101.4	101.6	104.5	103.3	103.3	85.0	104.6
STEAM FLOW, kg/s	400	386	381	382	368	364	274	409	411	417	355	416
DISTURBANCE FACTORS												
DRUM	3.44	3.46	3.13	3.11	3.09	2.99	2.92	3.39	3.41	3.54	3.16	3.99
SUPERHEATER OUTLET	2.31	2.29	2.18	2.20	2.21	2.14	2.12	2.23	2.18	2.28	1.92	1.89
TSV'S	2.88	2.89	2.76	2.80	2.78	2.68	2.73	2.90	2.84	2.93	2.46	2.57
COLD R/H PIPE INLET	1.79	1.77	1.73	1.79	1.75	1.68	1.77	1.89	1.85	1.83	1.51	1.49
REHEATER OUTLET	1.56	1.52	1.49	1.54	1.46	1.40	1.48	1.57	1.57	1.61	1.32	1.52
IP STEAM CHESTS	1.50	1.49	1.44	1.50	1.44	1.36	1.41	1.54	1.55	1.57	1.32	1.54

NOTES:
1. RESULTS ARE FOR PURGES WITH BOILER STOP VALVES FULLY OPEN
2. READINGS ARE AT TIME OF MAXIMUM DISTURBANCE FACTOR UNLESS OTHERWISE STATED
3. ALL PRESSURES ARE ABSOLUTE

13 Floating of boiler safety valves

13.1 Introduction

Protection against steam pressures being generated in the boiler in excess of design limitations is provided by a series of pressure relief valves (or safety valves) fitted to the boiler drum, final superheater outlet, cold reheater inlet and hot reheater outlets. The safety valves are required to relieve the pressure whenever the heat input to the boiler is in excess of the heat requirement of the turbine. This can occur with sudden reduction in electrical load, or an uncontrolled increase in firing rate.

Up to 20 safety valves are installed on a 500 MW unit and, in order to ensure protection of the superheater when the pressure becomes excessive, the safety valve settings are chosen such that the superheater safety valves will lift before the drum safety valves. If both sets of valves should lift, the choice of blowdown or reseat settings ensures that the drum valves reseat before the superheater safety valves, thus maintaining circulation in the superheater and preventing any chance of overheating. Likewise the reheater outlet (hot) safety valves lift before the reheater inlet (cold) safety valves to ensure that there is a steam flow through the reheater at all times.

The *blowdown* is the difference between the lifting and reseating pressures.

To conform with the Factories Acts and insurance company requirements, the safety valves have to be demonstrated to operate at the correct pressures by operating the boiler at excess pressure under controlled conditions and testing each valve in turn. This operation is known as *floating safety valves*. In some CEGB stations, the reheater safety valves are not floated but are set before delivery to site in the manufacturer's works, but in all the more recent 500 MW and 660 MW units the reheat valves are floated. They can be checked at the end of the steam purge when all debris

PURGE NUMBER	R31	R32	R33	R34	R35	R36	R37	R38	R39	R40	R41	R42
DRUM PRESSURE, bar	84.4	89.0	77.9	81.5	72.1	88.7	83.5	82.6	80.6	69.2	78.4	77.6
S/H OUTLET PRESSURE, bar	53.7	55.8	48.0	51.9	44.1	51.6	52.3	49.0	40.7	41.3	48.4	50.1
S/H OUTLET TEMPERATURE, °C	339	346	353	359	364	272	289	409	413	403	386	343
TSV INLET PRESSURE, bar	40.4	41.0	35.8	38.0	32.1	39.2	38.6	36.1	36.2	30.6	35.8	38.4
TSV INLET TEMPERATURE, °C	318	320	321	333	331	262	271	360	377	382	372	335
10 in PIPE ENTRY PRESSURE, bar	34.2	34.1	30.8	31.9	27.0	33.2	33.5	30.4	30.5	26.1	30.2	32.5
10 in PIPE ENTRY TEMPERATURE, °C	293	303	308	314	313	243	247	328	347	352	345	297
COLD R/H PIPE ENTRY PRES. bar	31.3	31.0	28.2	28.9	24.5	31.0	30.5	27.7	27.6	23.8	27.7	29.9
COLD R/H PIPE ENTRY TEMP. °C	288	296	302	309	304	239	240	325	342	350	344	276
REHEATER OUTLET PRESSURE, bar	28.1	27.7	25.5	25.9	22.0	27.7	27.3	24.7	24.6	21.3	24.7	26.5
REHEATER OUTLET TEMP. °C	375	371	356	373	373	324	334	387	401	400	391	394
IP STEAM CHEST PRESSURE, bar	26.6	26.2	24.0	24.6	21.1	26.5	25.8	23.6	23.4	20.3	23.5	25.1
IP STEAM CHEST TEMP. °C	369	366	356	362	361	332	331	355	374	382	384	388
16 in PIPE ENTRY PRESSURE, bar	23.1	22.4	21.0	21.0	18.1	22.6	22.2	20.1	20.1	17.6	20.0	21.6
16 in PIPE ENTRY TEMP. °C	356	353	342	349	348	318	312	337	357	366	368	371
24 in PIPE ENTRY PRESSURE, bar	17.00	16.86	14.97	15.20	13.60	16.10	15.90	14.40	14.30	13.0	14.9	15.3
24 in PIPE ENTRY TEMP. °C	345	340	334	338	337	306	307	329	347	359	359	364
24 in PIPE EXIT PRESSURE, bar	8.74	8.53	7.91	8.0	6.45	8.58	8.51	7.62	7.76	6.46	7.70	8.34
24 in PIPE EXIT TEMPERATURE, °C	315	292	306	302	289	234	271	284	303	314	305	334
TARGET PLATE FITTED	NO	NO	NO	NO	SS12A	NO	NO	NO	NO	SS12B	NO	NO
PERIOD STOP VALVES FULLY OPEN, S	30	45	30	45	40	30	30	30	30	30	30	30
DRUM PRESSURE BEFORE PURGE, bar	105.0	101.2	98.7	99.5	85.4	102.9	99.4	100.4	98.0	87.7	96.0	97.0
STEAM FLOW, kg/s	459	458	408	414	353	450	442	395	385	332	398	408
DISTURBANCE FACTORS												
DRUM	4.31	4.06	3.69	3.62	2.98	3.93	4.04	3.26	3.18	2.74	3.48	3.70
SUPERHEATER OUTLET	2.44	2.36	2.26	2.16	1.90	1.93	2.00	2.30	2.20	1.95	2.29	2.11
TSV'S	3.06	3.01	2.80	2.76	2.40	2.60	2.61	2.82	2.27	2.48	2.95	2.67
COLD R/H PIPE INLET	1.89	1.95	1.74	1.77	1.52	1.58	1.56	1.75	1.73	1.53	1.85	1.52
REHEATER OUTLET	1.78	1.78	1.50	1.56	1.35	1.57	1.57	1.53	1.50	1.30	1.57	1.54
IP STEAM CHESTS	1.76	1.76	1.51	1.53	1.31	1.58	1.57	1.44	1.43	1.25	1.54	1.53

NOTES:
1. RESULTS ARE FOR PURGES WITH BOILER STOP VALVES FULLY OPEN
2. READINGS ARE AT TIME OF MAXIMUM DISTURBANCE FACTOR UNLESS OTHERWISE STATED
3. ALL PRESSURES ARE ABSOLUTE

FIG. 4.71 (*cont.*) Ince *B* Power Station, unit 6: reheater steam purge, summary of results

has been removed and the IP steam chest has temporary valves, subject of course to the temporary pipework and valves being able to withstand the test pressures. It is more usual in the present-day units (1986) to 'float' the reheat valves with the unit on-load. This chapter covers the procedure for floating the reheater safety valves with the unit on-load.

13.2 Safety valve system description

13.2.1 Valve locations

Safety valves are fitted on the drum, superheater and reheater and Fig 4.73 shows a typical layout for a 500 MW oil-fired boiler.

It can be seen that there are six valves on the boiler drum, three at each end; four on the superheater outlet pipework, one on each leg; and ten on the reheater, one in each of the rehéat outlet-pipework legs and six on the reheater inlet legs.

13.2.2 Valve description

The types of safety valve used on CEGB boilers vary, the main types being the Dewrance Consolidated Maxiflow with compensated back-pressure closing and the Hopkinson Hylif torsion bar safety valve. The principle of operation of both types of valve is similar but the Dewrance valve uses a conventional spring to hold the valve closed, whereas the Hopkinson Hylif valve uses torsion bars. Figure 4.74 shows a Dewrance Consolidated Maxiflow safety valve and Fig 4.75 shows a Hopkinson Hylif torsion bar safety valve.

Since their principle of operation is similar, the following describes the Dewrance Consolidated Maxiflow safety valve with compensated back-pressure closing.

Back-pressure closing valves with pneumatic assistance are now widely used on all new CEGB units. This type of valve has downward flow, so that the sealing load on the valve disc is maintained by the steam

FIG. 4.72 Photographs of first and final target plates from Ince *B*, unit 6 steam purge.
(a) Overall view of plates from superheater and reheater purges
(b) Enlargement of reheater plates, showing improvement

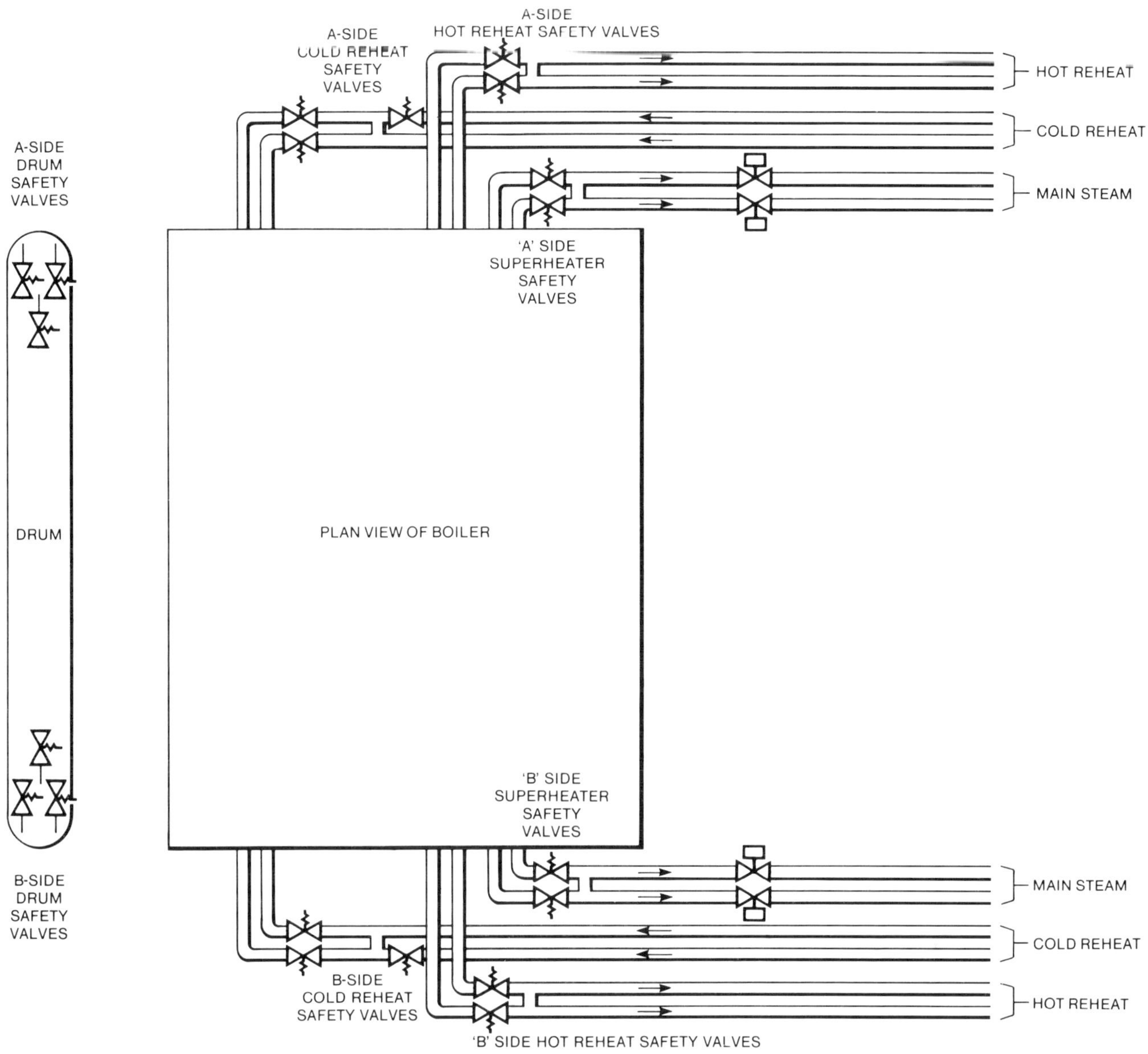

FIG. 4.73 Location of boiler safety valves on a 500 MW oil-fired boiler

pressure up to the instant of operation and therefore positive operation and repeatability is obtained. The valve operates during closing with a back-pressure from steam trapped on the upper side of the disc holder assisting the spring to return the valve back down onto its seat.

Figure 4.76 shows the valve internals and its working principle.

When the valve is blowing, a small amount of steam is bled from the huddle chamber (A) through two connecting holes (B) drilled in the disc holder (J) into chamber (C), escaping to atmosphere between floating washer (D) and overlap shoulder (E), and also through orifice plug (F).

As the steam pressure reduces, the overlap shoulder (E), upon reaching the top of the floating washer (D), traps steam in chamber (C), setting up pressure on top of the disc holder, and assists the spring in closing the valve.

With the valve in the full-lift position, the lift stop (G) resets against the cover plate (H), eliminating valve hunting and adding stability of operation to the valve.

With the valve blowing, the area between the floating washer (D) and the spindle is thus increased by the difference of the two diameters on the overlap collar. Under this condition, steam in chamber (C) escapes to atmosphere through the secondary area formed by the floating washer (D) and overlap collar (E), and by the hole in orifice plug (F).

Upon closing, the overlap collar (E) moves down onto the floating washer (D) reducing opportunity for escape of steam from chamber (C). Steam pressure then builds up in chamber (C) at a rate regulated by the size of the hole in orifice plug (F), the pressure

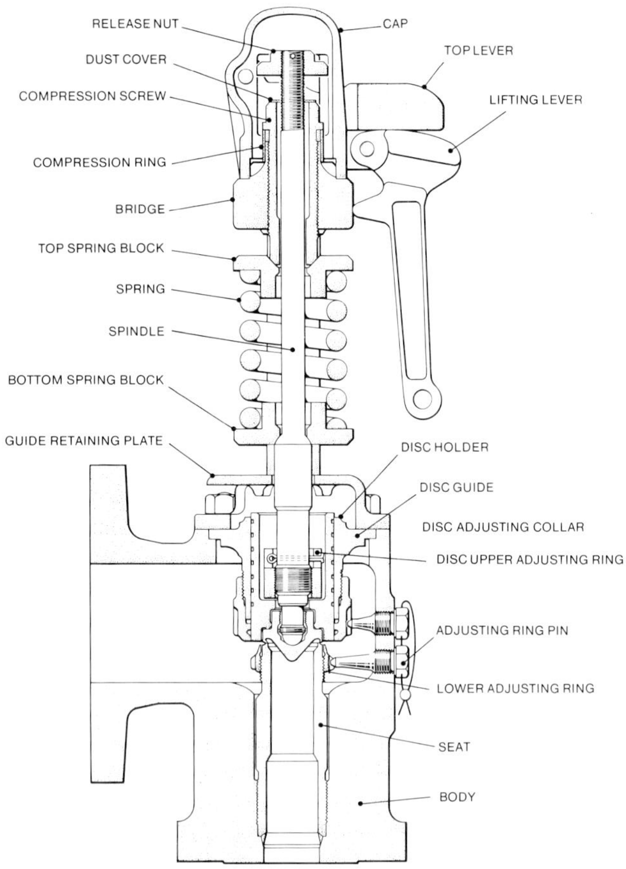

FIG. 4.74 Typical spring-type safety valve

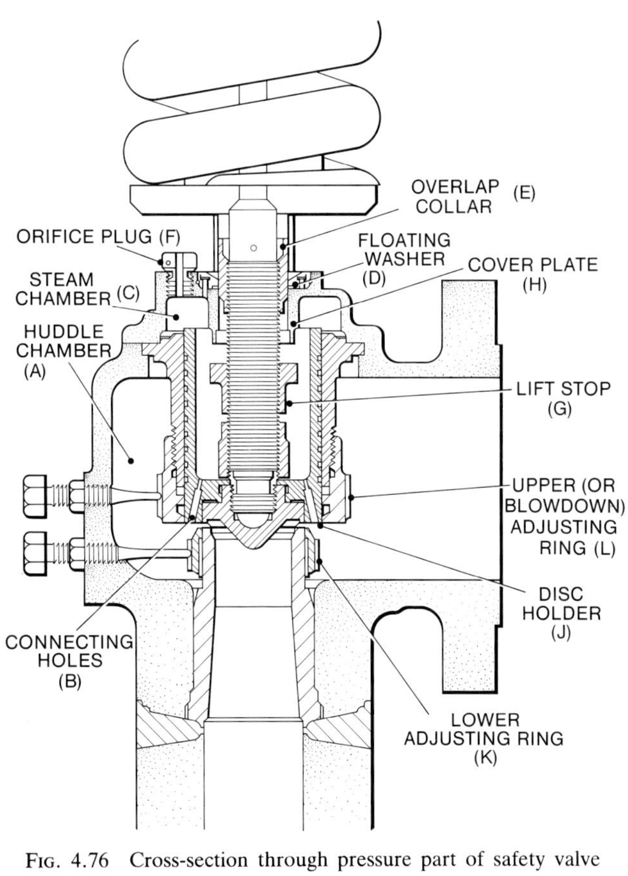

FIG. 4.76 Cross-section through pressure part of safety valve

generated producing a downward thrust in the direction of the spring loading. The combined thrust of the locked-in pressure and spring loading results in positive and precise closing. Cushioning of the closing action is controlled by position of the lower adjusting ring (K).

Regulation of valve blowdown is effected by adjustment of the upper (or blowdown) ring (L). Each ring is notched and can be adjusted *in situ* by removing the appropriate locking pin, inserting a screwdriver blade through the exposed hole and rotating the ring in the required direction.

For the drum and superheater, this type of safety valve is then fitted with a pneumatic servo (anti-simmer device) to assist the valve operation. 'Simmer', in this context, is where the valve just begins to pass steam in a spitting-type action.

13.2.3 Anti-simmer device

The anti-simmer device fitted to the Consolidated Maxiflow valve consists of a single-acting pneumatic actuator mounted on the bridge of the main valve. The actuator acts upon the main valve stem by means of a plunger not connected to the stem and clear of the stem when air pressure is removed. (This setting allows periodic checking of the control units and actuators without operating the safety valve.)

The diaphragm actuator is fitted with an internal return spring and is glandless, thus eliminating friction and reducing the possibility of variation of applied load. Figure 4.77 shows a typical anti-simmer device as fitted to a valve.

The object of the anti-simmer device is basically to set the safety valve mechanically about 2% below the desired blow-off pressure, and at all times of normal operation to apply an additional load of about 7% by the pneumatic diaphragm. Thus, if the boiler is operated at 2–3% below the 'set' pressure, there is still a substantial loading on the valve, keeping it tightly shut and avoiding 'simmer' under extreme circumstances; if, for instance, the pneumatic loading fails to be removed, the valve will still blow off at approximately 5% above design pressure.

It is not normally required to fit anti-simmer devices to the reheater safety valves, the normal compensated back-pressure closing valve being sufficient, but the following design stipulation is made:

> 'The safety valve of the reheater shall have a discharge capacity equal to 100% MCR evaporation of the boiler and 20% of this discharge capacity shall be situated at the reheater outlet. The blowing pressure of the reheater safety valves at the inlets and at the outlets will permit a pressure drop across the reheater of 5% at MCR.'

The anti-simmer actuator for each drum and superheater safety valve is provided with a self-contained air control system mounted on a steel plate. The system comprises an air filter, a diaphragm-operated air

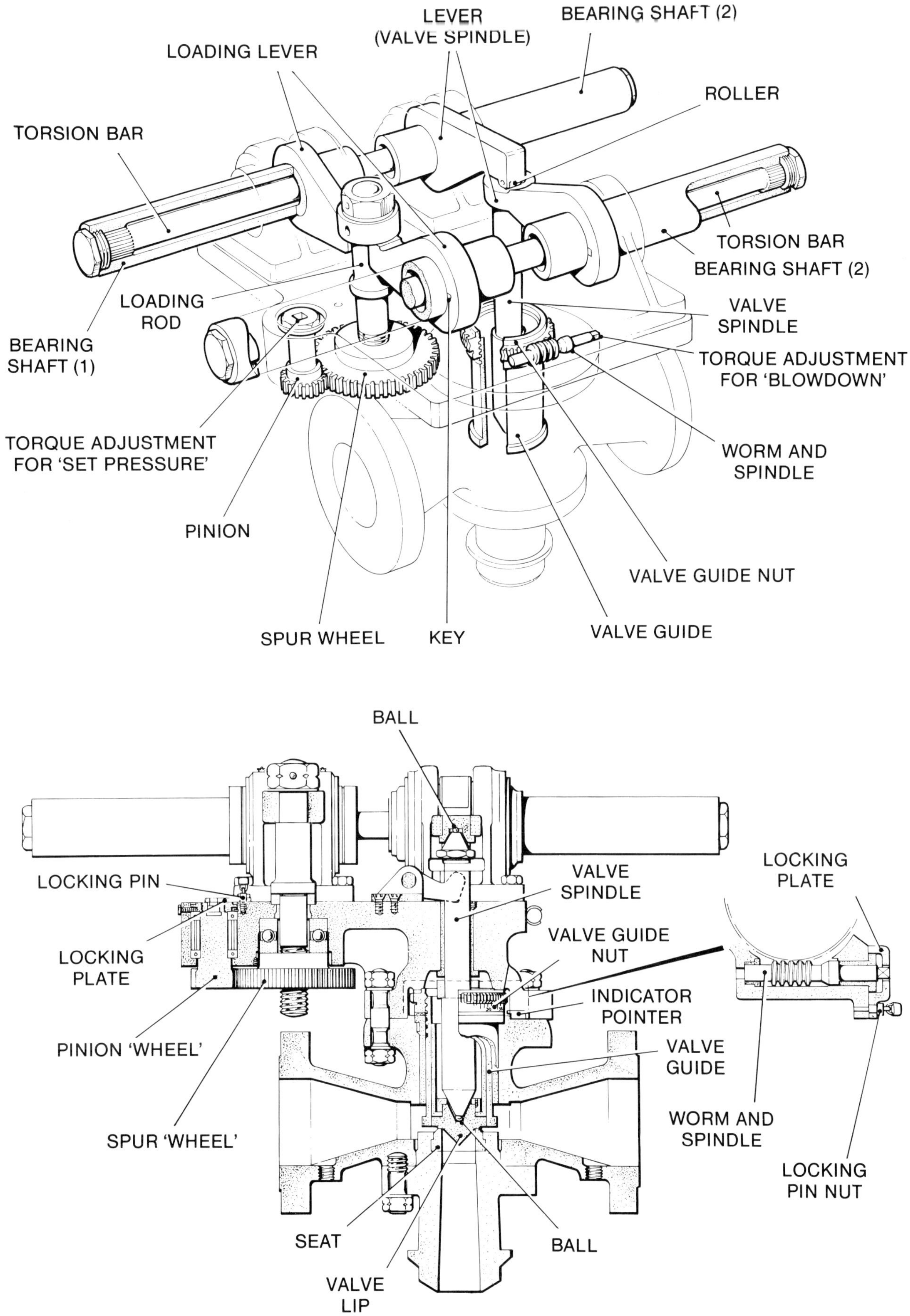

FIG. 4.75 Typical torsion bar safety valve

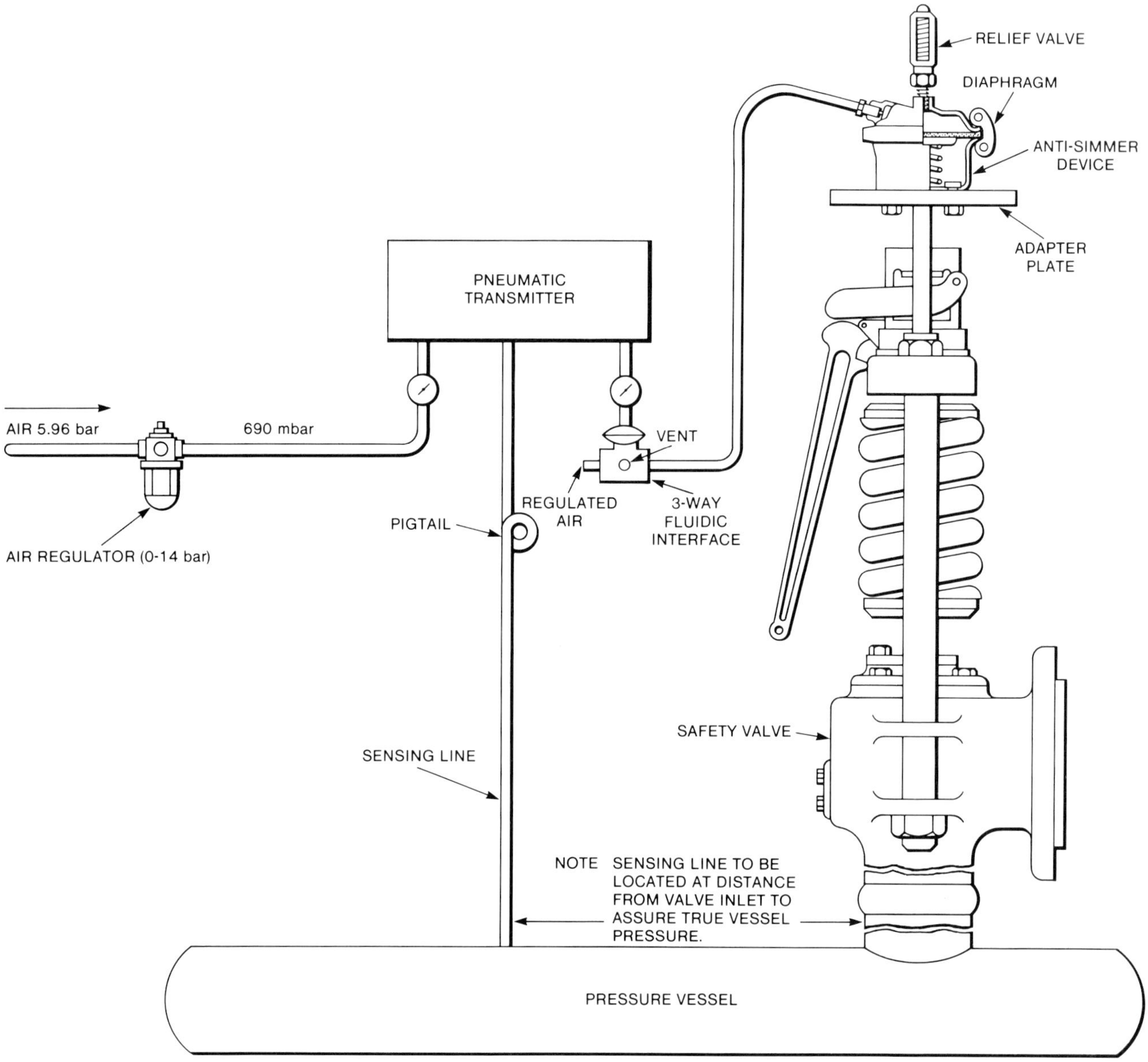

FIG. 4.77 Physical arrangement of anti-simmer pneumatic loading system

regulator and pressure gauge, and a 110 V AC solenoid-operated three-way control valve. Air supply to the control station is provided by the station instrument-air system. Operation of the three-way solenoid valve is initiated by a series of microswitches and latching or delatching relays, the microswitches being operated by a Bourdon tube pressure transmitter, sensing drum or superheater outlet pressure, dependent upon the application.

The pressure transmitter, relays and microswitches are mounted in a fabricated steel box, located adjacent to the boiler such as to require minimum length of impulse pipe. Figure 4.77 shows the detailed layout and Fig 4.78 shows the electrical control unit.

In normal operation, the latch relay is energised by the normally-closed contacts of the low pressure microswitch; the latch relay in turn energises the solenoid controlling the three-way control valve. With the solenoid energised, air is supplied to the pneumatic actuator, thus applying supplementary loading to the safety valve spindle.

If boiler pressure starts to rise above normal, the pressure transmitter opens the normally-closed contacts of the low pressure microswitch, thus removing supply from the latch relay. The latter, however, remains mechanically latched, retaining the supply to the solenoid controlling the three-way air valve and consequently the supplementary loading.

In event of boiler pressure rising to the 'lift' or high pressure setting, the normally-open contacts of the high pressure microswitch close, energising the delatch relay which removes the mechanical latching and releases supplies to the solenoid controlling the three-way valve. The valve reverses position under action of its integral return-spring — isolating the air supply and dumping air from the top of the pneumatic actuator to atmosphere, thus removing the supplementary loading and allowing the safety valve to move rapidly to full lift.

The 'valve open' lamp on the control box would be illuminated by auxiliary contacts (Fig 4.78).

In the event of a reduction in boiler pressure, the high pressure microswitch opens, removing supplies from the delatch relay. Once boiler pressure falls to the desired valve reset pressure, the low pressure microswitch normally-closed contacts make, energising the latch relay which in turn energises the solenoid-operated three-way valve, allowing air into the pneumatic actuator, which then applies the supplementary loading to effect rapid shutdown of the safety valve; the 'valve open' lamp extinguishes once the latch relay energises. Facilities are provided on the pressure transmitter to adjust the actuating points of the microswitches in order to establish 'lift' and 'reseat' settings.

A test pushbutton is fitted to the control box cover, and wired into the electrical circuitry, to facilitate in-service testing of the pneumatic actuator.

Faults which can occur to the system and resulting actions are as follows:

- *Failure of electrical supply and/or failure of air supply* Either failure results in the supplementary air loading being removed from the actuator, and the valve should lift at the mechanical setting established in the safety valve setting exercise.
- *Failure of pressure switches to operate* If the high pressure microswitch fails to operate, the pneumatic loading will not be removed and the valve will not operate until the boiler has overpressurised by approximately 5%. If the low pressure microswitch fails to operate, the supplementary pneumatic loading will not be reapplied and the valve will not shut until it reaches its mechanical reseat setting at about 5% blowdown.
- *Failure of the solenoid valve* Dependent upon the position in which the valve seizes, the safety valve will either lift at approximately 2% below normal set pressure, or will not lift until the boiler overpressurises by approximately 5%.
- *Failure of latched relay latching mechanism* With a boiler pressure rise above normal and a failure of the latching relay mechanism, the 110 V AC supply to the solenoid valve is removed and hence the safety valve would lift at the setting established in the safety valve setting exercise. On fall of boiler pressure, the latched relay will be energised without being latched, allowing a rapid shutdown of the safety valve.

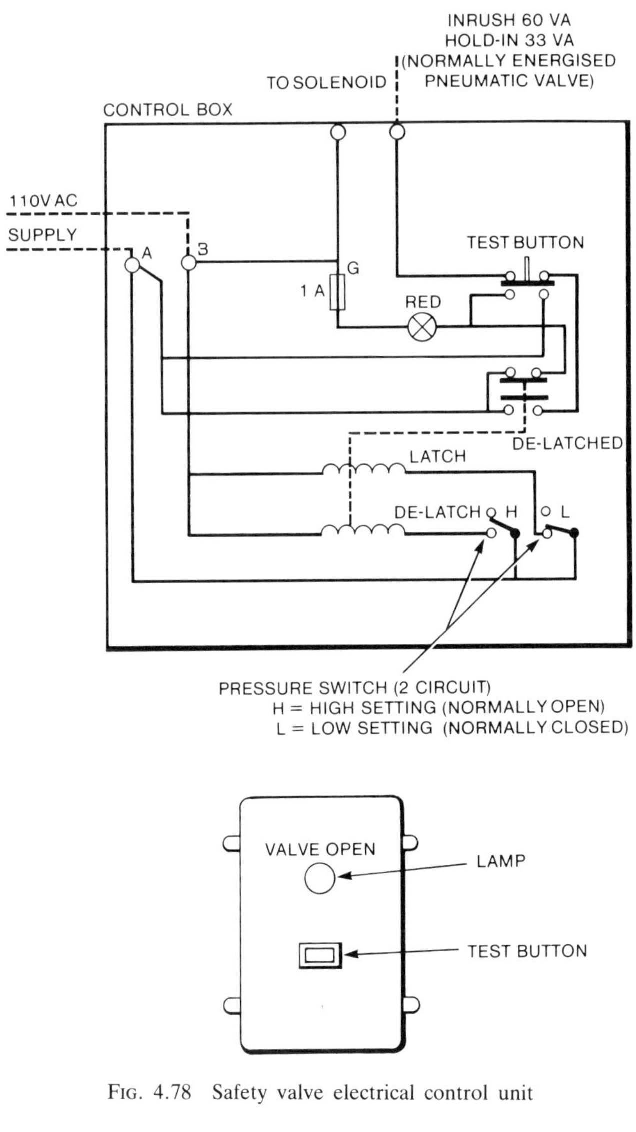

FIG. 4.78 Safety valve electrical control unit

13.2.4 Hydroset testing device

Setting of safety valves under steam pressure can present problems:

- Superheaters can be damaged due to starvation of the tubes
- Valve seats can be cut by steam erosion
- The boiler has to be operated in the overpressure condition for long periods.
- Costs of fuel, feedwater and manpower are expensive.
- On once-through boilers, which have to be tested with the turbine on-load, large load changes occur with the rapid pressure changes that are needed to float safety valves.

Although these problems cannot be eliminated entirely, they can be lessened by using a hydraulic device developed by the Dewrance/Dresser Company to assist safety valve floating. The hydraulic device is known as a 'Hydroset' and is fitted to the safety valve under test to assist the steam pressure under the valve seat to be in balance with the spring compression above the valve seat. To enable the Hydroset to be fitted, the pneumatic anti-simmer device is taken off the valve and the Hydroset unit is fitted to the compression screw of the valve. Figure 4.79 shows a typical device. The operating plunger of the hydraulic actuator is connected to the safety valve spindle by a turnbuckle, containing right- and left-hand threads. The hydraulic

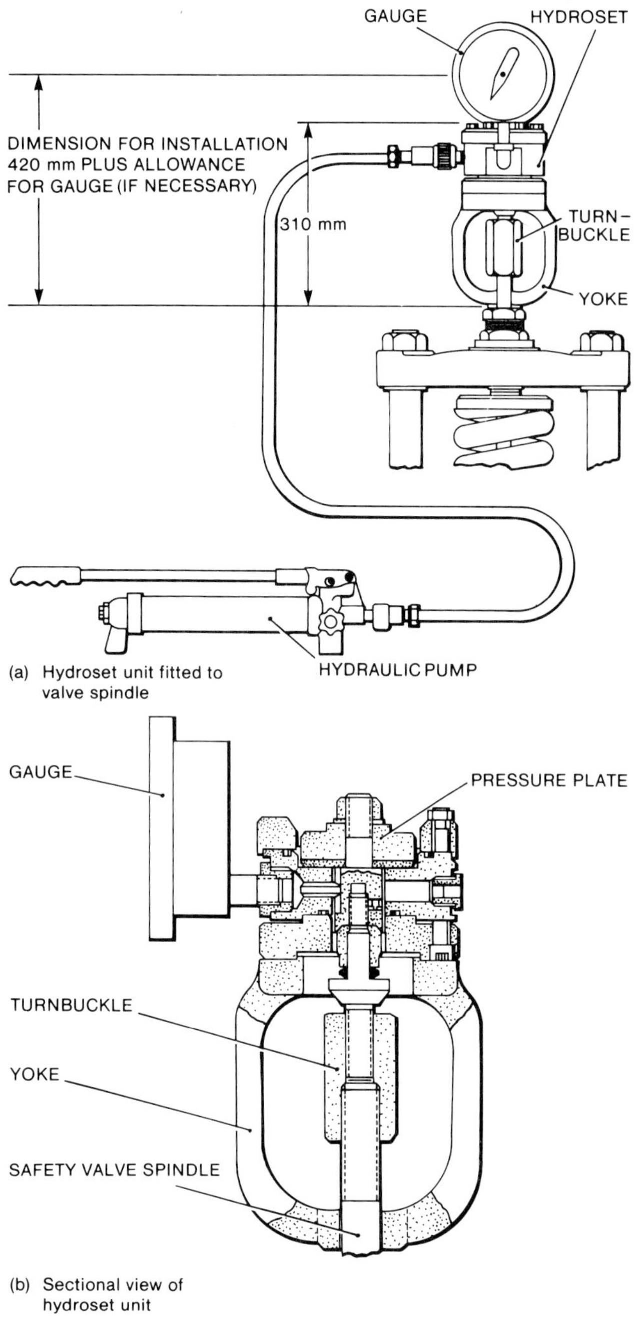

FIG. 4.79 Hydroset testing device

actuator is connected by a flexible hose to the discharge of a single-ram hydraulic pump. A calibrated pressure gauge is fitted to the Hydroset actuator.

The Hydroset device is used to set the 'blow off' (safety valve lift pressure) of all the boiler safety valves with the boiler at a pressure of approximately 60% MCR prior to the actual exercise of floating the valves and setting the blowdown range. The Hydroset is also used to set the reheater safety valve 'blow off' points once the unit has reached a load where the steam pressure in the reheater is 60% to 80% of the design mechanical lift-pressure (or the pressure is artificially raised to this level by throttling the IP turbine interceptor valves). The Hydroset unit does not facilitate the establishment of reseat pressures.

13.2.5 Safety valve gags

The drum and superheater safety valves are set by gagging (restraining) all valves except the one to be set. The highest-pressure valve is set first.

Care should be taken not to overtighten the gag, as damage to valve seats and/or spindles can occur.

Before any gags are removed or adjustments made, the boiler pressure must be reduced to a safe and steady pressure. A typical gag is shown in Fig 4.80.

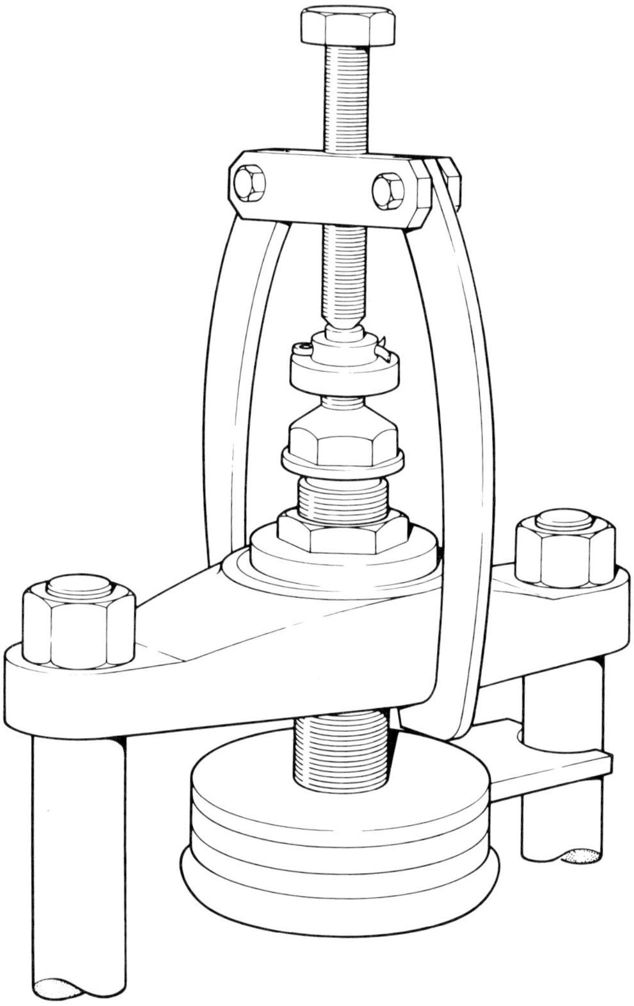

FIG. 4.80 Typical gag

13.2.6 Safety valve settings

Drum and superheater The settings of safety valves depends on the type of valve used and on whether the valve has anti-simmer devices. The following settings are for the Ince *B* 500 MW oil-fired boilers, which have Dewrance compensated back-pressure closing Maxiflow valves, with anti-simmer devices on the drum and superheater, and Dewrance compensated back-pressure closing Maxiflow valves without anti-simmer devices on the reheater.

The boiler is designed to operate at a drum working pressure of 181.7 bar, with 166.5 bar at the superheater outlet. The maximum drum design pressure is 191.7 bar.

Figure 4.81 shows the corresponding valve settings on the drum and superheater and it can be clearly seen that the blowdown pressure here is kept very low due to the anti-simmer devices.

- Drum valves

4 off	lift pressure	191.7 bar	reseat 185.5 bar
2 off	lift pressure	190.3 bar	reseat 184.1 bar

- Superheater valves

2 off	lift pressure	174.8 bar	reseat 169.6 bar
2 off	lift pressure	173.4 bar	reseat 168.2 bar

Reheater The reheater valves are not fitted with anti-simmer devices.

2 off	lift pressure	50.54 bar	reseat 48.54 bar
4 off	lift pressure	51.04 bar	reseat 49.02 bar
4 off	lift pressure	47.59 bar	reseat 45.71 bar

13.2.7 Safety valve escape pipework

The design requirements for the safety valve escape pipes are dealt with in detail in Design Memorandum 068/15 'Safe Discharge of High Temperature Steam Exhausts', and noise elimination in Design Memorandum 068/17 'Elimination of Noise Due to Steam Emission from Station Exhaust Systems'.

The alternative arrangements used for safety valve escape pipes are shown in Figs 4.82, 4.83, 4.84 and 4.85.

The sealed-type escape pipes shown in Figs 4.82 and 4.83 are generally preferred where noise elimination is important, since the slip joint (shown in Fig 4.84) has a noise-generating jet at the joint which is difficult to silence adequately.

Safety valve silencers, when fitted, must be inspected to check that the baffles are securely mounted and also that the mountings are able to withstand the high reaction-forces. Diffuser-absorptive-type silencers are usually fitted to the valves that are most likely to lift

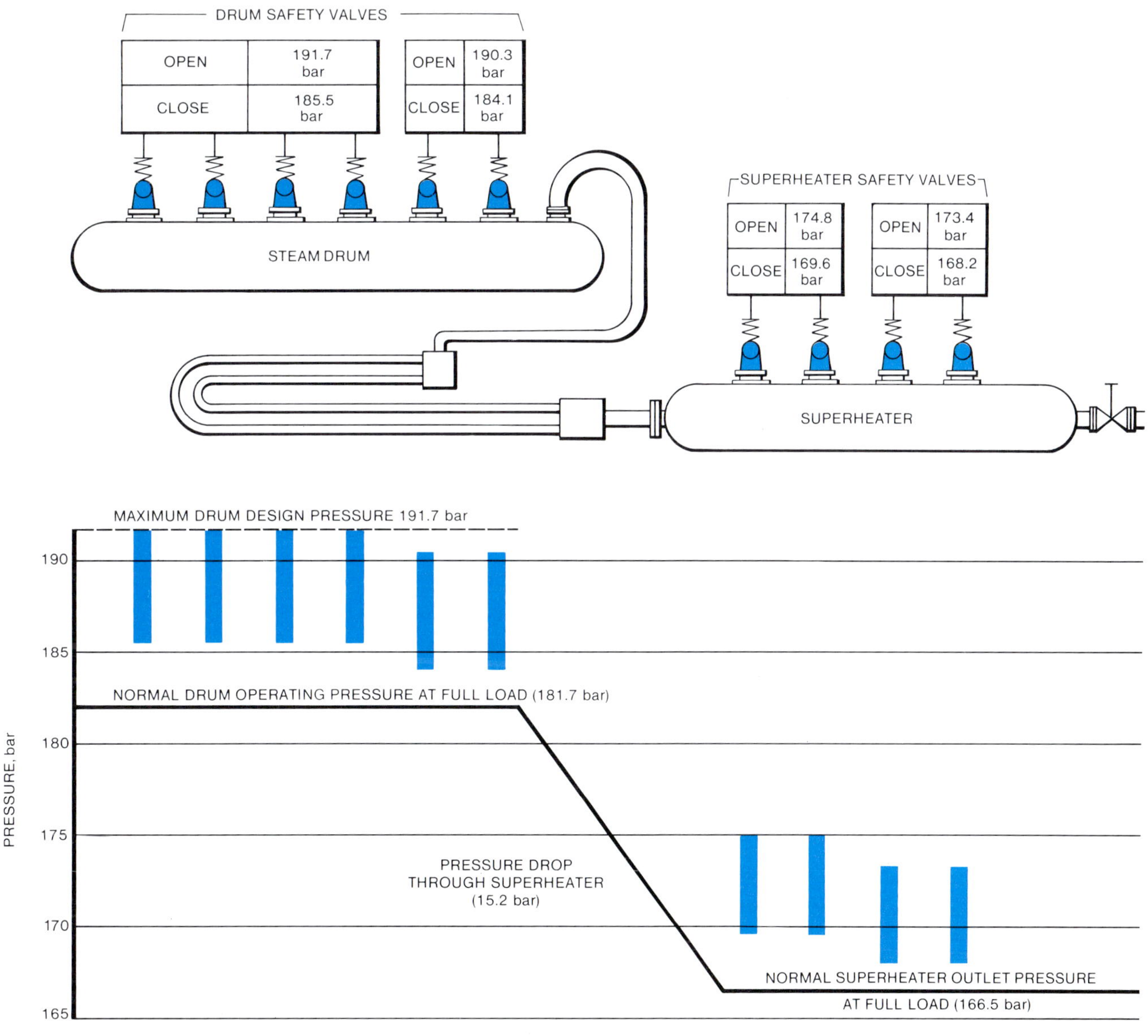

FIG. 4.81 Settings of drum and superheater safety valves with anti-simmer devices

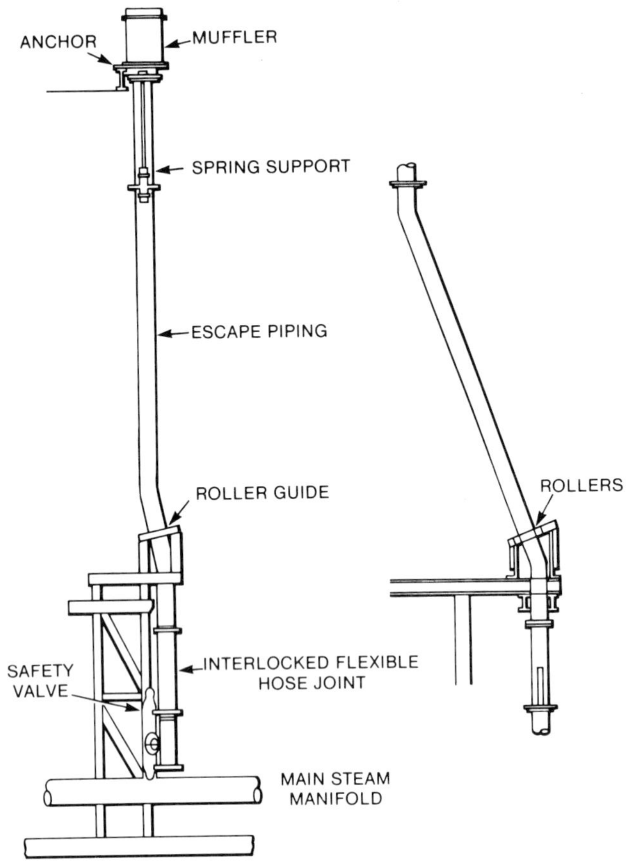

FIG. 4.82 Safety valve escape piping with interlocked flexible hose joints

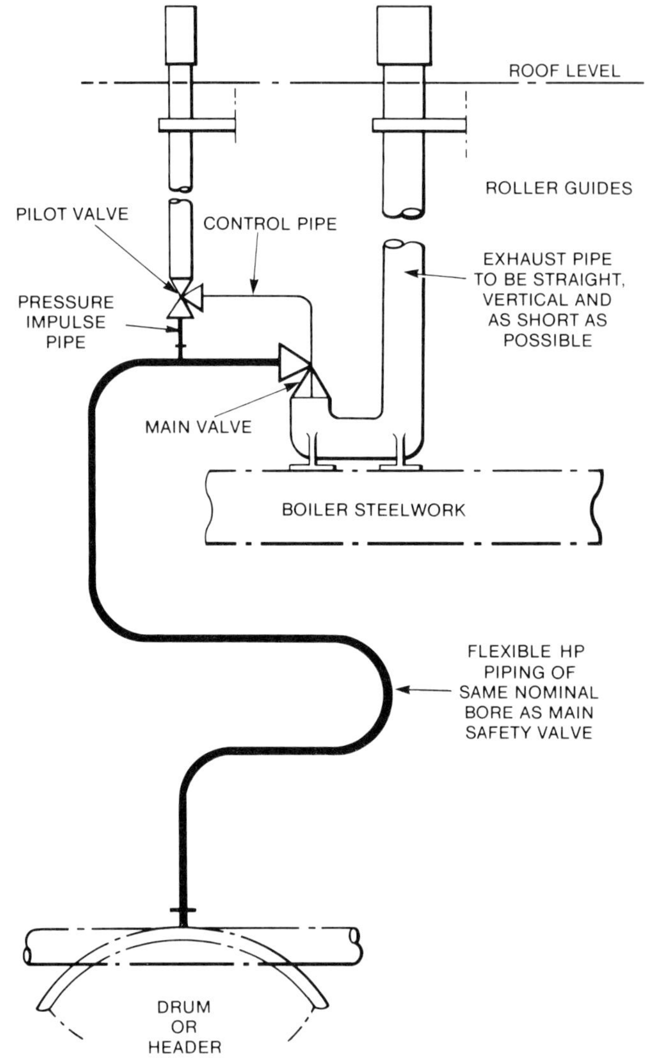

FIG. 4.83 Arrangement of pilot-operated safety relief valve

during station operation and therefore must be capable of the highest performance (approximately 40 dB sound attenuation). Ordinary diffuser-type silencers are capable of approximately 15–20 dB sound attenuation and are fitted to prevent nuisance on valves less likely to blow.

13.3 Initial preparations

Prior to commencing the floating exercise, the following preparations must be made:

- *Instrumentation* The test pressure gauges to be used in the various stages of the safety valve setting exercise must be calibrated in accordance with the requirements of the SCL, using a dead-weight tester. This procedure is described in Chapter 3, Section 1.

  All commercial boiler instrumentation necessary to operate the boiler to full design pressure must be commissioned and available.

- *Signalling systems* Adequate telephone communications between the boiler and the CCR must be commissioned and, in addition, a signalling system must be set up to inform the unit operator of the engineer's requirements for floating the valves. This usually consists of a set of coloured lights with which the engineer can signal to the unit operator. A normal signalling system is:

  Yellow light HOLD PRESSURE STEADY
  Green light RAISE PRESSURE
  Red light LOWER PRESSURE

  It must be made clear that the light must be ON for the unit operator to respond to the signal from the engineer in charge of floating the safety valves.

  A klaxon should also be available, to be sounded as a warning to site personnel before each valve lift is started.

- *Boiler steam pressure control* The superheater and reheater purge will have been completed and a full inspection made of all safety valves, control stations and escape pipework. Adequate provision must be made to reduce steam pressure on the unit boiler, by the use of steam dump valves, if fitted, or by the use of the final superheater outlet header drains and boiler blowdown drains.

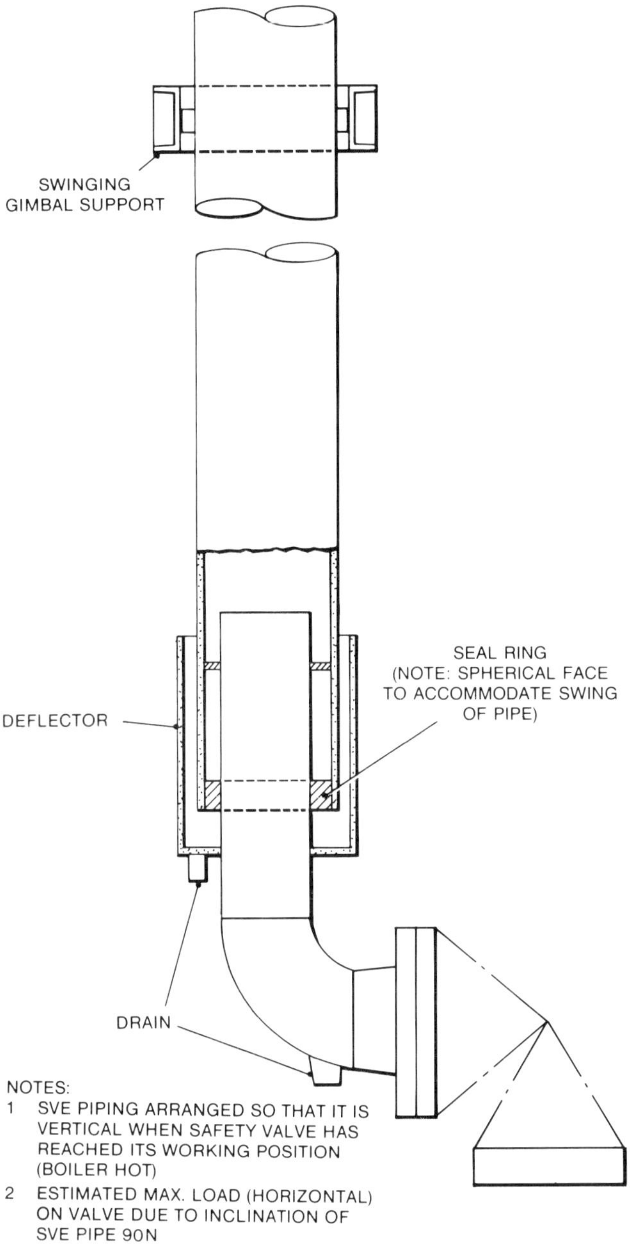

FIG. 4.84 Safe slip joint for safety valve escape (SVE) pipe

- *Boiler water level and chemical control* The boiler water level control must be available, using start-up and main feedwater regulating valves on manual control, and both of the starting and standby feed pumps should be fully available to supply the boiler.

  A long-range water level gauge glass should be available with remote TV viewing in the CCR, as for the steam purge and chemical clean operations.

- *Chemical control* This should be exercised to the Station Chemist's requirements and will be as laid down in Generation Operation Memorandum 75.

  To minimise the risk of stress corrosion, caustic soda should not be added to the boiler water during safety valve floating and the water conditions during this operation should be maintained at a minimum of 50 PPM ammonia, with a reserve of hydrazine. A value of 10 PPM of hydrazine should be aimed at.

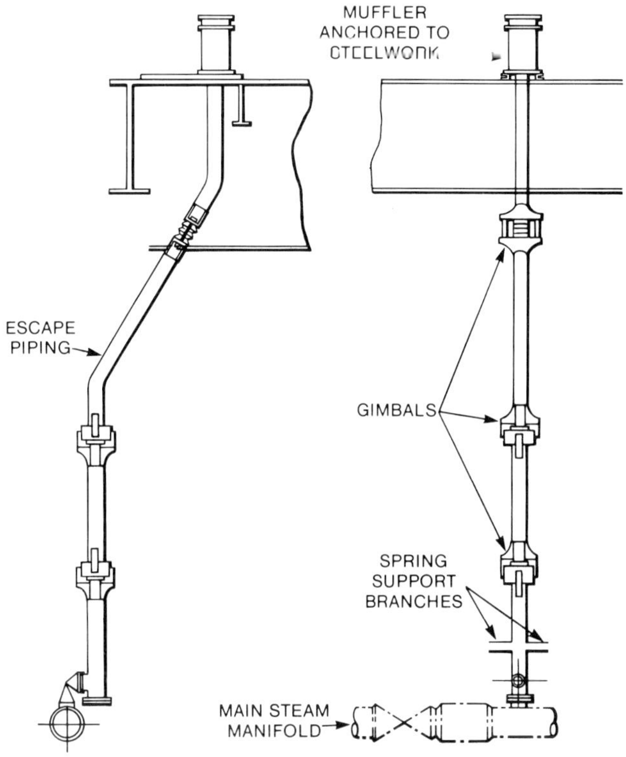

FIG. 4.85 Safety valve escape piping using gimbal expansion joints

Before commencing the safety valve floating programme, it should be proved that all cleaning chemicals have been thoroughly flushed from the boiler water circuit. The boiler should contain not more than 0.1 PPM of chloride ions or 0.2 PPM of sodium ions.

- *Boiler firing equipment* The float safety valve programme will normally take place following the steam purge operation and so boiler firing equipment will be available, but it must be ensured that sufficient oil burners are available on both oil- and coal-fired units to enable the boiler to be fired at a high rate to enable the steam pressures needed to float the valves to be obtained as quickly as possible. Due regard must be made during the firing of the boiler to operating parameters; in particular, drum metal temperature-differentials and superheater metal temperatures must be carefully monitored.

13.4 Safety and resources

13.4.1 Safety

Personnel safety

- Floating safety valves, particularly on prototype units, can be hazardous and therefore only essential personnel should be in the valve vicinity. The area around the safety valves and escape pipework must be barriered off and adequate warning notices displayed. Access to the roof area of steam discharge must be strictly controlled and restricted to test team personnel.

- Access platforms to the safety valves may be required and these should be built with alternative access/escape paths for emergency purposes.
- High noise levels will be generated during the safety valve floating and therefore ear defenders must be provided and worn by test team personnel.
- A campaign should be carried out prior to safety valve floating to warm site personnel and the general public local to the site of the procedure that higher-than-normal noise levels will be generated. Testing should be carried out during normal day hours, be suspended at night and should not be carried out on Sundays, whenever possible.

Plant emergencies

Adequate procedures should be available to deal with any emergencies that may occur during the testing of the valves. Likely emergencies that could occur are:

- Failure of communications
- Malfunction of the electromechanical control system
- Safety valve sticking in the 'shut' or 'lifted' position
- Safety valve bouncing
- Leakage or failure of safety valve escape pipework, supports, silencers, etc.

Emergency actions

- If a drum or superheater safety valve sticks open, the boiler pressure should be lowered in an attempt to reseat the valve. Feeding of the boiler will be adjusted to match the valve discharge and drum level, the overriding consideration being to keep within the allowable drum metal temperature-limits. Burners will continue to be fired with the object of controlling the pressure fall rate.

  The economiser recirculation valve should be used to reduce the risk of thermal shock to the drum.

  If a valve fails to reseat by the time the boiler pressure approaches 100 bar, or water make-up becomes an embarrassment, then the fire will be extinguished, the furnace purged and the boiler boxed up. The boiler will then have to be depressurised to inspect and repair the valve.
- If a reheater valve sticks open, the boiler pressure should be lowered in an attempt to reseat the valve. Depending on the load, the drop in boiler pressure may result in a load drop. With due regard to drum metal temperature-limits, the load can, if necessary, be dropped until the reheater is just above atmospheric pressure. The subsequent action to be taken will then be based on an assessment of the situation at the valve.
- Because the boiler is now to be taken to full design pressure, particular vigilance will be required to ensure free expansion of the boiler and particularly to the safety valve exhaust system. Procedures should be available to monitor these expansions and to notify the test team leader should problems occur during the testing.

13.4.2 Resources

Equipment

- The initial preparations described in Section 13.3 of this chapter
- Electrical supplies to anti-simmer devices
- Pneumatic air available for anti-simmer devices
- Signalling system available and tested
- Hydroset unit and test pressure gauges
- Hydraulic rig for adjusting anti-simmer devices limit switch settings
- Ear defenders
- Multimeter
- 500 V Megger
- Temporary barriers and warning notices
- Gags
- Test pressure gauge, complete with test certificates

Staffing

A team of six people will require the following time to complete the floating safety valve exercise:

- Set high and low pressure microswitch settings on the anti-simmer device with hydraulic pump rig — seven days.
- Inspection of valves — two days.
- Estimation of mechanical lift with Hydroset — two days.
- Setting of calculated mechanical lift — two days.
- Setting of design lift — two days.

The reheat safety valves will take approximately two days to complete, using the Hydroset unit.

13.5 Float safety valve procedure

13.5.1 Pre-commissioning checks

- All mechanical inspections completed and the relevant SCLs completed on the safety valve escape pipework, silencers and servo assistance equipment.
- All the checks and precautions in Sections 13.3 and 13.4 of this chapter completed and the procedures ready.
- Boiler ready for pressure raising, or at pressure following the steam purge exercise.

13.5.2 Commissioning procedure

Note: The following procedure is written for the floating of the Dewrance compensated back-pressure closing Maxiflow safety valve. The procedure will be similar for other types of valves, particularly the mechanical lifting procedure to set the superheater and drum valves. There will be some difference with other types of servo assistance. A simplified commissioning network diagram is shown in Fig 4.86.

Set the high and low pressure limit switches of safety valve anti-simmer device control boxes

The limit switches controlling the operation of each of the drum and superheater safety valve's anti-simmer devices are set, using a Dewrance hydraulic test pump. This necessitates disconnection of the impulse pipe into each control box and the connection of the hydraulic test pump by suitable adaptor; the electrical supply to each control box must be isolated.

Two test meters (such as an Avometer) are first connected across the terminals of the high and low pressure limit switches. The pump is operated so that pressure shown on the test pressure gauge (associated with the test pump) is set to the requisite lift pressure for the selected valve; for the highest set drum safety valve this is 193.5 bar, and for the highest set superheater safety valve this is 176 bar. Once settled on the desired pressure, the high pressure limit switch can be carefully adjusted such that the normally-open limit switch closes, shown by the test meter indicating continuity.

Note: During the period that pressure is being raised to the lift pressure, the normally-closed contacts of the low pressure limit switch must be 'Open', shown by the test meter indicating loss of continuity.

Hydraulic pressure should then be carefully reduced to the reseat pressure level, and the low pressure limit switch adjustment regulated until the contacts close and the test meter again indicates continuity.

Once high and low pressure limit switch settings are established, they should be checked for repeatability by raising and lowering the pressure in the hydraulic test unit to lift and reseat pressures, respectively.

Upon completion of limit switch setting, operation of the electrical control circuitry should then be checked, so that correct operation of the pressure switches operates the solenoid valves in the air pipework to the anti-simmer device actuator.

Once satisfactory operation of the electrical control system is established, operation of the anti-simmer actuator should then be checked after reconnection of the impulse pipe to the control pressure switch.

The anti-simmer device safety valve should then be carefully adjusted until the valve lifts. The accurate setting of the safety valve is particularly important, in that it is vital that the safety valve relieves pressure on top of the actuator diaphragm if the control system fails and a situation develops such that the main safety valve is required to lift.

The electrical supply to the selected valve should next be energised. In this instance, with no boiler pressure, the low pressure limit switch should be in its normally-closed position, supplying the latch relay and consequently energising the solenoid of the three-way valve. The valve then should open and switch the regulated air supply onto the anti-simmer device actuator. Observation at the valve should indicate travel of the anti-simmer actuator operating-rod towards the valve spindle. The system should then be checked for any evidence of air leakage and that the air pressure is being correctly regulated.

Reverse action of the anti-simmer actuator should then be checked by depressing the 'Test' pushbutton: this removes supplies from the three-way solenoid valve, exhausting air to atmosphere, with resultant retraction of the actuator operating-rod. The 'valve open' indicating lamp should also be illuminated.

Upon release of the pushbutton, supplies are reconnected to the solenoid valve, with subsequent operation of the anti-simmer actuator — the latch relay being retained in circuit through the normally-closed contacts of the low pressure limit switch.

Upon completion of operational checks on all drum and superheater safety valve anti-simmer actuators, the safety valves are ready for setting the mechanical lift points, using the Hydroset test unit.

Set mechanical lift-points of drum safety valves, using the Hydroset test unit

To set the mechanical lift point of the drum and superheater safety valves a Dewrance/Dresser Hydroset test unit is needed. The anti-simmer actuator is removed and a Hydroset unit fitted, to the compression screw of the valve to be set, by the specially constructed yoke. The operating plunger of the Hydroset actuator is then connected to the safety valve spindle by a turnbuckle having right-hand and left-hand threads which facilitates connection by turning clockwise. The Hydroset actuator is connected by a flexible hose to the discharge of a single-ram hydraulic pump. A calibrated pressure gauge is fitted to the Hydroset actuator.

For this method of valve setting, the boiler pressure must be standing between 60% and 80% of the required mechanical lift pressure: this is to avoid the possible inaccuracy in setting, if the boiler pressure were too low; if too high, the valve could be sent into the full-lift position. An ideal time to programme this is after the completion of the steam purge, when the boiler is at pressure.

Graphs will be available from the manufacturer to indicate the differential between the desired set-pressure and boiler pressure, and the hydraulic pressure to be applied to the Hydroset actuator.

Separate graphs should be provided for reheater, superheater, and drum safety valves.

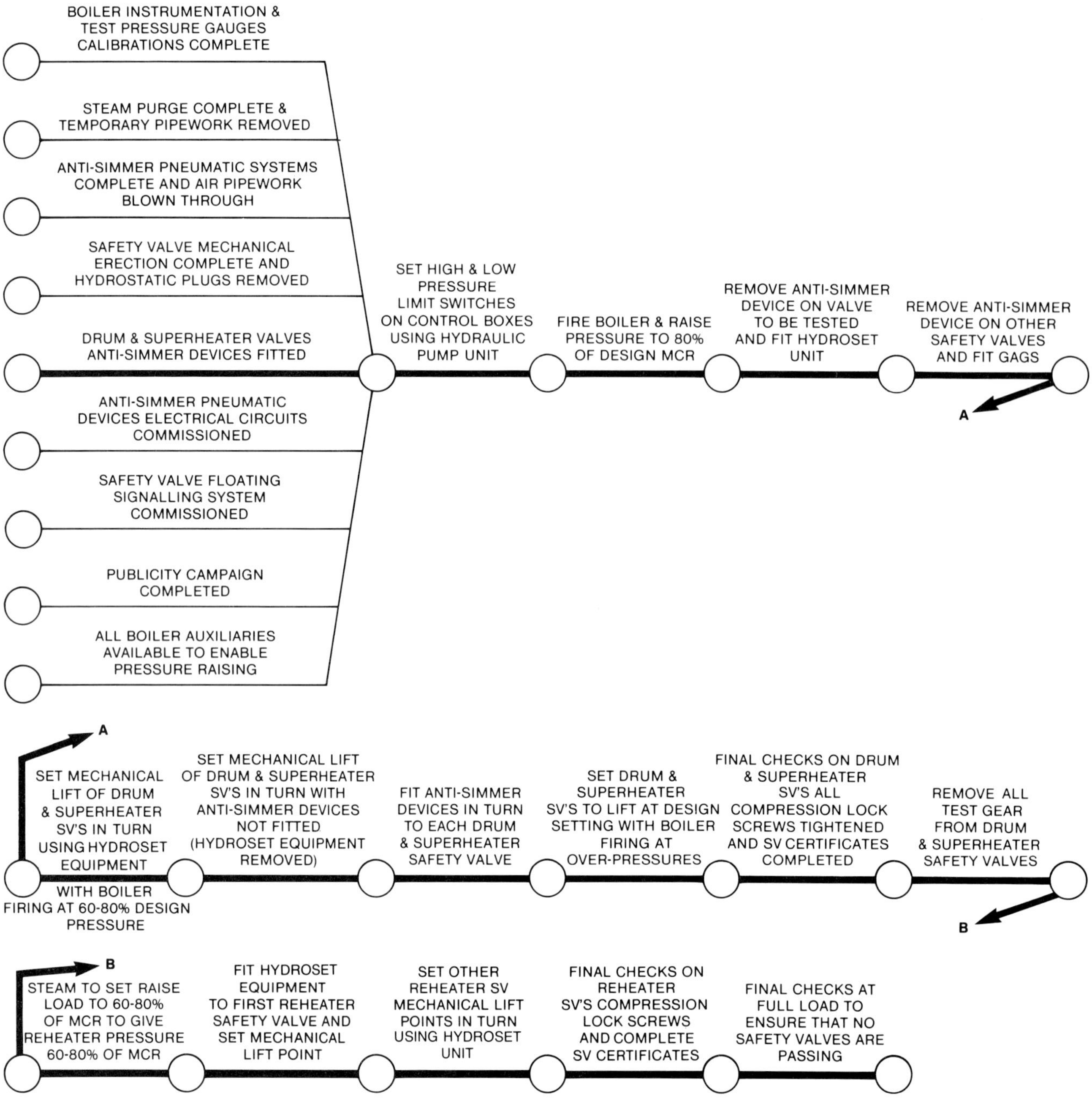

FIG. 4.86 Floating drum, superheater and reheater safety valves commissioning network

A typical graph for a drum safety valve is shown in Fig 4.87. An example of how the graph is used to establish the required hydraulic pressure is as follows:

- Boiler pressure 140 bar, i.e., approximately 70% of safety valve lift pressure.
- Safety valve lift pressure is 191.7 bar (see Fig 4.81).
- Difference between pressures is 51.7 bar.
- On graph, differential pressure of 51.7 bar gives reading on the horizontal axis of 44 bar.
- 44 bar should be hydraulic pressure required to put the valve into the simmer position (lift point).
- If the hydraulic pressure to achieve valve simmer is greater than 44 bar, it will be necessary to reduce spring compression; if less, increase spring compression.

The calculated compression spring adjustment, tabulated in adjustment of compression screw per flat against change in pressure, will be provided by the valve manufacturer for each valve.

Access to the drum and superheater valve stations must be complete and safe, including emergency escape scaffolding from the boiler drum and platforms and superheater safety valve access galleries; this facilitates rapid access to and from the safety valves if emergency action is needed at the valves.

Access to the boiler house roof should be prevented for the duration of the exercise, after a final inspection that all personnel are withdrawn from the area.

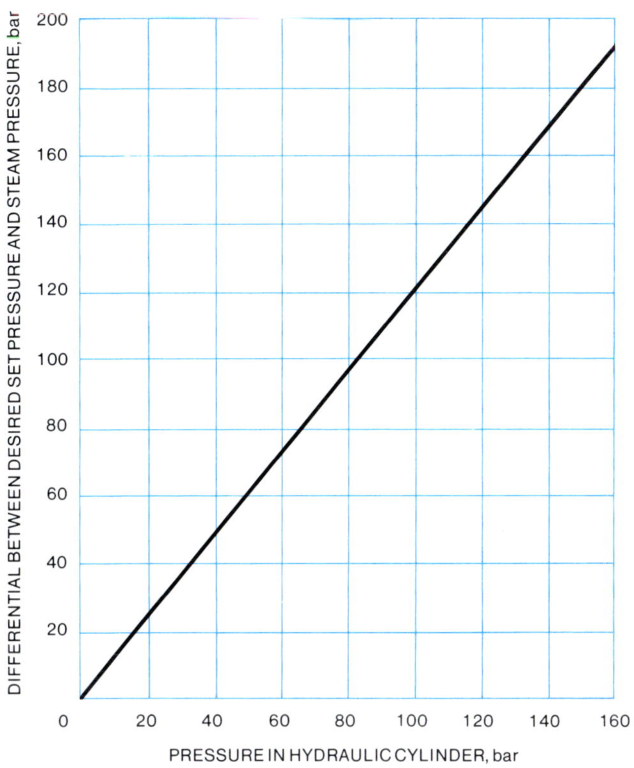

FIG. 4.87 Pressure diagram for use with hydraulic setting device (Hydroset)

The order of valve setting will proceed such that the designed highest-set drum safety valves are set first.

With all gags fitted, except for the valve under test, communications established, inspector's test gauge fitted and in service, and Hydroset actuator fitted to the chosen highest set-point drum safety valve, the boiler pressure is gradually raised towards the desired system pressure of, say, 140 bar. Once achieved, it is important to maintain the pressure steady, using a combination of firing rate and regulation of dump steam control valves, or the final superheater drain valves, as required.

Once steady boiler pressure is established, the hydraulic system pressure can be raised progressively towards the desired pressure, obtained from the graph.

The hydraulic system should be checked carefully for any sign of leakage and for the correct level of hydraulic fluid in the pump: too much fluid in the pump and hose can cause malfunction of the Hydroset, due to excessive expansion of the fluid under the hot environment of the boiler. The hose should essentially be free of any kinks, loops or knots which could impede the flow of the hydraulic fluid. With the hydraulic system checked as satisfactory, hydraulic pressure should be gently raised to the desired level, carefully maintaining observation for the start of the valve simmer-point. If the desired hydraulic pressure is reached without the simmer point being achieved, the hydraulic pressure should be reduced to zero; then, after ascertaining that the boiler pressure is still steady, the hydraulic system is again raised to the desired pressure.

If simmer point is again not reached, the hydraulic pressure should be gently increased until the valve starts to simmer, as detected by the valve 'spitting' and/or downward deflection of hydraulic system pressure gauge. The deflection pressure should be noted, the hydraulic pressure removed and the valve compression reduced in accordance with the compression-spring adjustment chart. Alternatively, if the valve simmers at a hydraulic pressure lower than desired value, compression should be increased.

Once the desired Hydroset pressure and valve simmer-point occur simultaneously, hydraulic pressure should be reduced to zero, and the repeatability of the valve lift-setting established by once again raising the Hydroset pressure to desired value.

After satisfactory setting of the highest-lift drum safety valve, the gag should be carefully removed from the next chosen high-set valve and the previous procedure repeated.

It is repeated until all the drum safety valve lift-points are set at the desired values.

Once all the drum valves are set, the boiler pressure should be reduced to approximately 100 bar, the test pressure gauge then transferred to the superheater connection, and the setting procedure repeated for the four superheater valves, using a boiler pressure of approximately 120 bar.

With all the drum and superheater lift-points established, the mechanical lift and reseat pressure settings now need to be checked, without the Hydroset unit.

Check mechanical lift and reseat settings of drum and superheater safety valves (without anti-simmer devices fitted)

This exercise is effectively that of conventional safety valve floating, except that here the lift and reseat pressures will be 2% below the design points, because of the removal of the anti-simmer devices.

With the boiler pressure at approximately 100 bar, preparations should commence for the start of the operation, i.e., communications between valve stations and control room as described in Section 13.3 of this chapter established, test pressure gauge fitted to the drum connection, and safety precautions regarding the general area and boiler house roof established.

All the drum and superheater safety valves — other than the highest set drum valve — must be gagged, and obviously the anti-simmer actuators must still be left unfitted.

When all these conditions are satisfactory, the engineer in control of operations at the boiler-top issues instructions to raise pressure, using the coloured light signalling system.

Pressure will then be progressively raised at a rate within the allowable rates of saturation pressure rise, together with drum and tube metal-temperatures: the reheater at this time is receiving no steam circulation, so particular attention must be paid to ensure that metal-temperature limits are not exceeded. To ensure a good clean valve lift, it is important that the last 15 bar pressure rise is achieved rapidly, using burners and by throttling the final superheater drain or steam dump valves. Also, control of water level is important, to avoid the possibility of abruptly lifting water from the drum into the superheater, as the valve lifts.

The point at which the valve lifts is noted and recorded, then the engineer at the boiler top initiates the 'Lower pressure' signal. The unit operator then reduces firing and rapidly reduces pressure by use of the final steam dump valves, until the 'Hold' signal is given from the boiler top.

Note In the event of the valve not lifting at the required mechanical lift pressure, boiler pressure should be allowed to rise to a maximum of approximately 3.5 bar above design pressure in order to try to establish the lift point. If the valve has not lifted at +3.5 bar, however, the engineer will give the 'Lower pressure' signal, and reduce the pressure to approximately 125 bar, to allow the engineer to reduce spring compression in preparation for a further attempt.

The point at which the valve reseats is noted and recorded. A decision will be made as to whether the valve settings are satisfactory for use as the basis for final design settings, using the anti-simmer facility. If not satisfactory, the spring compression or blowdown ring will be adjusted, as required by the engineer, and a further attempt will then be made, using the above procedures. Dependent upon the circumstances under which the original lift occurred, the engineer may request that the valve be left to cool for a period; the valve will then be gagged once again and preparations made to test the other high-set valve on the chosen side of the drum. This procedure continues until all the mechanical lift and reseat pressures of the drum safety valves are established.

Once achieved, the boiler pressure is reduced to approximately 100 bar, the test pressure gauge transferred over to the superheater connection, and communications systems re-established. The foregoing procedure is then repeated to set the four superheater safety valves; again, the setting sequence starts from the highest set valve.

Once all the drum and superheater valves are set for mechanical lift and reseat points, preparations should be made to fit the anti-simmer actuator and set the valves to operate at design lift and reseat points. Boiler pressure should be reduced to approximately 100 bar, to facilitate the fitting of the pneumatic actuators to the heads of the valves.

Set design lift and reseat points at drum and superheater safety valves (with anti-simmer devices fitted)

The procedure is virtually the same as that used for previous activities. Starting with the highest-set drum safety valve (to which the anti-simmer actuator will now have been fitted), air and electrical supplies are made available, and checked for operation by use of the test button. The boiler pressure should be standing at approximately 100 bar. The remaining drum and superheater safety valves should all remain gagged. The test pressure gauge will be refitted to the drum connection.

Once all safety precautions are established as satisfactory, with communications between valve stations and control desk checked, the engineer in control of operations at the boiler-top signals to 'Raise pressure'.

It is expected that the efforts taken to set the mechanical lift and reseat pressures, together with design high and low set-points of the anti-simmer device pressure switches, should ensure that the safety valve lifts at very close to design requirements.

Once again, the pressure rise over the last 15 bar should occur rapidly, to effect a sharp clean lift. The point at which the valve enters into full lift is noted and recorded; once this occurs, the engineer signals to reduce pressure.

The point at which the valve reseats should be noted and recorded. Recorded lift and reseat points will be scrutinised by the insurance inspector and CEGB engineer. If considered satisfactory, the lift and reseat pressures will be recorded on the Safety Valve Certificate. In the event that the valve does not lift at the required design pressure, the boiler pressure should be allowed to rise only to a maximum of 3.5 bar above the required design point. If the valve still fails to lift, the engineer will signal for the pressure to be reduced and held at approximately 125 bar.

The engineer will then check the operation of the high pressure limit switch to see whether it has failed to remove the supplementary loading being applied to the valve.

Similarly, if design reseat pressure is too low, operation of the low pressure limit switch, which might have failed to reapply the supplementary loading to the valve early enough, will be checked.

Note It is important to avoid having too low a blowdown range, i.e., 'reseat' pressure too close to 'lift' pressure. If this occurs, it is possible for the valve to start 'bouncing' quite violently.

Once design lift and reseat pressures are achieved, the gag is removed from the next high-set drum safety valve, the anti-simmer actuator fitted, the air and electric supplies made available, and actuator operation checked by use of the test button. With satisfactory operation achieved, the design lift and reseat points of the valve are set, as previously.

After successful setting of the six drum safety valves, the boiler pressure is lowered to permit fitting of the test pressure gauge to the superheater connection. The anti-simmer actuator is fitted to the highest-set superheater safety valve, electric and air supplies are made available and the actuator operation is checked using the test button facility. The design lift and reseat pressures are then set, in accordance with previously described procedures.

After successfully setting the four superheater safety valves, the boiler pressure should be reduced to approximately 100 bar, and the test pressure gauge removed.

The engineer then checks that all compression lock-screws are tightened, and that the measurement between the underside of the compression screw and topside of the locknut is taken at each drum and superheater safety valve and recorded on the specially prepared record sheet. From this point, the boiler can be steamed in accordance with the requirements of the overall commissioning programme.

Set mechanical lift point of reheater safety valves, using Hydroset test unit

The reheater safety valves cannot be set until the unit is on-load, thus entailing a separate exercise, which will now be described.

The mechanical lift and reseat pressures of all reheat safety valves are set in the manufacturer's works. However, the development of the Hydroset unit has facilitated dynamic testing of these valves in service, but it only establishes the mechanical lift point and not the reseat point. This limited facility has been welcomed and accepted by both the CEGB and Insurance Inspectorate.

To carry out this exercise, some method must be used to regulate the reheater system pressure to between 60% and 80% of design pressure, whilst the unit is operating at part-load, such that operating heat-fluxes are somewhat less than those at MCR conditions. It is therefore necessary to arrange for the turbine to be run at this rating: on a 500 MW unit, this would be approximately 350 MW.

The basic procedure is that used when setting drum and superheater safety valves, and the valves are be set in descending order from the four highest-set reheater inlet valves, then the two low-set reheater inlet safety valves, to the four similarly-set reheater outlet safety valves.

Obviously, with this type of exercise, a substantial amount of co-ordination is necessary between the valve station and the unit control desk. In this instance, overall control must rest with the Shift Charge Engineer, who will have as a priority maintenance of generation of target load commensurate with continued safety of the turbine/boiler unit, particularly if turbine interceptor valves are being used to control reheater system pressure. Close observation of reheat metal temperatures, together with reheater outlet steam temperature, is needed, in addition to observing various turbine parameters throughout the exercise.

The previously used light signalling system should be available, together with telephone communications at a suitable position adjacent to the valve stations at either side of the boiler. Temporary access must be provided between valve stations and galleries to enable the engineers to gain safe and speedy access to and from the valves, in case emergency action is needed. A suitable pressure tapping must be provided to fit a test pressure gauge at each communication control station.

The reheater safety valves are not fitted with anti-simmer devices, so only the mechanical lift pressure needs to be set at the design point, using the Hydroset unit.

Whilst the boiler will be operating under the automatic boiler control system, it is considered prudent to select the reheater control loops to 'hand' control, under which the unit operator will make any adjustments needed to maintain stable operating conditions throughout the safety valve setting exercise.

Note During this exercise, which is being carried out with the unit on-load, it is not permissible to have the nine remaining safety valves gagged, as would be the case with a static unit; to do so would obviously infringe insurance regulations. However, in the interests of personnel safety during the periods that test engineers are attending to the valve being set, it is considered prudent to fit gags to the immediately adjacent valve/valves. The gags will be removed as the engineers withdraw from the area.

If the reheater system pressure is being deliberately elevated by use of turbine interceptor valves, it is possible that any request for a reduction in system pressure made by the engineer at the valve station will be achieved by opening the interceptor valves, rather than by reducing generated load. This point is, however, perhaps speculative and the actual mode of operation has to be determined between the CEGB and turbine manufacturers.

The Hydroset hydraulic unit should be fitted to the selected highest-set reheater inlet valve and the operation tested, as previously.

With the unit settled at target load, preparations should be made for the safety valve setting. The engineer requests that the reheater system pressure be raised to a target level of 60–80% MCR pressure. Observation of operational parameters is maintained for a test period of, say, 30 minutes, to establish that stable operation is attainable with the throttled conditions.

Once it has been established that satisfactory unit conditions are attainable, the Shift Charge Engineer advises the engineer at the valve communication station that it is in order to proceed with valve setting.

The engineer now establishes the mechanical setting of the chosen valve by raising the hydraulic system pressure towards the required level using a graph simulation (Fig 4.87). Care must be taken to ensure that the valve does not enter into the full-lift position. Hydraulic pressure is then gently raised until the valve simmer-point is reached, as denoted by 'spitting', or by downward deflection of the hydraulic system pressure gauge.

As previously described, safety valve compression spring-loading is increased or decreased, as necessary, to accommodate the difference between design lift pressure and that actually determined by use of the Hydroset.

Once the design lift-point of the first valve is established, the lift pressure of the valve is recorded, the Hydroset unit transferred to the next valve to be set, and the operation of the hydraulic circuit checked in preparation for setting the design lift-point of this valve.

Upon completion of the exercise, the turbine interceptor valves will be regulated to normal operating position, if they have been used to regulate pressure. The engineers will ensure correct tightening of all compression lockscrews, and will measure the distance between the underside of the compression screw and the topside of the locknut at each valve, readings being entered on the specially prepared record sheet, to be kept for reference purposes for future work on the valves.

The test pressure gauge, Hydroset test unit, valve gags, etc., are then removed to a place of safe keeping. All temporary scaffolding facilities are dismantled.

The unit loading can now be adjusted to suit the requirements of Grid Control or of the continuing commissioning programme.

14 Boiler commissioning from completion of safety valve floating

This chapter has covered boiler commissioning from completion of erection up to the floating of boiler safety valves. This involved operating the boiler to design pressure on limited numbers of oil burners only, and a great deal remains to be commissioned when the boiler and turbine are operated as a unit on-load. This is covered in Chapter 6 and will involve operating oil-fired boilers with all oil burners in service: this often manifests problems not encountered with limited burner operation. It also involves the commissioning of milling plant and coal burners on coal-fired boilers.

Boiler automatic controls, involving single- and three-element drum level, combustion, superheater and reheater temperature, air-heater steam pressure and temperature, superheater and reheater and master pressure controls are also covered in Chapter 6.

CHAPTER 5

Turbine-generator and feedheating systems pre-steam-to-set commissioning

Introduction

The sections in this chapter describe the commissioning requirements for each system and major subsystem of a large modern turbine generator and its associated feedheating and condensate systems. The sequence in which the activities are described is not necessarily in order, although many of the activities are interdependent. It is therefore important to construct programmes to establish the relationships between the parallel and sequential activities, in line with modern planning practice.

Each section contains a particular example for commissioning purposes, together with a programme giving significant activity details for that system. In practice, these programmes must be integrated with overall erection and commissioning programmes for the turbine-generator and feed system. Similarly, these overall programmes would be further integrated with those for the boiler and common plant.

The interface between completion of construction and the commencement of commissioning may be difficult to define. This point must be clearly established to ensure a smooth transition between them in order that overlap and, perhaps more importantly, gaps are not created. The responsibilities for the two activities should also be clearly defined, since they can be organised in a number of different ways.

Most British manufacturers have their own commissioning organisation and may therefore take this into account when preparing erection and commissioning procedures. Such procedures should be carefully examined by the CEGB.

The following sections, therefore, contain references to a number of aspects of erection which will be important to the CEGB, not only for operational and maintenance purposes, but also in the event of a major plant shutdown and recommissioning activity. This information will include works test records, plant preservation during storage and erection, site test records and other data, such as final alignment and clearance readings.

Some commissioning activities are highly specialised. The sections dealing with electronic governing, excitation and automatic voltage regulators (AVR) are intended as a guide. The CEGB should provide or seek specialist advice to ensure that the supplier's information is sufficiently comprehensive, as well as to provide commissioning surveillance and expertise.

It is common practice in the CEGB to use the manufacturers' procedures and programmes including, where applicable, standard or specialised check lists, after they have been examined in detail and agreed by the principal parties involved. Some of these procedures may have to be expanded to include a number of different suppliers or to integrate certain activities. An example of the latter is the boiler steam purge activity, which will include both boiler and turbine high pressure (HP) pipework systems.

It is suggested that three commissioning Papers of Principle be constructed:

- For the main turbine and condenser
- For the generator, exciter and main electrical connections
- For the condensate, low pressure (LP) and high pressure (HP) feed systems

The preparation of these papers is described in Chapter 2. It is important that the Principles document is sufficiently comprehensive. Appendices A, B and C to this chapter show examples of contents lists relevant to the three Papers of Principle.

A comprehensive list of drawings is required. These will be of a general nature for the Papers of Principle, but detailed commissioning documents must have detailed lists of drawings relevant to the procedures.

Detailed procedures should also include comprehensive schedules of instruments, cables, valves, orifice plates, cabinets, cubicles and alarms.

It is important that plant conditions, including operating parameters and limitations, are specified carefully. These will be related to the particular plant to be commissioned and will not be general. Similarly, the services required for commissioning must be given considerable thought. Examples for the turbine and condenser are included in Appendices D and E to this chapter respectively.

Plant and personnel safety policy statements are another important aspect of commissioning and should be prepared for the particular activities to be undertaken. Examples for the turbine and condenser are included in Appendices F and G to this chapter respectively.

A policy statement on plant preservation should also be given in each Paper of Principle. Later sections of this chapter make reference to preservation details, where relevant.

Each section in this chapter includes a general description of plant items or systems to be commissioned. This description is intended to give the reader an account of plant likely to be encountered, together with sufficient detail of the specific examples upon which the discussion on detailed commissioning procedure is based. The description is followed by a list of commissioning objectives. One commissioning objective common to all sections is:

- To ensure the completion of erection in accordance with specifications and drawings.

1 Main steam turbine

The plant items covered by this section are the high pressure (HP), intermediate pressure (IP) and low pressure (LP) stages of the main steam turbine, including the HP and IP valve chest and connecting (i.e., 'loop') pipes to the turbines. The steam pipes between boiler stop valves and turbine valve chests are dealt with in Section 2.

1.1 General description

The detailed commissioning procedure relates to a large modern 660 MW turbine comprising an HP turbine, an IP turbine and three LP turbines.

The HP turbine (Fig 5.1) is of triple-shell design, comprising an inner casing formed with a horizontal joint, a sleeve casing which is a complete barrel without a horizontal joint and an outer casing having a horizontal joint. The HP turbine rotor shaft is a single-flow design and is machined from a chrome-molybdenum vanadium steel forging with eight discs machined from the forging.

The IP turbine (Fig 5.2) is a double-flow design having seven stages of blading situated each side of a central steam-inlet belt. The first three stages of each flow are located within an inner casing and the remaining four stages within a diaphragm carrier ring. The IP rotor is forged from the solid and gashed by machining.

The first stage of each flow is subjected to the maximum temperature of reheat steam used in the IP turbine. Therefore it is necessary to provide IP shaft cooling to these stages and the immediately adjacent areas.

The cooling steam is introduced to an annular chamber between the two first-stage nozzle plates. From this chamber, steam is directed along the rotor and up the inlet face of each first-stage disc, cooling the rotor and discs before combining with the main steam flow through the turbine.

The LP turbine (Fig 5.3) consists of three double-flow turbines having six expansion stages situated within one of three separate condensing sections which jointly constitute the condenser shell. The steam to drive this turbine comes from the IP turbine exhaust and is connected in parallel to the three LP turbines by the IP/LP crossover pipework.

Steam is bled from No. 1 LP turbine to No. 1 direct contact (DC) heater, from No. 2 LP turbine to No. 2 DC heater, and from No. 3 LP turbine to No. 3 DC heater (see Section 14 of this chapter).

Similarly steam is taken from the final stages of all three LP turbines to the moisture extraction condensers.

There are two HP valve chests (Fig 5.4), one on each side of the HP turbine, each comprising two main stop valves and two throttle valves. The valves are arranged so that a steam flow is maintained through all the loop pipes when one of the valves is shut. This facilitates on-

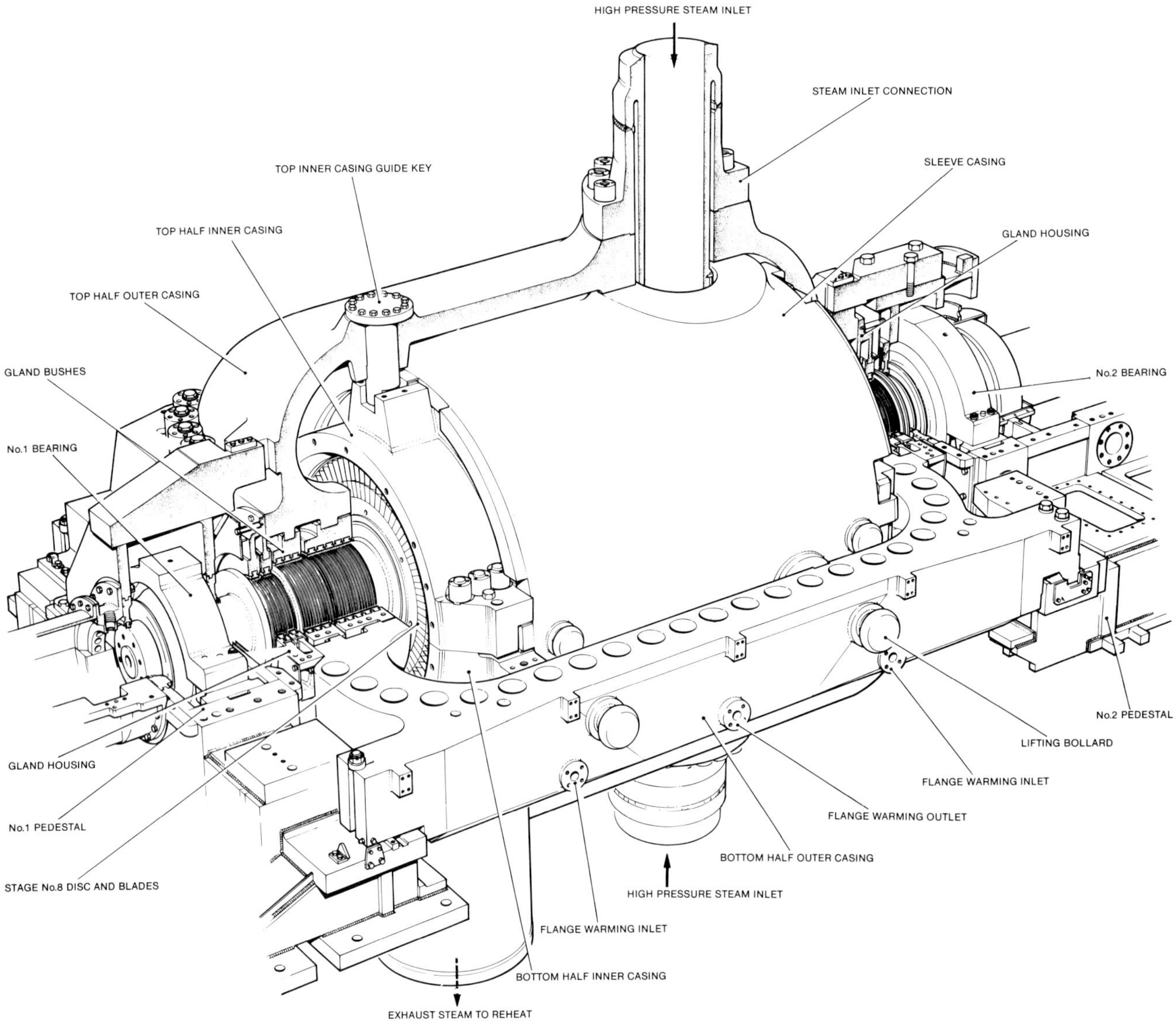

FIG. 5.1 HP turbine

load valve testing and also allows the turbines to be run up, kept on load, or shut down in the event of a valve control fault.

There are two IP valve chests (Fig 5.5), one on each side of the IP turbine. Each chest comprises two reheat stop valves and two interceptor valves. The valves are interconnected in a similar manner, and for the same reasons, as for the HP valves.

All valves are operated hydraulically by means of valve operating gear (Fig 5.6). The gear is controlled by signals from an electronic governing system.

1.2 Commissioning objectives

The commissioning objectives for these major components of the turbine are:

HP, IP and LP turbines, valve chests and loop pipes

- To ensure that all final data for alignment, clearances, etc., are recorded
- To ensure that instrumentation is complete
- To ensure that lagging and cleaning are complete

HP and IP valve chests

- To ensure that steam valve and associated relay settings and movements are recorded
- To check that operation is free from stiction and lost motion
- To check valve trip features

HP and IP loop pipes

- To check erection methods, including pipe closure and weld techniques.

1.3 Detailed procedures

The detailed procedures involve few functional operations, but serve as a very comprehensive post-erection

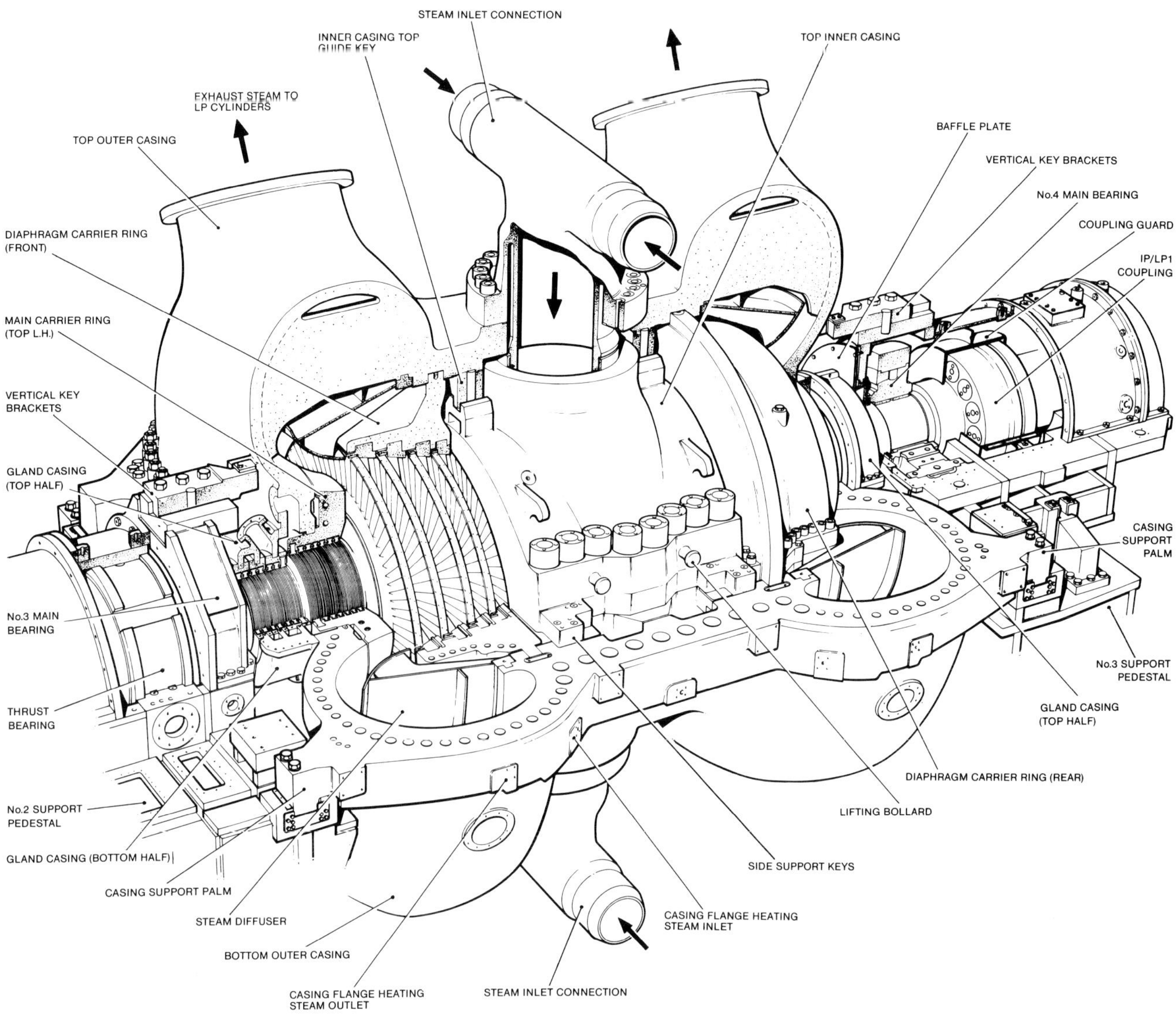

FIG. 5.2 IP turbine

check to ensure that these major components are complete to drawings and specifications. The checks must be completed before any of the associated systems, such as the lubricating oil system, are commissioned.

The Detailed Procedures Document should therefore comprise a very detailed inventory of:

- Erection records and data, including final clearance and alignment readings
- Instrument schedule of all pressure connections transmitters and thermocouples
- Alarm list
- Cabinets, cubicles
- Cabling
- Final inspection and boxing-up procedure
- Pre-operational activities procedures
- Instrument functional tests

Clean-condition working applied during the erection process must be maintained until such time as the final inspection and boxing-up procedures are complete. Some items may not be completely boxed up, so that other commissioning activities may proceed. For example, bearing covers may be left off and condenser steam spaces left open. These points must be integrated into the relevant procedures and programmes, and clean condition working maintained until the other activities are complete.

1.3.1 Erection records

These records should be available to the CEGB as part of the pre-commissioning activities. Data contained in the records, such as final clearance readings, should be examined and retained to form part of the history of the machine. Data is also important in establishing, for example, cold datums, which are used to assess differential expansions.

The erection records will contain much other information not directly associated with commissioning as such, but certain key points will act as final erection checks.

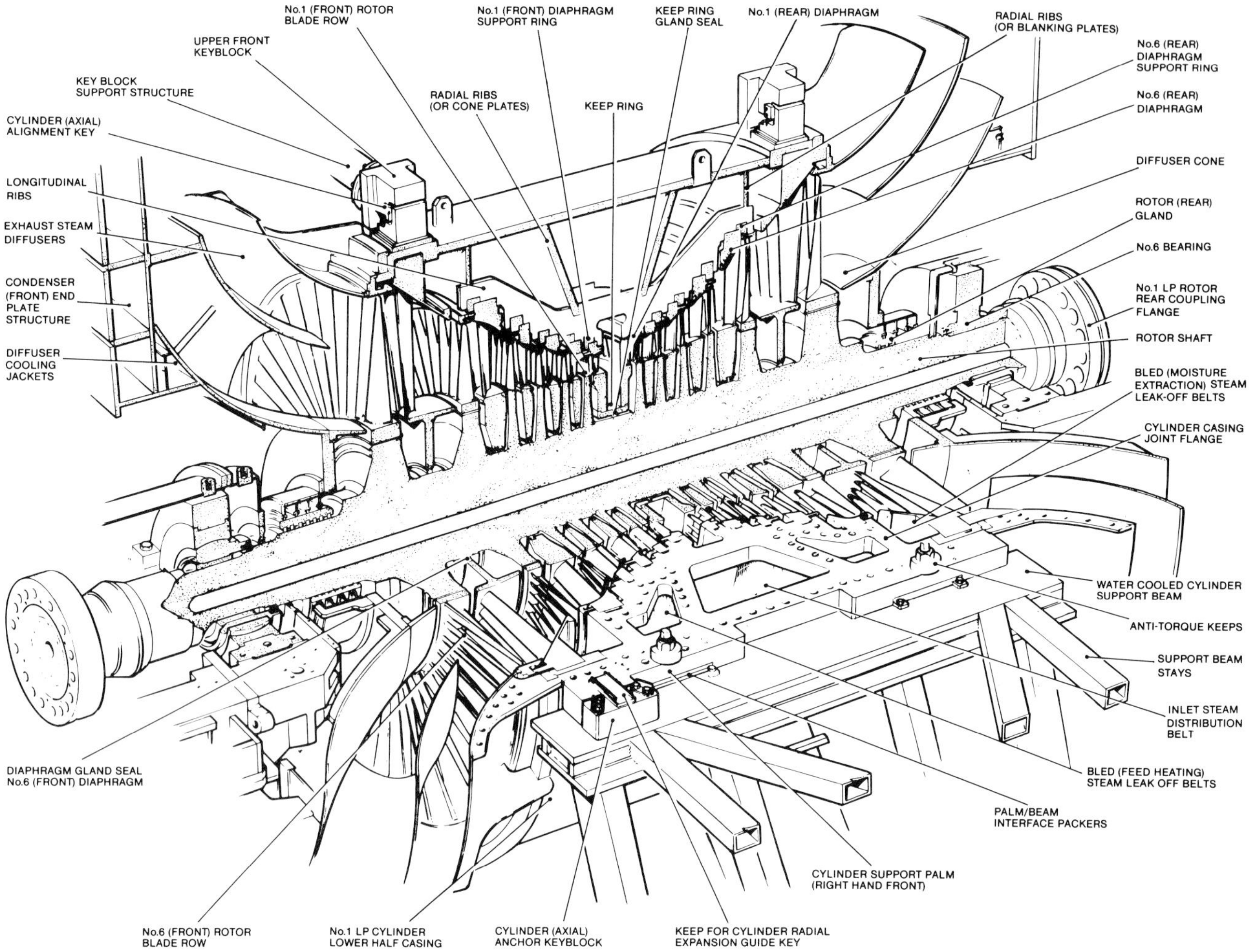

FIG. 5.3 LP turbine

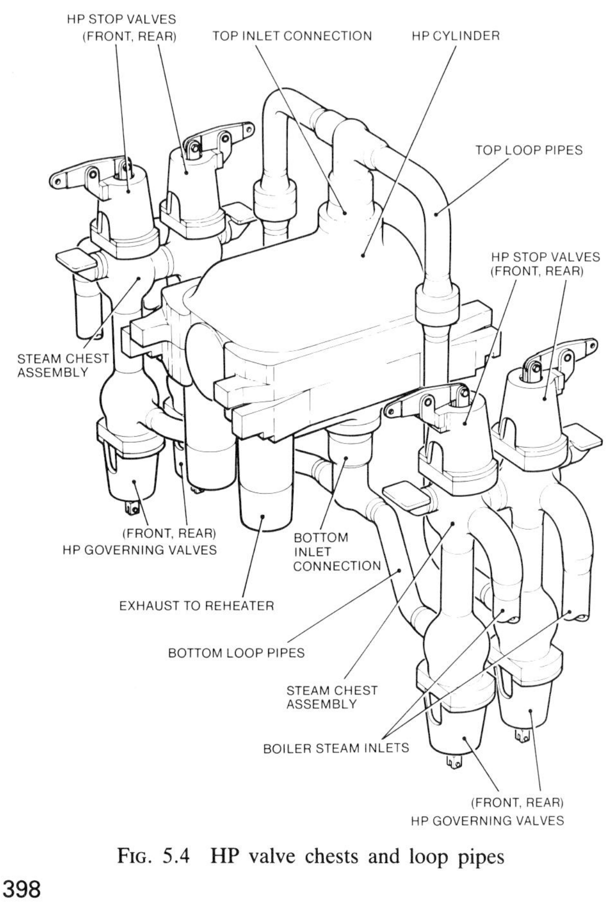

FIG. 5.4 HP valve chests and loop pipes

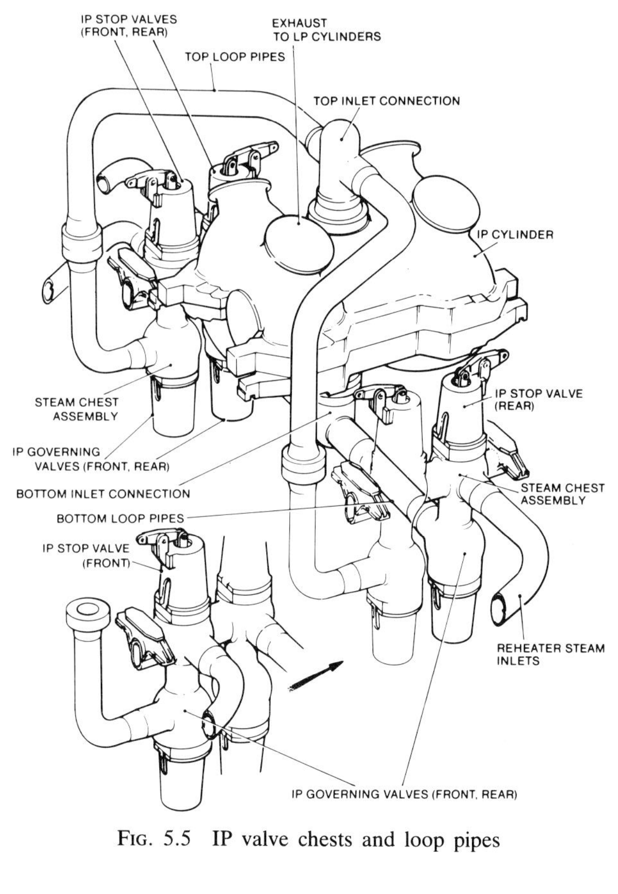

FIG. 5.5 IP valve chests and loop pipes

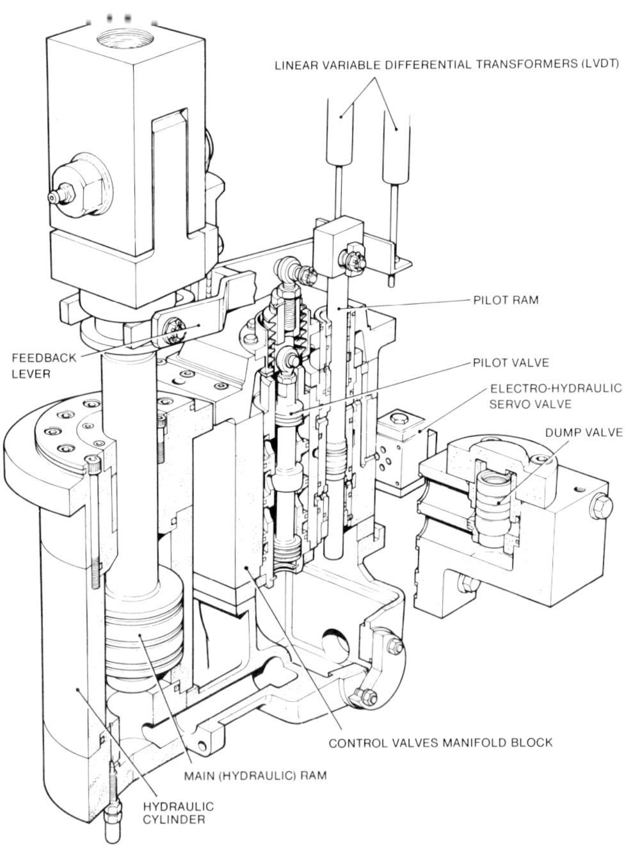

FIG. 5.6 Valve operating gear

A typical valve operating gear for all turbine steam admission valves, i.e., HP and IP emergency stop and governor valves. The steam valve (not shown) is opened against the action of a heavy duty spring, by means of the operating gear main ram.

The control gear comprises an electrohydraulic servo-valve, a hydraulic pilot ram and associated pilot valve, together with a solenoid-operated dump valve.

The servo-valve actuates the pilot ram to move the pilot valve, via the feedback link. The pilot valve then admits hydraulic fluid to, or releases it from, the main ram. The main ram movement is transmitted back to the pilot valve, via the feedback linkage, until the pilot valve is restored to a neutral position for the new position of the main ram. Pilot ram travel is signalled to the electronic valve control equipment by means of duplicate LVDT. This feedback signal instigates a further signal from the electronic control equipment to the electrohydraulic servo-valve to stop servo-valve movement when the main ram (and hence the steam valve) has reached the required position.

The dump valve is opened in an emergency to effect rapid release of power fluid from the main ram, thereby allowing the steam valve to close under the action of its spring.

1.3.2 Pre-operational activities — HP, IP and LP turbines

A thorough check should be made to ensure that fouls due to thermal expansion do not occur.

When flushing is complete (see Section 3 of this chapter), the mechanical aspects of turbine protective devices must be checked. The checks will be for freedom of movement and correct clearances, e.g., between trip latch mechanisms and overspeed rings, and between vacuum trip operating pistons and trip levers. The operation of the hand trip mechanism and duplicate trip solenoids may then be proven.

With lubricating oil supplies in service to supply the overspeed ring oil injection device, the device may be proven; for this, a temporary splash cover will be required.

Shaft earthing and voltage monitoring brushes must be checked to ensure that they are electrically connected and, where appropriate, that the specified series resistors are fitted. The electrical circuits are checked by lifting the brushes clear of the shaft. By means of a multimeter bridge and insulation test, check that brush resistance to earth is infinite, with the links or electrical connections disconnected. With links inserted and the series resistor shorted, check that the resistance to earth is zero (earthing brushes only). Then, with the shorts removed, check resistance to earth is correct (earthing brushes only — about 1 Ω).

Lagging must be checked to specification. The thickness is important and should be checked at numerous points over the whole surface of the turbine cylinders by means of a thin steel rod pushed through the lagging material. Lagging beneath cylinders must be tightly bonded. This can be checked by means of one or two larger holes to ensure that it has not sagged away from casings. Gaps between lagging and casing affect the heat distribution of the cylinders and create unnecessary stress and distortion. It should be noted that lagging is usually specified of greater thickness on the bottom casings in order to compensate for convective heating of the top casings, and therefore the need to lose more heat from the top in order to maintain uniform heat dissipation overall.

Where spray lagging has been used, this will be overlaid with a hard coating. A number of areas must be left off to allow for drainage when lagging is being dried during initial running of the turbine. Examples of erection records that would be examined as part of the pre-commissioning activity are:

(a) Complete inspection of cylinder support work

(b) Keys and palm packers to all keyways identified correctly and greased

(c) Internal casing drains clear and free from debris

(d) Thermal shielding securely fitted

(e) Internal piping and connections securely fitted and clean

(f) Final packing pieces fitted

(g) Rotor balance weights caulked and locked

(h) Dowels fitted

(i) Loop continuity checks on instrumentation prior to being boxed up

(j) Set up and record settings for rotor differential expansion equipment and ensure that all other supervisory equipment internal to pedestals is correct and complete.

Typical data required for plant history and maintenance is:

- Clearances for: Bearings
  Pedestals
  Glands (axial and radial)
  Keys
  Palms
  Diaphragm and wheels (axial and radial)
- Datums for: Glands to pedestals (axial)
  Cylinder palms to pedestals
  Rotors to cylinders (axial and radial)
  Rotors to bearing supports (axial)
- Relative heights and slopes for cylinder pedestal and half joints
- Shaft alignments, measured at each coupling
- Concentricity checks for each coupling, each pedestal and cylinder gland bore-to-rotor
- Shaft/casing 'bump' check records
- Coupling bolt extensions when tightened
- Cylinder bolt tightening records

1.3.3 Pre-operational activities — HP and IP valve chests

A thorough inspection must be made to ensure that construction is complete, temporary supports are removed and that fouls due to thermal expansion do not exist.

Erection records should be examined to ensure that the final settings and clearances associated with valves and valve chests are complete.

All external hydraulic pipework to valve operating gear will have been flushed and reinstalled (see Section 3 of this chapter).

Instruments, including valve position potentiometers, local indicators and valve limit switches, must be checked against a schedule.

Mechanical adjusting devices for pivots, links, etc., must be checked to be correctly set and locked in position in accordance with the detailed procedures. All points of lubrication must be checked and specified lubricants applied. Before valve chests are finally fitted with valves and covers, internal cleanliness must be verified. Modern equipment, using TV and fibre optics for inspection of inaccessible cavities within the chest, may be used.

For the system shown in Fig 5.6, the manufacturer will specify detailed functional checks to ensure that the valves and hydraulic operating mechanisms are set and operating correctly. These checks will ensure that, for each valve:

- Power piston travel is checked and recorded
- Valve is checked to be closed, confirmed by an external reference mark and recorded
- Electrical devices, such as hydraulic oil dump solenoids, are energised and proven to operate and reset
- Pilot valves and linkages are checked and recorded for freedom and travel.

The valves should be operated under hydraulic power over their full range. The hydraulic power system, which will have been flushed and commissioned, must therefore be put in service. Special electrical test equipment may be required in order to inject the required signals to enable the hydraulic valves to be operated, and hence the associated steam valves.

During this operation, the total travel of the steam valve must be recorded and a check made for freedom of movement. By means of micrometer clock gauges, movements of power and pilot pistons will also be checked and recorded.

The steam valve trip mechanisms must also be verified, but not by actual closure of the steam valve, as damage to valve and valve seals may occur. The steam valve is therefore driven open by its hydraulic relays and wedged in an approved manner by means of a packing piece under the power-piston steam valve linkage. The hydraulic oil dump valves may then be operated. The dump valves are then reset, the steam valve driven fully open and the packing removed.

It is important to check that there is no lost motion or friction in the valves and control systems. This may be verified by means of the valve position indicating device, a linear voltage differential transformer used in conjunction with a pressure transducer to monitor power-piston hydraulic oil pressure. The two signals are fed into an X–Y plotter and the power piston lift and pressure characteristic obtained for full opening and closing of the valve. A typical characteristic is shown in Fig 5.7.

On completion of these tests, test equipment must be removed. Final functional tests will require the main governing system to be in service and are therefore dealt with in Section 8 of this chapter.

1.3.4 Pre-operational activities — loop pipes

The HP and IP loop pipes are two separate, but complementary, systems of pipework designed to withstand pressures of about 155 bar and 40 bar, respectively, at approximately 540°C for the particular turbines described. The pipes interconnect the HP and IP valve chests with their respective cylinders.

The top loops on each side of the turbine are closed with a measured closure pipe containing a large flanged connection.

The pre-commissioning activity therefore comprises a detailed inspection of erection and weld procedures. The weld procedures will include specified hardness checks and butt-weld heat treatment. The welds will be inspected, using specified techniques such as ultrasonic testing, magnetic particle inspection and radiographic methods. Specifications will include weld electrode and filler materials.

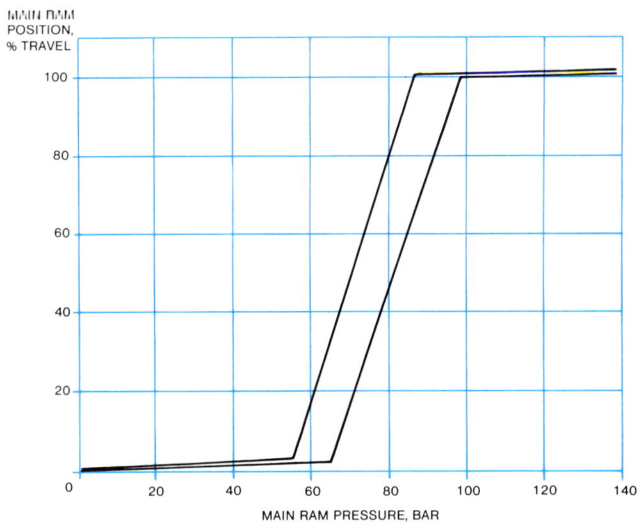

FIG. 5.7 Valve movement against motive pressure

Hydraulically-operated valve gear can be checked for lost motion or friction in the valves and control systems by plotting valve movement (i.e., operating ram position) against power piston hydraulic oil pressure (main ram pressure). The two signals can be fed into an *X-Y* plotter to give the characteristic shown. Friction and lost motion will be evident as non-linear aberrations.

The pipe closure procedure will be written so that the closure pieces can be manufactured to take into account weld gaps when setting up the piece and subsequent shrinkages of the butt weld.

1.4 Commissioning network and resources

No material resources are needed for this activity. Commissioning of the valve operating gear is covered by the network shown in Section 8 of this chapter.

The work covered by this section does not require a commissioning network. However, programmes dealing with the commencement of turbine commissioning should indicate the requirement for this work and make suitable allowance for the time required. The CEGB must decide either to accept the supplier's confirmations or to arrange for a double check. This would be carried out over a period of time towards the end of the erection programme with some time, say 2–3 days, allowed at the completion of erection.

2 Lubricating oil system

This section also includes:

- Main turbine turning gear (oil systems only)
- Oil storage, transfer and cleaning system
- Turbine greasing system

2.1 General description

A typical lubricating oil system (Fig 5.8) comprises a large main oil tank, a main oil pump directly mounted on the turbine-generator shaft at the exciter end, a booster pump, an AC motor-driven and a DC motor-driven ancillary lubricating oil pump (called flushing pumps) and lubricating oil coolers.

The main oil pump is a centrifugal type, supplying approximately 0.12 m^3/s of oil at 13.7 bar for the generator seal-oil system and the booster pump turbine. The booster pump, which is immersed in the oil tank, is driven by a Francis type single-stage turbine. Exhaust oil from the booster pump turbine flows through a duplex filter assembly, through selected oil coolers, to the bearing oil manifold supplying oil to the turbine-generator bearings. A portion of this oil is tapped off to the overspeed trip rings, via a test injection valve to allow on-load testing of these rings.

The AC and DC driven centrifugal-type flushing pumps are also mounted in the main oil tank. Both pumps assist in maintaining bearing oil pressure when starting and shutting down the main turbine. Also, the DC driven pump acts as a back-up in the event of an emergency, such as loss of AC supplies.

When the duty AC flushing oil pump is in operation, the main oil pump is pressurised through a priming connection. As the turbine speed rises, the main oil pump output increases until, at approximately 2500 r/min, a spring-loaded back pressure valve in the booster pump discharge line will open to allow the main oil pump suction to be pressurised, thus reducing the amount required from the duty AC flushing oil pump.

When the main turbine reaches a speed of 3000 r/min, the main oil pump booster system will be self-sustaining and as the duty AC flushing oil pump will still be in operation, the booster discharge will be overpressurised. The duty AC flushing oil pump is therefore arranged to stop automatically when the main oil pump discharge pressure rises to 11 bar.

Under normal operating conditions, two main oil coolers will be in service, with one on standby. A spring-loaded relief valve fitted in a by-pass line across the inlet and outlet sides of the coolers will open at a differential pressure of 1.4 bar, thus ensuring an oil supply to the bearings should there be a system failure, due to blockage or an inadvertently shut valve.

An additional oil cooler by-pass facility is provided by an air-operated valve between the oil cooler inlet and outlet pipes. This valve responds to a temperature controller sensitive to the oil temperature changes downstream of the coolers. The valve is open when the oil temperature is below 45°C and is fully closed above 45°C.

Oil from the main oil coolers is taken, via the bearing manifold, direct to the main bearings of the turbine and to those of the generator, exciter, turning gear, main shaft-driven oil pump bearings and its geared coupling. A tapping is taken off the line downstream of the coolers to supply the boiler feed pump turbine. Oil

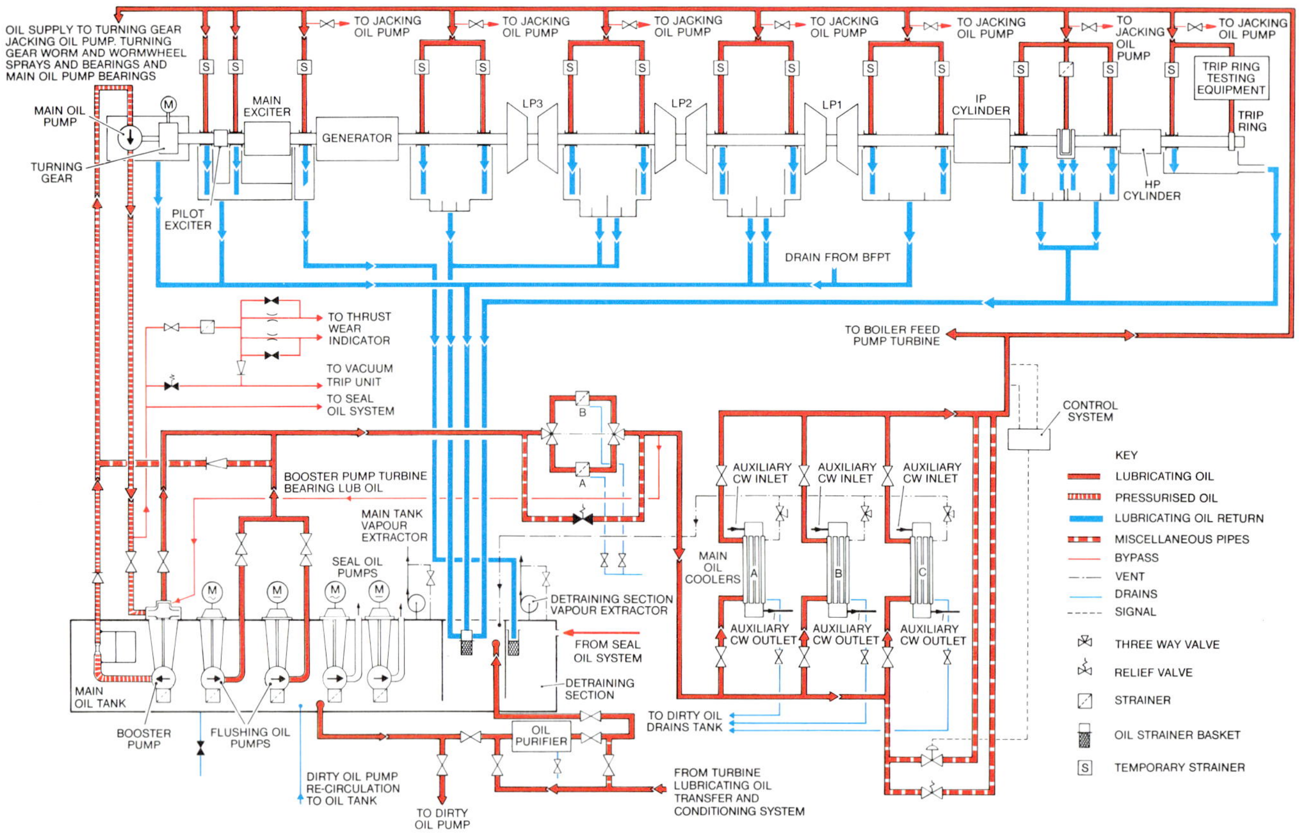

FIG. 5.8 Turbine lubricating oil system

In this typical lubricating oil system, lubricating oil is not used as a source of hydraulic fluid for the operation of turbine steam admission valves. Another feature of this example is that jacking oil supplies to the jacking oil pumps are tapped off from the lubricating oil supplies to the turbine and generator bearings. Seal-oil pumps are mounted directly on the main oil tank, but details of the seal-oil system are not shown.

returning from the lubricating and seal-oil system gravitates back to the main oil tank.

When the turbine is running down and the main oil pump pressure is decreasing with speed, the AC flushing oil pump is started automatically when the bearing oil pressure falls to 0.8 bar.

The DC flushing pump serves as a standby to the main oil pump should the AC flushing oil pump fail to start. It will be brought into operation automatically when the bearing oil pressure falls to 0.4 bar.

Two oil strainers are provided at main turbine pedestal No. 2, one in the oil line to the main turbine thrust and surge bearings and the other in the oil line to the thrust bearing wear indicator.

A high pressure jacking oil system, incorporating seven separate jacking oil pumps and a turning gear jacking oil pump, are integrated with, and supplied from, the lubricating oil system. These pumps enable the main bearing of the complete rotor assembly to be raised or floated in the bearings during turbine-generator start-up and shutdown, thus preventing damage to the bearings when shaft speeds are too low for hydrodynamic lubrication to take place.

The turning gear (see Section 6 of this chapter) is supplied with lubricating oil and also has its own jacking oil pump. It is interlocked with the jacking oil pumps when the turbine-generator shaft is started from rest. Once the shaft is at a turning gear speed of approximately 30 r/min, lubricating oil flowing to each bearing forms a hydrodynamic oil wedge.

The turning gear is also interlocked with the seal-oil system to ensure that a seal-oil supply has been established to the generator hydrogen seals. These seals would otherwise be damaged by metal-to-metal contact between their rotating and stationary elements.

The lubricating tank contents are maintained to a high standard of purity and low level of entrained moisture by means of a centrifugal oil purifier which takes its supply from a connection near the bottom of the oil tank and returns purified oil directly to the tank.

Two vapour extractors are provided on the main oil tank. One is connected to the main section of the tank, the other to a detraining compartment within the tank which receives oil returned from the generator hydrogen seals.

The system in Fig 5.8 shows that the lubricating oil for the main boiler feed pump and its steam turbine is supplied from, and returned to, the main oil tank.

A number of large modern turbines use lubricating oil at high pressure to power the main turbine steam admission valves. In such cases, the main power oil pump would be mounted on the turbine-generator shaft, with an electrically-driven oil pump mounted on the main oil tank. However, the very latest turbine installations use an entirely separate power oil system using a fire-resistant fluid (FRF) (see Section 3 of this chapter).

The oil storage, transfer and cleaning system (Fig 5.9) is a central facility which can recover oil from a turbine oil tank for regeneration. This system also acts as a central store for both clean and dirty oils. The regeneration unit (Fig 5.10) heats the oil to about 54°C and then sprays it into a vacuum chamber maintained at about 915 mbar. Entrained water boils off and passes to atmosphere. Water-free oil is then pumped via a filter unit to complete the regeneration process.

The greasing system is a simple manual system. Some modern plants are fitted with automatic greasing systems and the choice depends largely on the frequency of greasing and the materials of the sliding surfaces.

2.2 Commissioning objectives

The commissioning objectives for the lubricating oil system are:

- To ensure that the system is clean prior to the reception of lubricating oil
- To ensure that the system is flushed at velocities higher than service conditions
- To ensure that controls, indicators, alarm and interlocks are proven

2.3 Safety aspects

The Detailed Procedures Document should clearly state the safety requirements for lubricating oil system commissioning. Care must be exercised to ensure that construction operations are restricted in and above the immediate vicinity of the system to be flushed. Particular attention must be paid to ensure that no welding, heat treatment or other operations, which could cause ignition of oil vapour, or spilt oil, are carried out.

Barriers and notices should be enforced whilst flushing is in progress and it is advisable to ensure that unauthorised persons do not enter the restricted areas and that a careful watch is kept on activities which might cause a fire risk.

Where permanent fire protection is installed, this should be available for use. It may be advisable for automatic systems to be left available on hand control, particularly when work associated with flushing procedures is going on within the fire-protected zone.

A number of suitable portable fire appliances must be available at designated and well-identified points outside the flushing zones and, if necessary, adjacent to plant items within the zone where particular risks are identified.

During prolonged operations, staff should regularly patrol areas where oil systems are in service. Spillages must be dealt with promptly, because of the risk both of fire and of injury to personnel. Designated places must be provided for the disposal of materials used to clean up spillages.

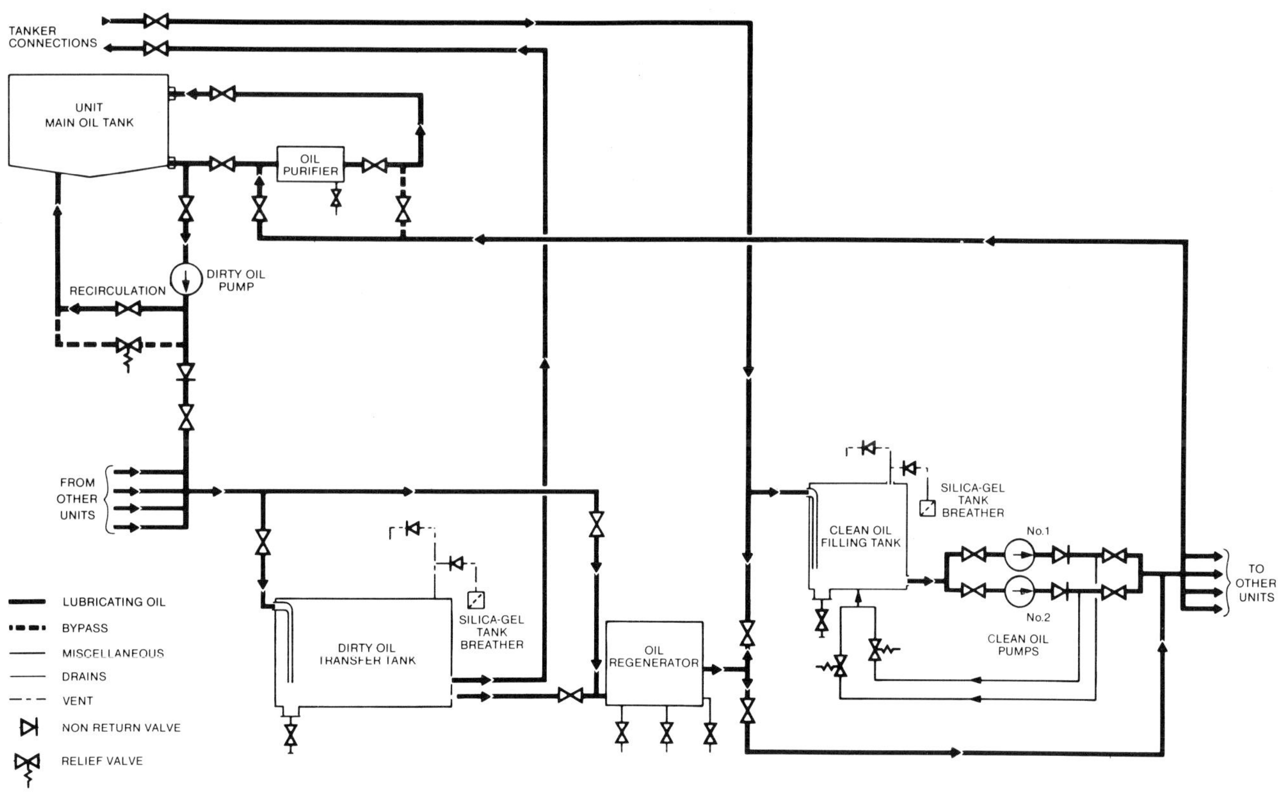

Fig. 5.9 Oil storage, transfer and cleaning system

This system can take clean oil from road tankers, and either store the oil until required to top up, or fill, the individual unit main oil tanks. Dirty oil can also be taken from the unit main oil tanks and regenerated for further unit use, or despatched to road tankers. In this example, the oil regeneration unit (Fig 5.10) operates on a different principle from the unit oil purifier (Fig 5.8).

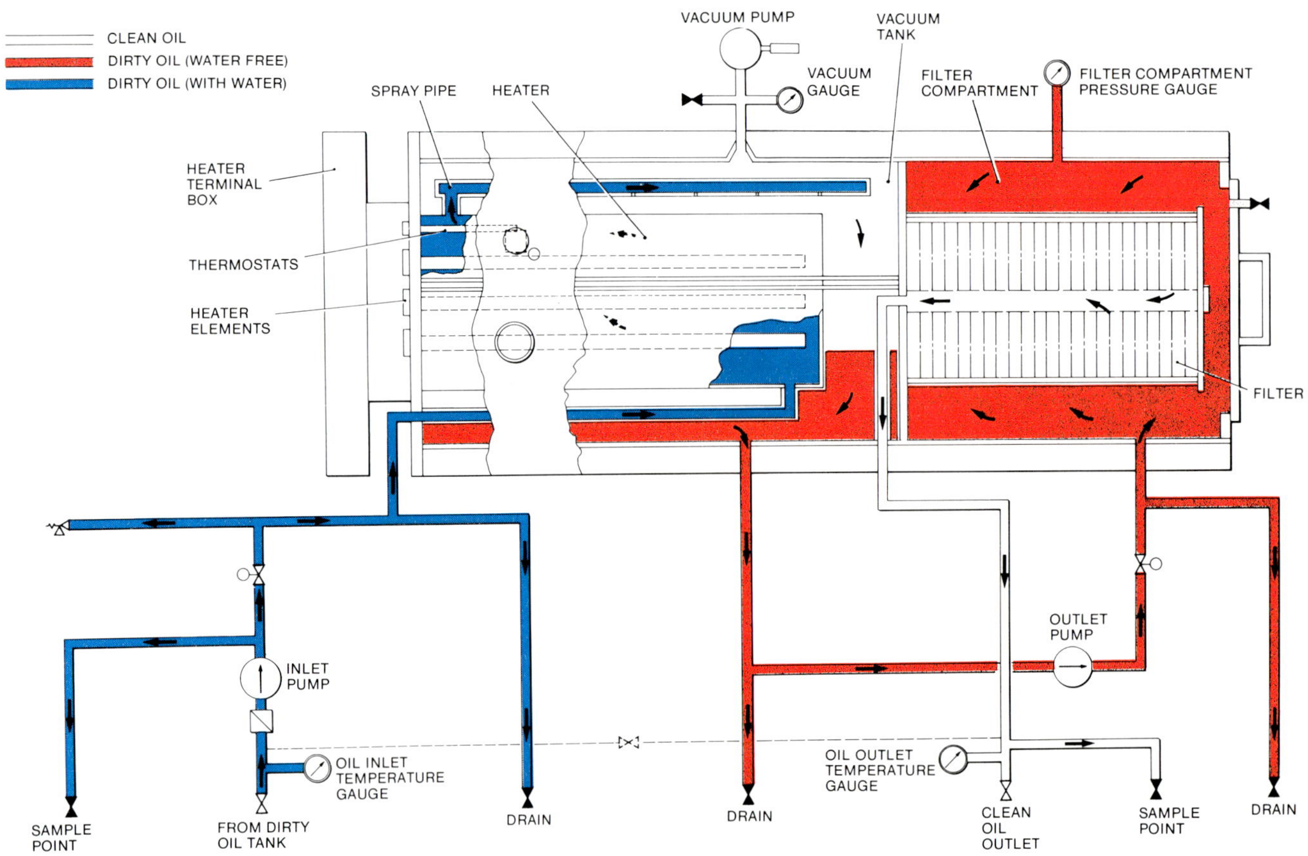

FIG. 5.10 Central oil storage system regenerating unit

2.4 Detailed procedures

The commencement of commissioning is preceded by a detailed inspection of items listed in the Detailed Procedures Document to ensure that they have been installed correctly to drawings and specifications.

For the system described, the Detailed Procedures Document should contain a complete inventory of the following items:

Main oil tank plinths and bund

Oil cooler plinths

Access platforms, ladders and handrails

Pipework supports

Main oil tank

Oil coolers A, B and C

AC oil pump

DC oil pump

Booster pump and oil turbine

Vent blowers (2)

Jacking oil pumps (7)

Turning gear oil pump

Centrifugal oil purifier

All filters, including duplex filter unit

All oil pipework and supports

Instruments, including instrument pipework, panel wiring, etc.

Orifice plates

Valves (all types)

Cabling

Manual greasing system

Main turbine turning gear (oil system only)

The checks should include a thorough internal inspection of the main oil tank for cleanliness and condition of internal surfaces, particularly where special finishes are provided. Personnel should wear suitable overshoes to protect internal surfaces. Access to the tank must be supervised at all times and particularly after final inspection has taken place.

Blanks and covers should also be in place. During erection, the condition of pipework should have been carefully monitored. However, the pre-commissioning inspection should ensure that the internal surface condition of pipework is sound. Where corrosion has taken place, a decision will have to be made for a chemical cleaning activity to be included.

2.4.1 Flushing procedure

The purpose of flushing any system is to remove loose materials such as debris and surface scale. This protects the components which are dependent upon the system from damage, unnecessary wear or deprivation of the system fluid.

The basic techniques of flushing are:

(a) To create a fluid velocity greater than that encountered when the system is in service, for all modes of operation.

(b) To disturb debris and promote its entrainment in the flushing medium, so that it can be trapped or discharged from the system.

High velocities can be achieved in a number of ways, either singly or in combination. These are:

- Sectionalising the system so that higher velocities are created in a smaller number of flow paths
- Reducing the system resistance by removal of components which restrict flow such as orifice plates, valve or filter internals
- Heating the flushing medium
- Using a medium less viscous than the working fluid

Debris can be disturbed in a number of ways which can be applied during the flushing process such as:

- Hammering pipework to dislodge internal scale and disturb debris
- Thermal cycling; this is used primarily to remove scale and other flaky particles from internal surfaces
- Sudden changes in the velocity of flushing medium
- Turbulence created by, say, the introduction of air into the flushing medium

When the lubricating oil system is clean and fit to receive oil, the main oil tank may be filled with oil. It is important to monitor this operation carefully to ensure no spillages or leaks.

Prior to reception of the initial oil charge, all drain valves should be locked shut, together with tank and pump outlet valves. The tank should be filled through the main oil strainer and a sample of each delivery taken and checked against specification.

This oil will be used to flush the system and may be to a specification different to that used for normal service conditions. In such cases, this initial, or flushing, charge will have a lower viscosity in order to assist the achievement of higher velocities within the lubricating oil system, so that debris and other particulate matter carried by the oil can be trapped on suitably placed fine mesh strainers. Instead of a special flushing oil charge, common practice is to use oil of normal specification and achieve lower viscosities by heating the oil. The lubricating oil system should be thermally cycled as part of the flushing technique, so that loose scale can be dislodged from internal surfaces. When the initial charge of oil has been supplied, the AC and DC flushing pumps will be commissioned against closed discharge valves. Vent blowers and the oil purifier will also be commissioned.

Main strainers, individual bearing strainers and orifice plates will be removed prior to flushing and the main oil tank return strainer checked as clean and in place. It is preferable to fit a strainer with a finer mesh than that used for normal service, provided the mesh will allow oil to pass without overflowing.

The oil should be heated by means of a temporary heat source, such as an electric heater or hot water supply to the cooling water side of an oil cooler. The oil temperature should be about 60°C, but must not approach the flash point. For normal turbine lubricating oils, the maximum oil temperature for flushing would be about 70°C.

The main turbine bearings and all jacking oil pumps must be by-passed. Also, other components with fine clearances, such as the main oil pump impeller, should be removed or by-passed.

Oil coolers must be in service and primed on the cooling water side. The oil purifier should also be in service.

The lubricating oil pipework system is then sectionalised by means of blanks fitted with identification tags and the jacking oil system excluded. Full faced joints must not be used with the blanks since they may accidentally be left in place when the blanks are removed. The initial flushing circuit should be via the main oil supply strainer by-pass and the main oil cooler by-pass. The internals of spring-loaded by-pass valves must be removed. Oil cooler oil inlet and outlet valves are also shut. The main oil supply strainer can be put in service during the by-pass flush. The fitting and removal of blanks should be strictly controlled by means of written authorisations. The sections should be chosen to give similar parallel flow paths of equal diameter. The flushing sequence should be such that no new section of pipework is introduced upstream of a section already flushed.

As each section is primed for the first time, the AC flushing pump must be put in service against a closed discharge and the discharge valve opened slowly.

Flushing is conducted using high flows through each section, together with periodic heating and cooling of the oil. The temperature cycle helps to dislodge debris and this can be assisted by hammering the pipework. As flushing proceeds, the AC pumps should be shut down periodically and the system pipework allowed to cool for several hours to aid the heating/cooling cycle. The DC flushing pump should be introduced periodically into the flushing circuit. The flushing process can also be enhanced by injecting air into the pipework at a convenient point close to the pump discharge valves. This creates turbulence and thus the entrainment of debris. Pump motor currents must be checked to avoid overload.

Debris collected in the main oil tank strainer is monitored and recorded until the section is deemed to be clean. The operation is repeated in turn on the other sections. When the pipework has been satisfactorily flushed, the temporary pipework blanks and by-pass pipework must be removed and the system reinstated.

When the oil pipework has been flushed, each bearing pedestal is prepared for flushing by removing the top and bottom bearing bushes and disconnecting jacking oil pipework at the last connection to each bearing bush. The pedestals are then thoroughly cleaned and covers, including oil wipers, fitted to ensure an oil-tight pedestal.

The system is again sectionalised by blanks at pedestals so that flushing of individual pedestals can take place. The flushing procedure is similar to that for the oil pipework. The oil flow should be the maximum capable of being handled by the pedestal drain whilst maintaining oil levels within the pedestals below the oil wiper. As each pedestal flush proceeds, the associated jacking oil line may be flushed by lowering the settings of the jacking oil pump relief valves, which will then allow the discharge of oil back to the main oil tank when the jacking oil pumps are put in service. The jacking pump discharge valves are throttled in and the relief valves set to their prescribed values. The discharge valve is then opened in order to flush the line to the bearing.

Pipework and pedestal flushing of the main boiler feed pump and turbine are included with the main turbine flushing procedure. The temporary strainers to each pedestal should be checked periodically and their condition recorded.

At completion of flushing, bearing pedestals should be thoroughly manually cleaned and inspected before reinstatement. Blanks must be removed and the jacking oil lines reconnected. Clean conditions must be maintained throughout this work. The temporary strainers at the bearing pedestals will be left in, but orifice plates may be replaced.

The AC flushing pump is then put in service together with the jacking oil pumps. Periodically the system must be shut down and individual bearing oil strainers inspected. This process is continued until no further debris is collected at the strainers over a period of at least 24 hours. The oil strainers may then be removed. A final sample of oil is then taken from the main oil tank and analysed. The fine oil strainer in the main oil tank should be inspected and cleaned. The strainer may be left in service and inspected from time to time during initial turbine operations. The temporary lubricating oil heater may be required for the final functional checks.

2.4.2 Final functional checks

Final functional checks are carried out after the flushing procedure. Those items not commissioned and items which have been commissioned to a limited extent for the purposes of flushing, e.g., local pump operation only, must now be completed in accordance with the check lists provided in the Detailed Procedures Document.

The first step of the functional check procedure is to ensure that electrical control circuits are complete, proven and energised. All alarm circuits specified in the Detailed Procedures Document alarm test will also be proven and operational, so that alarms can be set to indicate at prescribed settings. The electrical supplies for all auxiliary units must also be available. Instrument control air supplies, orifice plates, strainers, etc., must also be in place.

Subsequent steps depend upon the particular design of the lubricating oil system and it is important that the procedure proves all alarms, interlocks, sequences and controls. Thus, for any particular lubricating oil system, the procedures must be prepared and checked with a thorough understanding of the working and design features of that system.

The approach to the procedure should be to use actual signal inputs where possible such as an oil pressure, temperature or flow signal, auxiliary switchgear contact, etc. In cases where it is not practicable to use an actual plant signal, a simulated signal can be used, e.g., injection. Signals can also be derived from an intermediate source, such as an interposing relay. This is useful in setting up a system, but should be followed later by a full functional check. It may be convenient to check an interlock in stages by, for example, checking a pressure sensor to a relay, and then the relay to the device.

The functional testing procedure should also carry out negative checks, i.e., to check for errors in design and installation. This is done by devising checks that prove the system does not operate unless signalled to do so. For example, an automatic pump start feature should be proven to start correctly from, say, a low pressure signal, and then proven not to start from that signal when selected to 'manual' and 'off' modes.

For the system under consideration (Fig 5.8), the next step would be to check the AC flushing pump functionally, using standard check lists, whilst checking all associated alarms and indicators. The lubricating oil system temperature should be raised to working temperature, say 45°C, using the temporary oil heater.

- Put pump in service against closed discharge valve from local control station; check operation of stop button
- Restart pumps and slowly open discharge valve
- Record discharge pressure and motor current with oil at service temperature
- Record main duplex strainer differential pressure
- Record oil cooler outlet temperature
- Record thrust bearing oil filter differential pressure
- Record oil pressure at turbine centreline
- Change oil coolers (A and B; B and C; C and A), recording oil pressures at bearing centre line
- Check for no pressure fluctuation during changeover
- Stop the pump

When the system is primed, care must be taken to ensure that instrument lines and connections to control devices and valves are also primed.

The same procedure is then repeated for the DC flushing pump. Further key steps in the final functional procedures are:

- Select AC and DC flushing oil pumps to AUTO and check that both pumps start automatically.
- Use injected signals simulating main oil pump pressures/flows to check AC and DC flushing pump AUTO stop feature. Remove signals and prove that the pumps again start automatically.
- Prove the remote manual control of the AC and DC flushing pumps.
- With the AC pump in service, prove that the DC pump does not start when selected to AUTO. Stop the AC pump and prove that the DC pump then starts automatically.
- Prove the DC flushing pump AUTO start feature for loss of the electrical turning gear. Prove that the pump stays in service when the turning gear is restarted.
- Prove remote operation of oil tank vent blowers and record motor currents. Stop the blowers and check that blower fail alarms are initiated.
- Prove the automatic operation of the main bearing oil strainers by simulating high differential pressures. Check associated alarms.
- Jacking oil pumps are supplied from the lubricating oil system and interlocked not to start on low lubricating oil pressure. This should be proven by throttling the AC flushing oil pump discharge pressure below that of the interlock pressure, and then proving in turn that each jacking oil pump will not start.
- Prove that jacking oil pumps do not start on low seal-oil pressure.
- For each jacking oil pump in turn, with the exception of the turning gear jacking oil pump, check the lift of its bearing journal by means of a micrometer clock gauge, recording both transient and steady-state jacking oil pressures and jacking pump motor current.
- Jacking pumps are interlocked not to start when the turbine speed is greater than 100 r/min. This can be proven by injecting a speed signal just greater than 100 r/min and then attempting to start the pumps. The pumps are interlocked to stop when the speed rises above 100 r/min and can be proven in a similar manner.
- Prove that all jacking pumps start on AUTO when the lubricating oil pressure is normal and the turbine shaft speed is below 100 r/min.
- Prove all jacking oil supply pressure low alarms, one at each turbine bearing, by lowering the setting of associated alarm contacts. Ensure that the alarm sensors are correctly calibrated.
- The main oil tank high and low alarms should be proven by physically raising and lowering the level of oil within the tank. This will prove that the connections between tank and float switches or columns are free.

It is important to check the level within the main oil tank at which lubricating oil pumps lose their suction. This operation should be carried out by putting the oil pumps into service and then gradually lowering the tank oil level until there are signs that the pumps are losing suction. This may be indicated by noise, decreasing motor current or decreasing pump discharge pressure. Care must be taken to ensure that the pumps are not damaged by this operation. The level at which pumping performance is beginning to be affected must then be recorded accurately against a dip-stick or calibrated contents gauge. This activity may be combined with a similar test on the AC and DC seal-oil pumps (see Section 10 of this chapter).

At the completion of the final functional checks, the system is ready for operation. The temporary oil heater may be required for final functional checks of the seal-oil system.

2.5 Oil transfer, storage and cleaning system

This system (Figs 5.9 and 5.10) should be treated in a similar manner to that for the unit lubricating oil system.

As this is a central system, it may be necessary during construction to segregate the supply and return transfer pipework to other turbine-generator units by means of blanks and/or valves. Such points of segregation must be clearly identified and controlled jointly by the persons responsible for construction and for commissioning or operation on behalf of the CEGB.

2.6 Greasing system

In the plant described, this is a manual system; auto-greasing systems are also common and the choice is affected by the sliding surface materials and required frequency of greasing.

Commissioning is a simple procedure, following lines already described in this chapter. Care must be taken to ensure that the recommended lubricant is used and that grease is being pumped to the appropriate sliding surfaces. In the case of auto-greasing systems, care must be taken to ensure that greasing is not excessive and that the operating cycle intervals are adequate.

The application of grease to the point at which it is required can usually be checked visually, for example, by observing that it is being squeezed out from between sliding surfaces. Where this is not possible, grease should be applied during the final assembly stage of the component to be greased to ensure that it flows to the point of delivery.

2.7 Commissioning network and resources

A simplified commissioning network for the final erection checks, the flushing activity and final functional checks is illustrated in Fig 5.11.

The critical path is shown as a thick line. The time required to complete each critical path activity will depend upon a number of factors, including length of working day and manpower used. The flushing exercise for a large system, such as the lubricating oil system, can take several weeks to complete. The time is absorbed by the number of flushing circuits involved and the thermal cycling operations. It is advisable to allow plenty of time for the flushing operation, rather than make it a critical path activity within the overall commissioning programme.

Manpower requirements are small, these being mainly watchkeeping whilst circuits are being flushed. Even this requirement can be reduced to periodic inspections if alarms and indicators, particularly the main oil-tank low-level alarm, are commissioned and active at a manned location, such as a central control room.

A few personnel will be needed for the collection and inspection of samples, as well as for changing over spades and the fitting/removal of items, such as temporary pipework and strainer internals.

The main material resource is an initial oil charge, including oil retained in the primed lubricating oil system at approximately 70 m^3 for a modern 660 MW turbine-generator.

A large source of heat for the lubricating oil system is required. If an electric heater is used, it will need to have a capacity of approximately 300 kW for relatively cold climates. Alternative forms of heating, such as hot water, will have to be of equivalent heating capacity.

3 Turbine valve gear hydraulic fluid system

The plant items covered by this system are:

- Hydraulic power system package
- Interconnecting pipework to turbine steam valves
- Fluid conditioning unit

3.1 General description

A number of turbines use high pressure lubricating oil (termed 'power oil') as the hydraulic fluid to operate turbine steam admission valves. This power oil is supplied from a turbine-shaft-driven oil pump. For start-up and emergency conditions, an electrically-driven power oil pump is provided. Large modern units tend to use fire-resistant fluids (FRF) instead of lubricating oil, since the risk of fire due to high pressure lubricating oil in proximity to hot components is then eliminated.

The power oil supply to the steam admission valves of the turbine described is derived from a separate

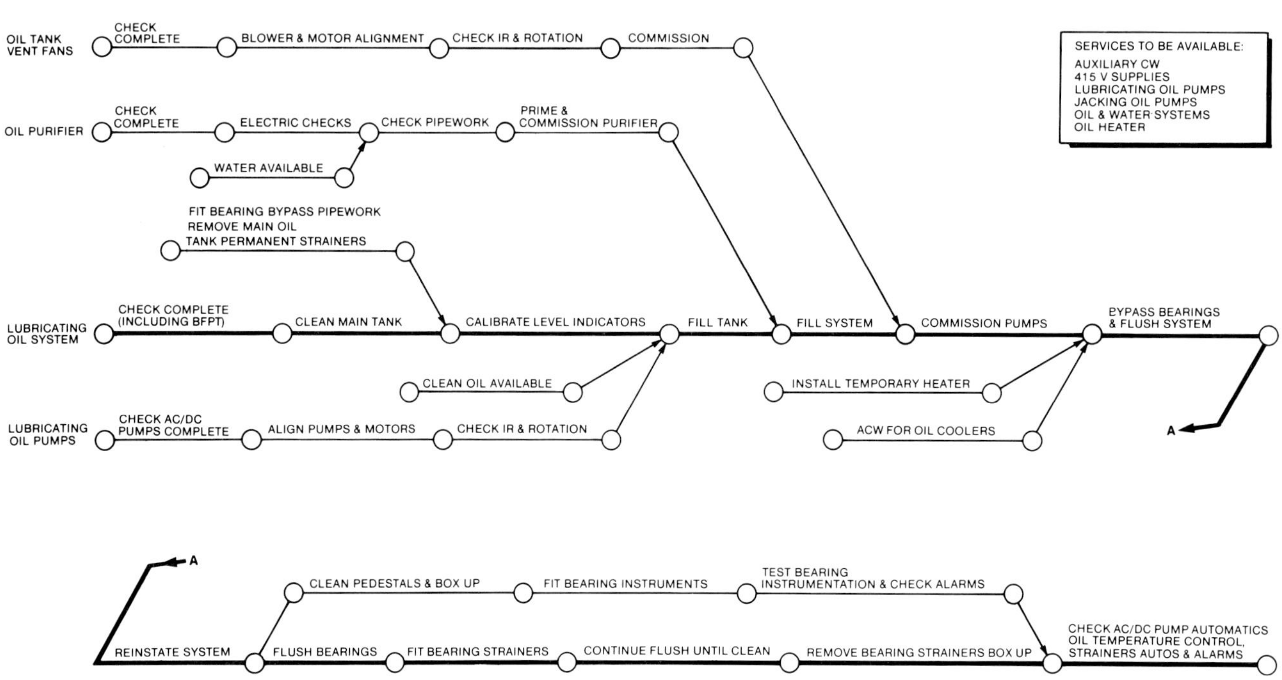

FIG. 5.11 Lubricating oil system — commissioning network

hydraulic package, using fire-resistant fluid as the operating medium (Fig 5.12). The package and connecting pipework are made in stainless steel. The package consists of two separate, identical pumping circuits which are interlocked against failure of either circuit. Each circuit has a main and boost pump, interlocked so that the main pump does not start until the boost pump provides about 2.7 bar. The interlock is operated by a pressure switch in the supply to the main pump suction. This pressure is regulated at about 3.3 bar by a relief valve.

The main pump is of axial piston, variable delivery type, operated by means of a variable-angle swash plate. The angle is controlled by hydraulic oil pressure from the main pump discharge.

Before the hydraulic fluid is suitable for normal service use, it must be heated. This is achieved by using the main pump at its full pumping capacity by causing the swash plate to be set at maximum pumping angle. This occurs when the fluid supplying the swash plate is cut off by means of a solenoid valve, which also vents the swash contents. The solenoid valve is controlled by a thermostat switch mounted in the main hydraulic oil tank. Fluid leaking past the main pump piston returns to the tank via fine filters (3 μm).

Each pumping circuit enters a common outlet manifold via a relief valve and a non-return valve. The relief valves are set at approximately 147 bar. A pressure switch effects automatic starting of the other pumping circuit.

The hydraulic power system supplies the steam admission valves to the main and boiler feed pump turbines. In order to maintain continuity of supply pressure when the pumping circuits change over, a pressure accumulator is fitted to the outlet manifold.

The system is provided with supply temperature and pressure indication, including high and low temperature alarms. Each pumping circuit has a water-cooled cooler, the cooling flow of which is controlled by a pneumatic valve, which closes on falling temperature sensed from the pumping circuit discharge pipe.

Each pumping circuit also has its own fluid tank of about 1.6 m^3 capacity. The tanks are filled by means of a gear pump fitted with a 10 μm filter in its delivery. The pump may also be used to transfer fluid from one circuit's tank to the other, and for emptying the tanks. On either side of the transfer pump are connections for a fluid conditioner. A five-way valve is provided, so that isolation from the fluid tanks can be obtained for the pump, filter and conditioner. The fluid tanks have a common vent to atmosphere via a silica gel breather, to prevent ingestion of moisture. A storage tank is supplied, which is used to dump fluid from one of the main tanks for conditioning.

The main tanks have high and low level detectors; the former is used to prevent a booster pump from starting and the latter to cause its respective booster and main pump to trip. Alarms are also initiated from these detectors.

The conditioning unit (Fig 5.13) comprises a vacuum unit and a Fuller's Earth filter. The conditioning unit takes its supply either from the bottom of the pumping circuit cooler, or from the storage tanks. The fluid passes through a strainer and a heating unit to a dirty-fluid pump, the discharge of which is in the form of spray into the vacuum unit. The large fluid surface area created by the spray allows entrained air to be dissipated to atmosphere via the vacuum pump. The air-free fluid is then pumped to the Fuller's Earth filter where residual moisture, acids and particulate matter are removed. The fluid then flows through a paper-element filter, which traps Fuller's Earth particles that might be carried over. The conditioned fluid then passes to each of the storage tanks.

3.2 Safety of personnel

The use of fire-resistant fluids will need careful consideration in respect of handling and disposal. A number of these fluids are of the phosphate ester type (tri-aryl phosphates) and their use in hydraulic systems has become widespread in recent years. Phosphate-ester fluids may readily be confused with mineral oil, which they are designed to replace. Plant using these fluids should be clearly identified and the means of handling and disposal defined.

The most serious risk is that of swallowing the fluid and this would include inhalation of the fluid if discharged into the atmosphere under high pressure as a mist. Inhalation of vaporised fluid and eye/skin contamination is a lesser hazard. However, vaporised fluid in the form of white fumes may be highly irritant to the respiratory tract. In such circumstances, breathing apparatus should be worn.

3.2.1 Handling precautions

Eating, drinking or smoking must be prohibited whilst the fluid is being handled, to prevent possible contamination of the lips and mouth.

Eye splashes may cause severe irritation and face visors and goggles must be worn at all times when the fluid is being handled, or when working on systems containing the fluid. If an eye splash does occur, then the affected eye should be irrigated with clean water as soon as possible and the individual referred for medical treatment.

To avoid prolonged contact of the fluid with skin, protective gloves made from an impervious material will be worn. After handling the fluid and before eating, drinking or smoking, the hands must be thoroughly washed.

All containers of fire-resistant hydraulic fluid must be clearly marked. Similarly, leakages should be collected in clearly marked containers.

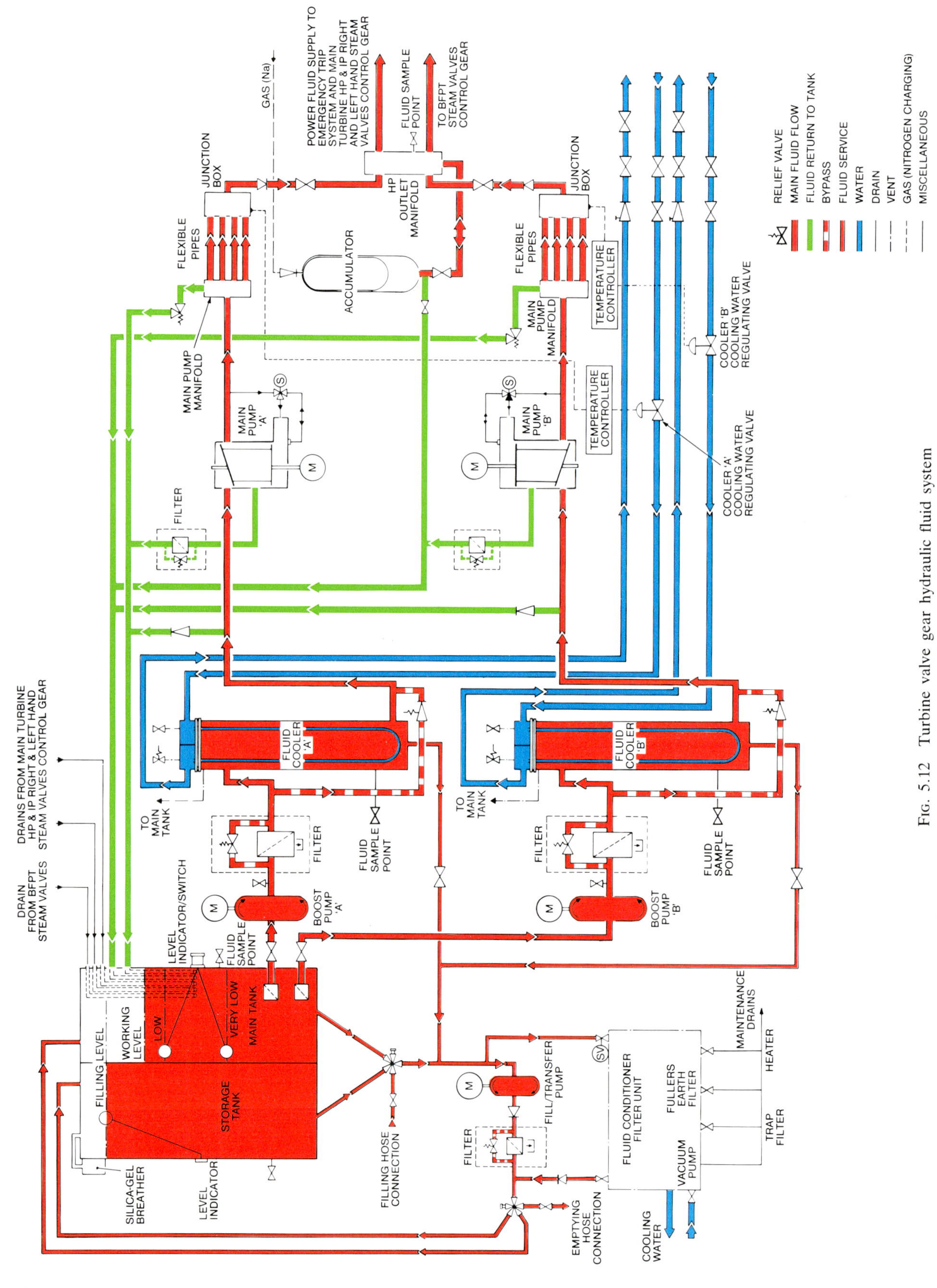

FIG. 5.12 Turbine valve gear hydraulic fluid system

The system comprises duplicate storage, filtering, cooling and high pressure pumping facilities, together with a fluid conditioning unit, using fire-resistant hydraulic fluid.

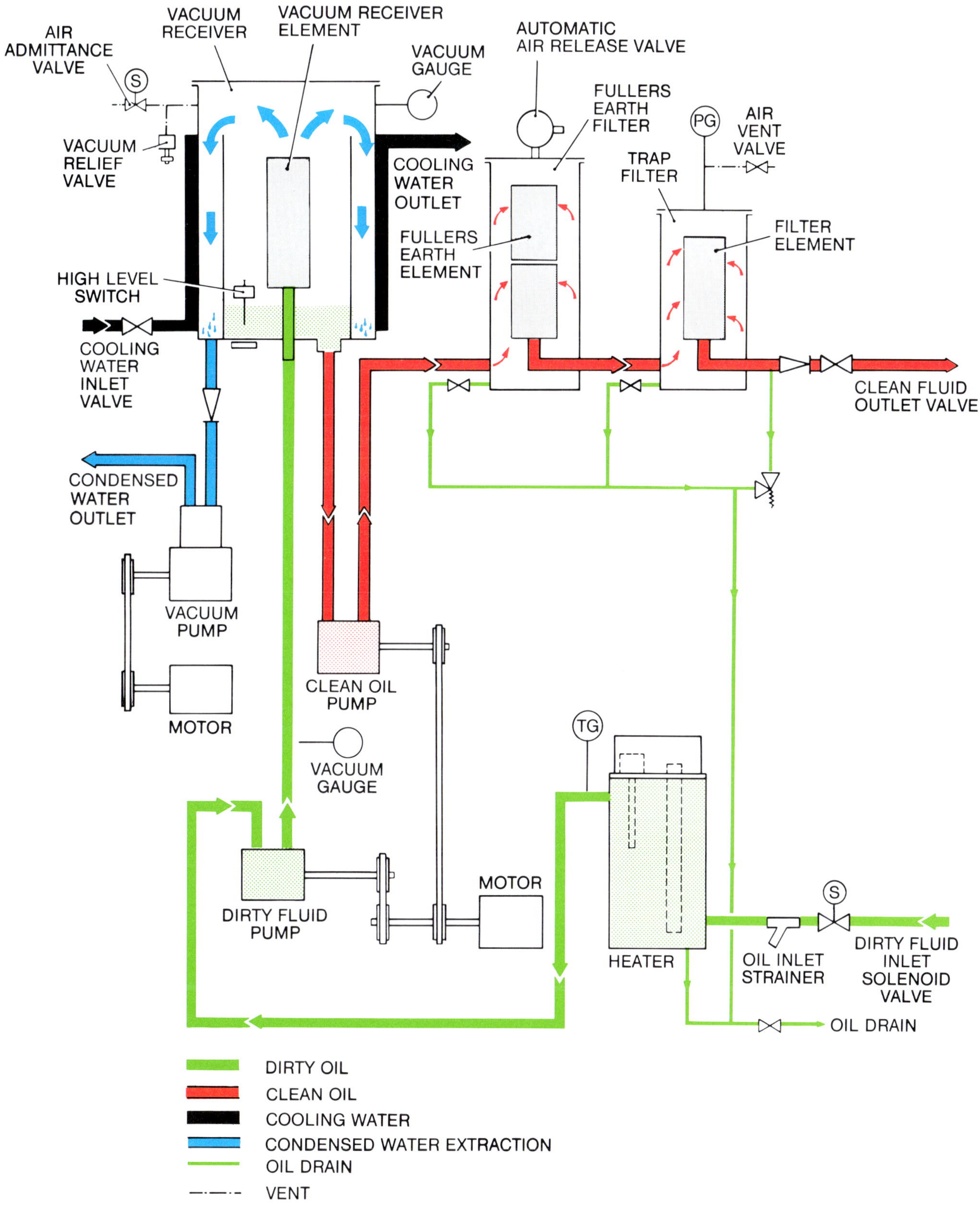

FIG. 5.13 Turbine valve gear hydraulic fluid system — conditioning unit

3.2.2 Action in the event of leakage

Phosphate-esters are classified as toxic materials and must be prevented from entering surface drainage channels.

All leakage or spillage must be contained locally, e.g., by the use of oil absorbent granules, sandbags, etc. These materials should be placed in polythene bags and clearly labelled, when being collected for disposal. Access to the area should be restricted by the erection of barriers and warning notices.

If leakage of phosphate ester occurs onto hot pipework, white choking fumes are formed which can cause irritation of the throat and chest. Under such circumstances, care should be taken to avoid inhaling the fumes.

3.2.3 Storage

It is essential that the fire-resistant fluids are stored inside a building and that the fluids are not subjected to extremes of temperature.

3.3 Commissioning objectives

- To ensure that the package has been installed correctly and connected to the turbine steam valve relays to specification and drawings.
- To ensure that the pipework system is clean and free from scale.
- To ensure that controls, indicators, alarms and interlocks are proven, including:
  - (a) Correct functioning of the hydraulic accumulator during hydraulic pumping circuit change-over.
  - (b) Correct operation of the main pump diverter valves on falling, and then rising, fluid temperature.
  - (c) Correct operation of pressure relief valves.
- To test the fluid conditioner unit for flow rates, conditioning performance under rated pressure and temperature conditions.
- To establish a sampling system for the hydraulic fluid.

3.4 Detailed procedures

The commencement of commissioning is preceded by a detailed inspection of items listed in the Detailed Procedures Document to ensure that they have been installed correctly and to drawing.

For the system described, the Detailed Procedures Document should contain a complete inventory of the following items:

Hydraulic package and connections to steam valve relays

Main storage tanks

Main pumps

Booster pumps

Transfer pump

Coolers

Filters — booster pump suction and discharge

Filters — main pump leak-off

Instruments, including instrument pipework and electrical connections

Orifice plates

Valves

Cabling

Pipework and supports

The checks on the erection activity should ensure that the components of the system have been stored in warm dry conditions. Motors should have IR tests conducted at regular intervals, say two-monthly, and results recorded.

Where off-load heaters are fitted, they should have been energised during the storage period.

All test certificates should be listed and checked for availability and completeness.

Appropriate lubricants should have been applied to bearings and all motor/pump units rotated one half turn every two weeks.

The hydraulic package and conditioner unit should have been thoroughly checked, including: direction of rotation of motors and pumps; identification of suction and discharge branches; fitting of dowels, gaps, coupling alignments recorded and correct; lagging and cleading complete or adequate for purpose; and coupling guards fitted. The internals of vessels should be inspected for cleanliness and all seals checked for compatibility with the hydraulic fluid to be used. All orifice plates should be listed and checked for correct size.

All pipework must be checked for correct erection, including such matters as jointing materials compatible with the hydraulic fluid, and welding carried out in accordance with an approved specification. Where compression fittings are used, these must be of an approved type for the system pressure and correctly fitted.

The pipework must be pressure tested in accordance with an appropriate specification. For the system described all the high pressure pipework would be tested to about 266 bar and the drains and low pressure pipework to 13 bar. Flexible connections must also be pressure tested and checked for correct fitting. Hoses should not have kinks in them.

3.4.1 Chemical clean

For the type of high pressure hydraulic system described, the internal surfaces of the pipework and tanks must be chemically cleaned prior to flushing to ensure that they are free from grease and that scale resulting from site welding of the stainless steel is detached. This procedure necessitates by-passes being fitted to all steam valve actuator and emergency trip valves.

Chemical cleaning is an activity often carried out by a specialist contractor and requiring chemical expertise. A suitable acid is used, which must be pure and have a very low chloride content, since chlorides will attack the base metal. All chemical solutions must be made with demineralised water.

Separate cleaning pumps are required and these must be fitted with fine discharge filters, so that debris is not introduced into the hydraulic system. Supply and drain lines will also be needed and these must be pressure tested with water to the same pressure as specified for the permanent installation.

A heating system is also required, using materials which will not be attacked by the acid used. The heating system could be a stainless steel coil, using steam or hot water as the heating medium. The whole system, including temporary pumps and pipework, is filled with demineralised water. It is important to ensure that all parts of the system are fully primed. The water is then heated to 70°C whilst being circulated.

A 0.2% solution of trisodium phosphate and a non-ionic surface wetting agent, such as a 0.5% solution of Berol 09, is added to the demineralised water and is circulated for about three or four hours, whilst maintaining temperature. The system is then completely drained, refilled with demineralised water and checked to be fully primed. The fluid is then circulated in order to flush all parts of the system, after which it is drained.

The system is again filled with demineralised water, as before, and a suitable acid solution introduced. The solution could be 7% solution of chloride-free sulphuric acid but, owing to the extreme care required in making up and handling the acid, it is preferable to use a 3% nitric acid solution.

For the sulphuric acid solution a temperature of 60°C is required and for the nitric acid a higher temperature of 90–95°C will be needed. A suitable inhibitor such as 0.15% stannine LTP must be added in order to prevent acid attack of the bare metal, thereby limiting chemical activity to the removal of oxides and scale. The system is then circulated for about eight hours at the appropriate temperature, after which it is drained, filled and flushed as before, until all traces of acid are removed.

The system must then be inspected, any loose debris being cleaned out from collecting points, such as tanks, and temporary filters inspected, cleaned and replaced. The system is then refilled with demineralised water and passivated by adding a solution of 3000 PPM hydrazine, 50 PPM ammonia. This solution is circulated and heated to 90°C, the circulation being maintained for about six hours, after which it is drained as quickly as possible whilst hot. As soon as this is complete and with the pipework still hot, the system should be blown through with nitrogen to purge it of moisture and maintain a non-reducing environment. It can then be stored under nitrogen until such time as it is required to be filled, all temporary by-passes and equipment having been removed and the system reinstated.

3.4.2 Flushing procedure

Before this procedure commences, the system must be checked for completeness. The flushing procedure is accomplished by circulating the service hydraulic fluid at high velocity, using the system pumps. Initially, the main tanks will be filled with the hydraulic fluid and the fluid then circulated via the pumps directly back to the tanks by means of a temporary connection between the pump outlet manifold and the tank, until that circuit is clean (Fig 5.14). The storage tank is flushed in a similar manner. The fluid must be sampled at the end of each flushing stage and analysed. The fluid conditioning unit must be commissioned and in service during the flushing of these tanks. During these operations, correct operation of relief valves and diverter valves is proven. Also, the system and its components must be checked for leakage. The system pipework is flushed by separating it into a number of flow paths in order to achieve very high velocities. For the system described, twelve flow paths are established, one being in service at each stage of the flushing programme (Fig 5.15). The flushing sequence of some of these flow paths is initially carried out without a filter, but filters are introduced for the later stages of the programme. The flushing stages are as follows:

(a) Power fluid to emergency trip valves, by-passing them to drain

(b) Power fluid to left hand (LH) and right hand (RH) main stop valves

(c) Power fluid to LH HP and IP governing valves

(d) Power fluid to RH HP and IP governing valves

(e) Protection fluid (i.e., power fluid to the protection device) to LH HP stop and IP stop valves

(f) Protection fluid to RH HP stop and IP stop valves

(g) Protection fluid to LH HP and IP governing valves

(h) Protection fluid to RH HP and IP governing valves

(j) Repeat of (e), (f), (g) and (h), but with filters in circuit

(k) Power fluid to RH HP and IP governing valves (local filter in circuit)

(l) Power fluid to LH HP and IP governing valves (local filter in circuit)

(m) Power fluid to LH and RH main stop valves (local filter in circuit)

(n) Boiler feed pump turbine power fluid pipework

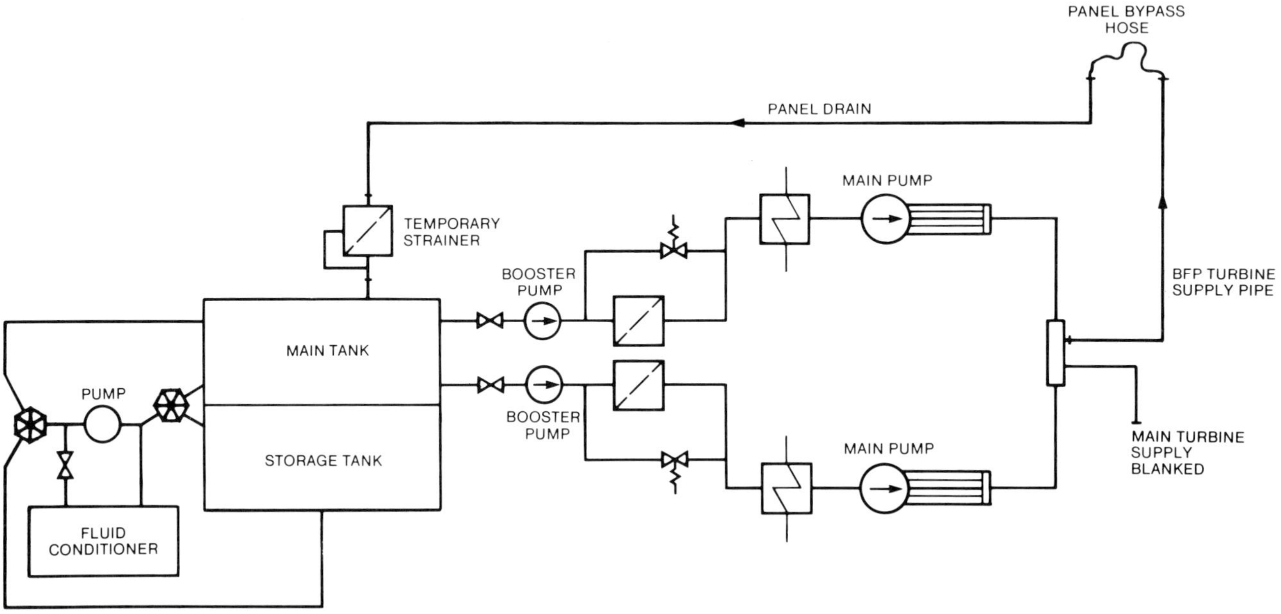

FIG. 5.14 Turbine valve gear hydraulic fluid system — schematic for flushing the pumping unit

Shows a typical arrangement for flushing the turbine valve gear hydraulic fluid package. Fluid is circulated at high velocities using the booster and main pumps. Fluid to the turbine valves is blanked off. Fluid from the main pumps is returned via a flexible connection and temporary strainers to the main fluid tank.

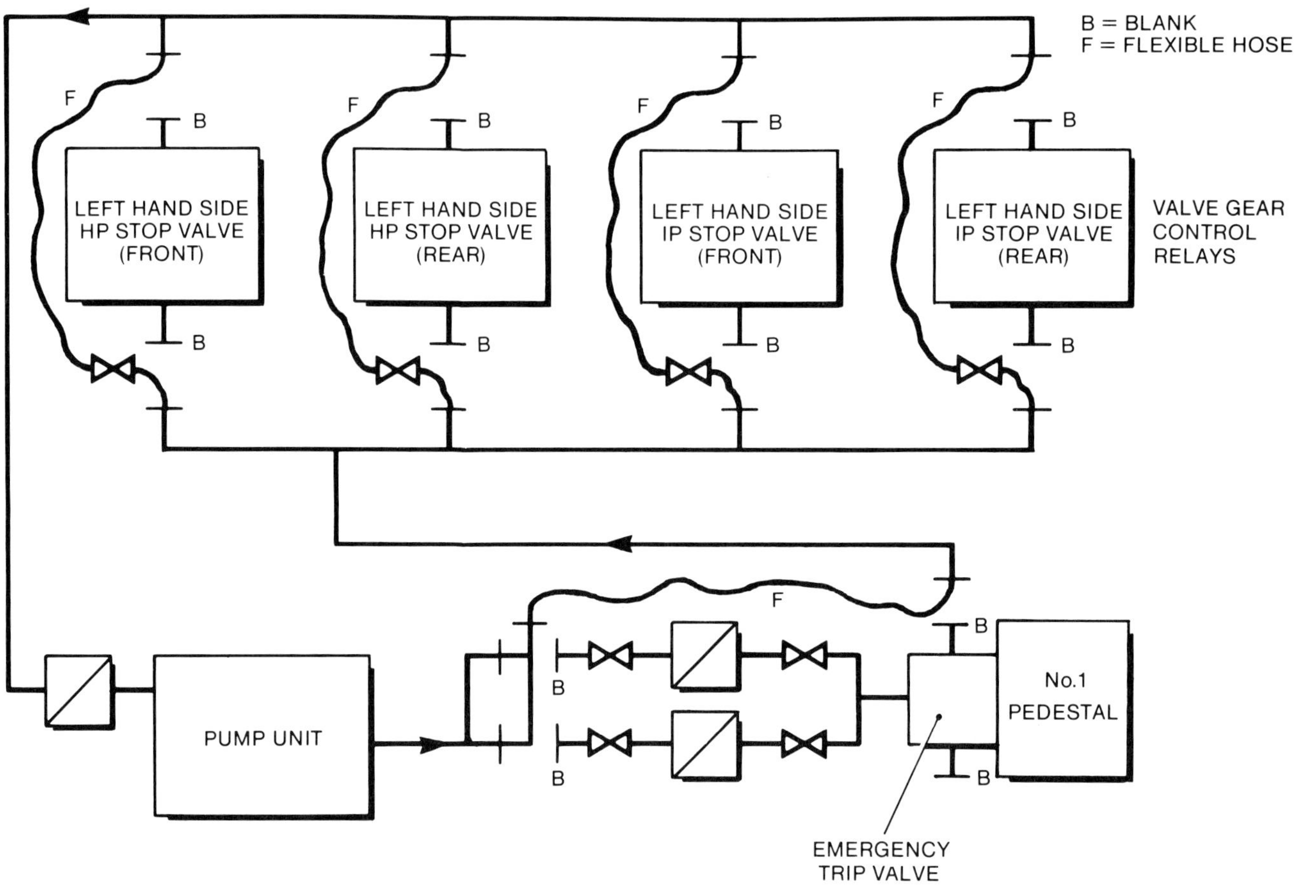

FIG. 5.15 Turbine valve gear hydraulic fluid system — pipework flushing schematic

One of the flushing circuits used to flush pipework connecting the hydraulic pumping unit with the turbine steam admission valve control relays. The pumping unit will have been flushed and commissioned. The unit is then used to flush, in stages, the remainder of the system. Here flexible connections are used to by-pass the emergency trip valve mechanism and its associated filters and also to by-pass the HP and IP stop valve control relays, so that the supply pipework to the relays can be flushed. The relays are flushed at a later stage.

Each group is fitted with a temporary by-pass around its steam valve servo and trip gear mechanisms.

During the flushing process for each circuit, the pumps will be stopped periodically and cooling water applied to the hydraulic fluid coolers. Thus, when the pumps are restarted, a slug of cold fluid followed immediately by hot fluid is pumped through the circuit to assist the removal of scale.

Blanking plates used for separating the hydraulic fluid system into its circuits must be carefully controlled in a manner similar to that for the lubricating oil system blanks.

At the conclusion of the flushing procedure, all temporary pipework and blanking plates must be removed and the permanent system reinstated, complete with new filter elements.

The flushing procedure will encompass the commissioning of the various duplicate pumping circuit components. The principal steps of the procedure are:

- Ensure that the main hydraulic fluid tank outlet strainer is clean.
- Fill the storage tank by using a hose connection at the filling pump suction and setting the five-way valves to the filling positions.
- Sample each drum of hydraulic fluid to be used for filling the system. The samples must be checked for DC resistivity and particulate contamination, and the results recorded.

Note Extreme care must be taken to ensure clean conditions when transferring hydraulic fluid from the drums. In particular, the outer surfaces of the drum, its filler cap and the hose which is put into the drum must be clean. The hose must be of a material compatible with the fluid.

- Fit a temporary strainer and return line to the main fluid tank.
- Fit a temporary connection from the hydraulic fluid unit outlet manifold to the main fluid tank.
- Fill the storage tank to the full mark and check the calibration of the tank level indicator.
- Transfer fluid from storage to main tank, via the conditioner unit.
- Circulate the fluid through the conditioner unit until fluid condition is acceptable.
- Commission a booster pump. This should be done by reference to a standard check sheet, with the main pump shut down. The booster pump discharge is returned to the main tank via the relief valve, which should be checked for correct operation.
- The main pump should then be commissioned by reference to a standard check sheet. This will be done with the main pump open to the outlet manifold and recirculated via the temporary connection, but with the accumulator isolated.
- As the main fluid temperature increases, the main pump thermostatically-controlled recirculation valve should be checked for opening at the correct temperature.
- At a fluid operating temperature of 40°C and with the recirculation valve closed, check the discharge relief valve operation by throttling the main pump discharge valve. Shut down both pumps.
- The booster and main pumps for the other pumping circuit are then commissioned in the same manner. This circuit is then left in service and the main tank thermostat set to 50°C for the flushing procedure. The temperature must not exceed 55°C
- With flow through the cooler and its water side drain open, check for no cooler tube leaks.
- Commission the fluid conditioner unit and leave in service throughout the flushing procedure. If contamination exceeds limits, flushing must be stopped until fluid conditions are acceptable.
- Remove the temporary connection from the outlet manifold to main fluid tank and reinstate normal pipework.
- Flush the system pipework progressively in accordance with the commissioning programme (Fig 5.17).

The pipework flushing procedure will require a number of fittings, such as adaptors and bends, to enable the various system components to be by-passed. Contract isolating valves will be removed and replaced by spares. The valves will be reinstated and flushed only after the associated pipework upstream of the valves has been flushed to an acceptable standard.

To achieve high velocities, the number of parallel paths is limited to four for both power fluid and protection fluid supply lines.

Sampling points are required and these are best incorporated at adaptor flanges upstream of isolating valve positions. Every flange not connected to a by-pass hose must be blanked and all sampling points must be plugged.

When contract valves have been reinstated and flushed, the contract filters must then be flushed.

The retention of debris by the main oil tank temporary return-strainers must be checked throughout the flushing activity. Booster and main pump differential pressures across their respective filters will also be recorded.

At completion of the flushing activity, remove all temporary fittings and reinstate system to normal.

Isolate the system and retain isolation until required for service.

3.4.3 Final functional checks

When the system has been reinstated, all those components of the system which were not commissioned for the flushing procedure must be checked against schedules in the Detailed Procedures Document, e.g., alarms, instruments and indicators.

The hydraulic fluid system should then be put in service and the system temperature raised to 40°C.

All relevant associated alarms and indicators must be proven at each step of the final functional checks, the principal steps being:

- Lower main tank fluid level and check boost pump trips. Prove pump will not start from its local/remote control station.
- Restore main tank level and prove boost pump start/stop operation.
- With boost pump stopped, operate main pump 'start' button and prove pump does not start.
- Start boost pump and then prove main pump start. Record motor current and discharge pressure.
- Lower main tank fluid level and check both pumps trip. Operate boost and main pump 'start' buttons and prove no response.
- Restore main tank level. Prove boost pump and main pump start/stop operations.
- Repeat these checks for the other pumping circuit. With one circuit in operation and the other selected for automatic start, prove standby cut-in on loss of the running circuit. This operation is then repeated for the other pumping circuit. Fluid pressure should be observed throughout the changeover.
- With a pumping circuit in service, reduce fluid temperature to below the normal value of 40°C until the system temperature is constant at 30°C. Observe operation of main pump thermostatically-operated diverter valve, which should cause the swash plate to move to maximum angle. The fluid temperature should then gradually recover to 40°C owing to the heating effect of the main pump.
- Repeat the operation for the other pumping circuit.
- Isolate the cooling water supplies to a cooler and prove the associated cooling water fail alarms. As main tank temperature rises to 50°C prove high fluid temperature alarm.
- With cooler fully in service, reduce fluid supply temperature to 35°C and prove low alarm. Reduce the fluid temperature to 20°C by preventing the thermostatically-controlled diverter valve from operating and prove the main tank low temperature alarm.
- For the installation described, the coolers control the hydraulic fluid temperature by means of pnuematically-operated cooling water valves. Alarms associated with loss of control air should be proven.
- Filter differential pressure alarms can be proven by adjusting the alarm contacts to initiate a high pressure alarm. The contacts are then reset. It is important that the calibration of the differential pressure device has previously been checked.
- The fluid conditioner unit pump controls and alarms should be checked in a similar manner.

3.5 Examination and maintenance of phosphate-ester fire-resistant hydraulic fluids

The CEGB has produced a Code of Practice for the examination, maintenance and handling aspects of phosphate-ester fire-resistant fluids.

These fire-resistant fluids are used in hydraulic systems operating up to pressures of 140 bar with valve clearances of the order of 2–4 μm. The passage of fluid through such fine gaps imposes a high rate of shear which produces local high temperatures accompanied by fluid degradation. The mechanical and chemical properties are therefore of great importance. The significant properties are:

- Viscosity
- Density
- Thermal and oxidation stability
- Air release
- Foaming
- Resistivity
- Cleanliness
- Hydrolytic stability
- Anti-rust
- Demulsibility

Degradation products, particularly in the presence of water, are corrosive and can erode finely machined parts. Particulate contaminants in high velocity fluid conditions are also erosive.

The air content of fluid passing from the main tank to the fluid pump is a meaningful indicator of fluid condition and should be checked during the commissioning process.

3.5.1 Examination of new fluid on delivery

Personnel responsible for sampling and testing must have the necessary experience and skill. Sampling is carried out in accordance with procedures described in BS5263.

3.5.2 Deterioration and important properties of fluids in service

A monitoring frequency must be established which will enable slow deterioration of the system during normal operation to be detected. During commissioning a high frequency of monitoring should be established from which a monitoring regime for normal service can be derived.

3.5.3 Parameters requiring monitoring — their causes and effects

Acidity

This can increase due to:

- Degradation of fluid through increase in moisture content or high fluid temperature
- Chemical contamination
- Ineffective operation of the Fuller's Earth filters

Acidity may cause corrosion, particularly in the presence of moisture.

Moisture content

This may be caused by:

- Abnormal ingress of water, e.g., cooler fault, ineffective reservoir breather or careless handling/transport/storage of the fluid
- Normal ingress of moisture but inadequate conditioning of the fluid
- Water entrained in the system after inadequate system flushing.
  (*Note:* The density of water is less than that of a phosphate-ester.)

Moisture may cause long-term system and fluid deterioration, corrosion, rust and seizure of sensitive control elements.

Particulate contamination

This may be caused by:

- Entrainment of particles generated from mechanical deterioration of system components
- Inadequate maintenance
- High fluid temperatures
- Carryover from Fuller's Earth filters

Particulate contaminants can cause valves to malfunction through silting and through component deterioration due to erosion.

DC resistivity

This may be related to:

- Increase in acidity or moisture content
- Topping up with an unsuitable fluid
- Contamination with electrically conducting materials

The effects of a decrease in DC resistivity can be acidity, fluid deterioration or erosion by electrochemical effects.

Mineral oil contamination

This may arise from:

- Topping up with the wrong fluid
- Residual contamination from components tested in mineral oil
- Entrainment of mineral oil in the system due to inadequate flushing.
  (*Note:* The density of mineral oil is less than that of a phosphate-ester.)

Mineral oils can give rise to deposits and seizure of control elements. Serious contamination will decrease the fire-resistant property of the hydraulic fluid.

Viscosity

Changes in viscosity may occur owing to degradation of fluid additives. This may affect the performance of system components.

Chlorine content

This can be caused by contamination with:

- Sea water
- Chlorinated solvents
- Protective treatments containing chlorine compounds

Chlorine can cause corrosion and contamination products causing erosion. Gross contamination will result in a need to change the fluid.

Aeration

This may be caused by:

- Faulty seals
- Poor conditioning
- Air released from the fluid owing to contamination or chemical changes

Aeration can cause pitting to pumps and system components where fluid pressure changes occur. It can also cause 'spongy' operation of the system.

3.5.4 Conditioning equipment

Conditioning equipment plays an important part in maintaining correct fluid conditions. Fuller's Earth filters are excellent for the extraction of acid, sediment and water. They must be replaced when differential pressures across them exceed 2 bar, or when acidity and resistivity levels cannot be contained.

A guide to the sampling and testing regime required for phosphate-ester fluids is shown in Table 5.1.

TABLE 5.1
Phosphate ester fire-resistant fluids — a guide for tests and testing frequencies

Test	Category of fluid duty	Frequency	Method	Test result action level	Action (on reaching limit, check indicated section on plant items as listed)
Acidity	Very severe	Monthly	IP 1A (BS 4705)	>0.3 mg KOH	Excessive pumping circulation Fuller's Earth filters (where provided)
	Severe	Monthly			
	Normal	Monthly			
Moisture content	Very severe	Weekly	Karl Fischer (BS 2511)	>200 PPM	Seals Coolers Dehydration equipment Reservoir breather
	Severe	Monthly			
	Normal	Monthly			
Particulate contamination	Very severe	Weekly		10 000 particles >5 μm per 100 ml or fluid downstream of main filters	Filter condition (where provided) Seals Wear of moving parts Corrosion due to acidity
Insolubles	Severe	6-monthly		>0.001%	
	Normal	Annually			
Mineral oil content	Very severe	Weekly		>3% mineral oil	Seals external contamination
	Severe	6-monthly			
	Normal	Annually			
DC Resistivity	Very severe	Weekly		<50 MΩm at 20°C	
	Severe & normal	Not required			

Figure 5.16 shows the relationships betwen DC resistivity, moisture content, acidity and the presence of chloride ions.

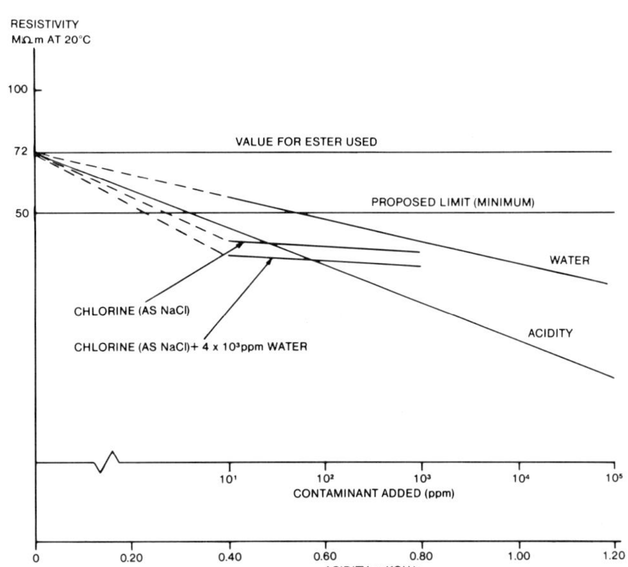

FIG. 5.16 Relationship between DC resistivity, moisture content, acidity and chloride ion for phosphate ester hydraulic fluids used for turbine steam valve control systems

Reference is made in Table 5.1 to Conpar testing, which is used to measure and categorise contamination levels of hydraulic fluid, where the fluid system components have clearances of less than 10 μm. The testing procedure must be clearly defined.

3.6 Commissioning network and resources

The network for commissioning the hydraulic package, flushing the complete system and initial operation of hydraulic servo units is illustrated in Fig 5.17.

A specialist will be required to carry out the necessary tests on the fluid. He should be consulted well in advance so that test equipment is specified and facilities provided.

An ample quantity of hydraulic fluid is required to charge the system, and also to replace losses. The quantity of fluid depends upon the initial cleanliness of the system and the care with which activities are carried out but, as a general guide, at least one complete spare change of fluid should be obtained.

The capacity of the system described is about 1.6 m^3 for each of the duplicate units.

4 Condenser

The plant items covered by this section are:

Condenser structure

Condenser cooling water (CW) tubes and tubeplates

Vacuum raising and maintaining plant

4.1 General description

4.1.1 Condenser structure

A number of different condenser designs exists for large modern turbines. The designs fall into two basic

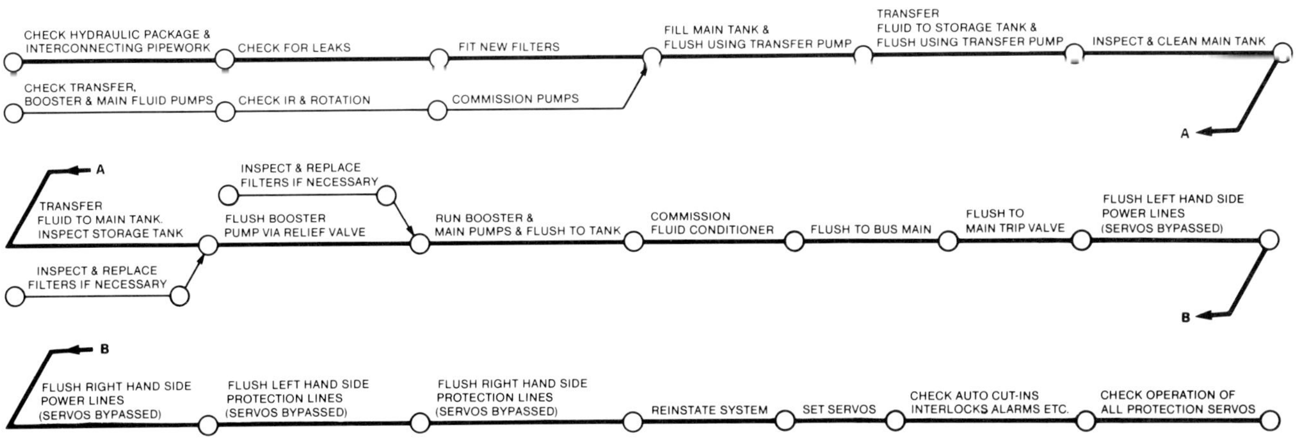

FIG. 5.17 Turbine valve gear hydraulic fluid system commissioning network

groups: underslung (with transverse or longitudinal CW flows) or pannier.

The pannier design may comprise individually constructed condensers mounted one on either side of the LP turbines and connected to their exhausts by ducts (Fig 5.18) or an integrated type, where the pannier condensers form part of the LP turbine outer casing (Fig 5.19). The integral pannier condenser is a more complex structure than the others and is therefore used as an example for commissioning purposes, since the structure depends upon a cooling system to maintain its geometry when the turbine is in service. The CW side of the condenser is similar to most other types.

The condenser embraces and supports the LP turbine casings. Each half (or pannier) condenser is divided

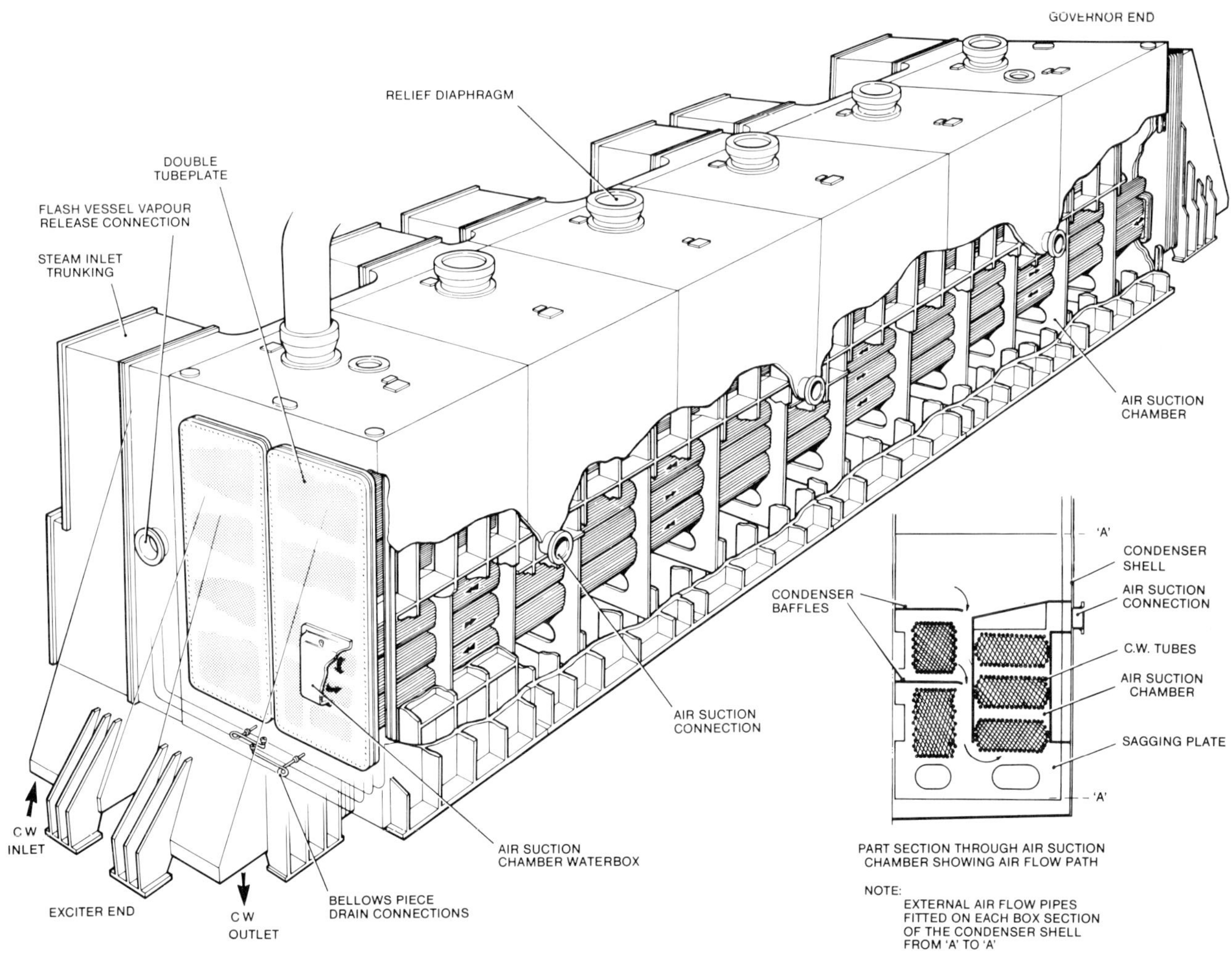

FIG. 5.18 Arrangement of individually constructed pannier condenser

One of a pair of individually constructed and separately supported pannier condensers. The condenser is connected to the LP turbines by means of steam ducts.

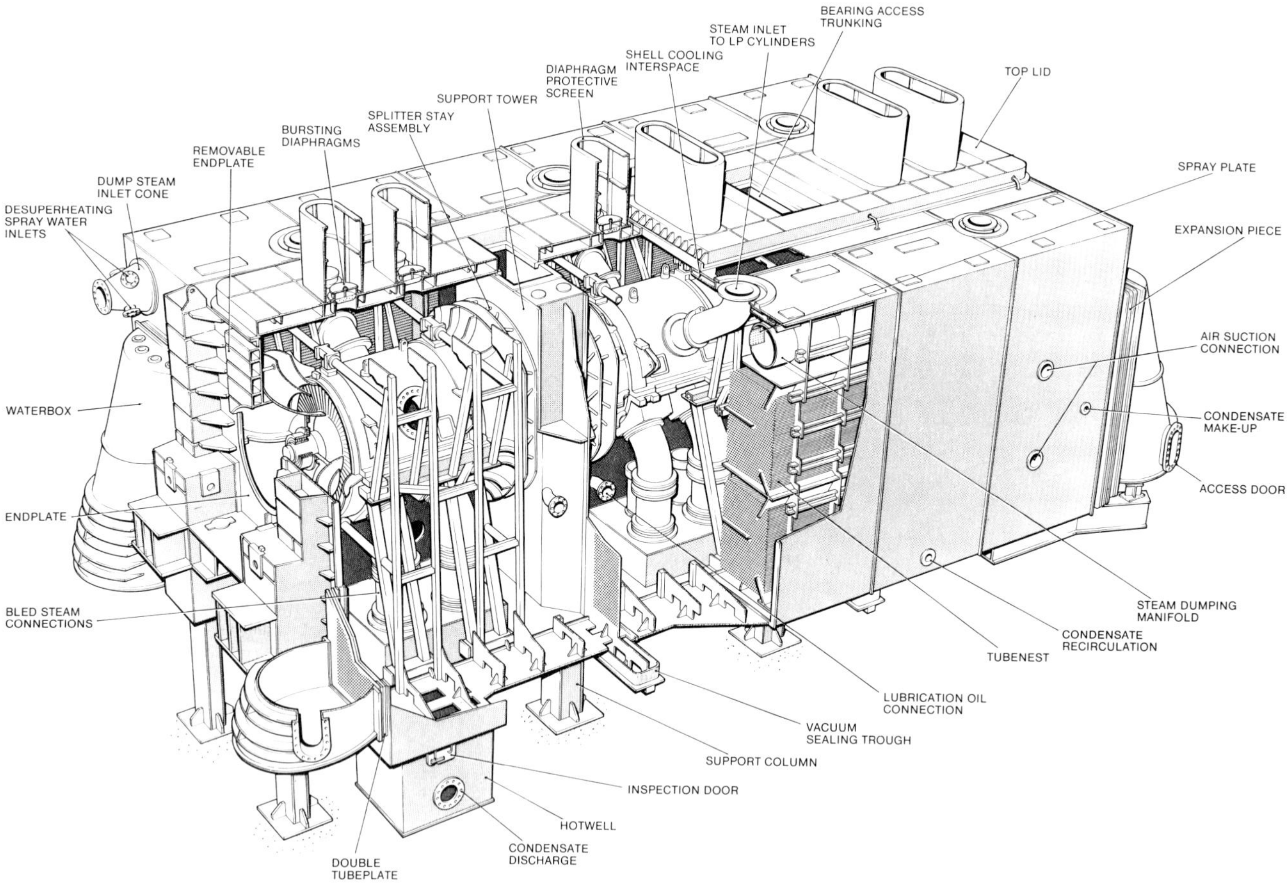

FIG. 5.19 Arrangement of integral pannier condenser

This completely integrated condenser embraces three LP turbines. Each half (or pannier) condenser is divided longitudinally into three sections, one for each turbine, each section having a different vacuum. The structure geometry is maintained by internal hollow struts which are cooled by condensate. Struts in the turbine exhaust areas are fitted with spray nozzles to cool exhaust steam.

longitudinally into three sections, one for each LP turbine, each section having a different design vacuum.

The structure is supported on steel columns which deflect under axial expansions.

The condenser structure is supported internally by means of hollow struts through which condensate is circulated in order to maintain thermal stability. The struts in the exhaust areas of the LP turbines are fitted with spray nozzles to prevent overheating in these areas.

Cooling water boxes are attached to the condenser, the outlet box having a large bellow piece which accommodates condenser expansion. The water boxes are protected against suctions by means of air admission valves. The suction within the water boxes can become excessive when, for example, the CW flow to the condenser is interrupted, the continuing momentum of the CW flow in the system causing high vacuums. The air admission valves are mounted on each condenser outlet water box. The valves have duplicate control systems and test facilities. For the system under consideration, each condenser has one air admission valve fitted at its topmost point which is air operated by means of a solenoid valve. The opening and closing times can be adjusted by means of restrictors in the air supplies to the air admission valves. The solenoid valves can be manually overridden for test purposes. The operating signal is derived from tapping points near the top of the outlet water boxes. Owing to the slightly different vacuums within each section of the condenser, the condensate level is higher in the higher vacuum section. The sections are interconnected by undershot weirs, so condensate flows through these weirs when the head of water is sufficient to overcome the vacuum differences between the sections, flow being from the cold (CW inlet) end to the hot (CW outlet) end of the condenser.

Bursting discs are fitted to the steam spaces at each section to protect the structure against overpressure.

The condenser has the ability to sustain a high part-load with one pannier out of service.

Each pannier has a single CW flow through titanium tubes which are expanded into double tubeplate assemblies. The tubeplate interspace is filled with condensate at a pressure greater than CW pressure in order to minimise the ingress of sea water under fault conditions. Some condensers are protected on the CW side by a protective coating. For such coatings, the differential pressure between tubeplate sealing water and CW in the condenser water box must be limited, otherwise leakage from the sealing space will tend to lift the coating. The tubeplate sealing system is therefore more complicated (Fig 5.20) than for non-coated water boxes, which have simple static head sealing systems (Fig 5.21).

4.1.2 Tubeplate sealing system

For coated condenser water boxes, the tubeplate sealing system comprises duplicate pumps in a closed loop system. The loops comprise four parallel flow paths, one for each of the two tubeplates for each pannier condenser.

A make-up tank fed from the condensate system feeds the tubeplate sealing system. The tank also caters for expansion and is fitted with a low level alarm.

Differential pressure between tubeplate and condenser water box is controlled and limited by means of an air-operated valve fitted in the common pump inlet lines to front and rear tubeplates.

Cathodic protection can be used to protect the internal surfaces of CW water boxes and tubeplates. This may be a simple system, using sacrificial anodes, or a more complicated system involving electrical controls.

4.1.3 Vacuum raising and maintaining plant

Vacuum raising plant is supplied to create an initial vacuum within the condenser steam space and then to sustain that vacuum and remove non-condensable gases. Vacuum raising plants therefore comprise start-up units and main units. A number of different principles are applied to vacuum raising and two typical vacuum pumps are illustrated in Figs 5.22 and 5.23.

The operating principle for the pump illustrated in Fig 5.22 is a liquid ring which is generated by the centrifugal action of an impeller mounted eccentrically in its impeller casing. The ring of water forms a seal between the tips of the impeller vanes and the casing. As the impeller rotates, the volume contained between a pair of impeller vanes and its water seal will cycle

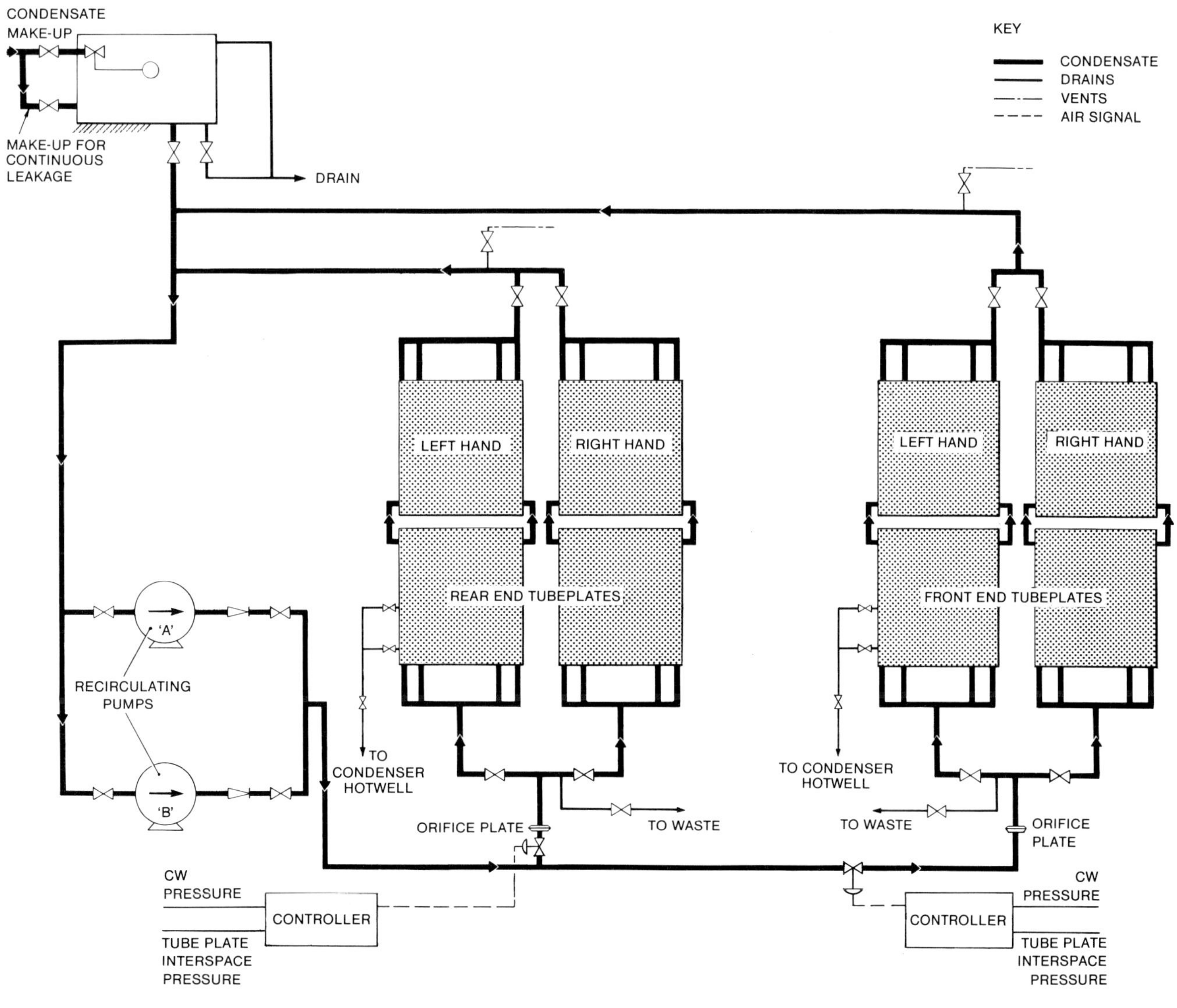

FIG. 5.20 Tubeplate sealing system for tubeplates covered by a protective coating

The coating can become detached if a leak exists and the tubeplate pressure is higher than the condenser water box pressure. Circulating pumps create a pressure head which is modified by diaphragm-operated valves. The valves close to reduce tubeplate pressure when tube plate/condenser CW differential pressure exceeds a prescribed value.

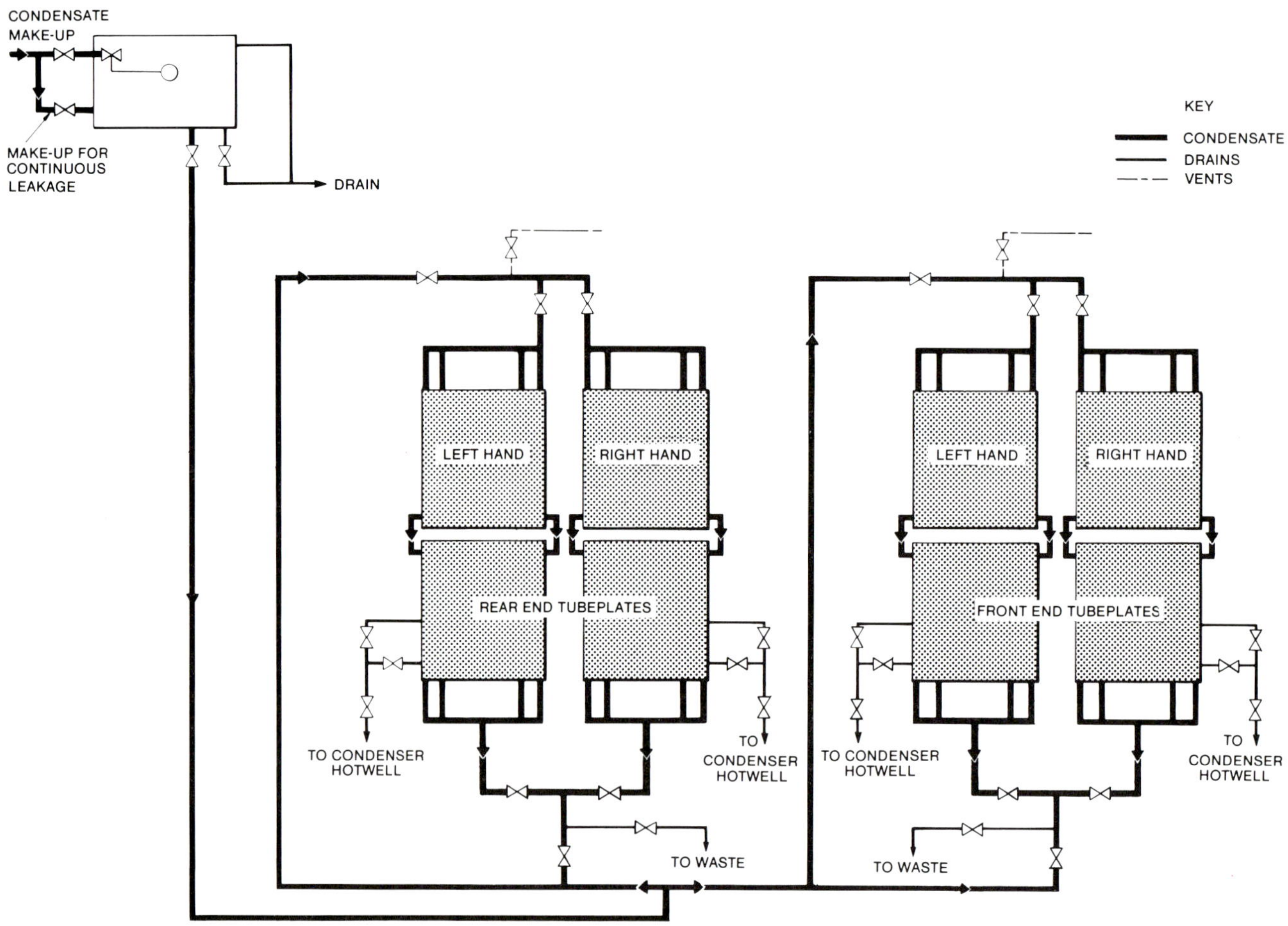

Fig. 5.21 Tubeplate sealing system for tubeplates without a protective coating

Shows a static head tubeplate sealing system for tubeplates without a protective coating. The static head is sufficient to maintain tubeplate pressure greater than the highest condenser CW pressure, thereby preventing an inflow of CW in the event of a tubeplate leak.

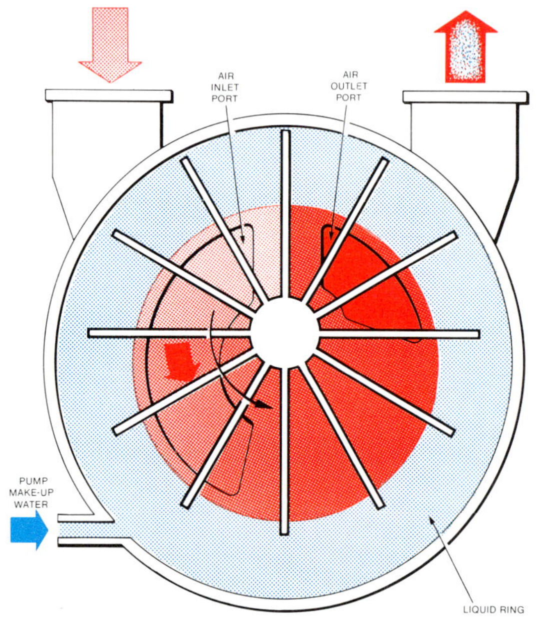

Fig. 5.22 Liquid ring vacuum pump

An eccentrically-mounted impeller rotates in a circular casing. A ring of water forms a seal between the tips of the impeller vanes and the casing. Air is drawn into the space between the vanes as the volume of the space increases. As the vane rotates, the space decreases, so the trapped air is compressed and ejected to atmosphere.

between maximum and minimum values for each revolution.

Inlet ports allow air and non-condensable gases to be drawn into the space between the vanes when the volume is increasing from minimum. As the vane rotates, the port is shut and further rotation acts to compress the air as the volume decreases. As the vane completes its revolution, an outlet port is opened and the compressed air/gases are discharged.

The operating principle of the Leblanc-type pump (Fig 5.23) is ejector action. Water is propelled into the divergent cone of an ejector which is open to the condenser. Air is sucked into the ejector and discharged to atmosphere via a separator.

As an example for commissioning purposes, a system of vacuum raising and maintaining plant deploying liquid ring pumps is described.

Figure 5.24 illustrates a vacuum raising system and Fig 5.25 illustrates a vacuum maintaining system.

The liquid ring pump draws directly from the condenser air suction busmain and discharges to atmosphere, via separators, which return entrained moisture to a tank. The tank acts as a reservoir, supplying the liquid ring pump seal by means of a free-suction sealwater pump and a cooler.

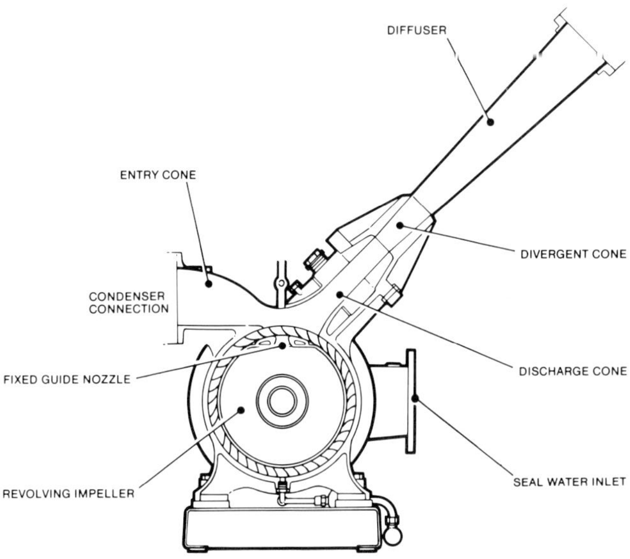

FIG. 5.23 Leblanc type vacuum pump

Water is propelled into the divergent cone of an ejector, which is open to the condenser. Air is sucked into the ejector and discharged to atmosphere, via a separator.

The main air pumps operate on the same principle, but using a two-stage liquid ring pump. The condenser air suction busmain feeds into a pre-condensing unit. Partial vacuum is created in this unit by means of a steam-operated vacuum augmentor. Superfluous vapour is removed from the flow stream by means of sprays within the pre- condensing units, the spray being derived under suction from a break tank. The level of accumulated spraywater and superfluous vapour collected within the pre-condenser is maintained by a pump and level control device.

The flow path continues from the pre-condenser, via the vacuum augmentor, to an auxiliary condenser operating on the same principle as the pre-condenser. The flow path from the auxiliary condenser is via an air-operated ejector, its motive power being air drawn in from atmosphere under suction from the liquid ring pump. The pump discharge and liquid seal arrangements are the same as for the start-up vacuum pump.

The two main air pumps are each of 100% duty and the two start-up pumps are each of 50% duty. The starting of each pump is sequence controlled, with automatic starting of the pump selected to standby duty.

Quick-closing valves in the air pump suction connections to the condenser air suction busmain prevent water being drawn from the air pumps when the pumps are shut down.

4.2 Commissioning objectives

The commissioning objectives are:

- To ensure that erection is complete to specification and drawings
- To ensure that the structure cooling system is flushed and its spray nozzles function satisfactorily
- To ensure that the tubeplate sealing system functions correctly
- To ensure that the tubeplate leak detection system operates satisfactorily
- To ensure that there is no passage of cooling water to the condenser steam space
- To ensure that cathodic protection systems (where fitted) for the cooling water box are commissioned satisfactorily
- To commission the start-up and main air pump packages
- To carry out a condenser vacuum test
- To check condenser level control systems and alarms
- To check the effect of CW flow with one condenser in service

4.3 Detailed procedures

The commencement of commissioning is preceded by a detailed inspection of items listed in the Detailed Procedures Document to ensure that they have been installed correctly to drawings and specification.

For the system described, the Detailed Procedures Document should contain a complete inventory of the following items:

Foundations

Panniers, including baseplates, keel plates and LP cylinder supports.

Pipework and supports, including cooling system, drains and vacuum grid system.

Instruments, including instrument pipework and electrical connections.

Valves

Cabinets and cubicles for vacuum panel, condenser level controls and tubeplate leak detection equipment.

Cabling

Access platforms and ladders

Vessels

Pumps

Orifice plates

Inspections internal to the condenser must ensure that internal access is complete, pipework complete and supported, pipework connections and entries free from debris and deflectors fitted. The internal spaces should also be clean and free from debris.

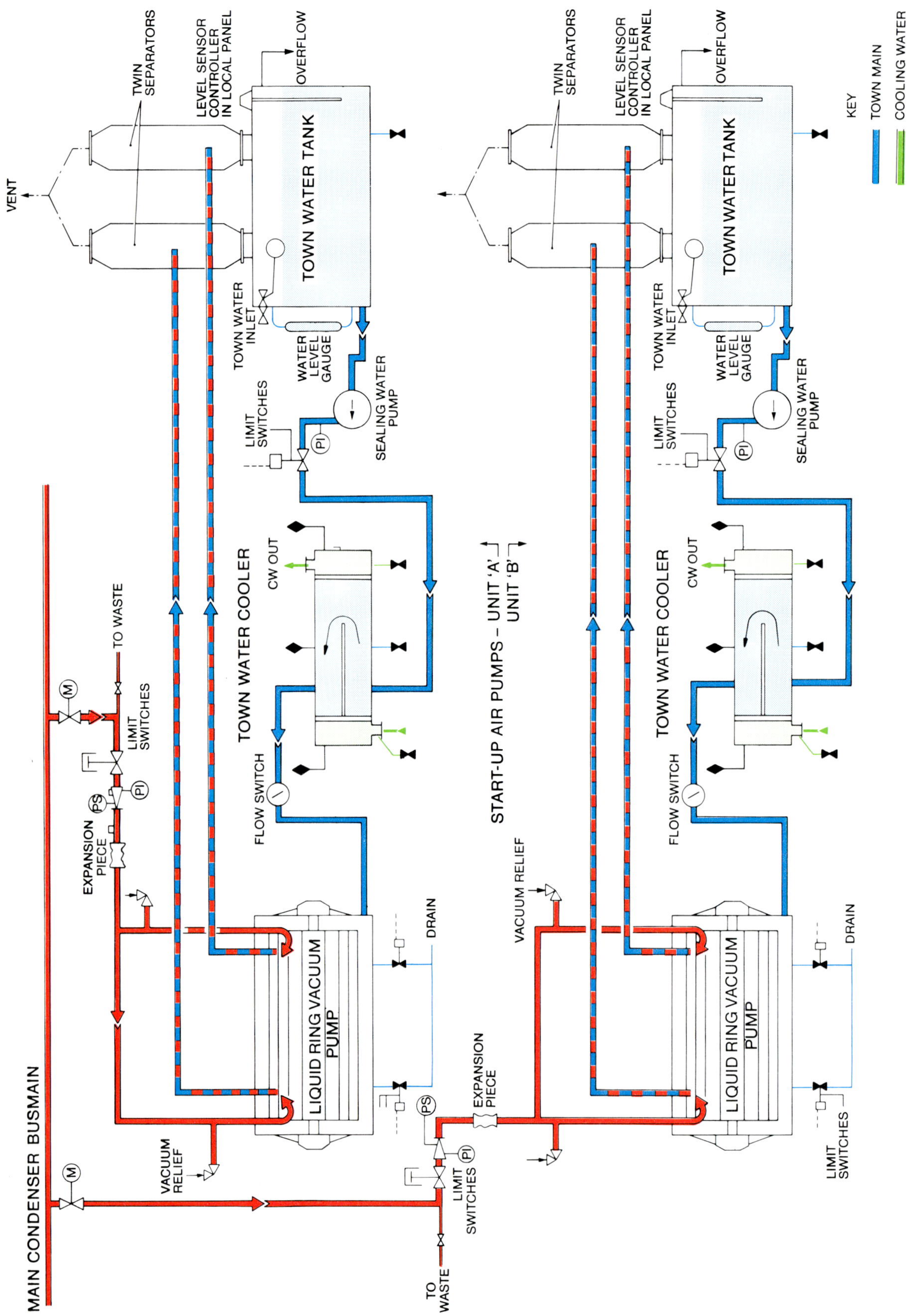

FIG. 5.24 Vacuum raising system using liquid ring pumps

This raises initial vacuum within the condenser steam space under unit start-up conditions over a short period of time. When an appropriate level of vacuum has been achieved, the vacuum maintaining system (Fig 5.25) takes over to exhaust non-condensable gases within the condenser.

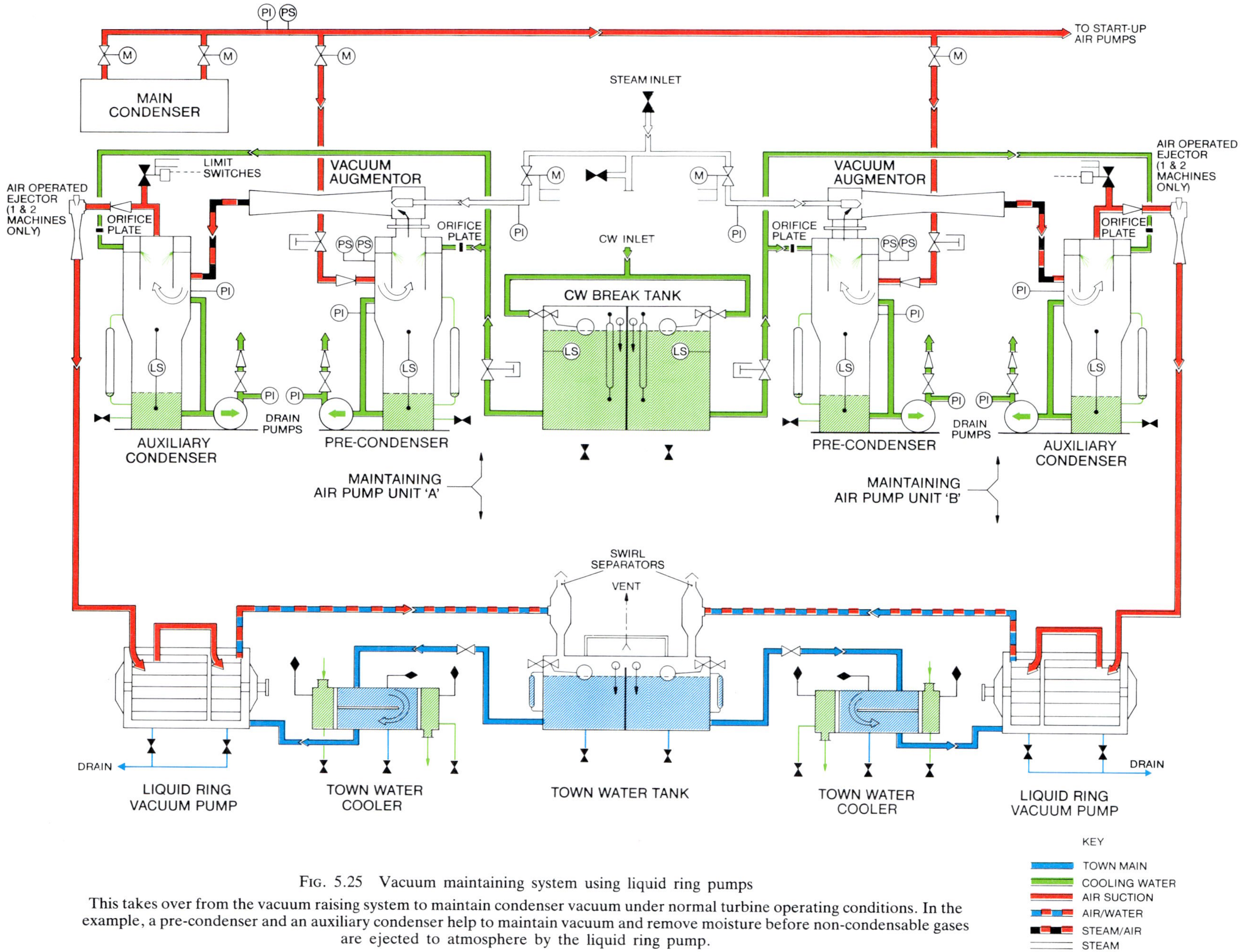

FIG. 5.25 Vacuum maintaining system using liquid ring pumps

This takes over from the vacuum raising system to maintain condenser vacuum under normal turbine operating conditions. In the example, a pre-condenser and an auxiliary condenser help to maintain vacuum and remove moisture before non-condensable gases are ejected to atmosphere by the liquid ring pump.

Some condensers are fitted with CW tube vibration damping inserts, comprising Butyl-rubber tubing threaded between each row or tubes mid-way between tube support plates. These must be checked for completeness.

If the condenser water boxes have been covered with a protective coating (such as Limpetite), each stage of the process will have been checked against specifications and a final inspection carried out.

The tubeplate interspace will have been pressure tested.

The condenser inlet and outlet CW valves will have been commissioned and used as part of the turbine erection and alignment activity, which requires the CW side of the condenser to be filled with water to simulate normal service load on the structure.

The final erection checks for the start-up and main air pumps should comprise a thorough check of each unit together with a check of their foundations, plinths, interconnections, air suction pipework and supports, service pipework and supports such as auxiliary steam, towns water and CW supplies, main and instrument electrical supplies and instrument air supplies.

The auxiliary steam lines to the main air pumps must be purged to atmosphere before final connection to their air pumps. Service water systems should similarly be flushed.

Records of main and start-up pump/motor coupling alignments should be available.

All connections to the condenser steam space should be complete before the air pumps are commissioned, and associated valves (bled-steam isolating, extraction pump suction, etc.) should be shut. Other connections to the condenser steam space, which cannot be isolated by valves, should be blanked off and a record kept of all blanks.

The outer glands must be fitted to all LP turbines; bearing pedestal covers must also be fitted to ensure that oil is not drawn through the glands into the condenser steam spaces, when vacuum raising.

The principal steps of the procedure for commissioning the condenser structure and associated systems are as follows.

4.3.1 Condenser structure

- Apply a pressure and vacuum test to the internal and external sections of the vacuum grid network, including associated instrument impulse lines. A pressure of approximately 2 bar should be applied and sustained for 24 hours, with a pressure loss not exceeding 0.1 bar. A vacuum of 950 mbar is similarly applied, the loss of vacuum not exceeding 15 mbar in 24 hours.
- Check that a condensate extraction pump is commissioned and available, together with a supply of demineralised water for flushing the condenser structure cooling system.
- Sectionalise the cooling system by means of blanks, recording all blanks used. Orifice plates must be removed. Spray nozzles must also be removed and blanked.
- Flush sections in turn until a satisfactory level of cleanliness has been achieved. This should be judged by the clarity of the water. The conductivity of the water can also be measured and compared with that of the supply. At the completion of the flush, little change in conductivity is to be expected.
- Ensure that all blanks are removed, and orifice plates and nozzles refitted.
- Using the condenser extraction pump, put the structure cooling system in service. Enter the condenser steam space and ensure that the spray pattern is satisfactory. Check that all nozzles are clear.
- Check condenser boxes. Functionally check the condenser level switches and associated alarms.
- Check condenser CW inlet and outlet valves for local and remote operation.

4.3.2 Tubeplate sealing system flushing procedure

- Check that header tank is clean and that its supply line is flushed.
- For tubeplate sealing systems with circulating pumps (tubeplates with protective coatings), disconnect pump suction pipes and flush system to waste from header tank with a supply of clean water. Reconnect pipework: isolate pump suction and discharge valves.
- Retract tubeplate conductivity-sensing probes. Disconnect the tubeplate sealing connection at the bottom of each tubeplate. Remove all system orifice plates. Flush to waste each tubeplate section in turn from the top to the bottom. Reconnect bottom tubeplate connections. Gravity flush rear tubeplate system via front tubeplate drain, and vice versa.
- Prime both circulating pumps and flush discharge pipework to waste at the nearest and lowest point. Commission both pumps, using standard check sheets, checking the automatic cut-in feature for each pump selected in turn to standby. To achieve a circulation whilst commissioning the pumps, jack open the differential pressure control valve.
- Circulate the system using each pump in turn. Shut down pumps and drain. Refill and repeat this operation until system is clean.
- Replace conductivity probes and orifice plates. Circulate and flush the system by means of the various drains until the conductivity of the water is satisfactory.
- Using the heater, and with its make-up supply isolated, check that the system is watertight.

4.3.3 Condenser water box air admission valves

- With CW system out of service and condenser water box admission valves isolated from their water boxes, prove that each valve will open when its operating solenoid valve is overridden.
- Repeat the override operations and set the air flow restrictors to each air admission valve to give the desired opening and closing times of approximately 1.5 s and 3 s, respectively.
- With the vacuum signal on, check that the solenoid switch operates at approximately 0.5 bar falling and that the air admission valve functions correctly. Remove the vacuum signal and check that the admission valve resets.
- Prove the 'low air pressure' alarm on the supply to each air admission valve by shutting off the air supply and allowing the pressure to decay.
- Prove the 'electrical supplies fail' alarm by switching off the solenoid control supplies.

4.3.4 Tubeplate sealing system final functional checks

- Recheck local and remote operation and indication of condenser CW inlet and outlet valves.
- Test each tubeplate conductivity and prove associated alarms, by removing and inserting in a solution having a known conductivity.
- Ensure that condenser CW inlet and outlet temperature indicators are calibrated and functional.
- Ensure that the CW air admission valves are commissioned.
- With the condenser CW system and tubeplate sealing system in service, isolate sealing system header tank and prove its low level alarm. Record the discharge pressures and motor currents of the tubeplate sealing system pumps and observe the functioning of the differential control valves. Recheck the auto cut-in feature of the standby pump and observe the response of the differential control valve.
- Reduce the condenser water box tubeplate sealing differential pressure by adjusting tapping point valves. Alternatively, pressure signals may have to be injected if sufficient adjustment is not obtainable. Check the various low differential pressure alarms for each tubeplate.
- Repeat the test for high differential pressures.
- For the anticipated CW water box maximum/minimum pressure range, check that the tubeplate sealing system differential pressure control valve maintains the prescribed differential pressure. This test should be carried out by injecting the appropriate signals. The performance of the system can subsequently be checked with the CW system in service. A chart recorder, with input pressure signals from the tubeplate interspace and from the inlet and outlet condenser water boxes, will enable the response of the tubeplate sealing system to be assessed for all modes of operation.

4.3.5 Condenser vacuum raising and maintaining system

The activities described relate primarily to the liquid-ring-type main and start-up air pumps, but the principles should be applicable to other installations.

- Ensure that seal-water supplies are available to seal the glands of valves subject to vacuum
- Shut the condenser air suction busmain isolating valves
- Check service water supplies, e.g., float valves to header tanks, and establish a level in the tank.

Start-up pumps

- Check that electrical supplies are available to liquid-ring pump drain solenoid and that it closes on energising.
- Commission the seal-water pump, using a standard check sheet.
- Commission the liquid ring pump, using a standard check sheet. This may be started empty, but under no circumstances must the water level be above the casing centre line. Close the liquid ring pump suction valve and ensure that its 'snifting' valve operates to admit air to the pump at a specified vacuum. Shut down the pump and de-energise its solenoid drain valve to open.
- Select a start-up air pump to remote control with its sequence control timer at maximum setting. Where the sequence is interlocked with availability of gland steam, inject a gland steam pressure signal.
- Initiate and time the sequence, ensuring that:
  (a) Liquid-ring pump drain valves close
  (b) Service water is supplied to the ring pump
  (c) Ring pump starts
  (d) Motor heater switches off
  (e) Suction valve to air-suction busmain opens at prescribed value (approximately 660 mbar).
- Shut down the start-up pump. Note sequence of operations and record sequence time. Set the sequence timer and prove by means of another start-up operation. Repeat the operations for the other start-up air pumps.

Main pumps

- Set up the pre-condenser and auxiliary condenser systems by initially ensuring that the CW break tank outlet valves are closed and by checking the float control to the break tank. Open the break tank outlet valve slowly and prove that it closes when pre-condenser and auxiliary condenser high level switches operate. This can be achieved by filling these condensers using a hose pipe.
- Commission the pre-condenser drain pump, using a standard test sheet. Leave pump in service and similarly commission the auxiliary condenser drain pump.
- With both drain pumps in service, check that the pre- and auxiliary condenser levels are maintained.
- Commission the liquid ring pump in the same manner as for the start-up air pumps, and then shut down the pumps.
- Shut the CW inlet valve to the CW break tank and allow the pre- and auxiliary condenser drain pumps to empty their respective vessels. Shut down both drain pumps.

At this point the pumps can be used to carry out an initial vacuum raising test, so that air leaks are identified and rectified. During or after this test, final functional tests can take place. A full vacuum raising test should then be carried out to check the capability of start-up and main pumps.

Initial vacuum raising test

Vacuum is raised initially using both the start-up pumps, the condenser air suction busmain being isolated at the condenser.

The condenser will have been prepared for vacuum raising by ensuring that all systems connected to it are isolated by means of valves or blanks. Associated vent and drain valves must be shut and all tapping points which cannot be isolated plugged. Bearing pedestals will be intact.

- Open air suction busmain isolating valve to condenser and check that the non-return valves fitted to the start of pump suction lines are open.
- Check for air leaks on all joints and connections to the condenser. This may be done using ultrasonic testing devices which detect noise above audible levels generated by leaking air. Alternatively, a freon gas detector can be positioned at the air pump discharge and freon gas discharged around all joints and connections.

  This procedure should be meticulously carried out, using detailed drawings showing joints and connections and/or by the use of detailed check lists. Where possible, identified leaks should be remedied immediately. Large leaks may have to be temporarily plugged. A log must be kept of all leaks, such as welds and flanges, that need repair. These repairs must then be rechecked.
- Leak checks must be continued until it is certain that all leaks have been accounted for. A vacuum of about 690 mbar should be achieved.
- Those parts of the turbine, HP heaters, LP heaters and de-aerator, bled-steam pipework and reheater which are connected to the condenser steam space should be introduced, where possible systematically, so that the vacuum test and leak check can be extended to them.
- With the entire system in circuit at atmospheric pressure and with the start-up air pumps shut down, put the pumps into service and establish the maximum vacuum achievable, which should be of the order of 680 mbar. Record the time taken to reach this value. Shut down the air pumps and record the time for the vacuum to decay to atmospheric pressure.

4.3.6 Final functional checks

Start-up air pumps

- Prove remote start-up operation.
- Shut down pump, set down sequence timer and prove start-up operation is blocked. Reset timer.
- Defeat liquid ring pump solenoid supplies and prove that pump start-up sequence is blocked. Restore solenoid supplies.
- Prove other sequence start features are blocked, as appropriate, e.g., failure of seal-water pump, liquid ring pump selected to local operation.
- Prove start-up pump trips when its suction fails to achieve a prescribed value. This may be achieved by venting the suction switch to atmosphere.
- Prove start-up pump stops on loss of seal-water pump and observe correct function of sequence steps on shutdown.
- Prove other sequence trips.

Main air pumps

As the main air pumps will not have been used for the vacuum raising and test exercise, they should be checked for correct remote sequence operation:

- Set the main air pump to remote control, with its sequence timer at maximum setting
- Initiate the start-up sequence, ensuring that the:
  - (a) Pre-condenser and auxiliary condenser drain pumps start
  - (b) Auxiliary condenser vacuum break valve closes
  - (c) Liquid ring pump starts

(d) Suction valve to air suction busmain opens at prescribed value (approximately 660 mbar vacuum)

(e) Steam-driven air-ejector steam valve opens at prescribed pre-condenser vacuum (approximately 900 mbar)

(f) Overall opening time is recorded.

- With sequence times still at maximum setting, shut down the main air pump. Ensure that the sequence of operations is correct and record the time taken.
- Adjust the sequence timer to a value just greater (say 120%) than the respective start-up or shutdown times (whichever is the greater) and prove by means of another start-up operation.
- Repeat the operations for the other main air pump.
- Set up fault conditions and prove that sequence shutdown is automatically initiated. Check all associated alarms. Examples are:

  (a) High pre-condenser level

  (b) High auxiliary condenser level

  (c) Loss of pre-condenser or main condenser pumps

  (d) Loss of liquid ring pump

- Prove that sequence start-up is inhibited at the appropriate stages by simulating failure of response from each of the sequence check requirements. Check all associated alarms. Examples are:

  (a) CW break tank outlet valve not open

  (b) Pre-condenser or auxiliary condenser pumps not running

  (c) Pre-condenser or auxiliary condenser levels high. Auxiliary condenser vacuum break valve open. Liquid ring pump not running.

  (d) Pre-condenser vacuum not greater than −660 mbar. Quick-closing air suction valve not open. Main condenser vacuum not greater than −900 mbar. Steam ejector steam valve not open.

Final vacuum raising test

The main turbine must be on turning gear and the glands sealed with steam. All systems, including the reheater, must be in circuit.

- Start both start-up air pumps on sequence control. Measure time taken to reach maximum vacuum and record the vacuum achieved. Record liquid ring pump motor currents on the start-up air pumps.
- Start both main air pumps on sequence control and similarly record details of time, vacuum and currents. Normal vacuum of 950 mbar should be obtained.
- Shut down each start-up air pump in turn, checking that the vacuum remains steady.
- Shut down a main air pump. Then shut down the second main air pump and prove automatic start of the other one. Repeat the operation with the roles of the two main air pumps exchanged. Measure the minimum vacuum values achieved during the changeover.
- Shut down all air pumps. If a vacuum break valve is fitted, prove the valve opening and closing operations whilst vacuum is high.

4.4 Commissioning network and resources

Figure 5.26 illustrates a commissioning network for the condenser. Figure 5.27 is a network for the condenser air pump final erection checks, commissioning of start-up and main air pumps, initial vacuum raising (termed the 'cold pull'), final vacuum raising with turbine shaft glands sealed, and functional tests.

Considerable man-hour resources are likely to be required to search out and rectify all leaks systematically and carefully. It is important that all leaks are identified and repaired in order to avoid vacuum raising and consequential delays later. Access will be required in order to check normally inaccessible flanges and other parts of the plant where leakage checks need to be made. Where possible, access provided during construction activity should be retained for the purpose of checking for leaks, particularly for joints which would otherwise be difficult to get to, or where it would be expensive and time consuming to provide such access. At least six to eight days should be allowed for this work.

The manpower required for rectification work should be relatively small for items such as pipework joints, but would be considerable for very large joints and where the dismantling of large components is necessary. It is essential, therefore, that erection of large items which will subsequently be under vacuum is carried out carefully and, wherever possible, tested for leakage at the construction stage.

5 Turbine gland steam system

5.1 General description

The quantities, pressures, and temperatures of the sealing steam to the various turbine glands are important and must be sufficient to seal the glands over the full operating range of the turbines.

A number of different designs of gland steam sealing systems are employed by the various turbine manufacturers. All of them use steam derived from the boiler for initial turbine start-up operations. As the turbines reach stable operating conditions, the boiler-derived steam can be substituted from other sources, such as the HP turbine gland annular pocket. The system

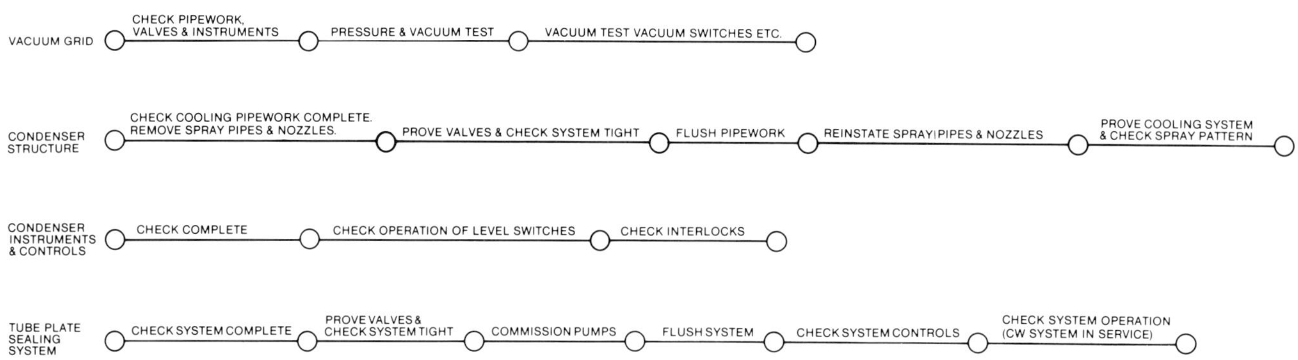

FIG. 5.26 Commissioning network for condenser structure

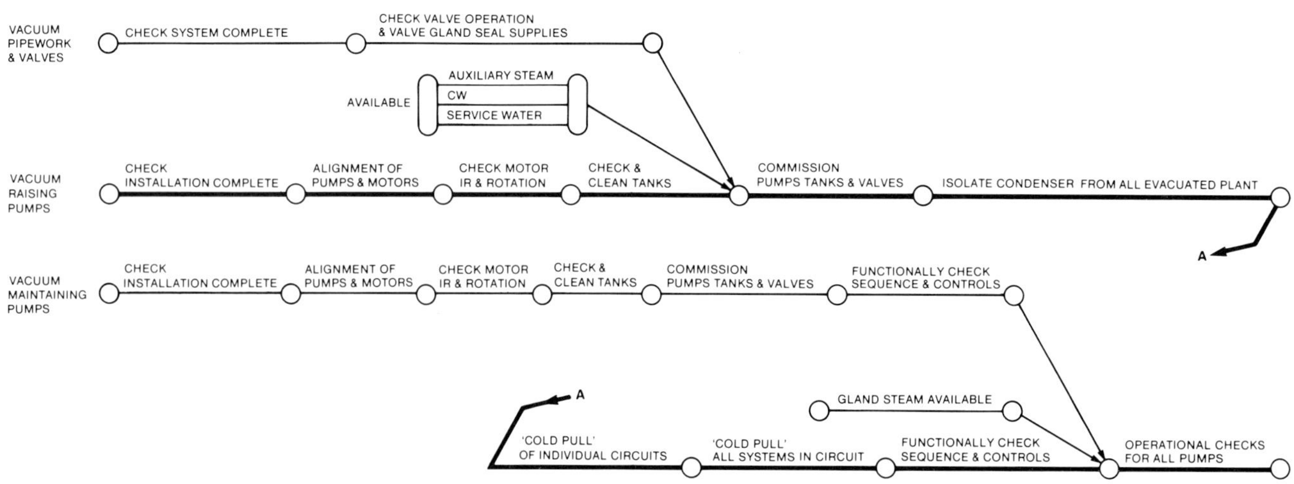

FIG. 5.27 Commissioning network for vacuum raising and maintaining systems

illustrated in Fig 5.28 provides full sealing automatically over the whole range of operating conditions, with remote control from the control room. The system comprises the following parts:

Sub-atmospheric line By which the outermost pockets of all glands are connected to the sub-atmospheric gland steam condenser, and are thus maintained at a slight vacuum by one of two full-duty blowers. The blower selected for standby duty is automatically brought into service when the vacuum falls to a prescribed value.

The atmospheric line By which the adjacent inner pockets of all the shaft glands, except those of the HP cylinder, are connected to the atmospheric gland steam condenser. The outer pockets of the turbine steam valve spindles are also connected into this line.

One inner pocket of each shaft gland is connected by means of a gland sealing line to the supply and dump valves. This sealing line is maintained at a fairly constant pressure by supplying live steam (if the pressure is low) or dumping excess steam (if the pressure is high). The pressure is normally controlled automatically in the LP manifold by a pneumatic controller with proportional integral action. The controller senses gland supply steam pressure and puts out an air signal to the pneumatic valves, which are operated on a split range via their respective positioners. The pressure in the HP manifold is slightly higher than that in the LP manifold. Motorised by-pass valves provide standby supplies, if the pneumatic valves fail whilst the turbine is in service.

Live steam is supplied from upstream of the turbine HP stop valves and, after being throttled through the supply control valve, it is reduced to a suitable temperature for the HP and IP cylinder glands by means of the first-stage desuperheating system, the required temperature being obtained by mixing uncooled steam with steam cooled in the first stage of the desuperheater. The cooling-coil valves may be set to suit starting conditions and the state of gland wear.

The steam supplied to the LP cylinder glands is cooled to about 120°C in the second stage of the desuperheater. Again, valves are set to control the amount of cooling, as required by the state of the glands.

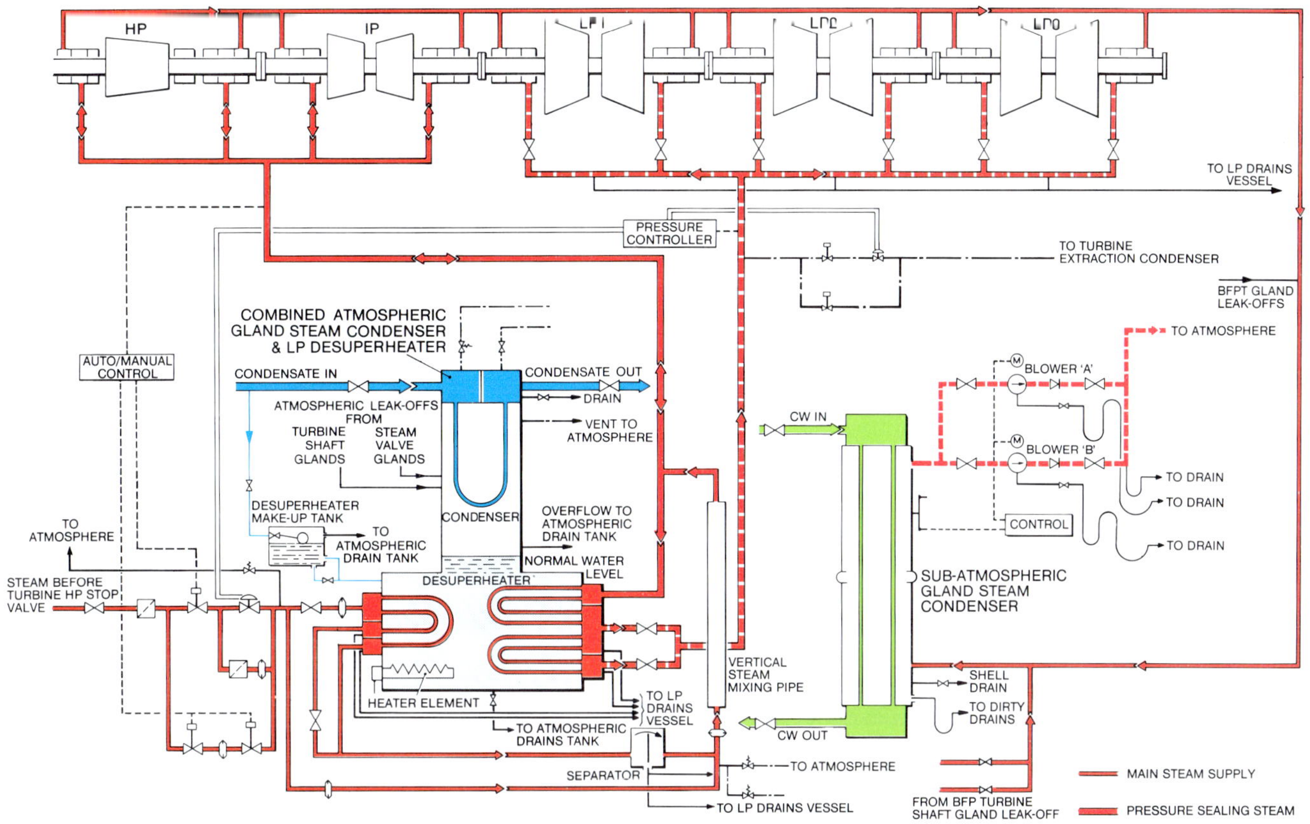

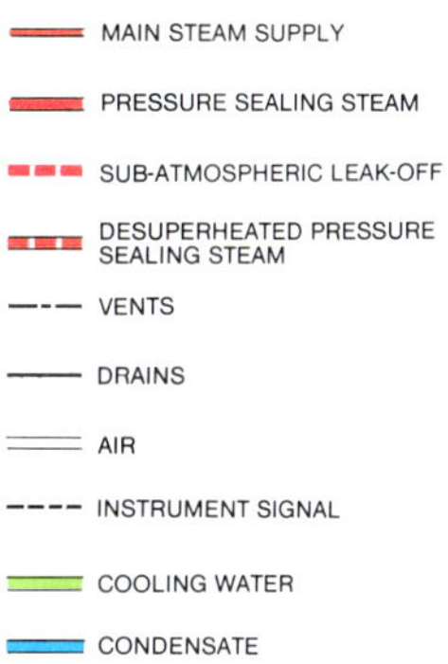

FIG. 5.28 Gland steam system

One of a number of different types of gland steam systems applied to large turbines using live steam as the source of sealing steam for all glands. It is designed to be self-sealing at a prescribed unit load, using steam bled off from stages within the HP rotor and IP rotor shaft glands. Pressure control is exercised by sensing the gland steam supply pressure to the LP rotor glands which is used to control either live steam supply or, when the system is self-sealing, steam bled off from the system to the turbine extraction condenser. Gland steam temperature control is non-regulatory and depends on water-immersed cooling elements and flow restricting orifices.

The pressures in the individual lines are indicated separately by local pressure gauges.

Additional leak-off lines return steam to suitable points in the system for optimum efficiency, namely:

HP cylinder Inner cylinder gland leak-off to cold reheat line, outer cylinder first-stage leak-offs to IP/LP crossover pipes.

Main steam admission valves First-stage leak-offs from HP stop and governing valves, and IP governing valves, to de-aerator.

The gland steam sealing system described also caters for gland leak-off from the turbine of a steam-driven boiler feed pump with connections from the valve spindle leak-offs on the feed pump turbine to the atmospheric line and de-aerator.

5.2 Commissioning objectives

The commissioning objectives are:

- To steam purge the gland steam system pipework
- To prove the correct operation of the gland steam control system, together with alarms and indicators, over the whole operating range
- To prove the correct operation and location of drains and drain traps
- To prove that the gland steam safety valves function correctly

5.3 Detailed procedures

The commencement of commissioning is preceded by a detailed inspection of items listed in the Detailed Procedures Document to ensure that the system has been installed correctly and to drawing.

For the system described, the Detailed Procedures Document should contain a complete inventory of the following items:

Gland steam separator

Combined atmospheric gland steam condenser and LP desuperheater

Sub-atmospheric gland steam condenser

Valve spindle leak-off chamber

Atmospheric drain tank

Sub-atmospheric condenser blowers (2)

Instruments, including instrument pipework and electrical connections

Orifice plates

Valves

Cabling

Pipework supports

Cabinets and cubicles

The first stage of the commissioning activity is to steam purge the gland steam system pipework. This is usually carried out at the same time as the unit steam purge activity.

The various components of the system are then commissioned and their function proved, including interlocks, indicators, alarms and controls.

Final commissioning of the system cannot take place until the turbine is operated over its entire range.

5.3.1 Steam purging

The principles of steam purging are similar to those for flushing, in that the system is purged with steam at higher velocities than those achieved in normal service.

The gland steam system is therefore separated into circuits, no circuit being introduced downstream of a system which has not previously been purged. The use of blanks must be carefully controlled in the same manner as described for the lubricating oil system flushing procedure. Blanks, temporary pipework and fittings must be designed and fitted to the same standard as the gland steam system, with particular attention to the support of components exhausting to atmosphere. Exhausts to atmosphere must be adequately guarded to protect personnel. Orifice plates and other restrictions must be removed.

For the system described, the steam purge circuits would be as follows:

- Live steam supply, control valve by-pass line and gland steam condenser by-pass line via drain connections to atmosphere. Drain valves must be removed and temporary connections to atmosphere fitted.
- Live steam supply, control valves and gland steam condenser via drain connections to atmosphere.
- Live steam supply, control valves, gland steam condenser by-pass and then the two parallel circuits of the HP/IP and LP connections to atmosphere.
- The LP section of the parallel circuits would then be blanked and the remaining circuit blown fully to atmosphere. Figure 5.29 illustrates one of the purge circuits.

For each stage of the purge, steam is applied slowly to the pipework to warm it and to check for leaks. The pressure is then gradually raised to ensure that the system is not subjected to violent thermal variations or to water-hammer effects. An adequate steam pressure is required to achieve a flow greater than that likely to be encountered in service. The flow should be maintained for about ten minutes. A minimum of two purges per stage is required and the pipework should be allowed to cool between each stage. This thermal cycling aids the removal of scale.

At the completion of the purge, all temporary fittings should be removed and the system reinstated, including orifice plates and valves.

5.3.2 Final functional checks

A check must be made to see that the system is fully reinstated after the steam purge, including all connections to the glands.

The principal steps for the system described are:

- Commission each gland steam blower using standard check sheets. Adjust each blower discharge valve in turn to give the specified suction (about 13 mbar) within the sub-atmospheric condenser. Check the starting and running motor currents for each blower.
- Prove automatic start for each blower in turn on low sub-atmospheric gland steam condenser vacuum, together with associated alarms.
- Check the functioning of all valves, including remote control and indication.
- The gland steam desuperheater only operates with the correct level of condensate. Fill the vessel and prove that the low level alarm clears. The condensate is protected against low temperature by means of an immersion heater, which should be proven to operate and also to switch off on low condensate level.

On completion of the functional checks, the gland steam sealing system is ready for unit operation, during which control of the system will be set up and proven over the complete operating range of the turbine.

5.4 Commissioning network and resources

A commissioning network for the gland steam system, including steam purge and final functional checks, is illustrated in Fig 5.30.

Access is required to the places where pipework is to be disconnected and blanked. The materials required must be prepared in advance of commissioning and manpower made available to carry out the work.

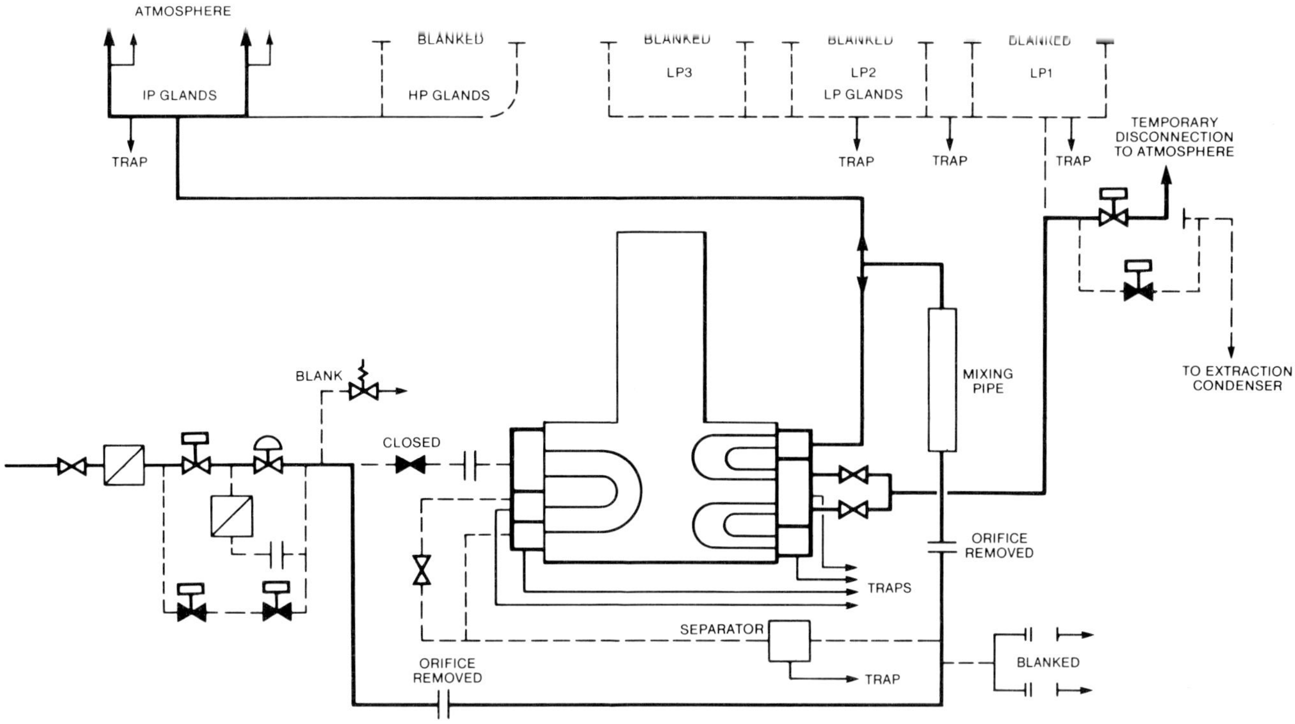

FIG. 5.29 Gland steam system — typical purge circuit

The circuit includes the live steam supply line and control valves, gland steam condenser by-pass and the two parallel circuits of the HP/IP and LP connections. The chain-dotted lines indicate those circuits which are isolated from the purge by means of blanks or valves. Pipework at the IP glands and in the tee-off to the extraction condenser has been removed in order to give a flow path to atmosphere. The traps are blown through during the purge process.

6 Miscellaneous turbine ancillary systems

There are a number of turbine ancillary systems which, although important, are relatively small or simple and can be dealt with by applying commissioning principles already established. The purpose of this section is to summarise the commissioning procedures for a number of these systems which are likely to be encountered. Examples of such systems are:

- Auxiliary cooling water (ACW) system
- HP and IP turbine flange heating system
- IP rotor cooling system
- Forced-air cooling system
- Electric turning gear

6.1 General description

6.1.1 Auxiliary CW system

Auxiliary cooling water systems are usually supplied from the main CW system. There are systems which are self-contained and supplied from a separate source, the commissioning of which is covered in Chapter 3. The supply taken from the main CW system may have its own filter. The flow through the ACW system will be a significant proportion of the total CW supply, of the order of 10%. The ACW system comprises a number of series/parallel paths, and is in parallel with the condenser CW system. A typical ACW system is shown in Fig 5.31.

The ACW system supplies cooling water to a number of turbine, generator and steam feed pump auxiliaries, such as turbine oil coolers, generator seal-oil coolers,

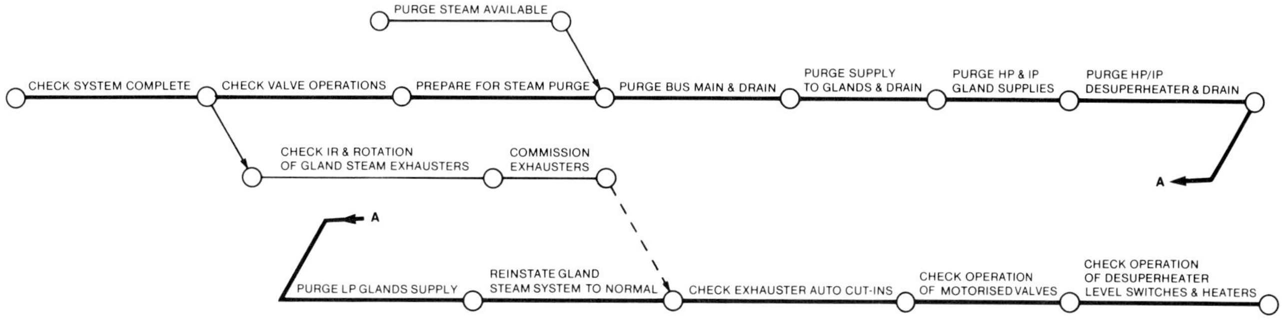

FIG. 5.30 Commissioning network for a gland steam system

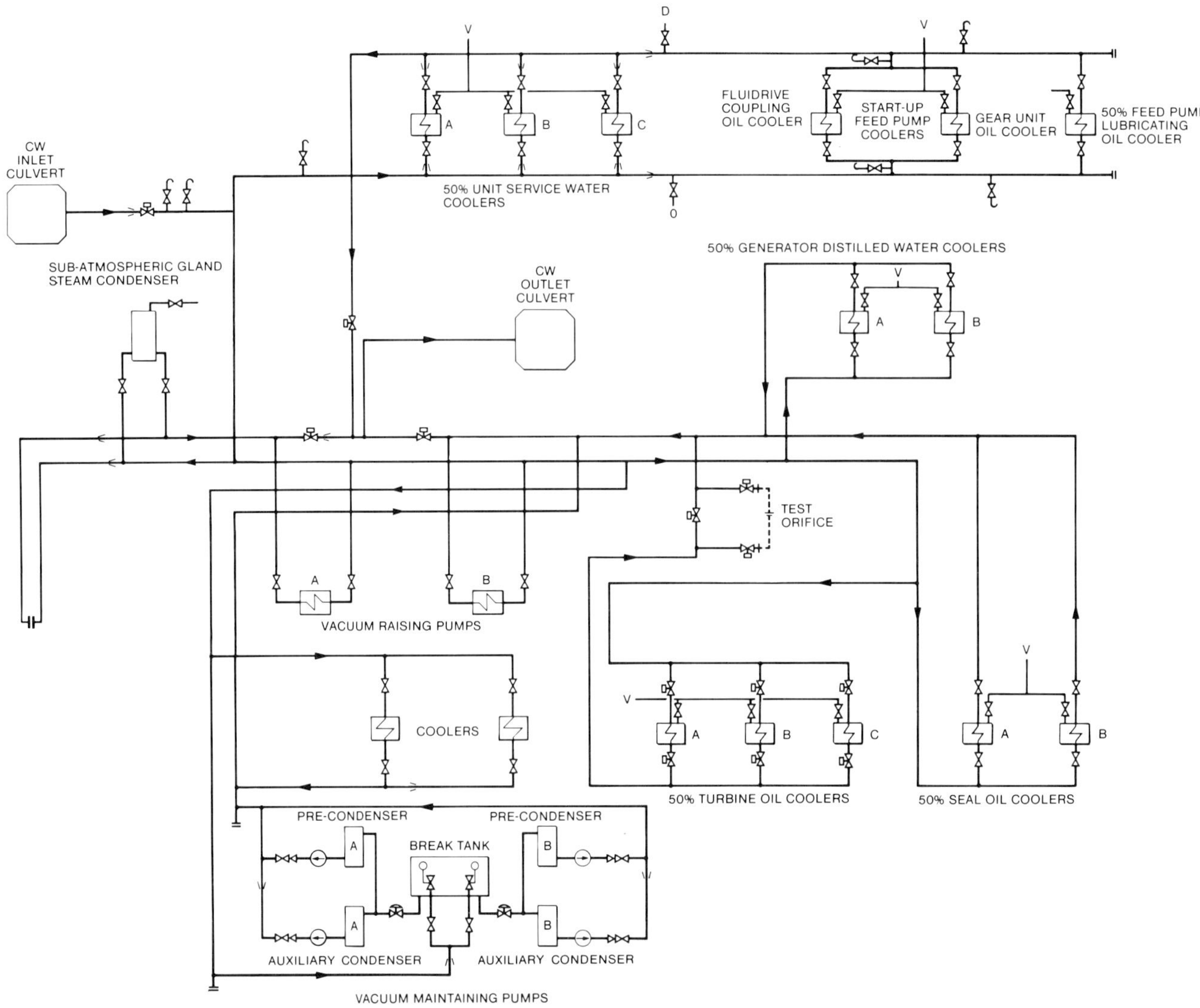

FIG. 5.31 Typical auxiliary cooling water system

A typical cooling water system used to cool unit auxiliaries. Cooling water, in this case seawater, is obtained from the unit cooling water inlet culvert at a point before the condenser inlet isolating valve. Water then flows in three parallel paths to a number of unit auxiliary coolers, before returning to the condenser outlet culvert. The example refers to certain unit auxiliaries, such as vacuum raising pumps, which are used in other sections of this chapter as examples for commissioning purposes.

vacuum raising plant, gland steam sealing systems, etc. Coolers require to be vented for the purposes of priming, as well as to maintain prime during service. Continuous venting systems may be connected to vacuum raising plant, similar to that described for the main condenser, or to some other convenient sub-atmospheric system, such as the top of the main condenser CW outlet water box. The ACW system may have a number of return connections to the main CW system downstream of the condenser outlet valves.

Where an ACW supply strainer is fitted, it will be of the duplex type, having automatic changeover on high differential pressure and automatic backflushing of the strainer when out of service.

6.1.2 HP and IP flange heating system

When large turbines are started up from cool or cold conditions, the thick metal flanges of the HP and IP turbine outer casing heat up more slowly than the thinner sections of the casing barrels. The controlled application of flange heating reduces temperature differences between the barrels and their flanges and bolts.

Passages are formed in the flanges which connect with annular spaces around the cylinder bolts, thus providing flow paths for warming steam. Steam is provided from a convenient point, usually from before HP turbine stop valves, and then via a steam strainer to HP and IP flange steam regulating valves.

Orifices may be incorporated both for regulating flange steam system warming and for controlling drains from the low points of the HP and IP flange heating systems.

Instrumentation for the HP and the IP systems will include steam pressure and temperature indication, regulating valve position, and turbine flange-to-barrel differential temperatures and alarms.

A typical system is shown in Fig 5.32.

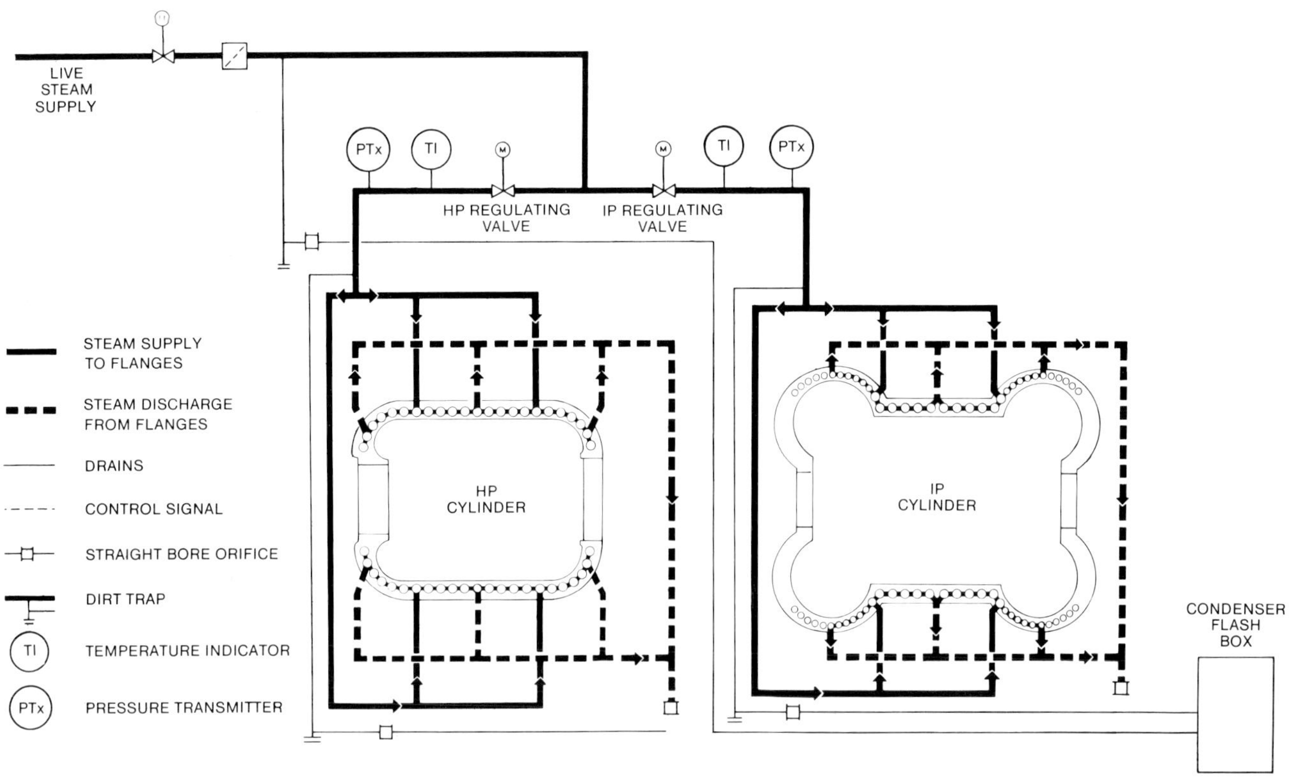

FIG. 5.32 HP and IP flange heating system

6.1.3 IP cylinder rotor cooling system

Large turbines, with steam entry at the mid-span of the turbine rotor and exhaust at either end, may need the centre portion of the rotor to be cooled and ventilated. Without this facility, the shaft temperature at this point may become excessive, reducing its creep life.

Figure 5.33 shows an IP rotor cooling system. Steam is introduced into the space between the two glands of the IP first-stage nozzle plate and flows across the faces of the rotor discs to mingle with the steam flowing through the two respective steam paths of the turbine. Steam is taken from the HP cylinder inner-gland leak-off. Orifice plates control the rotor cooling steam and fine adjustment is obtained by means of a valve, which is set and locked during the commissioning process.

A temperature indication is provided, which would show about 440°C under normal operating conditions for the system described. A marked drop in temperature indicates a reduction in steam supply, possibly due to a blocked orifice plate.

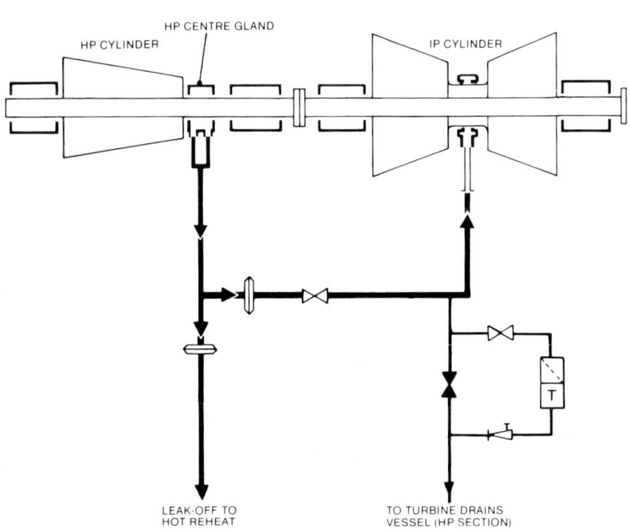

FIG. 5.33 IP cylinder rotor cooling system

6.1.4 Forced-air cooling system

Most modern large turbines are provided with systems which enable a reduction in the natural cooling times from hot shutdown to the point at which the turbine turning gear can be shut down. The natural cooling time will be of the order of 120 hours and this can be reduced to about 30 hours, thereby reducing outage times and costs.

There are a number of different systems, of which Figs 5.34 and 5.35 are typical. For the purpose of establishing a commissioning procedure, only Fig 5.35 is described here.

Air is introduced into the HP cylinder via inlets in the cold reheat pipework at the cylinder exhaust. The air flow is contrary to the steam flow path and the air discharges to atmosphere at the steam inlet end of the outer casing. A secondary circuit provides air flows via the HP valve chest drains.

Air is introduced into the mid-span of the IP cylinder between the inner and outer casings and discharges to atmosphere via outlets on the exhaust belt. A secon-

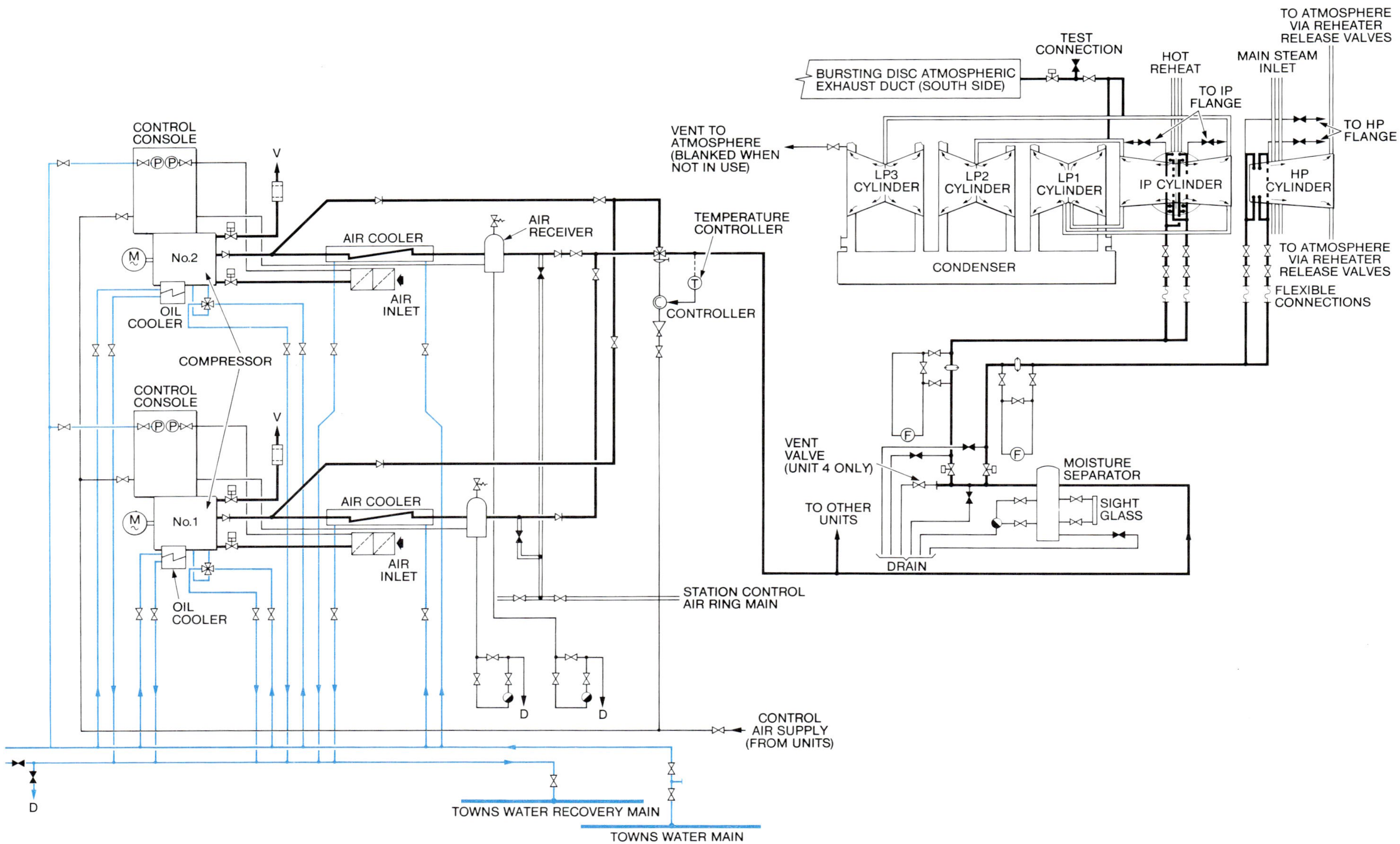

FIG. 5.34 Turbine forced-air cooling system

This forced-air cooling system uses a central compressed air unit, comprising duplicate compressors and coolers. The temperature of the air is controlled throughout the cooling process.

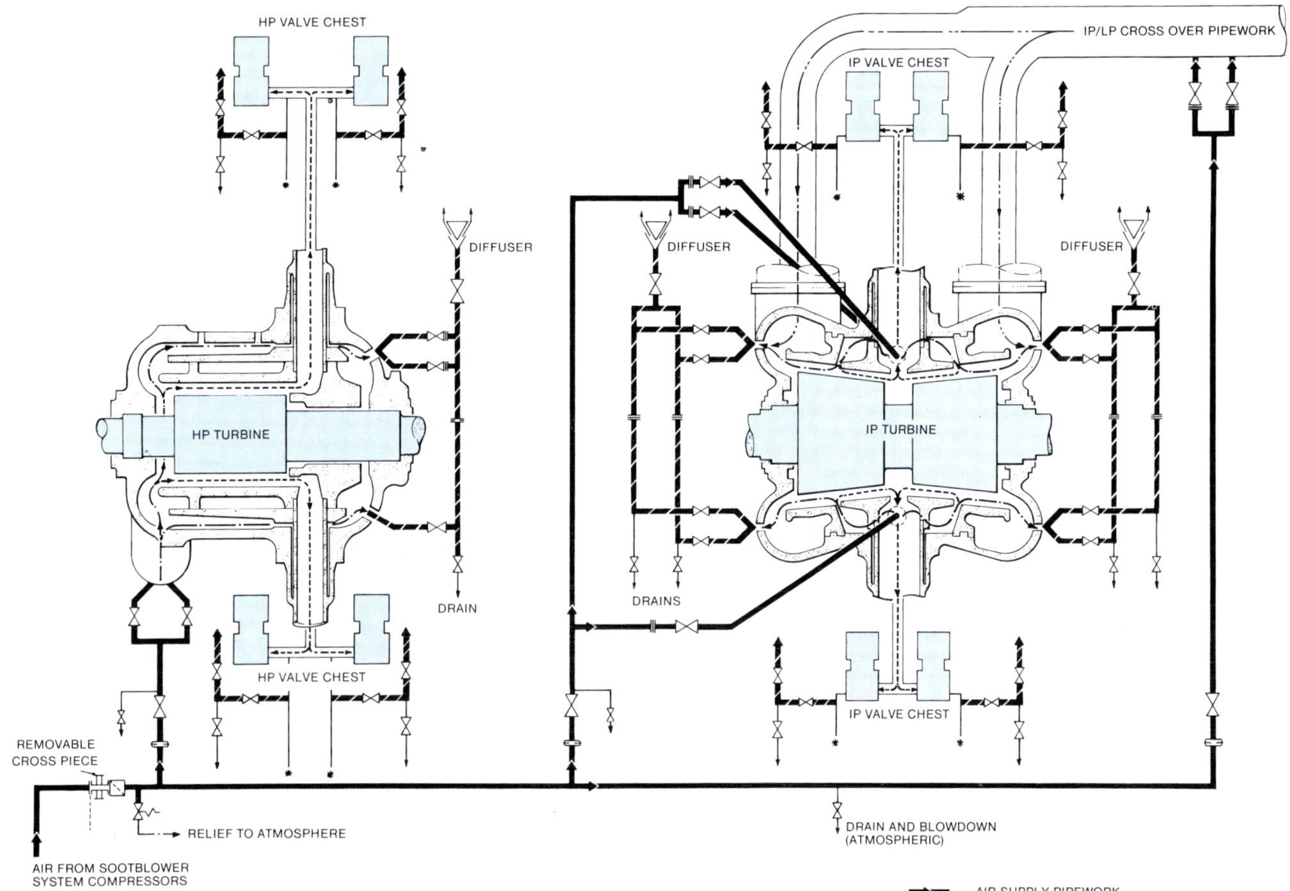

FIG. 5.35 Forced-air cooling system using air derived from an air sootblowing system

Another forced-air cooling system using air derived from an air sootblowing system. It is designed to give uniform cooling to the HP and IP cylinders and their associated valve chests. Individual air flows are preset by means of orifice plates sized during the commissioning phase, so that adjustments to valves are eliminated.

dary circuit provides air flows via the IP valve chest drains. Air is also introduced into each LP crossover pipe to act as a buffer to hot air flowing from the IP exhaust, thus keeping the LP turbine and condenser environment separated from the forced-air cooling conditions.

The forced-air cooling supply for the system described is obtained from the sootblowing compressed air system. The connections between the forced-air and the turbine cooling supply should have a means of physical disconnection and here a special cruciform device is used (Fig 5.35). The forced-air system is set up manually, the necessary control being achieved by means of orifice plates sized during commissioning tests. The progress of cooling is monitored by the normal turbine casing instrumentation. Temperature points are provided, together with alarms to check for steam leakage back into the forced-air cooling systems. These alarms are inhibited when the forced-air cooling supply is on.

6.1.5 Electric turning gear

Turbines are fitted with a means of rotating the shafts in order to prevent thermal bending, which would occur if the turbine came to rest in a hot condition.

A number of different designs are used, all incorporating a motor drive unit and a self-synchronising clutch mechanism, which permits automatic engagement and disengagement of the motor drive to the turbine shaft.

The turning gear speed is usually sufficient to maintain a hydrodynamic oil wedge between the turbine journals and their bearings. Thus, jacking oil pumps might not be essential but would normally be left in service in the event of a turning gear trip. Starting and stopping of the turning gear is achieved by speed-sensing devices. Interlocks are provided between the turning gear controls and the lubricating oil and seal-oil systems to prevent the turning gear being in service when these are not in service.

6.2 Commissioning objectives

6.2.1 Auxiliary CW system

- To apply a static pressure test to the system encompassed between the ACW supply and the return isolating valves, and then to flush the system
- To establish ACW flow from the main CW system and prove the ACW venting system
- To prove correct functioning of the ACW strainer (where fitted)
- To check that flows in individual paths are adequate under operational conditions

6.2.2 HP and IP flange heating system

- To purge the steam pipework
- To check valve operation
- To check pressure and temperature indicators and alarms
- To prove operation of warming steam supply and system drains
- To prove flange heating control for the turbine cold start operation

6.2.3 IP cylinder rotor cooling system

- To purge the steam pipework
- To check the sizing of orifice plates and to set the fine control device to limit the maximum cooling steam temperature

6.2.4 Forced-air cooling system

- To pressure test the cooling pipework
- To establish cooling air flow rates during a forced cooling operation for a fully heat-soaked turbine, using throttled air valves, and thereby estimate orifice plate sizes
- To repeat the forced cooling operation with orifice plates fitted and valves fully open, to check orifice plate sizes
- To prove effectiveness of the forced cooling operation by taking the turbine off turning gear and observing casing temperatures, and then observing rotor eccentricities with the turbine on turning gear

6.2.5 Electric turning gear

- To check for freedom of movement and to record the torque needed to turn the turbine shafts manually
- To check the manual and automatic operation of the electric turning gear, including interlock features, indicators, alarms and controls.

6.3 Detailed procedures

Completion of erection must be checked in a manner similar to that given for other systems.

6.3.1 Auxiliary CW system

- Vent pipework must be checked and outlet valves must be shut and vent connections isolated or blanked, as required. All drain valves must be shut.
- The ACW system is then subjected to a static head test using a suitable supply of water, e.g., from a head tank. The head tank should provide a head of water greater than the pressure expected when the system is in service. The water connection is made to a suitable drain point.
- Prime the system slowly, checking for free flow of water from the lowest vents in the system first. Shut each vent as priming progresses. For coolers not having vents to atmosphere close to the top of the cooler, disconnect vent pipework adjacent to the top of the coolers and fit temporary vent valves.
- When the system and all coolers have been primed, check for leaks and loss of water from head tank.
- Reconnect vent pipework and open the vent valves. Vent the highest point of the vent pipework until it is primed. Shut the vent and prove that the system is watertight.
- Check each vent to the priming system is free, by establishing in turn a flow from the cooler and through the vent pipework to atmosphere.
- Prove that the ACW strainer, where fitted, operates correctly. Use injected signals to check changeover and backflushing features and associated alarms and indicators.
- Prove all system power-operated ACW valves, including cooler temperature control valves where fitted.
- When the system has been tested, flush the system from its drains, operating each drain in turn to prove it is clear. Then isolate the water supply and drain the system down, including the drain valve used to prime the system.
- Ensure that the system is reinstated and blanks removed.

6.3.2 HP and IP flange heating system

Before the final closure of the HP and IP flange heating system pipework on to the turbine, it must be purged. This may be carried out using compressed air, or by using steam. If steam is used, the purge can be included as part of the unit steam purge activity and would follow a similar pattern to that described for the gland steam system purge.

For a purge using either air or steam, the flow paths will be limited to maximise the flow and no flow path will be introduced downstream of one which has not been purged. Components such as orifice plates and strainer internals must be removed. Blanks must be carefully controlled. The system should be purged progressively, using drains to obtain initial flow paths to atmosphere. The final stages of the purge will include the individual lines connecting with the various parts of the turbine. At completion of the purge, drain pots and vessels, such as the condenser flash vessel (Fig 5.32), will be opened and checked free from debris.

The system must then be reinstated, all blanks and temporary connections being removed, with filters and orifice plates refitted.

Final functional checks can then be carried out, in a manner similar to that described for other systems. Final commissioning of the system will take place after steam-to-set (Chapter 6).

6.3.3 IP cylinder rotor cooling system

This can be purged and final functional checks carried out in the same manner as for the HP and IP flange heating system.

Final commissioning will again take place after steam-to-set.

6.3.4 Forced-air cooling system

Before final closure of the forced-air cooling system pipework on to the turbine, it should be blown through with air in a similar manner to that described previously to achieve debris removal. It should be noted that the pipework back to the cruciform will be purged using steam from the turbine.

After the system has been reinstated, the cooling air supply is used to check for air flows to the turbine through each inlet point in turn and its associated outlet point.

Final functional checks will include calibration and alarm initiation for the temperature points used to sense steam leakage into the forced-air cooling system. The inhibition of the alarm when the forced-air cooling supply is in service should also be proven.

The performance of the forced-air cooling system is established after steam-to-set (Chapter 6). The initial test will be to establish a natural cooling operation after hot turbine shutdown. Subsequent tests will establish appropriate orifice plate sizes and achieve the remaining commissioning objectives.

The effectiveness of the forced-air cooling operation should be finally assessed by taking the turbine off turning gear at a temperature recommended by the manufacturer (typically 150°C hottest HP or IP casing temperature). The cooling rate of each measured point should continue smoothly. There should be no tendency for differential temperatures, such as casing top/bottom, to approach the limits set by the manufacturer.

The lubricating oil system which, when in service, aids the removal of heat from the shafts, should also be shut down. When the cooling trends of the casings are adequate and there is no tendency to high differential temperatures, the shafts should be manually turned and then put on electric turning gear. This will normally be several hours after the oil system has been shut down. When the shafts are on turning gear again, the eccentricities should be noted.

No significant change should be observed between these and the values recorded just before the shafts were taken off turning gear. High eccentricities indicate that the internal casing temperatures are not sufficiently uniform, thereby causing one or more shafts to hog or sag within the casings. Further forced-air cooling tests should then be carried out to determine the precise temperature at which the turbine can be safely taken off turning gear.

Causes of eccentricity other than those due to inadequate forced-air cooling should also be considered. For example, it has been known for thermal recirculation through the casings to take place via external pipework. In one instance, a manufacturing error was the cause of excessive thermal flows within the turbine casings, causing high casing differential temperatures.

6.3.5 Electric turning gear

The lubricating and jacking oil supplies will have been commissioned as described in Section 2 of this chapter.

With lubricating, jacking and seal-oil systems in service, the turning gear is operated by hand. This is usually achieved by means of a hand-ratchet attachment to the end of the motor drive shaft. A key-operated interlock system will be provided to ensure that the motor cannot be switched on whilst the hand-turning ratchet is engaged.

The effort required to turn the shafts, by means of a torque ratchet, should be recorded. This value is used as a basis of comparison for future hand-turning operations. Hand turning should always be applied before electric turning gear is used, when the turbine shafts are at rest.

Before commissioning the electric turning gear, the lubricating and seal-oil systems must be available and turbine-generator erection complete. A standard check sheet will be used for the turning gear motor. The interlock between motor supplies and the hand-turning facility must be proven. The starting and stopping of the electric turning gear is then proven, noting the start-up and steady-state motor current and the speed of the turbine shaft rotation.

Interlock features and associated alarms and indicators are then checked. This should be carried out to prove each permissive start interlock in turn, as well as for tripping interlocks. A simulated shaft speed signal (approximately 100 r/min) will be required to prove automatic start and stop of the turning gear for main turbine speed rising and falling.

On completion of these checks, the operation of the turning gear must be observed after steam-to-set.

6.4 Commissioning network and resources

Commissioning networks should be drawn for each of the systems described in this section.

7 Turbine supervisory equipment, turbine protection and electrical control

This section deals with the equipment associated with the main turbine-generator and a steam-driven boiler feed pump.

The list of equipment will include:

- Turbine supervisory (turbovisory) equipment for the measurement of:
  - Speed
  - Eccentricity
  - Differential expansion
  - Casing and pedestal movement
  - Valve positions
  - Bearing vibration
- Protection and electrical control
- Pressure and differential pressure switches for:
  - Hydraulic fluid system
  - Lubricating oil system
  - Turbine vacuum trip protection
  - Turbine low steam pressure protection
- Limit switches for:
  - Turbine vacuum unloading protection
  - Turbine steam valve position indication
- Solenoids for:
  - Protective devices such as valvegear tripping mechanisms

The systems and equipment used will vary considerably with the type of installation and the supplier. The purpose of this section is therefore to give a general approach to the commissioning of such equipment.

It is essential to establish a complete inventory of all the instruments and their functions, together with cable schedules, cabinets, cubicles and schematic drawings.

The turbine protection and electrical control systems will interact with equipment associated with more than one supplier. It is therefore important that the interfaces are clearly defined and the procedures are designed to prove complete systems.

7.1 General description

Turbovisory equipment is provided so that the condition of turbine shafts, casings, valve chests and similar components can be monitored and maintained in a safe condition over the whole operating range of the plant.

The equipment can be categorised into four main areas:

- Sensors for speed, eccentricity, differential expansion, casing and pedestal movement, valve position and bearing vibration.
- Cubicles for equipment providing energising supplies to the sensors, conditioning the various signals and providing outputs to operate alarms, drive indicators and driving data processing equipment.
- Data processing equipment, which provides computed information for signals and alarms, permanent records (including post-incident recording) and fast logging facilities.
- Termination equipment and interconnecting cables.

Examples of components used in turbovisory systems are:

- Power supplies with built-in monitoring facilities.
- Standard oscillators (e.g., 5 kHz).
- Current output units.
- Speed measurement circuits, e.g., 0–10 V DC output, proportional to input frequency.
- Speed monitor circuits, the output of which can be used, for example, to drive relays which change state when the frequency signals fall below prescribed levels.
- Rotation-sensing circuits, e.g., giving indication of reverse rotation.
- Eccentricity measurement circuits, e.g., measuring shaft eccentricity over speed ranges from, say, 20 to 3500 r/min: such circuits may give 0–10 V DC outputs proportional to peak-to-peak eccentricity.
- Position measurement circuits, e.g., converting the output of rectilinear potentiometers to 0–10 V DC proportional signals.
- Differential expansion measurement circuits which, when used with displacement detectors, produce 0–10 V DC outputs, corresponding to the displacement of the detector sensors from a steel target.

- Vibration measurement circuits to give an output proportional to the amplitude of vibration (independent of frequency).
- Signal level monitors, e.g., comparing input voltage signals with a reference signal and thereby generating an output to drive, say, an alarm.

Turbine protection and electrical control systems are best described by schematic drawings and descriptions of the particular systems employed. Such schematic drawings will show, for example:

- HP and LP heater bled-steam valve control
- Hydraulic fluid control system
- Steam feed pump turbine trip and control systems
- Main turbine protection scheme
- Generator and unit transformer protection schemes
- Generator protection scheme

The turbine-generator protection scheme will also include protection against overspeed and low vacuum, both of which act to trip the turbine steam admission valves. The steam valves can also be tripped by means of a mechanical trip lever and electrical solenoids. The solenoids are energised from a number of sources connected with unit electrical protection.

Trip protection is also provided for low lubricating oil pressure, low stator water flow, high condensate level in the condenser, turbine valve low hydraulic fluid pressure, low steam pressure at turbine steam admission valves, electronic governing malfunction, etc.

For devices such as turbine overspeed bolt rings, vacuum trip units and main trip solenoids, on-load test facilities with appropriate interlocks are provided.

7.2 Commissioning objectives

7.2.1 Turbine supervisory equipment

The objectives when commissioning the turbovisory equipment are:

- To verify that works testing of equipment cubicles has been carried out.
- To ensure that sensors, cubicles, interfaces and cabling have been installed in accordance with erection check sheets and that cable segregation and screening has been applied, where appropriate.
- To carry out calibration checks between sensor and turbovisory cubicle equipment to establish various settings such as linearity, span and zero. The measured parameters are then adjusted by known amounts and remote alarm settings, indicators, recorders and computer inputs checked.
- To prove turbine shaft axial position limits within the clearances of the thrust bearing and to compare measured movement against indicated movement.

After steam-to-set:

- To check speed measuring channel outputs for consistency and accuracy during turbine run-up and shutdown operations

7.2.2 Turbine mechanical protection devices

The objectives when commissioning these devices are:

- To prove satisfactory operation of trip latching mechanisms
- To prove operation of all mechanical tripping devices

After steam-to-set:

- To prove on-load testing of turbine overspeed rings and vacuum trip unit, with machine at speed
- To prove functional operation of overspeed trip rings (3300 r/min)
- To prove that the vacuum trip operates correctly with the unit at speed and condenser vacuum decaying

There may be other objectives to suit the particular installation.

7.2.3 Turbine electrical protection devices

The objectives when commissioning these devices are:

- To ensure that sensors, cubicles and cabling have been installed in accordance with erection check sheets.
- To operate all electrical devices and prove that they function in accordance with the protection system design.
- To prove that devices, such as relays and solenoids, operate over their specified voltage range, particularly at reduced voltage.
- To prove that electrical trip and reset devices, such as turbine trip and reset solenoids, function correctly when the mechanical trip systems are in service. This will include local and remote emergency trip push-buttons.
- To establish healthy conditions for each of the protective devices and systems, and then simulate a failure for each.

Some examples of electrical protection devices are:

Trip due to loss of turbine steam valve hydraulic pressure

Trip due to loss of lubricating oil pressure

Trip due to low vacuum

Load reduction due to reduced vacuum

Trip due to low steam pressure to turbine

Load reduction due to reduced steam pressure to turbine

Trip due to loss of stator winding water flow

Trip due to loss of (electronic) governing

Trip due to remote and local emergency pushbuttons

7.3 Detailed procedures

7.3.1 Turbine supervisory equipment

Final functional checking of turbovisory equipment will entail the use of special equipment, the exact nature of which will depend upon the particular turbovisory system installed. For the system quoted, the special equipment would be:

Digital voltmeter (AC/DC ranges)

Digital frequency counter

Sine/square wave oscillator (frequency range 0.6–10 Hz, amplitude 20 V peak-to-peak)

Double-beam oscilloscope

Decade resistance boxes (five decades, 0–10 kΩ)

Wheatstone bridge

Vibration test levels

Static displacement rig for rectilinear potentiometer calibration

0–36 V DC variable power supply

Degaussing coil (for eccentricity equipment)

In addition, good quality multimeters, variable load resistors and recorders will be needed.

Final erection checks will include inspection of dowels for sensor mounting brackets, records of final settings and gaps for detector heads, inspection of conduits and termination of screens, etc.

Some of the tests will be to ensure that the performance of works tested items has not changed during erection. Others will establish correct termination and calibration of items of equipment. Where standard components are used, such as amplifiers and oscillators, standard check sheets should be applied.

For most measurements it is possible to simulate the plant-mounted sensor, e.g., by use of a decade box. The sensor is disconnected and the simulator substituted to give a measure of gain and bias. These values are recorded and can be used for future checks. Thus, by reference to these records, components in the measuring units can be replaced with the unit on-load and circuit gains adjusted to accommodate the replacement item by using the simulated sensor.

The approach to calibration checks for complete circuits should be as follows.

Differential expansion

Sensors are often mounted on moveable carriages. Adjustment to the carriage can therefore be made and measured accurately, thereby simulating movement of the sensed item (e.g., shaft to casing, casing to casing). With sensing heads locked in position, it may be possible to move the sensed item, e.g., shaft part of the sensor.

Some differential measurements may be obtained by compounding two signals. For example, a rotor to inner-cylinder measurement can be obtained by compounding the rotor to pedestal value and the inner-cylinder to pedestal value. The signals are added electronically within the turbovisory equipment and the combined signal must be checked for polarity and calibration by adjustment of the two constituent signals.

Rectilinear potentiometers

By disconnecting the fixed end, the range and sensitivity can be set up.

Eccentricity

The gap measurement circuit can be calibrated by shimming the detector. The peak measurement circuit can be checked with an oscillator. Alarm settings may be checked by inserting a shim of appropriate thickness through the detector/shaft gap.

Vibration

Vibration detectors usually comprise a large mass carrying pick-up coils mounted on leaf springs. By injecting a DC current into the coils, the mass can be moved in either direction. Measuring the current indicates the condition of the detector, which can be checked against works test figures. Similarly, the amplifier sections of the vibration measurement can be checked by injection of a test frequency. A vibration table is suitable for checking the overall performance of all but the largest detectors.

Speed

An oscillator can be used to inject a test signal into speed measurement circuits to check their linearity. Also, the output of the detectors can be checked against shaft speed, measured by other means, when at turning gear speed.

Final checks are carried out with all rotors coupled and the jacking oil system in service. Rotor and cylinder temperatures must be cold and not affected by condensate recirculation, lubricating oil circulation, steam purging operation, etc. The shaft is jacked against the normal load face of the thrust bearing and a micrometer dial gauge put in contact with a suitable point on the shaft and set to zero. The jacking oil supply should be in service for the minimum period of time to minimise movement due to heating effects of the oil. All differential expansion indicators are then noted. The shaft is then jacked onto the surge face of the thrust bearing; the readings are again noted and compared with the clock gauge.

Similar procedures will be applied to boiler feed pump turbine turbovisory equipment.

7.3.2 Turbine protection

Final erection checks do not normally require any special equipment but, owing to the number of relay mechanisms involved, a relay test panel for the secondary injection of relays is advisable.

All equipment will be visually inspected and erection records examined to ensure that it is complete. The inspection will ensure that items such as links are fitted, that relays are in position and that time delay relays act in the right sense. With all fuses removed, the insulation resistance of the circuits should then be checked, using a megger. Incoming supplies must be isolated from the equipment and be checked to be within specified tolerances. The system should then have all its fuses removed and its supplies switched on. Using the relevant schematic or cable diagrams for each supervisory circuit, each supervisory relay is energised in turn by inserting its fuse. The fuse must have the correct rating. Every link in each leg of the circuit should then be disconnected separately and the de-energisation of the relay checked.

The checks should then proceed by proving that each item of the protection system operates according to drawing, including the initiation and clearance of alarms. For example, every contact must be proven to be in the circuit, as shown on the drawing, by operation of the contact and confirmation of the expected result. For parallel and series systems, every combination of contacts must be checked.

The contacts of pressure switches, differential pressure switches and vacuum switches must be checked using pressure/vacuum pumps and *not* by shorting contacts or by forcing relays.

Incoming equipment supplies must be adjusted over their specified ranges to prove correct equipment function.

Upon completion of the above checks, the mechanical aspects of devices, such as vacuum trip mechanisms, must be in their final state and set for operational service. The turbine protection and control systems may then be proven functionally by, for example, putting the turbine steam valve power oil system in service and operating the various valves and their trip mechanisms. This aspect of final functional checking becomes part of the individual system commissioning procedure and the various commissioning networks and procedures must clearly show the phasing and interactions of the activities.

Some examples of the protection and control activities which will be carried out in this way are:

- Steam valve open/shut indicators and controls
- Steam valve trip systems in both trip and reset modes
- Turbine trip mechanisms in trip and reset modes, including the operation of trip selection mechanisms: this will include hand, vacuum and solenoid trip features.

Similar examples can be given for the protection and control of other systems, e.g., lubricating oil system, stator cooling system, etc.

7.4 Commissioning network and resources

The number of man-hours required for commissioning is considerable.

Tight control must be experienced in phasing the work so that subsequent activities do not affect the integrity of the work at any stage. Thus, whilst the commissioning networks are simple, the detail of the work is highly interactive with other major activities such as final erection of bearings, availability of lubricating oil, etc. Overall networks must therefore clearly indicate these interactions and the time required to complete them. It is impracticable to construct programmes with the degree of flexibility needed to incorporate the fine detail of commissioning for this type of activity. Considerable importance must therefore be attached to routine commissioning meetings, where adjustments can be made and short-term programmes issued.

The network illustrated in Fig 5.36 is a simplified example for a turbine protection system. It comprises two main parallel activities for checking the protection equipment and the associated plant items, such as bled-

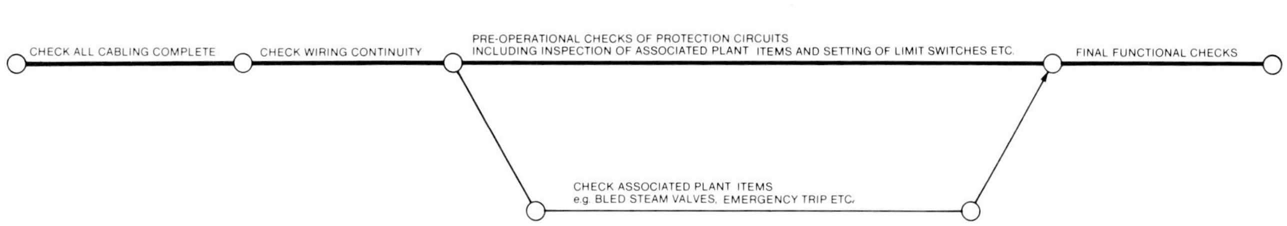

FIG. 5.36 Simplified commissioning network for turbine protection systems

Illustrates a typical, but simplified, network for turbine protection systems. The two main parallel activities for checking the protection equipment and the associated plant items will, in practice, be expanded in more detailed networks.

steam valves and emergency trip mechanisms. Each of the two parallel paths will, in practice, be expanded into more detailed networks. Many of the activities will form parallel paths and may therefore be supported by a detailed commissioning document, comprising numerous item check lists.

No specialist materials are needed for mechanical and electrical protection systems; those for turbovisory equipment are specified in Section 7.3.2 of this chapter.

8 Main turbine governor

This section deals with the commissioning of both mechanical and electronic governing systems. The procedures for commissioning steam admission valves and their hydraulic relays are dealt with in Sections 1 and 3 of this chapter.

8.1 General description — electronic governor

The modern trend is towards electronic governing systems. The following describes a typical system as an example of the approach to commissioning.

The control gear on the turbine steam admission valves operates the valves in direct response to signals received from the turbine governing equipment. Unlike mechanical governing systems, there are no mechanical or hydraulic connections between governor and valve control gear, although hydraulic systems are used to operate the relays which control all HP and IP emergency trip and governor valves (see Section 3 of this chapter).

Electronic governor systems incorporate monitoring and protection equipment to give controlled run-up and shutdown, as well as speed and load control. Emergency trip facilities associated with turbine and generator protection are also incorporated.

Control of the turbine is achieved by measurement of speed, the measured signals being transmitted to the control system from pick-up heads.

A high degree of monitoring and protection is incorporated in the control system. Fail-safe principles are a feature of the design: triple channels giving single-channel redundancy with cross-channel comparative monitoring are used throughout the system to ensure reliability. The primary function of the control system is to keep the turbine under control and safe under all conditions.

To achieve a high level of reliability in this control system, solid state electronic devices are used, with all components conservatively rated for their duties. Discrete-function modules containing printed circuit boards connected by plugs and sockets are used to permit removal for maintenance without affecting the operation of the turbine.

The valve actuating gear for all turbine steam valves is the same in principle. The gear incorporates a spring acting to close the valve opposed by a hydraulic ram acting to open it. Hydraulic fluid to operate the ram is supplied through the control gear which provides controlled steam valve operation over the full valve stroke, with synchronous operation of all valves having the same duty, e.g., all governing valves operate at the same speed to an equal amount of opening. The control gear also provides for rapid closing in an emergency.

The control gear comprises hydraulic units assembled together to pass power fluid to, or from, the steam valve actuators to operate the valves, as demanded by the turbine or external inputs.

Three types of control gear are used, each designed to control its valve in the manner best suited to the valve duty. One type serves the HP stop valves, the second serves the HP and IP governor valves and the third the IP stop valves. With the exception of the IP stop valve, the steam valves are opened to supply the required amount of steam to the respective turbine cylinders. The IP stop valve is fully opened and the turbine run-up controlled by the HP stop valve. Both stop valves are normally opened wide and the turbine is governed by the governor valves.

The principal areas of the electronic governing system are:

Speed/load control

Run-up system

Pay-off system

On-load valve test

Alarms

Arming and protection

Power and auxiliary power supplies

Figure 5.37 shows a simplified block diagram of the main governor system and Fig 5.38 shows a more comprehensive version.

Terminology

In dealing with a specific example of an electronic governing system, terminology may be different for different suppliers. The following glossary of terms is applicable to the example chosen:

Acceleration signal An analogue representation of the turbine acceleration.

Acceleration trip signal When the acceleration exceeds a preset value and duration, i.e., during load rejection, an acceleration trip signal is generated.

Alarm module Monitors the governor to give local and remote indications of failures and discrepancies in the system.

Analogue signal A continuously variable signal.

Arming bin Ensures that the governor is correctly set for run-up, when turbine protection is reset.

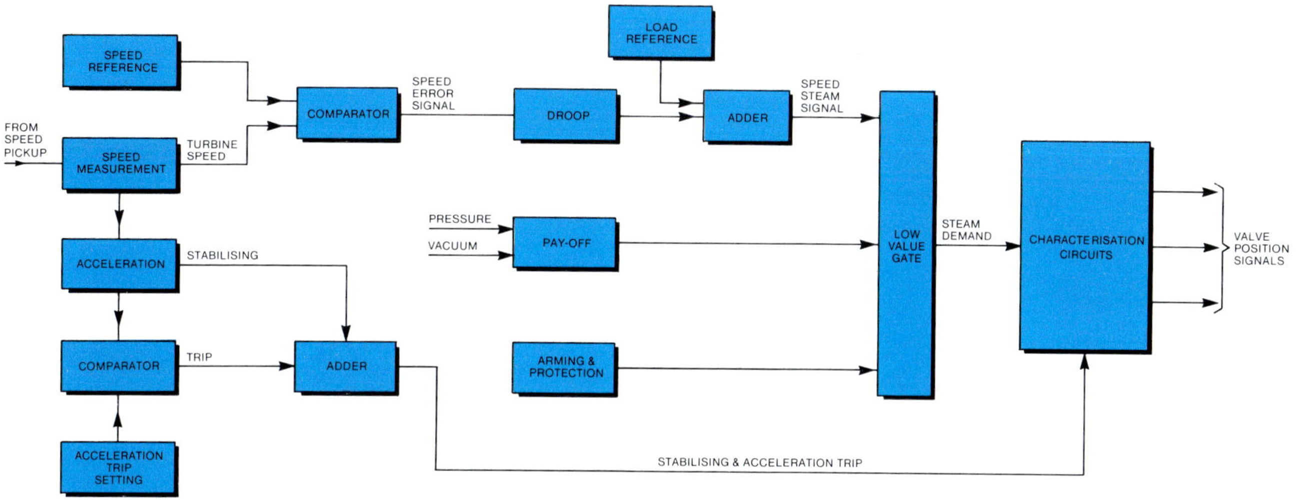

FIG. 5.37 Simplified electronic main governor system block diagram

Bias signal Modifies the steam demand signal to the IP governing valves when the HP + IP governing mode is selected.

Bin This consists of a framework which only contains a PCB rack and/or relay mounting facilities.

Digital signal An 'on' or 'off' signal.

Driver A higher power amplifier which drives a highway.

Gate (electronic) Generally a device which either allows or inhibits the passage of a signal. It can be a specialised device, e.g., the low value gate passes the lowest value of several inputs.

Gate (mechanical) Hinged framework in the cubicle, which carries standard panels, bins and modules.

Highway A line carrying a signal which goes to several destinations; similar to a busbar in power technology.

Load reference signal This is a speed error reference signal. With the generator synchronised, this signal is summated with the speed error signal resulting in a speed/steam signal.

Mode control signal This signal modifies the steam demand signal to the IP governing valves to follow either the HP ONLY governing mode or the HP + IP governing mode.

Module This comprises a framework with a power supply unit, together with two or three printed circuit board (PCB) racks and relay mounting facilities.

On-load valve test module Generates the signals necessary to close a valve for test purposes.

On-load valve test position signal selector bin Routes the selected valve position signal to the on-load valve test panel recorder during an on-load test sequence.

Pay-off setting bin Contains the controls for setting the pay-off droop and ramp rate.

Pay-off signal A signal, derived from either the steam pressure or vacuum transducers, which sets the maximum conditions. It feeds into the low value gate.

Protection relay panel Closes the valves in response to external protection signals.

Speed control module Acts as a buffer between the external controls and the governor.

Speed computing module Computes the required steam flow from turbine speed input signals and load reference.

Speed error signal This is a summation of measured and desired values of machine speed. An adjustable gain amplifier is used to provide the desired speed/load characteristic (droop).

Speed steam signal This is a summation of the speed error signal and the load reference signal.

Stabilising and acceleration trip signal A summation of the acceleration trip signal and the acceleration signal, which is summed with the steam demand signal in the highway drivers of the governing valves. This gives stabilisation of the turbine and closes the valves in the event of a load rejection.

Stop valves setting bin Synchronises the stop valves before run-up and limits their opening during run-up.

Valve control module Controls a valve in response to position signals from the valve highway control module.

Valve highway control module Computes the required valve positions from the required steam speed signal. The steam/speed signal is characterised by the module to give valve lift signals. The required steam flow signal is either the steam/speed signal or the pay-off signal.

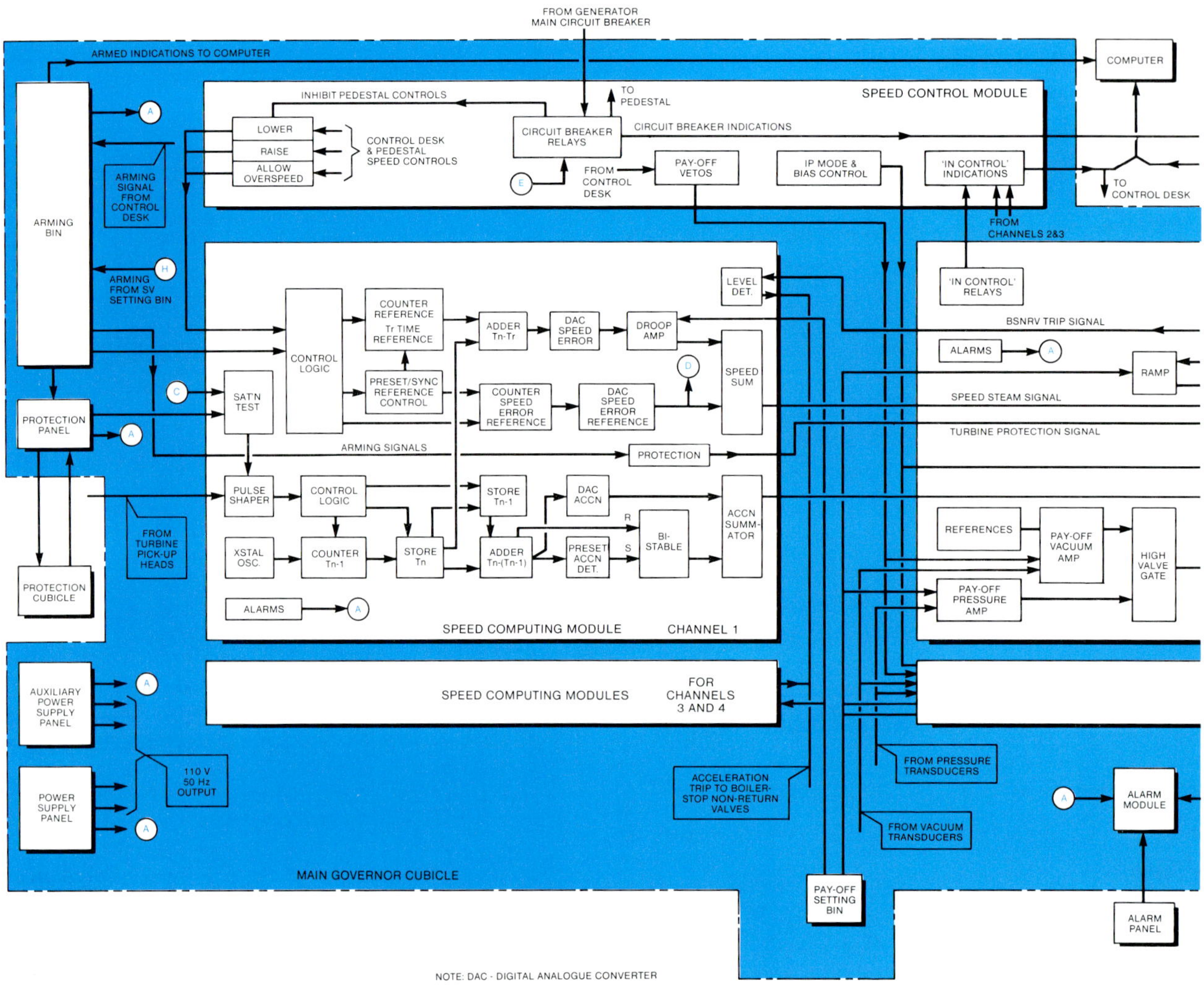

FIG. 5.38 Comprehensive electronic main governor system block diagram

Valve trip bin Trips the IP stop valves, either in response to signals from the on-load valve test module, or if the IP governing valve is tripped, or the turbine generator main protection trip is initiated.

8.1.1 Speed/load control

The complete electronic governor and hydraulic control system for the turbine valves provides a closed-loop speed control system for the narrow-range speed control of a turbine-generator. It will also allow controlled overspeed to a ceiling of 113% speed for the testing of mechanical overspeed protective devices. The speed is controlled by positioning the turbine valves, using valve position demand signals.

Turbine speed is measured by three magnetic pick-ups which generate a frequency proportional to turbine speed. Highway signals produced by the three channels are compared with each other on a '2 out of 3' voting system. If discrepancies are detected, the appropriate alarms are raised and the faulty channel disconnected. If a second highway discrepancy occurs, a main turbine trip is initiated. The frequency signal is computed digitally with a speed reference signal to give a speed error signal. To give the necessary speed/load (droop) characteristics, the speed error signal is modified by the droop amplifier. The output of the droop amplifier is added to the load reference (speed error reference) to give the speed/steam signal.

The load reference is used to provide a valve lift signal before synchronisation. Fine control of turbine speed is achieved by means of a motorised potentiometer operated from the unit control desk, or by computer. After synchronisation, the load reference signal is used to load the turbine which will then be operating under closed-loop control within its normal (i.e., narrow) governing range. In addition, a digital computation is carried out to produce an acceleration signal. If the average acceleration exceeds a preset rate and duration, an acceleration trip signal is given to close the HP and IP governor valves only. This signal can also be used to initiate the closure of other valves, such as bled-steam non-return valves, in order to limit turbine speed rise. The speed error signal, the acceleration signal and the acceleration trip signal are converted to analogue voltage levels.

Controls are provided at the pedestal for varying the speed reference; this is used for overspeed testing.

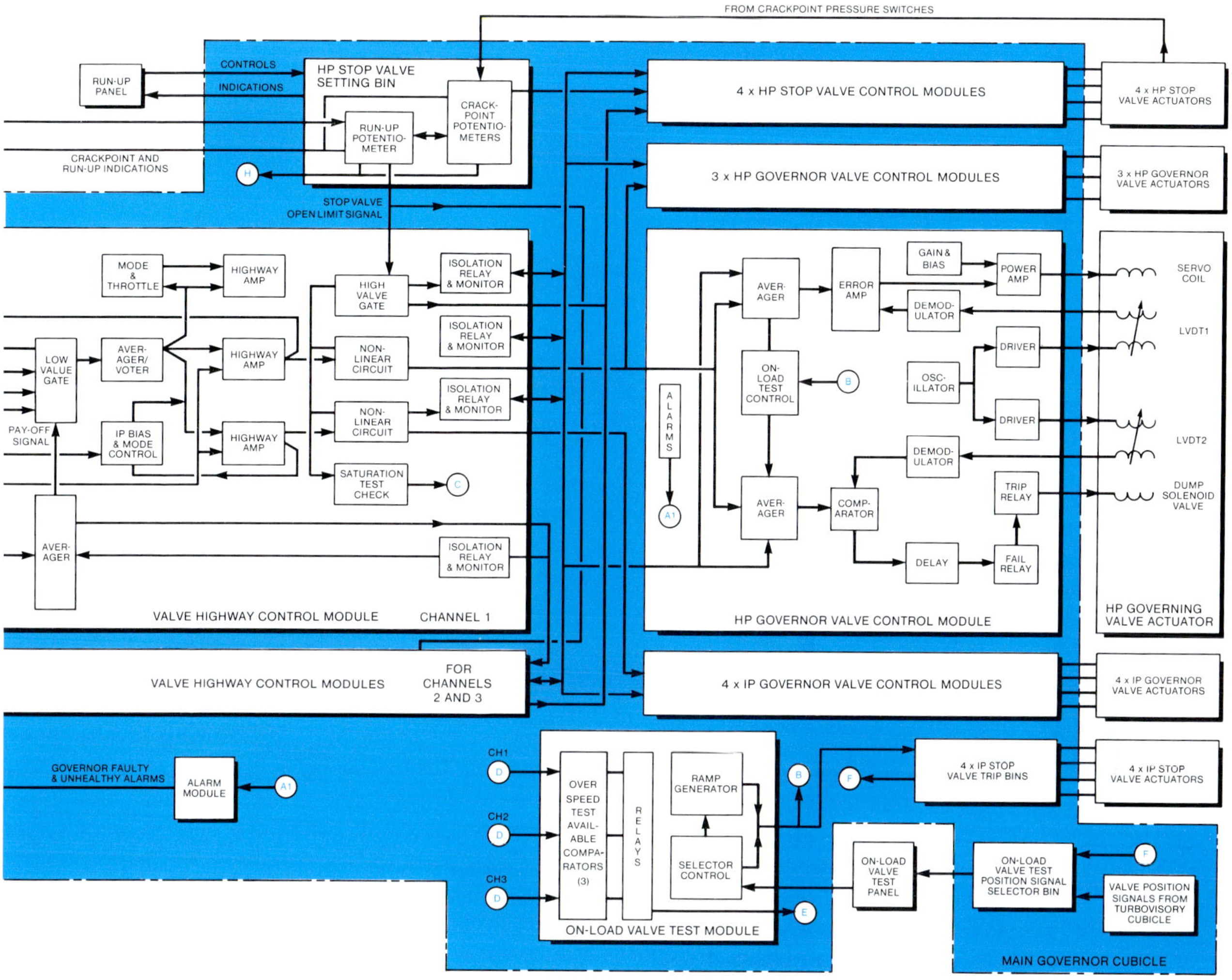

These controls are disconnected when the generator circuit-breaker is closed, i.e., when the unit is synchronised.

The speed/steam, the 'pay-off' and a protection signal derived from the arming system (see Section 8.1.6 of this chapter) are compared by the low value gate in each valve highway control module. The lowest value signal, the steam demand signal, is fed to characterisation circuits. The rate of increase of steam demand signal is limited by a ramp circuit which is connected across the low value gate. Decrease in steam demand signal is not limited and therefore acts instantaneously. The characterisation circuits generate the valve position signals from the steam demand signal.

For the HP governor valves, the characterisation circuits sum the steam demand, stabilisation and acceleration trip signals and the summated signal passes through a non-linear circuit, which converts the signal into the required valve position signal. For the IP governor valves, the steam demand signal is added to the stabilising and acceleration trip signal, and then further signals are added from the 'IP bias' potentiometer and a mode control signal. The mode control signal modifies the valve position to give either HP ONLY or HP + IP governing modes. The summated signal passes through a non-linear circuit, which converts it into the required valve position signal.

For the HP stop valves, the steam demand signal is compared with a valve opening limit signal and the lower of these two values is the valve position signal.

There are four valve control modules controlling four HP stop valves, four modules controlling four HP governor valves and four modules controlling four IP governor valves. The IP stop valves are controlled by the two valve trip bins.

Highway signals from the three channels are connected to each valve control module and are averaged and amplified. The averaged highway signal (valve position demand signal), and an actual valve position signal derived from a linear variable differential transformer (LVDT) associated with the valve, are compared and an error signal produced. The error signal controls the servo-coil associated with the valve actuator driven by the module and thereby positions the valve.

A second comparison between the averaged highway signal and a valve position signal, derived from a second LVDT, is carried out and, if a discrepancy is detected, an alarm is raised and the dump solenoid is operated to close the valve. Valves are designed to trip in pairs, this being achieved automatically by means of a valve intertrip signal.

When governor valves are wide open, they can no longer respond to genuine signals and are then said to be 'saturated'. An automatic saturation test facility is therefore provided which checks that the computed values are correct when governor valves are signalled to be fully open. If the computed values for a channel are incorrect, the channel will trip and initiate an alarm. For the example under consideration, the saturation test is started at approximately eight-minute intervals. Thus, when one channel has been tested successfully, the next channel's test is initiated (normally referred to as 'enabled'). Again, upon successful completion, the next channel is enabled, thus completing a saturation test for all three channels. The enabling process then recommences for the first channel, and so on. The enabling process is achieved by the injection of an additional speed pulse from one of the turbine speed pick-ups, so that its frequency is doubled. The speed computing channel which is fed from this pick-up will then see the turbine running at twice its nominal speed with a large acceleration. The channel will then generate a large signal (termed 'over close') and an acceleration trip signal provided the channel is fully saturated. The valve closing signals and acceleration trip signals are then detected and used to terminate the saturation test. The valve closing signal is applied to one channel at a time and therefore causes a discrepancy signal to be generated by the differences between the channels. The closing signal is then detected by discrepancy monitors and disconnected before it can affect the valves. This is achieved by means of a 'look ahead' pulse which trips the monitors before they can receive the saturation test signal pulse.

8.1.2 Run-up system

The facilities included within the governor provide an open-loop control of the valves; there is no speed control within the governor until the turbine approaches nominal speed.

Normally, the valve sequencing is such that the IP governor valves open first followed by the HP stop valves, followed by the HP governor valves. During run-up, it is preferable to open both sets of governor valves and then to control the steam flow by means of the HP stop valves only. This change in sequence is carried out in the stop valve setting bin. The opening of the HP stop valve is set by the run-up potentiometer within the bin. The potentiometer can only be adjusted during run-up or run-down, and is defeated when the generator is synchronised.

To obtain even heating of the turbine and valve chests, it is essential that all the stop valves begin to open at the same instant. Each stop valve is fitted with a hydraulic fluid pressure switch which detects the point at which the valve is about to open. This is called the 'crack point'. Each valve is driven independently to its crack point before run-up commences. The potentiometers which drive each valve to its crack point are located in the stop valve setting bin.

Interlocks are provided to ensure that all the conditions necessary for run-up are met before the system can be armed. Details of associated control are included in Section 3 of this chapter.

8.1.3 Pay-off system

The pressure and vacuum pay-off system detects adverse conditions in boiler steam pressure and condenser vacuum and then sets a limit to the maximum steam flow by means of the low value gates in the valve highway control modules, and also by reducing the speed error reference in the speed computing modules.

The boiler pressure and condenser vacuum are each measured by three electrical transducers. Boiler pressure is compared with a reference level in the pay-off setting bin and the difference fed to an amplifier whose bias and gain are adjusted by controls on the pay-off setting bin. These controls are designated 'start pay-off' and 'pay-off range', respectively.

Condenser vacuum is compared with a 'start pay-off' reference level in the valve highway control module and the difference fed to an amplifier. The highest positive value of the pressure and vacuum pay-off circuit outputs is fed to a channel-averaging circuit. A 10% minimum limit to the steam flow signal is applied in this circuit to prevent overheating of the LP turbine blades under low vacuum conditions. Switches are provided on the control desk so that the operator can render the pay-off systems inoperative, whilst seeking other ways to recover vacuum.

The three channel pay-off signals are compared and, if discrepancies are detected, the appropriate alarms are raised and the faulty channel disconnected.

If the pay-off system is operative and controlling steam flow, and one of the three channels is inoperative, then speed error references in the speed computing modules are automatically lowered by the speed control module. This continues until the 'speed/steam' signal from the speed computing module takes control of steam flow from the pay-off system at the low value gate in the valve highway control module.

8.1.4 On-load valve test

Groups of valves may be tested whilst the turbine-generator is on-load. A group of valves consists of an HP stop valve, an HP governor valve, an IP stop valve and IP governor valves: it can be selected for test by an on-load test panel pushbutton at the control desk. Operation of a pushbutton causes the first valve in the group to be connected to the on-load valve test module for testing. A ramp signal is then connected from the relevant valve control module, causing the selected valve to close in two seconds. The associated valve position signal from the turbovisory cubicle is routed via the on-load valve test position signal selection bin to an indicator and recorder on the on-load valve test panel and shows successful valve closure. The open or closed position of each IP stop valve is indicated by an associated limit switch. The valve is reopened after a successful test closure and the associated valve control module disconnected from the on-load valve test module.

The tested valve is now tripped and left closed whilst other valves in the group are similarly tested. When all valves in the group have been tested, each valve is in turn reset and brought back to governor control by means of the on-load valve test panel. When all valves are reset, the group is deselected and the on-load valve test module is reset ready for testing another group of valves.

8.1.5 Alarms

Two alarm modules monitor the system and give local and remote indication of failures and discrepancies in the system. 'Healthy' signals are normally connected to the alarm modules from monitor points in the system. When a failure or discrepancy is detected, the 'healthy' signal is removed and the relevant alarm is raised. All alarms also raise a 'governor fault' alarm. Certain groups of alarms requiring immediate attention also raise a 'governor unhealthy' alarm.

8.1.6 Arming and protection

The arming system ensures that the governor is in a safe condition for initiation of turbine run-up. The system consists of a circuit within the arming bin, speed computing modules and protection relay panel. The governor protection system closes all steam valves in order to protect the turbine, i.e., if the governor loses control of the turbine or on receipt of signals from the protection cubicle.

Before the turbine is run up, all valves are held closed by the protection system and the arming system. Before arming can commence, the protection trips must be reset.

If two out of three speed computing channels are armed, then the arming system releases the trips from the valves and the IP stop valves open. Other valves are held closed only by the speed/steam signal from the speed computing modules. (The arming action sets the speed error reference to -300% load equivalent, giving a large close speed steam signal.)

The turbine may now be run up by means of controls on the run-up panel and the run-up system takes control of steam flow.

8.1.7 Power and auxiliary power supplies

Two independent 110 V, 50 Hz, single-phase power supplies are connected to the power supply panel, which feeds all module power supply units and also the turbine protection trips. Motorised potentiometer units are supplied from a separate source.

One 110 V, 50 Hz, single-phase supply and a 240 V, 50 Hz single-phase supply are connected to the auxiliary power supply panel. The 240 V supply is fed to cubicle heaters and lights and the 110 V supply is connected to cubicle fans.

8.2 General description — mechanical governor

Mechanical governing systems are in common use for all sizes of turbine-generators. It is possible to replace a mechanical governing system with an electronic one, but the cost is high. An alternative is to enhance the response or modify the characteristics of a mechanical governor by electronic means. Thus, the commissioning and recommissioning of mechanical governing systems is still relevant to this volume.

The central component of a mechanical governing system is the centrifugal governor driven by gears from the turbine shafts. The governor acts on a system of hydraulically-operated relays, which in turn operate the hydraulic relays which control the turbine steam admission valves.

A typical system, using lubricating oil as the hydraulic fluid, is illustrated in Fig 5.39. The steam admission valve relays are operated by a main power oil system supplied by one of two pumps. During normal turbine operation, main power oil is supplied by a main oil pump driven by the exciter shaft. During turbine start-up and shutdown, when the main pump pressure is insufficient, an AC auxiliary power oil pump is used. The auxiliary oil pump, in this example, also supplies the lubricating oil system at low turbine speeds.

Both the main and the auxiliary power oil pumps draw oil from the main oil tank, the main pump being primed by means of a booster pump which is driven by an oil turbine supplied from the power oil system. Pressure in the main power oil system is maintained substantially constant (in this example, 14 bar) by a pressure control valve and by-pass. These valves return some oil to the lubricating oil system, but the bulk of it supplies a secondary power oil system. The secondary power oil system pressure is controlled (in this example at 7 bar) by means of a control valve which again discharges surplus oil into the lubricating oil system. The system acts as a source of oil at stable pressure to the mechanical governor relay oil system, termed the pilot oil system. The secondary power system also supplies oil for the operation of auxiliary relays associated with turbine governor valves, for the resetting of emergency trip plungers, the on-load testing of the mechanical overspeed trip mechanisms and an unloading oil system. The pilot oil system provides a hydraulic signal to control main and reheat governor valves. The signal pressure is regulated by means of the mechanical governor. The mechanical governor consists of two centrifugal governors mounted on a lay shaft driven by the HP turbine rotor. One governor, termed the main governor, controls the turbine speed

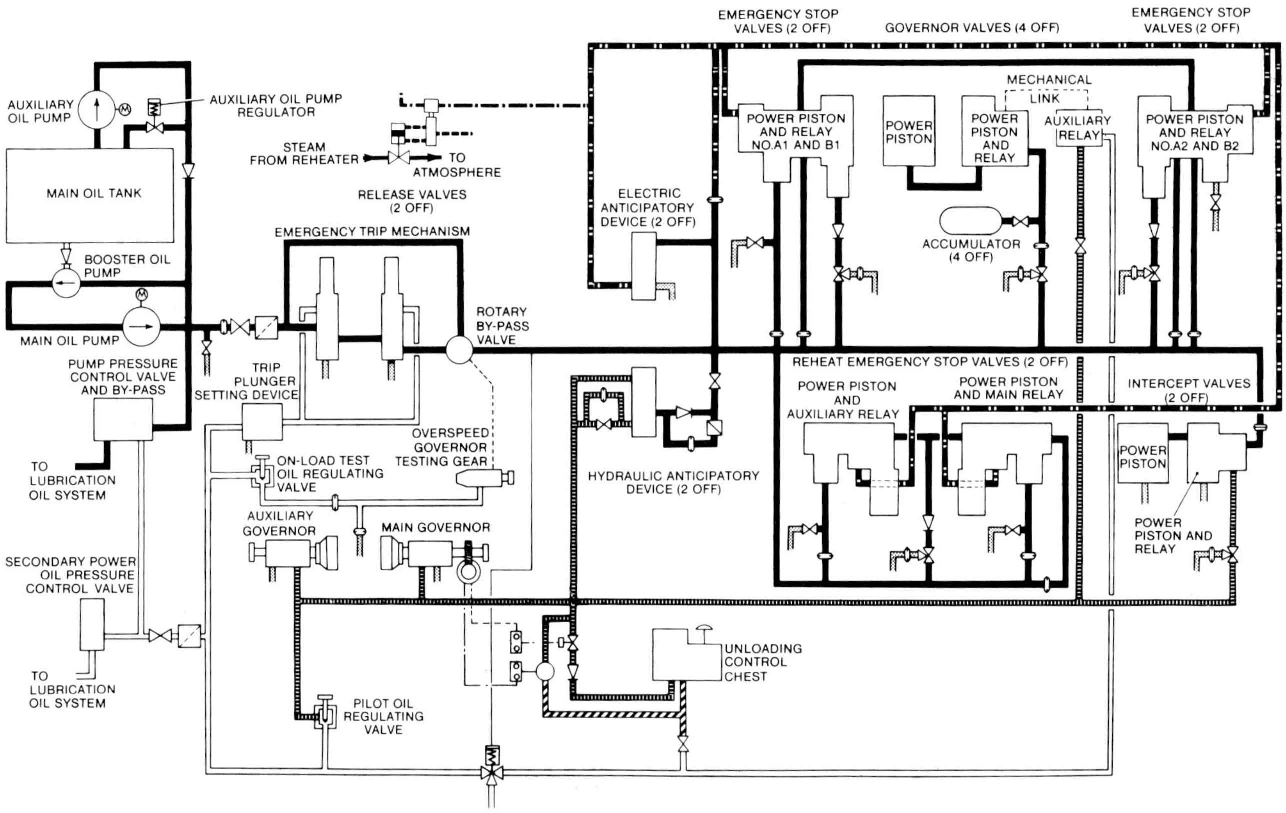

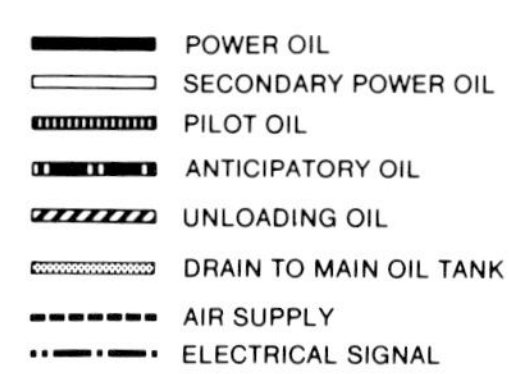

FIG. 5.39 Diagrammatic arrangement of a mechanical governing system

This mechanical governing system uses lubricating oil as the hydraulic fluid. The oil is supplied at high pressure, either by means of a DC motor-driven auxiliary power oil pump for start-up, shutdown and emergency use, or by means of a turbine shaft-mounted oil pump for normal operation. The principal use of the power oil is to operate power oil relays associated with each turbine emergency stop valve and governor valve. Power oil also provides the source of relay oil, called pilot oil, used in conjunction with the main and auxiliary centrifugal mechanical governors to control the position of HP and IP governor valves.

A number of devices, such as the hydraulic anticipatory device, electric anticipatory device and the unloading control chest, are incorporated to enhance governor action and provide turbine protection against undue acceleration or overspeed.

and load over the normal operating ranges. The second, auxiliary, governor is similar in construction, but comes into play at high turbine shaft speed in order to limit the degree of overspeed on, say, sudden load rejection.

The main governor is illustrated in Fig 5.40. At approximately 80% of normal running speed, the centrifugal force acting on the governor weights overcomes the governor spring tension. The movement is transmitted via bell cranks to a pilot oil piston which reduces pilot oil pressure to open the main and reheat governor valves. As speed increases, pilot oil pressure falls to a point where the governor valves control the flow of steam to the turbine. In order to raise speed further, pilot oil pressure must be increased. This is achieved by motorised movement of the governor gear, which adjusts the position of the pilot oil piston sleeve so as to reduce the flow of oil from the pilot oil system.

Normal practice is to set the motorised sleeve, termed the speeder gear, to the minimum governor-controlled speed setting. Thus, when the turbine speed is raised from turning gear speed by use of the emergency stop valves, a point will be reached when further opening of the emergency valves has no effect on turbine speed. The machine is then under governor control. At this point, the emergency stop valves would normally be opened fully, unless the turbine speed is to be raised to prove the action of the mechanical overspeed protection devices. The shaft speed at minimum governor setting (approximately 2850 r/min) is sufficient for the shaft-driven power oil pump to be self-sustaining. The auxiliary power oil pump will have provided sufficient pressure to drive the booster oil pump turbine and therefore provide pressure at the suction of the shaft-driven oil pump. This suction pressure is monitored. Thus, when the shaft reaches

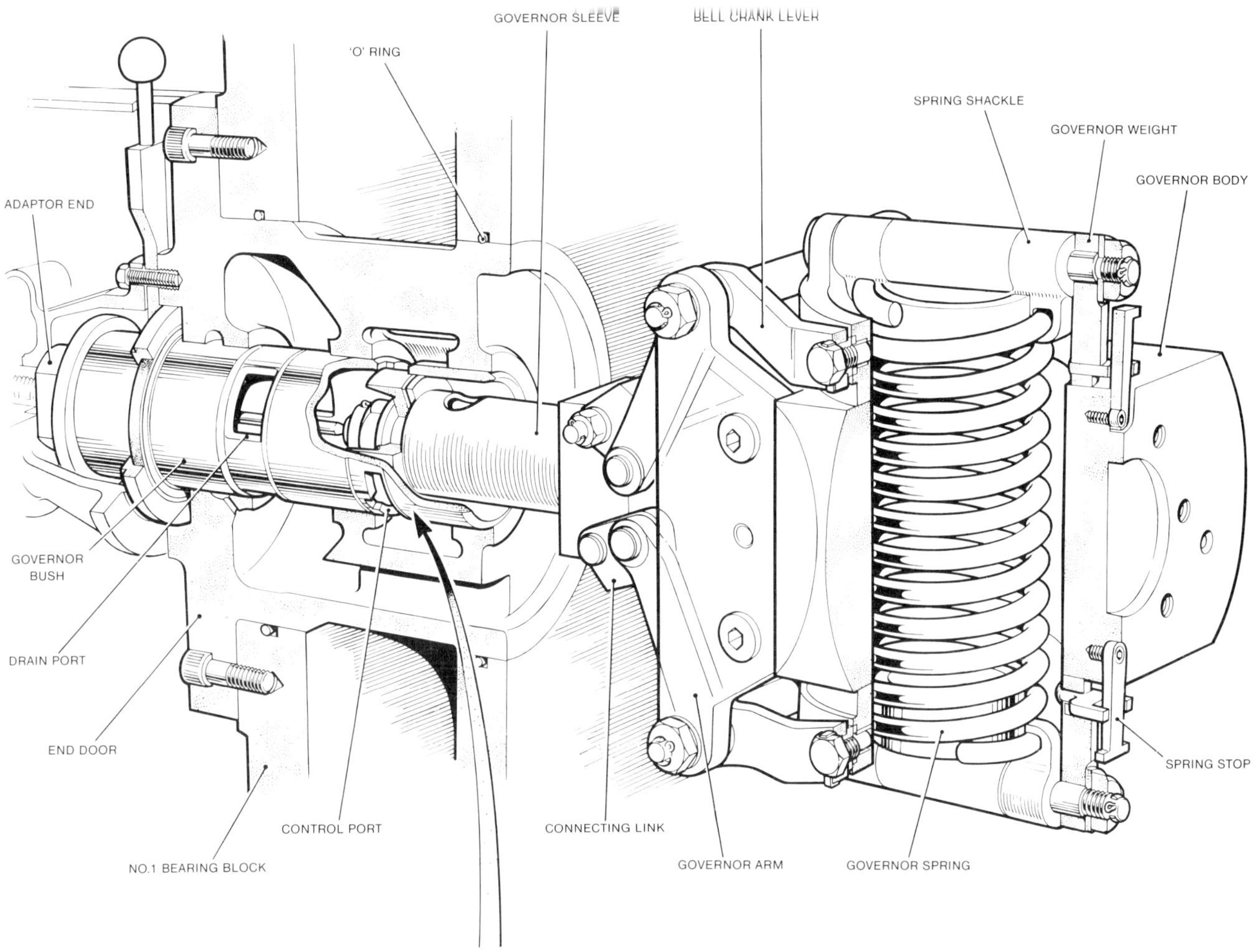

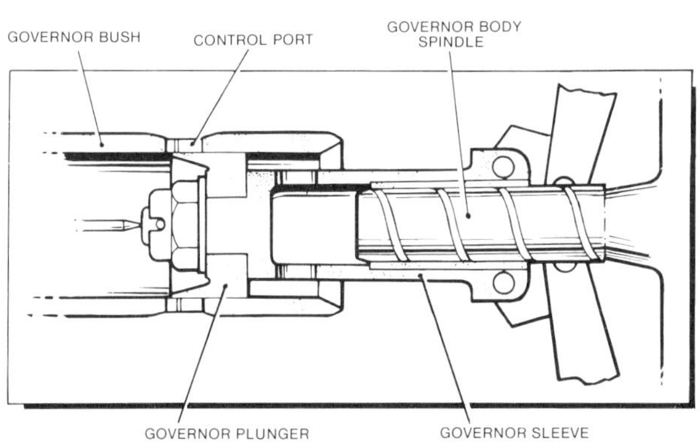

FIG. 5.40 View of a typical centrifugal mechanical governor

In this typical main centrifugal mechanical governor, the weights move apart under centrifugal force against the action of springs. In doing so, a plunger is caused to move within a sleeve, termed the governor bush, thereby controlling pilot oil pressure. In the main governor referred to in Fig 5.39, the position of the governor bush can be adjusted by means of an electric motor. This movement changes pilot oil pressure and is used to control turbine speed and loads by moving the HP governor valves. The auxiliary governor shown in Fig 5.39 is similar to the main governor shown here, the essential difference being that the control ports in the auxiliary governor bush do not come into operation until turbine shaft speed is above the normal operating range. The governor bush is preset and its ports shaped, so that proportionately more pilot oil is drained with increasing shaft speed.

minimum governor speed, a fall in suction pressure will be observed as the main power oil pump takes over from the electrically-driven auxiliary power oil pump. At this stage, the auxiliary power oil pump can be shut down.

When the generator has been synchronised to the grid system, movement of the motorised pilot valve sleeve varies pilot oil pressure to open or close the governor valves and hence control the generator output.

8.2.1 Unloading oil system

This is supplied from the secondary power system through an unloading oil regulating valve to control the turbine on partial loss of condenser vacuum, or of turbine stop valve pressure. The opening of the unloading system is controlled by condenser vacuum and stop valve steam pressure unloading devices. When vacuum or steam pressure falls below a predetermined setting, leak-off ports open and unloading oil pressure falls. A schematic diagram of an unloading oil system is illustrated in Fig 5.41.

8.2.2 Anticipatory system

For the mechanical governor system described, duplicate hydraulic anticipatory devices are provided to assist the normal mechanical governors in the event of a partial or complete rapid load rejection. Duplicate electric anticipatory devices are also provided, again to limit speed rise under electrical fault conditions, which cause the generator to trip, but not the turbine.

This particular system also prevents complete unloading and motoring of the generator on sustained reductions of stop valve pressure or vacuum. The unloading oil system is interconnected with the pilot oil system through a motor-operated isolating valve and non-return valve, which allows oil to flow from the pilot oil system only. This valve is closed on unit start-up, when pilot oil pressure is higher than unloading oil pressure owing to low stop valve pressure and low vacuum. When pressure and vacuum reach prescribed values, the isolating valve is caused to open. A differential pressure switch senses difference in pressure between the unloading and pilot oil systems. Thus, when low vacuum or pressure conditions exist, the difference in pressure is used to operate a switch and drive the governor motor towards its lower limit. Less steam is then admitted to the turbine, which tends to restore vacuum and pressure at a reduced load. If conditions do not restore, the governor motor will continue to reduce load until an anti-motoring device operates within the unloading equipment to stop the governor motor.

The unloading oil pressure control valve is positioned by means of an air signal derived from a pneumatic selector relay. This relay receives air signals from a stop valve pressure-sensing device and a vacuum-sensing device. In the unlikely event of pilot oil pressure not falling as rapidly as unloading oil pressure, the non-return valve between the two systems allows pilot oil to flow into the unloading oil system, so that a degree of unloading is achieved.

The hydraulic devices (Fig 5.42) are connected into the pilot oil system in parallel with the main and auxiliary centrifugal governors. A rapid fall in pilot oil pressure created by movement of the mechanical governors, due to a sudden load rejection, causes the hydraulic anticipatory devices to operate and drain oil from the pilot oil system, thereby assisting the governors to close the HP and IP governor valves rapidly.

The hydraulic anticipatory device is kept in its normal closed position by the action of pilot oil on an

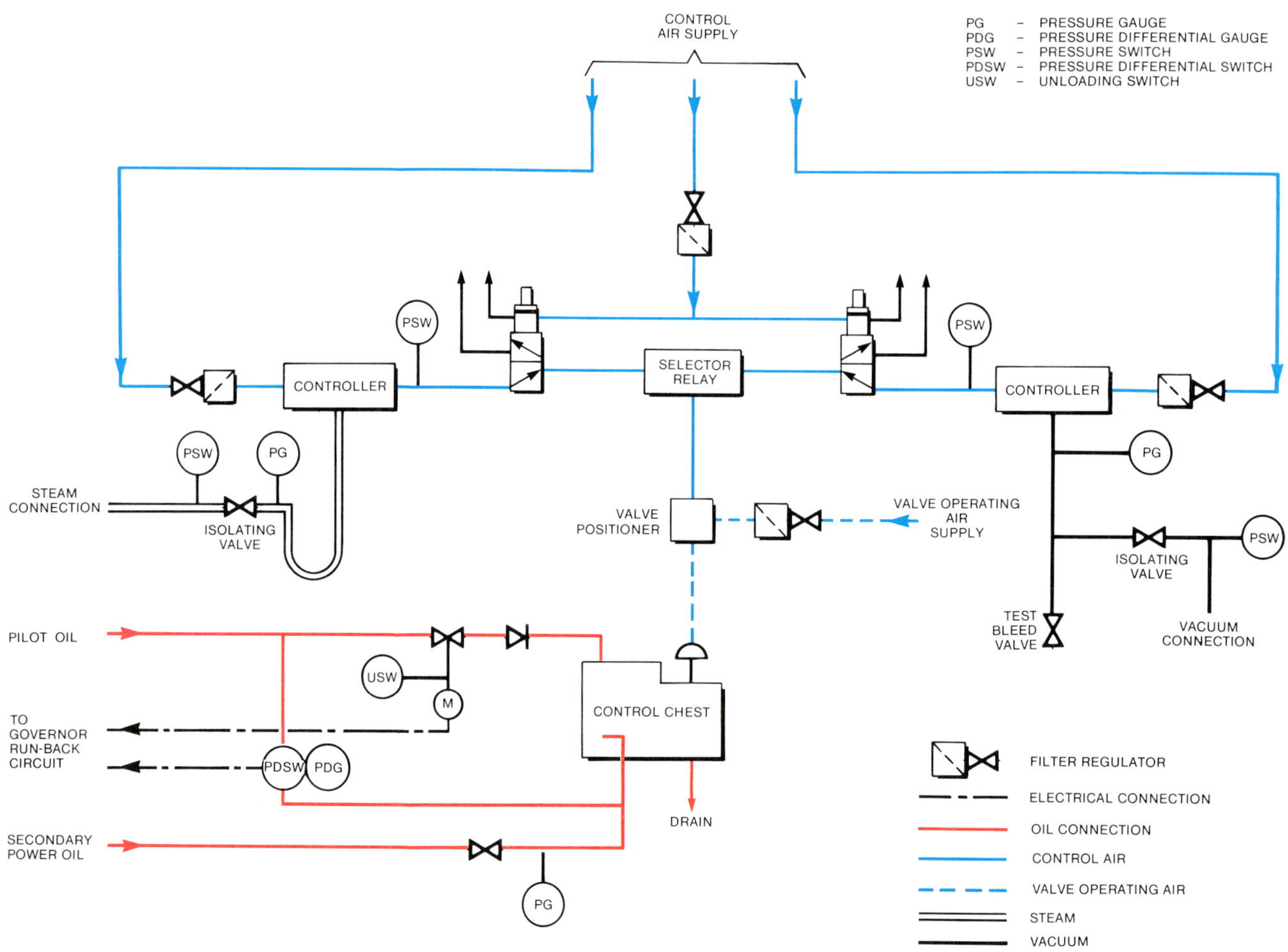

FIG. 5.41 Diagrammatic arrangement of an unloading system

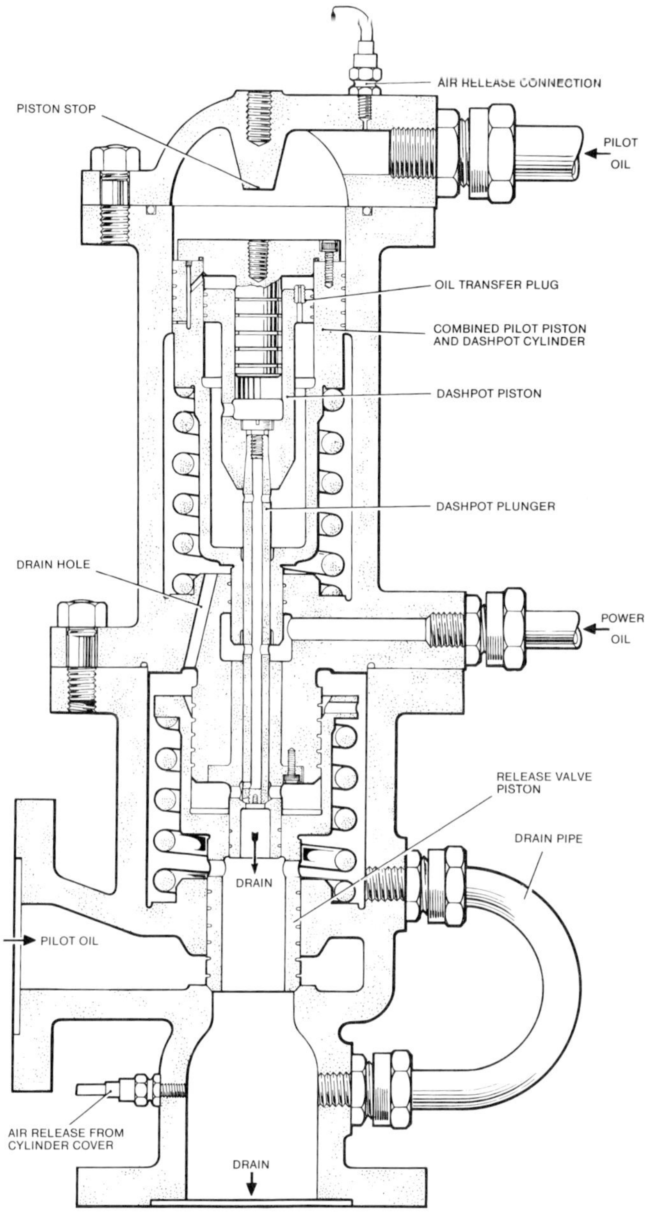

FIG. 5.42 Hydraulic anticipatory device

assembly of spring-loaded pistons. The assembly incorporates a dashpot open to power oil pressure. For normal variations in pilot oil pressure, the resultant forces on the assembly due to the combined effect of power oil acting via a transfer plug and pilot oil is downward.

For a rapid fall in pilot oil pressure, the transfer plug is unable to pass sufficient power oil for the system to remain in balance. The design is such that resultant upward forces are applied rapidly when a specific rate of pilot oil pressure decay has been reached, thereby causing the device to snap open and dump pilot oil to drain.

The electric anticipatory device (Fig 5.43) is a solenoid-operated valve which dumps the anticipatory oil system to drain when the solenoid is energised. The anticipatory oil system is supplied via an orifice from the power oil system.

Under normal operating conditions, power oil is supplied to a spring-loaded piston via a plunger attached to the solenoid. The lower end of the piston forms the dump valve, which is shut. In the event of the generator circuit-breaker opening, the solenoid is energised and moves the plunger upwards to cut off power oil to the spring-loaded piston, whilst putting power oil acting on the piston to drain. The piston therefore lifts under the action of its spring and dumps anticipatory oil. The supply of power oil to the anticipatory system is restricted by the fixed orifice and thus anticipating oil pressure decays. This causes the operation of spring-loaded dump valves associated with the power oil supply to the relays of each emergency stop valve. The stop valves therefore close, but the governing system remains active and the governor valves remain open. After a prescribed period of time, the solenoid is de-energised to restore anticipatory oil pressure and hence power oil supplies to the emergency stop valves. The stop valves therefore reopen.

The time delay is chosen so that the governing system can adjust to the loss of load, the final speed being a slight permanent rise above that which prevailed prior to the loss of generation.

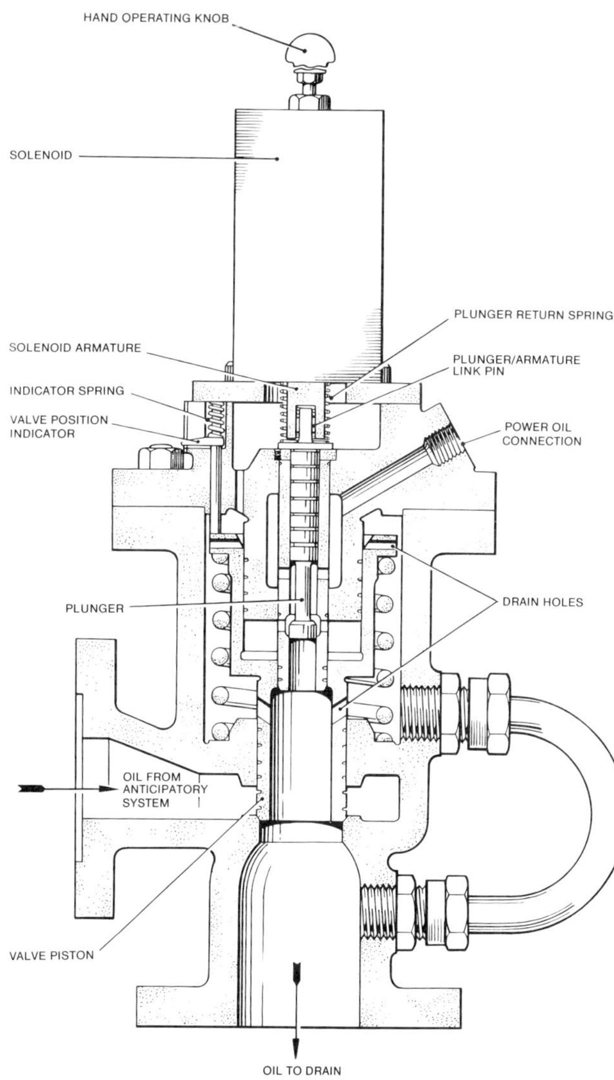

FIG. 5.43 Electric anticipatory device

8.3 Commissioning objectives

8.3.1 Electronic governor

The objectives when commissioning the electronic governor are:

- To verify appropriate works tests, where completed and with records available
- To verify appropriate erection checks have been completed and records available
- To ensure the correct opening and closing of all HP and IP steam valves for various 'speed error reference' and 'droop' settings of the governor
- To prove turbine 'run-up' and 'run-down' control stations
- To verify the performance of the governor with simulated speed and acceleration signals injected into the governor circuits and to ensure correct operation of the steam valves
- For the example chosen, to verify smooth changeover from HP to HP + IP governing modes and IP bias action
- To verify the integrity and settings of condenser low vacuum and low boiler pressure pay-off systems
- To determine certain parameters, such as valve gear opening and closing times, valve position and response for various signal changes.

A number of objectives can only be met when the turbine is at speed or on-load. They are:

- No-load speed range (+6% to −7% of rated speed) at the 4% droop setting
- On-load emergency overspeed bolt testing
- Overspeed testing
- After synchronising, to activate the trip of each protection channel in turn at various load increments, whilst observing any change in load for steam-derived signals and valve movements
- To determine the actual droop characteristics for the various droop selections available: measurements have to be taken to establish load and steam flow relationships, and speed/load characteristics for various speed error reference settings and system frequencies
- To carry out power supply switch tests to ensure non-corruption of governor circuits

When the generator is synchronised, other tests will be carried out, including on-load valve testing and dynamic valve gear response testing.

8.3.2 Mechanical governor

The objectives when commissioning the mechanical governor are:

- To verify that appropriate erection checks have been completed and records available
- To flush the governor relay oil systems
- To ensure correct opening and closing of governor valves with varying pilot oil pressure
- To prove turbine 'run-up' and 'run-down' control functions for emergency stop valves and governor valves
- To make and verify the initial settings for the governing system, including vacuum unloading, low boiler pressure unloading, hydraulic-anticipatory and electric-anticipatory systems before steam-to-set.
- To carry out functional checks on mechanical trip, trip by-pass and trip reset mechanisms.
- To determine certain parameters, such as valve gear opening and closing times, valve movement for various pilot oil pressure signals, etc.

As for the electronic governing system, a number of objectives can only be met when the turbine is at speed or on-load. They are:

- No-load speed range
- On-load overspeed bolt testing
- Overspeed testing
- To determine the actual droop characteristics of the governor
- To record data, such as the relationship between steam flow and valve openings
- To prove hydraulic and electric anticipatory devices
- To prove vacuum and low steam pressure unloading devices.

8.4 Detailed procedures — electronic governor

The detailed commissioning document will include a comprehensive list of all equipment supplied, including instruments, cabinets and cubicles, schematic and detailed drawings. A number of instruments associated with the electronic governor system are likely to be supplied under other contracts. These must be identified clearly and commissioning procedures integrated. The turbine valve gear will be erected and commissioned, including its hydraulic power supplies (see Sections 1 and 3 of this chapter).

All works tests and site erection tests must be listed and checked for completion. Special equipment will be required, dependent upon the particular installation.

The supplier will have developed test procedures and equipment which must be thoroughly understood and scrutinised on behalf of the CEGB. Typical equipment would include:

Governor test boxes

Oscillators

Function generators

Oscilloscope

Transistorised continuity/insulation tester

Shaft pulse simulator test box

Hydraulic pilot ram position simulator

Servo-valve injection unit (e.g., ±100 mA, 0–5 mA adjustable bias and stepped output)

A check on the completion of erection will comprise inspection of the following:

Power supplies and distribution system

Alarm modules

Speed control and computing modules

Valve highway control modules and valve control module

Protection panels

Arming, trip and setting bins

On-load test module

These checks will include, where appropriate, basic function checking of auxiliary equipment, such as valve gear, vacuum and pressure transducers, control switches, etc. Cabling must be checked for correct identification and termination, including earthing. Wherever the equipment items have been repeated, standard test sheets should be drawn up or modified to suit the particular item or module. The checking should not just repeat works testing, but test the equipment, as erected, using the installed wiring whenever possible. It should ensure that the plant is in good working order after erection and also that it has been erected correctly.

All remote equipment and cabling will be shown to operate into the cubicle/item, with power supplies on. Relays should be operated and change of state of contacts observed, when checking them for correct connections. Steady, short-circuit or open-circuit indication is not sufficient proof of correct contact function.

Much of this preliminary checking may be carried out before all erection work, such as cabling, is completed. It may be necessary to work this way when constraints on erection and commissioning programmes are tight: the two programmes then have to be tightly integrated and co-ordinated to ensure that checks are not invalidated by subsequent work. Considerable care also has to be taken to ensure that any rectification work required after tests is carried out promptly and the test repeated, where necessary. Also, where modifications are needed, an agreed procedure must be established, which must include an appropriate level of approval before they are implemented, after thorough assessment to ensure that the integrity of the system has not been affected. Detailed procedures are required to prove the modification.

Before commencing these checks for each cubicle or plant item, care must be taken to ensure that personnel working on remote items, such as valve gear, are not endangered. A control procedure with personal responsibilities must be established to ensure that such dangers cannot arise. Physical disconnection, control of supplies, the use of barriers, the recording of safety precautions and the appropriate level of supervision must be considered. In general, for modular systems, the modules should be initially unplugged and only plugged in as required for testing.

Before electrical supplies are put into a cubicle, the insulation of all power wiring to earth must be tested.

The insulation resistance of all fans, heaters and interconnecting cables must be tested using, say, a 500 V megger. Results of IR tests should be greater than 50 MΩ.

Supplies should be checked to be within specification, including the setting of undervoltage relays. Fuses must be checked to be correctly rated and properly identified. For ring circuits, the ring should be disconnected at a suitable point and the continuity of each part checked. Continuity should be checked progressively by, for example, opening and closing links to prove that the circuit is in fact being energised and de-energised from one source.

Fans, heaters and lighting must be checked to operate at an early stage of these checks and the circuits suitably identified as being live. Heaters must be checked to be free from obstruction which might cause fire or damage, e.g., loose wiring resting on the element.

Particular attention must be paid to the preliminary checking of alarm modules in respect of power supplies, test facilities and indicators to ensure that they are available from the outset of commissioning. Alarm modules are usually complex items of equipment, which need to be checked in a progressive logical manner to prove both that they function correctly and that they do not function incorrectly for combinations of events. Each module should be tested in accordance with an approved test procedure. For the system described, alarm module outputs are in the form of positive pulses. These pulses are checked by means of an oscilloscope, to verify the duration of the pulses, which must be of sufficient length to activate computer-scanned alarms.

Other modules, such as the speed control module, must also have well-defined commissioning procedures designed to ensure that equipment operates correctly for normal situations, as well as for all reasonable combinations of malfunction.

Much of this testing may be achieved by the application of shorts to simulate a particular function, such as the operation of a relay, a relay contact, an indicator lamp, etc. Certain veto devices are incorporated in the speed module, e.g., vacuum and pressure pay-off: the correct functioning of these devices must be checked.

Manual and motorised potentiometers must be checked for correct value over the whole range of operation and, where relevant, must agree with position indicators. Limit switches must also be checked for correct settings. All connections to remote controls and indicators must be proven, such as governor raise and lower switches, and various indicators, such as run-up and pay-off indicator lamps.

All remote inputs from the plant, such as valve 'crack-point' (i.e., point at which valve physically just lifts off its seat) switches, must be proven, either by actual operation of the valve or by simulation. Where simulation is used, the actual initiation of the plant signal must be proven at a later date.

For the system described, steam valves are controlled by 'bins' and modules. As well as checking availability of remote inputs, all external circuit drawings must be checked to be of latest issue. The drawing can be used as a check list, particularly for contacts, and should be marked up and signed as checking proceeds. In particular, every contact must be proven to be in the circuit shown on the drawing by opening and closing it and observing that the required result has occurred. For parallel and series contacts, every combination must be tried and shown to be correct with results as expected.

It may be convenient at this stage to check that the operation of the setting bins interacts correctly with the plant, but extreme care is required to ensure that steam valve movement does not endanger personnel. It may be practical to disconnect the the steam valve linkages and operate their relays alone. Plant items, such as crack-point bias potentiometers, should be driven through their range from the bin, whilst noting the energising/de-energising of associated relays when limits are reached.

Interlocking facilities which, for example, require potentiometers to be in set positions before subsequent actions can take place, must be checked. This would apply, for example, to the 'run-up in control' interlock with the three valve highway control modules when the four crack-points and the run-up potentiometer are in the minimum positions. Where '2 out of 3' facilities allow for normal operations or a tripping function, all combinations should be checked.

When the preliminary commissioning checks have taken place for the steam valve control modules, steam valve position checks can take place. It is important to check that the valves are complete and can be operated safely without danger to personnel or equipment, and that other mechanisms, which may operate when hydraulic fluid is available, are in a safe condition. Where possible, steam valves should not be tripped shut, as this can cause damage to valves when they are cold and not cushioned by steam flow. The number of trips from the fully open position should be kept to a minimum.

Initial checks would be carried out for control in the HP governing mode at 4% droop setting, the emergency trip valves being in the reset position and the governor system armed. A temporary instrument will be required to monitor a steam flow demand signal which is required for these checks.

Preliminary alignment settings are then applied to the hydraulic pilot relay of each valve. Datums are taken, using micrometer dial gauges to record and validate the position of the pilot and main rams with respect to a simulated valve control signal. The valve is capable of moving open to a position greater than the fully-open position required to admit full steam flow to the turbine at rated conditions of pressure and temperature. This 'over-open' position should be recorded to ensure that full valve travel to its mechanical limits is obtainable. Figure 5.44 illustrates main ram movement and pilot ram movement for a signal range

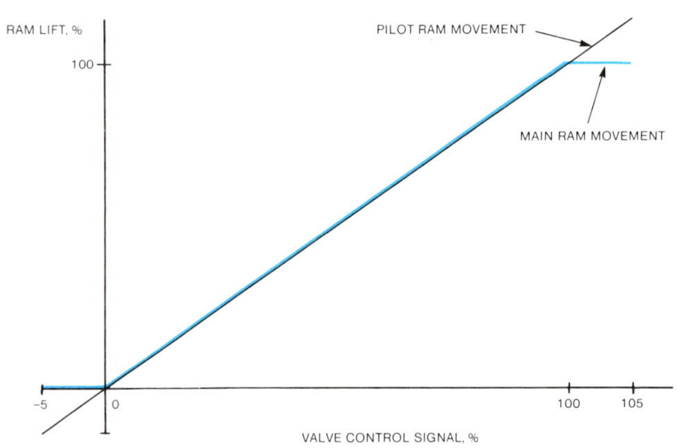

FIG. 5.44 Graph showing main and pilot ram lift against a base of % control signal

of approximately −5% to 105%. It can be seen that the main ram is mechanically limited over its range of travel, which should equate to 0–100% control signal. The pilot ram will have a slightly greater range of travel and will move a small amount (−5% to 0% signal value) before the main ram starts to lift: similarly, it will move a further small amount (100% to 105% signal value) when the main ram has reached the limit of its travel.

When each valve has been treated as above, the openings of the HP stop valves relative to a valve lift signal are checked to ensure that the openings are matched and in unison. All four valves would therefore be shut initially, with the micrometer dial gauges fitted to measure valve travel. The requirements for the system described is to operate a 'governor raise' control until 'run-up in control' is indicated. At this point, the 'crack-point raise' control is operated until 'crack-point' is indicated; the micrometer dial gauges are inspected to ensure that no movement has taken place. The 'run-up raise' control may then be operated until appropriate indicators show that the governor is in narrow-range control and that the run-up potentiometer has reached its maximum position.

By using governor control, the steam flow demand signal can be varied. By lowering the signal to a specified value, all four stop valves should be checked and adjusted to be just open. By raising and lowering the signal, equal operation of all valves is checked, including complete closure and the point of opening. Inspection of readings obtained should show valve movements to be within about 0.1 mm of each other. A typical plot of steam valve main ram movement against demand signal (in volts) is shown in Fig 5.45, for all four valves.

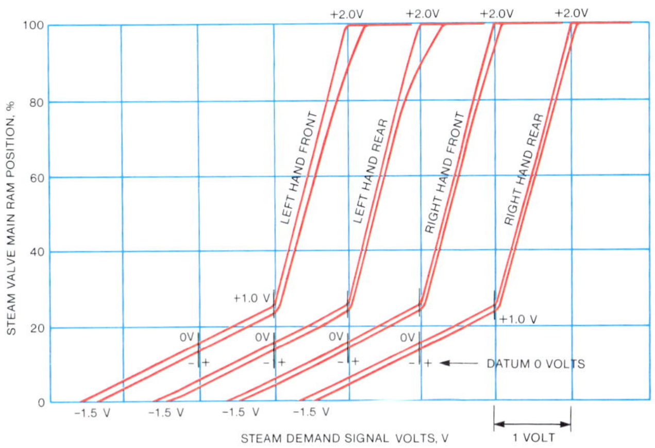

FIG. 5.45 Graph of turbine HP stop valve (main ram) movement against steam demand signal (volts)

The movement of a turbine main steam valve (i.e., main ram) is plotted against the steam demand signal, measured in volts. The figure shows the graphs obtained for each of four main HP stop valves, so that direct comparison can be made to ensure that the results match each other.

A similar operation can then be carried out for the HP and IP governor valves. For the example under consideration, the IP governor valves will lead the HP governor valves and must be fully open at 50% steam flow signal, the HP valves being fully open at 100% signal.

With all valves shut, operate the hand trip lever and check the feature of this design which causes the electronic governor to disarm and the run-up and crack-point potentiometers to go to their minimum settings.

Commissioning of the various other bins, such as protection and arming bins, should then proceed along similar lines. For the example under consideration, these bins are closely interlocked with valve highway modules, speed control module and computing channel digital modules, through relay interconnections or power supplies. All the devices would therefore have to be available at the same time and the detailed commissioning activities closely integrated.

For the on-load test system, a similar approach is again applied. The sequence of valve operations must be checked in detail. The voltage signal required to stroke each valve (or the actual valve position) should be recorded against a base of time, and checked to be smooth (Fig 5.46). Incorrect operation of sequences must be shown not to have any effect, for all possible

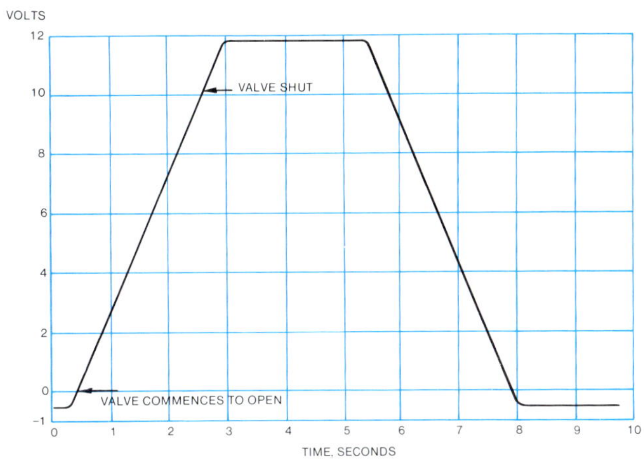

FIG. 5.46 Graph of voltage signal required to stroke a turbine steam admission valve plotted against time

A typical graph obtained for a turbine steam admission valve, when using the on-load test facility, showing the output, in volts, from the on-load valve test module against a base of time. A smooth, linear profile indicates a satisfactory control signal.

combinations. For the system under consideration, correct operations would be:

- A valve will only trip when it is being stroked
- Not more than one valve group can be selected for test at any time
- Direction of test (open/close) has to be reselected after each valve is tested
- No step change in valve position can occur unless the valve is correctly tripped shut.

8.4.1 Response tests

When commissioning of the electronic governor system has progressed to the point where the system is functional, as previously described, a number of response tests should be carried out. The response tests are the equivalent of final functional tests, described in other sections of this chapter.

The electronic governor system must be fully available, including all the associated turbine steam valve gear, with hydraulic supplies in service. The boiler will be depressurised and isolated from the turbine. The micrometer dial gauges will be in position on the steam valves, as previously described. The vacuum and steam pressure trip function must be isolated, the turbine emergency trip relays reset and the electronic governor armed. For the system described, the IP stop valves will travel to the full open position and the HP stop, IP and HP governor valves will remain closed under governor control.

Static response tests

The governing mode will be selected to HP control mode, with governor droop at 4% setting. A temporary test instrument will indicate the governor speed error reference signals, which should be set initially to a value causing the HP and IP governor valves to open fully, with HP stop valves remaining closed. During these tests, all indicators must be checked to operate correctly.

The HP stop valves should then be opened to their crack-point positions, the time taken being recorded. Also, the relay oil pressure under the power pistons at the crack-point position should be recorded. The micrometer dial gauges should confirm that the valves have not lifted off their seats. At this point, crack-point failure should be simulated by by-passing the relevant pressure switches. The respective HP stop valves will then lift a small amount and this is recorded.

The valves may then be driven open in increments, using the 'run-up' control, and valve lifts recorded until all valves are fully open. The closing and opening of valves by lowering and raising the speed error reference signal is then carried out and readings taken as before.

The 'run-down' control can then be initiated to check synchronised closure of all four HP stop valves. The 'crack-point lower' control is activated to ensure that all four valve actuators move to their shutdown positions.

A number of interlock features will need to be proven, to ensure:

- That the 'crack-point lower' feature is only operative when the run-up potentiometer is at a minimum setting
- That the 'run-up' control is only operative when all four HP stop valves are at crack-point positions
- That the 'run-down' control is only operative when generator main circuit-breaker is open.

The HP stop valves and the HP and IP governor valves can then be run shut. The IP stop valves are not under operator control and will remain open. The trip lever can be used to shut the IP stop valves.

Some governing systems are interlocked to disarm on loss of hydraulic fluid pressure, usually on a '2 out of 3' pressure switch configuration. This feature should be proven not to trip on loss of one pressure signal, and also to disarm on all combinations on loss of two pressure signals.

Operations similar to those described should be carried out for the other droop settings.

Response checks at simulated speed

A simulated shaft speed signal of 3000 r/min is required, the run-up potentiometer being in its maximum position. With HP governing mode and 4% droop settings selected, the speed error reference signal should be increased and the simultaneous opening of IP governor valves noted. Further raising of the signal will confirm simultaneous opening of the HP stop valves. The speed error reference signal is then increased by suitable increments and records of all valve lifts taken. This operation is then repeated for the HP and IP governor mode. For the commissioning example chosen, the HP and IP governing mode can be selected for a number of IP percentage bias settings. Readings should be obtained for all the HP and IP bias settings and examined to validate the IP bias limits.

The HP governor mode should then be selected and the shaft speed signal gradually increased: the HP and IP governor valves will start to close. The speed signal at which they become fully closed must be recorded. This check should be repeated for different speed error reference signals. A similar set of readings is then taken for each droop setting. With the final droop setting selected to 1% and the speed signal restored to 3000 r/min, the speed error reference signal should be reduced until the HP governor valve lift is in mid-range. The droop selector should then be switched from 1% through to 4%, noting that there is no change to the positions of the governing valves. A typical response curve is illustrated in Fig 5.47.

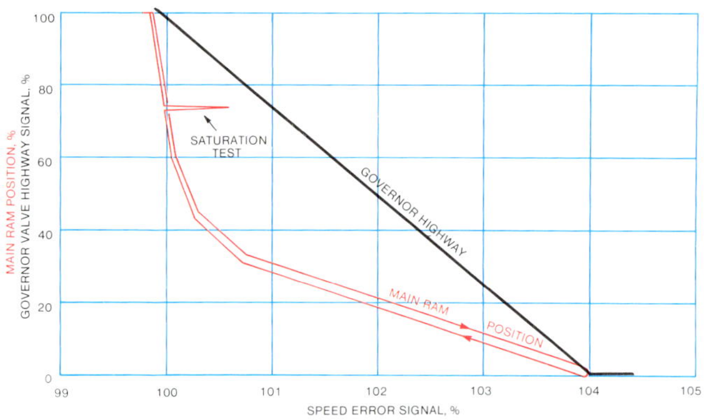

FIG. 5.47 Graph of HP governor valve highway signal and main ram position against % speed error signal

The figure illustrates a typical response curve obtained for an HP governor valve when the turbine-generator is at speed. A simulated shaft speed signal is applied to the electronic governor. The response of HP governor valve ram position is then obtained for a range of simulated speed error reference signals. The curve shows the effect of a saturation test signal. The graph also shows the corresponding governor highway signal associated with the speed error reference signal. The governor highway signal should be linear.

With the governor selected to HP + IP mode for all IP bias selections, the above checks should be repeated for all droop settings.

Tests can then be carried out to check governor response for various speed signal rates of change for which a speed signal ramp generator is required. The tests are conducted with speed signal initially set to 3000 r/min at 4% droop setting and in the HP governing mode. The speed error reference signal will have to be raised to its maximum value, which is approximately

107% of synchronous speed. All steam valves will be observed to be open. The simulated acceleration signal is then applied at the ceiling value for which the system is designed to respond (for the example chosen, 3.3 r/s^2) and the speed limited to, say, 3180 r/min. The HP and IP governor valves should be observed not to close.

A simulated acceleration signal at which the governing response to acceleration is specified to operate (3.7 r/s^2) is then applied. The valves should immediately shut, and reopen fully when acceleration ceases. The acceleration signal is often used to initiate power-assisted closing of bled-steam valves and this function must also be noted.

8.4.2 Determination of governor speed/load characteristic

The governor should be selected for the HP governing mode. By reference to the valve lift against steam flow signal characteristic obtained earlier, the HP governor valves should be set open to a position equivalent to, say, 25% load. The valve position must then be accurately recorded, together with the 'steam/speed' signal generated by the governor.

The steam/speed signal is then varied over its range and the movements of the governor valves and corresponding steam/speed signals recorded. The speed signal variation should be limited, so that movement of the governor valves does not exceed about ±5% load change.

This process is repeated for load values of approximately 50%, 80% and 100%. These results will be referred to during on-load tests in order to determine the governor incremental droop characteristics.

8.4.3 Vacuum pay-off

A vacuum pump is connected to the impulse pipework of the three pay-off transducers and the vacuum monitored by means of a mercury column. The vacuum pay-off device must not be in the gagged position.

The governing system must be set up with HP and IP steam valves open and the governor in HP control mode. Each transducer vacuum is allowed to decay in turn; no vacuum unloading action should occur. The exercise is then repeated for all combinations of '2 out of 3'. Vacuum unloading should then take place over a specified range of vacuum. All associated alarms and indicators must be verified.

The steam valves will drive to a position equating with a specified steam flow (approximately 10% MCR flow) at the low value of specified vacuum. Further vacuum reduction should not affect the steam valve position. By raising the governor load setting, no change of steam valve position should occur whilst vacuum remains low. As vacuum is slowly restored to the pay-off transducers, the HP and IP valve positions should remain unaffected, unless the speed error reference signal is increased.

A final test can then be carried out by rapidly reducing vacuum to all three transducers simultaneously, noting the fast response of the steam valves.

8.4.4 Pressure pay-off

This is dealt with in a similar manner to vacuum pay-off.

8.4.5 On-load valve testing

With all valves fully open, the full valve test sequence should be operated and the time for valve closure noted, together with observation of indicators. With each valve fully shut, the valve trip should be operated and an attempt made to open the valve sequentially. This attempt should fail.

8.5 Detailed procedures — mechanical governor

From Fig 5.39, it can be seen that a mechanical governing system is a complex hydraulic system. Here, lubricating oil is used as the hydraulic fluid, so the commissioning activity must be integrated with that of the lubricating oil system.

When the lubricating oil system is to be flushed, the relay oil system should be included in the flushing programme. Preparation of the relay oil system for flushing must be carried out with considerable care and under clean conditions. The principles of flushing are the same for the lubricating oil system, necessitating numbers of parallel circuits and the removal of components which would be damaged by debris or which would restrict the flow of the flushing medium.

Components to be removed and stored safely are:

- Internals of emergency stop valve and governor valve relays
- Internals of non-return valves and strainers
- Hydraulic accumulators
- Hydraulic anticipatory devices
- Electric anticipatory devices
- Vacuum and pressure unloading devices
- Orifices

Certain relay components can be completely removed, if this is more convenient than removal of their internal components.

Temporary pipework is then installed and blanks fitted in order to complete the flushing circuits. Temporary fine-mesh strainers will be required at the lubricating oil return to the main oil tank. The auxiliary power oil pump will also be required to circulate oil for flushing purposes. It is important that all parts of the system are flushed in each of the flushing circuits. Thus a number of circuits will be flushed with, for example, their trip mechanisms in the trip and reset positions and with relay and other valves in both open and shut positions. The number of flushing circuits depends upon the particular system, but about ten individual circuits will be required for the system described and each circuit requires at least six hours of flushing. When the flushing operation is complete, the relay system should be reinstated under clean conditions working, all temporary pipework and blanks being removed. The relay oil system should then be isolated from the lubricating oil system until it is needed for subsequent commissioning activities.

Before the commissioning activities can take place, preliminary settings of governor gear relay-linkages must be checked against relevant detailed drawings. These will be carried out at the same time as similar settings for the steam valve relay linkages.

Various aspects of governor control can then be proven, as follows, having ensured that the various electrical supplies are available for the governor motor and associated control and indicator equipment:

- Ensure that the relay system is in the tripped condition.
- Ensure that the governor motor control and indicators have been proven in accordance with standard check lists. The motor should be checked for freedom of movement and gearbox lubrication. The freedom of movement of the manual governor control over its entire range must be checked at the same time. Electrical checks, such as motor field resistance and the continuity of motor field and armature, should also be recorded.
- Prove the contacts of all limit switches, using a multimeter. Actual operation of each limit switch is desirable, such as those which limit the travel of the main governor pilot piston sleeve.
- Set the governor sleeve to mid-position and prove that the governor motor operates correctly in the raise and lower directions from the unit control desk. The governor can then be operated to its maximum raise and minimum lower limits to prove that the motor stops when the limit switches operate.
- Where computer control signals are provided for speed and load control, the operation of the governor motor by these signals should also be proven. It is important to verify that, when under computer control, manual control from the unit control desk is inoperative. Similarly, when under manual control, the computer signals will not control.
- All electrical relays associated with the governor motor should be checked to energise and de-energise correctly, together with the correct functioning of all contacts associated with each relay.
- All relays, limit switches, controls and indicators associated with the turbine trip and trip reset mechanisms can also be proven in a similar manner at this stage.

When all the above steps are completed, the commissioning of turbine emergency stop valve controls should be carried out in a similar manner to that described above. This will include operation from the unit control desk and, if applicable, computer control of motorised stop valve relays. The relay oil system must be left in the tripped position.

The next stage of commissioning can then take place with the relay oil system in service. The boiler will be at atmospheric pressure, with its firing equipment isolated, before further commissioning takes place. Alternatively, the boiler stop valves may be isolated and the steam connections to the turbine depressurised and vented to atmosphere.

Before oil is admitted to the relay oil system, the system should be set up for initial priming, in accordance with the manufacturer's instructions, the isolating valves to the main power oil and secondary power oil systems being open. The emergency trip mechanism will remain in the tripped position.

Items such as governor valve accumulators will be charged to pressure with their isolating valves shut. On-load test valves will also be in the untripped operating condition. Hydraulic and electric anticipatory devices and unloading devices will also be isolated. The auxilary power oil pump is then put into service and the pump pressure control valve checked to ensure that correct main power oil pressure is achieved. Similarly, the secondary power oil control valve will be observed and the pressure noted.

The commissioning activity can then proceed as follows:

- Reset the emergency trip plungers
- Observe the correct opening of steam admission valves where applicable. (In this example, the IP emergency stop valves should open fully automatically.)
- Check the correct operation of all pressure switches, including the correct sense of pressure switch outputs
- Check operation of the main (HP) emergency stop valves (ESVs). (In the example, the valve relays reset when their motorised controls are operated in the lower direction, after which they can be opened. The valves are ganged in pairs of motorised master ESV and slave ESV. The slave is checked to open after the prescribed opening of its master valve.)
- Set the mechanical governor to minimum position and then adjust the pilot oil regulating valve to give a prescribed value of pilot oil pressure (for this example, 3.5 bar). The pilot oil supply valves to the auxiliary relays will have been shut in accordance with the manufacturer's instructions for initial priming of the system. These valves may now be opened, causing their associated relays to shut and hence the governor valves to open partially.

- Operate the governor over its entire range to prove full opening and closing of the governor valves.

During the above operations, the initial settings applied to turbine steam admission valves should be checked and adjusted as necessary.

Some adjustments may be required so that groups of valves operate in unison.

'Friction loop' characteristics should be taken for HP and IP governor valves by plotting a graph of readings of pilot oil pressure against valve movement for each valve over its entire operating range. The graph should be smooth, thereby indicating that the valve and its relays are operating smoothly. Also, the curve obtained for valve opening will be close to that for valve closing, thereby indicating negligible lost motion between valves, linkages and relays.

The governor valve accumulator can be opened after the friction loop characteristics have been obtained.

The electric anticipatory units can then be commissioned. Some preliminary checks (which do not require the relay oil system in service) may already have been carried out:

- Ensure that all fuses are installed and of correct rating.
- Check solenoid resistance and prove circuits earth-free, using a multimeter.
- Check mechanical freedom of the solenoid plungers. These are often fitted with an extension for hand operation.
- Apply prescribed settings to oil pressure switches and prove operation by injecting oil pressure signals.
- Prove all associated controls, alarms and indicators, including supply fail alarms.
- Simulate the opening of generator main circuit-breaker by means of an appropriate auxiliary relay contact. The energising and subsequent de-energising after a set time interval of the anticipatory solenoid is then proven. For this, a detailed understanding of the system is needed, so that appropriate simulated oil signals can be applied.

The relay oil system can then be established, with trips reset, and HP and IP emergency stop valves closed. The main generator circuit-breaker can then be simulated closed and the governing gear operated to open HP and IP governor valves.

The generator circuit-breaker is then simulated to be open, at which point the electric anticipatory devices should function to open the HP and IP emergency stop valve anticipatory oil pistons. Since the emergency stop valves are closed, they will be protected against damage that might otherwise occur if the valves were tripped shut whilst cold and without the cushioning effect of steam flow. After the appropriate time, the solenoids will de-energise.

For the example chosen, reheat release valves are provided to release pressure in the reheater in the event of, say, a unit trip. These release valves will also be actuated to open via the anticipatory oil system when the emergency stop valves are caused to close, and to close when they are open.

When the solenoids are de-energised and in their normal operating condition, the governing gear is then returned to its minimum position, after which the relay oil system may be tripped.

The hydraulic anticipatory devices are commissioned with the relay oil system in service, the HP and IP emergency stop valves closed, and the governing gear set with HP and IP governor valves open. Initially, power oil supplies to the device will be isolated. The power oil supply is then de-isolated and the device observed not to operate, ensuring that there are no fluctuations in pilot oil pressure, and thus no movement of governor valves. The device is fitted with a test facility which should then be proven in accordance with the manufacturer's instructions, without causing any fluctuations in pilot oil pressure.

Commissioning of the unloading oil system will follow a pattern similar to that described for similar hydraulic devices. A vacuum test rig will be required to check and set the vacuum controls. When the initial commissioning of vacuum and pressure switches, relays, solenoids, alarms, controls and indicators has taken place, the relay oil system can be put into service and the trip relays reset. The isolating valve for the secondary power oil supply to the unloading gear is then opened, and also the motorised valve between the unloading and pilot oil systems. The effectiveness of the non-return valve between the two systems is proven by rapidly closing the secondary power oil supply to the unloading gear. Pilot oil and unloading oil pressures should fall at the same rate, indicating that the non-return valve has opened. A simulated steam signal is then applied and reduced as required to prove unloading operation. A vacuum signal is applied in a similar manner. The anti-motoring device will be proven to function at prescribed settings at the same time.

When the various components of the governing system have been commissioned, the on-load test facility for HP and IP emergency stop valves and governor valves may be commissioned. The approach to the commissioning of the test facility relays and associated equipment has been described in Section 1 of this chapter. When initial tests are complete and functional testing can take place with the relay oil system in service and steam valves open, the correct operation of limit switches should be checked at each step of the test sequence by means of a multimeter.

All valve position indicators must also be proven. Certain large machines are fitted with valve timing equipment, which allows an assessment of correct valve operation to be made. The equipment can be commissioned at this stage and proven to operate when on-load test commissioning has taken place. The timing equipment may be initiated from a variety of sources, such as emergency trip plunger operation, opening of the generator main electrical breaker or low pilot oil pressure. Actual proof of these initiating devices should not take place unless the steam valve chests are at or near service temperature.

8.6 Commissioning network and resources

Special equipment required to commission an electronic governing system is listed in Section 8.4 of this chapter.

People carrying out this activity will be specialists and fully conversant with the equipment.

The commissioning network (Fig 5.48) comprises a number of parallel paths and is relatively simple, but involves a considerable number of man-hours. The interrelation with other activities should be specified with care to avoid conflicts. Because of the parallel paths, it may be easier to use check lists. Commissioning meetings again play an important part in correlating various activities and interests and in issuing short-term programmes to ensure that conflicts are avoided. This is particularly important when the governing system requires steam valves to be operated.

The mechanical governing system activities are illustrated in Fig 5.49. This network will, in practice, be integrated with the lubricating oil system network (Fig 5.11) for the flushing activities. Operation of steam valve relays and their associated commissioning activities must again be carefully co-ordinated.

Resources for flushing will be similar to those required for the lubricating oil system. No specialist skill or equipment is required for a mechanical governing system, other than the use of persons competent to understand and perform the tasks required. However, the application of preliminary settings and the need to prove each component of the governing system in a careful and systematic manner requires considerable time (about one to two weeks).

When the system has been fully commissioned, it is prudent to repeat functional tests of the governing system, such as the opening and closing of valves, proving remote and local controls, operating trip and trip reset devices (taking care to trip the steam valves from a partially open position when cold), etc., prior to admitting steam-to-set.

9 Generator, exciter and main connections

This section deals with the final erection checks associated with the construction of the generator, exciter and main connections. The detailed description and commissioning of the generator sub-systems, such as stator water, gas, seal-oil, etc., are dealt with in later sections.

9.1 General description

Modern generators are usually conventional two-pole, three-phase hydrogen and water cooled machines, solidly coupled to the main turbine at one end and to the generator exciters at the other. British practice is for the machines to run at 3000 r/min (Fig 5.50).

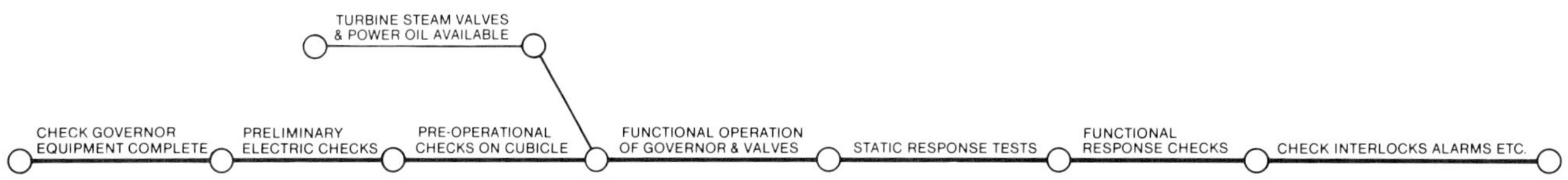

FIG. 5.48 Commissioning network for an electronic governing system

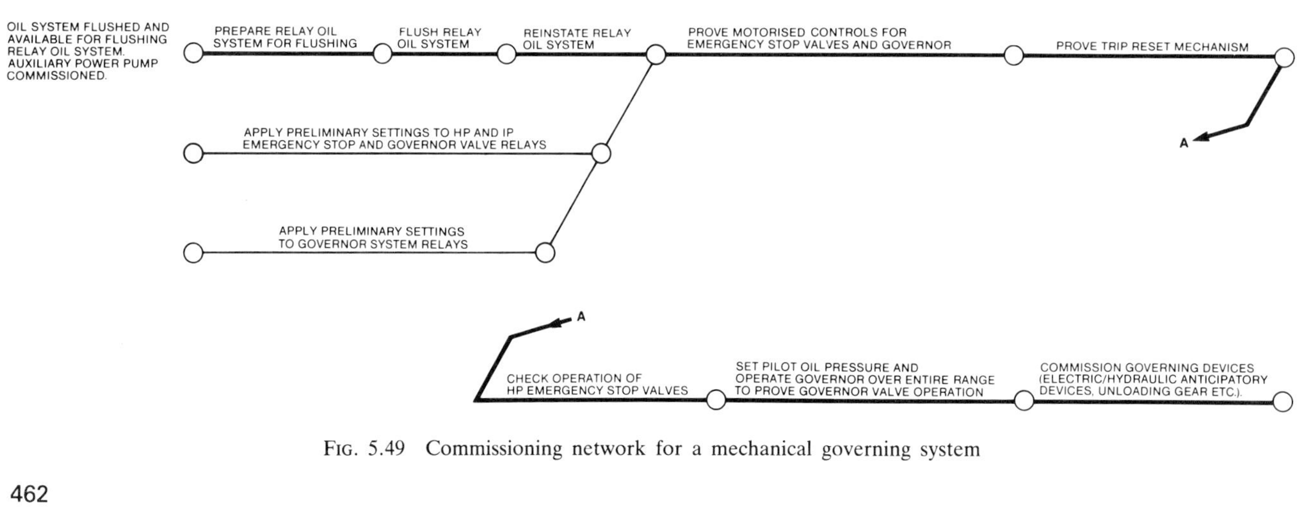

FIG. 5.49 Commissioning network for a mechanical governing system

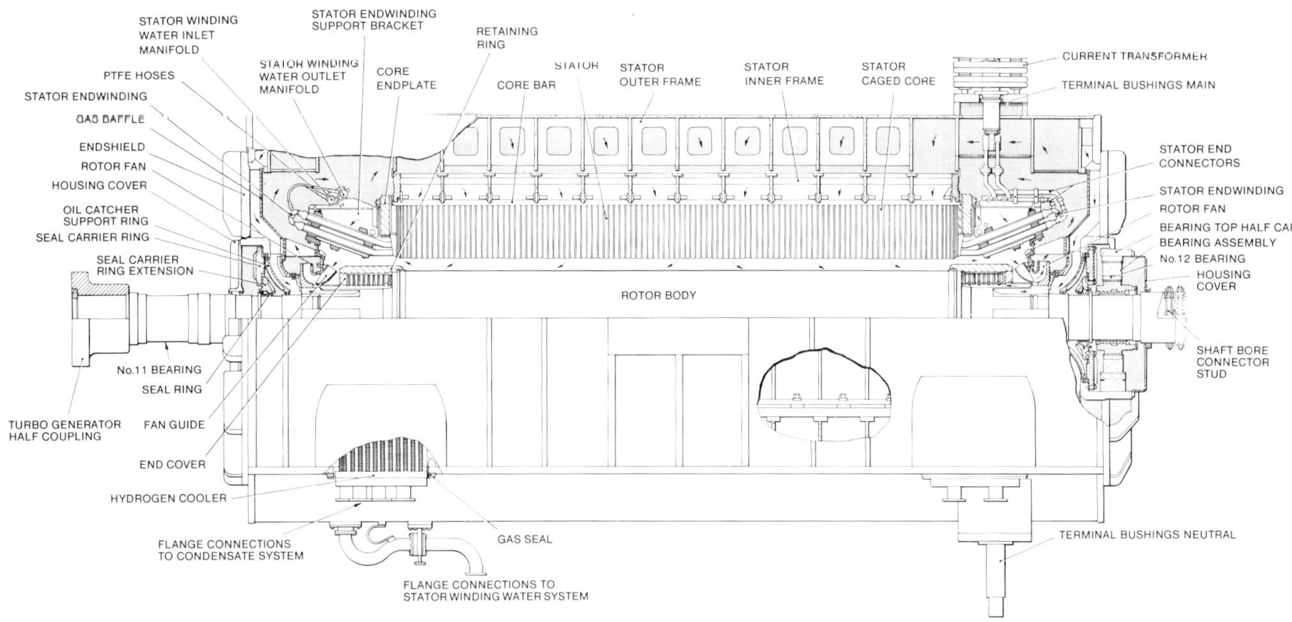

FIG. 5.50 Typical sectional arrangement of a 660 MW generator

Brushless excitation systems have an advantage over the slipring type, since they eliminate sliprings and the problems associated with brushes due to sparking and carbon dust. Excitation is by means of a permanent magnet generator electrically connected and coupled to the main exciter unit. The rectifier unit is a rotating assembly. However, excitation systems using sliprings are in common use.

With the brushless system, the instrumentation tapping points for monitoring generator rotor voltage, current, etc., are all contained in the rotating elements, so it is necessary to transfer signals from these points to the stationary instrumentation. Instrument sliprings and brushes could be used for this, but would introduce the very elements the brushless system was designed to eliminate. In the given example, a telemetry system is used to transfer the signals from the rotor across an air gap to the stationary equipment by means of a multi-winding rotating transformer, with one set of coils on the rotor and the other set of coils in a stationary housing.

To date, the complexity of telemetering systems has reduced equipment reliability and the future trend is likely to be towards the use of sliprings for this purpose, despite their known shortcomings.

In the example, the generator and exciter ratings are as follows:

Generator

Maximum continuous output (MW)	660
Maximum continuous output (MVA)	776
Terminal voltage (kV)	23
Speed (r/min)	3000
Frequency (Hz)	50
Hydrogen pressure (bar)	4.14

Exciter

Output (kVA)	3140
Voltage (V)	470
Power factor	0.95 (lagging)
Speed (r/min)	3000
Frequency (Hz)	150
Coolant	air

Usually, a generator of this type will operate at rated MVA, frequency and power factor, with voltage not more than 5% above or below rated voltage, but not necessarily in accordance with the standards of performance established for operation at rated speed.

If the rotor excitation is too low to sustain the load, the generator may become unstable. At low leading power factors, the generator must be operated within the specified limits of MW and MVA to maintain a margin of stability.

The excitation system is designed to ensure that the controlled generator will always work within its stable range.

A water system circulates cooled, filtered, demineralised water through the generator stator winding. Provision is made to monitor water flow, temperature, conductivity and entrained gas or air in the water (see Section 12 of this chapter).

A hydrogen gas system ensures that heat is removed from the generator rotor. The generator casing is therefore filled with hydrogen under pressure (see Section 11 of this chapter). Carbon dioxide is used as a scavenging gas, so that an explosive mixture of hydrogen and air is never formed in the casing. Instrumentation is provided to monitor the temperature, pressure and purity of the gas in the casing.

A seal-oil system enables the hydrogen to be contained within the generator casing by supplying cooled and filtered oil to the radial-type shaft seals (see Section 10 of this chapter). Seal-oil is taken from the shaft-driven main oil pump or from an AC or DC motor-driven oil pump when the main oil supply is not available. The seal-oil pressure is automatically controlled so that it is always higher than the casing pressure. The seal-oil system is equipped with instrumentation to monitor the condition of the seal-oil at all times.

In the example chosen here, the main generator connections (Figs 5.51 and 5.52) are of the phase-segregated type, with aluminium busbars within aluminium enclosures. The system is cooled by natural air circulation. Connections between busbars and generator terminals are made by means of flexible copper braids to prevent transfer of stress from busbar to generator connections. Provision is made for access to the braids and to the polytetrafluoroethylene (PTFE) hoses connecting with the stator water system. The hoses are fitted with test electrodes so that generator insulation measurements can be carried out.

The excitation system commissioning procedures are described in Section 13 of this chapter. The major components for a brushless system are the pilot exciter, the main exciter and the rotating rectifier (Fig 5.53).

The pilot exciter is an inductor alternator, using permanent magnets mounted within the six stator-pole assemblies. The magnets are made from high energy permanent magnet material.

The main pole is constructed from laminated electrical steel sheet and is formed in two halves with a horizontal joint at the generator shaft centre line. It also carries the compensating winding coils in holes punched in the pole face. These coils have brazed pole-to-pole connections, except those crossing the centre line which use insulated cables connected to two terminal boards, one on either side of the joint. Each pole is fitted with a separate short-rated coil to permit initial magnetisation of the permanent magnet, and for subsequent remagnetisation should this be necessary.

The pilot exciter rotor is directly coupled to the main exciter and carries the rotating elements of the telemetry aerials. Electrical components associated with the aerials are encapsulated in holes machined in the shaft. The rotor core is constructed from low loss electrical sheet steel laminations, with eight large teeth formed at the periphery.

The telemetry aerial comprises a rotating and a stationary assembly. The rotating assembly has four elements and the stationary assembly, which surrounds it, has three elements. The coils of the elements are positioned side-by-side axially along the shaft. Connections from the ends of the four rotating coils are taken to electronic and resistor pods, located in holes machined radially through the shaft.

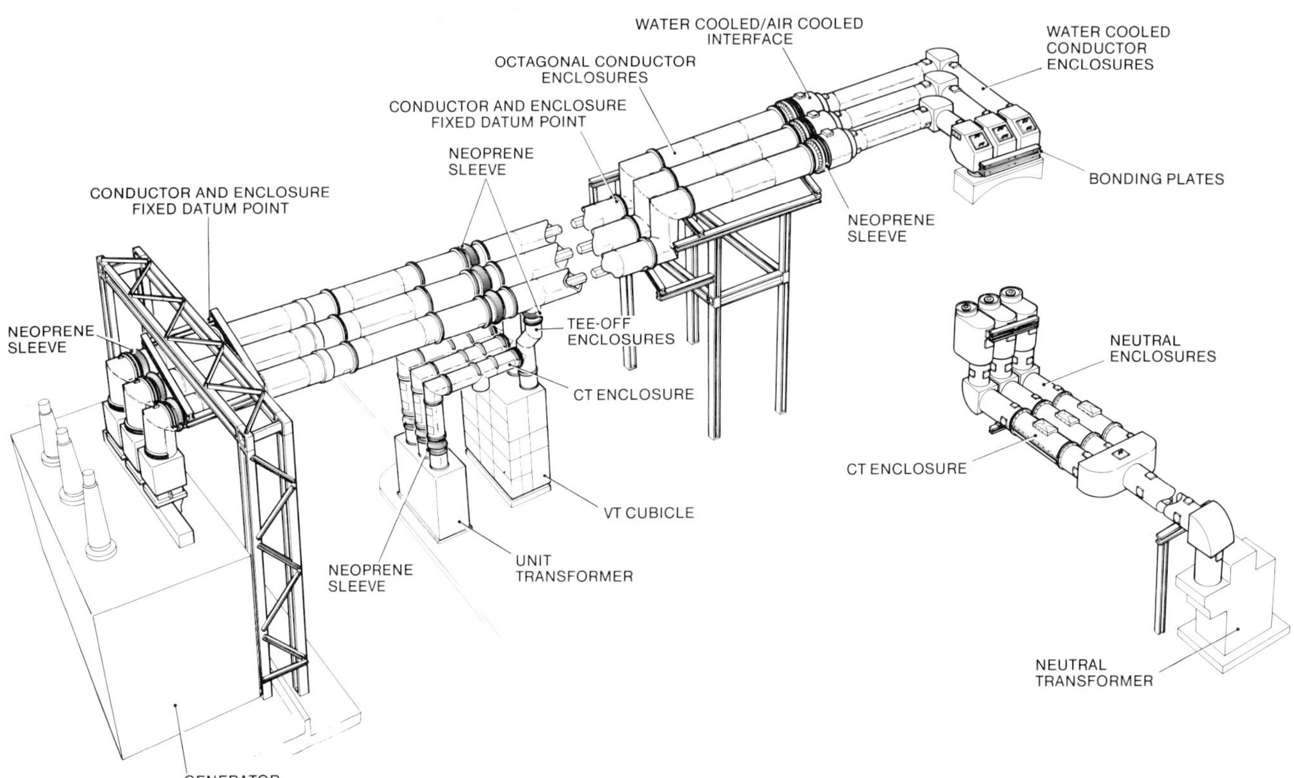

FIG. 5.51 Generator main connections

A typical phase-segregated main generator connection system, comprising aluminium busbars housed within air-cooled aluminium enclosures.

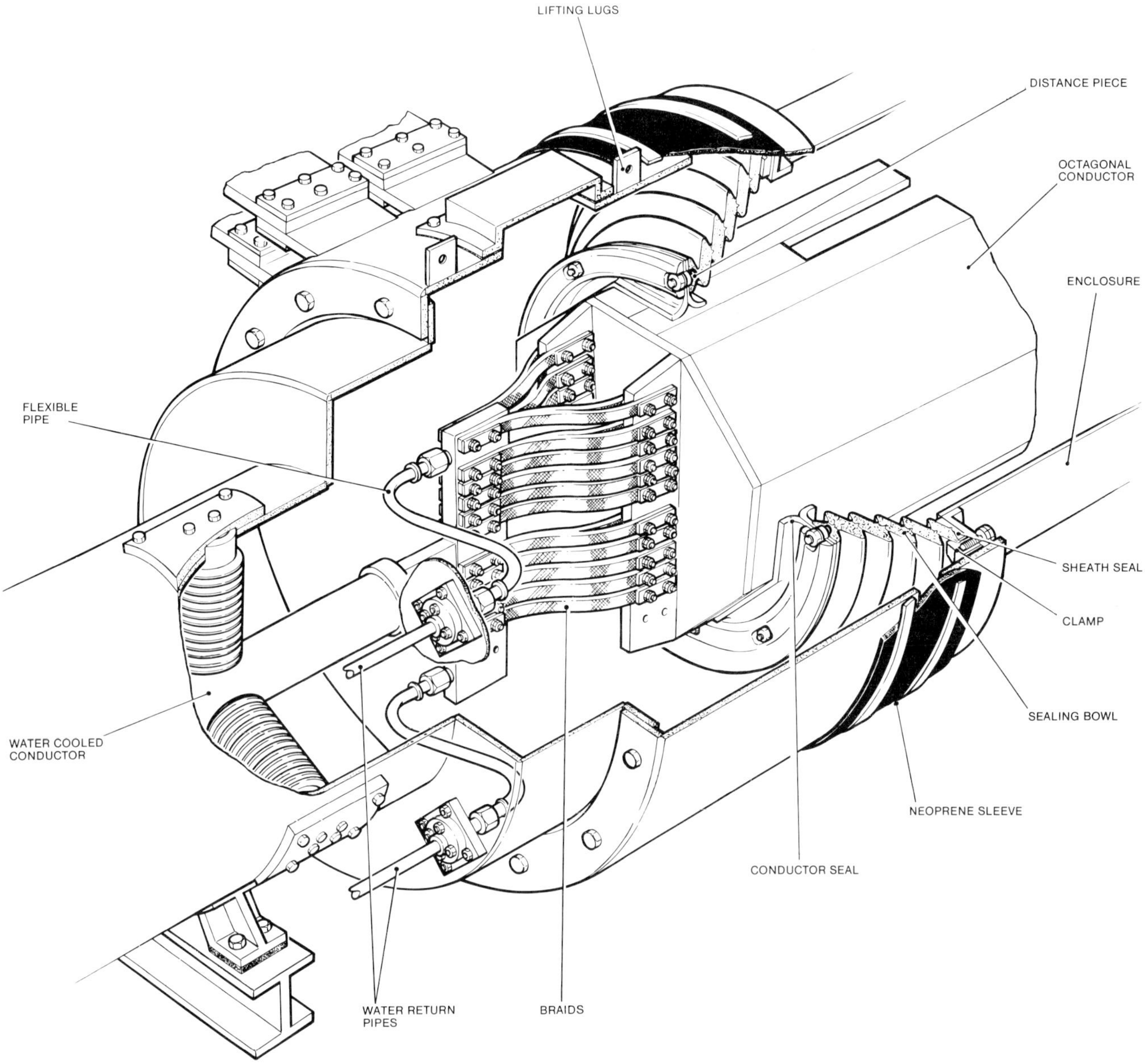

FIG. 5.52 Detail of generator main connections

A detail of the interface between generator main connection water-cooled conductor and phase-segregated busbar.

The main exciter (Fig 5.53) is a brushless three-phase rotating armature AC generator directly coupled to the generator. The DC field windings are mounted in the stator. The AC output from the rotor is rectified by diodes mounted on the rotating rectifier. In order to reduce diode commutation reactance, a fully interconnected damper winding is fitted to the exciter pole faces. The exciter is air cooled by its own closed-circuit ventilation system, which in turn is cooled by condensate.

The DC supply for the field coils is connected to a terminal board on either side of the exciter bedplate. Copper busbars from the terminal board pass through the bedplate and bolt onto laminated copper input supply bars, which in turn connect with the coil-connecting stages.

The rotating rectifier (Fig 5.53) is permanently coupled to the end of the exciter shaft and supported by a Michell-type tilting-pad bearing. The three-phase AC supplies to the rectifier are taken from the main exciter rotating armature, axially along the surface of the shaft.

The rectifier components are contained against centrifugal force by a steel retaining ring. The rectifier carries a number of diodes in parallel, giving about 20% standby capacity in the rectifier bridge circuits. Two circular rows of rectifier sub-assemblies are accommodated within the retaining ring. Each arm comprises nine parallel diodes and one series-connected diode. The rectifier bridge circuits have anode-based diodes in the positive arms and cathode-based diodes in the negative arms of the bridge.

Each diode is protected by two high breaking capacity (HBC) fuses connected in parallel, which isolate the diode in the rare event of it becoming faulty. The remaining diodes can carry full excitation current, up to

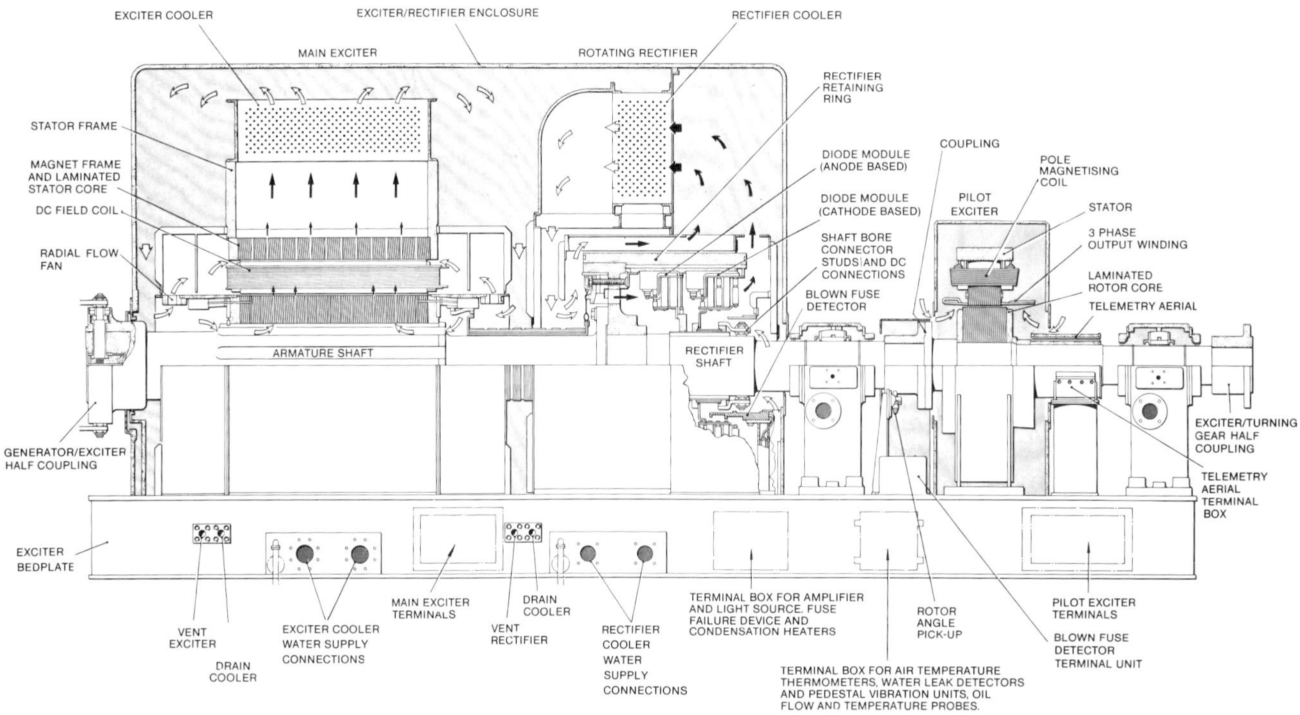

FIG. 5.53 Typical section of pilot and main AC exciters with rotating rectifier for a 660 MW generator

a loss of two of the nine diodes in each arm of the rectifier.

Each diode module also has a resistance-capacitance (RC) spike-voltage suppression circuit and indicator fuse. A fuse safeguards each RC circuit in the event of a capacitor failure and also acts as a damping resistance.

The indicator fuse is also used in conjunction with fuse failure equipment designed to detect the failure of main fuses used for diode protection. The fuse failure equipment provides alarms and indications, allowing the plant operator to decide on appropriate action for any combination of fuse failures.

9.2 Commissioning objectives

The commissioning objectives for these major components of the generator are:

- To ensure that the generator is maintained in good condition throughout the storage, erection and commissioning process
- To ensure that all final data for alignments, clearances, etc., are recorded
- To ensure that instrumentation is complete
- To ensure that type tests for prototype generators or repeat designs which have been significantly modified are accounted for, either by means of works tests and/or by site testing during the commissioning process
- To identify and assess details of works testing
- To carry out insulation and high voltage (HV) tests
- To ensure that the components of the pilot and main exciter assemblies function correctly. Where rotating rectifiers are concerned, this will include the fuse failure detection system
- To measure the resistance of flexible braid connections for future reference purposes.

9.3 Detailed procedures

The detailed procedures involve few functional checks, but serve as a comprehensive post-erection check to ensure that the generator, exciter and main electrical connections are complete to drawing and to specification. This part of the commissioning procedure will include all the generator auxiliary systems, such as seal-oil and stator water, the detailed description and commissioning of which is dealt with in later sections.

The Detailed Procedures Document should therefore comprise a comprehensive inventory of:

- Erection records and data, including rotor/stator air gap and alignment readings
- Instrument schedule of all pressure connections, transmitters and thermocouples
- Alarm list
- Cabinets, cubicles
- Cabling
- Final inspection and boxing up procedure
- Pre-operational activities procedure
- Instrument functional tests

Erection records will include a check on any process, such as welding or brazing, used during the erection process. The type of checks for welding were discussed in Section 1.3.4 of this chapter. In respect of brazed copper components, the procedure should be carefully defined. Joints must be cleaned, visually examined for cracks, porosity or other defects, and checked by means of dye-penetrant methods, where possible.

All locating keys and dowels must be checked to be in position and temporary packers removed. Some packers will be specified for welding. Locking tabs and plates must be checked.

Where insulated components have been erected, such as oil wipers carried on insulating supports, IR values will be specified consistent with clean, dry insulation. A 500 V megger suffices to check this. Care must be taken to ensure that rotating parts are not fouled, for example, by cables and clips.

Before the generator and its components, such as bearing pedestals, are boxed up, final inspections should be carried out to ensure that internal equipment is complete and clean.

All records must be checked for satisfactory completion. Typical records are:

Bearing and wiper assembly clearances and settings

Hydrogen seal clearances and settings

Generator and exciter air gaps

Generator rotor-to-pedestal axial datum

Generator and exciter rotor-to-stator axial and radial datums

Locating key clearances, packers and dowels

Generator seal, bearing and oil wipers IR readings

Pedestal half-joint slopes

Instrumentation check and calibration records

Whilst the generator is in storage, and also during the erection and post-erection period, its temperature should be held at 5–10°C by means of fan heaters to keep it in good electrical condition. Routine inspections should be made and records kept, including polarisation index (PI) testing. The polarisation index is equal to the IR reading after ten minutes divided by the IR reading after one minute. The ratio should be greater than 2 and the IR after one minute should be greater than 400 MΩ at 20°C. This should be measured using a 1000 V motor-driven megger. Values obtained are plotted on a graph to indicate IR trends and should be within specified values. Care must be taken in megger testing to ensure meaningful results. For example, the megger should be connected to the inlet and outlet stator water manifolds, with the live terminal to the phase or phases being measured and the earth terminal to the frame. This would also require the removal of manifold earth straps. Electrical leakage paths from manifolds to frame, and between manifolds and windings, will have a serious effect on the measurements.

These resistances should be checked, using a low voltage multimeter, before the IR test takes place to ensure that no insulation damage results from tracking.

If the insulation resistance is low, this may be due to wetness in the stator water system connections, such as the PTFE hose connections. These should be blown dry with air by pressurising the windings to, say, 5 bar and blowing the system down through drain valves, and also through temporary valves fitted to the manifold connections.

The generator rotor IR should also be measured at the same time as the stator, using a 500 V megger. The rotor windings should be shorted, using fuse wire. If the connections between generator and exciter are made, so that rectifiers are in circuit, the rectifiers must be shorted out. Also, earth links on shaft contacts must be removed.

The preservation of generators is important, not only for initial erection and commissioning activities, but also for recommissioning and long-term storage.

When large hydrogen-cooled generators are out of service for short periods of a few days and not open to atmosphere, they should be maintained with hydrogen in the frame. Seal-oil will have a certain moisture content and there will be a certain release of water to gas. Hydrogen dewpoints must therefore be maintained by purging with dry gas or by means of off-load hydrogen circulators (see Section 11 of this chapter). For longer-term outages, hydrogen should be replaced by a dry inert gas, such as nitrogen or air. The gas must be replaced if either the dewpoint increases or the insulation resistance falls. If the generator is not gas-tight, the gases may be replaced continuously. Dry carbon monoxide is not inert and will react with any moisture in the casing to form an acid, which would give rise to steel corrosion-cracking on rotor end-rings. Thus CO_2 is normally used only for purging.

The stator winding water system should be checked during erection, by means of regular air pressure and vacuum tests. These require water inlets, outlets, vents and drain connections to be blanked off and the water space within the windings pressurised with air to a suitable pressure (say 5.5 bar). The pressure drop over about twelve hours is then noted. The leakage should be very small (say 0.02 bar). If the pressure drop is greater than this and the leaks cannot be identified, the test should be repeated using helium gas, and a helium-sensitive detector used to search for leaks. The process is repeated until all leaks are cured. The vacuum test should be carried out at, say, 700 mbar, with decay no greater than 10 mbar over twelve hours.

For hydrogen coolers which are fitted into the generator frame, cooler packs should be pressurised with water at, say, 14 bar and checked to be leak free before being fitted into the generator.

Other aspects of generator commissioning, such as stator water system flushing and generator air pressure tests, are dealt with in subsequent sections.

Generator main electrical connections of the type described will have been erected in accordance with certain process specifications, such as Cadweld and aluminium welding techniques. Post-erection checks must therefore ensure that these processes have been carried out and recorded correctly. Clearance between conductors and casings must be thoroughly checked. The Cadweld process is used for making copper/copper connections: each weld must be free from slag and pressure tested at about 6–7 bar water pressure for one hour.

The resistance value of each flexible braid joint should be measured for specified bolt torque settings. Some judgement must be exercised in assessing joint resistance, but ducter readings in the range of 3 to 7 MΩ can be expected. Braids may have to be disconnected at a later stage, e.g., when commissioning the stator water system (see Section 12 of this chapter). The resistance value of each braid must be rechecked when the joint is remade.

Where single-point earths have been specified, these should be checked. All earths must be checked for continuity.

Where current transformers are fitted, their carriers should be marked uniquely to indicate their correct position in the event of subsequent removal. Position and numbers should be recorded.

Voltage transformers (VT) should be subject to a number of tests, viz:

LV magnetising current	LV pressure
Induced voltage	Earth continuity
Ratio and phase angle (at rated voltage and rated burden)	Polarity
	Partial discharge
Working burden	

VT isolating, earthing and interlock mechanisms must be thoroughly checked. Where the busbars are fitted with earthing mechanisms, these must also be checked. High voltage testing of the busbars should be to BS159 (1975). The test is carried out with the busbars and neutrals disconnected at the generator. Typical tests are as follows:

IR tests

Phase to phase:	greater than 100 MΩ
Phase to earth and neutral to earth:	greater than 100 MΩ

HV test at 66 kV DC for 15 minutes

Each phase independently:	leakage 10 mA to 5 mA
Neutral independently:	leakage 10 mA to 5 mA

For the generator, measurements of IR and PI serve two functions:

(a) They indicate whether there are any gross defects in the windings, and hence whether it is safe to apply an AC test voltage.

(b) They indicate whether there is moisture present in the bulk, or on the surface, of the insulation. Such moisture could affect the validity of AC test results.

Capacitance and loss tangent (tan δ) tests may also be applied to assess bulk insulation conditions of either individual phases or the whole winding, at voltages up to the line value, using a Schering or ampere-turn bridge. The best method of assessing the quality of insulation is to test the stator windings individually. The tests also provide a comparison, from which ageing can be assessed during the life of the generator.

The tests should be applied at a stage in the commissioning programme just prior to steam-to-set. Before the high voltage test, the stator water system must be drained. Water in the PTFE hose connections should be thoroughly dried out using dry air. Further IR readings are then taken as previously described. The PI value is then calculated and should be greater than 2 before the high voltage test is applied. The stator water system should then be returned to service and circulated for some five hours before the test to ensure a stable conductivity of not more than 400 μS/m. The stator water temperature should also be stable and close to 20°C. Water manifolds should be disconnected from earth and their resistance to earth tested, using a 1 kV megger. The value should not be less than 5 kΩ.

The generator should be filled with an inert gas such as carbon dioxide or hydrogen with the casing pressure raised to its normal operating value (4.1 bar in this example). The generator stator terminals will be electrically isolated from the main generator connections which will be earthed. The generator stator neutral will also be disconnected and the neutral side earthed. The rotor windings will also be earthed.

As a precautionary measure, thermocouples should be earthed and casing drains checked for oil or water ingress. The gas or air dewpoint should be recorded.

From the IR and polarisation index obtained, it can be determined if a generator dry out is required. Although the criteria are not absolute and some judgement may be exercised after consulting the manufacturer, a dry out will not be needed if the IR test values given earlier in this section are achieved, together with a PI index $\geqslant 2$. Another factor to be considered is known damage to insulation, which will encourage the take up of moisture into the insulation. Causes of such damage may be overheating, electrical tracking, discharge, erosion or delamination.

Also, when an overpotential test is envisaged and the gas humidity is worse than −10°C dewpoint, there is a risk of flashover on the end-windings. Thus, the generator must be dried out.

Methods of drying out a generator are:

(a) External heating using, for example, heated water in the stator cooling circuits and hot dry air circulated through the stator

(b) Internal heating by means of an external generator connected to the stator windings. The current flowing through the windings generates heat.

(c) Vacuum drying by the application of vacuum to the generator frame, in conjunction with some form of external heating.

(d) Internal heating by means of circulating currents within the stator windings, created by shorting the stator connections and rotating the turbine generator under steam, with controlled excitation (Chapter 6).

(e) Windage dry out, achieved by running the generator at a speed high enough to raise generator winding temperature (Chapter 6).

The aim of any dry out method is to remove moisture from all surfaces subject to electrical potential. The only effective method of removing moisture from the surface is to replace existing gas with dry gas, and to continue this operation until equilibrium is reached between dry gas and the insulation surface. A measure of the effectiveness of dry out is when the dewpoint of the gas leaving the generator is more than 10°C higher than gas entering the casing. Alternatively, the dewpoint of gas leaving the generator should not exceed −20°C.

The application of heat to the conductor on its own does not remove moisture, but merely drives it to the surface of the insulation. Similarly, windage dry out on its own merely circulates and heats the gas, whether it is damp or dry. Thus the moisture pushed up and entrained in the gas must be displaced or dried.

Vacuum techniques carry the danger of delamination of the insulation and damage to seals if the vacuum is too high. However, the dry out is effected very quickly. This technique has had limited application in the CEGB. It is advisable to choose the level of vacuum with care, in consultation with the supplier.

Heat applied to conductors should only be considered an aid to dry out. Where conductor heating is effected by DC injection there is a risk of shaft magnetisation, so this method is basically undesirable.

The choice of method depends upon a number of factors. Whenever possible, the need for a dry out should be avoided by preserving the generator in good electrical condition throughout the storage, erection and commissioning process. Methods (d) and (e) are both good choices from the point of view of control, but they are in-line commissioning activities. As the cost of fuel required to operate the plant is high and the time for the activity can be long, these aspects should be taken into account when making the choice.

Exciter storage and erection preservation is similar to that described for the generator. Temporary meggering leads should be provided for both rotor and stator. Where a protective cocoon is applied to the whole exciter assembly, these leads will be brought outside the cocoon. For rotating rectifiers, the diodes must be shorted out (probably at the supplier's works). This may be done by using fuse wire. Monthly IR checks should be taken and recorded, with any earth links lifted or removed. It is essential that all shorts are removed after the last IR check and before running the machine.

Off-load heaters should be used to maintain an air temperature of about 10°C above ambient temperature. Dessicant bags may be put inside the cocoon in a position where they may be easily inspected and retrieved. The bags should have indicators which can be observed from outside the cocoon.

On completion of construction and before boxing up components, a thorough inspection for cleanliness must be carried out, particularly with respect to magnetic particles. Vacuum cleaning should be used. Blowing with compressed air must definitely not be used.

Electrical connections must be checked for both mechanical and electrical integrity, including correct phase 'rise' or 'rotation'. External cooling services, using water, should be pressure tested and leak free. Leakage detectors should be commissioned prior to the application of water cooling services.

Services such as fuse detection, rotor angle magnetic pick-up, voltage monitoring, etc., must be checked and final clearance readings recorded.

Equipment enclosures, such as fibreglass hoods, should be fitted and checked to be clean and secure; their air seals must be examined to ensure that there is no ingress of unfiltered air. Filters must be checked to ensure that they are installed. Some sealing devices use an external air supply, which should be connected and in service at the earliest opportunity. All fittings (such as thermometers) should be checked, covers for terminal blocks fitted and permanent blanking plates in position and secure.

The alarm circuits, as listed in the detailed commissioning document, will be commissioned in the same manner as described in other sections.

9.4 Commissioning network and resources

A commissioning network should not be needed for the activities discussed in this section. However, commissioning programmes must indicate clearly, and make provision for, electrical testing. This must be carried out in an approved manner with physical disconnections, barriers and earthing leads being controlled and maintained by an authorised person. It is important that detailed discussion takes place at relevant commissioning meetings to ensure that, where electrical testing is required, work in the vicinity is suspended or completed before testing takes place.

Persons involved in electrical testing must have sufficient competence and experience both to carry out the tests and to interpret the results obtained.

10 Seal-oil system

This section deals with the seal-oil system, including pumps and coolers, and explains its relationship with the lubricating oil system.

10.1 General description

Figure 5.54 illustrates a typical seal-oil system for a large modern generator.

The purpose of a seal-oil system is to maintain a seal between the generator rotor as it passes through the stator casings, thereby containing the high pressure stator gas.

There are a number of different seal designs, but they all operate on a similar principle, comprising a fixed element and a rotating element attached to the shaft. The rotating element acts as a thrust bearing withstanding the axial load imposed by generating casing pressure. Lubricating oil is introduced between the faces of the rotating and stationary elements, acting as both a lubricant and sealing medium. The seal is designed so that most of the oil flows towards the atmospheric side of the seal. A small proportion flows towards the gas side of the generator; this is collected and returned to a special compartment of the main lubricating oil tank, via a detraining chamber. In the event of oil spilling into the generator, liquid detector alarms are provided. Thermal and differential expansion of rotor to casing is catered for in the seal design by allowing movement of the seal-oil housing within the end plates of the stator. Leakage between seal-oil housing and stator is contained by conventional means, such as O-rings and lip-seal devices.

The temperature of seal faces is usually indicated by means of thermocouples embedded into the stationary element of the seal close to the seal face. The seal faces are of the 'white metal' type.

Under normal operating conditions, seal-oil would be supplied from the turbine-generator shaft-mounted oil pump. For start-up, shutdown and emergency conditions, AC and DC motor-driven pumps are provided.

The oil supplied from the main oil pump is at high pressure. In the example given, the pressure would be about 8.25 bar. The supply pressure is controlled to give the required seal-oil/gas differential pressure, which is of the order of 1.4 bar. Individual manual trimming valves are fitted in the supply at each seal.

Seal-oil coolers are provided and act in a manner similar to the lubricating oil coolers. Self-cleaning strainers are fitted after the coolers.

The bulk of the oil flowing from the seals is returned to a compartment within the main turbine lubricating oil tank. The oil is allowed to settle, so that any small amount of hydrogen in suspension can separate out from the oil before the oil passes through a submerged weir into the main lubricating oil compartment. Hydrogen is vented from the settling compartment by means of a fan, which discharges to atmosphere outside the plant buildings.

Oil from the gas side of the seals passes to a detraining chamber, via 'tell-tale' sight glasses. Also, oil level within the detraining chamber may be seen by means of sight glasses. The oil flow will be very small and is usually observed as it passes over a V-notch weir. Hydrogen detrained from this oil passes back to the generator frame. Detrained oil passes to the hydrogen section of the main oil tank via a float-controlled sealing chamber, which prevents hydrogen escaping from the system into the oil tank. Alarms are provided for high/low levels in the sealing chamber. The separate lubricating oil section of the main oil tank also has its own vent fan, which discharges to outside atmosphere.

AC and DC driven pumps are mounted on the main oil tank in a similar manner to that described for the lubricating oil system. The AC pump is essentially a standby to the main seal-oil supply, as well as providing start-up and shutdown services. The DC pump acts in a similar manner, but provides emergency seal-oil supplies on loss of AC seal-oil pump electrical supply. The DC pump has its own control system to maintain oil/gas differential pressure. The DC pump supply enters the seal-oil main after the coolers and is therefore not at risk in the event of interruption of supply at the coolers.

Pressure accumulators are installed to ensure uniform pressure of seal-oil supply during pump start-up, shutdown and changeover. The accumulators also provide an instant supply of pressure energy in the event of a system disturbance caused, for example, by stiction or radial misalignment at a seal.

The seal-oil system is interlocked with the electric turning gear and jacking oil pumps, so that the shaft cannot be rotated unless the seal-oil system is in service. These interlocks are dealt with in the relevant sections of Chapter 6.

10.2 Commissioning objectives

The objectives when commissioning a seal-oil system are:

- To flush the system
- To set the system up for both AC and DC seal-oil pumps separately
- To ensure that seal-oil/gas differential pressure is controlled over the specified operating range
- To prove alarm indicator interlocks and controls, including standby pump cut-in
- To assess seal-oil supply transient pressures during pump changeover with accumulators in service and out of service

- To ensure that seal-oil supply flows and detraining flows are within specifications
- To prove the integrity of the seal-oil detraining system
- To set AC and DC seal-oil pump relief valves.

10.3 Detailed procedures

The Detailed Procedures Document for the seal-oil system should contain a complete inventory of its components. These will include:

Detraining chamber

Seal-oil coolers

AC and DC pumps

Hydrogen vent fans

Seal-oil strainers

Pipework and supports

Valves

Cabling

Orifice plates

Instruments including pipework, panels, wiring, etc.

Cabinets and cubicles

Final erection checks include a thorough check on the completion of the system, including vent pipework.

Care must be taken to ensure that blanks used to segregate the seal-oil system from the lubricating oil system during the erection process have been removed. Blanks may also be required to segregate the seal-oil system from the lubricating oil system during the commissioning process.

During the erection stage, the internal condition of seal-oil pipework will have been noted. If the condition of internal surfaces is poor, a decision to carry out chemical cleaning before, or after, erection will have to be made. This would be part of the overall activity for both seal-oil and lubricating oil systems, so that chemical cleaning can be carried out for both systems simultaneously, if found to be necessary.

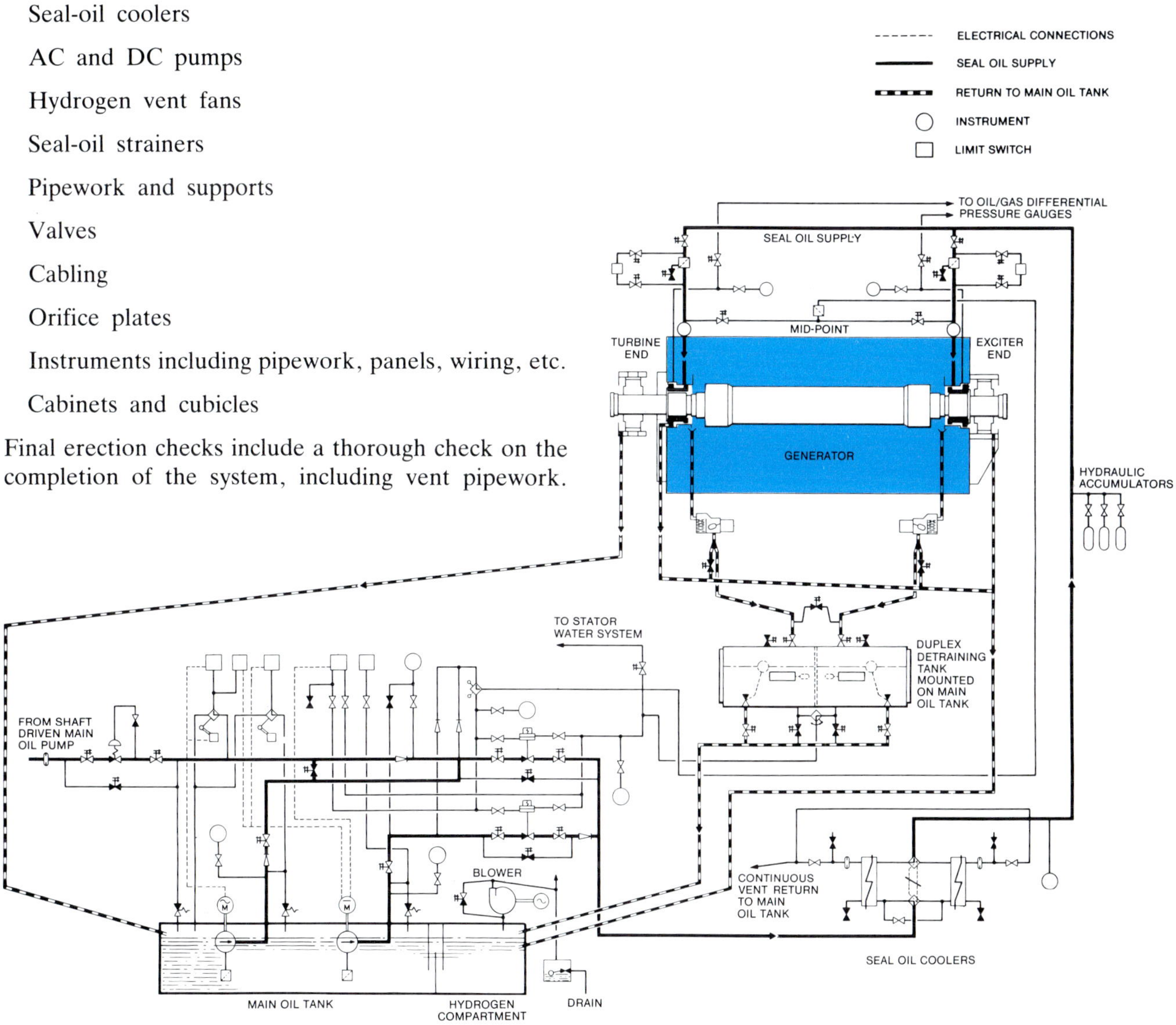

FIG. 5.54 Seal-oil system

Under normal operating conditions in this typical seal-oil system configuration, seal-oil is obtained from the turbine-generator lubricating oil system. For start-up and emergency conditions, AC and DC motor-driven pumps are also provided. Both the main and AC pump seal-oil supplies are pressure regulated to give the correct seal-oil to hydrogen differential pressure. The DC pump is equipped with its own pressure regulator.

The majority of the seal-oil supplied to the seal faces returns from the atmospheric side of the seal to a self-contained compartment within the main lubricating oil tank. The very small quantity of oil from the hydrogen side of the seal passes, via alarm flood chambers, to a hydrogen detraining chamber. The alarm chambers detect high seal-oil flow. The detraining chamber vents hydrogen back to the frame and acts as a pressure seal between the generator frame and the oil tank.

The seal-oil return flow compartment within the main oil tank is fitted with a vent fan which discharges residual hydrogen to atmosphere.

The lubricating oil system must be commissioned to the point where flushing has been completed. Owing to the integrated nature of the lubricating and seal-oil systems, commissioning of the seal-oil system should ideally follow on from completed commissioning of the lubricating oil system.

10.3.1 Flushing procedure

The initial commissioning requirement is to prepare the seal-oil system for flushing. To this end, the AC and DC seal-oil pumps will be commissioned in accordance with standard check sheets. Similarly, the main oil tank vent fans for both the seal-oil return section and the main sections of the oil tank will be commissioned in accordance with standard check sheets. Both vent fans must be proven to discharge freely to atmosphere. Vent pipework must be checked to rise throughout its length and drain traps and drain valves, where fitted, proven to operate.

The seal-oil strainer elements, orifice plates and detraining tank float valve, which are likely to restrict the high oil flows required for flushing, will be removed, as for flushing procedures described for other sections. Where access can be gained to the system, e.g., to the strainers and detraining tank, the internal condition should be checked for cleanliness. The seal-oil return strainer located within the main oil tank must be checked to be clean and in position.

As for the lubricating oil flushing procedure, a temporary heat source is required. Flushing of both systems can be integrated, as previously suggested. The seal-oil coolers need a cooling supply, which would normally be auxiliary cooling water. Before cooling supplies are applied, the ACW side of the cooler must be pressure tested, as described in Section 6.3.1 of this chapter.

Some parts of the seal-oil system will need to be by-passed. Flexible hoses should not be used. By-pass pipework should be constructed and supported to the same standards as the system pipework. By-passes will be fitted between the inlet and return connections to each seal.

Where seal-oil supplies are taken from the lubricating oil system, that part of the seal-oil system should have been flushed as part of the lubricating oil flushing procedure.

The circuits chosen for flushing must ensure high velocities. No dirty circuit must be introduced upstream of one to be flushed. Permanent by-pass connections should form part of the first circuits to be flushed.

Blanks must be controlled in the same manner as described for the lubricating oil system. Safety precautions will be similarly applied.

The AC seal-oil pump is used to carry out the flushing procedure. Before oil is admitted to the first circuit to be flushed, the pump discharge relief valve is set for low pressure with the pump discharge valve just 'cracked' open. The AC pump is then started and the circuit slowly filled and checked for leakage. The relief valve is then set to a higher pressure and the discharge valve gradually opened until full flow is established. The temporary seal-oil heating source is then applied. Flushing then follows in a manner similar to that described for the lubricating oil system, including periodic shutting down of the AC pump, removal of heating source in order to achieve thermal cycling, the use of the DC oil pump from time to time, the inspection of the seal-oil return strainer, etc.

Flushing of the seals should then be carried out; this will depend upon the particular supplier's recommendations. It may be possible to move seals back from their thrust faces in order to achieve a free flow across the faces of the seals. Clean supply filters can be used to ensure that debris, which might be introduced during final reinstatement of the system, is not swept into the seals. Also, seal-oil return filters should be in place.

The isolating valves or flow trimming valves to each of the seals must initially be shut. The seal-oil system is then put into service using the AC seal-oil pump, with oil at the temperature previously used for flushing. The isolating or trimming valves are then opened slowly and the flow across the seals adjusted to ensure that flows are not excessive. Seal-oil flow-return sight glasses must not be flooded.

The generator frame liquid detection traps must be commissioned and in service, and be physically checked at frequent intervals. At a suitable stage temporary supply filters may be removed and the flushing operation continued without them.

After about eight hours of continuous flushing and when strainers show no sign of debris, the operation is complete. It may be possible to inspect the seals by removal, say, of covers. This should be done under strictly controlled clean conditions. When the seals have been inspected and found to be satisfactory, they should be reinstated.

10.3.2 Final functional checks

The main oil tank contents should be raised to working temperature, say 45°C, for the purpose of carrying out the final functional checks. This may be achieved by means of the temporary oil heater suggested as a heat source for commissioning the lubricating oil system (see Section 2 of this chapter). Also, ensure that the main oil tank is at its specified working level.

The seals will be checked to ensure that they are fully reinstated and that the faces are not jacked apart. The generator should be coupled to the main turbine shaft, or alternatively it must be adequately restrained from axial movement within its normal axial running position.

Electrical control circuits must be proven and energised. All alarm circuits specified in the Detailed Procedures Document alarm list will also be initially commissioned and available, so that alarms can be proven and checked that they initiate at prescribed settings. The electrical supplies for all auxiliary units must also be available.

The seal-oil system valves will be set in their normal operating positions for priming the system, with the exception of certain valves which relate to other systems — such as the seal-oil/gas differential control and the main lubricating oil supply to the seal-oil system. The seal-oil accumulators will also be isolated.

Relief valves, such as those situated at the AC and DC seal-oil pump discharge, will initially be set to operate at low pressure. Both pumps will have been at least partially commissioned from the local control station for the seal-oil flushing procedure. The pumps may therefore be put in service in turn and their relief valves adjusted to give the specified delivery pressure. The AC and DC pump motor currents are then recorded.

Where a relief valve is fitted on the seal-oil side of the lubricating oil system supply, the relief valve may also be set in a similar manner, the supply from the lubricating oil system being isolated. If the relief valve is upstream of a non-return valve, the internals of the valve must be removed first and replaced after the relief valve has been set.

The seal-oil system is primed very carefully, using the AC seal-oil pump. For the system described, this would be achieved by opening the by-pass valve to the seal-oil/gas differential pressure control valve. Whilst this is being carried out, the seal-oil flow rates, detraining tank level and generator casing liquid detectors will be monitored continuously. The seal-oil pressure must not be allowed to exceed specified operational values at the seals for the casing under atmospheric pressure.

When the system is primed, care must be taken to ensure that instrument lines and connections to control devices and valves are also primed. Examples are the signal lines to the seal-oil/gas differential pressure control valves, which connect with the oil supply to the seals and to the seal-oil detraining chamber (thereby sensing gas pressure). Priming may be achieved by 'cracking' the last connection to the device until oil flows freely. Care must be taken to ensure that spillages are contained or cleaned up.

The seal-oil/gas differential pressure control valve is then initially checked by slowly opening the oil inlet valve, which should cause the former to close fully. The oil pressure at the inlet to the seals should remain fairly constant. The by-pass valve is then gradually shut, causing the differential control valve to open and assume control of the seal-oil supply pressure. By opening and closing the by-pass valve, the overall response of the differential valve can be observed, as the flow is varied.

The DC seal-oil pump is then started with the inlet valve to the seal-oil/gas differential pressure control valve shut. The control valve by-pass valve is then opened slowly; the seal-oil system pressure should rise slightly, thereby indicating that the DC seal-oil pump has taken over from the AC seal-oil pump. The AC seal-oil pump is then stopped. Having checked that the system remains stable, the inlet to the seal-oil/gas differential pressure control valve on the DC seal-oil pump is opened slowly. The control valve should remain static. The by-pass valve is then slowly shut: the differential valve should close and assume control of seal-oil pressure. By opening and closing the by-pass valve, the overall response of this differential control valve can also be observed.

The AC seal-oil pump is then started, the system again checked to be stable and the DC pump stopped. If the system remains stable, remote and local indicators should be recorded.

A number of readings should be recorded for future use:

- AC seal-oil pump discharge pressure
- AC seal-oil pump motor current
- Seal-oil cooler outlet temperature
- Seal-oil supply strainers differential pressure
- Seal-oil pressures adjacent to seals
- Oil/gas controller differential pressure
- Seal-oil drain flows
- Detraining tank level
- Seal-oil supply flows to seals

The supply pressure at the seals should also be noted for each cooler in service and any fluctuations during changeover of the coolers observed.

The DC seal-oil pump should then be started and its seal-oil/gas differential pressure control valve again observed to respond. A similar set of readings to those taken for the AC seal-oil pump should be recorded.

The automatic start features for both the AC and DC seal-oil pumps should be proven. Sensing lines to pressure switches associated with this feature will have been primed, as previously described. The manual and auto-selections for the AC seal-oil pump should be selected in turn and associated stable indicators checked. The pump should not start when selected to 'auto'. Where an auto-start test facility is provided, this is selected to start the AC seal-oil pump, and alarms and indicators checked. The test facility may then be deselected and the AC seal-oil pump left in service. A similar operation is then carried out for the DC seal-oil pump.

When the test auto-start facilities have been proven, a functional auto-start should be carried out for each pump in turn by switching off the running pump, when the other pump is selected to 'auto'. This should then start automatically. A number of readings should be recorded, including system conditions before and after the auto-start, together with minimum seal-oil pressure during the changeover and maximum pump motor current. The time to complete the sequence must also be recorded and the correct operation of indicators and alarms observed.

Alarms and indicators for the vent blowers should also be proven by stopping and starting them.

All outstanding alarms which have not been functionally proven during the above operations should now be checked. They may previously have been set using injected signals and they should be checked to function correctly by varying the system conditions, as required. A limited amount of simulation may be required. This may be achieved by isolating a sensing line to the device and then 'cracking' a union, or bleeding the line slowly by some other means.

For the system described, a number of important alarms can be checked as described above. These include:

- *Oil/gas differential abnormal* Isolate the oil-side sensing line and depressurise.
- *Seal-oil strainer differential pressure high* Isolate the low pressure sensing line and depressurise.
- *Seal-oil drain flooded* Close detraining tank inlet valve and allow the alarm float switches to operate. Ensure that the line does not fully flood back to the seals.
- *Detraining tank low level* Open tank body drain and allow level to fall sufficiently for float switch to initiate alarm.
- *Seal-oil cooler outlet temperature high* Set down contacts and then reset to correct operating value. The alarm should have been calibrated previously.
- *Seal-oil flow low* The flow sensor valves should be throttled to initiate the alarms, the flow devices and indicators having been calibrated previously. The alarms may be inhibited via an interlock with turbine speed below a certain level (say 2250 r/min) and this can be checked via the appropriate contact in the speed sensing circuit. The correct operation of the contact at the prescribed speed will have been checked by the injection of a speed signal.
- *Seal-oil pressure low* The sensing line can be isolated and depressurised. Check also that the DC seal-oil pump starts.

The seal-oil accumulators may now be charged to their specified pressures (e.g., nitrogen at about 3.5 bar). With all accumulators in service, the readings taken previously for automatic cut-in of the AC and DC oil pumps are repeated. They are further repeated for both one accumulator out of service and then two. With only one accumulator out of service, there should be no significant change in system pressure during changeover.

It is important to prove the level within the main oil tank at which the seal-oil pumps lose their suctions. This operation should be carried out by putting the oil pumps in service and then gradually lowering the tank oil level until there are signs that the pumps are losing suction, which may be indicated by noise, decreasing motor current or decreasing pump discharge pressure. Care must be taken to ensure that the pumps are not damaged by this operation. The level at which pumping performance is beginning to be affected must then be recorded accurately against a dipstick or calibrated contents gauge.

This activity may be combined with a similar test on the lubricating oil pumps (see Section 2 of this chapter). The check should be carried out with the hydrogen vent fan shut down, as this may affect the level within that part of the oil tank.

The level at which the seal-oil pumps begin to lose suction must be compared with, and comparable to, those recorded for the lubricating oil pumps, particularly for those systems which derive seal-oil from an auxiliary lubricating oil pump. A low tank level would result in failure of seal-oil supplies, which in turn would lead to loss of gas from the generator, both to atmosphere and to the main oil tank, with consequential risk of explosion or fire. Ideally, the seal-oil pump suctions will be below those of the lubricating oil pumps.

Upon completion of the final functional checks, the system, with all accumulators in service, is ready for the commissioning of the stator gas system. The performance of the system, including automatic changeover operations, will be observed and recorded during the generator frame gassing and degassing operations (see Section 11 of this chapter). Under no circumstances should the generator shaft be rotated on turning gear or by the main turbine without the accumulators in service, since damage to the seal faces can occur. The various interlocks between the lubricating oil system and turning gear must be proven, as indicated in the appropriate sections of this chapter.

For seal-oil systems with supplies from an auxiliary oil pump, similar system checks can be carried out as described in this section for the final functional checks. Where supplies are obtained from the shaft-driven main oil pump, the system must be proven to operate from this source of supply when the turbine is at speed. This is indicated in Chapter 6.

10.4 Commissioning network and resources

A commissioning network is illustrated in Fig 5.55. Resources are as specified for the lubricating oil system (Section 2.7 of this chapter).

11 Generator gas system

This system comprises the gas cooling circuit of the generator, including circulating, drying, filling and purging equipment.

11.1 General description

Figure 5.56 illustrates a typical gas cooling system for a large modern generator.

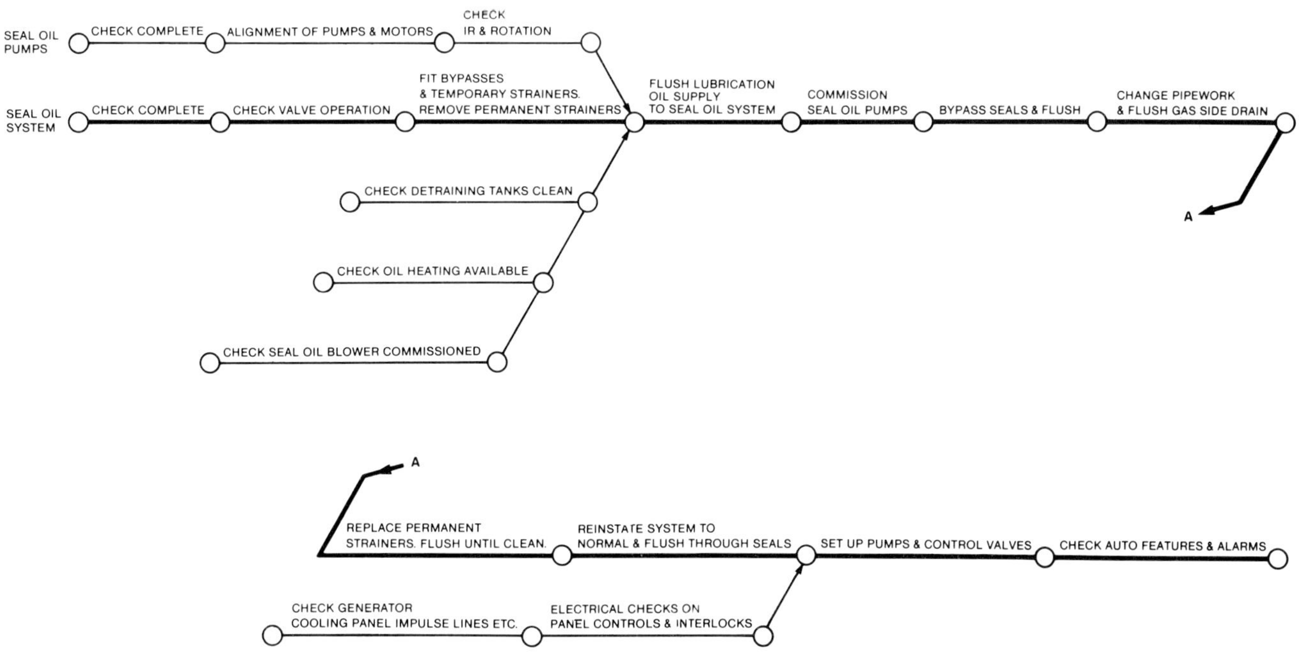

FIG. 5.55 Seal-oil system commissioning network

Hydrogen gas under pressure is used to cool the rotor windings and the stator core, the stator windings being cooled by water (see Section 12 of this chapter). The gas circuit is contained within the stator frame and gas is prevented from escaping by means of the seal-oil system.

The stator is therefore a pressure vessel, capable of withstanding an internal explosion.

Hydrogen is circulated around the rotor and stator core by means of fans, one at each end of the rotor. Hydrogen coolers, using condensate as the cooling medium, are contained within the stator frame. Hot hydrogen passes across the coolers before being recirculated by the rotor fans. Differential pressure created by the fans causes a small proportion of gas to pass through an external dryer which maintains the dewpoint within the generator at a low value. Free moisture within the frame collects in liquid leakage detectors, which initiate alarms. The detectors will also indicate the presence of free oil. It is important that gas pressure is maintained at a higher level than stator winding water pressure. An alarm is provided for this purpose. Also, the hydrogen seals will normally only be effective when gas pressure is greater than a specified value.

Hydrogen is explosive at atmospheric pressure when mixed with air, the explosive limits being between 4% and 74% hydrogen by volume. A hydrogen purity meter is fitted, which uses a small gas flow to atmosphere. The gas flow is normally indicated by means of a rotameter device and alarms are initiated for abnormal purity conditions. Pure hydrogen can be applied to the purity meter in order to check its calibration. An alarm is initiated if purity falls below 95%, but a figure of 99% is usually achieved by routine topping up of frame losses.

Hydrogen is supplied to the generator from central storage facilities. Automatic make-up valves are sometimes fitted and the pressure within the frame is controlled by a regulator. A relief valve is provided to limit casing pressure. The generator is purged of hydrogen by the introduction of a buffer gas, usually carbon dioxide. The carbon dioxide is then displaced with air. The reverse applies when the frame is to be charged with hydrogen. Limits are specified for hydrogen purity during both the purging and charging operations.

The hydrogen dryer comprises two identical cylinders containing a drying agent, such as activated alumina. The alumina acts as a dessicating reagent and therefore removes moisture from hydrogen when the activated cylinder is in service. The other cylinder will be off-line, during which time its contents are reactivated by a heating element. Moisture is purged by means of a bleed flow of hydrogen, which then passes through a cooler and separator before rejoining the gas flow to the in-service cylinder. When moisture reaches a given level in the separator vessel, an alarm is initiated.

Some large machines are fitted with off-load hydrogen circulating facilities. The need for these depends upon a number of factors, such as the kind of electrical insulation, together with general dewpoint condition of the gas and the type of operation (e.g.,

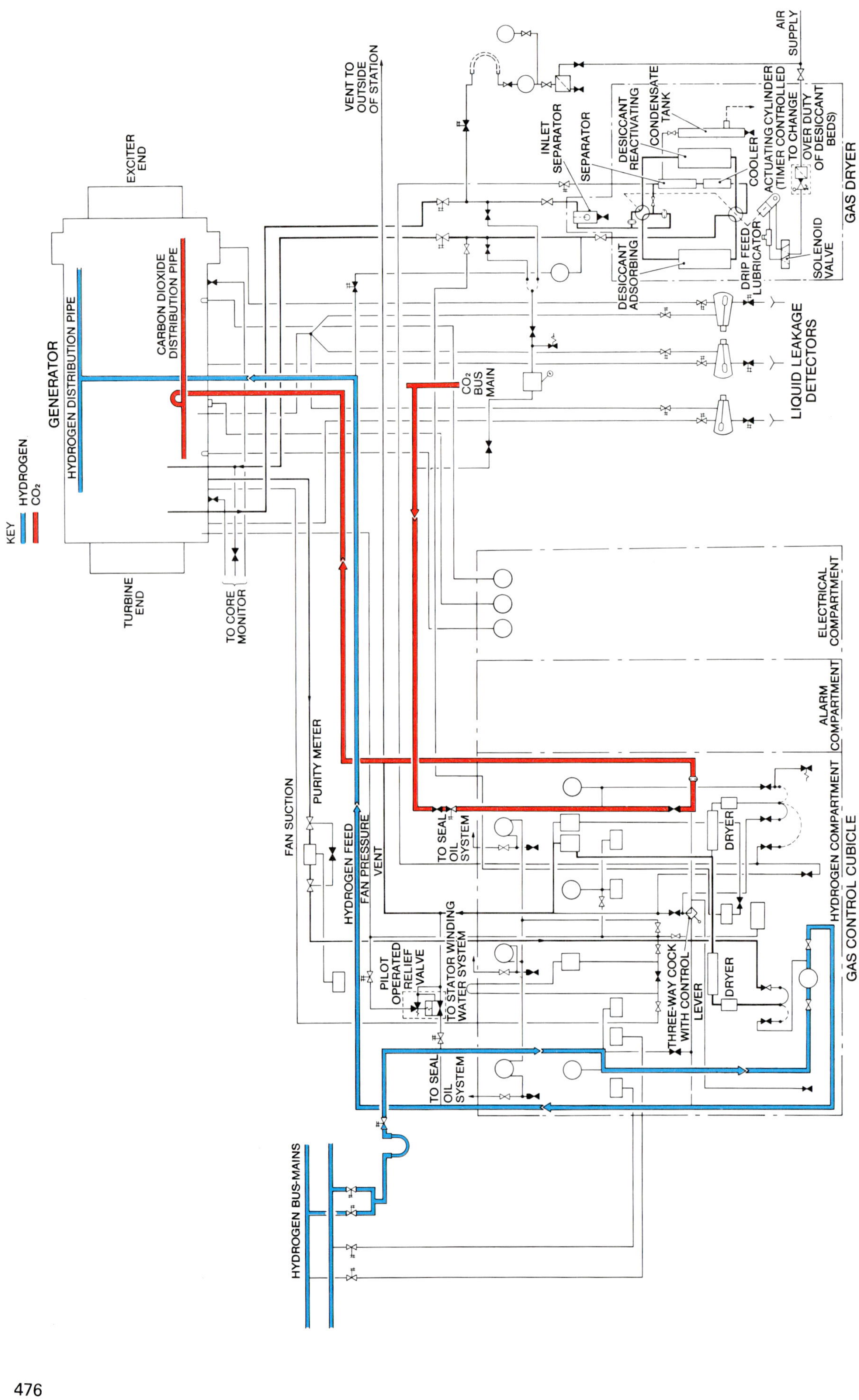

FIG. 5.56 Generator gas system

In this typical generator gas cooling and drying system, hydrogen gas is supplied from duplicate busmains to a hydrogen control cubicle and then passes to a distribution pipe fitted inside the top of the stator. Carbon dioxide, used for purging air or hydrogen, is also supplied through a busmain to a distribution pipe inside the bottom of the generator frame. On-load hydrogen driers are circulated with hydrogen by the differential pressure created by shaft-mounted circulating fans within the generator. Off-load hydrogen driers (not shown) are sometimes fitted.

two-shift, base load) encountered. Under certain conditions, when the dewpoint levels are high, moisture can settle out on electrical insulation, particularly when the generator is shut down and becomes cool. An off-load hydrogen circulator provides continuous drying of hydrogen, since circulation through the normal dryers is only possible when the generator rotor is turning at speed. A typical installation is shown in Fig 5.57. Such units are generally a complete package connected into the hydrogen dryer inlet line in parallel with an automatic by-pass line, which opens to allow normal flow of hydrogen to the dryer when the generator is at speed.

The package is connected to the hydrogen vent system and to the carbon dioxide purge system, the latter by a flexible hose. The automatic by-pass valve opens or closes when the gas circulator motor supply switch is open or shut.

The circulator will allow either hydrogen or air to be circulated. It is designed to trip if its lubricating oil level is low, if its seal-oil level is low, if the differential pressure across the circulator is high or low, or if the circulator motor is overloaded.

Interlocks are provided so that the circulator will not start unless lubricating oil and seal-oil levels are above a minimum value and unless the low differential pressure trip is inhibited by a time device.

11.2 Commissioning objectives

The objectives when commissioning the generator gas system are:

- To ensure that the generator is essentially gas-tight (air pressure test)
- To prove all alarms
- To ensure that purity monitoring and calibration are satisfactory
- To prove the generator charging and purging facilities and to record times taken and quantities used: also, to monitor frame temperature near to the point of CO_2 gas entry
- To check control interaction between seal-oil, stator water and generator gas systems
- To ensure that hydrogen drying equipment operates satisfactorily
- To assess the daily hydrogen consumption of the generator
- To prove off-load hydrogen recirculation equipment (when fitted).

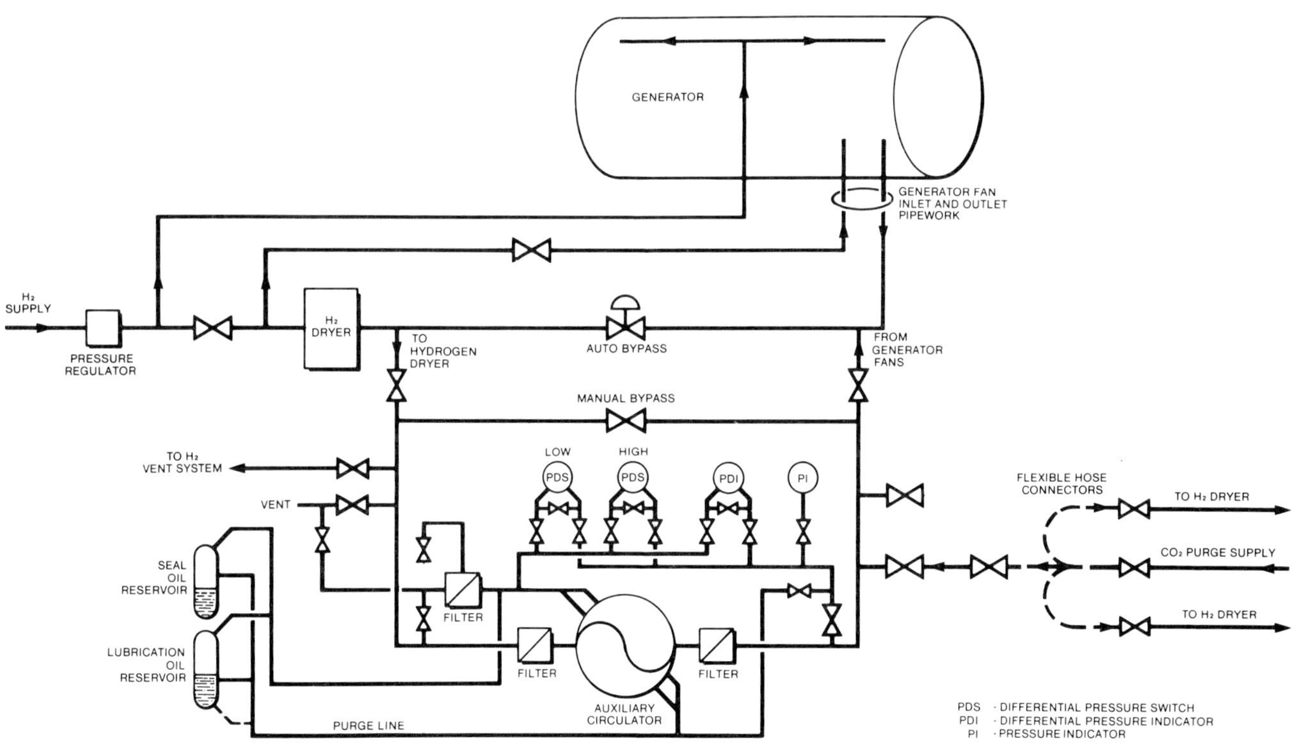

FIG. 5.57 Off-load hydrogen circulating system

The illustration shows a typical off-load hydrogen circulating system. When the generator is shut down, or at turning gear speed, the auxiliary circulator circulates hydrogen, via the hydrogen drier, to the generator.

11.3 Detailed procedures

The detailed procedures document for the generator gas system should contain a complete inventory of its components. For the example given, the list would include:

Vessels: gas dryer

Pumps: off-load circulating blower, where provided

Pipework and supports for air, hydrogen and carbon monoxide systems

Valves

Cabling

Orifice plates

Instruments, including pipework, panels, wiring, etc.

Filters

Cabinets and cubicles, e.g., generator gas control cubicle.

Final erection checks should include a thorough check on the completion of the system, including supplies of hydrogen, carbon dioxide and air. Special attention must be paid to insulated flanges and earthing straps as specified on relevant drawings. For hydrogen systems, particular care must be taken to ensure segregation of electrical components in accordance with design specifications. Vents to atmosphere must be checked to ensure that they are erected correctly. This includes small venting systems, such as the hydrogen purity meter, to ensure that pockets of gas cannot accumulate in the hydrogen cubicle.

Instrument and air supplies will be required. Where hydrogen dryer automatic changeover equipment is operated by air, the supply will normally be provided with air filters.

A number of orifice plates is likely to be used and their correct sizing and installation should be checked.

The first item to be commissioned will normally be the gas dryer. Its cylinders should be checked to ensure that they are charged with dry desiccant and that the filters which prevent desiccant carryover are in place.

For the dryer illustrated in Fig 5.56, satisfactory operation of the changeover valve operating cylinder is checked with the actuator disconnected from the valves. The actuator is then reconnected and the full operation proven. Certain valves have specific greasing arrangements and these should be checked (e.g., Audco valves). The changeover sequence may be tested initially by manual operation of the switching drive system and indicator lamps proven for the in-service and regenerating cylinder. The dryer condensate tank alarm can be proven by back-filling with water from a hose connected to its drain.

When the gas dryer has been commissioned, it may need to have its desiccant dried out. Even if this is not required, the following operation will prove the drying function.

- Ensure that generator air supply is disconnected from the dryer. This is usually facilitated by means of a unique U-pipe.
- Ensure that dryer outlet to the generator is isolated.
- Check the dewpoint of the supply air and ensure that it is reasonably dry (say –40°C at atmospheric pressure).
- Connect the air supply U-pipe with the dryer set to 'auto' and, at the commencement of a drying cycle, admit air to the dryer at a predetermined flow rate.
- Measure the dewpoint leaving the in-service dryer and regulate the air purge flow to the reactivating cylinder to enable it to heat up as quickly as possible to a prescribed value. Repeat this operation for the other cylinder.

11.3.1 Air pressure test

The seal-oil system must be commissioned and in service for the air pressure test.

It is advisable that the liquid leakage detectors and alarms are proven. This may be done by back-filling the respective float chambers with water by means of a hose connected to the drains. The stator water system, hydrogen coolers and lubricating oil systems should not be in service, since they may introduce heat to the stator, thereby heating the air and distorting the air pressure drop test readings. The seal-oil system coolers must be in service to control the seal-oil temperature to a constant value throughout the test.

Hydrogen and carbon dioxide supply lines should be disconnected and isolated to ensure that air under pressure does not leak into them. The air pressure test will create deflections of the stator end-plates and this can be measured by means of strategically placed micrometer dial gauges. The object is to ascertain that deflection is uniform and hence does not move the seal housings out-of-square to the rotor.

The generator is then charged with dry air, via the hydrogen dryers, and the pressure raised slowly. Seal-oil flows from the gas side of the seal should be recorded at increments of about 0.5 bar hydrogen pressure. The seal-oil system should be observed to function correctly and to maintain the prescribed oil/gas differential pressure.

A thorough check of all joints, flanges, welds, etc., is made for leaks, using soap solutions or approved branded products. If leaks are found, the generator should be depressurised and the leaks made good. When the generator has reached maximum rated pressure (about 6 bar for the system described), the area around the generator should be evacuated of all personnel. Casing relief valves are then isolated and the pressure raised slowly to the maximum prescribed value. The pressure is then maintained for about one minute, after which the relief valve isolating valves are opened slowly to reduce frame pressure, and their lift and reset settings checked and adjusted by raising and lowering the frame pressure.

The carbon dioxide supply limiting valve can also be checked under air pressure in a similar manner, if valve arrangements permit.

The valve will have been isolated, as previously mentioned. The limiting valve may then be back-pressurised from the frame and its lift and reset operation checked and adjusted to a specified value.

The pressure may then be dropped to working pressure and the loss of air monitored by recording pressure drops hourly over, say, a 24-hour period. It may prove useful to estimate actual leakage rates by means of a manometer connected between an air pressure reference cell and the generator frame.

A systematic leakage check should be made of the complete system. Soap solution, ultrasonic detection equipment or freon gas injection and detection gear may be used.

Air leakage into the stator winding water system and into the rotor bore should be checked, the latter at the rotor bore end-plugs.

When the leakage rate is deemed to be satisfactory, a 24-hour pressure drop test should be carried out. A standard test gauge must be connected to the system and the air pressure raised to rated pressure. The temperature at the beginning and end of the test is recorded and the pressure drop corrected for temperature. A corrected pressure drop of about 0.035 bar would be expected.

An air leakage rate of about 5 m^3/day is acceptable. On completion of the air test, all vent lines should be blown through to prove that they are clear. The generator may then be depressurised and the amount of liquid collected in the liquid detectors should be checked. If the detector alarm is initiated during the air test, the liquid should be drained off, measured and the quantity recorded, so that the total quantity can be assessed. If water is present, check for stator water or hydrogen cooler leaks. If oil is found, check for stiction at seals or seal-oil system controls.

11.3.2 Final functional checks

Upon completion of the air test, the remainder of the generator gas system may be commissioned, commencing with the frame depressurised but set up for initial filling.

All pressure initiated alarms must be proven and set correctly by injecting the relevant pressure switches. The values at which alarms are initiated and reset should be noted. For alarms initiated from more than one source, all inputs should be proven, including re-initiation of the common alarm, if one in the group has already been accepted.

Alarms involving loss of electrical supplies, such as gas analyser equipment, should be checked together with indicators of 'supply healthy' states. Fuse fail alarms can be proven simply by removing the fuse or inserting a failed fuse. Other supply monitoring alarms should be checked in a similar manner, including alarms monitoring the alarm supply.

The purity indicator for hydrogen in air must have its setting checked (normally 95% hydrogen in air). A pure hydrogen sample from the hydrogen mains is passed through the indicator and its calibration checked. The flow rate through the meter should be adjusted to give the specified value. Low purity alarms should be seen to clear. The hydrogen supply should then be disconnected from the purity indicator and the meter inlet valve opened to atmosphere. Hydrogen will slowly dissipate and the low purity alarms should initiate at the appropriate setting.

The purity indicator used for purging the generator (air in carbon dioxide, hydrogen in carbon dioxide), where fitted, should then be checked. The indicator for air in carbon dioxide should be checked first. Since carbon dioxide is heavier than air, the check will be made with the carbon dioxide tapping point selected at the top of the generator frame. The initial value, air being in the frame, will be close to 100%. The indicator should then be calibrated on pure carbon dioxide obtained from the supply main, the indicator showing 0% air in carbon dioxide. The flow rate through the meter should be adjusted to give the specified value.

The generator frame may then be purged of air with carbon dioxide by following normal operating instructions. The purity indicator for air in carbon dioxide should be observed throughout the operation, until the specified level of purity has been reached. The quantity of carbon dioxide used should be recorded for future reference. The temperature of the carbon dioxide pipework and stator casing at the point where they connect should be measured to ensure that the metal is not overchilled by the expansion of the gas as it enters the stator frame. Undue chilling has been known to cause brittle failure in this area. The carbon dioxide may then be displaced by hydrogen in accordance with normal operating instructions, but with the hydrogen supply limiting valve set to about half open. The purity indicator will be selected to read hydrogen in carbon dioxide, with the initial value before admitting hydrogen checked to be 0%.

At a purity of 85% hydrogen in carbon dioxide, the purity indicator should be switched to hydrogen in air and purity increased beyond 95%, noting that low purity alarms clear. The purity indicator may also be selected to the gas dryer outlet and this should give the same result as that from the frame. The quantity of hydrogen used should be recorded for future reference.

The generator hygrometer (dewpoint meter) cannot be effective until the generator shaft is rotating at speed, since it depends upon rotor fan differential pressure to supply the meter. However, it is normally possible to pass a sample through the meter from the frame by venting to atmosphere. Alarms can be proven by adjustment of set points. Similarly a flow can be established through the gas dryer hygrometer and alarms proven in the same manner.

Upon achieving casing purity above 95%, generator frame pressure may be increased to normal working value, the performance of the seal-oil system being noted.

The hydrogen pressure may then be raised to casing relief valve pressure, all non-essential personnel having been excluded from the vicinity. The lift and reset values are checked and compared with those previously achieved under air pressure. The casing supply limiting valve can then be adjusted to maintain a casing pressure just greater than relief valve setting. The casing pressure regulating valve can be set up by reducing the frame pressure to normal working value and adjusting the regulator.

A thorough final search for hydrogen leakage should then be carried out and the performance of frame pressure control and hydrogen dryers noted. Generator hydrogen gas loss can be calculated as follows:

gas loss per day (m³/day) at normal temperature and pressure (NTP)

$$= 6800 \left(\frac{P_{gs} + P_{bs}}{T_s + 273} - \frac{P_{gf} + P_{bf}}{T_f + 273} \right) \times \frac{V_g F}{t}$$

where:

P_{gs} = gas pressure at start of test, bar gauge
P_{gf} = gas pressure at finish of test, bar gauge
P_{bs} = barometric pressure at start of test, bar gauge
P_{bf} = barometric pressure at finish of test, bar gauge
T_s = absolute temperature at start of test, K
T_f = absolute temperature at finish of test, K
t = duration of test, hours
V_g = volume of generator voids and associated pipework, m³
F = correction factor to 99% purity (Fig 5.58)

Note For practical purposes, estimates of gas loss per day in the order of ±20% of actual value are acceptable, so factor F can normally be ignored. If a more accurate result is required, for example, for commissioning and recommissioning, F should be derived from Fig 5.58.

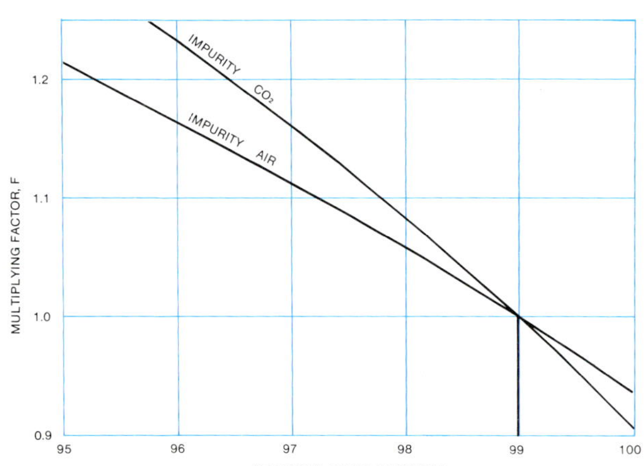

FIG. 5.58 Multiplying factor (F) to obtain loss rate for 99% pure hydrogen

The hydrogen purity indicator may be compared with chemical sampling methods. Care must be taken to ensure that the sample is obtained from gas within the frame and that it is not contaminated with air.

11.3.3 Off-load hydrogen circulation

Where off-load hydrogen circulation equipment is installed, this should be commissioned at an early stage since, when the seal-oil system has been commissioned, it can be used to maintain the generator in good electrical condition.

Prior to commissioning the off-load hydrogen circulation equipment, relevant hydraulic pressure test certificates must be available, together with material test and works test reports, as appropriate.

When the installation is completed but still physically isolated from the generator circuit, particularly if the generator is charged with hydrogen, an air pressure test should be carried out. The test pressure will be the same as the main generator frame test pressure and no loss of pressure would be expected over, say, a six-hour period.

The package will also have been proven on a closed circuit test run, using helium gas at the frame test pressure.

All joints and brazed or welded connections should be 100% tested by radiographic and magnetic particle inspection techniques.

A thorough inspection of the installation should be carried out, similar to procedures described for packaged systems elsewhere in this chapter, using standard check lists for items such as valves, cables, instruments and the hydrogen circulator. All safety devices, alarm and control settings must be checked to operate correctly at specified settings.

When these items have been proved and are available for operation, the package should be operated on by-pass when charged with air at atmospheric pressure. Pressure is then raised to frame test pressure. These two tests are then repeated with the package in circuit with the generator. Correct function must be proven, particularly of protection devices, along lines described in other sections of this chapter.

The package can then be gassed with hydrogen along with the generator. It must then be isolated from the generator and degassed, checks being carried out to ensure that isolating valves do not pass gas. A purging operation to air should then be proven. The package should then be recharged with hydrogen in accordance with approved operating instructions and tested on by-pass at frame test pressure. When this test has been carried out successfully and the system remains gas-tight, the test must be repeated with the package in circuit with the generator. The rate of removal of moisture can then be monitored over a prolonged period to check the effectiveness of the drying system.

11.4 Commissioning network and resources

Figures 5.59 and 5.60 illustrate commissioning networks for the generator gas system and the auxiliary hydrogen circulator equipment. Reference should be made to Section 12 of this chapter for commissioning network details of the air test and the relationship between the stator winding water system and gas system networks.

Adequate stocks of hydrogen and carbon dioxide must be available before the commencement of commissioning. This will require central storage facilities and distribution pipework to the unit to be commissioned. It is important that distribution pipework to other uncommissioned units is properly controlled by means of disconnections, blanks, etc.

The quantities of hydrogen and carbon dioxide required will depend upon the size of generator to be commissioned. It is advisable to allow for at least two filling and purging operations, plus an estimate for losses. A reserve of carbon dioxide must always be retained for emergency purging of the generator being commissioned, as well as for other operational units.

For a typical 660 MW machine the quantity of hydrogen required to fill the generator is about 850 m^3 at NTP. Losses can be assumed to be 25 m^3/day.

The quantity of carbon dioxide required to purge the generator of air or hydrogen is approximately 0.4 tonnes. It should be noted that more hydrogen and carbon dioxide will be needed if the generator is on turning gear, since the gas (or air) in the generator frame tends to mix with the gas (or air) being applied.

A source of dry air is required to charge the generator initially. This can be taken from an instrument air system, care being taken to ensure that the dewpoint is continuously monitored and acceptable. The air can be applied directly to the generator or, preferably, via the hydrogen dryer.

A bottle supply of helium gas is required for commissioning the auxiliary circulator.

12 Stator winding water system

This system comprises the stator windings, their inlet and outlet connections and the external components, such as coolers and pumps.

12.1 General description

A typical stator winding water system is shown in Fig 5.61. It consists of a closed circuit incorporating two 100% circulating pumps, two 50% duty coolers, a duplex strainer and a hydrogen detraining tank. The pumps can be selected for duty or standby, the standby pump cutting in when cooling water flow falls below 90% of rated flow. The coolers are arranged in series; each cooler can be by-passed, but this may result in a load reduction on the generator, dependent upon stator winding temperatures.

The stator water is derived from the condensate extraction pump discharge, which feeds a float-controlled header tank, via a non-regenerative demineralising unit. This unit is also in continuous service in parallel with the main cooling circuit. The cooling medium for

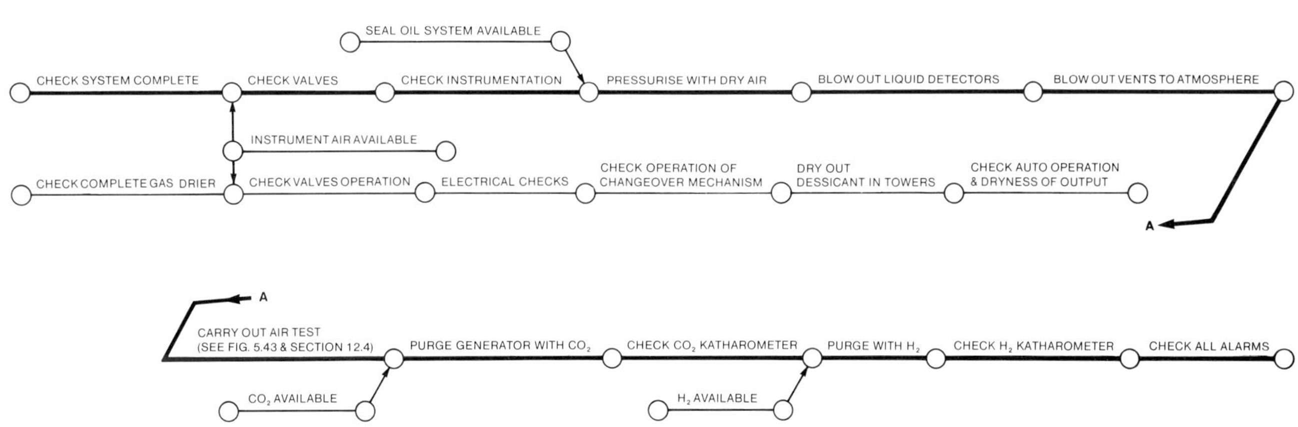

FIG. 5.59 Generator gas system commissioning network

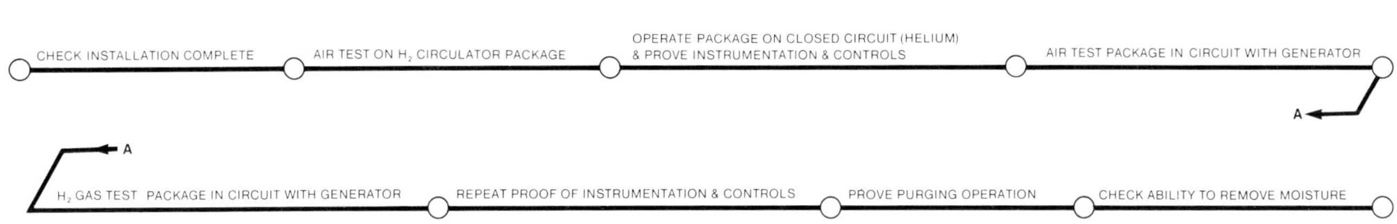

FIG. 5.60 Generator gas system — auxiliary hydrogen circulator commissioning network

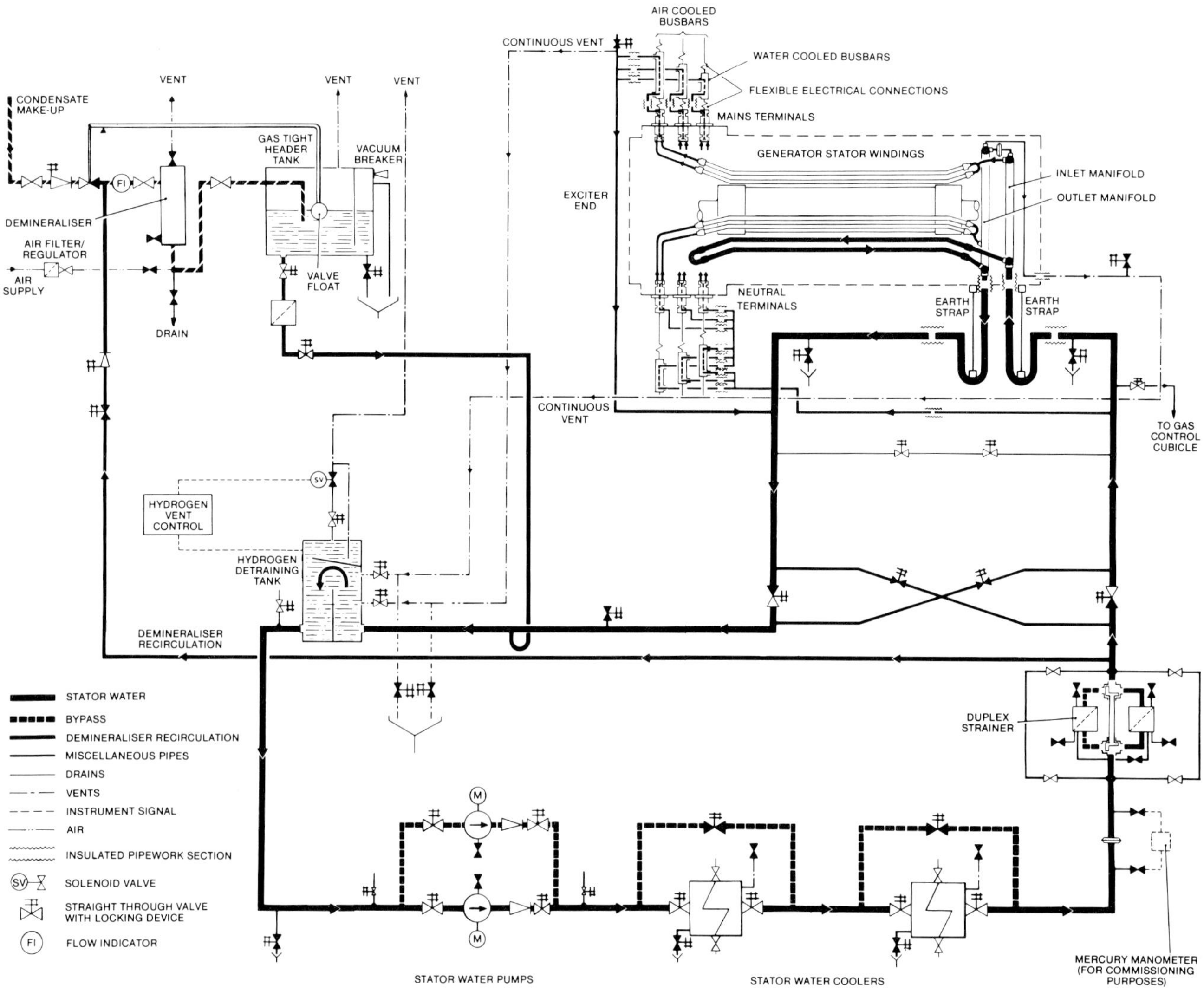

FIG. 5.61 Generator water system

A typical generator stator water cooling system. The closed circuit uses pumps to provide circulation to the stator conductors and coolers, using condensate as the cooling medium. A hydrogen detraining tank allows hydrogen to accumulate and be automatically vented to atmosphere.

the stator water coolers is derived from the condensate system.

Cooling water enters the generator frame via an insulated pipe, which connects with a winding water inlet manifold. Stator water is distributed from this manifold through PTFE hoses in a number of parallel paths to stator winding coil header boxes. After passing through the windings, the stator water leaves the coils via a further set of PTFE hoses to an outlet manifold, and thence to the detraining tank.

The system also provides cooling both for the main terminal connections and their copper busbars, which connect with the aluminium busbars, and also for the neutral terminal bushings and their busbars, connecting with the neutral earthing transformer.

The parallel flow paths described above are arranged so that they have the same pressure drops.

The hydrogen detraining tank automatically discharges detrained gases by means of a timer-operated solenoid valve. If the accumulation of detrained gas exceeds a certain volume, an alarm operates.

The system described has the ability to maintain the pressure derived from the pumps to a value approximately 0.35 bar below casing gas pressure by non-regulatory means. Gas/water differential pressure indicators and alarms are provided. The unit will trip if water flow drops below 80% of rated flow.

High stator water temperature (typically 75°C) is indicated by an alarm. For the example quoted, a trip setting of 90°C would be used.

12.2 Commissioning objectives

The objectives when commissioning the stator winding water system are:

- To ensure that windings have been maintained in good condition during the storage and erection process

- To flush the system
- To carry out stator water flow distribution checks (bubble tests)
- To carry out final vacuum, pressure and IR tests
- To prove that the stator water system performs according to design, including all alarms, indicators, controls and interlocks
- To measure the total conductance of the generator connections for future reference purposes.

12.3 Detailed procedures

The Detailed Procedure Document for the stator water system should contain a complete inventory of its components. For the example given, the list would include:

Foundations: coolers, pumps, detraining vessel, etc.

Vessels: stator water coolers, stator water head tank and hydrogen detraining tank

Pumps: Stator water pumps

Strainers: Duplex strainer and head tank outlet

Demineralising unit

Pipework and supports

Valves

Cabling

Orifice plates

Instruments, including pipework, panels, wiring, etc.

Final erection checks should include a thorough check on the completion of the system and on the records of processes carried out, such as 'Cadwelding' and brazing; where access can be gained, such as head tank and filters, an inspection for cleanliness should be made. Records of air pressure, vacuum and IR tests should be inspected.

12.3.1 Flushing procedure

The water circuits use small bore tube and are highly convoluted. Since even the smallest obstruction can cause generator overheating, extra care is taken in the flushing.

It is similar in principle to flushing procedures previously described, but the process is now carried out in both forward and reverse directions. This reversal of flow flushes out particles which might otherwise be lodged in the system. As part of the procedure, an air bubble test (described later in this section) is used to ensure that the windings are free from obstruction.

The external parts of the stator water system are fitted with a by-pass and the connections to the windings are blanked off. Similarly, a by-pass would be fitted across the HV and neutral connections. The high points of the by-pass pipework should be fitted with vents.

All orifice plates are removed. Also, components which might impede flow, such as the differential water/gas pressure control valves and duplex filter elements, are removed. Temporary filters should be fitted to the suctions of the stator water pumps.

The stator water system supply line is disconnected and flushed to waste, after which it is reinstated and the demineralising resins put into the demineralising unit. The stator water system is then primed and flushed to waste at suitable points, such as pump suctions, discharges, vents and drains. The flush should be carried out progressively, starting from points near to the header tank, e.g., pump suctions.

Valves can be used to limit the system as flushing progresses. Disconnections can be made to provide temporary points where the system can be flushed to waste, thereby preventing debris being carried into enclosed items, such as vessels.

The stator water pump suction filters should then be inspected and the pumps commissioned in accordance with standard check sheets. The pumps can then be used for high velocity flushing of the stator water external system.

The flushing circuits must be planned such that no unclean pipework is introduced into the circuit upstream of a section already flushed. Flushing then proceeds circuit by circuit, using both pumps and ensuring that there is an adequate level in the head tank. Each section being flushed will be completely drained down periodically during the flush, until the section is declared clean.

When the entire external system has been flushed, components which can be inspected, such as filters, vessels and traps, should be opened and checked as free from debris.

The external system, including the by-pass pipework, is then circulated for a prolonged period with regular inspection of pump suction filters until the system is deemed to be free of debris. It is important to ensure that the external system is thoroughly flushed. A minimum period of four to five days should be allowed, but full advantage should be taken of further programme time if available. The attached system is then drained and refilled via the demineralising unit. A chemical test is carried out to determine the conductivity of the water entering the system, which should be of the order of 0.01 μS/m.

The system is then circulated continuously for 24 hours, after which samples are taken from at least two different points of the system. The conductivity should not have changed significantly from the original value.

The main by-pass pipework and other temporary connections are then removed and the system reconnected to the stator windings. The entire system is then primed with water with a conductivity of about 0.01 μS/m. Circulation through the windings is carried out in both forward and reverse directions, using permanent connections provided for this purpose. Each circulation period should be approximately twelve hours. During each circulation, the level of the head tank should be monitored and the vents operated to ensure full priming and flow. The strainers should also be inspected at regularly defined intervals, the observations being recorded.

When the system is deemed clean, it is then drained, refilled with water of the same conductivity and circulated for a further 24 hours. Samples are then taken as before, but there should now be no significant deterioration in conductivity from that of the make-up.

The flushing procedure is concluded with an air bubble test, which is performed by injecting small quantities of air into the common inlet to the stator water system with the water in circulation. The injected air causes a continuous stream of bubbles to pass through the manifolds, hoses and windings which can be observed at the PTFE hose connections. Every outlet hose connection from the windings should have a similar stream of bubbles, thereby indicating that the windings are unobstructed. The correct direction of flow should also be noted. It may be necessary to provide access to the inside of the generator in order to observe the PTFE hoses. For the example chosen, access would be gained by removing a hydrogen cooler element. Upon completion of the bubble test, the stator water system may be drained and air pressure, vacuum and IR tests carried out.

12.3.2 Final functional checks

The first step of the functional checks procedure is to ensure completion of electrical control circuits and alarm circuits, having made certain that the stator water system is complete, with blanks and temporary filters removed and permanent items, such as control valves and filter internals, refitted.

The stator water system will be primed and the seal-oil system then put in service. The casing is then pressurised with air to the normal operating hydrogen pressure.

The actual steps of the procedure will depend upon the system to be commissioned. For the system described, each stator water pump will be put into service in turn, and the winding inlet and outlet valves adjusted to give a specified value of stator water/gas pressure differential. The stator water pressure will be below the gas pressure. The stator water flow is measured by means of an orifice plate and manometer.

For each pump in service, pump discharge pressure, winding inlet and outlet pressures, and pump motor currents are measured. The duplex strainer differential pressure (in the clean state) should also be noted.

For systems with water/gas differential pressure controls, the gas pressure should be varied over a prescribed range and the control system adjusted and verified to function correctly. The range of gas pressures should include enhanced pressures, where specified, for generator overload conditions. The foregoing pressure readings will then be repeated for all possible combinations of stator water coolers out of service.

Automatic standby duty for both stator water pumps must also be proven, including associated alarms and indicators. During pump changeover, the foregoing pressures should be noted. Stator water low flow alarms may be initiated and the alarm should have previously been set correctly, using the orifice plate and manometer. Other alarms should also be checked, such as duplex filter differential pressure and stator head tank high and low levels.

The hydrogen detraining tank can be tested by injecting clean air at an appropriate point. The automatic venting feature should be inhibited and the gas in water alarm proven. With the automatic venting feature in service, its operation should be verified. Test buttons may be provided to open the vent and these should also be checked. Where automatic venting is by time switch, the correct time interval must be set and confirmed.

Ensure that all other alarms are proven, particularly for alarms with more than one initiating source.

12.4 Commissioning network and resources

Figure 5.62 illustrates the commissioning network for the stator water system.

Other than the requirement for a supply of demineralised water, there is no significant material resource. Some chemical analysis is required and the use of a chemist is advisable.

The network illustrated for the generator gas system (Fig 5.59) and the stator winding water system (Fig 5.62) will need to be integrated. This is illustrated in Fig 5.63, which also gives more detail of the generator air test, together with the commissioning requirements for the hydrogen coolers.

13 Automatic voltage regulator (AVR) and excitation systems

This section is concerned mainly with the AVR. Excitation power for the main generator is derived from the (brushless) excitation system, described in Section 9 of this chapter. The excitation output, rectified by silicon diodes, is fed to the main generator rotor.

An AVR is a highly complex system. This section deals with the approach to commissioning such a system but, of necessity, is based upon a specific example. The detailed commissioning procedures dealt with in Section 13.3 of this chapter are therefore a balance between a general approach to commissioning all electronic AVR systems and the specific requirements for the example chosen. As in any specialised activity, specialist advice should be sought by the customer to ensure that commissioning procedures are

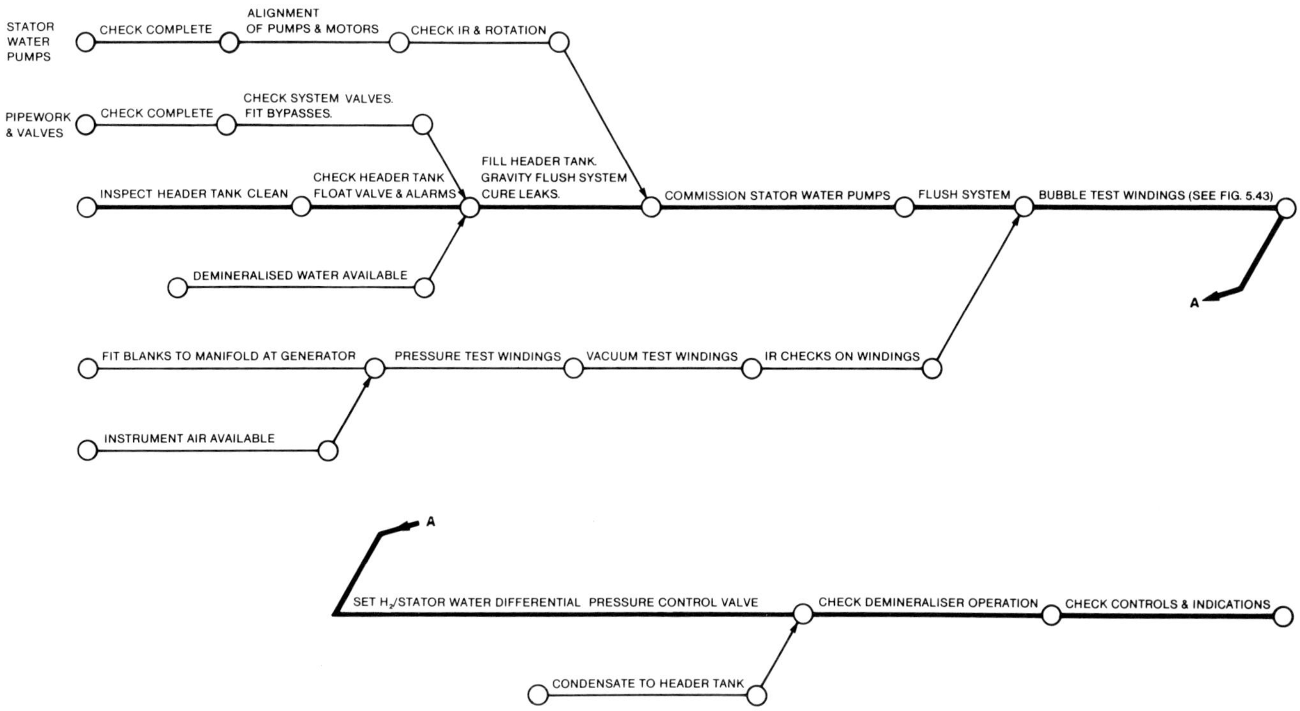

FIG. 5.62 Generator stator winding water system — commissioning network

thorough and well documented. Care must be taken to ensure that tests are designed to meet the needs of commissioning objectives and are controlled so that damage does not occur.

13.1 General description

A number of excitation control systems of different designs and responses have been and are still in use. Modern systems, often designed for use with high output turbine-driven generators, must be accurate, highly responsive and capable of establishing excitation automatically, in order to maintain constant generator voltage when changes occur due to load disturbances. A simplified schematic drawing of a typical system (Fig 5.64) will be used as an example for commissioning purposes.

This system is of dual-channel type comprising two identical channels, either of which is capable of controlling generator excitation independently. However, most systems are single channel, and trip from automatic to manual control in the event of a fault. In the dual-channel system, if both channels become unserviceable, control is automatically switched from automatic to manual operation. The three-phase output from the shaft-driven pilot exciter is fed into a thyristor power convertor, the DC output of which is supplied to the field of the main exciter at a level dependent upon excitation control signals. The converter is capable of

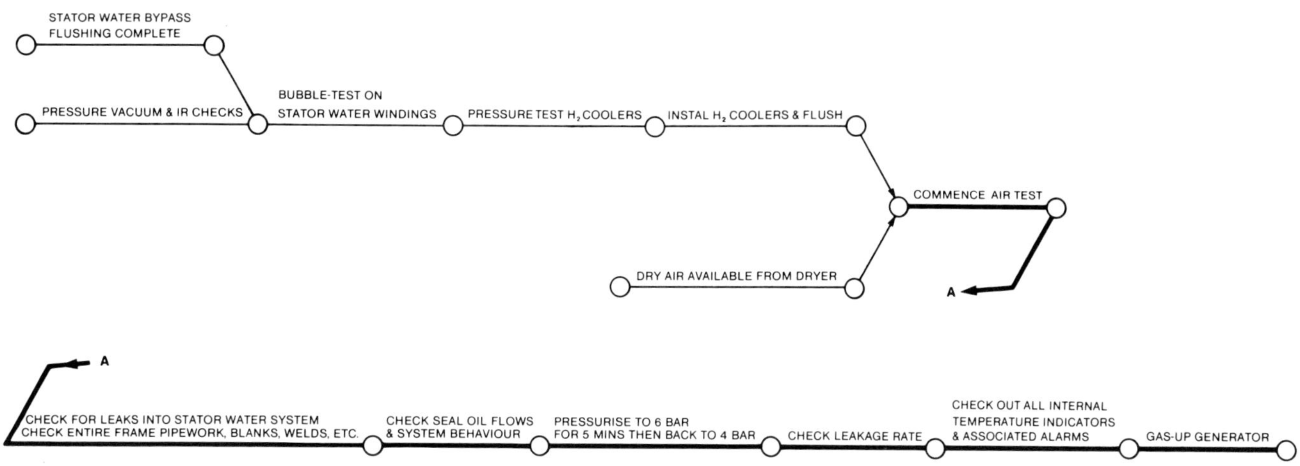

FIG. 5.63 Integrated network for generator gas and stator water systems

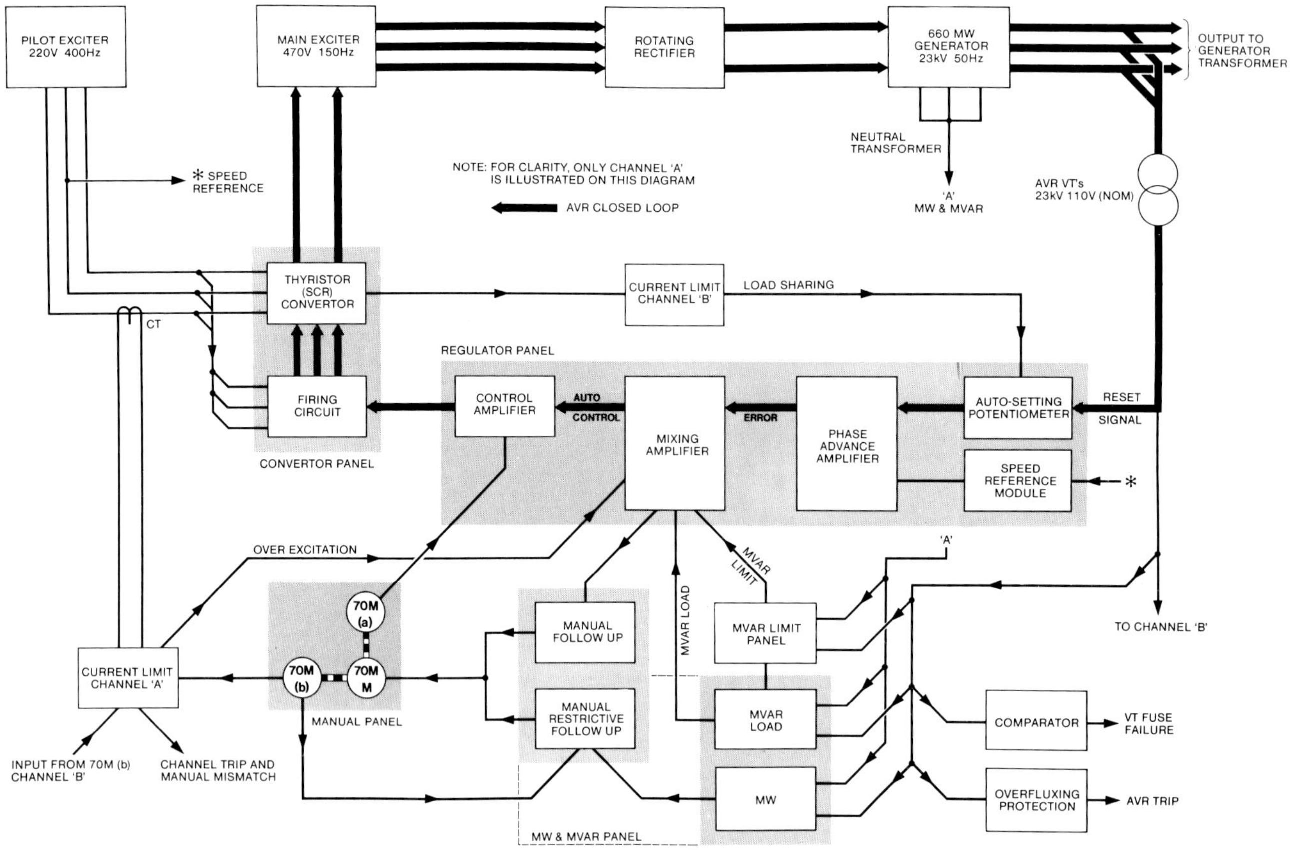

FIG. 5.64 Simplified schematic diagram of a brushless excitation system

inversion in order to suppress the exciter field current quickly. The main exciter output, a function of its field excitation, is rectified by the rotating rectifier diodes and fed to the generator rotor, thus determining main generator excitation. Additional control facilities are provided, by which the AVR controls current limits, under and over excitation limits, and parallel running compensation.

The telemetry equipment associated with the brushless design of excitation provides contactless measurement or indications, as follows:

Main generator field voltage

Main generator field current

Failure of rectifier bridge arm

Main generator field earth leakage

Average field winding temperature

The equipment is shown schematically in Fig 5.65.

When the AVR is in automatic control, each channel compares a proportion of the generator output voltage with a stable reference and, if the generator voltage changes, produces an error signal of the appropriate polarity to alter the existing input signal to the converter. Converter load sharing is controlled by the automatic adjustment of the output of one channel with respect to the other. Stability circuits are incorporated and limiting signals for excitation and converter current are generated, if the preset limits are reached.

Fault conditions also produce automatic responses. If the machine protection operates for any reason, both channels are tripped out and the exciter field is suppressed. If one channel develops a serious fault, that channel is switched out of service and the remaining channel continues to control in 'auto' mode.

If the remaining channel fails, it is switched to manual control. In the manual mode, the 'auto' control signal is switched out and the excitation controlled by adjustment of the standing input to the converter.

The remote 'Raise and Lower' switch is operative in both 'auto' and 'manual' modes.

The AVR maintains the generator output voltage within +0.5% of the set value from no-load to maximum continuous rating.

13.1.1 AVR system description

The AVR output stage is a fully controlled, full-wave, three-phase thyristor bridge which can supply an ample margin above the normal output current for quick response to a demand for increased excitation. The bridge also allows for a rapid reduction of exciter field current by inverted operation of the thyristors, i.e., by firing the thyristors in that part of the AC cycle so that energy stored in the field is fed back to the DC supply.

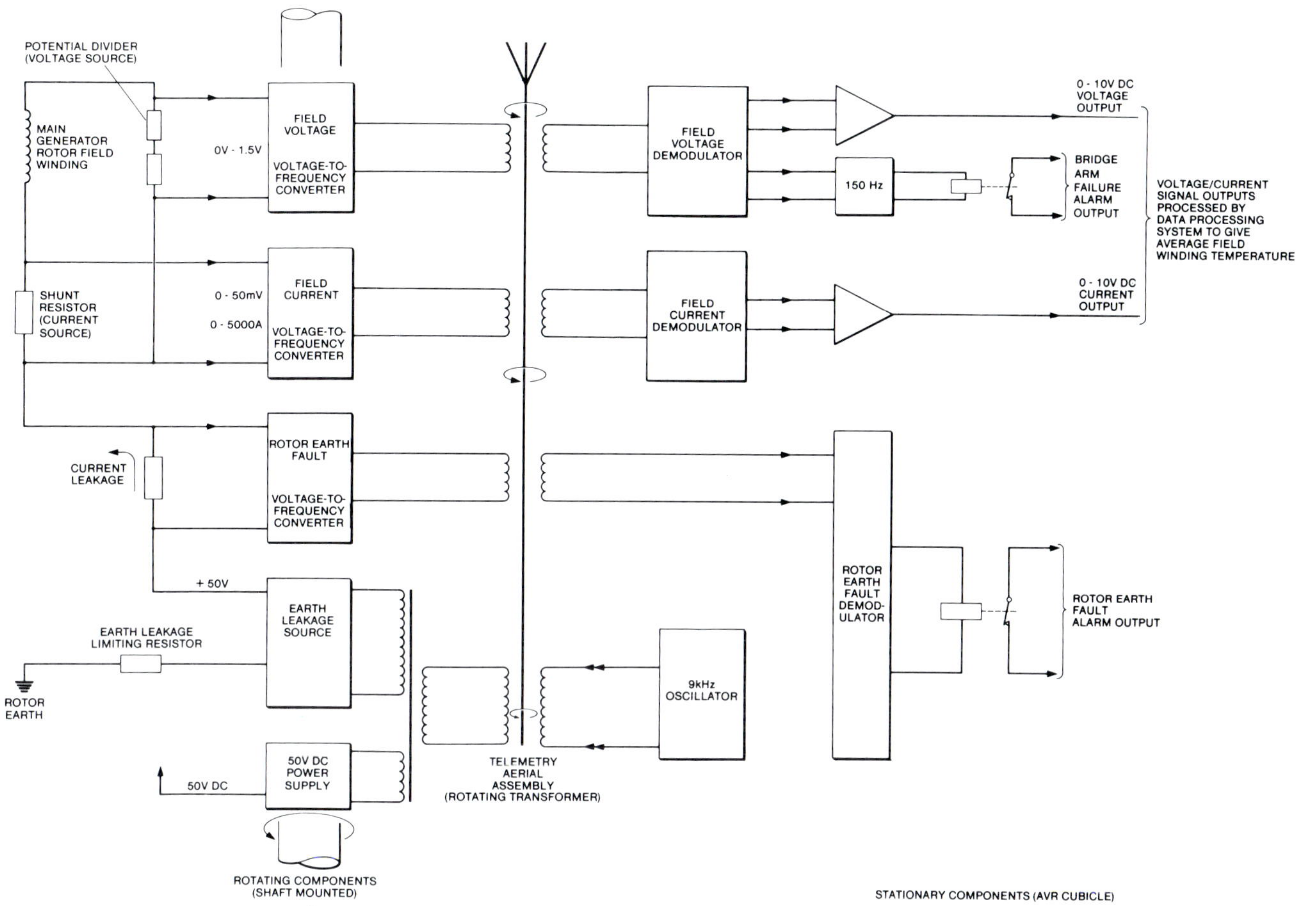

FIG. 5.65 Telemetry system for brushless excitation scheme

The system transmits information on voltage, temperature and current, as well as alarms. It comprises rotating elements which sense the required signals and transmits them to stationary elements via an aerial assembly.

This system is unique to one British supplier. Future developments for brushless excitation systems could use sliprings to transmit this type of information.

A control signal derived within the AVR equipment controls the timing of the thyristor gating pulses, relative to the anode AC supply, thus determining the output of the converter.

When the generator speed is below a specified value on run-up and run-down, or in the event of a trip, an inhibit signal suppresses the gating pulses, reducing the output to zero.

Manual control of a single channel

Manual control of a single channel of the AVR (Fig 5.66) is used only on commissioning, or in exceptional circumstances during unit operation. The manual circuit, comprising a motorised potentiometer, also performs a basic function in the 'auto' control system. The potential at the potentiometer slider provides the 'manual' control signal, which is fed via a control amplifier to the thyristor converter. Thus the excitation level is set by the position of the potentiometer slider, which can be controlled from the unit desk, or locally at the equipment cubicle.

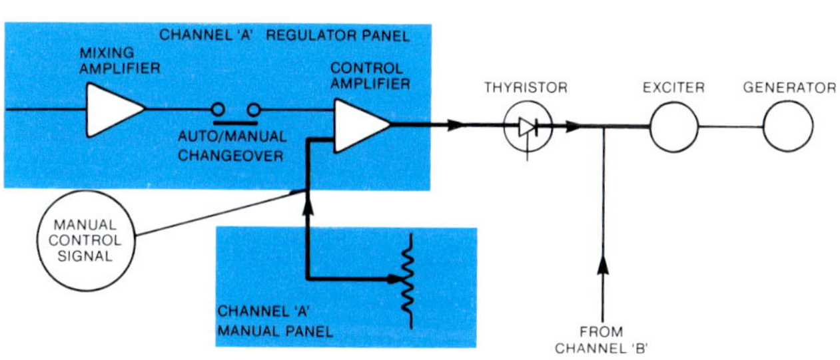

FIG. 5.66 AVR system — manual control block diagram

Overvoltage protection on open circuit causes the excitation control to switch off, thereby suppressing the exciter field.

Automatic control of a single channel

Automatic control of a single channel (Fig 5.67) is achieved by varying the input to the thyristor converter. The input is derived from the sum of the manual control signal plus an auto control signal, the latter being zero during steady state conditions. During a

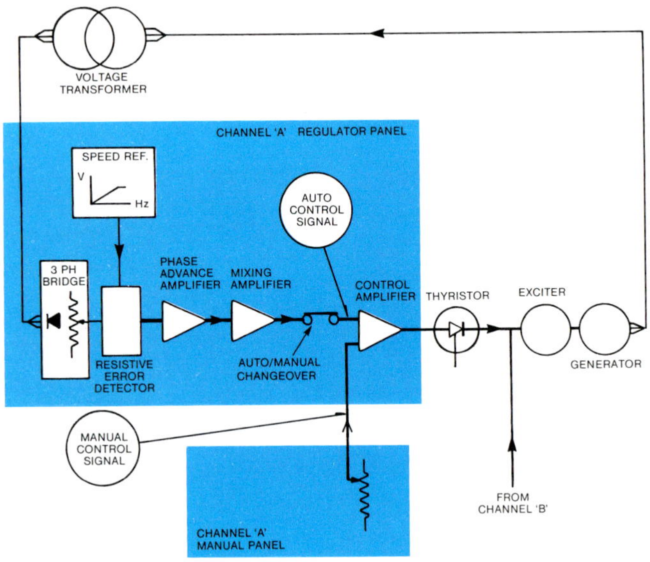

FIG. 5.67 AVR system — auto control block diagram

change in operating conditions, the auto control signal has a finite value of appropriate polarity and the sum is more or less than the value of the manual control signal. If new conditions are sustained, the manual control potentiometer is automatically moved to provide a new standing input, thereby reducing the auto control signal to zero again.

Manual restrictive circuit

The rapid response of the automatic control system enables the generator to be under-excited beyond the limit at which it can remain stable if switched to manual control. The manual control potentiometer travel is therefore restricted such that, in the event of a trip to manual control, the standing signal level from the potentiometer is at or above the safe limit.

Under-excitation limit

Two measuring circuits compare the phase and amplitude of the generator output current with the phase of the generator output voltage. One circuit produces a signal proportional to generator output reactive current and the other a signal proportional to output power (MW). These two signals are fed to an amplifier which produces an output signal which equates to a reset limit, which is related to the limit line of the generator capability diagram. The output signal acts to oppose further reduction of excitation beyond the limit line.

Overfluxing protection on manual and automatic control

To prevent the generator transformer overfluxing when the generator is running up to, or down from, rated speed, a sample of the generator output voltage is taken at the frequency of the output. If the voltage/frequency ratio tends towards an unsafe value, a predetermined threshold in the circuit is overcome and current is drawn from the manual control circuitry. This reduces the manual control signal, which in turn reduces the output of the thyristor converter.

Current limits

The output of a current transformer in the AC input supply lines to the thyristor converter applies a limiting signal to the thyristor converter, hence limiting current.

'Auto' control of dual channel

In normal operation, both channels are switched to 'auto' control and the exciter field is supplied by both thyristor converters in parallel. With the exception of a converter load-sharing device, both channels operate independently. Load sharing is achieved using the rectified output of the current transformers in the AC input to each of the converters. The outputs are compared and if they differ significantly, then one of the channels is reset to achieve sharing.

13.2 Commissioning objectives

The objectives when commissioning the AVR and the excitation system are:

- To verify that appropriate works tests were completed and their records are available
- To verify that appropriate erection checks have been completed and that their records are available
- To check the polarity of AVR current transformers
- To carry out static tests on AVR and excitation control equipment
- To obtain static characteristics of the equipment and preset the various control functions and limiting devices
- To check AVR and excitation system protective devices, including generator excitation suppression.

The objectives associated with unit operation are all dealt with in Chapter 6.

13.3 Detailed procedures

Most of the pre-steam-to-set activities can be carried out using temporary supplies and dummy electrical loads. For this reason, it is possible to carry out preliminary AVR tests before cabling has been completed and before erection of the associated generators and exciters is complete.

Care must be exercised to ensure that no work is permitted on the AVR equipment or cables connected with the equipment for the duration of the preliminary tests. Cables connected to terminals which are required for temporary supplies must be removed and insulated in a safe manner. The Detailed Procedures Document should state the exact requirements and responsibilities required to achieve this objective.

For the system described, the temporary supply and specialised equipment requirements are:

Motor generator (MG) frequency converter to provide a 220 V, 400 Hz, 120 A source

A three-phase, phase shifting transformer, rated at 200 VA in the range 415/100 V

A 0–300 V variable DC supply

A current transformer (CT) primary injection source, capable of providing 200 A

Ultraviolet (UV) recorder

Transistorised continuity/insulation tester

Three-phase variac transformer

Rectifiers for temporary supplies, as required

Low voltage DC injection source

Other equipment that will be required is:

Digital voltmeter
Oscillators
Oscilloscope
Diode tester
Multimeters

The 400 Hz supply is used to simulate the output from the heteropolar pilot exciter.

Initial commissioning activities will ensure that all components have been erected correctly in accordance with latest drawings and specifications. The activities must be designed to ensure that:

- Panels and modules are connected correctly
- Panel and module components, such as relays, transformers, potentiometers, rectifier bridges, relays, regulators, measuring modules, limiting modules, restrictive circuits, thyristors, etc., have been connected and are in good working order
- Insulation resistances are acceptable
- Continuity of connections is satisfactory
- Power supplies are within specified limits

Drawings can be used as a record of checks as they progress. Every contact must be proved to be in circuit, as shown on the drawings, by operation of the contact and confirmation that the required result is achieved. For parallel and series contacts, all combinations must be tried and shown to give the correct operation. All relays and contactor trip coils should be energised in turn and proven to operate at minimum specified voltage. Where normal supplies are supported by batteries, ensure that both battery operation alone and the 'normal' supply, as provided by a temporary source, will operate the equipment. This can be done by checking the operation of a selected number of relays and ensuring that the control circuits would function.

Power supplies must be checked to be within the specified range.

Insulation resistance of circuits should be checked by means of a multimeter, with fuses removed. Care must be taken to ensure that these checks do not cause damage to equipment. Where necessary, ensure that items likely to be damaged are disconnected for the purpose of the checks, e.g., printed circuit boards.

All fuses must be correctly rated, identified and proven to be healthy.

All links must be checked to be in place, and all control and protection relays plugged in.

Overfluxing relays and potentiometers will be set to initial prescribed values.

Remote features, such as raise/lower controls, will be proven to operate their appropriate relays. Similarly, main generator circuit-breaker auxiliary contacts associated with the AVR must be proven. This would normally be achieved by actual operation of the breaker under dead bar conditions. Machine protection circuits which connect with the AVR will similarly be proven.

Items having specific polarity, such as CTs, will each be checked using a standard check list.

A final careful check must then be made to ensure that jumpers, shorts, etc., used during these tests have been removed, that cores have been replaced and that the equipment is ready for functional testing.

It is advisable to ensure that standard check lists are produced for all similar items of equipment. Procedures relating to very specific items of plant such as the AVR regulator panel, speed reference modules, AVR follow-up modules, AVR manual restrictive modules, etc., may also be suited to standard check list format.

13.3.1 Final checks

AVR regulator panel

External cabling should be disconnected. Individual components within the panel such as speed reference boards and motorised potentiometers and relays will be checked in accordance with standard check lists. The test supply, in this case the three-phase, 50 Hz, 0–150 V Variac, will be connected and set to zero output. The voltage may then be raised slowly. As the voltage is increased, voltages at the secondary terminals of transformers can be checked. Also, rectifier DC levels and wave shape can be verified for prescribed input values. Rectifier ripple frequency should also be checked to be within specified limits. A graph should be drawn of rectifier input RMS voltage against the DC output voltage, from which smooth rectifier performance can be judged.

The characteristics of other components, such as mixing amplifiers and phase advance amplifiers, should also be checked in a similar manner. These amplifiers need a UV recorder to measure the response of step changes of input. Their response should remain stable over their range of travel.

Control amplifier gain should be plotted for various levels of input.

AVR speed reference

A similar approach to testing should be applied. The output will be checked to be proportional to speed and the point at which the module becomes saturated established.

Motorised potentiometers

A mechanical check for freedom of gear trains can be made by disconnecting the motor output drive. Associated microswitches operated by the gear train should be proven. The uncoupled motor can be energised and demonstrated to operate in both directions. It should then be coupled to the gear train and full operation checked. The current drawn by the motor is measured and compared with nameplate rating. The rating of each potentiometer must be checked. Position indicators should agree with wiper position. The cams can then be set up to specified limits. It is inadvisable to operate motor-driven potentiometers by hand, as the drive mechanism may be damaged.

AVR MVAR limiting panel

Individual components such as relays and MW measuring modules will be checked, using standard check sheets. Transformer tap settings must also be set to specified values, their IR values verified by means of multimeter and secondary terminations checked to be correct.

Panel characteristics will again be checked by connection of appropriate test supplies. Power supplies will be required for certain modules (e.g., ±24 V). For these tests, it is necessary to vary the phase of the current with respect to the test supply voltage. The curves produced should be consistent with the operating curve derived from the generator capability diagram.

AVR comparator relay

The example chosen for commissioning operates for frequencies between 40 Hz and 480 Hz. The choice of test supply frequencies should be determined in advance in agreement with the supplier. Fuses should be checked for correct rating. Where plug-in boards are used, associated supply fuses must be removed before a board is unplugged and only replaced after a board is plugged in, otherwise board components can be damaged. Tests should commence with fuses and associated boards removed.

The test supply is connected to the panel through a Variac, which is set at zero when the supply is switched on. Tests can then proceed sequentially and voltages raised and lowered to obtain various component characteristics.

The voltage level at which thyristor input fuses will fail when the thyristor is caused to fire when unloaded should be ascertained. The voltage can then be reduced by adjusting to the appropriate input limiting resistor. With a correctly rated new fuse in position and a dummy load connected across the thyristor output, the load can be reduced until the current limit operates. The value of output current when limited, measured by means of an ammeter in series with the load, should be recorded.

AVR thyristor converter cubicle

It is important that transformer taps on thyristor converters are selected to match the actual medium frequency test supply being used. Unless taps are correct, the required gain cannot be set and some components may be overloaded, or run at too low a voltage. Correct selection can be established by setting the test source to nominal machine speed and measuring the terminal volts at no-load. The transformer voltage taps are then set to the nearest suitable value, which will depend on the particular function of the transformer. For example, it may be sufficient to set the tap to the nearest value above or below the test value. For a power transformer, it might be prudent to choose the tap above the test value, since a primary tap set too low will cause excessive surplus power to be dissipated in the converter power supply.

The phase rotation of supplies to thyristor converters should be checked by using an oscilloscope: also, the pulse shape should agree with specification.

The converter should give a constant DC output voltage for various input voltages and frequencies. The converter firing angle should be shown to change over the specified range for the given input voltage change: this is established by measuring the firing angles, as seen by block pulses. Thyristors may be tested using a standard 'diode' test unit. Failed thyristors may have been damaged by the passage of large fault currents and show up as a short circuit, or near short circuit, at low test voltages (10–30 V). An ohmmeter, giving 1.5 V maximum (open circuit), is not reliable for short-circuit testing owing to forward and reverse non-linearity of both good and damaged thyristors when tested at these low voltage levels. Thus, a realistic test voltage must be applied. Most slightly damaged thyristors will show high leakage or breakdown at voltages of 100–500 V even if their actual rating is as high as 1200–1800 peak transient voltage (PTV).

The thyristors may be checked for forward and reverse leakage and gating sensitivity. A positive test voltage, no greater than the PTV rating of the thyristor, should be applied.

Gating sensitivity may be ascertained by applying a selection of gate currents in order to find the minimum current required to gate the thyristor fully, as indicated by full-scale deflection of the test unit. Partial or inadequate gating is indicated by a lower scale reading. Acceptable leakage and gating limits should be judged against data sheets supplied by the manufacturer.

With the converter switched on, all thyristors can be checked to be firing normally by observing the output wave form with the oscilloscope and establishing that all phases are similar and that ripple is within specified levels.

13.4 Commissioning network and resources

As stated previously, the AVR is a highly complex system and requires people with specialised knowledge and experience. Since the AVR is essentially a package, much of the commissioning work can be carried out without interference with other activities. A complex commissioning network is not needed, provided that detailed procedures are laid down in a logical and orderly manner. A simple network (Fig 5.68) indicates the relationship between erection and commissioning requirements and should be the subject of routine commissioning and erection planning meetings.

Specialised equipment will depend upon the particular installation and the equipment needed for the system described is listed in the previous subsection.

14 Condensate extraction and LP feedheating system

The list of plant items covered by this section is:

- Condenser extraction pumps
- Condensate pipework and associated coolers
- LP heaters and associated bled-steam pipework
- LP heater lift pumps
- De-aerator/storage vessel
- Condensate make-up/surplus system
- Vents and drain systems

For commissioning purposes, this system is normally taken from the condenser extraction pump suction valves to the de-aerator outlet isolating valves. The make-up system terminates at the valve immediately upstream of the make-up control valve, and the surplus system terminates at the isolating valve downstream of the surplus control valve (Fig 5.69).

14.1 General description

The condensate extraction and LP feedheating system are part of the overall feedwater system, which recovers condensed steam from the condenser hot wells and returns it to the boiler via the feed-heating system and the economiser. In the first stage of this process, the condensate is used to recover useful heat from a number of sources, such as the generator hydrogen and stator water systems, generator main exciter, gland steam system, etc. In the second stage, condensate is pumped from the condensate system into the LP feedheating system through an isolating valve. The condensate system is designed to operate on a closed cycle for unit start-up conditions and is equipped with make-up and surplus facilities. During start-up, the condensate quality may be outside prescribed limits. Closed circuit operation allows this to be monitored and controlled, using the make-up and surplus facilities. When the quality is under control, condensate may then be pumped forward to the LP feedheating system. Condensate may be pumped from the closed circuit system, either to reserve feedwater tanks, or to waste, depending on quality. The recirculation loop provides protection for generator coolers and gland steam condensers, by sustaining the removal of heat at low condensate flows.

In the particular system under consideration (Fig 5.69), condensate is extracted by one of two 100% condenser extraction pumps and pumped to the stator cooler, hydrogen cooler and exciter air coolers, through three parallel paths. Condensate then combines and flows to a parallel combination of boiler feed pump seal-water cooler and atmospheric gland steam condenser. A further parallel combination of three turbine moisture extraction condensers completes the condensate system.

Condensate provides a source of water to the condenser tubeplate sealing system, condenser structure cooling, main feed pump suction filter (secondary filter) flushing, sealing water head tank for sealing valve glands subject to vacuum, etc.

The condensate system is fitted with a priming tank fed directly from the condensate system, which allows the system to remain primed when condenser extraction pumps are shut down. Low level in the tank

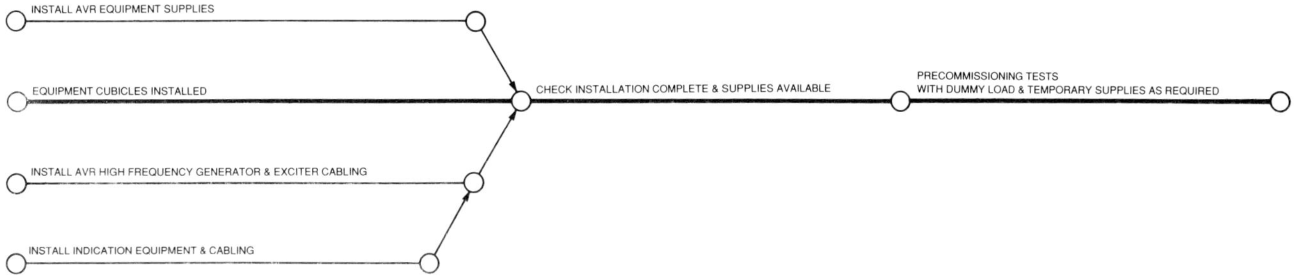

Fig. 5.68 Simple network relating key AVR erection and commissioning activities

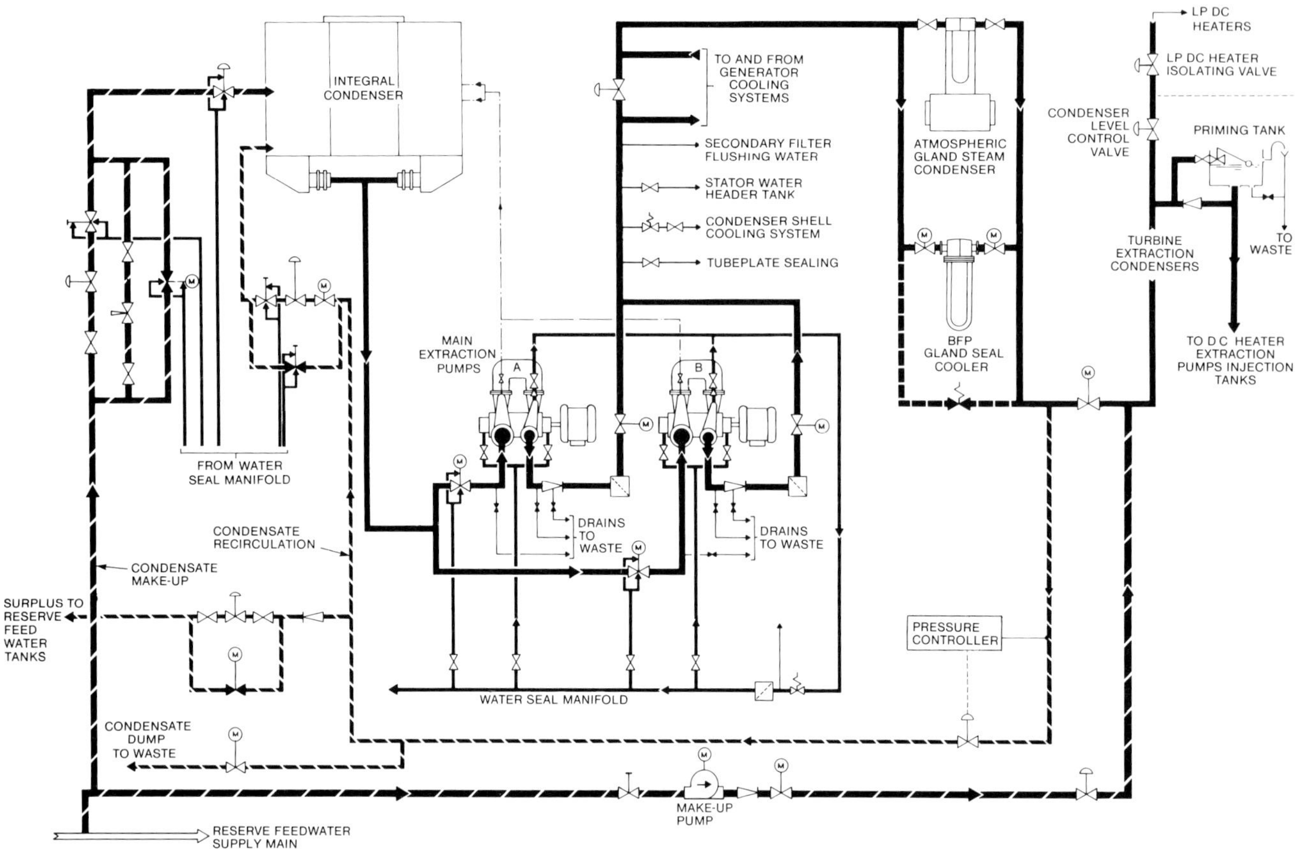

FIG. 5.69 Condensate extraction, recirculation, make-up and surplus system

This typical condensate extraction system extracts condensate from a condenser for delivery to the LP feedheating system. In this example, heat is recovered to the condensate system from a number of turbine auxiliaries, such as a boiler feed pump gland seal cooler, an atmospheric gland steam condenser and turbine moisture extraction condensers. The extraction pumps can also be used to recirculate water back to the condenser during start-up and low load operating conditions, when the condensate quantity is small. Also, when the condenser level is high, water can be pumped out of the system either to reserve feedwater tanks or, if contaminated, to waste. Make-up water can either be admitted direct into the condenser or pumped into the system at a point before the low pressure feedheating system.

initiates an alarm. In the event of the system losing prime, it is possible for hydraulic shock to occur when an extraction pump is put into service or reinitiated after a trip. Thus, vulnerable components, such as hydrogen coolers located within the generator body, are protected from damage, thereby protecting the generator from water ingress.

Figure 5.70 shows a distilled water (DW) system forming a parallel path with a selected number of hydrogen and exciter coolers. Under normal circumstances, these coolers form part of the condensate system parallel flow path. If, however, condensate temperature rises to a level such that ineffective cooling occurs, the DW system provides for extra cooling from the auxiliary cooling water (ACW) system. The DW pumps cut in on high condensate temperature, thereby providing a closed circuit cooling system owing to the presence of the non-return valve between the inlet manifold to the selected coolers and the other parallel-connected hydrogen and stator water coolers.

The LP feed system (Fig 5.71) receives feedwater from the condensate system via an isolating valve. Provision may also be made for the LP feed system to receive a proportion of make-up water, so that the unit can generate at partial load, whilst the condensate system is operating on closed circuit.

LP feedwater is then pumped to a high level LP heater, where feed temperature is raised by means of LP steam drawn from the turbine (Fig 5.72).

Feedwater then passes by gravity to one or more similar LP heaters arranged in a vertical cascade, termed an LP heater train. Some heater trains are arranged in two vertical cascades, feedwater being pumped from the bottom of the first cascade to the top of the second by means of intermediate lift pumps.

Two types of LP heater are in common use: the direct contact (DC) heater and the indirect, or tubed, heater.

DC heaters enable bled-steam to mix with the LP system feedwater without heat transfer losses. However, this needs careful design to ensure that feedwater cannot be carried over, or ingested into, the rotating parts of the turbine. LP DC heaters are therefore equipped with comprehensive protection

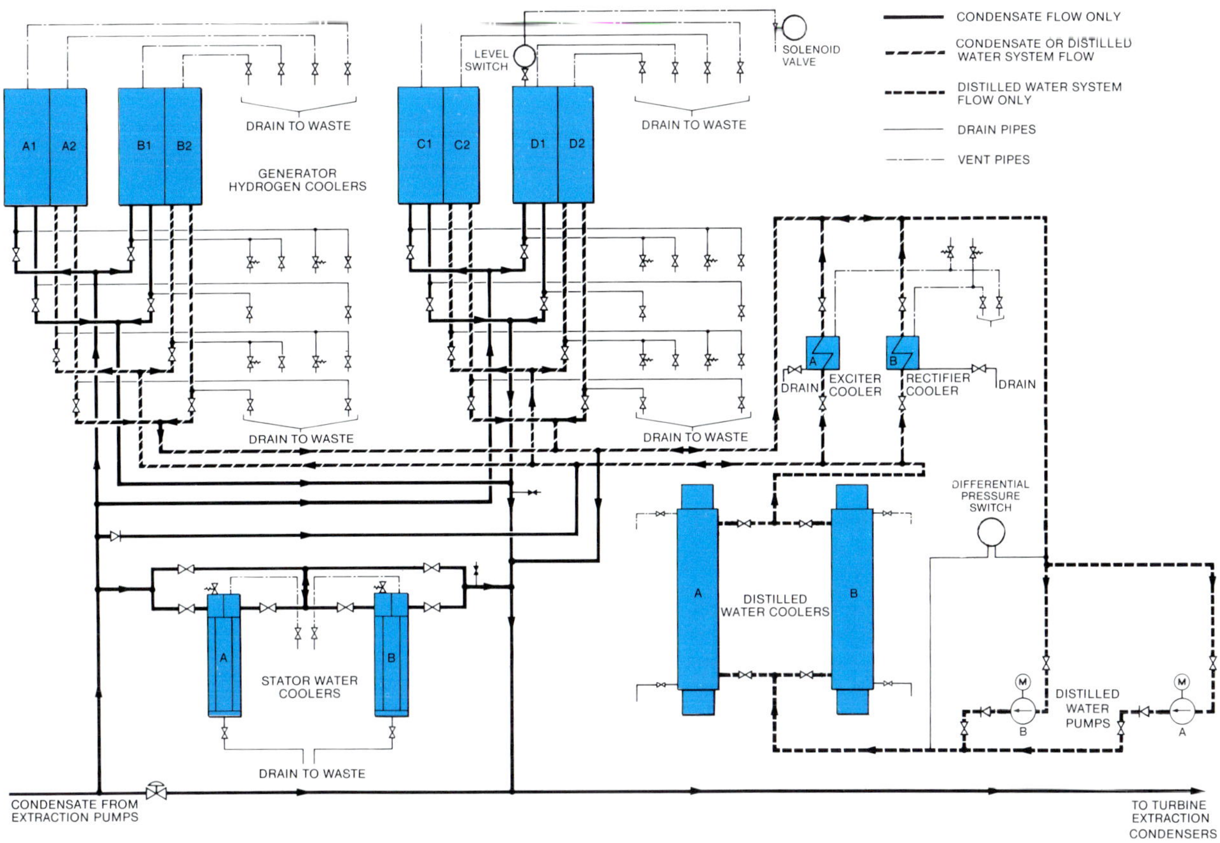

FIG. 5.70 Hydrogen and stator water cooling water system

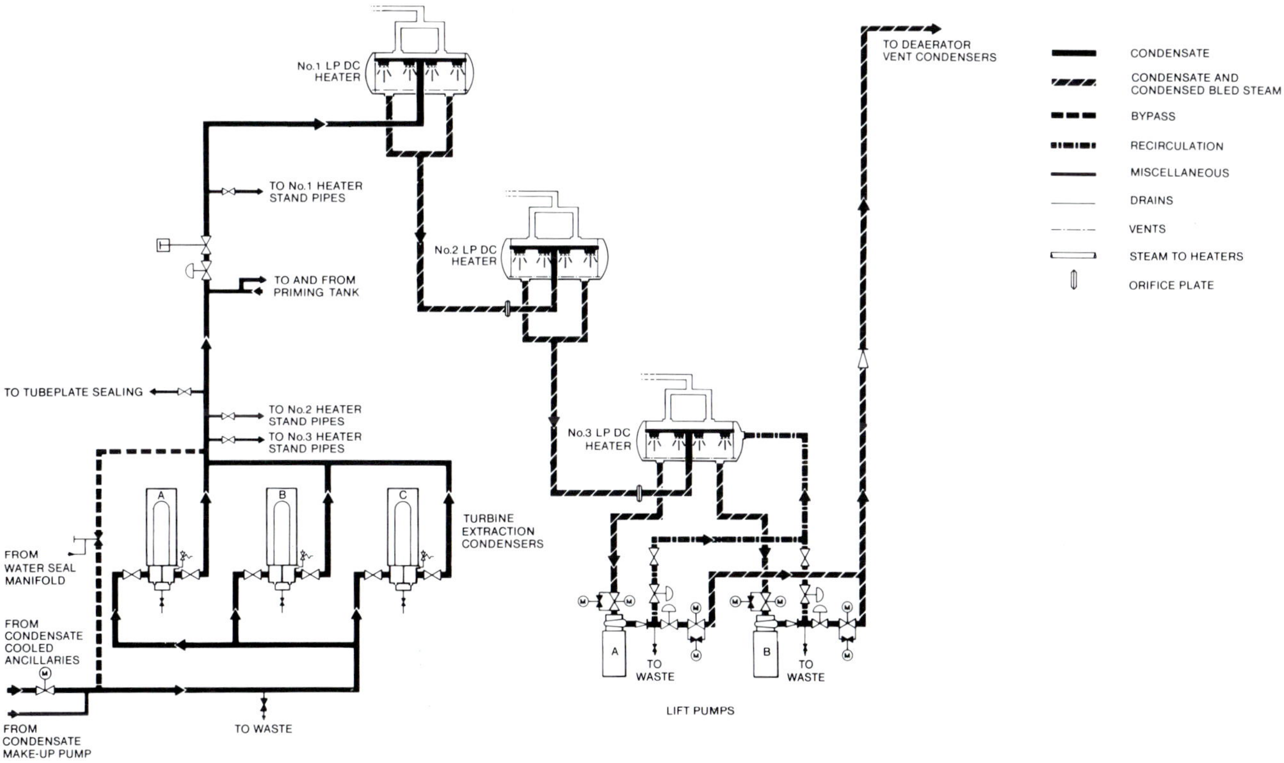

FIG. 5.71 LP feedheating system — feedwater

A low pressure feedheating feedwater system, comprising a vertical cascade of three LP direct contact heaters. Water is pumped from the lowest heater to the de-aerator by means of duplicate lift pumps.

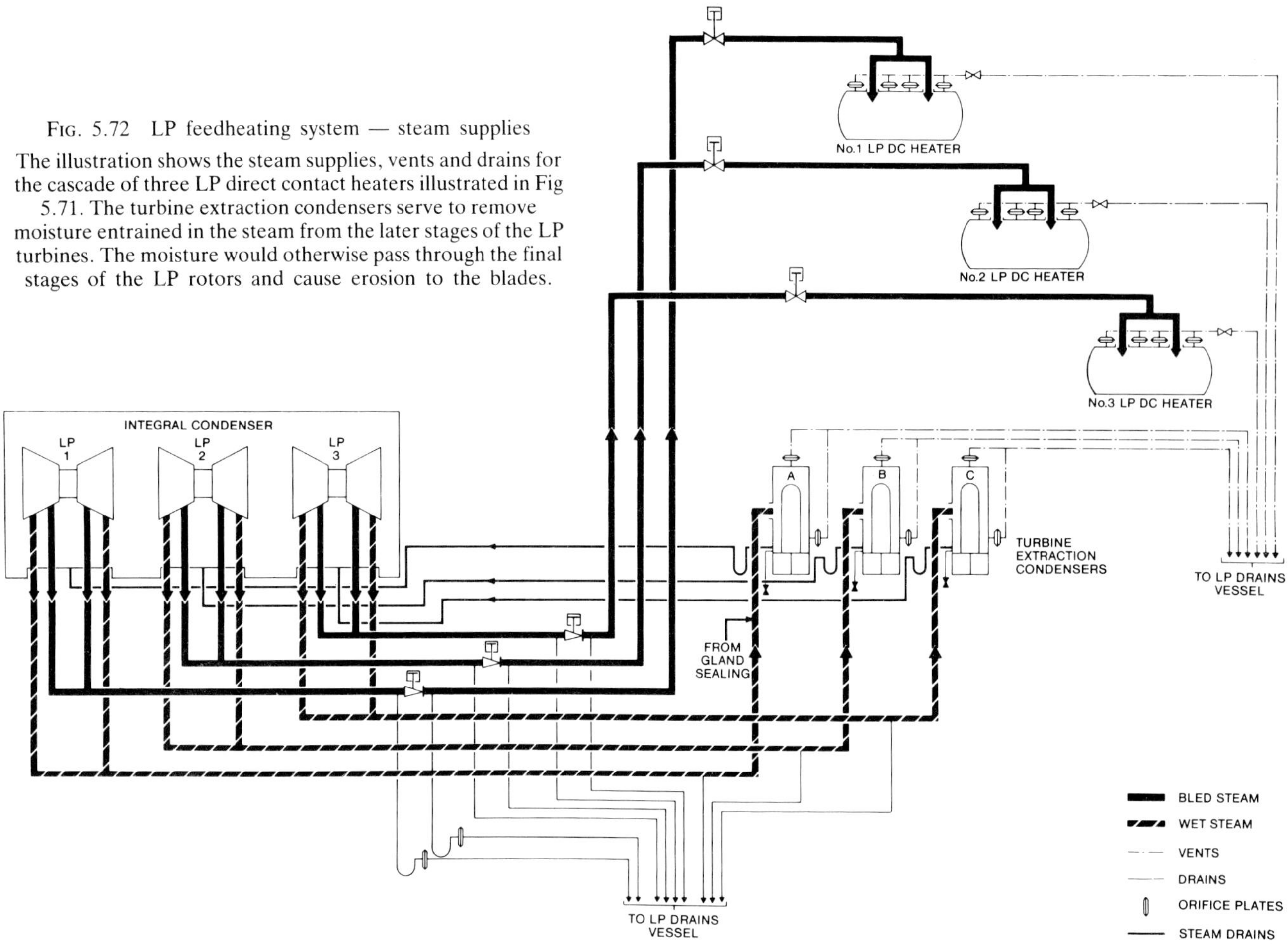

FIG. 5.72 LP feedheating system — steam supplies

The illustration shows the steam supplies, vents and drains for the cascade of three LP direct contact heaters illustrated in Fig 5.71. The turbine extraction condensers serve to remove moisture entrained in the steam from the later stages of the LP turbines. The moisture would otherwise pass through the final stages of the LP rotors and cause erosion to the blades.

systems which, for example, would cause lift pumps to trip, the inlet valve to the LP feed train to close, and valves in the bled-steam supplies to the LP heater bodies to close. Heater by-pass systems may be incorporated to enable the main unit to remain in service at reduced load. The LP heaters are physically segregated by means of seal water loops arranged between the heaters, thereby preventing steam flowing between them. Water-level control within the heater train is achieved by valves in the feed-flow to the train, and by recirculation control at lift pumps. The commissioning procedure for prototype LP DC heaters must be extremely comprehensive to ensure that the system dynamics are not such that seal water loops fail or become unstable, with consequent risk to the turbine, under all operating conditions, including routine testing. Current designs within the CEGB have been proven to be completely stable, although earlier designs required certain modifications.

Tubed LP heaters operate at less efficiency than DC heaters and provision has to be made to transport condensed bled-steam to lower pressure parts of the system.

LP feedwater systems terminate at a high level de-aeration/storage vessel. The vessel acts as a reservoir, supplying the high pressure feedwater system. Steam and water admission to the vessel is designed so that the LP feed is de-aerated, i.e., the gaseous products are removed and vented to the condenser for disposal to atmosphere via the condenser vacuum maintaining plant. The LP feed system shown in Figs 5.71 and 5.72 has DC heaters in a single cascade. The heater extraction pumps (Fig 5.73) are of the totally-immersed electrical rotor type, which are similar in principle to those used on assisted circulation boilers.

14.2 Commissioning objectives

The objectives when commissioning the condensate and LP feedheating systems are:

- To ensure that commissioning boundaries and activities are properly controlled and integrated with other commissioning activities
- To ensure that controls, indicators, alarms, interlocks and protective devices are proven, initially by means of dry functional checks
- To carry out internal inspections for cleanliness and condition of surfaces prior to the admission of water
- To flush the system in a controlled manner
- To commission condensate and LP feed system pumps in a sequential and controlled manner, whilst ensuring that plant safety aspects associated with immersed electrical-rotor lift pumps are taken into account

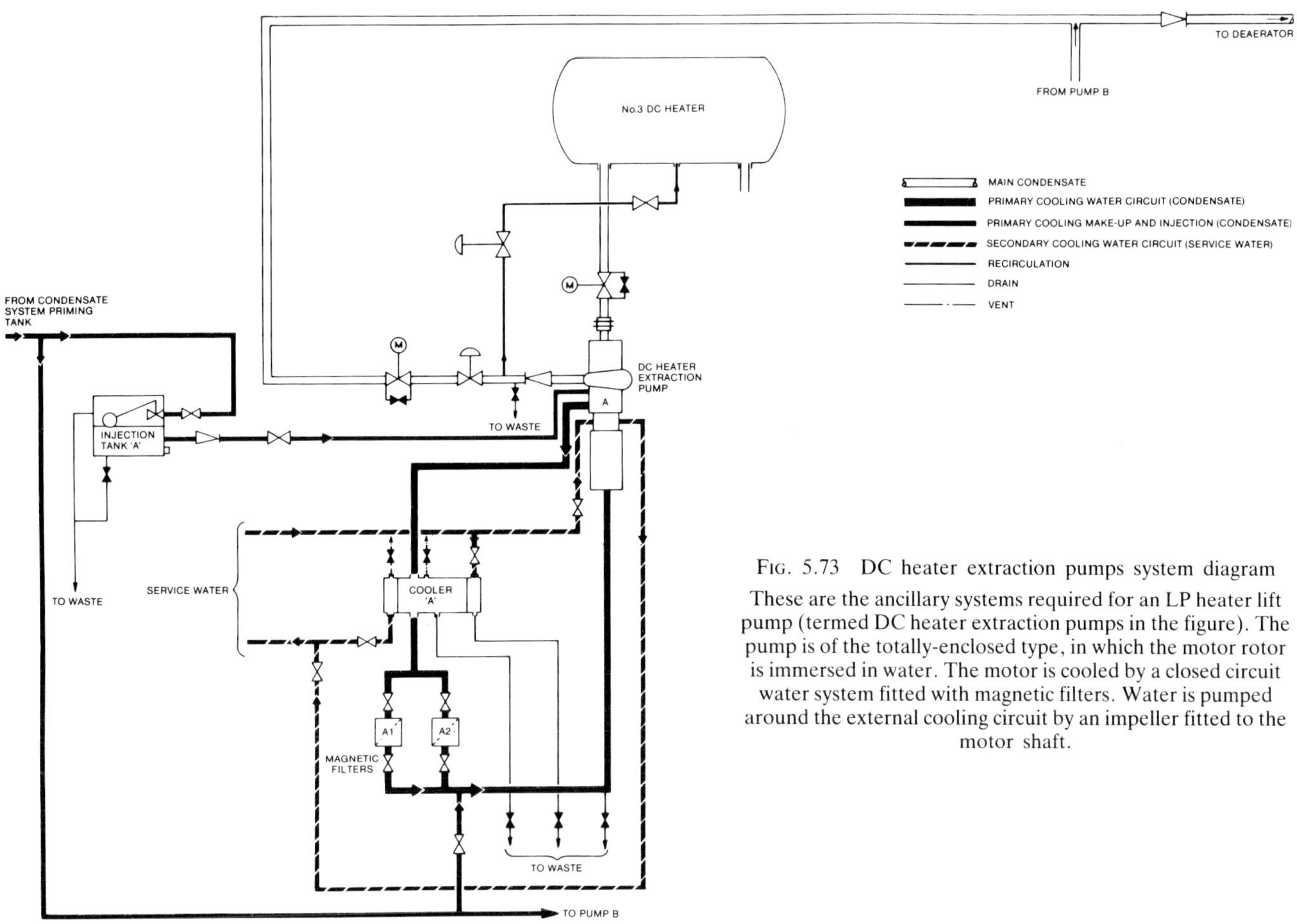

FIG. 5.73 DC heater extraction pumps system diagram

These are the ancillary systems required for an LP heater lift pump (termed DC heater extraction pumps in the figure). The pump is of the totally-enclosed type, in which the motor rotor is immersed in water. The motor is cooled by a closed circuit water system fitted with magnetic filters. Water is pumped around the external cooling circuit by an impeller fitted to the motor shaft.

- To carry out wet functional checks.

Other tests will be carried out after steam-to-set and are covered in Chapter 6. Such tests are:

- Stability tests, where appropriate
- On-load testing of protection systems
- Performance tests.

14.3 Detailed procedures

Commissioning commences with a complete inspection of the condensate and LP feed system to ensure that it has been installed correctly to drawings and specification, including the presence of items such as orifice plates, filter elements and dowels.

For the system described, the Detailed Procedures Document should contain a complete inventory of the following items:

Foundations: LP heater, other vessel support steelwork, plinths for motors and other components

Vessels: LP heaters, turbine moisture extraction condensers
Condensate priming tank
Hydrogen, stator water, and exciter coolers
De-aerator and associated vessels

Pumps: condensate extraction pumps and LP heater lift pumps

Filters: e.g., condenser extraction pump discharge filters

Pipework and supports, including water, bled-steam, vents and drains

Instruments, including pipework and electrical connections

Valves

Cabinets and cubicles, including chemical analysis panels, instrument and control cubicles

Cabling

Access ladders, platforms and handrails

Orifice plates

Standard check lists should be used, where applicable, for all items listed in the inventory.

It is important to ensure that interconnections with other systems are defined in the Detailed Procedures Document and controlled in an appropriate manner by the use of valves or blanks.

Preservation records should be inspected. The LP heater lift pump, since it is of the totally-immersed electrical rotor type, will have been transported and stored full of water, usually in a vertical position. The rotor should be turned slightly more than one full revolution, say 1.25 turns, every three months. The motor unit cooling water level must be topped up, as required, with demineralised water. Insulation resistance should also be checked and recorded every three months.

Records of weld checks will be maintained, as indicated in previous sections, together with other information which might form part of the customer's history and maintenance system. The de-aerator is the largest vessel in the LP feed system and final inspection will include checks for the removal of temporary stiffeners and also that all wedges, clamps, etc., are removed. Where temporary items have been welded to the parent metal, the surfaces must be made good, ground flush and the area checked by means of crack detection techniques. When erection of the de-aerator is complete, including associated vent vessels and integral pipework, and the internals inspected and cleaned out, a hydraulic test will be done as part of the erection activity. Open-ended connections are blanked and the vessel filled and vented free of air before raising its pressure to the specified test value. Pressure is applied gradually and the vessel checked for leaks. For the example described, a pressure test at about 11 bar sustained for 30 minutes would be applicable. The test pressure must not be exceeded and a suitable preset relief valve and test pressure gauge should be fitted. A minimum water temperature may also be quoted and this will be typically 7°C. After the test, the de-aerator should be fully drained and dried out, using warm dry air.

Preservation conditions should then be maintained, as required, the vessel being treated with a suitable conditioning agent, such as a vapour phase inhibitor (VPI), and adequately sealed. Internal inspection routines should then be established, say every eight weeks, and details logged. Where welded blanks have been used, these will subsequently be removed. Surfaces will again be ground flush and checked free from laminations and cracks.

Prior to actual commissioning activities, final inspection for cleanliness and complete installation of the deaerator internals should take place, as well as for all relevant external fittings.

The various instrument and electrical supplies should be available but isolated from the devices to be commissioned.

14.3.1 Flushing procedure — condensate system

The first part of the condensate system to be flushed will include the make-up system, condenser extraction pumps and those parts of the condensate system which do not form an integral part of the condensate circuit, such as the dump and surplus connections. All orifice plates should be removed. Initially, the condenser steam spaces will have been checked clean and boxed up ready to receive water. The condensate make-up system must be available and flushed to waste at an appropriate point. The connection to the make-up pump will be isolated at this stage. The condenser can then be filled to normal working level.

Each condensate extraction pump is then commissioned in accordance with standard check lists against closed discharge valves. The rest of the circuit to be flushed should then be checked to ensure that various supplies to items, such as the tubeplate sealing system, condenser shell cooling, etc., are isolated. Coolers should be by-passed and the isolating valve to the LP feed system closed. The system may then be primed whilst the condenser level is maintained at working level by means of the make-up connection, but with instrument connections remaining isolated.

An extraction pump is then put into service and the condensate system operated on recirculation back to the condenser. Water is rejected from the system, using the dump connection, at the same time maintaining a supply of make-up to the condenser until the system is clean. The second extraction pump should be put into service during this operation. Water cleanliness is achieved when its conductivity is below 500 μS/m and its iron content below 0.1 PPM. With the system shut down, all the associated coolers are then carefully primed and the system rechecked to be fully primed.

With the coolers initially isolated, an extraction pump is again put in service and each cooler flushed through in turn. When the system is deemed clean, the various instrument and small bore services can be disconnected at a suitable point downstream of an isolating valve, and each connection flushed to waste. During, and at the completion of, the flushing procedure, various strainers in the system must be opened and inspected.

The system described incorporates a closed circuit cooling system, the distilled water (DW) system, which provides extra cooling to selected hydrogen and exciter coolers. This system is flushed initially by disconnecting the DW pump suction and discharge pipework. This part of the flushing operation can be incorporated with the condensate flushing programme and the extraction pump is used to flush the DW pipework to waste with the DW coolers isolated. The DW pump discharge connection can then be flushed by putting each DW cooler into circuit in turn.

The DW system is then segregated from the rest of the condensate system by isolating the inlet valves to the gland steam condenser and the boiler feed pump gland-seal water cooler (Figs 5.69 and 5.70). The spring-loaded by-pass valve across these coolers will also be blanked. The DW pump suction and discharge connections are then replaced and a temporary filter with differential pressure indicator fitted to the suctions of the DW pumps. The system is then primed using the condenser extraction pump. Each DW pump can then be commissioned in accordance with standard check lists. The DW system should be run on closed circuit with one DW cooler in service for a prolonged period of, say, eight hours, with all coolers primed but isolated. The system is then drained to waste, the temporary strainers inspected and replaced, and the system reprimed for further circulation until it is clean. The hydrogen and exciter coolers forming the DW system are put into circuit in turn as part of the flushing procedure and the process repeated until the entire circuit is proven to be clean. The coolers should be included in subsequent draining and filling operations.

14.3.2 Flushing procedure — LP feed system

The internals of the LP heaters and de-aerator will have been thoroughly inspected and cleaned by hand. The de-aerator can be filled separately and drained to waste, this operation being repeated, as necessary, until it is considered free of debris.

The LP feed train is flushed by using the condenser extraction pump to achieve as high a velocity of water as possible. The LP heater lift pumps and motors will be removed from their casings and the casing used as a drain to waste. The heater seal drains will be left open during the flushing operation and checked free. The LP feed system make-up pump should be commissioned as part of the flushing operation and can, if required, be used for flushing instead of the extraction pump. The pipework to the make-up pump should initially be flushed to waste and then reconnected. The make-up pump is then commissioned in accordance with a standard check list.

Before flushing takes place, the inlet isolating valve to the high level LP heater will be isolated to ensure that water does not pass to the LP heater train. Coolers, such as the turbine moisture extraction condenser, and tanks, such as the condensate head tank, can then be filled and flushed to waste. The isolating valve is then opened and the LP train flushed to waste. When the LP heater train has been flushed, the system should be reinstated. Before the lift pumps are fitted, water supplies to pumps and motor cooling circuits must be thoroughly flushed to waste.

This procedure must be carried out very carefully, since particulate matter within the water systems of the pump can cause damage. The pump (Fig 5.73) is fitted with coolers to prevent heat transfer from the pump to the motor. These 'motor neck' coolers have a service water supply. The supply and return water services must be flushed to waste by disconnecting the pipework, the cooler being flushed in both directions during this process. Similarly, the service water supplies to the motor heat exchanger will be flushed. Flushing of the primary cooling water circuit of the motor and the motor heat exchanger requires great care. This circuit will include a filter with magnetic elements and these must be removed. Water connections to the lift pump motor will be disconnected and blanked. A temporary flushing source of high quality water is connected to the heat exchanger, which will then be connected in series to the filter. The system is then flushed to waste, until clean to the standards previously specified. If temporary pipework is used, it must be removed under clean conditions, ensuring that openings are not left exposed. Permanent connections which are removed and not included in the flush must be hand cleaned and inspected before reinstatement. Filter elements must be checked clean before refitting.

The motor and pump priming pipework must also be flushed to waste in a similar manner, thereby ensuring that every part of the motor water circuit and associated priming system has been flushed.

The bled-steam lines to de-aerator and LP heaters should be thoroughly inspected, wherever possible, as there will be insufficient steam flow available from the turbine until high loads are achieved. The inspection can form part of the final inspection of heater and de-aerator internals after the flushing procedure.

The cold vacuum test, which forms part of the condenser commissioning programme described in Section 4 of this chapter, will be phased in with the condensate and LP feed system commissioning programme.

14.3.3 Final functional checks

Final functional checks should be carried out in two distinct phases:

- Dry checks, with the condensate and LP feed system available for operation
- Wet checks, with the condensate and LP feed system in service.

Dry checks

The purpose of the dry checks is to ensure that all the control, interlock, protection, alarm and indicator functions are set and operating correctly, including the operation and control devices and valves. There is a need to have a thorough understanding of the system and its control design philosophy. The checks should prove that the control system functions as designed, including all possible failure modes, such as loss of electrical and air power supplies, loss of electrical or air control signals.

The condensate and LP feed system should be set, as specified in relevant operating instructions for start-up conditions. All services must be available, including those required for instrumentation and control. The condensate extraction pump-motor switchgear will be selected to 'test' or 'on jumpers', such that main auxiliary contacts operate without the motor being switched on. The condensate start-up controls will also be set in the 'dump' mode, i.e., the operating mode in which condensate is rejected from the condensate system.

The condensate priming tank low level alarm and associated interlocks are then proven, as follows:

- Check that condensate priming tank low level alarm is working initially, and then clears when the tank is filled.
- Check hydrogen cooler loss-of-prime alarm by partially draining the coolers.
- With both hydrogen cooler loss-of-prime alarm and priming tank low alarm working, prove that neither condenser extraction pump will start when initiated to do so, in both 'automatic' and 'manual' start modes. Repeat the operation, with only one alarm working in turn.

The next step in the final functional test procedure is to carry out appropriate dry interlock tests on controls, such as the LP heater and de-aerator level switches. These switches will operate to close certain valves, e.g., bled-steam isolating valves and condensate isolating valves. Also, level control devices will act to cause recirculation of water to heater lift pump suctions and operation of condensate make-up and surplus valves. The lift pump must be prevented from starting during these tests, but the initiation and tripping of its contactor will be proven.

LP DC heaters and de-aerator will be equipped with duplicate test columns to enable the level switches to be tested. These test facilities are incorporated into the dry interlock tests to check the correct function of each column in turn. The tests should be constructed so that normal operation will occur at correct levels but, for abnormal levels, pump and valve trips will be initiated and pump starts inhibited, in accordance with the design philosophy.

It is particularly important that there is sufficient water in the lowest LP heater, such that the lift pump is able to start safely and thereafter maintain either a forward pumping or a recirculation flow. The dry functional checks will therefore ensure that the pump cannot start unless there is adequate water level above the pump suction, that recirculation takes place at the appropriate heater level, and that the pump stops when the de-aerator level reaches a given high level and cannot be restarted unless the de-aerator level falls to a lower prescribed value.

The condensate/LP feed isolation valve is interlocked, not only with the LP feed train protection, but also with the condenser recirculation valve and the condenser extraction pumps. These features must also be proven during the dry tests.

All valves having automatic control and override features must be checked in a similar manner. The detail of these checks will need careful consideration to prove that valves operate correctly in accordance with their control signals and that valve failure modes on, say, loss of electrical supplies, power air, and pneumatic or electrical signals are safe for all failure combinations. The interaction of certain valves with each other, and also with certain pumps, must also be proven for all modes of operation, interlock and protection.

There will be specified values of operating signals relating, for example, to condenser water level or condensate system pressure and valves will be checked to operate to these values. Other valves will be set to operate through their range to specified opening and closing times.

It is particularly important that valves associated with protection systems are proven to operate from all signal sources. For example, the condensate/LP feed isolating valve will be caused to shut for high level in any one of the LP heaters, or in the de-aerator. It is also caused to shut on turbine trip, or if tripped from the control desk. Equally, this valve is permitted to open when heater levels are satisfactory, when the LP feed train by-pass valve is shut and when turbine-trip valves are reset, as detected by hydraulic fluid pressure at the turbine-trip valves.

Wet checks

On completion of the dry checks, wet functional checks can proceed. During this process, outstanding commissioning details for plant used during the flushing operation will be completed, e.g., remote control and indicators associated with the condenser extraction pumps. The LP heater lift pumps will not have been commissioned at this stage. The pumps will therefore be primed in accordance with operating instructions. Water must be present to the appropriate working level in the correct LP heater. The condensate system is therefore put in service on recirculation and water admitted to the LP feed train until the level in the lowest LP heater is adequate.

Alternatively, the LP feed system make-up pump can be used. The lift pump discharge valves must be shut and operating supplies isolated. The lift pump motor switchgear must be set to local operation and communication established between a person at the pump and another at the motor switchgear. The motor may then be started and, if motor current levels and all other aspects are normal, the pump is left in service on closed circuit operation. A number of readings, such as differential pressure across the motor cooler, motor temperature and pump discharge pressure, are taken until steady conditions prevail. Confirmation of correct rotation is established by noting that pump discharge pressure is greater than 60% rated discharge head. The other lift pump is then commissioned in the same way.

When the condensate and LP feed systems are ready for operational checking, a circulation route from condenser to de-aerator is used. The extraction pumps will be fully proven in turn, including automatic start of the pump selected to stand by. Readings of extraction pump discharge-strainer differential pressure, motor current, etc., will be taken. The response of the pressure-sustaining system during start-up, shutdown and extraction pump changeover will be noted. Also, the feed system priming tank will be observed to maintain an adequate level. The inhibition of extraction pump start for various combinations of low level in the priming tank and hydrogen coolers will be proven. Similarly, the pumps will be proven to start when these levels are normal.

An alarm 'double extraction pump failure' is often provided for the circumstances when both pumps are out of service, and this may be proven at this stage. The make-up pump should be checked in a similar manner.

With forward flow established to the LP feed train and with the LP heater lift pump in service pumping to the de-aerator, the lift pumps should be tripped simultaneously with the condensate/LP feed isolating valve. This can be initiated by simulation of, say, the de-aerator high level trip. The accumulation of water in the bottom of the vertical cascade of heaters is then checked to ensure that the lower heater is of sufficient capacity to accommodate the contents of the other two heaters. A check should also be made to see that water has not spilled over into the heater bled-steam lines. The time taken from trip initiation to the point where water reaches the high level in the low level heater should be noted.

Normal mode

A low forward flow condition is set up. The de-aerator level can then be varied over its control range to prove operation of the surplus and make-up valves. Rates of response and signal values will be noted and compared with those achieved for the dry checks.

The condenser, LP heater and de-aerator levels should also be varied over their control ranges to prove features such as the automatic starting and stopping of LP heater lift pumps.

All other aspects of control valve operation, system protection and interlocks as set up during the dry functional checks will be proven.

Where appropriate, surge tests can be carried out with cold water, using various combinations of pumps and vessel water levels so that the stability of the system is verified.

Dump mode

In the dump mode, the condensate/LP feed system isolating valve will be closed, the condensate system being on closed circuit recirculation, with make-up to the LP feed system. The make-up control to the LP feed system, using the 30% make-up pump, should be proven by varying the de-aerator over its control range and observing the pump discharge control valve. Tests similar to those described for the normal mode are carried out.

When the normal and dump modes of operation have been proven, all the on-load test facilities must be operated in turn from each of the duplicate devices

Wet functional testing is concluded by carrying out a full functional trip test for the entire condensate and LP heater system, the tests being repeated from every initiating source.

14.4 Commissioning network and resources

Figure 5.74 illustrates a commissioning network for the condensate and LP feed system described.

The closed circuit cooling system, the DW system, which features in the system described, has a separate commissioning network (Fig 5.75). In practice, it can be integrated with Fig 5.74.

Most systems are provided with off-load de-aerator heating facilities, which may use electric or steam heating. The test does not include a description or details of commissioning procedures. In practice, off-load heating systems may be commissioned separately, with final functional checks being incorporated into the LP feed system programme at a convenient point, when water is available in the de-aerator.

15 HP feed system

The list of plant covered by this section is:

- Feed pump suction filters
- Main and standby feed pumps, including leak-off system
- High pressure heaters and associated bled-steam pipework
- High pressure feed pipework

The extent of this system would normally be from the de-aerator outlet valves to the economiser inlet valves. Drain and vent systems interconnect with the lower pressure systems and the precise commissioning interface should be determined in advance.

15.1 General description

High pressure feed-heating systems generally comprise electric motor-driven feed pumps, as well as a steam turbine-driven main feed pump, together with a number of high pressure heaters. The feed pumps derive their suctions from a high level de-aerator/storage vessel via fine mesh suction strainers. Typical duplicate suction strainers, which may serve all pumps, are shown in Fig 5.76. The filter medium is a rotating microwire element made of stainless steel wire, wound screw-fashion around a drum.

Backflushing is achieved by means of a stationary synthetic-rubber-faced scraper, which is spring-loaded against the element. The scraper is supplied with flushing water diverted from the filter water via a valve which opens when the element drive motor starts. This

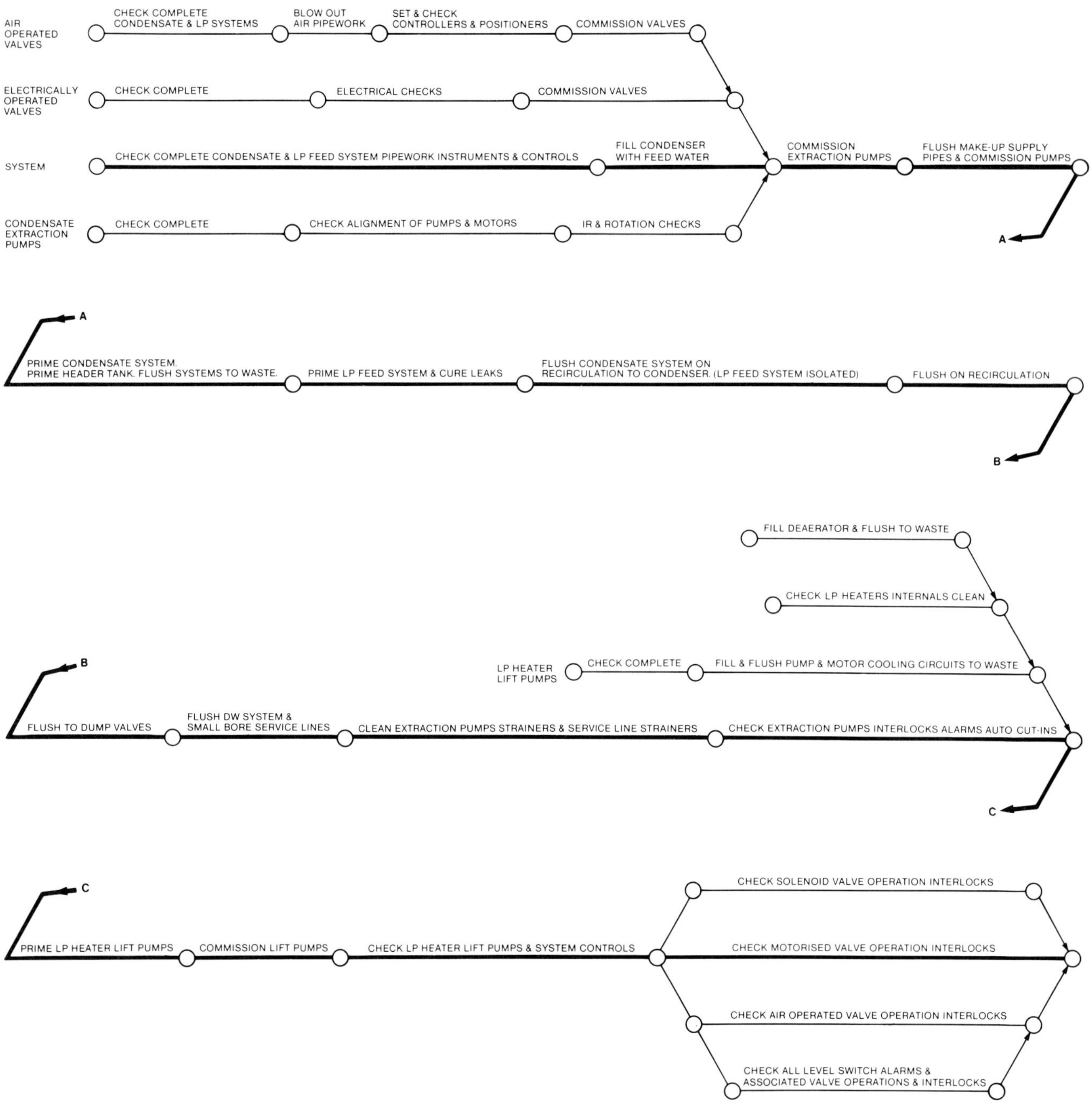

FIG. 5.74 Condensate and LP feed system commissioning network

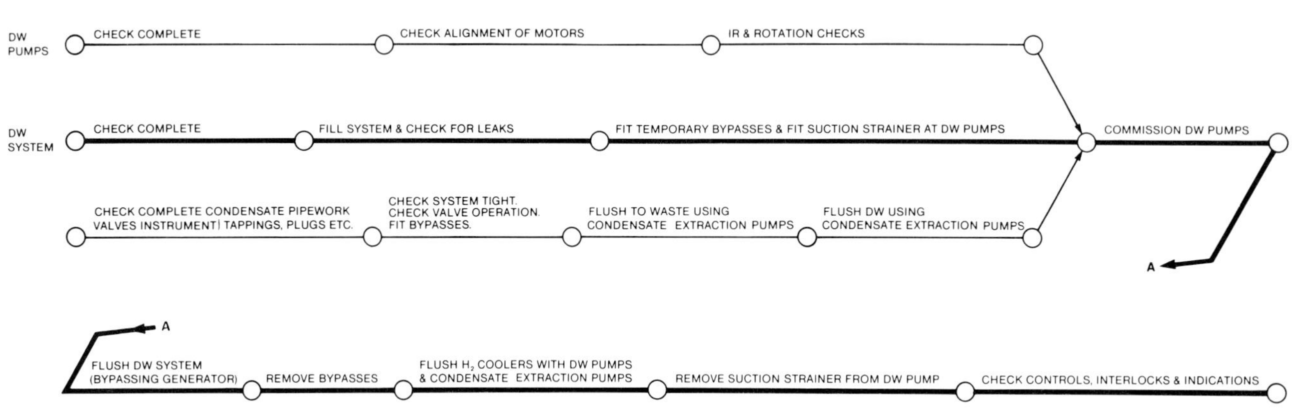

FIG. 5.75 Distilled water system commissioning network

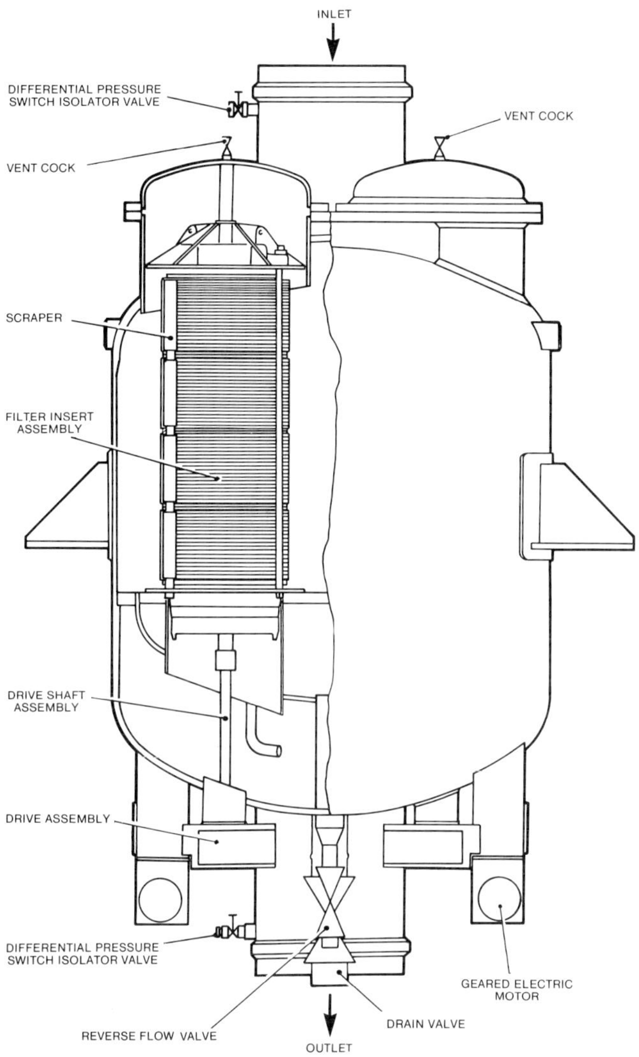

FIG. 5.76 Typical feed pump suction strainer

This feed pump suction strainer comprises a filter assembly which is rotated by a motor against a fixed scraper. The filter element consists of a stainless steel wire, spirally wound to give a very small clearance between each turn.

is initiated by high differential pressure across the filter. The system is provided with a secondary filter, the purpose of which is to recover flushing water.

The motor-driven feed pumps in the example chosen for commissioning purposes comprise a 50% duty and a 15% duty pump (Figs 5.77 and 5.78). The purpose of the 15% pump is to enable the steam-driven feed pump to be run up to minimum speed in parallel with the main turbine before the steam feed pump can take over pumping duty.

Other systems have two 50% duty electric feed pumps, which can completely substitute for the main steam-driven pump. The speed control systems for the 50% and 15% pumps differ, in that the larger one is controlled by means of a variable rotor resistor and the smaller one by means of a variable speed hydraulic coupling between a fixed speed motor and the high pressure pump.

All the feed pumps consist of tandem pumping units, one being the low pressure or 'booster' stage which delivers into a high pressure stage. The design of both booster and high pressure pumps is similar for all pumps.

For the example illustrated, all the high pressure stages have external thrust bearings, which absorb residual thrusts acting on the rotating elements. Some pump designs incorporate internal thrust bearings, using feedwater pressure from the high pressure stage discharge to counteract residual thrust. Thus, on loss of water to the pump, the internal thrust bearing can fail. For the design illustrated, the external oil-lubricated thrust bearing enables the pumps to run dry. However, it is not recommended that this feature be proven as part of the commissioning activities or at any other time, since it is intended to avoid short-term damage.

A common feature of modern feed pumps, particularly the cartridge-type high pressure stages, is that they can be removed and replaced quickly. Thus the alignment and datums for both pumps and foundations must be to a very high degree of accuracy and fixing arrangements standard between pumps of the same rating.

The main feed-pump used as an example is driven by a steam turbine (Fig 5.79). The turbine is a single-cylinder machine, directly coupled at one end via a flexible coupling to the high pressure pump. The booster pump is connected to a reduction gearbox, which is in turn flexibly coupled with the other end of the turbine. The turbine is designed to operate on bled-steam from the main turbine cold reheat line or 'live' boiler steam. The live steam supply enables the pump turbine to be operated at times when the main turbine load is between 0% and 30%. It can also be used, if necessary, over the whole operating range, but results in an overall reduction in unit thermal performance.

A small quantity of live steam is admitted to the turbine at all times in order to keep the live steam valve chests, pipes and turbine stages hot and available for instant use.

The turbine also provides a source of bled-steam to one pair of high pressure heaters.

Steam exhausts from the turbine at a pressure corresponding to the transfer pressure between the main turbine IP and LP cylinders. The bulk of the steam goes to the de-aerator and the remainder to the LP turbines. When the main turbine is not in operation, the exhaust route to the IP/LP crossover connection is isolated, so that steam can only pass to the de-aerator. However, if the flow should exceed the demands of the de-aerator the surplus is diverted to the main condenser flash vessel.

The example chosen has electronic governing. Other types have the more conventional mechanical governing systems. In both, as in the main turbines, on-load valve test facilities are provided. Steam feed pump turbines cannot normally overspeed owing to the hydraulic load of the feed pump. For electronic governing, the overspeed protection is checked by lowering the overspeed signal setting into the normal speed

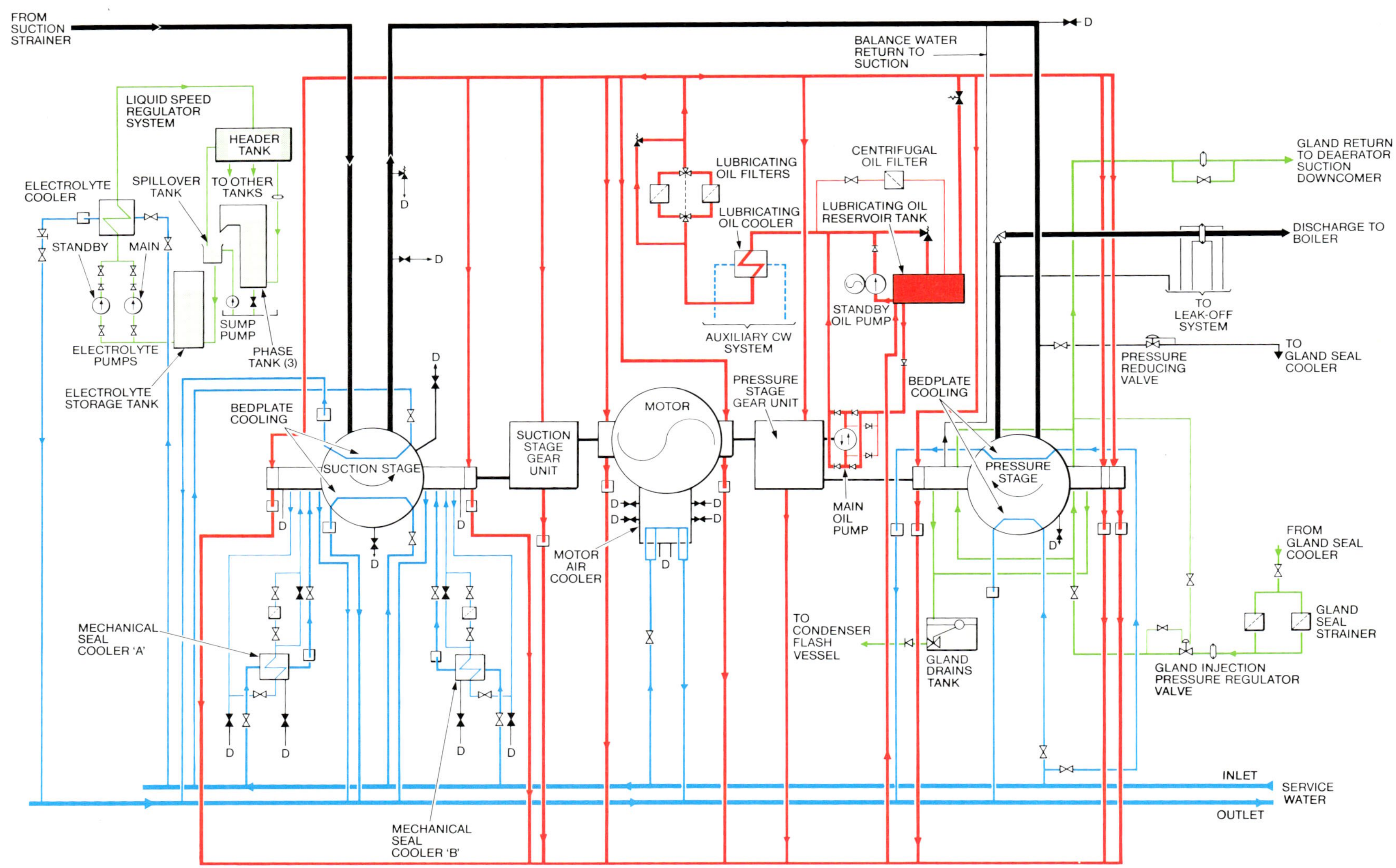

KEY
CONDENSATE
LUBRICATING OIL
SERVICE WATER
ELECTROLYTE
MECHANICAL SEAL SUPPLY
GLAND SEAL SUPPLY
ACW

FIG. 5.77 50% feed pump

The illustration shows a 50% duty motor-driven feed pump. The motor is connected at one end to the suction stage pump via a reduction gearbox and at the other end to the pressure stage pump via a step-up gearbox. Motor speed control is achieved by means of a liquid regulator, which can vary the resistance of the electrical rotor.

The HP pump glands are sealed by abstracting water from the pressure stage discharge and applying the water to the glands, via a regulating valve and strainers.

The suction stage pump glands are sealed by service water, which also provides bedplate and liquid regulator electrolyte cooling. Bearings for suction and pressure stage pumps and the motor are lubricated by means of a self-contained lubricating oil system, comprising a motor-driven oil pump and a pump shaft-driven oil pump (which is capable of pumping in both normal and reverse rotation). The oil system also supplies the gearboxes.

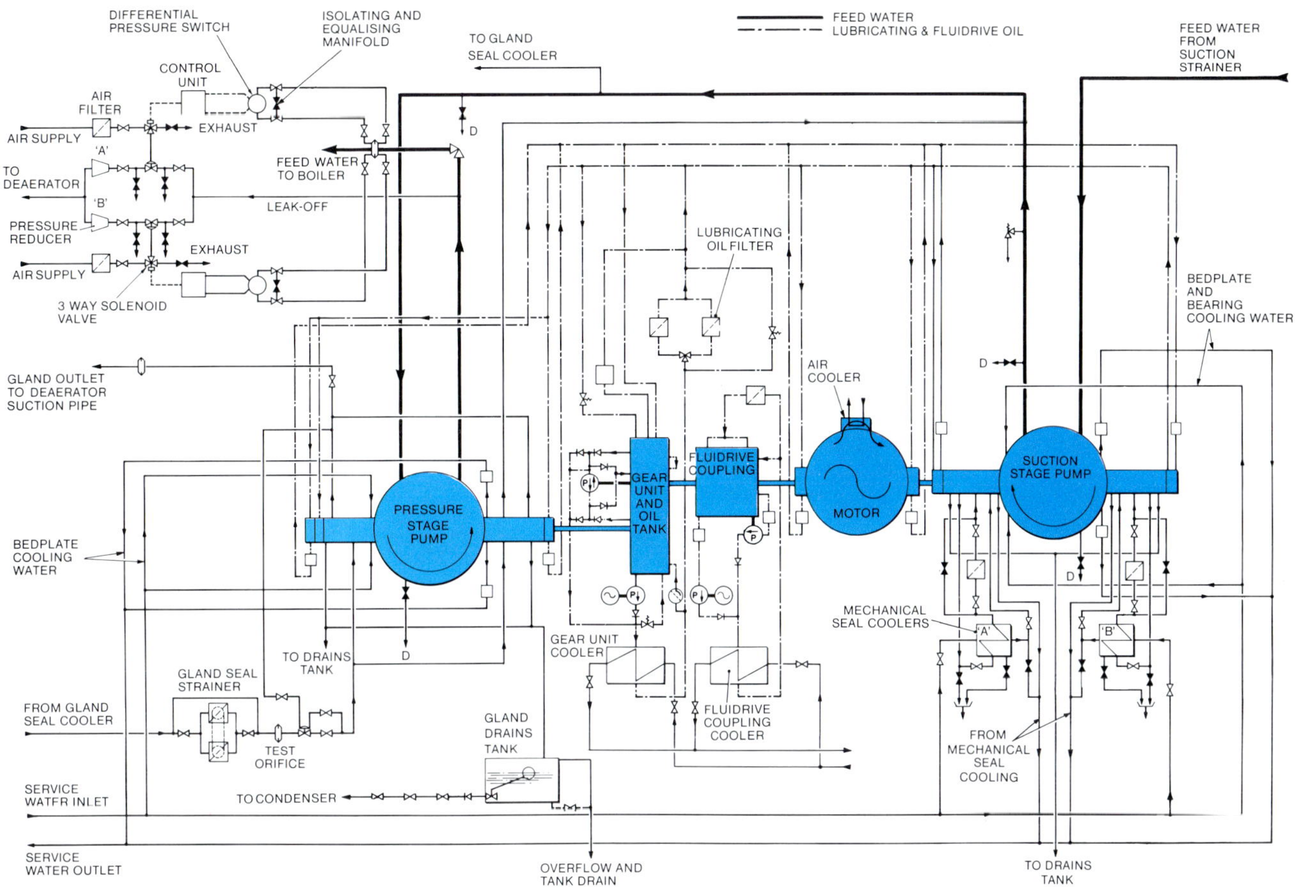

FIG. 5.78 15% feed pump

The 15% duty feed pump motor is directly coupled at one end to the suction stage pump, and at the other end to the pressure stage pump via a fluid coupling and step-up gearbox. Pump speed control is applied only to the pressure stage pump by means of the variable-speed fluid coupling. The ancillary services are similar to those for the 50% feed pump (Fig 5.77).

range. Actual overspeed for all governing systems can only be proven functionally by disconnecting the turbine from its pumps.

The feed regulating valves are not included in this section, as they are part of the boiler commissioning activity. However, the availability of the feed valves and the boiler must be considered in relation to feed system commissioning.

A typical HP feedheating system is illustrated in Figure 5.80. Most modern high pressure heaters are mounted vertically with feed entry and exit at the bottom. Alternatively, HP heaters can be installed horizontally. They are usually arranged in parallel banks of two or three heaters each. The example used shows a system comprising two heaters in each bank. The higher pressure pair draws steam from the main turbine, whilst the lower pressure pair draws steam from the steam feed pump turbine. Drainage is arranged such that each lower pressure heater derives steam from the drains of the adjacent higher pressure heater.

HP heaters are protected from high water level in their steam spaces by means of inlet and outlet isolating valves. When the valves close, they either open an integral by-pass line or divert the flow through a separate by-pass line fitted with spring-loaded non-return valves. The isolating valves are often motor operated, but another type is also in service, using the high pressure feed-water as the source of motive power. The latter has a more complicated control system, incorporating heater bank priming and depressurising facilities, as illustrated in Fig 5.81.

Heater drainage for each bank is separate from, and controlled independently of, the other bank. It consists of starting drains and main drains. The starting drains discharge through motorised valves to a low pressure drains vessel, and hence to the condenser. The drains open when the heater body pressure is less than that equivalent to about 10% load. The starting drains also ensure that heater bodies are drained when shut down, thus preventing waterline tube-shell corrosion.

Main drainage from the lower pressure heater is normally by heater body pressure to the de-aerator. However, for pressures below about 30% load, the drains are diverted to the low pressure drains vessel. This can also occur at high unit load, when the main boiler feed pump is out of service.

For the higher pressure heaters, drainage is normally cascaded to the lower pressure heaters. Drains can also be diverted to the de-aerator by a signal which indicates that the lower pressure heater body pressure is below 30% load. The higher pressure heater also diverts

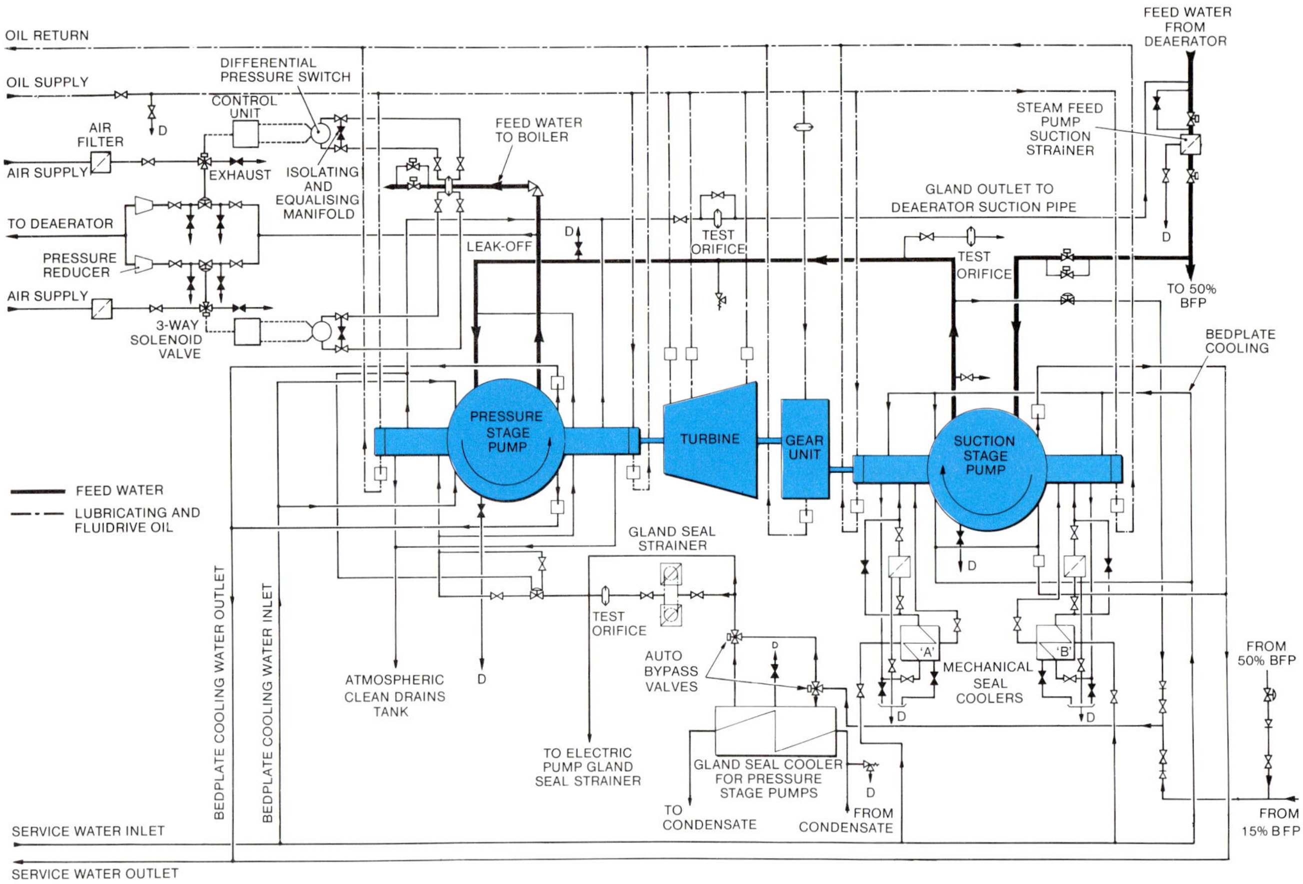

FIG. 5.79 Main boiler feed pump

The main 100% duty boiler feed pump is driven by a steam turbine. The suction stage pump is directly coupled at one end to the steam turbine and the high pressure stage is coupled to the other end of the turbine by means of a gear box. Ancillary services are similar to those for the 50% feed pump (Fig 5.77), but the lubricating oil system is supplied from the main turbine.

drains direct to the LP drains vessel when below 30% load.

Bled-steam non-return valves are similar to those for the LP feed system and are assisted to close in the event of a unit trip. The lower pressure heater non-return valves are also assisted to close when the steam feed pump turbine trips.

15.2 Commissioning objectives

The objectives when commissioning the HP feed system are:

- To verify that designed interchangeability of certain components has been taken into account
- To ensure that commissioning boundaries are properly controlled and integrated with other commissioning activities
- To ensure that controls, indicators, alarms, interlocks and protective devices are proven, initially by means of dry functional checks
- To carry out internal inspection for cleanliness and condition of surfaces prior to the admission of water
- To flush the systems in a controlled manner
- To commission the components of the HP feed system in a sequential and controlled manner, whilst ensuring plant safety aspects are taken into account
- To carry out wet functional checks.

Other tests will be carried out during the early commissioning and operational activities of the boiler, and also after steam-to-set; such tests are:

- Stability tests, where appropriate
- On-load testing of protective systems
- Performance tests, including verification of feed-pump characteristics and automatic feed control.

Steam-driven feed pumps

Additional commissioning objectives are:

- To operate the feed pump turbine uncoupled in order to verify governor control and characteristics, including overspeed and other protection features.
- To operate the coupled turbine/pump unit and to verify that operational parameters are within specification over the entire operating range.

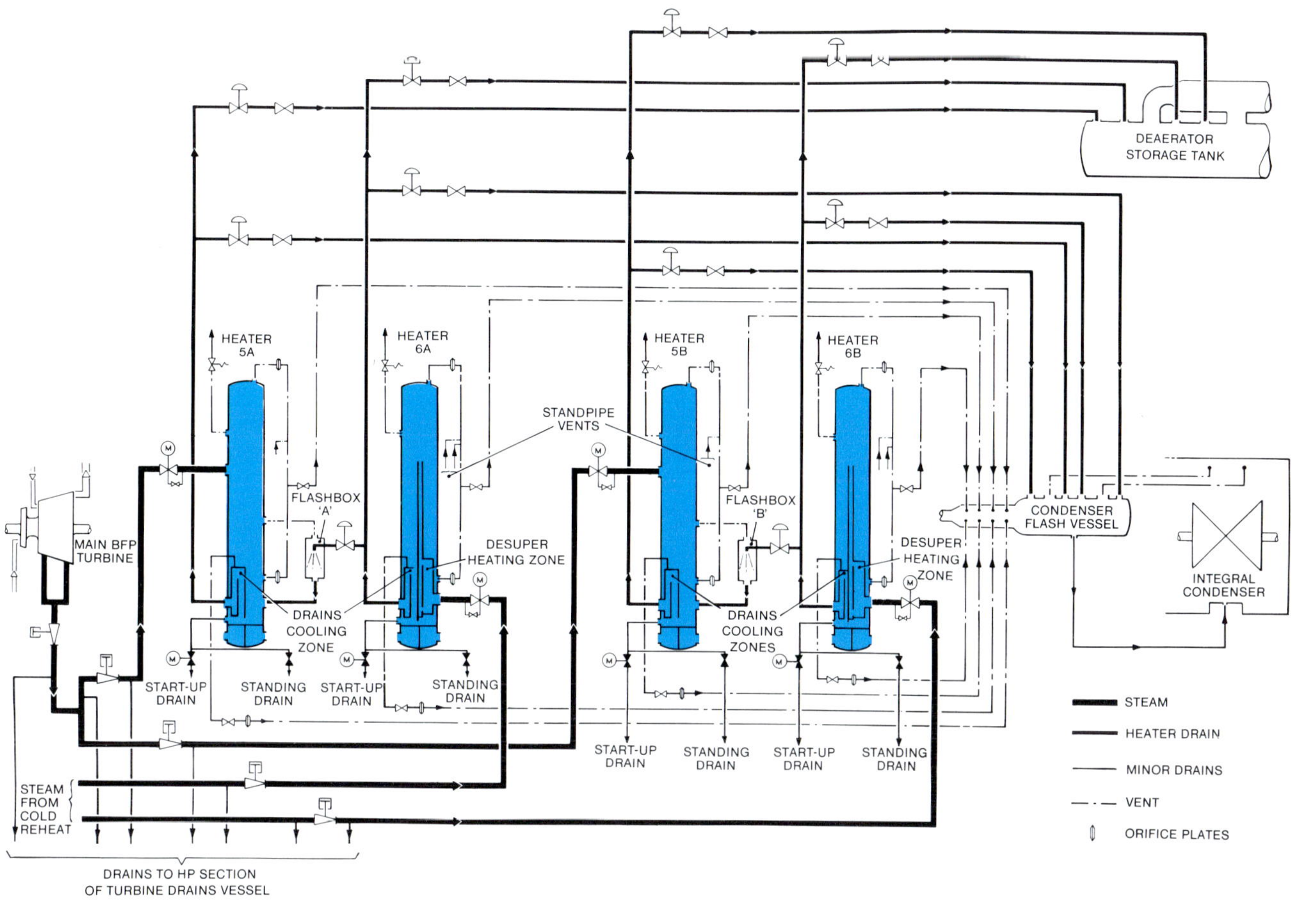

FIG. 5.80 HP heater system bled-steam supplies, vents and drains

The illustration shows a typical HP heater bled-steam system, comprising two parallel banks of heaters, with two heaters in each. The source of steam to one pair of heaters is derived from steam bled-off from stages of the steam feed pump turbine. The other pair of heaters derives steam from the exhaust, i.e., cold reheat, of the main turbine. Provision is made for steam from the higher pressure heater (Number 6) in each bank to pass to the lower pressure heater in that bank. Thus, if the steam feed pump turbine is out of service, the Number 5 heaters will derive all their steam from the Number 6 heaters. Provision is made to vent the heater bodies, including the drain cooling zones, to a condenser flash vessel, via flow restricting orifices. Start-up drains are also provided. Owing to the wide variation in heater body pressure encountered over their entire operating range, controls are provided so that steam can be vented to either the de-aerator or the condenser flash vessels.

- To operate the main feed pump in conjunction with other feed pumps and unit operation to demonstrate control of bled-steam, exhaust steam and steam supplies to the boiler feed pump turbine, within the context of overall feedwater control.

15.3 Detailed procedures

The Detailed Procedures Document should contain a complete inventory of items, as follows:

Foundations: HP heaters, feed pumps, filters and other vessel support steelwork

Vessels: HP heaters and heater drains vessel

Pumps: steam- and electric-driven feed pumps and ancillary equipment

Filters: feed pump suction filters

Pipework and supports, including vents and drains

Instruments, including instrument pipework and electrical connections

Valves, including special valves such as HP heater isolating valves, where applicable

Cabinets and cubicles, including chemical analysis panels and instrument cubicles

Cabling

Access ladders, platforms and handrails

Orifice plates

Standard check lists should be used, where applicable, for all items listed in the inventory. Some items, such as HP heater quick-acting isolating valves, will require special check lists.

As for the condensate and LP feed system, it is again important that interconnections with other systems are defined in the Detailed Procedures Document and controlled in an appropriate manner by the use of valves or blanks. This is particularly important where

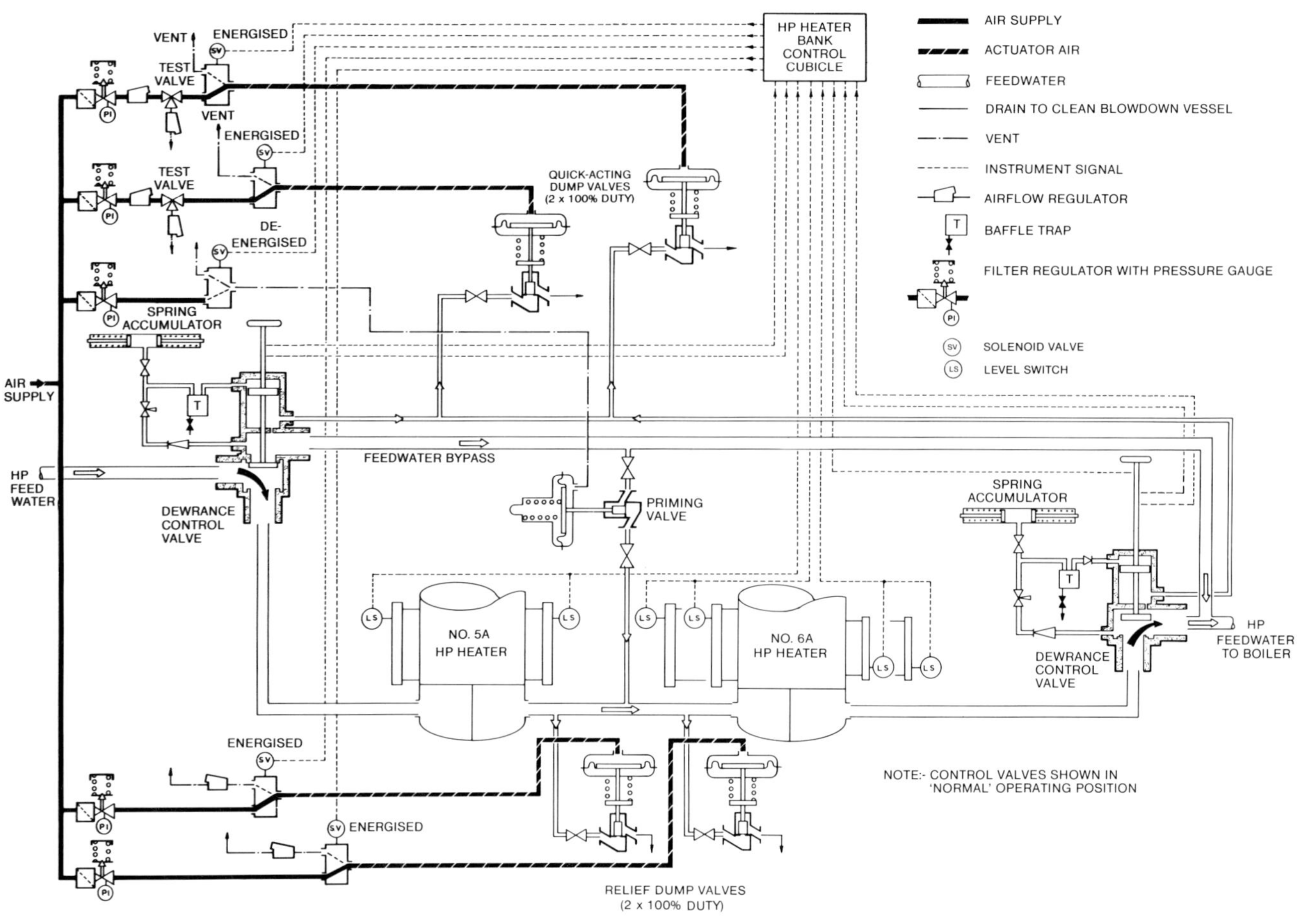

FIG. 5.81 HP heater feedwater system using feed operated isolating valves

The HP heater feedwater system for the heaters illustrated in Fig 5.80 applies to a particular manufacturer and is not typical of systems used in Britain, in that the feedheater inlet and outlet isolating valves are not motorised but are operated by means of feedwater pressure. Duplicate quick-acting dump valves open to dump feedwater from one side of the valve-operating piston, thereby closing the valve. The inlet valves have internal feedheater feed by-pass connections. A further feed by-pass connection fitted with a spring-loaded non-return valve is also incorporated into the system, but not shown in the illustration. In order to release the pressure contained between the feedheater inlet and outlet valves, which would otherwise tend to hold these valves open, duplicate relief dump valves are provided.

drains and vent systems are concerned. For example, many of these connect either directly or indirectly with the de-aerator and condenser. Similarly boiler feed pump leak-offs connect with the de-aerator. Services such as feed pump gland-seal water supplies are often derived from adjacent feed pumps, each pump delivering into a common supply manifold from which their gland-seal water supplies are drawn. Also services, such as lubricating oil and power oil, are often common with the main unit.

Preservation records will be maintained, as described elsewhere. Items such as the main boiler feed pump turbine will be carefully monitored during storage and erection.

Records of weld checks and other processes must be inspected. Information required by the CEGB such as final alignment, clearance and cold datum records, together with checks for keys, dowels, etc., will be collated, as described for the main turbine-generator.

Initial HP feed system commissioning can be segregated into discrete parts, commencing with feed pump suction filters. It is normal practice to commission at least one electric-driven feed pump, as it will be needed for subsequent activities, such as the boiler steam purge. These pumps can be functionally commissioned on closed circuit leak-off to the de-aerator. The HP feedwater circuit, excluding HP feed heaters, is the next commissioning priority. Where steam-driven main feed pumps are installed, commissioning activities can be dealt with separately. However, it should be an objective for main turbine and steam-driven boiler feed pump commissioning activities to be integrated with steam-to-set operations. Certain parts of the boiler feed pump turbine commissioning activities which are common with the main turbine, such as the lubricating oil system flush, should be integrated with the main activity wherever possible.

15.3.1 Feed pump suction filters

Each valve, vessel, electric motor, valve actuator and section of pipework will be checked to the appropriate item check list. Standard check lists will be used, where appropriate. Insulation resistance of all circuits will be checked with fuses removed, care being taken to ensure that equipment which might be damaged by the IR checks has been disconnected. Incoming supplies will be available and within specified voltage limits. Fuses must be of correct rating and clearly identified. Each pressure switch, level switch, local and remote instrumentation indicator and alarm must be checked to function correctly, in accordance with the filter control scheme and to principles described in other sections.

The pipework should be flushed before the filters are functionally checked. This can be done at the end of the erection process by high pressure water-jetting. The filter elements should be removed and the flush water diverted to waste, by disconnecting the filter pipework adjacent to each feed pump inlet valve. After flushing the pipework, the feed pump connection must be inspected internally before reinstatement, and the filter elements then replaced under clean conditions.

The suction filters must be full of water before their motors run, the correct direction of rotation being confirmed.

A high filter differential pressure can be simulated by setting down the differential pressure switch to ensure that filter drive motors cut in and out. For the system described, this will apply to both main and secondary filters. The secondary filter has an automatic backwash facility derived from the condensate system and this should be proven.

Correct differential pressure settings should be applied from either known calibration of the devices or by means of an injected signal. Timers in the control circuit must also be checked for correct setting. The timers ensure positive operation on receipt of a differential pressure signal.

15.3.2 Motor-driven feed pumps

These are large plant items and final erection checks, including the inspection of preservation and erection records, will be similar to those described in Section 1 of this chapter. Motors will be stored in a manner similar to that described for the generator and main exciters.

Gearbox preservation will be specified by the supplier. Typically, the gearbox will be filled with a suitable protective oil, as specified by a reliable oil supplier. It should be compatible with the recommended service oil, so that it can simply be drained and replaced with the normal oil when required. Bearings will be preserved in a similar manner, drain and other connections being blanked off. A routine should be established to turn shafts at regular intervals, say twice per month. The number of revolutions will depend on the gear ratio but should ensure that the high speed shaft rotates in excess of one full turn and that both input and output shafts rotate by fractions more than complete revolutions. This ensures that the components are not left in the same position during storage.

For the commissioning examples chosen, the 15% and 50% electric motor-driven feed pumps are similar in layout, but differ essentially in respect of pump speed control.

The 50% pump is coupled to its suction and high pressure stages by reduction and step-up gear boxes, respectively. Pump speed is controlled by means of variable speed motor. The 15% pump motor is directly coupled to its suction stage, but the high pressure stage is driven via a variable-speed fluid drive unit and gearbox.

Preliminary checks are carried out to ensure that main feed pipework and valves are correctly supported and that pump alignment is not affected by the attachment of the pipework.

Flushing procedures

Pump glands and seals must be by-passed, restrictions to flow removed and the associated pipework flushed. High pressure water-jetting is advisable, as this is a practical way to ensure that loose scale and debris is thoroughly removed. Care must be taken with systems common to several feed pumps to ensure that debris cannot accumulate in dead legs or in parts of the system temporarily segregated by blanks. The water used for flushing must be clean and capable of producing a flow greater than normal system flow for periods of about 30 minutes. Flushing is deemed complete when water conductivity is below 500 μS/m and its iron content below 0.1 PPM. Where possible, strainers should be used to collect flushing water debris, so that an assessment can be made of progress. When flushing is complete, the system must be fully reinstated, primed and checked for leaks under static pressure.

Other pump services, such as bedplate cooling, should be flushed using the normal sources of cooling water, but again with orifice plates and other flow restrictions removed.

The lubricating oil system will be flushed in a manner similar to that previously described for other lubricating oil systems. A temporary heating source and oil centrifuge is desirable. Certain components in the system, such as non-return valves, may require the removal of internals, or the complete valve to be substituted by spool pieces.

Temporary mesh strainers will be needed for the suction and discharge connections to the oil pumps driven from the feed pump shaft. Strainers will also be needed in, or at, the oil tank return connections. Relief valves will initially be adjusted to lift at lowest possible pressure. Pump and gearbox bearing connections will be by-passed and open ends blanked. Permanent flexible connections may have to be removed. These will be cleaned manually, capped and stored safely. The general condition of the oil tank system should be inspected, and the system should be cleaned manually before being charged with oil. The components required for the flushing programme will be commissioned to standard commissioning check lists.

Centrifugal oil pumps will be commissioned against closed discharge valves but, for positive displacement pumps, a flow path is established via their discharge relief valves set to lift at minimum pressure.

For the 15% feed pump fluid-drive system, flushing will follow a similar pattern.

Functional checks

After flushing procedures are complete and feed pump systems reinstated the pump leak-off valves should be commissioned to open above a specified flow rate, and associated alarms and indicators proven. The flow rate will equate to the valve operating signal.

The feed pump ancillary systems can then be made available, or put in service, to prove such features as:

- AC and mechanically-driven oil pump pressure switches, flow indicators, sequence controls, alarms and indicators
- Service water system flow rates and indicators
- Pump gear unit and fluid coupling cooling systems

These tests will be carried out with the feed pump motor switch 'on jumpers', i.e., in a position where main and auxiliary contacts can be operated without the motor being switched on.

Pump suction systems are often equipped with a start-inhibit feature, or an alarm, if the flow path between de-aerator and pump is not healthy. This would be achieved by 'valve open' limit switches on relevant valves. All combinations of valves in open and shut positions should be tried to prove both healthy and unhealthy flow path states.

Other applicable features which inhibit pump start or cause the pump to trip must be proven, such as:

- Reverse rotation
- Low lubricating oil pressure
- Feed pump balance-water temperature high
- Low fluid coupling lubricating oil pressure
- Fluid coupling actuator not in start position
- Liquid controller electrolyte temperature high
- Liquid regulator not in minimum position

Similarly, all 'start healthy' features should be proven.

Simulated signals may be required to prove differential pressure alarms, such as:

- Feed pump packing leakage high
- Feed pump gland-seal filter
- Feed pump lubricating oil filter
- Fluid drive coupling oil filter

Other examples of alarms to be proven are:

- Bearing temperature high
- Lubricating oil tank level low
- Gear-unit oil tank level low
- Fluid drive coupling oil tank level low
- Mechanical seal-water temperature high
- Lubricating oil cooler and fluid coupling oil cooler outlet temperature high

Pumps are protected against reverse rotation by means of a shaft sensor, with contacts that close when the shaft reverse rotation exceeds 1800 r/min; they should be proven by physically rotating the shaft backwards. An alarm will be initiated and signals passed to start the AC lubricating oil pump and, where applicable, fluid coupling lubricating oil pump. Finally the pump stop/emergency-stop features must be proven.

Operation and performance testing

When functional tests are complete, a motor rotation check, if not previously carried out, must be done with the motor uncoupled. The coupling is then made, so that the pump can be put in service on leak-off flow only.

Appropriate operating instructions are therefore applied to prime and pre-start check the pump. It is advisable to start the pump on full sequence control, the pump discharge valve being shut and its supplies isolated.

Communications should be established between the remote control station, the pump and the pump emergency stop button. Pump speed will be set to minimum value. For electrolyte controllers, the electrolyte strength will have been verified.

As soon as the motor starts, there should be an immediate change in discharge pressure to a value corresponding to leak-off quantity commensurate with pump speed and water temperature. These values will be specified by the pump supplier's characteristic curves. The pump starting current and steady running current should be recorded, together with time taken to achieve steady state current. It is inadvisable to trip the motor manually whilst drawing starting current.

Operating parameters, including vibration levels, will be logged to ensure that they stabilise at satisfactory levels. A log of all parameters should be taken, initially at frequent intervals and then, as trends establish, at longer intervals.

After a suitable period of time at minimum speed, say one hour, speed may be slowly increased in suitable increments, establishing steady conditions at each step. Speed should be verified, where possible, by means of a portable tachometer and compared with the contract instrument.

It is important that all personnel in the vicinity of the pump are protected from noise by the use of ear defenders. Noise measurements should be taken and staff advised accordingly.

When the pump has operated satisfactorily at full speed for at least one hour, and providing stable conditions have been achieved, speed can then be reduced slowly and the pump shut down. As the circuit-breaker is opened, the motor heaters should switch on.

During pump operation, suction filter differential pressure and automatic operation should be observed. Where necessary, a manual start of filter cleaning sequence should be initiated and the effects on filter differential pressure noted.

The pump performance characteristic should be checked against the supplier's curves.

Full forward flow of the pump will be proven when the HP feed system and boiler are available. The pump will be started initially on leak-off flow against a closed discharge valve. The valve will be opened slowly to prove smooth operation of the discharge non-return valve.

As boiler operating conditions permit, the pump will be proven over its flow/pressure range, including correct closing and opening of the leak-off valves, and the pump output verified to establish duty point at maximum and minimum speeds.

Tests will also be carried out to demonstrate automatic control in response to feed demand signals. Performance tests will be carried out at a later stage.

15.3.3 HP feedheaters

The start of commissioning is similar to that for the LP feed system and consists of a thorough check of all components for completion. The items to be checked will be the feedwater system from feed pump suction filters via pumps and feedwater regulating valves at the HP heaters, including the HP heater by-pass pipework, together with all associated drains and vents.

The HP heaters will also be thoroughly checked, including all associated fittings, such as relief valves, water gauges and access doors. Before the heaters are boxed up, the internal spaces must be inspected and cleaned.

The steam connections and associated vent and drainage systems to condenser and de-aerator, together with cascade drainage between heaters, will be thoroughly checked, including control equipment and cubicles.

Heater inlet and outlet hydraulically-operated valves of the type described will have a complex control system, involving heater bank priming and depressurising systems and associated control equipment.

During and after erection, special precautions may have been taken to ensure that heater and pipework systems remain in good order by the circulation of warm dry air. The time between disconnection of these precautions and steam-to-set should therefore be kept to a minimum, during which the HP heater commissioning activities can take place.

These activities will commence with dry functional checks of all valves, controls and protective devices. Again, a thorough understanding of the control system philosophy is required. All electrical, pneumatic and other supplies must be checked and available. Associated alarms and indicators will be proven.

Heater control systems are designed to operate drains for start-up, shutdown and other modes of operation when the main turbine is on-load including, if applicable, the main steam-driven feed pump being in or out of service. These systems are usually dependent on heater pressures, which initiate the opening or closing of certain drain valves, thereby diverting drains to higher or lower pressure parts of the system. These controls are functionally checked by injecting appropriate signal pressures.

Heater protection is by high water level in the heater bodies detected by duplicate float columns. Operation of the level switch acts to cause the heater bank to be by-passed, whilst at the same time initiating closure of the associated bled-steam non-return valves. These protective devices must be proven from each initiating source. Similarly, the test facilities provided to test the float switches on load must be checked.

The bled-steam valves will be checked as described for the LP feed system. Non-return valves associated with bled-steam derived from a steam feed pump turbine will be initiated to close on loss of the turbine, whilst all bled-steam valves will similarly close on loss of the main unit.

The ability to put heater banks back in service when trip signals are removed should also be proven.

For the hydraulically-operated feedheater isolating valves, functional testing is limited. Full commissioning of the system can only take place when a feed pump is in service and the HP feed system pressurised.

When dry functional checks are complete, water should be admitted to the system so that a limited flush can take place. Precautions should be taken before flushing to ensure that instrument lines and heater level columns, etc., have been isolated. The flushing procedure is limited by the fact that the only available full flow path is to the boiler. Therefore use has to be made of vent and drain lines. An electrically-driven feed pump is used and this is initially put in service on leak-off low only, the de-aerator having been heated by means of the off-load heating facility.

The HP feed system must be segregated from the boiler and other feed pumps and it is then primed in accordance with operating instructions. The flush is carried out and the flow limited by partially opening the feed pump discharge valve. Instrument tapping points should be blown through when the flush is considered completed.

Dynamic tests of heater control and protection systems can only be carried out when the boiler is in service and under pressure. A convenient time is after the floating of boiler safety valves.

Heater protective devices should again be proven from all initiating sources. The consistency of opening and closing operations of feedheater isolating valves should be observed, and the time of operation recorded.

The remaining commissioning tests cannot be carried out until steam-to-set. This will include the cleaning of bled-steam lines and the bled-steam side of the heaters and is achieved by putting all the steam drains to waste (see Chapter 6).

A cold vacuum test will have been carried out on the HP heater system, as described in Section 4 of this chapter.

15.3.4 Steam-driven boiler feed pump

The steam-driven boiler feed pump is a major unit auxiliary. The steam turbine, whilst being very much smaller than the main turbine, will be commissioned in a similar manner, the general principles of which are stated in Section 1 of this chapter for cylinders and valve chests. Whilst there are differences between main turbine and feed pump turbine governing systems, whether they be of the electronic or mechanical type, the general principles are similar and reference should therefore be made to Section 8 of this chapter. The commissioning activities associated with the suction and pressure stages of the feed pump are similar to those for motor-driven feed pumps.

The oil supplies for both the steam feed pump turbine and its pumps are usually derived from the main unit system. The phasing of completion of erection and the flushing procedure must therefore be closely integrated, so that flushing of the main unit and the feed pump can be carried out at the same time.

Similarly, the power oil supplies to the steam admission valves on the steam feed pump turbine are also usually derived from the main unit system and the completion of erection and commissioning activities must again be integrated.

The main steam leads, loop pipes and valve chests will be steam purged as part of the main turbine activity. This requires the removal of valve chest internals and the fitting of specially designed temporary steam-blow pipework. Chests need lagging and therefore thermocouple and other chest instrumentation must be fitted.

All openings and tapping points which might be contaminated by lagging must be sealed and the lagging activity phased, so that other clean conditions required for, say, pump erection are not breached.

Commissioning summary

Final erection checks must be made to check the availability of instrumentation protection and control systems, and the inspection of preservation and erection records (see Section 7 of this chapter). Lagging must be checked all over for thickness and secure bonding to metal surfaces. Holes must be left to enable the lagging to drain during dry-out. All sliding surfaces and other greased points must be charged with the appropriate grade of grease and a thorough check made to ensure that nothing will prevent expansion. In the cold state, various readings must be checked and recorded, such as pedestal palm and keyway clearances, pedestal movement and cold datums. Also, with rotor on both running and surge thrust faces, the clearances for rotor to cylinder, rotor to pedestal, and cylinder to pedestal should be recorded.

The turning gear must not be energised until its motor has been uncoupled and checked for correct rotation.

A complete check will be made of all the ancillary systems, such as bedplate cooling, gland sealing, etc., as well as for turbine governing, valve gear, supervisory and protection equipment.

With the turbine uncoupled from its suction and pressure stages, a turning gear sequence start and stop should be initiated. When this has been proven successfully, a fault signal should be injected for each stage of the sequence to ensure that the sequence is inhibited and that associated alarms and indicators are correct. As before, alarms and other signals arising from more than one source must be proven from all sources. The sequence interlocks are usually associated with low lubricating oil pressure and open pump discharge valves. When the turbine is coupled to its pumps and the pumps primed, the shafts should be rotated by operation of the turning gear. If the shafts turn freely, the full sequence start and stop of the turning gear is then proven, note being taken of the time to reach steady motor current and the value of the current. Speed signals are used to simulate turbine speed, so that automatic start and disengagement can be checked. Safety interlocks which allow hand operation of the turning gear must also be demonstrated. If the turning gear has a self-contained gearbox, it can be expected that the supplier will recommend that the oil be drained and the gearbox flushed and then recharged with fresh oil after a period of time on turning gear, say 50 hours.

Operational checks on the pumps can only take place when steam supplies are available and in conjunction with main unit operation. The unit operation will be supported by an electric feed pump. The steam feed pump will be pre-warmed in accordance with operating instructions and will be gradually run to minimum speed with the pump on leak-off operation. This activity is dealt with in Chapter 6.

15.4 Commissioning network and resources

Figure 5.82 illustrates a simplified commissioning network for a motor-driven feed pump. The details for commissioning the lubrication systems are not shown, but can be derived from similar examples given previously. Similarly, details for the control system are not

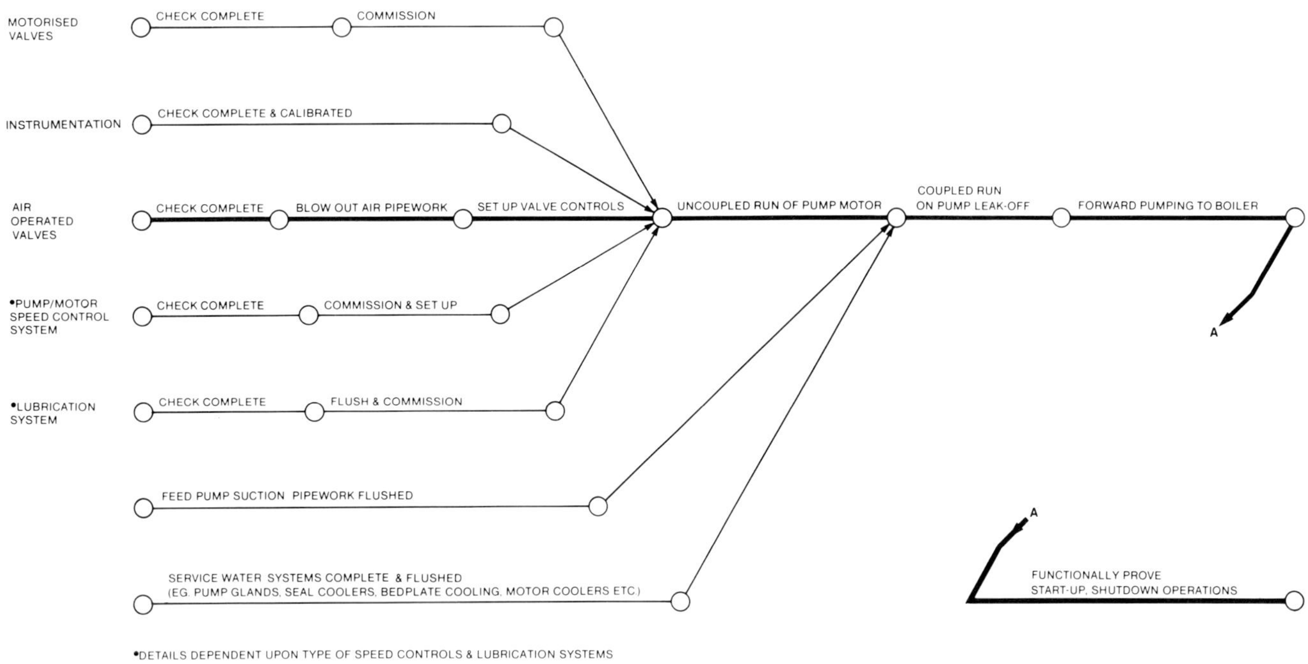

FIG. 5.82 Typical motor-driven feed pump commissioning network

indicated, since these will be dependent upon the type of system installed. However, the systems are relatively small and simple and the logic of a network can be derived from similar examples given elsewhere.

Figure 5.83 illustrates a commissioning network for the feed system pipework and heaters described in the text and shows the point at which a motor-driven feed pump is required. In practice, the commissioning activities for the feed system and feed pumps will be integrated and co-ordinated through the routine commissioning meetings.

Figure 5.84 illustrates the commissioning network for a steam feed pump turbine. This network does not show the activities for the pump, but these will be similar to those for the motor-driven pumps. The activities for the lubricating oil system will apply to both the steam turbine and the pump. As stated in the text, the lubricating oil system is usually integrated with that for the main turbine.

All of the networks cover both pre and post steam-to-set. The aspects of commissioning after steam-to-set are dealt with more comprehensively in Chapter 6.

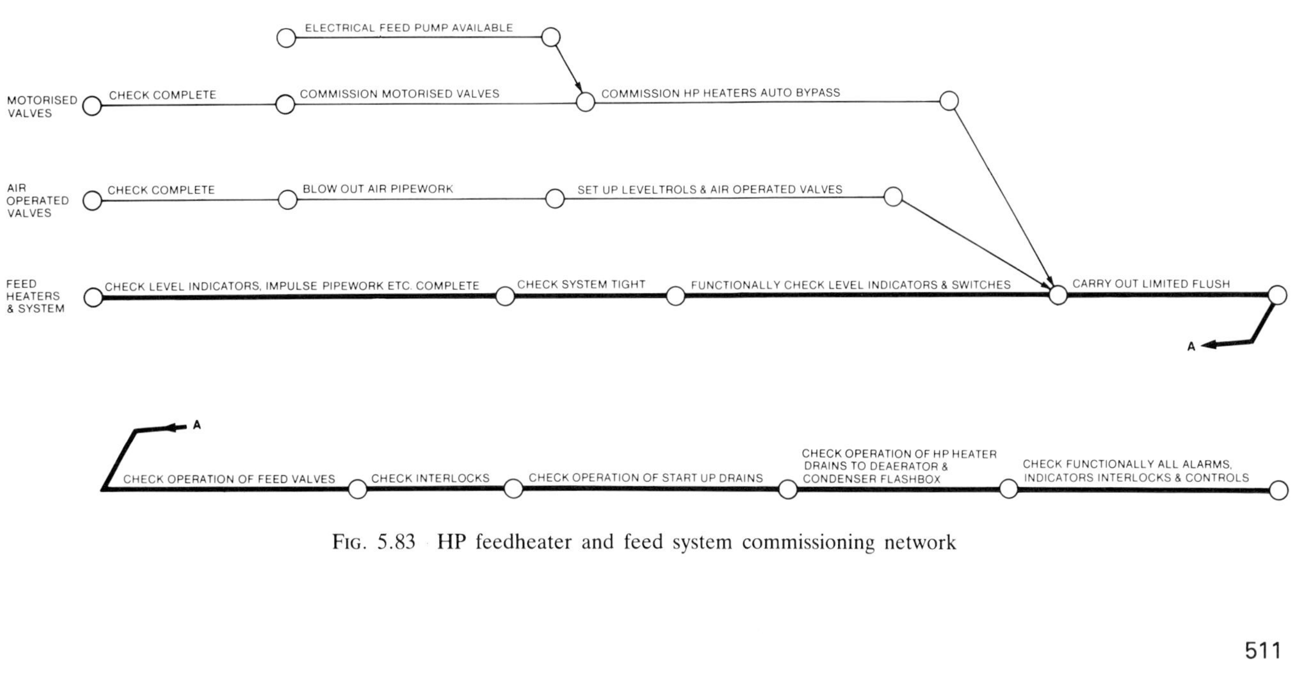

FIG. 5.83 HP feedheater and feed system commissioning network

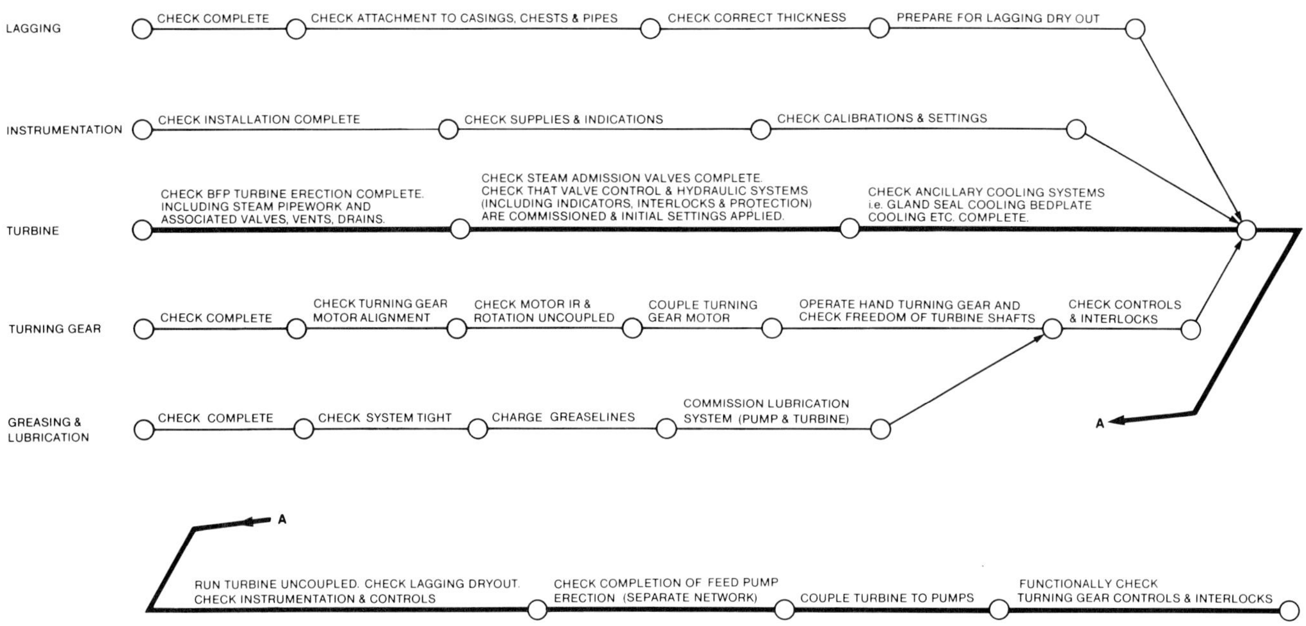

FIG. 5.84 Boiler feed pump turbine commissioning network

Appendix A

Paper of principle for turbine and condenser: typical list of contents

1 Introduction — plant covered by the paper of principle

2 Objectives and assessment points

3 General description of plant and systems

3.1 HP turbine
3.2 IP turbine
3.3 LP turbine
3.4 Forced-air cooling
3.5 Electric turning gear
3.6 Lubricating oil and greasing system
3.7 Emergency stop and throttle valves
3.8 Hydraulic fluid system
3.9 Flange heating system
3.10 Gland sealing system
3.11 IP rotor cooling system
3.12 Condenser
3.13 Tubeplate sealing system
3.14 Air extraction plant
3.15 Condensate system
3.16 Bled-steam system
3.17 Electronic governor
3.18 Turbovisory equipment
3.19 Turbine protection
3.20 Alarm system
3.21 Data processing system

4 Design intent

4.1 Overall parameters
4.2 Life of plant
4.3 Operating modes
4.4 Interchangeability of plant components
4.5 Main turbine cylinders and rotors
4.6 Stop and throttle valves and electronic governor
4.7 Turbine electrical and mechanical protection
4.8 Turbine unloading gear
4.9 Bled-steam system
4.10 IP rotor cooling system
4.11 Flange heating system
4.12 Gland sealing system
4.13 Turning gear
4.14 Lubricating oil system
4.15 Condenser
4.16 Condenser tubeplate system
4.17 Condenser high water level trip
4.18 Forced-air cooling system
4.19 Steel foundations
4.20 Lagging and cleading

5 Design data

5.1 General
5.2 HP, IP and LP turbines
5.3 Limit of operating conditions
5.4 Air extraction plant
5.5 Lubrication
5.6 Condenser
5.7 CW waterbox air extraction equipment
5.8 Condensate extraction pumps
5.9 Boiler feed pump (BFP) gland seal water cooler
5.10 Combined atmospheric gland steam condenser and LP desuperheater
5.11 Sub-atmospheric gland steam condenser
5.12 Condenser flash vessel
5.13 HP/LP drains vessel and drain pumps

6 Terminal points

7 Documentation

7.1 Associated documents
7.2 Manuals — operation, maintenance, plant description
7.3 Pre-commissioning and commissioning tests
7.4 Commissioning report

8 Pre-commissioning and commissioning tests

8.1 HP, IP and LP turbines including bearings and pedestals
8.2 HP and IP loop pipes, IP/LP crossover pipes, bled-steam pipework and valves
8.3 IP rotor cooling system and HP and IP cylinder flange heating systems
8.4 Turbine HP and IP stop valves, HP throttle valves, IP interceptor valves and steam chests
8.5 Turbine stop, throttle and interceptor valves and hydraulic fluid system
8.6 Electronic governor
8.7 Lubricating oil system and turning gear
8.8 Turbine gland steam system
8.9 Air extraction system
8.10 Condenser, structure tubeplate sealing system and CW waterbox vacuum raising and maintaining system
8.11 Condensate system, extraction pumps and distilled water system
8.12 Drains, HP/LP drains vessel, condenser flash vessel and drains pumps
8.13 Forced-air cooling system
8.14 Turbovisory equipment
8.15 Turbine mechanical protection devices
8.16 Turbine electrical protection devices
8.17 Foundation settlement measuring equipment
8.18 Special instrumentation
8.19 Enthalpy drop tests
8.20 72-hour run
8.21 Heat rate tests
8.22 Overload tests
8.23 Unit flexibility checks
8.24 Load rejection tests
8.25 Spinning reserve tests
8.26 Operational trials

9 Relationship with other tests

10 Plant conditions

11 Services required

12 Responsibilities

13 Resources

14 Safety of plant

15 Safety of personnel

16 Plant preservation

Appendix B

Paper of principle for generator and associated systems: typical list of contents

1 Introduction — plant covered by paper of principle

2 Objectives and assessment points

3 General description of plant and systems

4 Design intent

5 Design data

6 Terminal points

7 Documentation

8 Pre-commissioning and commissioning tests

8.1 Generator stator, rotor, exciter and pilot exciter (including rotating rectifiers)
8.2 Stator winding water system and generator water cooled connections
8.3 Air cooled connections including VTs
8.4 Gas system
8.5 Seal-oil system
8.6 Automatic voltage regulator (AVR)
8.7 Protection
8.8 Metering
8.9 Special instrumentation
8.10 Continuous full load operation
8.11 Generator heat run
8.12 Official heat rate
8.13 Overload test
8.14 Operational trials
8.15 Load rejection tests
8.16 Automatic synchronising equipment tests

9 Relationship with other tests

10 Plant conditions

11 Services required

12 Responsibilities

13 Resources

14 Safety of plant

15 Safety of personnel

16 Plant preservation

Appendix C

Paper of principle for condensate, LP and HP feedheating systems: typical list of contents

1 Introduction — plant covered by paper of principle

2 Objectives and assessment points

3 General description of plant and systems

3.1 Condensate dump, make-up system and priming systems
3.2 LP feedheaters
3.3 Heater lift pumps
3.4 De-aerator
3.5 Feed pump suction filters
3.6 Steam-driven feed pump
3.7 Motor-driven feed pump
3.8 Feedwater control valves
3.9 HP feedheaters
3.10 Bled-steam valves

4 Design intent

4.1 LP feed system
4.2 Boiler feed pumps
4.3 HP feed system

5 Design data

5.1 General
5.2 LP feed system
5.3 Boiler feed pumps
5.4 HP feed system
5.5 Feed regulating valves

6 Terminal points

7 Documentation

7.1 Associated documents
7.2 Manuals — operation, maintenance, plant description
7.3 Pre-commissioning and commissioning tests
7.4 Commissioning report

8 Pre-commissioning and commissioning tests

8.1 Condensate and LP feed system
8.2 Boiler feed pump suction filters
8.3 Steam-driven feed pump
8.4 Motor-driven feed pump
8.5 HP feed system
8.6 Performance tests

9 Relationship with other tests

10 Plant conditions

11 Services required

12 Responsibilities

13 Resources

14 Safety of plant

15 Safety of personnel

16 Plant preservation

Appendix D

Paper of principle for turbine and condenser: typical plant conditions (including operating parameters and limitations)

1 Steam cannot normally be put to the turbine until the boiler feed system, generator, turbine protective systems, turbovisory equipment and turbine auxiliary systems are complete to a stage suitable for running.

2 Steam will not be put to the set until the computer and alarm systems are in service.

3 Any temporary strainers or filters installed during flushing exercises must be removed before the unit is run unless they are specifically designed for on-load conditions.

4 Before attempts are made to apply electrical turning gear to the turbine after a prolonged shutdown the shaft must be turned by hand for one complete revolution.

5 Prior to turning the turbine-generator the generator seal-oil system must be in service to prevent seal damage.

6 The turbine must not be left on the electrical turning gear if a high eccentricity develops. Hand turning must be applied and continued until the eccentricity falls below the limiting value (typically 100 μm).

7 The lubricating oil busmain pressure must not fall below the minimum value when the turbine shaft is rotating.

8 The lubricating oil outlet temperature from the turbine bearings should not exceed the limiting value (typically about 80°C maximum).

9 The tripping and interlock facilities associated with all turbine bled steam valves will be commissioned and fully proven before steam is applied to main and boiler feed pump turbines.

10 Before putting steam to the main turbine the condenser structure and shell cooling system must be in service.

11 When applying steam to the unit for any run-up the specified rates of rise of valve chest loop pipe and turbine cylinder temperature must not be exceeded. Also differential temperature limits for valve chest and turbine cylinders must not be exceeded.

12 If one half of the condenser is isolated on the CW side of the unit, load must be restricted in accordance with the manufacturer's instructions.

13 The commissioning of the foundation settlement equipment should be carried out before the sections of the main turbine generator shaft are aligned, so that a set of readings can be taken immediately after alignment. The weight on the supporting columns should be such as to simulate actual operating conditions.

Appendix E

Paper of principle for turbine and condenser: typical services required

1 Water treatment plant fully available

2 Unit control air system in service

3 Cooling water (CW) and auxiliary CW systems in service: Two CW pumps to be available for service

4 Towns main water system in service

5 General service cooling water system in service

6 All fire fighting services associated with the turbine

7 Auxiliary steam available where required for heating services, condenser air pump air ejection, main boiler burner purging, atomising and cooling, etc.

8 Steam to be available from the main boiler

9 Unit computer and alarm system, including station services computing and alarm facilities, if provided

10 All plant telephones to be commissioned and radio communications available as required

11 All electrical switchboards and supplies to be commissioned and available.

11.1 A station transformer
11.2 Station switchboard and auxiliary boards
11.3 Unit board and associated unit auxiliary boards
11.4 240 V DC station and unit supplies
11.5 110 V DC station and unit supplies
11.6 50 V DC station and unit supplies
11.7 110 V AC instrument supplies

12 Unit gas turbine where appropriate

13 General service air system

14 Reserve feedwater supplies

Appendix F

Paper of principle for turbine and condenser: typical safety of plant policy statement

1 All cableway covers and fire barriers must be erected before plant commissioning commences. Portable and fixed fire protection equipment must be available.

2 All HV electrical motors will have their starts limited to two consecutive starts or three equally spaced starts in any one hour.

3 The initial run and subsequent runs of the turbine will not be carried out until all protection devices, safety interlocks, emergency pushbuttons and on-load testing facilities have been proven and witnessed initially by simulation and, thereafter, functionally.

4 The turbine or its associated plant will not be run if oil leaks occur in the regions of hot pipework or casings.

5 Special areas of safety risk should be noted and the appropriate action taken. For example:

Forced-air cooling

(a) A special watch will be required on metal temperatures and differential expansions to ensure that there is no possibility of a rub taking place between turbine fixed and rotating parts.

(b) If the turning gear motor trips and cannot be restarted the forced air cooling system must be shut down and natural cooling will be relied upon for the set to cool down, with the rotor being turned by hand.

(c) When the cooling tests have been completed the forced-air cooling system must be isolated from the turbine casing before the turbine is run up.

Turbine valves

The boiler stop valves must be isolated and downstream vents/drains opened for pre-operational activities on the steam chests and valves of the main boiler.

Turbovisory equipment

Clean conditions should be maintained at the detector heads throughout any tests and no unauthorised access allowed following test completion.

Turbine turning gear

(a) The turbine unit must be turned by hand before the first attempt to electrically turn the machine.

(b) Electrical supplies to the turning gear must be isolated before hand turning is attempted unless an adequate interlock system exists.

Appendix G

Paper of principle for turbine and condenser: typical safety of personnel policy statement

Plant identification

All plant identification and labelling must be complete before commissioning commences.

Communications

Methods of communications must be established.

Fire protection

Where permanent fire protection systems are installed they must be commissioned.

A suitable selection of portable fire fighting equipment will be provided at strategic points around the plant appropriate to the fire hazards anticipated.

Where areas of particular hazards occur, such as the hydrogen panel for the generator, notices will be posted prohibiting smoking or naked lights.

All fire and emergency drills and procedures will be finalised and operative. All personnel concerned in operating and commissioning will be fully conversant with these procedures and appropriate persons trained in the use of fire equipment.

Fluid spillages

Adequate supplies of absorbent materials will be available to mop up or contain any spillages of oil.

Lagging

All pipework and casings will be lagged in accordance with the specification, or temporary lagging applied to a standard which will protect personnel who require access.

Lighting

Adequate lighting levels must exist in all areas required for access and operation.

Lubricating oil flush

Precautions will be taken to ensure that during the oil flushing procedures no sources of heat will be used in the vicinity of the turbine unless absolutely essential.

Flexible connections will not be used during the oil flushing operation.

CHAPTER 6

Unit commissioning and post-commissioning activities

1 Introduction

1.1 Scope of chapter

The contents of this chapter outline the commissioning activities that are necessary to commission a 660 MW or similar large unit from steam-to-set through to full commercial operation, plus post-commissioning activities, including tests of a contractual nature which are carried out after achieving full load.

Unit commissioning activities start from the time steam is first admitted through the main steam pipes to the turbine. The steam-to-set activity brings together for the first time all the various subsystems of the turbine, boiler and common plant systems to function as a complete unit. These subsystems would have previously been individually commissioned, as described in Chapters 3, 4 and 5 of this volume.

1.2 Commissioning Paper of Principle

A single unit commissioning Paper of Principle must be constructed which brings together all the tests identified in the other plant-area Papers of Principle, which can be identified as 'unit tests'. The preparation of Papers of Principle is dealt with in Chapter 2. It is important that the document is comprehensive and Appendix A shows the contents for a unit Paper of Principle.

Once the tests have been identified, it is necessary to agree the order of the tests to enable a programme of commissioning activities to be produced. The programmes should be structured to achieve three main activities:

- From steam-to-set (STS) to synchronisation
- From synchronisation to full-load and completion of a 72-hour run
- From completion of the 72-hour run to completion of other agreed tests

A typical programme to meet this objective is shown in Fig 6.1.

At the time when unit commissioning starts, fresh aspects of safety arise and these are contained in item 12 of Appendix A.

1.3 Detailed unit commissioning procedure

From the programme, detailed procedures can be prepared for each identified activity in the critical path. These are then combined into a single book which is the reference document for all unit commissioning activities. Copies of this document are distributed to the various parties who are either involved in or affected by the activities. Figure 6.2 gives an example of a typical contents list. Apart from the main critical path activities, there will be a requirement to carry out parallel commissioning activities at specific points in the programme.

As indicated in Chapter 2, it is important when these critical activities are taking place, to be absolutely clear as to individual responsibilities. It is normal to name one person to be responsible for the specific test being undertaken: that person is also responsible for the operation of the plant covered by the test. If plant outside his control malfunctions or conditions require changing, his prime responsibility is to contact the person in charge of that plant and to protect his own plant and personnel. In practice, the Shift Charge Engineer is in overall charge of all plant, except when items of plant are specifically excluded during commissioning and testing.

To ensure that there is no ambiguity, the names of persons responsible for the various activities are displayed in the Central Control Room (CCR) and are changed when either personnel or duties change. The person in charge of the commissioning activity would be either a Test Team Leader or a Test Controller.

The main turbine and boiler contractors have their own commissioning organisation, headed by an experienced commissioning engineer. The contractor's commissioning engineers, together with the Site Commissioning and Construction Engineers, agree and formulate the commissioning programme. The order of any programme is generally dictated by the rate of con-

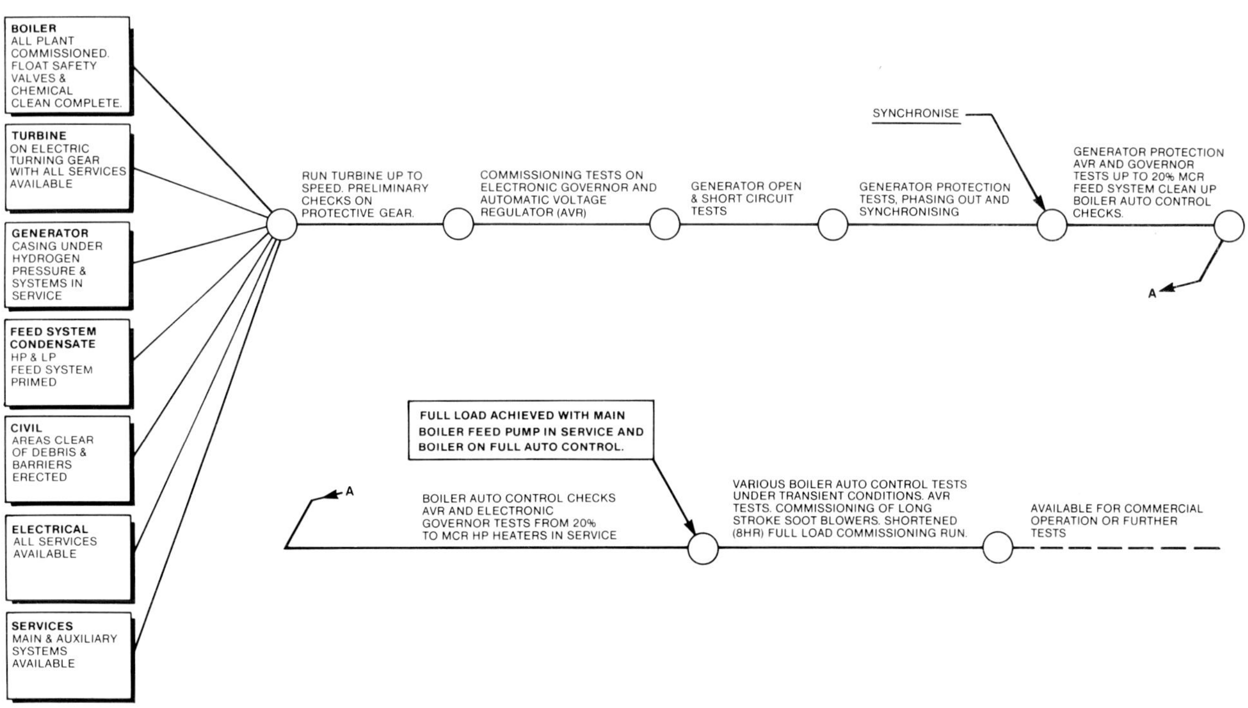

FIG. 6.1 Typical unit programme

struction progress, but once the STS commences, the options for varying the commissioning programme are limited. It is desirable to start STS with the knowledge that all the plant required has been commissioned and tested to enable the programme to proceed without planned stoppages from STS through to full-load. In reality, there are often activities which are incomplete at the start of the STS activity but which can be allowed to continue as parallel activities to a later completion. An example of this on a coal-fired unit is late completion of coal-firing equipment, which does not impede synchronisation on oil burners.

The acceptance of such a construction delay would lead to changing the objective and to proceed along the commissioning programme up to synchronisation only. It is likely that plant problems will be identified during this activity that can be resolved in parallel with the construction delays, hence improving the overall time for putting the unit into commercial operation. It is always necessary, when considering such actions, to consider the effect on working practices and progress by the contractor when plant is in operation in the construction area, for example, noise levels may be higher, it may be warmer, etc.

1.4 Plant and system assessment

In order to control the final stages of the commissioning activity, two assessments are carried out: the first covers the activities planned from STS to synchronisation, the second from synchronisation to full-load.

Prior to the commencement of the STS activity, an assessment is carried out on all the plant and equipment associated with the activity to identify the following:

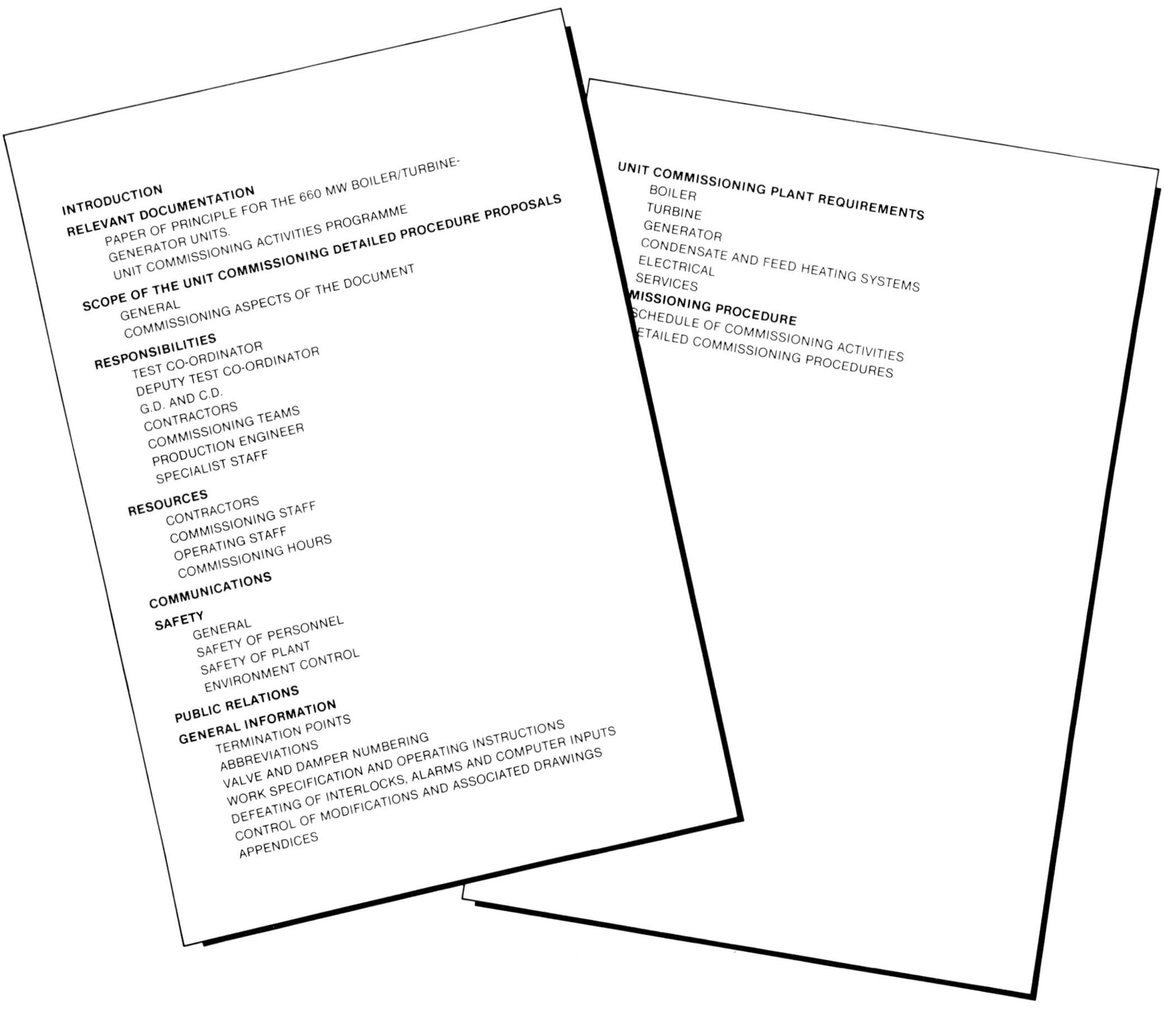

FIG. 6.2 Schedule of contents for a typical unit commissioning detail procedure

- Plant limitations or outstanding defects that may effect operation.
- Availability of adequate plant protection systems.
- Availability of all commissioning documentation needed for the activity.

TABLE 6.1

Commissioning assessment point — schedule of documents, tests and services

Relevant document	'Source'
Erection specification	
Papers of Principle:	
660 MW boiler/turbine unit	Approved by Commissioning Panel
Main boiler plant	Approved by Commissioning Panel
Main turbine and condenser	Approved by Commissioning Panel
Feed system	Approved by Commissioning Panel
Generator	Approved by Commissioning Panel
Generator transformer	Approved by Commissioning Panel
Unit transformer	Approved by Commissioning Panel
Detail commissioning procedures including test sheets	Prepared and agreed as applicable by the following commissioning teams:
	Boiler
	Turbine
	Feed heating and feed pumps
	Generator/transformer
	Unit tests
	Unit detail procedures from steam-to-set approved by Commissioning Panel, other procedures approved by site Commissioning Engineer

- Availability of up-to-date operating instructions.
- Availability of adequate resources to meet the objectives of the activity.
- Availability of the required management information, control and audit procedures including maintenance procedures.
- Status of the programmed/commissioning testing activities to date.

The responsibility for carrying out the assessment rests with the Test Controller, acting on behalf of the Commissioning Panel. The assessment is contained within a document which is presented to the Commissioning Panel for approval. It includes a schedule of documentation.

The assessment identifies any items of plant that will not be available at the STS date, together with a list of all mechanical and electrical services that are required. The status of the fire protection and detecting system on the unit is recorded; where such systems are not available, this fact should be recorded in the Commissioning Log and on the appropriate fire and unit panel in the CCR.

Examples of assessment schedules are given in Tables 6.1 and 6.2, showing typical contents lists.

1.5 Defeating interlocks, alarms and computer inputs

As a further precaution prior to the commencement of the STS activity, the Test Controller instigates a total review of the interlock, alarm and computer input inhibit schedules for the individual plant items and systems that are to be commissioned. This information must be recorded as part of the Commissioning Detail

TABLE 6.2

Commissioning assessment point, 660 MW main boiler and associated systems — commissioning documentation and test schedule

Section	System check procedure	Prepared and approved	Programmed test completion date	Remarks
Boiler dynamic tests	Boiler air test			
	Chemical clean			
	Float safety valves			
Pressure parts	Pressure parts			
Draught plant	ID fans A and B			
	FD fans A and B			
	Gas recycling fans A and B			
	Steam air heaters A and B			
	Gas air heaters A and B			
	Air and gas damper actuators			
	Boiler gas passes and casings			
	Air and gas ducts			
	Boiler interlocks and protection functional checks			
	Draught plant proving tests			
Combustion system	Oil burners			
	PF mills			

Procedure. There should be no further defeating of interlocks or inhibiting alarms, unless such action has the agreement of the Test Controller, who will be responsible for recording the facts and advising all other interested parties.

1.6 Control of modifications and drawings

At the same time, the Test Controller should review the drawing schedules of each plant item and system that has been commissioned to ensure that the schedules carry the correct revision reference to which the plant was commissioned, plus details of any 'on-site' modification awaiting further drawing revision. In addition, the drawing schedules of plant items which are to be commissioned as part of the STS activity are reviewed.

From commencement of the STS activity, it is necessary to have a method of controlling further modifications that become necessary due to experience of plant operation. Whether these modifications take the form of a major contractual change, found necessary to enable the plant to meet its design specification, or a small modification, it is necessary to set up a procedure to control and approve all such modifications, including the updating of drawings and operating instructions, as applicable.

Major modifications have to be agreed between the Station Manager/Project Manager and appropriate contractor, whereas smaller modifications can generally be approved and actioned by the Test Controller.

1.7 Schedule of permissible operating conditions and limits

When commissioning a new unit for the first time, it is essential that all staff involved are made fully aware of the operating limits. The numerous limits are recorded in a schedule which will be available at the unit desk. Table 6.3 shows typical values in the turbine and boiler areas. In addition, the computer printout for both pre-critical and critical alarm settings should also be available. Action on receipt of alarms schedule must also be complete and fully available before commencing the STS activity.

1.8 Plant limitations

Once the STS activity commences, there will be periods when steady state is reached: it is essential, if the plant is to be left under the control of the normal station operating staff, that they are quite clear under what conditions the plant is left in their care. Therefore, when commissioning activities terminate at the end of a day, the Test Controller records the state of the plant at that time and identifies any actions required overnight, for example, cleaning of strainers and repairing defects.

TABLE 6.3
Permissible operating conditions and limits

Plant parameter	Typical values
Boiler	
Final steam outlet pressure	165 bar
Final steam temperature	541°C
Reheat steam temperature	541°C
Final feed temperature	254°C
Primary superheater metal temperature	556°C
Secondary superheater metal temperature	520°C
Reheater metal temperature	526°C
Drum water low level alarm	250 mm below NWL
Drum water high level alarm	100 mm above NWL
Drum water low level trip	330 mm below NWL
ID, FD, GR fans low lubricating oil pressure	1.0 bar
Turbine	
HP cylinder metal differential temp. limits — alarm	−25°C to +25°C
HP cylinder metal differential temp. limits — maximum	−25°C to +25°C
IP cylinder metal differential temp. limits — alarm	−25°C to +25°C
IP cylinder metal differential temp. limits — maximum	−25°C to +25°C
Steam conditions — steam pressure at TSV	159 bar
Steam conditions — steam temperature at TSV	538°C
Turbine differential expansions:	
HP rotor/cylinder — alarm	−4.0 to +9.1 mm
HP rotor/cylinder — advise trip	−4.9 to +9.9 mm
IP rotor/pedestal — alarm	−9.0 to +12.7 mm
IP rotor/pedestal — advise trip	−9.6 to +13.9 mm
LPA rotor/pedestal — alarm	−16.0 to +35.5 mm
LPA rotor/pedestal — advise trip	−17.5 to +39.9 mm
Maximum bearing vibration — alarm	50 μm
Maximum bearing vibration — advise trip	90 μm
Maximum turbine eccentricities — on bearing	25 μm
Maximum turbine eccentricities outside critical speed	75 μm: speed hold
Maximum during turbine run-up	125 μm: advise trip
Maximum turbine on load — receipt of eccentricity alarm	Immediate investigation
Bearing oil temperatures: maximum	77°C
advise trip	82°C running
Note: Consult manufacturers operating manuals and design documentation for a complete list	
Condenser	
Condenser high water level alarm	0.8 metre rising
Condenser vacuum low alarm	−925 mbar
Condenser vacuum deloading sequence	
Unloading commences	−893 mbar
Progressive unloading to 10% MCR	−893 to −860 mbar
Load at 10% MCR	−860 mbar
Trip operates	−840 mbar
Generator	
Maximum permissible rotor winding temperature	130°C
Maximum permissible stator winding temperature	90°C

TABLE 6.3 (*continued*)

Plant parameter	Typical values
Maximum permissible stator core temperature	105°C
H_2 seal face temperature normal	80°C
H_2 system — normal operating pressure	4.1 bar
H_2 system — maximum rated coring pressure	5.2 bar
H_2 system — frame purity	95% H_2 in air
Generator transformer winding temperature alarm	100°C
Generator transformer winding temperature trip	120°C
Unit transformer winding temperature alarm	105°C
Main boiler feed pump	
Rotor eccentricity	70 μm
Bearing vibration — alarm	50 μm
Differential expansion alarm	−1.9 to +2.5 mm
Differential expansion trip	−2.4 to +3.0 mm
Lubrication oil system low alarm	0.76 bar
De-aerator (D/A)	
D/A normal working level	3.17 mm
D/A high level alarm	3.55 mm
D/A low level alarm	1.83 mm

In addition, the planned activities for the next 24 hours should be written down.

Under these conditions, standard documentation is used to enable the plant state to be recorded on a daily basis. In this way, the Unit Test Team are made aware of any problems which may have developed in their absence.

2 Preparation for steam-to-set

2.1 Scope

The STS activity will normally commence approximately 8–12 weeks after completion of the float safety valve (FSV) activity. During these 8–12 weeks, plant will be isolated whilst work is completed on the unit and therefore not available for rechecking.

Prior to filling the boiler in preparation for pressure raising, all electrically-operated drain and control valves on the boiler, including the main boiler stop valves and by-passes, should be operated over their full range and all modifications checked. Particular attention must be paid to boiler stop and by-pass valves which may have been removed after the steam purging of main and reheat pipework for remedial work to the valve discs. These valves have an important role in the STS activity and any malfunction later could delay the programme. In addition, all manually-operated valves should be rechecked for freedom of operation. Checks should be extended to include both hand and electrically-operated drain valves associated with the main steam pipes, hot and cold reheat pipes, HP and IP cylinder drains. It is important where valves use extended spindles for remote operation, that sufficient allowance for expansion has been made to prevent either disengagement or distortion. For example, on a 660 MW oil-fired unit operating at 160 bar 538°C, the main steampipes expand 75 mm downwards, with a similar amount for the hot reheat steampipes. The cold reheat pipes expand 50 mm downwards.

Depending on the length of time spent on chemical clean, steam purge and FSV, it is prudent to inspect the condenser tubeplate prior to the commencement of the activity for any fouling, and clean as necessary. The auxiliary cooling system will not have been fully in service prior to this activity but, if strainers are fitted, they should be checked for cleanliness.

Prior to commencement of boiler pressure raising, the large motors associated with ID, FD and GR fans, and boiler feed pumps, etc., are subjected to final insulation-resistance check, using a 2.5 kV motorised megger; a value in excess of 100 MΩ should be obtained. As part of the precheck, plant that is sequence controlled should be subjected to thorough proving with the auxiliaries system in service, and the sequences proved through to the simulated closure of the control circuit-breakers, isolated at the switchgear from the busbars and the outgoing feeder circuit.

In addition, certain checks must be carried out on the turbine, associated with the speed governor, its hydraulic control system and the turbine trip solenoids, prior to charging the main steampipes.

Once the state of the plant is confirmed, boiler pressure raising can commence. The pressure raising activity is aimed at achieving certain terminal conditions at the turbine stop valves (TSV) prior to 'rolling' the turbine on steam. Typical initial conditions required for the first run-up are shown in Table 6.4. These show minimum steam conditions at the TSV of 70 bar 360°C, which will give sufficient pressure for the gland steam system to be commissioned.

This activity is initially carried out with the boiler stop and by-pass valves locked shut, and with main steampipe drains in the fully-open position. The opening of the main steampipe drains is to stop any premature admission of steam to the turbine due to possible let-by from the main stop and/or by-pass valves. For better control, it is recommended that pressure raising continues up to 30 bar before any attempt is made to charge the main steampipes.

TABLE 6.4

Initial conditions required for first turbine run-up

Steam temperature before HP turbine stop valve	360°C
Steam pressure before HP turbine stop valve	70 bar
Boiler reheat steam temperature	220°C
Condenser vacuum	−915 mbar
Turbine barring speed	36 r/min
Bearing oil pressures	1.0 bar
Bearing oil temperature	>20°C
Governor hydraulic fluid temperature	40°C
Turbine bearing eccentricity	<25 μm

2.2 Boiler filling, gas and air plant start-up

Prior to firing the first burners on a large, natural circulation oil-fired boiler, the following preliminary actions will have been carried out:

- Drum and superheater air release valves open
- Reheater air release valves open
- Enclosure floor drain valves open
- Superheater circulating control main and by-pass valves open
- Superheater and reheater spray control valves shut
- Boiler emergency drain valve shut
- Electrical operation of the drum blowdown isolating and control valves checked and the valves shut
- Electrical operation of the drum lowering emergency isolating and control valves checked and the valves shut
- Adequate clearance between the boiler and galleries, etc., checked, to allow for expansion
- Confirm that the expansion reference pointers are fitted to the boiler and set to zero
- Confirm that all access and inspection doors are closed

The boiler can then be filled to normal working level (NWL), using either the boiler filling pump or an electrically-driven feed pump. If time permits, the boiler should be filled via the boiler filling pump; in this way, water enters at the bottom of the boiler and all air is vented into the steam drum. However, this operation takes approximately four hours (on a 660 MW unit) and then requires the disconnection of the filling hose before pressure raising can commence. Filling from the feed pump is much faster, with the water entering through the economiser, typically taking 45 minutes to one hour. The normal precautions concerning temperature and quality of water must be observed.

Pressure raising commences by starting the gas air heaters and, where fitted, checking that the bled steam air heaters are available to take steam under manual control.

It is then necessary to establish both an air and a gas path through the boiler by opening dampers, including the following: air heater gas by-pass, air heater gas inlet and outlets, air heater air inlet and outlet, hot air ducts to windbox, burner air doors, hot air recirculating, etc.

The next operation is to start the induced draught (ID) and forced draught (FD) fans on sequence; it is recommended at this stage that all four fan groups should be put into service in preparation for a 'boiler furnace purge'. This is required to ensure that gas passes are clear of combustible gases prior to the ignition of the first burner and lessens the risk of furnace explosions on start-up. The previous checks will have ensured that the dampers associated with the fans are in their correct position for fan starting. A typical starting sequence is as follows: start 'A' ID fan; allow starting current to decrease to normal fan running value (approximate time 100 seconds); then start 'A' FD fan. Subsequent fans are started in a similar manner, which allows closer control of combustion chamber conditions and places less stress on the works auxiliary power systems. All fans are started in this manner until the running régime defined by the manufacturer is achieved..

The furnace pressure controller should be adjusted to obtain a desired value of −0.25 mbar suction. At this stage, it is desirable to put the ID fans on auto control; this reduces the possibility of furnace overpressurisation, as well as ensuring more stable operation. The gas path through the boiler is established through the burner air doors on large oil-fired units or through secondary air dampers on coal-fired units: for a boiler furnace purge, all these doors should be open. However, dependent on the number of burners fitted and the sequence requirements of the furnace purge, it is possible to achieve a satisfactory purge with only half of the air doors open, provided that the disposition of these doors across the width of the furnace ensures no furnace and ductwork 'dead spots'. The furnace purge is carried out at an airflow equivalent to 50% MCR, which would be typically 300 kg/s for a large oil-fired boiler. This is achieved by opening the FD outlet control dampers on both fans to obtain the correct airflow, ensuring that the furnace pressure was still being maintained at −0.25 mbar suction. Once the correct airflow has been established, the purge can commence and continue until the furnace purge sequence is shown to be complete. Depending on the system and stages of purge, this may take from five to ten minutes. If a two-phase purge is installed, this will require adjustments to the gas recirculating (GR) fan output whilst still maintaining the desired airflow. Once the purge is complete, both FD fans are switched to auto control and air doors or secondary air dampers progressively closed, leaving open only those where it is intended to fire the oil burners. By use of the windbox/furnace pressure differential mode of auto control, the output of the FD fans reduces automatically as the air doors are progressively closed, allowing the furnace pressure auto control to function and maintain correct furnace conditions.

With the completion of the purge, it is normal practice to ensure that the gas air heaters are by-passed on the gas side, with the gas inlet dampers shut and the gas by-pass damper open. The inlet dampers are not opened again until the air heater air outlet temperatures reach 110°C. The air heater by-pass damper is closed to ensure that all the air flows through the air side of the gas air heaters.

Many manufacturers recommend continuing after the furnace purge with all fan groups in service, but this has been found to cause control problems at low flows with small control damper openings. A more controllable situation is achieved by using only one ID and FD fan. During pressure raising this enables the rate-of-rise of superheater steam temperatures to be kept under close control by more accurate control of excess air in the furnace. With few oil burners in service in a large furnace, no accurate readings of oxygen, or CO, in the flue gas will be available to give this guidance.

Using only one fan group has the obvious disadvantage that, in the event of a trip on either fan, a boiler firing trip results, giving a break in operation. However, the benefit of better superheater temperature control and combustion outweighs this risk. As the second fan group is still available, any such trip should not cause more than, say, a 20-minute delay.

Irrespective of whether the boiler is pulverised-fuel-fired (PF) or oil-fired (OF), the initial pressure raising is carried out using oil burners. The number of oil burners selected for this operation will depend on the capacity of the individual oil burners fitted and their turndown ability. The oil burners of a PF boiler are of small capacity, with one burner for each PF burner, and possibly up to six PF burners associated with one coal mill. It is normal under these circumstances to choose burners situated towards the bottom of the furnace and situated on either side of the furnace centre. The choice of oil burners may be constrained by milling plant availability and whether the later stages of pressure raising make use of coal firing, which is now rare in UK practice.

In an oil-fired boiler, the oil burners will have a greater individual capacity and a better turndown ratio. Typically, a large oil burner may be capable of a maximum flow of 6 t/h but can operate between 2 t/h and the maximum, together with the ability to control oil flow. Here, the choice of burners is easier and only needs the selection of four bottom row burners, two in each half of the furnace, where the boiler comprises 24 burners situated in two symmetrical opposed firing furnaces. For a 36-burner frontwall-fired boiler with the burners arranged in three rows, the selection of burners is increased to six on the bottom row.

A PF boiler comprising 60 burners arranged for opposed firing in five rows with ten PF mills and six burners/mill in a split furnace needs, depending on oil burner size, from six to twelve burners in service, equally divided in both halves of the furnace, for initial pressure raising. The burners chosen should be the bottom row of burners.

2.3 Pressure raising

With the furnace purge completed and the draught plant set up to fire the selected burners, it now only requires a check on the fuel oil conditions before burner firing commences. Depending on the system, methods of ignition and atomising, the following should be checked:

- Fuel oil pressure at minimum value.
- Fuel oil temperature at correct value, depending on type of oil.
- Propane gas pressure at correct value.
- Control air pressure at correct value.

Having established these services are available at the desired value, initiation of the first oil burner can commence. As most of these should have been tested earlier during the chemical clean and float safety valve activities, difficulties do not normally arise, as a level of skill, both in operation and fault finding, will have been achieved. It is important that safety interlocks and equipment are checked to work and are not overridden in an attempt to speed up the programme. Further oil burners are established up to the number required, with due regard to the boiler temperature limitations, of which the following are typical:

- Maximum pressure raising rate not to exceed 67°C/h of saturated steam temperature.
- Drum heating temperature differential limits must not exceed those given in the drum heating cycle curve.
- Limit on gas temperature at primary superheater (540°C is a typical limit for a large oil-fired unit).
- Limit on gas temperature at reheater inlet (480°C is a typical limit for a large oil-fired unit).

The last two limits are measured by either installed gas-side thermocouples or retractable thermoprobes. Pressure raising continues within the constraints of the boiler temperature limits until a superheater pressure of 30 bar is obtained. During this period, air releases will have been closed and the operation of all superheater drains checked. Whilst carrying out this activity, equal numbers of burners should be in service in both halves of the furnace. In addition, strict monitoring of combustion conditions must be maintained including checks on smoke levels and airflows. Apart from the obvious concern for the environment, too much excess air can cause difficulties in controlling the rate-of-rise of saturated steam temperatures.

During the steam purge and chemical clean, the drum water level single-element control system should have been commissioned, using preliminary settings. As pressure raising commences, this control should be in service but, if the correct drum level was already established at the start of pressure raising, the water level will swell and the boiler may require no further supply of water during the pressure raising to 30 bar.

2.4 Turbine valve gear testing

2.4.1 Governor prestart checks

Prior to pressurising the main turbine HP steam pipes with steam, the valves of the turbine valve gear must be checked over their full range of movement. Before this operation, it is necessary to confirm that the turbine is already turning on electrical-turning gear and that no let-by exists from the main boiler stop and by-pass valves. This ensures that any inadvertent admission of steam via the steam pipes or drain vessels does not cause shaft/cylinder distortion. As most modern governors are electronically controlled, the following pre-start checks should be carried out:

- All power supplies on and all monitor alarms clear.
- All plug and socket connections on steam valves and at the front-end pedestal are made secure and wire locked.
- Inputs from vacuum pay-off lines clear and the associated instrument isolating valves to the transducers open.
- Governor set for 'HP governing mode' with the droop setting at 4%.
- Governor hydraulic fluid system fully in service.

With the prestart checks completed, the checks on the valve operations can commence. Initially, all the HP and IP governor and stop valves will be in the tripped condition. The first check is to reset the main trip valve and confirm that this is correctly indicated. After arming the governor and opening the HP and IP stop valves, the error reference signal is checked over the full range of the governor. During this operation the HP and IP governor valves should open fully, and they should be fully closed on completion.

Tests are then carried out to confirm that, for a three-channel governor control system, the tripping of one channel does not initiate a governor failure or a trip signal to the main turbine protection. Two channels are then tripped to confirm that, under this condition, a 'governor failure' is recorded and a trip signal is sent to the turbine main protection. These single-channel and double-channel tests are then repeated to confirm that the same results are obtained for all combinations of the three channels: the tests require the individual tripping of channels A, B and C to confirm that the combinations BC, AB, AC have no spurious circuitry. Tripping the channel combinations BC, AC and AB confirms the two-channel tripping requirements. This may have been done previously but, if any work has been carried out on valve gear and/or governor cubicle since, it is essential that the tests are repeated.

Having completed the governor channel-trip tests, the only action remaining before carrying out the turbine valve-gear test is to trip the hydraulic fluid system to confirm that the loss of this fluid trips the electronic governor. If a standard mechanical governor is fitted, it is likely that the preliminary checks will proceed directly to the valve gear testing.

2.4.2 Turbine valve gear tests

The valve gear normally comprises four HP stop valves, four HP governor valves, four IP stop valves and four IP governor valves. The tests are to confirm that each valve moves to the fully-open position and then to the closed position without any irregular movement. A trip test confirms that the valves close simultaneously and instantly. Modern large turbine governor systems are normally equipped with an on-load test facility, which enables one group of valves to be opened and tripped from the unit desk with the unit on full-load. The on-load test panel has indicators for each of the 16 valves, with limit switches indicating whether the valve is 'fully open' or 'fully closed'. The actual valve test is best carried out using the on-load test panel, which allows valve movement, test panel operation and indication to be confirmed in one operation. The test panel can be grouped into four sections. The first section operates the following:

- *A* HP turbine stop valve
- *A* HP turbine governor valve
- *A* IP turbine interceptor stop valve
- *A* IP turbine interceptor governor valve

The remaining sections operate the *B*, *C* and *D* streams of stop and governor valves.

The on-load test commences on a valve stream after arming the governor and opening the valves in the sequence stated previously. The stroke of each valve is observed locally to confirm its freedom of operation; in addition, the 'open' indications in the CCR are checked and adjustments carried out, where necessary. Having moved the four valves to the 'open' position, the valves are then closed/tripped; once again, correct operation of the valves is checked both locally and on the 'on-load' test panel in the CCR.

The tests are then repeated for the *B*, *C* and *D* streams of the stop and governor valves. This completes the turbine valve test; the governor is then isolated to ensure that inadvertent operation of the turbine valve gear does not occur during steampipe warming.

2.4.3 Turbine trip solenoids

Facilities are provided on modern turbine-generators to enable turbine trip solenoids to be individually tested on-load without tripping the turbine. It is necessary to demonstrate this facility prior to the first turbine run.

Figure 6.3 shows an example of this type of equipment. The test is carried out by selecting the front or rear trip valves for test, confirming on the selected valve that the isolator sleeve has moved across. In addition, pressure indication is normally provided which, on moving the isolator sleeve, will reduce from its normal operating pressure to zero. When the solenoid test pushbutton is operated on the valve selected, tripping can be confirmed by reading the pressure indication which would reduce from normal operating pressure to zero.

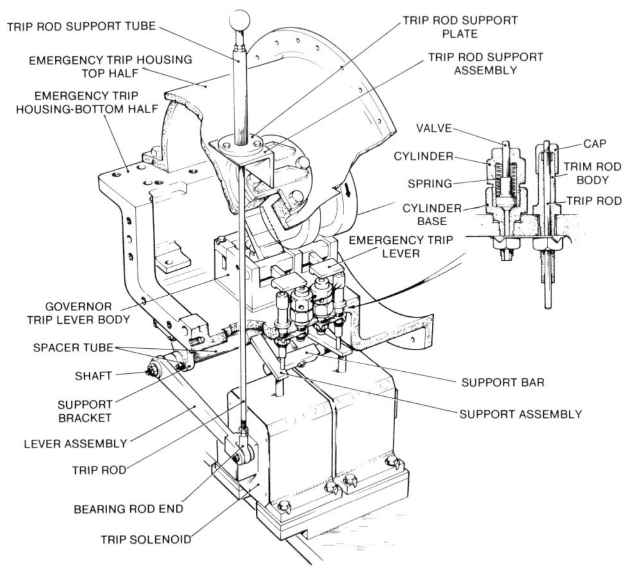

FIG. 6.3 Solenoid and hand trip mechanism

The reset mechanism is then operated to reset the valve pedestal; this is confirmed by the pressure being restored to its previous operating value.

The selector lever is then returned to the neutral position, checks being made that the isolator sleeve has also returned to the neutral position and that the pressure indication of the sleeve records its previous value.

Having now completed the pretesting of the turbine governor valves and trips, charging of the main steam pipes can commence.

2.5 Main steampipe pressurisation

The high steam flow in 660 MW units needs four main steampipes from the boiler to the turbine HP stop valves. Tappings are taken from these pipes for the live steam supply to the main boiler feed pump (MBFP) and the gland steam system (see Fig 6.4). These

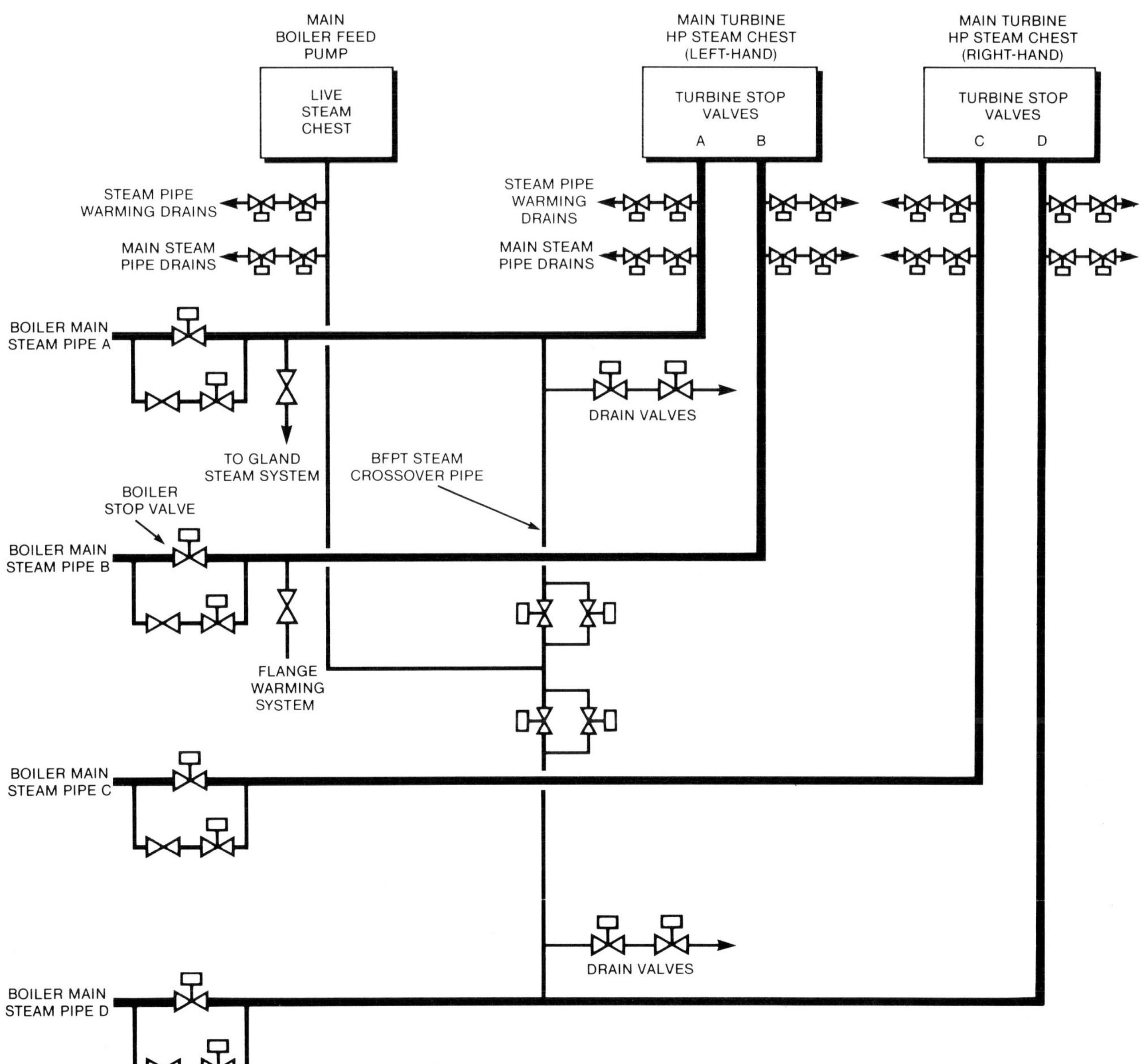

FIG. 6.4 Main steam pipe connections

tappings incorporate electrically-operated drain valve sets, each comprising two valves in series to ensure that any wear is taken on one valve and that the other will continue to give a tight shut-off.

In addition, drain valves are fitted on the live steam line to the MBFP which takes its supplies from two of the four main steampipes.

When commissioning or even recommissioning a turbine, it is desirable to monitor its expansions. To enable this to be carried out on the HP and IP cylinders, reference points must be fitted on the cylinders. Typical positions of these reference points on the HP cylinder are shown in Fig 6.5; similar positions are also adopted for the IP cylinder. These references must be established when the cylinders are cold and after construction work is complete, but prior to any steam admission to the cylinders. When they are fitted the cold set values are measured, using either a vernier calliper or a clock gauge. These readings form the basis for all future checks. Figure 6.6, apart from showing the reference positions, also records typical cold-set values. Expansions are then arrived at by subtracting or adding the latest figure from/to the original cold datum value.

2.5.1 Procedure for main steampipe pressurisation

In Section 2.3 of this chapter, the boiler was taken up to a pressure of 30 bar at 300°C. This should be confirmed and re-established, if necessary. The turbine shafts must be rotating on the electrical turning gear. All electrically-operated drain valves on the main steam pipes and those associated with the MBFP are set to the 'open' position.

Check that the turbine main gland steam HP inlet isolating valve is closed.

Once the drain system has been set up, the four boiler stop valve by-pass valves are opened to the 50% position. Each by-pass valve set normally comprises a hand-operated valve in series with an electrically-operated valve: all four electrically-operated valves open together when operated from the control desk. This ensures equal admission to each steam pipe and thus

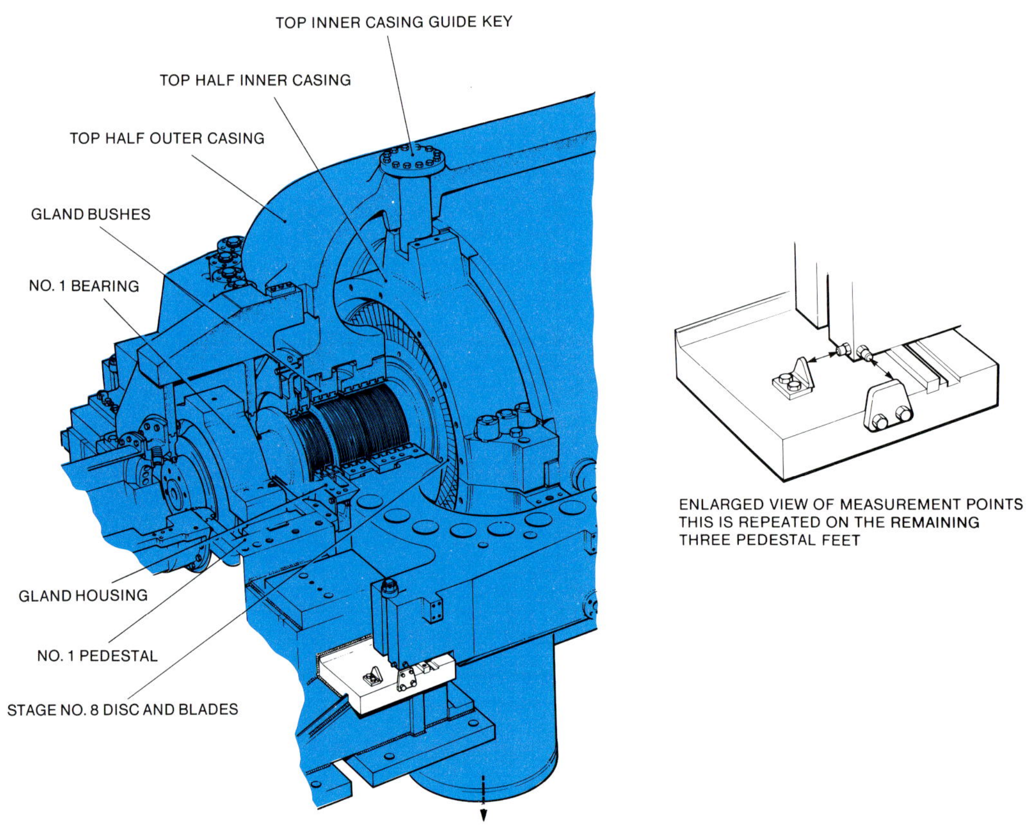

FIG. 6.5 HP cylinder showing reference measuring points

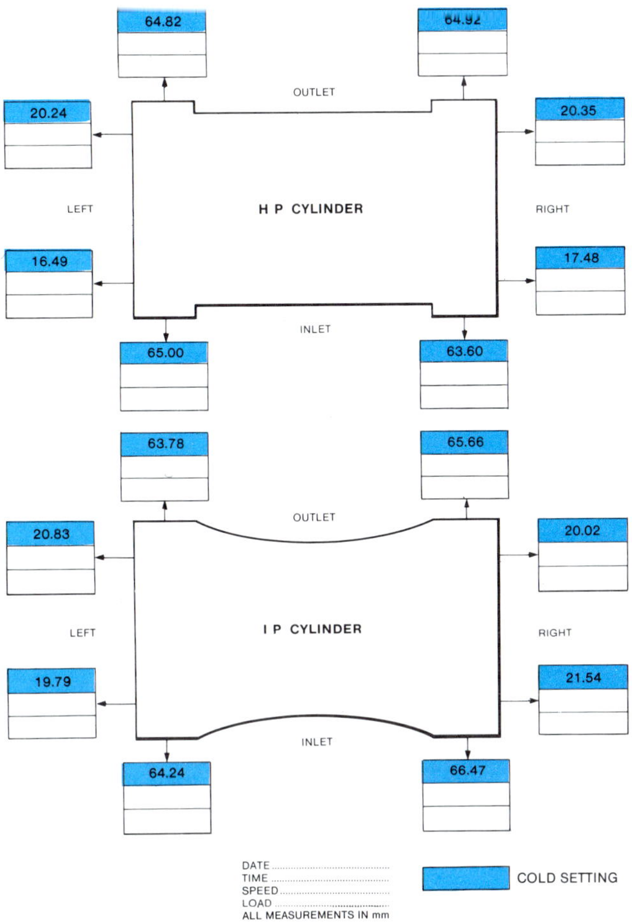

FIG. 6.6 HP and IP cylinder reference point expansion checks (machine cold)

even heating. At this stage a local check on the valve state is worthwhile to confirm that the manually-operated valves are fully open and that the electrically-operated valves are in the correct position.

During steampipe warming, continual reference to the following indications must be made:

- Boiler steam pressure and temperatures
- Steam temperatures before turbine stop valve
- Metal temperatures of each turbine stop valve
- Differential metal temperatures for each turbine stop valve

The rate of warming (see Fig 6.7) should be limited to ensure that saturation temperatures reached on the initial pressurisation of the steampipes do not exceed the turbine stop valve chest metal temperature by more than 100°C. The maximum metal temperature differentials are stated in the schedule of permissible operating conditions and limits. Examples of the rate of increase of steampipe metal temperatures are 100°C/h when condensate heating, and 135°C/h whilst super-heating. The former condition relates to the fact that the early temperature rise in the steampipe is a function of the pressure within the pipe and hence the rate-of-rise of saturation temperature, whereas the latter reflects a greater response to the steam temperature from the boiler. Careful observation of trends will indicate the transition. Pipe temperatures are maintained within 14°C of each other during the warming process, with adjustments to drain valves and boiler stop by-pass valve openings to achieve this.

The opening of the main steam by-pass valves continues until they are fully open, with the boiler pressure still being maintained at 30 bar. These by-pass valves are usually of small bore and, with the number of drain valves on the system that are fully open, it is not normally possible to obtain pressure in the steam pipes. As this is the first time the steampipes have been charged, no indication of pressure will be available until the measuring devices are commissioned and put into service.

Control of steam chest heating depends upon the individual design in this area, but one method used on a design with four steampipes involved shutting off the steam supply to one pipe on each side of the turbine and establishing a warming flowpath down the remaining pipe on each side. This flow is through two stop valve bodies on each side with a reversal of steam flow in the isolated pipes by use of drain valves near the steam chest. More conventionally the steam chests are heated by conduction but, if problems are experienced, alternatives such as that above should be considered.

Once steam chest metal temperatures are within 50°C of the indicated steam temperature at the turbine, the drain valves near the turbine are partially closed and the steampipe drains shut. This transfers drainage control to the turbine end of the main steampipes.

The main boiler stop valves can then be slowly opened whilst maintaining the required limits by use of the drain valves at the turbine to pressurise the main steampipes to the same value as the boiler pressure.

Pressure raising of the boiler then restarts to obtain the typical conditions for commissioning the gland steam system:

- Steam pressure 40 bar
- Steam temperature 320°C

The rate-of-rise in steam saturation temperature must not exceed 67°C/h. Dependent upon the type of boiler, drain valves associated with various parts of the boiler may need to be operated at regular intervals.

Having pressurised the main steampipes, the steam pressure indicators before the turbine stop valves can be put into service. With this activity complete and the readings confirmed as correct, the boiler superheater drain valves may be closed, leaving all boiler and main steampipe drainage to be controlled by the drain valves near the steam chests. It must be recognised, however, that any incident needing the boiler stop valves to be closed will require the immediate opening of the

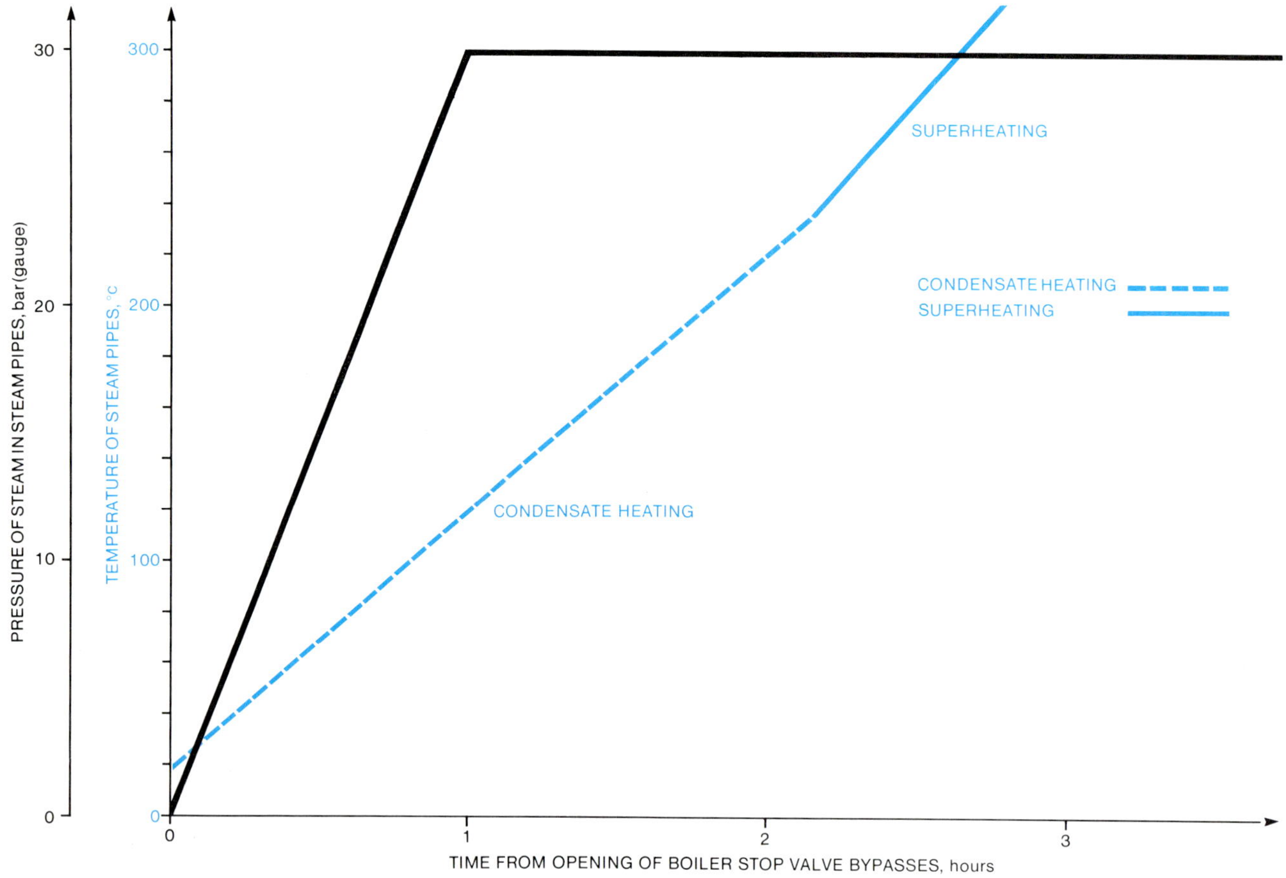

Fig. 6.7 Main steampipe heating curve

superheater drain valves to ensure the safety of the superheater.

Once satisfied that there is no external evidence of steam leakage from the main steampipes and HP steam chests, the boiler stop valves are closed to approximately 30% open, with the by-pass valves shut, in preparation for putting the turbine gland steam system into operation prior to the first turbine run-up. The restricted opening of the boiler stop valves is a precaution in case the turbine HP steam stop valves fail to close in an emergency. Then total closure of the boiler stop valves can be achieved in a shorter time (e.g., 12 s, as opposed to 2 min).

With the steam chests charged to 40 bar and 320°C, a further complete set of reference position readings is taken (see Fig 6.8).

The boiler firing conditions are adjusted to maintain steam conditions of 40 bar and 320°C whilst the gland steam system is put into service for the first time.

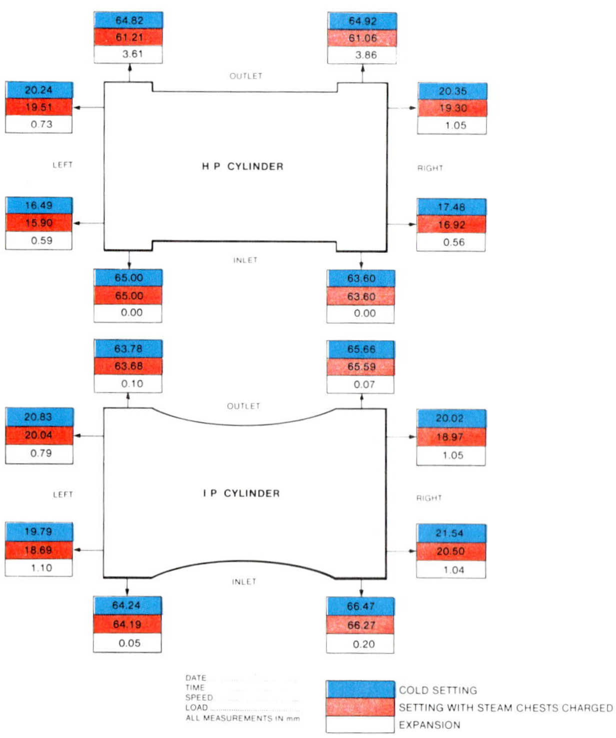

Fig. 6.8 HP and IP cylinder reference point expansion checks (steam chests charged)

2.6 Setting up the gland steam system

The description of the gland steam system and the pre-STS activities associated with it is contained within Chapter 5 of this volume, but Fig 6.9 refers to a typical system. The general principle of operation is that the live steam supply is throttled by the supply control valve to the correct pressure and is then reduced to an acceptable temperature for the HP and IP cylinder glands by the first stage of the desuperheating system. The steam supply for the LP cylinder glands is further cooled by the second stage desuperheater.

The system would have been initially set up prior to the availability of steam. The pressurisation of the main steampipes allows the gland steam system to be checked for the first time. The gland steam system warming process commences by slightly opening the HP gland steam isolator. The rate of warming of the pipework should not exceed a maximum value of 100°C/h. Once the pipework has reached 100°C, the gland steam exhauster fan should be started. It is then necessary to ensure that the steam temperature is kept above saturation for the pressure in the HP manifolds.

After warming is complete, the pressures and temperatures for the HP, IP and LP glands should be checked. Typical values being:

	Pressure	*Temperature*
HP/IP glands	0.15 bar	300°C
LP glands	0.05 bar	120°C

If adjustments are necessary, they should be made on the appropriate control valve. The steam conditions at the sub-atmospheric gland steam condenser, gland steam condenser and LP desuperheater should also be checked.

Having confirmed that the correct pressures and temperatures exist at the various glands, vacuum raising can now be started.

2.7 Vacuum raising

With the turbine glands sealed, the boiler pressure is raised whilst vacuum raising is in process to achieve the

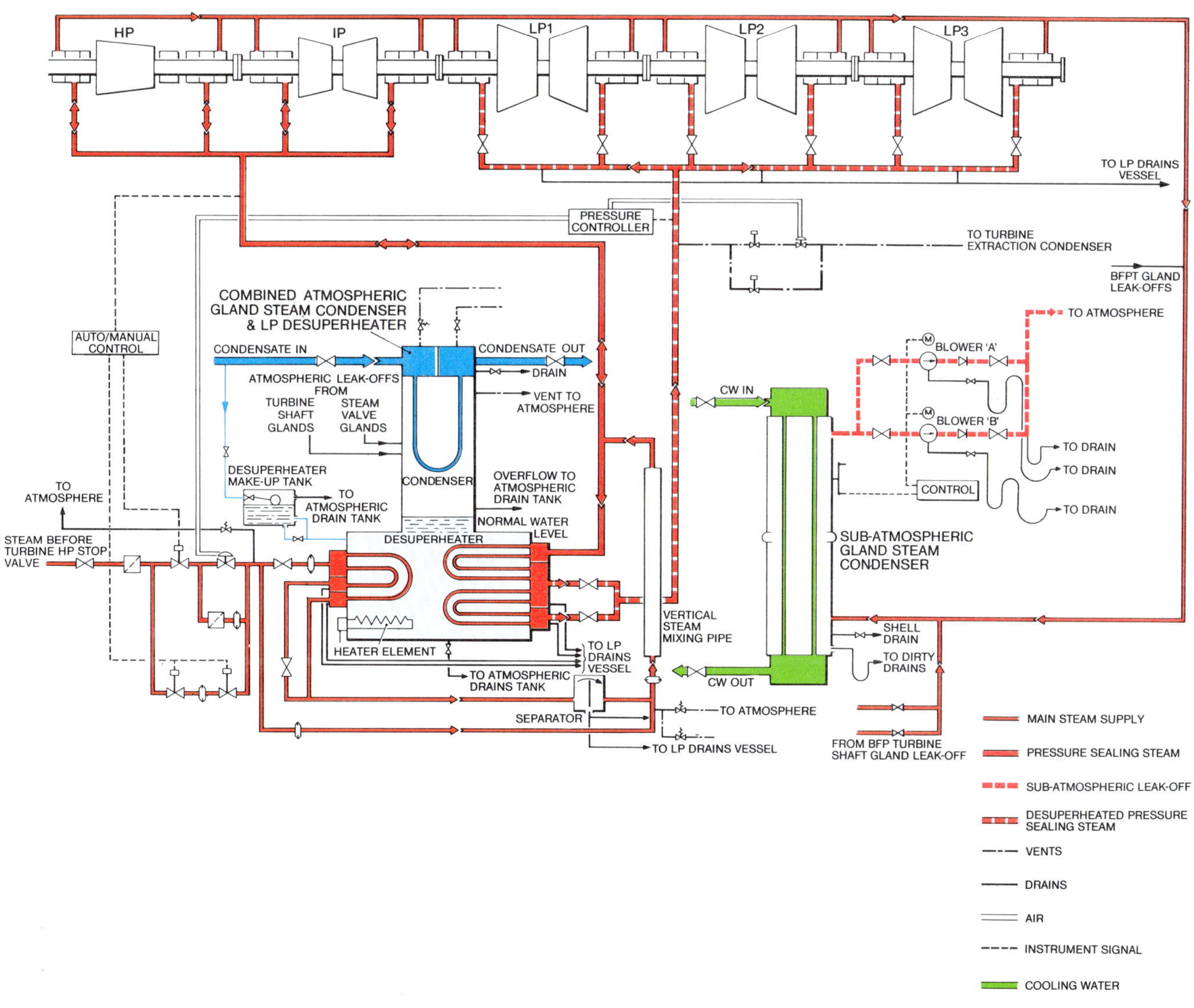

FIG. 6.9 Typical gland steam sealing system

terminal conditions necessary to roll the turbine on steam, typical boiler conditions being 70 bar 360°C.

A cold vacuum raising exercise has been carried out as part of the pre-steam-to-set activities and is described in Chapter 4 of this volume. The vacuum raising activity about to commence differs from the cold vacuum pull in only one respect, namely that the HP, IP and LP glands are sealed. A value in excess of 700 mbar, achieved during the cold vacuum raising with the full vacuum system in service, will usually ensure that the present activity will be successful. The hot vacuum raising comprises the following activities on a particular plant, using two types of air pump.

2.7.1 Test procedure

- HP, IP and LP glands sealed at correct pressure
- Condenser vacuum breaking valve closed
- Main condenser air suction valves open
- Start-up air pumps air suction main isolating valve open
- Main air pumps air suction main isolating valve open
- Both start-up air pumps to be placed into service
- When condenser vacuum has reached approximately 800 mbar, start both main air pumps
- Shut down start-up air pumps at a vacuum of approximately 900 mbar. Vacuum raising to the desired value should be achieved using both main air pumps. A typical value for a 660 MW turbine with six exhausts would be 915 mbar and it should take 20 minutes to achieve from the initial starting of the start-up air pumps.

Having reached a hot vacuum condition for the first time, a test can be carried out to measure the quantity of air being handled by the main air pumps, together with a check on the performance of the plant. This can be carried out if measuring devices have been fitted to the main air pumps. It is recommended that this be done and also that consideration be given to fitting these permanently. The knowledge of the quantity of air being handled by the plant relative to the vacuum obtained identifies the quantity of air ingress into the system. This information will confirm if the conditions can be maintained, possibly on one air pump, as the design air quantities are known. Also, if the correct vacuum cannot be obtained, the test will confirm whether it is a vacuum raising plant problem or a system air leakage problem.

As the commissioning programme proceeds and the pressure line moves down from the HP cylinder, the tests are repeated. In this way, the air ingress is known at different points in the operation from steam-to-set to full-load and can assist with the location of air ingress.

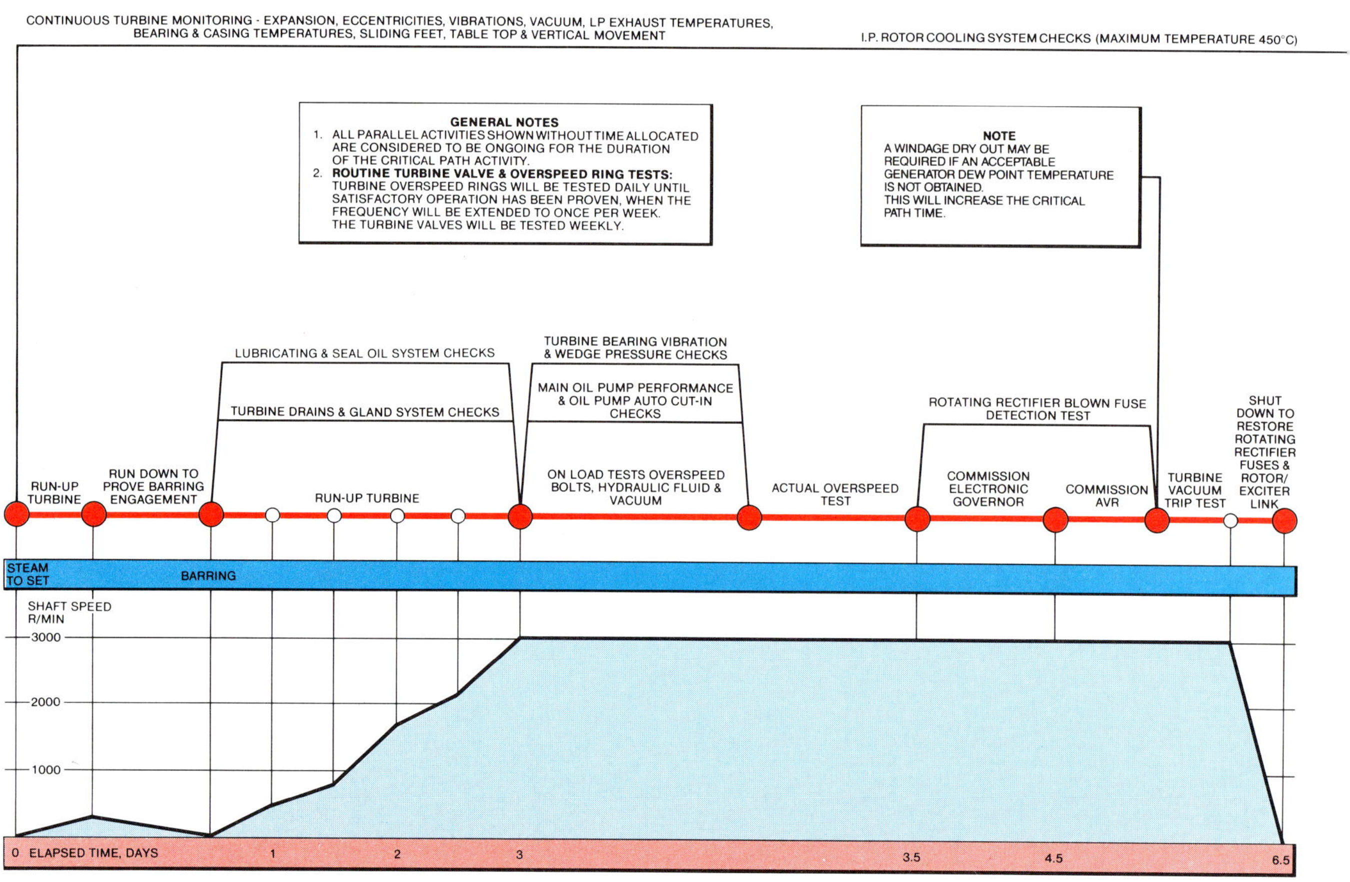

With the condenser vacuum at its desired value and the boiler conditions at 70 bar/360°C, the turbine can be turned on steam for the first time.

3 Initial steam-to-set to first synchronisation

For these operations, a typical 660 MW five-cylinder steam turbine will be considered and appropriate values quoted; where necessary, the more general underlying principles will be discussed. A typical programme for this period is shown in Fig 6.10. If the lubricating oil system does not supply power oil and, in consequence, only high volume, low pressure oil pumps are provided, it may be necessary to supply auxiliary heating to enable correct oil temperature to be achieved at turbine start-up. A system which has a separate non-mineral hydraulic fluid system for turbine governor control is discussed in Chapter 5 of this volume. The lubricating oil system which complements such a system is set up with the AC-driven pumps in service on manual control; any DC-driven pumps are also set for manual control. All lubricating oil system alarms and cut-ins will have been checked previously and confirmed prior to initial STS.

3.1 Initial turbine run-up to synchronous speed

3.1.1 Initial checks

Modern turbine-generator electrical turning gear is designed to engage automatically on run-down, such that the turbine shaft speed does not fall below the electrical turning gear speed of approximately 40 r/min. The electrical turning gear motor automatically starts at 100 r/min turbine speed on run-down and meshes to become fully engaged at 40 r/min. The reverse occurs on run-up, where de-meshing commences as soon as the shaft speed is greater than the turning gear speed, with the electrical turning-gear motor stopping automatically at 100 r/min. In addition, the heavy bearings of the turbine-generator need a high pressure 'jacking' oil supply to assist with lifting the shaft. The pumps which supply this oil are arranged to start and stop in phase with the electrical turning gear equipment.

The turbine will already have been put on electrical turning gear from a stationary position, but it is essential to confirm the correct operation of this turning gear when the turbine shaft accelerates and decelerates. The first objective in the commissioning programme must be to turn the shaft on steam up to 150 r/min and then to reduce speed to check the correct operation of the electrical turning gear. Doing this early in the programme reduces the risk of shaft damage should it be

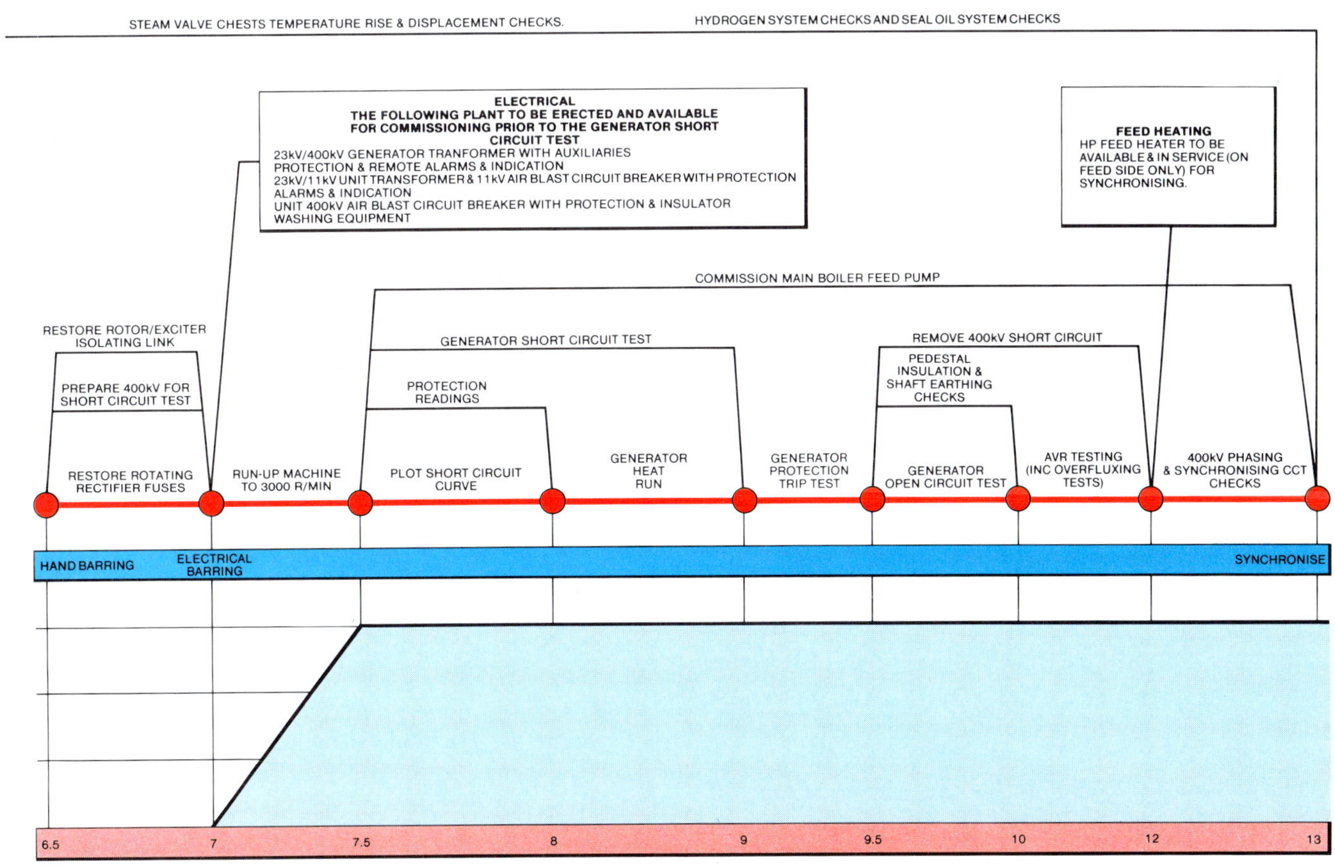

FIG. 6.10 Commissioning programme for a 660 MW unit — from steam-to-set to synchronisation

necessary later to put the machine on electrical turning gear due to some unforeseen circumstances. Failure of the turning gear later in the commissioning programme can, at best, cause a delay and, at worst, cause minor shaft damage.

3.1.2 Electrical turning gear checks

The object is to apply steam to the turbine for the first time to check the operation of the turning gear in all modes. The procedure to be adopted is as follows:

- Confirm that steam terminal conditions are as specified in the turbine pre-run requirements (70 bar, 360°C).
- Confirm that no turbine eccentricity level is greater than 25 μm when the turbine shaft is being turned electrically.
- Check that the condensate system is in the 'dump' mode.
- Confirm that the boiler auxiliaries are functioning correctly.
- Reset the turbine front and rear trip valves to allow power oil to reach the turbine stop and governor valves.
- Set the boiler stop valves to a maximum opening of 30%. This is purely a safety precaution to allow quicker closing of these valves should difficulties be experienced with closing the turbine stop valves.
- Set up the turbine electronic governor.
- Time-mark all recorders to enable co-ordinated information to be obtained.
- Check that the current taken by the electrical turning gear motor under these conditions is similar to that previously noted.
- Operate the governor controls in order to reach a point where the IP governor valves are fully open and the HP governor valves are 20% open. Confirm that the HP stop valves remain closed.
- Further operation of the governor controls should be carried out to commence opening the turbine HP stop valves, thereby accelerating the turbine at the rate of 30 $r/(min)^2$ up to a speed of 150 r/min.
- Check that the electrical turning gear current starts to fall immediately the shaft rotates on steam and record the shaft speeds at which the electrical turning gear and the jacking oil pump motor trip.
- With the machine at 150 r/min, carry out initial checks on bearing vibration and eccentricities, which should show little change from their standstill values. The accuracy of the turbine speed measurement should be checked at this point.
- Confirm by visual inspection instrumentation checks that there are no obvious plant problems.

Having established that the turbine is running satisfactorily, it is then necessary to take action to prove the cut-in of the electrical turning gear and jacking oil pumps.

- Close the turbine HP stop valves and, as speed falls, check that the electrical turning gear motor and the jacking oil pumps start automatically at 100 r/min.
- As the speed falls further, check that the electrical turning gear engages and maintains the shaft speed at approximately 40 r/min.
- Check the electrical turning gear motor current against previously recorded values.

If this activity has been carried out satisfactorily and the plant is performing normally, then the next stage, the run up to 400 r/min, can commence. The figure of 400 r/min is chosen because it is clear of the lowest shaft critical speeds (usually associated with the generator) but is a speed where an oil wedge is formed. This gives an early indication of bearing operation and vibration characteristics. With the proving of the electrical turning gear, the main turbine run-up can now commence.

3.1.3 Run-up to 400 r/min

At this stage, both the HP heaters and main steam-driven boiler feed pump are isolated and the LP feed system is in service on the water side with the bled-steam valves open. Feed flow to the boiler will be via the HP heater by-pass line utilising an electric feed pump. The de-aerator should contain water at 100°C and the boiler feed pump suction filters should be in service. Seal oil for the hydrogen seals is supplied by the AC seal oil pump with the connection from the main turbine oil system isolated.

With the above conditions satisfied, it is necessary to check again that the various terminal conditions are satisfied before admitting steam to the turbine. Any lengthy delay between completion of the electrical turning gear test and the next stage requires careful examination of all conditions, including the state of plant and auxiliaries.

At this stage, it is desirable to put both oil pumps in service to ensure that oil pressure is always available at the turbine bearings. Operate the governor controls to open the HP turbine stop valves and accelerate the turbine at a rate of 30 $r/(min)^2$ up to a speed of 400 r/min and hold. During this acceleration, check that the electrical turning gear motor and jacking oil pumps automatically shut down at 100 r/min turbine shaft speed.

Operate the HP and IP flange warming regulating valves as soon as the turbine rolls on steam. The regulating valves should be set to maintain HP and LP flange/barrel differential temperatures within the permissible limits which are stated within the manufacturer's operating instructions.

At 400 r/min, check around the turbine and take a full set of readings, including the cylinder reference points. This will be the first time that steam has passed through the HP cylinder and has provided a heat source for removing any moisture present in the insulation. If the HP and LP cylinders have been thermally insulated with dry material then the only moisture removed will be atmospheric. However, if a wet-setting insulation has been used, then a considerable drying out period will be required. To assist with removal of moisture a series of approximately 6-mm diameter holes should be made in the lagging penetrating from the outer skin to the cylinder. This is particularly important if a proprietary sealing finish has been applied to the outer layer of insulation as the holes will be the only path available for moisture to escape. The quantity of vapour escaping from the lagging holes will identify the rate of insulation dry-out. Throughout this period, regular checks of cylinder expansions are made by measurements at the cylinder reference points.

It is essential that information is recorded covering many aspects of the operation during the whole commissioning process. This should have been defined in advance and arrangements made to obtain it. This can be done automatically, using the computer at prescribed intervals or, if this facility is not available, by manual intervention at the relevant times. If built-in computer facilities are not available, then the use of special data loggers, chart recorders and/or log sheets should be considered. What is fairly certain is that, if an incident occurs, you will almost certainly not have enough information, or you will miss it by 10 seconds, or the equipment will break down!

The items which should be recorded under steady state conditions are identified in Table 6.5. This list represents all the parameters to be monitored during the commissioning activity up to full-load. As commissioning progresses, the stated parameters are gradually added to the recording process, until eventually all items stated are being recorded.

The time spent at 400 r/min is controlled by the rate of heat distribution around the HP cylinder, which in turn is dependent upon the wetness of the insulation. The cylinder reference point readings indicate how expansion in the different planes is taking place. Unequal expansions need to be examined to see whether they are critical and to determine the possible causes. Typically, uneven steam temperatures, badly applied insulation and faulty cylinder sliding-feet lubrication can cause this condition. Further indication of the cause can be obtained from monitoring the various HP and IP cylinder differential temperatures.

At this low speed, all turbine differential expansions and eccentricities should indicate values which are well below the specified operating values: if this is not so, then an investigation should take place before any further increase in speed is considered.

The time spent at 400 r/min varies from machine to machine, but is likely to be of the order of 8 hours for a first commissioning unit. It takes time to carry out and record all the necessary checks in addition to confirming that the plant is in a state to allow the run-up to proceed. Whilst holding the turbine at 400 r/min for approximately 8 hours, it is essential to control the boiler combustion airflow distribution during this low steam flow condition, thus ensuring good combustion and preventing excessive steam temperature due to high excess air levels.

TABLE 6.5
Items to be recorded at steady state conditions

Boiler
Steam and water conditions
Steam and water chemical purities
Steam pressures and temperatures
Tube metal temperatures
Draught plant performance
Air and gas temperatures
Oil burner/PF mill performance
Fuel oil operating conditions, i.e., pressure, temperature
Environmental checks on smoke, dust and noise levels
Turbine
Levels of vibration and eccentricity at each bearing
Lubricating oil pressures, temperature and chemical conditions
HP cylinder steam and metal temperatures
IP cylinder steam and metal temperatures
Gland steam pressures and temperatures
Condenser and condensate performance, including condenser vacuum
CW and ACW systems performance
Stator water system temperature and pressures
Hydrogen cooling system temperature and pressures
HP feed system: steam pressure and temperature of HP heaters, drain temperature and final feed temperature
Main boiler feed pump
Bearing vibrations
Rotor eccentricities
Differential expansions
Lubricating oil pressures and temperatures
Steam and feed system pressure and temperature
Electrolyte temperature and condition
Pump speed

3.1.4 Run-up to 800 r/min

When conditions are considered satisfactory at 400 r/min, the governor controls are operated to accelerate the turbine up to 800 r/min at the rate of 30 r/(min)^2. As it is likely that the first major critical-speed band will be around 850 r/min, it is essential to choose a speed point away from this to prevent critical frequency induced vibration.

Once established at 800 r/min, the speed is held constant, all HP and IP cylinder differential temperatures are monitored, flange-warming is controlled and a full set of plant data is recorded. In addition, a visual check is carried out around the complete unit for any leaks or obvious pipe fouls.

Also, a full set of readings from the cylinder reference points is taken which will influence the decision as to when the turbine run-up programme can proceed. Apart from continuing the drying out process and, because of the increased steam flow, ensuring that changing turbine differential temperatures are controlled, this operation at 800 r/min is similar to that at 400 r/min.

The time spent at 800 r/min on the first commissioning unit is again likely to be about eight hours. The slightly increased turbine steam flow will not decrease the need for careful boiler operation and the need to maintain the original terminal steam conditions.

3.1.5 Run-up to 1800 r/min

When conditions are considered satisfactory at 800 r/min, operate the governor control gear to accelerate the turbine up to 1800 r/min at the rate of 30 $r/(min)^2$. As the second major critical speed band is likely to be around the 1800–1900 r/min range, it is first necessary to confirm the bottom end of this band, using vibration instrumentation, before choosing the 'hold' speed. The choice of hold speed must ensure that there is sufficient margin to prevent the turbine speed drifting into the critical speed band area, but that sufficient speed increase has been obtained to give a significant increase in steam flow. With good boiler control, a margin of 50–100 r/min should be sufficient.

During the speed increase from 800 r/min to 1800 r/min, the eccentricity and vibration levels must be closely supervised. If a limit of either is reached, the turbine speed is reduced to a known safe level. It is essential for future information, and for safe operation, to record any vibration or eccentricity peaks that occur during the speed rise.

Having stabilised the turbine speed at approximately 1800 r/min, all plant operating parameters must be checked and recorded. This will be the first time during the run-up programme that changes in turbine oil temperature, vibration, eccentricity, differential expansion, etc., begin to show significant changes and require close monitoring. Most turbines during normal cold starts will require a hold at this speed to ensure that the HP and IP turbines, reheater and associated pipework are warmed through. Examples of the conditions and limits that are required to be satisfied for a 660 MW turbine are given in Table 6.6. Figure 6.11 shows how certain turbine thermal parameters vary with time under soak conditions at 1800 r/min. The early commissioning runs of the unit establish the best speed to hold the turbine to satisfy the nominal 1800 r/min soak period, both for critical speed and differential-temperature control requirements. It is often a balance of increased steam flow to achieve temperature differentials against increased speed and vibration problems. The theoretical prediction of critical frequencies for multi-shaft configurations is not perfect: this is an area to be approached practically, and with caution.

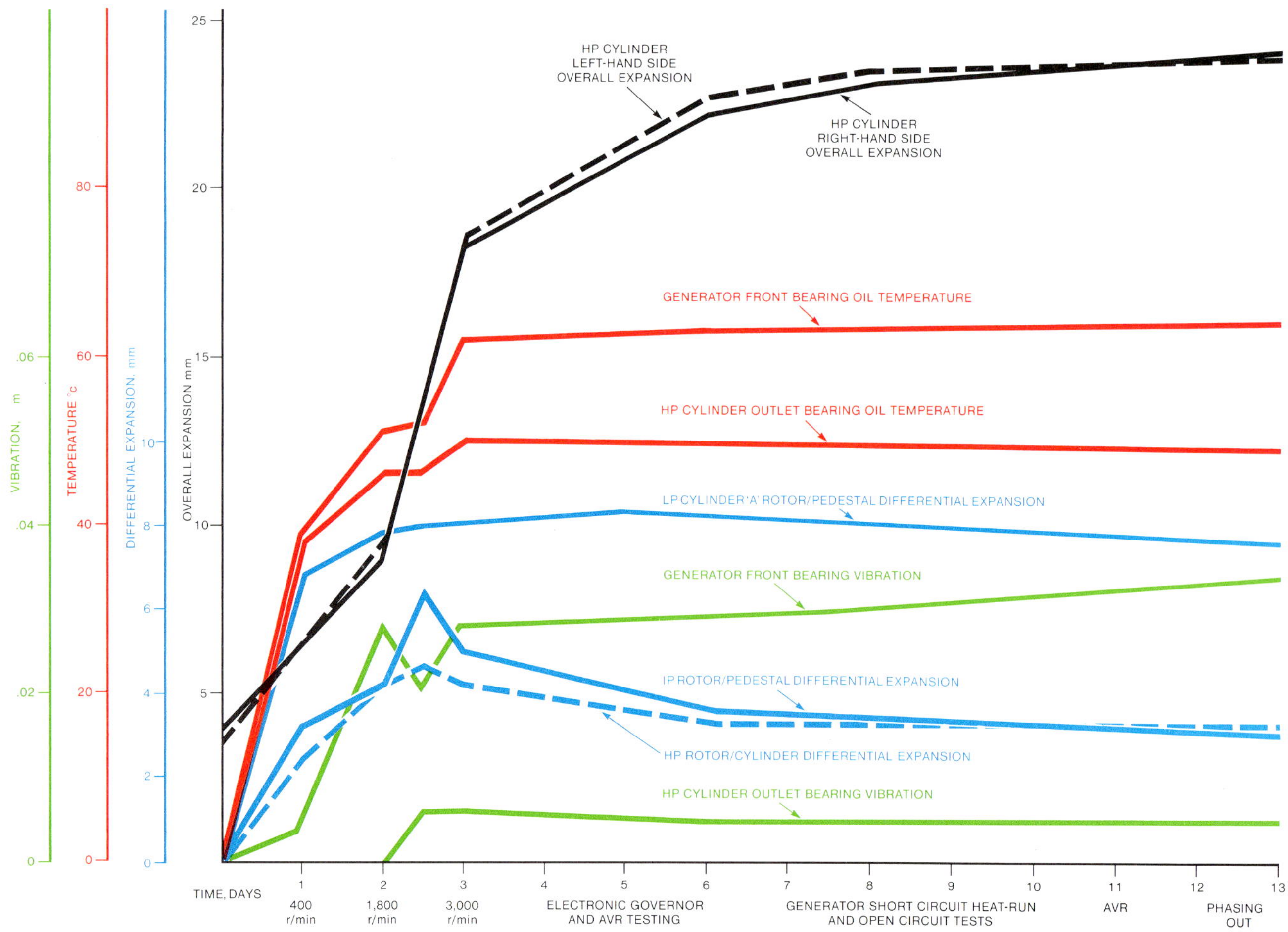

FIG. 6.11 Typical turbine parameters from steam-to-set to synchronisation

TABLE 6.6
Typical turbine soak conditions at 1800 r/min from a cold start

Measurement	Maximum value, °C
HP S/V after R/H front inner/mid differential	36
HP S/V after L/H front inner/mid differential	36
HP S/V after R/H rear inner/mid differential	36
HP S/V after L/H rear inner/mid differential	36
HP G/V after R/H front valve inner/mid differential	36
HP G/V after L/H front valve inner/mid differential	36
HP G/V after R/H rear valve inner/mid differential	36
HP G/V after L/H rear valve inner/mid differential	36
HP inner cylinder/inlet inner differential	>150
HP cylinder inlet inner/mid differential	28
IP G/V before R/H front valve inner/mid differential	28
IP G/V before L/H front valve inner/mid differential	28
IP G/V before R/H rear valve inner/mid differential	28
IP G/V before L/H rear valve inner/mid differential	28
HP rotor surface	>150
HP rotor surface/bore	34
IP rotor surface	>100
IP rotor surface/bore	44
IP inner cylinder inlet flange inner/mid differential	>100
IP cylinder inlet inner/mid differential	48

During the hold period, checks are carried out on the turbine cylinder drains system to confirm that the drain lines are clear, either by temperature indication or by touch. If the drain system comprises steam traps fitted with by-pass valves, it is normal practice to isolate the traps, leaving the steam drainage to pass through the by-pass line. In this way, any pipe debris will not enter the trap and cause it to malfunction; it also ensures a positive drainable path. A typical example of a turbine drainage system is shown in Fig 6.12 (a) and (b).

Checks are also carried out on the generator hydrogen and seal-oil system to monitor leakage and fan differential pressure.

With the increase in bearing temperatures, the oil coolers must be in service on the water side. Figure 6.13 shows a turbine lubricating oil system for a 660 MW turbine. Oil temperature control is achieved by opening and closing the oil temperature control valve fitted in parallel with a pressure relief valve which responds to temperature changes downstream of the oil cooler. The valve opening and closing is controlled by a temperature measurement in the oil return pipe to the coolers. With this valve closed, no oil passes through the coolers: this is so during the initial part of the turbine run-up. It is fully closed up to an oil temperature of 40°C and fully open at temperatures over 50°C. In this operating mode, oil will by-pass the main coolers when return oil temperatures are below 40°C and, as the temperature rises, control will be achieved by oil flow through both the by-pass line and the coolers. If the oil return temperature exceeds 50°C, all the return oil will pass through the oil coolers. As a protection against malfunction of the temperature control valve, a pressure relief valve is fitted in the by-pass line which opens when a predetermined pressure drop occurs across the coolers, thus providing an uninterrupted supply of oil to the bearings.

The hold at 1800 r/min continues until all the conditions specified in the turbine thermal parameters have been satisfied. In addition, further readings are taken from the HP and IP turbine cylinder reference points. If unequal expansions are indicated, the hold time at this speed point will need to be extended until such time as the expansions are under control. Checks on the amount of moisture being vented from the turbine cylinder lagging will give a good indication of the cause of unequal expansions.

The time spent at 1800 r/min on the first unit commissioned is likely to be approximately twelve hours. The time taken depends upon what plant problems develop, their investigation and their solution. Examples of these are receipt of abnormal alarms, instrumentation readings which differ from those expected, and general plant malfunctions. The run-up should not proceed unless the person in charge is completely satisfied with the state of the plant.

3.1.6 Run-up to 2150 r/min

Once conditions are considered satisfactory at 1800 r/min, the governor control is operated to accelerate the turbine up to 2150 r/min at the rate of 60 r/(min)^2. During the speed increase from 1800 to 2150 r/min, close supervision of the eccentricity and vibration levels should be carried out. This speed ramp takes the turbine through the second major critical speed band, the bottom end of this band having been previously identified in the speed ramp up to 1800 r/min. The turbine speed should not be held in the critical band but should be accelerated through it. The actual critical speed band is recorded for future reference and for use in the plant operating instructions. The main objective of the increase to 2150 r/min is to pass through an important critical speed band and to allow a further increase in steam flow which enables further information to be gathered on matters such as expansion, auxiliary plant operation and general plant operation.

The following are continuously monitored: turbine overall and differential expansions, HP and IP shaft eccentricities, bearing vibration, condenser vacuum, LP exhaust temperatures, bearing temperatures, and steam valve chest and cylinder temperatures. In addition, further checks are made on the HP turbine drain systems.

The time spent at 2150 r/min is likely to be in the order of four to six hours, dependent upon the same factors noted on the 1800 r/min hold.

3.1.7 Run-up to lower end of governor range

Once conditions are satisfactory at 2150 r/min, the next part of the turbine run-up will take the machine for the

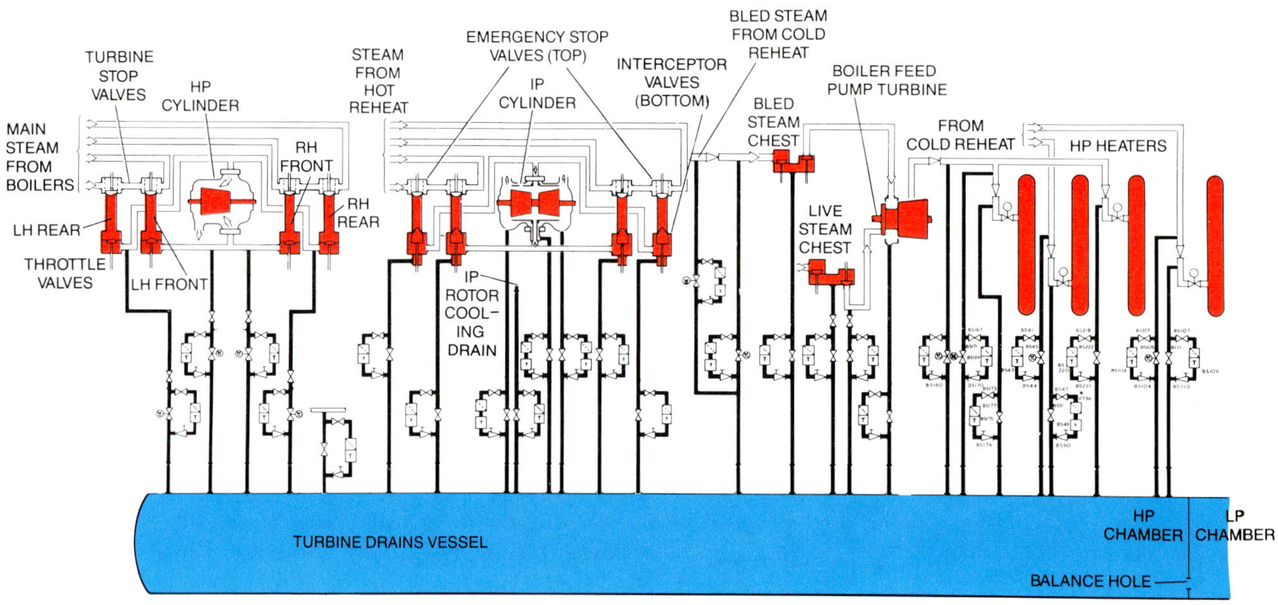

(a) HP Drainage System

first time on to the bottom of the governor range, at a speed of approximately 2790 r/min. At this speed, the main shaft-driven turbine oil pump should become available. However, before this pump can be allowed to deliver oil, it must be checked to ensure that it is generating a positive supply of oil to meet the requirements of both the bearings and seal-oil systems. During the speed ramp, checks are made to confirm that the interconnecting oil valve between the main and seal-oil systems is shut, and that the AC and DC flushing oil pumps are running and selected to a non-automatic mode of operation. This should prevent inadvertent shutdown of the electrically-driven oil pumps when the main oil pump starts to deliver oil.

Having confirmed that the oil system is set up as required, the governor control gear is operated to accelerate the turbine at the rate of 300 r/(min)^2 up to the bottom of the governor range, 2790 r/min. Check that the governor takes control and record the speed at which this occurs. It is at this point that, if an electronic governor is fitted and its speed reference probes have been set incorrectly, no signal will be sent to the governor, and hence the governor control valves will stay open instead of closing. In these circumstances, speed control of the turbine will still be achieved by throttling the turbine stop valves. It is necessary during the run-up, as the speed approaches the governor range, to keep a close check on the rate of turbine acceleration to ensure that the turbine speed will not increase beyond 3000 r/min, if the governor malfunctions. This is to allow a speed margin between the controlled speed and the potential operating speed of the overspeed protection. At this time, the overspeed protection has only been rig tested and its exact speed of operation under operational conditions is not known.

During the speed rise from 2150 to 2850 r/min, close supervision is maintained on the shaft eccentricity and vibration levels. This speed increase will take the machine through the next critical speed band at approximately 2500 r/min: the vibration levels experienced when passing through the critical, together with actual critical speed band, should be recorded. The turbine is accelerated quickly through the critical speed band, the speed never being held in this range.

If governor control has not been established by the time the turbine speed has reached 2950 r/min, the run-up programme will have to be abandoned if checks show that no speed error reference signal is being received from the speed probes. Other governor faults would need to be investigated to see if the machine could be left running at speed whilst remedial work was carried out. If the speed probes are at fault, the only course of action is to run the turbine down to electrical turning gear speed and to readjust their gaps. Electronic governors use at least three speed probes and require a healthy signal from two of them to enable the governor to function.

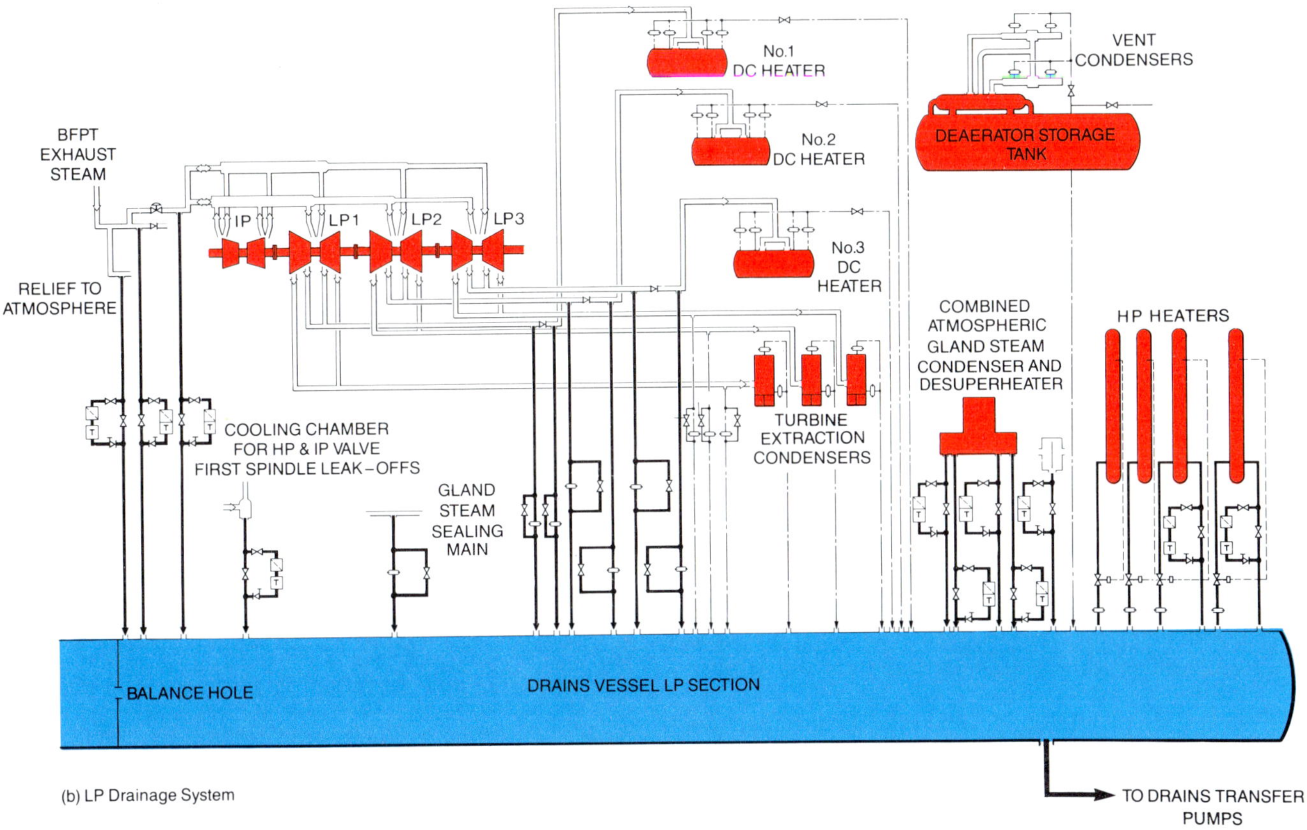

FIG. 6.12 Typical turbine drainage system (HP and LP drains)

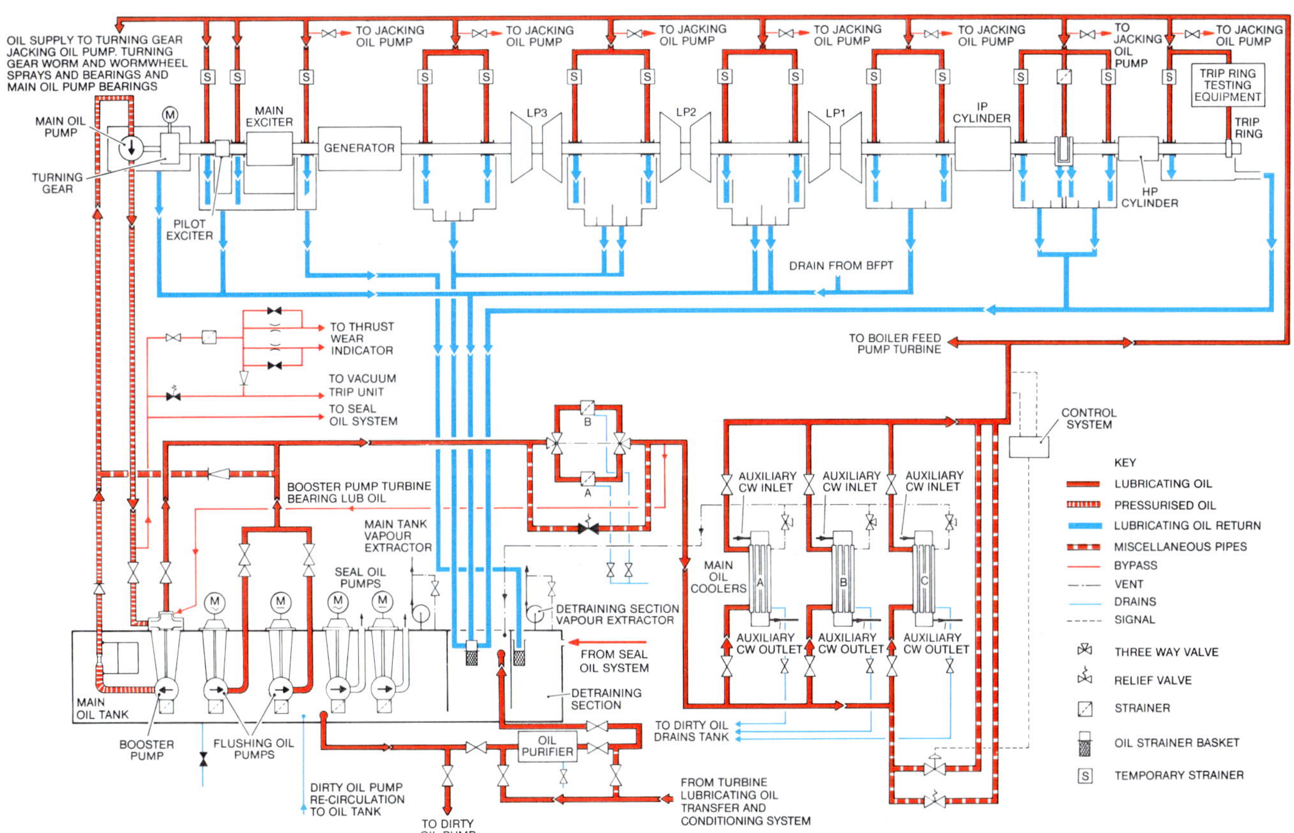

FIG. 6.13 Typical turbine lubricating oil system

Having stabilised the turbine at the minimum governor control speed and confirmed that the governing system is in fact in control, all plant operating parameters should be again checked and recorded. At the higher speed, the additional volume of steam required to drive the turbine will, amongst other things, result in higher bearing temperatures. Hence the need to check the operation of the main lubricating oil temperature control system to ensure that the oil return temperature is being controlled between 40 and 50°C.

In addition to repeating the checks previously stated, the following operations/checks are carried out:

- Check that the main oil pump discharge pressure is building up.
- Check the seal-oil temperatures and the hydrogen fan differential pressure.
- Check the HP and IP cylinder drains, confirming that the lines are still hot and, if steam-trapping is fitted, confirm that by-pass operation is still in use.
- Confirm that conditions are being maintained on boiler, and check steam chest temperatures.

The time spent at 2850 r/min is likely to be of the order of four to six hours, dependent upon the previously indicated factors.

3.1.8 Run-up to 3000 r/min

Once conditions are considered satisfactory at 2850 r/min, operate the governor controls to accelerate the turbine under governor control to 3000 r/min at the rate of 60 r/(min)^2. Confirm the machine is at 3000 r/min, checking signals from the electronic governor against the normal speed indicators fitted both in the control room and locally at the turbine HP pedestal.

Having stabilised the turbine for the first time at 3000 r/min and again confirmed that the electronic governor is controlling, all plant operating parameters are checked and recorded.

The checks include the following:

Boiler

- Steam pressure
- Superheater steam temperature
- Reheater steam temperature
- Primary superheater metal temperatures
- Boiler expansions
- Fuel oil temperatures and pressure control
- ID and FD fans
- Air heater performance
- Boiler water conductivity

Turbine

- Turbine expansion
- Shaft eccentricities
- Bearings — vibration, temperatures and wedge pressures
- Condenser vacuum and performance of air plant
- LP turbine exhaust temperatures
- HP and IP cylinder temperatures
- IP rotor cooling system temperature
- Hydrogen system leakage, fan differential pressure
- Seal-oil system temperature and pressure
- Gland steam system — pressures and temperatures at the glands
- Turbine drains
- Condensate conductivity
- Electric boiler feed pump performance and control

With the turbine at 3000 r/min for the first time, the hydrogen drier is put into service and the generator dewpoint taken and recorded. The dewpoint is likely to be positive, irrespective of how dry the generator was kept during the construction period, being closer to zero for a well-stored generator than one which has had poor storage. Whilst other activities proceed, with the turbine at 3000 r/min, it is essential that the hydrogen drier remains constantly in service (prior to first exciting the generator) and that the moisture extracted is measured and recorded together with the dewpoint value.

The maximum accepted generator dewpoint temperature prior to exciting is 20°C below the cool gas temperature. Where the cool gas temperature exceeds 40°C, the maximum acceptable dewpoint is +20°C. If a dewpoint value within the limits specified has not been obtained at the end of the AVR commissioning with the rotor/exciter isolating link removed, it is necessary to carry out a windage dry out.

3.2 Turbine lubricating oil pump checks

Having confirmed that plant operation is satisfactory, the next part of the programme sets up the turbine lubricating oil system to its normal operating state.

The first operation is to check that the shaft-driven main oil pump is delivering its design discharge pressure; once this is confirmed, first the DC flushing oil pump and then the AC flushing oil pump are shut down and selected to 'auto start'. The shutting down of these pumps is staggered to allow time for the close monitoring of the performance of the oil system, which is being supplied for the first time by the shaft-driven oil pump. Once stable conditions have been obtained, the pressure of the oil system is checked and recorded to confirm that design conditions have been obtained.

With the lubricating oil system operating normally, the seal-oil system can now be introduced onto the main oil pump supply by gradually opening the interconnecting valve between the seal-oil and main oil systems and checking that the seal-oil system pressure remains satisfactory, whilst confirming the correct operation of the pressure regulator.

With the interconnecting valve open, the discharge valve of the AC seal-oil pump is closed gradually, checking that the seal-oil is being maintained at the correct pressure by the differential pressure regulator. Once satisfied that the system is under control and operating to the design conditions, the AC seal-oil pump is shut down, the discharge valve reopened and pump selected to 'auto start'; this is also the selection position for the DC seal-oil pump. With the lubricating and seal-oil systems now in normal service, the system is checked for correct pressure and temperature operation.

The 'auto start' facility of both the AC and DC flushing oil pumps will have been initially set and tested without the main oil pump in service (see Chapter 5 of this volume). Before proceeding with the overspeed tests, it is essential to prove the integrity of the AC and DC flushing pump 'auto starts' by initially reducing the turbine speed, recording the speed and minimum bearing oil pressure at which the AC and DC flushing pumps cut in, and confirming that changeover takes place without pressure fluctuations.

Typical turbine speeds corresponding to the required oil pressures for pump cut-in are:

- AC flushing oil pump, 2500 r/min
- DC flushing oil pump, 2490 r/min
- AC seal-oil pump, 2000 r/min
- DC seal-oil pump, 1900 r/min

Having confirmed the correct 'auto starting' of the oil pumps, the turbine speed should then be increased to 3000 r/min to check the cut-out speeds of the AC and DC flushing oil pumps. It is normal practice not to fit auto shutdown sequences to the AC and DC seal-oil pumps, this being left as an operator action once the turbine is running correctly at 3000 r/min. The automatic shutdown of the flushing oil pumps is set to occur at a rising bearing oil supply pressure of 550 mbar.

3.3 On-load testing of protective devices

Once it is considered that the turbine is operating satisfactorily at 3000 r/min within the specified operating limits and that the boiler and auxiliary systems are all functioning correctly, the next part of the commissioning programme is *on-load protection testing* to check the functioning of certain protective features with the turbine-generator on-load. This ensures that protective devices will work, when called upon in operational service. For example, it is important that valve gear and overspeed rings are checked for movement when operating at full load for long periods of time. This is achieved without significant load drop or risk of trip. Examples of items which can be checked in this manner are given below, together with cross-references to the appropriate texts in this chapter:

- Overspeed rings by oil injection (Section 3.3.1)
- Turbine valve gear (Section 3.3.2)
- Governor range checks (Section 3.3.3)
- Turbine trip solenoids (Section 3.3.4)
- Governor hydraulic fluid pressure switches (Section 3.3.5)
- Direct vacuum trip (Section 3.3.6)

An example of a trip mechanism which isolates the power hydraulic fluid supply to the main steam turbine valves under normal operating conditions, as well as rapidly depressurising the system under trip conditions, is shown in Fig 6.14. This system is common to all aspects of the on-load protection scheme.

Having indicated that the next stage is that of testing the on-load protection, it has to be remembered that the following checks must be carried out regularly to ensure the continued safe operation of the unit during these activities:

- Turbine drainage and gland steam system
- Continuous turbine monitoring
- IP rotor cooling system
- Hydrogen and seal-oil system
- Steam valve chest temperature rise and movement.

3.3.1 Overspeed rings by oil injection

The turbine is generally equipped with two overspeed rings, and the objective of this test is to ensure that each ring is free to move, with the set at normal speed. This test only confirms this aspect, as the actual speed(s) of the shaft at which the rings trip can only be recorded during an overspeed test. Reference to Fig 6.15 will assist with understanding the steps taken whilst carrying out the test:

- Operate the selector mechanism to select the ring to be tested (front or rear). This operation ensures that the trip valve associated with the selected ring will not operate to trip the turbine.
- Observe that the reading on the pressure gauge associated with the trip valve selected falls to zero, indicating that the isolating sleeve has moved into position.
- Depress the testing valve to admit oil to the rear of the ring, causing it to become unbalanced and move out to strike the trip lever.

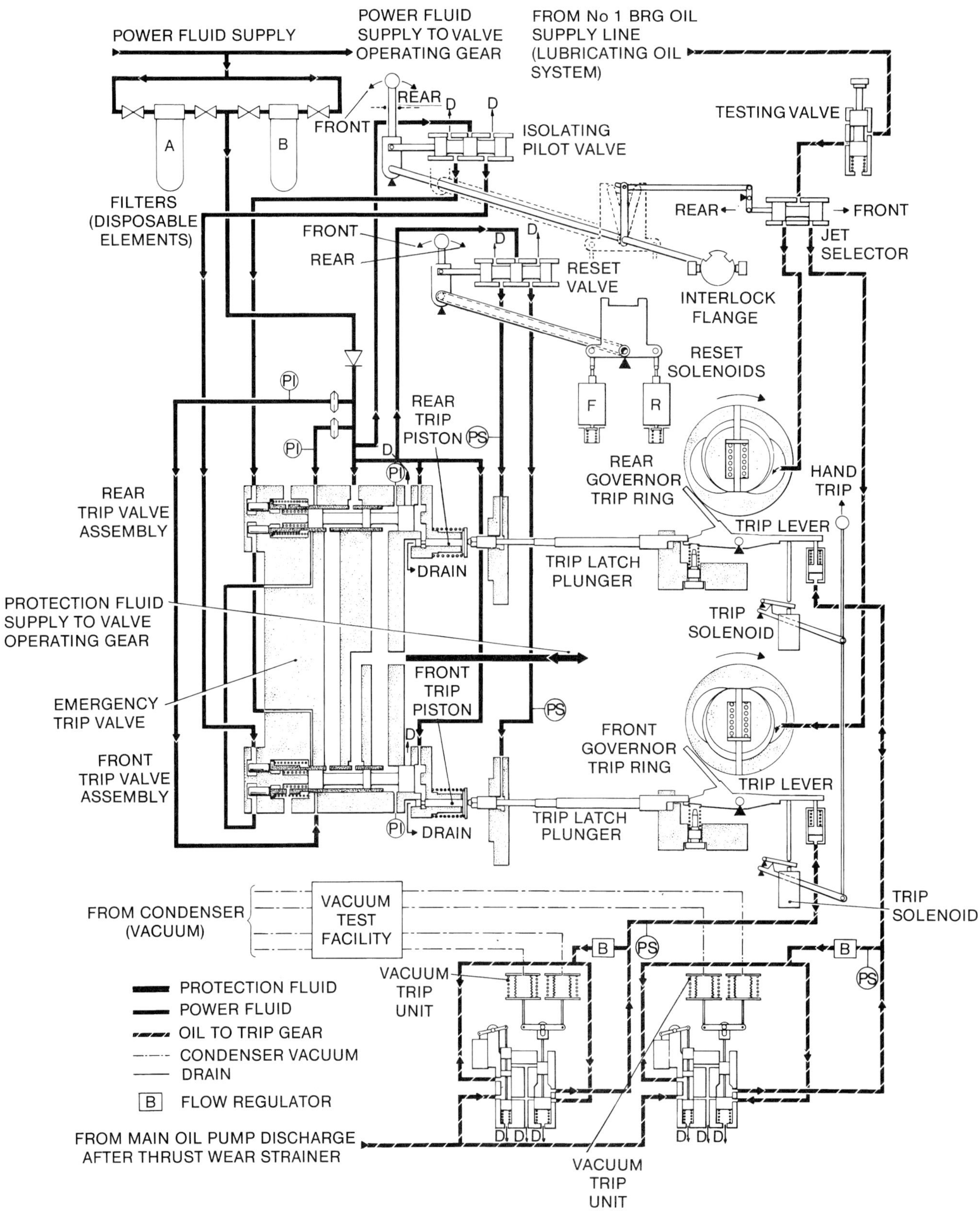

Fig. 6.14 Emergency trip system

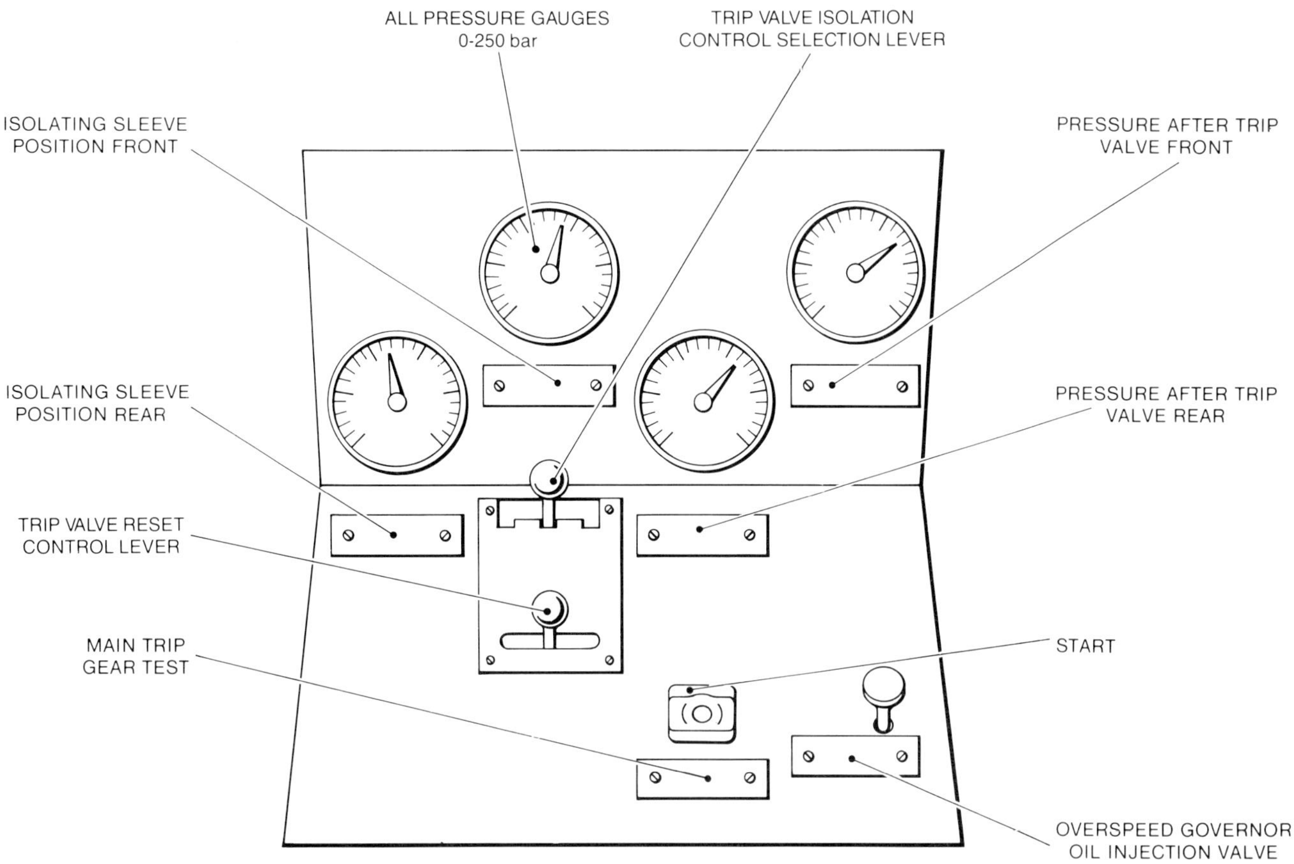

FIG. 6.15 Overspeed ring test panel

- Observe that the reading on the 'pressure after trip valve' gauge has reduced to zero, indicating that the overspeed ring has operated and tripped the trip valve.
- Release the test valve.
- Operate the reset mechanism by selecting the ring tested in order to reset the trip valve.
- Observe the reading on the 'pressure after trip valve' gauge and check that it has returned to its normal value of 135 bar. This confirms that the trip valve has reset.
- Ensure that both selector levers have been returned to neutral and observe that the reading on the 'isolator sleeve position' pressure gauge has returned to normal.

The activity to check the operation by oil injection of the selected overspeed ring is now complete. The same procedure is adopted to test the other overspeed ring.

3.3.2 Turbine valve gear

As part of the pre-STS activities, the turbine valve gear will have been operated to confirm correct operation and indication at the on-load test panel. That test was carried out without steam present, so this test is to confirm that the heating of valve chests and components has not affected the ability of the valve gear to operate normally. Reference to Fig 6.16 will assist in understanding the actions carried out during this procedure:

- Check groups *A*, *B*, *C* and *D* test panel lamps by operating lamp test button. Replace any defective bulbs.
- Select Group *A* valves for testing by operating selector pushbutton and select 'throttle valve' for test.
- Depress the valve test 'stroke' pushbutton and simultaneously operate the valve test 'trip' pushbutton; confirm that the governor valve closes and remains in a tripped position.
- Check that the on-load test panel indication confirms that the valve is shut.
- Operate the valve sequence selector, selecting 'main stop valve' for test.
- Depress the valve test 'stroke' pushbutton and simultaneously operate the valve 'trip' pushbutton; confirm that the stop valve closes and remains in a tripped position.

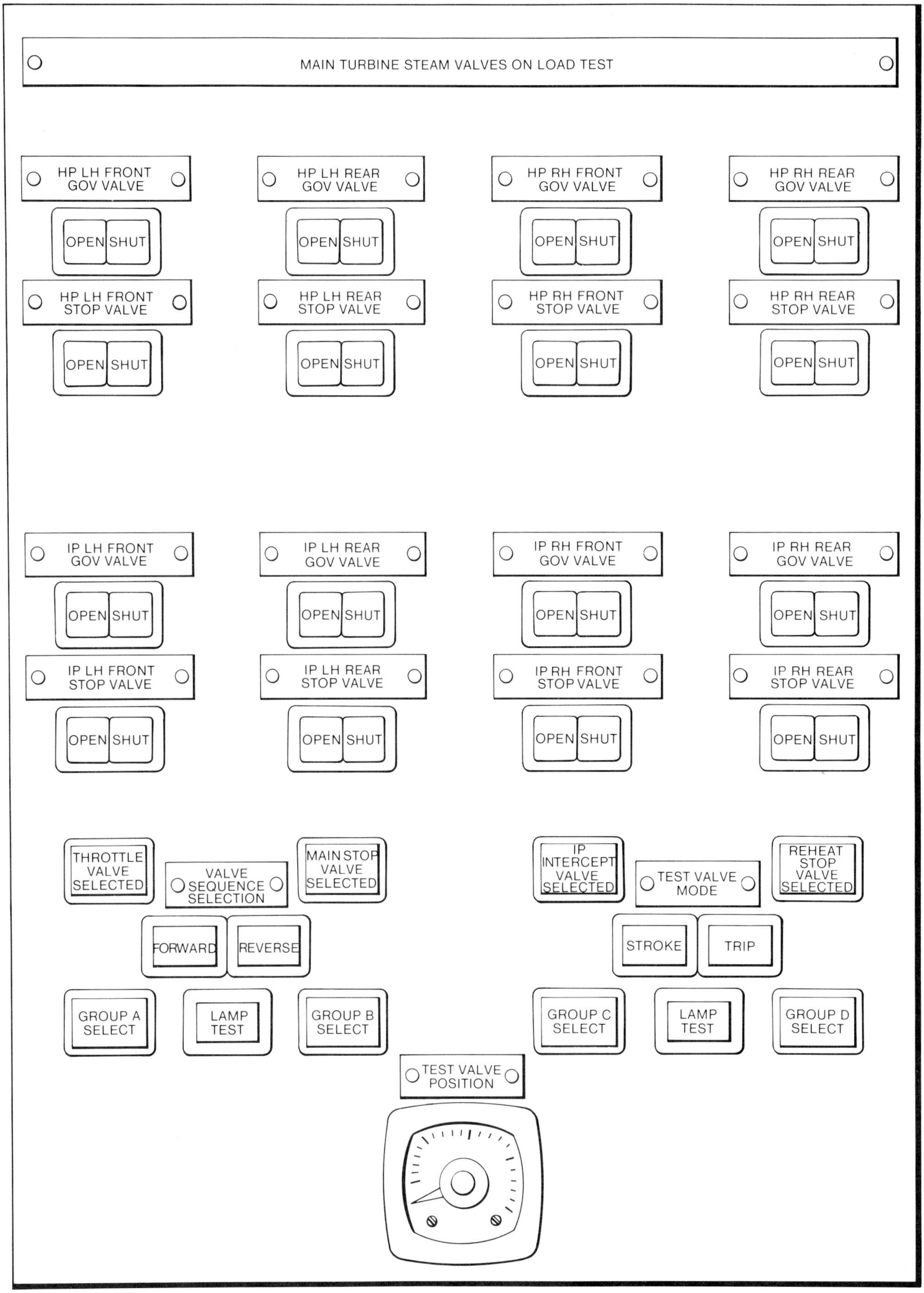

FIG. 6.16 On-load turbine valve gear test panel

- Check that the on-load test panel indication confirms that the valve is shut.
- Operate the valve sequence selector, selecting 'IP governor valve' for test.
- Depress the valve test 'stroke' pushbutton and simultaneously operate the valve 'trip' pushbutton; confirm that the governor valve closes and remains in a tripped position.
- Check that the on-load test panel indication confirms that the valve is shut.
- Operate the valve sequence selector, selecting 'IP stop valve' for test.
- Depress the valve test 'stroke' pushbutton and simultaneously operate the valve 'trip' pushbutton; confirm that the stop valve closes and remains in the trip position.
- Check that the on-load test panel indication confirms that the valve is shut.

This now completes the closing/tripping of the *A* stream of turbine stop and governor valves. The turbine is now running at 3000 r/min with the following valves closed:

- *A* HP turbine stop valve
- *A* HP turbine governor valve
- *A* IP turbine stop valve
- *A* IP turbine governor valve

Before reversing the valve test procedure, the effect on turbine bearing vibration and eccentricities under this mode of operation should be checked and recorded. No problems should arise, as the design permits operation under this condition, but any abnormalities must be investigated before proceeding.

With group *A* valve testing complete, the 'reverse' selector is operated to open the valves in the following order:

IP governor valve

IP stop valve

HP governor valve

HP stop valve

- Depress the valve test 'stroke' pushbutton to reset the trip and check that the IP stop valve opens.
- Check that the on-load test panel indication confirms that the valve is open.
- Repeat the reverse actions and select operations for the remaining three valves until all group *A* valves are open.
- Deselect the *A* group of valves after completion of test.
- Select groups *B*, *C* and *D* valves in turn for test, adopting the same procedure and checks as outlined for the *A* stream of valves.

3.3.3 Governor range check

Having completed both the on-load valve testing and the overspeed oil injection test, the top of the governor range can now be checked with the knowledge that governor valve operation is correct.

This requires the turbine speed to be raised to the top of the governor range by operating the speed control to a level where no further speed increase takes place. This occurs at approximately 3180 r/min.

During the time that the machine speed exceeds 3000 r/min, the following are checked:

- Turbine vibrations and eccentricities
- Hydrogen seals and seal-oil system

Having recorded the top of the governor speed range, reduce the turbine speed to 3000 r/min and check that the operating parameters which increased during the activity have returned to the previous levels.

3.3.4 Turbine trip solenoids

A turbine is normally equipped with one or more trip solenoids and this test is to confirm their operation. The test procedure is similar to that adopted for the on-load overspeed ring test, except that oil injection is not used. Instead, the solenoid test button is depressed and correct operation is denoted by the reading on the 'pressure after trip valve' gauge falling to zero. Reference to Fig 6.15 will assist with understanding the steps taken.

3.3.5 Governor hydraulic fluid pressure switches

The object of this test is to demonstrate that loss of governor hydraulic fluid pressure operates the pressure switches and gives correct alarm indications.

A typical installation uses six pressure switches arranged in two groups of three. A governor trip occurs when two out of the three pressure switches in a group indicate loss of pressure.

The system is designed to enable one group at a time to be taken out of the system for testing, whilst still retaining the full protection against loss of hydraulic fluid pressure on the other group. Figures 6.17 and 6.18 show test panels associated with this equipment and the procedure outlined below refers to equipment on such a panel:

- Turbine to be at 3000 r/min.
- Operate the 'test selector' switch normally fitted onto the unit control desk to either 'Test *A* Group' or 'Test *B* Group'. The selection automatically isolates the group selected for test electrically. Hence the protection is retained on the other group, requiring a '2 out of 3' operation of the pressure switches to confirm loss of hydraulic pressure.

- Having completed the selection on the unit control panel, a master key is released for use in the local hydraulic fluid test panel, which is located adjacent to the HP end of the turbine.
- The *A* and *B* groups each have their own local test panels (Fig 6.18). On the selected panel, operate the lamp test switch, check that all panel lamps light and that all panel lamps extinguish on deselection.
- Select the 'test' position of the lamp isolating switch.
- Open the hydraulic fluid test line valve and confirm that the test gauge reading records the fluid system pressure, normally 138 bar.
- Remove the key from the test selector and insert it into a key exchange box fitted on the local panel, which allows two channel keys in the exchange box to be removed. The panel is arranged such that the test is carried out with '2 out of 3' combinations, i.e., A_1 with A_2, A_2 with A_3, and A_3 with A_1.
- These two channel keys can now be inserted into two locks, allowing the hydraulic fluid test valves to be operated for the first combinations.

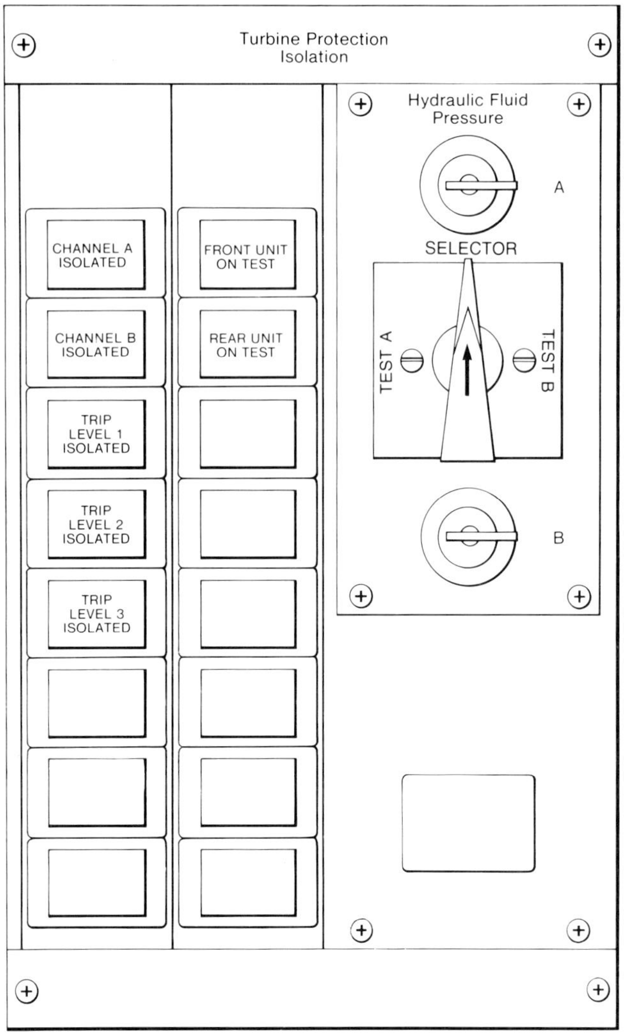

FIG. 6.17 Control room hydraulic fluid pressure test selector panel

- After opening the test valves and confirming the correct pressure, the drain valve is opened slowly to decrease the pressure.
- Operation of the corresponding pressure switches will be indicated by the lamp indication. The pressure indicated on the test gauge when the pressure switch operates must be recorded.
- Confirm that all associated alarms and indications have been received in the CCR.
- The test drain valve is now closed slowly, confirming by indication that, as the pressure increases, the switches reset and that all the alarms and indications received during the test are cleared.
- The hydraulic fluid test valves are then returned to their normal operating positions and the two channel keys removed.
- The same procedure is then completed for the other two combinations of the *A* group.
- Having completed the *A* group combination testing, the two keys removed from the key exchange box associated with the *A* group are returned to the box, allowing release of the master key.
- The master key returned to the test selector allows this switch in the first instance to be selected to 'Normal'. Before proceeding to the *B* group, a quick check should be carried out to ensure that no problems have developed during this activity.
- Operate the test selector switch and select 'Test *B* group'.
- Repeat the same procedure for the *B* group.
- On completion of this test, a final check should be made to confirm that all valves associated with the test panels are in their correct position. Test keys must be refitted into the key exchange box and the master key returned to the 'test selector' on the unit desk, which in turn should be switched to 'Normal'.

3.3.6 Vacuum trip test

The vacuum trip system shown in Fig 6.14 is used to trip the turbine before the condenser vacuum falls to the level at which the possibility arises of LP blading overheating due to windage friction. A typical trip value for a 660 MW turbine-generator is −840 mbar.

Facilities are provided to enable individual testing of both the vacuum trips to be carried out on-load without tripping the turbine, whilst at the same time recording the vacuum setting. A typical test procedure is outlined below and reference to Fig 6.18 will assist with understanding the procedure. The vacuum test panel is normally located adjacent to the turbine HP cylinder:

- Confirm that the panel lamps on the vacuum test panel are functioning by operating the lamp test

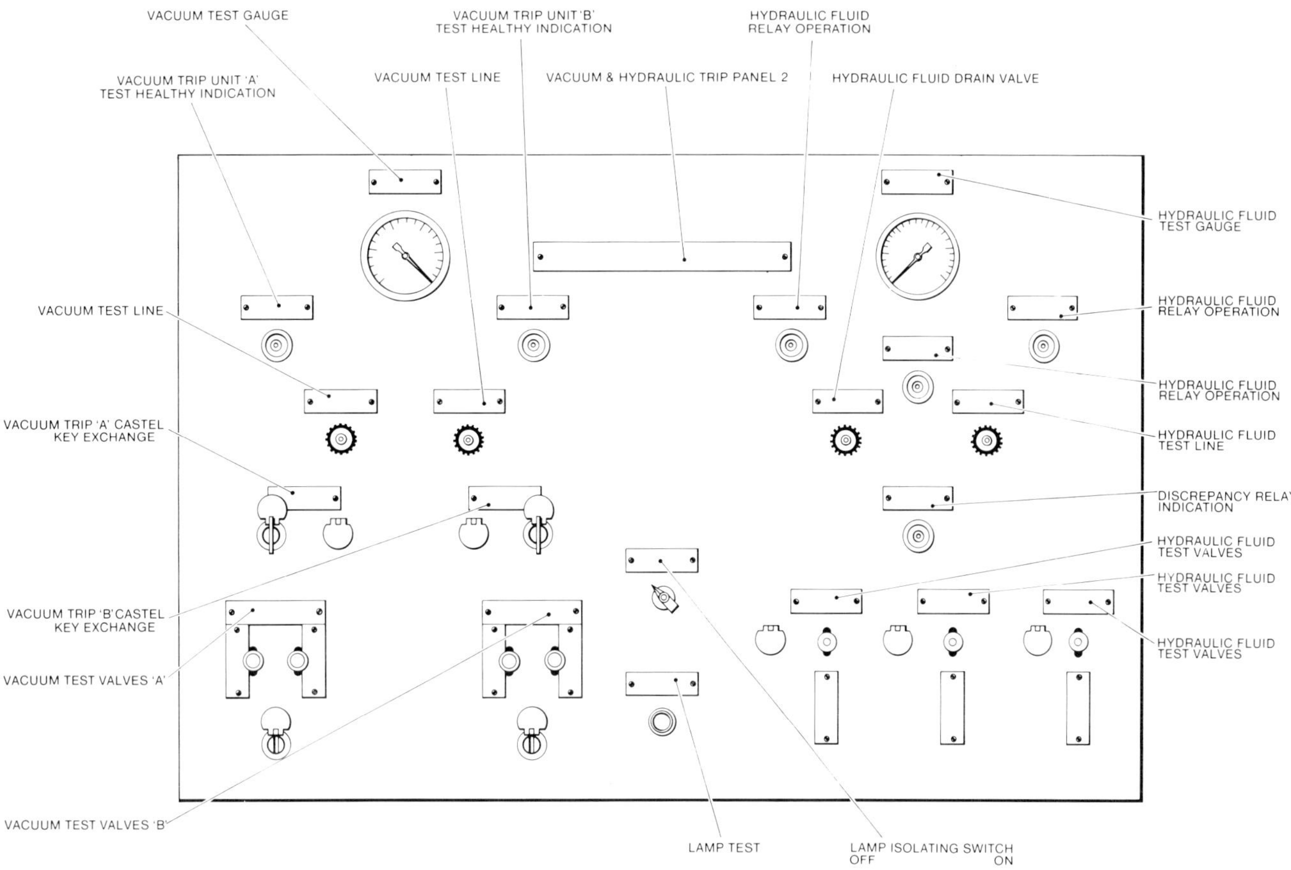

FIG. 6.18 Turbine hydraulic fluid pressure switch test panel

button. Check that all lamps extinguish on completion of test.

- On the test panel, open the vacuum isolating valve and check that the reading obtained on the gauge corresponds to the known condenser vacuum.
- Select the front or rear trip valves for testing and confirm that the isolating sleeve has moved into position by the pressure gauge indication reducing from 135 bar to zero.
- Having selected the appropriate trip valve, this will release a key that can then be used in the on-load test panel. The keys are of the mechanical interlock type with a different code for each of the two trip valves.
- The insertion of this key in the vacuum trip on the local test panel isolates the tripping circuit.
- Having isolated the tripping circuit, a further key is released which is then inserted in the vacuum test panel. This enables the selected vacuum trip unit test valve to be operated.
- The test valve is operated and, at the same time, the vacuum vent valve is opened to release the vacuum in the line, indicated by the test gauge reading reducing to zero. Once confirmed, the release valve is shut. This is repeated for the second test valve
- With the two test valves in their 'normal' positions, both valves are operated simultaneously to the 'test' position and the vacuum released by operating the vacuum release valve. This expands both bellows units simultaneously, causing the trip lever to be actuated.
- At an indicated value of −840 mbar, the alarm light should illuminate, indicating that a trip has occurred: the vacuum release valve is then shut. Further evidence of a trip is that the 'pressure after trip valve' indication for the selected valve reads zero.
- Reset the selected trip valve and check that the indications are correct by carrying out a lamp test.
- Having confirmed that all test valves on the vacuum test panel are in their correct position, return the keys to their normal positions and deselect the trip valve selected for the test.
- Select the other trip valve and repeat the procedure.
- On completion of this test, a final check should be made to confirm that all valves associated with the vacuum test panel are in their correct positions. All test keys should be refitted and both trip valves checked that they are in their normal positions.

3.4 Turbine overspeed testing

Having completed the 'on-load protection' testing, the next activity is to carry out a turbine overspeed test to confirm that the overspeed protection operates and initiates a trip. The speed at which the trip occurs for each ring is recorded. It is a requirement in the CEGB that an overspeed test is carried out at least every six months and that the overspeed rings are 'on-load' tested monthly. The 'on-load' testing of the overspeed rings has already confirmed that the rings are free to operate: these tests confirm the speed at which they operate. Figure 6.19 shows a typical assembly of an overspeed trip mechanism. The rings are normally adjusted to operate near to 3300 r/min. The reason for this is that 10% overspeed is the operating maximum allowed, but to have a ring operating considerably lower than this could involve a turbine trip if the system frequency were to rise, for example, if it tripped at 3100 r/min, then a system frequency of 51.7 Hz would cause a trip.

The following conditions should be monitored during the overspeed test:

Turbine speed

Turbine bearing vibration and eccentricity

Turbine differential expansions

Hydrogen and seal-oil system

The information obtained on the speed ramp, with particular regard to turbine bearing vibrations and eccentricities, will be useful information for consideration in load rejection testing. A typical test procedure is as follows:

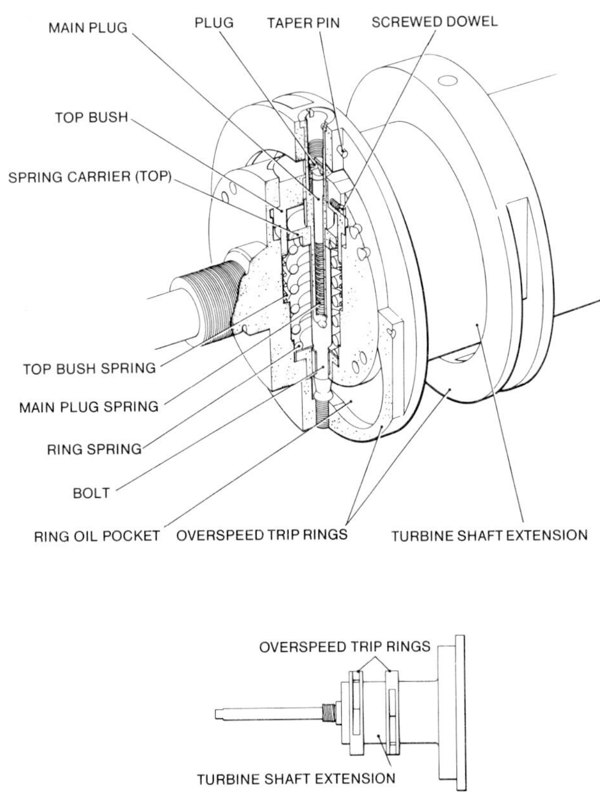

FIG. 6.19 Overspeed trip mechanism

- Turbine at 3000 r/min, boiler stop valves still only 30% open.
- As the exciter link is not yet fitted for this first overspeed test of the turbine, no voltage can be induced into the rotor. Under normal turbine operation this will not be so and hence isolations should be carried out on the field circuit prior to overspeed tests. This is particularly important where generators are connected directly to step-up transformers, as overfluxing can occur.
- Select the overspeed ring to be tested using the procedure outlined in Section 3.3.1 of this chapter.
- Selection will be confirmed by the reading on the pressure gauge associated with the selected trip valve falling to zero.
- With one type of electronic governor, the turbine has to be within the 'overspeed allow' range (typical minimum value 2980 r/min). This is confirmed by a coloured light illuminating to indicate that control for an overspeed test is available. The method of achieving this state varies with manufacturer and the operating instructions for the particular turbine must be consulted.
- With the turbine within the overspeed test range, the turbine speed can be increased after freeing the overspeed range release button.
- Operation locally of the speed button will signal the governor valve to open and the machine speed to increase at a rate of approximately 10 r/s.
- Record the speed at which the trip operates. This can be displayed on a digital tachometer on the local test panel, which is fed from the speed signal being received by the governor from the speed probes. To obtain a permanent record of the overspeed value at which the trip occurred it is necessary to install a high speed recorder.

Check that the indication on the gauge 'pressure after trip valve' falls to zero on the local test panel when a trip has occurred. This is a useful cross-check.

If a trip has not occurred by the time the turbine has reached a maximum speed of 3350 r/min, the test on the selected overspeed ring must be abandoned. The second ring should be selected and, if a similar failure to trip is experienced, the test should be discontinued, the turbine run down in speed and put on electrical turning gear prior to adjusting the overspeed ring settings:

- When a trip occurs, confirm that the HP and IP governor and stop valves have tripped and that the appropriate indication has been received on the unit control desk.

- Allow the turbine to run back to 3000 r/min to enable the overspeed ring to reset.
- Reset the turbine trip valves. This action should not be attempted until the overspeed rings have reset automatically at 3000 r/min.

If the machine trips cannot be reset before the speed reaches 2850 r/min, the turbine should be allowed to run back to a known safe speed band outside any critical speeds. This speed has been established during the first run-up. During this activity, the 'auto cut-in' facility for both the AC and DC flushing oil pumps and AC seal-oil pump can be rechecked:

- Move the 'overspeed ring' selector to the normal position.
- Operate the governor controls to ensure that it will allow normal valve movements for speed increase to take place. At this stage the IP stop valves should be fully open.
- Operate the governor to open all four HP governor valves to approximately 20% opening and the IP governor valves to 80% opening. Open the HP stop valves and raise the speed at the rate of 300 r/(min)^2. Confirm that governor control is re-established at approximately 2840 r/min.
- Raise the turbine speed to 3000 r/min by operating the governor control.
- With the turbine speed at 3000 r/min, the levels of turbine bearing vibration and eccentricity are recorded and checked against the levels experienced prior to the first overspeed test.
- Select the other overspeed ring for test and carry out a second overspeed test.
- On the satisfactory completion of the overspeed trip test, the overspeed selection is returned to normal and the turbine re-established at 3000 r/min.

Speed values within the band 3250 to 3320 r/min are generally acceptable for overspeed tripping. If trips occur outside this band, it is necessary to adjust the setting. This can only be carried out when the turbine shaft is stationary.

The decision of when to adjust the overspeed ring setting depends on the actual value obtained. However, if both overspeed rings operate below the top of the governor range or in excess of 3320 r/min, the test programme should be delayed to enable the settings to be adjusted, followed by a repeat test.

3.5 Commissioning tests on electronic governor

3.5.1 Objective

The pre-commissioning test and checks, described in Chapter 5 of this volume, will have initially proved the electronic governor equipment. However, with the turbine at 3000 r/min, tests are carried out to demonstrate:

Governor steady-state control This includes:

Speed range checks

Speed error reference signal checks

Speed stability checks under steady state conditions

Governor droop characteristic setting and checks

Governor transient control These tests demonstrate the stability of the governor control system under transient conditions by inducing valve gear trips.

3.5.2 General description

The electrohydraulic governor system provides a closed-loop speed control system for the narrow range speed control of a turbine-generator. A simplified block diagram of such a system is shown in Fig 6.20. Speed is measured by magnetic pick-ups which generate a frequency proportional to turbine speed. The frequency signal is compared with the speed reference to produce the speed error signal. To obtain the speed/band (droop) characteristics, the speed error signal is modified by the droop amplifier, the output of the droop amplifier being added to the load reference to produce the speed/steam signal. The load reference is used prior to synchronising to produce a fine speed control and then after synchronisation to load the turbine. The characterisation circuits generate the valve position signals from the steam demand signals.

For the HP governing valves, the characterisation circuits sum the steam demand, the stabilising and acceleration trip signals; the resultant signal passes through a non-linear circuit which converts it into a valve position signal.

For the HP stop valves, the steam demand signal is compared with a valve opening limit signal from each stop valve and the lower of the two values is used as the valve position signal.

The IP governor valves are controlled as a result of the steam demand signal being added to the stabilising and acceleration trip signals which is, in turn, added to the signal from the IP bias potentiometer.

The HP and IP valves are controlled from separate modules, with four modules controlling the four HP stop valves, four controlling the four HP governing valves and a further four modules controlling the IP governing valves. The IP stop valves have only two positions, i.e., fully-open or fully-closed, and are tripped in response to signals from the on-load valve test equipment or by the tripping of the corresponding IP governing valve. Figure 6.21 shows a typical 660 MW turbine control system.

As the governor has three separate channels, highway signals produced by each of these are compared and, if discrepancies are detected, the appropriate alarms are raised and the faulty channel disconnected.

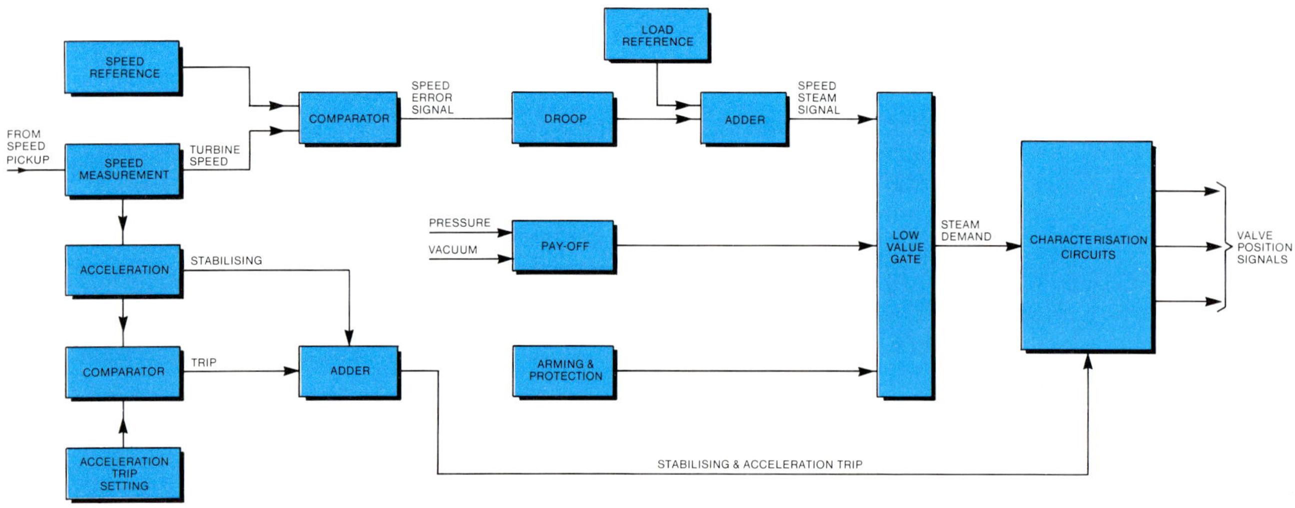

FIG. 6.20 Main governor system block diagram

If a second discrepancy occurs, the dump solenoid operates and trips the turbine.

Highway signals from the three channels are connected to each valve control module and are averaged and amplified (see Fig 6.22). The averaged highway signal (valve position demand signal), and a valve position signal derived from a variable differential transformer associated with the valve, are compared and an error signal produced. The error signal is used, via a power amplifier, to control the servo coil associated with the valve actuator and thereby to position the valve.

A second comparison between the averaged highway signal and a valve position signal derived from a second

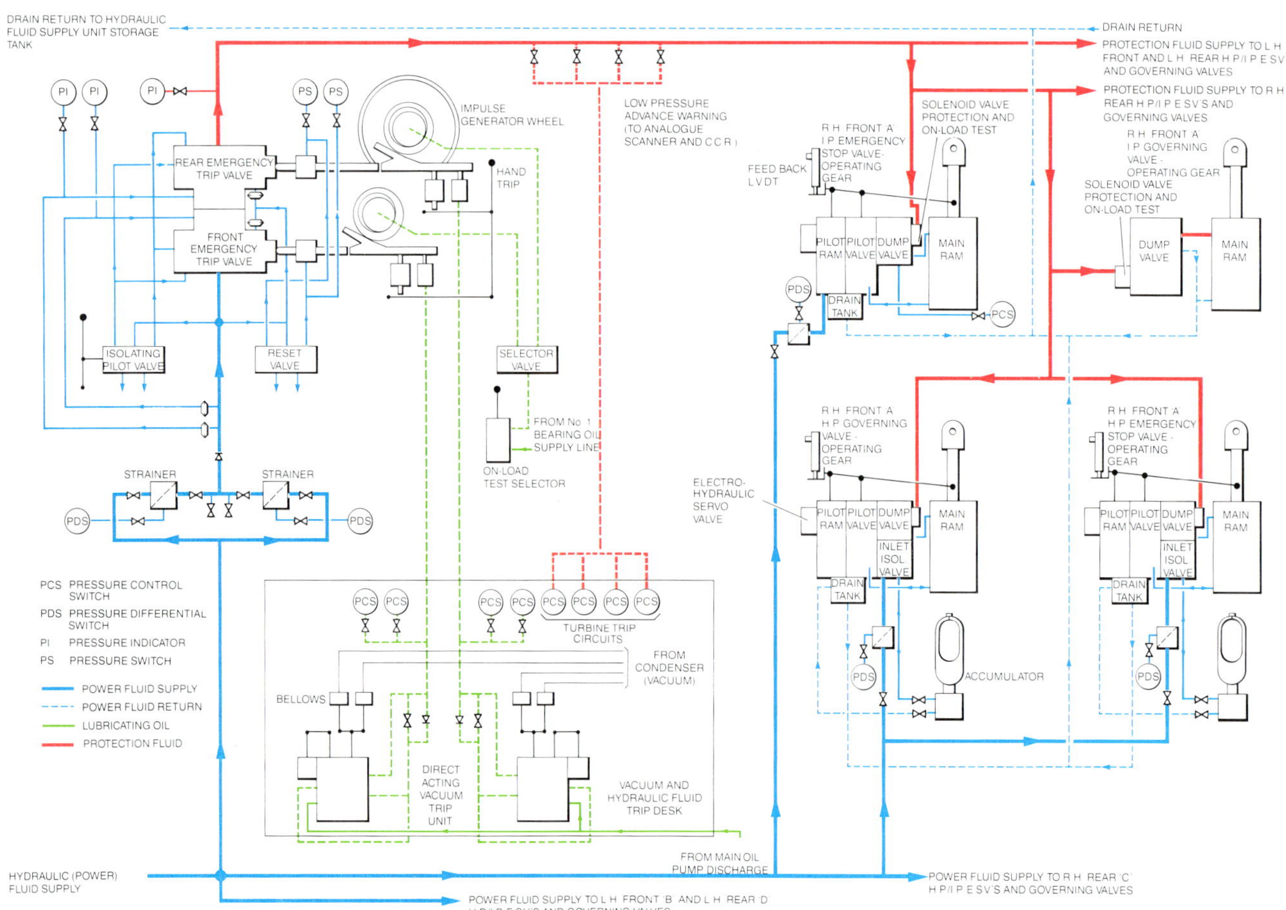

FIG. 6.21 Typical turbine valve control and protection, hydraulic system

linear variable differential transformer checks the signal; if a discrepancy is detected, the dump solenoid operates to close the valve. This safety feature ensures that failure of equipment, such as position indicators, does not cause erroneous governor action.

3.5.3 Commissioning tests

Governor control checks

The first check on the governing system took place during the initial turbine run-up to synchronous speed (Section 3.1.8 of this chapter) and identified the speed at which full governor control commences (expected value 2790 r/min). With the turbine speed held at the lower end of the governor range, checks are carried out to confirm the following:

- Turbine speed
- Turbine governor valve openings
- Steam pressure and temperature at turbine stop valves
- Condenser vacuum

It is important that this information is obtained regularly during this activity and a 30-second update is recommended. The outputs from the three magnetic pick-ups must be checked to be producing the correct level of speed signal.

The stability of the governors must be checked at synchronous speed under steady state conditions to confirm that turbine speed is held constant.

The range of speed control should be checked by adjusting the speed error reference and seeing that the turbine speed is steady over the complete governing range. The method of testing the governor range is as follows.

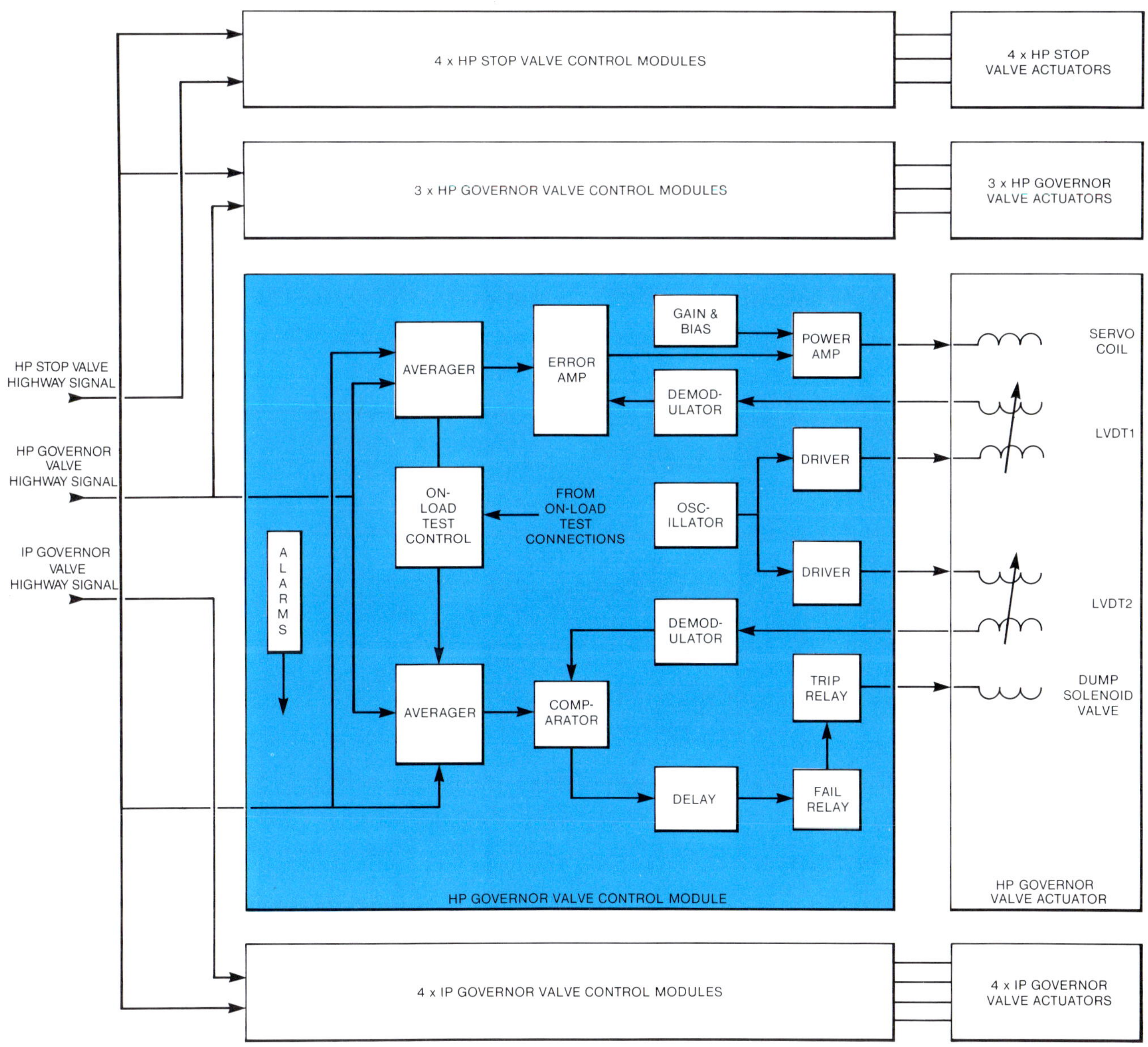

FIG. 6.22 Main governor valve control modules

Lower the speed error reference signal until a minimum speed is reached at which the turbine is under the control of the governor. Lowering the speed error reference further will prevent the governing system from controlling at that speed point and the turbine will begin to run down.

This minimum speed which can be sustained by the governor is recorded at the 'bottom of the governor range'. Having established the minimum speed, the speed error reference signal is increased to its maximum value, recording the speed of the turbine at the 'top of the governor range', the expected value of this speed being approximately 3150 r/min.

Where the governing system has a facility for selecting the governor droop characteristic, the tests to establish the bottom and top governor range levels should be carried out at each of the available droop settings. Figure 6.23 shows a governing system with 5 droop settings, namely 1%, 1.5%, 2%, 3% and 4%, with a speed limit of +6% for the upper speed error reference and −7% for the lower speed error reference.

Table 6.7 indicates the predicted maximum and minimum governor speeds for various governor droop settings.

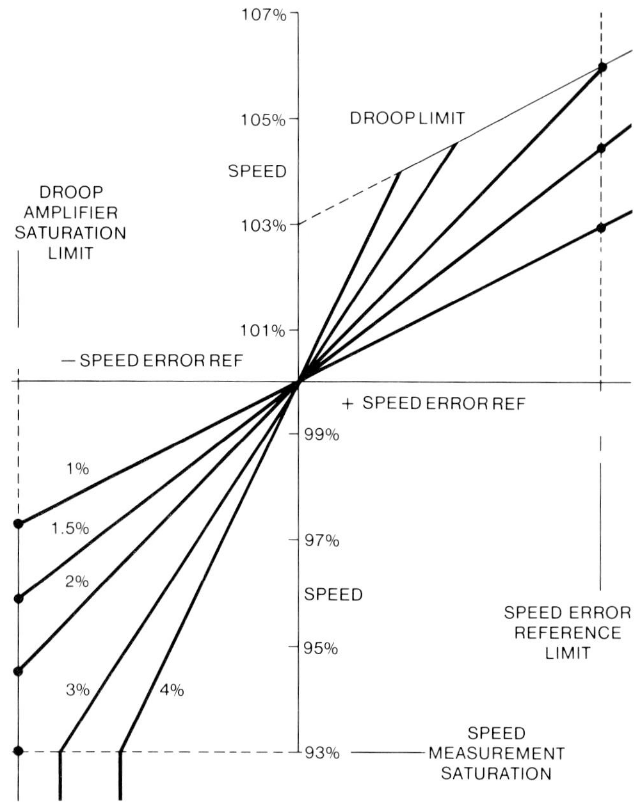

FIG. 6.23 Main governor droop circuit characteristics

TABLE 6.7

Typical maximum and minimum governor speeds for various governor droop settings

Governor droop settings	Governor range, r/min	
	Maximum	Minimum
4%	3180	2790
3%	3180	2790
2%	3175	2830
1.5%	3125	2870
1.0%	3090	2910

Governor system stability checks

With a governing system using three channels, the loss of any one channel should not affect either the governing system or the control of turbine speed. Although it will have been checked as part of the pretesting work, it must be retested with the turbine at speed.

To test this, a speed computing module must be switched out, thus tripping the associated governor channel. The alarms received are checked for correct operation and any disturbance in turbine speed noted. Having confirmed correct operation with no channel interaction, the tripped module should be reset after realigning the references. The test should be repeated for the other two channels.

Valve tripping

The rest of the test programme involves tripping each of the four HP and IP stop and governor valves in turn, checking that the turbine speed remains constant and that no interaction is induced into the governing system by the trips. The method to be adopted is as follows:

- Trip one HP stop valve control module and confirm that the selected valve has tripped.
- Reset the module and confirm that the valve reopens.
- Repeat this procedure for the other HP stop valves, HP governor valves, IP stop valves and IP governor valves.

The final part of the governor commissioning tests is to monitor the acceleration trip circuitry. This requires the installation of an oscilloscope connected to the governor system. The monitoring should be carried out over at least 10 minutes, during which time the turbine speed is varied, using speed error references. The check should confirm that the acceleration trip setting selected will protect the turbine without causing instability. A typical trip level is $7\%/s^2$.

The electronic governing systems for modern units are complex and specialised. This section only outlines the main parameters to be checked, the detailed knowledge for these tests to take place being obtained from the equipment manufacturers and other specialist engineers. Without the availability of specialists or allowing sufficient time to train staff, delays in the commissioning programme could be experienced as faults occur. It is essential that as much pre-testing as possible is carried out prior to steam-to-set, as equipment faults can cause repeated start-ups and shutdowns of the plant if they are not detected at an early stage.

3.6 Commissioning tests on the automatic voltage regulator (AVR)

3.6.1 Introduction

The commissioning of the AVR and all its associated equipment is carried out in three distinct stages:

- Pre-commissioning tests, which are static tests. These aspects are described in Chapter 5, Section 13 of this volume.
- Commissioning tests with the links between the exciter and generator field removed, with the turbine at 3000 r/min. The static testing used a motor-generator set to provide the 400 Hz supply, but with the turbine at 3000 r/min the 400 Hz supply can be derived from its normal source, the permanent magnet generator. With the link between the exciter and generator field removed, the AVR equipment can be energised from its permanent supply point without exciting the generator field.
- AVR testing, including overfluxing tests. This testing is carried out with the links between the exciter and the generator in the service position, with the turbine at 3000 r/min on open-circuit. This activity will be described in Section 3.14 of this chapter.

3.6.2 General description of an AVR system

Two three-phase thyristor converters, operating in parallel, rectify and control the current supplied from the 400 Hz permanent magnet pilot exciter to the main exciter field. Each converter is controlled by its own AVR, with an associated manual follow-up, manual restrictive circuit and importing MVAr limiter. A current limit switch circuit in each channel (converter and AVR combination) prevents overloading of the converter. Overvoltage protection is also provided, which is initiated by a spark gap across the converter output. Balanced loading of the two channels is maintained by means of a current-sharing circuit. Figures 6.24 and 6.25 will assist with understanding the text.

During generator run-up, the excitation is activated at a preset speed, with the AVR selected to 'auto' control. During generator run-down, the excitation is inhibited at approximately the same speed by means of an inhibiting voltage signal fed into the thyristor firing circuits.

In order to reduce the possibility of overfluxing the generator transformer, the AVR speed reference reduces proportionally with frequency below 47.5 Hz. This ensures that the generator terminal voltage reduces proportionally with speed below 2850 r/min.

The manual follow-up maintains the manual rheostat in the correct position to give a smooth changeover

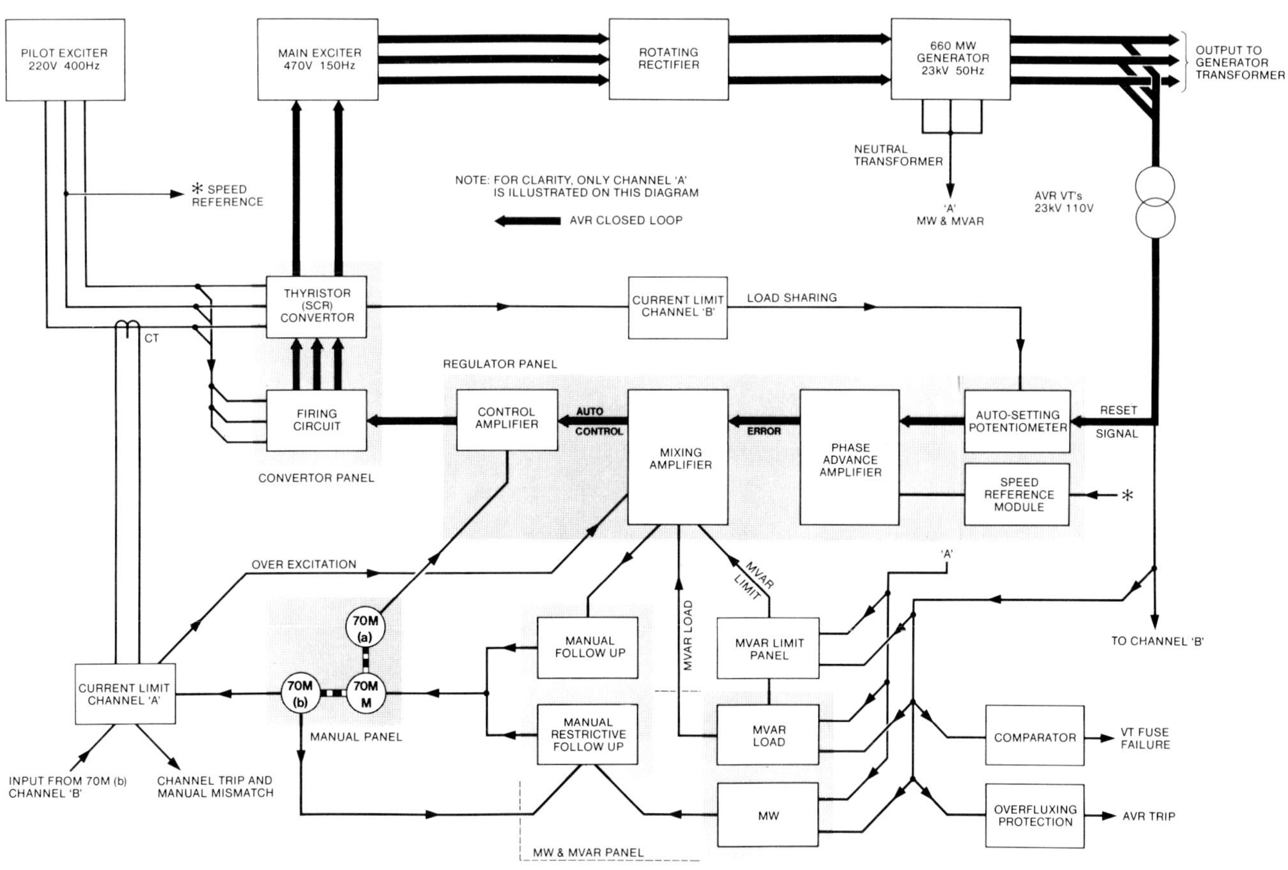

FIG. 6.24 AVR functional flow diagram

(minimal excitation change) should the AVR trip to 'manual' control due to a fault. The action of the manual restrictive circuit is to inhibit the follow-up of the manual control potentiometer by placing a restriction on its travel, such that, should a trip occur, the standing level of signal from the potentiometer is at or above the safe limit. Without this inhibition, there would be a risk of generator instability if the AVR tripped under these conditions. With the manual restrictive circuit operating, a smooth changeover is not possible, but the disturbance is minimised by selecting the 10% limit rather than the normal 20% limit for manual control. Following a trip to manual control, it would be necessary for the operator to increase the excitation until the generator operates within the 20% limit.

The following protective devices are provided for each channel:

- VT fuse failure
- Thyristor supply fuse failure
- Converter cooling failure
- DC voltage supply failure
- Over-excitation: after a preset time delay.
- Under-excitation: after a preset time delay

With two channels in operation, the initiation of any of these devices would cause the faulty channel to trip, leaving the healthy channel controlling the excitation in auto control. Should a fault occur on the remaining channel, the AVR on that channel would trip to manual control.

Two more protective devices operate only when the generator main circuit-breaker is open. The first is an overvoltage relay on each channel, which monitors the generator terminal voltage. When an excess voltage occurs on the generator, its excitation is suppressed by removing the control signal to the firing circuit of each converter and shorting out the control signal input. At the same time, the three-phase supply contactors to each channel are tripped.

The second protective device is the transformer overfluxing protection relay in each channel which operates when a preset voltage/frequency ratio is exceeded. Initially, it operates on the AVR and manual rheostats to reduce this ratio. If this is not successful, after a preset time delay it suppresses the generator excitation in the same manner as the overvoltage relay.

A fault on the generator which causes the main protection trip to operate also suppresses the excitation in the same manner.

3.6.3 General design criteria

The AVR should be capable of maintaining the generator terminal voltage to within +0.5% of the set voltage over the whole of the generator operating range. An MVAr limiter should be included to prevent the generator from operating beyond stability limits of 4% at full-load to 10% at no-load. A manual follow-up device maintains the manual control at the setting taken up by the automatic control. This is inhibited when the generator operates beyond a 10% full-load to 10% no-load stability limit in order to reduce the risk of generator instability if the AVR should trip to manual control when operating beyond these limits. The 10% limit is chosen for this setting, in preference to the manual stability limit of 20% at full-load to 10% at no-load, to avoid an excessive rise in the generator terminal voltage following an AVR trip. The operator is expected to increase the generator excitation until the machine is operating within the manual stability limit by tap changing on the generator transformer.

The MVAr limiter settings will be made during the static pre-commissioning tests and checked with the generator on-load.

The VAr limiter settings will be made during AVR commissioning.

3.6.4 Commissioning tests

The static testing of the AVR and the excitation control equipment will have confirmed the following:

- The static characteristics of the equipment.

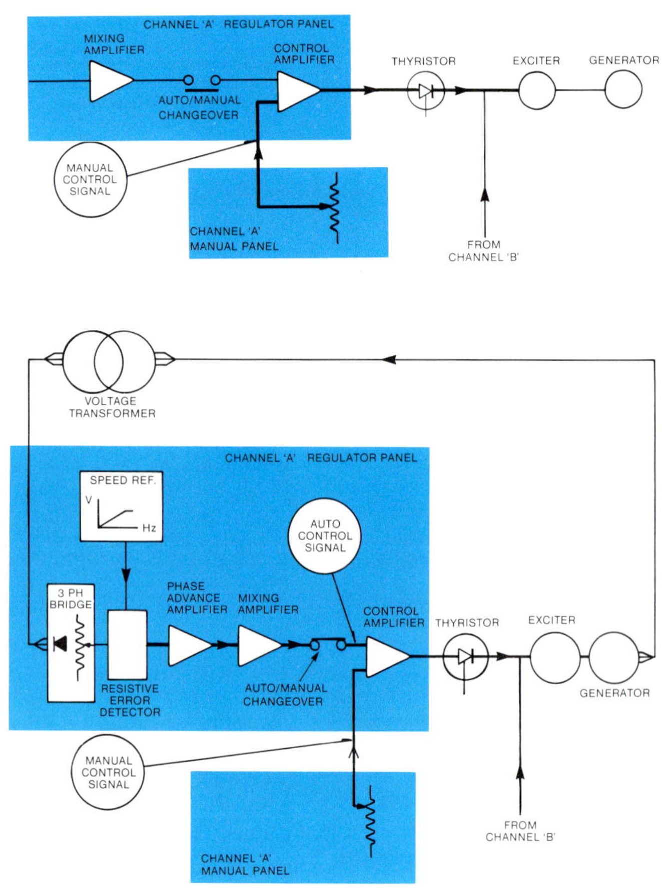

FIG. 6.25 AVR auto and manual control block diagram

- The correct preset of various control functions, including the operating points of the MVAr limits.
- The correct operation of all relays and contactors.
- Correct operation of all remote controls, indications and alarms.
- Correct operation of all AVR and excitation system protective devices, including suppression of the generator excitation.
- An acceptance value of insulation resistance of the equipment.

The tests to be carried out at 3000 r/min are as follows:

- Proving the output of the 400 Hz permanent magnet pilot exciter and its phase rotations.
- Measuring the 400 Hz permanent magnet pilot exciter output voltage and selecting the converter transformer tap position for both the *A* and *B* channels.

Independent channel checks

The following checks are carried out on each of the two channels independently:

- Check the supply available from the supply module of the thyristor converter.
- Check that the DC output from the thyristor bridge is smoothly controlled.
- Confirm that the voltage output collapses on tripping the thyristor bridge.
- Confirm that the firing angle of the converter bridge changes from 90° to 30° by adjusting the gain of the converter to produce a voltage change in the order of 3 V.

The MCR excitation range should be set such that 95% of the converter output voltage is equivalent to that necessary for MCR excitation.

The minimum voltage output is also set, subject to the following:

- The minimum generator voltage in 'manual' mode is not critical but must be less than the minimum voltage obtained in 'auto', the converter gain control being set to obtain the correct voltages.
- Confirm that, when set to MCR excitation in the 'auto' control position, the converter is driven into saturation after a predetermined time interval and that a trip to 'manual' results. The permanent magnet generator and converter output voltage should be measured during this test.
- Having completed the checks on the selected channel, they should be repeated for the other channel, isolating the channel which has already been tested.

Dual-channel checks

With both *A* and *B* converter isolators closed, adjust the voltage control to ensure that both AVR channels are sharing the load equally and that the converter output voltage is equivalent to that required for MCR operation.

The AVR is then switched from 'manual' to 'auto' which should cause the converter output to rise rapidly to the saturation level and remain at this condition until the second channel trips to 'manual', at which point it should return to the previous levels. Both channels should not trip out completely, only tripping to 'manual'.

The output voltage of the PMG and converter is recorded during this test.

3.7 Rotating rectifier blown-fuse detection test

3.7.1 Objective

With two dummy fuses fitted, the objective is to confirm that the rotating rectifier unit indicates the faults by generating two negative pulses and to locate the faulty fuses by analysing these pulses.

3.7.2 General description

The blown-fuse detector equipment is installed as part of the protective systems where rectifiers are installed and mounted on the exciter shaft. Figure 6.26 shows a rotating rectifier assembly. Two circular rows of rectifier sub-assemblies or modules are connected and accommodated in the retaining ring, the complete rectifier being a '3–2–1–9' connection of 54 diodes. The notation signifies three AC connections, two DC connections, one diode in series per arm and nine parallel diodes per arm. The number of diodes in parallel allows a 22% standby capacity in the rectifier bridge. Figure 6.27 shows the rectifier electrical connections. Each diode is protected by two high breaking capacity (HBC) fuses connected in parallel which isolate the diode in the event of it becoming faulty, leaving the remaining diodes healthy to carry full excitation current. Satisfactory operation would be expected with two of the nine diodes in each arm out of service.

In addition to the two main HRC fuses, each diode module has a resistance-capacitance spike voltage suppression circuit and indicator fuse. The indicator fuses, in conjunction with the blown-fuse detector equipment, are designed to detect the failure of the main HBC fuses used to protect the diodes.

The indicating fuse consists of a ceramic body connected between screwed metal ends which are used to attach the fuse to the rectifier module and adjacent component. The ceramic body contains a small explosive charge connected internally to the metal ends and surrounded by a sand filling.

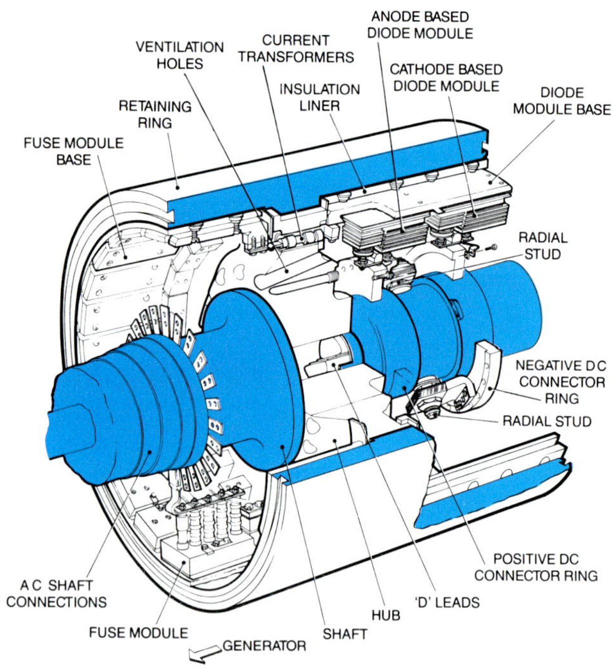

FIG. 6.26 Rotating rectifier, sectional view

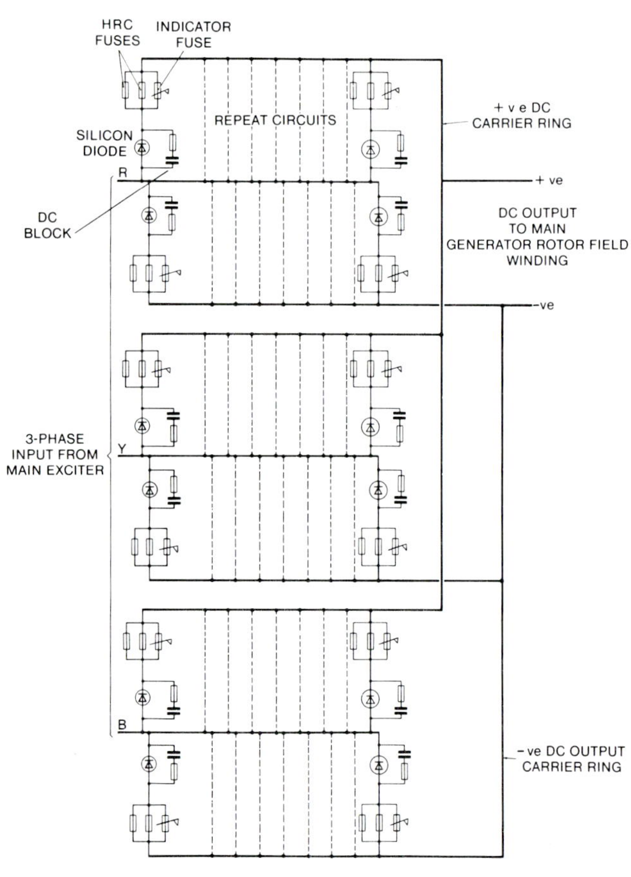

FIG. 6.27 Rotating rectifier electrical connections

Normally, each indicating fuse is short-circuited by the main fuse with which it is associated and carries no current. When the main fuse ruptures, due to a diode failure, a small current will flow through the indicating fuse which detonates the explosive charge and drives out the indicator pin through the centre of the screwed metal end of the fuse. This pin then projects approximately 16 mm and interrupts a beam of light focused on a photoelectric cell once per revolution, thereby giving an indication of a blown fuse on the blown-fuse detector equipment. The indicator pin remains extended by a spring and will not return to its normal position.

The blown-fuse detector equipment (Fig 6.28) comprises two main units the optic head: and the terminal unit containing the electrical equipment. The optic head is mounted on the exciter bedplate adjacent to the rotating rectifier. A vertical fabricated steel support column is bolted to the top plate of the bedplate outboard of the rectifier retaining ring and on the optical centre line of the shaft. The optic detector head is mounted horizontally on the top of this support column, projecting into the space between the rectifier modules and rectifier shaft.

The body of the head is machined from phenolic laminated cloth board and has mounted on it three photoelectric cells (photo-cells) with separate light sources, two on the bottom face and one on the top face. In each, a light beam is projected across a space onto the photo-cell. The light beam from the inboard bottom unit passes over the path traversed by the tops of the negative bridge-arm indicating fuses as the shaft rotates. The light beam from the outboard bottom unit passes over the path traversed by the tops of the positive bridge-arm indicating fuses. Under normal operation, the paths of the light beams are uninterrupted and the light shines continuously on the photo-cell.

If a diode becomes faulty and its main fuses rupture, the indicating fuse associated with that diode will also rupture, driving out the indicator pin. Depending on which arm of the rectifier the faulty diode is in, the pin on the indicating fuse will break the light beam to one or other of the two photo-cells. The beam will be interrupted momentarily once every revolution of the rectifier shaft; this will cause an electrical pulse to be generated in the photo-cell circuit. This is the simplest form of detection and indicates a faulty diode, but it does not indicate the location of the faulty diode. To obtain this information, a reference marker is required and this is provided by the third photo-cell unit. By using a suitable oscilloscope, the signal from the reference marker and from the blown fuse photo-cell circuit can be compared and the angular position of the faulty diode determined.

Light for the three beams is provided by three separate fibre-optic light guides. These are taken from the terminal unit mounted on the top of the bedplate

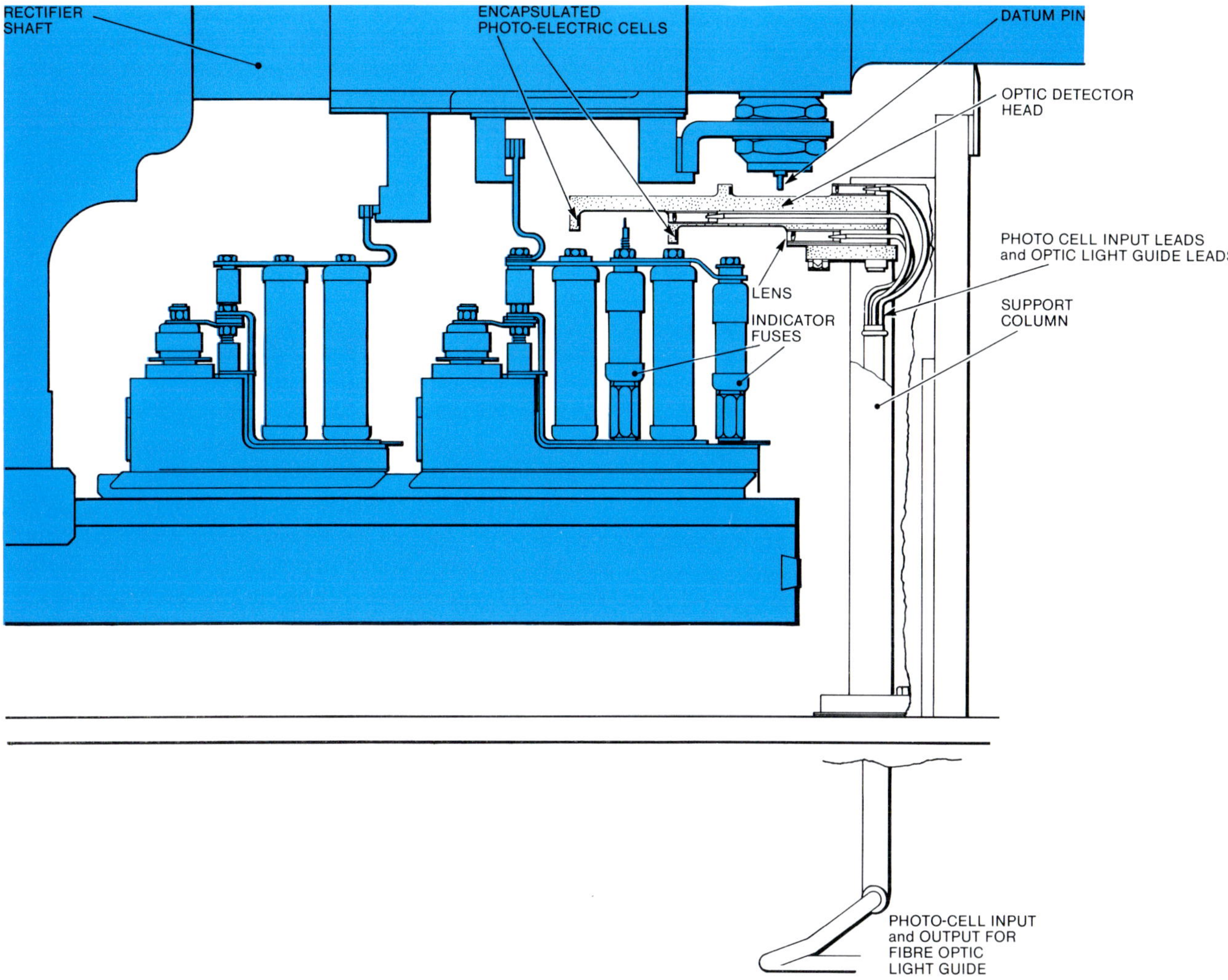

FIG. 6.28 Blown fuse detector

and pass through a conduit tube in the bedplate into the base of the optic detector head support columns and hence up the back of the column into the detector head.

3.7.3 State of plant for test

- Turbine-generator running unloaded and not synchronised at 3000 r/min.
- Two rotating rectifier dummy fuses fitted to simulate failed fuses and to check location equipment.
- Rotor/exciter isolating link removed.
- Blown-fuse alarm computer input inhibited.

3.7.4 Test procedure

- An oscilloscope is connected to the output of the datum photoelectric cell which will confirm that negative pulses of approximately 5 V of 0.5 ms pulse width are present.
- These pulses are used to trigger the oscilloscope. When connected to the 'fuse failure' photoelectric cells, two negative pulses should occur.
- Note and record the time delay between the start of the scan and the occurrence of these two pulses.
- The location of the blown fuses is obtained by dividing the time delay by a factor supplied by the equipment manufacturer. This should be checked against the known location of the failed fuses and any errors corrected.
- Remove the inhibit on the computer input and check that the alarm has initiated correctly.

3.8 Turbine vacuum trip testing

3.8.1 Introduction

The commissioning programme, having now advanced to a point where a turbine shutdown is required to refit the exciter rotating rectifier fuses and links, provides an ideal opportunity to carry out the turbine vacuum trip test without affecting the overall programme time. This test is carried out with the turbine at 3000 r/min by allowing the condenser vacuum to fall until the loss-of-vacuum protection trip operates.

3.8.2 General description

The direct acting vacuum trip mechanism is shown in Fig 6.14 and comprises duplicated units, each of which consists of two spring-loaded metal bellows mounted alongside each other and coupled by a pivoting link, pin-jointed at its centre to a trip valve plunger.

3.8.3 Test procedure

The test is carried out by taking one of the vacuum maintaining units (air pumps) out of service, allowing the condenser vacuum to decay slowly. No attempt should be made to break vacuum by opening the vacuum break valve until after the operation of the trip. If vacuum decay is very slow due to a tightly sealed unit, the second vacuum maintaining unit should be shut down.

As the condenser vacuum falls, the operation of the low vacuum alarm is checked and recorded, a typical value being −910 mbar. Vacuum will continue to fall until the trip operates at −840 mbar. Whilst carrying out this controlled test, steps should be taken to avoid the unnecessary tripping of the boiler firing equipment, which would normally occur following the operation of the vacuum protection trip. The isolation of the boiler firing trip would normally be achieved either by the operation of the normal/isolated test switch or by the removal of trip links. In either, the action taken should be recorded.

After the operation of the trip, the following should be checked:

- HP and IP governor stop valves tripped
- Electronic governor tripped
- Correct alarm operation
- As the turbine runs down, confirm the cut-in pressures of the following oil pumps:
  AC flushing oil
  DC flushing oil
  AC seal-oil
  Compare the results with previous values.
- Continue monitoring turbovisory conditions during run-down.
- Check that electrical turning gear motor and all bearing jacking oil pumps have automatically started at about 100 r/min.
- Check that the electrical turning gear has engaged correctly and that the electrical turning gear motor current is at its normal value.

Once satisfied that the turbine is on electrical turning gear, the boiler firing should be shut down and the main steampipes isolated by closing the boiler stop valves.

At this point of the activity, some six days into the commissioning programme, maintaining the boiler firing during the rundown is a recommended precaution as, in the event of an electrical turning gear failure, the turbine could be run-up again, thus preventing any damage or programme delay whilst waiting for stationary shafts to straighten.

3.9 Turbine shutdown in preparation for generator short-circuit testing

3.9.1 Restoring rotating rectifier fuses and links

Prior to turning the turbine shaft on steam, the isolating link between the generator and exciter was removed, to prevent any possibility of the generator being excited during the early part of the commissioning programme, especially during the AVR testing.

In addition, if rotating rectifier fuses are used for excitation protection the testing of the rotating rectifier blown-fuse detection equipment (Section 3.7 of this chapter) requires the fitting of two dummy fuses in place of normal rectifer fuses, which would have been fitted prior to the commencement of the STS activity.

Having now completed the AVR tests that required the exciter link removed, and the rotating rectifier blown-fuse detection test, these fuses and links must be refitted before proceeding to the next part of the programme. The rotating rectifiers are located between the main exciter and pilot exciter, enclosed with the main exciter within the main enclosure. Figure 6.26 shows a typical installation.

The position of both the fuses and the links in the exciter requires the turbine to be off-load and off electrical turning gear before the refitting of the fuses and links can commence. In addition, steps must be taken to provide safety from the system by applying the necessary isolating and earthing arrangements. As the work is around the exciter area, it would not normally be necessary to purge the generator gas frame of its hydrogen.

If during the initial turbine runs any leakage from the hydrogen seals has been noted, it would be prudent to discharge the hydrogen with CO_2, and initiate 'no smoking or naked light' working practices. It would be normal practice for electrical turning of the shaft to be in service for approximately 12 hours after the shutdown before commencing hand turning operations.

The procedure to be adopted is as follows:

- Turbine shutdown.
- Maintain the turbine on electrical turning gear with all jacking oil pumps in service.
- Isolate the turbine from the system in preparation for issue of the Permit for Work.
- Remove as many covers as possible prior to the issue of the Permit for Work whilst the turbine shaft is still rotating on electrical turning gear.

- Ensure that the replacement links and fuses are available and that staff carrying out the work are in possession of the correct tools and are competent to carry out the work.
- Having established that the work can commence after the 12-hour electrical shaft-turning time, isolate the electrical turning gear and issue the Permit for Work. Leave oil pumps running.
- Ensure that the turbine hand turning equipment is available at the appropriate point on the turbine shaft and that resources are continuously available to carry out the operation.
- Hand turning of the turbine shaft should be carried out at least once in every 30 seconds, rotating the shaft through a half turn.
- The work of refitting links and fuses should proceed on a continuous basis. Hand turning must continue whilst this work is in progress.
- On completion of work, cancel the Permit for Work and return turbine to normal electrical turning gear.
- Once on electrical turning gear, check turbine shaft for eccentricity levels and let it run on electrical turning gear until an acceptable level has been reached before attempting a turbine run-up.

3.9.2 Erecting 400 kV short-circuit

The next part of the commissioning programme is the generator short-circuit test, and the shutdown following the vacuum trip test presents an opportunity for installing the short-circuit. The short-circuit could be fitted without the turbine shut down but, from a safety and control aspect, when commissioning a new turbine-generator, it is probably best done under shutdown conditions.

The short-circuit link comprises sheathed insulated cable strapped and connected across the three phases. Figure 6.29 shows a typical system with the short-circuit erected across the 400 kV generator phases between the main generator circuit-breaker and busbar isolators. First, it is necessary to calculate the current which will flow at 400 kV through the short-circuit. This is normal full-load current and, for a generator which has a full-load current of 20 000 A at 23 kV, current flow at 400 kV will be approximately 1200 A. The insulated cable must therefore be capable of passing this current safely.

This may require either two or three shorting cables across the phases, dependent on the cabling and generator rating. The shorting cable must be prepared in advance and fitted with approved fixing clamps, which are also capable of passing the required current.

The fitting of the short-circuit must be a controlled operation, with the agreement of all parties involved including 'System Operation' who are responsible for the control and operation of the EHV transmission system. On the CEGB system, a Permit for Work would be usual in order to provide safety from the system by applying isolations at all points of supply.

Once the short-circuit has been erected, it is essential that strict control is exercised over the operation of the 400 kV busbar isolators. These must remain locked in the isolated position for the time that the short-circuit remains fitted. The nominated Test Controller is responsible for ensuring that operation of the busbar isolators does not take place until after the short-circuit has been removed or rendered ineffective.

3.10 Generator short-circuit test

3.10.1 Objective

This activity is carried out after a shutdown for replacement of the rotating rectifier link and the fitting of a short-circuit connection capable of passing full stator current for several hours. The objectives of this test are:

- To determine the generator short-circuit curve.

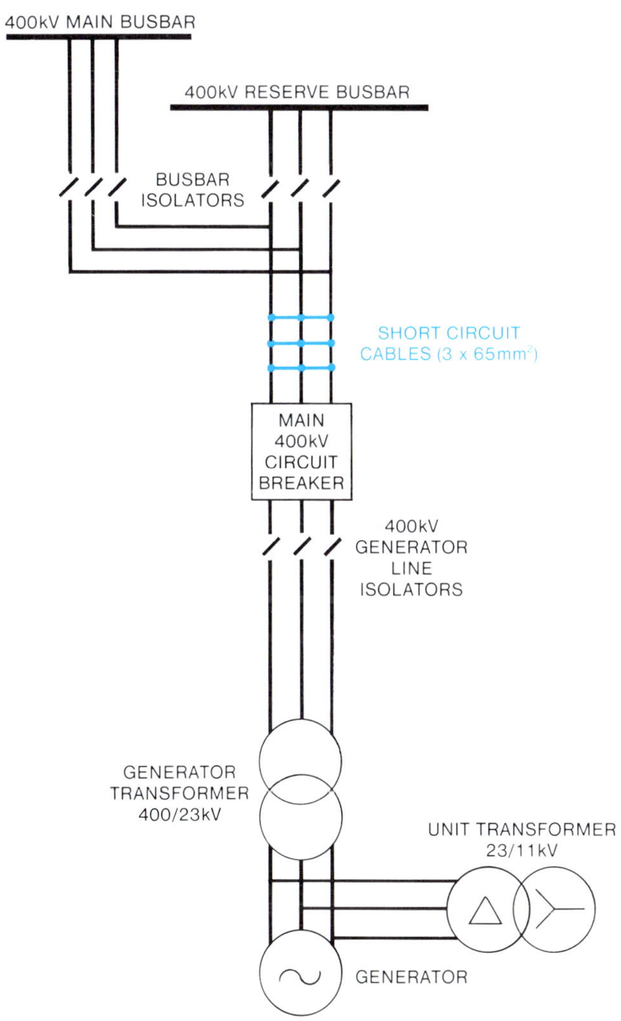

FIG. 6.29 Position of 400 kV short-circuit

- To determine the excitation decay curve of the field winding.
- To take readings in the protection circuits at 10% and 100% full-load current values.
- To conduct a short-circuit heat run on the generator, associated busbar and step-up transformer connections.

3.10.2 Short-circuit curve and protection readings

Figure 6.29 shows a typical arrangement of step-up transformer and 400 kV system. Note that the three-phase short-circuit is applied in such a manner that it is possible to move it clear of conductors without shutting the machine down. This allows certain other tests, such as the open-circuit test, to take place without the need for a shutdown.

It must be remembered that this is the first time that voltage has been applied to the windings and substantial current circulated through them. Consequently, the approach is cautious and the test proceeds with small increases in excitation. Prior to the start of the 'short-circuit test programme', it is essential to take a final generator dewpoint check; this should be carried out with the turbine at 3000 r/min. The maximum acceptable dewpoint for exciting the generator is 20°C below the cool gas temperature.

Where the cool gas temperature exceeds 40°C, the maximum acceptable dewpoint is +20°C.

3.10.3 Initial state of plant

- The pre-commissioning/test documents for the generator must be complete and verified.
- Overall trip and alarm tests must have been successfully carried out.
- Safety Rules Clearance Certificates must have been issued for the generator, generator transformers, unit transformer, and associated 400 kV, 23 kV and 11 kV connections, including voltage transformers (VTs).

On the generator and unit transformers, the following must be completed:

- Waterspray fire protection systems. Satisfactory deluge tests must have been carried out.
- All transformer oil levels must be correct, with the transformer breathers operating satisfactorily and the gas/oil relays vented.
- All oil valves must be set for correct operation and final oil samples taken. Oil pumps will have been run for at least 24 hours prior to the test for circulating the oil and are now shut down for a manual start.
- The HV compound fencing must be checked to be complete, with the compound secure and free from debris.
- Appropriate danger notices are positioned and displayed.
- The earthing system is complete.
- On the unit transformer, the off-load tap selector should be locked in mid-tap position.
- All specified protection must be checked by secondary injection and appropriate settings applied to relays.
- Reserve and main busbar 400 kV isolators open. Generator line isolator open.
- 400 kV earth switches open.
- Generator VT fuses removed.
- Temporary three-phase short-circuit erected.
- All protection and control links associated with the 23 kV and 11 kV equipment in position.
- Installation and commissioning of the instantaneous earth fault relay complete. This is fitted to trip the generator main circuit-breaker.
- Unit transformer 11 kV circuit-breaker and VT isolated.
- 23 kV earth switch locked open. 23 kV VTs in their operational position.
- Both AVR channels, A and B thyristor converter AC and DC switches to be open.
- AVR 400 Hz supply isolator to be open.
- Temporary pushbutton to be installed at the control desk, wired to trip the field suppression relay, with an additional pushbutton also installed in the AVR cubicle. Both pushbuttons are wired in parallel with the machine protection circuits.
- Auto-synchroniser isolated.
- Generator transformer on mid-tap position and the tapchanger selected for remote operation.
- Turbine at 3000 r/min.

The following systems must be in service and operating correctly:

Generator stator water circulating and cooling systems

Hydrogen cooling system

Condensate system

Auxiliary and main CW systems

The test procedure commences by carrying out switching to set up the 400 kV electrical system. Figure 6.29, showing the position of the temporary three-phase short-circuit, shows all the circuit isolators and the circuit-breaker in the open position. For the test, the main and reserve busbar isolators, together with the earth switches, are required to remain open. To ensure complete safety against inadvertent operation, these isolators should be locked open and the keys controlled by the person in charge of the test. The following procedure is typical for carrying out a short-circuit test of a large generator under site conditions:

- Replace generator voltage transformer fuses.
- Close line isolator switch.
- Close the main circuit-breaker and remove the generator protection control links.
- The turbine is now operating closed onto the electrical short-circuit with no excitation. In this condition, a small value of stator current would be expected due to residual magnetism. The value is recorded; typically, 250 A is normal for the type of generator being considered.
- Check that both channels of the AVR are in the minimum excitation position.
- Close both AVR channel thyristor converter AC breakers and operate the temporary AVR field suppression pushbutton to confirm that the tripping facility is operative. Reset the protection.
- Close the AVR 400 Hz isolator and lock it closed.
- Close one AVR channel thyristor converter DC isolator.
- Close selected channel thyristor converter AC breaker.

These actions will produce a stator current with minimum excitation. The magnitude of this current, together with the rotor voltage and current, are recorded. Typical values at this stage are:

Stator current at minimum excitation 800 A

Rotor voltage at minimum excitation 55 V

Rotor current at minimum excitation 220 A

- Increase the excitation slowly to circulate 10% full-load current; a typical value for a 660 MW generator is 2000 A. In parallel with this activity, if negative phase sequence protection is fitted, the output voltage of the protection relay should be recorded, a typical value being 120 mV. With 10% full-load current circulating, preliminary readings on the protection are taken and recorded.
- Table 6.8 shows a typical list of protection equipment where readings would be taken. The list is extensive due to the need to cover all the protection fitted and the recording of this information takes approximately 1½ hours, with four engineers working in two pairs.
- With the turbine running at 3000 r/min and with 200 A circulating in the rotor, checks are carried out to assess the performance of the stator water and hydrogen cooling: the coil and core slot temperatures are noted. Temperatures will gradually increase during this activity, and it is essential that a base line set of readings is recorded prior to closing onto the short-circuit. These temperatures must be closely monitored at each increase in stator current and time allowed for stabilisation of temperatures and the recording of the readings. Table 6.9 shows typical parameters to be recorded from zero current to full-load current and during the period of the generator heat run.

TABLE 6.8

List of protection readings taken during the short-circuit test at 10% and 100% full-load current

Generator transformer winding temperature indicator HV winding
Generator transformer winding temperature indicator LV winding
AVR interposing CT secondary
Generator transformer biased differential protection
First feeder protection relay
Stator earth fault relay (system 1)
Stator earth fault relay (system 2)
Loss of excitation protection relay
Negative phase sequence protection relay
Low power relay 1
Instrument transducers interposing CT secondaries
Low power relay 2
Generator transformer HV REF protection relay
Second feeder protection relay
Efficiency test interposing secondary CTs
Tariff metering interposing secondary CTs
Stator ammeter
Generator ammeter 400 kV
First feeder protection relay
Second feeder protection relay
Breaker fail protection relay
Overcurrent protection relay
Phases unbalanced relay
Feeder flow/instruments CTs
Generator commissioning instruments earth fault relay
Circuit ammeter

On completion of the protection readings, the following activities should be carried out, if the protection equipment is fitted:

- Prove the operation of the negative phase sequence system.
- Check and record the voltage at the HV restricted earth fault relay when CT leads are shorted and disconnected.
- Confirm the value of stator current at which the generator transformer biased differential protection operated.
- Increase the excitation on the generator in 10% increments up to full-rated current to plot the AVR short-circuit curve (see Fig 6.30). At each step, record and take action as necessary, depending on the performance of the stator winding and its cooling systems.

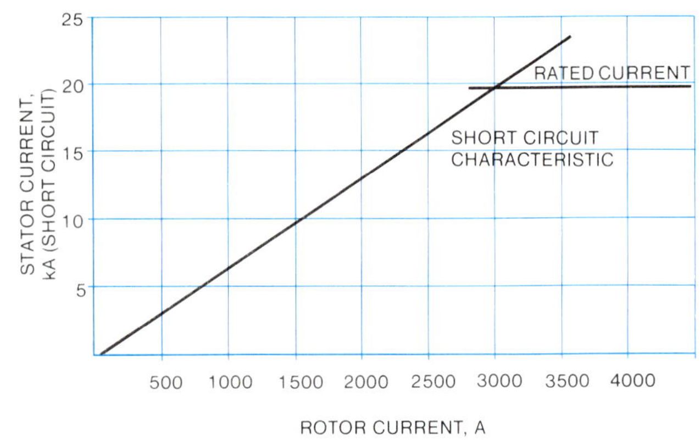

FIG. 6.30 Typical short-circuit curve

- Whilst circulating full-load current, repeat the readings taken on the protection at 10% current flow.
- A visual examination of the three-phase short-circuit should be carried out to confirm that there are no problems.
- Having completed the short-circuit plot and protection readings at full-load current, the excitation decay performance on field suppression can be carried out before the commencement of the generator heat run.
- The excitation decay test requires that the single channel of the AVR that is in service be tripped, confirming that the AC thyristor converter AC breaker opens.
- The test is repeated with the second AVR channel in service. Both trips take place from full-load current.
- Having completed the excitation decay test on both AVR channels, the excitation should be increased to obtain full-load current in preparation for the full-load heat run.

3.10.4 Generator heat run

With the completion of the protection readings at 10% and 100% full-load current and the establishment of the excitation decay characteristic, the next test is the full-load heat run. This involves circulating full-load current until the generator, busbars and generator transformers reach their maximum working temperatures. The normal accepted time for temperature stabilisation to occur is 5–8 hours, but this depends on whether the heat run directly follows the protection readings.

The continued circulation of full-load current increases the temperatures of the stator core and stator water system and demonstrates the ability of the condensate system to control and dissipate the heat generated. It is therefore necessary to ensure that optimum turbine vacuum conditions are obtained, and that close attention is paid during the heat run to certain plant parameters. These are given in Table 6.10, which shows the stabilisation of temperatures and is produced on the basis that the heat run directly followed the protection readings.

In addition to monitoring the operation of the generator, the temperatures of the air and water cooled connections are checked. The method of checking uses a combination of temperature-sensitive tape and an infra-red camera. The infra-red camera produces immediate results, normally by a digital readout, whereas the temperature-sensitive tape information is

TABLE 6.9

Typical parameters to be recorded from zero to full-load current during the short-circuit test

STATOR CURRENT	BEARING VIBRATIONS: GENERATOR FRONT	BEARING VIBRATIONS: GENERATOR REAR	BEARING VIBRATIONS: EXCITER FRONT	BEARING VIBRATIONS: EXCITER REAR	CORE TEMPERATURES: BETWEEN TOP AND BOTTOM COILS, PHASES A	B	C	CORE TEMPERATURES: IN CORE BETWEEN SLOTS	STATOR WATER SYSTEM: PRESSURE IN	PRESSURE OUT	TEMPERATURE IN	TEMPERATURE OUT	CONDUCTIVITY	WATER OUTLET TEMPERATURES, PHASES A	B	C	HYDROGEN SYSTEM: GENERATOR CASING PRESS	GAS TEMPERATURE FROM COOLER	GAS TEMPERATURE TO COOLER	SEAL FAN TEMPERATURE TE	SEAL FAN TEMPERATURE EE	ROTOR: CURRENT	ROTOR: VOLTAGE	CONDENSER VACUUM A	B	C
AMPS	mm	mm	mm	mm	°c	°c	°c	°c	bar	bar	°c	°c	ms/m	°c	°c	°c	bar	°c	°c	°c	°c	amps	volts	mbar	mbar	mbar
2000	0.025	0.025	0.020	0.018	29	30	29	32	3.79	1.46	32	30	1.86	32	33	32	4.10	31	33	56	51	556	47.9	44.9	47.3	43.8
4000	0.023	0.024	0.026	0.020	30	30	30	32	3.77	1.46	32	31	1.65	32	33	33	4.08	31	33	55	51	723	80.7	43.4	45.8	42.3
6000	0.024	0.025	0.028	0.021	31	31	31	33	3.77	i.46	32	33	1.76	33	34	34	4.09	31	34	55	51	1026	111.4	43.3	45.7	42.2
8000	0.021	0.024	0.028	0.021	31	32	31	33	3.77	1.46	33	36	1.85	35	36	35	4.10	31	35	55	51	1356	145.1	43.3	45.7	42.3
10000	0.020	0.025	0.025	0.019	34	34	33	33	3.77	1.46	34	40	2.06	38	38	39	4.12	32	36	54	51	1666	177.8	43.8	46.2	42.8
12000	0.021	0.025	0.028	0.020	36	36	36	34	3.76	1.46	35	45	2.18	42	41	40	4.14	32	38	55	51	1965	210.0	44.7	47.1	43.7
14000	0.028	0.025	0.031	0.023	38	38	38	34	3.75	1.46	36	50	2.45	46	44	46	4.16	32	40	55	51	2296	246.5	44.0	46.3	42.9
16000	0.029	0.025	0.029	0.024	41	41	41	36	3.74	1.46	38	56	2.68	50	47	50	4.18	33	43	55	51	2584	280.0	43.9	46.3	42.9
18000	0.028	0.025	0.030	0.023	45	45	45	37	3.71	1.46	39	64	3.08	55	51	55	4.21	33	46	55	51	2885	316.5	43.6	46.0	42.7
19500	0.032	0.025	0.032	0.025	48	48	48	39	3.69	1.46	41	70	3.25	60	55	59	4.23	33	48	55	51	3139	348.1	43.5	45.9	42.5

obtained by inspection after shutdown to confirm the maximum temperature reached. The temperature-sensitive tapes should cover a range of temperature values based on the design temperature criteria, including tapes to cover the possibility of higher temperatures than predicted.

Once temperatures have stabilised for at least one hour, the heat run can be considered complete. The excitation should be decreased to minimum, followed by the opening of the selected AVR channel thyristor converter AC and DC circuit-breaker and isolator; this removes all residual excitation. Opening the main circuit-breaker and replacing the generator protection and control links terminates the activity.

The three-phase short-circuit, especially erected for this activity, must be removed before synchronising; however, the open-circuit and overall trip test can be completed without the removal of the short-circuit. If this policy is adopted, the main circuit-breaker must be rendered inoperative and any isolation keys held under the control of the Test Controller.

3.10.5 Generator windage dry out

Objective

To obtain an acceptable generator dewpoint temperature prior to applying a voltage to the generator. The vacuum trip test concludes the first part of running the turbine at 3000 r/min and leaves the machine on the electrical turning gear. If an acceptable dewpoint has not been obtained during the period of 3000 r/min operation, the following dry out procedure must be adopted.

Introduction

During the construction period, regular monthly checks are carried out on the generator to establish the insulation resistance (IR) and polarisation index (PI) of the stator winding, together with the insulation resistance of the rotor winding, the generator frame being kept sealed with temporary end-covers, and with dehumidified air being circulated. On completion of the generator construction, flushing of the stator water system commences, initially on the external pipework and finally through the stator windings. Just prior to admitting water to the stator for the first time, a final IR and PI test is carried out, producing a PI value in excess of 2. Once water is present in the stator any further IR testing will produce low values. Providing a PI value in excess of 2 was established prior to water entering the stator, it is unlikely that a windage dry out will become necessary.

The final check occurs just prior to the commencement of the STS activity, when a high voltage test is carried out. A prerequisite of this test is to drain the stator windings of water and to purge all the water out of the many flexible connections between the conductors and the inlet and outlet stator manifold headers, using dry air. Prior to carrying out the HV test a further IR test is done, which must produce a PI greater than 2. After successfully completing the HV test, the stator windings are filled with water. Normally, the STS activity commences within two weeks of completion of the HV test.

If the steps outlined have been adhered to there is no reason to expect that it will be necessary to carry out a windage dry out. The decision for this depends solely on the value of the generator frame dewpoint. Section 3.1.8 of this chapter gives the value of dewpoint required and outlines the requirement to ensure that the frame dryer is put into service once the turbine has been established at 3000 r/min.

The following test procedure is included as a method of obtaining the correct dewpoint if the procedures identified have not been carried out correctly, or if long delays have occurred in the commissioning programme.

TABLE 6.10
Typical parameters to be recorded during the generator heat-run

TIME	STATOR WATER SYSTEM								HYDROGEN SYSTEM			BEARING VIBRATIONS				CONDENSER VACUUM AVERAGE	CORE TEMPERATURES			
	PRESSURE		TEMPERATURE		CONDUCIVITY	WATER OUTLET TEMPERATURES, PHASES			GAS PRESSURE	GAS TEMPERATURE		GENERATOR		EXCITER			BETWEEN TOP AND BOTTOM COILS, PHASES			IN CORE BETWEEN SLOTS
	IN	OUT	IN	OUT		A	B	C		FROM COOLERS	TO COOLERS	FRONT	REAR	FRONT	REAR		A	B	C	
HOURS	bar	bar	°C	°C	ms/m	°C	°C	°C	bar	°C	°C	mm	mm	mm	mm	mbar	°C	°C	°C	°C
0	3.69	1.46	41	70	3.25	60	55	59	4.23	33	48	0.032	0.025	0.032	0.025	45	48	48	48	39
1	3.69	1.46	41	71	3.78	61	57	61	4.26	34	50	0.024	0.024	0.037	0.022	44	52	51	51	42
2	3.70	1.46	42	71	4.00	62	57	61	4.27	34	50	0.028	0.024	0.035	0.022	45	52	51	52	43
3	3.68	1.46	41	71	4.20	62	57	61	4.27	34	50	0.031	0.024	0.031	0.021	45	52	52	51	43
4	3.71	1.46	41	71	4.50	62	57	62	4.27	34	50	0.027	0.023	0.027	0.020	45	52	52	52	43
5	3.73	1.46	41	71	4.60	62	57	61	4.27	34	50	0.026	0.023	0.029	0.021	44	52	52	51	43

Test procedure

Conditions prior to run-up:

- With stator water system drained, blown down and purged, carry out insulation readings. Typical values are manifold-to-earth 250 kΩ and manifold-to-winding 1 MΩ.
- Generator isolated from all points of supply.
- Generator hydrogen system purged and replaced with dry air at atmospheric pressure, with a continuous supply of clean dry air available.
- Seal-oil system in service.
- All instrumentation and alarms in service, particularly the generator thermocouples.
- Links and fuses between the generator field and exciter removed.
- AVR isolated.
- Hydrogen coolers primed on the water side, with no flow.
- Additional dewpoint monitoring instrumentation fitted to measure the dewpoint of the air entering and leaving the generator.

Dry out procedure

The dry out procedure is carried out with the stator water system fully primed and in service, and the generator frame filled with clean dry air at the normal pressure, thus ensuring that correct stator water/gas pressure differential is maintained:

- Run turbine up to dry out speed. The speed should be high enough to produce sufficient windage heating to raise the generator winding temperature without undue delay. Operating at 3000 r/min should be the objective.
- Attempt to obtain a temperature throughout the generator of between 70 and 80°C. The turbine speed controls the temperatures and these are maintained as steady as possible throughout the dry out after the initial heating period.
- The moisture in the stator water inlet and outlet manifolds is flashed off and extracted by connecting the manifold drains to a vacuum source.
- Logging of all temperatures, dewpoints and turbine speed is carried out at least hourly.
- Liquid leakage detectors are drained periodically, the quantity of water removed being measured and recorded.
- The dryout can be considered complete once the dewpoint of the casing air is less than 0°C at casing pressure.
- Once an acceptable dewpoint is reached, an IR reading of the stator in a fully-operational state is taken. With all the parallel paths of water present an acceptable IR reading would be about 0.5 to 1 MΩ.
- When the dry out is complete, the turbine is run down, put on electrical turning gear and the generator is allowed to cool. Cooling can be assisted by the controlled opening of hydrogen cooler vents, and by allowing water to pass through the coolers. Core-to-gas differential temperatures should not exceed 25°C, and the generator should be allowed to cool to 40°C before purging with CO_2 prior to filling the frame with hydrogen.
- Any temporary additional instrumentation is removed and preparations made to refit exciter links and fuses, after carrying out a HV test now that a satisfactory generator dewpoint has been achieved.

If it is necessary to carry out a windage dry out, then this will increase the overall commissioning programme by several days, dependent on the amount of moisture to be removed.

An alternative method to be considered would be to heat the stator water to 65–70°C, and to circulate it through the windings. This can only be carried out if the stator water system design included a permanently fitted heater. This heating, combined with the continuous purging of the hydrogen in the generator frame and coupled with effective moisture removal by the hydrogen drier, should reduce the overall time, compared with a windage dry out, and may be carried out at a more convenient time.

3.11 Generator overall trip test

3.11.1 Objective

During the short-circuit test, the links and fuses associated with the main electrical protection circuitry were disturbed. The proving of all electrical protection was completed prior to the commencement of the steam-to-set activity; however, it is now necessary to check that all circuitry is sound by carrying out a trip test to prove that the protection circuits are healthy. The trip testing is carried out with the turbine at synchronous speed, with measures taken to prevent the inadvertent operation of the boiler firing trip: the removal of links in the boiler firing trip circuit prevents the unnecessary tripping of the boiler firing system and unnecessary delays in the commissioning programme.

3.11.2 General description

The electrical protection of modern large turbine-generators is normally divided between first and second protection panels with separate DC supplies. Trip testing comprises two tests; in the first instance operating a protection relay connected to the first protection panel, followed by a test on a second protection panel circuit. The test should trip the main

400 kV generator circuit-breakers and the 11 kV unit transformer ACB, as well as tripping the turbine steam valves. After the trip, record which trips have operated, then reset all protection relays (including flags).

The turbine valve gear will need resetting and, after arresting the turbine loss of speed, a run-up back to 3000 r/min is started. The turbine run-up should not recommence in a critical speed band, but speed should be allowed to fall to a known safe speed band before acceleration takes place.

Having returned the turbine to 3000 r/min, a second trip is carried out from a relay circuit in the second protection panel. The procedure of trip and alarm resetting together with the recording relay operation, adopted for the first trip, is repeated for the second trip.

Figure 6.31, a typical 660 MW turbine-generator/boiler electrical protection system block diagram, identifies the extent of the protective system and nominates the protective devices to be operated as part of this test.

3.11.3 Initial state of plant prior to the commencement of generator overall trip test

- The turbine-generator short-circuit test activity has been completed.
- Reserve and main busbar 400 kV isolators locked open, with control selected to 'local'.
- Main generator circuit-breaker locked open.
- 400 kV earth switches are locked open.
- 23 kV earth switches are locked open.
- Generator VT fuses are in service.
- All protection/control links and fuses are in the service position, except the boiler firing trip links.
- Unit transformer 11 kV circuit-breaker is open and isolated.
- Unit transformer VTs are in service.
- 23 kV VTs are in service.
- Both AVR channels, A and B thyristor converter AC and DC circuit-breakers open.
- All AVR protection, control and power supply fuses and links in position.
- AVR 400 Hz supply isolator is closed.
- Temporary pushbutton installed for the short-circuit remains installed at the unit control desk.
- Auto synchroniser is isolated, if fitted.
- Turbine at 3000 r/min with the following systems in service and operating correctly: hydrogen cooling system, DW cooling system, stator water system, condensate systems, auxiliary and main CW systems, vacuum maintaining systems.
- The relevant station electrical auxiliary switchboards are in service supplied from the station transformers.

3.11.4 Overall trip test procedure

- Put the unit transformer 11 kV ACB into the test position.
- Close the main generator circuit-breaker.
- Operate the selected protective relay in the test protection circuit.
- Confirm that:

  The main generator circuit-breaker has opened.

  The unit transformer ACB has opened.

  The generator field suppression relay has operated.

  The turbine stop and governor valves have tripped.
- Check that all appropriate alarms and indications have been correctly received.
- Reset all alarms and cancel all discrepancy indications.
- Reset the protection relay flags, in both the protection panel and the AVR.
- Reset the turbine trip and arm the governor. Run the turbine back up to 3000 r/min.
- Check synchronising and reclose the main generator circuit-breaker.
- Check synchronising and reclose the 11 kV unit transformer ACB into the test position.
- Operate the selected protective relay, this time in the second protection circuit.
- Repeat the actions of checking results, resetting and recording alarm, etc., as identified for the first trip-test.
- Having confirmed that the test is complete, the 11 kV unit transformer ACB should be isolated. The turbine valve gear is reset, the governor rearmed and the machine is brought back to 3000 r/min in readiness for the commencement of the open-circuit test.

3.12 Generator open-circuit test

3.12.1 Objective

Having completed the generator protection trip test and checked that all plant has been restored to its normal state, the turbine can be run-up to speed and the generator open-circuit test carried out (see Fig 6.32). The objectives of this test are:

- To obtain the open-circuit curve up to rated voltage and to measure the magnitude of voltage present in

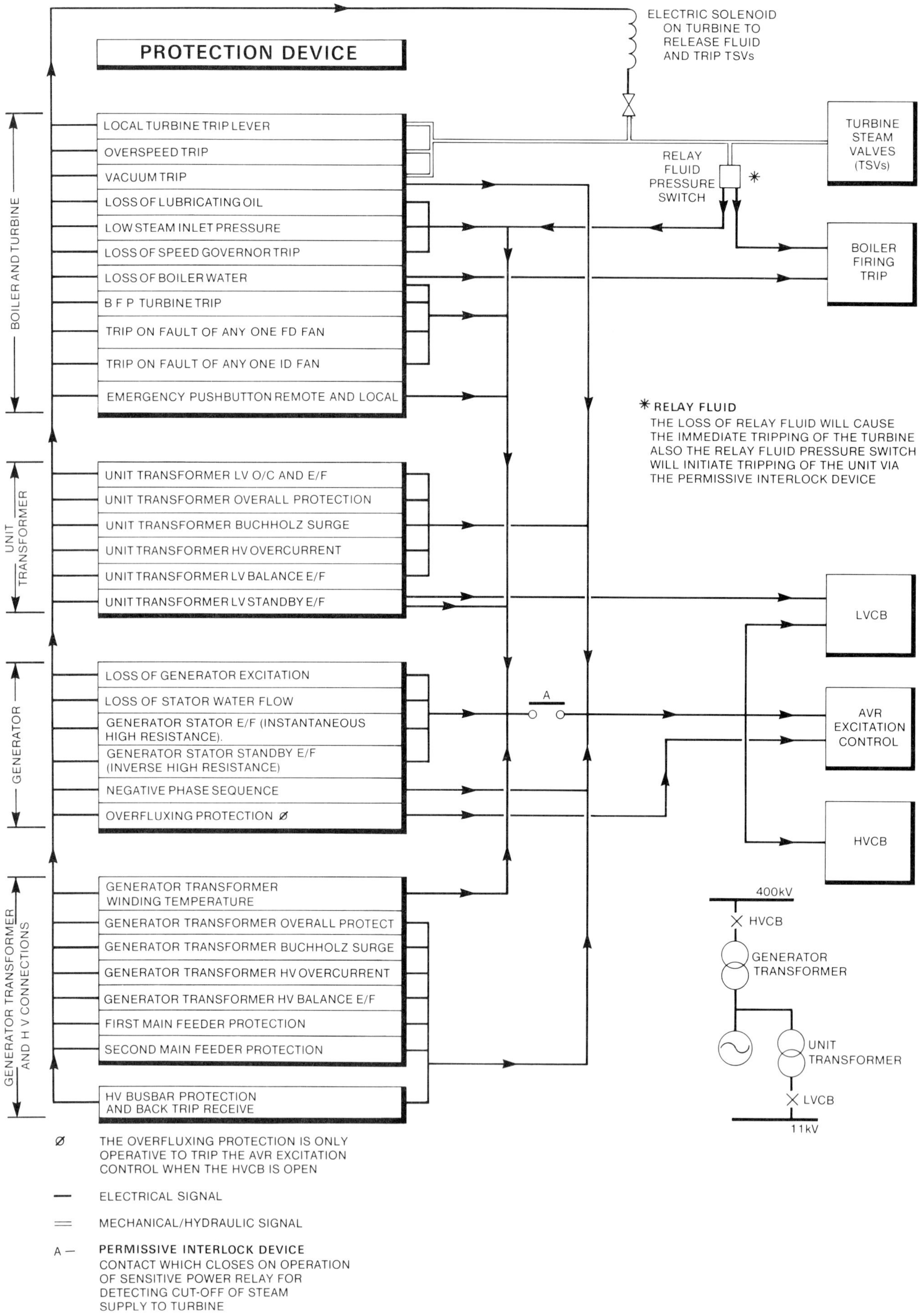

FIG. 6.31 Typical 660 MW turbine-generator/boiler electrical protection system block diagram

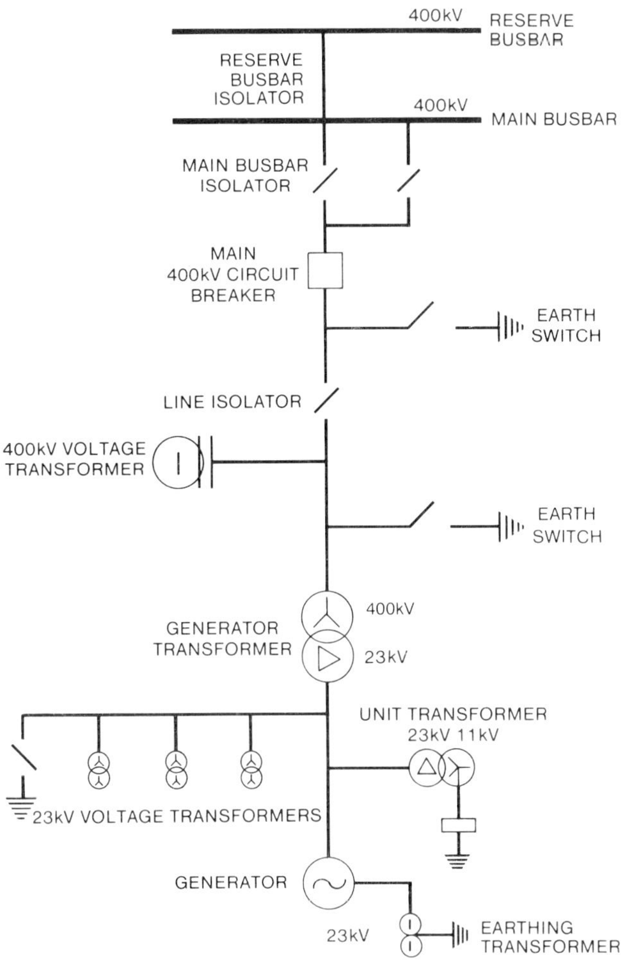

FIG. 6.32 Generator open-circuit test electrical connections

the protection circuits by increasing the excitation manually in steps.

- To carry out an overvoltage test at 104% of normal voltage to check the connections associated with the stator, generator, transformer and unit transformer.
- To determine the field decay characteristic on open-circuit.

3.12.2 Initial state of plant

Before commencing this test, the state of the plant for a generator solidly coupled to a step-up transformer and unit transformer is given in the following list. Whilst this is primarily an electrical state of plant list, the general running of all plant must continue to be monitored, as indicated previously:

- The generator overall trip test has been completed.
- The generator main and reserve busbar isolators must be open.
- The main circuit-breaker and its associated isolator must be open and rendered inoperative.
- The earth switches on both the high and low voltage sides of the generator transformer must be locked open.
- Generator voltage transformer (VT) fuses to be in service.
- All protection and control fuses and links to be in service.
- Unit transformer LV circuit-breaker to be open and isolated.
- Unit transformer VTs to be in service.
- Generator LV VTs to be in service.
- All AVR protection, control and power supply fuses and links to be in position.
- Both AVR channels A and B thyristor converter AC and DC circuit-breakers to be open.
- AVR 400 Hz supply isolator to be closed.
- Generator transformer to be on mid-tap position with tapchanger selected for remote operation.
- Unit transformer off-load tap selector to be locked in mid-position.
- Temporary pushbutton remains installed at the unit control desk.
- Any auto-synchronising equipment fitted should be isolated.
- Generator transformer coolers to be available for operation but selected for manual operation at the control unit desk.
- Turbine running at 3000 r/min, with the following systems in service and operating correctly:

  Hydrogen cooling system

  Generator stator water circulating and cooling systems

  Condensate systems

  Auxiliary and main CW systems

3.12.3 Test procedure

Having achieved these plant conditions, the open-circuit test can be carried out, as follows:

- Close the thyristor converter AC circuit-breaker and DC isolator for the AVR channel selected.
- Raise the excitation manually in 10% steps up to 23 kV to plot the open-circuit characteristic of the generator. Figure 6.33 shows a typical open-circuit curve.
- With the stator excited to 23 kV, the generator, generator transformer, unit transformer and generator VTs to be inspected for any sign of stress whilst being subjected to full rated volts for the first time.
- Carry out voltage measurements on protection and indication equipment. A typical list of equipment is shown in Fig 6.34.

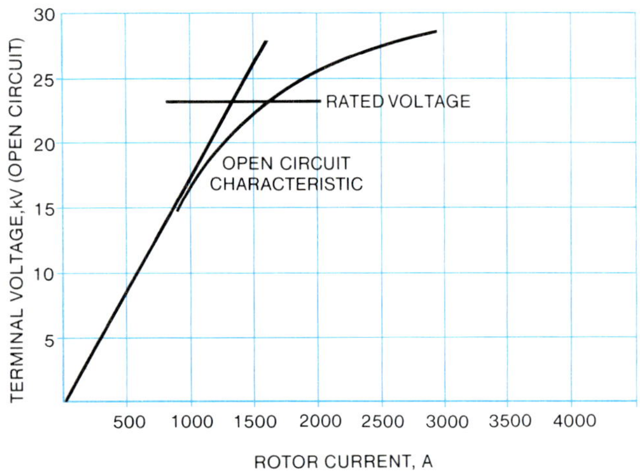

FIG. 6.33 Typical open-circuit curve

LOSS OF EXCITATION RELAY VT SUPPLY

LOW LOWER RELAY 1 VT SUPPLY

LOW LOWER RELAY 3 VT SUPPLY

VT SUPPLY FAULTY RELAY SUPPLY

GENERATOR METERING AT THE METERING PANEL VT SUPPLY

STATOR VOLTMETER AT STATION SUPERVISOR'S DESK

GENERATOR 400kV VOLTS AT THE TRANSMISSION DP

COOLER CONTROL SUPPLY

INSTRUMENTATION SUPPLY

EFFICIENCY/TEST SUPPLY

AVR1 SUPPLY

AVR2 SUPPLY

FIG. 6.34 Open-circuit test — equipment where voltage readings have to be taken

- Inhibit the operation of the generator transformer overfluxing protection in preparation for the over-voltage test.
- Increase the excitation to obtain 104% of rated voltage (23.92 kV for a nominal 23 kV system).
- Inspect the generator transformer, unit transformer and generator VT during the overvoltage period.
- Operate the generator transformer tapchanger through its range, ensuring during this activity that the 104% overvoltage figure is not exceeded. This ensures that the equipment works whilst energised.

The open-circuit test is now considered complete and the following procedure is then carried out:

- Decrease the excitation to obtain normal stator voltage and then carry out field suppression to determine the field decay characteristic on open-circuit.
- Confirm that the excitation has reduced to its minimum value and that the AC thyristor breaker has opened. Open the DC thyristor converter isolator.
- Restore the generator overfluxing protection to normal.

The total time to complete this activity is approximately 3–4 hours, but the preparatory work takes longer.

3.13 Bearing pedestal insulation checks

3.13.1 Objective

The generator bearings are insulated to ensure that, if small voltages are generated in the length of the rotor shaft, they cause no damage. If a small voltage were generated in the rotor shaft, it could cause current to flow around a circuit formed by the shaft bearings, seals, oil wipers and the stator outer frame. Such a current would result in pitting of the bearing surfaces, shaft journals and seal faces. To avoid this, the bearings, seals and oil wipers are insulated from the outer frame.

The pre-commissioning checks carried out on a bearing pedestal during assembly will have confirmed the existence and integrity of the insulation between the bearing bush and its support. Figure 6.35 shows a generator bearing assembly: the insulation resistance between each support pad and the bush to which it is attached and between each support pad and the support housing is measured, using a 500 V insulation tester (megger). A resistance of at least 1 MΩ must be obtained for each reading. Once the bearing is fully assembled, it is only possible to take one reading by use of the probe contact plate. One of the pre-steam-to-set activities for the turbine should be regular insulation checks, carried out initially with the shaft stationary with no oil pumps in service and repeated with the turbine shaft rotating on electrical turning gear with oil pumps in service. This test is now repeated with the shaft running at 3000 r/min.

3.13.2 Procedure for checking bearing pedestal insulation

The test is carried out using a 500 V insulation resistance tester (megger). A reading is obtained via an access hole provided in the bearing end cover through which an insulated test probe is inserted to make contact with the probe contact plate bolted to the top-

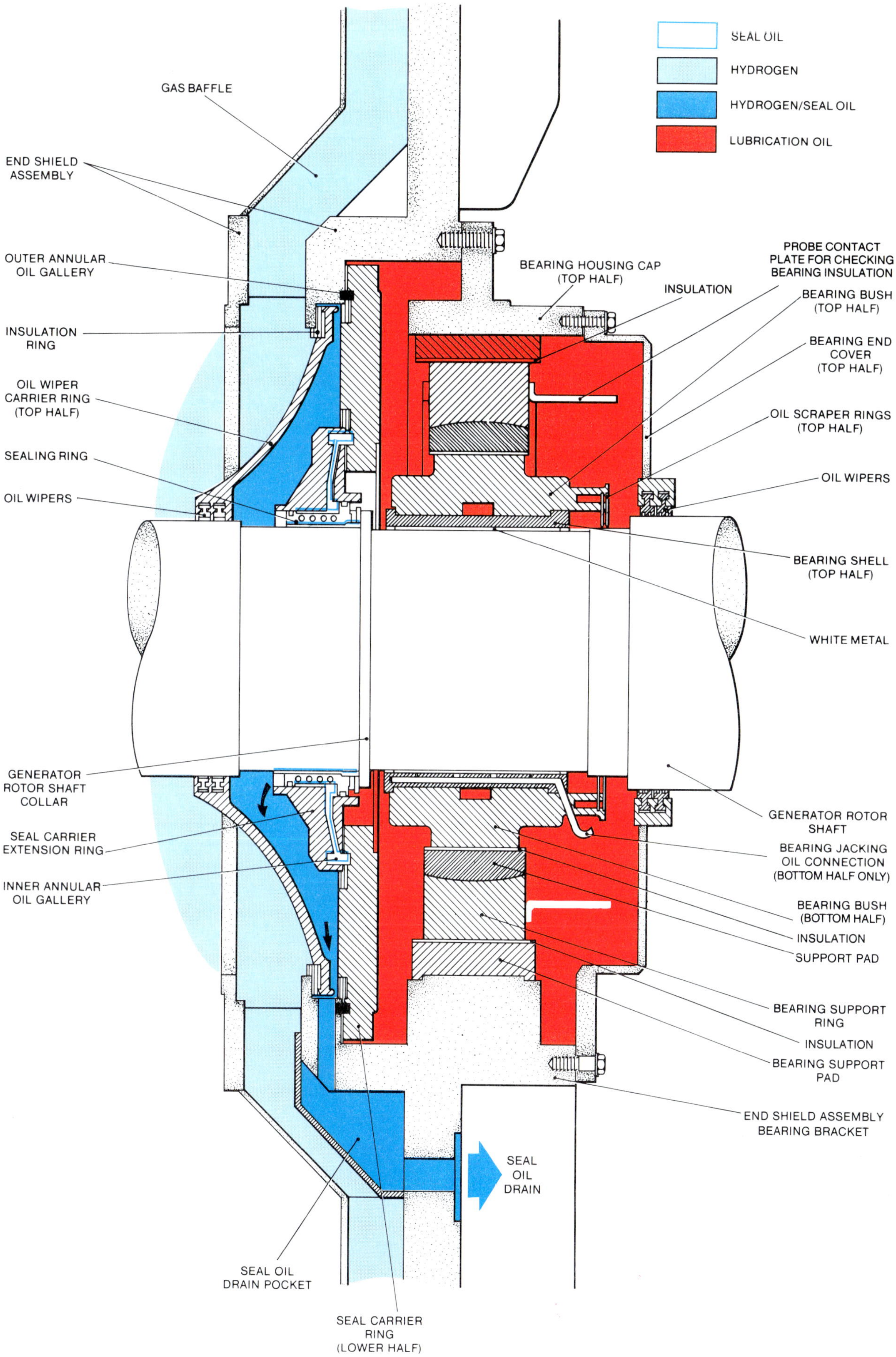

FIG. 6.35 Generator bearing assembly

half bearing support ring. The other lead of the megger is attached to the bearing cap. The insulation resistance is checked between these two points, and should give a result of at least 1 MΩ. If values lower than this are obtained, it is possible that ingress of moisture and/or debris has taken place. To check for this requires bearing dismantling and delays in the programme. It is a matter of past history and judgement as to whether the programme can continue without this defect being rectified. If a shutdown is to take place shortly, then it is preferable to keep running, but if there is going to be considerable delay before a shutdown occurs, it may be preferable to shut down to minimise the risk of bearing damage.

3.14 AVR tests including overfluxing

3.14.1 Tests with exciter link in position

As previously stated, the tests on the automatic voltage regulator (AVR) equipment fall into three distinct stages. The first two, namely the pre-commissioning tests and commissioning tests at 3000 r/min with the exciter link removed, have been dealt with in Chapter 5, Section 13 of this volume and in this chapter, Section 3.6, respectively.

The third stage of testing, with the exciter link in position and the turbine at 3000 r/min, is covered in this section.

3.14.2 Test programme

The tests on the AVR are carried out on each channel separately and then with both channels operating together. Checking and setting the voltage range and voltage step changes are carried out in order to set the AVR feedback controls for optimum response and current sharing. The increase in voltage should be carried out in steps of 5% of the generator terminal voltage in areas of the operating limits curve which are clear of limiting MVArs. This checks the on-load response of the AVR and should be carried out at various outputs.

The overfluxing protection settings are checked on ‘auto’ control, both for single-channel and dual-channel operation. The under-excitation limit, which comprises two measuring circuits comparing the output current with the phase of the generator output voltage, is checked at 0%, 25%, 50%, 75% and 100% MCR load points.

3.14.3 Test procedure

Voltage range

With one channel isolated and the AVR selected to ‘manual’, slowly increase the generator terminal voltage to its design value, noting the position of the voltage control potentiometer on the channel selected to manual operation:

- Check that the voltage limit switches prevent the generator voltage being increased beyond 104% of nominal voltage.
- Check that the terminal voltage is correct and adjust as necessary.
- Check that the normal mains frequency voltage is present in the right areas of the AVR and similarly that any higher frequency voltage derived from the permanent magnet pilot exciter is also present.
- Plot the AVR loop gain by varying the voltage and noting the response.
- Confirm that the balance indicator is giving a zero reading, indicating that there is no difference between manual and auto state. Adjust controls to achieve this, if necessary.
- Switch the AVR to ‘auto’ control.
- Check that the AVR controls the generator terminal voltage between 85% and 104% of nominal voltage over the full range of the control potentiometer.
- Check that the manual follow-up circuit is operating.
- Set the terminal voltage to its nominal value and trip the channel to ‘manual’. This should cause no disturbance to AVR operation.
- With an ultra-violet recorder connected into the AVR system, record the response during step changes. The voltage changes are obtained by the installation of a temporary switch unit which is used to give, initially, a 1% reduction in voltage transformer output voltage, and finally a 5% reduction. These tests are carried out at rated machine voltage.
- Repeat the above test at 85% of machine terminal voltage. It is not unusual to obtain a response under these conditions which is more oscillatory than that at nominal voltage.
- During the tests, observe the field converter output and record its value, checking also that the converter is not misfiring.
- Repeat the voltage range checks with the isolated channel in service, after isolating the tested channel.

Overfluxing protection

- With one channel isolated, set the machine terminal volts to the nominal value on the channel under test.
- Measure input voltages to enable the setting value of the input voltage for overfluxing protection to be calculated.
- Set the calculated value.

- Reduce the input voltage to nominal and recheck the setting accurately by increasing the input voltage again to the setting value.
- Repeat the overfluxing setting checks with the isolated channel in service, isolating the channel that has already been set.

Under-excitation limit

This comprises two measuring circuits which compare the phase and amplitude of the generator output voltage. One circuit produces a signal proportional to generator output reactive current, the other signal being proportional to the active load (MW). These signals are fed into an amplifier, producing an output signal which is used to prevent the limit line of the generator stability diagram being exceeded. This is achieved by opposing signals which would cause a further reduction of excitation beyond the limit line.

The operating point of the limiter is set just prior to synchronising and then checked at loads of 25%, 50%, 75% and 100% output. Stability will be checked by operating just in the exporting MVAr zone and then at the limit points. This will be achieved initially by operating the generator transformer tapchanger and, when the unit is on-load, by introducing voltage steps into the AVR of up to 5% of terminal voltage.

Dual-channel operation

- With both AVR channels in service, bring terminal volts up to nominal value.
- Confirm balance indicator is at zero on both channels.
- Check that AVRs will control voltages between 85 and 110%.
- Check the response to 1% and 5% step voltage changes in a similar manner to that on the single-channel tests. The response should be similar to the single-channel test and should be repeated at nominal and 85% machine voltage.
- On completion of these tests, record the decay time of the generator terminal voltage by open-circuiting the AVR.
- Check the operation of the voltage relay which automatically switches the AVR in and out of service during run-up and run-down of the turbine. This test should be carried out in both dual- and single-channel operation.

The checks on the AVR system have now been completed sufficiently to permit 'auto' operation of the AVR dual channels at first synchronising, although further tests and checks recur during the loading programme up to 100% MCR.

3.15 Phasing and synchronising circuit checks

3.15.1 Objective

The objective is to determine the phase order of the generator 400 kV connections by comparing them with a known phase order. Figure 6.36 shows a generator and two feeder circuits which will be used in this example of the phasing and synchronising. Initially, the three-phase VTs of feeder 2 are energised from the generator circuit with the feeder isolated from the 400 kV system. Comparisons are then made between the VTs of feeders 1 and 2. When these are complete, checks will then be carried out on the generator synchronising circuits to allow the turbine-generator to be synchronised to the 400 kV grid system for the first time.

3.15.2 Initial state of plant

The following are the plant requirements prior to the commencement of this activity:

- The station, unit, auxiliary and other electrical service boards necessary to operate the generator are in service.
- The turbine-generator waterspray protection system is complete, a successful deluge test has been carried out and the system is in service.
- Generator unit transformer oil levels are correct and the transformer breathers are in service and satisfactory.
- All oil valves on the generator and unit transformers are in the position for normal operation.
- Generator transformer oil pumps are operational and selected for remote control.
- Generator open and short-circuit tests are completed.
- The 400 kV short-circuit has been removed.
- Bearing pedestal insulation has been checked and a value of at least 1 MΩ achieved.
- The generator earth switches are locked open on both the 400 kV system and the generator.
- Unit transformer 11 kV ACB is open and isolated.
- All voltage transformers associated with the turbine-generator are in service.
- The automatic voltage regulator is set for manual operation with all AVR protection, control and power supply fuses and links in the operational position.
- The A and B AVR channels, DC isolators and AC circuit-breakers are in the open position.

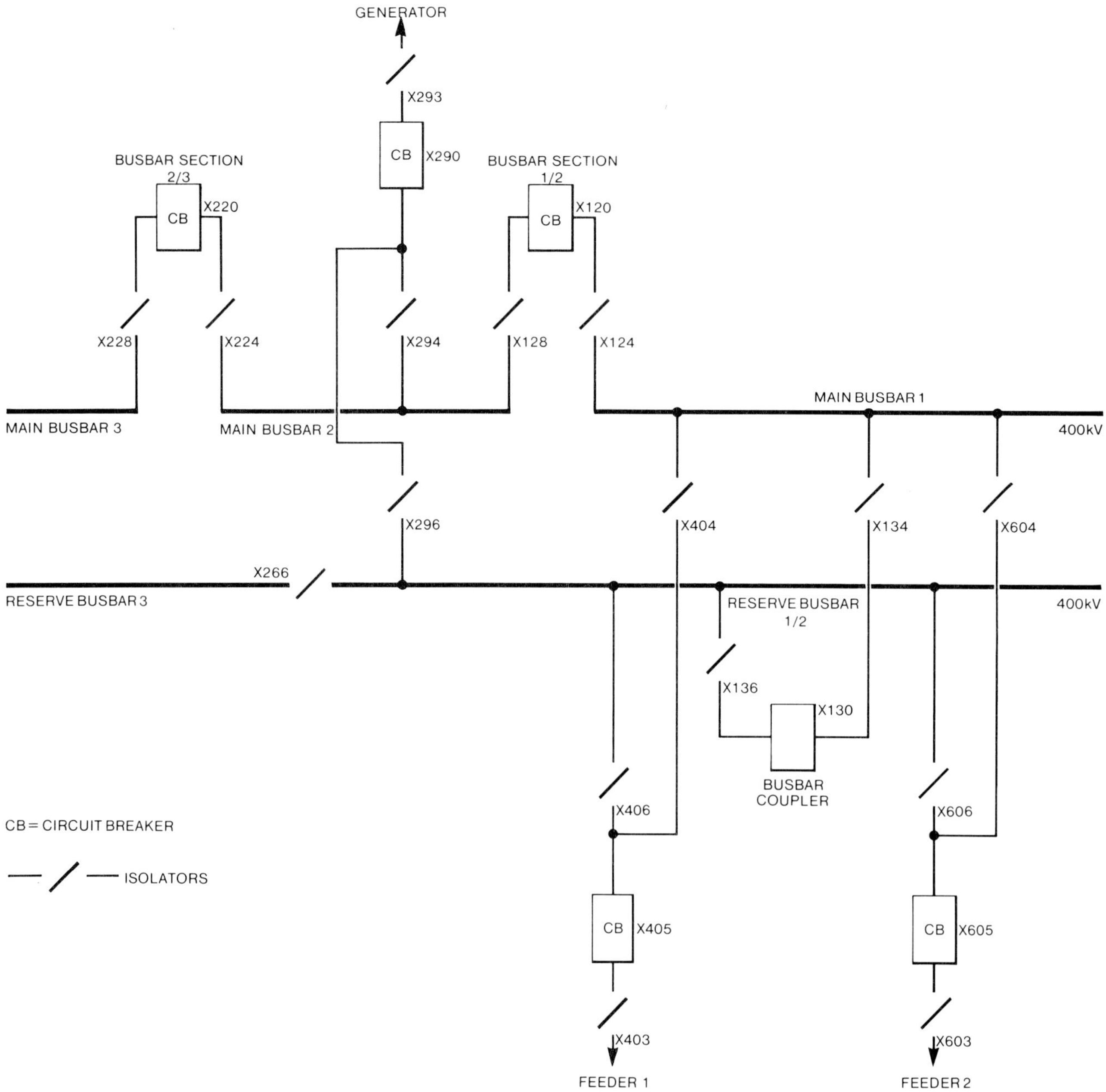

FIG. 6.36 Simplified 400 kV system for phasing and synchronising circuit checks

- All protection and control fuses associated with main generator 400 kV, unit transformer 11 kV and generator relay panel are in service.
- The generator and unit transformer tapchangers are in mid-position.
- If fitted, remove the boiler firing trip link to prevent unnecessary tripping of the boiler during this activity.
- The turbine-generator is at 3000 r/min.
- The generator 400 kV main and reserve busbar isolators are open (X296, X294). Line isolator open (X293). Circuit-breaker open (X290).
- The line VT fuses and links are in their normal position.
- The commissioning instantaneous earth fault relay is fitted to trip the main generator circuit-breaker (X290).

Prior to the commencement of this phasing and synchronising procedure the generator VTs have been energised from the 400 kV busbars via the generator circuit-breaker (X290), with the connections to the generator transformer disconnected. During this testing, checks have been carried out to confirm the phasing of the VT with respect to feeder 2 circuit yellow phase VT. In addition, feeder 1 VTs would have been checked against the VTs of feeder 2, with the synchronising wiring on both feeders proved. Checks on the generator synchronising circuitry would already be completed. Feeder 2 is a dead-end circuit, being isolated at its remote end for these tests.

3.15.3 Test procedure

Switching operations to isolate main busbars 1–2 and feeder 2 from the 400 kV system

- Initial setting up to be:

  Bus sections 1/2 and 2/3 circuit-breakers closed (X120 and X220)

  Bus coupler 1/2 circuit-breaker closed (X130)

  Feeder 1 reserve busbar isolator closed (X406)

  Feeder 1 main busbar isolator open (X404)

  Feeder 2 reserve busbar isolator open (X606)

  Feeder 2 main busbar isolator closed (X604)

- Open feeder 2 circuit-breaker (X603).
- Open bus section 2/3 circuit-breaker (X220).
- Open bus coupler 1/2 circuit-breaker (X130) discharging main busbar sections 1 and 2.
- Open busbar isolators (X136, X134 and X224).

Energising feeder 2 circuitry VTs from the generator, via main busbars 1 and 2:

- Put selected AVR channel into service on manual operation and set at minimum excitation.
- Close generator line isolator (X293).
- Close generator main busbar isolator (X294).
- Close generator circuit-breaker (X290).
- Close feeder 2 busbar isolator (X604) and line isolator (X603).
- Close feeder 2 circuit-breaker (X605) — dead circuit.
- Increase generator excitation manually to raise the stator voltage to the normal value and the feeder line voltage to approximately 400 kV. This step is likely to induce a charging current, thus causing a leading MVAr load on the generator.

To determine the phase order of the generator by energising feeder 2 VTs from the generator, via main busbar sections 1 and 2:

- Plug in the synchronising trolley, select 'synchronising' on the bus coupler circuit-breaker (X130) and check that 'running volts' only are present.
- The application of a short will allow an 'incoming volts' signal to be obtained even though the appropriate isolators are open. Check on synchronising trolley that both 'incoming' and 'running' volts are present.
- Adjust the turbine speed to slightly less than synchronous speed in order to obtain a slow slip speed.
- Remove the 'running volts' fuse on feeder 2, confirming on the synchroscope that 'running volts' have disappeared and that 'incoming volts' remain.
- Replace 'running volts' fuse on feeder 2 and confirm on synchroscope that 'running volts' reappear.
- With the generator at slow slip speed, observe the phase order between feeder 2 and feeder 1 VTs. This is done by phasing out between the VTs at the instant that the synchroscope across the bus coupler (X130) indicates (a) 12 o'clock and (b) 6 o'clock in accordance with Table 6.11.

TABLE 6.11

Phasing and synchronising circuit checks — phase order checks

	R–N Volts	Y–N Volts	B–N Volts	R–Y Volts	Y–B Volts	B–R Volts
Feeder 1 circuit VTs						
Feeder 2 circuit VTs						

Feeder 1 circuit VTs	Feeder 2 VTs			Synchroscope Accrom B/Comper (XBo)
	R, Volts	Y, Volts	B, Volts	
R Y B				12 o'clock
R Y B				6 o'clock

Notes R = Red phase, Y = Yellow phase, B = Blue phase, N = Neutral

- Deselect synchronisation on the bus coupler (X130).
- At feeder 2, open circuit-breaker (X605).
- At generator, open main circuit-breaker (X290).
- Open the following isolators: feeder 2 (X603 and X604) and generator (X293 and X294)
- Reduce generator excitation to minimum.

Switching to restore main busbars 1 and 2 to service and to isolate reserve busbars

- Close bus section 2/3 isolators (X224 and X228).
- Close bus section 2/3 breaker (X220).
- Close bus coupler 1/2 isolators (X134 and X136).
- Close bus coupler 1, charging main busbar sections 1, 2 and 3 (X130).
- Close feeder 1 busbar isolator (X404).
- Open feeder 1 busbar isolator (X406).
- Open bus coupler 1 circuit-breaker (X130 discharging).
- Open bus coupler 1 isolator (X136 and X134).

Energising feeder 2 circuit VT from generator, via reserve busbar

- Close generator line isolator (X293).
- Increase excitation manually on the generator to raise stator volts to design value and feeder 2 line voltage to 400 kV.

To determine the phase order of the generator, energising feeder 2 VTs from the generator, via reserve busbar section 1/2

- Plug in synchronising trolley and select 'synchronising' on the bus coupler circuit-breaker (X130). Observe the presence of both running and incoming volts. Set the turbine speed such that the synchroscope indicates a slow slip speed.
- Remove the running volt fuse from feeder 1 and check on the synchroscope that the 'running volts' indication disappears. The 'incoming volts' indication should remain.
- Replace the running volts fuse in feeder 1 and confirm on synchroscope that the 'running volts' indication has returned.
- With the generator at slow slip speed, observe whether the phase order between feeder 2 circuit VTs and feeder 1 circuit VTs is correct, by phasing out between the VTs at the instant that the synchroscope across bus coupler (X130) indicates (a) 12 o'clock and (b) 6 o'clock in accordance with Table 6.11.
- At the 12 o'clock position corresponding phase volts (e.g., red to red) should read zero and, at the 6 o'clock position, full line voltage should be obtained.

To check the phasing of the generator VTs by comparison with feeder 1 circuit yellow phase VT:

- Using the synchronising facilities across bus coupler 1 (X130), remove feeder 2 voltage fuse and confirm on the synchronising trolley that the 'incoming volts' indication disappears and that the 'running volts' indication remains.
- Replace voltage fuse on feeder 2 and confirm that the 'incoming volts' indication has returned.
- Remove the generator synchronising fuse and check on the synchronising trolley that the 'incoming volts' indication has disappeared. The 'running volts' indication should remain.
- Replace generator synchronising fuse and confirm that 'incoming volts' indication returns.
- Phase out between feeder 2 VTs and feeder 1 VTs at the same instant as the synchroscope indicates (a) 12 o'clock and (b) 6 o'clock in accordance with Table 6.12.

Table 6.12

Generator circuit and feeder 1 circuit yellow phase VT voltage comparison

Generator VT Feeder 1 VT Yellow phase	Synchroscope
Volts	12 o'clock
Volts	6 o'clock

- Simultaneously phase out between the generator VTs and feeder 1 circuit yellow phase VT at the same instant as the synchroscope indicates (a) 12 o'clock and (b) 6 o'clock in accordance with Table 6.12.
- Deselect 'synchronising' on bus coupler 1 circuit-breaker (X130).

Prove synchronising circuitry for generator main circuit-breaker (X290)

- Plug in synchronising trolley, select 'synchronising' on generator main circuit-breaker (X290). Confirm both running and incoming volts are present with the synchroscope at 12 o'clock.
- Remove generator synchronising fuse and confirm that the 'incoming volts' indication disappears on the synchronising trolley.
- Replace generator synchronising fuse and confirm that the 'incoming volts' indication reappears.
- Remove feeder 2 synchronising fuse and confirm that 'running volts' indication disappears on the synchronising trolley.
- Replace feeder 2 synchronising fuse and confirm that the 'running volts' indication reappears.
- Open feeder 2 circuit-breaker (X605) to discharge the feeder. Open line isolator (X603).
- Observe on the synchronising trolley that the 'running volts' indication has disappeared.
- Deselect 'synchronising' of generator main circuit-breaker (X290).
- Open main generator circuit-breaker (X290).
- Open main generator line isolator (X293).

Setting up 400 kV busbars and energising reserve busbar 1/2 for final synchronising checks on the generator

- Open feeder 2 reserve busbar isolator (X606)
- Open generator reserve busbar isolator (X296).
- Close feeder 2 main busbar isolator (X604).
- Close feeder 2 line isolator (X603).

- Close feeder 2 main circuit-breaker (X605)
- Close bus coupler 1/2 isolators (X134 and X136).
- Close bus coupler 1/2 circuit-breaker (X130), charging the reserve busbars.

Final checks on the generator before synchronising and paralleling generator with the 400 kV system, via bus coupler 1/2 and reserve busbars:

- Plug in the synchronising trolley and select 'synchronising' on generator circuit-breaker (X290).
- Connect a test meter to read the phase angle voltage difference between the generator incoming volts and the 400 kV system running volts. Observe that the test meter reads zero volts.
- Close generator reserve busbar isolator (X296).
- Close generator line isolator (X293).
- Check that the generator synchronising selector relay has energised. Check that the test meter indicates a fluctuating voltage as the generator runs with a small slip speed.
- Check that the test meter reads zero volts at the 12 o'clock position and maximum voltage at the 6 o'clock position.
- Raise the speed of the generator to slightly above 3000 r/min; the synchroscope must indicate that it is running fast.
- Reduce generator speed to below 3000 r/min; the synchroscope must indicate that it is running slow.
- Check 'synchronising' on the generator main circuit-breaker (X290).
- Synchronise and close generator main circuit-breaker (X290), paralleling the generator with the 400 kV system, via bus coupler 1.

Final checks on generator before synchronising and paralleling the generator with the 400 kV system, via bus section 1/2 and main busbar section 2

- Open generator main circuit-breaker (X290).
- Open generator line isolator (X293).
- Open generator reserve busbar isolator (X296).
- Close feeder 2 reserve busbar isolator (X606).
- Open feeder 2 main busbar isolator (X604).
- Close feeder 1 reserve busbar isolator (X406).
- Open feeder 1 main busbar isolator (X404).
- Open bus section 2/3 circuit-breaker (X220).
- Open bus section 2/3 isolators (X224 and X228).
- Close bus section 1/2 isolators (X124 and X128).
- Close bus section 1/2 circuit-breaker (X120).
- Plug in synchronising trolley, select 'synchronising' on generator circuit-breaker (X290).
- With a test meter connected to read the phase angle voltage difference between generator incoming volts and the 400 kV system running volts, observe that the test meter reads zero volts.
- Close generator line isolator (X293).
- Close generator main busbar isolator (X294).
- Check that generator 'synchronising selected' relay has energised. Check that the test meter indicates fluctuating voltage as generator runs with small slip speed. Check that the test meter reads zero volts at 12 o'clock and maximum voltage at 6 o'clock.
- Check 'synchronising' on the generator main circuit-breaker (X290).
- Synchronise and close the generator circuit-breaker (X290), paralleling the generator with the 400 kV system via bus section 1/2 and main busbars 2.
- On synchronising, apply a low load of approximately 10 MW. Reduce the generated output slowly to observe and record the operation of the low power relays (normal operation approximately 2–3 MW).

3.15.4 HP and IP cylinder reference point expansion check

Turbine expansions are continuously monitored during all operations up to the time the turbine is fully loaded at rated temperatures and pressures, and hence fully expanded. The reference point checks on the HP and IP cylinder against the cold set values should be taken on a daily basis during the complete commissioning programme. Figures 6.6 and 6.8 show the initial cold datum check and the amount of expansion that has occurred following charging the HP steam pipes. Figure 6.37 shows the expansion that has occurred after several days of turbine operation at 3000 r/min prior to first synchronisation. It enables an assessment to be made of the relative rates of expansion of the HP and IP cylinders together with consideration of the degree of lagging dry out that has occurred prior to increasing turbine load for the first time.

4 Commissioning of main boiler feed pump turbine

4.1 Introduction

A modern 660 MW unit feed pump arrangement is likely to comprise two 50% duty electric-driven feed pumps plus a main 100% steam-driven feed pump. The role of the 50% feed pumps is for initial start-up of the boiler and to maintain the unit on full-load in the event of a failure of the main feed pump. The commissioning of the electric feed pumps is described in Chapter 5 of this volume and is completed prior to the commencement of the steam-to-set activity. The electric feed pumps are fully tested during the steam purge and float safety valve activities.

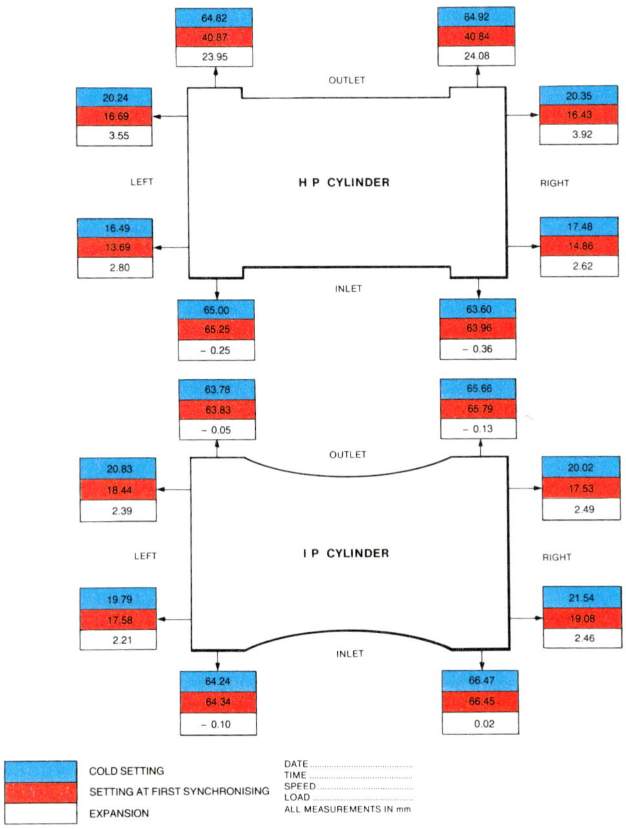

Fig. 6.37 HP and IP cylinder reference point expansion checks (first synchronising)

The main boiler steam-driven feed pump can only be fully commissioned when steam pressure is available from the boiler and the turbine is at 3000 r/min. The pre-commissioning checks described in Chapter 5 of this volume would have been completed prior to STS, adopting the general commissioning principles of the main turbine and cylinders for the steam-driven turbine.

The commissioning of the steam-driven feed pump is integrated into the overall commissioning programme with the objective of having this pump available and fully in service at synchronisation. Apart from the financial benefit over the use of electric feed pumps, it has the added advantage of enabling steam to flow at low pressure to HP heaters, enabling steam cleaning of the pipework and checks for steam leakages to be carried out early. Depending on the feed system installed and the quality of control exercised during construction, the 'iron' level of the condensate may restrict the progress of the commissioning programme if the main feed pump commissioning is left until later in the programme. It also has the advantage of achieving the designed feed outlet temperature relative to load.

The commissioning programme (Fig 6.10) shows the feed pump commissioning activity commencing during the generator short-circuit test and continuing over a five-day period in parallel with the short-circuit, open-circuit, AVR and phasing and synchronising checks.

During this period, apart from the generator trip tests, the turbine is operating at 3000 r/min whilst a series of electrical tests is taking place.

During these electrical tests, mechanical engineers will be available to commence the feed pump activity.

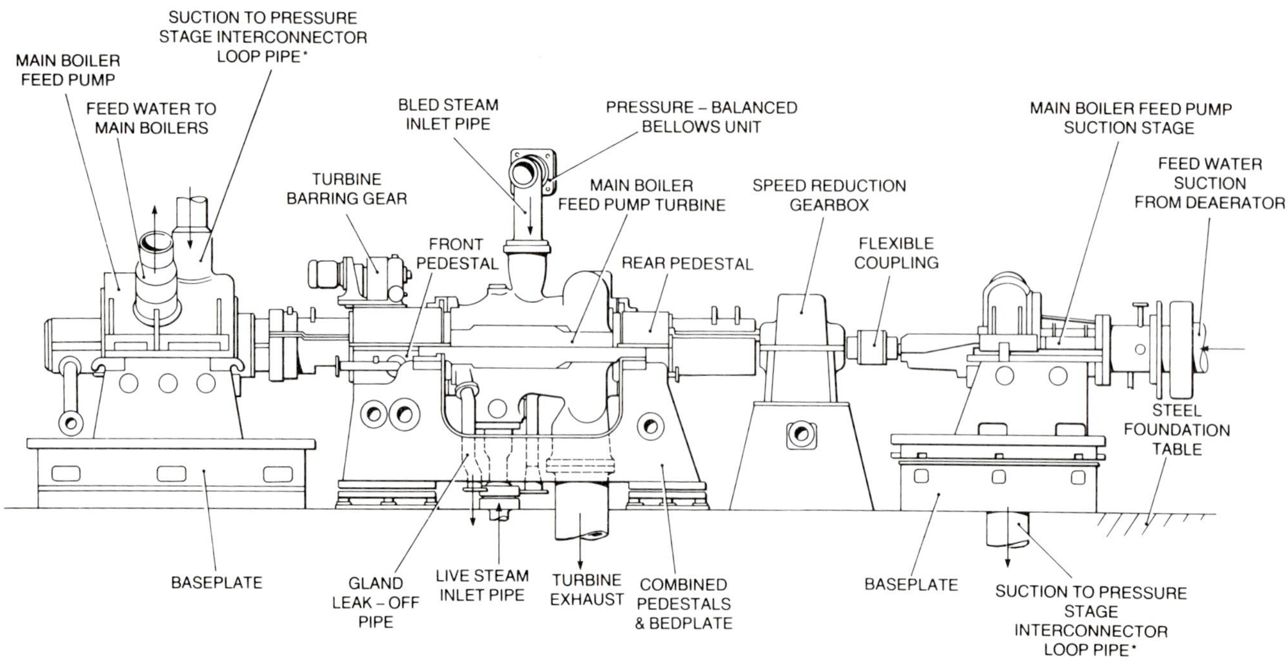

Fig. 6.38 Main boiler feed pump — general layout

4.2 General description

The main boiler feed pump comprises tandem pumping units (Fig 6.38), consisting of a suction booster stage driven through a speed reducing gear unit and a pressure stage pump driven direct by the main boiler feed pump turbine.

The bearing lubricating oil system for the feed pump is part of the main turbine oil system. The system is pumped under the head developed from the main shaft-driven oil pump with provision for pressure control if the turbine and boiler feed pump requirements are not compatible. A typical pressure for a feed pump oil system is 1 bar. The bearing return oil is cooled by the main turbine oil coolers.

4.2.1 Main boiler feed pump turbine

The boiler feed pump turbine is a single-cylinder, horizontal, variable-speed impulse reaction type. The turbine comprises five expansion stages, each stage consisting of a row of stationary blades and a row of moving blades. The boiler feed pump is designed to operate on either bled-steam from the reheat inlet pipework or live steam from the boiler. The live steam supply enables the feed pump to operate when the main turbine load is less than 30% MCR.

Live steam is admitted continuously at all times through a small pilot hole in the governor valves to keep the live steam chest and loop pipes at temperature when operating on bled-steam, thus avoiding thermal shock resulting from a changeover from bled steam to live steam. The boiler feed pump turbine is therefore designed to accept large and sudden changes of load over the complete load range of the main turbine without excessive thermal stress, fatigue or risk of cracking.

Separate steam chests and inlets on the boiler feed pump casing are used for the live/bled-steam supplies. Each steam chest houses a combined isolating and emergency stop (CIES) valve and a single-seated vertical governing valve. A permanent steam strainer of standard fine mesh, with 5 mm holes, is mounted with both the live and bled-steam CIES valves to prevent the ingress of foreign matter which could cause damage to the turbine.

Operation of the live and bled-steam stop and governing valves is via a hydraulic fluid control system.

Steam exhausts from the boiler feed pump turbine at a pressure corresponding to the transfer pressure between the IP and LP cylinders of the main turbine. The majority of this exhaust steam is used for de-aerator and boiler air heater steam preheating, any surplus steam being discharged to the IP/LP crossover. In addition, steam is taken from the appropriate boiler feed pump turbine stage for the bled-steam supply to the first stage of the HP heaters. A typical boiler feed pump exhaust steam system is shown in Fig 6.39, with Fig 6.40 showing the HP steam supply from the main boiler feed pump turbine.

4.2.2 Main boiler feed pump

The suction booster stage of the pump is necessary to increase the suction pressure to the high pressure stage. It receives feedwater from the de-aerator storage tank via the LP feedwater filtering system and discharges the feedwater at higher pressure via a loop transfer pipe to the pressure stage (see Table 6.13 for technical data).

The speed reducing gear unit is driven by the boiler feed pump turbine and the reduced speed output from this gearbox is used to drive the suction booster stage. The pressure stage raises the feedwater pressure to its required value and discharges through HP heaters and feed regulating valves to the boiler.

A dual fail-safe leak-off system operates automatically to prevent the feed pump from overheating when

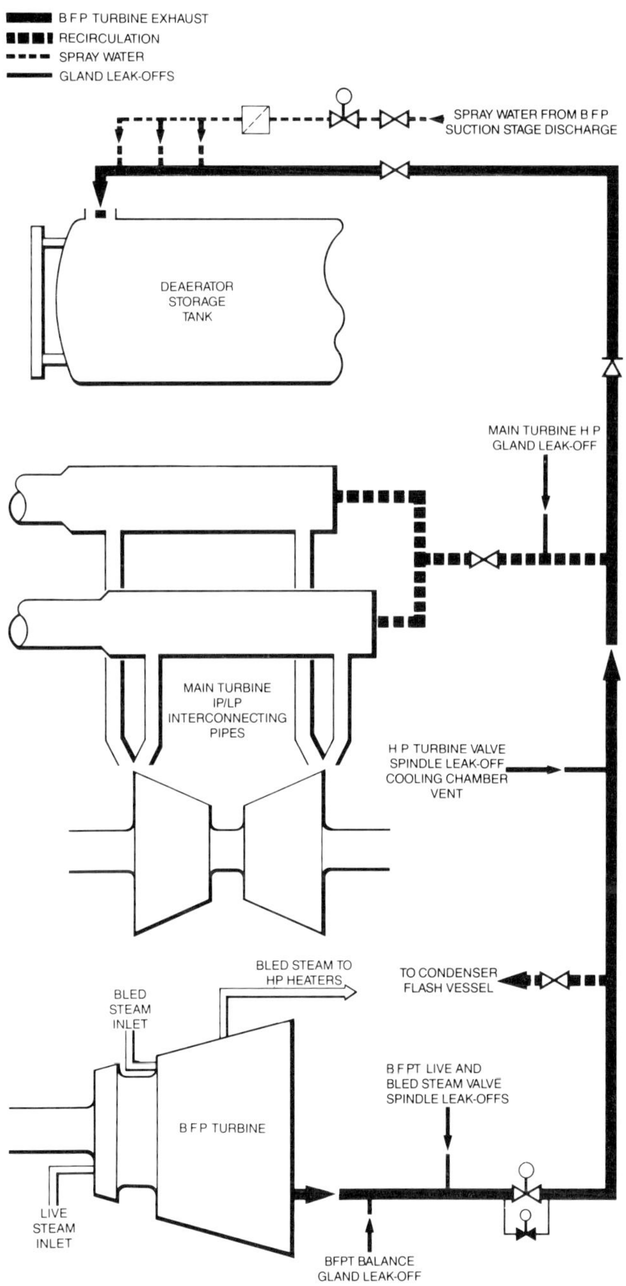

Fig. 6.39 Boiler feed pump exhaust system

TABLE 6.13

660 MW boiler feed-pump turbine — technical data

Suction booster stage	
Suction temperature	160°C
Vapour pressure	6.2 bar
Capacity	608 kg/s
Generated head	213 m (18.9 bar)
Suction pressure	8.3 bar
Discharge pressure	27.2 bar
Speed	2160 r/min
Power absorbed	1700 kW
Pressure stage	
Suction temperature	160°C
Vapour pressure	6.2 bar
Capacity	608 kg/s
Discharge pressure	202 bar
Suction pressure	27.2 bar
Speed	6450 r/min
Power absorbed	15,150 kW
Gear unit	
Power transmitted	1864 kW (2500 HP)
Input speed	6500 r/min
Output speed	2182 r/min
Leak-off system	
Leak-off quantity	76 l/s
Normal working pressure	202 bar
Normal operating temperature	160°C

the flow of boiler feedwater drops below a predetermined minimum rate. It is normal practice for each leak-off valve to pass 12.5% of MCR flow; with both leak-off valves operating together, the pump leak-off will be fully open up to approximately 28% MCR feed flow. The difference between 25% and 28% flow is to reduce the possibility of the leak-off valves hunting.

Mechanical seal units are keyed to the rotor shaft one on either side of the impeller, to prevent loss of pump water. Each unit contained within a seal housing consists of a rotating tungsten-carbide coated seal ring and a stationary carbon seal ring. Both rings make physical contact, forming a seal to prevent loss of pump water. Each unit has its own self-contained cooling system. The pressure stage glands require to be cooled via a gland seal cooling system. This cooling water supply must be reliable and of feedwater quality and is likely to be obtained from between the suction booster and the pump pressure stages. It is cooled, filtered and passed to the pressure stage glands, the gland seal cooler being cooled by condensate.

A typical main boiler feed pump system (Fig 6.41) shows a two-stage pumping unit identifying steam and water circuits, together with the supplementary services of gland seal water, control air, service water, etc.

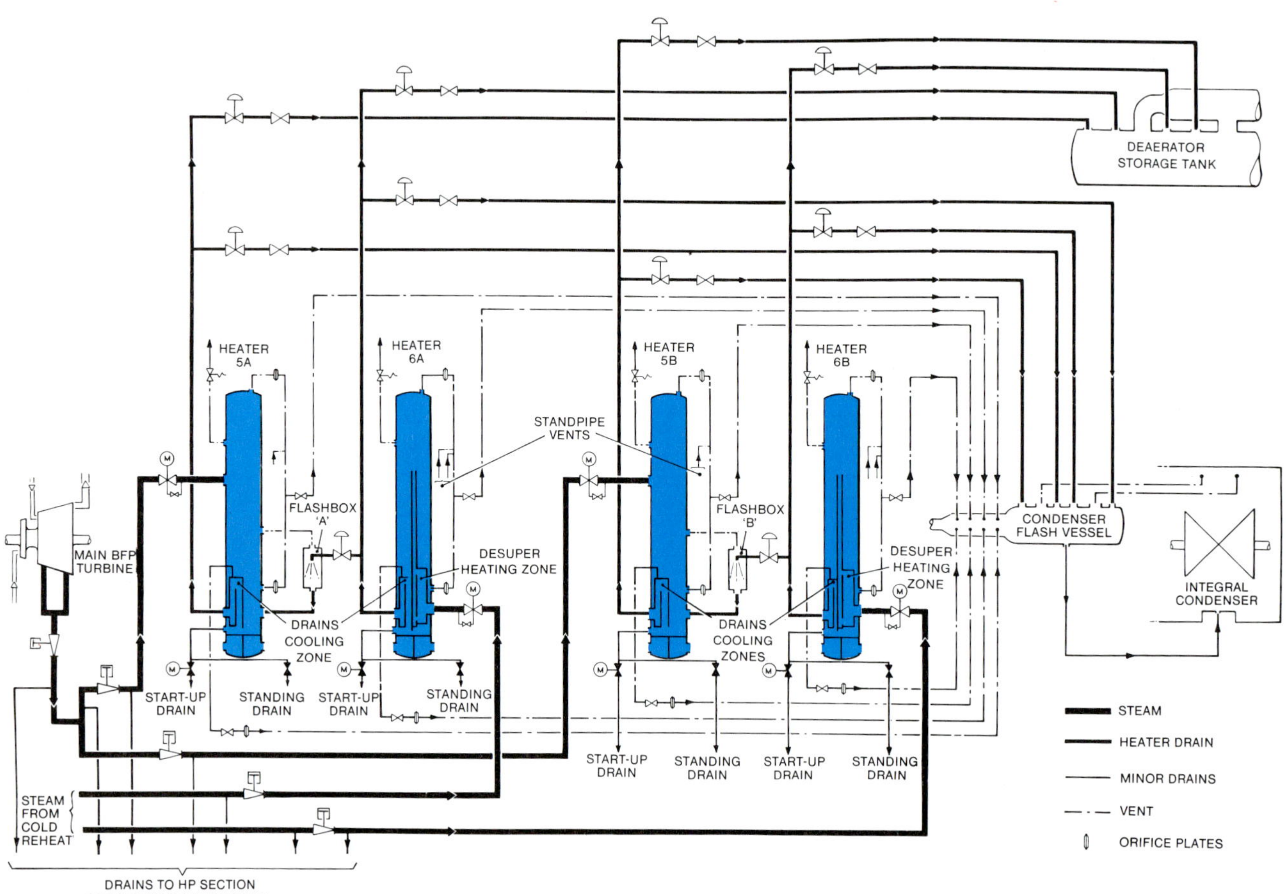

FIG. 6.40 HP heater steam system

4.2.3 Turbine governing and control

The speed control of the main boiler feed pump turbine, and hence the flow of feedwater to the boiler, is achieved by electronic governor controls. The governor system comprises three identical speed computing channels and two valve controllers for the live and bled-steam valves. In practice, the method adopted controls the live and bled-steam governor valves, utilising a speed error computing system and valve positioning loop. The control of the valves is such that the bled-steam valve is fully open before the live steam valve starts to open.

Primarily, the electronic governor provides a closed-loop speed control system during run-up, which controls the opening of the live and bled-steam valves. Secondary control for these valves is provided from the feedwater control system.

The turbine speed is measured by three separate magnetic transducers which generate a frequency proportional to turbine speed. This output frequency is converted into a voltage which is applied to the speed-sealing section, where the turbine speed is calibrated in terms of operating limits. The speed signal output from the speed-sealing section is compared with a speed reference (derived from a motorised potentiometer) to produce a speed error output from the comparator circuit. The speed error signal is fed to a low value gate, where it is compared with the load reference. The lowest of these three signals then becomes the controlling signal, determining the valve opening signal for either the live steam or bled-steam valves. A 3% speed error will open the bled-steam valve; a further 3% speed error produces a direct output to control the live steam valve.

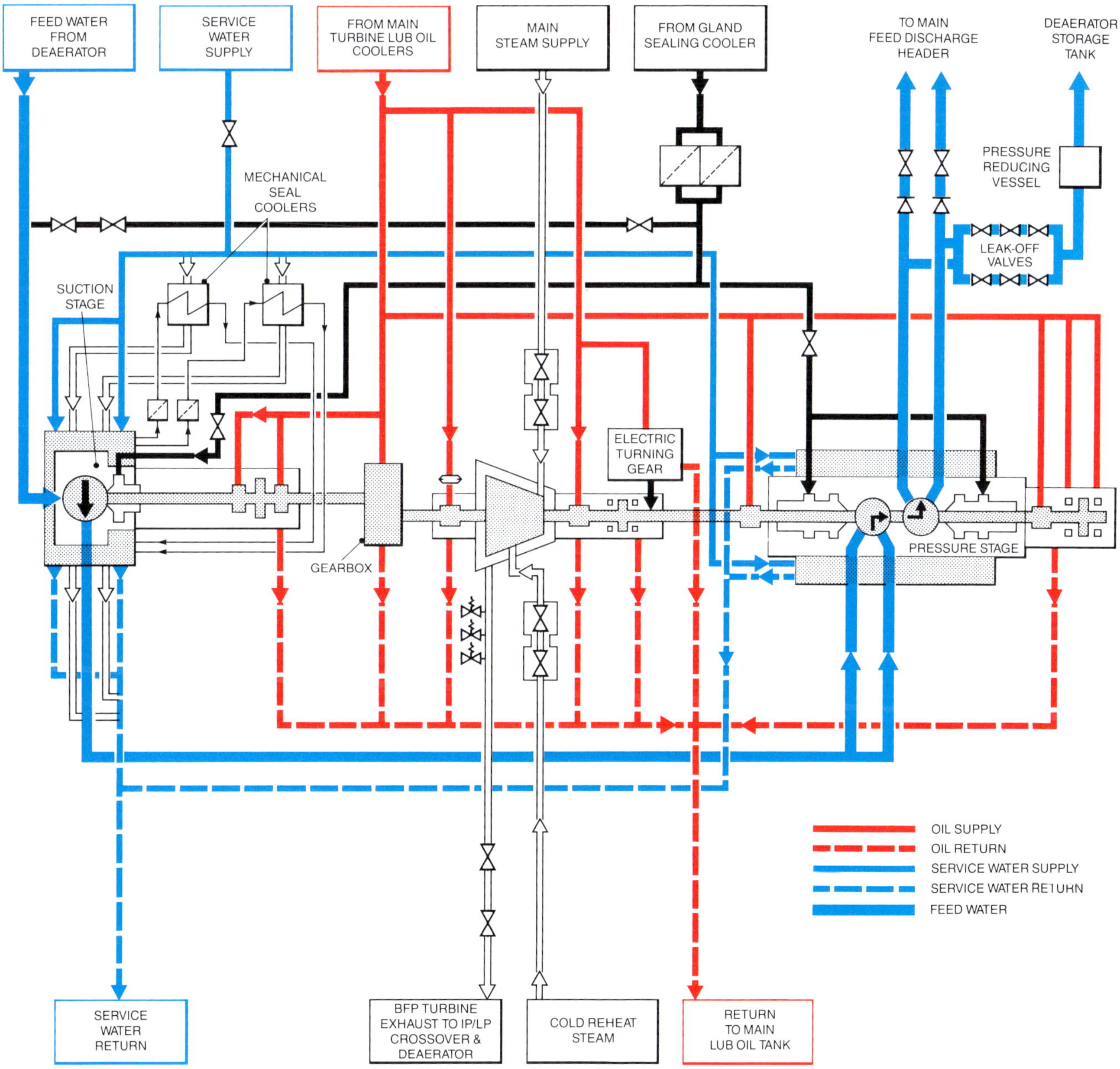

FIG. 6.41 Typical boiler feed pump system

The resultant valve position signals are used to control the speed/load of the turbine, depending on the mode of operation. A typical boiler feed pump governing system is shown in block diagram form in Fig 6.42.

4.2.4 Turbine protective devices

Protective devices are installed in the boiler feed pump control system to trip the turbine in the event of a condition arising that will hazard the turbine or pump. Automatic protective devices are installed to protect against the following:

- Turbine overspeed
- Bearing oil pressure low
- Feed pump balance water temperature high
- Loss of electronic governor

To prevent spurious tripping of a single protective device, the principle of 'two out of three' operation is employed for a number of protective devices, i.e., bearing oil pressure and feed pump balance water. In the case of the electronic governor, a trip is initiated resulting from any two of three speed channels tripping together. Hence the loss of one speed channel will not trip the governor system.

4.3 Pre-commissioning activities

The main boiler feed pump turbine can be divided into the following sections:

- Bearings, pedestals and foundations
- Live and bled-steam stop and governor valves
- Live and bled-steam exhaust systems
- Turning gear and lubrication system
- Governing and control
- Supervisory system
- Protective systems
- Boiler feed pump

Prior to the commencement of the commissioning activities for first running the pump, it is necessary to ensure that construction is complete in all the areas identified above and that all the pre-commissioning tests and checks have been done, using the standard check lists and construction records. Chapter 5, Section 15.3.2 of this volume deals specifically with the pre-commissioning checks and tests associated with a motor-driven feed pump; these checks are also applied to the feed pump side of the main boiler feed pump. However, the main feed pump, which utilises steam for its driving power, is also fitted with steam supply stop and governor valves required for the purpose of speed control. These aspects are not part of a normal electric feed pump so these pre-commissioning checks, which are not identified in Chapter 5 of this volume, must be specified.

4.3.1 Live and bled-steam stop and governor valve pre-commissioning checks and tests

- Check stroke of live and bled-steam stop and governor valves.
- Check the operating equipment of all the valves for any lost motion.
- The operation of live and bled-steam governor valves to be checked to ensure that the electro-

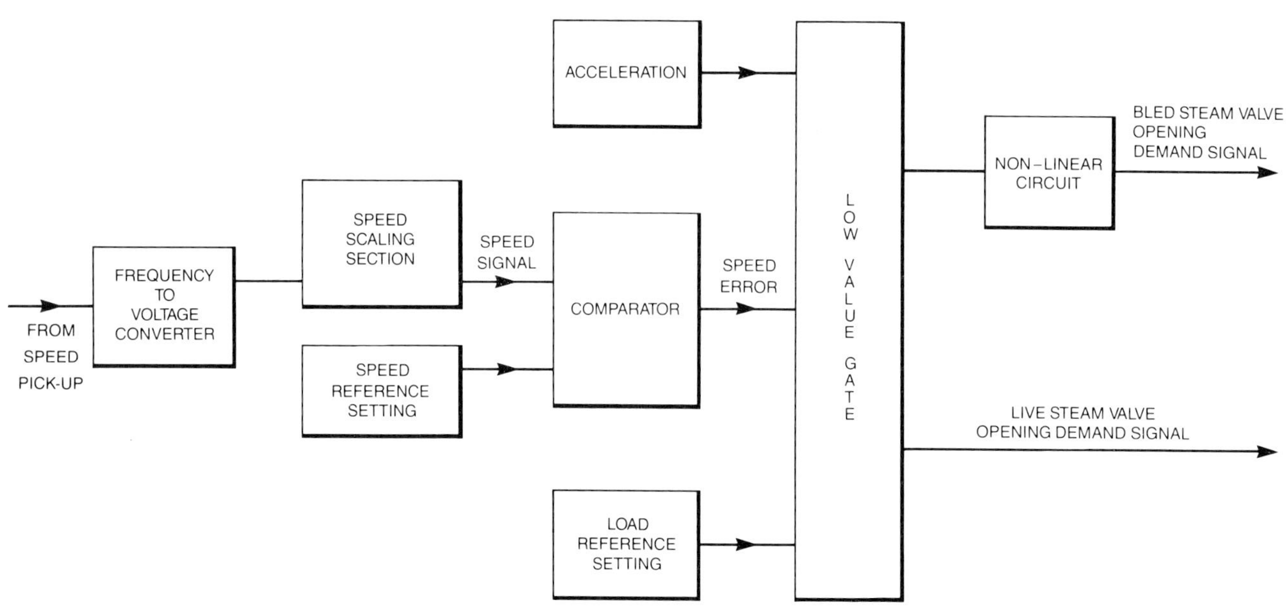

FIG. 6.42 Boiler feed pump electronic governor system — block diagram

hydraulic servo-valve operates correctly. This should be proven from 'remote', using the electronic governor.

- Check the calibration of the 'remote' valve position indicator on the control desk against the actual valve position for the stop and governor valves.
- Check the control of both the live and bled-steam governor valves. This is normally done by injecting a signal into the electronic governor representing the shaft speed.
- Check all steam valves individually for fast-closing operation.
- Check all steam valves to confirm that the on-load test equipment is functioning correctly.
- The stability of the live and bled-steam governor valves is to be checked to ensure that hydraulic fluid pressure surges can be absorbed by the accumulators fitted. This test is carried out by switching out the duty hydraulic fluid pump and allowing the standby pump to automatically start on falling pressure.
- Shut down the hydraulic fluid and test to ensure that at least three rapid strokes of the live and bled-steam governor valves can be achieved before there is insufficient pressure in the hydraulic system to operate these valves.

4.3.2 Live and bled-steam exhaust systems

- Check that all pipework, valves, instrumentation and cabling has been installed in accordance with the construction checklists.
- Air-operated valves must be 'stroked' to confirm free and correct operation. Any on-load test facilities must be checked and proved for correct operation.
- Where practicable, non-return valves movement should be confirmed free, ensuring that the valve is sufficiently sensitive to react to a reversal of steam flow.

4.3.3 Electrical turning gear

- Confirm that the lubricating oil system has been flushed clean, and that the correct lubricating oil pressures are being obtained at the bearings and gearbox.
- Confirm the freedom of shaft movement by manual turning.
- Confirm the starting performance of the shaft electrical turning gear and compare it with the design curve.
- Check that an electrical turning gear start is inhibited when low lubricating oil pressure is indicated. This can be achieved by simulating a low pressure by venting the impulse line.
- Confirm the correct operation of bearing high temperature alarms.

4.3.4 Protective devices

- Confirm that all cubicles are equipped and installed to the latest drawing revision.
- Check the electrical supplies.
- Check the operation of all protection devices and confirm that the required result is obtained. This check is made against the latest issue of the protection diagram.
- Electrical supplies must be lowered to the specified minimum voltage and a check made to confirm that relays and solenoids still operate at the reduced voltage.
- Test with hydraulic fluid system in operation to demonstrate the correct operation of the trip valves and reset solenoids.
- Satisfactory operation of the protective devices to be proven by establishing normal plant conditions and then simulating a failure.
- Local and remote emergency stop pushbuttons to be checked for correct operation.

4.3.5 Supervisory system

- The cubicle(s) and any cabling interface boxes to be checked for completion against construction checklists.
- Check the cabling associated with the measuring points for segregation and screening from other local cabling.
- Carry out calibration checks between the sensor and turbine supervisory cubicle amplifier to establish zero span and linearity settings. The measured parameter must then be adjusted by known amounts and remote alarm settings, indicators, recorders and computer inputs checked, the calibrations being carried out on the following: differential expansion, eccentricity, speed measurements and vibration measurements.

4.3.6 The feed pump

- Check that all pipework, valves, vessels, instrumentation and electrical equipment, including cabling, has been installed and checked against the construction records.
- Pressure stage pump glands and mechanical seals and associated pipework to be flushed until clean.

- The gear unit and all pump oil bearings and their associated pipework systems must be flushed clean, with cleanliness confirmed after temperature cycling the oil, with no debris being picked up in the fine-gauge strainers fitted for flushing.
- Confirm that the pump clearances and axial pump shaft movement are correct.
- Check the turbine/gearbox/pump alignment figures prior to first pump run.
- Establish the cooling water flows to all the applicable cooling systems; this will include bedplate cooling and seal water coolers. The appropriate flow switches, controls and alarms associated with the cooling system must be checked for correct operation.
- Check all the bearing high temperature alarms for correct operation.
- Check the mechanical 'seal water' high temperature alarms for correct operation.
- Prime both suction and pressure stages of the pump to check for evidence of leakage. This activity should be carried out early in the overall commissioning plan, at a time when the de-aerator is at ambient temperature. Under this condition, it is not essential for the feed pump to be on turning gear. However, if the activity is carried out with the de-aerator heated, the pump charging should only be carried out with the feed pump on turning gear. The proving of the complete pump and its pipework should be undertaken as early as possible, since any leakage is likely to be from a large diameter joint that will require time to repair. This could affect the later commissioning programme, if not started early.
- Where automatic sequence starting equipment is fitted to the various auxiliaries of the pump, the instrumentation and its associated alarms should be proved to be in working order, both 'local' and 'remote'.
- Check the operation of the electronic overspeed trip device.
- Manually rotate the rotor in the reverse direction to confirm the correct operation of the reverse rotation relay.

4.4 Preparations for first pump run

Having completed all the pre-commissioning checks with all results documented, preparations can now be made to put all the pump sub-systems into service in readiness for the first run. This will include carrying out the following actions:

- Check that all drains associated with the lubricating oil system are locked closed. Put the lubricating oil system into service and check that oil flows at the gear unit and at all bearings.
- Put the mechanical seal and bedplate cooling systems into service. Confirm that flow rates are correct.
- Ensure that the feed pump discharge valve is closed.
- Confirm that the de-aerator is filled to normal working level and that make-up water to the de-aerator is available.
- Prime the 'suction filters', which are situated on the suction line from the de-aerator to the LP stage of the pump. Initially, it is recommended that the filter backflush be left permanently in service with the filter controls on manual. The initial runs of the pump on leak-off will cause a considerable amount of mill scale to be scavenged from the pipework, especially from the leak-off lines. Since this debris is transported direct to the de-aerator, running the filters continuously on back flush ensures its removal from the system.
- Having primed the suction filters, it is good practice to lock the filter inlet valves open.
- Confirm that both the suction and pressure stages of the feed pump have been primed.
- Confirm that the mechanical seal is primed, with valves in their normal operating positions.
- Once the feed pump has been primed, check that the indicated suction pressure at the low stage pump under static conditions equates to the designed head figure.
- Ensure that the 'gland cooling' system is in service and that the system and coolers are primed. Confirm that the gland outlet suction valve is locked open.
- Check that all power supplies to the supervisory equipment are available and that the following equipment is available for service: bearing vibrations, eccentricities and differential expansions.
- Turn the boiler feed pump shaft by hand and check for freedom of rotation.
- Having confirmed the pump to be free, start the electrical shaft-turning gear, check the oil flows and that the shaft is being turned satisfactorily without high motor current. Confirm that there is no evidence of rubbing on the shaft.
- Confirm that the valve gear hydraulic system which operates the main steam stop and governor valves is in service.
- Carry out an off-load test of the on-load valve test equipment. The procedure for carrying out this test is given in Section 4.8 of this chapter.
- Check, and set up as required, the steam exhaust paths from the turbine, gland steam and spindle leak-offs.

- Check that all the feed pump steam drains are correctly set, and that the main turbine is at 3000 r/min with the condenser operating under normal vacuum conditions.
- Check that the feed requirements for the boiler are being supplied from the standby electrical feed pump, with the HP heaters in service on the water side only.
- Charge the 'live system' supply leg after warming-through the pipework and chest. The warming rates should be the same as stated for the main turbine, i.e., 100°C/hour when condensate heating, and 135°C when superheating.
- Reset the live and bled-steam stop valves.
- Set the speed governor to minimum setting.

4.5 First pump run

Having now completed all the pre-commissioning checks and ensured that all subsystems associated with the feed pump are set for correct operation and that the live steam chest is at the required temperature, the pump run-up activity can commence. The pump commissioning operations are in two distinct phases.

Phase 1 Consists of the pump operating on leak-off only in a closed loop comprising the de-aerator storage vessel, feed suction pipework and suction strainers.

Phase 2 Follows Phase 1 immediately, providing that the feed path to the boiler is established. The checking of pump performance will be continuous until the main boiler/turbine unit has reached its maximum design rating, i.e., 660 MW.

- During the complete commissioning operation the pump should be operated within its operating envelope. A typical curve for a 660 MW feed pump is shown in Fig 6.43.
- Throughout the running operation, the following parameters should be continuously monitored:

  Rotor differential expansion
  Rotor eccentricity
  Pedestal vibration
  Levels of bearing vibration
  Bearing temperatures and pressures
  Steam inlet and outlet pressures and temperatures
  Casing metal differential temperatures
  Gland seal pressures
  Turbine and pump speed
  Movement of the feed pump table top

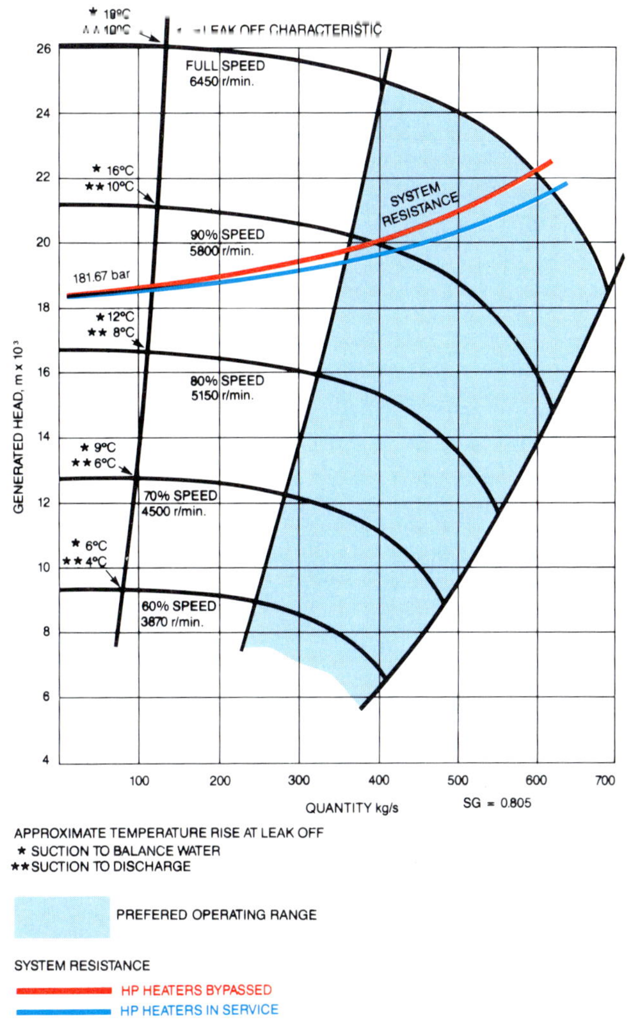

FIG. 6.43 Typical 660 MW boiler feed pump operating envelope curve

- Check that all expansion joints on the turbine are adequately greased on the following items:

  Dowel post and axial keys
  Pedestal keys
  Stop and governor valve pedestal feet

- Throughout the commissioning operation, continuous monitoring should be carried out to check the degree of speed control exercised by the electronic governing system over the complete speed range of the feed pump.
- Closely monitor the cylinder steam drains during the run to ensure that all lines are hot. Steam traps should be initially isolated and by-passed and not put into service until it is considered that all debris has been removed from the system. In practice, this will not occur until all the commissioning tests are complete.
- Back-warm the boiler feed pump turbine until the outer flange temperature reaches a minimum value of 90°C.

- Roll the turbine on live steam for a cold start by the operation of the governor. Increase the pump speed to 1200 r/min and hold at this speed. At this condition check that the electric turning gear motor stops at 25 r/min, and that pump discharge pressure rises immediately on start-up and continues to rise to the level corresponding approximately to the leak-off quantity at the applicable pump speed. If the pressure does not increase, the pump must be shut down and the system revented.
- Carry out all necessary monitoring at 1200 r/min, and hold at this speed until the cylinder flange temperature has reached 160°C.
- When satisfied with conditions at 1200 r/min, increase speed to 2000 r/min for further heat soaking and monitoring.
- Once conditions have stabilised at 2000 r/min, increase speed to 3000 r/min and change over control from live steam stop valve to governor valve. Confirm that the turbine exhaust steam path is to the main turbine condenser.
- Complete all monitoring and recording at 3000 r/min.
- With the pump at this speed, recheck that the cooling water and gland sealing water flows are still correct. Adjust if necessary.
- Check filter differentials. Change over and clean, as required.
- Carry out a complete check of the feed pump for evidence of any form of leakage.
- Check leak-off pipework to ensure that pipework hangers are all functioning.
- Check that speed control of the pump is satisfactory under the operation of the electronic governor.
- When satisfied with conditions at 3000 r/min, the pump speed is increased in two stages up to a maximum speed of 70% full speed (4500 r/min). The monitoring exercises previously stated are to be continued.

  Note: It is not recommended to run a steam feed pump at a higher speed than 4500 r/min against a closed discharge valve, as this is likely to induce excessive pump vibrations.
- Carry out on-load testing of the turbine overspeed protection device.

4.6 Taking up pumping duty

- When satisfied with the pump operation at 70% speed on closed discharge, the pump is now ready to be brought into pumping duty.
- With the HP heaters already in service fully primed on the water side, open the steam isolating valves to the HP heaters fully after first setting all the valves associated with the drainage control to their operating positions. Prior to the commencement of this activity, the operation of the drainage system and its associated alarms and protection system will have been fully tested.

  The advantage of bringing the HP heaters in service on the steam side at this early stage is that it gives better control over the rate of bled-steam pressure increase, as well as allowing early cleaning of the associated steam pipework.
- With the pump speed at approximately 70%, open the feed pump discharge valve and increase pump speed until the feed non-return valve lifts and feedwater flows.
- Increase pump speed to meet the boiler feed flow requirements, maintaining its operation within the operating envelope; check the operation of the electric feed pump to confirm that its output flow has reduced and that the leak-off valve has opened.
- Carry out a check on the operation of the leak-off valves, confirming that the valves close at design flow (in this case, 182 kg/s rising) and reopen at 152 kg/s falling. This requires adjustment of the feed pump speed and hence of its output.
- Continuously monitor the parameters stated and carry out a complete inspection of the pump and its subsystems. Check the feed discharge pipework, including confirmation that the pipe supports are not showing signs of being overloaded. This will be recognised either by the supports operating on their stops or by damage to the supports.
- With a pump flow greater than the leak-off flow requirements and with the leak-off valves shut, carry out an on-load test of the leak-off valves. Confirm that the valves open fast in a smooth operation and that they close fully when commanded.
- As the commissioning programme proceeds past 25% MCR, the bled-steam supply from the cold reheat will become available. Change over progressively to bled-steam and check that the live steam governor valve closes and that the bled-steam governor valve takes over speed control.
- As the pump output increases, measure and record the flow and head values at constant speeds and check the results against predicted performance curves.
- Carry out on-load testing of the live and bled-steam valves once the pump operation is solely on bled-steam. The procedure for this is described in Section 4.8 of this chapter.
- During the complete commissioning tests, from initially proving the pump up to full-load, the plant

should always be operated within the defined operating parameters and limits. If limits are exceeded, action must take place immediately to control the problem.

4.7 Pump shutdown

A planned shutdown of the pump from a heated condition should be carried out early in the programme, prior to the unit reaching approximately 25% MCR.

- The objective of this early shutdown is to confirm that, under hot conditions, the boiler feed pump steam valves trip correctly, the leak-off valves have opened, the pump run-down is smooth, and the electrical turning gear motor starts and successfully turns the shaft for 4 hours.
- Having established that the electric feed pump is either still in service or is put into service, trip the main boiler feed pump from the emergency trip pushbutton. Confirm that all valves have tripped and correct alarm indications have been received.
- Confirm that the turbine exhaust-to-condenser valve has shut.
- Check that the pump runs down smoothly, recording the time of run-down to turning gear speed and, at approximately 25 r/min, confirm that the turning gear motor has started.
- Maintain the lubricating oil system in service for the 4-hour shutdown period. Leave all cooling water supplies in service until the feed pump has cooled.
- Close the feed pump discharge valve.
- Having completed a 4-hour shutdown, a further pump run should take place. After this shutdown period, the run-up will be in a 'hot' category, unlike the initial run which was a cold start. The same parameters require monitoring, the run-up being, on this occasion, to the manufacturer's standard operating instructions.

4.8 Pump protection system testing

This section details the procedure for carrying out the following tests:

- Pre-steam to boiler feed pump turbine, test of the on-load valve test equipment
- On-load valve test at 330 MW
- On-load valve test at 660 MW
- On-load test of electrical overspeeds
- On-load testing of overspeed trip channels

4.8.1 Testing boiler feed pump turbine and governor valves

The main boiler feed pump governing system should be considered identical to the main turbine governing system, both for testing and the need to confirm that the valve gear is free over its full range of movement. The on-load testing of the boiler feed pump governing gear is different from that of the main turbine in as much that the test can only be carried out when the pump is operating entirely on the bled-steam supply. This requires the live and bled-steam valves to be fully open, with the live steam governor valve closed and the bled-steam governor valve controlling steam admission.

The valve testing on-load basically comprises the following actions:

- Tripping the bled-steam stop valve
- Confirming that the feed pump live steam governor valve is in control
- Resetting valves
- Closing live steam stop valves
- Confirming that the feed pump bled-steam governor valve is in control
- Resetting valves

The valve test carried out on the main boiler feed pump prior to charging the steam supply pipes is only demonstrating, under off-load conditions, that the on-load valve test operates correctly without the ability to check the control aspects.

During the commissioning phase, the effects on the main boiler feed pump operation of carrying out the on-load testing are unknown. It is sound commissioning practice to carry out the on-load valve test as soon as is practicable which, for a 660 MW unit, would occur at approximately 330 MW when the turbine is on bled-steam operation.

The test programme at 330 MW is an information test to indicate that the test at full-load can be carried out satisfactorily, so the test procedure written for the 330 MW operation is identical to that for 660 MW.

4.8.2 Off-load testing of the on-load valve test equipment

Initial state of plant

- Live and bled-steam pipes at atmospheric pressure
- Governor hydraulic fluid system in service
- Boiler feed pump turbine on electrical turning gear
- Feed pump filled with water and, if necessary, pressurised

Test procedure

- At the unit control desk, select the BFPT valve gear test switch to the 'test permissive' position; confirm that indication lights are correct.

- Confirm that all indication lamps are intact by operating the test switch.
- Test trip the live steam stop valve and confirm that the valve is closed (Fig 6.44).
- Operate the live steam stop valve operating gear; confirm that here the operating gear motor runs to the closed position and then, with a further operation, runs the valve back to its fully-open position.
- Operate the governor control test switch and confirm that the bled-steam governor valve closes and that the live steam governor valve opens. Check the smooth operation of both valves.
- Test trip the bled-steam stop valve and confirm that the valve is closed.
- Operate the bled-steam stop valve operating gear; confirm that the operating gear motor runs to the closed position and then, with a further operation, runs the valve back to its fully-open position.
- Reset the governor valve and check that the bled-steam governor valve opens and that the live steam governor valve closes.
- Deselect the valve gear 'test permissive' to 'normal' operation.
- Test complete.

4.8.3 On-load valve testing, 330 MW to 660 MW

- Initial state of plant satisfactory
- Load on unit at least 330 MW
- Feed pump control on bled-steam only, with live steam governor valve closed
- Live and bled-steam stop valves fully open
- Feed pump on 'auto' control
- Standby electric feed pump running on leak-off

Test procedure

The procedure adopted is identical to the off-load test procedure, as stated in Section 4.8.2 of this chapter. However, it is necessary to monitor the control aspects closely as the bled-steam governor valve closes and the live steam governor valve opens. This part of the test

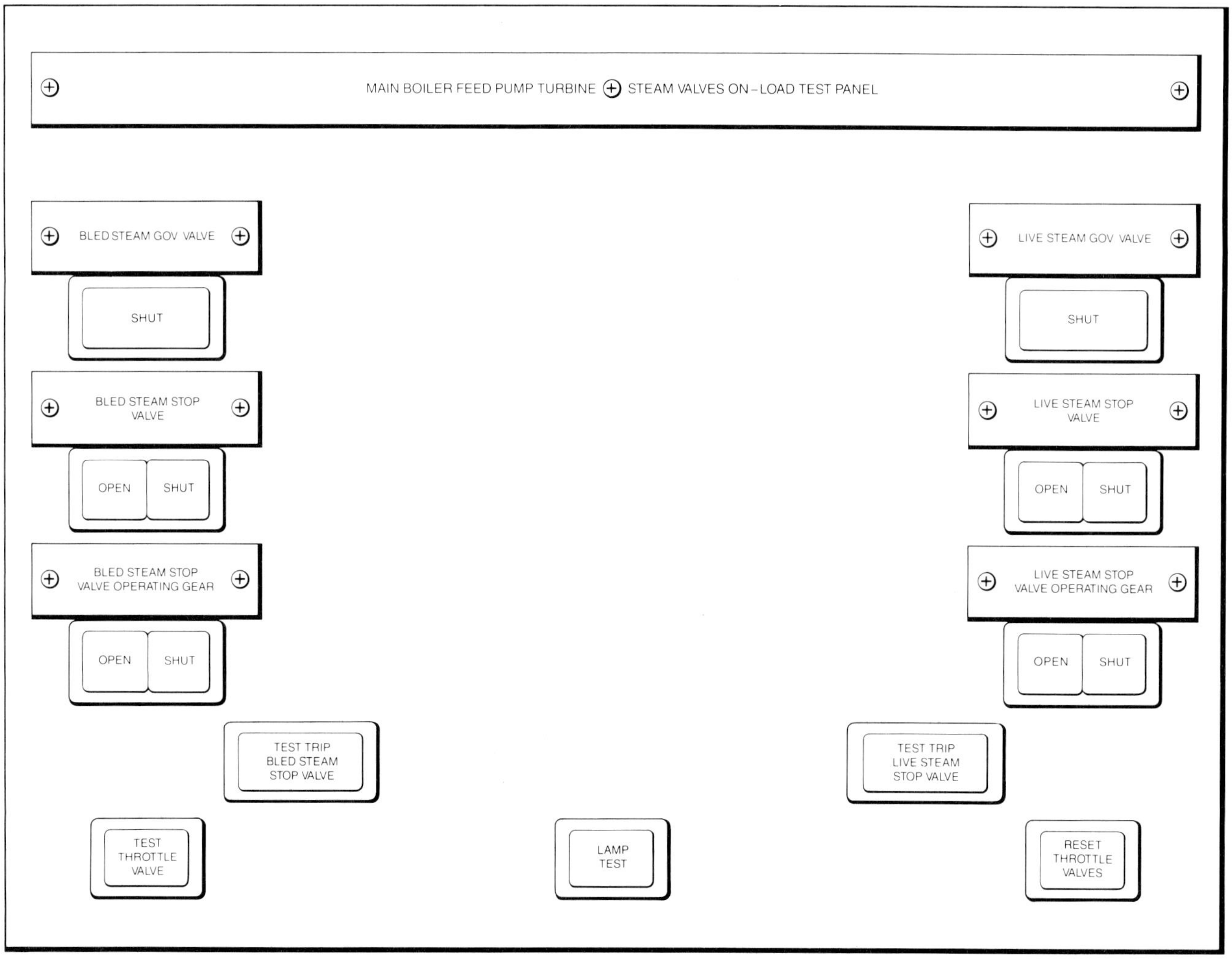

Fig. 6.44 Boiler feed pump on-load valve test panel

does not trip the valves but closes/opens the valves under the control of the hydraulic governor. Of necessity, their changeover needs to be a slow operation under sequence control and can be expected to take approximately seven minutes. During this period, control that was initially with the bled-steam governor valve is transferred progressively to the live steam governor valve. The changeover should be smooth and have no effect on the operating performance of the feed pump.

During the live and bled-steam changeover, it is necessary to monitor bearing vibrations and eccentricities closely, in addition to the pump performance.

4.8.4 On-load testing of electrical overspeeds

The turbine speed is measured by four sensors mounted in the same area as the governor sensors. These sensors transmit signals representing the turbine speed to the speed measuring equipment associated with the overspeed protection equipment.

If the turbine reaches 7100 r/min, i.e., 10% in excess of normal maximum operating speed, relay operation will trip the electronic governor providing two speed channels have been tripped. In addition the live and bled-steam valves will be tripped, together with the HP heater steam inlets and non-return valves.

The testing of this equipment is carried out on-load normally from the 'overspeed protection' cubicle (Fig 6.45) which would house annunciators, alarms, controls, digital speed indications and individual speed indicating meters used for the on-load overspeed testing.

Initial state of plant

Feed pump at 100% flow on auto control. Supporting electric feed pump in service on leak-off to cover possibility of malfunctioning during the test.

Test procedure

- At the test panel, confirm that all annunciator lamps are operating.
- Confirm that overspeed trip dial settings are correct. Note that the normal range of settings is 105% to 111% of normal maximum speed.
- Confirm that hydraulic fluid pressure indication is correct (65 bar).
- Confirm that speed signal input to the channel to be tested is in the range 3300 to 6650 r/min.
- Switch on the on-load test facility and select the channel to be tested.
- Slowly rotate the datum depression and check the datum on the selected channel module to achieve a trip condition. Cease rotating datum depression at instant of trip.
- Confirm that the channel has tripped and that correct annunciator indications have been received.
- Note the datum depression dial setting and speed on the digital speed indicator.
- Confirm by hydraulic fluid pressure indication that the trip valve has operated.
- Reset the datum depression set point to original set value and initiate the 'trip reset'; confirm the correct operation of the annunciators, alarms and reset.
- Confirm that hydraulic fluid pressure has returned to its previous value of 65 bar.
- Having completed the test on one channel, repeat in turn for the other three channels.
- On completion of testing of the four channels, the on-load test facility should be switched to the 'isolated' position and any test key removed and placed in a safe position.

4.8.5 On-load testing of overspeed trip channels

Initial state of plant

Main boiler feed pump operating on leak-off at a speed between 3300 and 4500 r/min. (*Note:* it is not recommended to operate the feed pump on leak-off flow above a pump speed of 4500 r/min.)

Test procedure

- Confirm the operation of all annunciator lamps.
- Confirm that the overspeed signal settings are correctly adjusted to enable the on-load test to be carried out. The setting must be within the speed range 105% to 111% of normal maximum speed.
- Confirm the setting of datum depression and correct hydraulic fluid pressure.
- Confirm that the speed signal inputs to the channels under test are in the range 3300 to 6650 r/min.
- Insert test key into the 'test trip' panel to energise the test facility.
- Select the two channels to be tested. (Note that the testing requires two channels at a time to be test tripped in the following combinations: A and C; B and C; A and D; B and D.)
- Rotate datum depression on channel A to induce a trip condition. Note the dial setting at point of trip.
- Check that channel A has tripped and that correct operation of annunciator alarms has been received.
- Confirm by hydraulic pressure indication that the trip valve has operated.
- Rotate datum depression on channel C to induce a trip condition.

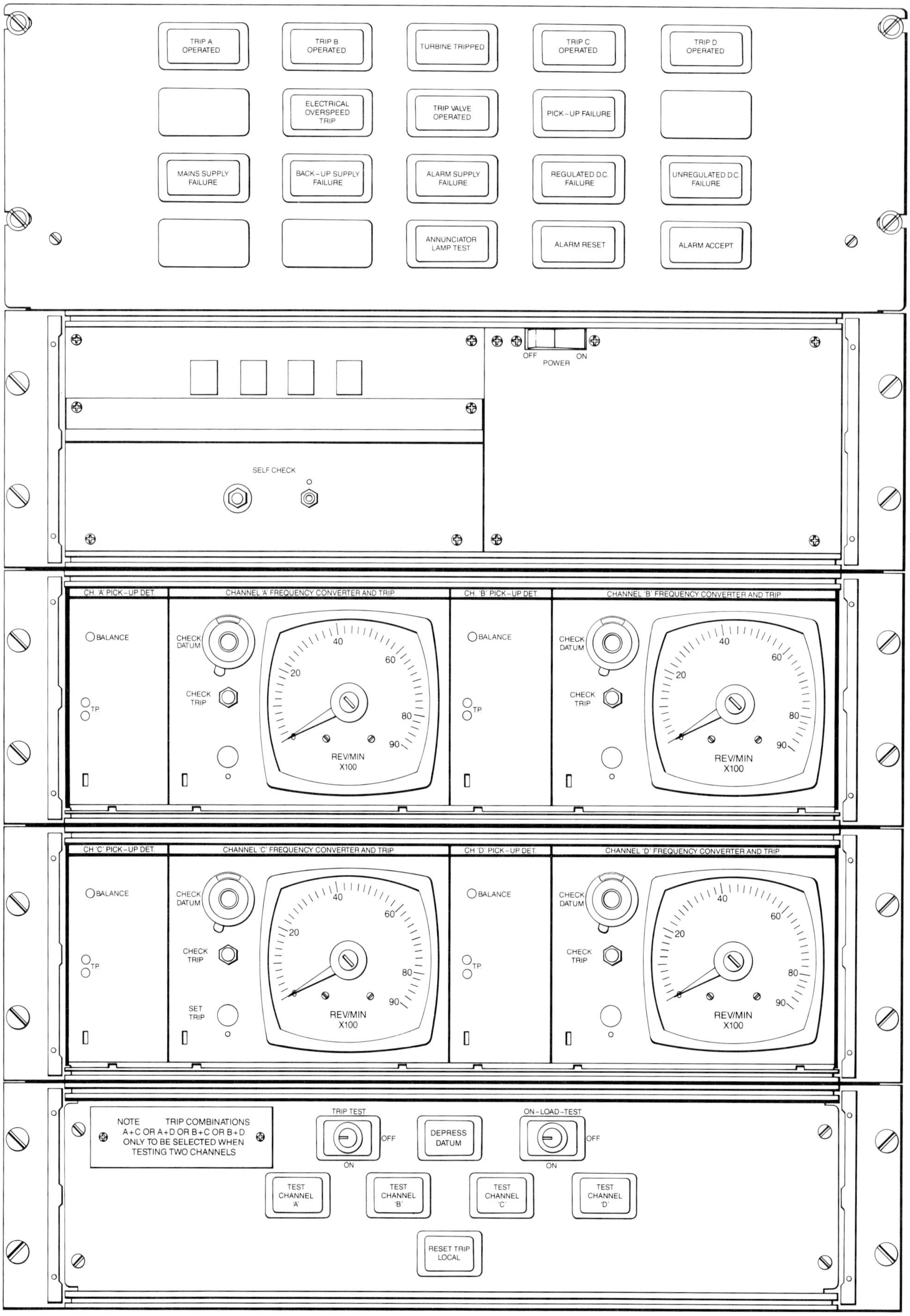

FIG. 6.45 Boiler feed pump turbine overspeed protection panel

- Observe that the boiler feed pump turbine has tripped, that the valve gear has closed, that the electrical overspeed has operated and that the turbine speed is falling.
- Reset the datum depression dials on channels A and C to their previous reference levels.
- Reset the trip and all alarms, checking on reset that all alarms operated have cleared.
- Confirm that the hydraulic fluid pressure has been re-established to its previous value.

The procedure stated for the A and C channel combinations is repeated for the other three combinations, i.e., B and C; A and D; B and D.

The timing of this commissioning test should be dictated by the overall requirements and objectives of the commissioning programme. However, the actual tests should be carried out as early as possible, thus confirming the correct operation of the protection devices.

On the completion of the test, the 'test trip' should be isolated by the removal of the test key.

5 Synchronisation to full-load and completion of 72-hour run

5.1 Increasing load in steps to 100% MCR

5.1.1 Objective

Having completed the first of the three main objectives of the commissioning programme stated in Section 1.2 of this chapter, i.e., steam-to-set to synchronisation, the second main objective of synchronisation to full-load and completion of the 72-hour run can now commence. This objective, for which a typical programme is given in Fig 6.46, is to commission the unit up to its full-load rating, including completion of all checks and tests necessary to ensure that, at the termination of this activity, the unit can be handed over at full rating for commercial operation. It will still be necessary to demonstrate its two-shift capability and performance guarantees which will take place within the third and final stage of the commissioning programme.

The programme in Fig 6.46 shows an elapsed time of 15.5 days from synchronisation to full-load, during which period all the boiler automatic control systems (auto systems) will have been commissioned and left in service. It must be remembered, however, that the times quoted for the auto systems in the programme assume that all pre-checks have been completed, that all the signals have been checked from the source points and that some initial settings have been applied to those auto systems already utilised during boiler firing activities. If this preparatory work has not been completed, the commissioning programme will be extended.

For safe and efficient operation of a 660 MW unit, it is essential that the commissioning of all auto systems is integrated into the overall commissioning programme and not left until some indeterminate date in the future. Confidence in the operation of the auto systems needs to be established from the commencement of the steam-to-set activity, with the single-element feedwater control and furnace pressure controls in service under auto control, the remaining control loops being commissioned at the appropriate load points.

5.1.2 Plant assessment

At synchronisation, the second and final plant/system assessment is carried out. The first assessment covered the steam-to-set to synchronisation activity and was made prior to the commencement of the turbine run-up and the harnessing of all the subsystems and auxiliary systems that form the unit. This second assessment, made with a greater knowledge of the plant, has the main objective of considering plant limitations or defects that may affect operation or prevent the unit reaching full output. The assessment should also check that the necessary commissioning documentation required to complete this second objective is available and approved. The responsibility for carrying out this final assessment rests with the Test Controller, acting on behalf of the Commissioning Panel.

5.1.3 Increasing load

The commissioning programme for the rate-of-increase of loading and the dwell times at the various loads is structured to fit the outstanding plant and system checks and auto system commissioning rather than just considering the output of the unit. The load hold points allow plant checking at a steady state condition, as well as incorporating all the necessary commissioning activities. The programme shows the first hold point at 15% and, after increasing to 20%, at 10% steps thereafter up to full-load.

Full design conditions of boiler steam pressures and temperatures are only obtained at load points in excess of 70% MCR. It is particularly important to exercise control over the pressure and temperature, as there will obviously be a chemical limitation on boiler pressure and, if the superheater sprays have not been commissioned, temperature control will not be available. It is evident that the mode of boiler firing must be strictly controlled, firing lower level oil or PF burners, with the excess air levels reduced to a minimum compatible with obtaining satisfactory combustion conditions.

Oil-fired boilers

Increasing the firing rate on an oil-fired boiler is achieved by firing the burners on maximum turndown until all burners are in service, and then by increasing the oil flow to each burner by the operation of a single control valve.

For normal operation that would be the best procedure and would be as recommended by the manufacturer. However, from a commissioning viewpoint, this would not be the correct approach as the unit will be operating at low loads for longer periods of time than envisaged for normal operation. The commissioning programme (Fig 6.46) identifies a period of eight days before 50% MCR is achieved. Since there is a need to control boiler temperatures and pressures during this period, the lower level burners should always be in service. At all times, depending on the furnace design, balanced numbers of burners in both halves of the furnace should be in service. Clean, spare oil burner lances should be readily available at all times, as it is inevitable that debris will cause burner blockage, and hence bad combustion, which should not be overcome by increasing the excess air levels.

PF-fired boilers

For a PF-fired boiler, the milling plant and its associated fans will all have been checked and tested without actually passing coal through to the PF burners. The checks and tests on the PF milling plant prior to synchronisation are contained in Section 4 of this chapter. The activity remaining after synchronisation is to demonstrate the ability of the air and coal systems to control the stability of the coal mill, its

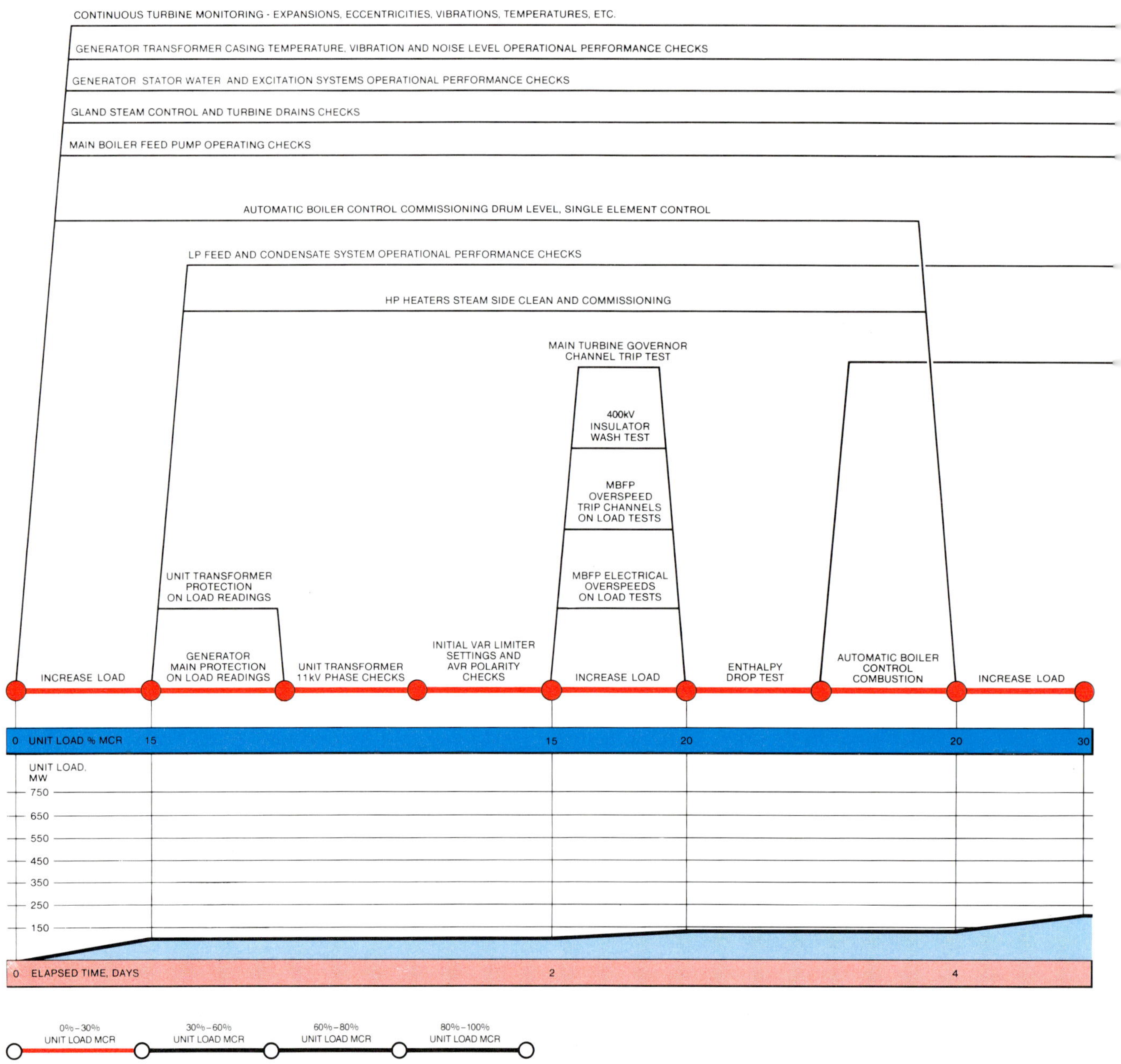

throughput capability and quality, and the coal distribution to individual burners.

The grinding performance and the PF distribution checks are carried out as part of the boiler efficiency and steam temperature control tests (see Section 6.1 of this chapter).

A large modern PF boiler may comprise up to 10 PF mills and 60 PF burners, each PF burner housing its own individual oil burner. It is easier to control boiler temperature using oil burners rather than PF burners, so the low load commissioning runs, up to 15% MCR, are best achieved with the oil burners because of the problem of PF stability. As in the oil-fired boiler, the low level oil burners in the PF furnace should be fired, as required, to control steam pressure, temperature and combustion performance. The milling plant comprises up to 10 mill groups, each containing a mill with a classifier, a coal feeder and a primary air fan, together with a seal-air system if the PF mill is designed for pressurised operation. With this design, the raw coal chute and coal feeders are all under pressure from the primary air fan. The speed of the mill grinding ring is approximately 40 r/min.

Having successfully stabilised unit operation at 15% MCR and completed the test activities at this load point, further load increases are carried out, using the PF milling equipment. For temperature control purposes, the lower level PF burners should be put into

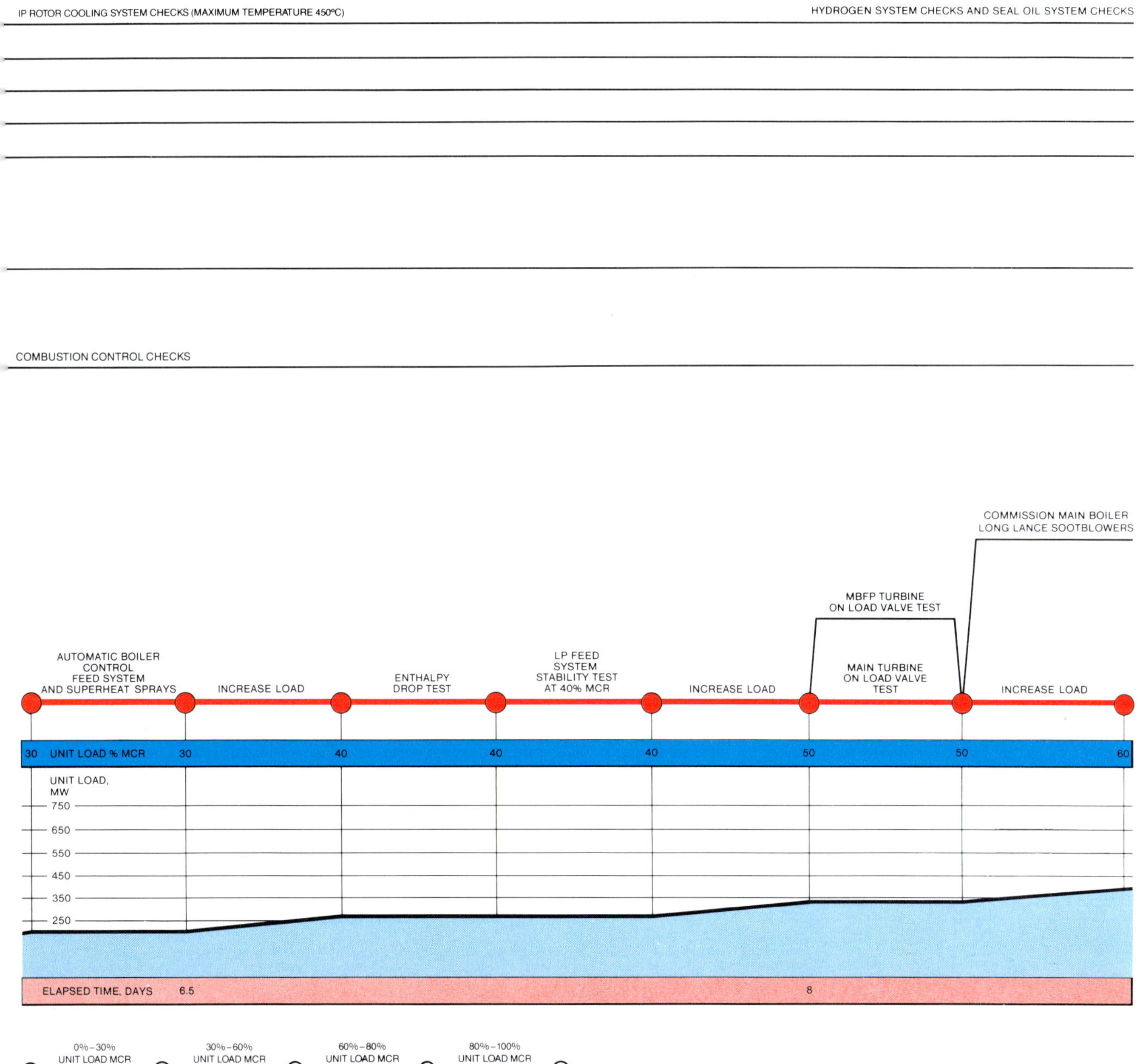

FIG. 6.46 Commissioning programme for a 660 MW unit — from synchronisation to completion of full load 72-hour run

service first. The order of establishing additional PF burners must take note of the burner distribution in the furnace. In order to ensure that PF firing stability is maintained, mill group burners should be fired in a pattern that enables burners to support each other. Firing an isolated group of PF burners should be avoided.

Prior to first firing from a mill, it is essential to ensure that all the oil burners associated with the mill group are in service to provide stabilised flames of sufficient quality to enable a burner flame detector to indicate a healthy condition. Having confirmed that the mill group pre-commissioning checks have all been completed and that coal is available in the bunker, the primary air (PA) fan is put into service on the chosen mill with a path through to the mill outlet. The output of the PA fan is controlled by inlet vanes, the temperature being controlled by mixing hot air and cold tempering air to obtain the desired air temperature to the mill. The PA fan is run to obtain a minimum mill temperature of 65°C before coal is admitted to the mill. Having established the correct mill temperature, the mill is started, followed by the coal feeder on minimum speed. The coal/air mixture is then delivered to the individual burners via the distribution pipework and flow-equalising devices to try to achieve equal distribution between burners. There will be an individual air register for each burner where swirl is imparted to the secondary air by fixed or adjustable vanes. The burners associated with a single mill are normally grouped in a

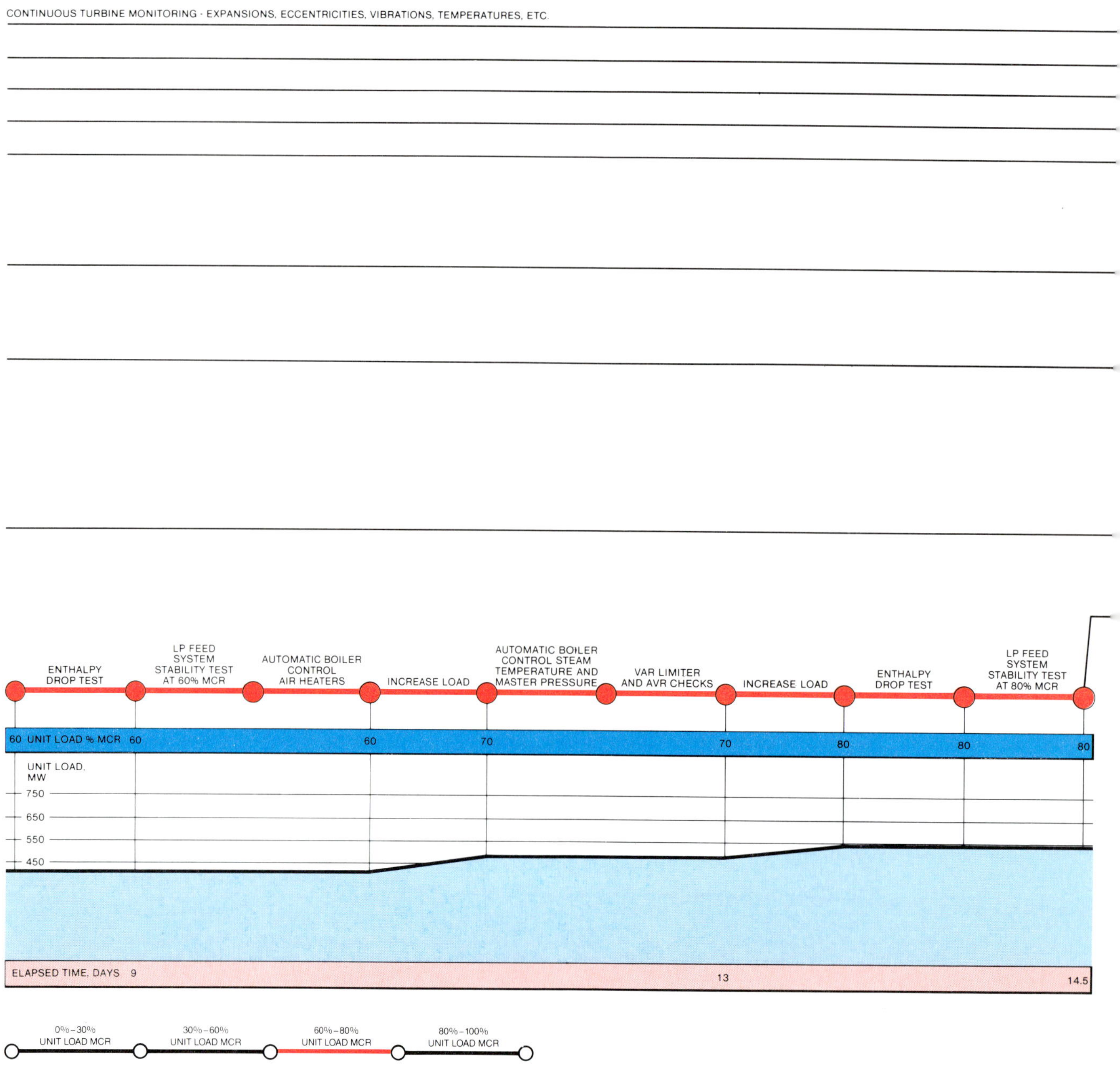

common windbox with its own isolating dampers. Each PF burner is provided with an oil burner and means of igniting the oil. Once the PF burners are firing, visual checks should be carried out for flame stability and combustion performance, and adjustments carried out as necessary. The oil burner should remain in service as a stabilising agent until several other PF mill groups have been established and confidence in flame stability achieved. During early commissioning, removing the oil burners from service below 50% MCR should not be attempted.

Subsequent firing of other mills should continue up to full load to meet the requirements of the commissioning programme, closely monitoring mill temperatures and differentials, primary air fan differentials, flame stability and combustion performance.

Plant monitoring parameters

As load is increased on the unit, it is essential to monitor all the operating parameters carefully and regularly to ensure that design limitations are not exceeded. At each load hold point on the commissioning programme, time should be allowed to check and record the information applicable to the plant conditions listed in Table 6.5. Figure 6.47 illustrates some of the turbine parameters measured from synchronisation to full load.

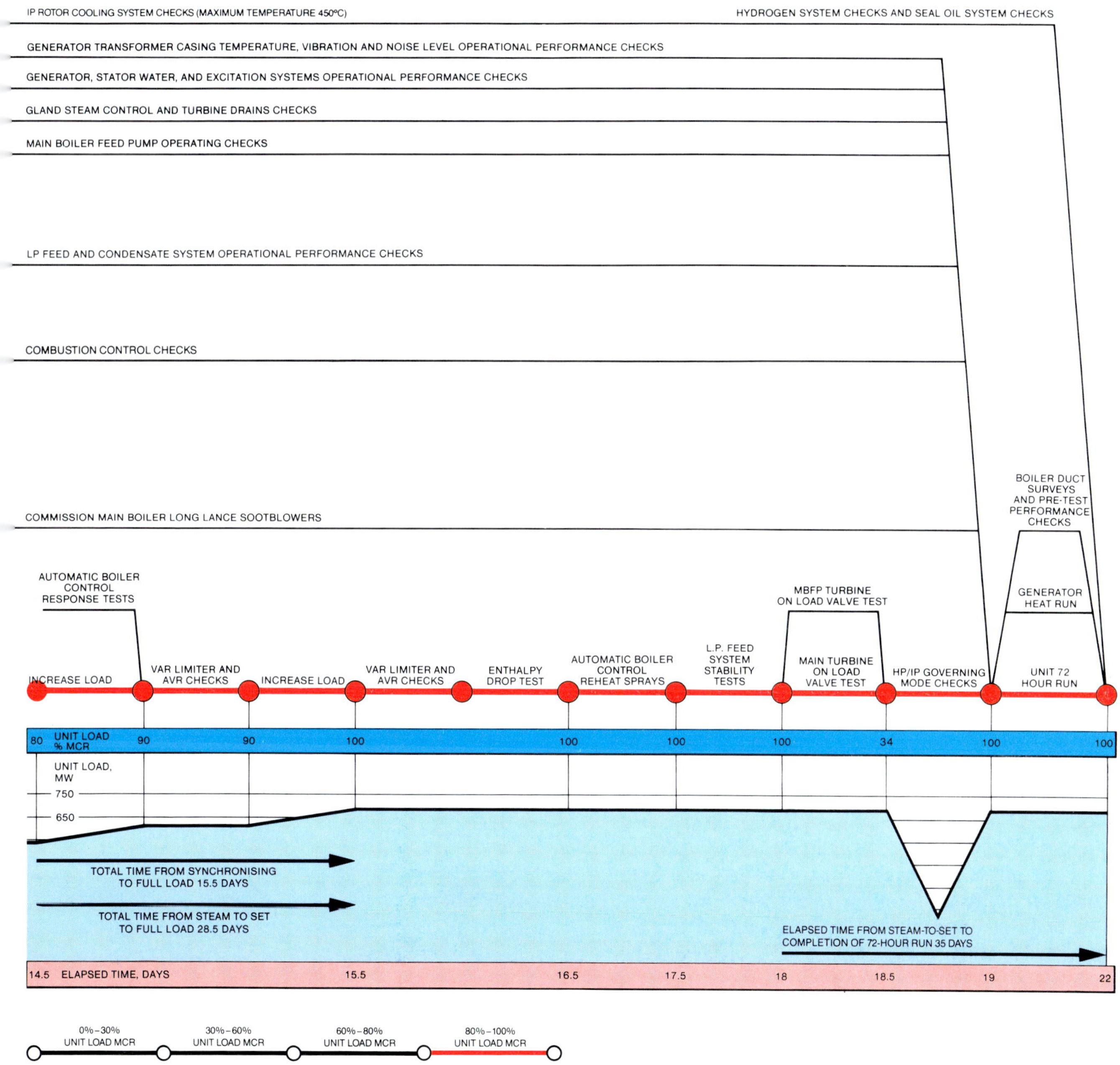

FIG. 6.46 (*cont.*) Commissioning programme for a 600 MW unit — from synchronisation to completion of full load 72-hour run

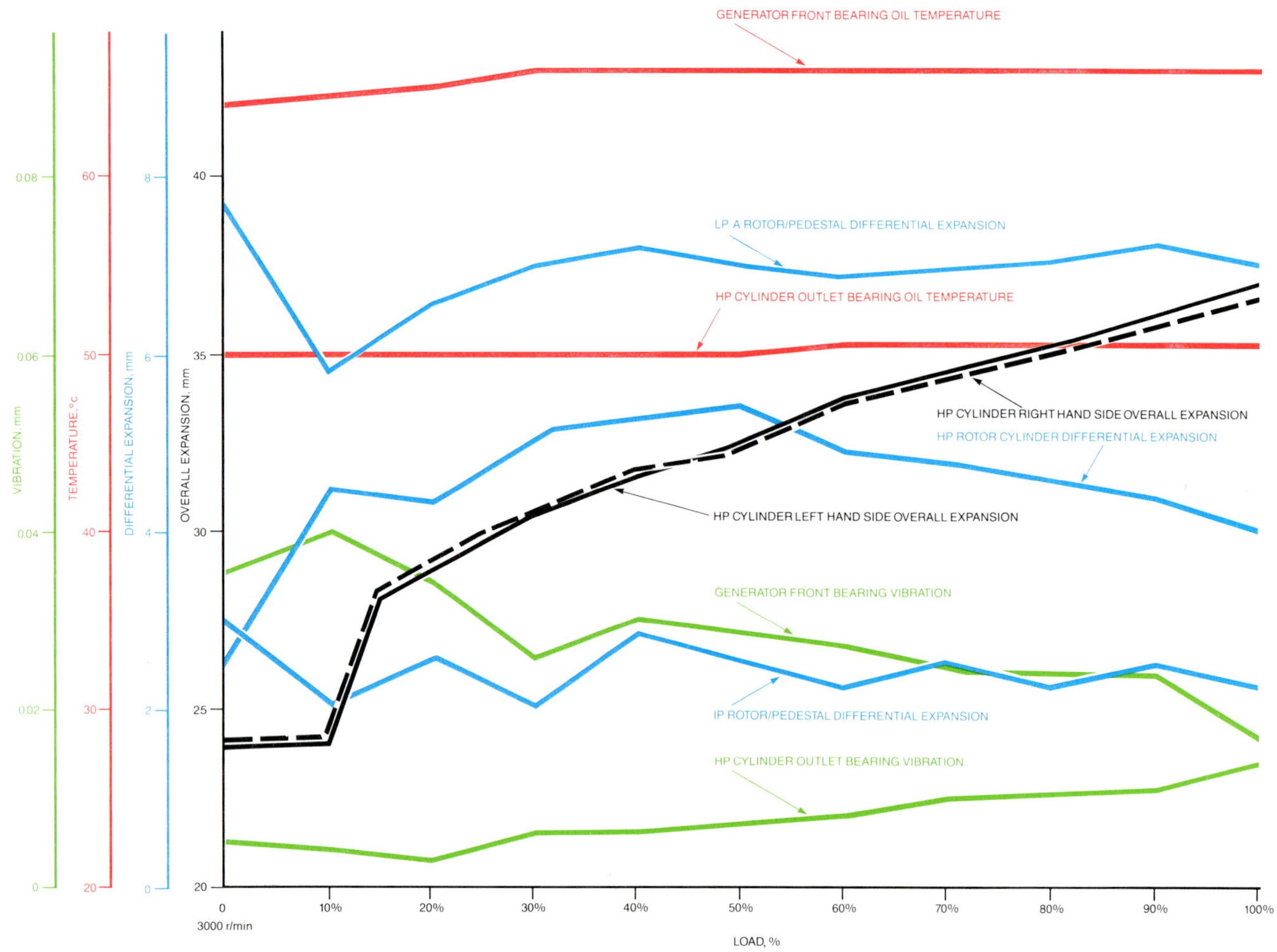

FIG. 6.47 Typical turbine parameters from synchronisation to 660 MW

In addition, the HP and IP cylinder reference point expansion checks should continue at each load hold point up to 660 MW. Figure 6.48 shows the final set of cylinder expansion figures at 660 MW. If these figures are compared with the original cold datum figures, it can be seen that the overall expansion of the turbine measured at the HP outlet is 37.01 mm on one side of the cylinder compared with 36.65 mm on the other side. This represents a differential between the two sides of 0.36 mm, which is too insignificant to affect turbine operation. It shows that the cylinder has been completely dried out and that the HP cylinder sliding feet have expanded freely. The other expansion checks confirm that all expansions are normal. It is worth remembering that, during the initial inspections prior to steam being applied to the turbine, sufficient clearance between pipes, housing, etc., to accept an expansion of up to 40 mm must be confirmed.

As full-load approaches, it will be the first time that the boiler has operated at full pressure and temperature. The expansion of the boiler will have been checked during the 'float safety valves' activity; however, it should be rechecked once full design operating conditions have been reached. A 660 MW boiler expands down from its supports by 175 mm between cold and hot conditions.

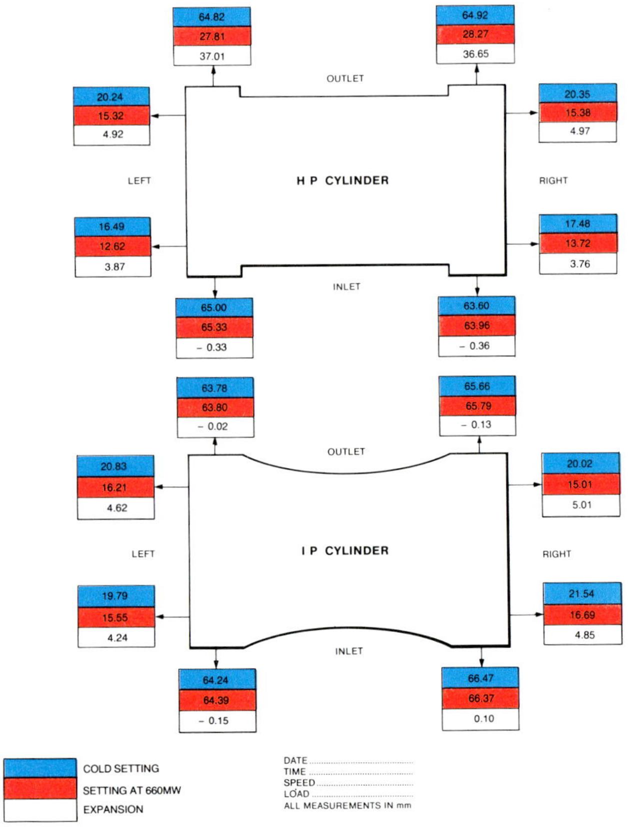

FIG. 6.48 HP/IP cylinder reference point expansion checks at 660 MW

In addition to recording and checking the operating paramcters as stated, the operational performance of the following should be regularly checked:

- Main boiler feed pump
- LP and HP feed systems
- Generator stator water and excitation systems
- Generator transformer casing temperatures, vibration and noise levels
- Ancillary coolers performance
- Operation of auxiliary plant, i.e., air systems, cooling water systems, fuel oil systems and auxiliary boiler systems

Chemical control

During the complete commissioning activity, strict chemical control is essential to ensure that the designed life of the plant is met. As will be discussed in Section 5.5 of this chapter on HP feedheaters, the early commissioning of the boiler feed pump and of the HP heaters steam pipework means that pipework is cleaned early. The debris scavenged by the increasing steam flow causes high levels of iron in the condensate, requiring excessive water dumping. If on-load 'polishing plant' capable of taking full flow is installed, then this debris in the system will not cause problems. It will, however, be necessary to ensure that the polishing plant is performing correctly. If this facility is not available, the load achieved may be affected by the capacity of the water dumping system.

The most likely cause of commissioning delays resulting from chemical requirements is the silica concentration in the boiler water. The boiler plant should be operated to ensure that $\ngtr$ 0.02 PPM of silica be present in the steam. Figure 6.49 shows the permissible silica levels against boiler pressure; at 130 bar boiler pressure, the maximum permitted silica level is 0.6 PPM. This is a high silica level which would ultimately limit the load carried by the turbine if it was not reduced. Typically, 450–500 MW could be achieved at this boiler pressure if the turbine were operated with its governor valves wide open; this risks possible superheater damage if a loss of load occurred that failed to trip the boiler firing because of the time required to reach the lift pressure of the boiler safety valves.

Providing that a good chemical clean has been carried out and the time interval between that activity and steam-to-set is short, or (alternatively) that the boiler has been continuously preserved under nitrogen, the levels of silica should be controllable without affecting the loading programme. This needs continuous maximum blowdown to the Station Chemist's requirements from the start of the steam-to-set activity. If a plant problem develops requiring a shutdown of known duration, the opportunity should be taken to depressurise the boiler by blowing it empty from 15 bar pressure and refilling with clean water: this procedure

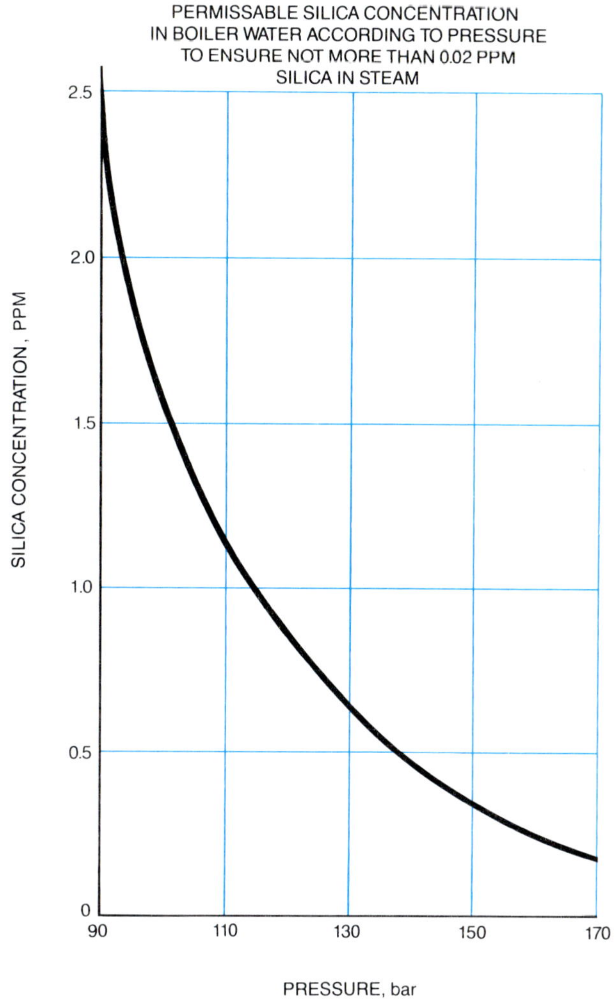

FIG. 6.49 Permissible silica levels

will assist in achieving early control of silica levels. Failure to respond to blowdown requirements for chemical control will only cause delays in the commissioning programme. The control of the boiler water pH and the monitoring of correct hydrazine levels in the boiler water should not present any problems.

The rest of this section describes the tests that are integrated into the overall commissioning programme with the object of achieving early and successful commercial operation.

5.2 On-load protection readings in the main generator circuit

5.2.1 Objective

With the generator operating at approximately 15% MCR, on-load protection readings are taken to confirm that all of these signals are correct. Changes in reactive power are needed during this test to enable certain protective devices to be checked. The test also confirms that all of the metering signals are correct.

During the short-circuit test, readings of protection were taken at 10% and 100% of full-load current. Although this confirmed the current transformer (CT) signals to the protective relays, the present test also incorporates voltage and metering checks.

5.2.2 Test procedure — main generator

With the unit at a steady output of 15% MCR (approximately 100 MW), carry out on-load protection readings as specified in Table 6.14.

TABLE 6.14

On-load protection readings in main generator circuit at 15% MCR

Current measurements
First feeder protection relay
Generator transformer biased differential protection relay
Stator earth fault relay
Loss of excitation fault relay
Negative phase sequence protection relay
Low power relay 1
Second feeder protection relay
Generator transformer HV restricted earth fault protection relay
Low power relay 2
Overcurrent protection relay
Generator circuit-breaker fail current check relay
Generator phases unbalanced relay
Voltage measurements
Loss of excitation protection relay
Low power relays 1 and 2
Protection VT supply faulty relay
Metering
Generator metering — main and check system
Generator MW reading on unit panel (and at any other point where this information is displayed)
Total generated output signals
Generated MVAr at unit desk
Stator voltages
Stator current
Generator 400 kV voltage

During the test, the turbine-generator must operate at a leading power factor, so that operation of the loss-of-excitation protection can be checked.

To check the actual value at which the low power relay operates, the turbine-generator load is reduced to approximately 1 MW.

5.3 Unit transformer — 11 kV phase checking

5.3.1 Objective

The objective is to check that the phase relationship between the unit transformer 11 kV connections and the unit/station 11 kV busbars is correct (see Fig 6.50). The phasing out is performed at the unit transformer switch cubicle, using the incoming transformer connections and the busbar connections in conjunction with an approved 11 kV phasing device.

The synchronising circuitry of the unit transformer is also checked and, if correct, the transformer can be synchronised with the 11 kV system for the first time.

5.3.2 Initial state of plant

- The generator protection on-load readings are completed.
- The generator is operating at 15% MCR (approximately 100 MW).
- The generator main circuit-breaker overcurrent protection is adjusted to a low setting.
- Station transformer(s) have been commissioned and are on-load supplying the 11 kV station switchboard.
- The unit transformer 11 kV air circuit-breaker (ACB) is in the *open* position and its associated voltage transformer is in the *service* position with its secondary fuses inserted.
- All protection and control fuses associated with the unit transformer ACB are in the *service* position.
- The unit transformer off-load tap selector is locked in the minimum tap position (tap 1). Note that the tap position may require changing. If so, this is normally performed with the turbine-generator off-load.
- The unit transformer overcurrent and earth fault protection devices are at a low setting.
- Consideration should be given to rendering the boiler firing protection inoperative for the duration of this test to prevent unnecessary tripping of the boiler in the event of errors during the test. For example, care must be taken when resetting the protective relays after the test, to avoid tripping.

5.3.3 Test procedure

- Unit transformer 11 kV ACB is to be removed from its switch cubicle to permit access to the incoming transformer and busbar connections.
- Phase out between both sets of connections, using approved phasing equipment. Record readings as in Table 6.15.

TABLE 6.15

Phasing checks between feeder and busbar spouts at the unit transformer 11 kV switchbay

Comparison check

Feeder spouts	Busbar spouts Red kV	Yellow kV	Blue kV
Red	12.8	6.4	6.4
Yellow	6.4	12.8	6.4
Blue	6.4	6.4	12.8

Magnitude check

	Red kV	Yellow kV	Blue kV
Busbar spouts	6.4	6.4	6.4
Feeder spouts	6.4	6.4	6.4

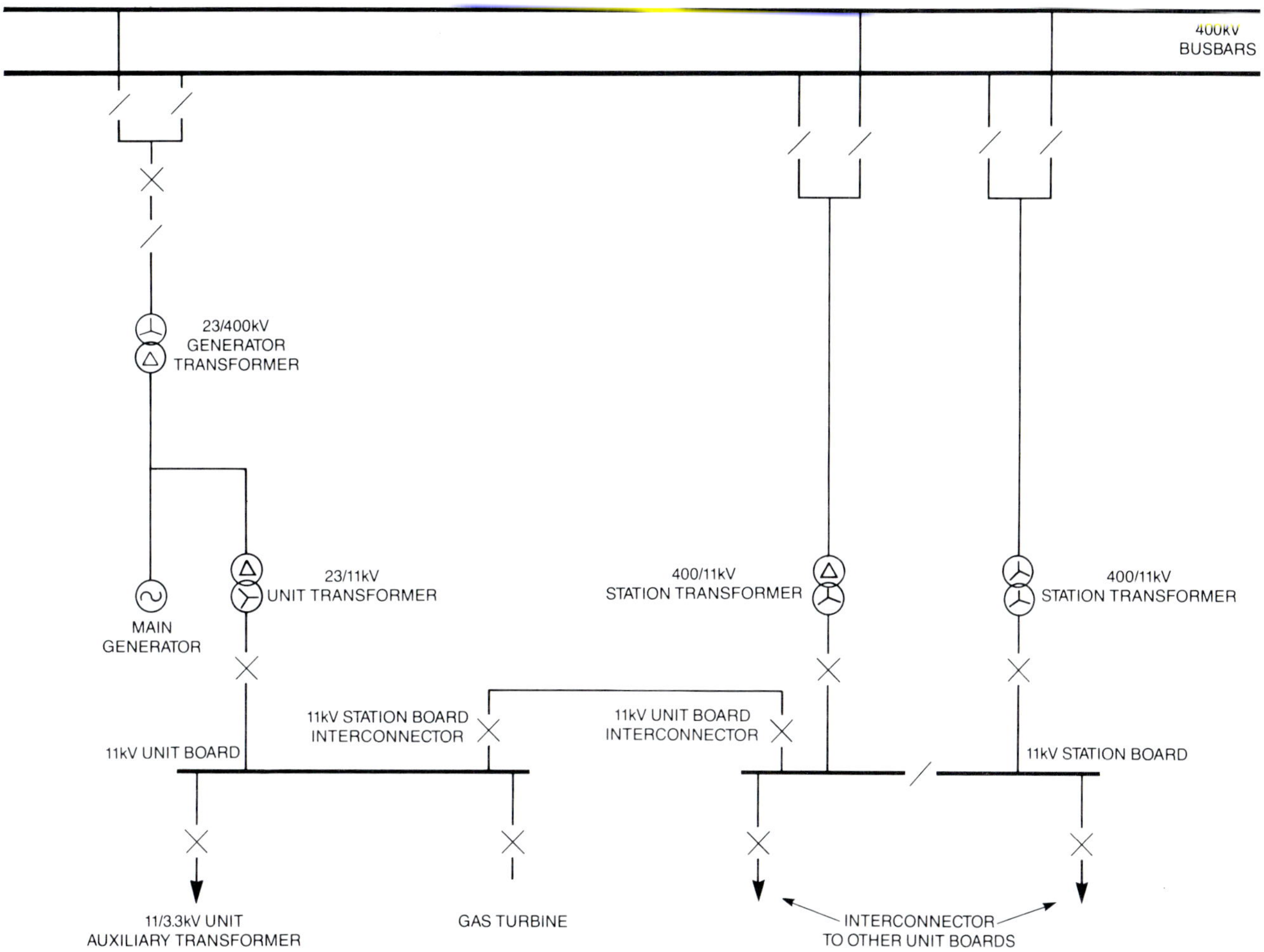

FIG. 6.50 11 kV unit and station switchboard distribution system

- Phase out between the secondary sides of the unit transformer VT and the station board interconnector to unit board VT. Record readings as in Table 6.16.

TABLE 6.16

(a) *Phasing between secondary sides of unit transformer VT and station switchboard VT*

Unit transformer 11 kV VT secondary	Station board interconnector 11 kV VT secondary, volts			
	R	Y	B	N
Red (R)	11.0	11.0	10.9	64.2
Yellow (Y)	10.9	10.9	10.9	64.1
Blue (N)	11.1	11.1	11.0	64.1
Neutral (N)	64.5	64.5	64.0	

(b) *Unit transformer and station board interconnection 11 kV VT secondary voltage comparisons*

	R–N	Y–N	B–N	R–Y	Y–B	B–R
Interconnector VT secondary	63.5	63.5	63.5	10.9	10.9	10.9
Unit transformer XVT secondary	64.0	64.0	64.0	11.0	11.0	11.0

- Check the synchronising circuit secondary wiring voltage measurements between the unit transformer and the station transformer interconnecting circuits. Record readings as in Table 6.17.

TABLE 6.17

Synchronising circuit secondary wiring voltage measurements between unit and station transformer

Unit transformer		Station board interconnector (SB intercon) feeder	
Unit trans.	*to*	SB intercon	64.5 volts
Unit trans.	*to*	SB intercon	64.1 volts
Unit trans.	*to*	SB intercon	64.4 volts

- Put the unit transformer 11 kV ACB into the *service* position. With the synchronising facilities in service, confirm that 'running volts' only are available and then, after inserting the synchronising interposing secondary incoming volts fuse, that 'incoming volts' indication is also available at the synchronising facility.
- At the station board interconnector on the 11 kV unit board, remove the running volts fuse. Check

that the 'running volts' indication disappears and that the 'incoming volts' indication remains.

- Replace the running volts fuse and confirm that the 'running volts' indication is re-established.
- At the 11 kV control panel, check synchronism. Confirm that the two systems are in parallel with no large voltage difference and then close the unit transformer 11 kV ACB, paralleling the unit transformer with the station 11 kV system for the first time.
- The unit transformer is now in parallel with the station board, with the station board interconnecting feeder still in circuit. This arrangement should continue for at least 1 hour to establish that the unit transformer can support its own unit load. During this period, the unit transformer should be inspected for any evidence of oil leaks or obvious signs of overheating.
- Once satisfied with the operation of the unit transformer, the station board interconnector is opened, leaving the unit transformer supporting the unit auxiliaries.
- The protection settings remain set down until the unit transformer on-load protection readings are completed.

5.3.4 Unit transformer — on-load protection readings

Objective

With the generator operating at approximately 15% MCR and the unit transformer supplying the unit auxiliaries, protection readings are taken to confirm that the protection signals are correct. Check that metering signals, voltage and current indications are correct.

In addition, readings are taken of the generator biased differential protection, if fitted, as it is likely to have one set of current transformers fitted in the unit transformer connections, teed-off the main generator connections.

Test procedure

- Take on-load protection readings on the unit transformer in accordance with the items specified in Table 6.18.
- On completion of the test readings, check the synchronisation and close the station board interconnector.
- Open the unit transformer 11 kV ACB.
- Reset the unit transformer overcurrent and earth fault protection to their original settings.
- Reset the generator main circuit-breaker overcurrent protection to its original setting.

TABLE 6.18

On-load protection readings on unit transformer at 15% MCR

Protective relays
Readings to be taken on the following protective relays:
Unit transformer biased differential relay
Unit transformer standby earth fault relay
Unit transformer LV restricted earth fault relay
Unit transformer HV overcurrent relay
Unit transformer 11 kV overcurrent earth fault relay
Generator transformer biased differential relay
Voltage/current measurements
Unit transformer 11 kV ammeter and voltage readings at the control panel
Unit transformer HV and LV current
Current indication at the 11 kV switchgear panel
Unit transformer HV and LV winding temperature indications
Metering
Unit transformer metering for MW and MVAr

- Reconfirm that all generator unit transformer protection devices are set to normal.
- Return the 11 kV unit transformer to service after checking synchronism.
- Open the station board 11 kV interconnector, leaving the unit transformer supplying its own auxiliaries.

5.4 AVR response tests and reactive power limiter settings

5.4.1 Introduction

When the turbine-generator is first synchronised, the automatic voltage regulator (AVR) should have been completely proved, by the tests identified in Sections 3.6 and 3.14 of this chapter, to be capable of controlling the voltage output on 'manual' and the automatic circuits checked by simulation. Closed-loop response will have been proved during the open-circuit test for dual-channel operation and under automatic control. As soon as possible after first synchronisation, the following need to be checked:

- Phasing of the VT and CT circuits to the active (MW) and reactive (MVAr) power measuring panels.
- Step response tests and single- and dual-channel operation.
- Setting of MVAr reactor.
- Check operation of the manual restrictive circuit.

5.4.2 Objective

The objective is to check the output of items associated with the AVR at different loads and, similarly, to check the performance of a number of circuits at different loads. These checks include the following:

- That the outputs of the MW and MVAr units on both channels are consistent with the load on the turbine.
- That the output of the MW transmitter is consistent with the MW load on the turbine.
- That the rotor angle meter functions correctly. This requires the use of a stroboscope.
- That the terminal voltage varies with the MVAr load in accordance with the droop setting after compounding control has been set.

These activities are carried out with each channel operating individually and with both channels operating in parallel.

5.4.3 Summary of proposed tests

At various loads, calibrate:

- MW and MVAr units
- MW transmitter
- Rotor angle meter
- Manual follow-up circuit, with channels operating individually and in parallel

Single channel:	Set compounding control
	Check response to step change
	Set minimum excitation limit
	Set manual restrictive circuit
	Check excitation range
	Check operation of over-excitation relay
Dual channel:	Check load sharing
	Check current limit
	Check response to step changes
	Check MVAr limit
	Check manual restrictive circuit

All of these checks are carried out immediately after synchronising and then at 15%, 50%, 75% and 100% MCR load points.

5.4.4 Test procedure

Select one AVR channel and isolate the second channel:

- With turbine synchronised and excited under manual control, check that the MVAr load can be changed smoothly.
- Select 'auto' and check that MVAr load is smoothly controlled.
- At zero MVAr load, set the compounding control.

Response

With a high speed (UV) recorder connected to measure machine terminal voltage, excitation and rotor angle, record the response at different MW loads at minimum and maximum excitation by introducing a step change.

At suitable loads with excitation on *manual*, introduce a 3% voltage step change. Then change to *auto* and record results.

Increase the excitation to obtain maximum lagging MVAr load. Decrease the excitation to minimum and check performance.

Under-excitation limit and stability check

- Check that a step response switch is fitted
- Adjust the excitation to operate at various points 20% outside the practical stability limit line.
- Close the step response switch and further increase excitation. Adjust the voltage bias, as necessary, until it begins to show a negative voltage.
- Open the step response switch and check that the MVAr limit alarm operates, that the change in MVAr is not more than 4% to 5% of rated MVA and that the generator voltage remains steady.
- Leave the step response switch open and disconnect it at the first opportunity.

Setting minimum excitation limit

- Check that the MW bias is set to minimum.
- With the turbine at nominal voltage (zero reactive load) and generating less than 20% of full load, lower the excitation by tapchanging on the generator transformer. The indicated MVAr should stop decreasing and remain in the safe area resulting from the checks carried out during the under-excitation limit and stability check.
- Adjust the control rheostat and tapchanger successively and progressively until the MVAr limit alarm is energised at a point just beyond the limit line.
- Record the turbine terminal voltage at the limit point. This voltage should be consistent with the set percentage droop.
- Reduce the excitation from the limit.
- With the turbine at nominal voltage (zero reactive load) and generating at a load in excess of 50% MCR tap, change the generator transformer until the MVAr load is again at the limit line. Repeat the above procedure at this load point, and at 75% and 100% MCR loads to check the minimum excitation limit over the full load range of the turbine-generator.

Manual restrictive circuit

- At the lowest possible generated load, trip the AVR to 'manual' control and reduce the excitation to bring the turbine-generator to the restrictive VAr limit line, with the terminal voltage maintained at the design value. Record the MW output and note the position on the VAr limit line.

- Repeat this test at the load points given in Section 5.4.3 of this chapter.
- Adjust the restrictive line as required, such that the inputs to the restrictive circuit are correct over the full MW operating range.

Excitation range

With the turbine-generator at full MCR load, confirm that the excitation circuit is in control by varying the excitation from maximum to minimum levels.

Over-excitation circuit

This check confirms the operation of the over-excitation relay. The relay should have been preset to give a value of 1.25% MCR. Check the actual operation point for the relay by increasing excitation.

Second AVR channel

Having completed the test and checks stated for the selected channel, they must be carried out with the other channel in operation.

Load sharing

With the turbine-generator on load and both AVR channels in service, the load-sharing capabilities of the channels need to be checked. This test is carried out at the load points given in Section 5.4.3 of this chapter.

- Check the current sharing between the two channels by measuring the voltage as current across the voltage control potentiometers. With high current output, the current sharing should be ±10% of MCR current.
- If the current sharing is not within tolerance, adjustments will be required.
- If the current sharing circuit tends to hunt, make systematic adjustments to achieve a sensitive current balance control.
- Confirm that the balance indicator of each channel indicates zero and that each manual potentiometer follows the control signal from its own channel and that the two are in synchronism, so that no mismatch occurs.
- Trip the AVR channels in turn to check that the trip of one channel does not induce any disturbances into the system.
- On reinstating the tripped channel, check that this does not induce any disturbances in the voltage control.
- Check that the current limiter for each channel does not trip at less than 110% MCR.
- Check the response of the AVR resulting from a 1% voltage change.
- Check that the manual mismatch alarm is not displayed; adjust, if necessary.
- Check that the MVAr limit circuit functions properly with both channels in parallel.
- Check that the manual restrictive circuit functions properly with both channels in parallel.

5.4.5 Generator stability diagram

A generator stability diagram (Fig 6.51) shows the practical stability line that the manual restrictive circuit follows. This diagram, for a 660 MW 0.85 PF rated turbine-generator, also shows the stator and rotor heating limits.

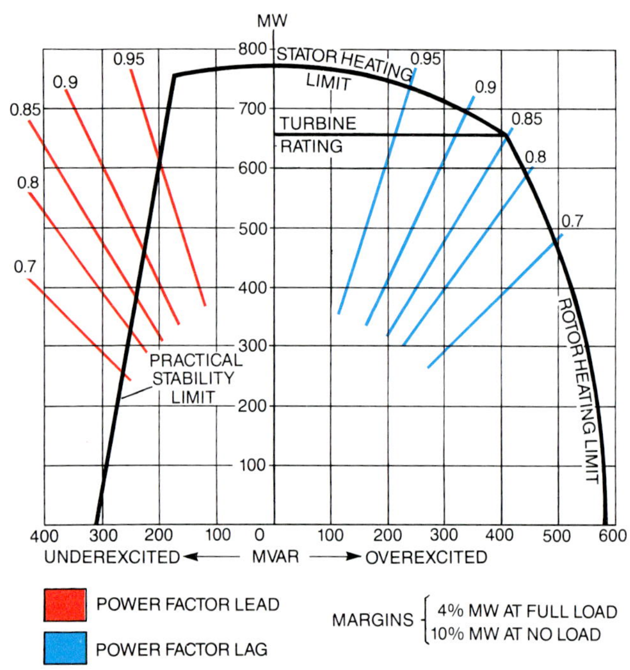

FIG. 6.51 Generator stability diagram

5.5 HP feedwater heaters

5.5.1 General description

A typical system is shown in Fig 6.40 and consists of two heater banks in parallel, each having two heaters connected in series. The heaters are mounted vertically with each heater being of the two-pass surface type, with internal baffles to form drain cooling sections near the drains outlet and, in the case of the two highest pressure heaters, i.e., the last stage in each bank of desuperheating zones, at the steam inlet. A typical final stage HP heater showing the heater internal construction, including baffles, desuperheating, drains cooling and condenser zones is shown in Fig 6.52.

The water side of the heaters is at feed pump discharge pressure with isolating valves at the feed inlet to each heater bank and at both final heater outlets. Operation of these valves is automatically initiated in

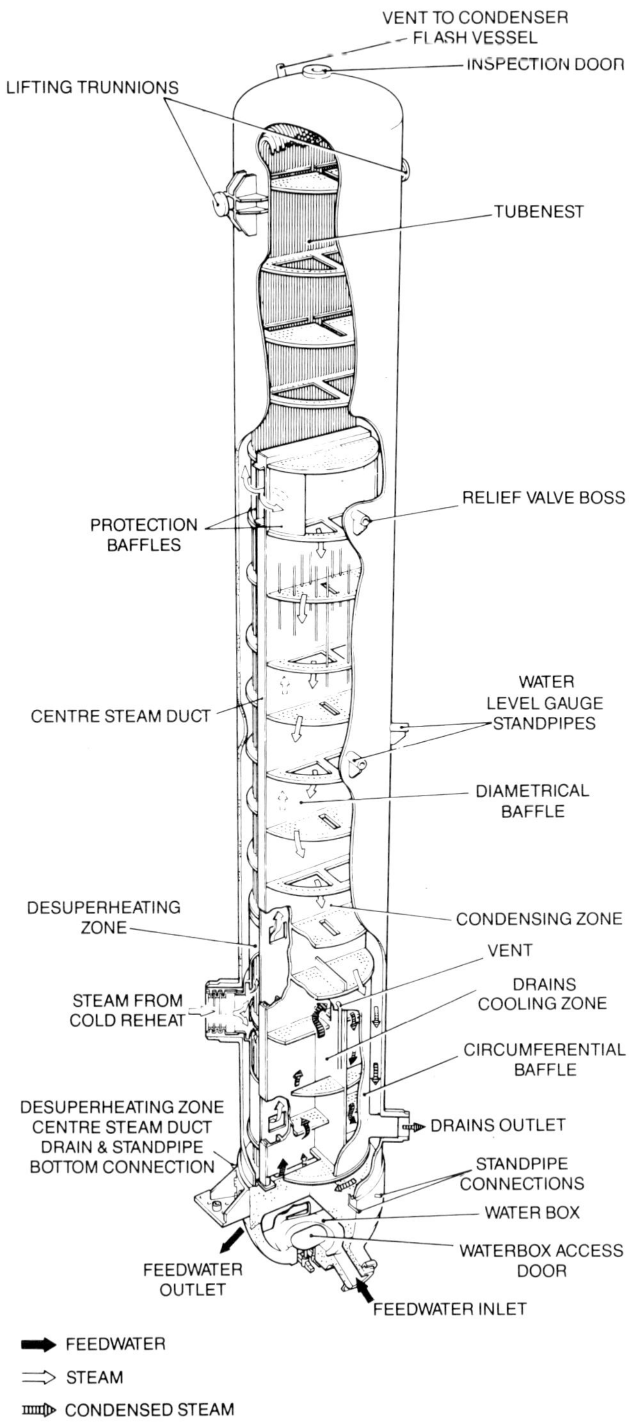

FIG. 6.52 Internal arrangement of a final stage HP heater

the event of a high water level being experienced on the steam side of the heaters. A high water level would also cause the associated steam isolating valve to close and an orificed by-pass to open, thus enabling full feedflow at reduced temperature to be maintained.

In the system shown, the steam supply to the No. 5 heaters is from the main boiler feed pump and to the No 6. heaters from the main turbine HP exhaust. Each steam supply line is fitted with a non-return valve to reduce the quantity of entrained steam available to overspeed the main turbine in the event of a turbine trip, as well as reducing the possibility of reversed steam flow from a quickly depressurised HP heater containing water.

The main steam drains from the heaters are finally discharged either to the turbine drains vessel, to the condenser flashbox or to the de-aerator, depending on the level of water and/or the pressures existing in the heater. In addition, drains from the No. 6 heater in each bank can flow to the No. 5 heater via a flashbox in which the flashed steam is used for heating. The distribution of the drains is automatically controlled by pneumatically-actuated valves, which are operated in response to signals from pressure and level switches mounted on the heaters.

In addition to the main drainage system, start-up drains are provided on each heater to ensure that the heater bodies are drained on the steam side at the time of shutdown. These drains discharge through motorised valves to the low pressure condenser flashbox which in turn discharges into the condenser. The start-up drains are open at all loads below approximately 10% MCR. A typical HP heater drainage system showing both the main and start-up drains can be seen in Fig 6.40.

The technical data associated with the feedheating system is given in Table 6.19.

TABLE 6.19
Technical data for an HP feedheating system

Steam side	No. 6 HP heater	No. 5 HP heater
Shell pressure, bar	49	24
Steam temperature, °C	340	246
Flow, t/h	120	69
Drain temperature, °C	200	160
Feed side		
Feed flow, t/h	999	999
Inlet feed temperature, °C	194	143
Outlet feed temperature, °C	254	194
Tube dimensions		
External diameter, mm	19	19
Thickness, mm (145 SWG)	145	145
Total heating surface, m^2	1395	1395
Tubenest		
Working pressure, bar	200	200

5.5.2 Commissioning tests

The tests associated with commissioning an HP feedheating system are divided into four sections:

- Pre-commissioning tests
- Wet functional tests (pre-STS)
- Dynamic functional tests (pre-STS)
- Commissioning and performance tests of system (post-STS)

5.5.3 Pre-commissioning tests

Mechanical checks Each vessel, valve, valve actuator and section of pipework is checked against the appropriate item checklist. The detailed commissioning procedure includes a complete itemised schedule of the checklists.

Electrical checks This includes motor rotation checks and the operation of all electrically-operated valves. The checks are recorded on the appropriate item checklists, and these are treated similarly to the mechanical checks.

Control and instrumentation checks Each pressure switch, level switch, local and remote instrumentation, and alarm is checked against the appropriate item checklist, treated similarly to the mechanical checks.

Dry interlock tests Dry interlock and sequence checks are carried out on all relevant plant.

5.5.4 Wet functional tests

Water side

Prior to finally closing and securing the HP heater waterbox access door, the tubeplate should be inspected to ensure that no debris is present. Having confirmed this, the feedheaters can then be charged on the water side to check for leaks, both internally and externally, together with limited flushing.

Priming the heaters is initially carried out under the head of the de-aerator. The pressurisation of the heaters then continues; this requires a feed pump in service.

The amount of flushing that can be carried out is generally very limited and amounts to flushing water through all the feed side drains. This is a low velocity flush but should be sufficient if strict control has been exercised during construction activities. An alternative method to obtain a high velocity flush would be to pump through the heaters to the boiler and then to drain the boiler: the disadvantage of this method is that any debris left in the heaters may lodge in the boiler tubing, so it is not recommended.

During these flushing operations the feedwater regulating valves should have their internals removed, if they are in the flushing circuit, to prevent damage to the control valve cage assembly.

Steam side

The bled-steam lines from both the turbine and boiler feed pumps will have been dry stored and capped during construction to preserve their internal surfaces. The cleaning of these surfaces is carried out after steam-to-set utilising the normal bled-steam supplies.

Cold vacuum pull

In carrying out the turbine cold vacuum test, the HP heater system is set for service operation. The cold vacuum pull provides an opportunity to check for leakage, especially on the drains system.

Final functional tests

The HP feed system drain control air-solenoid valves must be checked for operation from all initiating signals. Any motorised start-up drain valves should also be checked from their initiating signals.

The control actions initiated by the heater level switches must be checked independently. The operation of all valves associated with feedheater protection, their alarms and indications must be individually checked, the alarms also being checked for correct indication. Stroke time must also be checked and recorded for the main feedwater inlet and outlet valves and for the HP heater bled-steam isolating valves.

5.5.5 Dynamic functional tests

Functional tests of HP heater level columns on-load test facility

Duplicate level control columns are now fitted, equipped with a water-filling facility to carry out on-load tests by positively creating a high heater level without tripping the heater. A test is carried out to prove this facility, to confirm that the high level is indicated and, when switched to 'test', that tripping of the heater does not occur.

Full functional trip tests of the complete HP heater protection system

Full functional trip tests of the complete HP heater tripping scheme must be carried out from every initiating source.

This involves setting the HP heaters up for normal operation and checking that the correct sequence of valve tripping and changeover (as applicable) occurs in the right sequence.

5.5.6 Commissioning and performance test of system

The commissioning of the HP feedheaters after completion of all the various tests and checks carried out prior to the commencement of the steam-to-set activity is closely integrated with the commissioning of the main boiler feed pump. Section 4 of this chapter, covering the commissioning of the main boiler feed pump, states that the HP feed heaters should be in service on the water side as part of the main feed pump commissioning, prior to first synchronisation. The commissioning of the main feed pump provides the steam supply to the lowest pressure HP heaters and it is at this point that cleaning of the steam pipework commences, initially with a low steam flow at zero pressure, which increases as the turbine load increases. With the HP heaters set for operation, the bled-steam valves to the highest pressure heaters should be fully open. With the boiler feed pump in service and the turbine-generator at 3000 r/min but not on load, there will be a low steam flow to all the HP heaters. The condensed steam drainage is likely to contain high levels of iron that have been scavenged off the internal surfaces of the bled-steam pipework by the steam flow. As the load increases on the turbine, peaks in the level of iron content in the condensate system will be noted until finally the pipework has been cleaned up. It is likely that the iron levels experienced will require dumping of the condensate to waste.

Following satisfactory completion of the initial checks after steam admission to the HP heater and as the turbine load increases above 10% MCR, the operating point at which the main and start-up drains system transfer should be recorded and checked against initial settings.

A full functional trip test of the heater banks by operation of the heater protection system should be carried out during the loading programme to prove its integrity, this being the first occasion when the system is operating with steam and water present. The trip tests should be repeated at various loads including, finally, full-load. The actual tripping test is initiated on each occasion by flooding a level column without switching it to the test position.

As the load on the turbine is increased, the performance of the HP heaters should be continuously monitored with specific reference to the operation of the automatic drains control system. Initial checks of pressures and temperatures should be recorded at various load points and compared against the design values. The full turbine acceptance test will include sufficient test instrumentation to enable the heater performance to be checked. For a two-bank HP heater system, the main feed by-pass valve should be checked as being shut when both heater banks are in service.

A single-bank HP heater test must confirm that, under this mode of operation, satisfactory flow distributions are achieved through the single heater bank and the by-pass line.

5.6 Enthalpy drop test

5.6.1 Introduction

Deterioration of the cylinder efficiency of turbine-generators generally occurs most rapidly during the initial period of operation. It is recognised that it is difficult to conduct an acceptance test as soon as the turbine-generator commences commercial operation, so enthalpy drop tests should be carried out as part of the planned commissioning programme. This establishes basic data relating to the HP and IP cylinder efficiencies which can be used to assess deterioration. Deterioration will continue during the subsequent life of the machine, its rate being monitored by routine enthalpy drop tests.

5.6.2 Test equipment

Calibrated test instruments are used for all pressure and temperature measurements required for the cylinder efficiency calculations. Test instrumentation is installed to record LP cylinder exhaust pressure, extraction steam conditions and feedwater temperatures. The generator output is obtained from temporary test instrumentation until the accuracy of the permanently installed metering has been proven. In later tests, the permanent instrumentation can be used, if considered to be sufficiently accurate.

The accuracy of the cylinder efficiency determination will depend largely upon the accuracy of the steam temperature and pressure measurements. Care should be taken in selecting the location of the measurements and in the choice and accuracy of the equipment used. Ideally, the same equipment should be used in the same location for each test of a series to minimise changes in systematic errors.

5.6.3 Measurements

The measurements required for enthalpy drop tests are the pressures and temperatures at the inlet and outlets of the HP and IP cylinders. The LP exhaust pressure and the generator output are recorded for reference. In addition, some assessment of the gland leak-off quantities and the pressures and temperatures at the bled-steam tappings are useful, if they can be obtained accurately.

Enthalpy drop tests will only be comparable if steam conditions at the start of the HP and IP cylinder expansion are similar for each test. The reference conditions that should be the same for each test of an enthalpy drop series are the after-throttle pressure and temperature and the reheat temperature. In the early stages of operation, when load may be restricted, reference conditions should be chosen at the maximum after-throttle pressure which can be reliably repeated over a series of tests.

The boiler feed pump (BFP) turbine should be treated in a similar manner to the main turbine, the measurements required being steam pressures and temperatures at the inlet, outlet, exhaust and bled-steam tapping points of the turbine cylinder, together with pump speed. A BFP turbine enthalpy drop test should be carried out concurrently with those on the main turbine. Table 6.20 gives design and test results and indicates the measurements to be taken.

5.6.4 Test procedure

Care should be taken to ensure that steady conditions prevail during each test. The main boiler feed pump should preferably be in service at a steady output with the boiler feed control loops commissioned and functioning correctly.

The collection of the basic data should be an integral part of the overall commissioning programme at the following load points: 20%, 40%, 60%, 80% and 100% MCR values. The readings to be taken are those identified in Table 6.20, which shows the design and actual results obtained at the 660 MW load point.

TABLE 6.20

Examples of enthalpy drop test results for a 660 MW turbine-generator

Power station cylinder data	Turbine generator No.		
	Units	Design	Test
Test number			
Date			
Total running hours to date	Hours	—	2000
Generator output	MW	660	660
HP and IP cylinder steam pressures			
Before HP stop valves	bar	159.6	156.0
After HP governor valves	bar	155.7	150.2
At HP cylinder exhaust	bar	42.0	41.4
Before IP interceptor valves	bar	38.9	38.6
At IP cylinder inlet	bar	38.5	37.3
At IP cylinder exhaust	bar	5.0	4.8
Steam temperature			
Before HP stop valves	°C	538.0	536.0
At HP cylinder exhaust	°C	341.0	341.0
Before IP interceptor valves	°C	538.0	530.0
At IP cylinder exhaust	°C	259.0	255.0
HP cylinder			
Inlet enthalpy before SV	kJ/kg	3405.0	3404.1
Inlet entropy before SV	kJ/kgK	6.5	6.5
Actual exhaust enthalpy	kJ/kg	3065.5	3064.5
Isentropic exhaust enthalpy	kJ/kg	3030.0	3030.7
Cylinder efficiency	%	89.5	89.7
IP cylinder			
Inlet enthalpy before SV	kJ/kg	3532.0	3514.0
Inlet entropy before SV	kJ/kgK	7.2	7.2
Actual exhaust enthalpy	kJ/kg	2971.0	2972.0
Isentropic exhaust enthalpy	kJ/kg	2920.0	2918.5
Cylinder efficiency	%	92.5	91.0

Table 6.21 shows the tested cylinder efficiency results at the various load points during initial commissioning.

Having established the enthalpy drop test readings at the intermediate load points and at full-load, the rate of deterioration of cylinder efficiency is measured by repeat testing at full-load during the subsequent life of the turbine-generator.

Figure 6.53 shows the deterioration of the cylinder efficiencies of two identical 660 MW units. The HP cylinder design efficiency was 89.72% and one of the turbines initially showed an efficiency greater than this value; even after some 10 000 h of operation the efficiency, although showing a slight deterioration, was still better than design. In the second turbine, the initial cylinder efficiency was lower than design by approximately 0.1%, but this then deteriorated rapidly, showing an efficiency loss of approximately 2.5% over 10 000 h of operation. This illustration is presented to show the value of regular enthalpy drop testing. Here, a genuine plant problem existed with the HP cylinder which eventually necessitated opening up the cylinder to repair damage to the nozzle box and fixed blading. The normal rate of deterioration resulting from wear, as opposed to major cylinder problems, would follow the efficiency graph for turbine 1. The particular problem associated with turbine 2 would not have been obvious during normal plant operation, although calculated efficiency returns would indicate a fall off in performance without identifying the problem area.

With regard to the IP cylinder efficiency (Fig 6.54), the example quoted had a design efficiency of 92.5%, where the test results for both turbines give lower values of approximately 91%. Comparing the two turbines after 10 000 h operation, turbine 2 exhibited a fairly marked deterioration over the first 3000 h when compared with the other turbine. There were no obvious problems with either machine to account for the different deterioration rates: this highlights how similar plant can exhibit differences, especially on clearances, which result in approximately 1% difference in efficiency over the same time interval. Figure 6.55 shows the method of calculation.

The time taken during the commissioning programme for the enthalpy drop test is approximately 2 hours at each of the stated load points. This assumes that the

TABLE 6.21

Enthalpy drop test results at 20–100% MCR for a 600 MW turbine-generator

Nominal load % MCR	Test values of load MW	MW	HP cylinder efficiency, %	Governor valve opening, %	IP cylinder efficiency, %	Governor valve opening, %	Other plant conditions		
20	132	140	90.1	5	90.0	37	Electric feed pump in service		
							MBFP in service		
							L/S %	B/S %	Press bar
40	264	270	90.3	21	90.5	100	5	87	141
60	396	330	90.6	25	91.0	100	2	60	140
80	528	470	90.4	35	90.8	100	3	70	138
100	660	660	89.7	95	91.0	100	4	83	165
Official acceptance test 2000 hours later									
100	698		89.7	100	90.3	100	0	67	168

Notes MBFP = Main boiler feed pump
L/S = Live steam governor valve opening
B/S = Bled-steam governor valve opening

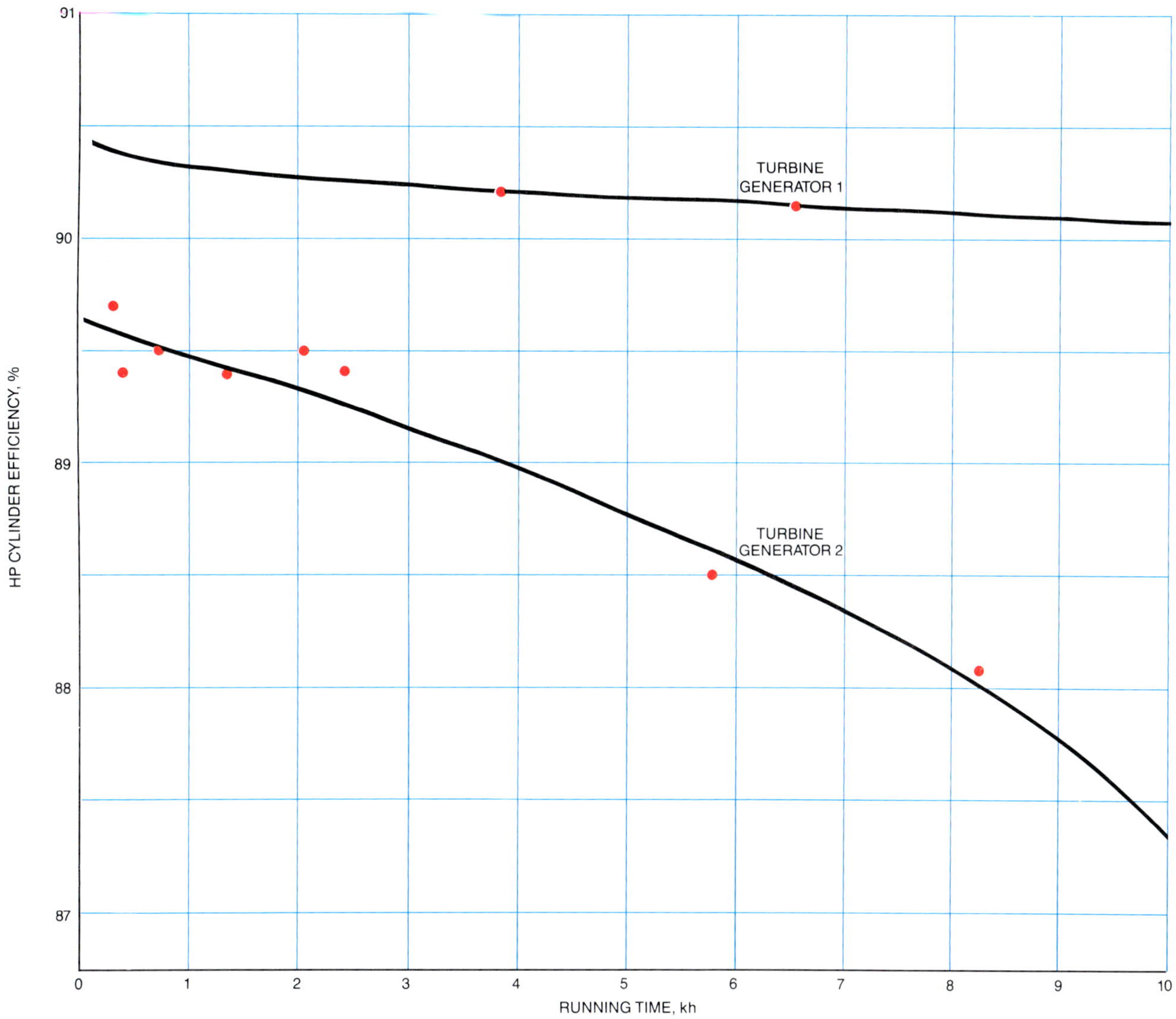

FIG. 6.53 HP cylinder efficiency performance plotted against time

equipment has previously been checked and calibrated, and is functioning correctly.

The full method of calculation is contained in Volume G, Chapter 7. The subject is included briefly in this volume to emphasise the importance of carrying out these tests as part of initial commissioning, as it is normal for the turbine acceptance test to take place some time later, possibly after the turbine has been in operation for some thousands of hours.

5.7 On-load turbine tests

Prior to first synchronisation of a turbine-generator, the numerous protection devices will have been tested and checked, as detailed in Section 3.3 of this chapter.

The typical unit programme (Fig 6.1), under the general notes, required routine turbine valve and overspeed ring testing by fluid injection daily until satisfactory operation was proven. In reality, satisfactory operation should be confirmed during the seven days prior to synchronisation.

The following tests must be incorporated in the routine test procedures — weekly for the overspeed and turbine valve tests, and monthly for the other tests:

- Overspeed rings by oil injection
- Turbine trip solenoids
- Governor hydraulic fluid pressure switches
- Direct vacuum trip
- Turbine valve gear

All of them, with the exception of the turbine valve test, can be carried out at any turbine load as they do not affect the magnitude of the turbine load at the time of the test. The on-load turbine valve test will, however, cause load reductions during the test period. Table 6.22 shows how turbine load reduces during tests at 260 MW and 500 MW. At each stage of the test, the closing of one valve is not compensated by a greater opening of the remaining valves. The only other factor affecting valve opening is the system speed: a fall in frequency coincident with the valve test will cause the

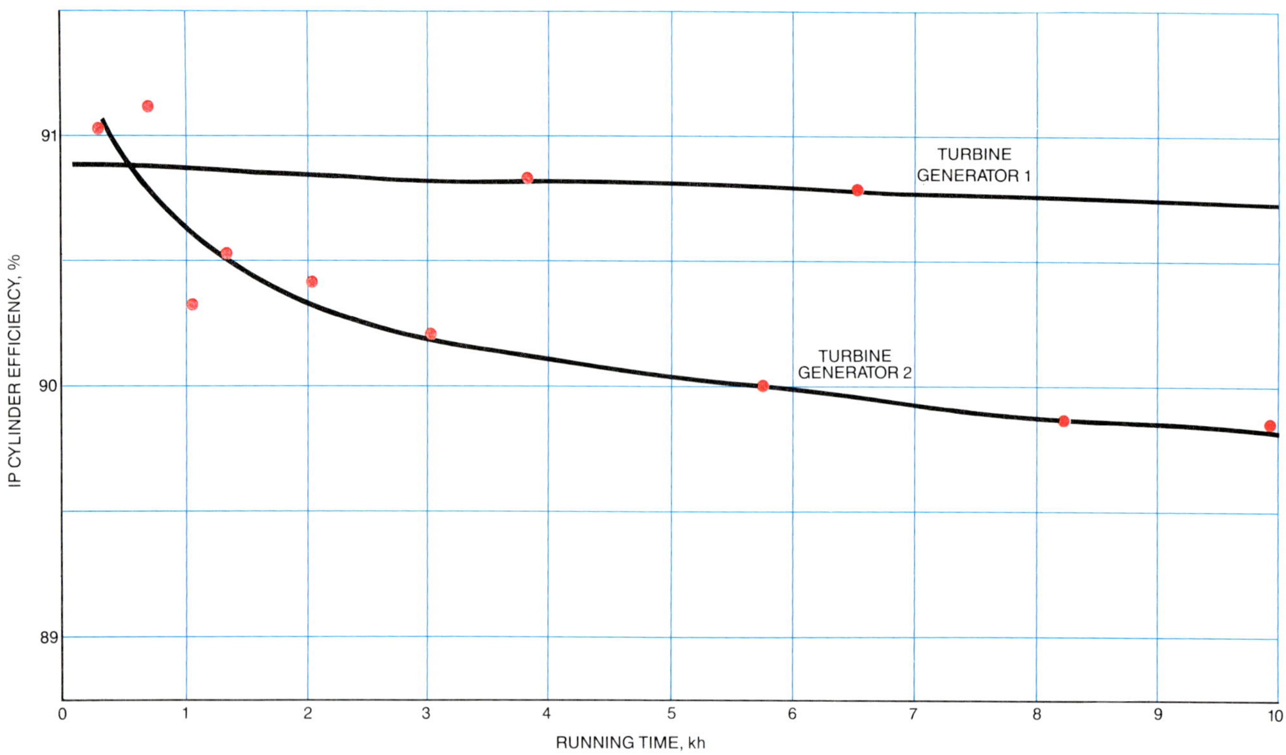

FIG. 6.54 IP cylinder efficiency performance plotted against time

speed governor to signal the valves to open further. Boiler pressure will tend to rise during the period during which a valve is closed for the test. This will increase the output but not to the original value.

5.7.1 Turbine governor channel trip test

Before the unit reaches 50% MCR, a governor channel trip test should be carried out to demonstrate that the tripping of one governor channel does not affect the operation of the other two channels or induce any trips in the turbine governing system. This will have been demonstrated statically, but it must be confirmed during commissioning that no stray circuits exist and that the turbine does not trip if one channel trips.

The test involves the tripping of a governor channel, checking results, rebalancing the circuit and returning it to service. This test is repeated for each of the three governor channels. These operations must not affect the output or the stability of the governor control system.

5.8 Sootblowers

5.8.1 Introduction

In order to retain maximum heat transfer efficiency in the boiler gas passes and associated gas air heaters, provision is made to remove combustion deposits from the superheater, reheater and economiser tubes as well as the elements of the gas air heater. The combustion deposits are removed, using compressed air for cleaning.

The number of sootblowers installed depends upon the boiler rating and the type of fuel burned. Typical numbers are 30 for an oil fuel boiler and 150 for a coal-fired boiler. The boiler design should incorporate provision for the later installation of additional soot-blowers which may become necessary dependent upon the analysis of future fuel burnt.

The control system for sootblowing is normally electronic, the panel housing the control equipment being equipped to accommodate additional soot-blowers. The control system will normally provide for a full sequence controlled sootblower operation to a selected programme.

Traditionally, the CEGB used sootblower systems with steam as the cleaning medium. Although effective for cleaning the boiler heating surfaces, these have high maintenance costs, plus the costs of generating the steam. The steam requirement for sootblowing was approximately 5 MW on a 660 MW unit. Air soot-blowing has now been developed and, providing the correct air pressure and flows are achieved, the cleaning effectiveness of the two media is very similar. The advantage of the air sootblowing system is reduced maintenance and running costs, though the capital cost of installing the additional air compressor systems may be slightly greater when compared with the steam system. Another advantage of the air system is that any sootblower compressor can be used for sootblowing on any boiler.

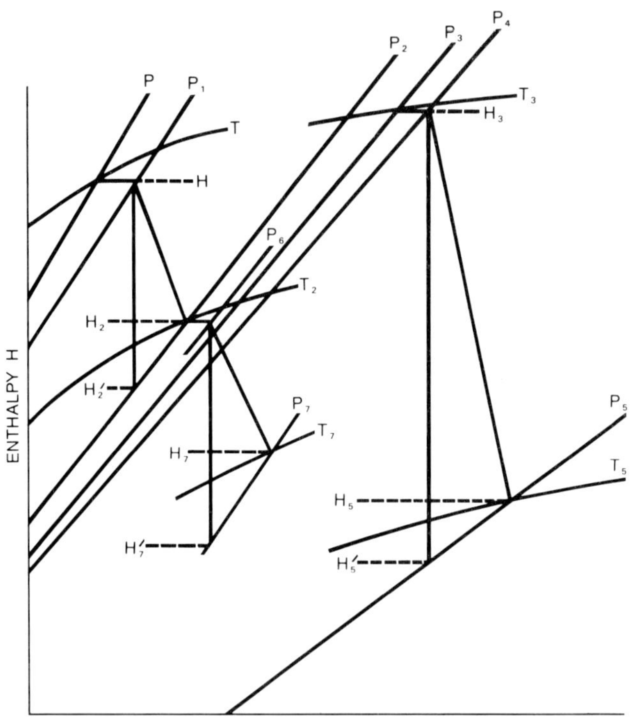

	MEASURED		DERIVED	
	PRESS	TEMP.	ENTHALPY	ENTROPY
BEFORE TSV	P	T	H	
AFTER GOVERNOR VALVES	P_1		H	S_1
AT HP CYLINDER EXHAUST	P_2	T_2	H_2	
ISENTROPIC HP EXHAUST	P_2		H_2'	S_1
BEFORE INTERCEPT VALVES	P_3	T_3	H_3	
AT IP CYLINDER INLET	P_4		H_3	S_4
AT IP CYLINDER EXHAUST	P_5	T_5	H_5	
ISENTROPIC IP EXHAUST	P_5		H_5'	S_4
AFTER BFP TURBINE GOVERNOR VALVES	P_6		H_2	S_6
AT BFP TURBINE EXHAUST	P_7	T_7	H_7	
ISENTROPIC BFP TURBINE EXHAUST	P_7		H_7'	S_6

$$\text{HP CYLINDER EFFICIENCY} = \frac{(H - H_2)}{(H - H_2')}\ \%$$

$$\text{IP CYLINDER EFFICIENCY} = \frac{(H_3 - H_5)}{(H_3 - H_5')}\ \%$$

$$\text{BFP TURBINE CYLINDER EFFICIENCY} = \frac{(H_2 - H_7)}{(H_2 - H_7')}\ \%$$

FIG. 6.55 Enthalpy drop tests — HP and IP cylinder efficiencies

A compressed-air sootblower system can be fully commissioned and operated (stroked) in a cold boiler furnace prior to the first firing of the boiler, provided the air supply is available. As this involves stroking the various sootblower lances into the boiler gas passes and blowing out through the blowers, control needs to be exercised relative to other construction activities. The sequence control system can be fully commissioned and correct operation proven before the unit is commissioned. The only subsequent commissioning activity is to operate the sootblowers under hot conditions with the boiler at pressure to prove freedom of operation without any interference due to boiler expansions, lubricant dry out, etc.

For the steam-operated sootblower system, sequence dry runs, including stroking the sootblowers into the cold boiler without the steam cleaning medium, are the only activities that can be completed without the unit

TABLE 6.22
On-load turbine valve test results

Time	MW	HP governor valve 1 opening %	HP governor valve 2 opening %	HP governor valve 3 opening %	HP governor valve 4 opening %	Steam pressure before turbine stop valve bar	Boiler outlet pressure bar	Remarks
14.08.10	284.2	20.3	19.9	20.7	20.9	136	143.9	
14.09.10	237.3	—	19.8	20.8	21.1	139.2	146.2	Valve 1 closed
14.14.10	214.8	—	20.0	20.2	21.0	142.6	149.5	Valve 1 still closed
14.15.10	268.0	19.9	20.0	20.2	21.0	138.9	146.5	Valve 1 opened
14.17.10	219.3	20.4	—	20.5	20.9	138.7	145.8	Valve 2 closed
14.21.10	246.7	22.5	—	22.9	22.9	137.5	144.7	Valve 1 still closed
14.22.10	295.9	21.2	20.9	21.8	22.8	133.4	141.4	Valve 2 opened
14.23.10	235.7	21.1	20.4	—	21.5	135.1	142.7	Valve 3 closed
14.28.10	211.5	19.6	19.7	—	20.5	138.5	146.7	Valve 3 still closed
14.29.10	275.6	20.0	19.7	20.4	20.6	136.3	144.5	Valve 3 opened
14.32.10	219.5	20.9	19.8	21.2	—	137.0	144.9	Valve 4 closed
14.38.10	216.0	20.2	19.8	20.5	—	138.2	145.9	Valve 4 still closed
14.39.10	271.3	20.0	19.8	20.4	20.6	133.9	141.5	Valve 4 opened
14.26.20	500.6	36.6	37.3	37.5	38.3	139.6	143.5	
14.27.20	462.7	—	38.2	39.9	38.4	140.1	143.8	Valve 1 closed
14.32.50	563.9	41.7	41.9	42.4	42.6	154.9	161.2	Valve 1 opened

being on load. Steam supplies are taken from a high pressure point associated with the primary superheater, enabling gas air heater blowing to take place at low loads, and from the reheater inlet pressure point. The following sections outline the commissioning checks for air and steam sootblowing systems with the boiler on load.

5.8.2 Steam sootblowers

A typical sootblower steam supply system is shown in Fig 6.56. The steam supply from the reheater is at approximately 42 bar, this being the same pressure obtained after the pressure reducing valve from the HP system supply point (165 bar). Each of the sootblower sections is supplied with steam through section valves. Automatic drain valves are fitted to ensure that the pipework remains free of condensate. The drain valves operate on temperature and normally close once the drain temperature has reached 150°C.

The steam section valves are controlled in the slightly open (or 'cracked') position as well as the fully-open position, this facility being used initially to warm and pressurise the sootblower system. The provision of the HP steam system enables air heater sootblowing to take place at low loads and at times when the boiler is off-load but at pressure, but this high pressure steam is more costly than the reheater steam supply and its use should be restricted.

Different types of sootblower are needed, depending on the section of the boiler in which they are to be used. The various types are:

Non-retractable sootblowers which are used in the furnace area and are designed to give a blowing arc of 70°.

Short retractable sootblowers which are also used in the furnace area and in the platen superheater, but may only traverse approximately 300 mm.

Long lance sootblowers which are used in the high temperature zones, such as the secondary superheater. The length of the 'lance' is approximately equal to just under half the width of the boiler. Steam must pass through this type of sootblower whenever it is entered into the boiler. If left in the boiler without steam passing through, it would deflect to such an extent that it would be impossible to retract and it would need to

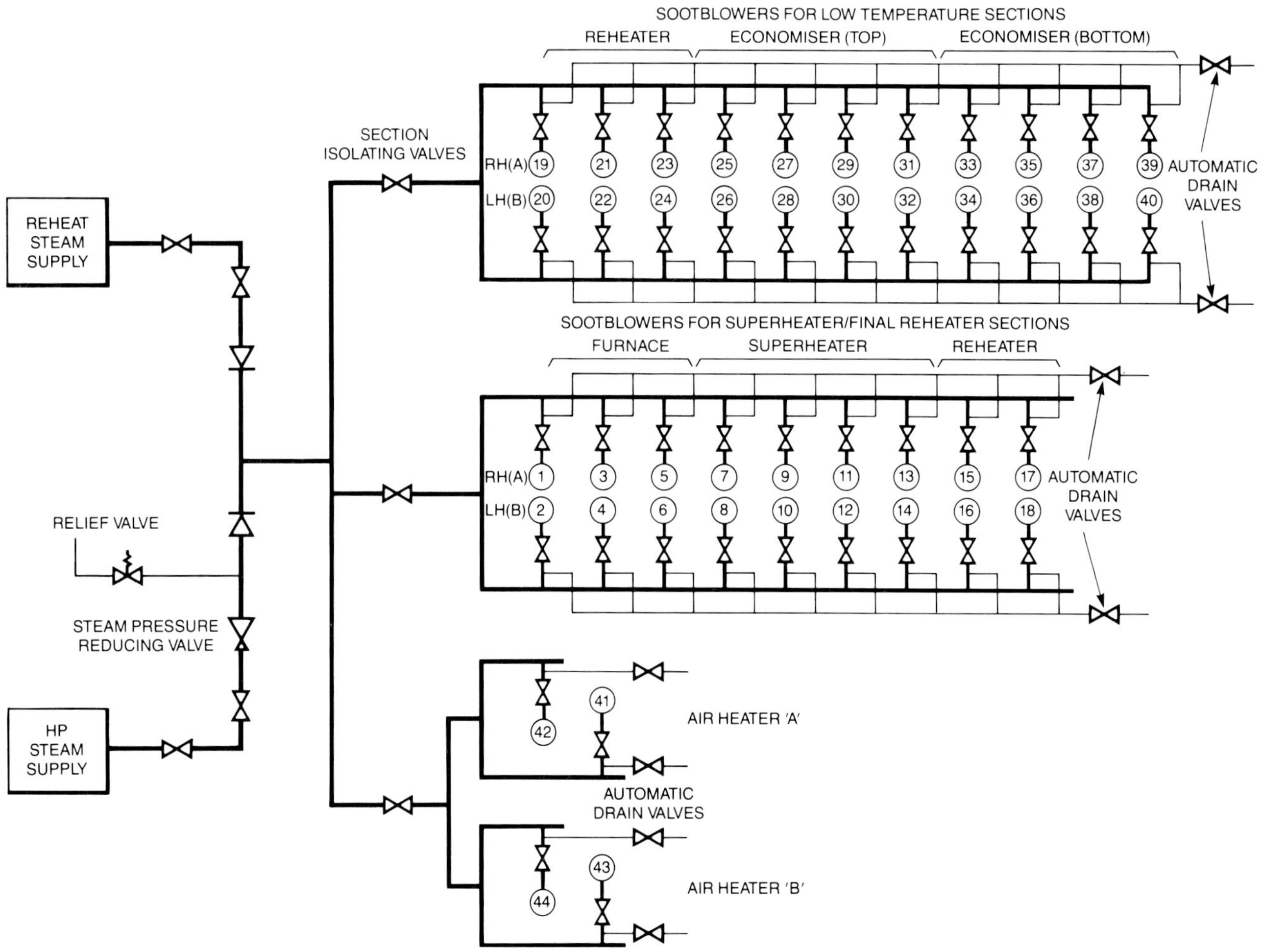

FIG. 6.56 Typical sootblower steam supply system

be cut for removal. In the event of low steam pressure, the sootblower should be retracted immediately.

Air heater sootblowers The blowing time of these sootblowers is set relative to the speed of the rotating air heater, each blowing for at least two full revolutions of the air heater.

The steam pressure requirements are different for each type of sootblower, varying from 14 to 20 bar. The steam pressure is controlled by orifice plates.

5.8.3 Compressed-air sootblowers

The air supply for sootblowing comes from compressors operating at 10 bar. These compressors are supplied specifically for sootblowing and are generally installed on the basis of one per unit, feeding into an air ring main serving all of the installed boilers. The sootblower compressors can also be used for other purposes, such as the turbine forced-air cooling system. Similar to the steam system, the compressed-air sootblowers installed in an oil-fired boiler would be arranged in three distinct sections:

- Superheater and reheater
- Economiser
- Gas air heater

In oil-fired boilers, it is not usual to provide sootblowers in the furnace.

An operational limitation may become necessary because of the air heater requirement for two compressors in service when two sootblowers are to be operated simultaneously, but for other operations only one compressor is necessary.

The types of sootblower are as follows:

- Long retractable for the superheater and reheater
- Half-retractable for the economiser
- Multi-nozzle sootblowers for the air heater

The blowing time of the long retractable sootblowers is approximately 12 minutes and for the half-retractable sootblower 6 minutes.

A typical sootblower layout for a 660 MW oil-fired boiler is shown in Fig 6.57.

A typical long retractable air sootblower is shown in Fig 6.58.

5.8.4 Sootblower commissioning

Compressed-air sootblowers

This type of sootblower can be commissioned prior to the first firing of the boiler. Chapter 4 of this volume describes the basic checks and method of setting up the sootblowers, including the air system. The final commissioning of the system can only be achieved with the boiler under pressure and carrying its full-load of 660 MW.

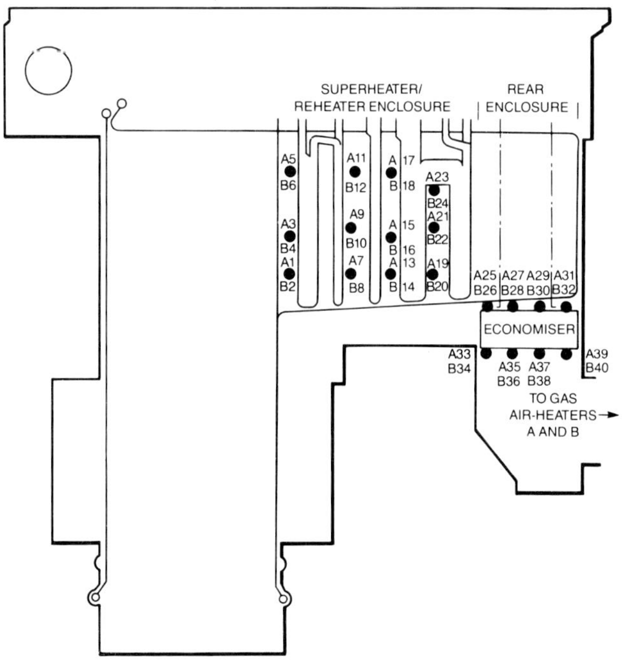

FIG. 6.57 Typical layout of sootblowers for a 660 MW oil-fired boiler

The checks at full-load are necessary to demonstrate that correct sootblowing operations can still be carried out under hot conditions, with the boiler fully expanded.

The following equipment has to be available prior to the commissioning checks:

- Portable air motor.
- Portable compressor or air bottle to check the pressure switch.
- Pressure gauges for checking the blowing pressure of each sootblower.

Prior to carrying out any sootblowing:

- Check that the sootblower is adequately supported and guarded.
- Check that, with the boiler expanded, the sootblower assembly is not fouling any supports.
- Check that the gearbox is filled to the correct oil level.
- Grease all grease nipples, where fitted, and ensure that they are fully charged with grease.
- Confirm that the gear rack is free from debris and obstruction.
- Check that the flexible air hose is not damaged.
- Check that the cable carrier is tensioned.
- Confirm that the refractory filling has been fitted in the wallbox.
- Confirm the availability of at least two sootblower air compressors, operating to the designed air requirements.

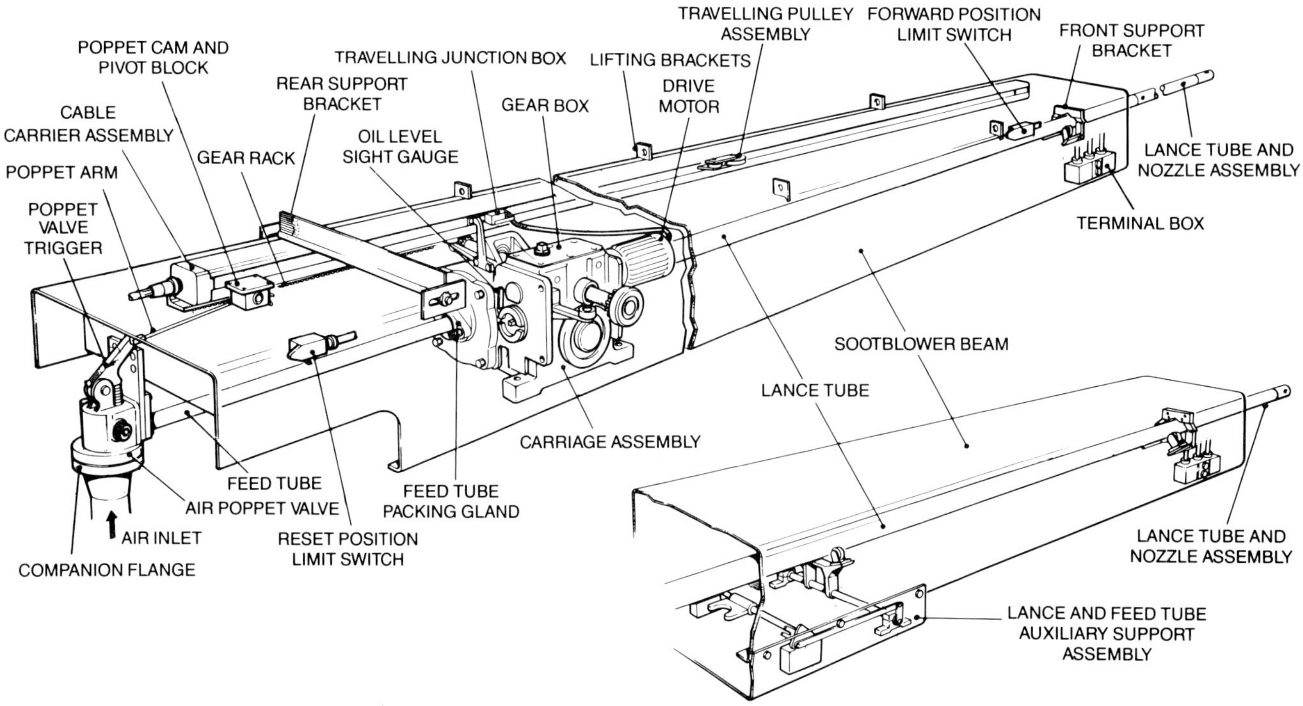

FIG. 6.58 Long retractable air sootblower

- Charge the air ring main, after completing the preceding checks, in preparation for sootblowing.

Compressed-air sootblower operation

- Select initial air heater section

- Initiate sootblowing sequence for air heater sootblowing

During sootblowing, check the following:

- Individual sootblower air pressure. Compare with original cold-set values.

- Check and record blowing time for each sootblower.

- Confirm that, on a single sootblowing test with one sootblower air compressor in service, the correct air pressures are obtained.

- Repeat this test with a pair of sootblowers operating with a single air compressor in service, and reconfirm correct air pressures.

- Observe operation of sootblowers, checking for any evidence of stiction.

  Note The air pressure settings obtained with the sootblowers on the initial setting will be found to increase slightly under hot conditions.

- Select the other two sections in turn and repeat the checks, as stated. As these blowers are in the higher temperature zones, this activity should not be attempted until the boiler loading has reached at least 50% MCR.

- In the event of reduced compressed-air pressure or total loss of pressure, the sootblowing operation should be immediately suspended. Any blowers still in the furnace must be withdrawn, using the portable air motor.

Steam sootblowers

The basic initial checks for this type of sootblower are described in Chapter 4 of this volume. In that chapter, low pressure (LP) steam from an auxiliary boiler system is available for early sootblower commissioning. This LP steam source may not be available and, if available, it may be at the wrong steam pressure and is not likely to be capable of supplying the correct steam flows. However, if available, it should be used for the initial setting of the sootblowing system. The steam supply system, as indicated in Fig 6.56, will have been checked as part of the pre-commissioning checks for completeness, for the correct operation of all valves and that it is adequately supported.

Prior to carrying out sootblowing under hot boiler conditions, it is necessary to carry out checks on the sootblowers similar to those for the compressed-air sootblowers. In addition, the steam supply system requires checking to ensure that all piping is lagged and that all electrically-operated valves operate over their full range of travel.

Steam sootblower operation

The first operation is to charge the steam system up to the steam section valves from the high pressure steam point.

- Select at the sootblowing control panel for 'steam warming' and initiate.
- Gradually open the high pressure steam supply valve to charge the line up to the pressure reducing isolating valve.
- Charge the pressure reducing valve. Record pressures.
- Confirm that the system relief valve is not lifting.
- Check that the non-return valve on the reheat steam supply is not passing steam.
- Proceed to the section valve and initiate charging of the sections.
- Check that the section valve initially goes to the 'cracked' position.
- Check that all drain valves are open and check the temperature at which they close.
- Confirm that the section valve is wide open and that all drain valves are shut, the steam system is at the correct pressure and that there is no evidence of pressure steam leakage.
- Select initially the air heater section and proceed as identified earlier for compressed air sootblowing.
- Check and recheck the steam pressure at each sootblower.
- Select the other two sections and repeat the checks. This part of the commissioning should not be carried out unless the boiler loading has reached at least 50% MCR.

Having confirmed correct operation of the sootblowers and associated systems from the HP steam connections, the test should be repeated from the reheater steam supply source. The test is carried out at 100% MCR to ensure that the design steam pressure is available. All results should be recorded and compared with those obtained from the HP steam supply. The steam pressures and stroke time should be identical, regardless of the steam supply source.

5.9 Automatic boiler control systems

The continuous, safe and efficient operation of any large modern unit requires the auto control systems to be in service and functioning correctly. It is therefore essential that the commissioning of the boiler auto control systems is integrated into the overall commissioning programme. Figure 6.46 showed how these tests are incorporated into the programme at the appropriate load points. This section covers the commissioning of the auto control systems (in general terms): a full technical description of these systems is available in Volume F, Chapter 2. The boiler automatic control loops which require integration into the commissioning activities are:

- Combustion control (PF or oil)
- Feedwater control
- Superheater temperature control
- Air heater steam pressure control
- Boiler master pressure control
- Reheater temperature control

The commissioning programme (Fig 6.46) indicates the need for load variations during the auto control systems tests. The basic setting up is carried out at steady state load; load variations are then introduced which allow performance of the controls to be closely monitored. High accuracy, high speed recorders are used to monitor parameters associated with the auto control systems, for example, the monitoring of the opening or closing of a control valve relative to the pressure on the system being controlled by that control valve. This high accuracy monitoring equipment is in addition to the standard installed instrumentation.

5.9.1 Combustion control of oil-fired plant

The combustion controls are adjusted so that they function from 20% MCR load upwards. Below this figure, auto control is difficult without special equipment and is usually not justified.

Volume F, Chapter 2 describes the pulverised-fuel combustion control.

Brief description of the loop

The control scheme operates in a differential pressure mode for start-up and low loads, and in airflow modes for loads in excess of 20% MCR.

In the start-up mode, the oil pressure to burners is measured by a pressure transmitter and compared with the desired value signal set by the operator. The resultant difference is converted to pulse signals to control the start-up oil valve.

The differential pressure between the windbox and furnace is measured by differential pressure transmitters which are temperature compensated, using the windbox air temperature transmitters. The output signals are averaged and compared with a desired value of furnace/windbox differential. The resultant difference is converted to pulse signals for the control of the FD fan vanes and/or the inlet damper. A balancing circuit maintains equal signals to each fan and a biasing circuit enables the proportion of load on each fan to be varied by the operator.

The boiler output steam pressure, which is measured by more than one transmitter producing an average signal, is compared with the desired value of steam pressure selected by the operator.

The measured value of oil flow to the burners is derived by subtraction of the recirculating oil flow from the boiler total oil flow. At approximately 15% MCR, oil flow control transfers automatically from the start-up valve to the main oil valve. Once the oil flow exceeds 25% MCR, the control scheme is changed to the airflow mode. Figure 6.59 shows a simplified combustion control system.

Commissioning checks — start-up valve with furnace windbox differential

The number of burners in service must require less than 25% MCR airflow in order to establish initial conditions. During this activity, the following parameters are monitored:

- Total furnace windbox differential pressure
- FD fan(s) damper and vane position
- Oil pressure to burners
- Start-up oil valve position

The commissioning checks include the following:

- Adjust the start-up oil valve to obtain the desired oil pressure.
- Adjust the controls to obtain desired furnace/windbox differential pressure.
- Adjust the pulse and analogue controllers, as required, to obtain 100% settings.
- Select the start-up valve to 'auto'.
- Adjust the pulse controller to obtain a stable signal between desired and measured values of the start-up oil valve position.
- Test until optimum settings are achieved.
- Initiate a small change in the furnace windbox desired value and check the response of the FD fan

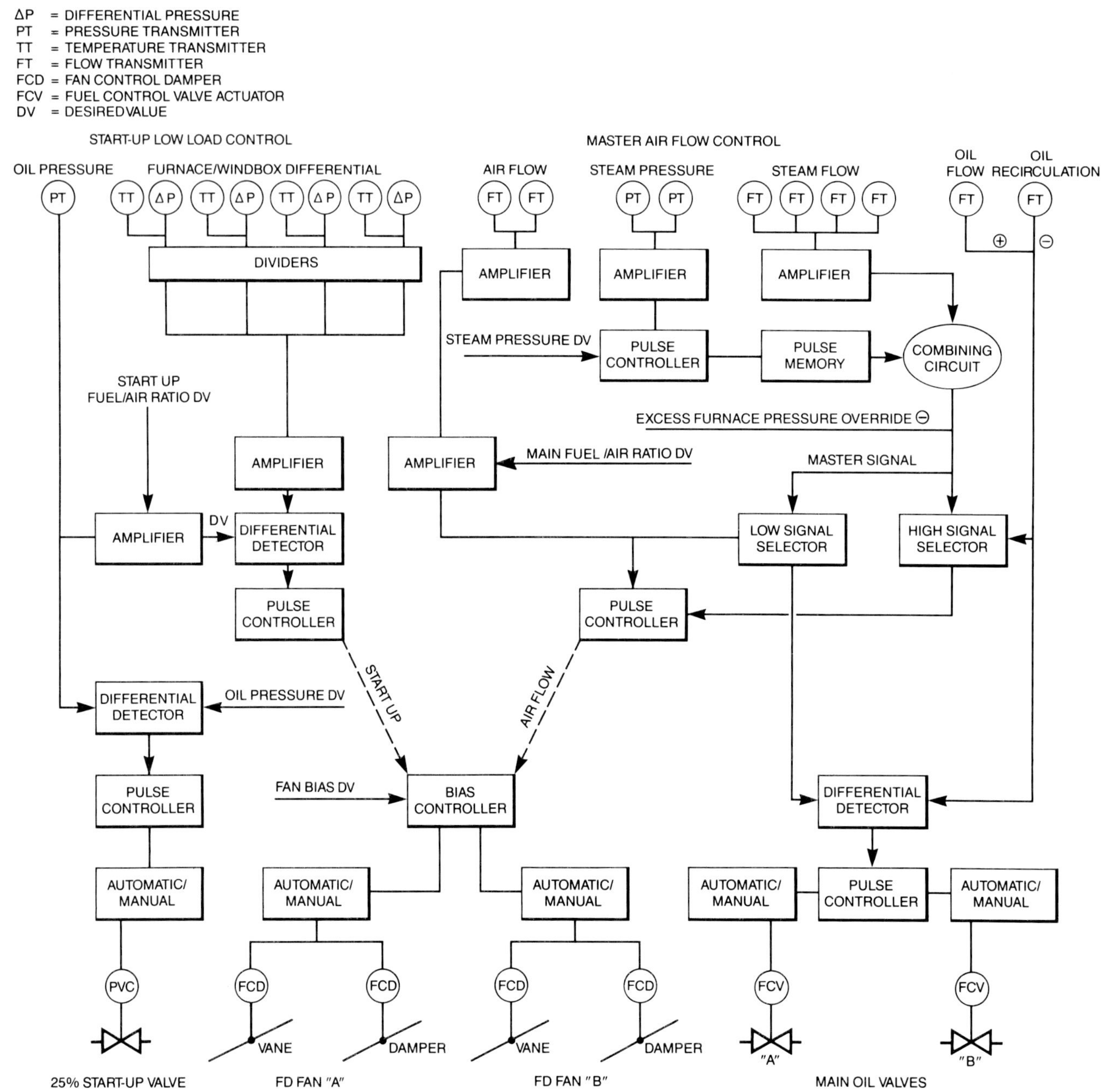

FIG. 6.59 Simplified combustion control system including master steam pressure control

vanes and inlet damper. Adjust the pulse controller to obtain the desired results.

- Continue the tests until optimum settings are achieved.

Commissioning checks — main oil valve with total airflow

This part of the commissioning activity takes place once the airflow exceeds 25% MCR with the ID fans on 'auto' control. During this part of the activity, the following parameters must be monitored:

- Oil flow
- Main oil control valve position
- Airflow
- FD fan(s) damper and vane positions

The commissioning checks include the following:

- Adjust the gain setting to the amplifier to obtain satisfactory oil/airflow ratio.
- Increase the oil input, and in consequence the airflow, until approximately 40% MCR airflow is reached; re-establish the correct oil/airflow ratio.
- Adjust the fuel/air ratio and then select the oil valve to 'auto' control.
- Increase/decrease airflow by operation of the FD fan vanes. Check the response of the oil valve and determine its correct setting for attaining the fuel/air relationship.
- Test until optimum settings have been obtained.
- Adjust the boiler master control station until the oil valve just starts to close. At this point, the airflow should be in balance.
- Select the FD fan vanes for 'auto' operation.
- Vary the steam pressure demand in both directions and check the movement of the FD vanes against the oil valve.
- Adjust the pulse controller until 'stable oscillations' are obtained.

  Note In this context, 'stable oscillations' mean that the system is damped so that the oscillations are of small magnitude, approximating to a constant value.

5.9.2 Feedwater control

The system comprises two loops, single-element control for start-up and three-element control for use above 25% MCR.

The start-up single-element loop should have been commissioned during early boiler firing checks and used during the chemical clean and steam purge activities; it is not commissioned as part of this programme. A load point of 30% MCR is specified for commissioning the three-element control.

Single-element control

The single-element control utilises a '25% feedflow control valve', i.e., a valve passing a quantity of feedwater corresponding to 25% MCR, to maintain working drum level. The pump control signal is derived from a differential pressure measurement across the start-up feed valve. The start-up scheme also incorporates feedflow orifice plates and transmitters. The output signals from the drum level transmitters are summed, density-compensated and averaged. This average signal is compared with the feedflow transmitter signal, and the resultant signal is converted to operate the feed valve.

The differential pressure signal from the feedwater control valve is compared with the desired value set by the operator. Changing the desired value alters the water flow to the boiler.

Three-element control

The three-element control scheme uses the drum level transmitter signals to modify the steam flow signals derived from orifice plates in the main steamlines.

The resultant signal is compared with the measured feedflow and any difference is converted into a pulse signal to control the speed of the main feed pump. The differential pressure across the main feed valves depends on feedflow and valve opening. A desired value is set by the operator and the valve moves to achieve this setting. A differential pressure transmitter is used for this purpose. Figure 6.60 shows a simplified feedwater control system.

Commissioning checks for the three-element drum control

The main boiler feed pump should be in service on single-element drum level control duty.

During this activity, the following parameters must be monitored:

- Drum level
- Feedflow
- Steam flow
- Boiler feed pump performance
- Differential pressures across main feed regulating valves
- Main feed valve position

The following procedure is used:

- Initially, the system should be operating with approximately 10 bar differential pressure across the main feed valve, with the start-up feed regulating valve maintaining the drum level.
- Adjust the pulse controller to obtain stable oscillations between the two main feed valves.

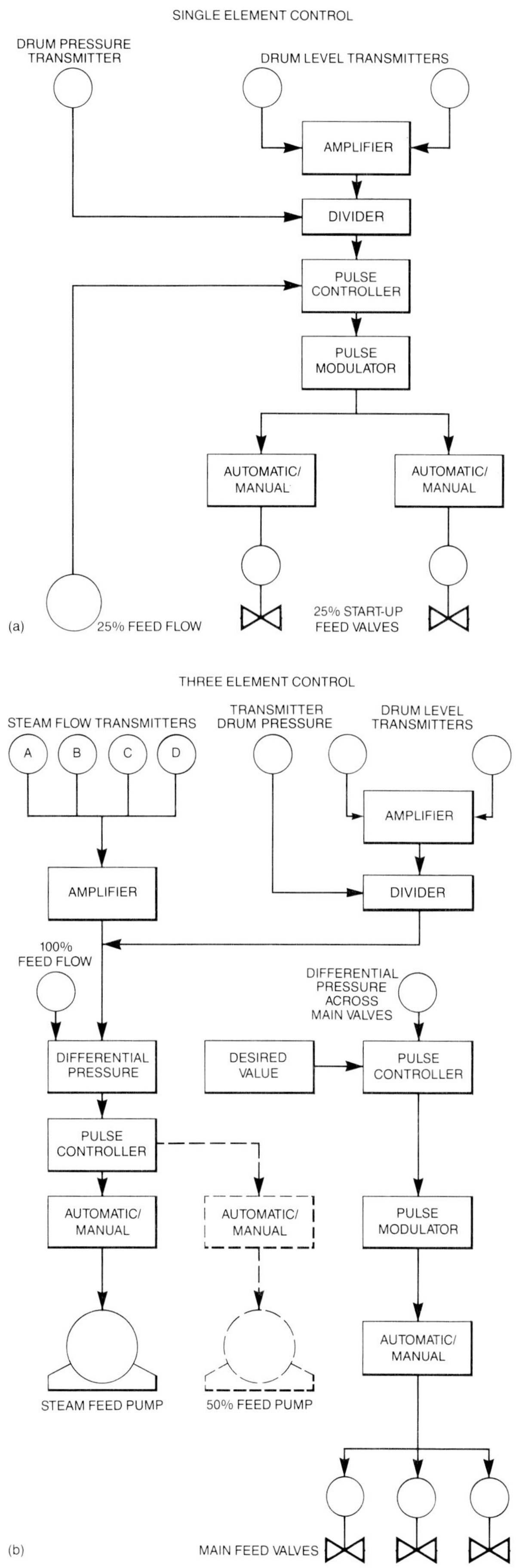

FIG. 6.60 Simplified feedwater control system

- Close the start-up feed valve slowly, whilst maintaining the desired value of pressure differential across the main control valve. During this operation, manual speed control of the main boiler pump will be needed to maintain the drum level.
- Adjust the pulse controller, making small changes in desired value settings.
- Obtain stable oscillations and test until optimum settings have been obtained.
- Switch the main feed pump to 'auto' control, making small changes in steam flow requirements. Adjust the controllers and test until optimum response is obtained.
- Adjust the drum level in both directions and adjust the controllers, as necessary, to obtain satisfactory operation.

5.9.3 Superheater steam temperature control

The superheater temperature control system is commissioned at the 30% load point. This ensures that steam temperature control is available early in the programme, so that strict control of boiler steam temperatures can be maintained as unit load increases. An interlock is normally installed to prevent operation of the superheater sprays below 25% MCR in order to prevent possible waterlogging of the superheater tubes, so the 30% load point represents the earliest opportunity on the commissioning programme to carry out this activity.

Brief description of the control loop

The superheater steam temperature control loop is divided into four individual loops to match the four streams in the secondary superheater circuit. The flow of steam through the primary and secondary superheaters is divided between the four parallel streams which are arranged in the gas pass to distribute heat absorption uniformly. The control equipment associated with each stream is independent except for a common desired value of steam outlet temperature. The control across the superheater is used so that the attemperator in each stream can fulfil two functions:

- Reduce the temperature fluctuations in the primary superheater caused by firing patterns.
- Control the steam temperatures to maintain outlet temperatures within the prescribed limits.

Figure 6.61 shows superheater and reheater steam flows for a 660 MW boiler. The pipework and valve system associated with a four-stream superheater and reheater spray system is shown in Fig 6.62.

The secondary superheater outlet temperature is measured by four resistance thermometers in each stream, mounted in different parts of the stream. These

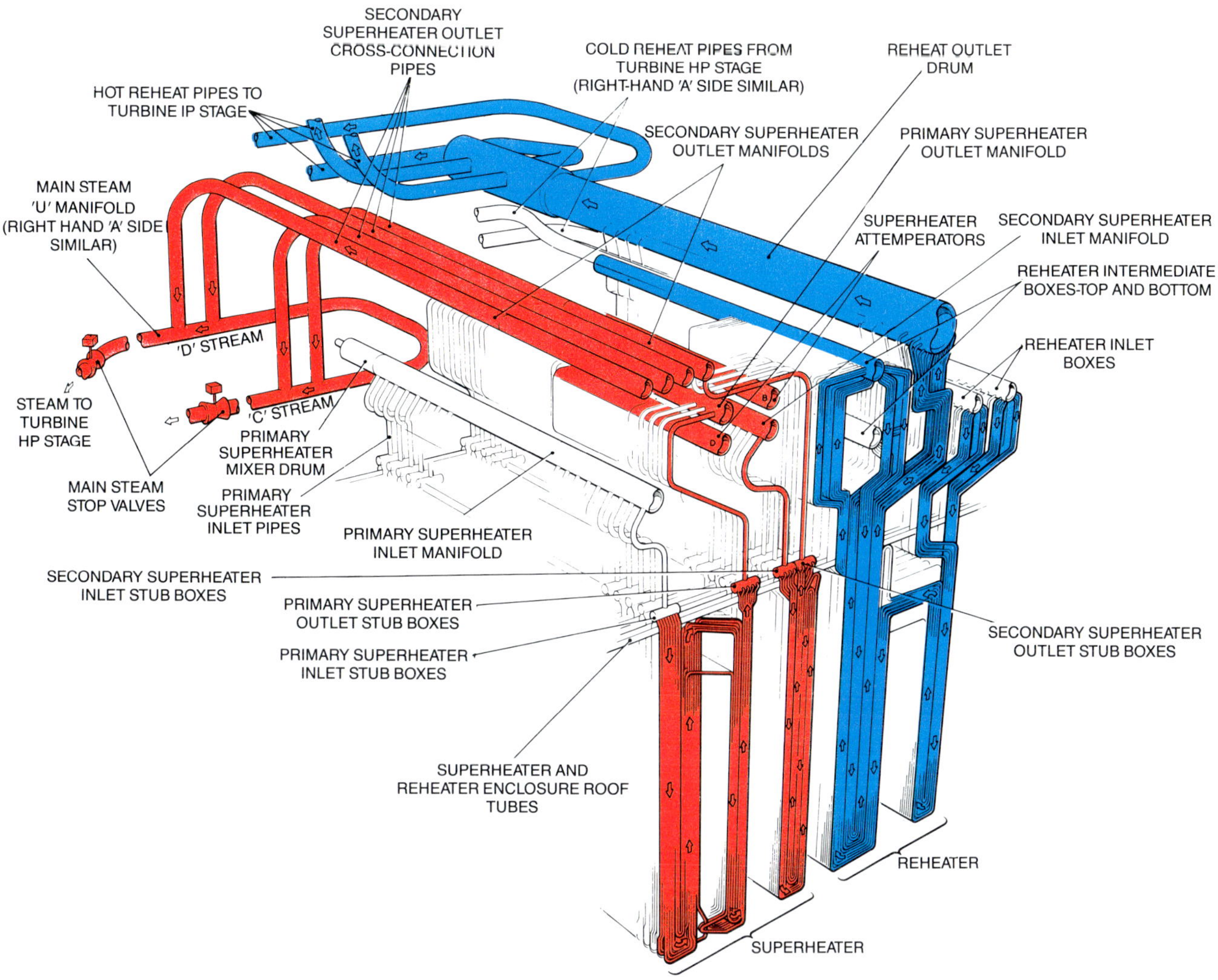

Fig. 6.61 Superheater and reheater steam flows for a 660 MW boiler

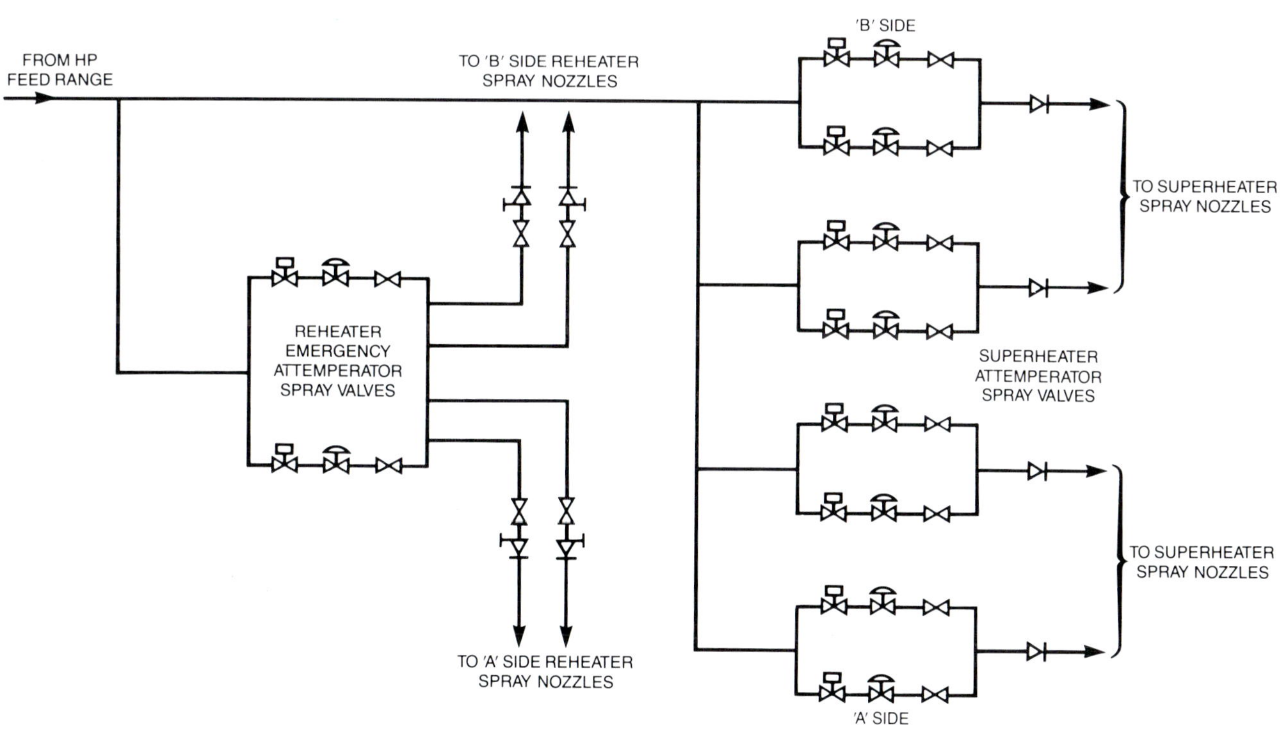

Fig. 6.62 Superheater and reheater attemperator spray valve arrangement

thermometers are connected in pairs, such that two average steam temperature signals can be obtained at the two temperature transmitters. Either of these signals can be used in the control loop by operating the selector switch. Figure 6.63 shows the system for one stream only, identical systems being installed on the other three streams.

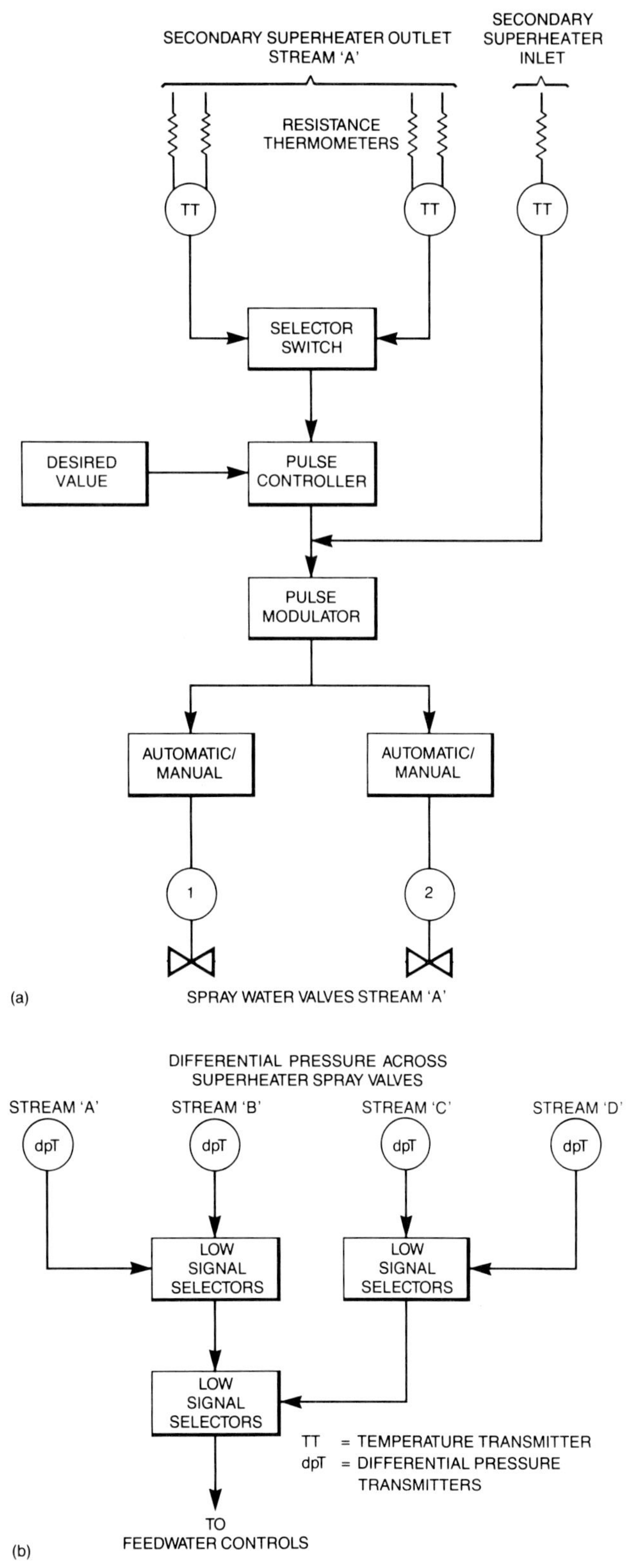

FIG. 6.63 Simplified superheater steam temperature control system

The measured value (MV) of steam temperature is compared with the desired value (DV) of temperature set by either the unit operator or, if fitted, by the unit computer. The resultant difference signal becomes the desired value for the secondary superheater inlet temperature. The measured value of the secondary superheater inlet temperature is derived from its associated resistance thermometer and transmitter.

The measured value of secondary superheater inlet temperature is compared with the desired value signal and any difference is converted to a pulse signal to operate the superheater spray control valve. As each stream contains two control valves, only one spray valve can be selected to 'automatic' at any one time.

Commissioning checks

The pipework and valve arrangements for the superheater spray temperature controls should have been flushed through, using the electrically-driven feed pump, just prior to the commencement of the boiler steam purge. This ensures that any debris left in the spray pipework is flushed into the boiler and then purged out to atmosphere during the steam purge. During flushing, checks are made on the pipework supports for movement and suspension, and to confirm that no pipework leaks exist. Prior to the flushing operations, commissioning checks must have been completed on all valves, including their means of operation, and on the pipework.

Prior to commencing activity:

- Confirm that the electrically- and hand-operated valves are set for correct operation, with electrically-controlled valves operating from the unit control desk.
- Confirm that the air supplies to pneumatic control valve exist.
- Confirm that the spraywater pipework is charged up to the control valves, and that the valves are not passing.

During this activity, the following parameters must be monitored:

- Superheater inlet and outlet temperatures
- Superheater spray valve position

The commissioning checks include the following: commencing the initial checks at the lower end of the spray control range, select the superheater spray control valve to 'manual', obtaining the desired temperature at the superheater outlet:

- Adjust pulse controllers.
- Adjust the pushbuttons on the control panel to achieve equal superheater inlet temperature indications.
- Transfer the selected spray valve to 'auto'.

- Change the superheater temperature and adjust to obtain the optimum response.
- Adjust to obtain stable oscillations at the superheater inlet temperature.
- Test and adjust until optimum settings have been obtained.
- Vary the superheater outlet temperature and check the loop response.
- Adjust pulse controller until stable oscillations occur.
- Apply the same procedure at the higher end of the superheater spray control range.
- Fully test until optimum settings have been obtained.

Apply the same settings to the remaining three spray control streams initially and repeat the same tests, making minor adjustments, as necessary.

At a suitable opportunity, the 25% MCR inhibit of superheater valve opening and the corresponding valve closing, following a trip, should be proven, i.e., that spraywater can only be used for superheater temperature control at loads above 25% MCR.

5.9.4 Air heater steam pressure control

The commissioning of the bled-steam controls for the bled-steam air heater should be integrated into the overall unit programme for this activity to commence at the 60% MCR load point. At loads below this level, the bled-steam pressure will be too low for operation. The steam air heater has two supplies, one from the auxiliary steam system and the other from the boiler feed pump exhaust line to the LP cylinder. The supply from this auxiliary steam system has commenced on manual control prior to steam-to-set. This activity is to recheck the auxiliary steam control valves and to commission automatic control for both the bled-steam and auxiliary steam supplies.

Brief description of the control loop

The control scheme uses temperature transmitters to measure the gas air heater inlet and outlet gas temperatures. Figure 6.64 shows a typical scheme. The air inlet and air outlet temperatures are summed to give a combined temperature at the bottom of the heater which is controlled by varying the steam flow through the steam air heater. The measured combined temperatures are compared with the desired value for the heater in two controllers. The resultant difference is converted to pulse signals in both controllers and fed through an 'auto/manual' station to the auxiliary steam valve and to the bled-steam valve. The unit sequence control transfers from one valve to the other. It causes the valve not in service to close by reducing the desired value to that controller.

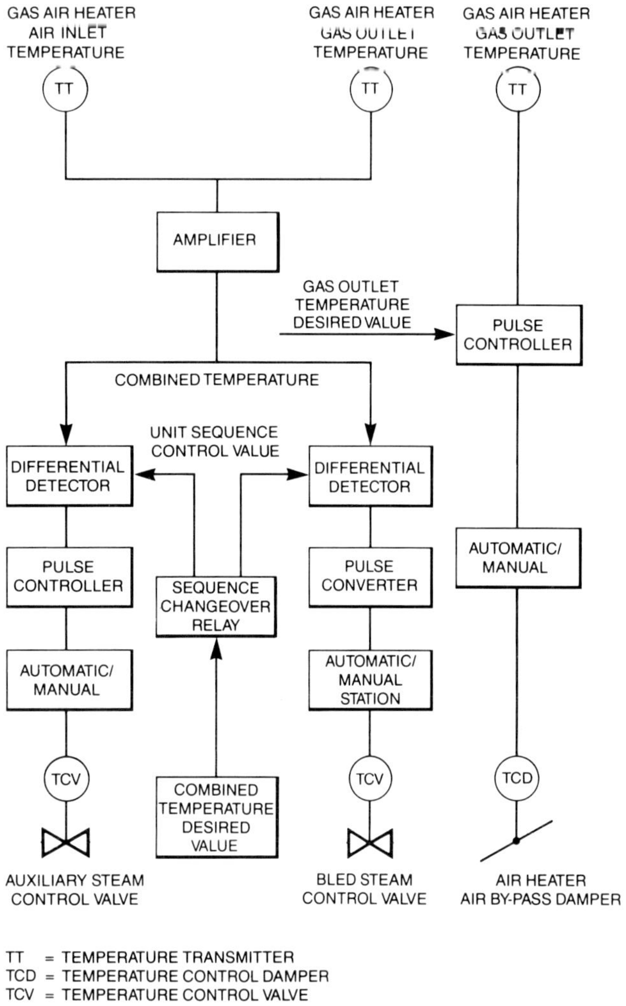

FIG. 6.64 Simplified air heater control system

A separate temperature transmitter is used to measure the gas air heater gas outlet temperature, which is prevented from falling below the desired value by opening the air by-pass damper. The output from the transmitter is compared with the desired value for the heater, with any difference being converted to pulse signals for controlling the position of the air by-pass damper.

Commissioning checks

During the activity the following parameters must be monitored:

- Steam air heater air outlet temperature.
- Gas air heater air outlet temperature.
- Auxiliary steam control valve position.
- Bled-steam control valve position.
- Air heater air by-pass damper position.

Auxiliary steam control valve

- With the auxiliary steam control valve on 'manual', adjust the valve opening to obtain a combined temperature difference of 208°C between the steam air heater air outlet and the gas air heater gas outlet.

- With the air heater air by-pass damper on 'manual' adjust the damper opening to obtain a temperature of 140°C at the air heater gas outlet.
- Adjust the combined temperature desired value to obtain identical readings of temperature on the fitted indicators.
- Select the auxiliary steam control valve to 'auto'.
- Adjust the desired value by small amounts and then adjust the pulse controller until the correct response has been obtained.
- Adjust the controller until stable oscillations occur.
- Test until optimum response has been obtained.

Bled-steam control valve

The initial conditions are that the desired temperatures are being controlled by the auxiliary steam control valve, with the air heater air by-pass damper and both the auxiliary steam and bled-steam valves being on 'manual':

- Start closing the auxiliary steam control valve, whilst opening the bled-steam valve, ensuring that the desired temperatures are still being monitored.
- With the auxiliary steam valve fully closed, adjust the desired value (if necessary) to obtain equal temperature indications.
- Select bled-steam valve to 'auto'.
- Change the desired value by small amounts and then adjust the pulse controller until the correct response has been obtained.
- Adjust controller until stable oscillations occur.
- Test until optimum response is obtained.

The condensate drains will normally return to the condenser flashbox. As this is the first time that a bled-steam flow has been established through the pipework, initially all condensate drains must be dumped to waste. Regular checks must be carried out to establish the condensate purity before it is allowed to discharge into the condenser flashbox.

A typical maximum value for conductivity before recovering the condensate drains is 10 μS/m. This water purity must be monitored during the rest of the load programme and the following shutdowns until the system has been confirmed as clean.

5.9.5 Master steam pressure control

The commissioning of the master steam pressure control occurs in the overall commissioning programme at the 70% MCR load point. The master steam pressure controller can only be commissioned if all the associated combustion control loops are functioning correctly on 'auto' control.

Brief description of the control loop

The master steam pressure control loop has an overriding effect upon the whole of the system and is used as a measure of turbine load. Its function is to maintain the boiler outlet pressure constant under all load conditions; for this reason the boiler outlet pressure is measured, together with an anticipatory loop measuring boiler steam flow. Figure 6.59 includes a typical master steam pressure control loop.

The boiler stop valve pressure is measured by two pressure transmitters, each producing an analogue signal proportional to steam pressure. These signals are fed to amplifiers, each of which produces two identical, but isolated, signals for further processing.

The difference detector compares the values of the signals and initiates an alarm if an excessive difference exists.

The desired value (DV) of the boiler steam outlet pressure is determined by the operator or computer. The pulse controller compares this DV with the measured value (MV) from the amplifier module. The steam flow is measured by four flow nozzles, one in each main steam pipe. Each flow nozzle has an associated differential pressure transmitted which produces an analogue signal. These signals are converted to a flow value by use of a square root extractor module. The amplifier module sums the four steam flow signals and produces two identical signals, each equivalent to the average steam flow.

If the stop valve pressure differs from the desired value, an error signal is generated which is not equal to the steam flow signal. This results in a signal which is fed to the selector module and is used as the output signal.

The signals from the master pressure controller initiate action in all the combustion control loops, thus maintaining correct combustion conditions as the boiler pressure changes are corrected.

Commissioning checks

During this activity, the following parameters must be monitored:

- FD fans damper/vane positions
- Oil pressure to burners
- Airflow to boiler
- Furnace/windbox differential pressure
- Drum level
- Feedflow
- Steamflow
- Superheater and reheater steam temperatures

- Steam pressure
- Electrical output

The commissioning procedure is as follows:

- Check that all combustion control loops are operating correctly on 'auto' control.
- Select the boiler master steam pressure controller to 'manual'.
- Adjust the steam pressure desired value to obtain identical readings of pressure on the indicators.
- Using the boiler master control, adjust the boiler pressure requirements under 'manual' control and confirm that all systems follow and function correctly.
- Once satisfied on 'manual' control that the systems all follow the master signal, select the master steam pressure controller to 'auto'.
- Make small changes in the desired value setting of the steam pressure in both directions and monitor that correct performance is being maintained.
- Repeat, inducing larger changes of pressure, until satisfied with the ability of the master steam signal to maintain the correct boiler pressure within the defined tolerances, without pressure excursions.
- Make adjustments to controllers, as required.

5.9.6 Reheater temperature control

The commissioning of the reheater temperature control system occurs in the overall commissioning programme at the 100% MCR load point. The control comprises two distinct loops, initially using the gas recycling (GR) fans to recycle a proportion of the boiler flue gases. If sufficient control is not obtained by this method, a spraywater attemperator is used. The GR fan is commissioned prior to steam-to-set and is available, on manual control, for reheater temperature control. This activity sets up the reheater temperature controls for 'auto' operation over the full range of the boiler, utilising either system.

Brief description of the control loop

The reheater steam temperature is controlled initially by recirculating hot boiler exhaust gases into the furnace hopper to maintain steam temperatures. If temperatures increase such that, at minimum GR fan damper opening, control is not being achieved, the water sprays are used. The role of the GR is basically to increase the temperature to ensure that the design conditions of reheater steam temperature are achieved.

The secondary reheater outlet steam temperature is measured by two pairs of resistance thermometers (Fig 6.65). Either of these signals is used, via a selector switch, and is compared with the desired value of reheater steam temperature set manually by the operator. Any difference between measured and desired value signal is fed to the controller which forms part of the DV signal for the gas recirculating inlet damper control. This DV signal can be overridden to 'manual' by a protection circuit on excessive primary superheater outlet temperatures. Four resistance thermometers measure this temperature. The operator can adjust the DV signal to reduce the temperature or the protection circuit will generate lowering pulses to reduce it.

The DV signal for controlling the GR fan inlet dampers is derived from oil flow transmitters associated with fuel oil flows. The other variant comes from the primary reheater outlet, where the temperatures are measured by four resistance thermometers which are combined in an amplifier to provide an average temperature signal. This is used as an anticipatory signal, operative only when the measured steam temperature is changing. The resultant DV signal for the inlet damper control is compared in the difference detector with the measured value obtained from the GR fan balancing control circuit. This balancing circuit measures the position of both the GR fan inlet dampers, passing these signals through amplifiers to the difference detector. The signals also feed via a controller to the pulse controller which receives any difference signal output from the difference detector. The resultant output signal from the pulse controller is fed to the gas recycling inlet damper actuator which controls the quantity of flue gas being recycled, and hence the reheat temperatures.

The signal selected by the selector switch in the secondary reheater outlet temperature-measuring equipment is used as the initial measured value (MV) signal for the reheater spray control. Similarly, the DV signal for this control is the same as that used for the GR fan inlet damper control loop, but modified by an offset signal to prevent the spray valves acting in unison with the inlet damper control. The combination of these two signals becomes the desired value signal that is compared with an MV signal generated from the reheater spraywater flow transmitter. The resultant difference in signal is then converted to a pulse signal and routed to the reheat spray control valves. Figure 6.62 shows the reheater spray valve arrangement: in this example only one spray valve can be selected to 'auto' at any time.

The speed of the GR fans is regulated by a hydraulic coupling which is adjusted to maintain a constant differential pressure across the fan inlet damper. The differential pressure is measured by a transmitter and compared with a DV differential pressure signal set by the operator. The resultant signal difference is converted into pulse signals and fed into the hydraulic coupling actuator. Figure 6.66 shows the control loop.

Commissioning checks — gas recycling fans and damper control

During this activity, the following parameters must be monitored:

- Primary and secondary reheater outlet temperatures.

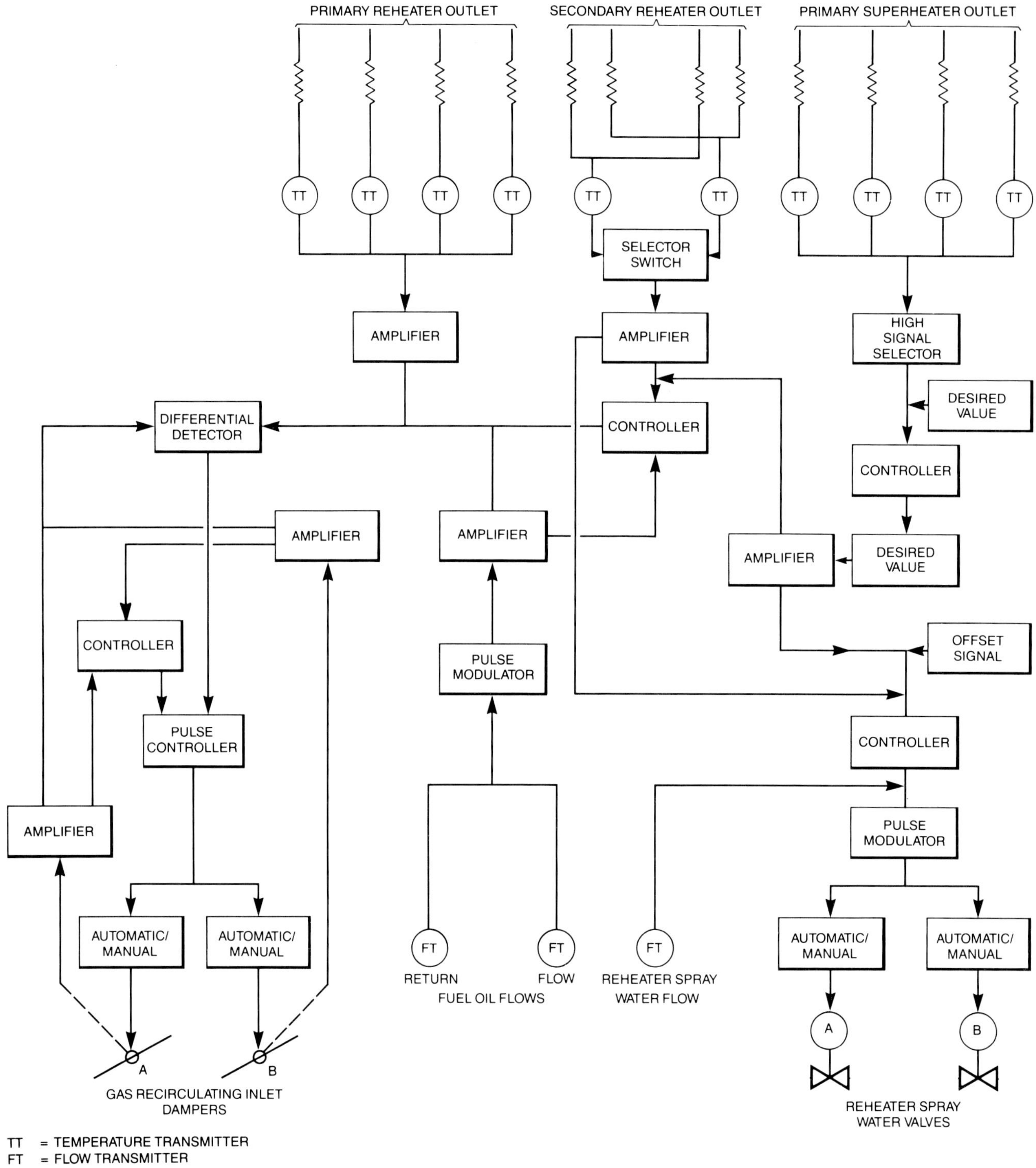

FIG. 6.65 Simplified reheater temperature control system

- Gas recirculation differential pressure across inlet dampers.
- Gas recirculation inlet damper position.
- Gas recirculation hydraulic coupling speed.

The commissioning procedure is as follows:

- Set the reheater steam temperature control to 'manual'.
- Adjust gas recirculation to obtain desired temperature and differential pressure across the GR fan inlet dampers.
- Transfer the gas recirculating hydraulic coupling control to 'auto'.
- Adjust the desired value to create small changes (approximately 5%) and adjust the controllers to obtain the desired response. Continue adjustment until 'stable oscillations' are obtained.

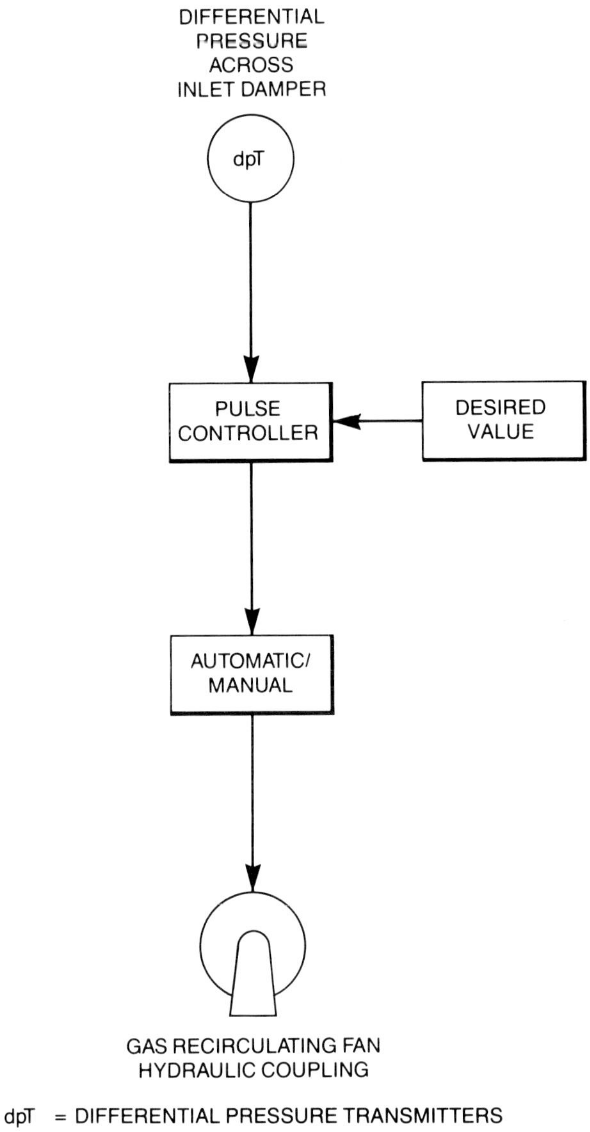

FIG. 6.66 Simplified gas recirculating fan speed control system

- Select a secondary reheater outlet temperature transmitter.
- Adjust the desired value of the reheater outlet temperature until the temperature measurements are indicating equal values.
- With the gas recirculation on 'auto' control, transfer the inlet damper status to 'auto' control.
- Introduce a temperature step-change at the reheater outlet and note the response between the demand signal and the damper position signal.
- Check for correct response and adjust pulse controller until 'stable oscillations' are obtained.
- Continue testing until optimum responses are obtained.
- Return the GR fan inlet dampers to 'manual' and adjust to create a minimum differential damper opening of 5%.
- Transfer back to 'auto' and confirm that the control loop returns the dampers to a balanced position.
- Adjust the pulse controller, if necessary.
- The test on the differential vane opening is carried out at both ends of the temperature range, i.e., low reheat and maximum temperatures.
- With the GR fan dampers and fans on 'auto', change the desired value of the reheater outlet temperature by 5% and note the response in obtaining the new desired value.
- Adjust the controller, if necessary, until stable oscillations occur.
- Carry out further tests by inducing temperature changes.
- Adjust the primary superheater high outlet-temperature protection control, and set to the desired value.
- Reduce source output until GR fan dampers start to close.
- Adjust the controller to obtain a smooth readjustment of the desired value.

Commissioning checks — reheater sprays

During this activity the following parameters must be monitored:

- Reheater outlet temperatures
- Reheater spray flow
- Reheater spray control valve position

The commissioning procedure is as follows:

- Confirm that electrically- and hand-operated valves are set for correct operation, with the electrically-controlled valves operating from the unit control desk.
- Confirm the air supply to the pneumatic control valve.
- Confirm that spraywater pipework is charged up to the control valves and that the valves are not passing.
- Adjust the reheater outlet temperature to check that the three indicators are all reading the same value.
- Adjust the additional reheater outlet desired value by approximately 5°C to create a difference in the indicated temperature readings.
- Select the chosen reheater spray valve to 'auto'.
- Adjust the additional desired value to initiate opening of the spray control valve.
- Adjust the controller until stable oscillations take place between reheater spray control valve and the reheater spray flow.

- Continue to adjust temperature control to obtain a value where the sprays open again and are adjusted to confirm that overspraying is not taking place.
- Test fully by making temperature changes in both directions until optimum responses are obtained.

5.10 LP feed system stability tests

5.10.1 Introduction

Direct contact (DC) feedheaters are used in low pressure feedheating systems because of their superior thermal efficiency compared with a tubed type of heater. The bled steam and condensate are mixed within the DC heater, such that all the bled steam is transferred to the condensate. The condensate will then leave the heater at the saturation temperature corresponding to the bled-steam inlet pressure. In the tubed type of heater, a temperature difference will always exist between the condensate and the bled steam due to the intervention of the tube wall.

The absence of any physical mechanical barrier between the condensate and bled steam in DC heaters could lead, under certain operating conditions, to reverse condensate flows along the bled-steam lines, with consequent risk of damage to the turbine blading. This reverse flow can occur due either to malfunction or to basic inherent instability created by the design.

This section describes the tests to be carried out at varying loads to establish the performance and stability of a DC feedheating system, particularly if it is the prototype of a new design.

5.10.2 General description

A typical LP feed system comprises three low pressure heaters, plus the other heat exchangers for turbine moisture extraction, hydrogen coolers and gland steam condensers, together with the de-aerator. Figure 6.67 shows, in simplified form, the basic constituents of an LP feed system.

Using DC heaters for improved thermal efficiency, the LP feed system would be as indicated but, even if the installation comprised tubed-type LP heaters, the LP system would be basically as shown.

Plant mode of operation

The condensate path, after passing through the condensate-cooled ancillaries and the turbine moisture extraction condensers, enters the first and lowest pressure DC heater. The three DC heaters are arranged in vertical cascade. The elevation of each heater above the next in the train is sufficient to allow condensate to gravitate from heater to heater, overcoming the heater pressure differential and the frictional pressure drops within the system. For the system shown, the centre lines of the heaters are DC1 at +44.66 m, DC2 at +30.0 m, DC3 at +17.8 m, and the de-aerator at +39.7 m.

Isolation between heaters is achieved by U-shaped connecting pipework, in which the condensate forms a manometric seal, providing a hydraulic head to support the pressure difference between the upper and lower heaters.

After passing through the DC heater train, condensate is then pumped by a DC heater extraction pump into the de-aerator storage vessel via control valves.

The flow of condensate to DC1 is controlled by a condenser level control valve which maintains the condenser level within specified control limits. In the event of high condensate levels being experienced in any of the DC heaters, high level control switches mounted on the centre line of each heater will initiate a trip. The operation of the trip will cause rapid closing (1 s) of the bled-steam valves, plus the closure of the condensate isolating valve in approximately 15 s. The closing time for this valve is set at 15 s to prevent hydraulic shock.

A DC heater by-pass valve is provided which enables a condensate flow rate of up to 50% MCR to be passed to the de-aerator in the event of any DC heater being out of service.

The bled-steam supplies are taken from the LP turbine stages and, if the main turbine consists of three LP cylinders, the bled-steam supplies are taken from each cylinder as follows:

DC1 — 4th stage No. 1 LP

DC2 — 3rd stage No. 2 LP

DC3 — 2nd stage No. 3 LP

The latest CEGB design of DC heater incorporates extensive modifications from the early design to improve system stability. These include deepening the seal loops between DC1/DC2 and DC2/DC3 to enable the system to accept larger disturbances without heater flooding. Orifice plates have been introduced in each seal to increase line resistance for forward and reverse flow to improve system damping.

The capacity of DC3 vessel has been increased so as to contain all condensate draining from DC1 and DC2, together with that entering DC1 as the condensate isolating valve closes.

Stabilising tanks have been fitted in DC2 and DC3 in place of the upper condensate dispersal trays to provide additional condensate storage and improve system stability.

Stability has been further improved by taking bled-steam tappings for the DC heaters from sources whose pressure cannot vary independently, i.e., from the LP cylinders. The provision of a condensate recirculation line to DC3 ensures that the impellers of the DC heater extraction pump are protected against cavitation at low loads by retaining a minimum water level in the heater. Such cavitation could result in significant heater level fluctuations, as well as damage to the pump.

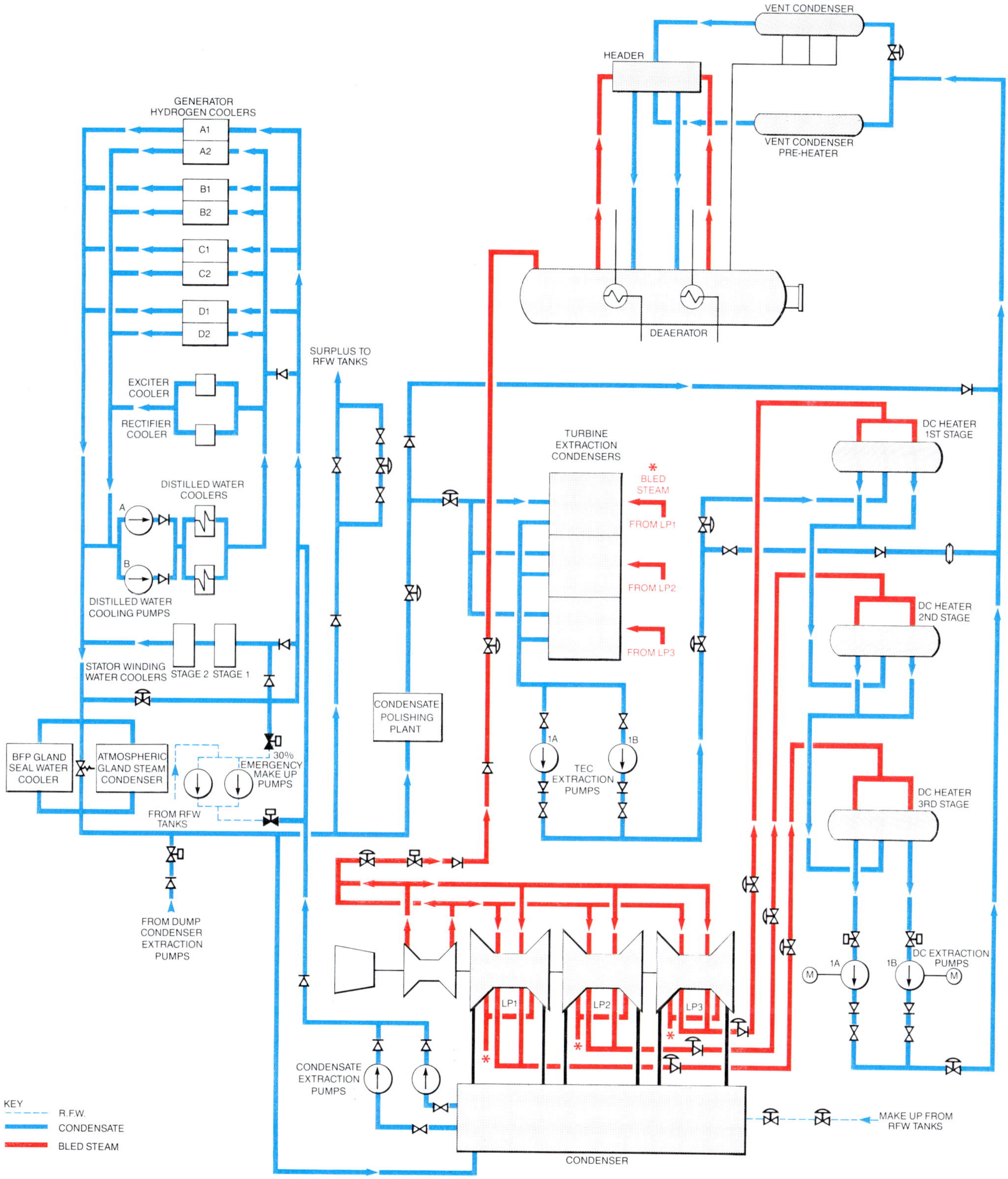

FIG. 6.67 Typical LP feedwater system

In the event of a turbine trip, the condensate isolating valves, the bled-steam isolating valves, together with the bled-steam non-return valves for each DC heater, are commanded to close, thus reducing the quantity of steam entrained in the system to a minimum.

5.10.3 Initial state of plant

The plant should be in the following operating condition:

- Turbine load at the selected value.
- Unit transformer out of service. Electrical supplies to be taken from the station transformer (this is a sensible precaution but is not essential).
- HP and LP feed systems fully in service.
- All standby plant selected and available for duty.
- The necessary test equipment available and in service, including the liquid level detection system on the bled-steam lines and UV recorders.
- During the actual testing, the valves associated with the bled-steam drain lines to be open.

5.10.4 Test procedure

The tests for the LP feed system are divided into three sections:

(a) Wet functional tests of individual plant items

(b) Dynamic functional tests of system

(c) Performance and system stability tests

Items (a) and (b) are described in Chapter 5 of this volume. The procedure for item (c) is described in the following test sequence:

- Checking individual bled-steam valve closing operation on each heater in turn.
- Reopening of bled-steam valves in reverse order.
- Overfeeding test on DC1 heater.
- Checking the operation of the standby DC heater extraction pump.
- Tripping of the DC heaters and operation of the plant on DC heater by-pass.

As the overall objective of the test is to confirm stability, it is necessary to carry out the above tests in graduated steps commencing with lower loads, analysing results before proceeding to the next load point, and finally carrying out the tests at 100% MCR. To enable the correct assessment to be made, the tests should be carried out at 40%, 60%, 80% and 100% MCR.

Bled-steam valve closure tests: bled-steam valve for DC1

- Carry out on-load test of all heater protection systems, including confirming the operation of DC heater by-passes.
- Close the DC heater 1 (DC1) bled-steam isolating valve.
- Observe closely and record the maximum and minimum water levels, together with pressures in DC2 and DC3.
- Confirm that the bled-steam valve has fully closed.
- Check that all bled-steam drain lines are free of water by information obtained from liquid level detection system.
- Check that all relevant alarms and trips have been correctly received and can be reset before proceeding.
- Once conditions have stabilised, reopen the bled-steam valve.
- Observe closely and record the maximum and minimum water levels, together with pressures in DC2 and DC3, as DC1 is reintroduced on the steam side.

The test is then repeated for DC2 and DC3, recording the water levels and pressures in DC1 and DC3 for the DC2 test, and in DC1 and DC2 for the DC3 test.

Bled-steam valve — reopening in reverse order

- Close each DC heater bled-steam valve in order DC1, DC2, DC3.
- Confirm that each valve is fully closed.
- Open DC3 bled-steam valve.
- Repeat checks on maximum and minimum water levels in DC heaters following the opening of DC3 bled-steam valve.
- Check that alarms and trips have operated correctly and can be reset.
- Repeat the above test opening DC2 and DC1 valves.

Overfeeding of DC1 heater

- Establish a higher than normal level in the condenser.
- Start the standby condenser extraction pump.
- Check the rise in water level in DC1, DC2 and DC3 and record the maximum levels reached.
- Check whether the standby DC heater extraction pump cuts in.
- Check that all alarms have operated correctly and can be reset.
- Shut down the condenser extraction pump and return it to its standby mode.

Operation of standby DC heater extraction pump

- Confirm that on-load testing of the heater protection system at this load point has been carried out; if not, carry out the test.
- Select the standby DC heater extraction pump to 'manual' and start the pump.
- Observe and record the maximum and minimum water levels experienced in each of the three DC heaters.
- Observe the operation of the pump discharge valves and recirculating control valves to ensure that no undue oscillations are set up in the DC heater train.
- When conditions have stabilised, check the steady state conditions with two DC heater extraction pumps running. Observe any detrimental effects of pipework vibration or movement.
- Shut down the standby DC heater extraction pump, reselect to 'standby'. Confirm that stable conditions return following the shutdown of the second pump.

Tripping of DC heaters and operation of the plant on DC heater by-pass

- Confirm that on-load testing of the heater protection system at this load point has been carried out; if not, carry out the test.
- Operate the DC heater high water level group 'close' button to trip DC heaters.
- Confirm that DC heaters have tripped and that DC heater feedwater by-pass valve has opened.
- Check that all DC heater bled-steam isolating valves and non-return valves have closed, together with DC heater condensate isolating valve.
- Monitor condenser and de-aerator water levels and ensure that the control system is operating satisfactorily in the by-pass mode and that the turbine can be maintained at the test load level.
- When steady state conditions have been established, reinstate the DC heater system and close the by-pass valve.

5.10.5 Establishment of operating techniques

It would be normal practice to expect that the maximum condensate flow capability of the DC heater by-pass would only support a 50% MCR output from the turbine.

For tests carried out at loads higher than 40% MCR, it will be necessary to de-load the turbine to a load within the by-pass limit. The commissioning tests, apart from proving the stability of the system, should also identify the longer term operational procedures for dealing with genuine DC heater trips.

A DC heater trip at all load levels above 50% MCR will leave the condensate system 'over watered' and, generally, with high condenser water level and low de-aerator level. This will be a transient condition that will require very close monitoring of both of these levels and the need to establish the correct de-loading rate with due regard to the capability of the DC heater by-pass flow limit. If the condensate system fails to cope with the de-loading rate, the unit can be expected to trip on low water level in the de-aerator.

5.10.6 Test results

Typical stability results at 100% MCR are shown in Fig 6.68. These results show not only the levels following from the trip in DC1, DC2 and DC3 but also the condenser and de-aerator levels, together with de-aerator temperature, pressure and load curves. The graphs also show a trip on high de-aerator level when reinstating heaters after trip, when full condensate stability had not been achieved.

The stability tests performed show that the DC feedheating system is stable and that operational procedures incorporated in the stability test programme were not considered to impose any severe disturbance on the system.

The test results shown relate to the 100% MCR test. Similar results will have been taken at the other test loads and analysed before finally proceeding to the 100% test.

The successful completion of this test clearly demonstrates the stability of the system and gives confidence in dealing with any genuine DC heater trips in the longer term operation of the plant.

5.11 HP/IP governor mode checks

5.11.1 Introduction

Modern power plant may have fitted two distinct modes of governor control:

(a) Conventional HP valve governing mode in which load is controlled by HP governing valves, with the IP valves providing a protective function.

(b) The HP/IP governing mode in which both the HP and IP governor valves participate in turbine control. This provides additional stored energy in the reheater to improve the emergency response, which is achieved by the IP governor valves operating partially closed.

This test procedure describes how the IP governing mode should be commissioned; the effect on IP throttle governing for the stored energy test is described later in this chapter under that heading.

5.11.2 Initial state of plant

- Turbine at full rated load under steady state conditions

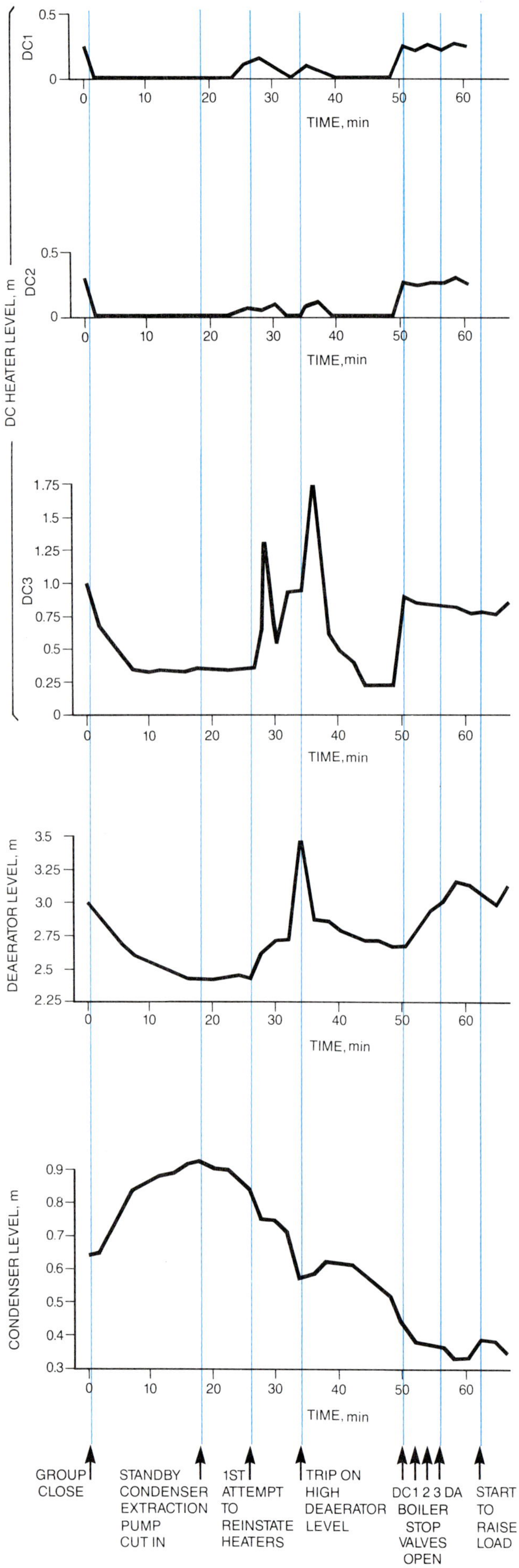

- HP feedheating to be operating to design conditions
- Main steam-driven boiler feed pump operating as designed
- Turbine operating on HP governor control

5.11.3 Test procedure

- Record the reheater steam pressure at full-load under HP governor control.

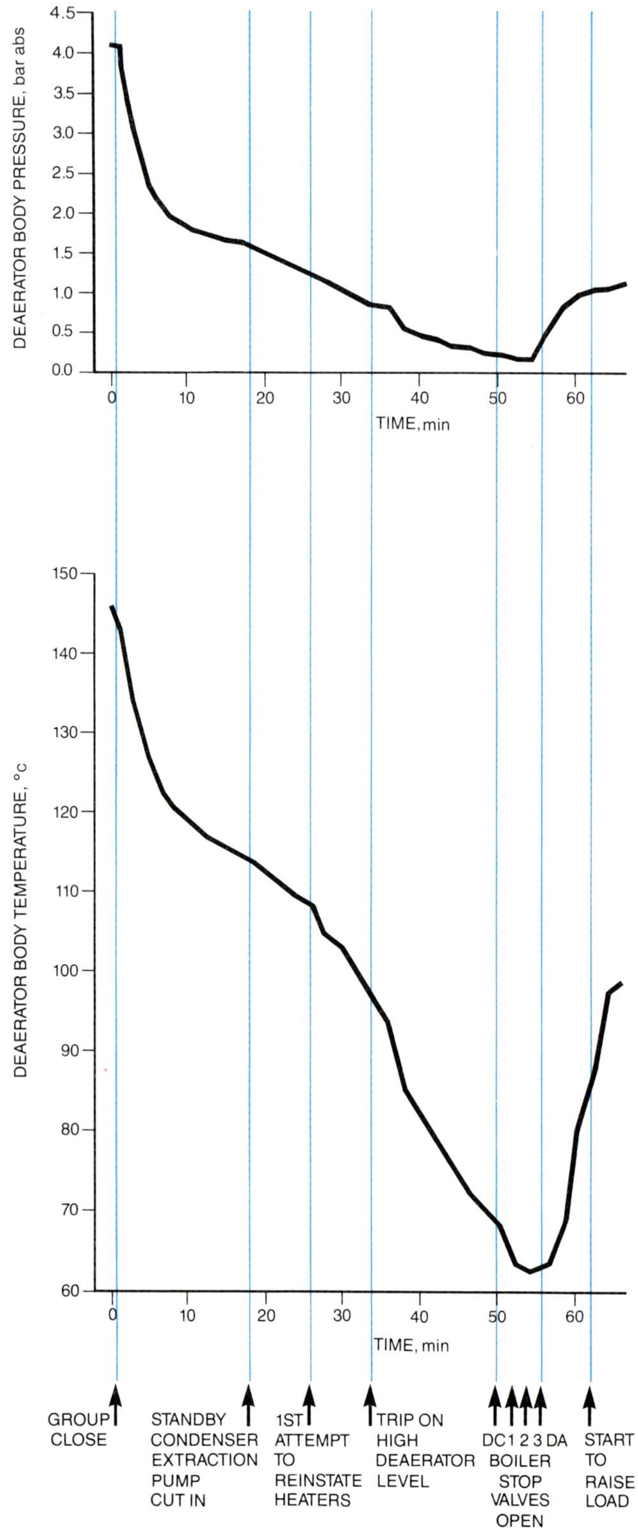

- Select the governors to HP/IP with the bias on the IP governor valve set to +5% (note that, under this condition, the IP governor valve will remain fully open and no change should be observed in the turbine load).
- Reduce the load on the turbine in 10% steps from 100% to 30% MCR. Record the following at each graduated load point:

  HP governor valve positions

  IP governor valve positions

  Reheater steam pressure

  Unit load
- Increase the load to 50% MCR and adjust the bias on the IP governor valves until the required reheater pressure is obtained.
- Change the load in both directions from 50% MCR in 10% steps to confirm that, for loads below 50% MCR, the reheater pressure falls and that, for loads above 50% MCR, the reheater steam pressure is sensibly maintained, with the IP governor valve fully open at 90% MCR.

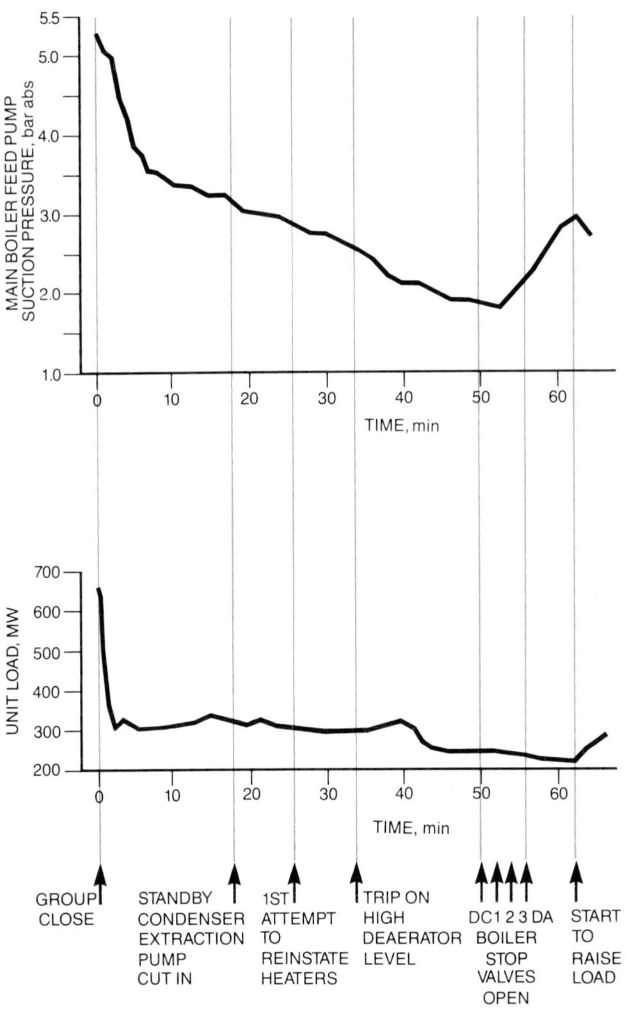

FIG. 6.68 LP feedwater system 100% MCR stability test

- Whilst checking the reheater pressure and load changing 50%–0–100%, record the valve positions of both the HP and IP governors.
- Whilst carrying out this test programme, it is essential to ensure that rated terminal conditions are maintained at each load point.

5.12 Unit 72-hour run

5.12.1 Introduction

The objective of the unit 72-hour run is to prove the capability of the boiler/turbine unit and its associated auxiliary plant to run at its maximum continuous rating for 72 hours.

Any reduction or cessation in unit output during the test will be considered by the appropriate parties to decide whether the loss of output was outside the plant supplier's responsibility and whether the plant may otherwise have met the test requirement.

Whilst this may seem to be a short period of running at full-load for a new unit, it is a defined point after which the unit is accepted for commercial operation, if successful.

5.12.2 State of plant

- Unit to be operating at 660 MW at rated power factor with all steam pressures and temperatures at design values.
- The boiler to be substantially leak-free and operating at its design condition.
- Plant to be operating normally, with all automatic control loops functioning correctly.
- Plant operation to be carried out with the normal complement of operating staff.
- No other testing of plant to be carried out during the 72-hour run.
- All auxiliary plant systems to be confirmed as operating correctly and any known defect liable to affect or hinder the test to be rectified prior to the commencement of the test.
- Nominated responsible persons to be identified by name, one from each main group, i.e., Turbine Contractor, Boiler Contractor, Project Engineer and Site Commissioning Engineer. They must agree in advance the ground rules for the test, as well as agreeing that the plant is in a condition for the test to commence.

5.12.3 Test procedure

The actual start time of the test is to be agreed by the nominated persons acting on behalf of the contractors and customer.

- The unit to be confirmed to be in a stable condition and operating at the test load by continuously monitoring the following parameters during the 72-hour period:

  Generated MW

  Reactive power

  HP steam inlet temperature and pressure

  IP steam inlet temperature and pressure

  Boiler superheater steam temperature and pressure

  Boiler reheater steam temperature and pressure

  Printout from the station computer system of all other plant parameters

- Continue generating at the agreed test conditions for 72 hours.
- On completion of the test, collect logs and other test data for analysis.

5.12.4 Rules to be applied to the 72-hour run

- The 72-hour run output to be judged on the examination of the half-hour summation metering and the adjustment carried out to correct small amounts of under-generation by over-generation. The limits of this correction are ±20 MW.
- The 72-hour run to be continuously monitored and facts recorded on a special log sheet. Reasons for any shortfalls and the actions taken must be noted.
- Any plant problems which occur during the test but do not cause a loss of load should also be recorded.
- Any plant problems which cause a large loss of load for periods in excess of 1 hour, but do not require the unit to be taken off load, should, in the first instance, be recorded together with the reason. The decision relative to the effect on the test will be decided by the nominated group. The likely decision would be to extend the test for a period of time equal to the shortfall.
- Any plant problem which causes the unit to be taken off load requires a decision from the nominated group. The decision will be influenced by the cause and the time interval since the test commenced. This could result in a decision that the 72-hour run has been completed or has to be restarted.

6 Commissioning tests after completion of the 72-hour full-load test

6.1 Introduction

Having completed the second main objective of the commissioning programme, namely synchronisation to full-load and completion of the 72-hour full-load test, the unit can now be delivered for commercial operation. The various tests carried out during the commissioning programme may have indicated problems which affect plant operation. When handing plant over for commercial operation, it is essential to identify any plant limitations, together with any specific temporary operating procedures, and defects which, although requiring correction, do not at that time affect plant operation. This information must be conveyed both in writing and verbally to the operating staff.

The completion of the 72-hour full-load test still leaves the third main commissioning phase to be completed as identified in Section 1.2 of this chapter. This final phase completes the outstanding contractual performance and operating type tests and ends by demonstrating the ability of the station to bring a unit onto load with the external grid system de-energised, using the installed gas turbine or diesel generators. This is known as a 'dead station start'.

The test programme for this final phase incorporates the following:

- Boiler efficiency and steam temperature control tests
- Unit guaranteed heat rate and overload test
- Unit flexibility tests
- Load rejection tests
- Automatic synchronisation checks
- Operating trials
- Dead station start

6.2 Boiler efficiency and steam temperature control tests

6.2.1 Introduction

The boiler efficiency and steam temperature control tests form the basis of the contractual plant acceptance tests and occur as soon as possible after the completion of the full-load 72-hour test. Figure 6.69 shows a programme for the proposed tests, which includes seven days of plant cleaning and preparation. It also identifies the start-up and shutdown tests which are discussed in Section 6.6.6 of this chapter.

The programme comprises:

- Boiler efficiency tests at 100% and 80% MCR
- Steam temperature control tests

This section describes the steps necessary to set up the boiler plant for the test, but does not include the method of calculation which is described in Volume G, Chapter 7.

6.2.2 Scope of the tests

The tests have the following aims:

- To demonstrate the guaranteed boiler thermal efficiency at boiler maximum continuous rating

(BMCR). This test may be carried out at any load within the range 90% → 100% BMCR.

- To demonstrate that the boiler is capable of supplying steam at the guaranteed BMCR: a typical value of BMCR for a 660 MW unit is 579 kg/s.
- To demonstrate the attainment of full superheater and reheater steam temperature at a load below BMCR, normally 70% BMCR.
- To demonstrate the automatic superheater and reheater steam temperature control, as load is varied at a rate of 5% full-load per minute over the range 70% → 100% → 70% BMCR under normal operating conditions.
- To demonstrate the performance of the steam temperature control equipment under overload conditions whilst maintaining the BMCR of 579 kg/s but with the economiser feed inlet temperature depressed by approximately 60°C. This test is best carried out as part of the unit guaranteed heat rate programme (Fig 6.70), which includes an overload test, following the completion of the boiler test programme. By adopting this approach, time will be saved in the overall programme.
- To demonstrate the performance of the combustion control equipment at 100% BMCR and during load changes.
- To check dust emission tests at 100% BMCR and 70% BMCR.
- For coal-fired plant, to test precipitator efficiency at 100% BMCR.

6.2.3 Method of testing

The boiler efficiency tests are conducted in accordance with the recommendations contained in BS2885 'Code for acceptance tests on stationary steam generators of the power station type, relating to the determination of thermal efficiency by the heat loss method'.

The gross thermal efficiency of the boiler is determined by summing the measured losses and the specified values for radiation and unmeasured losses. Typical values at 100% MCR are 0.25% for the unmeasured losses and 0.3% for the radiation and convection losses.

The guaranteed steam output and thermal efficiency of the boiler are based on certain operating parameters. Typical values are as follows:

Feedwater temperature to the economiser	254°C
Steam pressure at boiler stop valve	167 bar (absolute)
Reheater and superheater steam temperature	541°C
Air temperature at the inlet to the regenerative air heater, with the bled-steam air heater in service	32°C

The fuel being burned must represent a typical basic design fuel.

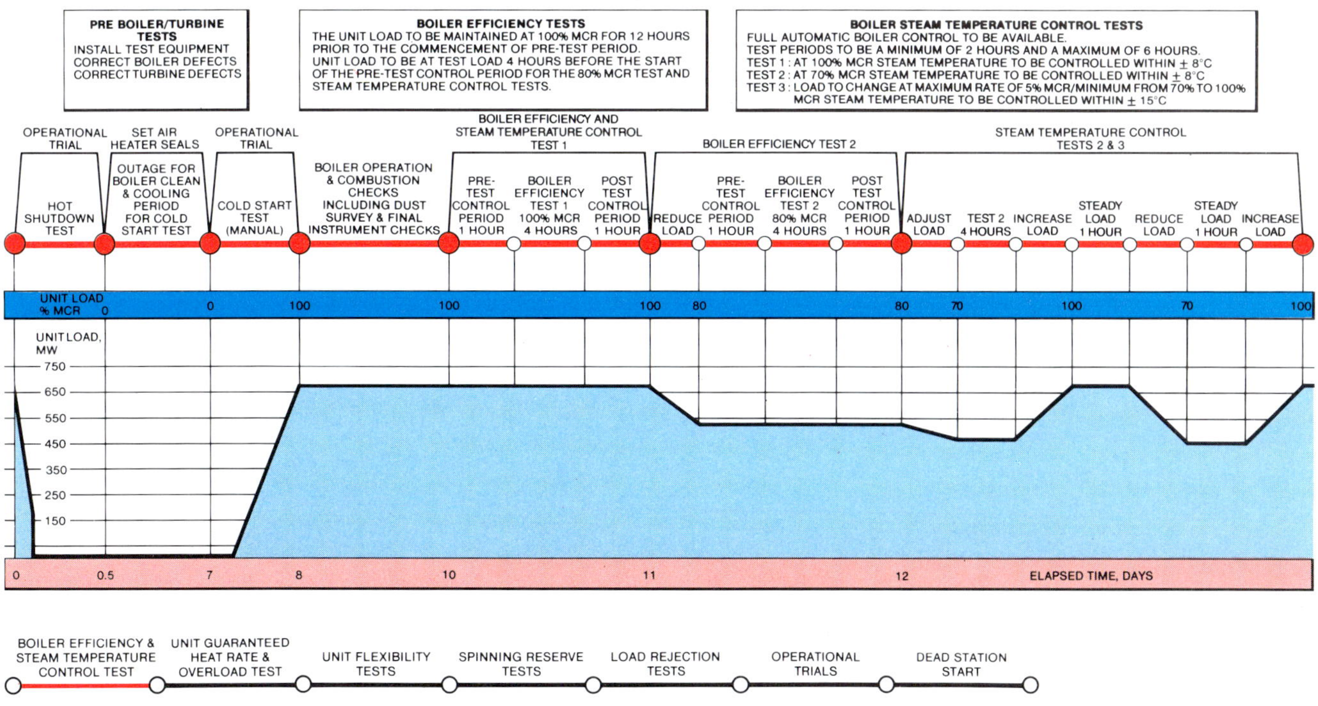

FIG. 6.69 Boiler efficiency and steam temperature control tests

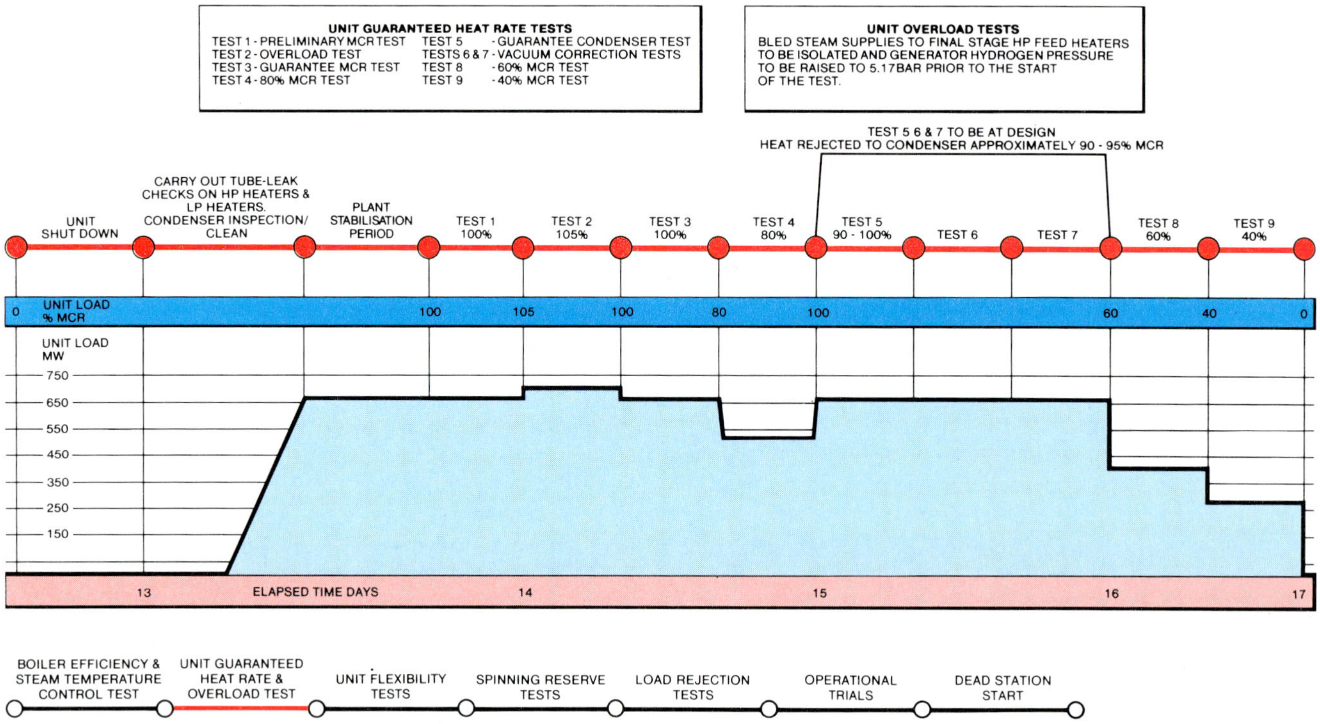

Fig. 6.70 Unit guaranteed heat rate and overload tests

6.2.4 Pre-test activities

Prior to the commencement of the 72-hour full-load test, the test equipment necessary for the boiler performance tests should have been installed as far as is practicable and its performance monitored during the full-load operation.

During the 72-hour run, the boiler performance and the accuracy of the commercially fitted instrumentation is checked. An ID fan duct probe survey should also be carried out to determine representative measuring positions for flue gas sampling.

The gas loss across the air heater seals is checked during the 72-hour run to enable any necessary adjustments to be completed during the pre-test shutdown period.

In a PF boiler, coal and air distribution checks are completed on each PF burner by the end of the 72-hour run to enable any necessary adjustments in the PF distribution to take place.

During the period of the 72-hour run, agreement should be reached between the contractors and the customer that the plant is in a satisfactory condition for the test to take place, subject to an agreed work programme during the boiler efficiency pre-test outage. A similar agreement is needed after the outage, during the two-day period of boiler operation and combustion checks prior to the commencement of the first pre-test period.

During the programmed shutdown identified in Fig 6.69, the following operations/checks should be carried out:

- Joint inspection by customer and contractor of the boiler gas passes to determine the state of cleanliness and to agree a cleaning programme, if necessary.
- To set, or adjust, the gas air heater seals in line with the gas leakage rate checks to enable the leakage rate to be controlled at approximately 4%, or the manufacturer's design figure.
- Cleaning of air heater elements, if necessary.
- To review the overall condition of the plant.
- To complete the installation of all the test instrumentation.
- To rectify all the defects associated with the boiler and turbine that would affect plant performance.
- For an oil-fired boiler, to clean and calibrate all burners. Carry out rig flow checks to confirm that each burner is within tolerance.
- For a PF boiler, to rectify known PF leaks and any other problems associated with operation of the milling plant.
- Agreement should be reached between the contractor and customer that the fuel to be used for the test falls within the general specification for the design.

6.2.5 Test programme

The boiler efficiency and steam temperature control test programme consists of the following:

- Boiler outage for approximately seven days to clean, rectify defects and install test equipment.
- With the unit returned to full-load, a period of two days checking boiler operation and combustion, including dust emission tests, a final survey and the checking of all instrumentation.
- Boiler efficiency Test 1 at 100% MCR.
- Steam temperature control Test 1 at 100% MCR, confirming that designed steam temperature can be maintained at design values within ±8°C.
- Boiler efficiency Test 2 at 80% MCR.
- Steam temperature control Test 2 at 70% MCR, confirming that full designed steam temperature can be maintained within ±8°C.
- Steam temperature control Test 3. Load changed at 5% MCR/minute from 70% → 100% → 70% BMCR to confirm that the steam temperature can be controlled within ±15°C.
- During all the tests, the boiler should be operating on full automatic control. During the steam temperature control Test 3 load changes, the combustion control capability of the auto systems should be checked.

6.2.6 State of plant

Before the pre-test control period for boiler efficiency Test 1, the unit should be maintained at 100% MCR for 12 hours to establish stable plant conditions.

The state of the plant prior to the commencement of the control period for each test is as follows:

- The boiler to be operating at the test load in a steady state.
- Any programme of boiler sootblowing must be completed.
- Any programme of boiler water blowdown for chemical control must be completed.
- Any steam supply to other services should be isolated.
- For an oil-fired boiler, all oil burners in service.
- For a PF-fired boiler, the designed number of mill groups for full-load to be in service.
- The turbine governor control to be selected for 'HP governing' only.
- All LP and HP heaters and the main boiler feed pump to be in service.
- All the unit standby auxiliaries to be available but not in service.
- The unit computer to be in operation.
- The boiler automatic control systems to be fully in service.
- If computer loading control is installed, it should be available for service during the steam temperature control Test 3.
- Any flue gas additives injected into the flue gases as part of normal operation to be shut down at least 1 hour before tests involving dust emission measurements.
- The bled-steam air heater to be in service, with steam supplied from the auxiliary boilers.

6.2.7 Boiler efficiency Test 1 at 100% MCR and steam temperature control Test 1

At the commencement of the pre-test control period:

- Ensure that the boiler has been at the test load for the stated time.
- Ensure that the state of the plant is as listed in Section 6.2.6 of this chapter.
- Ensure that the unit is operating in a stable condition.
- Ensure that the necessary test personnel are stationed and ready to take readings or any other duties, as detailed.

At a time agreed between the parties, commence the 1-hour pre-test control period.

- Time-mark all relevant chart recorders.
- Take an initial set of readings.
- Maintain the boiler in a steady operating condition at the test load.
- Repeat the readings every 15 minutes.
- After 1 hour, providing operating conditions have remained stable and by agreement with the parties concerned, the 4-hour boiler efficiency Test 1 and the steam temperature control Test 1 at 100% MCR can commence.
- Take the first set of readings.
- Repeat the readings every 15 minutes throughout test.
- After 4 hours at steady load, the test will be considered complete.
- On completion of the test, commence the 1-hour post-test control period. This allows a quick check to be made of the various results obtained whilst the boiler is still operating within its test limits. Any problems with results can be rechecked. If necessary, the post-test control period can be extended.

- Continue to take readings at 15-minute intervals throughout the post-test control period.
- At an agreed time, terminate the control period and notify all staff involved in the test.
- Collect all log sheets, charts, etc., produced during the overall test period.

6.2.8 Boiler efficiency Test 2 at 80% MCR

On completion of the 100% boiler efficiency and steam temperature control tests, reduce load to 80% MCR. Maintain the boiler at the 80% MCR load point for a minimum of 4 hours prior to the commencement of the pre-test control period for this test.

The state of plant for this test is identical to that specified in Section 6.2.6 of this chapter.

The test programme used is the same as that stated in Section 6.2.7 of this chapter for the 100% boiler efficiency test, but at 80% MCR.

The completion of the 80% boiler efficiency test concludes the efficiency testing.

6.2.9 Steam temperature control Test 2 at 70% MCR

On completion of the 80% boiler efficiency test, reduce the load to 70% MCR for the purpose of checking that the superheater and reheater steam temperatures can be maintained within the tolerance of ±8°C.

The state of plant specified in Section 6.2.6 of this chapter also applies to this test.

Maintain the boiler at the test conditions for a period of four hours, during which period both the superheater and reheater steam temperatures should be continuously monitored to confirm that full design temperatures are being maintained within the specified tolerance.

6.2.10 Steam temperature control Test 3

Load changing from 100% to 70% BMCR

- On completion of the steam temperature control Test 2, increase unit load to 100% BMCR and maintain stable conditions for a minimum period of one hour.
- The state of plant for this test is as specified in Section 6.2.6 of this chapter.
- Confirm that all the boiler automatic control systems are functioning correctly and that the superheater and reheater steam temperatures are being maintained at the design values.
- Ensure that test personnel are in position, mark all relevant recorders and take an initial set of readings.
- Commence the test programme by setting the computer load control program to achieve a deloading rate of 5% MCR/minute, from 100% to 70% BMCR.
- Initiate the deloading program and monitor the rate of deloading, together with the degree of control exercised by the automatic temperature control systems over the superheater and reheater steam temperatures. These temperatures should be controlled within a tolerance of ±15°C during the load changing programme. Accurate monitoring of the steam temperatures is required, so the temporary installation of high speed, high accuracy test instrumentation is essential.
- Maintain the boiler at the 70% BMCR load point for a minimum of one hour to enable stable conditions to be obtained prior to the commencement of the load ramp 70–100% BMCR.

Load changing from 70% to 100% BMCR

- On completion of the 1-hour steady state test period at 70% BMCR, set the computer load control program to achieve a loading rate of 5% MCR/minute from 70% to 100% BMCR.
- Initiate the loading program and monitor the same parameters as those specified for the deloading test.
- On completion of the load ramp, maintain stable conditions at 100% BMCR and monitor boiler performance during the period.

Overload to 70% BMCR

As already stated, this part of the test should be incorporated into the unit guaranteed heat rate and overload test on the completion of the overload test. By adopting this policy, the unit is already operating on overload without the HP heaters in service; hence there is a financial saving through not having to operate unnecessarily at a lower thermal efficiency.

The overload test requires the boiler to operate at 100% BMCR with the feed outlet temperature reduced by approximately 60°C from its normal design value. With a typical design value of 254°C, the economiser inlet feedwater temperature would be 194°C for this test. The other effect of taking the final stage HP heaters out of service is to increase the output of the unit from 660 MW to approximately 705 MW.

The increase of 45 MW represents the increase in steam flow through the reheater, IP and LP cylinders.

The unit will have stabilised during turbine testing at the overload condition. During this time, the superheater and reheater steam temperature should be checked to be controlled within ±8°C.

On completion of overload test, set the computer deloading program to deload at 5% MCR/minute from 705 MW to 70% BMCR (i.e., 462 MW).

Proceed as previously stated for the deloading tests.

Monitor the steam temperature to confirm that control within ±15°C has been maintained during deloading.

6.2.11 Combustion control

During the 100% and 80% boiler efficiency tests, the boiler is set for optimum conditions, using the minimum amount of excess air compatible with good combustion. The efficiency test includes dust emission tests which helps to assess the quality of combustion. During the steam temperature control tests, the opportunity should be taken to monitor the level of combustion control exercised by the boiler automatic systems during load changing. Monitoring the dust emission levels during the load changes, together with the levels of O_2 and CO, permits further assessment of the combustion control being exercised by the boiler automatic controls. The test may indicate inadequate response in the control and hence the need to re-time the automatic controls and repeat the tests.

The dust emission tests also enable the dust emission level recorder to be calibrated.

6.3 Unit guaranteed heat rate and overload test

6.3.1 Introduction

The unit guaranteed heat rate and overload tests form the basis of the contractual turbine plant acceptance tests and take place as soon as possible after completion of the full-load 72-hour test. As an ageing deterioration allowance of 0.1%/month is included in UK contracts, starting 3 months after joint synchronising, it is essential that this test programme should be achieved as soon as possible.

Figure 6.70 shows the programme following immediately after the boiler efficiency and steam temperature control tests. The programme for a 660 MW unit comprises the following series of tests:

Test 1 — preliminary guaranteed heat rate test at 100% MCR

Test 2 — overload test at 705 MW

Test 3 — guaranteed heat rate test at 100% MCR

Test 4 — guaranteed heat rate test at 80% MCR

Test 5 — guaranteed condenser test

Tests 6 and 7 — vacuum correction tests

Test 8 — guaranteed heat rate test at 60% MCR

Test 9 — guaranteed heat rate test at 40% MCR

Tests 2, 3 and 5 are contractual requirements, the other tests being necessary to obtain basic data relevant to the longer term operation of the plant. If the main turbine governor has the facility of HP/IP governing control, i.e., the IP governor valves throttle at loads below 90% MCR, a test should be carried out at a suitable load point to obtain comparative efficiency results.

6.3.2 Pre-test activities

Prior to the shutdown for boiler cleaning and defect rectification, and in preparation for the boiler efficiency and steam temperature control tests, the plant is checked for internal and external leakage on the HP feed and condensate systems so that they can be rectified during the shutdown. These checks include:

- With the de-aerator at maximum working level and isolated from all external sources of make-up water, monitor the rate of fall and calculate the make-up loss requirement. The normal duration of each guaranteed heat rate test is 4 hours and during the test the make-up water supply is isolated, so it is essential to confirm, prior to the start of the test programme, that the make-up losses are low enough to enable the unit to operate for a 4-hour period under test conditions.
- Check the test measuring devices to ensure that there are no by-pass or recirculating flows. If these are detected, repairs should be effected.
- The operating point of the feed-pump 'leak-off' valves must be established and the valve leakage in the closed position assessed.
- Check the HP heaters for any evidence of tube leaks, and repair if necessary.
- Check the cleanliness of the condenser tube plate, and clean if necessary.
- Check that the HP heater drains control system is functioning correctly for all loads in the test programme.
- Draw up a comprehensive list of plant defects in sufficient time to allow rectification to take place.
- During the shutdown prior to the boiler efficiency tests, install the special measuring devices for the turbine efficiency test. These include test thermocouples, orifice plates and monitors. All test equipment should have its calibration certificates checked prior to installation.
- All flow paths not included in the formulation of the guaranteed heat rate must be isolated during the actual tests unless they are essential for the safe operation of the plant. Where flows cannot be isolated, a flow rate and necessary corrections is agreed by all parties connected with the test.
- An agreed 'test isolation' schedule is prepared in advance of the tests and successful operation of the unit with these test isolations in service demonstrated before the test programme commences. Table 6.23 identifies the necessary test isolations.

TABLE 6.23
Unit guaranteed heat rate test isolations

The isolations are divided into four geographical routes:
- HP heaters and turbine
- De-aerator and feed pump suction
- Boiler feed pumps
- Make-up

HP heater/turbine
- Flange warming valves
- HP heater standpipe vents
- HP heater baffle plate drains
- HP heater vents (steam)
- HP heater start-up drains
- HP heater drains vessel drains
- HP heater inlet air release
- HP heater condensate inlet/outlet drains
- HP heater standing drains
- HP/LP drains vessel drains
- HP heater drains to condenser and condenser flashbox
- Hot reheat warming drains
- Main steam warming drains on each steam leg
- Main steam/feed pump leg drains
- Hot reheat drains
- Cold reheat drains
- IP steam chest drains
- HP loop pipe drains
- Gland steam drains
- HP and IP chest trap by-pass valves
- Condenser standby vents
- Bled-steam drain valves on LP feed system
- Condenser standpipe vents
- Condensate dump valve
- Extraction pump suction and discharge drains
- Make-up line drain
- Hydrogen cooler inlet/outlet vents and drains
- DW cooler drains
- Stator water cooler vents

De-aerator feed pump suction
- Spraywater to de-aerator bled-steam pipe
- DC heater extraction pump air release
- De-aerator standpipe vents
- De-aerator filling valves
- De-aerator draining valves
- De-aerator vent to condenser
- Auxiliary steam to de-aerator control
- Tubeplate sealing tank make-up, drain and overflow
- Boiler feed pump suction filter drains and by-pass
- Condensate to de-aerator drain
- DC heater extraction pump discharge drains
- H_2 cooler by-pass valve
- Tubeplate sealing make-up, by-pass, drain and overflow
- BFPT exhaust to condenser
- Injection tank drains
- Cooler shell drains
- Cooler air vents

Boiler feed pumps
- Boiler feed pump pressure stage drain
- Boiler feed pump loop pipe vent
- Spraywater to de-aerator bled-steam pipe
- Feed regulating valves, vents and drains
- Main steam feed pump leak-off isolations to de-aerator and drain vent
- Boiler feed pump pipe drain
- Boiler feed pump pipe air release
- 50% boiler feed pump suction stage drain and vents
- 50% boiler feed pump loop pipe vent
- 50% boiler feed pump pressure stage drain and vents
- 50% boiler feed pump gland drain tank overflow
- 50% boiler feed pump gland seal supply vent
- 50% boiler feed pump leak-off isolators and drains
- Boiler feed pump live steam and bled-steam chest by-pass and drains
- Gland steam by-pass valves
- BFPT exhaust by-pass

Boilers
- Superheater attemperator spray air release valves
- Reheater attemperator spray air release valves
- Drum, superheater and reheater air release valves
- Secondary superheater drain valves
- Air heater valves
- Hydrostep drains and vents
- Economiser inlet air release valves
- Boiler filling and emptying valves

Make-up
- Make-up inlet and control isolating valves
- Make-up by-pass valves
- Surplus to RFT control, isolating and drain valves
- Condenser recirculating isolator, by-pass and drain

This table gives guidance to areas which must be considered when drawing up a test isolation schedule. It is not exhaustive.

- The checks and tests identified in this section are part of the overall plan to ensure that the plant is operating correctly without any unmeasured or unknown leakage paths. A considerable number of these items will require to be checked again just prior to the commencement of the actual test programme.

In addition, at a suitable point in the overall commissioning programme and in advance of the efficiency test programme, the overload capability of the unit must be demonstrated. This is achieved with the unit generating at its full-load rating of 660 MW by closing the bled-steam isolating valve on each of the final stages of HP heating in turn (Fig 6.40 shows a typical HP heater steam system) and recording the increase in output for each heater. With the overload conditions established, monitor the following:

- Boiler steam and water temperatures
- Boiler metal temperatures
- Stator winding temperatures
- Stator cooling water temperatures
- Turbovisory conditions
- Turbine steam and metal temperatures
- Generator output
- Generator stator current
- Generator stator core temperatures

During the overload test, for periods not exceeding 2 hours, the maximum stator current may be permitted to increase to a higher value, if system conditions demand this, providing that the generator frame hydrogen pressure is also increased. Typical values for a 660 MW unit are:

Stator current increased from 19 500 A to 20 700 A.

Stator hydrogen pressure increased from 4.1 bar to 5.2 bar.

The overload should be maintained for a minimum of 2 hours and any operating parameter limits that are exceeded should be recorded and their effects considered when preparing operating procedures for overload operation.

6.3.3 Methods of measurement

The various items of data for the test are obtained as follows:

Boiler feed Boiler feedwater flow is measured by an orifice assembly located in the line from the de-aerator outlet. This is fitted specially for the test and removed immediately on its completion for recalibration checks.

Steam flow to the main boiler feed pump Nozzle flow is measured by an assembly located in the boiler feed pump steam supply pipework from the cold reheat line.

HP live steam flow to the main boiler feed pump Tests are carried out with the feed pump operating on reheat steam; however, it is normal practice to keep the live steam chest at temperature by maintaining a small steam flow at all times. This flow needs to be calculated and the value of flow agreed by the parties involved in the test.

Gland leakage steam Measured by orifice plates which may be fitted permanently.

HP exhaust steam to IP rotor cooling Orifice plates are installed in the pipeline to the IP rotor cooling.

Spindle leakage of HP governor valves When the governor valves are not fully back-seated, there will be leakage from the valve spindles. Orifice plates are installed to measure this leakage.

Boiler feed pump gland drains flow A flowmeter by-pass loop is installed between the gland drains tank and the condenser.

Pressure measurement The inlet steam pressure to the HP/IP cylinders is measured by test dead-weight gauges. Other pressures are obtained from calibrated pressure transducers and standard gauges. For vacuum measurements, mercury-filled U-tube manometers fitted with verniers are used.

The turbine LP exhaust pressure is recorded by a mercury-filled precision U-tube manometer.

Temperature measurements These measurements are obtained by using calibrated iron/constantan or chromel/constantan thermocouples inputting to multi-channel data loggers with typewriters and other print-out facilities.

The main cooling water temperatures are measured by installing sampling tubes connected to a mixing chamber where the water temperature can be measured using precision grade mercury-in-glass thermometers.

Electrical measurements The generator electrical output is measured using the two-wattmeter method. For confirmation of accuracy, two single-element and one double-element class 1 watt-hour meters are used, backed up by indicating wattmeters.

Calibrated class IAL current and voltage transformers are used with precision grade ammeters and voltmeters.

6.3.4 Test programme

The unit guaranteed heat rate and overload test programme comprises the following series of tests, which take 5 days to complete. Before starting Test 1, check the HP heaters and the LP heaters (if tubular heaters) for evidence of tube leakage. Check the gland seal cooler for leakage and the condenser tubeplate for cleanliness.

- *Test 1: preliminary test at 100% MCR* The objective of this 2-hour test is to confirm that all the test equipment is functioning correctly and that plant conditions are satisfactory for the test to commence.

- *Test 2: overload test* This demonstrates the capability of the unit to sustain a load higher than its normal maximum rated output for short periods of time. During this test the plant must be closely monitored to ensure that permissible operating conditions and limits are not exceeded and that any factors limiting the level of overload or its duration are highlighted.

- *Tests 3, 4, 8 and 9: unit guaranteed heat rate tests at 100%, 80%, 60% and 40% MCR* The tests below 100% are used to establish the slope of the Willans Line (see Volume G, Chapter 7).

- *Test 5: guaranteed condenser test* This measures the heat rejected to the condenser with the condenser CW outlet valves throttled to obtain the designed CW temperature rise across the condenser.

- *Tests 6 and 7: vacuum correction tests* These are carried out with the designed heat rejected to the condenser, (a) with the condenser valves fully open and (b) with the condenser vacuum reduced by approximately 25 mbar by admitting a quantity of air into the condenser. These tests confirm the vacuum correction curves.

6.3.5 State of plant

Prior to the start of Test 1, the unit should be kept at its first test load condition for four hours to establish stable plant conditions.

The state of the plant prior to the start of the test programme should be as follows:

- Any programme of boiler sootblowing completed.
- Any programme of boiler water blowdown for chemical control completed.
- Any steam supply to other services must be isolated.
- Bled-steam air heater should be operating on auxiliary steam.
- Any mechanical filters on the feed pump suction lines are to be in service with automatic backflushing isolated.
- Standby feed pumps isolated and drained.
- Main boiler feed pump leak-off valves to be isolated *only* if previous checks have indicated that the valves pass. They should be isolated at the last possible moment and opened immediately the test is complete.
- The 'test isolation' schedule applied, the final phase of isolation being to isolate the make-up and surplus valves.
- For an oil-fired boiler, all oil burners in service.
- For a PF-fired boiler, the designed number of mill groups for full-load to be in service.
- The turbine governor control to be set for 'HP governing'.
- All LP and HP heaters to be in service, together with main boiler feed pump confirmed to be operating on reheat steam only.
- All the unit standby auxiliaries to be available but not in service.
- The unit computer to be in service.
- The boiler automatic control systems fully in service.
- If computer loading control is installed, it should be in service to maintain constant load irrespective of frequency changes.

6.3.6 Test procedure

The test programme identifies a series of seven tests, the basic procedure for each being similar, varying only by the load and the objective of the test. The general procedure for all the tests is:

- Confirm that all pre-test preparations are complete and that the test isolation schedule (Table 6.23) has been applied.
- Confirm that the state of plant conforms to Section 6.3.5 of this chapter.
- Ensure that test personnel are in position.
- Confirm with all parties concerned that the plant is running steadily and that it is in a condition for the heat rate tests to commence.
- Commence pre-test period, taking all readings at 5 minute intervals.
- On completion of pre-test period, confirm with appropriate parties that the full test can commence, which will last for 2 hours.
- On completion of test maintain test conditions for a further hour.
- On completion of test period, collect all test log sheets.

A summary of typical test results is given in Table 6.24 which includes the test results for both HP and HP/IP governing modes. From this, the loss of efficiency when operating in the HP/IP mode can be seen.

6.4 Unit flexibility tests

6.4.1 Introduction

The unit flexibility tests are designed to establish confidence in the operational flexibility of the unit prior to the two-shifting trials. The tests are designed to impose a sequence of increasingly onerous tasks upon the unit, providing information and experience for the succeeding tests.

The test programme comprises the following:

- Stored energy tests
- Firing rate response test
- Ramp loading tests
- Step loading tests
- Spinning reserve test

Figure 6.71 shows the programme for the stored energy and firing rate response tests.

Figure 6.74 shows the programme for the ramp loading and step loading tests.

Figure 6.77 shows the programme for the spinning reserve test.

This series of tests follows the completion of the boiler and turbine efficiency tests and is preceded by a shutdown for the removal of the test instrumentation, specifically those flow assemblies which need to have their calibration checked before the efficiency test results can be finalised.

The plant conditions specifically required for each test are identified under the 'State of plant' heading for the test. In addition, the general plant conditions that are common to all tests, include the following:

- HP/LP feedheaters in service

TABLE 6.24

Typical efficiency test results for a 660 MW water cooled generator

	Governing modes							
	HP	HP/IP	HP	HP/IP	HP	HP/IP	HP	HP/IP
Load at generator terminals, less electrical auxiliaries, MW	660	660	528	528	396	396	264	264
Boiler feed pump input, MW	14.05	14.05	11.36	11.36	9.54	9.54	8.62	8.62
Heat loading of bled-steam air heater, MW	21.54	21.54	19.49	19.49	17.59	17.59	13.75	13.75
Steam pressure to bled-steam air heater, bar	4.89	4.89	3.93	4.14	3.00	3.17	2.10	2.34
Heat consumption reheating and feed heating, kg/kWh	8027	8027	8138	8261	8525	8712	9086	9286
Steam consumption reheating and feed heating, kg/kWh	3.02	3.02	2.97	3.13	3.04	3.27	3.16	3.61
Feedwater temperature at final feed outlet, °C	252	252	238	250.5	222	249	204	240
Turbine exhaust pressures, mbar absolute	35/43/60	35/43/60	30/36/47	31/37/47	27/31/39	27/32/40	24/28/33	25/28/33
Generator efficiency, %	98.6	98.6	98.65	98.65	98.63	98.63	98.42	98.42
Cycle efficiency, %	44.85	44.85	44.23	43.58	42.23	41.32	39.17	38.33

- Main boiler feed pump in service
- All oil burners or the full-load complement of PF burners in service
- Unit computer and data processor in service
- Governor speed droop setting at 4%
- Both AVR channels in service

6.4.2 Stored energy test

This test enables the stored energy of the HP and IP steam pipework to be assessed and demonstrates the short-term response to a 'spinning spare' demand. Boiler drum level swell may be detectable and the quality of the feed system controls to a demand disturbance will be demonstrated. Temperature deviations should be small, but the rapidly changing HP steam and feed pressure will exercise the ability of the attemperator outlet temperature control system in maintaining the correct spraywater flow.

The test programme comprises four separate tests for a turbine with HP/IP governing mode control. On each test, the load is increased at a loading rate equivalent to 5% MCR without any adjustment to the boiler firing rate (see Fig 6.71).

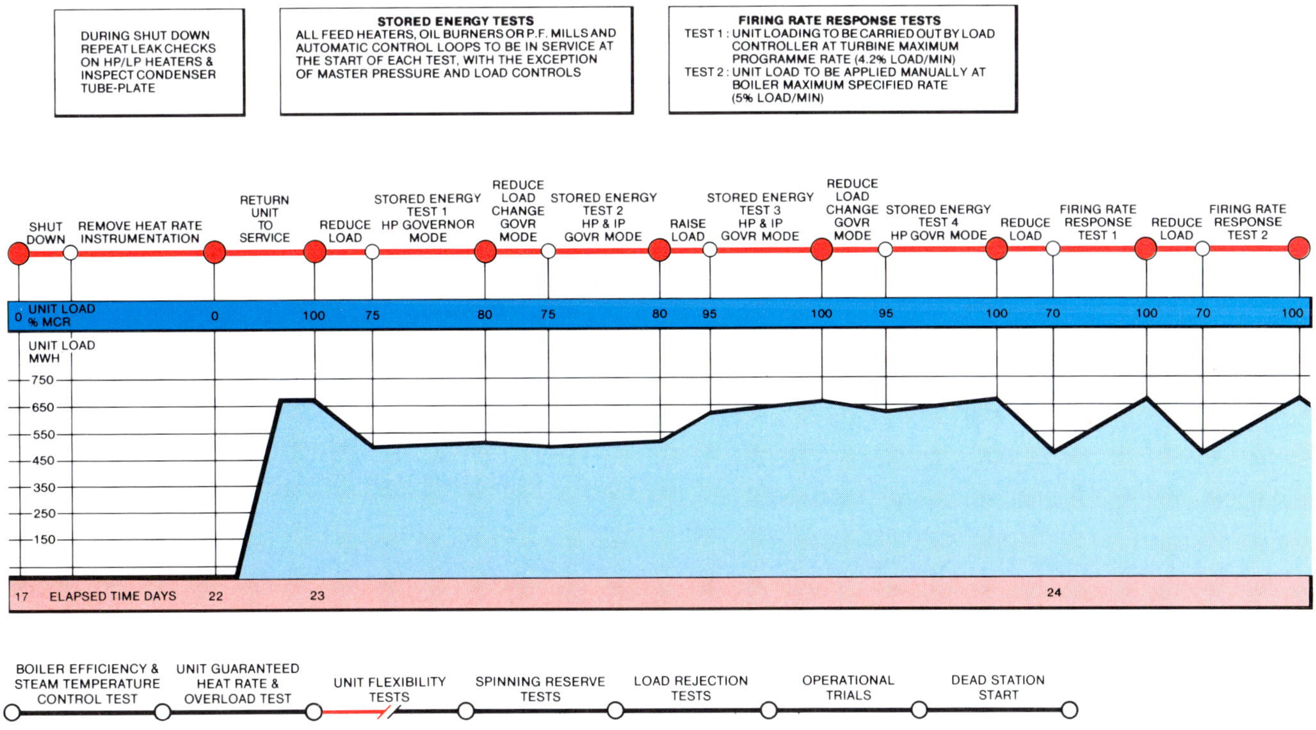

FIG. 6.71 Unit flexibility programme — stored energy and firing rate response tests

State of plant

- The initial load starting points for the four tests are:
  1. Unit load at 75% MCR with the turbine governor in HP mode.
  2. Unit load at 75% MCR with the turbine governor in HP/IP mode.
  3. Unit load at 95% MCR with the turbine governor in HP mode.
  4. Unit load at 95% MCR with the turbine governor in HP/IP mode.
- Boiler to be operating at rated pressure and temperature.
- All boiler automatic controls to be in service, with the exception of the boiler master pressure controller.
- Computer control of unit load to be out of service.
- Additional chart recorders may require fitting to measure systems frequency and turbine HP/IP governor valve positions.

Test readings

The following are continuously monitored and recorded.

Generator output MW and MVAr

System frequency

Superheater outlet steam temperature

Primary superheater outlet steam temperature

Attemperator outlet steam temperature

Reheater outlet steam temperature

Drum level

Total steam flow

Feed flow

Total oil flow to burners

Furnace pressure

PF mill loadings

Turbine differential expansions

Turbine metal temperatures

Superheater spray valve differential pressure

GR fan inlet damper differential pressure

Together with:

HP and IP governor valve positions

Spraywater flow

Boiler combustion oxygen and smoke density

Test procedure

Stored energy Test 1

- Confirm that unit is operating at 75% MCR with the turbine governor control in the HP mode.
- Confirm that the state of plant conforms to that described earlier in this section.
- Apply a load demand signal to the turbine governor equivalent to 5% MCR (33 MW).
- Allow the unit to pick up load without any manual adjustment to the firing rate.
- Carefully monitor the effects of the demand on the main boiler and turbine parameters and observe the control exercised by the boiler automatic systems.

The same procedure is adopted for Tests 2, 3 and 4, selecting the HP/IP governing mode for Tests 2 and 4. Tests 3 and 4 are carried out at 95% MCR.

Test results

The two main parameters, namely MW output and boiler steam pressure, for each of the four tests are shown in Fig 6.72. At 75% MCR in HP governor mode, the increased load is maintained for approximately two minutes at the additional 33 MW requirement before gradually falling over a total time of seven minutes to the original load point. Comparing these with the 75% MCR HP/IP governor control mode shows the increased load maintained for approximately 5 minutes before gradually falling over a total time of 9 minutes to the original load. A similar pattern can be seen for the 95% MCR test.

It must be remembered that this short-term increase of MW output is achieved without increasing the boiler firing rate.

6.4.3 Firing rate response tests

These tests show the response of the boiler combustion and furnace pressure controls when subjected to heavy demands. The tests also demonstrate the ability of the boiler heat transfer surfaces to undergo rapid changes in heat flux without danger of dryout or excessive metal temperature.

The tests commence with the boiler at reduced steam pressure controlling the turbine load at 70% MCR. The computer control of the boiler master pressure desired value (DV) should be set to 'manual'.

The programme comprises two tests. In Test 1, the boiler master steam pressure DV is pulsed manually to increase the unit load at the rate of 28 MW/minute.

In Test 2, the boiler master steam pressure DV is manually pulsed at the maximum rate, i.e., a continuous signal.

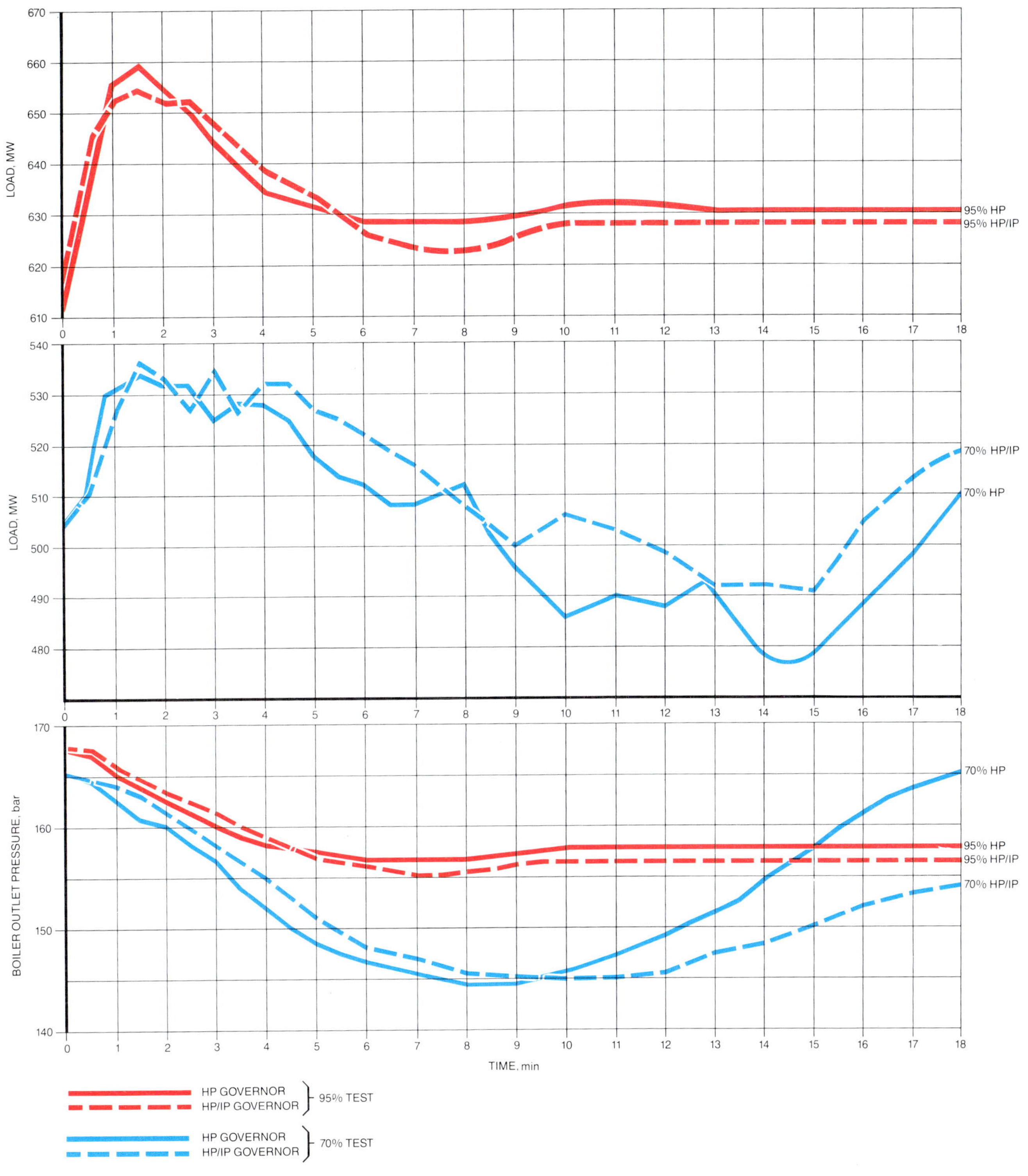

FIG. 6.72 Stored energy test results

State of plant

The following conditions exist for both tests:

- Unit operating initially at 462 MW (70% MCR).
- HP and IP turbine governor valves fully open.
- Boiler pressure adjusted to meet load and governor valve opening requirement.
- Boiler master pressure control selected to 'manual'.
- Computer loading program switched out.
- Computer control of boiler master pressure selected to 'manual'.
- Computer control of superheater and reheater steam temperature set to 'manual'.
- All other automatic boiler-control loops in service on 'auto'.
- Turbine governing mode set to 'HP'.

Additional instrumentation is fitted in the following areas:

Boiler furnace, superheaters and reheater metal temperatures.

Boiler furnace heat-flux meters.

Boiler downcomer pitot tubes and density meters.

Test readings

In addition to the test readings shown in Section 6.4.2 of this chapter, the following are continuously monitored and recorded for a minimum time of approximately 10 minutes, by which time full boiler pressure and full unit load should be obtained:

Total air flow

Boiler furnace, superheater and reheater metal temperatures

Boiler furnace heat-flux meters

Boiler downcomers pitot tubes and density meters

Boiler combustion oxygen and smoke density

Test procedure — Test 1

- Ensure operating conditions are stable at the initial test load of 70% MCR (462 MW).
- Confirm plant operating as specified under '*State of plant*'.
- Take initial set of test readings as previously specified.
- Pulse-operate the boiler master steam pressure DV (desired value) to raise boiler pressure and consequently unit load at a rate equivalent to 28 MW/minute.
- Carefully monitor the effects of the demand on the boiler automatic controls.
- Continue raising boiler pressure at the desired rate until full boiler pressure and full unit load are obtained.
- After approximately 10 minutes, stabilise unit operation.
- Restore unit load to 70% MCR and stabilise conditions in preparation for Test 2.

Test procedure — Test 2

- The same procedure is adopted for this test, except that the boiler master steam pressure controller DV has a continuous raise signal applied to it.
- This forcing of the boiler master pressure control may cause the furnace to pressurise.
- Carefully monitor the effects of this load forcing, especially on the drum level and furnace pressure.
- After approximately 10 minutes, stabilise unit operation.

Test results

The main parameters, namely MW output, boiler steam pressure and drum level, for each of the two tests are shown in Fig 6.73. From these test results it can be seen that, for Test 1, the rate of load increase was only 19.5 MW/min against the desired rate of 28 MW/min. For Test 2, a rate of unit load increase of 28 MW/min was achieved.

The drum level control can be seen to be adequate; the minimum drum level occurred in Test 2 at the higher loading rate. However, this was still in excess of the low drum water trip level of −330 mm.

There is no doubt that this type of firing response test is more arduous than straightforward load ramping with the boiler at normal full operating pressure.

6.4.4 Ramp loading tests

The ramp loading tests demonstrate the ability of the boiler automatic control systems to respond as they are subjected to increasing rates of load raising. The programme comprises three tests, each commencing from the 70% MCR load point (see Fig 6.74):

Test 1 Increase load at the rate of 10 MW/minute to 100% MCR

Test 2 Increase load at the rate of 12 MW/minute to 100% MCR

Test 3 Increase load at the rate of 28 MW/minute to 100% MCR

This test programme exercises the automatic boiler controls in a sequence which approaches the severity of the step loading tests (Section 6.4.5 of this chapter).

Although the short-term demand on the feed system will not be as severe as in the stored energy tests (Section 6.4.2 of this chapter) and similarly for the combustion controls when compared with the firing rate response tests (Section 6.4.3 of this chapter), this test provides the opportunity to check the ability of the automatic controls to function correctly without undue deviation.

State of plant

The following conditions must exist for each of the tests:

- Unit operation initially at 70% MCR (462 MW).
- Turbine governor control in 'HP governing' mode.
- Boiler operating at full rated pressure and temperature.

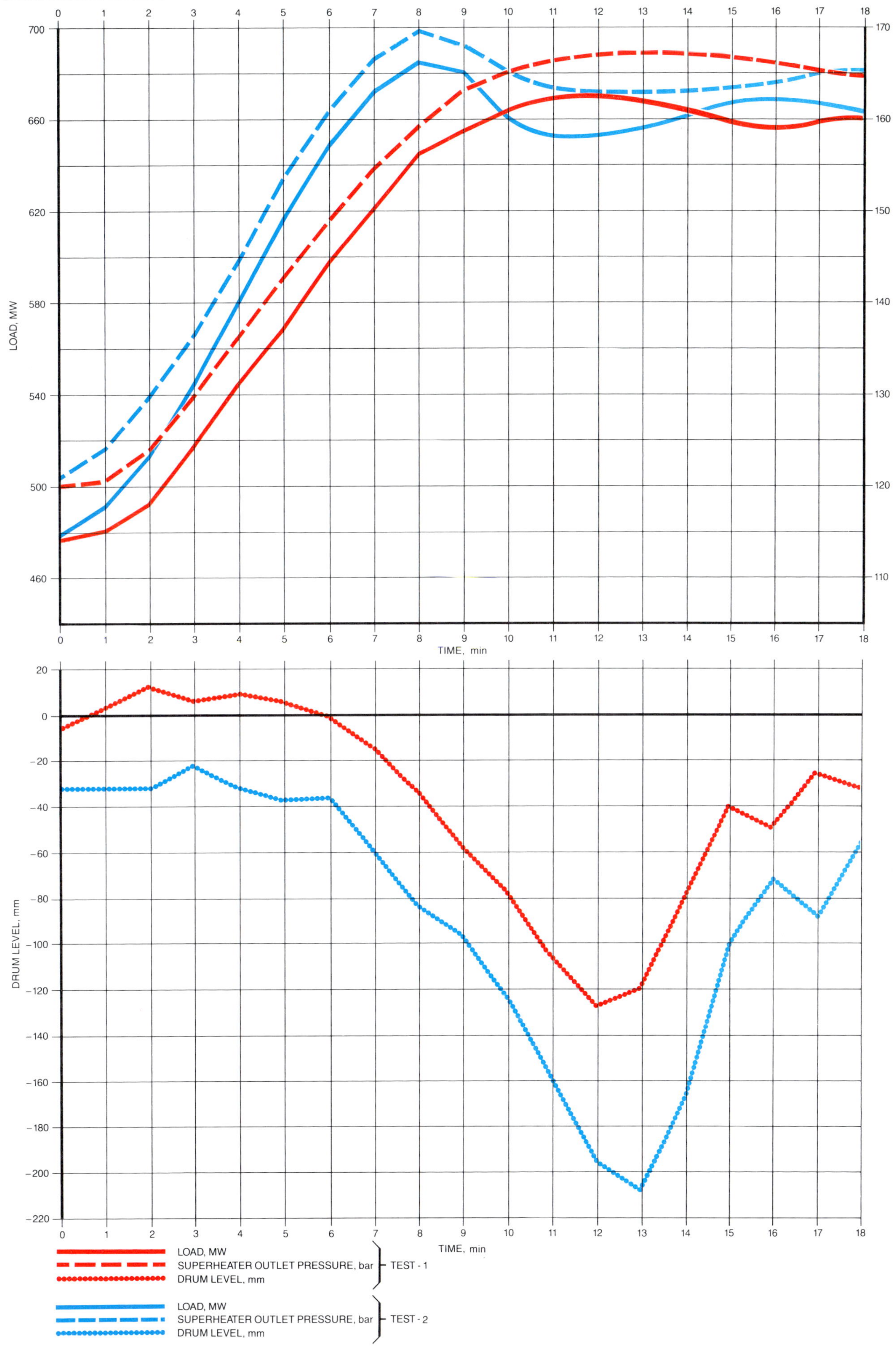

FIG. 6.73 Firing rate response test results

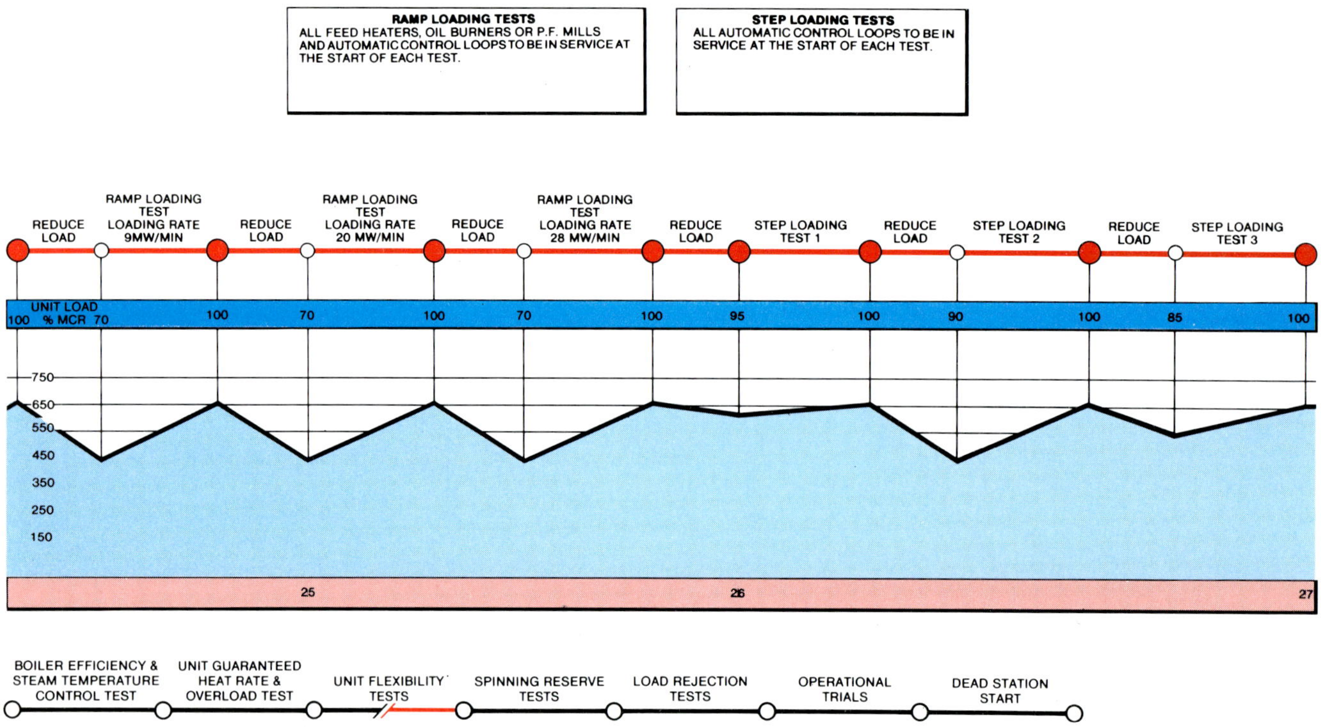

FIG. 6.74 Unit flexibility programme — ramp loading and step loading tests

- Computer control of unit loading in service.
- All boiler automatic control systems operating on 'auto'.
- Additional instrumentation fitted to measure boiler furnace, superheater and reheater metal temperature, and boiler furnace heat fluxes.

Test readings

In addition to the test readings shown in Section 6.4.2 of this chapter, the following are continuously monitored and recorded for a minimum of 25 minutes, by which time full-load with stable conditions should have been achieved. This time is less for Test 3, because of its high rate of load increase.

Total air flow

Spraywater flow

Turbine HP governor valve positions

Test procedure — Test 1

- Ensure that operating conditions are stable at the initial test load requirement of 70% MCR (462 MW).
- Confirm that the boiler is operating at full rated pressure and temperature and that the plant is operating as specified under '*State of plant*'.
- Take an initial set of readings.
- Select the computer loading program to achieve a load ramp of 10 MW/minute with a target load setting of 660 MW.
- Initiate the loading program and closely monitor the unit operating conditions as the load increases, paying particular attention to the automatic boiler control response.
- Stabilise the unit at full-load, continue monitoring until this has been achieved.
- When satisfied that the test is complete, reduce unit load to 70% MCR (462 MW) and stabilise conditions in preparation for Test 2.

Test procedure — Test 2

As Test 1, except that the computer loading program is set for a load ramp rate of 12 MW/minute with a target load setting of 660 MW.

- Closely monitor the plant parameters, as specified.
- Stabilise the unit at full load; continue monitoring until this has been achieved.
- When satisfied that the test is complete, reduce unit load to 70% MCR (462 MW) and stabilise conditions in preparation for Test 3.

Test procedure — Test 3

The last test is identical to the previous two tests, but with computer ramp loading rate set for 28 MW/min with a target load setting of 660 MW.

Test results

The main parameters, namely MW output, boiler steam pressure and drum level, for each of the three tests are shown in Fig 6.75.

The test results show that the selected loading ramps were basically achieved, although in Test 1 the actual load rate achieved was slightly lower than the desired rate of 8.75 MW/min. Boiler pressure during the loading ramps shows that reasonable control is being exercised over the boiler firing rate. Control of drum level is successful in each test.

The effect of Test 3 when ramping at 28 MW/minute on furnace heat flux and boiler metal temperature requires full analysis of all the test results. For the examples given, the boiler automatic control functioned very well.

6.4.5 Step loading tests

These tests exercise all the boiler automatic control loops over a wide range. They demonstrate the capability of the control systems of providing a fast but stable response to the unit within the constraints bounded by the largest expected system frequency reduction.

The step loading is achieved in an electronic governor system by injecting a signal into the speed error reference circuit which will signal a demand for increased output. The time for a full load change (0–100% MCR) for an electronic governor valve opening is of the order of 40 s. A conventional hydraulic governor system is much slower in response. The test procedure described is applicable to an electronically-controlled governing system only.

The test programme consists of three tests: each increases the severity of the demand on the auto systems.

State of plant

The following conditions must exist for each of the tests:

- The initial load starting points for the three tests are:

  Test 1: unit load 95% MCR (627 MW)

  Test 2: unit load 90% MCR (594 MW)

  Test 3: unit load 85% MCR (561 MW)

- Turbine governor control in 'HP governing' mode.
- Boiler operating at full rated pressure and temperature.
- All boiler automatic control systems operating on 'auto'.
- Additional instrumentation fitted to measure boiler furnace, superheater, and reheater metal temperatures, and boiler furnace heat-fluxes.

Test readings

The same quantities as those listed in Section 6.4.4 of this chapter are continuously monitored and recorded for a minimum time of 8 minutes.

Test procedure — Test 1

- Ensure that the operating conditions are stable at the starting load 95% MCR (627 MW).
- Confirm that the plant is operating as specified in the '*State of plant*'.
- Confirm that the boiler is operating at full rated pressure and temperature.
- Take an initial set of readings.
- Subject the turbine governor to a step demand to increase load immediately following the signal for full throttle valve opening.
- Stabilise the unit at full load and continue monitoring for approximately 8 minutes until this has been achieved.
- When satisfied that the test is complete, reduce unit load to 90% MCR (594 MW) and stabilise conditions in preparation for Test 2.

Test procedure — Tests 2 and 3

Use the same procedure as described for Test 1 except that the initial load is 90% MCR (594 MW) for Test 2, and 85% MCR (561 MW) for Test 3.

Test results

The main parameters of MW, steam pressure and oil flow for each of the three tests are shown in Fig 6.76. In addition, the variation of drum level control is shown for Test 3 only, the most demanding test of the series.

For each test, there is an instant increase in load immediately following the full opening of the governor valves. In each test, the instantaneous demand increase took the unit load up to about 640 MW approximately 5 minutes after the step demand was applied. The results clearly show the fall in superheater pressure following the load step and the resulting increase of fuel input to restore the boiler pressure to its normal value corresponding to the full-load output of the turbine.

The drum level results for Test 3 show that control of drum level was maintained following an instant demand increase of some 80 MW.

The effect of the step loading increases on furnace heat-flux and boiler metal temperature requires full analysis of all the test results. From the examples given, the boiler automatic control loops all functioned very well.

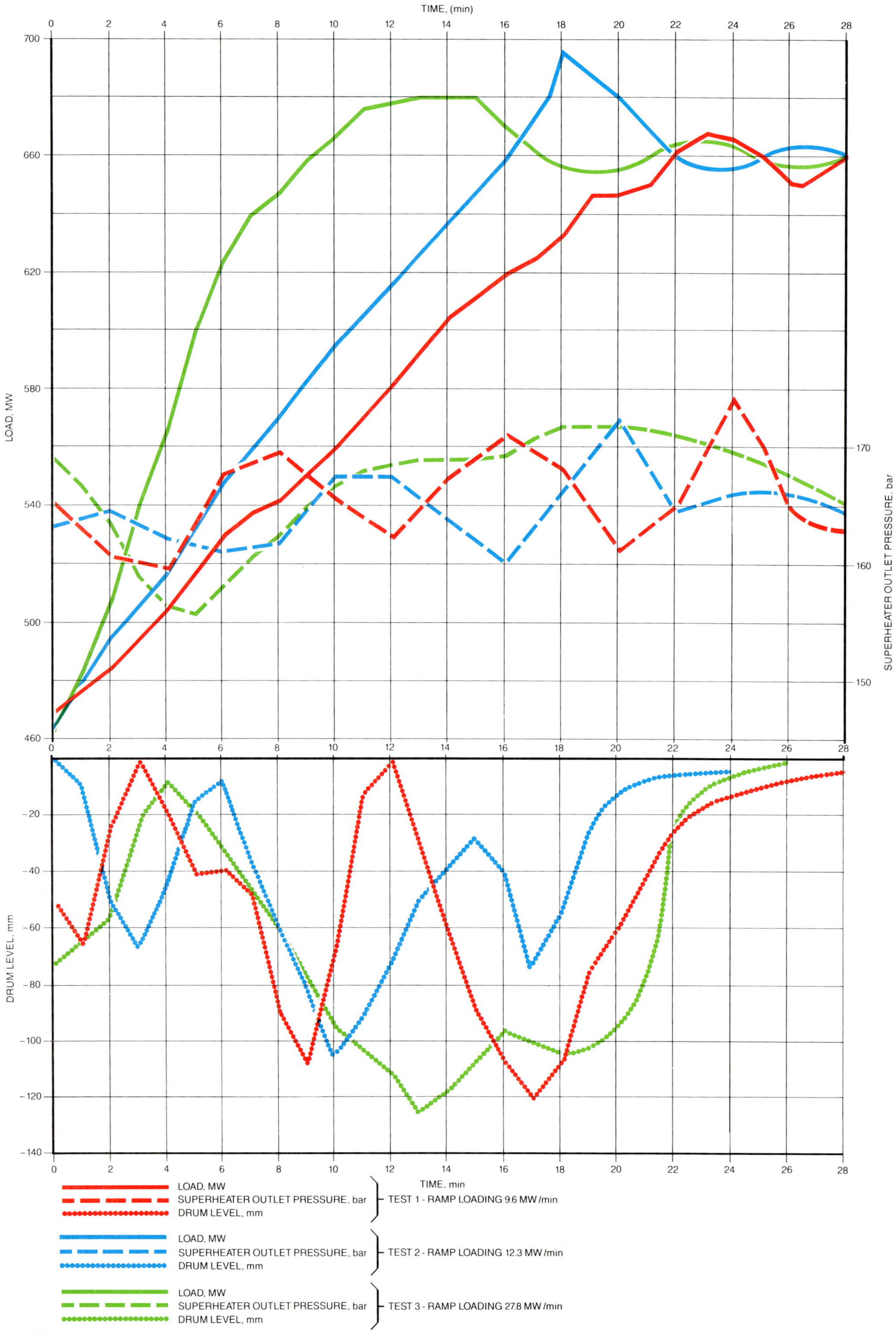

FIG. 6.75 Ramp loading test results

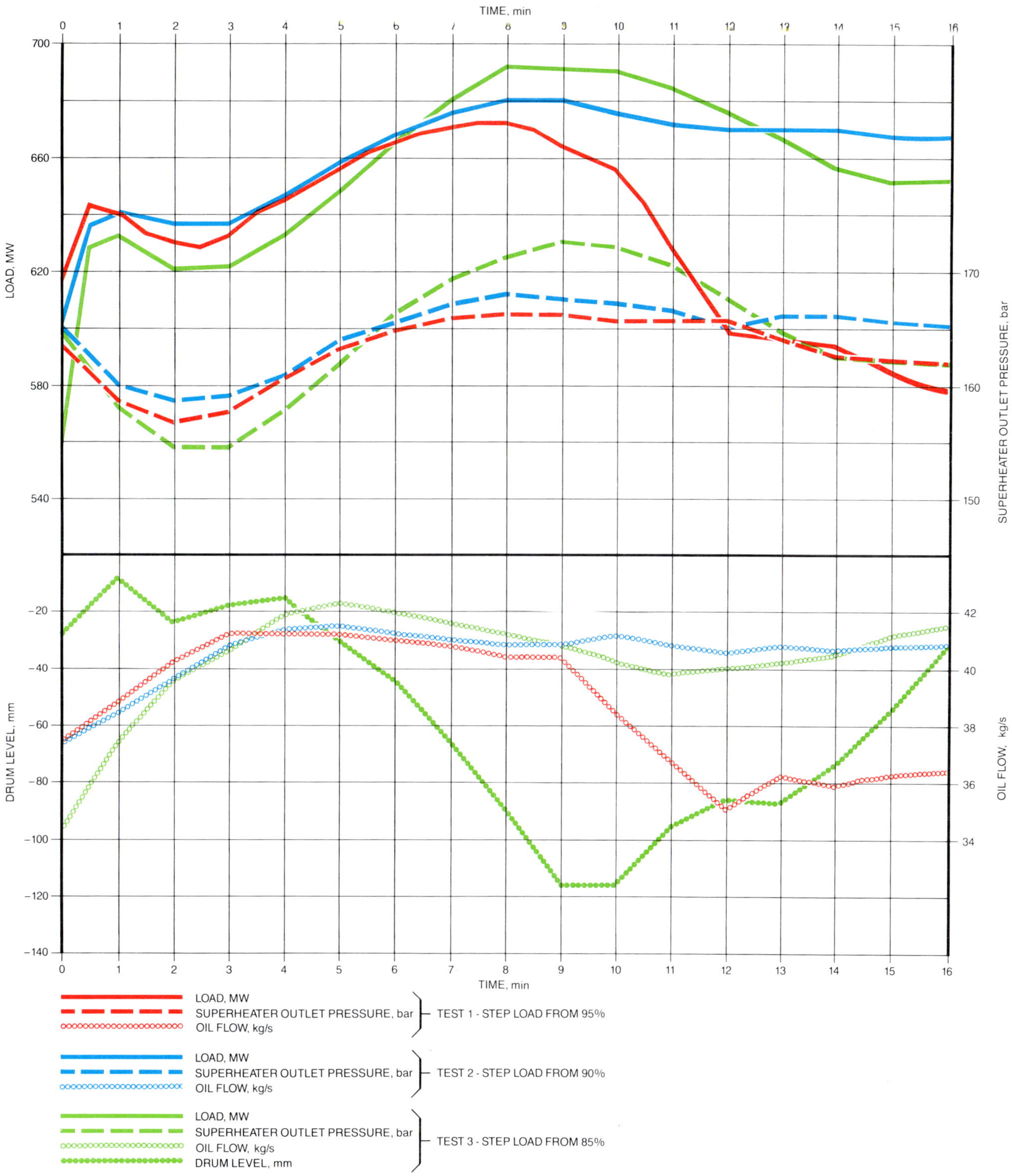

FIG. 6.76 Step loading test results

6.4.6 Spinning reserve test

The objective of the spinning reserve test is to prove the specified spinning reserve capability of the boiler and the ability of the turbine to respond to step changes in demand.

The test is arranged such that whilst operating at 75% MCR (495 MW) an instantaneous demand for the remaining 25% load is initiated, with the first 100 MW of this increase being met within 3 seconds and the remainder over a period of minutes (see Fig 6.77).

The test is carried out for both the HP/IP and HP governor modes and is only applicable to an electronic governor control system. The initial step demand for the first 100 MW is equivalent to the system frequency falling by approximately 0.2 Hz and is applied by injecting a simulated reduced frequency signal into the governor speed measuring circuit.

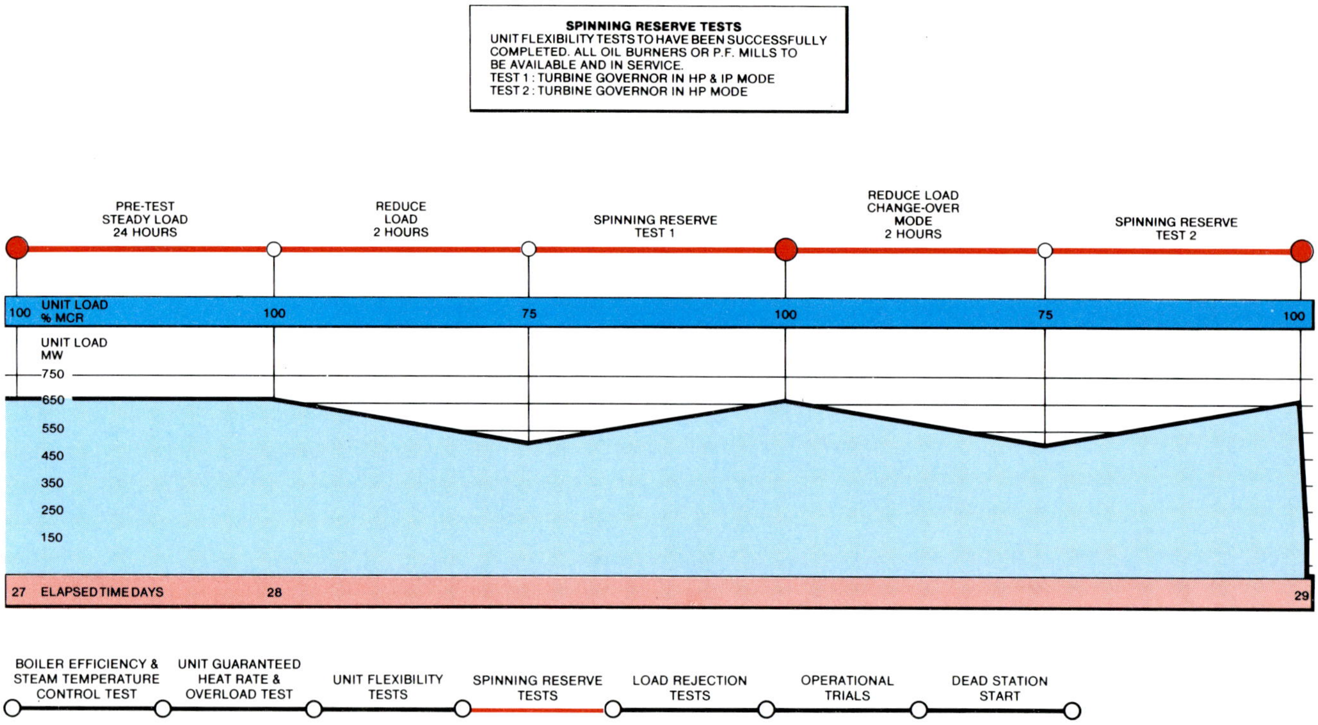

FIG. 6.77 Unit flexibility programme — spinning reserve test

State of plant

The following conditions must exist for the test:

- The unit to be at the initial test load of 75% MCR (495 MW).
- Governor control in HP/IP or HP mode.
- The controller for the rate of steam demand to be set at 100%.
- The governor pressure pay-off controls to be set at full steam pressure conditions at the turbine stop valve.
- Boiler operating at full rated pressure and temperature.
- All boiler automatic control systems operating on auto.

Test readings

In addition to the test readings shown in Section 6.4.2 of this chapter, the following are continuously monitored during the test:

Superheater outlet steam pressure

HP and IP governor valve positions

Superheater and reheater spray valve positions

Oil flow to burners

Main feed pump discharge pressure

Main feed pump speed

Boiler combustion oxygen and smoke density

Test procedure — HP/IP mode

- Confirm unit operating conditions are stable at the initial test load requirement of 75% MCR (495 MW).
- Confirm boiler is operating at full rated pressure and temperature.
- Confirm plant is operating to the requirements specified under the '*State of plant*'.
- Take initial set of readings as identified in the test readings.
- Inject into the electronic governor speed measuring circuit a simulated reduced frequency signal equivalent to 0.2 Hz.
- After the initial pick-up of 100 MW, set the computer program for a load increase rate of 28 MW/min up to 660 MW.
- Stabilise the unit at full load and continue monitoring until this has been achieved.
- When satisfied that the test is complete, reduce load to 75% MCR (495 MW), change the governor control to HP only and stabilise conditions for next test.

Test procedure — HP mode

A repeat of the HP/IP mode test with governor control selected for HP only.

Test results

The results of tests showing the increase in MW demand are given in Fig 6.78. From these results it can be seen that the test objective of obtaining an instantaneous load increase from 495 MW to 660 MW in 63 seconds was not achieved. From Fig 6.78 it can be seen that an immediate load increase of 60 MW in less than one minute was obtained in the HP/IP mode and 40 MW in the HP mode. Load increased fairly quickly with 610 MW being achieved in four minutes. The rate of load increase caused the boiler pressure to fall and a small time delay became evident as the automatic control system increased the firing rate of the boiler. This is demonstrated by the load falling off before full load is finally achieved in 13 minutes. The shape of the load curves for both tests are almost identical, the only difference being the load level which, as expected, produced a higher load for the HP/IP test, thus confirming that operating in this mode provides a slightly better spinning reserve capability. Fine tuning of the boiler automatic controls to improve response time may result in a faster rate of load increase.

6.5 Load rejection tests

6.5.1 Introduction

This test demonstrates the ability of the turbine speed governing system to control the turbine speed to a value less than the overspeed bolt setting, following the complete loss of load. The load rejection test simulates the total loss of electrical load without any prior or subsequent signals to the turbine protective tripping devices.

As no protective devices operate, the speed of the turbine is controlled by the response of the governor system to a rise in speed. These tests prove the basic design of the governing system as well as confirming that no stiction exists in the governor valves and that the bled-steam non-return valves operate correctly. In addition, the tests demonstrate the ability of the automatic voltage regulator to control the stator voltage during the speed rise immediately following the load rejection. A typical programme for the load rejection tests is shown in Fig 6.79 and covers a series of five tests as follows:

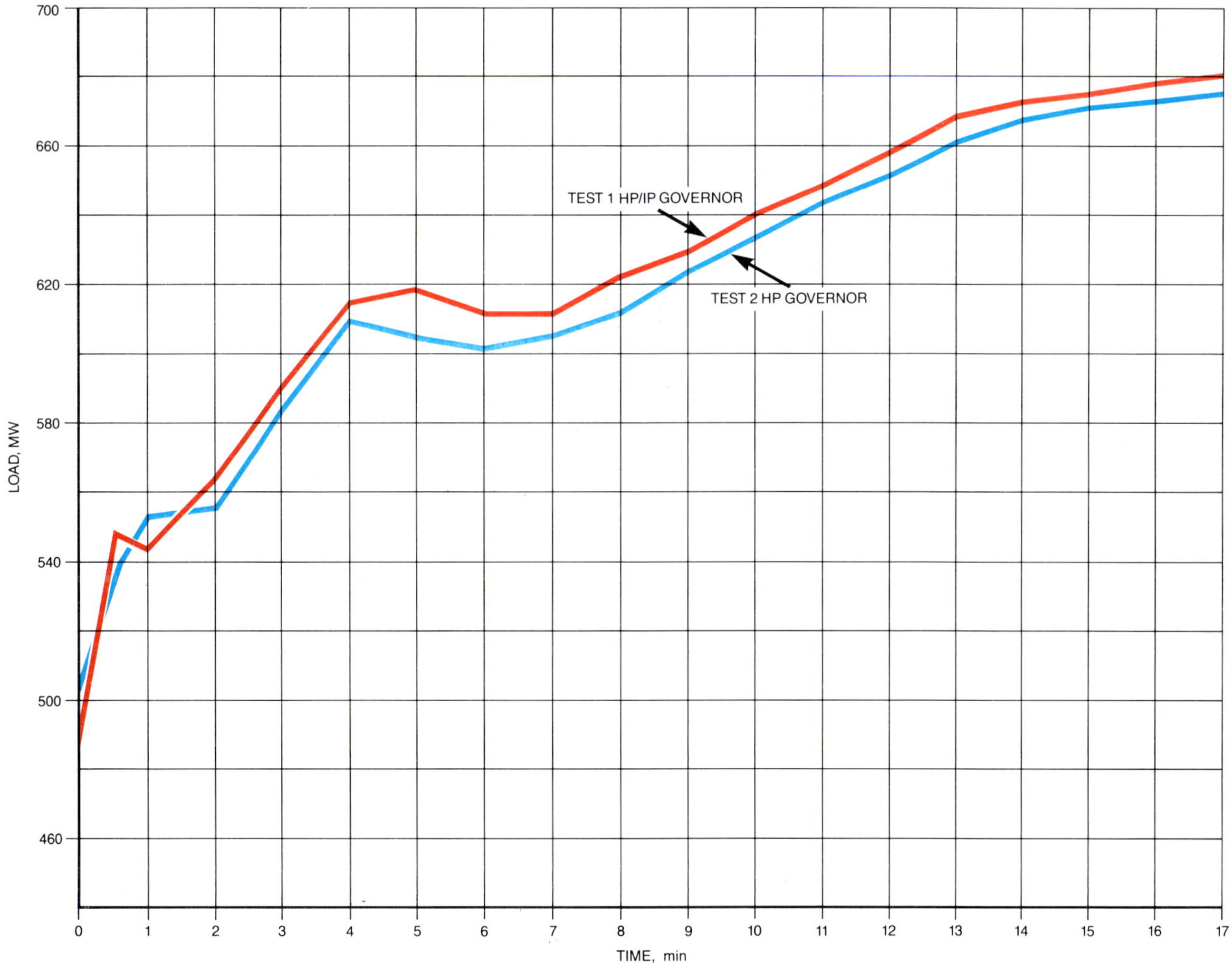

FIG. 6.78 Spinning reserve test results

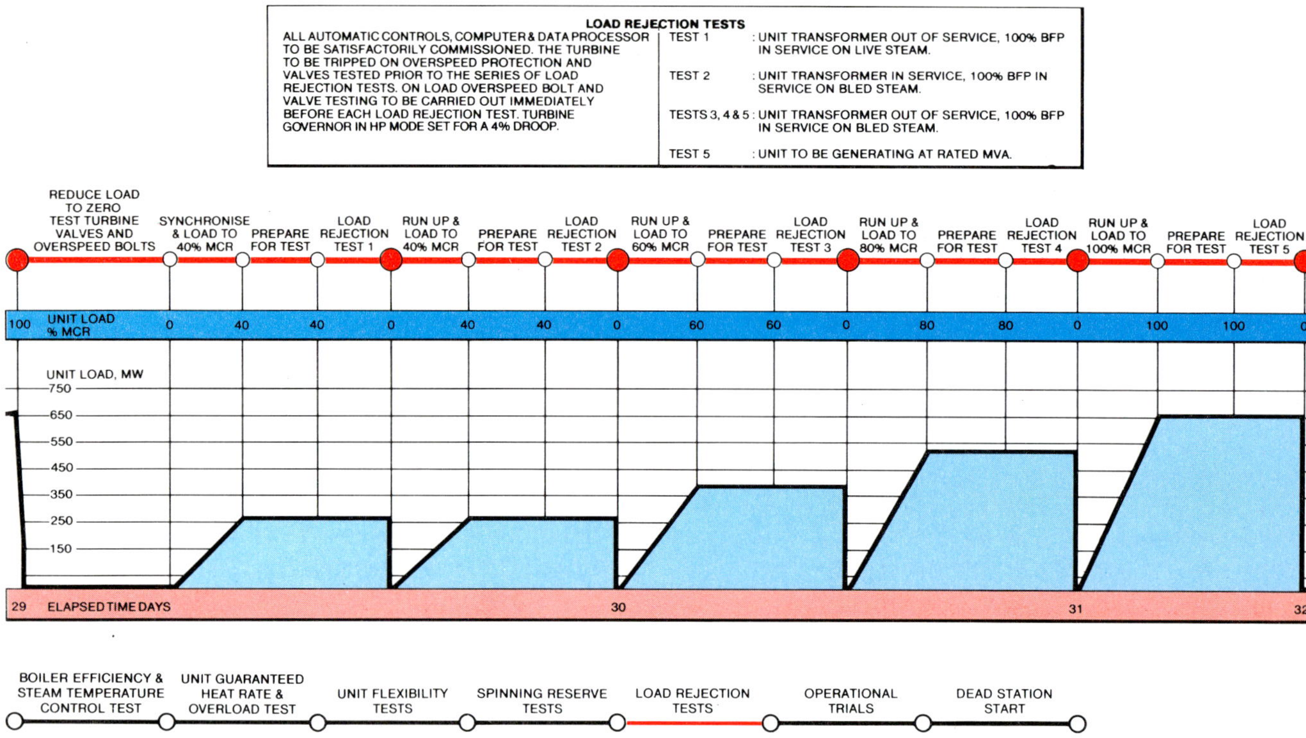

FIG. 6.79 Load rejection test programme

Test 1 40% load rejection

Test 2 40% load rejection with unit transformer remaining in service (for this test only).

Test 3 60% load rejection

Test 4 80% load rejection

Test 5 100% load rejection at rated MVA

Tests 1 to 4 are basically information tests, the results being analysed and compared with the predicted values before proceeding to the next scheduled test. In Test 2, where the unit transformer remains in service, there is the opportunity to monitor the unit auxiliary power system voltage and frequency, especially during the temporary speed rise. Test 5 is the most important but the preceding steps are necessary to minimise risk to the unit. This test, at full load and rated MVA, also has contractual implications. The programme identified in Fig 6.79 should be adopted for the load rejection tests for the first unit in a multiple-unit station where the units are of identical rating, manufacturer and design. For later units, the information tests can be reduced to one single test at 50% MCR before carrying out the full-load rejection. The full-load rejection test should be carried out under conditions that promote the highest temporary speed rise. This is achieved by ensuring that the feedheating system is fully in service and that the unit transformer is out of service. By adopting these conditions, the unit load will reduce to zero and the quantity of steam entrained in the bled-steam pipework at the time of the rejection will be at its maximum value. Any genuine rejections that occur later will always give a lower level of temporary speed rise with the unit transformer in service as the load would reduce to approximately 10 MW and not zero. Whilst carrying out these tests, steps should be taken, where possible, to prevent the unnecessary lifting of the boiler safety valves. There are two main reasons for this. The first is environmental, i.e., loud noises should be avoided whenever possible, particularly if there is a local resident population, and the second is to avoid damage to safety valve seats with the subsequent risk of leakage.

The tests must avoid risk to plant and personnel, so they need to be carried out to a co-ordinated programme.

The load rejection tests should not be attempted until all control loops, boiler automatic control systems and data processors have been satisfactorily commissioned and until major plant defects or design limitations have been eliminated.

6.5.2 Pre-test preparation

At a unit shutdown, immediately prior to the running period within which it is intended to carry out the load-rejection test programme, the following checks should be carried out, irrespective of whether they have been carried out previously:

- Carry out a full overspeed test on both overspeed rings, confirming correct operation at +110% speed with a tolerance of ±1%. If the test values are outside those specified, adjustments to the settings of the rings must be made, with further retesting on the first run-up after the adjustment.

- Move both the HP and IP governor valves over their full range to confirm that they are free from stiction and that the valve closing times are not greater than 0.2 s.
- The operation of the speed acceleration relays should be checked by the injection of a signal equivalent to that which would occur in the event of a full-load rejection. This must be seen to operate the relays and send a closing signal to the HP and IP governor valves.
- In an electronic governing system, the three speed measuring channels should be checked to ensure that a faulty signal channel will not invalidate operation of the governors, and similarly that loss of either of the two AC supplies does not trip the governors.

In addition to these checks, the unit should be inspected prior to the shutdown and plant defects that need rectification identified before the commencement of the load rejection programme.

6.5.3 State of plant

The following conditions must exist for the five tests:

- All boiler automatic control loops in service.
- All HP and LP feedheating systems in service.
- Main boiler feed pump in service. For Test 1 it should be fed from the live steam supply, and for the other four tests from the bled-steam supply.
- Turbine vacuum unloading and trip protection devices in service.
- Turbine governor control in HP mode.
- Automatic voltage regulator in service on 'auto' control, with both channels functioning.
- Unit computer in service.
- Unit transformer in service for Test 2 only and out of service for the other four tests.
- A special communications and signalling system to be installed and proved to be operating satisfactorily. Normal direct wire and PAX systems proved to be functioning, as well as personal radios.
- The area around the turbine to be barriered and notices posted warning of the tests.
- High speed and accurate multi-channel tape recorders to be installed to measure the quantities identified in Section 6.5.4 of this chapter.

6.5.4 Test readings

The following parameters should be recorded on high accuracy tape recording equipment.

Turbine load

Turbine speed

HP governor valve positions — each valve

IP governor valve positions — each valve

HP stop valve positions — each valve

IP stop valve positions — each valve

Stator current

Stator voltage

Exciter current

Exciter voltage

Automatic voltage regulator output current

Automatic voltage regulator output volts

Steam pressure before HP governor valves

Steam pressure before IP governor valves

Steam pressure after HP governor valves

Steam pressure after IP governor valves

Acceleration relay operation

Speed error signal

Timing pulse

HP governor highway

IP governor highway

In addition, the following are recorded:

Boiler steam outlet pressure

Boiler steam outlet temperature

Boiler reheater outlet pressure

Boiler reheater outlet temperature

Turbine HP cylinder exhaust pressure

Turbine condenser vacuum

HP heater bled-steam pressure

Hydrogen casing pressure

Stator water pressure and temperature

Main BFP steam pressure before bled/live steam stop valve

Main BFP steam temperature before bled/live steam stop valve

Main BFP exhaust pressure

Turbine hydraulic fluid control system pressure

Turbine differential expansions, eccentricities and bearing vibrations

6.5.5 Pre-load rejection checks

The following should be checked just prior to the actual load rejection test:

- Turbine overspeed test to be carried out at the beginning of the first day's test programme and on each successive day's testing.
- Immediately prior to each test, test the turbine overspeed by oil injection.
- Carry out on-load valve testing at the load of the rejection test. Ensure that all valves operate correctly.
- Confirm the correct operation of the turbine electronic governors immediately prior to the test by varying the unit load.
- Confirm the correct operation of the boiler feed pump electronic speed governor immediately prior to the test by varying pump speed.
- Exercise the boiler feed pump regulating valves immediately prior to commencing the tests.
- Confirm that all bled-steam non-return valves are operating satisfactorily.
- Prior to each separate test, confirm that the signalling and communication systems are functioning. The test personnel must be fully briefed on the test countdown procedure and the countdown must be clearly displayed in defined areas.
- On the test days, confirm that system conditions are such that the load rejection can take place at a specified time.

6.5.6 Load rejection countdown

Typical countdown procedure

T−60 min	Ensure that the initial conditions are correct for the relevant test. Confirm with Systems Control that the test will take place in 1 hour.
T−30 min	Confirm that the tests/checks identified in pre-load rejection checks have been completed. Commence the countdown by advising the test personnel that the countdown programme has started. This signal initiates the posting of various notices giving warning of the test and advises all other staff to keep outside the restricted area.
T−10 min	Increase the boiler firing rate to increase boiler pressure to a value higher than normal operating pressure but below safety valve level. Reconfirm with Systems Control that the load rejection will take place in 10 minutes.
T−8 min	Increase drum level to +100 mm above drum centre line.
T−4 min	Shut the superheater and reheater spray isolating valves. Reduce the oil flow or stop the coal feeders supply. The objective is to have the fuel input cease at the instant of load rejection. The timing of this will vary with the type of fuel and the size of the PF milling equipment.
T−2.5 min	Open the superheater outlet and main steampipe drain valves to control steam pressure to the desired value.
T	Open the generator circuit-breaker to reject the load being carried.
T+2 min	Stop the countdown.

Signalling arrangements

A typical signal arrangement is:

T−30 min	Switch on steady red light
T−2 min	Commence flashing red light
T−1 min	Switch on green light
T−10 s	Commence flashing green light
T	Switch on both green and red lights to steady illumination
T+2 min	Extinguish both red and green lights

Following load rejection, the turbine speed will rise to a value that should be less than the overspeed setting. The boiler pressure will rise rapidly but, if the firing rate has been controlled correctly and with the utilisation of the drain valves, unnecessary lifting of the boiler safety valves should have been avoided. The boiler drum level will fall very quickly and is likely to go below the minimum level of the gauge glass unless the enhanced drum level as indicated at T−8 min had been established and maintained up to the time of the rejection.

Careful monitoring of both the boiler and turbine parameters should continue until steady state is achieved. Particular attention should be paid to turbine differential expansions, eccentricities and bearing vibrations whilst the turbine is running at speeds in excess of 3000 r/min.

Once the maximum temporary speed rise has been recorded, the turbine will continue to run at a speed in excess of 3000 r/min dependent on the governor setting at the time of the load rejection. Operation of its governor will reduce the turbine speed back to 3000 r/min in preparation for resynchronising.

6.5.7 Test results

Figure 6.80 shows graphs derived for a 660 MW turbine generator from a full-load rejection test on an electronic-governor-controlled turbine, operating with a 4% governor droop setting.

From curve (b) it can be seen that the 'maximum temporary' speed rise to 3220 r/min was obtained within 4 seconds, this being well inside the overspeed setting of 3300 r/min. Although outside the timescale of the graph, a speed reduction to 3140 r/min (permanent speed rise) occurs approximately 10 minutes after load rejection and is dependent on the rate-of-decay of steam pressure after the HP governor valves. The steam pressure after the IP governor valve (curve (m)) took approximately 5 seconds to decay to zero.

The HP governor valves (curve (c)) and the IP governor valves (curve (e)) were fully closed within 0.25 seconds, whilst the HP stop valves (curve (d)) were fully closed in 0.4 seconds.

The performance of the automatic voltage regulator (curves (n) and (p)) was excellent. Curves (f) and (g) show how the stator current reduced to zero within 0.1 s and that the stator voltage was maintained at 23 kV after the load rejection.

6.6 Automatic synchroniser

6.6.1 Introduction

The objective of this test is to confirm that the turbine-generator can be synchronised to the system by the use of automatic synchronising equipment.

The programme has three stages:

- Pre-commissioning checks on the automatic synchronising equipment
- Dead busbar tests
- Live busbar tests

6.6.2 Principle of operation of the automatic synchroniser

With the turbine-generator running at 2850 r/min with the generator excited to normal stator voltage, the automatic synchroniser is switched into service.

Voltage matching then takes place by varying the position of the generator transformer tapchanger. When the 'incoming' and 'running' voltages are matched, signals are sent to the electronic governor to alter the speed error reference signal. The turbine-generator speed increases to a value slightly above synchronous speed (approximately 10 r/min). At this point, the synchroniser sends a signal to the governor control to lower turbine-generator speed towards synchronous speed. When the slip frequency is within the range set in the automatic synchroniser, the relative phase between the two voltages is measured and, when the phase angle difference is < 5%, closure of the generator circuit-breaker is initiated.

The automatic synchroniser incorporates an override circuit, set to produce a load pick-up of approximately 5% MCR on synchronising.

The pulsing rate of the governor control is set to ensure that synchronising takes place with the governor datum above synchronous speed and, in addition, allows one phase coincidence within the synchronising bandwidth.

6.6.3 Pre-commissioning tests of the automatic synchroniser

The automatic synchroniser unit will be preset for the switching disconnector/circuit-breaker closing time (including any interposing relays and contactors), the setting normally being between 75–350 ms.

The voltage matching limits are set and checks are made to ensure that the voltage transformer (VT) supplies have the correct polarity and phase. During manual synchronising, the VT supplies are measured at the input to the synchroniser unit and any discrepancy is taken up by adjustment on the internal resistor.

6.6.4 State of plant for live and dead busbar tests

- The 400 kV system, as shown in Fig 6.81, is solidly coupled with bus section 1/2 and 2/3 and bus coupler 1 is closed.
- The relevant station boards, auxiliary boards and services boards required to run the turbine-generator are in service.
- The gas turbine is electrically isolated.
- The generator circuit-breaker line disconnector and busbar selector disconnectors are open.
- The turbine-generator is running at 3000 r/min.
- The unit transformer 11 kV ACB is open.
- The AVR is in service, selected for 'automatic' operation on dual-channel control.
- All protection, control fuses and links are in the service position.
- Boiler firing trip links, if fitted, are out of service.
- Automatic synchroniser isolated.
- Recent timing tests on the generator circuit-breaker have given consistent closing cycle times from initiation to closing of contacts. The time for the closing of the interposing relay has been measured.
- Calibration tests on the automatic synchroniser are complete.
- Output signals from the automatic synchroniser to the governor speed error reference raise/lower, the generator transformer tapchanger and generator

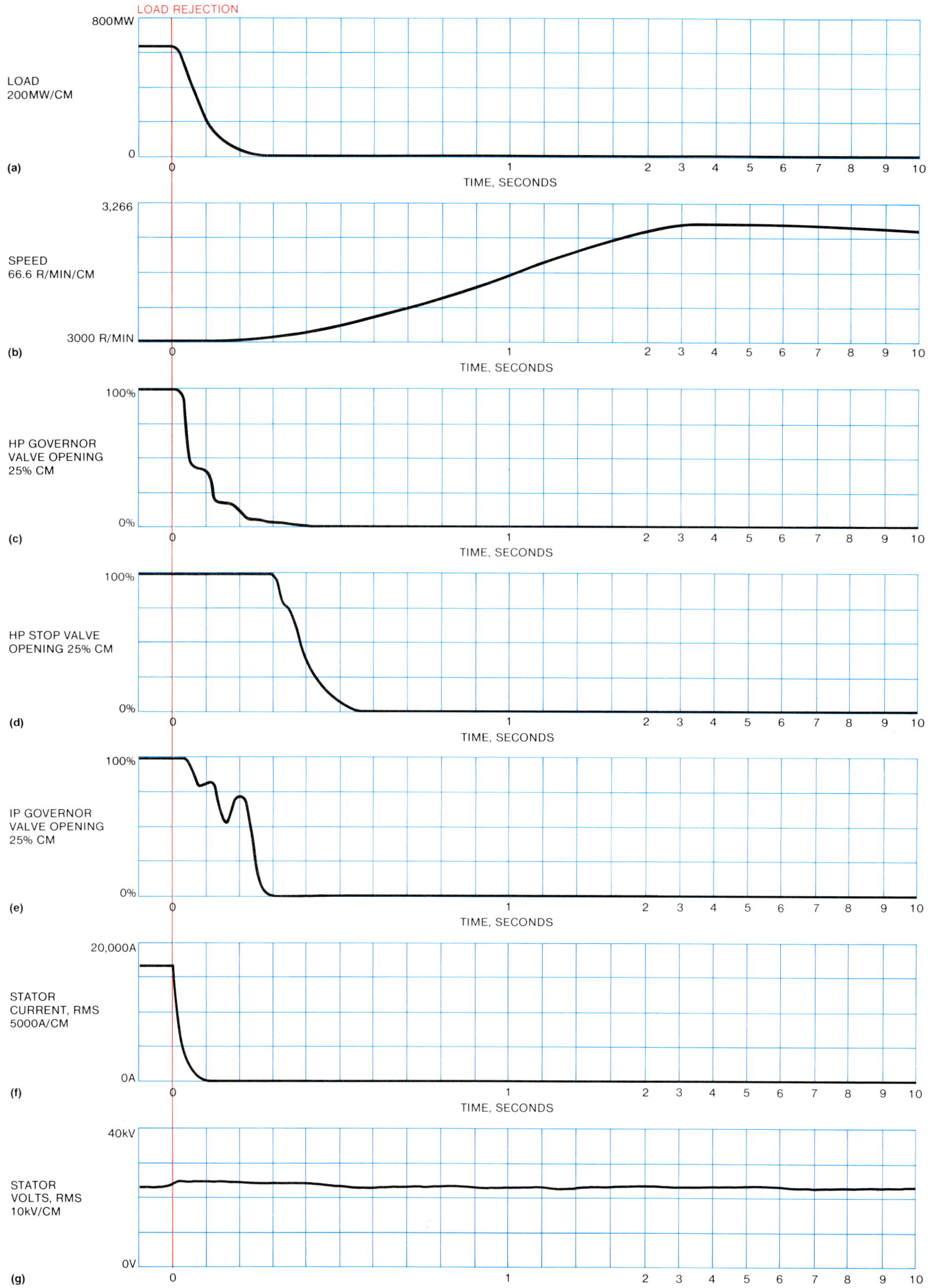
LOAD REJECTION
800MW
LOAD
200MW/CM
0
(a)
TIME, SECONDS
3,266
SPEED
66.6 R/MIN/CM
3000 R/MIN
(b)
TIME, SECONDS
100%
HP GOVERNOR
VALVE OPENING
25% CM
0%
(c)
TIME, SECONDS
100%
HP STOP VALVE
OPENING 25% CM
0%
(d)
TIME, SECONDS
100%
IP GOVERNOR
VALVE OPENING
25% CM
0%
(e)
TIME, SECONDS
20,000A
STATOR
CURRENT, RMS
5000A/CM
0A
(f)
TIME, SECONDS
40kV
STATOR
VOLTS, RMS
10kV/CM
0V
(g)
0 1 2 3 4 5 6 7 8 9 10
TIME, SECONDS

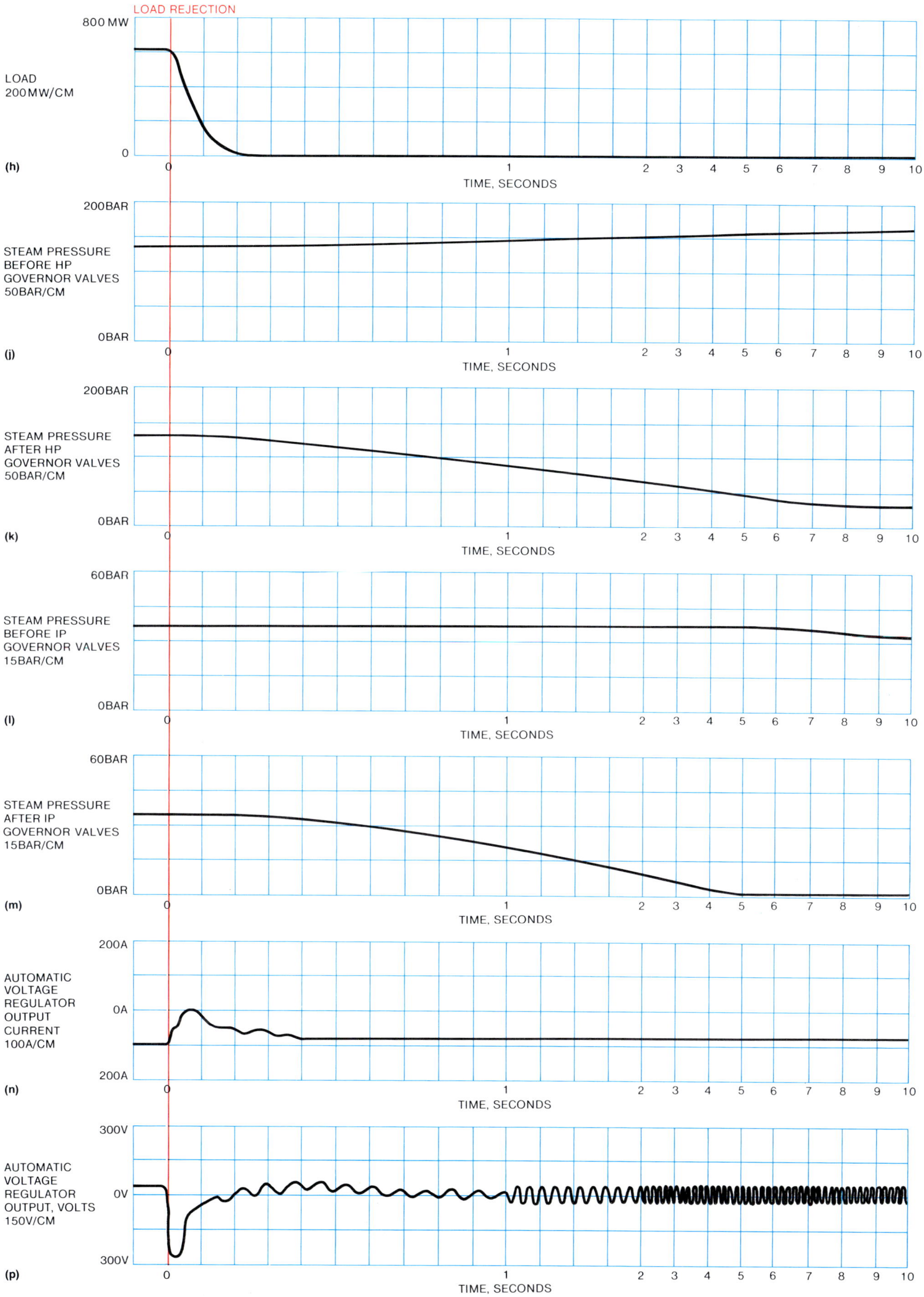

FIG. 6.80 Typical full load rejection curves for a 660 MW turbine generator

circuit-breaker are checked for correct 'sense'. Indications at the control desk and synchronising selector relay must also be checked.

- A recorder is fitted into the automatic synchroniser circuit to monitor the incoming and running volts, governor pulses and circuit-breaker closing.
- Initial settings applied to the automatic synchroniser are:

400 kV circuit-breaker closing time	75 ms
Voltage matching setting	2%
Slip override setting (maximum % difference frequency)	0.4%
Governor pulsing: pause length	2.4 s
pulse length	520 ms

6.6.5 Dead busbar tests

General

Figure 6.81 shows a simplified 400 kV dual busbar system. Phasing and synchronising circuit checks have been completed (Section 3.15 of this chapter) and subsequently numerous manual synchronising operations have been carried out.

For this part of the commissioning of the automatic synchroniser, the 400 kV generator line disconnector and the main and reserve busbar disconnectors must be locked in the 'open' position. The keys associated with the locking must be in the possession of the Test Controller.

This part of the test programme requires a minimum of five successful 'dead bar' shots. If the automatic synchroniser needs resetting before the test is successful, then the total number of dead busbar auto-sequence checks will be greater than five. The starting point for Test 1 commenced with the turbine-generator at 3040 r/min, and for Test 2, at the lower end of the governor range, nominally at 2840 r/min.

Test procedure — Test 1

- Apply a by-pass in the auxiliary circuit wiring to allow an 'incoming volts' signal to be obtained, even though the appropriate generator line disconnector switch is open.
- At the automatic synchroniser, insert the 'running', 'incoming' and auxiliary power fuses.

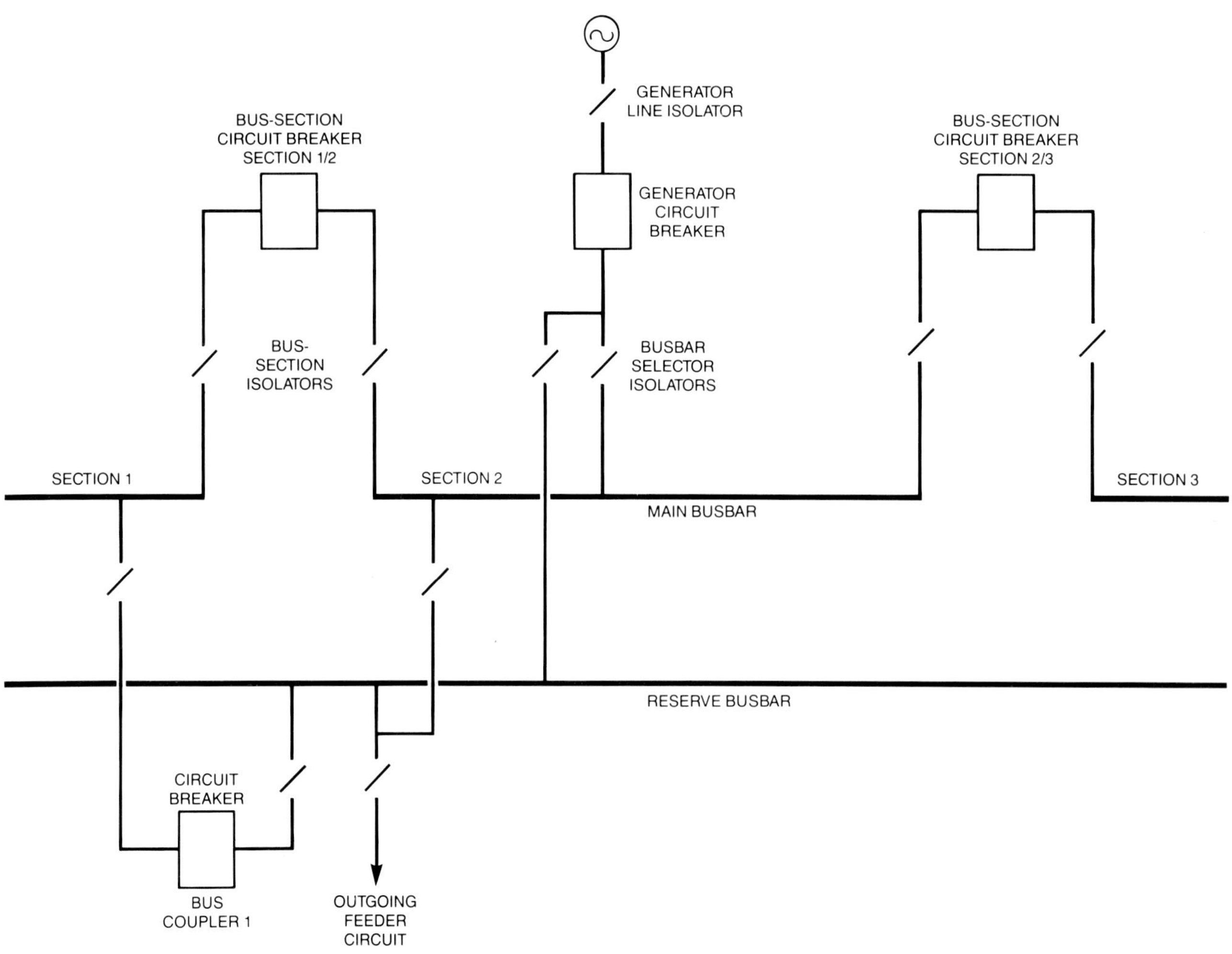

FIG. 6.81 Simplified 400 kV system for automatic synchroniser checks

- Plug in the synchronising trolley and select 'synchronising' on the generator circuit-breaker and confirm the presence of both 'running' and 'incoming' volts.
- Measure the values of the 'running' and 'incoming' volts at the automatic synchroniser.
- Adjust the generator excitation slightly, simultaneously observing a change to the 'incoming' voltage as measured at the automatic synchroniser. Return the excitation to its previous level.
- With the turbine running slightly above synchronous speed, observe (by the use of a test voltmeter) that the synchroscope voltage records zero at 12 o'clock and 127 V at approximately 6 o'clock (this is associated with 110 V secondaries on the voltage transformer).
- Adjust turbine speed to 3040 r/min.
- Operate the generator transformer tapchanger to obtain a system voltage just in excess of the nominal operating level of 400 kV.
- Set the synchronising to 'auto synchroniser' and initiate.
- Observe that the generator transformer tapchanger operates to match the voltage.
- Confirm that the turbine speed decreases.
- Start monitoring the performance of the automatic synchroniser, using the test recorder. Table 6.25 shows the parameters to be measured.
- Check that the generator circuit-breaker closes and that the correct indications are displayed on the 'auto synchroniser' selector.
- Stop the test recorder connected to the automatic synchroniser and examine the recording of the dead busbar automatic synchronising sequence.
- Open the generator circuit-breaker.

If the recording shows that the automatic synchroniser settings need adjustment, these should be agreed and the new settings recorded. A further dead busbar automatic synchronising sequence check should then be carried out.

Test procedure — Test 2

- Adjust the turbine-generator speed to 2840 r/min.
- Set the generator transformer tapchanger to a high value.
- Initiate the automatic synchroniser.
- Observe that the generator transformer tapchanger reduces voltage for matching purposes and that the turbine generator speed increases from 2840 r/min to a value in excess of synchronous speed and then decreases towards synchronous speed.
- Start the test recorder in the automatic synchroniser circuit.
- Check that the generator circuit-breaker closes and that correct indications are displayed on the automatic synchroniser selector.
- Stop the test recorder and examine the results, which should be recorded as in Table 6.25.
- Open the generator circuit-breaker.
- If the recordings show that adjustments to the initial settings of the automatic synchroniser are needed, agree the changes and reset. Repeat the dead busbar synchronising checks.
- When satisfied that no further adjustments to the automatic synchroniser settings are necessary, carry out three more dead busbar automatic synchronising checks to confirm that the settings are correct.
- On completion of the dead busbar checks, open the generator circuit-breaker and remove the by-pass wiring used to obtain the incoming voltage signal.

TABLE 6.25

Record of dead and live busbar automatic synchroniser sequence tests

Trace No.	Automatic synchroniser relay settings					Closing error degrees	Steady state load picked up (for live busbar tests only)
	Generator circuit-breaker total close time	Voltage matching setting	Slip override setting	Governor pulsing			
				Pulse length	Pause length		

6.6.6 Live busbar tests

General

These tests directly follow the dead busbar tests.

Test procedure

- Open bus section 2/3 circuit-breaker (Fig 6.81) and ensure that the busbar disconnectors are open and made inoperative.
- On bus section 2/3, adjust the overcurrent protection relay to its minimum setting.
- Set the generator overcurrent protection relay to its minimum plug and time multiplier setting.
- Close the generator main busbar selector.
- Close the generator line disconnector.
- Confirm that turbine-generator speed is at 3000 r/min.
- Plug in the synchronising trolley on the generator circuit and select for 'manual synchronising'. Confirm the presence of both 'incoming' and 'running' volts.
- Adjust the turbine speed to indicate a slight slip speed.
- Deselect 'manual synchronising' and unplug the synchronising trolley.
- Adjust the turbine speed to 2840 r/min.
- Adjust the generator transformer tapchanger to a high tap number.
- Set to 'synchronising' on the automatic synchroniser and initiate.
- Check that the generator transformer tapchanger reduces the voltage to obtain matching and that the turbine speed increases to just above synchronous speed and then decreases towards synchronous speed.
- Start the recorder to monitor the performance of the automatic synchroniser.
- Confirm that the generator circuit-breaker has closed and that the generator output immediately following closure is not greater than 10% full load.
- Stop the recorder and examine the results of the live busbar automatic tests. If the load following synchronising is $> 10\%$, decrease the slip setting and increase the run-down rate. If the load increase is $< 2\%$, decrease the run-down rate. A typical acceptable value is 5 $\pm 1\%$.
- Open the generator circuit-breaker.
- If further adjustments are found necessary to the automatic synchroniser settings, repeat the live busbar tests.
- When satisfied that no further adjustments to the automatic synchroniser settings are needed, carry out three more live busbar automatic synchronising checks to confirm that the settings are correct.
- The final condition is with the generator circuit-breaker closed and the generator supplying power.
- Disconnect the recorder fitted to the automatic synchroniser.
- On bus section 2/3 circuit-breaker, reset the overcurrent protection to its normal value.
- On the generator circuit, reset the overcurrent protection to its normal values.

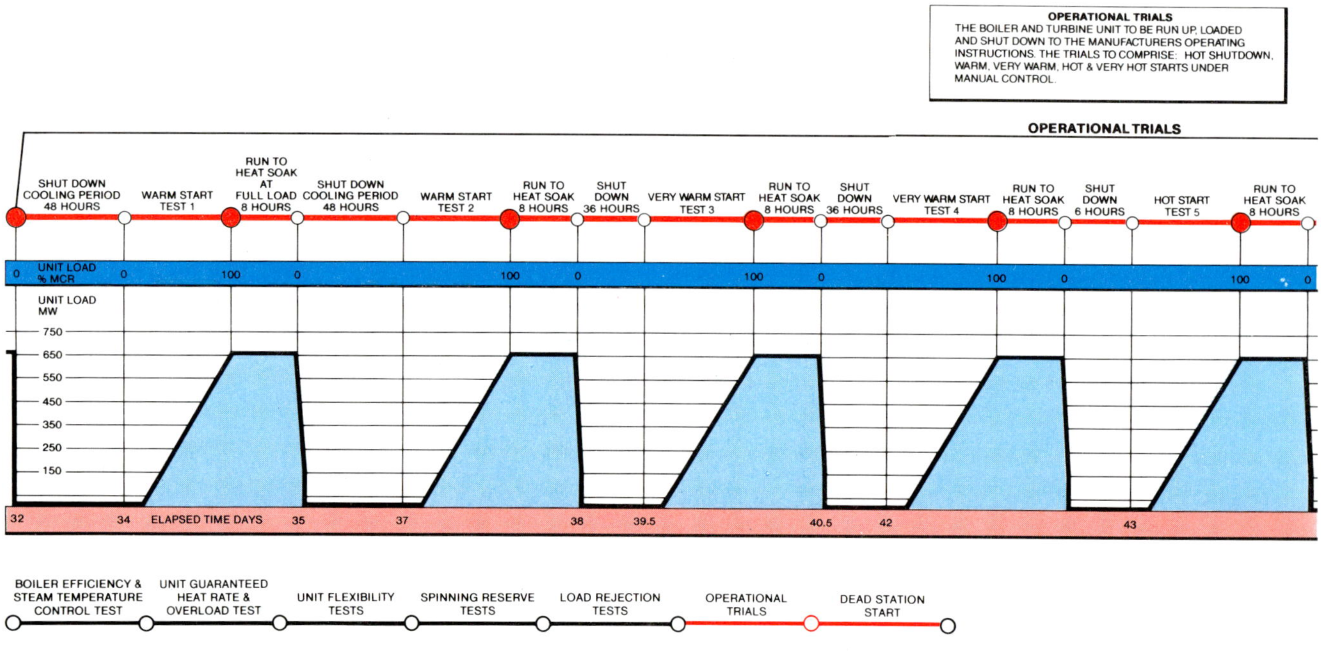

- Close bus section 2/3 disconnectors and circuit-breaker, thus making the 400 kV system solid again.

6.7 Operational trials

6.7.1 Introduction

These trials demonstrate that the boiler/turbine unit can be satisfactorily started, loaded and shut down as stated in the plant specification. They indicate any inadequacies or weaknesses in the plant which will reduce its operational reliability. The ability of the various auxiliary systems to cope with sequential hot starting of more than one unit will be established at a later date. The demand and consumption of these services such as auxiliary steam, make-up water, auxiliary power and fuel are recorded during the operational trials on the first unit. It is normal CEGB practice to carry out the operational trials on the first unit commissioned, the information obtained being applied to all subsequent units. These trials also establish the minimum operating load at which the unit can run successfully for several hours. To ensure adequate turbine life, it is essential to demonstrate that the following conditions are met for all starting conditions and loading rates:

- There are no unacceptable levels of turbine bearing vibration and eccentricities.
- There are no problems of differential expansion in any cylinder.
- There are no unacceptable temperature differentials which would reduce the thermal fatigue life.
- There is no undue differential expansion of the steel foundations.

For the boiler it is essential to demonstrate that the tube metal temperatures do not exceed the design limits, and that the maximum limit of boiler drum metal temperature heating differentials are not exceeded during all the starting conditions and varying loading rates. In addition, to demonstrate that the rate of steam temperature increase can be achieved without exceeding any metal temperature limitations.

The test also allows an assessment of the number of operators required for the various starting modes, particularly for the hot and very hot starts.

During the commissioning programme, the unit will have experienced, in the first instance, a slow run-up procedure to full turbine speed and subsequently during the programme will have been exposed to a number of cold starts, in addition to other kinds of start. Before the boiler efficiency and steam temperature control tests and after completion of the unit guaranteed heat rate and overload test, the programme presents the opportunity to monitor shutdowns and to carry out two cold start tests. The knowledge obtained from the initial run of the turbine and subsequently from the cold starts and other operations is assessed to determine any plant deficiencies before starting the series of operational trials. Figure 6.82 shows a programme of operational trials incorporating nine unit starts and shutdowns; it does not include cold start tests, as operational requirements are likely to dictate that the plant will operate within 'warm and hot start' modes.

660 MW units normally use automatic computer facilities for turbine run-up and load control. Initially, this equipment should not be used until sufficient knowledge of the main plant behaviour during the various categories of turbine run-up and loading have been established. A test programme similar to that shown in Fig 6.82 is then decided, using the computer run-up and loading program.

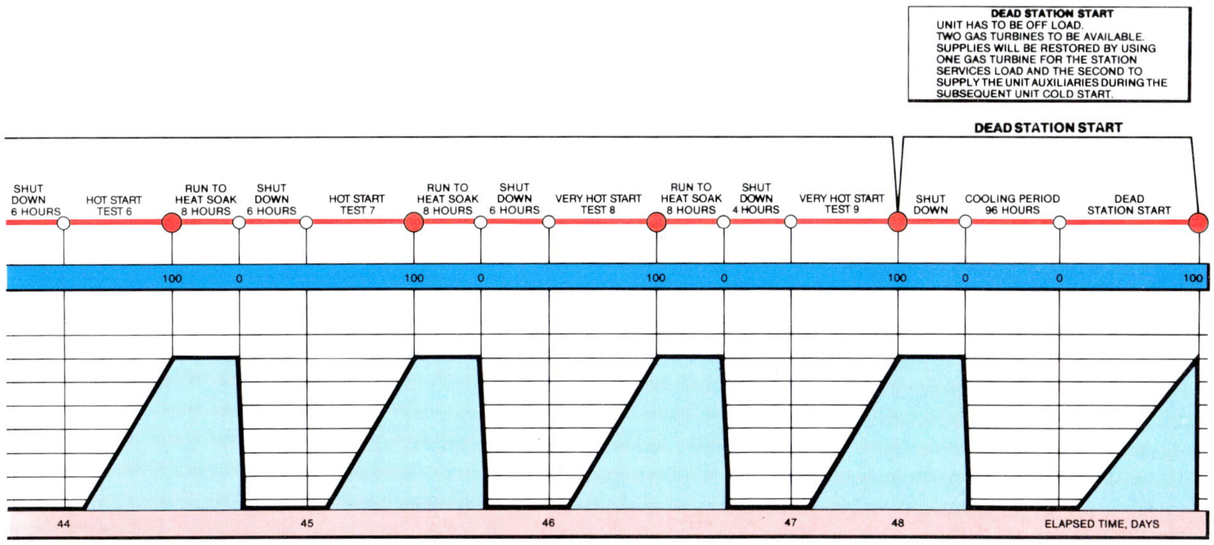

FIG. 6.82 Operational trials including dead station start — test programme

6.7.2 Test programme

The test programme contains the following series of tests under manual control:

Test 1	Warm start	Category 2
Test 2	Warm start	Category 2
Test 3	Very warm start	Category 3
Test 4	Very warm start	Category 3
Test 5	Hot start	Category 4
Test 6	Hot start	Category 4
Test 7	Hot start	Category 4
Test 8	Hot start	Category 4
Test 9	Very hot start	Category 5

The starting category is based on the temperature in the HP and IP cylinders. Table 6.26 identifies the temperature bands relating to the selection of the starting category. The effectiveness of the cylinder lagging, coupled with the method of shutting down the plant and the ambient temperatures, influence the cooling rates of the cylinders. As a rough guide to starting categories, the length of shutdown gives the following results:

Category 5	Very hot start after 4 to 6 hours shutdown
Category 4	Hot start after 6 to 18 hours shutdown
Category 3	Very warm start after 18 to 30 hours shutdown
Category 2	Warm start after 30 to 54 hours shutdown
Category 1	Cold start over 54 hours shutdown

6.7.3 Tests 1 and 2 — warm start tests, Category 2

The conditions in the boiler for this category depend on how the boiler was shut down. Full steam temperatures and pressures are maintained for as long as possible during the deloading period and, after furnace purging following the shutdown, all dampers are closed to reduce cooling paths through the boiler. By these actions, the boiler is kept at pressure during the shutdown period prior to the start of the turbine run-up. Table 6.27 identifies the plant conditions required for the commencement of the turbine run-up. The first requirement is to identify the boiler operating conditions for the turbine run-up and to fire the boiler to meet these conditions. For Category 2, Table 6.27 identifies a steam inlet pressure of 69 bar and a temperature of 375°C. The boiler firing rate is controlled to satisfy the following conditions:

- The gas temperature at the primary superheater is limited to 540°C and at the reheater inlet to 480°C until an adequate steam flow is established, the gas temperature being monitored by retractable probes.
- The maximum pressure raising rate must produce a saturated steam temperature rise rate which does not exceed 1.12°C/minute.
- The drum heating temperature differential limits must not exceed the values given in Fig 6.83.

During the charging of the boiler/turbine steampipes, the pressure here must be limited such that the saturation temperature reached on the initial pressurisation does not exceed the turbine stop valve chest metal temperature by more than 100°C. During steampipe warming, the rate of increase of the metal temperature should be 100°C/h when condensate

TABLE 6.26

Definition of temperature for turbine starting — categories 1–5

Point of measurement	Starting category, °C				
	1 Cold	2 Warm	3 Very Warm	4 Hot	5 Very Hot
HP SV, after valve, RH front inner HP SV, after valve, LH front inner HP SV, after valve, RH rear inner HP SV, after valve, LH rear inner HP GV, after valve, RH front inner HP GV, after valve, LH front inner HP GV, after valve, RH rear inner HP GV, after valve, LH rear inner	All points < 120	All points 120–230	All points 230–350	All points 350–450	All points > 450
HP loop pipe probe surface IP SV, before valve, RH front inner IP SV, before valve, LH front inner IP SV, before valve, RH rear inner IP SV, before valve, LH rear inner	All points < 70	All points 70–180	All points 180–300	All points 300–400	All points > 400
IP loop pipe probe surface	< 205	205–305	305–370	370–410	> 420

SV = Stop valve GV = Governor valve

TABLE 6.27
Start category acceptance levels

Check	Point of measurement	Acceptance level for starting category				
		1 Cold	2 Warm	3 Very Warm	4 Hot	5 Very Hot
Condenser vacuum, mbar	Condenser master vacuum ring main	−914	−914	−914	−914	−914
Turbine barring, r/min	Two analogue inputs of speed	36	36	36	36	36
Bearing oil pressure, bar	Digital input from barring interlock relay	0.7	0.7	0.7	0.7	0.7
Bearing oil temperature, °C	Analogue signal from main oil manifold	32	32	32	32	32
Control fluid temperature, °C	Analogue signal from oil cooler discharge	32	32	32	32	32
Generator frame hydrogen pressure, bar	Main generator frame	4.1	4.1	4.1	4.1	4.1
Generator hydrogen cooling water, bar	Pressure switch	2	2	2	2	2
Steam inlet temperature, °C	Before stop valves (accept 1 out of 2 for both LH and RH)	360 360	375 375	420 420	460 460	460 460
Steam inlet pressure, bar	Before stop valves (accept 1 out of 2 for both LH and RH)	69	69	76	103	103
Eccentricity, μm	At each bearing	<25	<25	<25	<25	
Differential expansion, mm	At each cylinder	within alarm levels				

heating and 135°C/h whilst superheating. Steampipe temperatures should be maintained within 14°C of each other during the warming process. The boiler pressure and temperature should be raised to 69 bar and 375°C, as specified in Table 6.27. The tabulated items are monitored throughout this period.

The rates of increase of boiler pressure and temperature are given in Fig 6.84. This also shows gas temperatures, feed/steam flow, firing rate and turbine run-up and loading rate.

The main boiler feed pump is introduced into the system in time to be available for pumping duty at synchronisation. A constraint on the boiler feed pump is that it cannot be started until turbine vacuum has been raised. To raise the feed pump speed at the optimum rate requires that the steam chests are first warmed and expanded in a manner similar to the main turbine. Control of differential temperatures must be maintained within defined limits. Vacuum raising cannot take place until the main steampipes are pressurised and the turbine glands sealed. Vacuum raising takes approximately 20 minutes and hence it is essential to pressurise the main steampipes as evenly as possible to enable the vacuum raising and feed pump operation to commence.

The turbine run-up should start once the conditions identified in Table 6.27 have all been satisfied and should continue at the rates specified in Table 6.28.

Once synchronised, the loading rates should follow those stated in Table 6.29; for Category 2, full load is achieved in 140 minutes after synchronisation. During

TABLE 6.28
Turbine run-up rates for the five start categories

Speed range, r/min	Acceleration rate in r/min^2 for various starting categories				
	1 Cold	2 Warm	3 Very Warm	4 Hot	5 Very Hot
Electrical turning gear speed up to 800	30	150	All ranges	All ranges	All ranges
800 to 1750 (critical speed band)	300	300	300	300	300
1750 to 2150	60	150			
2150 to 2800 (critical speed band)	300	300			
2800 to 3000	60	150			

the period of the turbine run-up, the following parameters associated with the boiler, turbine-generator and main boiler feed pump should be continuously checked and recorded:

Boiler

Gas pass temperatures

Superheater and reheater steam temperatures

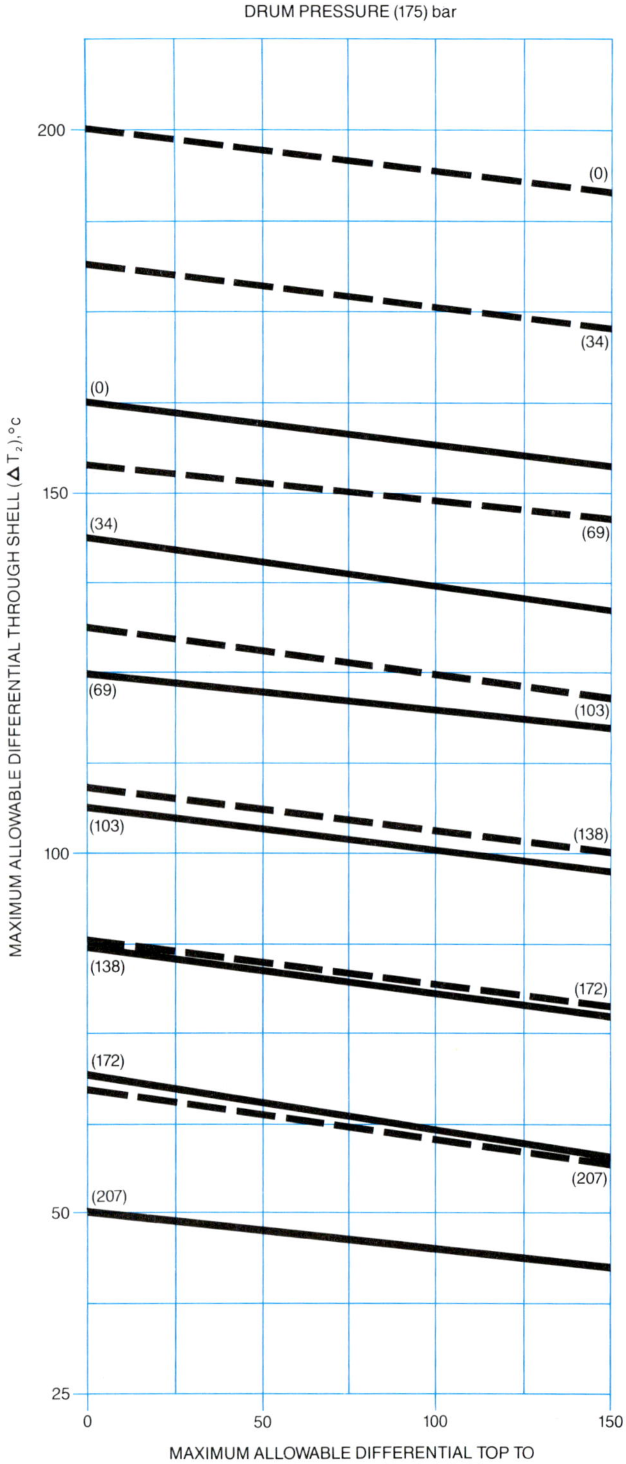

FIG. 6.83 Limits of permissible drum metal temperature differentials

Primary and secondary superheater metal temperatures

Boiler drain flow temperatures

Boiler drum heating and cooling differential temperatures

Boiler pressure

Combustion performance, i.e., O_2 levels, smoke emission and dust emission levels

Turbine-generator and main boiler feed pump

Steam-to-metal temperature differentials

Casing and steam chest temperatures

Turbine differential expansion

Turbine rotor eccentricity

Turbine bearing vibrations

Turbine bearing oil drain temperatures and pressures

Turbine HP and IP cylinder steam chest metal differential temperature

Generator seal-oil pressure

Generator gas temperatures

Generator hydrogen purity

Generator stator core and slot temperatures

Stator water pressures and temperatures

Once the unit has been established at full load, the results require analysing to see how closely the machine performs to the manufacturer's specifications, and to identify any problems or improvements in operational technique before further trials in this category are commenced.

6.7.4 Tests 3 and 4 — very warm start tests, Category 3

The warm start tests identified in Fig 6.84 show a total time from initial boiler firing to full-load of 340 min. In Category 3 tests, the shutdown time is shorter than for Category 2 tests and hence the turbine does not cool

TABLE 6.29
Turbine loading rates for the five start categories

Loading range, MW	Loading rate for start category									
	1		2		3		4		5	
	Rate MW/min	Time min	Rate MW/min	Time min	Rate MW/min	Time min	Rate MW/min	Time min	Rate MW/min	Time min
0 to 33	33	1	33	1	33	1	) 132 (0 to 132)	1	) 132 (0 to 132)	1
33 to 132	1.1	90	1.65	60	2	50	)		)	
132 to 660	6.6	80	6.6	80	9.6	55	12.3	43	27.8	19

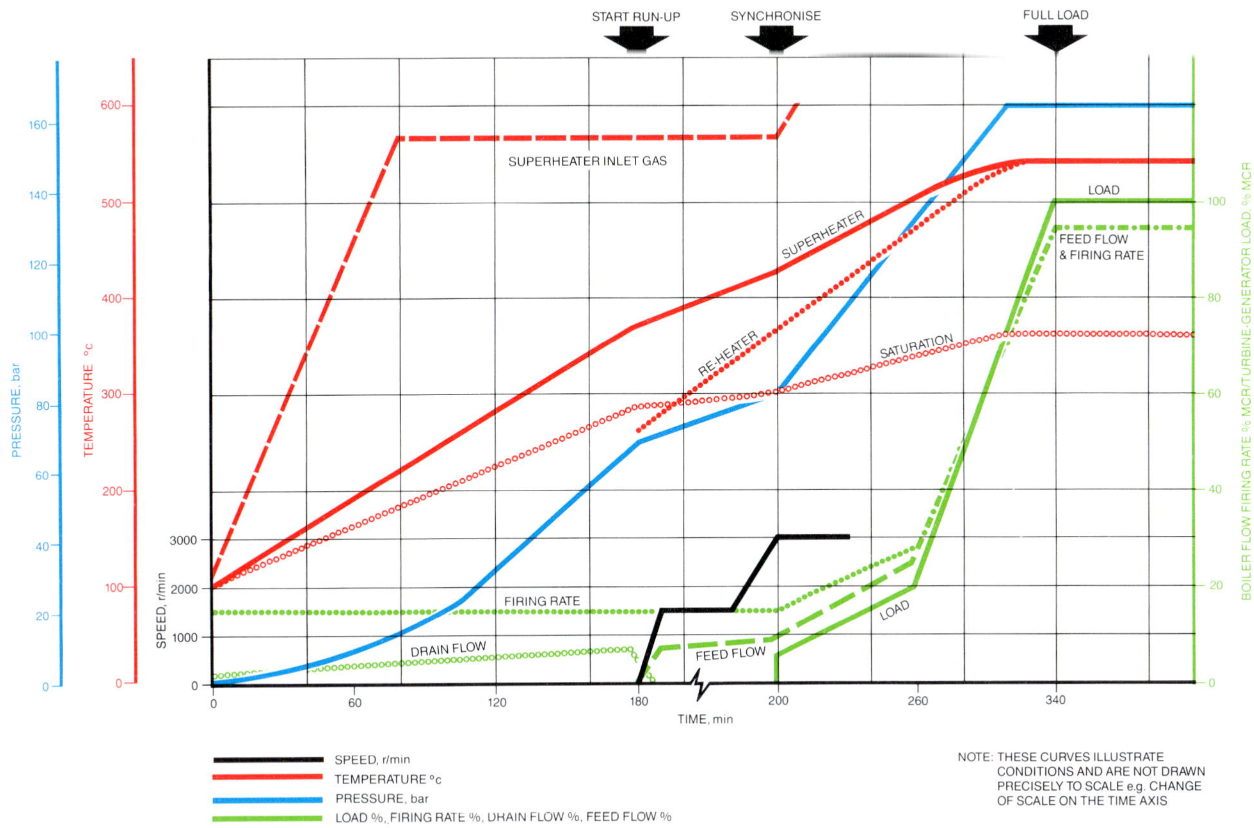

Fig. 6.84 Category 2 — warm start operating parameters

down to the same extent. Table 6.27 gives boiler conditions of 76 bar, 420°C superheater temperature, based on a HP cylinder temperature of between 230 and 350°C (Table 6.26). Reference to Fig 6.85 Category 3 shows that the initial boiler conditions are 50 bar/200°C. The time from this state to achieving the desired terminal conditions is reduced and the overall time from initial boiler firing to full-load is 240 minutes, 100 minutes shorter than the warm start time.

The reduced time available for this starting category means that the unit operation must be closely controlled. With the boiler already at a significant pressure at the start of the operation, increased firing will be necessary to obtain the higher temperatures required for the commencement of the turbine run-up. This demands close monitoring of the boiler gas and drain temperatures. A check that the specified superheater temperature of 420°C is obtained in the specified time of boiler firing prior to commencement of turbine run-up is also required. Both primary and secondary superheater temperatures will need to be checked.

Even though the time is reduced, the monitoring requirements are identical to the lower Category 2 start.

With the enhanced loading rate, particular attention is paid to turbine bearing vibration, eccentricities, differential expansions, turbine cylinder metal temperature differentials and boiler metal temperature, including drum differential temperatures.

Once the unit has been established at full-load, the results require analysing to see how closely the performance was to the manufacturer's specifications, to identify any problems or improvements in operating technique that could be applied for further trials in this category of start.

6.7.5 Tests 5 to 9 — hot and very hot start tests, Categories 4 and 5

The hot and very hot start categories are the most frequent types of starts to be carried out by the unit. They are the most demanding in the performance both of the operator and of the plant. The difference between the Category 4 and 5 starts is only in the time taken to full-load after synchronisation. Table 6.30 shows that a Category 5 start results in full-load 20 minutes after synchronisation, compared with 40 minutes for a Category 4 start.

Table 6.30

Predicted performance times for various start categories

	Category				
	1	2	3	4	5
From boiler firing to start of run-up, minutes	240	130	100	60	60
From start of turbine run-up to synchronisation, minutes	160	70	35	10	10
From synchronisation to full-load, minutes	170	140	105	40	20
Total time, minutes	570	340	240	110	90

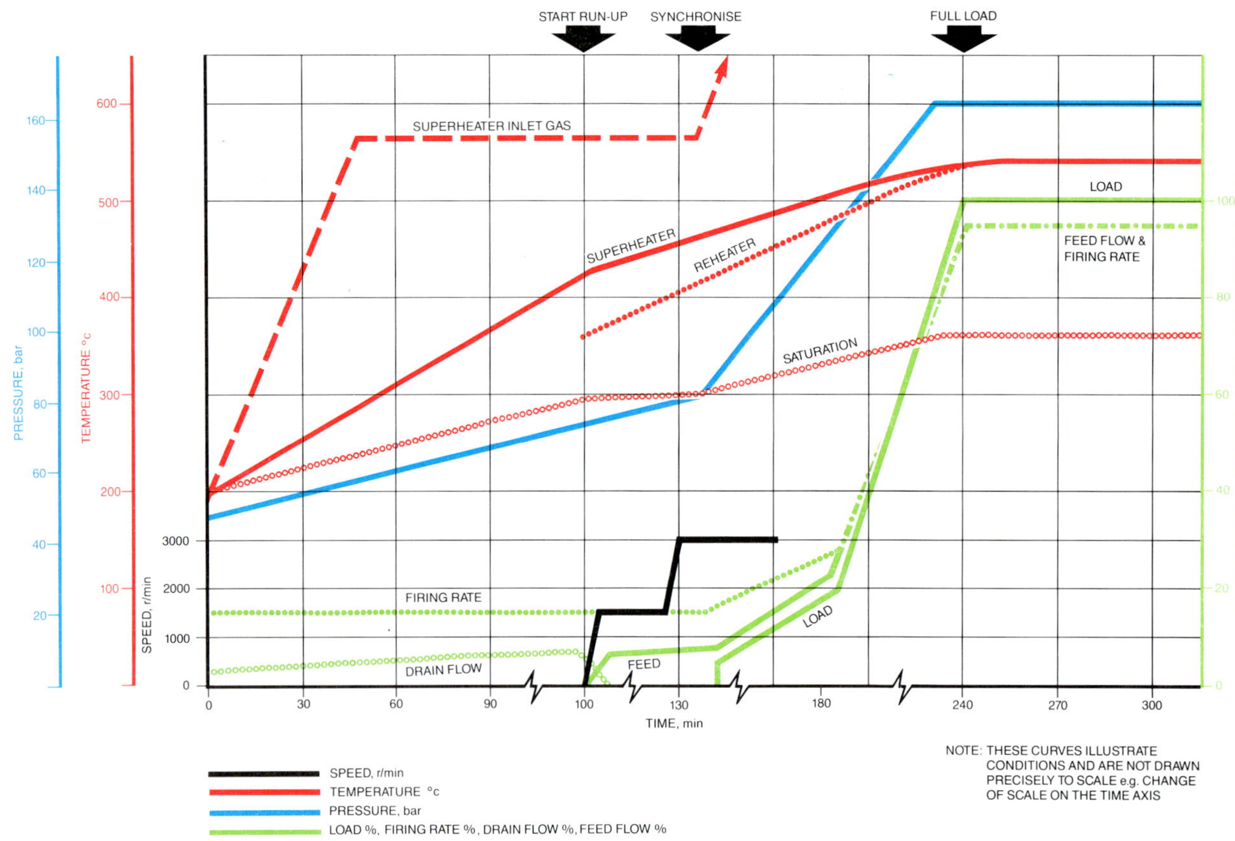

FIG. 6.85 Category 3 — very warm start operating parameters

The first part of a Category 4 or 5 test programme is to establish the shutdown procedure. The objective during the shutdown is to maintain steam temperatures at the design values for as long as possible to prevent unnecessary cooling of the HP and IP cylinders. Dependent upon boiler type, the selection and use of burners in the upper part of the furnace will assist with this objective. The shutdown time from full load to zero load should be completed within 20 minutes.

Typically, deloading rates are from 100% to 40% in 5 minutes and from 40% to 20% in 15 minutes. The remaining 20% of load is removed in approximately 10 s by continuous operation of the governor.

The test needs to establish the pressure at which the boiler should be shut down to ensure that, after extinguishing all fires and purging the furnace, the pressure does not rise to a value sufficient to operate the superheater safety valves.

For a 660 MW oil-fired boiler operating at 165 bar, a typical value at the time boiler fires are extinguished would be 130–140 bar. It is essential to maintain the highest possible pressure because, even though pressure loss will be small during the shutdown on the type of boiler being considered, a fall in pressure of some 30 bar will occur during the furnace purge prior to relighting. Figure 6.86 shows the rate of load reduction and boiler pressure and temperature values for the shutdown.

For Category 4 and 5 starts, the initial start conditions of the boiler are 103 bar and 460°C. These conditions are necessary to avoid cooling the HP cylinder, which will be in a temperature range 350–450°C for a Category 4 start and over 450°C for Category 5. Pressurising of the main steampipes should take place with temperature differentials within the range −50°C to +100°C.

Figure 6.87 shows initial boiler conditions of 80 bar and 300°C, requiring a steam pressure increase of 23 bar and a steam temperature increase of 160°C.

The furnace purge cools the superheater elements in addition to reducing boiler pressure. These tests have to establish the number and locations of the burners needed for a suitable firing rate to give the required temperatures and pressures without exceeding any boiler metal temperature limits. They also establish the point at which additional burners are required prior to synchronising, to enable the boiler to respond to the initial 20% load applied within a minute of synchronising. The hot starts will result in faster run-up and loading rates, so it is necessary to establish the boiler firing to ensure that the boiler output at rated pressure and temperature is available to meet the turbine load demand. This should be achieved without the need to resort to over-firing which can result from a mismatch between loading and boiler firing rates.

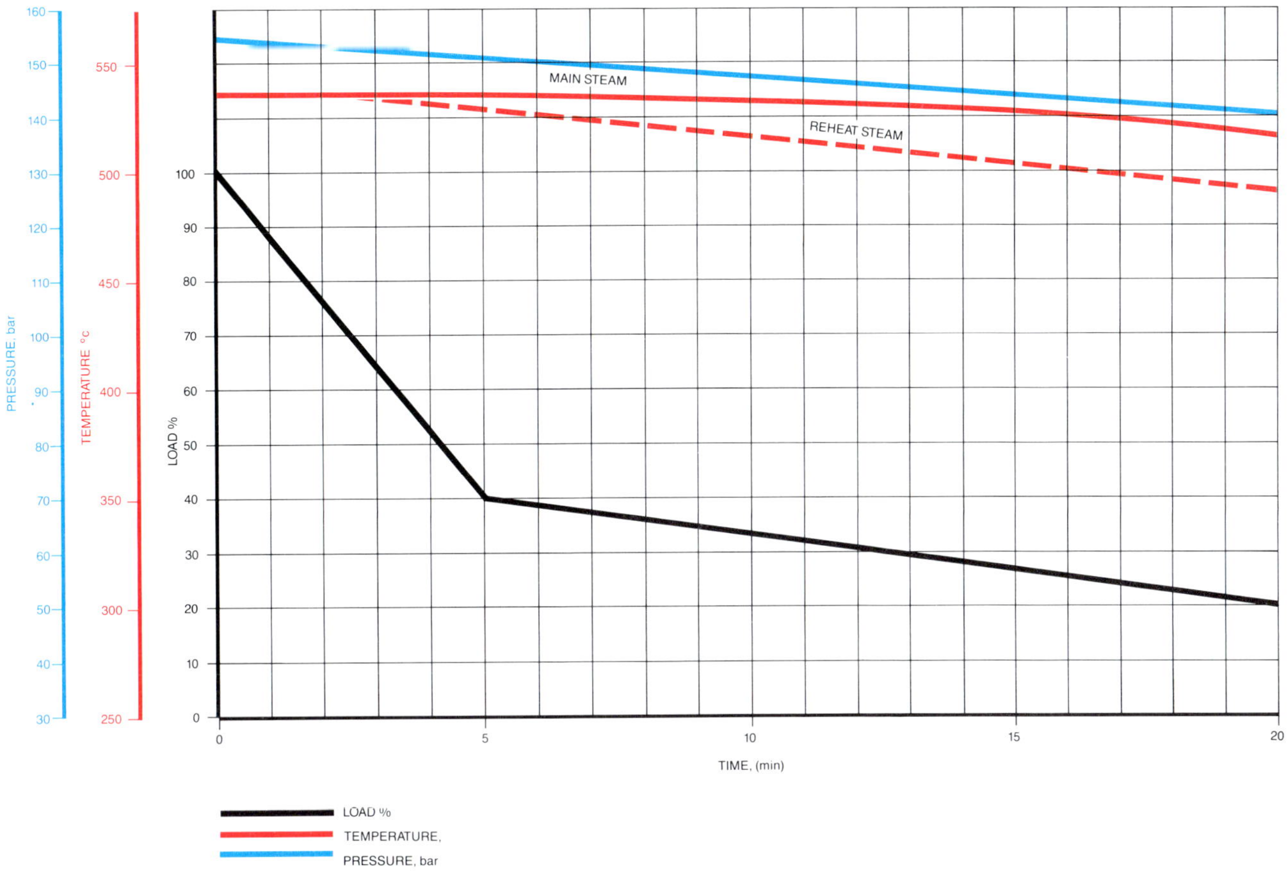

FIG. 6.86 Categories 4 and 5 — hot start shutdown

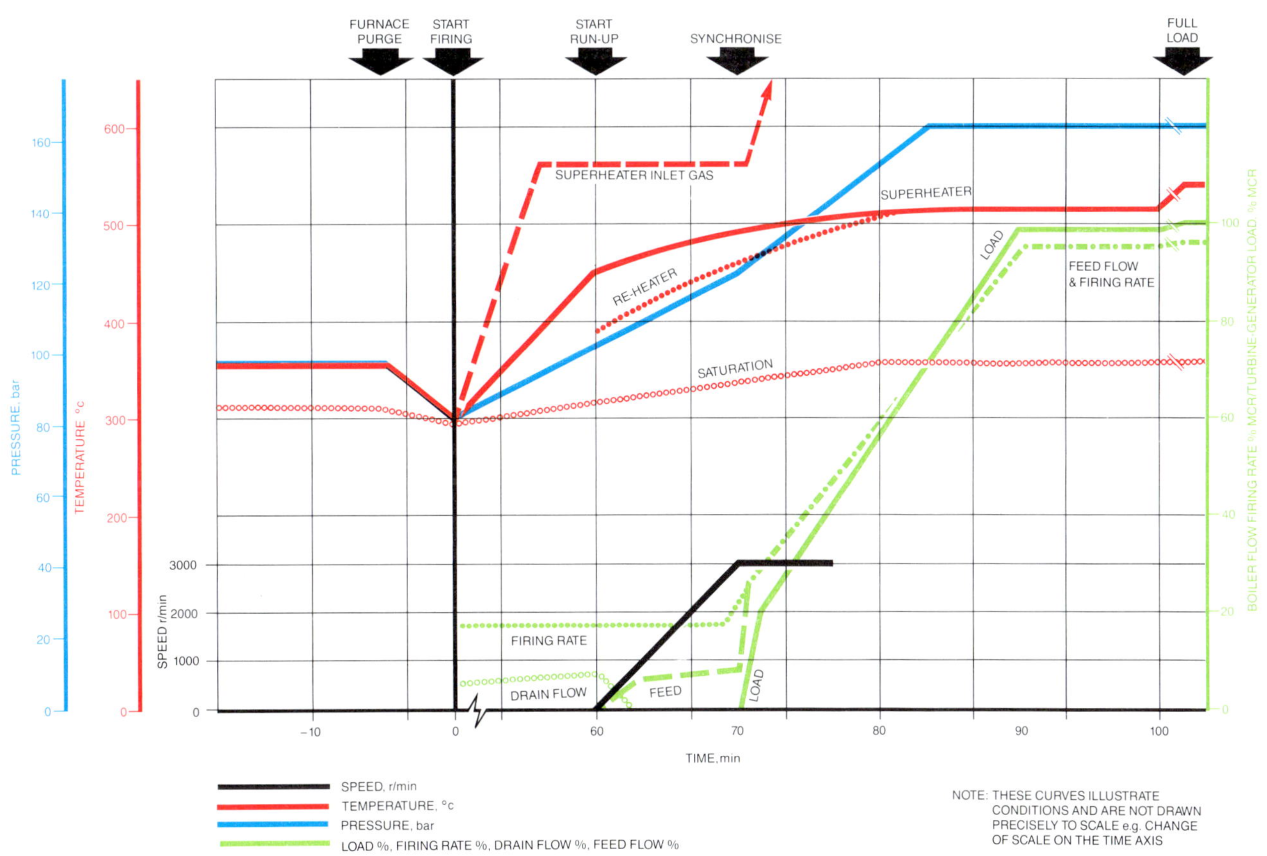

FIG. 6.87 Category 4 — hot start operating parameters

With the enhanced loading rates, close attention must be paid to the monitoring of the operating parameters, as previously stated.

6.7.6 Cold start, Category 1

For comparison, Fig 6.88 shows the cold start conditions when the boiler is at zero pressure and the HP turbine cylinder temperatures are less than 120°C.

6.7.7 Future operation

The operational trials check compliance with the contract specification and with the manufacturer's operating instructions. On completion of the test series, all the results are to be closely studied with a view to establishing that plant operation has been achieved within the contract specification and without exceeding any plant limits. Any plant problems resulting from the tests should be fully investigated and engineering solutions proposed.

Further training will be necessary to ensure that all staff are capable of operating the plant in a two-shifting mode to meet the stated time of operation. The results should also be analysed with a view to a further reduction in operation time, without exceeding limits, and further tests carried out to demonstrate this. Reducing the time for bringing back units on load improves the flexibility of the plant as well as reducing the cost in fuel and works energy of returning a unit to service.

6.8 Dead station start

6.8.1 Objective

To demonstrate that a shutdown power station can be restarted, synchronised and loaded, using gas turbines and station batteries, even when totally disconnected from the high voltage grid system and other external power supplies.

6.8.2 Test programme

(a) Perform pre-test trials to check the consequences of total loss of station AC supplies, including the effects on batteries resulting from loss of charger supplies. In addition, the adequacy of air-operated fire systems, when no charging air compressors are available, is tested. A single gas turbine is started and isolated from other sources of supply, and its ability to withstand and control the effects of starting some large 11 kV motors is demonstrated.

(b) Shut down the works power system and bring the station to a 'dead' condition.

(c) Re-energise the works power system, using batteries and gas turbines.

(d) Start a unit from cold Category 1 condition, to synchronise and load. The Category 1 condition assumes that the turbine has no need for electrically-driven shaft turning gear. It is therefore less risky to carry out the test under this condition than under another start category.

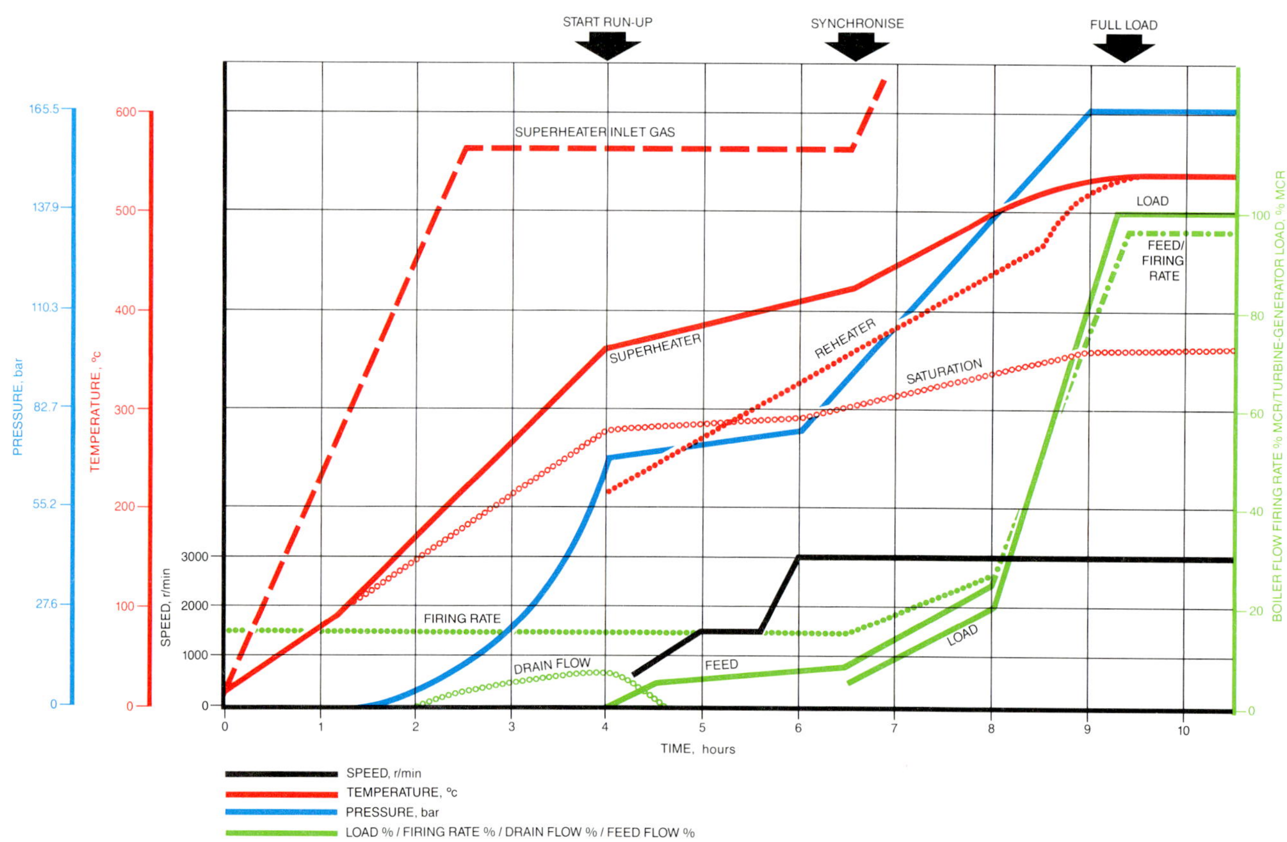

FIG. 6.88 Category 1 — cold start operating parameters

6.8.3 State of plant

- All turbines off load and in a cold start condition.
- All electrical timing gear off.
- At least two gas turbines with design ratings of 30 MW to be available for service, with one preferably being the gas turbine associated with the unit being started.
- Emergency lighting checked and proved available.
- All unit DC auxiliaries checked and proven available (it is expected that generator cooling systems are being maintained in service).
- Communication systems checked and proven for operation.
- The unit computer and data processing systems to be in service, supplied from inverters.
- The common service plant computer also to be in service on inverter supplies.
- The station transformer to be in service, with the HV grid system operating normally.

6.8.4 Pre-test trials

The pre-test trials identified in Section 6.8.2 of this chapter are carried out in advance of the dead station start test and the results carefully analysed to study the effect of having no station electrical supplies for approximately 30 minutes. The re-establishment of station supplies from the gas turbine should not take this time, but it is prudent to allow some extra time to enable any problems encountered to be overcome safely.

After switching out the station works power system, it should take approximately 15 minutes to restore station supplies using a gas turbine.

6.8.5 Shutting down the works power system

The first time this operation is carried out, the works power system should be systematically switched out to a written procedure. This involves switching out a number of transformers to prevent large inrush currents affecting the operation of the gas turbine following re-energising, when it is supplying the works power system. Having switched out the works power system, the final act is to switch out the station transformers, thus removing the source of incoming power.

6.8.6 Re-energising the works power system

With the auxiliary power system shut down, one gas turbine is selected for a remote no-load start. The signal to start causes the gas turbine to run-up to a generator shaft speed of 2800 r/min on its DC

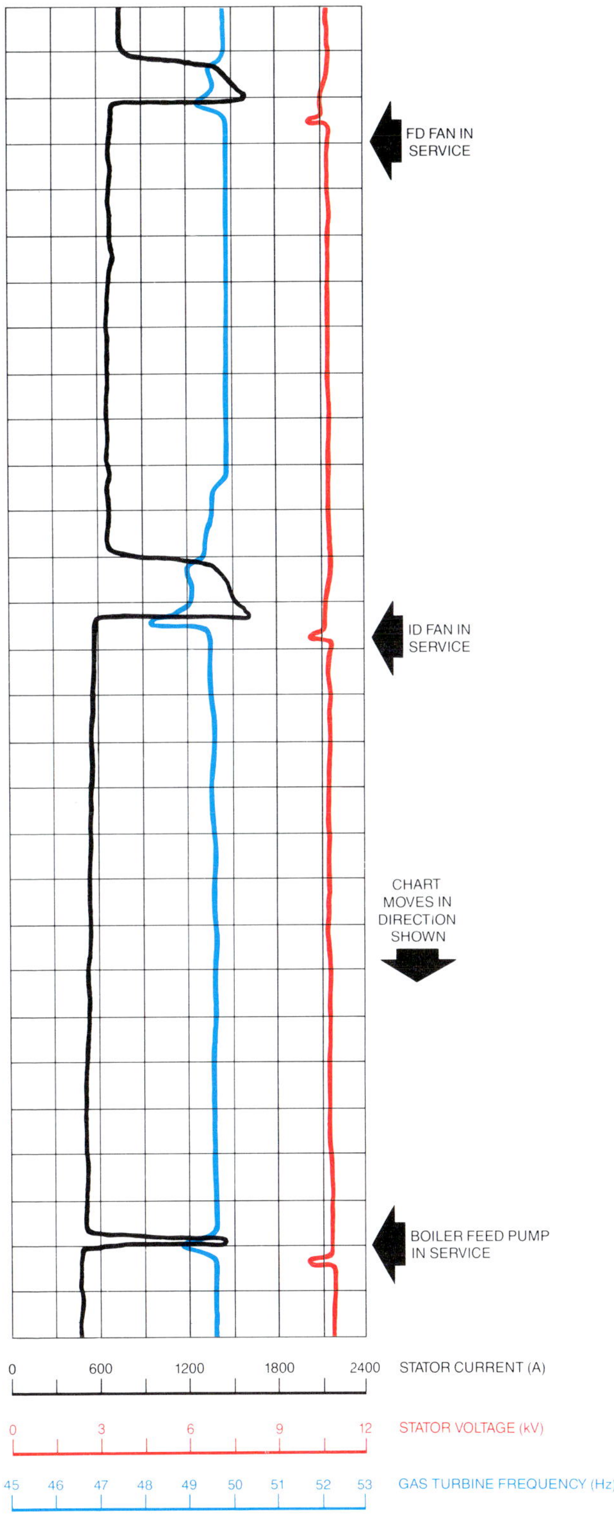

FIG. 6.89 Dead station start.

The chart shows the variation of gas turbine speed, stator volts and stator current, whilst starting a feed pump. ID fan and FD fan. Load prior to starts was 6 MW which included the CW pump and condensate extraction pump.

auxiliaries. The gas turbine speed is then increased to 3000 r/min with its generator voltage set to its nominal value of 11 kV and the automatic voltage regulator in service.

The gas turbine 11 kV switch is then closed onto the 11 kV unit board. Re-energise each station auxiliary switchboard shut down as part of the de-energising procedure. With AC supplies available, the AC seal-oil and flushing oil pumps on the turbine can be restarted and the DC pumps shut down.

During the re-energising procedure, the output voltage and frequency of the gas turbine are continuously monitored.

Once the works power system is fully energised and under the control of the gas turbine, a second gas turbine should be started and synchronised to its own unit board, running at zero load as a back-up in the event of failure of the first gas turbine.

6.8.7 Test results

With the auxiliary power system now re-energised, the unit can be returned to service in the normal manner for a cold start. The gas turbines must run for approximately 5–7 hours until synchronisation is achieved. Prior to the start of unit operation, the gas turbine will be supplying a base load resulting from lighting, heating, air conditioning, sump pumps, air compressors and battery chargers plus the turbine oil and seal-oil pumps. A typical figure for this load on a 4 × 660 MW oil-fired station would lie in the range 2–5 MW.

Chart recording facilities are needed to record gas turbine stator current, stator voltage and generator frequency (speed), especially during start-up of the large 11 kV motors. As these motors are progressively started, the ability of the gas turbine governor and voltage regulator to control frequency and voltage will be demonstrated. This is particularly important during the high levels of current present when starting large motors.

Figure 6.89 shows traces of the variation of gas turbine frequency, stator volts and current taken from a dead station start on a 660 MW unit. With a base load of approximately 2 MW, the starting of a 9.7 MW electric feed pump reduced the frequency by 1 Hz. This pump is speed controlled by the use of an external rotor resistance, giving a starting current of 1.3 times normal full-load current. The starting of the ID fan, which is a high inertia direct-on-line start 3.4 MW motor, produced the largest frequency drop of 1.4 Hz. For the example quoted, the CW pump was already in service and the base load on the gas turbine at the time of starting the items listed in Fig 6.89 was between 6 and 7 MW.

Table 6.31 shows the loadings on a gas turbine as the major items of plant are established in service, resulting in a maximum requirement of 14 MW. This demonstrates that a single gas turbine of output has sufficient capacity to enable a unit start to be carried out as well as supplying essential auxiliaries for other shutdown plant. Once the turbine-generator is synchronised, the unit transformer can be put in service, followed by the station transformer, thus allowing the gas turbine to be shut down.

TABLE 6.31

Gas turbine loadings during a dead station start

Gas turbine loadings	Time of day						
	16.12	16.37	16.52	16.58	17.03	17.10	18.50
	(a)	(b)	(c)	(d)	(e)	(f)	(g)
Load, MW	6	7	8	11	12	12	14
Stator voltage, V	11	11	11	11	11	11	11
Stator current, A	400	420	530	700	800	800	900
Governor, %	60	60	60	63	63	63	63
Frequency, Hz	50	50	49.9	50.2	50.2	50.2	50

(a) CW pump in service
(b) Fuel oil pump in service
(c) Boiler feed pump in service
(d) First ID fan in service
(e) First FD fan in service
(f) Both GR fans in service
(g) Second ID fan in service

Appendix A

Paper of principle for 660 MW boiler/turbine unit: typical list of contents

Section

1 Introduction

2 Objectives and assessment points

3 Plant description

3.1 General
3.2 Unit service
3.3 Special design aspects

4 Design intent

4.1 General control concept
4.2 Operational modes
4.3 Specified operational times
4.4 Target thermal efficiencies

5 Terminal points

5.1 Electrical
5.2 Mechanical

6 Documentation

7 Schedule of tests

8 Services required

8.1 Station auxiliary system — mechanical and general
8.2 Electrical system and equipment

9 Responsibilities

10 Resources

11 Safety assessment of plant

12 Plant and environmental safety policy

Note: Typical contents for Section 12 are included on the following pages

12.1 General

Emphasis that each organisation must ensure that its employees work to the various rational legislation, CEGB Safety Rules, Local Safety Instructions and Plant Operating Instructions.

12.2 Personal safety

Access This is to be complete to permanent finish or to equal standard temporary finish.

Protective equipment Where this is required, it is to be provided and used.

Section

Noise Noise measurements are to be taken, areas defined and appropriate protective equipment worn.

Lighting Permanent lighting to be installed if possible or temporary lighting to give the requested level of illumination for operational work.

12.3 Plant safety

Identification To be clearly identified to prevent mistakes and malfunction.

Lagging To be completed to protect personnel and minimise the risk of fire.

State All plant to be at a known state of completion and commissioning. This will usually be achieved by the documentation system and physical inspection.

Access In addition to the above, access may need to be restricted, either completely to certain areas or only certain personnel allowed to enter. This may require physical barriers and warning notices.

Controls Consideration should be given to the use of automatic controls and, unless already proven, they may need to be left in the manual mode initially.

Monitoring Enough permanent or temporary instrumentation must be provided to operate the plant safety for the work being undertaken.

Programmes The programme of testing to be agreed in advance, and communicated to all parties, with responsibilities clearly defined.

Emergency procedures All fire fighting, first aid and evacuation procedures must be defined, tested and understood. Particular emphasis must be placed on plant fire fighting equipment and procedures.

12.4 Environmental safety

Fluid spillage All hazardous fluids and their characteristics should be known in advance and plans drawn up to contain spillages. The policy and practice concerning spillage of environmentally-sensitive fluids which go outside the station boundary should be known.

Emissions The operating parameters should be defined to minimise this occurrence.

Complaints A procedure should exist on site for dealing with complaints on environmental matters, by whomsoever they may be raised.

SUBJECT INDEX